9th International Scientific – Research and Development of Mechanical Elements and Systems (IRMES 2019)

IOP Conference Series: Materials Science and Engineering
Volume 659

Kragujevac, Serbia
5 - 7 September 2019

Editor:

Nenad Marjanovic

ISBN: 978-1-7138-0914-2
ISSN: 1757-8981

Printed from e-media with permission by:

Curran Associates, Inc.
57 Morehouse Lane
Red Hook, NY 12571

Some format issues inherent in the e-media version may also appear in this print version.

Institute of Physics
Dirac House, Temple Back
Bristol BS1 6BE UK

Phone: 44 1 17 929 7481
Fax: 44 1 17 920 0979

techtracking@iop.org

Additional copies of this publication are available from:

Curran Associates, Inc.
57 Morehouse Lane
Red Hook, NY 12571 USA
Phone: 845-758-0400
Fax: 845-758-2633
Email: curran@proceedings.com
Web: www.proceedings.com

TABLE OF CONTENTS

MECHANICAL ELEMENTS AND SYSTEMS

PRODUCT DEVELOPMENT PROCESS

NEW TECHNOLOGIES AND MATERIALS

POWER AND MOTION TRANSMISSION SYSTEMS

TOWARD A SUSTAINABLE MOBILITY

IOP Conf. Series: Materials Science and Engineering **659** (2019) 011001 doi:10.1088/1757-899X/659/1/011001

9th INTERNATIONAL SCIENTIFIC CONFERENCE - IRMES 2019

RESEARCH AND DEVELOPMENT OF MECHANICAL ELEMENTS AND SYSTEMS

5-7 September 2019, Kragujevac

Organized by the Department for Mechanical Constructions and Mechanization at the University of Kragujevac, Faculty of Engineering

with support of:

ADEKO – Association for Design, Elements and Constructions

IFToMM - International Federation for the Promotion of Mechanism and Machine Science

Accepted, peer reviewed papers from the 9th IRMES 2019

Edited by

Nenad Marjanović, Kragujevac

IOP Conf. Series: Materials Science and Engineering **659** (2019) 011001 doi:10.1088/1757-899X/659/1/011001

Preface

The 9th International Scientific Conference - IRMES 2019 - Research and Development of Mechanical Elements and Systems was organized by the Department for Mechanical Constructions and Mechanization of the Faculty of Engineering at the University of Kragujevac and the Association for Design, Elements and Constructions – ADEKO from September 5th to the 7th 2019 at the Faculty of Engineering in Kragujevac.

On the previous eight IRMES Conferences (the first in 1995, the last in 2017), around a thousand papers have been presented, and there were over a thousand participants from all over the world. A long and successful tradition is a stable basis for organizing this and future IRMES Conferences.

The mission of IRMES Conferences is to serve the global community by improving, spreading and applying new engineering knowledge, with the goal of being used as a source of the newest and most relevant information for mechanical engineers and experts in related fields – on a local, regional and global level.

Specific goals, themes and fields of the IRMES 2019 Conference are defined in cooperation with the ADEKO association, and in accordance with current topics and problems. Thematic units of the conference are: Mechanical Elements and Systems (modeling and simulation, loading and stress conditions, tribology, noise and vibrations, maintenance and monitoring, safety, quality, reliability), Power and Motion Transmission Systems (development of new concepts, modeling and simulations, noise and vibrations, testing, safety, quality, reliability), Product Development Process (technology transfer, creativity and innovations, development and design, innovative product development, smart systems, industry 4.0, knowledge economy), New Technologies and Materials (CAD/ CAM/ CAE technology, intelligent production systems, robotics and mechatronics, rapid prototyping, new materials), Research and Development in Field of Vehicles and Transport, Power and Motion Transmission Systems and a Special Section "Toward a Sustainable Mobility".

We have ensured a wide international participation, in order to have as many high quality research papers as possible and in order to increase the significance and influence of IRMES Conferences on a global level. We would like to thank the reviewers on their hard work and dedication, which have increased the quality of the IRMES 2019 Conference.

We would like to thank all keynote speakers, authors, committee members, reviewers, sponsors and others who have helped this Conference and attributed to its quality and importance.

Conference Chair of IRMES 2019

Professor Nenad Marjanović, PhD

Kragujevac, 7th September 2019

IOP Conf. Series: Materials Science and Engineering **659** (2019) 011002 doi:10.1088/1757-899X/659/1/011002

Peer review statement

All papers published in this volume of *IOP Conference Series: Materials Science and Engineering* have been peer reviewed through processes administered by the proceedings Editors. Reviews were conducted by expert referees to the professional and scientific standards expected of a proceedings journal published by IOP Publishing.

IOP Conf. Series: Materials Science and Engineering **659** (2019) 012001 doi:10.1088/1757-899X/659/1/012001

Research on designing a multiloop planar linkage

D Badoiu[1] and G Toma[1]

[1]Petroleum-Gas University of Ploiesti, Faculty of Mechanical and Electrical Engineering, 39 Bucuresti Blvd., 100680, Ploiesti, Romania

E-mail: dorin.badoiu@gmail.com

Abstract. The paper presents the way of determining the dimensions of the elements of a planar linkage when imposing certain conditions regarding the movement of a component piston, aiming at obtaining a mechanism whose total mass is minimal. A computer program that simulates the kinematics of the analyzed mechanism has been developed by the authors using Maple programming environment. The NLPSolve function included in the Optimization Package of Maple has been used for establishing the dimensions of the elements that assure the minimal mass of the linkage. Finally, some simulation results are presented.

1. Introduction

The design of linkages is a complex process that aims at establishing an optimal functional constructive solution that fulfils the conditions imposed by the task to be accomplished. Many times in such a process it starts from a certain structural configuration of the linkage and the dimensions of the component elements are determined by solving a problem of estimating the extreme values of a function that expresses certain characteristic parameters [1-6]. From a mathematical point of view, this implies calculating the minimum or maximum value of this function in the presence of certain constraints expressing the conditions imposed by the functioning of the mechanism.

In the paper is presented the way of determining the dimensions of the elements of a planar linkage, consisting of seven links and three independent contours, when imposing certain conditions regarding the movement of a component piston, aiming at obtaining a mechanism whose total mass is minimal. A computer program that simulates the kinematics of the analyzed mechanism has been developed by the authors using Maple programming environment [7-8]. The NLPSolve function included in the Optimization Package of Maple has been used for establishing the dimensions of the elements that assure the minimal mass of the linkage.

2. Theoretical considerations

Figure 1 shows the mechanism to be analyzed. As it can be seen from the associated graph, shown in figure 2, the mechanism consists of three independent contours: 0-1-2-3-0, 0-1-2-4-5-0 and 0-5-6-7-0. The kinematics of the mechanism has been studied by projecting the vector circuits corresponding to the independent contours on the axes of the (Oxy) coordinate system [9-10].

Thus, by projecting the vector equation: $\overrightarrow{OA} + \overrightarrow{AB} + \overrightarrow{BO} = 0$, corresponding to the first independent contour 0-1-2-3-0, on the axes (Ox) and (Oy), we obtain the following system of equations:

$$\begin{cases} l_1 \cos\varphi_1 + l_2 \cos\varphi_2 = 0 \\ l_1 \sin\varphi_1 + l_2 \sin\varphi_2 - s_3 = 0 \end{cases} \quad (1)$$

where: $l_1 = OA$, $l_2 = AB$ and $s_3 = OB$.

From the system of equations (1), the angle φ_2 may be calculated from the equation:

$$\cos\varphi_2 = -\frac{l_1}{l_2}\cos\varphi_1 \tag{2}$$

Then, the displacement s_3 may be determined with the relationship:

$$s_3 = l_1 \sin\varphi_1 + l_2 \sin\varphi_2 \tag{3}$$

For the second contour 0-1-2-4-5-0, by projecting the corresponding vector equation: $\overrightarrow{OA} + \overrightarrow{AC} + \overrightarrow{CD} + \overrightarrow{DE} + \overrightarrow{EO} = 0$ on the axes (Ox) and (Oy), the following system of equations is obtained:

$$\begin{cases} l_1 \cos\varphi_1 + l_{21}\cos\varphi_2 + l_4 \cos\varphi_4 + l_{51}\cos\varphi_5 - x_E = 0 \\ l_1 \sin\varphi_1 + l_{21}\sin\varphi_2 + l_4 \sin\varphi_4 + l_{51}\sin\varphi_5 - y_E = 0 \end{cases} \tag{4}$$

where: $l_{21} = AC$, $l_4 = CD$, $l_{51} = DE$ and x_E , and y_E are the coordinates of point *E*.

By solving the system of equations (4), the angle φ_4 may be calculated from the equation:

$$C \cdot \cos\varphi_4 + D \cdot \sin\varphi_4 = E \tag{5}$$

where:

$$\begin{cases} C = 2l_4(l_1 \cos\varphi_1 + l_{21}\cos\varphi_2 - x_E) \\ D = 2l_4(l_1 \sin\varphi_1 + l_{21}\sin\varphi_2 - y_E) \\ E = l_{51}^2 - l_4^2 - (l_1 \cos\varphi_1 + l_{21}\cos\varphi_2 - x_E)^2 - (l_1 \sin\varphi_1 + l_{21}\sin\varphi_2 - y_E)^2 \end{cases} \tag{6}$$

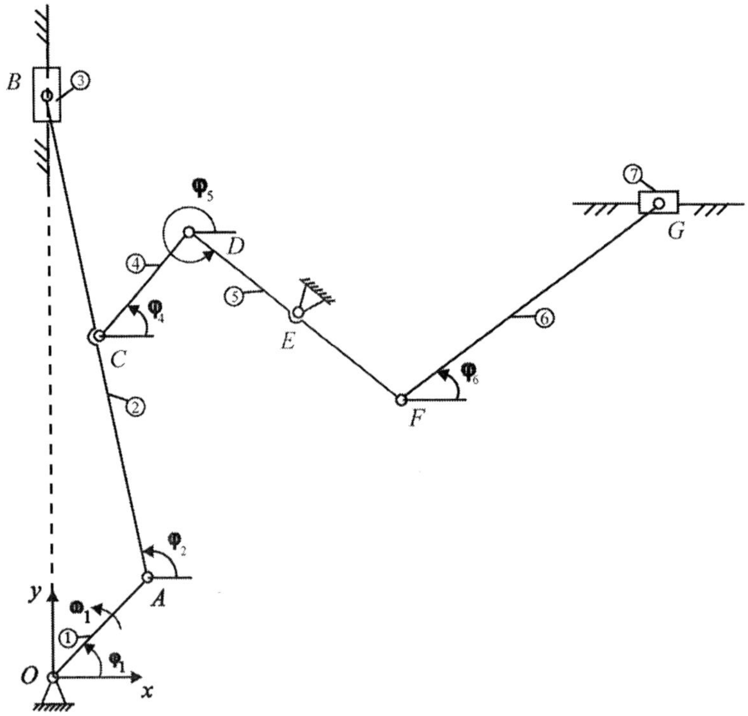

Figure 1. Planar mechanism.

The angle φ_5 may be determined with the relationship:

$$\varphi_5 = \text{ATAN2}(-(l_4 \sin\varphi_4 + l_1 \sin\varphi_1 + l_{21} \sin\varphi_2 - y_E), -(l_4 \cos\varphi_4 + l_1 \cos\varphi_1 + l_{21} \cos\varphi_2 - x_E)) \quad (7)$$

where ATAN2(*y*,*x*) calculates arctan(*y/x*) by taking into account the signs of *x* and *y* [9].

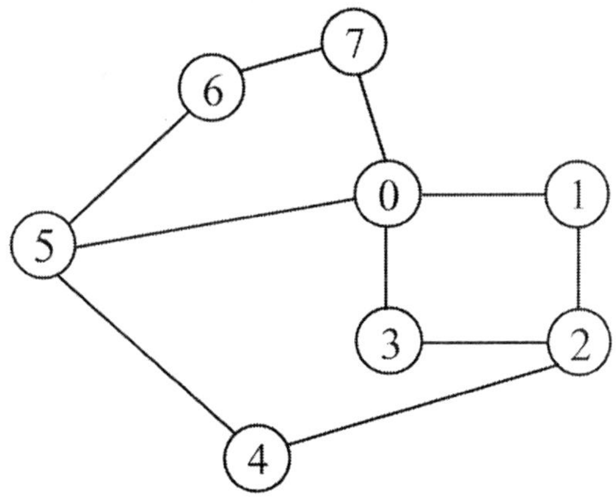

Figure 2. The associated graph.

The vector equation: $\overrightarrow{EF} + \overrightarrow{FG} + \overrightarrow{GE} = 0$ corresponding to the third contour 0-5-6-7-0, projected on the axes (*Ox*) and (*Oy*), leads to obtaining the following system of equations:

$$\begin{cases} l_{52} \cos\varphi_5 + l_6 \cos\varphi_6 - s_7 = 0 \\ l_{52} \sin\varphi_5 + l_6 \sin\varphi_6 - (y_G - y_E) = 0 \end{cases} \quad (8)$$

where: $l_{52} = EF$, $l_6 = FG$ and $s_7 = x_G - x_E$ (x_G and y_G are the coordinates of point *G*).

By solving the system of equations (8), the angle φ_6 may be calculated from the equation:

$$\sin\varphi_6 = -\frac{1}{l_6}(l_{52} \sin\varphi_5 - y_G + y_E) \quad (9)$$

and the displacement s_7 may be determined with the relationship:

$$s_7 = l_{52} \cos\varphi_5 + l_6 \cos\varphi_6 \quad (10)$$

3. Simulation results

Using Maple programming environment [7-8] a computer program that simulates the kinematics of the analyzed mechanism has been developed. The initial dimensions of the component elements have the following values: $l_1 = 0.15$ m; $l_2 = 1.05$ m; $l_4 = 0.5$ m; $l_5 = 0.5$ m; $l_6 = 0.8$ m; $l_{21} = f_2 l_2$, where $f_2 = 0.2$ and $l_{51} = f_{51} l_5$, where $f_{51} = 0.5$. The coordinates of point *E* are: $x_E = 0.55$ m and $y_E = 0.35$ m and the *y* coordinate of point *G* is $y_G = 0.5$ m.

The angular speed ω_1 is considered to have the value of 15 rad/s. The elements 1, 2, 4, 5 and 6 (figure 1) are made of steel bars with circular section having the following radii: $r_1 = 0.04$ m; $r_2 = 0.025$ m; $r_4 = 0.04$ m; $r_5 = 0.03$ m and $r_6 = 0.025$ m. The mass of the piston *3* is 5 kg and the mass of the piston 7 is also 5 kg.

We analyzed several cases in which it has been imposed the value of the stroke of the piston 7 and the values of the crank angle φ_1 for which the speed of the piston 7 is null, provided the total mass of the mechanism be minimal.

IOP Conf. Series: Materials Science and Engineering **659** (2019) 012001 doi:10.1088/1757-899X/659/1/012001

Further are presented some simulation results in the case when it was sought to determine the lengths of the bar-shaped components such that the piston *7* have zero velocity for the positions of the mechanism corresponding to the crank angle $\varphi_1 = 30^\circ$ and $\varphi_1 = 230^\circ$ and its stroke ($s_7(230^\circ) - s_7(30^\circ)$) be equal to 0.3 m, provided that the total mass of the linkage be minimal.

In this scope has been used the NLPSolve function included in the Optimization Package of Maple. The NLPSolve function solves problems involving the minimization or maximization of a continuous real nonlinear objective function possibly subject to constraints.

The NLPSolve command uses various methods implemented in a built-in library and the solvers are iterative in nature and require an initial point. In this case, the Sequential Quadratic Programming method has been used and the iterative calculus has been started with the initial values of the lengths of the component elements.

The NLPsolve function interrupts the calculation when one of the constraints is not fulfilled or complex numbers appear during the calculations. When solution of the problem is not identified after the calculations, a message appears specifying that no improved point could be found.

To avoid the occurrence of complex numbers during computations, the variation of the lengths of the component elements, less than l_4, was considered to be between 0.98 and 1.04 of the values corresponding to the previous step of the iterative process.

For l_4 it was considered that its value may vary between 0.98 and 1.3 of the value corresponding to the previous step.

After five iterations the program displayed the message that no improved point could be found. The values obtained for the lengths of the component elements are: $l_1 = 0.138\ \text{m}$; $l_2 = 0.968\ \text{m}$; $l_4 = 0.595\ \text{m}$; $l_5 = 0.461\ \text{m}$; $l_6 = 0.738\ \text{m}$; $l_{21} = f_2 l_2$, where $f_2 = 0.228$ and $l_{51} = f_{51} l_5$, where $f_{51} = 0.475$. For the coordinates of point E have been obtained the following values: $x_E = 0.511\ \text{m}$ and $y_E = 0.346\ \text{m}$ and for the y coordinate of point G the value: $y_G = 0.463\ \text{m}$. The total mass of the mechanism is 75.568 kg.

In figures 3, 4 and 5 are represented the variation of the displacement s_7, of the speed and of the acceleration of the piston 7, during a cinematic cycle, beginning with $\varphi_1 = 30^\circ$. The curves noted with 1 correspond to the initial mechanism and the curves 2 correspond to the mechanism obtained after the analyzed case.

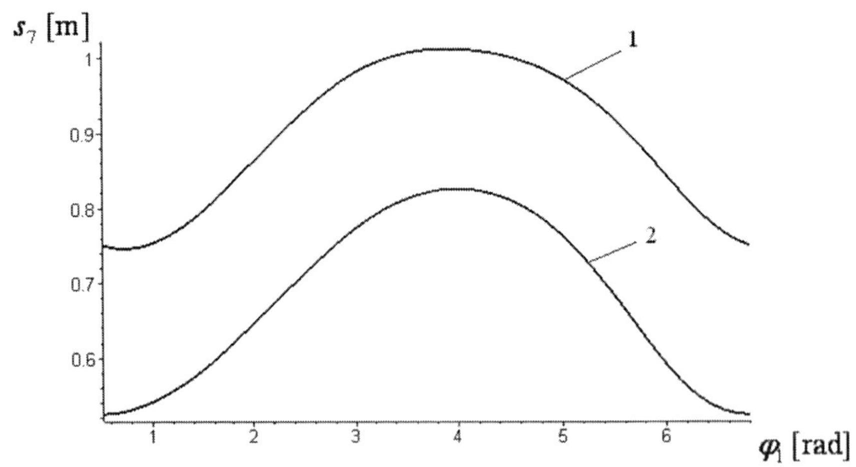

Figure 3. The variation of the displacement of the piston 7: curve 1 for the initial mechanism and curve 2 for the mechanism obtained after analysis.

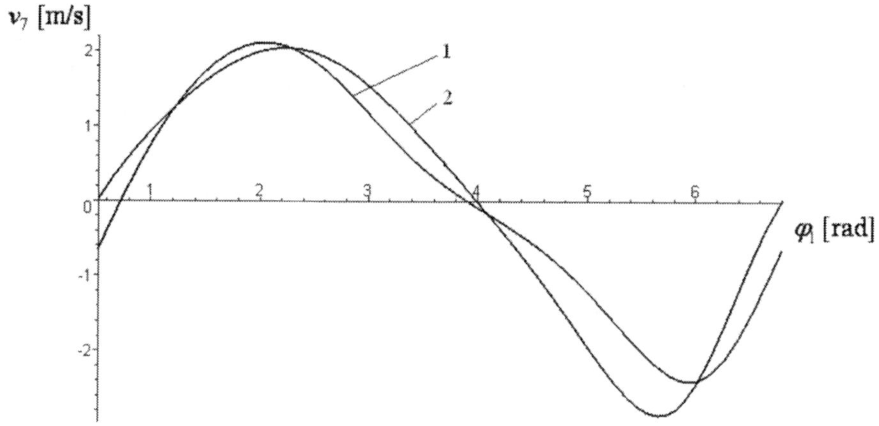

Figure 4. The variation of the speed of the piston 7: curve 1 for the initial mechanism and curve 2 for the mechanism obtained after analysis.

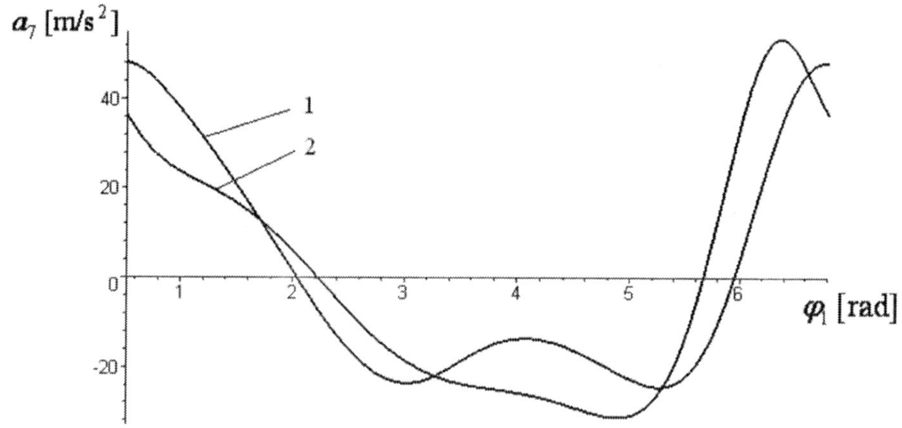

Figure 5. The variation of the acceleration of the piston 7: curve 1 for the initial mechanism and curve 2 for the mechanism obtained after analysis.

Figure 3 and figure 4 show that the imposed conditions for the stroke and for the speed of the piston 7 are fulfilled. Also, figure 4 and figure 5 show that the extreme values of the speed and of the acceleration of the piston 7 have not undergone significant changes as against the original situation.

4. Conclusions

In the paper a series of results concerning the synthesis of a planar linkage have been presented. In the development of the kinematics simulation program, the Maple programming environment has been used, which offers besides symbolic computing facilities also a package with extremely useful functions in optimization problems. In this sense, the NLPSolve function has proved to be extremely useful and easy to implement in the simulation program to solve the problem of optimizing the functioning of the mechanism studied in the paper. It is obvious that the way of realizing the synthesis of the analyzed linkage can be used in the case of any other planar linkage with different configuration. However, particular attention must be paid to choosing the intervals of variation of the values of the lengths of the component elements, as well as to the choice of the initial values of these lengths on each iteration.

IOP Conf. Series: Materials Science and Engineering **659** (2019) 012001 doi:10.1088/1757-899X/659/1/012001

References

[1] Sandor G N and Erdman A G 1984 *Advanced Mechanism Design: Analysis and Synthesis (Vol. 2),* Prentice-Hall Inc.

[2] Shigley J E and Uicker Jr J J 1981 *Theory of Machines and Mechanisms,* McGraw-Hill

[3] Badoiu D and Toma G 2016 On a Dynamic Optimisation Problem of the Quadrilateral Mechanism, *Journal of the Balkan Tribological Association* **22**(1) 250-260

[4] Toma G 2019 Research Concerning the Optimization of the Mechanism of the Conventional Sucker Rod Pumping Units, *Revista de Chimie* **70**(5) 1795-1799

[5] Badoiu D and Toma G 2019 Research Concerning the Predictive Evaluation of the Motor Moment at the Crankshaft of the Conventional Sucker Rod Pumping Units, *Revista de Chimie* **70**(2) 378-381

[6] Badoiu D and Toma G 2017 Research Concerning the Identification of Some Parameters of a Sucker Rod Pumping Unit, *Revista de Chimie* **68**(10) 2289-2292

[7] Monagan M B, Geddes K O, Heal K M, Labahn G, Vorkoetter S M, Mccarron J and Demarco P 2005 *Maple Introductory Programming Guide*, Maplesoft, a division of Waterloo Maple Inc.

[8] Bernardin L *et al.* 2014 *Maple Programming Guide*, Maplesoft, a division of Waterloo Maple Inc.

[9] Badoiu D 2001 *Structural and Cinematic Analysis of the Mechanisms* (in Romanian), Technical Publishing House, Bucharest

[10] Badoiu D 2003 *Dynamic Analysis of Mechanisms and Machines* (in Romanian), Didactical and Pedagogical Publishing House, Bucharest

IOP Conf. Series: Materials Science and Engineering **659** (2019) 012002 doi:10.1088/1757-899X/659/1/012002

Simulation of ejector for vacuum generation

L Macia[1], R Castilla[1] and P J Gámez[1]

[1]Universitat Politecnica de Catalunya, Faculty of Fluid Mechanics, Carrer Colom 11 08222 Terrassa, Catalunya, Spain

E-mail: llorenc.macia@upc.edu, robert.castilla@upc.edu, pedro.javier.gamez@upc.edu

Abstract. Supersonic ejectors are used in a wide range of applications such as compression of refrigerants in cooling systems, pumping of volatile fluids, or vacuum generation. The objective of the present paper is to mesh and simulate, in an OpenFOAM environment with an open access implicit density- based solver HiSA, the physics of the vacuum ejector, and, later, compare the results with experimental measurements. In order to achieve this a 2D axisymmetric mesh made by hexahedral cells has been created. Steady solutions have been obtained, with prescribed total pressure in primary and secondary inlets. Secondary total pressure ranges from 1 to around 0.2 bar in which the secondary flow is zero. Numerical results are compared with experimental measurement, with two flowmeter sizes for small flow rate accuracy. Two regimes are encountered. In supercritic regime the secondary is chocked and sonic flow is reached in the second nozzle. In subcritic regime, the secondary flow is subsonic. The agreement is good, although simulation tends to slightly overestimate flow rate for large values region.

1. Introduction

Supersonic ejectors are used in a wide range of applications such as compression of refrigerants in cooling systems, pumping of volatile fluids, or vacuum generation. In this latter case, also known as zero-secondary flow, physics in the vacuum ejector is more complicated than in steady cases, since recirculation bubbles existing in the diffuser exhibit transient behaviour during the start-up period [1].

In the supersonic ejectors the main concern is to achieve the maximum vacuum level and to increase the flow rate of suction. The induced flow, or secondary flow, is the air that is being carried across the second nozzle.

In Figure 1 a scheme of the primary and secondary flow, as well as the mixing chamber where both flows join, is presented. Figure 2 shows the model of the ejector simulated.

The purpose of the present paper is to numerically simulate an ejector and compare results with experimental measurements.

IOP Conf. Series: Materials Science and Engineering **659** (2019) 012002 doi:10.1088/1757-899X/659/1/012002

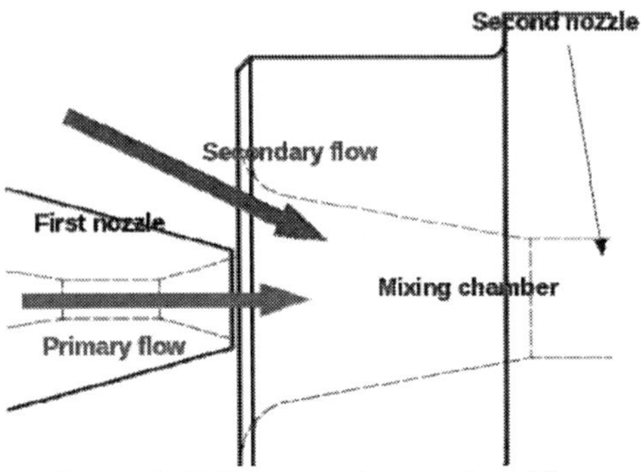

Figure 1. Primary and secondary Flow.

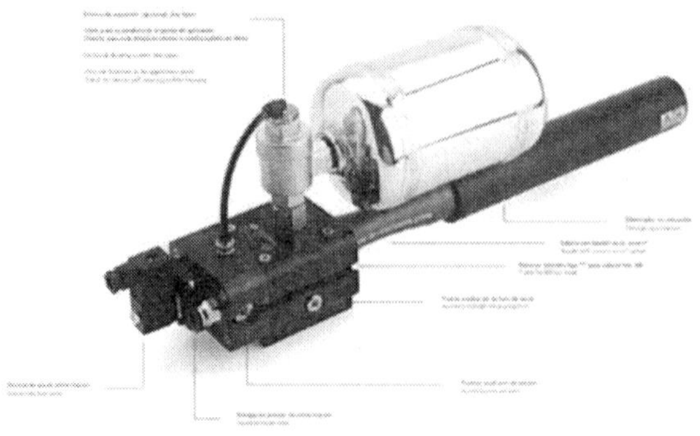
Figure 2. Ejector simulated.

1.1. State of art
According to the Del Valle [2] ejectors are used for vacuum generation since 1901. In this thesis the effect of sub-critical and critical function modes of an ejector is discussed.

The present work is based on the work of Jafarian [2] which is the first to study the transient phenomena in vacuum ejectors. According to Mazzelli [3], the best turbulence model for this kind of numerical simulation is the K-w SST, and, thus, it has been used in the present paper.

It has been used in this work the HiSA [4] implicit solver but some years before Xiang Gao et al. [5] developed a parallel density-based implicit solver with OpenFOAM, although it has not been used.

Arun Kumar et al. [6] wanted to show the physics of vacuum generation in ejectors when they are in zero-secondary flow such as bubbles on the secondary flow that obstruct the entrainment of air.

2. Experimental set up
In the Figure 3, the experimental setup is schematically presented.

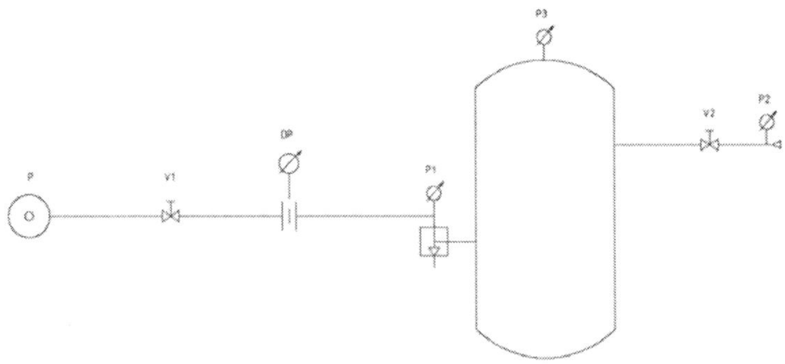

Figure 3. Schematic of experimental lab.

The primary pressure, denoted by P in Figure 3, has been kept steady and at a value of 6 bar for all the measurements. This pressure feeds the ejector, which sucks out the air from the vessel. The vacuum level achieved in the vessel is measured in the manometer P3.

The ball valve, in the scheme V2, controls the flow rate in and out the vessel when the system reaches a steady state. The flow rate is measured with a nozzle flow meter, shown in the picture as P2.

The performance curve obtained gives the values of the induced flow rate for each operating condition. The secondary flow is maximum for atmospheric pressure in the vessel and it decreases with low values of this pressure. Figure 4 shows a theoretical characteristic curve with a decreasing flow rate while increase the vacuum level.

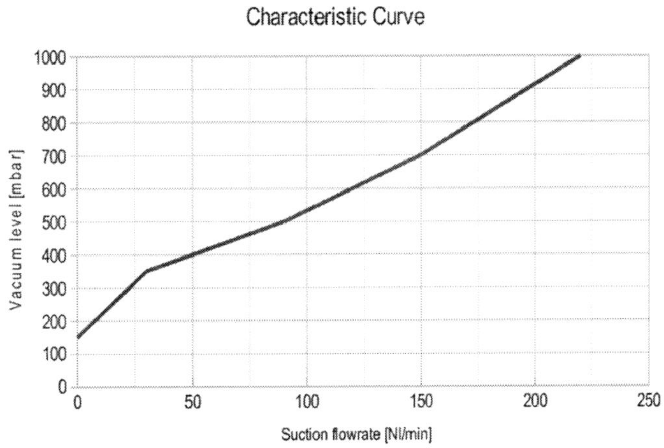

Figure 4. Standard curve of performance.

3. Numerical results

For a supersonic compressible flow, density based models are the favourite because of its capacity of capture discontinuities created by the shock waves. The equations for unsteady compressible flow are, for mass, momentum and energy,

$$\frac{\partial \rho}{\partial t} + \frac{\partial}{\partial x_i}\rho u_i = 0 \tag{1}$$

$$\frac{\partial}{\partial t}\rho u_i + \frac{\partial}{\partial x_j}\rho u_i u_j = -\frac{\partial p}{\partial x_i} + \mu\left(\frac{\partial u_i}{\partial x_j} + \frac{\partial u_j}{\partial x_i} - \frac{2}{3}\frac{\partial u_k}{\partial x_k}\delta_ij\right) \tag{2}$$

$$\frac{\partial}{\partial t}\rho E + \frac{\partial}{\partial x_j}\rho E u_j = -\frac{\partial u_i p}{\partial x_i} + \mu u_i\left(\frac{\partial u_i}{\partial x_j} + \frac{\partial u_j}{\partial x_i} - \frac{2}{3}\frac{\partial u_k}{\partial x_k}\delta_{ij}\right) + k\frac{\partial T}{\partial x_i} \tag{3}$$

with the perfect gases equation,

$$p = \rho R' T \tag{4}$$

have been solved on the OpenFOAM simulation toolbox. In the present paper the open access implicit density-based solver HiSA [4], that implements the AUSM+up upwind scheme for face fluxes, has been used. This solver allows to solve unsteady flows with larger Courant numbers than explicits solvers. According to Mazzelli [3], it has been proved that the best turbulence model are the k-w SST, and thus, it has been used in the present simulations.

The 2d axisymmetric hexahedra-dominated mesh has been generated with blockMesh, in order to get a suitable good quality. A python code that, eventually, creates the required dictionary, blockMeshDict, has been also used. This mesh is composed of 20300 cells. Figure 5 and Figure 6 show the total view of the mesh and the detail of the mixing chamber.

The boundary condition in primary inlet is 6 bar (relative) of total pressure, and Neumann condition for velocity. In the secondary inlet also the pressure is prescribed, and the flow rate is given by the simulation. In outlet the standard atmospheric pressure is set but, in order to avoid reflections, the waveTransmissive boundary condition is used.

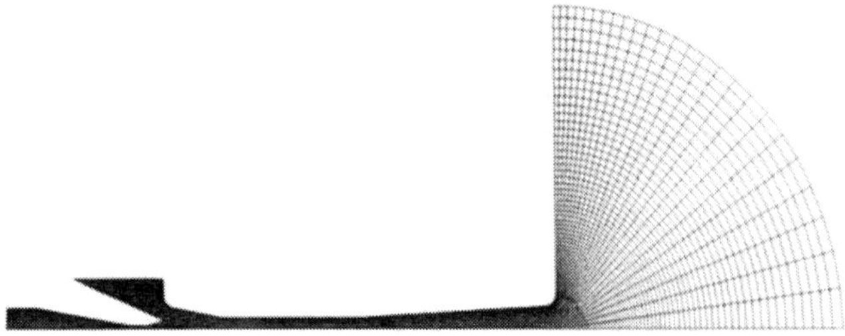

Figure 5. The mesh created by blockMesh.

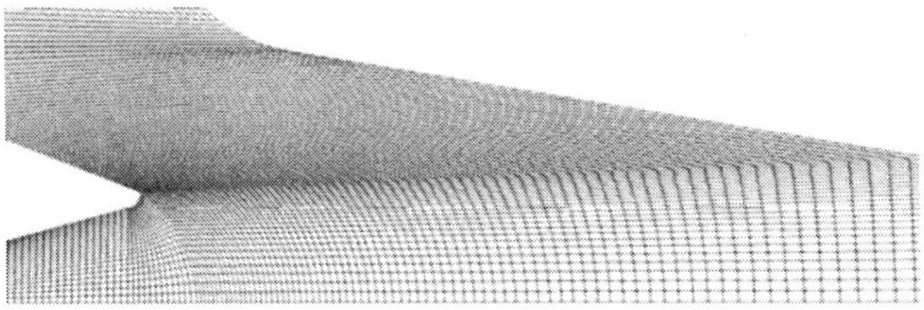

Figure 6. Detail of the critical part of the mesh.

A series of simulations, where the pressure at the secondary inlet has been changed, have been performed. The secondary pressure has ranged from atmospheric pressure (1 bar) to 0.2 bar. For each of this pressures the flow rate has been calculated.

4. Discussion

4.1. Experimental and numerical data

Figure 7 shows the numerical and the experimental results for the same geometry and operating conditions. A good agreement is found, although it seems that the numerical simulations overestimate the flow rate, about 10%, at the same vacuum pressure for small vacuum level of Ps = 0.28 up to Ps = 1 (large values of pressure in vacuum level) while underestimate the flow rate, about less than 5%, in the low values of vacuum pressure, at Ps = 0.21 up to Ps = 0.28.

It also tends to slightly overestimate the value of maximum pressure level: zero flow performance at Ps = 0.2 in the experimental results whereas Ps = 0.217 in the numerical ones.

4.2. Subcritical and critical modes

According to the thesis of Del Valle [2], it should be possible to observe a two regimens of secondary flow rate. So, at the beginning this flow is shocked at the very exit of the second nozzle. When at the vessel is at some vacuum level, the flow rate becomes sub-critical, and, finally, gets stuck so it becomes a zero-secondary flow ejector. In the fig, 8 is possible to see how the secondary flow is retained before the mixing chamber, due to the flow expansion of the primary flow.

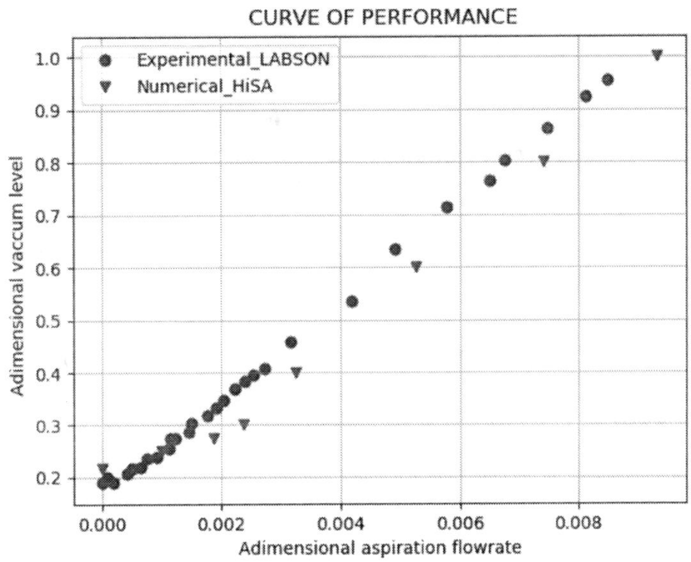

Figure 7. Curve of performance Experimental vs Numerical.

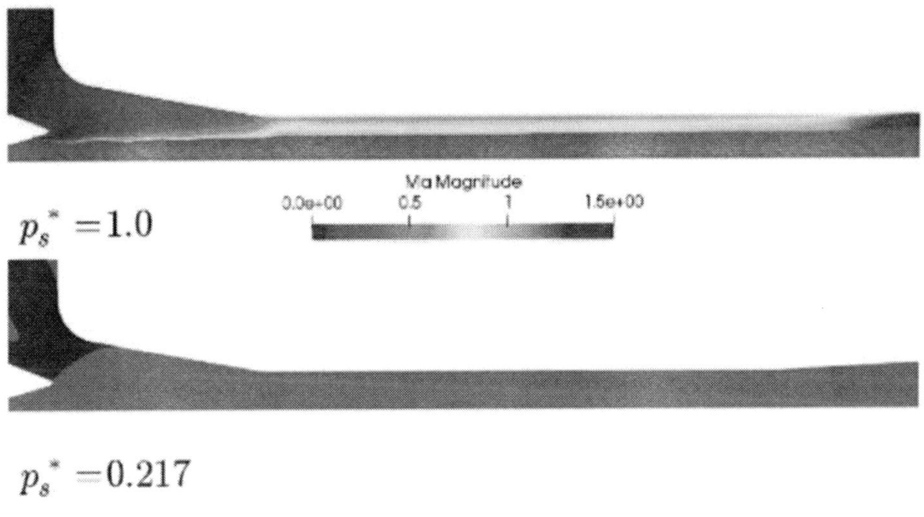

Figure 8. At 1.0 bar in the secondary field works in critical mode to work as zero-secondary flow at 0.217 bar.

5. Conclusion

Simulations of a supersonic ejector for vacuum generation have been presented. Simulations have been performed with the open source implicit density-based HiSA in the frame of the OpenFoam toolbox. Numerical results have been compared with experimental measurements, with an acceptable agreement, although CFD tend to slightly overestimate flow rate for the high values range.

It has been proved that the HiSA is a suitable implicit compressible gas based on density solver, and that it has been able to obtain results in less time expected than the explicit solvers.

The ejector simulated goes from a critical mode to a zero-secondary flow at 0,217 bar, which is approximately 80% of vacuum level.

Acknowledgments
Support of Industrial Doctorate (2018 DI 025) from Generalitat de Catalunya is acknowledged.
Many thanks to laboratory technicians Jaume Bonastre, father and son, for their help. Special thanks to Oliver Oxtoby, from CSIR (South Africa) for his invaluable comments.

References

[1] Jafarian A, Azizi M, and Forghani P 2016 Experimental and numerical investigation of transient phenomena in vacuum ejectors, *Energy* **102** 528–536
[2] García V 2014 *Eyectores para aplicaciones frigoríficas*, PhD thesis, Valladolid
[3] Mazzelli F, Little A B, Garimella S, and Bartosiewicz Y 2015 Computational and experimental analysis of supersonic air ejector: Turbulence modeling and assessment of 3D effects *Int. J. Heat Fluid Flow*, **56** 305–316
[4] Snedde G, Heyns J and Oxtoby O 2018 *HISA,* https://hisa.gitlab.io (accessed on Jan, 2019)
[5] Gao X 2018 Developing a parallel density-based implicit solver with mesh deformation in OpenFOAM, *J. Comput. Sci.* **28** 59–69
[6] Kumar R A and Rajesh G 2018 Physics of vacuum generation in zero-secondary flow ejectors, *Phys. Fluids* **30** 100-121

IOP Conf. Series: Materials Science and Engineering 659 (2019) 012003 doi:10.1088/1757-899X/659/1/012003

Reliability of cylindrical tank exposed to fire

M Đelošević[1], S Sremac[1] and G Tepić[1]

[1]University of Novi Sad, Faculty of Technical Sciences, Trg Dositeja Obradovića 6, 21101 Novi Sad, Serbia

E-mail: djelosevic.m@uns.ac.rs

Abstract. The paper presents a methodology for assessment the reliability of tank due to typical accidents. The aim of the research is to identify fracture lines and shapes of fragments generated by tank explosions. Accident scenarios and their probabilities are defined with Event Tree Analysis (ETA). Static structural analysis of the tank is realized by the software package ANSYS 15. The probabilistic mass method (PMM) was used to assessment the shape of the fragments. The assessment of the reliability of the tank affected by the fire was carried out according to Fault Tree Analysis (FTA). Verification of the results obtained was made according to available accidents. It has been found that the construction of the tank and especially the type of end caps affects the fragmentation pattern, i.e. the shape of fragments created by explosion. The results of the research are practically usable during the design of the tank because they provide information that is not contained in EN 13445-3.

1. Introduction

Explosion of process equipment is one of the most common causes of chemical accidents [1]. Explosion of process equipment is an inevitable event in the accident chain - a domino effect [2]. Every rapid pressure build up in a confined space (thus a tank) causes an explosion [3]. Releasing of flammable substances is the most common cause of fire in chemical processing facilities, while fires in the vicinity of process equipment cause the appearance of hot BLEVE (Boiling Liquid Expanding Vapour Explosion) [4]. The primary accidents scenarios that accompany the explosion of process equipment include fire BLEVE, unfire BLEVE, mechanical explosion (ME), confined explosion (CE) and runaway reaction (RR) [5]. The root causes of rapid pressure increase within the process equipment can be due to evaporation of the liquid (BLEVE), overpressure (ME), combustion of gases, vapours and dust (CE) and uncontrolled chemical reactions (RR). The thermal influence results in a reduction in the critical pressure that causes the explosion of process equipment [6]. Explosion of pressurized vessels is accompanied by the effects of fragmentation, blast wave and thermal radiation. The fragmentation effect of the tank is characterized by every explosion, regardless of the type of dangerous substance [7]. The fragmentation effect is much more pronounced than the blast effect and thermal radiation (fireballs) and can manifest at distances over 1.2 *km* [8]. A fragmentation risk analysis can be used to assess the reliability of process equipment. In this paper, the assessment of the probability of fault of the tank is carried out through fragmentation probabilities using PMM [9]. The starting hypothesis of this study is based on the attitude that a greater number of generated fragments is an indicator of greater reliability of the pressure vessel. The research in this paper should show that the prevention of the hot BLEVE effect directly affects the higher reliability of the process equipment.

2. Case study: Horizontal cylindrical tank

The tank reliability assessment procedure is presented through a case study of the cylindrical storage tank (Figure 1) for LPG (liquid petroleum gas). The study is based on the following assumptions:

1) The probability of fault (unreliability) of the tank is conditioned by the effect of fragmentation,
2) The influence of secondary fragments is negligible compared to primary fragments,
3) The tank is designed according to EN 13445-3 and
4) The tank is directly exposed to fire.

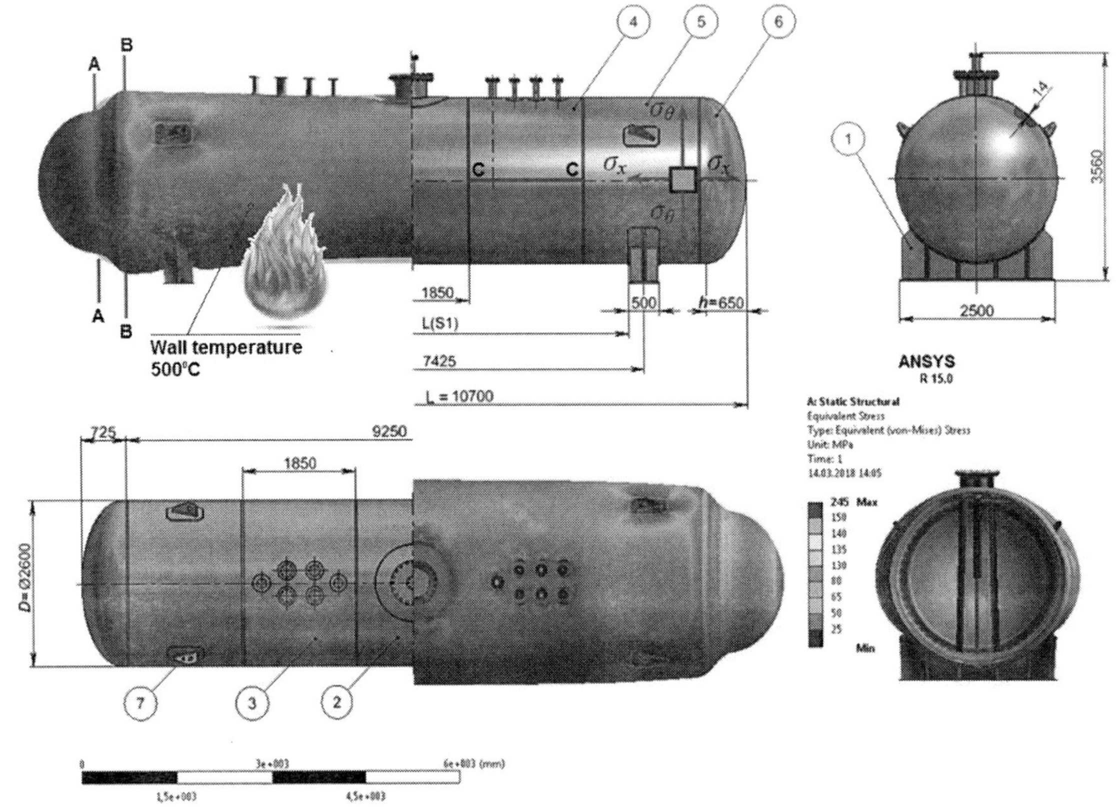

Figure 1. Dimensions and constructional type of tank with static structural analysis.

2.1. Critical zones of the tank

Cylindrical tanks are characterized by three critical cross sections: A-A, B-B and C-C. A critical cross-section depends to a large extent on the type of the end caps. Typical forms of end caps for horizontal cylindrical tank are spherical, ellipsoidal and torispherical. Spherical tanks contain a critical zone in the cross-section C-C, while the ellipsoidal end cap is characterized by a critical cross-section B-B. Critical zone A-A occurs only in tanks with torispherical end caps. In the given case study, a tank with ellipsoidal end caps according to DIN 28013 is considered, whose stress state can be represented by:

$$\sigma_x = \left[1 + \frac{3}{2}\frac{1}{\sqrt{3(1-\nu^2)}}\left(\frac{D}{2h}\right)^2 e^{-\lambda x}\sin(\lambda x)\right]\cdot\left.\left(\frac{Dp}{4\delta}\right)\right|_{x=\frac{\pi}{4\lambda}} = 46.5p \tag{1}$$

$$\sigma_\theta = \left[1 + \frac{1}{4}\left(\frac{3\nu\sin(\lambda x)}{\sqrt{3(1-\nu^2)}} - \cos(\lambda x)\right)\left(\frac{D}{2h}\right)^2 e^{-\lambda x}\right]\cdot\left.\left(\frac{Dp}{2\delta}\right)\right|_{x=\frac{\pi}{4\lambda}} = 92.5p \tag{2}$$

$$\sigma_{\max} = \sqrt{\sigma_x^2 + \sigma_\theta^2 - \sigma_x\sigma_\theta + \frac{3}{2}(\sigma_x - \sigma_\theta)^2} = 98p \tag{3}$$

Where is the tank diameter (D = 2600 mm), the height of the end cap (h = 650 mm), wall thickness of the tank (δ = 14 mm), the Poisson's ratio (ν = 0.3), the coefficient $\lambda = [12(1-\nu^2)/(D\delta)^2]^{-4} = 9.53\ m^{-1}$ and the tank pressure (p). The maximum stress of the tank with an ellipsoidal end cap is in cross-section B-B. The position of this cross-section is defined with $x = \pi/4\lambda$ measured from the beginning of the cylinder. For the given parameters is $x \approx 82\ mm$, which means that the critical cross-section is close to the zone of the joint of the cylinder and the end caps. The maximum operating pressure of the tank is determined on the basis of the allowable stress (f) for the tank material (S235J2G3) which according to EN 13445-3 amounts $f = R_m/2.4$ = 195.8 MPa (the tensile strength of the material is R_m = 470 MPa). The operating pressure inside the storage tank of the LPG is from 16.4 to 16.9 bar (average 16.7 bar). Equivalent (von Mises) stress according to (3) for the average pressure is $\sigma_{theory} = 98p$ = 163.7 MPa. This stress according to EN 13445-3 is $\sigma_{EN13445\text{-}3} = pD_m/(2ze_a)$ = 154.2 MPa, where the mean diameter of the tank $D_m = (D\text{-}e_a)$ = 2586 mm, wall thickness $e_a = \delta$ = 14 mm and coefficient of welded seam z = 1 (full control of welded seams). Equivalent stress in accordance with the FEM model (ANSYS 15) is 150.2 MPa. Comparative analysis shows that the mathematical model and EN 13445-3 give several percent higher eqivalent stress than the FEM model. Theoretical analysis and EN 13445-3 consider the cylinder of the tank without the influence of other structural elements (supports, lifting lugs, etc.).

2.2. Mechanism of explosion of the tank

The explosion is caused by a rapid increase in pressure inside the tank and always manifested by the fragmentation effect. Explosion of the tank is a result of extremely dynamic pressure changes. The intensity of the fragmentation effect is conditioned by the rate of pressure change. Fracture of the tank under the influence of quasi-static pressure is not accompanied by a fragmentation effect or poorly expressed. Explosion of the tank follows the detonation process whose velocity of the spherical longitudinal wave is up to several km/s. The velocity of the wave propagation (c) through the wall of the steel tank is $c = (E/\rho)^{0,5}$ = 163,6 m/s. This velocity is most common in the whole order of magnitude (10 times) less than the detonation velocity. The detonation velocity depends on the type of combustible substance within the pressure vessel and for the LPG is about v = 1800 m/s. The time required to propagate the wave through the wall of the tank is $t = \delta/c = 8{,}55\cdot10^{-5}\ s$. The impact wave due to detonation for the same time crosses the path $s = vt$ = 154 mm. This means that during the explosion of the tank in which the LPG is located at time t, radial deformation will be achieved $\varepsilon_\theta = \Delta D/D$ = 11,8%. This situation is not even theoretically possible, since the fracture of the tank occurs at a much less radial deformation $\varepsilon_{\theta,failure} = [(2-\nu)/392]\cdot(R_m/E)\cdot(D/\delta)$ = 0.18%. Numerical illustration has the role to explain the mechanism of the explosion of the tank. The tank fracture due to the quasistatic pressure always runs along the high stress zone and starts at the stress concentration points. The quasi-static pressure change allows the stress lines to find a critical zone of the tank from which the cracks spread. Due to the dynamic change in pressure, the stress lines do not have enough time to localize in the external critical zone of the tank. In this case, the crack propagation starts from internal sources of stress concentration. The expansion of the pressurized vessels takes place at a velocity of v, which affects the interruption of the stress lines transmitted by the velocity of c. Then the ductile materials of the tank behave like brittle and generate a larger number of fragments. Therefore, a higher detonation velocity causes a greater number of generated fragments. The previously described explosion mechanism is illustrated in Figure 2.

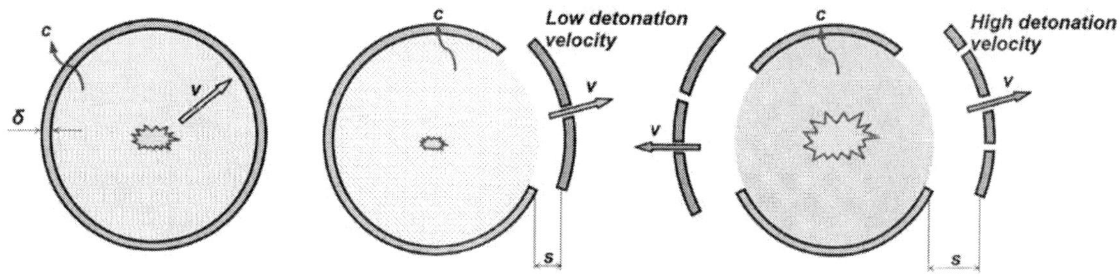

Figure 2. Mechanism of explosion of the tank depending on the detonation velocity.

3. Fracture lines and fragmentation patterns

The assessment of fracture lines and fragmentation patterns was performed on the basis of structural analysis and Monte Carlo simulation. Critical zones of the tank are defined by structural analysis, while the Monte Carlo simulation estimates the most probable masses of fragments (Figure 3). The fragmentation analysis of the tank is limited to two fragments, which is a typical in the BLEVE effect.

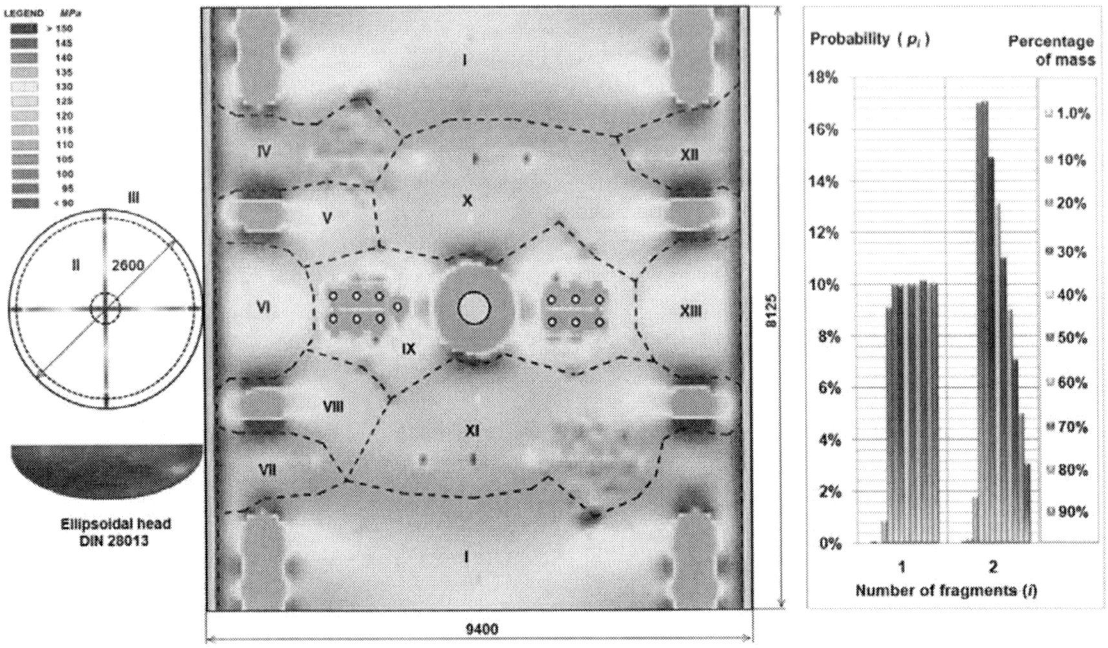

Figure 3. Potential fracture lines and mass distribution for the first two fragments.

The most likely fragmentation patterns due to the BLEVE effect are shown in Table 1. The masses of the fragments are expressed through a percentage share in the tank mass, according to the histogram of Figure 3. The mass of the tank is 12300 *kg*. The maximum probability of generating two fragments (≈17%) occurs if the aggregate mass of the fragments is between 1300 *kg* and 2225 *kg*.

Table 1. Characteristic forms of fragments due to tank explosion.

ONE FRAGMENT	TWO FRAGMENTS	
Fragment mass: 1350 *kg* Fracture zone: II+III+VI Generation probability: 9.10%	Fragment mass: 820 *kg* and 480 *kg* Fracture zone: II+III and IV+V Generation probability: 17.04%	Fragment mass: 1590 *kg* and 2050 *kg* Fracture zone: II...VII and VIII+IX Generation probability: 14.85%
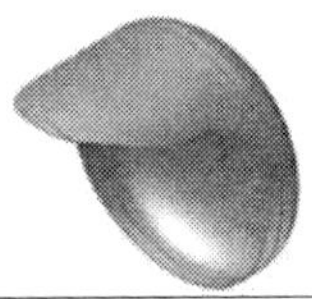	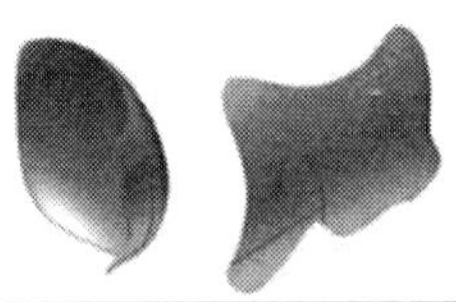	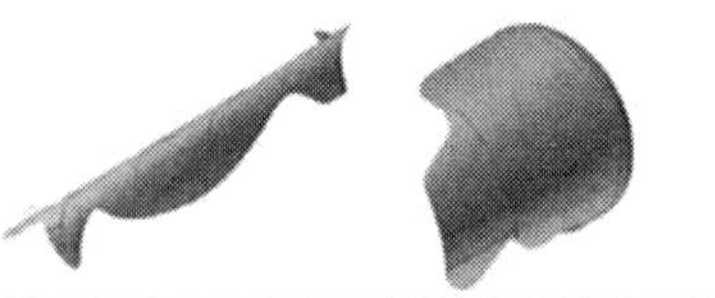
Fragment mass: 2050 *kg* Fracture zone: II+III+IV+V+VI Generation probability: 10.00%	Fragment mass: 205 *kg* and 2050 *kg* Fracture zone: VII+VIII and II ...VI Generation probability: 17.06%	Fragment mass: 1365 *kg* and 3545 *kg* Fracture zone: XIII...XV and II...IX Generation probability: 13.00%
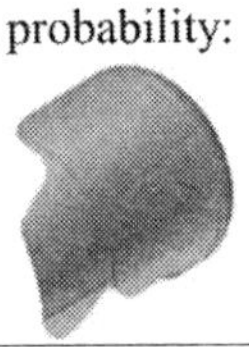	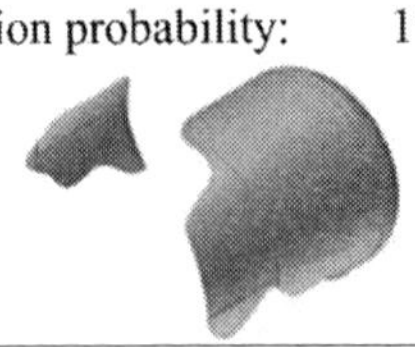	

4. Accident scenarios due to tank explosion

Explosion of the tank caused by critical pressure includes more accidents scenario, which are caused by the fracture of the segments of the tank. The division of the tank into segments was performed for easier assessment of fragmentation probabilities [9-10]. The first segment of the tank (S1) belongs to the cylinder in the length L (S1) according to Figure 1. The remaining part of the tank belongs to the segments S2 and S3. Propagation of fracture lines can be carried out individually or combined along segments S1, S2 and S3. The Event Tree Analysis (ETA) provides a total of 6 potential scenarios for the explosion of the tank (Figure 4).

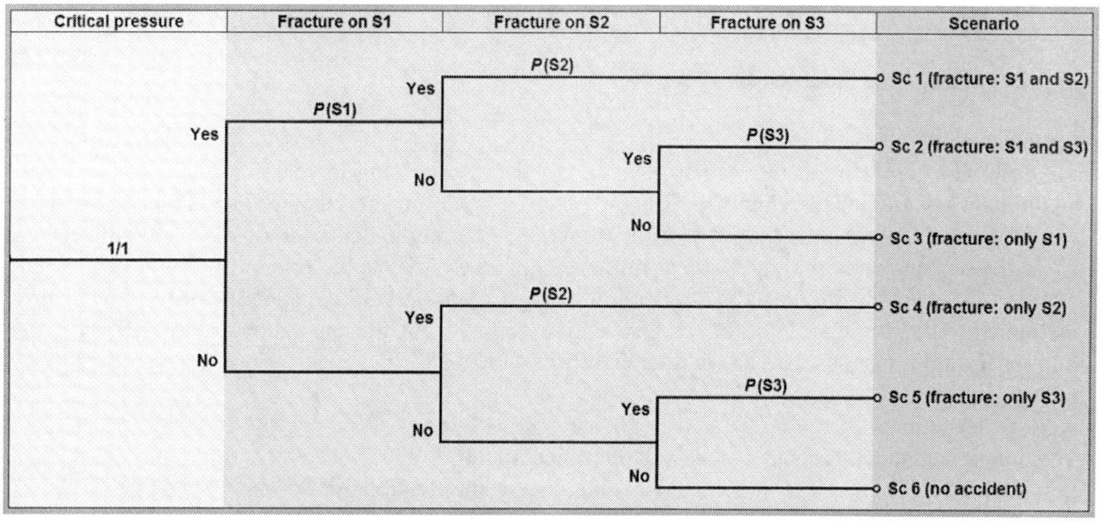

Figure 4. The definition of accident scenarios using Event Tree Analysis – ETA.

Accident scenarios Sc1 and Sc2 include a fracture of the tank in at least two segments S1 and S2/S3. Accident scenarios Sc3, Sc4 and Sc5 include a fracture only in one of the segments S1, S2 and S3. The last outcome (Sc6) is not an accident, as there is no fracture of any segment of the tank. The fragmentation probabilities of the segments were obtained by the implementation of PMM [9]. The limit fragmentation probabilities by segments and generated fragments are given in Table 2. The maximum values for S1 correspond to the minimum values for S2 and S3.

Table 2. Probability of fragmentation in the segments and number of fragments.

Segment	Probability %	Number of fragments 1	2	3	4	5	6	7	≥ 8
S1	*max*	35.500	25.000	9.375	2.500	0.500	0.094	0.013	0.019
	min	27.500	18.333	6.875	1.833	0.367	0.069	0.009	0.014
S2 or S3	*max*	11.250	11.791	5.165	1.467	0.299	0.056	0.007	0.011
	min	6.250	6.120	2.810	0.811	0.166	0.031	0.004	0.006

5. Assessment of tank reliability

Assessment of tank reliability includes qualitative and quantitative analysis using Fault Tree Analysis – FTA. The loss of the operating function of the tank is conditioned by the fragmentation effect and represents the top event. Realizing this event involves generating a number of fragments (1,2, ...). Basic events include accident scenarios (Sc1 ... Sc5) whose probability of realization can be determined on the basis of Table 2. The minimum fragmentation probability corresponds to the thermal influence due to the BLEVE effect (the temperature of the tank wall is about 500^0C). Maximum reliability values are achieved when the tank is not exposed to fire. The failure tree and the limit values of the tank reliability are given in Figure 5. The probability of tank failure is defined by $T = P(Sc1)+...+P(Sc5)$. Tanks of reduced reliability are characterized by a smaller number of fragments.

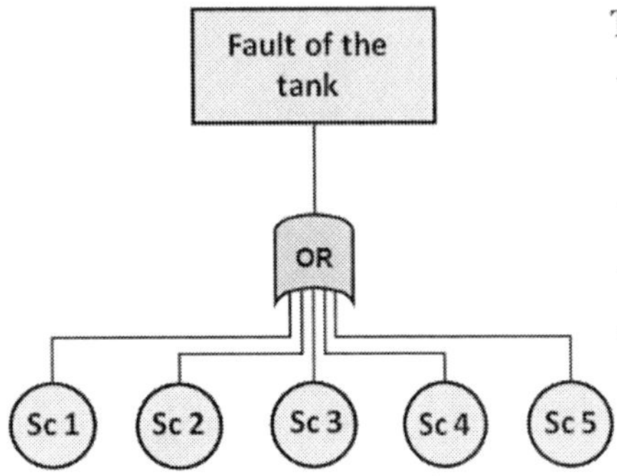

Figure 5. Fault Tree Analysis for tank.

Table 3. Reliability of the tank relative to number of fragments

Reliability [%]	Number of fragments 1	2	3	4	5	≥ 6
max	57.10	69.87	87.96	96.70	99.78	99.87
min	56.69	68.88	87.81	96.68	99.33	

6. Conclusion

The reliability assessment in this paper was carried out according to real exploitation conditions. A qualitative assessment of reliability (Table 3) can be made on the basis of available accidents for a known type of tank. A quantitative assessment requires the implementation of PMM [9] with ETA and FTA. It was concluded that the thermal influence adversely affects the reliability of the tank. A larger number of generated fragments is a reflection of higher tank reliability. The reliability analysis proves the starting assumption of the paper. Preventive action from the emergence of the BLEVE effect implies the simultaneous increase in the reliability of the tank.

Acknowledgments

This paper is made as a result of research on project of TR 36030, financially supported by Ministry of Education, Science and Technological Development of the Republic of Serbia.

References

[1] Hemmatian B, Abdolhamidzadeh B, Dabra R M and Casal J 2014 The significance of domino effect in chemical accidents, *Journal of Loss Prevention in the Process Industries* **29** 30-38

[2] Abdolhamidzadeh B, Abbasi T, Rashtchian D and Abbasi S A 2011 Domino effect in process-industry accidents – An inventory of past events and identification of some patterns, *Journal of Loss Prevention in the Process Industries* **24** 575-593

[3] Tugnoli A, Gubinelli G, Landucci G and Cozzani V 2014 Assessment of fragment projection hazard: Probability distributions for the initial direction of fragments, *Journal of Hazardous Materials* **279** 418-427 (https://doi.org/10.1016/j.jhazmat.2014.07.034)

[4] Eckhoff R K 2014 Boiling liquid expanding vapor explosions (BLEVEs): A brief review, *Journal of Loss Prevention in the Process Industries* **32** 30-43

[5] Gubinelli G and Cozzani V 2009 Assessment of missile hazards: Identification of reference fragmentation patterns. *Journal of Hazardous Materials* **163**(2-3) 1008-1018

[6] Plans E, Pastor E, Casal J and Bonilla J M 2015 Analysis of the boiling expanding vapor explosion (BLEVE) of a liquefied natural gas road tanker: The Zarzalico accident, *Journal of Loss Prevention in the Process Industries* **34** 127-138

[7] Mébarki A, Mercier F, Nguyen Q B and Saada R A 2009 Structural fragments and explosions in industrial facilities. Part I: Probabilistic description of the source terms, *Journal of Loss Prevention in the Process Industries* **22** 408-416 (https://doi.org/10.1016/j.jlp.2009.02.006)

[8] Mébarki A, Nguyen Q B and Mercier F 2009 Structural fragments and explosions in industrial facilities. Part II: Projectile trajectory and probability of impact, *Journal of Loss Prevention in the Process Industries* **22** 417-425 (https://doi.org/10.1016/j.jlp.2009.02.005)

[9] Djelosevic M and Tepic G 2019 Identification of fragmentation mechanism and risk analysis due to explosion of cylindrical tank, *Journal of Hazardous Materials* **362** 17-35

[10] Djelosevic M and Tepic G 2019 Probabilistic simulation model of fragmentation risk, *Journal of Loss Prevention in the Process Industries* **60** 53-75

Effect of friction on nominal stress results in a single tooth bending fatigue test

M Vukić[1], I Čular[1], R Mašović[1] and K Vučković[1]

[1]University of Zagreb, Faculty of Mechanical Engineering and Naval Architecture, Ivana Lučića 5, 10000 Zagreb, Croatia

E-mail: ivan.cular@fsb.hr

Abstract. One of the main reasons for gear failure is bending fatigue, which occurs due to cyclic stresses at the tooth root region. Bending fatigue can cause crack initiation which, in turn, may propagate and result in catastrophic tooth breakage. Single tooth bending fatigue tests are frequently employed to generate statistically significant fatigue data at relatively low price. Hence, they are often employed as screening tests for bending fatigue behaviour of actual running gear pair. Since relatively small alteration of nominal tooth root stress values can produce significant change in bending fatigue lives, it is important to adequately consider the effect of friction and test fixture deformation when conducting single tooth bending fatigue tests. In this paper, numerical investigation of tooth flank friction effect on spur gear load capacity during single tooth bending fatigue test is carried out. Two different test fixtures are considered. Linear elastic finite element analysis is carried out to obtain nominal tooth root stresses. The results show that friction at gear tooth flank as well as fixture deformation can affect nominal tooth root stress results.

1. Introduction

Single-tooth bending fatigue (STBF) test is often employed for bending fatigue testing of a gear tooth by applying a load normal to the flank at a fixed point. Main goal of this test, which is used as a screening test for operating gears, is to generate a statistically relevant quantity of bending fatigue data. The test itself is usually done via electrohydraulic universal test machine at a relatively low price. The overall cost of an experiment can further be reduced by conducting multiple tests on a single gear specimen.

Various types of STBF test fixture arrangements are currently being used in the field of mechanical gear testing and many studies have been conducted on them. Gasparini et al. [1] studied the effects of material, design and manufacturing parameters on bending fatigue life of helicopter gears. They employed a fixture design in which both test tooth and the reaction tooth are loaded at the same time with two linearly balanced forces, eliminating the need for the gear's support. Therefore, following the positioning of the gear, the support pin can be removed. Same fixture was used by Gorla et al. [2] for bending fatigue strength study of case carburized and nitride gear steels for aeronautical applications. In both investigations, the test teeth were loaded by a force applied at the highest point of a single-tooth contact (HPSTC). Handschuh et al. [3] conducted STBF tests on steel spur gears. The fixture arrangement used in this study consists of one actuating arm that loads the test tooth and one fixed anvil that supports the reaction tooth, while the specimen gear is supported in the fixture casing. Here, the load is also applied at the HPSTC. In the Boeing flexural design [4], which appears to have found favor in the aerospace sector, the specimen gear is rigidly supported on a shaft while the load is applied through a carbide block contacting the test tooth at the HPSTC. The loading block is held in the fixed orientation by a flexural loading arm and the reaction is carried through a block contacting the reaction tooth. This flexural design ensures accurate loading of the gear tooth with minimal migration of the loading point. Akata et al. [5] investigated performances of differently manufactured spur gears using a single-tooth bending principle in combination with three-point bending loading. They used a fixture design in which the specimen gear is placed on a lower test fixture and bending loads are applied via its upper part by means of needle elements placed between the tooth flank and the upper part. The distance between the loading centers of the needle elements is the diameter of HPSTC. In a similar setup, without using needle elements while the specimen gear sits on his own shaft rather than the lower test fixture, Bian et al. [6] conducted STBF tests on 38SiMnMo alloy steel. The last fixture design considered here, more commonly known as SAE (Society of Automotive Engineers) test fixture [7], primarily consists of a specimen gear, actuating arm and support anvil, all of which are mounted on a common shaft. The actuating arm is L-shaped, and it rotates about the shaft axis, contacting the test tooth at its tip.

According to Vučković et al. [8], friction can significantly affect the bending fatigue life of a gear. To investigate the effect of friction at the results of a single-tooth bending fatigue tests, two of the abovementioned fixture designs are considered. The first one is the three-point bending fixture according to Bian [6] and the second one is the SAE fixture design according to the official standard [7]. Furthermore, a computational model based on the elastic FEM analysis is proposed. The proposed model simulates contact between gear tooth flank and the fixture surface to obtain nominal tooth root stresses. Predicted behavior of the specimen gear tooth with regard to the fixture rigidity is compared to the numerical simulations given by the finite element method (FEM).

2. Materials and methods

Finite element analysis is carried out on AISI 8620H specimen gear, whose geometry is based on the SAE standard [7]. Geometrical and material parameters employed for FEM simulations of both fixtures are presented in Table 1.

Table 1. Geometric and material parameters of the specimen gear.

Parameter	Value
Number of teeth, z	34
Module, m	4,23 mm
Normal pressure angle of basic rack, α	20°
Pitch diameter, d	143,934 mm
Base diameter, d_b	135,252 mm
Root diameter, d_f	133,245 mm
Tip diameter, d_a	152,28 mm
Modul of Elasticity, E	210 GPa
Poisson coefficient, v	0,3
Density, ρ	7850 kg/m^3
Thickness of the gear, b	For three-point loading fixture: $b = 10$ mm
	For SAE fixture [7]: $b = 25.4$ mm

As shown in Table 1, thickness of the specimen gear is varied according to the type of simulated fixture. For the three- point loading fixture, the thickness is determined according to the estimated fixture/gear thickness ratio of 3:1 [6]. For both simulations, same load-time history based on sinusoidal load amplitude variation is applied (Figure 1). The load-time history is divided into five steps with equal time periods. The initial loading is increased from zero to the maximum load, followed by two cycles of unloading from maximum to minimum load (taken as 10% of maximum load) and then loading back to the maximum load. For three-point loading fixture maximum load is equal to F_{max} = 5 kN, while for the SAE fixture maximum load is F_{max} = 10 kN.

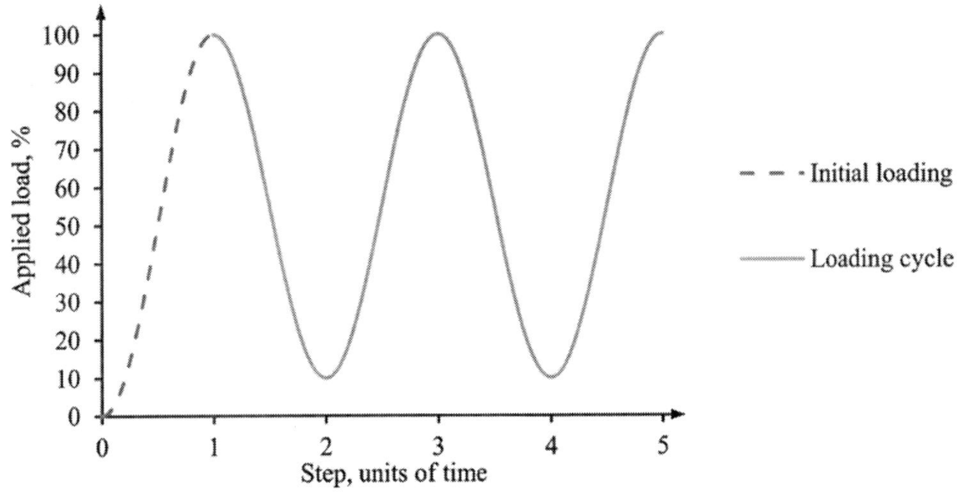

Figure 1. Applied load-time history.

2.1. Three-point loading model

The three-point loading model consists of two main parts: the loading anvil and the specimen gear (Figure 2a). Boundary conditions and the loading point are assigned to a numerical model according to Figure 2b. Geometry of the loading anvil is determined to ensure the loading force is vertically tangent to the base diameter of the gear.

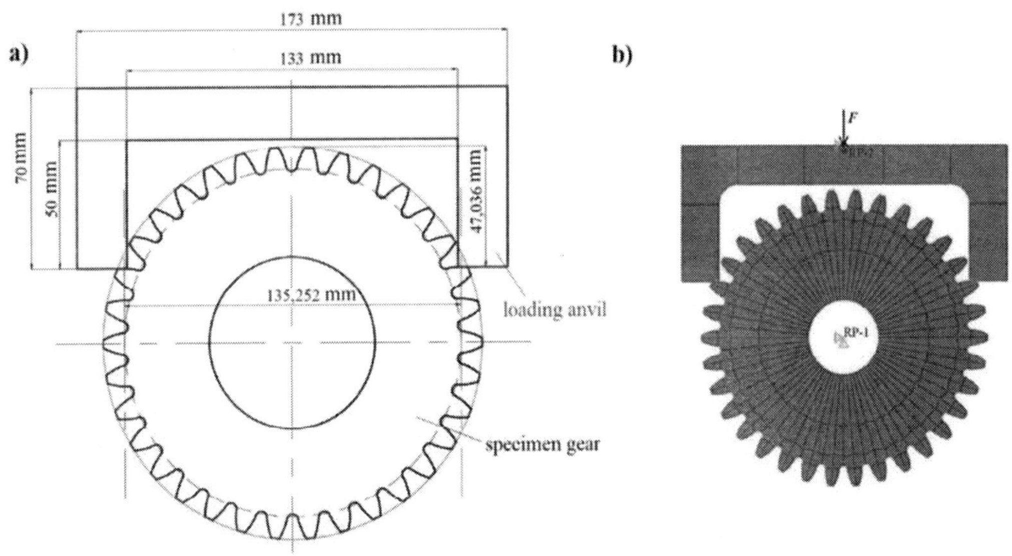

Figure 2. Three-point loading model: (a) geometric model, (b) numerical model with boundary conditions and loading point.

Load is applied at the center of the loading anvil. Two cases are considered. First case assumes elastic behavior of the loading anvil, while the second one assumes ideally rigid behavior. When the loading anvil behaves as an elastic body (Figure 3a) while in contact with the tooth flank, horizontal plane of the anvil should slide in the outwards motion, meaning friction force F_f should act in the opposite direction on the anvil. The friction force acting upon the tooth flank will then act in the outwards direction and consequently increase the nominal stresses at the tooth root fillet.

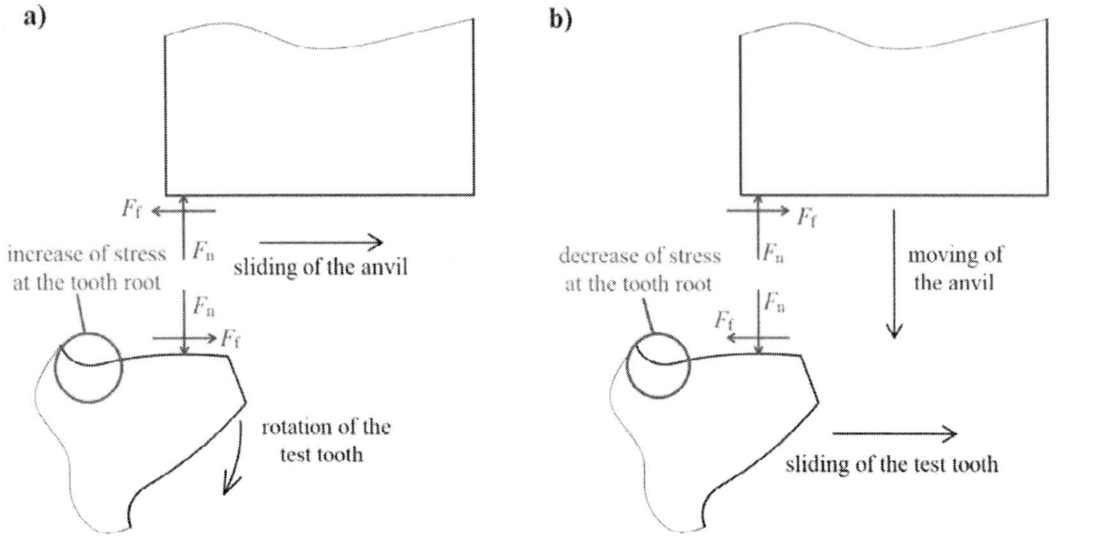

Figure 3. Predictions for the three-point loading model: (a) elastic behaviour, (b) ideally rigid behaviour of the loading anvil.

For the second case, depicted in the Figure 3b, the rigidity of the anvil should cause the sliding motion of the tooth flank in the outwards direction, meaning friction force on the tooth flank will act in the opposite direction. This should lead to an overall decrease of principal stress at the tooth root fillet.

2.2. SAE fixture model

Second load model, based on the SAE fixture design, consists of three main parts depicted in the Figure 4b. Specimen gear is loaded via an upper anvil; at the same time the gear is supported by a fixed lower anvil.

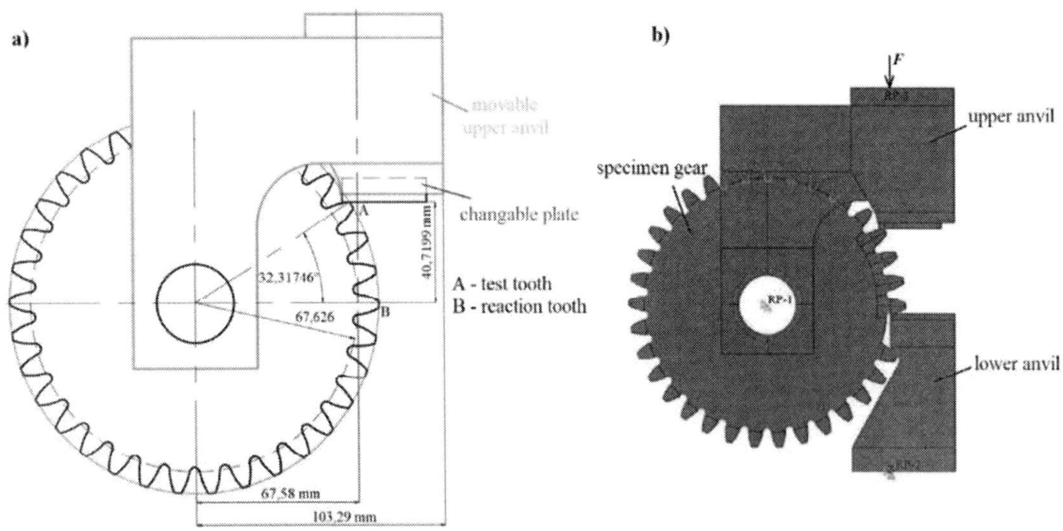

Figure 4. SAE fixture loading model: (a) geometric model, (b) numerical model with boundary conditions and loading point.

Load is applied at the top of the L-shaped upper anvil which rotates about the center S_1 (Figure 5) while the test tooth bends around its root. Since upper anvil rotates about greater radius than the tooth itself, it causes the friction force on the tooth flank to act against its direction of rotation, thereby decreasing the principal stress at the tooth root fillet. Directions of the corresponding forces at the contact point between the tooth flank and the changeable plate on the upper anvil are depicted in the Figure 5b.

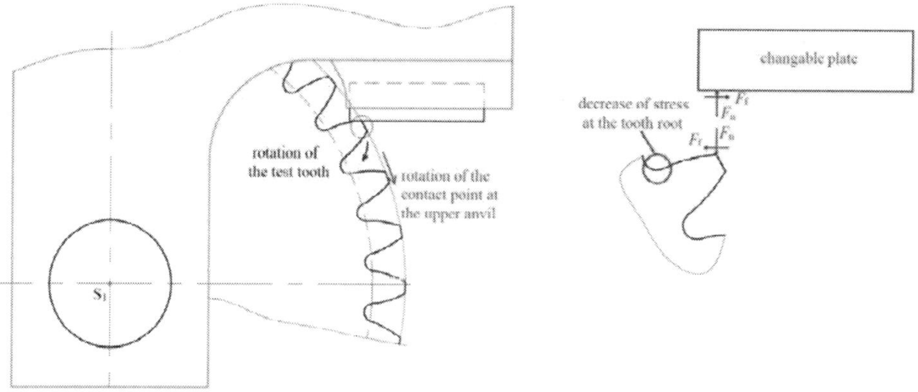

Figure 5. SAE loading model: behaviour prediction upon loading and forces at the contact point.

Numerical analysis for both presented models is conducted within the commercially available FEM software Abaqus 6.13 [9]. According to Pehan et al. [10], plane stress can be assumed if the gear thickness is less than six times it's module. Therefore, two-dimensional FE analysis is carried out. Parts within both models are discretized by 8-node biquadratic plane stress quadrilateral elements, designated as CPS8 in Abaqus-Standard. For both FE models, convergence test is carried out to determine the required number of finite elements. The mesh is refined, and the number of finite elements is increased in the regions of contacting flanks and corresponding root fillets to detect the slightest changes and increase the accuracy of the results.

3. Results and discussion

Numerical simulations are conducted for both proposed loading models by varying friction coefficients and loading.

3.1. Three-point loading fixture model results

The effect of friction on the stress-time history for the three-point loading model with elastic behavior of the loading anvil is shown in Figure 6. As predicted, the principal stress at the tooth root fillet increases with the friction coefficient, which causes higher friction force at the tooth flank.

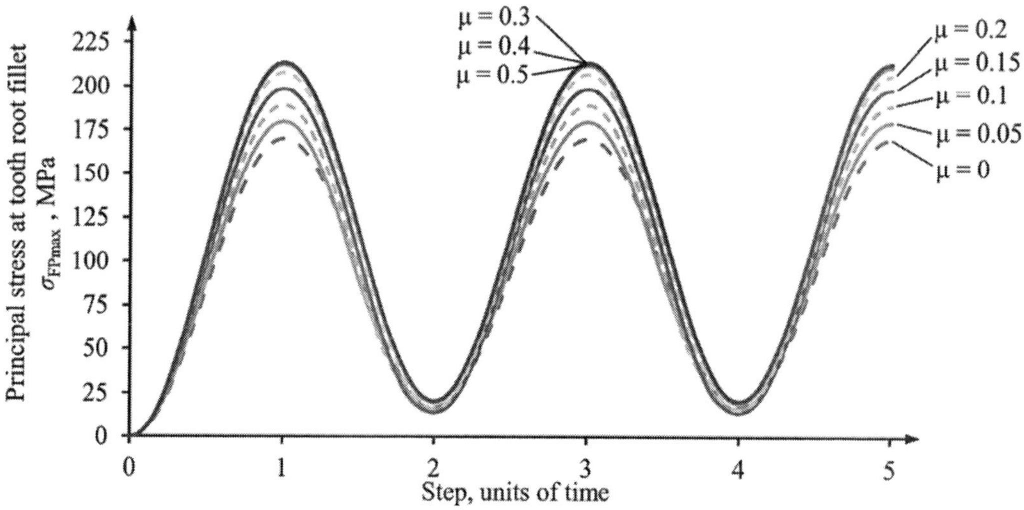

Figure 6. Principal stress-time history for the tooth root fillet and elastic behaviour of the loading anvil.

Contrary to the elastic behavior of the anvil, when ideal rigidity of the loading anvil is assumed, the principal stress at the tooth root fillet decreases with an increase of friction coefficient (Figure 7).

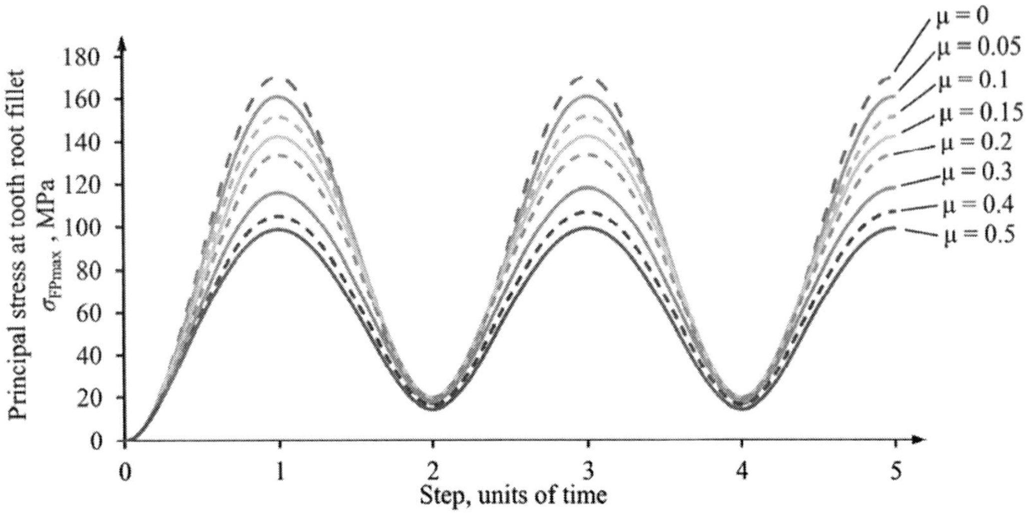

Figure 7. Principal stress-time history for the tooth root fillet and rigid behaviour of the loading anvil.

Since the elastic behavior of the loading anvil is more realistic, numerical simulation for various load values are conducted only for this case. As it can be observed from the Figure 8, with an increase in the friction coefficient, the maximum cyclic stress is increased (Figure 8a) and the minimum cyclic stress is decreased (Figure 8b), i.e. the stress range is enlarged. This will, in turn, reduce bending fatigue life of the gear.

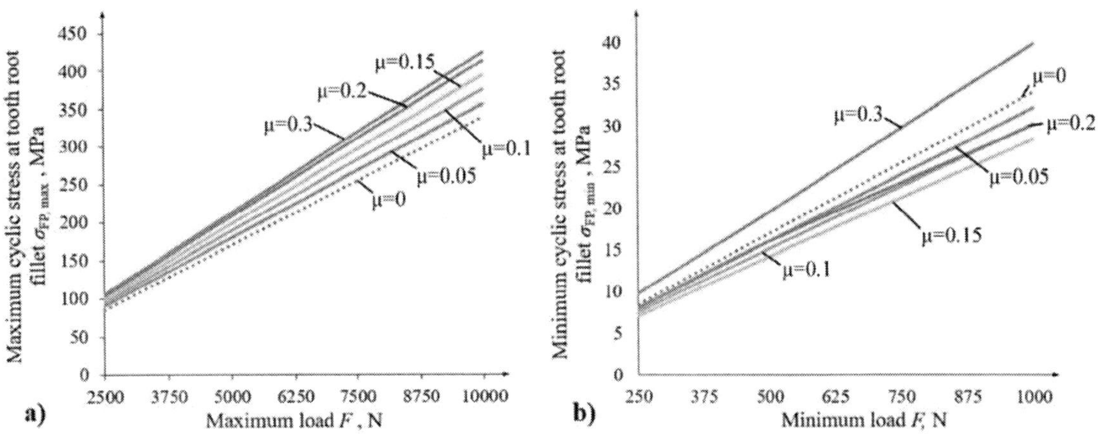

Figure 8. The effect of friction and different loads on cyclic stresses at tooth root fillet for elastic behaviour of the loading anvil: (a) maximum cyclic stress, (b) minimum cyclic stress.

3.2. SAE fixture model results

Stress-time history for the SAE fixture design model is provided in Figure 9. As shown in the diagram, the maximum principal stress at the tooth root fillet is decreased with an increase of the friction coefficient.

IOP Conf. Series: Materials Science and Engineering **659** (2019) 012004 doi:10.1088/1757-899X/659/1/012004

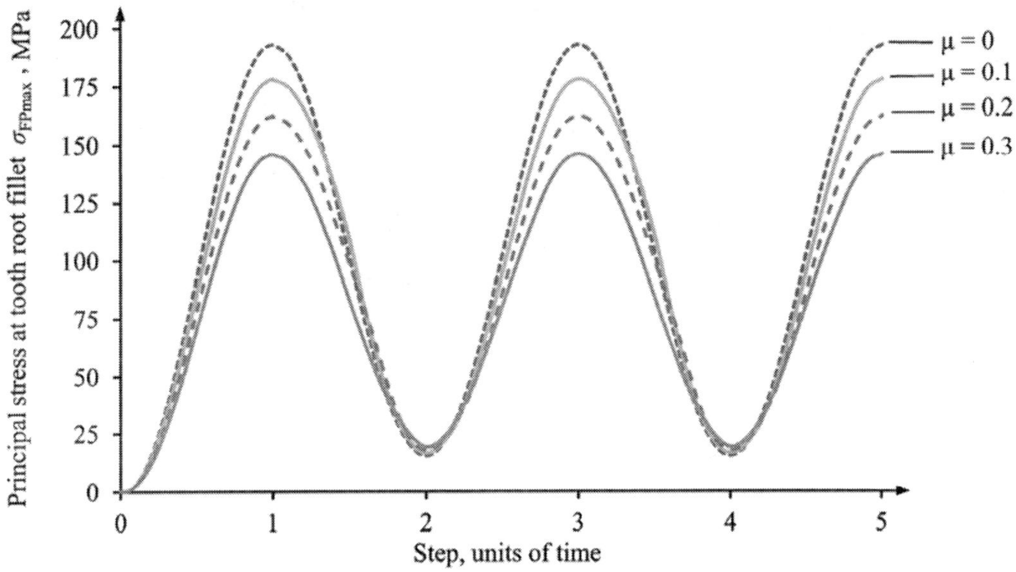

Figure 9. Principal stress-time history for the tooth root fillet of SAE fixture model.

Contrary the behavior for three-point loading model observed in Figure 8, with an increase in the friction coefficient, the maximum cyclic stress is decreased (Figure 10a) and the minimum cyclic stress is increased (Figure 10b), i.e. the stress range is reduced. This will, in turn, increase the bending fatigue life of the gear.

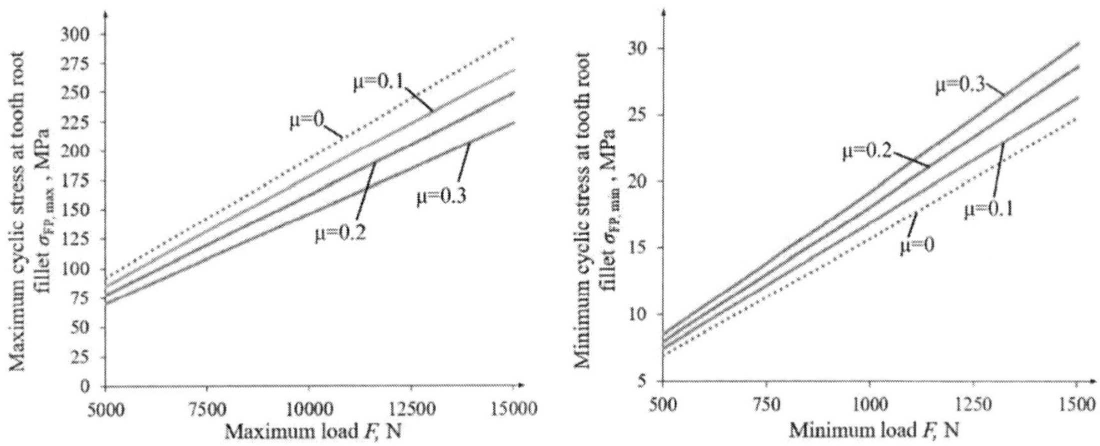

Figure 10. The effect of friction and different loads on cyclic stresses at tooth root fillet: (a) maximum cyclic stress, (b) minimum cyclic stress

For both load models, it can be observed that the friction force influences tooth root stress results. Even though variation of maximum principal stresses is relatively small, according to [11], amplitude change of 10% can cause an increase or a decrease in fatigue life by a factor of three.

4. Conclusion

In this paper, numerical analysis of the principal stresses at tooth root fillet while considering friction force and fixture rigidity was presented. By employing the elastic finite element analysis, contact between tooth flanks and fixture for two different load models was simulated and maximum principal stresses at the tooth root fillet were obtained. The results were compared with the initial predictions and were found to be in a good agreement. Furthermore, it was found that friction can impact nominal tooth root stress results, which can influence bending fatigue life of a gear. Since STBF tests are often used as screening tests for actual bending fatigue behavior of the gear, friction coefficient should be determined before the test and properly compensated for, or reduced to minimum for the most accurate results.

References

[1] Gasparini G, Mariani U, Gorla C, Filippini M and Rosa F 2009 Bending Fatigue Tests of Helicopter Case Carburized Gears, *Gear Technology*
[2] Gorla C, Rosa F and Conrado E 2017 Bending Fatigue Strength of Case Carburized and Nitrided Gear Steels for Aeronautical Applications, *International Journal of Applied Engineering Research* **12** 11306-11322
[3] Handschuh R F, Krantz T L, Lerch B A and Burke C S 2007 Investigation of Low-Cycle Bending Fatigue of AISI 9310 Steel Spur Gears NASA/TM-2017-214914
[4] Davis J R 2005 Gear Materials, Properties and Manufacture 2005, *ASM International*
[5] Akata E, Altinbalik M T and Çan Y 2004 Three-point load application in single tooth bending fatigue test for evaluation of gear blank manufacturing methods, *International Journal of Fatigue* **26**(7) 785-789
[6] Bian X X, Zhou G, Liwei, Tan J Z 2012 Investigation of bending fatigue strength limit of alloy steel gear teeth, *J. Mechanical Engineering Science* **226** 615-625
[7] Standard SAE J1619, Single Tooth Gear Bending Fatigue Test 2017, *SAE International*
[8] Vučković K, Galić I, Božić Z and Glodež S 2018 Effect of friction in a single-tooth fatigue test, *International Journal of Fatigue* **114** 148-158
[9] Abaqus 6.13, *Dassault Systems* 2012 Simulia
[10] Pehan S, Hellen T, Flašker J and Glodež S 1997 Numerical methods for determining stress intensity factors vs crack depth in gear tooth roots, *International Journal of Fatigue* **19** 677-685
[11] Stephens R I, Fatemi A, Stephens R R and Fuchs H O 2001 *Metal Fatigue in Engineering*, John Wiley & Sons, Inc.

IOP Conf. Series: Materials Science and Engineering **659** (2019) 012005 doi:10.1088/1757-899X/659/1/012005

Contact pressure analysis of slewing rings

R Skyba[1], S Hrček[1], L Smetánka[1] and M Majchrák[1]

[1]University of Žilina, Faculty of Mechanical Engineering, Univerzitná 8215/1, 01026 Žilina

E-mail: rudolf.skyba@fstroj.uniza.sk

Abstract. The paper deals with creation of functional parametric 3D model of slewing bearing. The slewing bearing is loaded by the radial force that acts from the centre of the bearing and it is also perpendicular to the axis of bearing. 3D fully parametric model of slewing ring was designed in Creo Parametric 3 and it is controlled by the ball diameter. To calculate several geometric bearing modifications, the 3D model from Creo was imported to the Ansys Workbench computational program. Based on many analysis in Ansys, it is possible to evaluate the size of contact pressures, determine the most loaded place and compare values between individual bearing modifications. By transferring the measured values of the contact pressures to the Microsoft office, it was possible to construct a graphical dependence between the values of measured stresses and the magnitude of the radial force acting on the slewing bearing.

1. Introduction

The well-known fact about large diameter slewing bearing is that they are primary designed for applications where large amount of axial load is transferred through thin-walled rolling rings to rolling elements, which usually are balls or rollers. Inner ring is evenly surrounded by rolling elements located on its orbits [1]. This also applies to the outer ring, with one difference and it is that the rolling elements are located on its inner side because there is the raceway located [2].

The stiffness of slewing ring is one of most important thing if we talk about them. They have significant influence on the dynamics of a rotating shaft and also on the precision of the machine systems [3]. Life of rolling elements and the rolling contact fatigue can be determined by full-scale endurance tests. This tests that are realized in bearing test rigs are expensive and time consuming [4]. Lin analyses the displacement and coefficient of radial stiffness with angular contact [5]. For the analysis John Harris's method based on Hertz-contact theory has been used. In order to compare the results obtained by the mentioned method and Palmgren's empirical relations the finite element method has been used [6]. Research about the radial stiffness of the radial bearing was realised by Mullick and it was also based on John Harris's finite element method [7]. For solving a nonlinear equations the Newton Raphson's method has been used, while in contact analysis uses finite elements method. In this study the gyroscopic moment and the influence of sliding was neglected [8]. The results shows that the relative displacement and stiffness of the bearing rings depend on the centrifugal force and combined loads [9]. Antoine et al. propose two new approximate, methods for determination the angle of contact depending on the preloaded and speed or in special cases for elastic preload [10]. Duval et al. concluded that the treatment thickness, in the investigated depth range of 3,2 mm to 4,5 mm, has no real influence because critical points from computations being at a depth of less than 3,2 mm. Increasing the initial contact angle from 45 degrees to 50 degrees does not affect the level of damage compared to the chosen reference case [11]. However, this increase of the angle makes appear a second critical zone, at the track border, with a non-negligible probability of failure also the reduction of conformity compared to the reference is penalizing in term of fatigue damage. On the other hand, in case a surface cracking is favoured, compared to the under layer cracks predicted in other cases [12]. The most influential parameter is undoubtedly the diametric gap. The introduction of a gap with respect to the references case (null gap), strongly reduces the fatigue strength and increases the risk of cracking in the track border. This parameter can be used as an indicator in the context of damage monitoring on the slewing ring [13].

Radial stiffness depends on the elastic deformation (deflection) of the bearing under load and can be expressed as a ratio of load to deflection. However, since the relationship between deflection and load is not linear, only guideline values can be provided [14].

The main task of this paper is to select concrete type of bearing then create virtual 3D model parameterize it and then to modify this model by several parameters direct from the environment of Ansys workbench calculation program while the model is under the load [15].

Methodology of solving this task includes selection of critical parameters for creation of the parametric bearing model. In this case we introduce two main parameters like ball diameter and contact angle. There are also minor parameters like bearing pitch diameter which has significant influence on the amount of used rolling elements. If we change number of rolling elements directly from Ansys then after we update model and value of elements will be increased we lost defined boundary conditions of bearing. So we will keep the same value of pitch diameter during all measurements.

A methodology based on finite element method has been developed to calculate accurately the value of ball forces around slewing bearing and its deformation. This methodology can use the non-linear springs to simulate contact. It is necessary to use this method when stiffness of support structure is not uniform.

In the end we need some calculations like calculation of semi-axis, contact pressures, and the depth at which the maximum tension is applied. If we have all of this we can move forward and based on many calculations of the deformations performed in the Ansys we can compare influence of ball diameter and contact angle. Based on analysis graphs were constructed.

2. Creation of basic geometry of four-point contact slewing bearing and its parameterization

Slewing bearing or ring, is large diameter bearing, usually loaded by axial force, small radial force and tilting moment at the same time see figure 1. Parts from which the bearing and rolling element consist are shown in figure 2. Dimensions that are needed for parameterization are showed on figure 3 and also table with dimensions and relations between them is shown on figure 4.

Slewing bearing considered in this paper is four point contact type. So if we introduce radial force only, the whole load will be transmitted through the 4 contact points see figure 5.

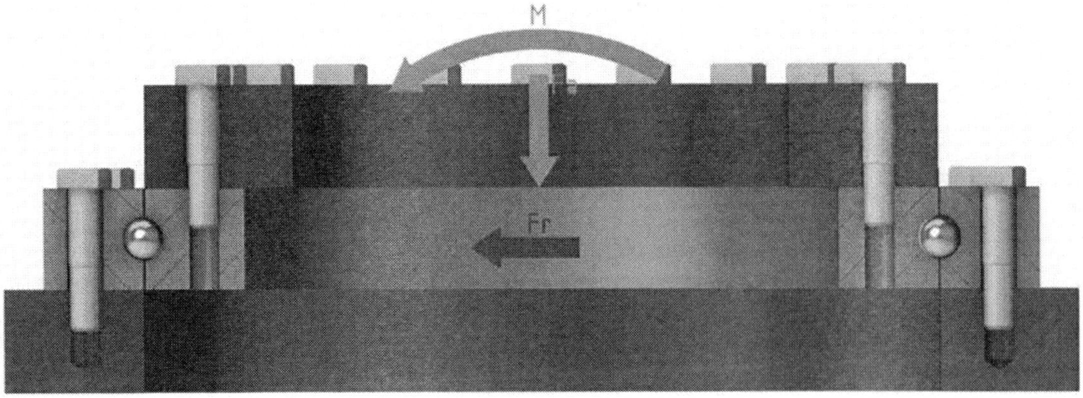

Figure 1. Slewing bearing loaded by tilting moment, radial and axial forces.

IOP Conf. Series: Materials Science and Engineering **659** (2019) 012005 doi:10.1088/1757-899X/659/1/012005

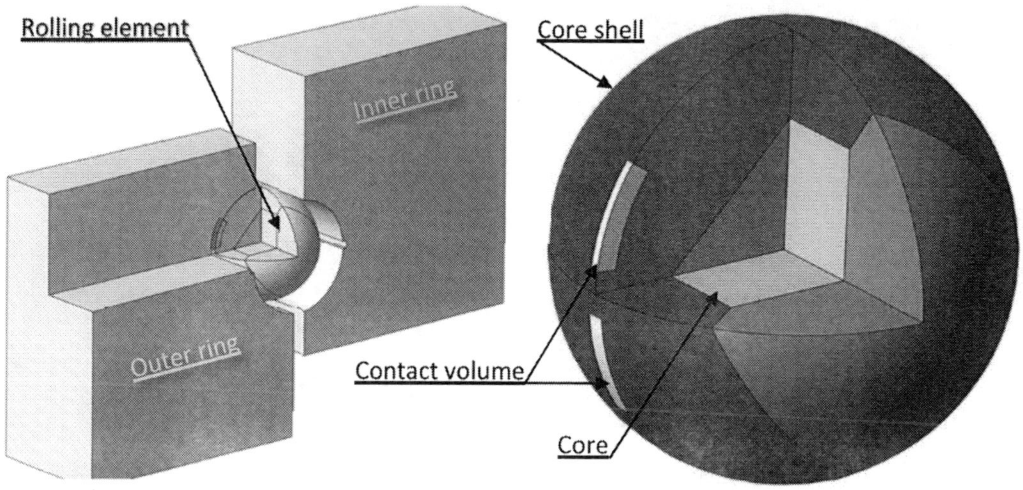

Figure 2. Parametric model of one bearing part (left), design of rolling element (right).

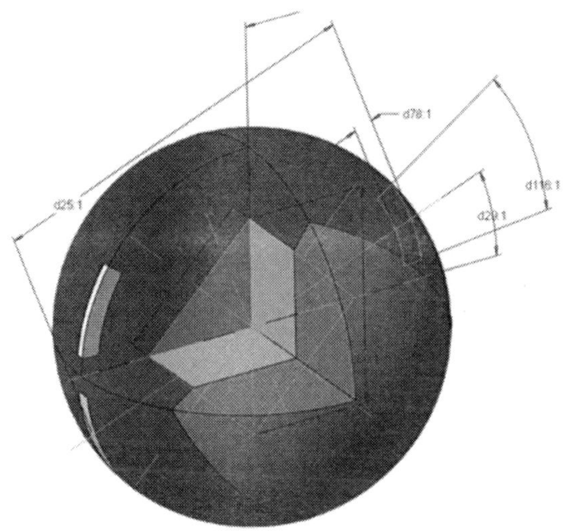

Figure 3. Parametric model of rolling element.

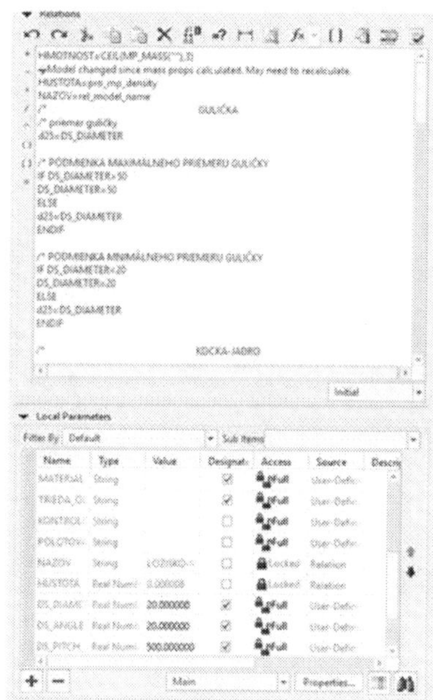

Figure 4. Relations between the geometry dimensions.

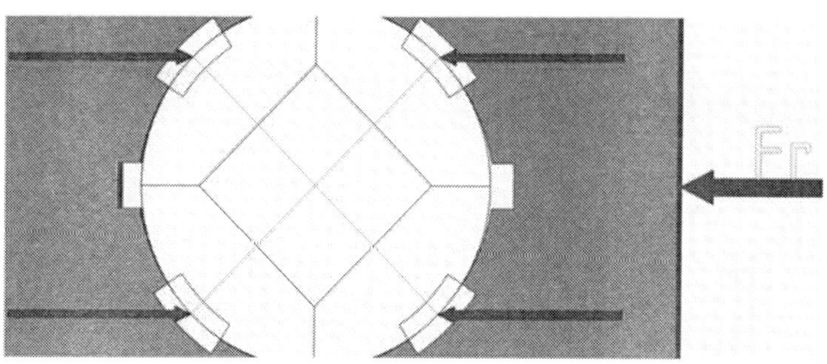

Figure 5. Distribution of load in four point contact.

IOP Conf. Series: Materials Science and Engineering **659** (2019) 012005 doi:10.1088/1757-899X/659/1/012005

3. Basic calculations

All calculation are based on Hertz-contact theory that is most suitable for this type of task. First of all we can find the curvature radii of inner and outer raceway.

$$\rho_{1I} = \frac{1}{r_{we}};\ \rho_{2I} = \frac{1}{r_{we}} \tag{1}$$

$$\rho_{1II} = -\frac{1}{r_i};\ \rho_{2II} = \frac{1}{r_{pw}} \tag{2}$$

From obtained values of radii of curvatures it is possible to obtain sum of values.

Then we calculated the value of cosτ.

$$cos_\tau = \left|\frac{\rho_{1I}-\rho_{1II}+\rho_{2I}-\rho_{2II}}{\sum\rho}\right| \tag{3}$$

In the next step we can calculate distance between the centres of raceways.

$$A = B \cdot D_{we} \tag{4}$$

After the distance between centres has been calculated, we move forward and we calculate the size of contact area.

$$2_a = 2 \cdot \mu \cdot \sqrt[3]{\frac{1}{E} \cdot (1 - m^2) \cdot \left(\frac{3Q}{\sum\rho}\right)} \tag{5}$$

$$2_a = 2 \cdot \vartheta \cdot \sqrt[3]{\frac{1}{E} \cdot (1 - m^2) \cdot \left(\frac{3Q}{\sum\rho}\right)} \tag{6}$$

After we obtain values of both contact zones, then it has been calculated the depth in which the maximum pressure acts.

$$2_{z0} = 2_b \cdot 0{,}25 \tag{7}$$

For calculation of stiffness between raceway of inner and outer ring and surface of rolling element it has been needed to calculate approximated parameter of ellipticity.

$$\kappa_{i,o} = a_{i,o}^{\frac{\pi}{2}} \tag{8}$$

Where the value of $a_{i,o}$ can be calculated by following equation.

$$a_{i,o} = \frac{2 \cdot f_{i,o} \cdot D_{pw}}{D_{we} \cdot (D_{pw} - D_{we}) \cdot (2 \cdot f_{i,o} - 1)} \tag{9}$$

Also in next step it was necessary to calculate stiffness between inner ring contact surface and ball. Then it was also calculated for outer ring and the ball.

$$\mathrm{K}_{i,o} = \pi \cdot \kappa_{i,o} \cdot E' \cdot \sqrt{\frac{2 \cdot \theta \cdot R_{i,o}}{9 \cdot \vartheta^3}} \tag{10}$$

After the calculation of both stiffness parameters we are able to calculate combined stiffness.

$$\mathrm{K}_n = \left[\left(\frac{1}{\mathrm{K}_i}\right)^{\frac{1}{n}} + \left(\frac{1}{\mathrm{K}_o}\right)^{\frac{1}{n}}\right]^{-n} \rightarrow n = \frac{3}{2} \tag{11}$$

Then we calculated the maximum of contact pressure by equation bellow.

$$p_0 = \frac{1{,}5}{\pi \cdot \mu \cdot v} \cdot \sqrt[3]{\left[\frac{E}{(1-m^2)} \cdot \frac{\sum\rho}{3}\right]^2 \cdot Q} \tag{12}$$

4. Boundary Conditions and Simulation

After introduction of 3d model from Creo parametric to Ansys Workbench boundary conditions has been set as you can see on figure 6. First of all, we set the conditions for point contacts because only by this way the right conditions can be achieved. For removing of movements and rotations there was used the boundary conditions like Displacement for movements and Remote control for movements and rotations.

Figure 6. Boundary conditions of bearing part.

Before calculations were performed the mesh method in contact zones has been set to hexahedral and their size to one tenth of millimeter. For achieving a fast calculations the size of elements in less important locations need to be set to highest value as possible. Mesh created on rolling element is shown on figure 7. For better computing the hexahedral fine mesh has been created on places where the maximum of contact pressure acts. On elements such as bearing rings was created coarse mesh because it has not a significant influence to the calculations see figure 8.

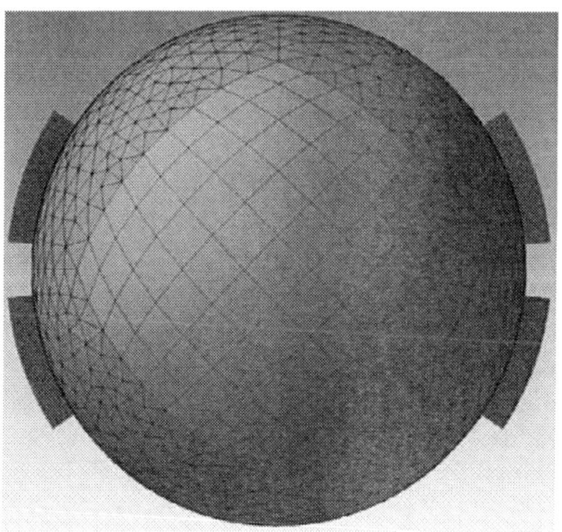

Figure 7. Mesh created on rolling element.

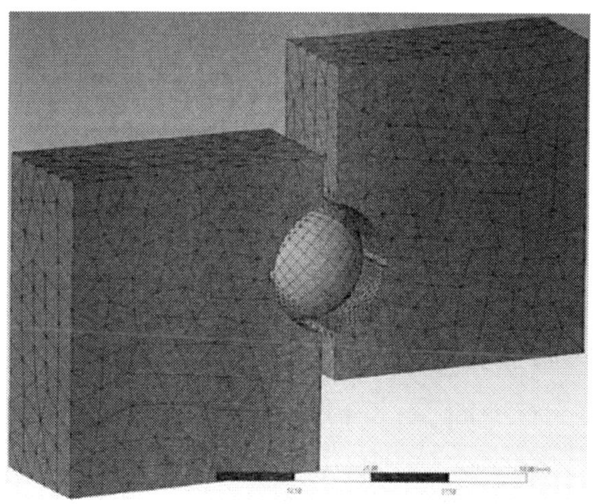

Figure 8. Mesh on bearing part.

After the boundary conditions has been set we start the solve mechanism in Ansys and after few hours we obtain the results. Concrete results and load distribution of contact pressure in contact surfaces see figure. 9.

The graph of dependence between force and the contact pressure has been created from the results obtained in Ansys Workbench figure 10.

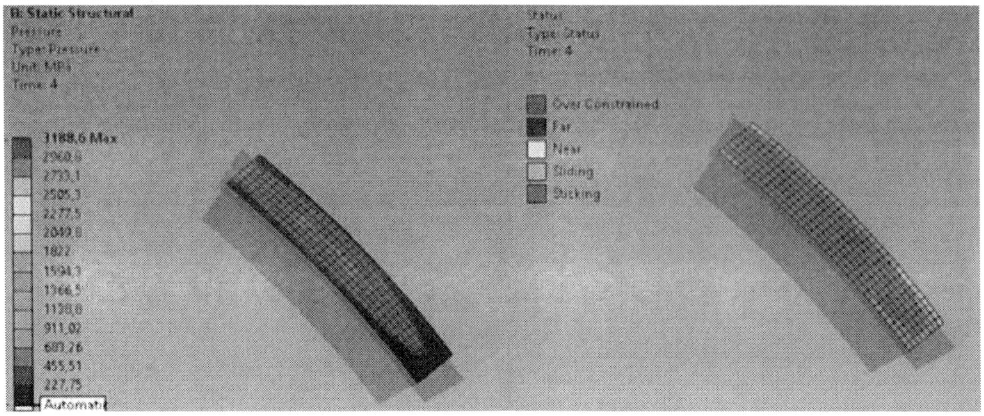

Figure 9. Distribution of contact pressure.

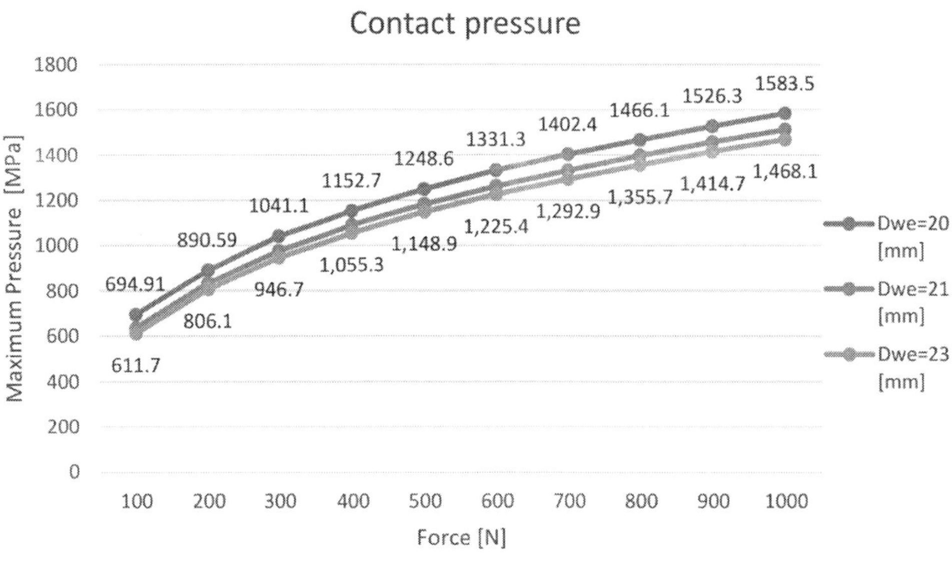

Figure 10. Graph of dependency between contact pressure and force.

5. Conclusions

Creation of 3d calculation model is very helpful if large number of calculations are needed for different geometric variants. By this method it is possible to change the geometry parameters and determine the relationships them. After the model of bearing is imported to the Ansys workbench it is necessary to set boundary conditions and create finite element mesh. Main rule for accurate calculations is that the mesh in most loaded location need to be created by hexahedral mesh method.

By measuring of contact pressure at different geometry parameters it has been proved that, the relationship between the applied force and maximum value of contact pressure is evenly growing, this means that when value of force is higher also the maximum of contact pressure will be greater, but there are also shear losses.

Acknowledgments
This study was supported by Ministry of Education, Science, Research and Sport under the contract No. 1/0595/18 – Optimising the internal geometry of roller bearings with line contact in order to increase their durability and reduce their structural weight. This study was supported by Slovak Research and Development Agency under the contract no. APVV-14-0508 – Development of new methods for the design of special large-size slewing rings.

References

[1] Bastovansky R, Tropp M, Lukac M and Brumercik F 2017 Molybdenum sheet metal test device, *Communications* **19** 124-127
[2] Gramblička S, Kohár R and Stopka M 2017 *Dynamic analysis of mechanical coveyor drive system*, In book: Procedia Engineering [elektronický zdroj] **192** 259-264
[3] Jedlinski L, Caban J, Krzywonos L, Wierzbicki S and Brumercik F 2015 Application of vibration signal in the diagnosis of IC engine valve clearance, *Journal of Vibroengineering* **17**(1) 175-187
[4] Hrcek S, Kohar R and Medvecky S 2012 Determination on the maximum roller bearing load with regards to durability thereof using FEM analysis, *Communications* **3**(14) 55-61
[5] Liu J, Shao Y and Lin C-M 2012 Vibration analysis of ball bearings with a localized defect applying piecewise response function, *Mechanism and Machine Theory* **56** 156–169
[6] Kohar R, Hrcek S and Medvecky S 2012 Usage of dynamic analysis to determine force interactions between components of rolling bearings, *Communications* **3**(14) 62-67
[7] Mullick A 1990 *An Investigation in Stiffness of Rolling Contact*, Chung Yuan Christian University, Department of Mechanical Engineering, Doctoral Thesis
[8] Kohar R and Hrcek S 2014 Dynamic analysis of a rolling bearing cage with respect to the elastic properties of the cage for the axial and radial load cases, *Communications* **3A**(16) 74-81
[9] Kohár R and Hrček S 2014 *Dynamic analysis of rolling bearings with an elastic cage*, In book: Modern methods of construction design, Cham: Springer (Lecture notes in mechanical engineering ISSN 2195-4356)
[10] Antoine J F, Visa, C and Sauvey C 2006 Approximate Analytical Model for Hertzian Elliptical Contact Problems, *Journal of Tribology, Transactions of the ASME* **128** 660-664
[11] Kučera Ľ and Gajdošík T 2014 *The vibrodiagnostics of gear*, In book: Modern methods of construction design – 54th ICMD 113-118, Springer
[12] Lukáč M, Brumerčík F, Krzywonos L and Krzysiak Z 2017 *Transmission system power flow model*, In book: Communications: scientific letters of the University of Žilina **19**(2) 27-31
[13] Martikan M, Brumercik F and Bastovansky R 2015 Development of mechatronic deformation system, *Applied Mechanics and Materials* **803** 173-178
[14] Medvecký Š, Kohár R and Hrček S 2012 *Dynamic analysis of cage behavior in a tapered roller bearing*, In book: KOD 2012 Konstruisanje - Oblikovanje - Dizajn: 7th international symposium machine and industrial design in mechanical engineering: proceedings: 24-26 May 2012, Balatonfüred, Hungary. - Novi Sad: Faculty of technical sciences 45-50
[15] Weis P, Kučera Ľ, Pecháč P and Močilán M 2017 Modal analysis of Gearbox Housing with Applied Load, *Procedia Engineering* **192** 953-958

Gears or rotors - three approaches to design of working units of hydraulic machines

J Stryczek [1], S Bednarczyk[1], E Codina[2], P J Gamez-Montero[2], L Ivanovic[3] and M Matejic[3]

[1]Wroclaw Fluid Power Research Group - Wroclaw University of Science and Technology, 27 Wybrzeże Wyspiańskiego st., 50-370 Wrocław, Poland
[2]IAFARG & LABSON - Dept. Fluid Mechanics - Campus Terrassa - Universitat Politecnica de Catalunya, Carrer de Colom, 11, 08222 Terrassa, Spain
[3]University of Kragujevac, Faculty of Engineering, Sestre Janjić 6, 34000 Kragujevac, Serbia

E-mail: lozica@kg.ac.rs

Abstract. The gerotor pump has many advantages over other hydraulic pumps, such as high reliability, compactness, lower noise in running conditions etc. Considering the increasingly stringent regulations in the field of environmental protection, one of the main goals in the industrial applications of gerotors is to improve their design by using light materials and optimize their performance. This paper will give an overview of modern research that focuses on improving the performance of gerotor pumps. A joint publication of three university centers, namely of Wroclaw, Terrassa and Kragujevac, presents their original approaches to the problem of designing gerotor and orbital systems. The presentations have been prepared in separate parts, but according to similar rules. The first part of the paper focuses on meshing between internal and external gear of gerotor. Analysis for both cases, moved axis and fixed axis, will be provided in accordance to their working characteristics and their application. The second part of the paper provides the design of a gerotor pump strikes a balance between volumetric efficiency, manufacturability and mechanical efficiency. The third part of the paper defines the methodology for selecting the best combination of geometric form parameters of the trochoidal gearing in relation to the reduction of contact stresses and improving other characteristics of the pump. All three parts provide manufacturing and application review in the fluid power machines. Paper concludes with presented theoretical and experimental review of the authors and their practice achievements.

1. Introduction

In hydraulic, pneumatic, mechanical and also internal combustion engines, gerotor or orbital systems are often used. Those words refer to systems composed of two elements, i.e. a rotor and a ring, which are internally engaged and in which the difference between the mating elements equals one. If both elements work at fixed axes and the rotor drives the ring, then the system is called a gerotor system. If, however, the rotor moves within a fixed ring by circular or orbital motion, then such a system is called an orbital system. Two fundamental problems related to gerotor and orbital systems need solving, namely: establishing a uniform terminology for this machine design area and formulating principles for the design of gerotor and orbital systems.

Among the authors who deal with the issues, there are two basic approaches to solving those problems. According to the first one, two co-operating elements are treated as gears so the problem is located in the field of gears [1, 2, 3]. According to the second one, those elements are treated as cams with a special outline so the problem is located in the field of the theory of machines and mechanisms [4, 5, 6]. In both cases, different terminology is used.

This paper presents research approaches to the described gearing of three university centres, namely of Wroclaw, Terrassa and Kragujevac. The methods presentation has been prepared in separate parts, but according to similar rules including geometry, kinematics, hydraulics, contact stress and technology. In the summary, all approach methods are compared.

2. Designing of gerotor and orbital gear systems for hydraulic machines according to FPRG WROCŁAW

2.1 Geometry

In the gerotor design process, two basic groups of curves, i.e. epicycloids and hypocycloids, are used. These types of tooting are used because of its very good meshing characteristics, resistance to shock load, low noise, etc. In the following text of this chapter much closer description of gerotor profile types will be presented.

By treating the rotor and the ring as gears, parameters characteristic for gears were adopted for their description. In Figure 1 are shown uncorrected and corrected gear profiles.

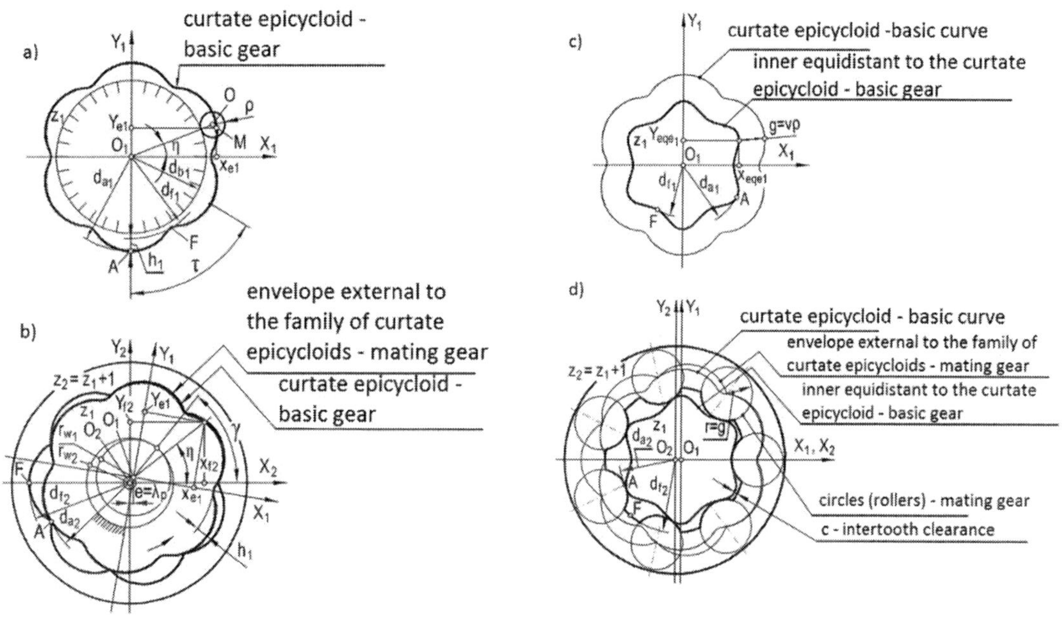

Figure 1. Designing epicycloid gears: a, b) uncorrected gears: internal and external; c, d) corrected gears: external and internal.

Figure 1a,b presents the principles of designing a system of uncorrected epicycloid gears. In Figure 1 the following parameters are shown:

z_1 - number of teeth identical to the number of the epicycloid's or hypocycloid's arcs present in the entire closed cycloidal curve,

m - a module which in accordance with the principle of constructing cycloidal curves is equal to the diameter of the moving wheel ρ, i.e. $m = 2\rho$

λ - tooth depth factor (λ=OM/ρ),

η - the angle of the cycloidal curve,

v - correction coefficient of the cycloidal profile,

g - shift of the profile ($g = v\rho$),
b - tooth width.

As shown further in this work, using those parameters, the geometry and kinematics of cycloidal gears can be described, and also hydraulic parameters of machines, such as delivery (absorption) and the pulsation of delivery (absorption) can be determined. It shows the versatility of the adopted parameters system and its usefulness in the entire designing process.

The system of parametric equations describing the family of wheels with an epicycloid outline in the X_2Y_2 system has the form:

$$x_{fe2} = \frac{m}{2}\left[\lambda\cos\gamma + (z_1+1)\cos\frac{z_1\eta-\gamma}{z_1} - \lambda\cos\frac{(z_1+1)z_1\eta-\gamma}{z_1}\right]$$
$$y_{fe2} = \frac{m}{2}\left[\lambda\sin\gamma + (z_1+1)\sin\frac{z_1\eta-\gamma}{z_1} - \lambda\sin\frac{(z_1+1)z_1\eta-\gamma}{z_1}\right] \tag{1}$$

Equations (1) should be supplemented with a condition of the envelope which connects angle of rotation γ of the basic wheel relative to the collaborating wheel with angle of the epicycloid η [3]:

$$\sin z_1\eta - \lambda\sin\left[z_1\eta + \frac{z_1(\eta-\gamma)}{z_1}\right] + \sin\frac{z_1(\eta-\gamma)-\gamma}{z_1} = 0 \tag{2}$$

Using equations (1) and (2) together, it is possible to determine the profile of the collaborating internal gear. The characteristic diameters of the collaborating gear are derived from formulas (3):

$$d_{a2} = 2(z_1\rho + \rho - \lambda\rho + \lambda\rho) = m(z_1+1) \tag{3a}$$

$$d_{f2} = 2(z_1\rho + \rho + \lambda\rho + \lambda\rho) = m(z_1+1+2\lambda) \tag{3b}$$

Analysing Figure 1b, it is noted that:
- formulas for determining the outline of an internal gear are difficult to use directly,
- the teeth of both gears have wrong proportions between the head and the foot, and the contact stresses may be too high,
- manufacturing of an internal gear of a complicated outline is difficult and expensive.

In such a situation, the gears and gear systems are corrected. First, the basic external gear is corrected. As shown in Figure 1c, the correction consists in the formation of the internal equidistant relative to the curtate epicycloid, and the shift of the equidistant equals $g = v\rho$. The system of parametric equations describing the equidistant takes the form:

$$x_{eqe1} = \frac{m}{2}\left[(z_1+1)\cos\eta - \lambda\cos(z_1+1)\eta - v\frac{\cos\eta - \lambda\cos(z_1+1)\eta}{\sqrt{1-2\lambda\cos z_1\eta + \lambda^2}}\right]$$
$$y_{eqe1} = \frac{m}{2}\left[(z_1+1)\sin\eta - \lambda\sin(z_1+1)\eta - v\frac{\sin\eta - \lambda\sin(z_1+1)\eta}{\sqrt{1-2\lambda\cos z_1\eta + \lambda^2}}\right] \tag{4}$$

The formulas for calculating the outside diameter and the root diameter in the corrected gear are:

$$d_{a1} = 2(z_1\rho + \rho + \lambda\rho - g) = m(z_1+1+\lambda-v) \tag{5a}$$

$$d_{f1} = 2(z_1\rho + \rho - \lambda\rho - g) = m(z_1+1-\lambda-v) \tag{5b}$$

Next, the collaborating internal gear is corrected. As shown in Figure 1d, circles of a radius equal to the equidistant shift of $r = g$ are drawn from the vertices of the envelope, and then those circles are connected to each other by arcs of the r_{f2} radius circle, which rotates from the centre of the collaborating gear O_2. The characteristic diameters of the collaborating gear after correction are derived from the formulas:

$$d_{a2} = 2(z_1\rho + \rho - \lambda\rho - g + \lambda\rho) = m(z_1 + 1 - \nu) \tag{6a}$$

$$d_{f2} = 2(z_1\rho + \rho + \lambda\rho - g + \lambda\rho) = m(z_1 + 1 + 2\lambda - \nu) \tag{6b}$$

The detailed derivation of the presented equations for corrected gerotor profile could be found in the literature [3, 13-15].

The correction of the epicycloidal gear system can be compared with the P-0 correction (without the axes distance change) in the classic involute mesh.

To sum up this part of the considerations, it can be stated that:
- the formulas for calculating the gear outlines are simple and easy to use in the design process,
- the teeth and gears are proportional,
- manufacturing of the gears is much simpler and cheaper compared to the uncorrected gears, particularly in the case of the collaborating internal gear.

Hypocycloidal gears and gear systems can be designed in a similar way to the one the epicycloidal gears and gear systems are designed. That design problem has been thoroughly explained in [13, 14, 15].

2.2 Kinematics

Principles of collaboration between epicycloidal gears in a system at fixed axes of rotation are shown in Figure 2a.

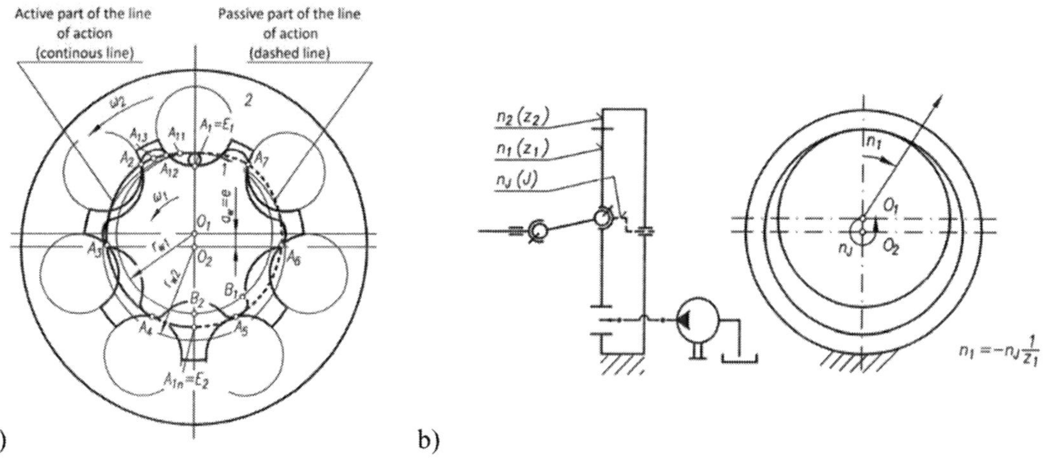

Figure 2. Collaboration (mating) of cycloidal gears; a) mating of the corrected epicycloidal gears with fixed axes O_1, O_2; b) mating of the epicycloidal gears with moving axes, where O_1 rotates around O_2.

The external gear drives the internal gear and all teeth of both gears are in constant contact with each other. The line of action has the shape of a loop (conchoid) which is divided into the active part carrying the load (continuous line) and the passive part not transferring the load (dashed line). The number of action is $\varepsilon = z_2/2$ and is much larger than in the involute gearing. That means that the teeth are much less loaded than in the involute gearing, and the mechanical cycloidal gear transmissions can transfer higher loads than involute gear transmissions. An extremely important feature of the collaboration is that between the gears, intertooth closed spaces are formed, which are necessary for transporting the working medium in hydraulic machines. The principles of the gears co-operation in the epicycloidal systems with moving axes are shown in Figure 2b. A case of a single hydraulic gearing was considered. An internal gear of O_2 centre remains motionless, and on that gear, an external gear of O_1 centre rolls with the planetary motion. It is a kind of gear transmission, in which the motion of the external gear is hydraulically forced and the transmission equals $n_1=n_j/z_1$. That principle of collaboration was used in orbital hydraulic engines [16, 17] and in control units [17].

2.3 Hydraulics - delivery (absorption) and pulsation of the delivery (absorption) of hydraulic machines

Using the characteristic parameters of the teeth and the mesh adopted at the beginning of this work, it is possible to make formulas for calculating of the hydraulic parameters of the machines. This applies to the delivery, or, alternatively, the absorption, as well as pulsation of the delivery, or, alternatively, absorption of the delivery of machines with fixed axes (gerotor) and delivery, or alternatively, absorption of machines with moving axes (orbital).

The delivery (absorption) of gerotor machines is determined by the formula:

$$\frac{q_{st}}{\pi bm} = \frac{1}{4}\left[(z_1+1+\lambda-\nu)^2 - \frac{z_1}{z_1+1}(z_1+1-\nu)^2 + z_1\lambda^2\right] \tag{7}$$

and the pulsation of delivery (absorption) from the formula:

$$\delta = f(z,\lambda,\nu) \tag{8}$$

In the case of orbital machines, which, by principle, work only as engines, the absorption is determined by the formula:

$$q_{or} = z_2 q_{st} \tag{9}$$

2.4 Design and manufacturing solutions for gears of cycloidal hydraulic machines and mechanical transmissions

Using the presented method of designing cycloidal gear systems, their construction documentation can be developed and manufacturing can be started. Figure 3 presents selected methods of manufacturing cycloidal gears.

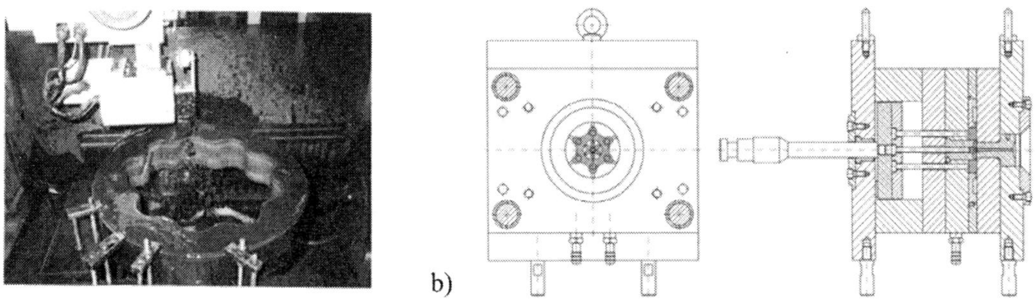

a) b)

Figure 3. Manufacturing of the cycloidal gears: a) spark erosion of the internal hypocycloidal gear, b) manufacturing of the internal epicycloid gear by injection moulding.

Individual gears, or small series, can be made by spark erosion [3] (Figure 3a), and the plastic cycloidal gears can be made by injection moulding [18, 19] (Figure 3b). The technological data, tools and instruments needed for the manufacturing have been developed by scientists of the FPRG WROCŁAW. As a result of the manufacturing process, various cycloidal gear systems have been made and used in the construction of hydraulic machines [3, 17] and mechanical transmissions [21].

3. Designing of gerotor for hydraulic machines according to IAFARG & LABSON TERRASSA

The design of a gerotor pump from scratch is drawn on existing research work and the own authors' experience and know-how. As a first step, the conceptual stage with the dimensional constraints lead to obtain a complete gear set by means of four basic parameters and simple formulae. The design process, a step forward, the most important characteristics of the pump such as theoretical performance indexes and the porting are presented.

The available gerotor gear sets by means of sheet commercial literature shows a great number of standard sets and each application has to be adapted to fit one of them. However, the selection of the design parameters that has led to the geometry of these standard sets is not clear, at least in the available information sheets. Despite of being the cost effective solution, it is advisable to choose the right selection of the design parameters that allows reaching a balance between contact stress, volumetric characteristics and flow irregularity. The methodology presented in this section is specifically applied to circular-tooth profiles (conventional-toothed gerotor).

3.1 Dimensional constraints parameters in a new-born gerotor pump

The characterization of the gerotor pump will be based on the volumetric capacity c_v in [cm^3/rev]. Then, the dimensional constraints will set up the corrected epicycloidal gear thickness H in [mm] and the external diameter of the external gear D_c in [mm]. The gear thickness is proportional to the volumetric capacity. However, the main influence of H is related with the filling capability of the chambers, and subsequently, on the volumetric performance and efficiency. The eccentricity e in [mm] of the gear set, which is the most difficult geometrical parameter to predict, can be calculated as,

$$e^2 - \left(\frac{D_c}{4} - \frac{wc}{2}\right) \cdot e + \frac{39.8 \cdot c_v}{H} = 0 \tag{10}$$

where wc in [mm] is the wall wide of the external gear (see Figure 4). The value of wc can be estimated by using the following expression,

$$wc = 0.075 \cdot D_c \tag{11}$$

being just an estimation because it is not taking into account the number of teeth of the external gear, working pressure, contact stress and gear material. Now, the arc radius of the external gear tooth g in [mm] can be calculated as:

$$g = \frac{z_2 \cdot e}{\lambda} - 0.995 \cdot (0.5 \cdot D_c - wc) + 2 \cdot e \tag{12}$$

where z_2 is the number of teeth of the external gear, z_1 is the number of teeth of the internal gear ($z_2 - z_1 = 1$) and λ is the tooth depth factor, with recommended values going from 0.6 (larger g) to 0.8 (shorter g) from Stryczek et al. [18].

Then, by knowing c_v, H and D_c from the specifications of the application, and by choosing z_2 the four basic parameters are defined and the designer can use the previous formulae, from equation (1) to (12), to determine the profiles of the gear set.

3.2 Theoretical performance indexes and porting

The theoretical torque T in [N·m/bar], theoretical power N in [W/bar·rpm], the tip velocity of the internal gear V_t in [m/s·rpm] and the theoretical radial load RL in [N/bar] can be estimated by using the following expressions:

$$T = 0.016 \cdot c_v \tag{13}$$

$$N = 0.0016 \cdot c_v \tag{14}$$

$$V_t = 0.0001 \cdot \left(\frac{D_c}{2} - wc - e \right) \tag{15}$$

$$RL = 0.1 \cdot H \cdot \left(D_c - 2 \cdot wc - 3 \cdot e \right) \tag{16}$$

RL value is a reference to the load on the shaft and its bearings, and V_t value must be taking into account regarding to the volumetric efficiency since the increase of V_t has a negative effect on the filling capability of the chambers.

The porting located at the housing on the casing is an important factor in the gerotor pump volumetric performance and efficiency and a proper porting design requires experience and knowledge by the pump designer. As a first step in the design process, the theoretical porting can be defined based on the contact points *P* where a chamber is comprised between two successive contact points and the end inlet port (*fa*), the beginning outlet port (*ii*) and the beginning inlet port (*ia*). The external porting radius *G* in [mm] and the internal porting radius *RiRi* in [mm] can be calculated as:

$$G = 0.5 \cdot D_c - wc \tag{17}$$

$$RiRi = 0.5 \cdot D_c - wc - 3 \cdot e \tag{18}$$

and the upper limit, the porting angle, and the lower limit, the porting length, can be simply obtained by calculating the corresponding contact points (P_{ia}, P_{fa}, P_{ii} and P_{fi}) by means of the line of contact [7]. The theoretical porting does not accomplish the minimum requisites, specially providing the best chamber filling as possible, and a former porting is then proposed in [5].

3.3 GeroLAB software

The specifications of the application configure three basic parameters c_v, *H* and D_c, and the fourth one by choosing z_2. These parameters can be introduced in specialized software named GeroLAB that will calculate and provide the designer with the technical drawings in CAD format (Figure 5), volumetric characteristics, contact stress, teeth clearance and former porting [22].

GEROLAB software is an integrated package system based on modules and written by using an open source code. Basically, the methodology consists of three phases, each of which is related to a basic module: design, volumetric characteristics and contact stress [23]. By consecutively following each phase, the software is capable of combining the design of a gear set for a gerotor pump. Then, as an integrated package, it is completed with two new modules. The minimum clearance module is the forth module that takes into account real effects regarding the geometry paths of the manufactured tolerances and clearances. The fifth module integrates the new approach of modelling relief grooves located at the porting, to make the transition from the theoretical to the former porting [24].

3.4 New gerotor technology: GeroMAG concept and the magnet-sleeve-sealed with polymer composite concept

GeroMAG concept is an innovative variable flow pump, sealed, compact and non shaft-drive with magnetic-driving outer rotor [25]. Up to date in the standard gerotor pump technology, the rotational movement was transmitted from the internal gear (driving) to the external gear (driven) by means of an exterior shaft-drive motor. The external gear takes the role of primary pump gear conducting the rotational movement as a driving outer rotor in the GeroMAG concept, without exterior shaft-drive, using electro-magnetic driving on the external gear through pole pieces (see Figure 5a).

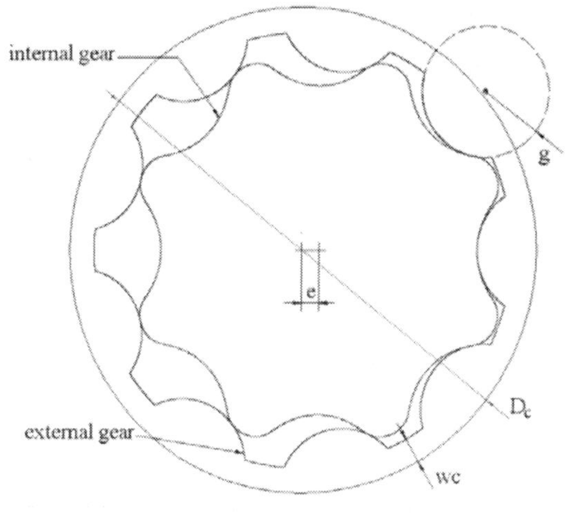

Figure 4. The basic geometrical parameters of the corrected epicycloidal gear.

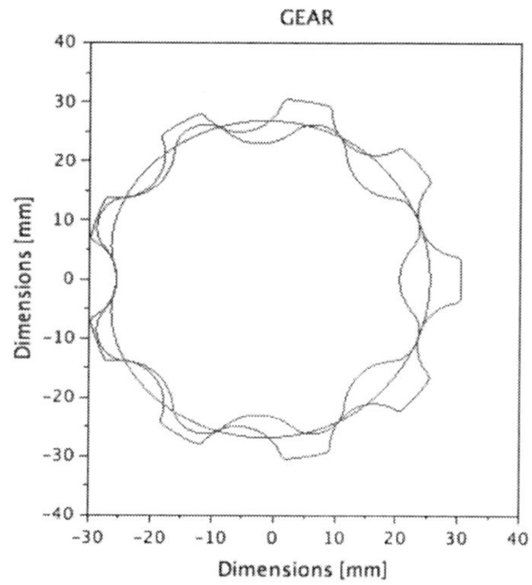

Figure 5. Gear set and line of contact by using GeroLAB [22].

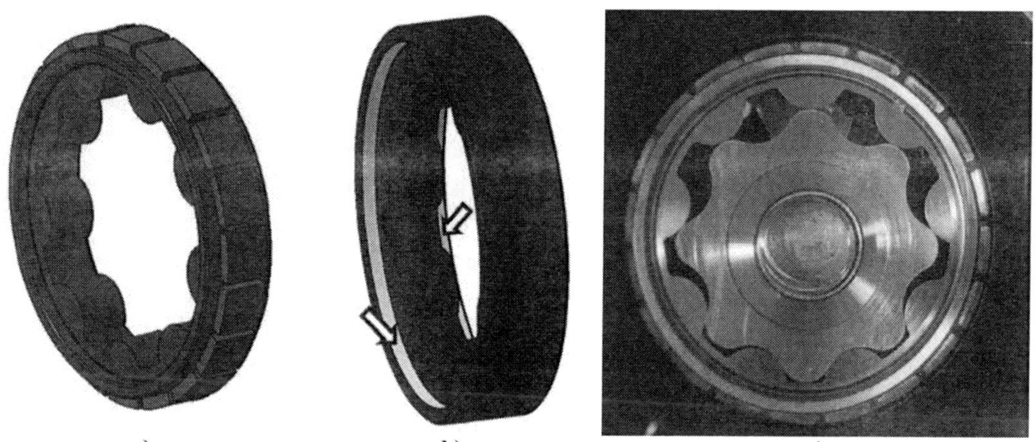

Figure 5. a) Pole pieces fixed on the external gear, in grey; b) guiding internal surface, in grey indicated with arrows; c) GeroMAG prototype with an in house manufactured transparent polymer pump cover.

In addition, the coaxially of each gear in its own rotation centre, as well as the eccentricity has to be guaranteed. A guiding internal surface is designed to guide the rotational movement of the outer gear. The guide is performed by the path of the inner surface in both sides of the external gear teeth faces (see Figure 5b).

This novel configuration accomplishes a standard volumetric flow rate at low rotational speed with satisfactory volumetric efficiency (see Figure 5c). GeroMAG concept is believed to lead to a cleaner and noiseless hydraulic technology to be part of a wider part of sectors. Moreover, if the configuration of the magnetic-driven outer rotor is also transferred to the internal gear, becoming likewise a magnetic-driven inner rotor, then the rotational movement could be performed with zero inter-teeth contact stress by controlling the electrical torque of both rotors.

Another example of new concept in gerotor technology is the magnet-sleeve-sealed mini gerotor pump [20]. This novel variable flow pump configuration, despite of the difficulties occurred and several disappointed fallouts obtained, such as frictional performance of the gear units, accomplishes a standard volumetric flow rate at low rotational speed with satisfactory volumetric efficiency. Experimental results prove feasibility as well as proof of concept.

4. Designing of gerotor for hydraulic machines according to Kragujevac

4.1 Mathematical description of gerotor tooting profile

Analysis takes into account the case where exterior gear of the trochoidal pump has one tooth more than the interior gear. The interior gear profile is described by peritrochoid equidistant, while the exterior one is described by the circular arc with radius r_c. The basic geometrical relations during unmodified and modified peritrochoid gearing generation are illustrated by Figure 6a. Peritrochoid is presented by the point *D,* fixed in a circle plane of radius r_a, which is rolling inward side on the outside of an immovable circle of the radius r_t. The equations of the peritrochoid are defined in the coordinate system of the trochoid $O_t x_t y_t$, according to Figure 6a [26].

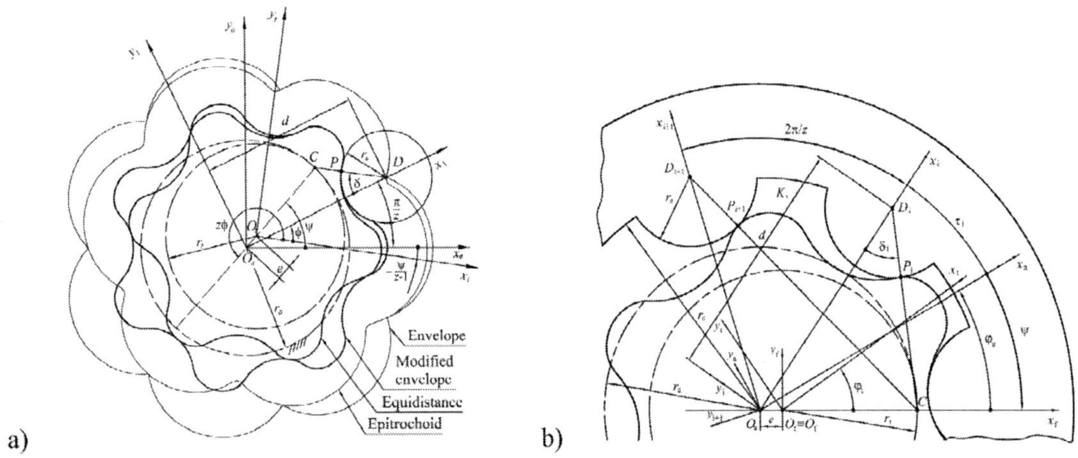

Figure 6. Gerotor gearing: a) generation of the trochoidal gearing; b) Schematic presentation of the gerotor pump gear pair with fixed shaft axes.

Figure 6.a illustrates that during the relative moving of kinematic circles, point *D* generates peritrochoid, while point *P* generates equidistant. The angle specified as δ is an angle between the normal *n-n* and radius vector of the point *D*, and can be defined as leaning angle.

Since at the trochoidal gearing is characterized with simultaneous meshing of all the teeth, it is necessary to defines general equations of the profile points coordinates applicable to all the teeth. The internal gear profile (Figure 6b) is defined by a position vector of the contact point P_i in the coordinate system of the trochoid $O_t x_t y_t$ using the following equations:

$$
\begin{aligned}
x_{P_i}^{(t)} &= e\left\{ z_2\lambda_c \cos\left[\tau_i + \frac{\psi}{z_2-1}\right] - \cos\frac{z_2\psi}{z_2-1} - c\cos\left[\tau_i + \frac{\psi}{z_2-1} + \delta_i\right]\right\} \\
y_{P_i}^{(t)} &= \left\{ z_2\lambda_c \sin\left[\tau_i + \frac{\psi}{z_2-1}\right] - \sin\frac{z_2\psi}{z_2-1} - c\sin\left[\tau_i + \frac{\psi}{z_2-1} + \delta_i\right]\right\}
\end{aligned} \tag{19}
$$

where $r_c = ce$, $d = \lambda_c e z_2$,, i is the ordinal number of the tooth, $\tau_i = \dfrac{\pi(2i-1)}{z_2}$,

$\delta_i = \arctan\dfrac{\sin(\tau_i - \psi)}{\lambda_c - \cos(\tau_i - \psi)}$ and $\psi = (z_2 - 1)\left(\phi - \dfrac{\pi}{z_2}\right)$.

The derived equations can be applied to different kinematic schemes of trochoidal pumps. They enable modelling of the gear pair meshing during the rotation, as well as determination of necessary parameters for any teeth pair at an arbitrary moment. In pumps with fixed shaft axes where the drive shaft is fixed to the internal gear (Figure 6b), the following expressions define the relations between the referent angle ψ and the shaft rotation angles $\varphi_a = -\psi$ and $d\varphi_t = \frac{z_2}{z_2 - 1} d\varphi_a$.

To unify the various geometric design of working units of hydraulic machines, parameters used in the chapter 2 can be related to variables used in this chapter as follows: $r_b = r_t d / r_a$, $\rho = \lambda_c e$, $g = r_c$, $\lambda = 1/\lambda_c$, $v = c/\lambda_c$ and $\eta = \phi$.

4.2 Gerotor functional characteristics

In order to obtain functional dependency which would provide projecting of the pump gear pair, based on the given starting data, a mathematical model of trochoidal gearing pump capacity characteristics has been developed. First to be considered was distribution of working fluid and definition of characteristic phases in the pump operating cycle, followed by description of the methods for defining working capacity and actual rate of delivery of a pump. Based on the analysis of results, relevant values were identified which have influence on pulsation of the rate of delivery and uneven flow.

Concerning of conditions which reduce contact forces creation, and thus reduce wear, analyzed were forces and moments which influence the gear pair of the trochoidal rotational pump. Starting with the specific conditions in which the load is transmitted simultaneously at a number of contact points, considered were fluid thrust force which affect the sides of the gear teeth, contact forces and contact stress [27, 28, 29]. To check analytical methods and determine error which occurs when applying above noted postulations, the finite element method was applied (Figure 7).

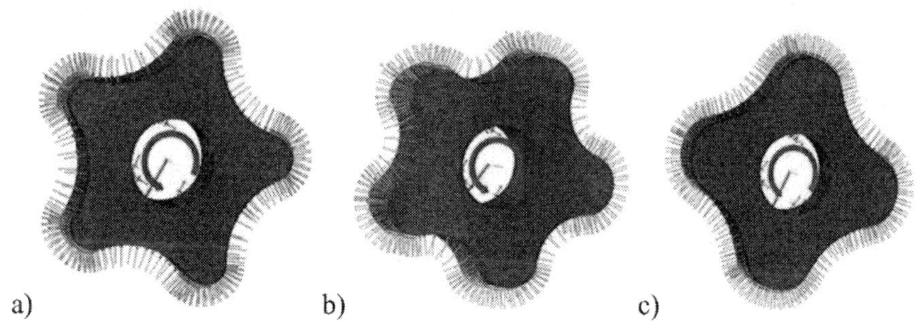

Figure 7. Models of the discrete gears simulation with forces defined and visualization of reaction support: a) z_2=6, λ_c=1.575 b) z_2=6, λ_c=1.375 c) a) z_2=5, λ_c=1.85.

Maximum contact stresses, forces and moments were analysed using analytical and numerical methods, taking into consideration the pressure variation in the pump chambers due to the fluid flow [26]. It should be pointed out that in the described investigations the theoretical trochoidal profile was considered, with presumptions on ideally precise geometrical characteristics. Starting from the fact that the real profiles are made with inevitable technological gaps, it is considered modelling of real meshing profiles [30].

4.3 Methodology for selection of the optimum teeth profile

Selection of the optimum profile should begin by defining the initial values of the designed gear pair. External load is one of the in advance given values, and it is defined through the working pressure and the pump volume. The geometric characteristics that remain constant and that are determined with the overall size of the gear pair are the following: the gear width *b*, the eccentricity *e* and the external gear root radius expressed by the meshing envelope parameter S_{fa} [27]. It is necessary to determine the external gear teeth number z_2 and the values of the coefficient λ_c and c, which define the optimum shape of the teeth profile. The convex-convex contact is taken into consideration. Since the convex part of the external gear profile is a circular arc of the radius r_c, the problem comes down to the analysis of the contact at the point of the internal gear profile with the smallest curvature radius. In that case, the equivalent curvature radius has the lowest value for the given profile and it can be calculated using the equation:

$$(\rho_{ekv})_{\min} = \frac{r_c\,\rho_{c\min}}{r_c + \rho_{c\min}}. \tag{20}$$

Introducing the expression for $\rho_{c\min}$

$$\rho_{c\min} = e\left(z_2 \sqrt{\left(\frac{3}{z_2+1}\right)^3 \left(\lambda_c^2 - 1\right)\left(z_2 - 1\right)} - c \right) \tag{21}$$

into the equation (22), taking into account the equation (21), obtaining the first derivate of the equation (20) per the parameter λ_c and then setting it equal to zero, we get the condition for the existence of the maximum equivalent radius of the curvature $(\rho_{ekv})_{\min}$ dependent on λ_c, in the form of:

$$\frac{1}{\left(z_2\lambda_c + 2 - S_{fa}\right)^2} + \frac{\left(\frac{3}{z_2+1}\right)^3 (z_2 - 1)\lambda_c \left[\left(\frac{3}{z_2+1}\right)^3 (z_2-1)\left(\lambda_c{}^2 - 1\right)\right]^{-\frac{1}{2}} - 1}{\left\{ z_2 \left[\left(\frac{3}{z_2+1}\right)^3 (z_2-1)\left(\lambda_c^2 - 1\right)\right]^{\frac{1}{2}} - z_2\lambda_c - 2 + S_{fa} \right\}^2} = 0. \tag{22}$$

As this equation is *transcendental* and the parameter λ_c cannot be expressed explicitly, the solution is obtained by an iterative procedure. Thus, chosen value is further checked against the complex criteria. It has been shown that gear pairs can be generated based on minimizing the equivalent curvature of the profile of the tooth in contact, when the condition that the specific sliding of the meshing profiles are equal is fulfilled. This procedure has been applied to a commercial pump, and the obtained solution has been proposed as a better solution for the given conditions. The analysis results have shown that, for the same working conditions, significant reductions in maximum contact stresses can be achieved. For the same teeth number, when the value of the parameter λ_c is decreased, the maximum contact stress can be reduced by about 16.8%, while for the smaller number of teeth, with an increase in the value of the parameter λ_c, approximately 35% lower values are obtained compared to the commercial pump [27]. The results obtained in this paper have wide and practical application especially in design of gerotor pumps with better performance characteristics.

4.4 Experimental verification

After the analytical check-up of the theoretical models, an experimental verification of results was conducted. Measuring of flow rate and volumetric efficiency was conducted on three different models of gear pairs (Figure 8) with simulation of real conditions of pump exploitation, in the same pump housing. One of them was a commercial gear pair, shown in Figure 8,a and the other two gear pairs, whose profiles were derived from calculations and proposed as optimal are made and shown in Figure 8,b and Figure 8,c.

a) 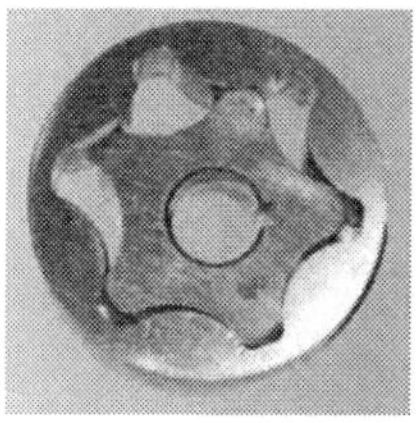 b) 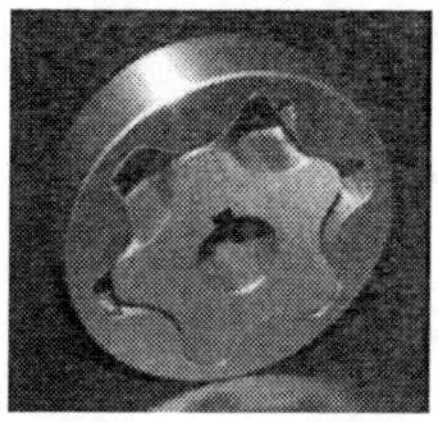 c) 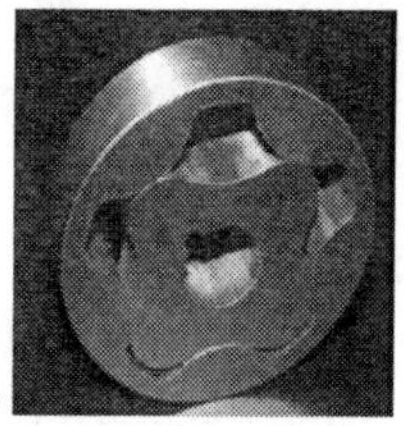

Figure 8. Models of three tested gear pairs: a) Commercial model b) Optimal model, z_2=6; c) Optimal model, z_2=5.

5. Conclusions

Wrocław Fluid Power Research Group assumes that gerotor and orbital systems should be treated as gear systems. Following that assumption, the scientists of the group has developed the knowledge of the fundamentals of designing and manufacturing of cycloidal gears and gear systems constituting gerotor and orbital systems. It is complex knowledge which includes geometry, kinematics, strength, hydraulics and cycloidal gears technology. For the description of those issues, an integrated system of technical parameters has been used, which is similar to the one applied for the classic involute gears.

The research group IAFARG & LABSON from Terrassa considers the gerotor pump as a hydraulic machine of which pumping action is carried out by means of a trochoidal-envelope toothed gear set. The research group unifies several approaches to characterize performance indexes either an existing gerotor pump or a new-born gerotor unit. These approaches mainly are dynamic simulation by using bond graph technique, computational fluid dynamics numerical simulation and experimental work.

Contribution of the Kragujevac group research lays in the detailed analysis of the effects of the geometric parameters of the trochoidal gearing on the reduction of contact stresses performed using the analytical-numerical method and in the new possibilities it opens up for further studies. The developed mathematical model and the obtained results can be of use to the constructors of the gerotor pump and motors for choosing the best constructive solutions that reaches higher coefficient of efficiency.

Acknowledgment

This research was funded by the Ministry of Education and Science of the Republic of Serbia under the contracts TR35033 and TR33015. We would like to thank to the Ministry of Education, Science and Technological Development of Republic of Serbia for their financial support during these investigations.

References

[1] Shannikov V 1948 *Planietarnyje rieduktory s viecentroidalnym cievochnym zacieplieniem*, Mashgiz
[2] Litvin F L 1968 *Tieoria zubcatyh zacieplenі*, Moskva, Nauka
[3] Stryczek J 2007 *Koła zębate maszyn hydraulicznych*, Oficyna Wydawnicza Politechniki Wrocławskiej, Wrocław
[4] Nang A and Maiti R 2012 Unification of epitrochoid origin profile design approaches for external lobed star member used in hydrostatic and gear units, *Journal Mechanical Science* **227**(2) 299-310
[5] Gamez-Montero P J, Castilla R and Codina E 2018 Methodology based on best practice rules to design a new-born trochoidal gear pump, *Proceedings of the Institution of Mechanical Engineers, Part C: Journal of Mechanical Engineering Science* **232**(6) 1057-1068
[6] Ivanovic L and Josifovic D 2006 Specific Sliding of Trochoidal Gearing Profile in the Gerotor Pumps, *FME Transactions* **34** 121-127
[7] Gamez-Montero P J, Castilla R, Khamashta M and Codina E 2006 Contact problems of a trochoidal gear pump, *International Journal of Mechanical Science* **48**(12) 1471-1480
[8] Maiti R and Sinha G L 1988 Kinematics of active contact in modified epitrochoidal generated rotary piston machines, *Mechanism and Machine Theory* **23**(1) 39-45

[9] Bronsztejn J N and Siemiendiajew K A 1970 *Matematyka, poradnik encyklopedyczny*, PWN, Warszawa
[10] Pham Duc Nhuan 1978 *O doborze parametrów konstrukcyjnych i modelowaniu pewnych maszyn hydraulicznych o zazębieniu cykloidalnym*, Politechnika Warszawska, Warszawa, Doctoral Thesis
[11] Lobastov V K 1978 *K tieorii Vniecenttoidnogo epicykloidalnego zacieplenia planietarnych miechanizmov*, Tieoria Miechanizmov i Mashin, Izdatielstvo Charkovskogo Gosudarstviennogo Univieristeta
[12] Colbourne J R 1974 The geometry of Trochoid Envelopes and Their Application in Rotary Pumps, *Mechanism and Machine Theory* **4**
[13] Stryczek J 1990 Projektieren der Zykloidenverzahnungen Hydraulischer Verdrangermaschinen, *Mechanism and Machine Theory* **25**(6) 597-610
[14] Stryczek J 2003 Gerotor und Orbitmaschinen. Projektiengsgrundlagen, *Oelhydraulik und Pneumatik* **10** 670-673
[15] Stryczek J 2005 Gerotor und Orbitmaschinen. Konstruktionen fuer neue technische Anvendungen, *Oelhydraulik und Pneumatik* **6** 386-390
[16] Wusthof P and Schneider J 1981 Grundlagen der Planetmotoren-Vorstchlung eines neuen Planetmotors, *Oelhydraulik und Pneumatik* **2**
[17] Stryczek J 2018 *Podstawy projektowania hydraulicznych mazyn zębatych*, Wydawnictwo Naukowe PWN SA, Warszawa
[18] Stryczek J, Bednarczyk S and Biernacki K 2014 Gerotor pump with POM gears: Design, production technology, research, *Archives of Civil and Mechanical Engineering* **14**(3) 391-397
[19] Stryczek J, Banaś M, Krawczyk J, Marciniak L and Stryczek P 2017 The fluid power elements and systems made of plastics, *Procedia Engineering* **176** 600-609
[20] Gamez-Montero P J, Antoniak P, Castilla R, Freire J, Krawczyk J, Stryczek J and Codina E 2017 Magnet-Sleeve-Sealed Mini Trochoidal-Gear Pump Prototype with Polymer Composite Gear, *Energies* **10** 1458
[21] Bednarczyk S 2015 Designing the Hypocycloidal Gearing in the Planetary Transmission, *Machine Dynamics Research* **39**(3) 5-23
[22] GeroLAB Package System, Available online: http://www.gerolab.es (accessed on 15 May 2019)
[23] Gamez-Montero P J, Castilla R, Mujal R, Khamashta M and Codina E 2009 GEROLAB package system: Innovative tool to design a trochoidal-gear pum, *Journal of Mechanical Design, Transactions of the ASME* **131**(7) 0745021-0745026
[24] Gamez-Montero P J, Garcia-Vilchez M, Raush G, Freire, J and Codina E 2012 Teeth clearance and relief grooves effects in a trochoidal-gear pump using new modules of GeroLAB, *Journal of Mechanical Design, Transactions of the ASME* **134**(5)
[25] Gamez-Montero P J, Castilla R, Codina E, Freire J, Morató J, Sanchez-Casas E and Flotats I 2017 GeroMAG: In-house prototype of an innovative sealed, compact and non-shaft-driven gerotor pump with magnetically-driving outer rotor, *Energies* **10**(4)
[26] Ivanović L, Devedžić G, Mirić N and Ćukovic S 2010 Analysis of forces and moments in the gerotor pumps, *Proceedings of the Institution of Mechanical Engineers. Part C: Journal of Mechanical Engineering Science* **1**(1) 1-13
[27] Ivanovic L 2016 Reduction of the maximum contact stresses by changing geometric parameters of the trochoidal gearing teeth profile, *Meccanica* **51**(9) 2243-2257
[28] Ivanović L and Matejić M 2018 FEM analysis of gerotor machines impeller with planetary motion, *IETI Transactions on Engineering Research and Practice* **2**(2) 1-8
[29] Ivanović L, Rakić B, Stojanović B and Matejić M 2016 Comparative analysis of analytical and numerical calculations of contact stresses at rotational elements of gerotor pumps, *Applied Engineering Letters* **1**(1) 1-7
[30] Ivanovic L, Devedzic G, Cukovic S and Miric N 2012 Modeling of the Meshing of Trochoidal Profiles With Clearances, *Journal of Mechanical Design* **134**(4) 041003-1 - 041003-9

Macro-roughness height determination of teeth surfaces obtained by gear generating method

D T Babichev[1], D A Babichev[2] and S Yu Lebedev[1]

[1]Industrial University of Tyumen, Institute of Transport, Melnikayte Street 72, 625039 Tyumen, Russia
[2]Sibur Holding, ZapSibNeftekhim LLC, East Industrial District, Block 1, No. 6, Building 30, 626150 Tobolsk, Russia

E-mail: lebedevsergey1995@gmail.com

Abstract. The article describes three types of gear teeth surfaces: smooth (envelope), striped (with combs) and scaled (with pyramids and combs connecting the pyramids). The method for combs and pyramids height (above an envelope) determination based on "velocity and acceleration of penetration" conceptions is introduced. Acceleration of penetration allows determining an actual workpiece surface deviation from a theoretical envelope without generating surface knowing. The computer simulations have been carried out, and the correlations between combs and pyramids height and gear machining parameters are exposed.

1. Introduction

Teeth surfaces produced by gear generating method can be classified into three following categories: smooth, stripped and scaled surfaces.

Smooth surfaces have no machining marks – i.e., macro-roughness (initial roughness). The smooth surface is an envelope $\sum_2$ of a one-parametric family of generating surfaces $\sum_1$ if $\sum_1$ (i.e., a cutting tool initial surface (CTIS)) is an actual surface. Teeth surface obtained by gear knurling is an example of a smooth surface structure.

Stripped surfaces have stripe-like (combs) machining marks, generally, of variable height and width (Fig.1). Stripped teeth surfaces can be obtained by generation through:

- ✓ a two-parametric family of generating surfaces when CTIS is an actual surface. The examples are: **1.** Involute gears grinding with cylinder and straight wheels (stripes are along the tooth line). **2.** Grinding with abrasive worm grinding wheel with a profile in axial section close to a wheel profile in normal section (stripes are along the tooth profile – i.e., the lay pattern coincides with the generating direction).
- ✓ of a one-parametric family of generating surfaces when CTIS is a cutting edge mark. The example is cylindrical gears shaping (stripes are along the tooth line).

Fig. 1 demonstrates a helical involute gear tooth grinded with a straight wheel. Spur involute gear side surface is composed of multitude plane surfaces touching calculated involute surface along the entire plane surface length. In these conditions, adjacent surfaces intersect at lines, creating certain surface texture. For teeth number z=18 and the number of wheel motions for teeth space grinding k_X =10 the value of the surface deviation angle is equal to $\Delta\alpha$=2^0.

Scaled surfaces are generally composed of parallelogram (upper image in Fig. 2) or hexagonal (lower image in Fig. 2) pits. Pit central points touch calculated tooth surface. Scaled surfaces were first systematized by B.K. Shunaev [1] while investigating coarse-pitch cylindrical gears machining by diagonal involute gear hobbing cutters.

Smooth surfaces are called *envelope* whereas rough surfaces (stripped and scaled) were called by G.I. Shvelyova [2, p.144] *wraparound surfaces* (Fig. 3).

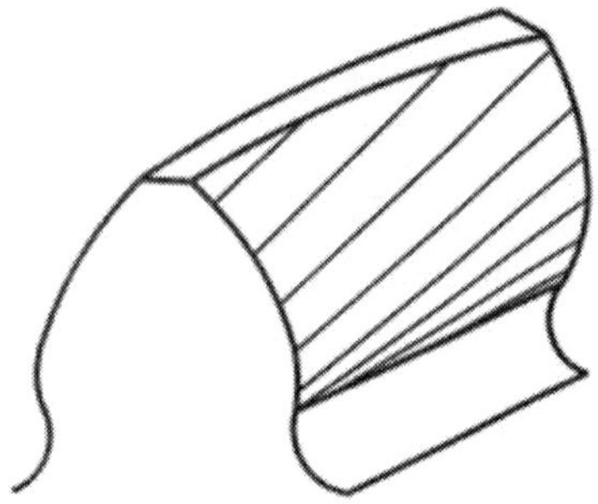

Figure 1. Stripped tooth surface.

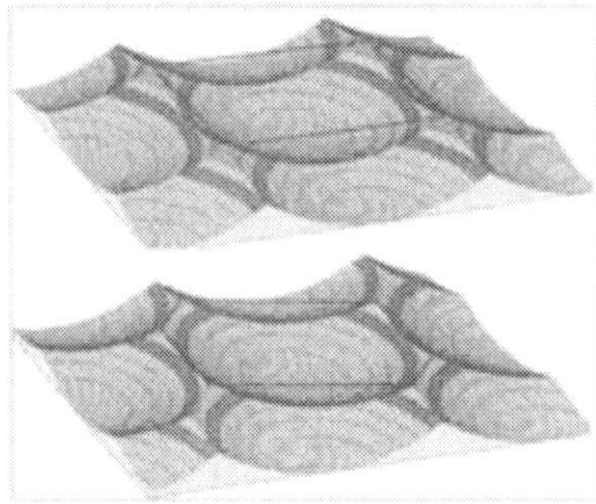

Figure 2. Scaled teeth surfaces.

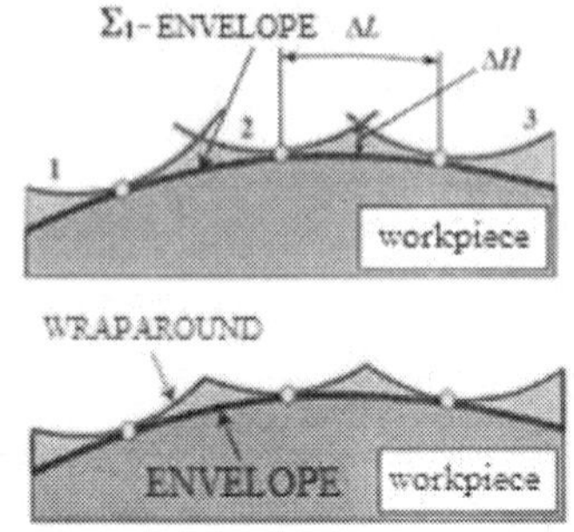

Figure 3. Envelope and wraparound.

The stripped surface is divided into a set of stripe-like surfaces each of which touches envelope$\sum_2$ along a line. Compared to the stripped surface each scaled surface pit touches envelope at a point. The wraparound surface is always above the envelope surface. Determining the heights difference – ΔH in fig. 3 – for particular machining types is considered a quite challenging task.

Typically to complete the task the envelope and points on the intersection line of complex adjacent pits or stripes surfaces should be found. In the article another method for determining maximum macro-roughness value ΔH_{max} is proposed:

- ✓ ΔH_{max} is calculated simultaneously with the envelope determination by means of a computer program.
- ✓ Quite simple formulae based on stripes, pits and combs forms and dimensions research are used. The research and formulae are presented in the article.
- ✓ Kinematic method for determining surfaces and lines produced by generating process is used. The method is described in [3-5] and is different from all known in the classical theory of gearing methods.

The abovementioned method is based on «velocity of penetration» V_N (proposed by V.A. Shyshkov [6]) and «acceleration of penetration» a_N [3-4] conceptions. The following method features can be denoted: 1) for the conjugated surface Σ_2 points determination, simple formulae of kinematics of a point are used (solving of meshing equation is not required); 2) for the surface Σ_2 point coordinates determination it is necessary to know: (x, y, z) coordinates of a point on a generating surface Σ_1; unit normal vector projections n_x, n_y, n_z on Σ_1; penetration velocity V_N and acceleration a_N of the point on Σ_1. The above-described approximate kinematic method serves for determining envelope surface points if wraparound points are found. The application example of the method including calculations accuracy estimation is considered in [4]. In the work [5] advanced method is presented: another new geometric-kinematic surface movement index at a point of space (a derivative of penetration acceleration a_N – i.e., the second derivative of penetration velocity V_N) was implemented for the first time. The new index allows increasing accuracy of the method by 2-3 orders of magnitude.

A brief account of «velocity and acceleration of penetration» terms that are fundamentals of the kinematic method described in [3] will be given further in the article.

2. Velocity of penetration

The velocity of penetration V_N is a relative velocity vector projection $\mathbf{V}_{12}$ on a normal $\mathbf{N}$ to a cutting tool initial surface (CTIS). V_N [6] is determined as a dot product of $\mathbf{V}_{12}$ with a unit normal vector $\mathbf{n}$ to CTIS. The unit normal vector *is directed away from CTIS:*

$$V_N = \mathbf{V}_{12} \cdot \mathbf{n} = V_{12x} \cdot n_x + V_{12y} \cdot n_y + V_{12z} \cdot n_z \quad (1)$$

IOP Conf. Series: Materials Science and Engineering **659** (2019) 012007 doi:10.1088/1757-899X/659/1/012007

if $V_N > 0$ – CTIS penetrates in workpiece; if $V_N < 0$ – CTIS moves off machining surface; if $V_N = 0$ – CTIS slides along machining surface. If $V_N = 0$, a cutting tool forms an envelope workpiece surface (at an envelope and a wraparound contact points).

3. Acceleration of penetration

The acceleration of penetration a_N is an acceleration of CTIS penetration in the machining workpiece, i.e., a_N – a time derivative of the penetration velocity V_N. Velocity and acceleration of penetration are *scalar values*: if $V_N>0$ – CTIS comes closer to machining surface at a set point; if $V_N< 0$ – CTIS moves off machining surface; if $a_N>0$ – penetration velocity increases. Differentiate (1) with respect to time t:

$$a_N = \frac{dV_N}{dt} = \frac{d}{dt}(\mathbf{V}_{12}\cdot\mathbf{n}) = \mathbf{a}_{12}\cdot\mathbf{n} + \mathbf{V}_{12}\cdot\dot{\mathbf{n}} \tag{2}$$

where $\mathbf{a}_{12}$ – relative acceleration of a point sliding along CTIS with "$-\mathbf{V}_{12}$" velocity.

The process of a_N design equations derivation from (2) is tedious. Therefore, final formulae for generalized meshing (fig. 4) are presented further in the article. In fig. 4 both meshing links are in helical motion around its axes while the axes are in helical motion around the interaxial perpendicular. The motion characteristics of the links are impact-free: motion parameters (ω_1, ω_2, ω_3, V_1, V_2, и V_3,) can be variable or constant in time, can be positive, negative or equal to zero. All the parameters are considered a one or two independent motion parameters function (t_1, t_2) which stands for a one- or two-parametric motion (generation). CTIS is on the link 1, the machining surface is on the link 2; the fixed coordinate system is $X_hY_hZ_h$. Any manufacturing meshing can be inserted in the described coordinate system.

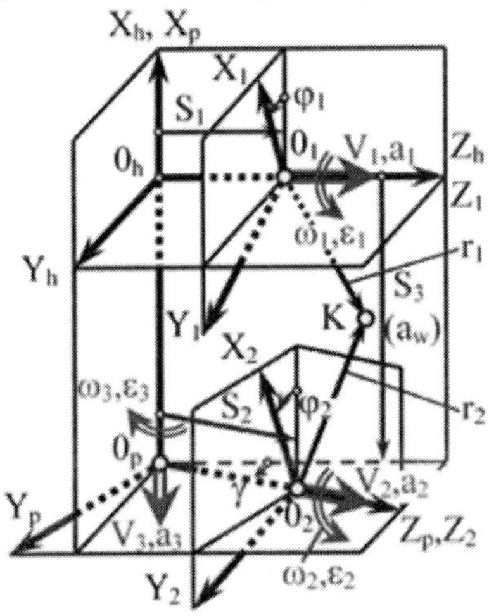

Figure 4. Generalized meshing.

Formulae for calculation of penetration acceleration a_N in generalized meshing (Fig. 4):

The relative angular velocity vector $\boldsymbol{\omega}_{12}$ in the system of coordinates $X_hY_hZ_h$:

$$\boldsymbol{\omega}_{12}^{(h)} = \{\omega_3\cdot\mathbf{i},\ -\omega_2\cdot\sin\gamma\cdot\mathbf{j},\ (\omega_1-\omega_2\cdot\cos\gamma)\cdot\mathbf{k}\} \tag{3}$$

- The relative linear velocity vector $\mathbf{V}_{12}$ in the system of coordinates $X_hY_hZ_h$:

$$\left.\begin{aligned} V_{12x}^{(h)} &= V_3 - \omega_1\cdot y_h + \omega_2\cdot y_p;\\ V_{12y}^{(h)} &= -V_2\cdot\sin\gamma + \omega_1\cdot x_h - \omega_2\cdot x_p\cdot\cos\gamma - \omega_3\cdot z_h;\\ V_{12z}^{(h)} &= V_1 - V_2\cdot\cos\gamma + \omega_2\cdot x_p\cdot\sin\gamma + \omega_3\cdot y_h. \end{aligned}\right\} \tag{4}$$

- The relative acceleration vector $\mathbf{a}_{12}$:

$$\left.\begin{aligned} \mathbf{a}_{12} = {} & [(\mathbf{a}_1-\mathbf{a}_2-\mathbf{a}_3)+(\boldsymbol{\varepsilon}_1-\boldsymbol{\varepsilon}_2-\boldsymbol{\varepsilon}_3)\times\mathbf{r}_1+\boldsymbol{\varepsilon}_2\times\mathbf{S}_3-(\boldsymbol{\varepsilon}_2+\boldsymbol{\varepsilon}_3)\times\mathbf{S}_1]+\\ & +[\boldsymbol{\omega}_1\times(\boldsymbol{\omega}_1\times\mathbf{r}_1) - \boldsymbol{\omega}_3\times(\boldsymbol{\omega}_2\times\mathbf{S}_2)-(\boldsymbol{\omega}_2+\boldsymbol{\omega}_3)\times((\boldsymbol{\omega}_2+\boldsymbol{\omega}_3)\times\mathbf{r}_2)]-\\ & -[2\cdot(\boldsymbol{\omega}_3\times\mathbf{V}_2) - (\boldsymbol{\omega}_3\times\boldsymbol{\omega}_2)\times\mathbf{r}_2]-2\cdot(\boldsymbol{\omega}_2+\boldsymbol{\omega}_3)\times\mathbf{V}_{12}-\boldsymbol{\omega}_{12}\times\mathbf{V}_{12} \end{aligned}\right\} \tag{5a}$$

Excluding from (5a) zero summands for cylindrical, bevel and worm gears we get:

$$\mathbf{a}_{12} = \boldsymbol{\omega}_1\times(\boldsymbol{\omega}_1\times\mathbf{r}_1)-\boldsymbol{\omega}_2\times(\boldsymbol{\omega}_2\times\mathbf{r}_2)+(\boldsymbol{\omega}_1-\boldsymbol{\omega}_2)\times\mathbf{V}_{12} \tag{5b}$$

- Acceleration of penetration a_N:

$$a_N = \mathbf{a}_{12}\cdot\mathbf{n}+\mathbf{V}_{12}\cdot\dot{\mathbf{n}} = \mathbf{a}_{12}\cdot\mathbf{n}+\mathbf{V}_{12}\cdot\mathbf{c}\cdot\omega_K+\mathbf{V}_{12}\cdot(\mathbf{c}\times\mathbf{n})\cdot\omega_K^{\perp} \tag{6}$$

where ω_K – angular rolling velocity of generating surface moving over machining surface towards the vector **c**; $\omega_K^{\perp}$ – angular rolling velocity of generating surface moving over machining surface in a plane perpendicular to the vector **c.** These parameters are calculated according to the following formulae:

$$\omega_K = (\mathbf{n}\times\mathbf{c})\cdot\boldsymbol{\omega}_{12} + \frac{\mathbf{V}_{12}\cdot\mathbf{c}}{R_R^1} + \frac{\mathbf{V}_{12}\cdot(\mathbf{n}\times\mathbf{c})}{R_K^1} \tag{7a}$$

$$\omega_K^{\perp} = \mathbf{c}\cdot\boldsymbol{\omega}_{12} + \frac{\mathbf{V}_{12}\cdot(\mathbf{n}\times\mathbf{c})}{R_R^2} - \frac{\mathbf{V}_{12}\cdot\mathbf{c}}{R_K^1} \tag{7b}$$

where **c** – unit tangent vector setting normal profile direction. R_R^1 и $R_R^{1\perp}$ – CTIS normal profiles radii of curvature corresponding to the profile along the tangent line **c** and the profile perpendicular to the unit normal vector **c** (if $R_R>0$ – body is convex). R_K – radius of geodesic torsion of a line on a surface along the tangent line **c** (if $R_K>0$ – twist is counterclockwise).

All the three formulae (7) components are rotating components in a normal plane section. The first component of (7a) is a relative angular velocity projection $\omega_{12} = \omega_1 - \omega_2$ on a normal to a cutting plane, i.e., rotation of the normal is caused by links relative motion. The second component of (7a) comprises rotation of the normal caused by bodies sliding *in a normal plane section* due to its surface curvature. The third component of (7a) comprises rotation of the normal caused by bodies sliding *toward perpendicular to a normal plane section* due to geodesic torsion of a line on a surface.

4. Macro-roughness height

"Combs" height ΔH above an envelope surface (see fig. 2) determination formulae were obtained in [3]. In the article, the formulae are given without derivation and proofs.

4.1. Height ΔH in case of one-parametric generation

The following two cases can be specified:

Case 1 – Generating by a cutting tool initial surface (CTIS) – an actual surface (gears grinding and honing) or a cutting edge mark (cylinder gear wheels cutting by shaping cutters, bevel wheels milling, etc.). Stripes and combs are directed *along a tooth line.*

Case 2 – Generating by a cutting tool initial surface (CTIS) – CTIS is a surface produced by a forming cutting edge motion in template-cutting method (gear wheels machining by involute side and end milling cutters and also toolheads). Stripes and combs are directed *along a tooth profile.*

In both cases "combs" height ΔH above an envelope surface can be determined according to the formula:

$$\Delta H = -0{,}125\cdot a_N\cdot \Delta t^2 \tag{8}$$

where a_N – *generating tool* penetration acceleration, calculated at a contact point of an instrument (its CTIS) with machining surface; Δt – period of time between two moments of adjacent stripes cutting determined according to one of the formulae:

$$\Delta t = \frac{2\cdot\pi}{\omega_0\cdot z_0\cdot k_X} = \frac{2\cdot\pi}{\omega_1\cdot z_1\cdot k_X} = \frac{2\cdot\pi}{\omega_0\cdot z_0\cdot k_Z} \tag{9}$$

where ω_0, ω_1 – angular tool and workpiece velocities respectively wherein acceleration of penetration a_N is calculated; z_0, z_1 – shaping cutter and workpiece teeth number respectively; k_X – number of shaping cutter or rack-type tool strokes per one teeth machining; k_Z – involute gear cutter chip grooves number or toolheads cutters number.

4.2. Height ΔH in case of two-parametric generation

The example of two-parametric generation is cylindrical gears machining by involute gear hobbing cutters. In this case, pits can be parallelogram (or rectangular, which is a particular case) or hexagonal in shape. Parallelogram pit shape can be only obtained when pits form *longitudinal and transverse rows* directed *precisely along generation and feed directions* respectively – see fig. 5 and 6. Even if a single one of the conditions is not fulfilled, pits will be hexagonal in shape (fig. 6).

Shapes and ***heights of all the corners*** of parallelogram pits (as a projection on a tangent plane) ***are equal***. The corners are tetrahedral pyramids (pyramidal asperities) in shape (fig. 5 and 6). In general case, pits and pyramids are "bevelled", i.e., angle $\theta \neq 90^\circ$ (see fig. 5). Tetrahedral pyramids height above an envelope surface – maximum faceting height – is determined as a sum of two heights (*$\Delta=\Delta_1+\Delta_2$ in fig.* 5) produced by generating and feeding respectively:

$$\Delta H_{MAX} = -0{,}125 \cdot (a_{NV} \cdot \Delta t_V{}^2 + a_{NS} \cdot \Delta t_S{}^2) \quad (10)$$

where a_{NV}, a_{NS} – generation and feed acceleration of penetration respectively;Δt_v – time required for shaping an adjacent pit set *along the generation direction*; ;Δt_S– time required for shaping an adjacent pit set *along the feed direction*.

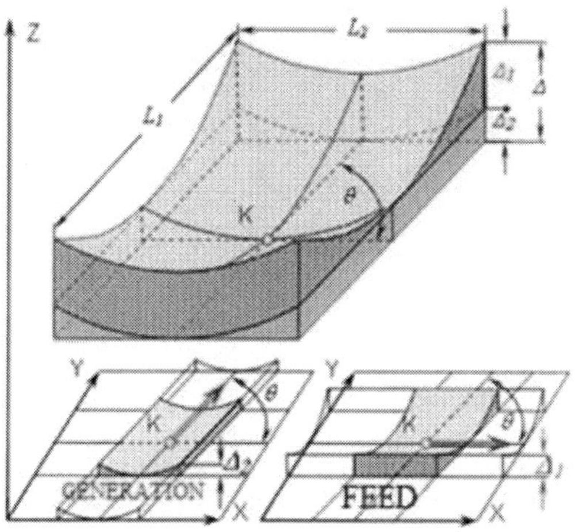

Figure 5. Parallelogram pit in case of two-parametric generation

Parameters ;Δt_v and ;Δt_S are determined according to the formulae:

$$\Delta t_V = \frac{60}{n_0 \cdot z_{0k}} \quad ; \quad \Delta t_S = \frac{60}{n_1 \cdot z_0} \quad (11)$$

where n_0, n_1 – instrument and workpiece numbers of revolutions per minute used for penetration acceleration calculation; z_0,z_{0k} –number of cutters and cutter chip grooves respectively.

Shapes of hexagonal pits (as a projection on a tangent plane) can vary significantly depending on the coefficient k_L – pits displacement ΔL (along a row) in adjacent rows to distance between pits L (in a row) ratio (shown on the right of fig. 6). In the case of hexagonal pits, pyramids are triangular. Figures 5 and 6 indicate that: three surfaces converge at all vertices of hexahedron angles, whereas at the vertices of parallelogram pits four surfaces converge.

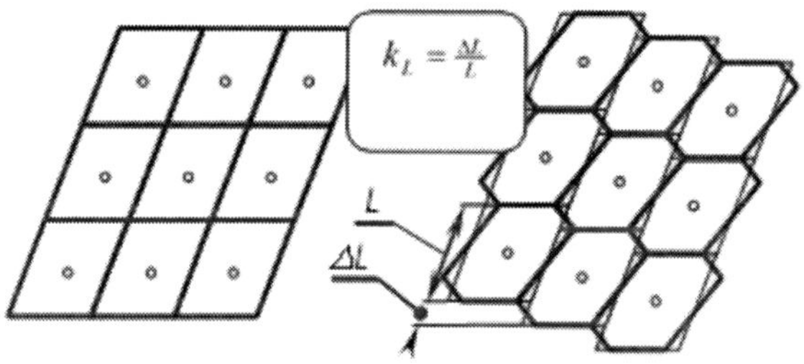

Figure 6. Pits shapes: parallelogram (if $\Delta L = 0$) and hexagonal (if $\Delta L \neq 0$)

For hexagonal pits obtained by a two-motions (generating and feeding) gear machining method it was analytically determined and computationally proven in [3] that: *all the six vertices heights are equal and comprise 75-100% of* Δ_{MAX} *calculated according to the formula* (10). It was also proven that macro-roughness height Δ depends only on the five following parameters: Δt_v and ;Δt_S – time periods required for shaping adjacent pits in a row and in two adjacent rows; a_{NV}, a_{NS} – generation and feed acceleration of penetration; k_L – coefficient of pits displacement in adjacent rows. Neither distances between adjacent pits ratio nor the angle θ between generating and feeding motion vectors nor cutting edge position on a generating surface have an impact on macro-roughness height.

For all pit types the following formula for pyramids height ΔH – i.e., maximum faceting height – determination was obtained:

$$K_Z = 1 - \frac{k_{HL}}{1+k_H} \cdot (2 - k_{HL}) \qquad (12)$$
$$\Delta H = K_Z \cdot \Delta H_{MAX}$$

where ΔH_{MAX} is calculated according to the formula (10), and

$$k_H = \frac{\Delta_1}{\Delta_2} = \frac{a_{NV} \cdot \Delta t_V^2}{a_{NS} \cdot \Delta t_S^2}; \qquad k_{HL} = k_H \cdot k_L \cdot (1 - k_L) \qquad (13)$$

Formulas (12-13) can be applied if $0 \leq k_H \leq 4$ and if if0 $\leq k_L \leq 1$.

ΔH to ΔH_{MAX} ratio denominated as K_Z we will call a macro-roughness height reduction coefficient. обозначенное через K_Z, назовем коэффициентом уменьшения высоты макронеровностей.

While studying scaled surfaces forms (rectangular, parallelogram and hexagonal), all adjacent pits are considered symmetrical and equal though it is possible only if the value of acceleration a_N is equal at all the points of machining surface. Each pit "central" point K is situated on a surface, which is a generating surfaces family envelope.

5. Pits shape and pyramidal asperities height in case of two-parametric generation

Algorithms and computer programs based on formulae (1) – (13) and generalized gearing (shown in fig. 4) formulae of point coordinates and vector projections transformation were developed to investigate the influence of machining type parameters on macro-roughness types and sizes.

5.1. Hexagonal pit shape

Fig. 7 and 8 illustrate how the pit transformation process is influenced by two key machining type parameters: the angle θ between generation and feed direction vectors (fig. 5) and $k_L = \Delta L/L$ coefficient which defines the value of pits displacement in adjacent rows along the generation direction (fig. 6). The result of computer simulations for pit dimensions ratio determination: $k_S = L_1/L_2$=2 (where L_1 – dimension along generation direction and L_2 – dimension along feed direction) is shown in fig. 7 and 8. In this case displacement coefficient $k_L = \Delta L/L$ changes its value from 0 to 0.5 as when k_L is equal to 0.5-1.0, the symmetrical hexagonal pit transforms into parallelogram pit. Accepting any value of k_L exceeding 1.0 is meaningless as the shape of adjacent pits does not change. Moreover, a new pit near the considered one will be formed not by the current cutting edge of a cutter, but one of the following edges. Fig. 7 illustrates the results of the computer modelling in case the angle θ =90^0 (fig. 5), fig. 8 – in case the angle θ =70^0 (pits and pyramids are "beveled").

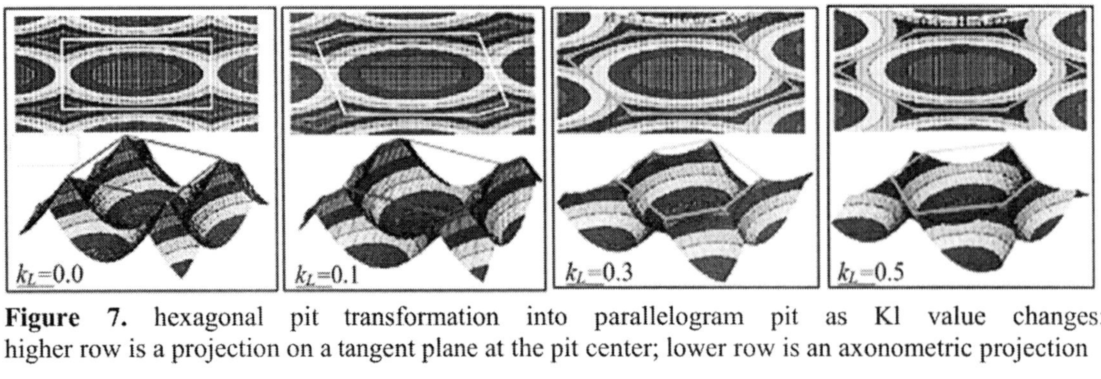

Figure 7. hexagonal pit transformation into parallelogram pit as Kl value changes: higher row is a projection on a tangent plane at the pit center; lower row is an axonometric projection

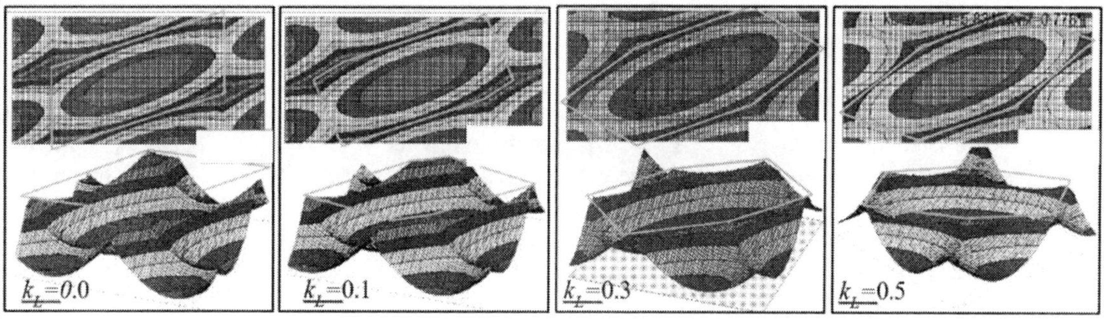

Figure 8. parallelogram pit transformation into hexagonal pit as Kl value changes: higher row is a projection on a tangent plane at the pit center; lower row is an axonometric projection

5.2. Dependence of pyramid height on generation parameters

After calculating pits and pyramids sizes and shapes it was estimated how the macro-roughness height Δ can decrease depending on generation parameters. The ratio of macro-roughness height Δ to maximum macro-roughness height Δ_{MAX} cannot be less than 0.75 which is the significant result of the macro-roughness height analytical research. It also was discovered that $k_\Delta = \Delta/\Delta_{MAX}$ ratio is a function of the two parameters: $k_\Delta = f(k_L, k_H)$– (fig. 9).

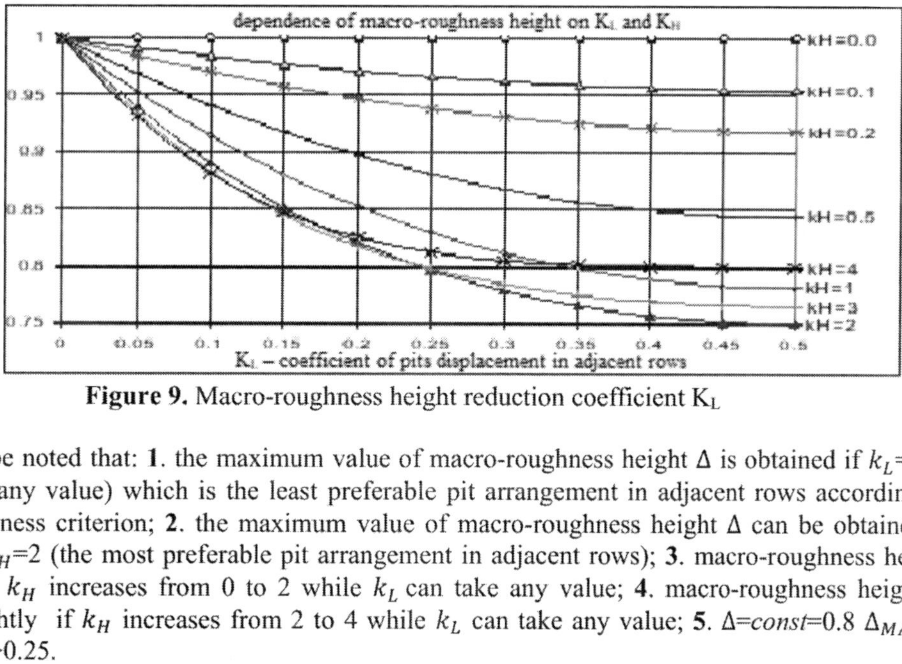

Figure 9. Macro-roughness height reduction coefficient K_L

It should be noted that: **1**. the maximum value of macro-roughness height Δ is obtained if k_L= 0.0 (k_H can take any value) which is the least preferable pit arrangement in adjacent rows according to surface roughness criterion; **2**. the maximum value of macro-roughness height Δ can be obtained if k_L=0.5 and k_H=2 (the most preferable pit arrangement in adjacent rows); **3**. macro-roughness height Δdecreases if k_H increases from 0 to 2 while k_L can take any value; **4**. macro-roughness height Δ increases slightly if k_H increases from 2 to 4 while k_L can take any value; **5**. Δ=*const*=0.8 Δ_{MAX} if k_H=4 and k_L>0.25.

Conclusion

The study of macro-roughness forms and sizes of surfaces machining by edge tools in case of two-parametric generation and by any tools in case of one-parametric generation with the use of "acceleration of penetration" term allows us to obtain:

- two general and simple formulae for determining maximum faceting height ("combs" height above an envelope surface) in case of one-parametric and two-parametric generation;
- the formula for calculating faceting height decrease coefficient *k* (depending on coefficient of pits displacement in adjacent rows). The *k* coeficient change its value from 1 to 0.75, i.e., generation parameters k_L и k_H have a slight impact on roughness.

The study also shows that:

- in case of two-motion gear machining method the shape of pits can be: parallelogram, rectangular or hexagonal. Moreover, heights of all 4 or 6 pyramids surrounding the pit are equal;
- the use of velocity and acceleration of penetration can simplify the process of resolving other gear generation analysis issues [3]: determining the curvature of surfaces formed by generating methods; estimating the cutting area and calculating the thickness of the cut layer; determining points coordinates on envelope surface through points coordinates on wraparound surface, etc. In other words, velocity and acceleration of penetration are the key parameters of generating process that should be calculated and used in computer programs along with $\mathbf{V}_{12}$, $\boldsymbol{\omega}_{12}$, **n**, **r** vectors. It is a crucial part of the current research conclusion.

Acknowledgments
This work was supported by grant (project № 9.6355.2017/БЧ) of government order of Ministry of education and science of Russia Federation for the period 2017–2019 in Tyumen Industrial University.

References

[1] Shunayev B K 1965 *Geometry of lateral surface with diagonal gear milling,* Design and technology in heavy engineering: proceedings of UPI, Sverdlovsk, USSR, Sat 14
[2] Sheveleva G I 1999 *Theory of formation and contact of moving bodies*, Mosstankin, 494 p
[3] Babichev D T 2005 *Development of theory of gearing and shaping of surfaces on basis of new geometric-kinematic representations,* Tyumen, Doctoral Thesis
[4] Babichev D T, Babichev D A, Pankov D N, Panfilova E B 2016 Kinematic method of finding points on envelope, knowing points on enveloping, *Bulletin of NTU "KhPI"* **26** 9-21
[5] Babichev D A, Babichev D T, Serebrennikov A A 2010 Using the derivative of acceleration of introduction when finding points on envelope, knowing points on enveloping, *News of NTU "KPI"* **27** 20-25
[6] Shishkov V A 1951 *Formation of surfaces by cutting by generating method,* Mashgiz

IOP Conf. Series: Materials Science and Engineering **659** (2019) 012008 doi:10.1088/1757-899X/659/1/012008

Use of indention velocity derivative for estimation of envelope points when the wraparound points are known method

S Yu Lebedev[1], D A Babichev[2] and D T Babichev[1]

[1]Industrial University of Tyumen, Institute of Transport, Melnikayte Street 72, 625039 Tyumen, Russia
[2]Sibur Holding, ZapSibNeftekhim LLC, East Industrial District, Block 1, No. 6, Building 30, 626150 Tobolsk, Russia

E-mail: lebedevsergey1995@gmail.com

Abstract. There is a modification to the undifferentiated surface points' estimation methods that are formed by envelope methods. This method: doesn't need a lot of additional calculations; easily fits the existing analysis systems of forming processes; essentially reduces the overall total of calculations at adequate accuracy.

1. Task definition

Reference paper [1] describes a kinematic method of envelope points estimation when the wraparound points [2] are known. The method is based on the usage of terms "indention velocity and acceleration", and as shown in [1] it lets reduce the value of the calculated points deflection from theoretical envelope by 1 to 3 orders of magnitude. Computer calculations which are partially provided in [1] showed that deflection behavior of the calculated points on Σ_3 from envelope Σ_2 (see figure 1) follows well defined laws: a) in the direction of envelope Σ_2 curvature radius growth, points Σ_3 are situated inside of the envelope surface Σ_2 field; b) deflections of points Σ_3 from Σ_2 are power low dependent. All this complies with physical sense and points at computational errors in the example under analysis being caused by third derivatives effect. Since indention velocity runs over first derivatives; indention acceleration - over second derivatives; and kinematic method [1] does not take account of higher order derivatives. It shall be noted that also earlier it was proposed to use high derivatives for analysis and synthesis of engagements. In such a way G.I. Shevelyova has developed the formal power series method [3]. D.T. Babichev wrote [4] about utility value of indention velocity high derivatives, but up to now this idea has neither been implemented in methods, nor in calculations.

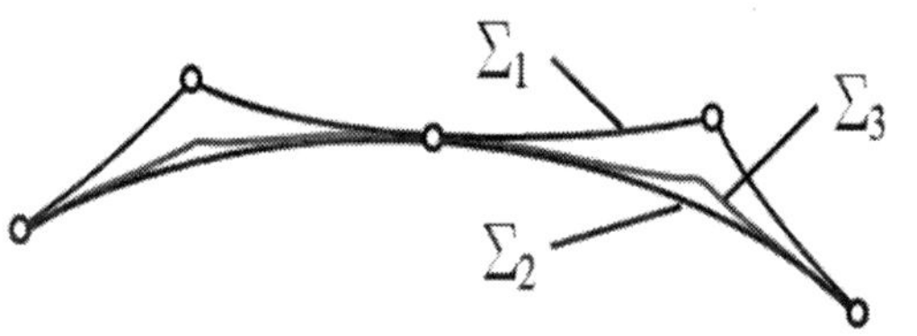

Figure 1. Types of surfaces: Σ_1 – wraparound; Σ_2 – envelope; Σ_3 – estimated using kinematic method.

2. Research objective

To develop envelope points estimation method over wraparound points coordinates using indention acceleration derivative for calculations. And to assess accuracy of the kinematic method provided in [1] when third derivatives are used in it.

Derivative of indention acceleration. It is found by differentiation of indention acceleration a_{BH} [1, formula (3)] with respect to time t:

$$\dot{a}_{BH} = \frac{da_{BH}}{dt} = \frac{d}{dt}(\mathbf{a}_{12} \cdot \mathbf{n} + \mathbf{V}_{12} \cdot \dot{\mathbf{n}}) = \mathbf{a}_{12} \cdot \mathbf{n} + 2 \cdot \mathbf{a}_{12} \cdot \dot{\mathbf{n}} + \mathbf{V}_{12} \cdot \ddot{\mathbf{n}} \quad (1)$$

where $\mathbf{V}_{12}$ –is relative velocity vector; $\mathbf{n}$ – unit normal vector to the surface, directed out of the generating element field; $\mathbf{a}_{12}$ – acceleration of the point situated on the generating surface Σ_1 and riding on it at a velocity "$-\mathbf{V}_{12}$"; $\dot{\mathbf{n}}$ – derivative of unit normal vector $\mathbf{n}$.

For flat rack-and-gear drive where generating profile is a straight line, using dependencies (3)-(5) and (6b)-(8b) from [1] and assuming in generic engagement [1, figure 1]:$\varepsilon_1 = \varepsilon_2 = \varepsilon_3 = a_1 = a_2 = a_3 = 0, \gamma = 90^0, \omega_2 = 1, V_1 = -a_W, \omega_1 = \omega_3 = V_2 = V_3 = 0, S_3 = a_w$, following simple formula was deduced for calculation of derivative $\dot{a}_{BH}$ of indention acceleration a_{BH}:

$$\dot{a}_{BH} = \omega_{12}^3 \cdot r_b = -\omega_2^3 \cdot r_b = -r_b \quad (2)$$

where r_b – base radius of the gear blank.

Using work formulae [1], along with dependence (2), plots were constructed (see figure 2) illustrating changes in: indention velocity V_{BH}, indention acceleration a_{BH} and its derivative $\dot{a}_{BH}$ along cutting edges in the rack gearing under consideration.

Indention velocity V_{BH} – see figures 2a and 2b – is equal to 0 at the action points K and K* of generating rack with envelope involute (not shown in the figure 2). At non-action points V_{BH} is numerically equal (under ω_{12}=1) to the arm h of the unit normal vector relative to the pitch point P. Here V_{BH}>0, $\mathbf{n}$ creates moment relative to the pitch point P, co-directing normalized angular velocity ω_{12} (reminder: $\boldsymbol{\omega}_{12} = \boldsymbol{\omega}_{TOOL} - \boldsymbol{\omega}_{BLANK}$). When the rack is moving, action points K and K* are shifted along the rack profile and due to it indention velocities V_{BH} are changed at all the profile points. Three material circumstances shall be noted: Firstly, thicknesses of the layers cut by the cutting edge are shown in the sketches of V_{BH} (under V_{BH}>0), figure 2b. It is one of the key ideas of V.A. Shishkov [5], who introduced the term V_{BH}. Secondly, in the points B and C of profiles' jag following rule is in action: If direction of V_{BH} is changed at the jag - this jag is shape-generating, i.e. the point is situated on the envelope (see point C, figure 2b); if direction of V_{BH} is unchanged at the jag - the jag is not shape-generating, i.e. the point is situated out of envelope (see point B, figure 2b). It is one of the ideas of the paper [5], where shape generation by jags was studied carefully by the author. Thirdly, when the generating motion direction is changed, velocity V_{BH} reverses the direction at all the points of the generating profile.

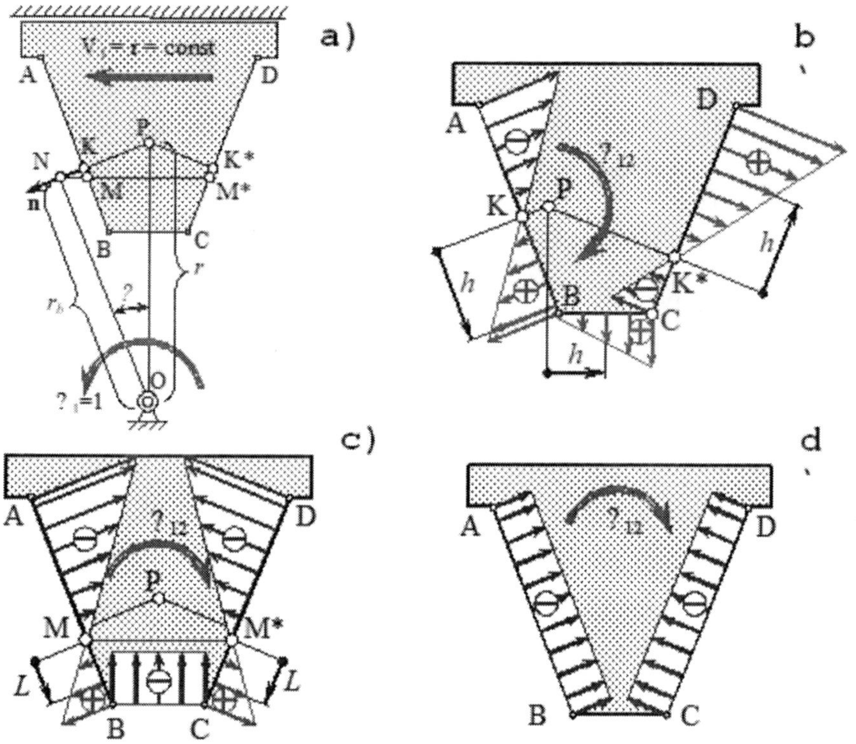

Figure 2. Indention parameters for straight intervals of a rack profile: a) engagement scheme; b) velocity V_{BH}; c) acceleration a_{BH}; d) its derivative.

<u>Indention acceleration</u> a_{BH} – see figures 2a и 2b – is equal to 0 at points M и M*. At other points of the side profiles AB and CD acceleration a_{BH} is in linear dependence on the distance e between current point and points M or M*: $a_{BH} = \omega_{12}^2 \cdot e \cdot ctg\alpha$, where α – is a rack pressure angle. In BC interval acceleration $a_{BH} = \omega_{12}^2 \cdot r_f$, where r_f – is a distance between BC interval and gear center - point O. Two material circumstances are to be noted. Firstly, value and direction of acceleration a_{BH} depend neither on rack movements during generation, nor on direction of this motion. Secondly, at the points under $a_{BH}>0$ *envelope is formed inside of the generating element field (see intervals MB and CM*), i.e. gearwheel tooth undercutting takes place.* [6]

<u>Derivative</u> $\dot{a}_{BH}$ <u>of indention acceleration</u> a_{BH}– see figure 2d: equal to $\dot{a}_{BH} = -\omega_{12}^3 \cdot r_b$; similar at all the points of side portions AB and CD, and does not depend on rack movements. If the generating motion direction is changed, then direction $\dot{a}_{BH}$ is changed as well. On the interval BC acceleration $\dot{a}_{BH} = 0$.

Principles of the kinematic method where indention acceleration derivative is used. Below is envelope points' estimation algorithm for the case of one-parameter envelope.

<u>*Stage 1*</u> is similar to the stage 1 described in [1]: estimate coordinates x_2, y_2, z_2 on the wraparound Σ_1. Meanwhile calculate and keep in mind: a) projections n_{x2}, n_{y2}, n_{z2} of unit normal vector $\mathbf{n}_1$ to the surface Σ_0; b) relative velocities vectors: angular $\boldsymbol{\omega}_{12}$ and linear $\mathbf{V}_{12}$; c) indention velocity V_{BH} and acceleration a_{BH} Additionally derivative of indention acceleration $\dot{a}_{BH}$ is to be found – for the involute gear teeth generation by a rack the formula is to be used (2).

<u>*Stage*</u> 2 is very alike with stage 2 of [1]:

- Calculate distances δ from the point on wraparound Σ_1 to envelope Σ_2 (along normals $\mathbf{n}_1$ to Σ_1) and time Δt of their transit by formulae:

$$\left.\begin{aligned}\Delta t &= \min\left\{abs\left(\frac{-a_{BH} \pm \sqrt{a_{BH}^2 + 2 \cdot \dot{a}_{BH} \cdot V_{BH}}}{\dot{a}_{BH}}\right)\right\} \\ \delta &= abs\left(V_{BH} \cdot \Delta t + \frac{a_{BH} \cdot \Delta t^2}{2} - sign(V_{BH}) \cdot \frac{\dot{a}_{BH} \cdot \Delta t^3}{6}\right)\end{aligned}\right\} \qquad (3)$$

- Estimate normal line **N** and its unit vector $\mathbf{n}^{(\Sigma 2)}$ to the envelope surface [1, formulae (13)].
- Calculate normal distance δ_n to the surface Σ_2 [1, formulae (14)].
- Estimate coordinates of the wraparound points Σ_2 ([1, formulae (15)].

3. Evaluation of the method offered and accuracy estimation

The software used for the research article [1] has been updated in accordance with the above mentioned algorithms and formulae. Using this software an involute spur gear forming process by counterpart rack has been simulated. Just like in [1], gear reference diameter was set to d=1000 mm, i.e. a_w=500 mm. See figure 3 for the results of this simulation. It shows graphically the deflections of calculated surface Σ_3 points (see figure 1) from the envelope Σ_2, i.e. from the involute. It is apparent that: a) deflections almost do not depend on point situation on the involute; b) deflections tend to reduce in a stable and significant way in case if the number of teeth z and number of cuts k increase when the gear blank is rotated by one tooth (if z or k is doubled, deflection is reduced by a factor of more than 30); c) deflection value is negligible: under z=10 and k=10 maximum involute deflection for the gear with reference diameter d=1,000 is equal to only 0.0001 micrometer; d) under $z*k$>1,500 method error $\Delta h<10^{-9}$ µm becomes commensurable with computer calculation accuracy which made up 15 to 16 decimal signs (wavy curves under z*k=1,600 and $z*k$=3,200 witness it clearly).

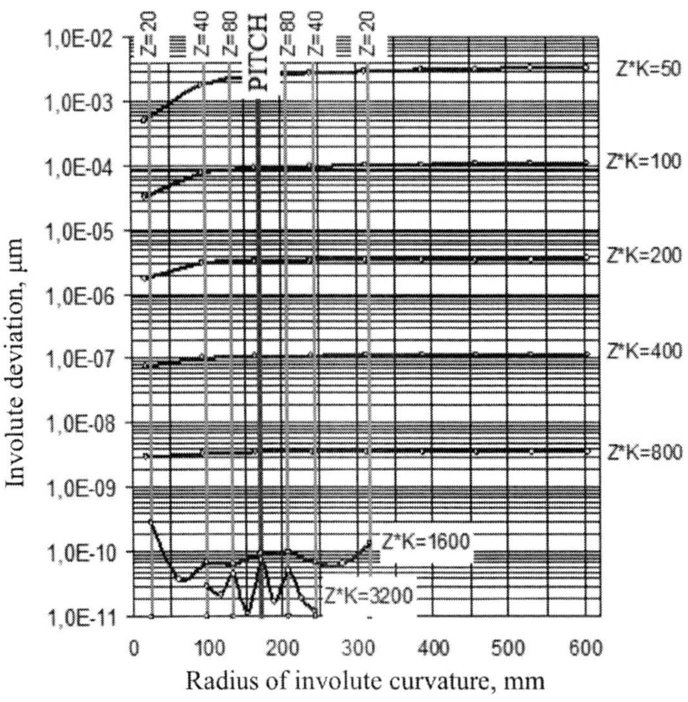

Figure 3. Kinematic method accuracy when indention acceleration derivatives are used.

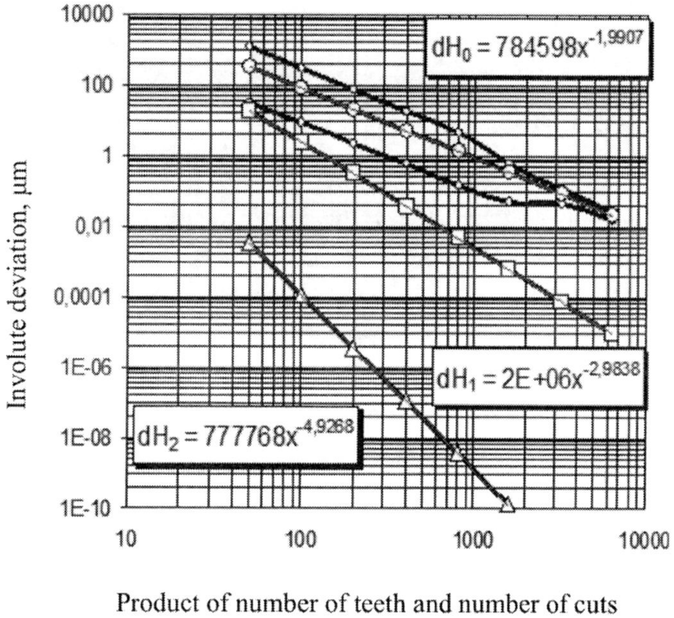

Figure 4. Maximum errors of the three envelope calculation methods.

Figure 4 shows accuracy comparison for three methods of envelope calculation depending upon the number of teeth z and number of cuts k *(by the example of rack treatment of an involute gear)* Three top curves (method 1) – maximum wraparound Σ_1 deflection from involute Σ_2 on the circles (from top to bottom): tip circle, reference circle, close to base circle. It is obvious that on the reference surface under z^*k=100 deflection makes up about 100 μm. Second-from-the-bottom curve (method 2) – maximum deflection for kinematic method described in [1]. One can see that on the reference circle under z^*k=100 deflection makes up more than 1 μm; i.e. reduced by almost two orders of magnitude. Bottom curve (method 3) – shows maximum deflection for kinematic method when indention velocity derivative $\dot{a}_{BH}$ is used. Apparently, deflection is equal to 0.0001 μm, i.e. less than initial one by six orders of magnitude and less by four orders of magnitude compared to method 2 results. Trends shown in figure 4 indicate that the errors dependencies on product z^*k are power-mode, but exponents are different:

$$\Delta_{method1} \approx \frac{8 \cdot 10^5}{(z \cdot k)^2} \quad ; \quad \Delta_{method2} \approx \frac{8 \cdot 10^6}{(z \cdot k)^3} \quad ; \quad \Delta_{method3} \approx \frac{8 \cdot 10^5}{(z \cdot k)^5} \qquad (4)$$

i.e. increase of cuts number k improves accuracy of method 3 in the most material way; and in the least material way (though, significant) – method 1 accuracy.

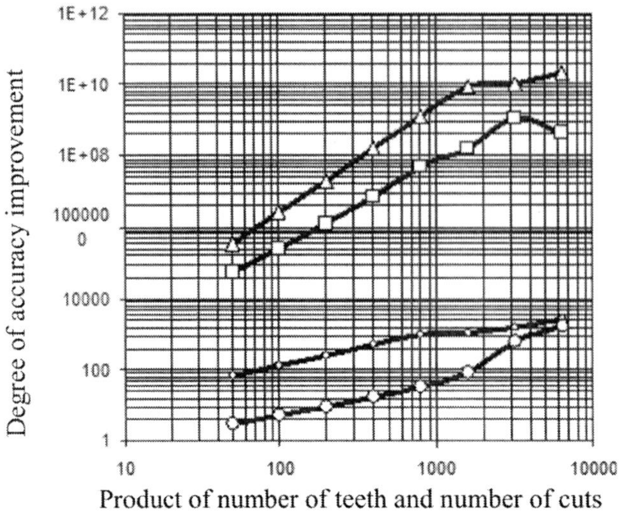

Figure 5. Comparative accuracy analysis of the two kinematic methods.

Figure 5 is more informative in terms of showing the degree of accuracy improvement for two kinematic points calculation methods on the Σ_3 (i.e. close to involute) compared to deflections of the wraparound jags (i.e. jagged curve tangent to involute). Two bottom curves – are minimum and maximum accuracy improvements of method 2. Two top curves – are minimum and maximum accuracy improvements of method 3, which is described above and includes use of derivative $\dot{\alpha}_{BH}$.

4. Conclusion

Indention derivative was used in kinematic method of envelope points lumping given the wraparound points. Simple example demonstrated that in this case accuracy of kinematic calculation of coordinates under the same number of cuts increases by 3 to 6 orders of magnitude. It is assumed that in the general case use of the derivative $\dot{\alpha}_{BH}$ improves accuracy not by 3 to 6 orders of magnitude but only by 2 to 3 orders of magnitude, since it was involute to be considered as envelope and in case of its formation second derivative is $\ddot{\alpha}_{BH}$=0. In other engagements, especially in space ones, forth derivative may be not zero, that would lead to calculation accuracy reduction.

Acknowledgment

This work was supported by grant (project № 9.6355.2017/БЧ) of government order of Ministry of education and science of Russia Federation for the period 2017–2019 in Tyumen Industrial University.

References

[1] Babichev D T, Babichev D A, Pankov D N and Panfilova E B 2016 Kinematic method of finding points on envelope, knowing points on enveloping, *Bulletin of NTU "KhPI"* **26** 9-21

[2] Sheveleva G I 1999 *Theory of formation and contact of moving bodies*, Mosstankin, 494 p.

[3] Sheveleva G I 1969 Method of power series in theory of gearing with point contact, *Mashinovedeniye* **4** 58-65

[4] Babichev D T 2005 *Development of theory of gearing and shaping of surfaces on basis of new geometric-kinematic representations*, Tyumen, Doctoral Thesis

[5] Shishkov V A 1951 *Formation of surfaces by cutting by generating method*, Mashgiz

[6] Babichev D T, Babichev D A, Pankov D N and Panfilova E B 2009 Analysis of formation of teeth by methods of bending around kinks on generating surfaces and lines, *Bulletin of NTU "KhPI"* **20** 32-44

Kinematic method of envelope points calculation when the wraparound points are known

D A Babichev[1], D T Babichev[2] and S Yu Lebedev[2]

[1]Sibur Holding, ZapSibNeftekhim LLC, East Industrial District, Block 1, No. 6, Building 30, 626150 Tobolsk, Russia
[2]Industrial University of Tyumen, Institute of Transport, Melnikayte Street 72, 625039 Tyumen, Russia

E-mail: lebedevsergey1995@gmail.com

Abstract. There are two kinds of surfaces that are formed by the movable bodies: wraparound and envelope surfaces. In this work the authors develop velocity (kinematic) envelope points' calculation method, when the envelope is known. The method is approximate; and based upon the usage of terms "speed and acceleration"; and lets reduce the amount of calculation using undifferentiated methods of envelope estimation

1. Problem statement

Theory of surfaces formation by movable bodies distinguishes [1] two types of formed surfaces: wraparound and envelope – see Figure 1. Wraparound Σ1 is a discrete family connected set of the reference gear tool surfaces pieces; envelope Σ2 is a surface tangent to all the elements of this discrete set. Envelope Σ2 is normally a smooth surface; wraparound Σ1 – is always a faceted surface (with jags). Envelope estimation Σ2 – one of the major objectives of gearing theory, and a large number of researches was focused on its resolution [2-11]. The above mentioned papers are based on differential methods: classic [2-3] and kinematic [4-11], which has been an essential method for gearing theory since the middle of 20-th century. Wraparound Σ1 is estimated [1,12-13] by direct tracking of generating surface points situation Σ0 relative to the blank – direct digital simulation method (PCM) as per [13]. Majority of specialists dealing with computer simulation of forming processes consider PCM as surer methods compared to kinematic ones and due to it they prefer to estimate wraparound rather than envelope. As a consequence, such profiling method is a basis of the of the most powerful Russian software solutions: a) EXPERT software for hypoid and spur gearing created by G.I. Shevelyova and her students; SPIDAL+ solution for worm type gearings created by V.I. Goldfarb and his students. As well as of many foreign solutions: a) PCM by S.V. Lunin (USA); b) Klingelnberg's software solution (Germany) developed with contribution from famous Russian bevel and hypoid gearing specialist M.S.Segal. Though, estimation of wraparound points Σ1 rather than envelope ones Σ2 has a number of

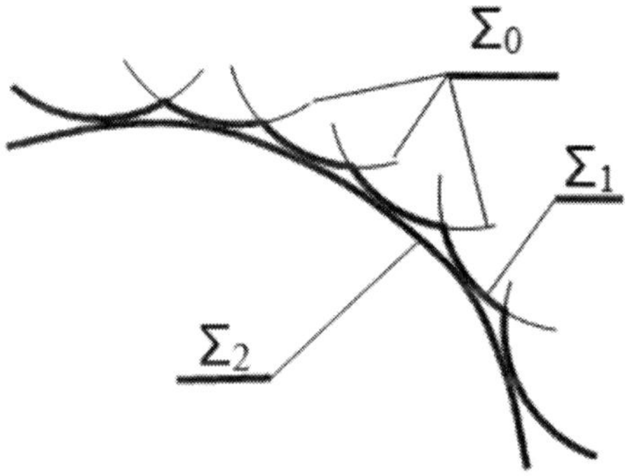

Figure 1. Figure 1 – Generating, wraparound, and envelope surfaces: Σ_0 – generating; Σ_1 – wraparound; Σ_2 – envelope.

disadvantages. Firstly, it requires more calculations; overall total is more by 1 to 2 orders of magnitude compared to envelope points calculations.

Secondly, difficulties arise with respect to evaluation of generating surface curvature required for, specifically, assessment of teeth contact strength and oil film load take-up. These difficulties are caused by the following:

- Curvature of band and rippled surfaces differs fundamentally and materially from envelope curvature. For instance, curvature radius of the involute, as an envelope rectilinear-face rack, gradually changes along this curve: close to the base circle , at the tips diameter . However, for wraparound involute formed by the same rectilinear-face rack the curvature radii at all the points (as well as at the jags). Since wraparound of such involute is a jagged line, tangent to involute at each and every of its straight intervals.
- The points estimated using non-differential methods are often situated on band and rippled surfaces at the nodes of some irregular uv-grid, which is almost impossible to correlate with hollow centers or band throats. That is why the estimated set of points is rather ill-placed relative to the hollows and bands: any of the points may turn out to be situated: either in the center of the hollow, or at the jag, or next to the pyramid apex. It leads to errors, if the curvature is estimated upon point approximation by curves or surfaces.

Thirdly, contact problem solving through numerical techniques gets more complicated: the points alleet warpage values due to random situation at the jags and in the facet hollows. It causes mistakes in estimated contact stresses and results in 'jagged' boundaries of the contact areas.

Commonly used means of 'approach' of wraparound Σ1 and envelope Σ2 is a very fine-feed task for computer simulation of shape generation process. Definitely, it reduces 'roughness' of the simulated surface, resulting in reduction of errors while solving a contact problem. However, it is not a solution of the problem with envelope curvature estimation, if for no other reason than envelope's curvature does not depend on the feed.

Work objective. To present a kinematic method of 'approach' of wraparound Σ1 and envelope Σ2, based on utilization of terms 'indention velocity VBH' and 'indention acceleration aBH '. And assess its accuracy, reliability and amount of calculations. This method can be used instead of fine feed method or along with it. Prior to presentation of the method itself, let us look upon indention velocity and acceleration, the basis of the method.

2. Indention velocity

VBH , suggested by V.A. Shishkov [4] – also known as velocity of mutual approach and separation [4] –is a velocity of reference gear tool surface (RGTS) penetration into the blank volume. Indention velocity is estimated as projection of a vector of gear tool's movement velocity relative to machined element onto direction of a normal vector to the RGTS at this point:

$$V_{BH} = \mathbf{V}_{12} \cdot \mathbf{n} = V_{12x} \cdot n_x + V_{12y} \cdot n_y + V_{12z} \cdot n_z \tag{1}$$

where $\mathbf{V}_{12}$ – relative velocity vector at the point on RGTS: $\mathbf{V}_{12} = \mathbf{V}_{tool} - \mathbf{V}_{blank}$;
n – unit normal vector to RGTS at this point, always directed out of the gear tool body, e.g. out of grinding disc, but never inwards.

Velocity V_{BH}, estimated through (1), is a scalar value which can be:

- positive – gear tool (Σ_1) indents into gear blank body (Σ_2), and removal of metal takes place, if it is available at this location of the gear blank;
- negative –Σ_1 moves off Σ_2, and machining at this location is impossible;
- zero – this point at the gear tool surface at the moment is forming the point at the surface which will become a result of machining: at the envelop of generating surfaces class.

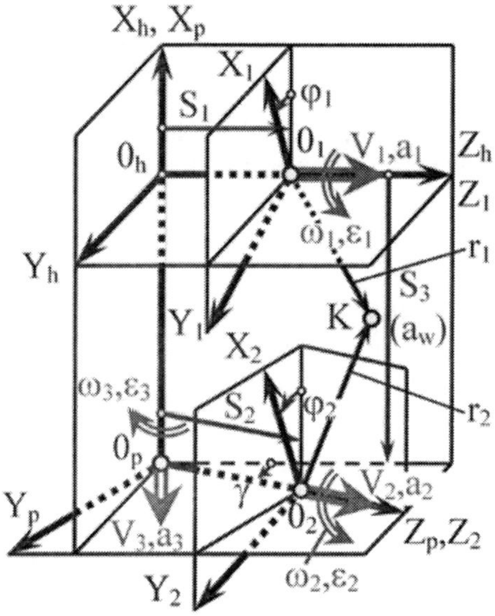

Figure 2. Generic gearing.

Term 'indention velocity' is not only applicable to technological gearings, but also to working ones. Moreover, in [10] kinematic value "indention velocity" extends to all the possible types of generating elements: surfaces' jags in the shape of edges or peaks; flat and space lines both with jags and smooth; points. And for all types of relative components' motion: one-parameter, two-parameter, and multi-parameter.

3. Indention acceleration

The a_{BH}, suggested by D.T. Babichev, is acceleration at which generating element (normally surface) indents ('sinks in') into the volume of the other body to form a mating element (normally a surface as well). I.e. a_{BH} indicates how fast indention velocity V_{BH} varies with time at this particular point of the space where the formed element is situated (formed surface in space gearings, and formed line - in flat ones). Indention acceleration is relative acceleration; as well as indention velocity is a scalar value. For the first time a_{BH} was used for estimation of cutting zones and calculation of thicknesses of the layers removed by cutting edges of generating gear tools [14]. Later on [10] it was found out that indention acceleration a_{BH} is an essential qualitative measure of shape-generation process: a) edge of regression at the surface formed by enveloping is a set of points where a_{BH} =0 during shape generation; b) if a_{BH} is *positive*, then *envelope is formed inside of generating part field*, that is impossible for metal and it means that undercutting takes place; c) contact of two mating surfaces is feasible for metal only in those points where a_{BH} <0; d) through indention acceleration a_{BH}, which ad hoc does not depend on section direction, a relative curvature is calculated [15] in any normal section and in one-parameter and multi-parameter gearings:

$$\frac{1}{R_p}=-\frac{\omega_k^2}{a_{BH}} \qquad \frac{1}{R_p}=-\sum_{i=1}^{n}\left(\frac{\omega_{k\,i}^2}{a_{BH\,i}}\right) \tag{2}$$

where ω_k – angular velocity of bodies roll motion in the plain of the section in which curvature is estimated; n – number of enveloping parameters. And this is far not complete list of the problems, where it is reasonable to use a_{BH} [10] for resolution.

Let's find a_{BH} upon differentiation of V_{BH} with respect to time t:

$$a_{BH}=\frac{dV_{BH}}{dt}=\frac{d}{dt}(\mathbf{V}_{12}\cdot\mathbf{n})=\mathbf{a}_{12}\cdot\mathbf{n}+\mathbf{V}_{12}\cdot\dot{\mathbf{n}} \tag{3}$$

where $\mathbf{V}_{12}$–is relative velocity vector; **n** – unit normal vector to the surface, directed out of the generating element field; $\mathbf{a}_{12}$ – acceleration of the point situated on the generating surface Σ_1 and riding on it at a velocity "$-\mathbf{V}_{12}$";

It shall be noted that indention acceleration differs from gearing equation derivative, normally [6] used for curvatures estimations. This difference is in calculation method of partial derivatives of curvilinear coordinates and of time parameter of enveloping; $\frac{\partial u}{\partial T},\frac{\partial v}{\partial T},\frac{\partial t}{\partial T}$ (below they will be marked as $\dot{u},\dot{v},\dot{t}$). In [6] they run over condition that the point remains action point (under gearing equation time derivative set to zero). Also here constraint equation $\dot{u},\dot{v},\dot{t}$ is presented differently:

$$\mathbf{C}_u\cdot\dot{u}+\mathbf{C}_v\cdot\dot{v}+\mathbf{V}_{12}\cdot\dot{t}=\mathbf{n}\cdot V_{BH} \tag{4}$$

i.e. point's sliding velocity over generating element shall be selected (by means of assigning $\dot{u},\dot{v},\dot{t}$) in such a way that this point's velocity vector would also be a vector of indention velocity. This is a key statement in a_{BH} calculation method. It can be calculated at any point of generating element surface, not only at action points.

One more material remark: unit normal vector derivative $\dot{\mathbf{n}}$ included in (3) sometimes called as unit normal vector end-point velocity [6], is also an angular velocity of roll motion. And ω_k and $\omega_k^{\perp}$ are projections of this vector $\dot{\mathbf{n}}$ on an axis of the right Darboux triad [16] with axes $\{\mathbf{c}, \mathbf{c} \times \mathbf{n}, \mathbf{n}\}$. I.e. formula (3) for indention acceleration a_{BH} calculation may be presented as:

$$a_{BH} = \mathbf{a}_{12} \cdot \mathbf{n} + \mathbf{V}_{12} \cdot \dot{\mathbf{n}} = \mathbf{a}_{12} \cdot \mathbf{n} + \mathbf{V}_{12} \cdot \mathbf{c} \cdot \omega_k + \mathbf{V}_{12} \cdot (\mathbf{c} \times \mathbf{n}) \cdot \omega_k^{\perp} \quad (5)$$

4. Calculation of parameters required for estimation of indention acceleration

The Estimation of vectorsω_k, $\omega_k^{\perp}$, and especially $\mathbf{a}_{12}$ included in (5), is a matter of considerable difficulties. Thus let's give formulae for their estimation in generic gearing shown in Figure 2. Actually any specific transmission or generation may be refined in the systems of coordinates and motions of this gearing. That is why below specified formulae are applicable to any specific gearings. In Figure 2 both members making up a gearing perform helical motions around their axes, and the axes take part in helical motion around center-to-center perpendicular. Generating element is at the member 1, generated one – at the member 2. Type of members' motion is not anyhow limited: motion parameters (ω_1, ω_2, ω_3, V_1, V_2 и V_3,) can be variable or constant in time, positive, negative or zero. All these parameters are assumed as curve of one, two, or n number of independent motion parameters $(t_1, t_2, \ldots t_n)$, that will correspond to one-, two-, or n-parameter motion (enveloping). It shall be noted that in [16, 10] it is shown, also via examples, that: cut surfaces sometimes resulted by advance/retract of gear tools are envelopes belonging to the group of multi-parameter generating surfaces. For this reason analysis of multi-parameter gearings represents a problem of practical importance, not just an abstract theoretical research. Supposing that for the gearing in Figure 2 following is known:

- displacements, velocities, and accelerations for all the six motions (□1, □1, □1, ...);
- coordinates of the point on the generating element (x_1, y_1, z_1; x_h, y_h, z_h; ... x_2, y_2, z_2);
- unit normal vector **n** to the generating element at this point (n_{x1},n_{y1}, n_{z1}; ... n_{x2},n_{y2}, n_{z2});
- curvatures of generating element at this point (R_R^1, $R_R^{1\perp}$, R_K^1, $\mathbf{c}=\{c_{x1}, c_{y1}, c_{z1}\}$).

Here: R_R^1, $R_R^{1\perp}$– generating surface's curvature radii in two mutually perpendicularnormal sections; R_K^1 – radius of geodesic torsion of the line along unit tangent line vector **c**; **c** – unit tangent line vector to the generating surface which sets first direction (curvature radius in it is equal to R_R^1). It shall be noted that if unit vector **c** follows one of the main directions at the generating surface, then $R_K^1 = \infty$, i.e. geodesic curvature $k_K^1 = 1/R_K^1 = 0$. Hence it is strongly recommended to define a tangent line **c** along one of the main directions at the generating surface: there will be no need to calculate R_K^1 or k_K^1.

Regarding direction of unit normal vector **n** and signs of curvature and torsion radii:

a) unit normal vector **n** is always directed out of the generating element field;

b) R_R^1 and $R_R^{1\perp}$ are positive for convex surfaces; c) $R_K^1 > 0$, if while moving along unit vector **c** unit normal vector **n** creates a left-handed screw.

Below are formulae for calculation of all the parameters required for indention acceleration calculation by formula (5). Though, the common known formulae for calculation of unit normal vector **n** and curvature radii R_R^1 and $R_R^{1\perp}$ through the given equation of generating surface are not shown. The coordinates conversion formulae for generic gearing in Figure 2 are not provided either.

4.1. Formulae dependent on the type of gearing

Below see them for Figure 2.

Relative angular velocity vector $\boldsymbol{\omega}_{12}$ in coordinates $X_hY_hZ_h$:

$$\boldsymbol{\omega}_{12}^{(h)} = \{\omega_3 \cdot \mathbf{i},\ -\omega_2 \cdot \sin\gamma \cdot \mathbf{j},\ (\omega_1 - \omega_2 \cdot \cos\gamma) \cdot \mathbf{k}\} \quad (6)$$

Relative linear velocity vector $\mathbf{V}_{12}$ in coordinates $X_hY_hZ_h$:

$$\left.\begin{aligned} V_{12x}^{(h)} &= V_3 - \omega_1 \cdot y_h + \omega_2 \cdot y_p; \\ V_{12y}^{(h)} &= -V_2 \cdot \sin\gamma + \omega_1 \cdot x_h - \omega_2 \cdot x_p \cdot \cos\gamma - \omega_3 \cdot z_h; \\ V_{12z}^{(h)} &= V_1 - V_2 \cdot \cos\gamma + \omega_2 \cdot x_p \cdot \sin\gamma + \omega_3 \cdot y_h. \end{aligned}\right\} \tag{7}$$

Relative acceleration vector $\mathbf{a}_{12}$:

$$\left.\begin{aligned} \mathbf{a}_{12} &= [(\mathbf{a}_1 - \mathbf{a}_2 - \mathbf{a}_3) + (\boldsymbol{\varepsilon}_1 - \boldsymbol{\varepsilon}_2 - \boldsymbol{\varepsilon}_3)\times\mathbf{r}_1 + \boldsymbol{\varepsilon}_2 \times \mathbf{S}_3 - (\boldsymbol{\varepsilon}_2 + \boldsymbol{\varepsilon}_3)\times\mathbf{S}_1] + \\ &+ [\boldsymbol{\omega}_1 \times (\boldsymbol{\omega}_1 \times \mathbf{r}_1) - \boldsymbol{\omega}_3 \times (\boldsymbol{\omega}_2 \times \mathbf{S}_2) - (\boldsymbol{\omega}_2 + \boldsymbol{\omega}_3)\times((\boldsymbol{\omega}_2 + \boldsymbol{\omega}_3)\times\mathbf{r}_2)] - \\ &- [2\cdot(\boldsymbol{\omega}_3 \times \mathbf{V}_2) - (\boldsymbol{\omega}_3 \times \boldsymbol{\omega}_2)\times\mathbf{r}_2] - 2\cdot(\boldsymbol{\omega}_2 + \boldsymbol{\omega}_3)\times\mathbf{V}_{12} - \boldsymbol{\omega}_{12}\times\mathbf{V}_{12} \end{aligned}\right\} \tag{8}$$

While analyzing two-parameter gearings, it is required to calculate twice by formulae (6)-(8), estimating velocities and accelerations upon each enveloping parameter. Accordingly, while analyzing **n**-parameter gearings one should estimate velocities and accelerations upon n number of values.

It is possible to apply formulae (6) – (8) for particular gearings having removed items containing zero accelerations, velocities and displacements. E.g. for all the right-angle gearings (hypoid, worm, globe, spiroid, etc), under: $\gamma = 90^0$, $\varepsilon_1 = \varepsilon_2 = \varepsilon_3 = a_1 = a_2 = a_3 = 0$, $\omega_2 = 1$, $\omega_1 = \pm i_{12}$ (here sign is "–", if direction of one of the velocities ω_1 or ω_2 does not follow the direction specified in Figure 2), $\omega_3 = V_1 = V_2 = V_3 = 0$, $S_3 = a_w$, we will obtain, instead of equations (6)-(8), a simpler formulae:

$$\boldsymbol{\omega}_1^{(h)} = \{0\cdot\mathbf{i},\ 0\cdot\mathbf{j},\ i_{12}\cdot\mathbf{k}\} \quad \boldsymbol{\omega}_2^{(h)} = \{0\cdot\mathbf{i},\ 1\cdot\mathbf{j},\ 0\cdot\mathbf{k}\} \quad \boldsymbol{\omega}_{12}^{(h)} = \{0\cdot\mathbf{i},\ -1\cdot\mathbf{j},\ i_{12}\cdot\mathbf{k}\} \tag{9}$$

$$\mathbf{V}_{12}^{(h)} = \{(-i_{12}\cdot y_h + y_p)\cdot\mathbf{i},\ \ i_{12}\cdot x_h\cdot\mathbf{j},\ \ x_p\cdot\mathbf{k}\} \tag{10}$$

$$\mathbf{a}_{12} = [\boldsymbol{\omega}_1 \times (\boldsymbol{\omega}_1 \times \mathbf{r}_1) - \boldsymbol{\omega}_2 \times (\boldsymbol{\omega}_2 \times \mathbf{r}_2)] - (\boldsymbol{\omega}_1 + \boldsymbol{\omega}_2)\times\mathbf{V}_{12} \tag{11}$$

here:

$$\mathbf{r}_1^{(h)} = \begin{bmatrix} \cos\varphi_1 & -\sin\varphi_1 & 0 \\ \sin\varphi_1 & \cos\varphi_1 & 0 \\ 0 & 0 & 1 \end{bmatrix} \cdot \begin{bmatrix} x_1 \\ y_1 \\ z_1 \end{bmatrix} \quad ; \quad \mathbf{r}_2^{(h)} = \begin{bmatrix} 1 & 0 & 0 \\ 0 & 0 & 1 \\ 0 & -1 & 0 \end{bmatrix} \cdot \begin{bmatrix} \cos\varphi_2 & -\sin\varphi_2 & 0 \\ \sin\varphi_2 & \cos\varphi_2 & 0 \\ 0 & 0 & 1 \end{bmatrix} \cdot \begin{bmatrix} x_2 \\ y_2 \\ z_2 \end{bmatrix} \tag{12}$$

under $\varphi_2 = \varphi_{20} + \varphi_1/i_{12}$; where φ_1 – enveloping parameter.

Similarly it is possible to obtain simplified formulae for spur, rack and other gearings, as well as for cam control assemblies and machine-tool gearings. Thus, for flat rack gearing, assuming: $\varepsilon_1 = \varepsilon_2 = \varepsilon_3 = a_1 = a_2 = a_3 = 0, \gamma = 90°, \omega_2 = 1, V_1 = -a_w, \omega_1 = \omega_3 = V_2 = V_3 = 0, S_3 = a_w$, we obtain:

$$\boldsymbol{\omega}_1^{(h)} = 0, \qquad \boldsymbol{\omega}_2^{(h)} = \{0\cdot\mathbf{i},\ 1\cdot\mathbf{j},\ 0\cdot\mathbf{k}\}, \qquad \boldsymbol{\omega}_{12}^{(h)} = \{0\cdot\mathbf{i},\ -1\cdot\mathbf{j},\ 0\cdot\mathbf{k}\} \tag{13}$$

$$\mathbf{V}_{12}^{(h)} = \{\ y_p\cdot\mathbf{i},\ \ 0\cdot\mathbf{j},\ \ (x_p - a_w)\cdot\mathbf{k}\ \} \tag{14}$$

$$\mathbf{a}_{12} = -\boldsymbol{\omega}_2 \times (\boldsymbol{\omega}_2 \times \mathbf{r}_2) - \boldsymbol{\omega}_2 \times \mathbf{V}_{12} \tag{15}$$

Also formulae (12) might be used assuming φ_1=0; and setting in formulae coordinates conversion $\varphi_2 = \varphi_{20} + \frac{S_1}{a_w}$, where S_1 – enveloping parameter.

4.2. Formulae non-dependent on the type of gearing

By them we calculate roll motion angular velocities ω_k, and $\omega_k^{\perp}$, parts of the equation (5).

Roll motion angular velocity ω_k is an angular velocity of unit normal vector **n** rotation to the generating surface relative to the generated surface measured in the plane set by vectors **n** and **c**. It is calculated by formula:

$$\omega_k = (\mathbf{n}\times\mathbf{c})\cdot\boldsymbol{\omega}_{12} + \frac{\mathbf{V}_{12}\cdot\mathbf{c}}{R_R^1} + \frac{\mathbf{V}_{12}\cdot(\mathbf{n}\times\mathbf{c})}{R_K^1} = \left(\boldsymbol{\omega}_{12}^{(i)} + \frac{\mathbf{V}_{12}^{(i)}}{R_K^1}\right)\cdot(\mathbf{c}\times\mathbf{n}) + \frac{\mathbf{V}_{12}^{(i)}}{R_R^1}\cdot\mathbf{c} \tag{16}$$

All three items in the central part of the formula (16) are rotations in the plane of the normal section passing vectors **n** and **c**. First item is a relative angular velocity vector projection ω_{12} onto the normal to the sectional plane, i.e. normal rotation in this plane defined by relative rotation of the components. Second item considers normal rotation towards the surface caused by sliding of the bodies in the normal section plane due to surface curvature at this section. Second item considers normal rotation towards the surface caused by sliding of the bodies in the direction perpendicular to the normal section plane and defined by geodesic torsion of the line at the surface.
Roll motion angular velocity in the plane perpendicular to the unit vector **c**:

$$\omega_k^{\perp} = \left(-\boldsymbol{\omega}_{12} + \frac{\mathbf{V}_{12}}{R_K^1}\right)\cdot\mathbf{c} + \frac{\mathbf{V}_{12}}{R_R^{1\perp}}\cdot(\mathbf{c}\times\mathbf{n}) \tag{17}$$

While analyzing two- and n-parameter gearings, it is required to calculate two or n times by formulae (16)-(17), estimating roll motion velocities upon each enveloping parameter.

5. Principle of envelope points estimation using kinematic method.
Indention Shape-generation is a process of *interaction between* generating element's *body surface* Σ_0 and *space* g_2 this body is moving in. If another body – "item" is placed in such space g_2, then moving generating element running through the space will form on this body a surface Σ_2 which is an envelope belonging to surfaces group Σ_0 – boundaries of the generating element.

The points contained in generated surface Σ_2 are the points of surface Σ_0 of the generating body, in which indention acceleration $V_{BH}=0$. Gearing equation has to be solved in order to estimate these points. If one of the curvilinear coordinate u or v is being estimated through gearing equation, it means calculation of the action point at the surface Σ_0 under the set value of enveloping parameter. If enveloping parameter t is being discovered, then it is estimation of surface Σ_0 displacement under which point given at Σ_0 would become action point. Nevertheless in any case estimation of the point at surface Σ_2 starts with solution of the gearing equation.

A markedly different method of envelope estimation, i.e. estimation of points at Σ_2, is considered in the current paper. Its specifics are as follows:

- Points at Σ_2 are estimated using direct method – without solving a gearing equation.
- For estimation of the "conjugate" point's coordinates at Σ_2 one needs to know only:a) point's coordinates at the surface Σ_1 (same as Σ_0); b) unit normal vector projections to Σ_1 at this point; c) indention velocity and acceleration (V_{BH} and a_{BH}) at this point.
- "Conjugate" points at the generated surface shall be found in the same system of coordinates where the parameters mentioned in the previous paragraph are set.
- The method is applicable to all the types of surface shape generations involving kinematic method: For one-, two-, and n-parameter motion of the generating element represented by surface, line or point.
- The method is approximate; the closer the datum point at the surface Σ_1 to the action point and to the generated surface Σ_2, the higher is accuracy of the method.

For the principle of point estimation at the surface Σ_2 formed by one-parameter enveloping see Figure 3. In this figure:

- points A_1 and B_1 are set at the generating surface Σ_1, and are situated close to the action point K;

  points A_2 and B_2 are the desired points at the generated surface situated, accordingly, "opposite to" points A_1 and B_1;
- $\mathbf{n}_1^{(A)}, \mathbf{n}_1^{(B)}$ – unit normal vectors to Σ_1;
- V_{BH} and a_{BH} – indention velocities and accelerations
- δ_A and δ_B – distances between the desired points A_2 and B_2 and datum points A_1 and B_1.

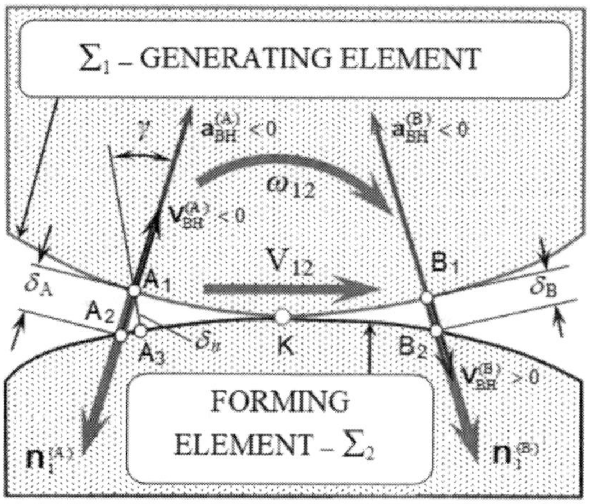

Figure 3. Principle of estimation at the envelope surface Σ_2 points.

Near the point B_1 shape generation of the surface Σ_2 is ongoing by means of surface Σ_0 "indenting" towards normal vector $\mathbf{n}_1^{(B)}$. Point A_2 at Σ_2 is already generated, and surface Σ_0 remoted from the point A_2 at distance δ_A is moving away from it at the velocity $V_{BH}^{(A)}$. Let us consider (due to small distances δ_A and δ_B) the "indenting" motion at the interval δ_B and separation motion at the interval δ_A as uniformly variable motions, i.e. $a_{BH} =$ const. These intervals δ, as well as time Δt of passing them, are estimated by formulae for uniformly variable motion:

$$\delta = -\frac{V_{BH}^{2}}{2 \cdot a_{BH}} \qquad \Delta t = -\frac{V_{BH}}{a_{BH}} \tag{18}$$

6. Algorithm of envelope points estimation using kinematic method.

Each In order to assess accuracy of the method, the process of spur gear involute profile generation by rectilinear-face rack was studied. Such gearing was selected as a study object because for involute it is not difficult to deduce formulae for estimation of distances measured from theoretical envelope represented by the involute itself. In the software engineered for simulation of this shape generation process the above mentioned algorithms and formulae have been implemented. Computations were made for the gear ring with reference diameter d= const =1000 mm. First stage involved analysis of three parameters' influence over the height of the wraparound jags above the involute. z – number of teeth; k – number of motions (cuts), performed by the rack when the gear wheel is rotated by one tooth; ρ_{EVOL} – involute curvature radius defining point's position at the gear wheel. The dependencies found are shown in the Figure 4.

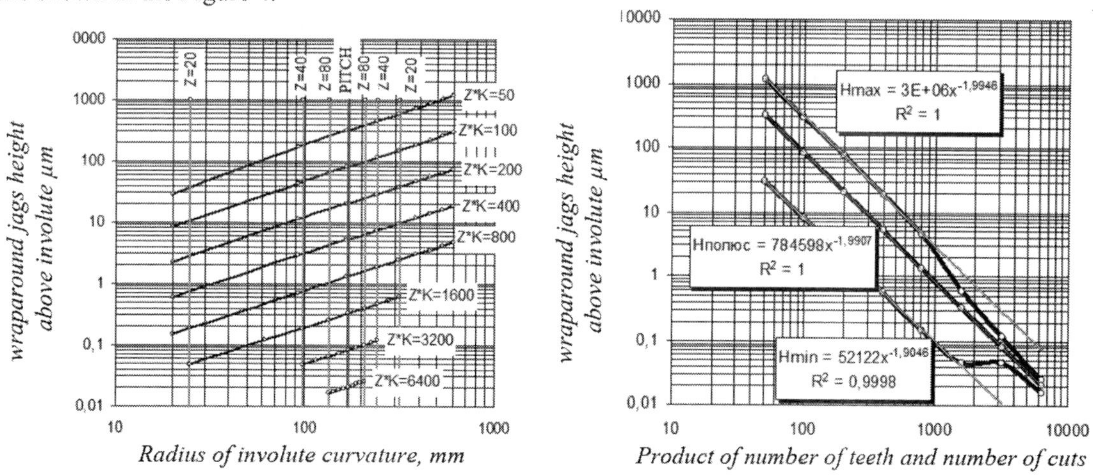

Figure 4. Wraparound Σ1 deflections from the involute Σ2

It is apparent that: a) at the reference cylinder wraparound jags deflections from the involute are defined by dependence μm, under d=1000mm; b) near the top circle under small teeth number it is four times more, and near the base circle it is 15 times less than at the reference circle.

Second stage involved study of the same three parameters (z, k, ρ_{EVOL})' influence to the deflection of the points estimated using kinematic method from the involute. See Figure 5 for the results of this analysis.

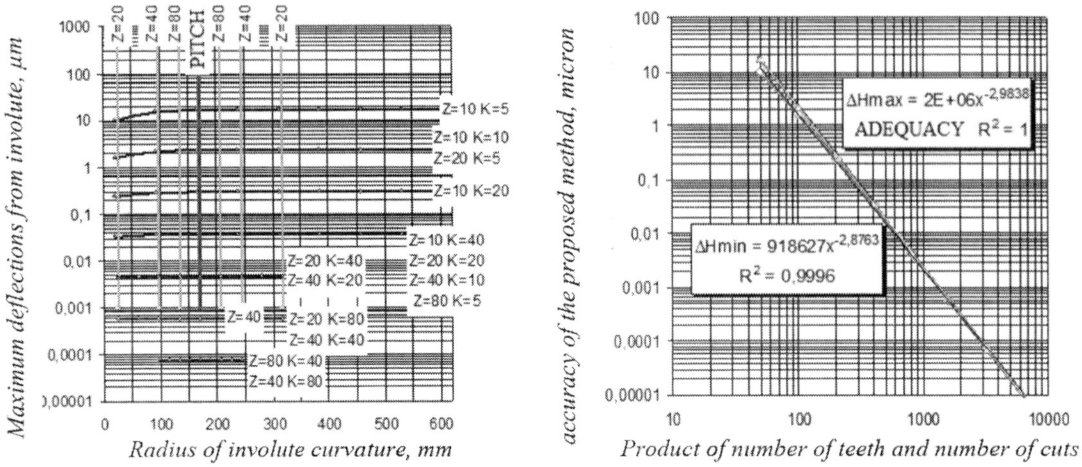

Figure 5. Wraparound Σ1 deflections from the involute Σ2

Following is concluded based on this figure: a) maximum involute deflections of the estimated points almost do not depend on point situation at it; b) maximum involute deflections follow the dependency – see trends in the right part of the Figure 5; c) application of the above described kinematic method as a "postprocessor" after non-differential methods of envelop estimation (through wraparound) enables re-duction of maximum envelope deflections of the estimated points by 1-3 orders of mag-nitude, i.e. 10-1000 fold.

Regarding reliability of the suggested method: It complies with reliability of non-differential methods if one condition is met: there shall not be any elements (strips or hollows) with very little curvature radius at the wraparound surface. An element with a very little curvature radius is the element at the wraparound (and generating), on which between two neighboring edges or peaks the angle between normal lines toward this ele-ment exceeds 3-4 degrees. Elements with a very little curvature radius are represented, namely, by all the jags at generating surface. For this reason the method is not applicable to analysis of fillets formed by jags. It shall be noted that the estimation method for wraparounds formed by jags is described in [15, 17]. That method has been tested by Doctor of engineering sciences E.S. Trubachyov in the system "SPDIAL+", and the re-sults confirmed high reliability with calculation time reduced by more than two orders of magnitude.

Acknowledgment
This work was supported by grant (project № 9.6355.2017/БЧ) of government order of Ministry of education and science of Russia Federation for the period 2017–2019 in Tyumen Industrial University.

References

[1] Sheveleva G I 1999 *Theory of formation and contact of moving bodies*, Mosstankin
[2] Gokhman H I 1886 *Theory of gearing, generalized and developed by analysis*, Diss. Master of Mechanics, Odessa
[3] Kolchin N I 1949 *Analytical calculation of flat and spatial links (with application to the profiling of the cutting tool and the calculation of errors in the links)*, Mashgiz, 210 p
[4] Shishkov V A 1951 *Formation of surfaces by cutting by generating method*, Mashgiz
[5] Davydova Ya S 1950 *Non-involute gearing*, Mashgiz
[6] Litvin F L 1986 *Theory of Gearing*, Nauka
[7] Erikhov M L 1972 *Principles of systematics, methods of analysis and issues of the synthesis of gearing schemes*. Doctoral Thesis
[8] Zalgaller V A 1975 *Envelope theory*, Nauka
[9] Goldfarb V I 2001 Some exercises with gearing equations, *Space of gears*. Collection of reports of the scientific seminar of the Educational and Scientific Center of Gears and Gear Design, Izhevsk – Elektrostal 20–24

[10] Babichev D T 2005 *Development of theory of gearing and shaping of surfaces on basis of new geometric-kinematic representations*, Tyumen, Doctoral Thesis

[11] Rama Krishna K and Dibakar S 2019 *Motion Space of Contacting Smooth Curves in Plane Using Screw Derivative*, In book: Advances in Mechanism and Machine Science: Proceedings of the 15th IFToMM World Congress on Mechanism and Machine Science, Mechanisms and Machine Science **73** 669-680

[12] Nesmelov I P, Goldfarb V I 1983 Non-differential approach to solving envelope problem, *Mechanics of machines* **61** 3–10

[13] Goldfarb V I, Lunin S, Trubachov E.S 2003 *Advanced computer modeling in gear engineering*, In book: Proceedings of ASME International Power Transmission and Gearing Conference, Chicago Illinois USA, September 2–6

[14] Langofer A R, Babichev D T, Raikhman G N, Shunaev B K 1986 Study on the computer load on the cutting edges of the gear cutting tool, *Machines and tools* **1** 18 -19

[15] Babichev D T 2008 Acceleration of penetration and curvature in gears, *Theory and practice of gears and engineering,* sat. report scientific and technical conference with international participation Izhevsk 157–161

[16] Dusev E E 1969 Connection between the curvatures of the mutually-bounded surfaces of the teeth of spatial links, *Mechanical Engineering* **3**

[17] Babichev D T 2004 On use of multiparameter envelopes in computer simulation of the processes of shaping in workers and technological gears *Theory and practice of gears*, Coll. report scientific and technical conference with international participation, Izhevsk 302–315

IOP Conf. Series: Materials Science and Engineering **659** (2019) 012010 doi:10.1088/1757-899X/659/1/012010

Design and performance simulation of TORVEastro three-link astronaut robot

F Samani[1] and M Ceccarelli[1]

[1]LARM2: Laboratory of Robot Mechatronics, University of Rome Tor Vergata, Rome, Italy

Email: francescosamani83@gmail.com

Abstract. In this article TORVEastro, a three-link astronaut robot, is presented with a conceptual design that is used also for a CAD design in performance simulations. This TORVEastro robot is designed for service applications in space stations. Requirements and characteristics are discussed with the aim to identify design problems and operation features. A study of feasibility is discussed through performance evaluation by using kinematics and dynamics simulations whose results show the feasibility of TORVEastro robot operation and its peculiarities.

1. Introduction

With the development of space technology, more complex space tasks will be carried out in the future. Compared with astronauts, it is advantageous to use space robots instead of human beings and therefore, space robots are increasingly playing a key role in on-orbit servicing operations. Since more than ten years, great attention has been paying to service robots, for the development of new robotic systems for applications even in non-technical areas [1]. Typical robots are already developed for medical care, space exploration, demining operation, surveillance, entertainment, museum guide. In some cases, the results have already come available on the market and in considerable literature about service robot is available not only on technical problems but it is reported in no-technical paper for space [2]. Advancements in interaction between a human and a robot, derives from the need for even more efficiency, flexibility and productivity, in a space contest, as well as for reduction in the human stress, risk and workload as pointed out for example in the paper of G. Michalos and his collaborators [3].

According to the International Federation of Robotics (IFR), "a service robot is a robot, which operates semi or fully autonomously to perform services useful to the well being of human and equipment, excluding manufacturing operations" [4]. According to ISO 8373 [5] service robots require “a degree of autonomy”, which is the “ability to perform intended tasks that are based on current state and sensing, with-out human intervention”. For service robots this ranges from partial autonomy - including human robot interaction - to full autonomy - without active human robot intervention. The International Federation of Robotics (IFR) statistics for service robots include systems based on some degree of human robot interaction or even full teleoperation as well as fully autonomous systems.

Currently an astronaut robot in-operation is Robonaut (NASA), a humanoid robot that has been developed project presented by the Robotics Laboratory at the Lyndon B. Johnson Space Center (JSC) of NASA in Houston, Texas [6]. Robonaut tasks re-quire to work alongside the astronauts (Robonaut can use space instruments and work in similar environments suitable for astronauts). The latest version of Robonaut, R2 is the first robot that is built in the US for the ISS (International Space Station), and it is was delivered by STS-133 in February 2011. It is a robotic torso that is designed to assist crew with EVA (Extra Vehicular Activity) and may contain tools that are used by the crew. However, Robonaut 2 does not have adequate protection to exist outside the space station and improvements and modifications would be needed to allow it to move around the station.

Another astronaut robot is Rollin' Justin from ESA, it has four-finger hands; Its mobile base allows autonomous operation over a long range [7]. This space robot consists of a system of light arm with light arm with four-fingered hands which makes it an ideal platform for operating for sensitive manipulations. The mobile base of this space robot allows autonomous operation over a long range. Motion detection sensors and cameras allow reconstructions of the robot environment. Unstructured, variable and dynamic environments require the space robot to act independently and without human support. The Justin robot is able to work safely with people. Its multiple degrees of freedom enabled Rollin 'Justin to pursue multiple goals simultaneously while respecting a hierarchy of tasks. For example, the robot can serve drinks by observing the environment, moving without singularity, avoiding collisions, responding in a manner consistent without collisions with the environment.

Kirobo is Japan's first robot astronaut (JAXA) [8] that was developed by University of Tokyo and Tomotaka Takahashi in cooperation with accompany Koichi Wakata, the first Japanese commander of the International Space Station. Kirobo arrived on the ISS on August 10, 2013 on JAXA's H-II Transfer Vehicle Kounotori 4, an un-manned resupply spacecraft launched August 4, 2013 from Japan's Tanegashima Space Center. Kirobo is about 13 inches tall, 7 inches wide and about 6 inches deep. He weighs about 2 lbs. and speaks Japanese [9]. The robot's capabilities include voice and speech recognition, natural language processing, speech synthesis and tele-communications, as well as facial recognition and video recording.

The characteristics of a space environment are related to low air pressure, large thermal excursions, high radiations and microgravity. In this regard, human beings in space require a life support system that is broad, flexible and complex. Therefore, it is important the implementation of an environmental control system and a rescue system, and this requires a lot of uninterrupted material supply [10].

In this paper the problem of designing a space service robot is attached so that it can support human being in space station. The analysis of a CAD modeling and a performance evaluation are carried out for a feasibility analysis.

2. 3-link astronaut robot design

The International Space Station (IIS) is a space station, as a habitable artificial satellite, in a low Earth orbits. Its first component was launched into orbit in 1998, with the first long-term residents arriving in November 2000 [11]. The ISS serves as a microgravity and space environment research laboratory in which crew members conduct experiments in biology, physics, astronomy, meteorology, and other fields [12]. IIS is equipped with handrails that are positioned on the outside surface to help the astronauts in the movement.

It is difficult to transport large or heavy objects from the ground into a space station [13]. The design of the space robot has to respect the requirements of the operations in the space station but, at the same time, its mass and volume should be as small as possible [14]. The characteristics of the proposed TORVEastro design are (fig. 1):

- the total degrees of freedom are 9;
- an occupied volume of 25 cm x 25 cm in the closed configuration;
- a total mass of the space robot less than 25 kg;
- robot arm length of 70 cm.

The TORVEastro space robot has a cylindrical body design with three legs, each of which is made up of two links and an end-effector. The symmetrical geometry of the legs makes them interchangeable thanks to the possibility of adopting a structure with multi-functional end-effectors. The conceptual design shows how one or two legs can be used for the most varied activities. The cylindrical body of the robot consists of curved surfaces and TORVEastro offers the possibility of facilitated movement in the outdoor space of orbital stations in order to be able to reach the localization objects. The service robot can move along, on the rods and handrails. In fig. 1 the conceptual design is shown with geometric and motion parameters.

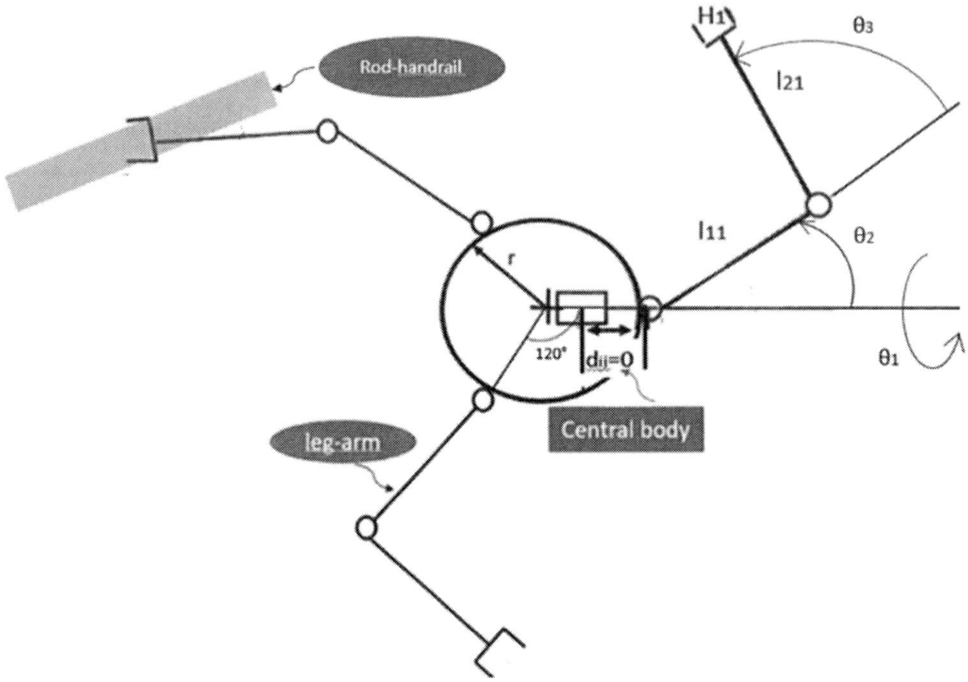

Figure 1. A conceptual design of TORVEastro.

3. Mechanical design and features

The proposed design of TORVEastro consists of a trunk and three legs. Control boards, sensors, battery and motors can be installed inside the cylindrical body (fig. 2). The moment of inertia of the robot in the center of mass of the TORVEastro is: Px = 0.84 kg m^2; Py = 1.1 kg m^2; Pz = 1.7 kg m^2. Calculations of the center of mass of the space robot have been performed in different configurations and the results show a proper dynamic capacity of TORVEastro. All electronic and sensitive parts can be repaired easily since the structure is easily accessible in each of its parts. The cylindrical body has a radius of 20 cm and a height of 15 cm. The max power of actuators is of 1 Watt, the max torque is of 1 N m, the number of actuators per end-effector is one, the number of the total actuators is twelve, the capacity of the battery is of 20 thou-sand mAh at the voltage is of 5V (valued consumption is of 0.2 J/sec, autonomy is 100 hours). The proposed TORVEastro configuration has the possibility to have symmetric three-link and the robot can be used with any posture that can be easily maintained because of a modular design with interchangeable parts. The end-effector can act either as a foot or as a gripper (fig. 1). Each link of the service robot consists of a 2,5-1 mm thick hollow cylinder with a circular cross-section arc of radius of 20 cm and an angular opening of 80 ° (fig. 2). Inside the link there is the possibility to pass the necessary wires of the sensors. The fig. 3 shows how the servomotors are positioned inside the central body. The Central Servomotor (CS) controls the rotation of θ1 by a drive motor shaft, the Left Servomotor (LS) controls the rotation of θ2 through a pair in mechanical tension wires. Finally, the Right Servomotor (RS) controls, through mechanical tension wires, the opening and closing of the arm 2 through a free disk wheel to rotate compared to the joint 2 and linked with another wheel fixed in the joint 3 (fig. 3). The connection between the links and the joints is designed according to have a symmetric geometry in order to have a proper balance of the masses compared to the link axis.

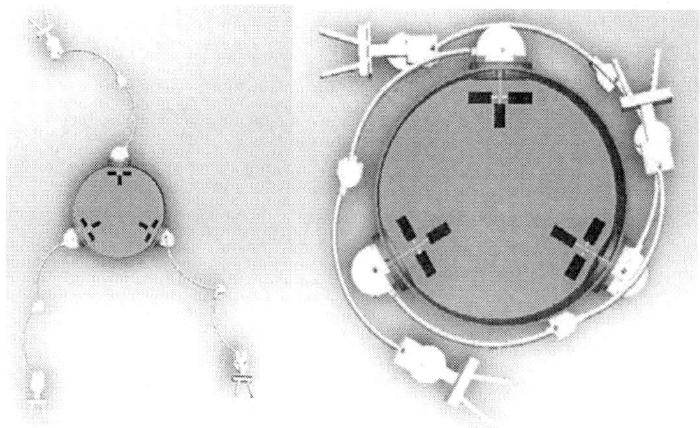

a) *b)*

Figure 2. A CAD model for mechanical design of TORVEastro in fig. 1:
a) Full open configuration; b) Full closed configuration.

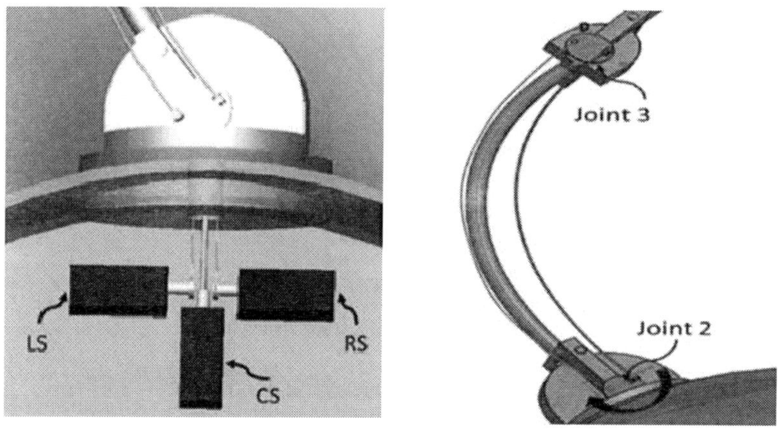

a) *b)*

Figure 3. Connection between servo-motors and joints by wires in fig. 2:
a) Internal position of the servomotors; b) Link between RS and the joint 3.

4. Performance evaluation

A space robot must have the ability to walk and simultaneously manipulate objects. In the walking mode TORVEastro moves by using grasp while the third foot does not participate to the walking and it can do some other tasks. In order to characterize the motion of the robot, simulations have been computed in SolidWorks with the characteristics data of micro gravity at 0.1 g to simulate micro gravity in the space and in addition there are low value of friction (because bearings are used to transform friction and sliding into rolling friction) and low damping area is assumed at the joints of robot mechanical design. In the first step, TORVEastro capability to move in the space environment has been evaluated by the using of the Torque Motor 1 (TM1), shown in fig. 3 as a central servomotor, it is able to produce a rotation in θ1. Velocity, angular velocity and acceleration of the end-effector are evaluated by the using of TM1. Reaction forces and moment are evaluated in each rotary joint with result in fig. 5. TM1 is used for a variable time of some seconds and its value (TM1) has a cubic pattern as a function of time operating by a maximum value of 1N m and a minimum value of -0,5 N m. In this motion mode TORVEastro is able to perform a rotation of 2π rad in 2.40 s. The maximum value of reaction force is of 9 N and the maximum value of the reaction moment is of 9.6 N m. The joints with largest reactions are joint 1 and joint 2 of link11. The simulation results show very slow value of joint reaction and proper value of mobility, due to the lightweight design of the space robot.

A wide range of values were analysed during the simulations of performance evaluation. The simulations were evaluated by the use of:

- Torque motor 1 (TM1);
- Torque motor 2 (TM2);
- Torque motor 3 (TM3);
- TM1+TM2;
- TM1+TM2+TM3.

The most critical situation in terms of relative speeds, accelerations and constraint reactions, it was shown during the simulation in which 1-2-3 actuators were used simultaneously as reported in fig. 5-6. TM1, TM2 and TM3 have a cubic pattern as a function of time. In this configuration the end-effector has reached the max value of velocity of 2.4 m s^{-1} and the max value of angular velocity compared to the joint 1 of 9.45 rad s^{-1}, the max value of linear acceleration has been of 15 m s^{-2} (fig. 7). The max value of reaction force has been of 11 N and the max value of reaction moment has been of 4,6 N m. The simulation results show a proper ability to move of TORVEastro. The kinematic analysis shows values and time histories that demonstrate the feasibility of this space robot for a practical implementation in service objectives in space stations. The constraint reactions show values that demonstrate solidity of the joint of the robot.

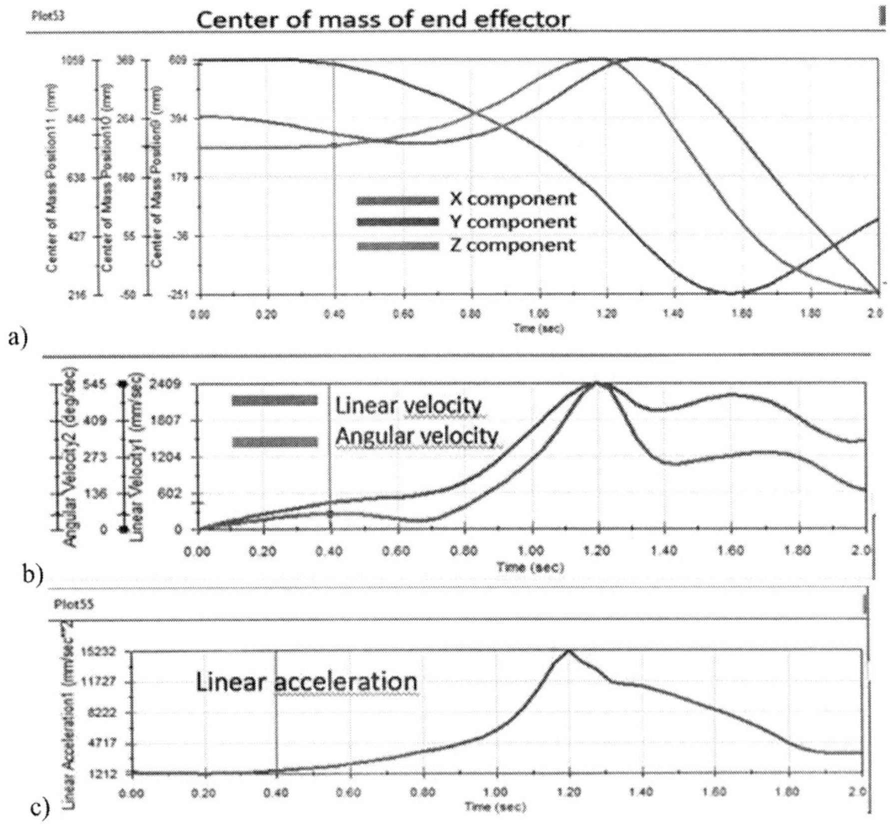

Figure 4. Plot of numerical results of the simulation:
a) Position of center of mass of end-effector (x, y, z components); b) Velocity and angular velocity of center of mass of end-effector; c) Linear acceleration of center of mass of end-effector.

IOP Conf. Series: Materials Science and Engineering **659** (2019) 012010 doi:10.1088/1757-899X/659/1/012010

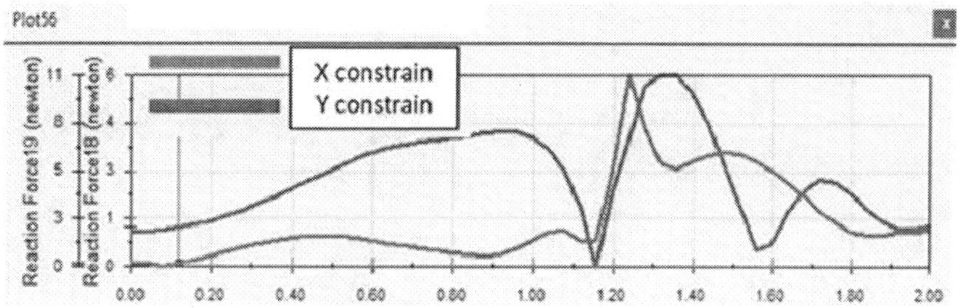

a)

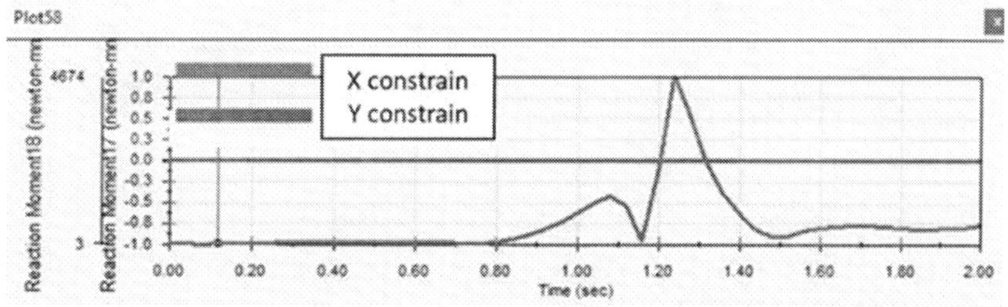

b)

Figure 5. Plot of numerical results of the simulation:
a) Max values of reaction force; b) Max values of reaction moment.

The links of the space robot have been studied by FEA (Finite Element Analysis) subjecting the end-effector to a load of 100 N. Assuming that the element size used for meshing an entity is (e), the average element size in layers radiating from the entity will be: e, e r, e r2, e r3, ..., e r^n (e is the element growth ratio). If the calculated average element size of a layer exceeds (E), where (E) is 3.21 mm the program uses (E) instead. The mesh radiates from vertices to edges, from edges to faces, from faces to components, and from a component to connected components (fig. 6). In the fig. 7 the red arrow represents the yield strength of the material. After several studies the thick of the two links of the robot is 2.5 mm for link1 and 1mm for link2 and, as shown in fig. 7, this configuration has the possibility to support the load of 100 N.

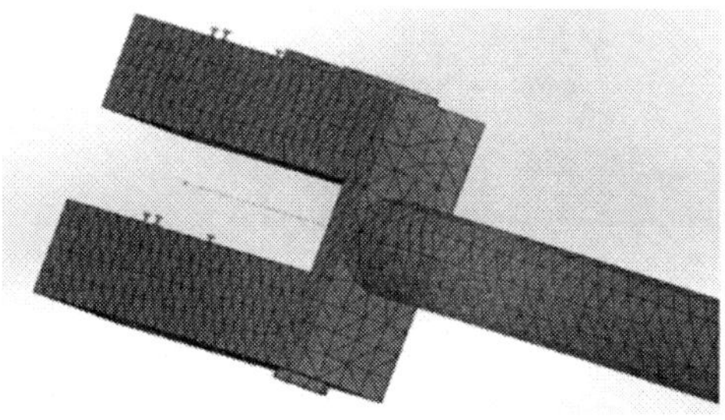

a)

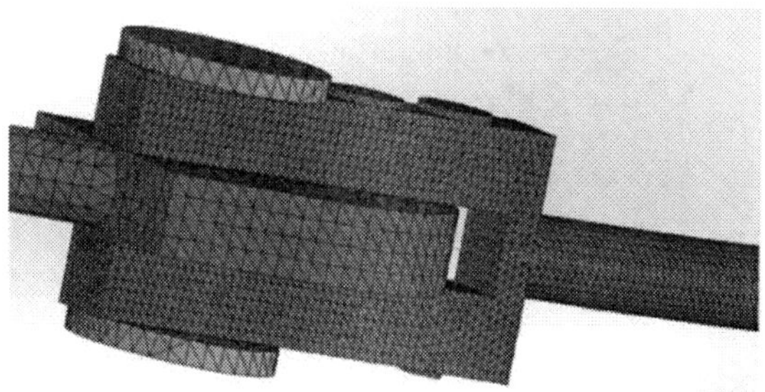

b)

Figure 6. Mesh density: a) Mesh density of joint 1; b) Mesh density of joint 2.

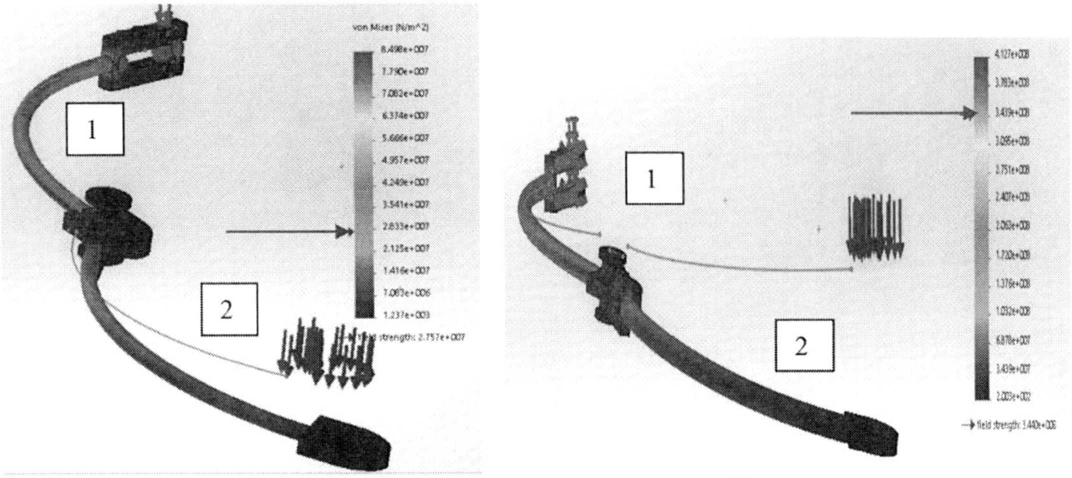

a) *b)*

Figure 7. Plot of results of FEA analysis:
a) Tick of link1 and link2 is 1 mm; b) Tick of link1 is 2.5 mm and tick of link2 is 1 mm.

5. Conclusions

TORVEastro robot is proposed and simulated for service applications to help astronauts to do assembling and to do some other tasks like repairing and monitoring works outside space orbital stations.

The proposed conceptual kinematic design is obtained as a result from an analysis of operation problems and requirements in space. The robot design is composed of one central body and three legs/arms with the purpose to have robust, versatile, compact, and light design. Simulations have been used to characterize the mobility performance to give first indications for a feasible design of a first prototype. Future work is planned for experimental validation and performance testing.

References

[1] Sun Z, Li H and Ceccarelli M 2018 *Prototype Design and Performance Tests of Beijing Astronaut Robot*, Applied Science

[2] Kawamura K, Rogers T E, Hambuchen K A and Erol D 2015 *Toward a Human–Robot Symbiotic System*, Household Service Robotics

[3] Michalos G, Makris S, Tsarouchi P, Guasch T, Kontovrakis D and Chryssolouris G 2015 Design considerations for safe human-robot collaborative workplaces, *Procedia CIRP* **37** 248–253

[4] IFR, *International Federation of Robotics*, https://ifr.org/

[5] ISO 8373:2012, *Robots and robotic devices*, https://www.iso.org/standard/55890.html

[6] Csorba K, Varga D, Tevesz G and Vajk I 2012 The RobonAUT autonomous mobile robot construction contest?, *IFAC Proc.* **9**(1) 360–365

[7] Christoph B, Wimböck T, Schmidt F, Fuchs M, Brunner B, et. al. 2009 *Rollin' Justin - Mobile Platform with Variable Base*, IEEE International Conference on Robotics and Automation

[8] Bogos S 2013 *Kirobo Will be Japan's First Robot Astronaut*, http://www.escapistmagazine.com/ news/view/125525

[9] Calderone L and Matthews K 2014 *Robot Astronaut 'Kirobo' Ready for Launch to International Space Station*, Robotics Tomorrow

[10] Yoshida K and Wilcox B 2008 *Space Robots and Systems*, Springer Handbook of Robotics 1031-1063

IOP Conf. Series: Materials Science and Engineering **659** (2019) 012010 doi:10.1088/1757-899X/659/1/012010

[11] Lambright W H 2019 Administrative Leadership and Long-Term Technology: NASA and the International Space Station, *Space Policy* **47** 85–93
[12] Hoppenbrouwers T, Ferra L, Markus M and Wolff M 2017 Acta Astronautica Operations Data Files, driving force behind International Space Station operations, *Acta Astronaut.* **138**(May) 255–261
[13] Thuot P J and Harbaugh G J 1995 Extravehicular activity training and hardware design consideration, *Acta Astronaut.* **36**(1) 13–26
[14] Ceccarelli M, Li H, Carbone G and Huang Q 2015 Conceptual kinematic design and performance evaluation of a chameleon-like service robot for space stations, *Int. J. Adv. Robot. Syst.* **12**

IOP Conf. Series: Materials Science and Engineering **659** (2019) 012011 doi:10.1088/1757-899X/659/1/012011

Delamination assessment of composite curved angles using simplified FEA models build-up by 2-D layered shell elements

M N Baba[1]

[1]Transilvania University of Braşov, Faculty of Mechanical Engineering, B-dul Eroilor no. 29, 500036 Braşov, Romania

E-mail: mariusbaba@unitbv.ro

Abstract. As widely known, an accurate analysis of the stress state across the thickness of a composite laminate requires the calculation of the interlaminar through-the-thickness stresses as well as the shear stress components at the layers' interfaces. These so-called out-of-plane stresses are the key parameters when defining the delamination strength of curved composite laminates, particularly when their radius of curvature and the thickness of the composite are of the same size order. Since the use of layered solid elements is far from a reasonable approach in practical design applications, based on the finite element analysis results obtained through the use of conventional layered shell elements, the purpose of this paper is to propose a simplified post processing approach that enable to predict the interlaminar stress components originated across the thickness of symmetrically balanced curved composite laminates under the combined action of axial forces, shear forces and bending moments applied in curvature plane. A benchmark analysis to provide a critical comparison between a reference finite element model constructed based on 3-D layered solid elements and a simplified model build-up by 2-D layered shell elements is also considered.

1. Introduction

The use of fiber-reinforced composite laminates across a variety of industrial fields is becoming increasingly common due to their weight saving potential with respect to conventional metallic materials. On the other hand, the fibres lying in the plane of a laminate do not provide reinforcement through-the-thickness direction and consequently this leads to interlaminar failures which significantly reduce the overall strength and integrity of composite structures. As well-known, one of the primary failure modes experienced by the fiber-reinforced composite laminates is the delamination induced due to interlaminar stresses caused by inherent factors such as the free-edge effects, structural discontinuities, impact damage by foreign objects, stiffness mismatch between adjacent layers of different orientation and in particular, curved-shape geometries when folding-unfolding under the applied external loads.

Besides the stress concentration effects near the laminate free edges as well as the impact damage failure mechanisms which have been extensively studied so far by analytical, experimental or numerical methods, few works were focused on the laminate curvature effects especially for highly curved composite laminates relative to their thickness.

When curved composite laminates are subjected to outward bending by unfolding, due to the combined effects of bending, axial and shear forces acting in plane of curvature the inner layers experience tension generating circumferential tensile stresses, whereas the outer layers undergo circumferential compression stresses. Otherwise, folding by inward bending generate compression of inner layers while tensile the outers. In addition, the radial stresses developed due to the internal bending moment tend to separate the interfaces while the shear stresses caused by the out-of-plane force components tend to shear apart the layers.

For design purposes, an efficient procedure to accurately predict the stress state across the thickness of a specific composite laminate requires the calculation of interlaminar through-the-thickness stresses as well as the shear stress components at the interfaces. These so-called out-of-plane stresses are the key parameters in failure analysis of curved composite laminates, particularly when their radius of curvature approaches the stacking thickness.

As concern the various strategies available at the moment to conduct the finite element analyses of highly curved composite laminates loaded in the curvature plane, there are actually two major complications that arise when trying to predict the interlaminar stresses. The first is that even the three-dimensional models built by layered solid elements allow a complete extraction of the interlaminar stress components; in general their practical use is far from a reasonable approach as it requires long computational time as well as the necessity to perform careful and well-controlled discretization in order to avoid the mesh dependencies of results. In fact, these are the main reasons why the finite element models used for structural stress analysis of lightweight composite structures are basically build based on layered shell elements in different depth of discretization. Nevertheless, the second issue is related to the limitations provided through the theoretical definition of layered shell elements that make them impracticable for the analysis of delamination failure of fiber-reinforced composite laminates. These are provided by the main assumptions of the classical laminate theory as it is built into the layered shell elements. Therefore, essentially, the structural response of the entire composite laminate is based on the definition of each individual lamina assuming the out-of-plane normal stress and the out-of-plane shear strains are vanishing so that, as mentioned above, the interlaminar stresses and the associated failures cannot be predicted by means of classical laminate theory. Moreover, as the curvature effects are not considered at all, when the ratio between the radius of curvature and the composite thickness tends to unity, the layered shell elements are becoming inadequate for predicting in-plane curvature stresses about circumferential direction.

Several computational models are available to calculate the closed-form analytical solutions for the out-of-plane stress components developing over the cross sections of curved layered composite plates or beams under plane strain or plane stress assumptions. However, their use as given in the original analytical form is quite difficult and often limited due to various geometries of parts involved in as well as some complex loading conditions. Lekhnitskii [1] has developed an analytical solution based on the theory of elasticity, for orthotropic curved beams loaded by concentrated shear forces and moments at the free ends. The results were extended by Ko and Jackson [2] to a multi-layered theory proposed for semi-circular laminated composite beams subjected to concentrated forces and moments applied at the free ends. O'Peck [3] derived closed-form two-dimensional solutions applicable both to sandwich composites as well as to fibre-reinforced laminates, to determine the displacements, strains and stress fields developed across the section of curved laminated orthotropic beams subjected to shear loads, moments and temperature changes. Martin [4] has investigated the delamination failure in unidirectional curved laminates. Closed form elasticity solutions and 2-D finite element analyses were conducted to locate the critical values of radial stress over the curvature region. The variation of the strain energy release rate with the delamination growth was then determined using the finite element analysis. Gonzales-Cantero et al. [5], [6], developed a complete analytical model to determine the radial, circumferential and shear stress components developed over the cross section of curved laminate composites subjected to simple bending in the curvature plane as well as surface pressures, assuming that each ply is composed by a number of fictitious laminas. In such a way the problem was simplified supposing a beam model in each lamina and thus the equilibrium and compatibility equations were solved taking the limit when the number of fictitious laminae is tending to infinity. The model does not consider the shear deformations as the displacements of every lamina are approximated by the displacements of a homogeneous equivalent material. Some studies in the literature have focused on the development of special layered finite elements [7], [8], or other numerical calculation techniques [9], [10], to predict the interlaminar failure of highly curved composite laminate structural parts. However, such methods are less efficient in the preliminary design phases when certain reliable, simplified and conservative analytical approaches in combination with some basic finite element analysis results are more effective.

Based on the finite element analysis results obtained through the use of conventional layered shell elements, the purpose of this paper is to propose a simplified post processing approach that enable to predict the interlaminar stress components originated across the thickness of symmetrically balanced curved composite laminates under the combined action of axial forces, shear forces and bending moments applied in curvature plane. A benchmark analysis to provide a critical comparison between a reference finite element model constructed based on 3-D layered solid elements and a simplified model build-up by 2-D layered shell elements is also considered.

2. Analysis

The numerical analyses were conducted on laminated curved angles made of a polyphenylene sulfide (PPS) thermoplastic resin reinforced with T300 JB continuous carbon fibers. The layout is virtually obtained by stacking a number of 13 plies of 0/90° bidirectional 5-harness satin fabric semi-impregnated with PPS polymer, following the sequence [(0,90)/(±45)6/(0,90)], which results in a 4 mm-thick laminate. The elastic properties of PPS lamina that were considered in the finite element analyses are reported in Table 1.

Table 1. The elastic properties of PPS lamina.

E_1 [*MPa*]	E_2 [*MPa*]	E_3 [*MPa*]	ν_{12} [-]	ν_{13} [-]	ν_{23} [-]	G_{12} [*MPa*]	G_{13} [*MPa*]	G_{23} [*MPa*]
55400	55400	8500	0.05	0.35	0.35	3510	3510	3510

The laminate angle under analysis is provided with a ratio between the radius of curvature and the composite thickness, R/h equal to 2. It has to be noted that this radius of curvature, denoted by R, refers to the centroidal axis that defines the middle plane of the laminate cross section.

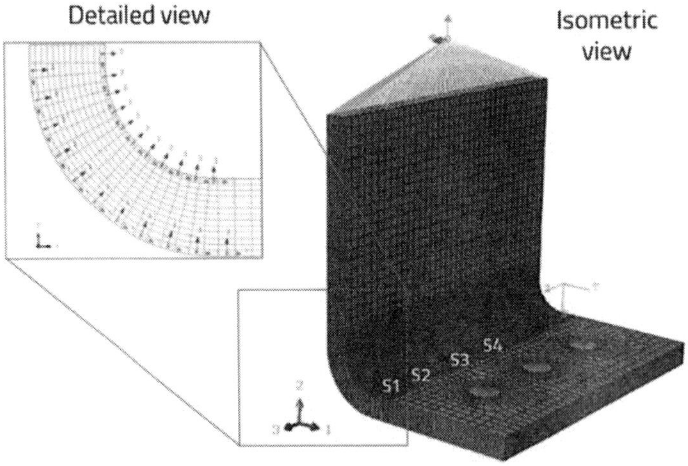

Figure 1. The reference FE model of layered solid elements.

Two types of computational models constructed in Abaqus/CAE finite element software were considered, namely a reference model build-up by 3-D layered solid elements (C3D8) and a simplified one based on layered shell elements (S4). Figure 1 shows a detailed isometric view of the reference model considered in the analysis. Within the same figure the consistent pattern of the orientations of the element coordinate systems is presented upon the lateral view of the curvature area.

The use of consistent element coordinate systems to define the material orientation in case of layered solid elements models enables a coherent definition of the laminate layout. This is particularly important when curved laminate angles are involved in, since the accuracy of the results can be significantly affected due to the anisotropic properties, in particular between in-plane as well as through-thickness direction. However, such a task requires an accurate control of meshing procedure.

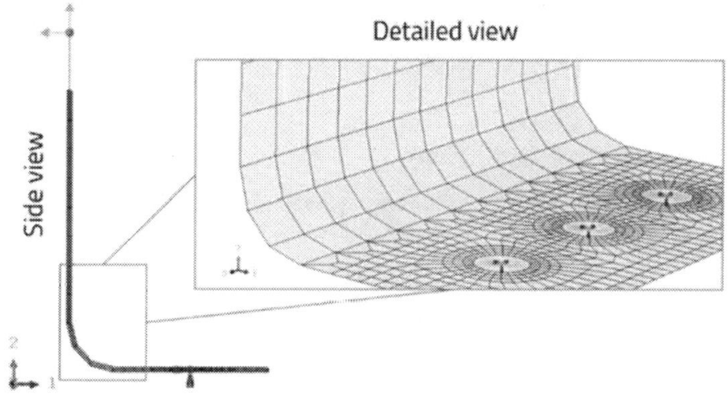

Figure 2. The simplified FE model of layered shell elements.

A side view containing an isometric detail of curvature area for the layered shell finite element model is presented in Figure 2. In this regard it has to be mention that both for the simplified model as well as for the reference model, the loading is virtually represented by two unit forces (F_1 = -1*kN*; F_2 = 1*kN*) and a unit couple (M_3 = 1*kNm*), applied on the free side of the angle in the curvature plane, with respect to the global reference coordinate system. Furthermore, as shown in Figure 1 and Figure 2, for both finite element models, the clamped boundary conditions at the bolt holes as well as the unit loads are applied by means of RBE2 elements.

As already explained in the introductory section, since the layered shell elements are inadequate for predicting the folding-unfolding interlaminar stresses, a simplified post-processing model based on shell forces and shell moments extracted at the elements located on the baseline of the curvature region is proposed. Such a model can be used to estimate the strength of highly curved quasi-isotropic laminates provided their symmetrically balanced layout configuration, as is the present case study of 0/90° bidirectional polyphenylene sulphide reinforced with continuous carbon fibers. Therefore, the overall equivalent elastic properties of the laminate may be considered in the analysis, as given in Table 2.

Table 2. The equivalent elastic properties of laminate.

E_x [*MPa*]	E_y [*MPa*]	E_z [*MPa*]	ν_{xy} [-]	ν_{xz} [-]	ν_{yz} [-]	G_{xz} [*MPa*]	G_{xy} [*MPa*]	G_{yz} [*MPa*]
41031	41031	8500	0.30	0.26	0.35	14066	3510	3510

Assuming a planar stress state, the radial stress component σ_r and the shear stress component $\tau_{r\theta}$ may be derived based on the general curved beam expressions:

$$\sigma_r(\theta) = \frac{s(\theta)}{h}\ ; \qquad \tau_{r\theta}(\theta) = \frac{3}{2}\frac{m(\theta)}{h\sqrt{r_i r_o}}\ , \tag{1}$$

where $s(\theta)$ is the flux of shear force (force per unit length, units [*N/mm*]) while $m(\theta)$ is the moment flux (bending moment per unit length, units [*Nmm/mm*]). In the above expressions the thickness and the width of the laminate are denoted by h and b while r_i and r_o indicate the inner radius and the outer radius respectively.

As Figure 3 shows, the curvature fluxes $s(\theta)$ and $m(\theta)$ can be easily derived through the translation of the element fluxes s_i, n_i and m_i, extracted at the baseline of the shell elements located on the ends of curvature. Therefore, the following transformations may apply:

$$s(\theta) = s_i \cdot \cos(\theta) - n_i \cdot \sin(\theta) \; ; \quad m(\theta) = m_i + s_i \cdot R\sin(\theta) - n_i \cdot [R - R\cos(\theta)] \, , \tag{2}$$

where R is the radius of centroidal axis of laminate:

$$R = r_i + \frac{h}{2} \, . \tag{3}$$

Within the above context it is important to note that the definition of the coordinate system is taken so that the positive s_i and m_i determine the angle opening (unfolding) while positive n_i causes angle closing (folding).

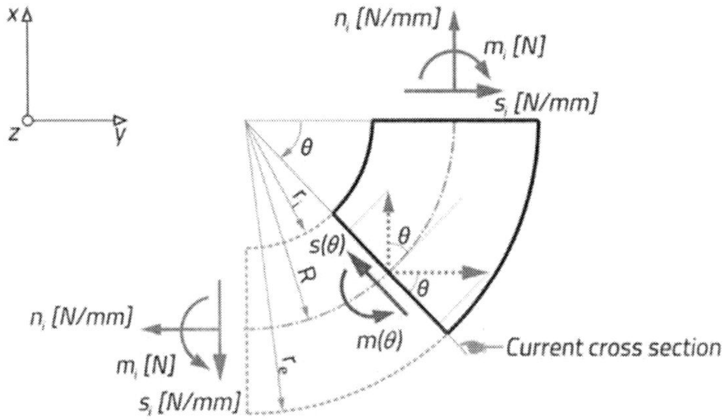

Figure 3. Data processing scheme based on the extracted fluxes.

Neglecting the contribution of in-plane circumferential stresses, according to Ref. [11] the quadratic failure criterion proposed by Brewer and Lagace may be used to predict the delamination onset:

$$\left(\frac{\tau_{13}}{f_{13}}\right)^2 + \left(\frac{\tau_{23}}{f_{23}}\right)^2 + \left(\frac{\sigma_1}{f_1}\right)^2 = 1 \, , \tag{4}$$

where τ_{13} and τ_{23} are the (average) interlaminar shear stresses, σ_1 is the (average) radial stress and f_{13}, f_{23} ,f_1 are the corresponding allowable (strength) values.

Based on the aforementioned criterion, for plane stress problems when positive radial stresses occur, the Factor of Safety may be determined as:

$$FoS = \frac{1}{\sqrt{\left(\frac{\sigma_1}{f_1}\right)^2 + \left(\frac{\tau_{13}}{f_{13}}\right)^2}} \, , \text{ if } \sigma_3 > 0 \tag{5}$$

while for regions with compressive radial stresses, the expression of Safety Factor reduces to:

$$FoS = \frac{f_{13}}{\tau_{13}} \, , \text{ if } \sigma_3 < 0 \, . \tag{6}$$

IOP Conf. Series: Materials Science and Engineering **659** (2019) 012011 doi:10.1088/1757-899X/659/1/012011

3. Results

The variations of the interlaminar radial and shear stresses with the curvature angle between 0 and 90 deg., obtained by means of the reference layered solid model are presented in Figure 4 and Figure 5. The stress components are evaluated with respect to a local cylindrical coordinate system whose radial direction is denoted by 1 while the tangential direction is denoted by 2.

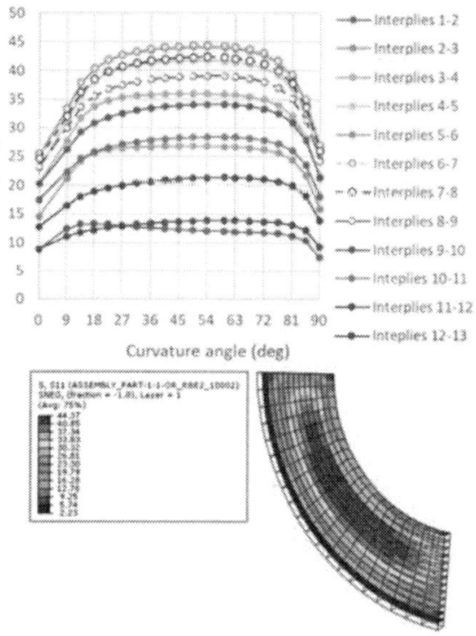

Figure 4. Variation of the interlaminar radial stresses with the curvature angle between [0-90] deg.

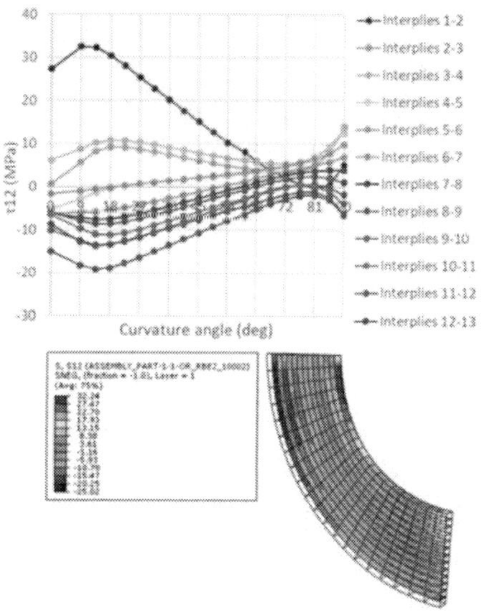

Figure 5. Variation of the interlaminar shear stress with the curvature angle between [0 - 90] deg.

The force and moment fluxes were extracted at both ends of the curved region. The local coordinate systems are defined for both angle sides so that their positive directions determine opening the angle. Figure 6 shows the selection of elements and the orientations of the local coordinate systems used for extraction of fluxes at the beginning and the end of curvature region.

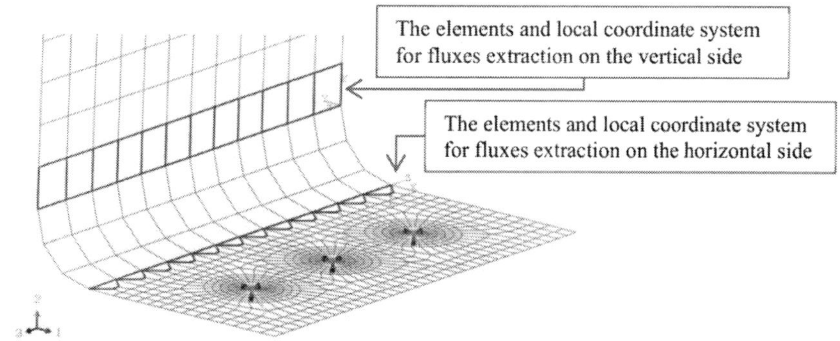

Figure 6. The elements selection and the local coordinate systems.

Figure 7 to Figure 12 present the fringe plots of component fluxes represented by the shell forces and shell moments at the baseline of curvature region, extracted both for horizontal and vertical sides of the angle. The corresponding local coordinate systems are considered as shown in Figure 6. It is to be noted that the sign of the shell moment is defined by the element in the sense that a positive moment produces compression on the positive element side. Thus, when post-processing data it is important to consider the element normal and change the shell moment signs accordingly.

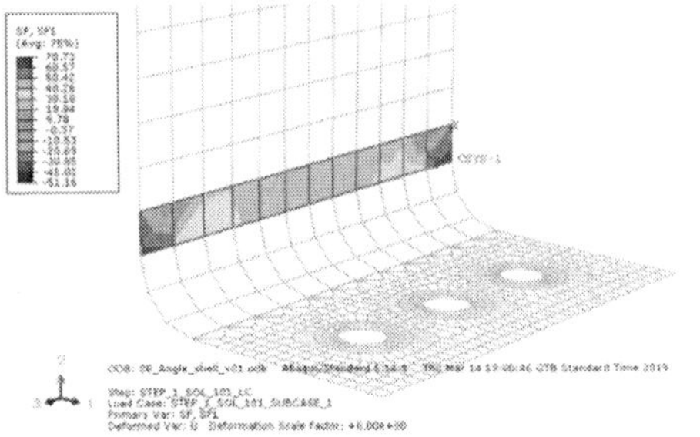

Figure 7. Fringe plots of shell forces n_i at the curvature base (vert. side).

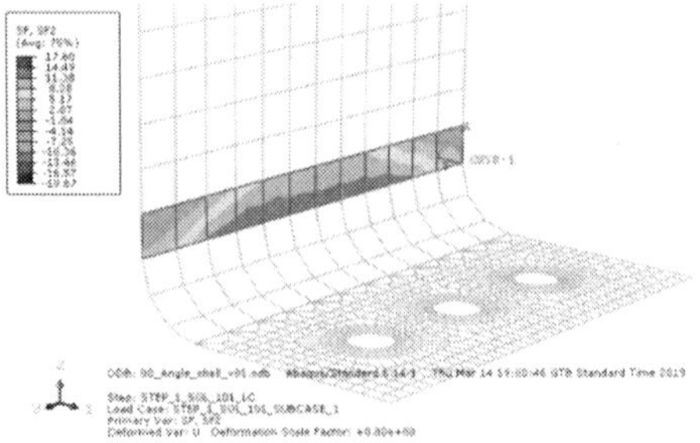

Figure 8. Fringe plots of shell forces s_i at the curvature base (vert. side).

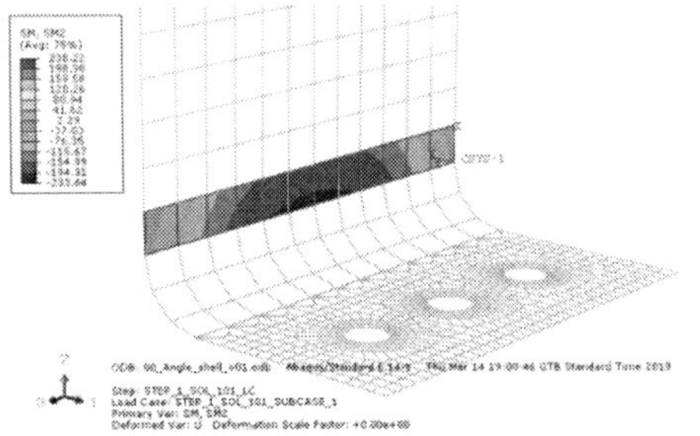

Figure 9. Fringe plots of shell forces m_i at the curvature base (vert. side).

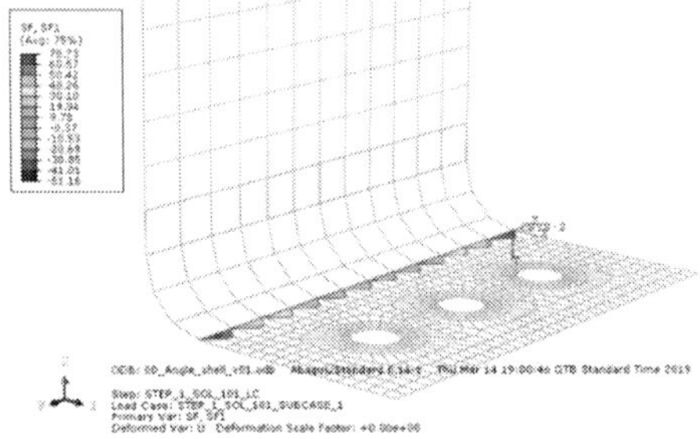

Figure 10. Fringe plots of shell forces n_i at the curvature base (horiz. side).

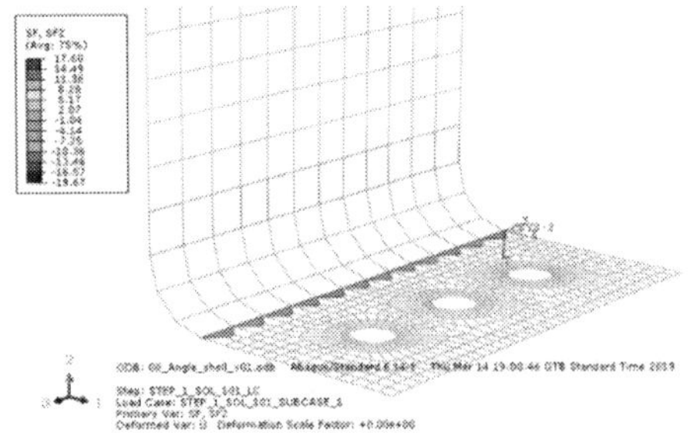

Figure 11. Fringe plots of shell forces s_i at the curvature base (horiz. side).

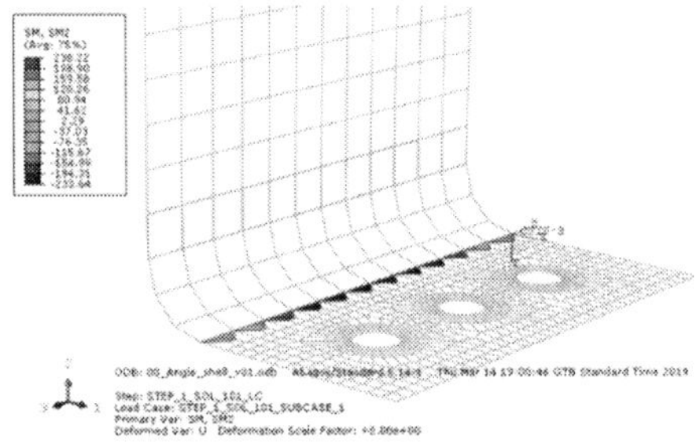

Figure 12. Fringe plots of shell forces m_i at the curvature base (horiz. side).

Figure 13 to Figure 16 show comparative plots of Safety Factor calculated for the reference solid model versus the simplified shell model, at the different sections of the laminate angle as denoted by S1, S2, S3 and S4 in Figure 1.

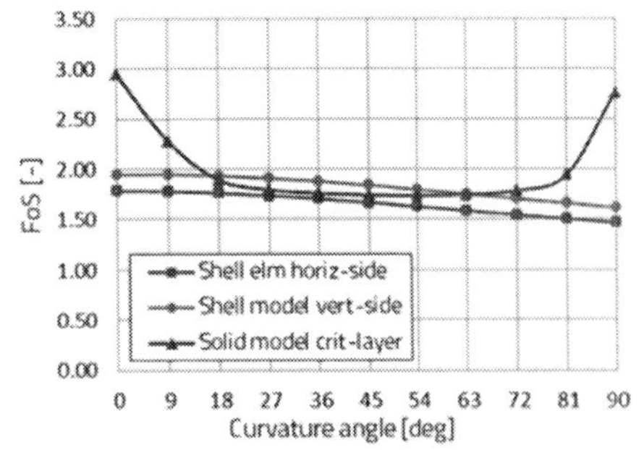

Figure 13. FoS comparison results of solid and shell model at section S4.

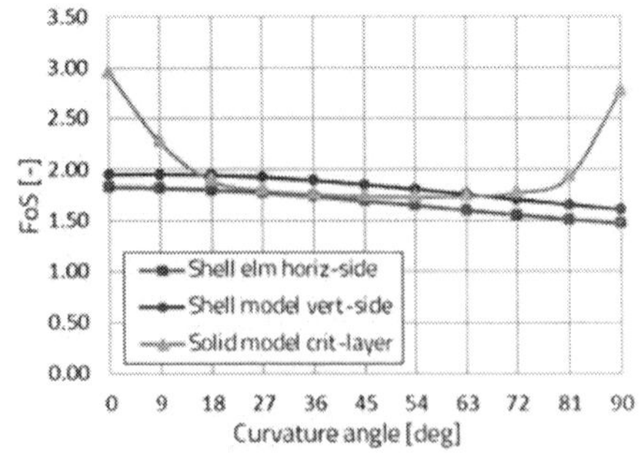

Figure 14. FoS comparison results of solid and shell model at section S3.

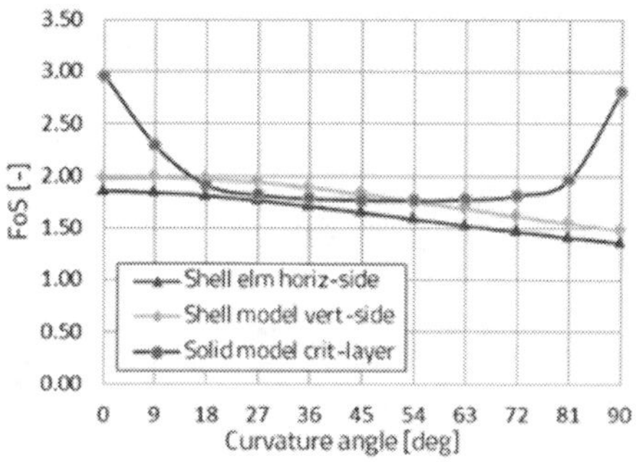

Figure 15. FoS comparison results of solid and shell model at section S2.

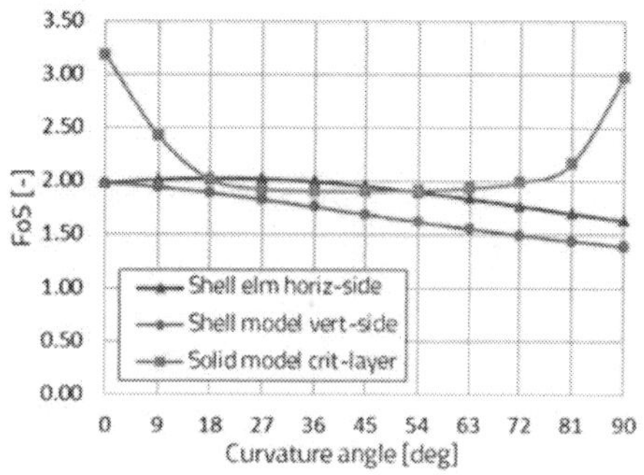

Figure 16. FoS comparison results of solid and shell model at section S1.

4. Conclusions

A simplified post-processing approach based on shell forces and shell moments extracted at the elements located on the baseline of the curvature region is proposed. The reliability of those input loads and thus the reliability of the method depends mainly on the level of discretisation of the shell-based finite element model. Hence, the approach described within the present paper might be successfully used in practical applications either as a preselection method that allows to identify the critical locations or as a pre-sizing method to calculate the reserve factors of highly curved angles made of quasi-isotropic laminates provided their symmetrically balanced layout configuration.

The base assumption of this approach is that the solution of the elastic problem is assumed to be bi-directional. It follows that the main components of stresses that are responsible for unfolding are through-the-thickness normal and shear stresses which may be easily calculated based on shear force fluxes and bending moment fluxes. Another conservative assumption is that the maximum stresses occur at the same location in the thickness direction.

The use of consistent element coordinate systems to define the material orientation in case of layered solid elements models enables a coherent definition of the laminate layout. This is particularly important when such curved laminate angles are involved in, since the accuracy of the results can be significantly affected due to the anisotropic properties, in particular between in-plane as well as through-thickness direction. However, an accurate control of meshing procedure is required. It is also to be noted that the sign of the shell moment is defined by the element in the sense that a positive moment produces compression on the positive element side. Thus, when post-processing data it is important to consider the element normal and change the shell moment signs accordingly.

The comparative critical analysis of the finite element results obtained by means of using layered solid elements with respect to the results computed based on processing the shell forces and shell moments extracted at the baseline of curvature region, show that the proposed simplified approach provides conservative results. Anyhow, further extension of the method is required to account for in-plane circumferential tensile/compression stresses as well as the use of an appropriate failure criterion.

References

[1] Lekhnitskii S G, Tsai SW and Cheron T 1968 *Anisotropic Plates*, Gordon and Breach Science Publishers

[2] Ko W L and Jackson R H 1989 *Multilayer theory for delamination analysis of a composite curved bar subjected to end forces and end moments*, NASA Technical Memorandum 4139

[3] Peck S O 2000 Curved Laminated Beams Subjected to Shear Loads, Moments and Temperature Changes, *Composite Structures: Theory and Practice*, ASTM STP **1383** 437-451

[4] Martin R H 1992 *Delamination Failure in a Unidirectional Curved Composite Laminate*, STP1120-EB *Composite Materials: Testing and Design (Tenth Volume)*, edited by GC Grimes 365-383

[5] González-Cantero J M, Graciani E, Blázquez A and Paris F 2016 A new analytical model for evaluating interlaminar stresses in the unfolding failure of composite laminates, *Composite Structures* **147**(2016) 260-272

[6] Gonzalez-Cantero J M, Graciani E, Blazquez A and Paris F 2014 *Analytic evaluation of radial stresses in unfolding failure of composite materials. Comparison with numerical solutions.* In: Proceedings of the 16th European Conference on Composite Materials ECCM16., 22-26 of June, Seville, Spain

[7] Vidal P, Gallimard L and Polit O 2014 Shell finite element based on the proper generalized decomposition for the modelling of cylindrical composite structures, *Computers and Structures* **132**(2014) 1-11

[8] Ferreira A J M, Carrera E, Cinefra M and Roque CMC 2011 Analysis of laminated doubly-curved shells by a layerwise theory and radial basis functions collocation, accounting for through-the-thickness deformations, *Comput Mech* **48**(2011) 13-25

[9] Matsunaga H 2003 Interlaminar stress analysis of laminated composite and sandwich circular arches subjected to thermal/mechanical loading, *Composite Structures* **60**(2003) 345-358

[10] Alibeigloo A and Shakeri M 2009 Elasticity solution for static analysis of laminated cylindrical panel using diferential quadrature method, *Engineering Structures* **2009**(31) 260-7

[11] Brewer J C, Lagace P A 1998 Quadratic stress criterion for initiation of delamination, *Journal of Composite Materials* **22**(1988) 1141-1155

IOP Conf. Series: Materials Science and Engineering **659** (2019) 012012 doi:10.1088/1757-899X/659/1/012012

Stress analysis of lifting table using finite element method

N Rašović[1], A Vučina[1] and M Obad[1]

[1]University of Mostar, Faculty of Mechanical Engineering, Computing and Electrical Engineering, Matice hrvatske bb, 88000 Mostar, Bosnia and Herzegovina

E-mail: nebojsa.rasovic@fsre.sum.ba

Abstract. Lifting table has been designed and developed through the concept of Learning Factory (LF) at the University of Mostar. The idea for lifting table design has come from the local industry needs for a lifting platform that should lift a man and/or load at a certain height. For safety reasons, design is checked under the loading using a method of finite element analysis. The paper predicts and explains methodology for structural analysis used in presented case study. Results of FEM analysis are basis for making ways and guidelines to optimize current design in order to get optimal parameters for weight, stability, capacity, mobility and layout of the lifting table.

1. Introduction

From the previous work [1], design of lifting platform has been realized through learning factory concept at Department of mechanical engineering University of Mostar. It is about mechanical lifting mechanism that is able to carry great loads by vertically hand moving. A need to use a lifting mechanism is very widespread across labs, workshops, departments etc. The goal of developed design is to lift some objects or man at a certain height in order to perform some operations. The required highest level of lifting was up to 550 mm. The technical requirements for design have been defined by previous work and lifting table has been achieved as represented (Table 1).

Table 1. Technical requirements of the mechanical lifting table.

Technical Requirements, Constraints		
Table dimensions: 600×300 mm	Max. costs: 150 €	
Max. lifting height: 550 mm	Easy assembly of parts: ✓	
Min. height of design: 200 mm	Easy to handle: ✓	
Max. load carrying: 300 kg	Easy dismantling: ✓	
Total structure mass: 30 kg	Adopt to available technology at LF: ✓	

Corrado et al. [2] has been focused on design and analysis of belt drive of lifting table scissor in only two characteristic positions, the highest and the lowest, where the stress was much lower than the yield stress of material (S275 and load carrying of 1000 kg). Dengiz et al. [3] has designed two-level scissor lifting system with hydraulic drive, where the stress was in the allowable limits (for St37, St50 and load carrying of 500 kg). Rani et al. [4] introduced even two hydraulic cylinders at two-level scissor for better operating and controlling. Besides stress analysis, some papers [5-8] have been dealing with design improvement by selecting some optimal criteria of topology optimization. In following sections, methodology of stress analysis and optimization of lifting table structure has been presented.

2. Objective and scope

Stress and strain analysis for the entire structure of lifting table represents one of the main tasks in the phase of embodiment design. This is the most important step because calculation results need to confirm design correctness. Correctness of design implies capability to carry out full predictable load, easy mechanical handling, safe lifting and smooth motion of the mechanism. Because of very low working speed, it will be observed three characteristic positions of the lifting table mechanism: the highest, the lowest and middle position under acting of maximal load. The objective of this paper is to check the stress and strain for all three characteristic positions and on that basis make ways and guidelines to optimize current design.

In accordance with previous mentioned, material selection has to be appropriate depending on component type and function it performs. The most of the components are made of structural steel (S275) like as two frames (upper and lower), four legs, longitudinally stiffener, transversely stiffener and lever. The four wheels are made of polyethylene. Slider and spindle screw nut are made of bronze, while spindle is made of stainless steel.

3. Methodology for structural analysis

Technical requirements (Table 1) will play main role in defining boundary conditions of parametrized model of lifting table. The Figure 1 shows workflow algorithm for structural analysis for previous defined requirements. Appropriate geometric mesh model that ensures low approximation errors and validity of obtained results has been created. This task can be done in few different derived mesh iterations in order to get congruence or confirmation of stress results.

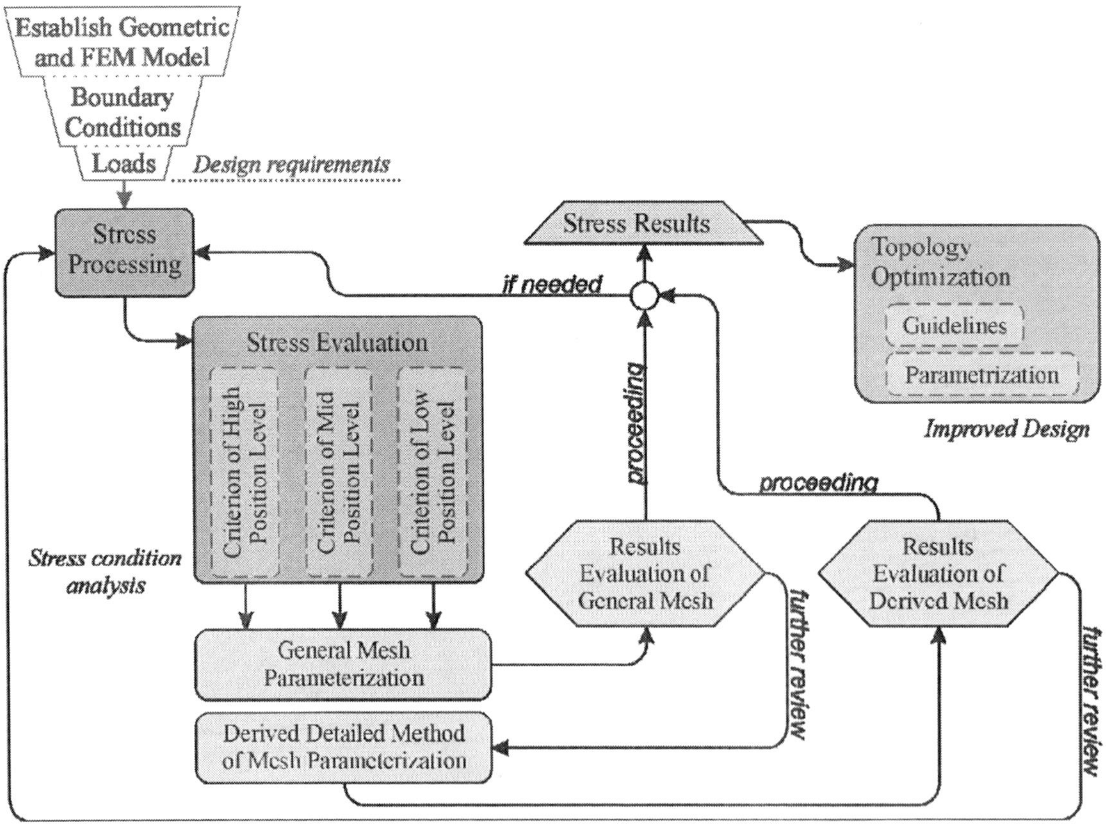

Figure 1. Workflow algorithm of stress analysis using FEM.

To achieve full validity of results, all relationships between parts in their hierarchal order within assembly have to be established. For purpose of relationships defining, Ansys® SpaceClaim has been used. The most important, all connections (joints) have to be defined from the motion type aspect and all of them are constrained in certain axes directions. From simplification and calculation aspect, the spindle has replaced by circular beam like cylindrical rigid body with same diameter and material. Friction coefficient has been added between polyethylene wheels and structural steel frames, which is 0.2. Support is defined as fixed type and it is located on the lower frame at the bottom face. Force value is 3000 N and it is applied on the upper frame at the top face.

4. Case study

For structural analysis implementation, Ansys® Workbench has been used [9, 10]. All three positions are analysed under full load of 3000 N, which is applied on top face at the upper frame. Entire configuration of the lifting table scissor is considered as a rigid structure, not as a mechanism. That means the components in their contacts have been defined as bonded connections. Equivalent (von-Mises) stress and total deformation are obtained by structural analysis and the criterion for decision is value of yield stress for structural steel (275 MPa) as maximal limit in defining stress condition of the structure (Figure 2).

Figure 2. Structural analysis for general mesh in the three characteristic positions under full load.

The maximal reached value of stress has been 224.11 MPa, related to low position and it is located at welded joint of mounting bracket and lower frame (Figure 2, c). In the middle position, maximal stress value has been 219.51 MPa and it is located at spindle (Figure 2, b). The minimal value of stress has been 121.49 MPa, related to high position and it is located at lever (Figure 2, a). These results represent spotted values that become a base for further analysis. The greatest value of total deformations for all mechanism positions does not exceed 1.7 mm (Figure 2), so strain analysis is not considered in further procedure.

Taking into account workflow algorithm (Figure 1), a few different meshes are created to confirm obtained results from the first mesh iteration (General Mesh$^{(1)}$). Each mesh has to be analysed, since each following mesh is generated with respect to the results derived from the previous one. The goal is to confirmed results from previous analysis (Figure 2), and the components with the greatest stress values have been taken to additional observing.

According to structural analysis (Derived Detailed Mesh$^{(2)}$, Table 2), three components had maximal stress values: lever (in high position), spindle (in middle position) and mounting bracket at welded joint (in low position). These components are taken to further analysis and mesh details are listed in the Table 2.

Table 2. Mesh details for multiple iterations of structural analysis.

Characteristics of Mesh	General Mesh$^{(1)}$	Derived Detailed Mesh$^{(2)}$	Derived Detailed Mesh$^{(3)}$
Mesh Method	Automatic	Hex Dominant	Hex Dominant
Element Size [mm]	6	3	1
Size Function	Adaptive	Curvature	Proximity and Curvature
Mesh Quality	Medium	Fine	Fine
Transition Ratio	0,272	0,272	0,272

Using results derived from previous mesh analysis, three component with the highest equivalent (von Missis) stresses are excluded into individual observing (Derived Detailed Mesh$^{(3)}$, Table 2) in order to get more precisely values. In this case, it should take into account that very small surface of elements could lead to enormously great and unreliable values of stresses. After the last iteration results have been represented (Figure 3).

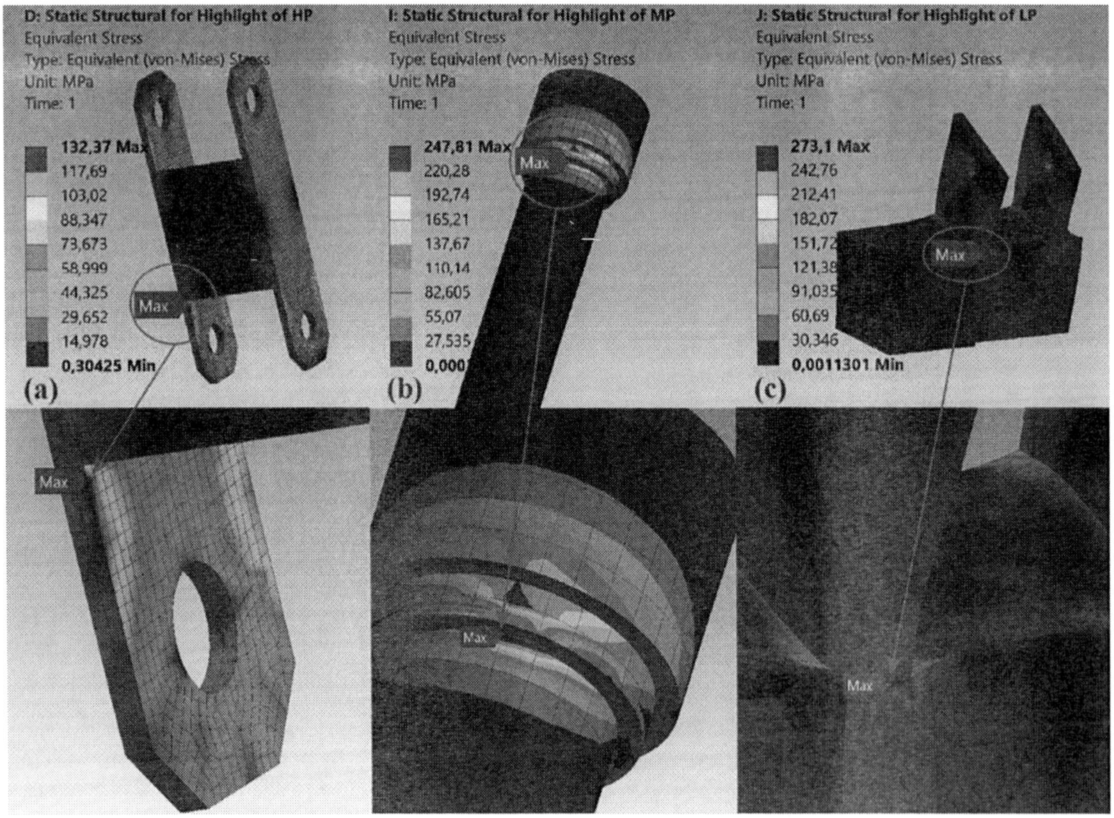

Figure 3. Derived detailed results of previous parameterization for excluded components.

As the Figure 3 shows, the maximal stress values of the excluded components have been increased for 10% approximately. This increasing has been related to the case of derived detailed mesh and these stress values represent purified results through few mesh iterations. It can be noted that obtained results are compromise in numerical evaluation of multiple model parameters. Stress results obtained in this section will serve in creating guidelines for design improvement.

5. Guidelines for design improvement

Improvement of lifting table design using method of topology optimization is presented in this section. The procedure of topology optimization is based on equivalent (von-Mises) stress results in order to generate optimal shape of part or structure. For a part subjected to a given constraint (equivalent stress) the intention was to increase its stiffness on the one hand and to reduce the weight on the other. Topology optimization results have to offer a potential geometry for removal of less efficient material from total part volume. To find the most precise geometry for topology optimization, the number of iterations have to be increased as much as possible, as well as number of elements. It means that procedure of FE analysis has to find out and suggest as much as possible of passive elements in each iteration step. To obtain more accurate and precise results of topology optimization process, it has been used parameters for fine quality of meshing (Derived Detailed Mesh[3], Table 2). To conduct topology optimization process, the objective goal is compliance minimization as specific criterion under static structural environment. The term compliance in topological terminology is opposite to term stiffness. Therefore, minimizing the compliance is equivalent to maximizing the global structural static stiffness.

From FE analysis, three characteristic positions of the scissor mechanism have given three components with the greatest stress values. Criterion of response constraint is volume reduction, and for all structure parts is 85% to retain. Thus, process of topology analysis has shown potential cutting contour of excess material (Figure 4).

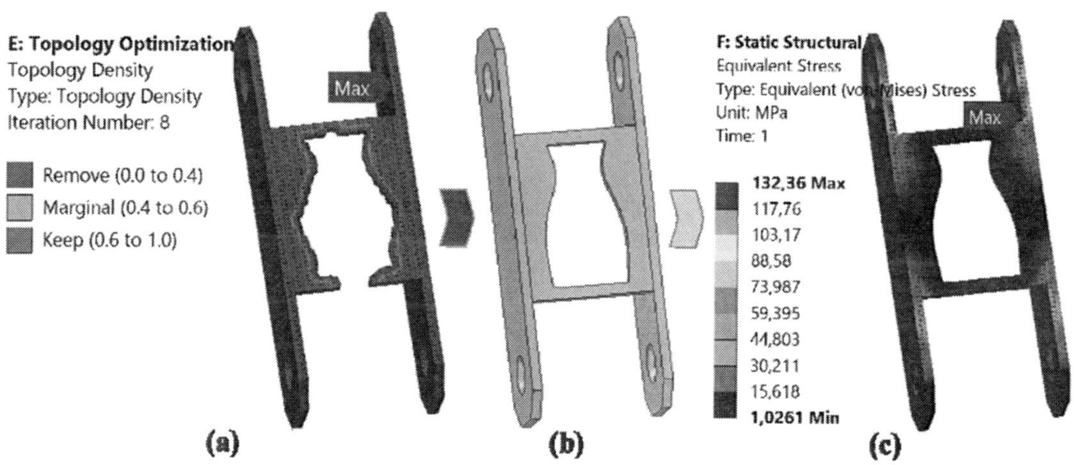

Figure 4. Recommendations of topology analysis (a), changing feature through geometry (b) and results confirmation by renew structural analysis (c).

Changes in geometry have been made according to proposed pattern in topology analysis (Figure 4, a), the shape of pattern is mostly retained and new feature has been created through SpaceClaim modeller (Figure 4, b). Renew structural analysis has confirmed previous results in stress, so validity of topology optimization is proven (Figure 4, c). Based on maximal stresses occurrence, some guidelines for potential topology optimization of the other two components, ordered by priority would be:

For spindle:

1. Add radius fillet at two transition edges of the spindle.

For welded joint of mounting bracket and lower frame:

1. Increase the thickness of mounting bracket for greater strength;
2. Add radius fillet at welded mounting bracket with lower frame;
3. Place mounting bracket on steel plate and make welded joint between plate and frame;
4. Strengthen mounting bracket with additional welded seam at frame;
5. Improve quality of welded joints.

These guidelines have been ordered by degree of priority, and they will be subject of further stress and topology analyses in order to further improve lifting table design. Note that each change in topology or design must be within constraints listed in the Table 1.

6. Conclusion

In this paper, the methodology for stress and topology analysis of lifting table has been presented. This methodology has given a way to calculate values of stresses and deformations taking into account all mentioned constraints. Results obtained by FEM analysis has shown that stress does not occur always at the same place. Its distribution is depending of scissor position and for three different positions, there is transiting over three different components. FEM results have confirmed that the maximum stress value does not exceed value of yield stress. It means that static principles have satisfied. This type of structural analysis has served as introduction to topology analysis based on criterion of volume reduction. The goal was to retain rigidity while reducing mass in lower stress area, dominant blue colour in stress analysis. Initial proposed pattern of boundary shape has been adjusted and applied in feature creation. Further structural analysis confirmed previous stress results. Maximal stress appeared at same place as previous, and topology optimization has been proven.

References

[1] Rašović N, Vučina A and Dedić R 2018 *Lifting Table Design in a Learning Factory at University of Mostar*, Proceedings of the 4th International Scientific Conference on Mechanical Engineering Technologies and Applications COMETa 2018, 374–379
[2] Corrado A, Polini W, Canale L and Cavaliere C 2016 To Design a Belt Drive Scissor Lifting Table, *International Journal of Engineering and Technology* **8**(1) 515–525
[3] Dengiz C, Şenel M, Yıldızlı K and Koç E 2018 Design and Analysis of Scissor Lifting System by Using Finite Elements Method, *Universal Journal of Materials Science* **6**(2) 58–63, https://doi.org/10.13189/ujms.2018.060202
[4] Rani D, Agarwal N and Tirth V 2015 Design and Fabrication of Hydraulic Scissor Lift, *MIT International Journal of Mechanical Engineering* **5**(2) 81–87
[5] Bankoti S, Jain N and Misra A 2015 Topological Optimization of 3D Structures by Optimality Criteria using ANSYS, *International Journal for Research in Emerging Science and Technology* **2**(2) 30–35
[6] Li D and Kim I Y 2018 Multi-material Topology Optimization for Practical Lightweight Design, *Struct. Multidiscip. Optim.* **58**(3) 1081–1094, https://doi.org/10.1007/s00158-018-1953-z
[7] Beňo P, Kozak D and Konjatić P 2014 Optimization of Thin-walled Constructions in CAE System ANSYS, *Tehnički vjesnik* **21**(5) 1051–1055
[8] Picelli R, Townsend S, Brampton C, Norato J and Kim H A 2018 Stress-based Shape and Topology Optimization with the Level Set Method, *Computer Methods in Applied Mechanics and Engineering* **329** 1–23, https://doi.org/10.1016/J.CMA.2017.09.001
[9] Rašović N, Vučina A and Dedić R 2019 Design and Analysis of Steel Reel Shaft by Using FEA, *Tehnički vjesnik* **26**(2) 527–532, https://doi.org/https://doi.org/10.17559/TV-20180116103950
[10] Dizdar S, Vucina A, Rasovic N and Tomovic R 2018 *Application of Pipestress and Ansys in Stress Analysis of Nuclear Pipe Support – Case Study*, IOP Conference Series: Materials Science and Engineering **393**(1) 012021, https://doi.org/10.1088/1757-899X/393/1/012021

IOP Conf. Series: Materials Science and Engineering **659** (2019) 012013 doi:10.1088/1757-899X/659/1/012013

Response statistics of single degree of nonlinear random structure with nonlinear damping characteristic and nonlinear elastic characteristic under white-noise excitations

P Stan[1] and M Stan[1]
[1]University of Pitești, Faculty of Mechanical Engineering and Technology, Romania

E-mail: petre_stan_marian@yahoo.com

Abstract. The paper investigates the applicability of the path integral solution method for calculating the response statistics of nonlinear dynamic systems whose equations of motion can be modelled by the use linearization differential equations. The present paper consists of discussion on dynamic response of structures under random load. They are random processes and commonly described by spectral density functions. An identification technique is proposed for a nonlinear oscillator excited by response-dependent white noise. Stiffness, damping and excitation are estimated from records of the stationary stochastic response. Assume that a single-degree of freedom structure is excited by a force which is a random process described by the spectral density function.

1. Introduction

Nonlinear, dynamic systems subject to random excitations are frequently met in engineering practice. The research goals are, firstly, the computation of stochastic, nonlinear response characteristics (with accuracy and efficiency as important criteria) and, secondly, the investigation and thorough understanding of stochastic, nonlinear response phenomena. The source of randomness can vary from surface randomness in vehicle motion and environmental changes, such as earthquakes or wind exciting high rise buildings or wave motions at sea exciting ofshore structures or ships, to electric or acoustic noise exciting mechanical structures. The desire to compute response characteristics, such as the statistical moments and the power spectral density of the response of these systems, leads to the development of methods that can be used to approximate this response. The excitations, that will be studied, are stationary, Gaussian processes [1,2]. They are random processes and commonly described by spectral density functions. Assume that a single-degree of freedom structure is excited by a force F which is a random process described by the spectral density function $S_F(\omega)$ [2]. If Gaussian assumption is adopted for the force and the structure is assumed to be nonlinear, the response is expected to be Gaussian distribution as well. As the force is not deterministic, the response of the structure is expected to be random.

2. System model

As a result, to obtain the standard deviation of the response, the velocity variance of the single-degree of freedom system the relative acceleration of this structure and the spectral density of the response.
A single-story building is modeled by four identical columns of Young's modulus E and height h and a rigid floor of weight m. The damping can be approximated by an equivalent damping constant c. The ground acceleration due to an earthquake is assumed to be a Gaussian white noise with a constant spectrum S_0. The columns have cilinder sections of diameter D.

IOP Conf. Series: Materials Science and Engineering **659** (2019) 012013 doi:10.1088/1757-899X/659/1/012013

The moment of inertia I for the columns and the total stiffness of the four columns k are:

$$I=\frac{\pi D^4}{64}, \tag{1}$$

$$k=\frac{3nEI}{h^3} \tag{2}$$

Note that as the ground acceleration is assumed to be a Gaussian white noise of constant spectral density S'_0 [3,4] the spectral density of the earthquake force that acts on the structure can be found to be $(0,5...1,1)m^2S'_0$.

The equation of motion for this single-degree of freedom structure under earthquake excitation can be written as

$$m\ddot{x}(t)+c\dot{x}(t)+g(x(t))=-m\ddot{x}_g(t), \tag{3}$$

where m is the mass, c is the viscous damping coefficient, $m\ddot{x}_g(t)$ is the external excitation signal with zero mean and $x(t)$ is the displacement response of the system.

Dividing the equation by m the equation of motion for this single-degree of freedom structure under earthquake excitation can be written as

$$\ddot{x}(t)+2\xi pc[\dot{x}(t)+\varepsilon r_1\dot{x}^3(t)+\varepsilon r_3\dot{x}^5(t)]+p^2x(t)+p^2\alpha x^3(t)=w(t) \tag{4}$$

where r_1, r_3 is the nonlinear factor to control the type and degree of nonlinearity in the system, ξ is the critical damping factor, p is the undamped natural frequency, for the linear system, α is the nonlinear factor to control the type and degree of nonlinearity in the system and $x(t)$ is the displacement response of the system. A positive value of α represents a hardening system while a negative value represents a softening system behavior.

In the general form, the previous equation is written

$$\ddot{x}(t)+h(x(t),\dot{x}(t))=w(t). \tag{5}$$

Determination of solutions is done considering the following linear equation

$$\ddot{x}(t)+\beta_{ech}\dot{x}(t)+\gamma_{ech}x(t)=w(t). \tag{6}$$

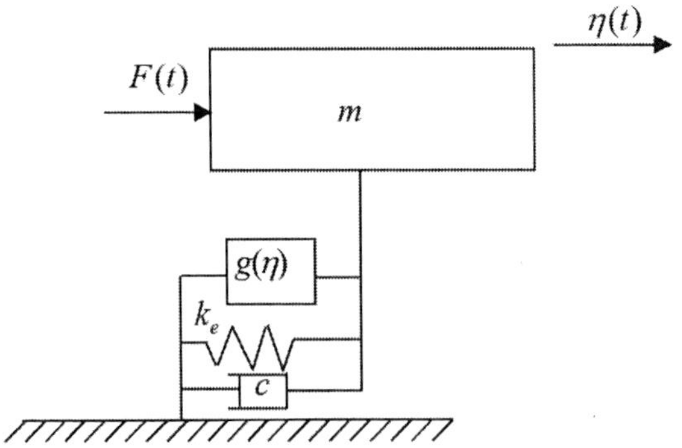

Figure 1. Spectral analysis of a SDOF system under random load.

The difference between the nonlinear stiffness and linear stiffness terms is

$$\varepsilon=h(x(t),\dot{x}(t))-\beta_{ech}\dot{x}(t)-\gamma_{ech}x(t). \tag{7}$$

It is required that the error is minimal, that is

$$\frac{\partial}{\partial \beta_{ech}} E[\varepsilon^2]=0\,, \tag{8}$$

respectively

$$\frac{\partial}{\partial \gamma_{ech}} E[\varepsilon^2]=0\,. \tag{9}$$

Because

$$E\{\varepsilon^2\}=E\{h^2\}+\beta_{ech}{}^2E\{\dot{x}^2\}+\gamma_{ech}{}^2E\{x^2\}-2\beta_{ech}E\{\dot{x}h\}+2\beta_{ech}\gamma_{ech}E\{x\dot{x}\}-2\gamma_{ech}E\{xh\}\,, \tag{10}$$

,we get a system of two equations with two unknowns

$$E\{\dot{x}h\}-\beta_{ech}E\{\dot{x}^2\}-\gamma_{ech}E\{x\dot{x}\}=0 \tag{11}$$

$$E\{xh\}-\beta_{ech}E\{x\dot{x}\}-\gamma_{ech}E\{x^2\}=0\,, \tag{12}$$

with solutions

$$\beta_{ech}=\frac{E\{x^2\}E\{\dot{x}h\}-E\{x\dot{x}\}E\{xh\}}{E\{x^2\}E\{\dot{x}^2\}-(E\{x\dot{x}\})^2} \tag{13}$$

$$\gamma_{ech}=\frac{E\{\dot{\eta}^2\}E\{\eta h\}-E\{\eta\dot{\eta}\}E\{\dot{\eta}h\}}{E\{\eta^2\}E\{\dot{\eta}^2\}-(E\{\eta\dot{\eta}\})^2}\,. \tag{14}$$

Since the average is zero, the solutions are written simplified

$$\beta_{ech}=\frac{E\{\dot{x}h\}}{E\{\dot{x}^2\}} \tag{15}$$

$$\gamma_{ech}=\frac{E\{xh\}}{E\{x^2\}}\,. \tag{16}$$

In case of equation (3), by linearization we get

$$\ddot{x}(t)+2\xi p\left\{\left[1+\varepsilon\left(r_1\frac{E\{\dot{x}^4\}}{E\{\dot{x}^2\}}+r_3\frac{E\{\dot{x}^6\}}{E\{\dot{x}^2\}}\right)\right]+p^2\frac{E\{\dot{x}x\}}{E\{\dot{x}^2\}}+\alpha p^2\frac{E\{\dot{x}x^3\}}{E\{\dot{x}^2\}}\right\}\dot{\eta}(t)+\frac{E\{xh\}}{E\{x^2\}}x(t)=w(t) \tag{17}$$

The damping feature is

$$\beta_{ech}=2\xi_e p_e=\frac{E\{\dot{x}h\}}{E\{\dot{x}^2\}}=2\xi p\left[1+\varepsilon\left(r_1\frac{E\{\dot{x}^4\}}{E\{\dot{x}^2\}}+r_3\frac{E\{\dot{x}^6\}}{E\{\dot{x}^2\}}\right)\right]+p^2\frac{E\{\dot{x}x\}}{E\{\dot{x}^2\}}+\alpha p^2\frac{E\{\dot{x}x^3\}}{E\{\dot{x}^2\}} \tag{18}$$

which can still be written

$$\beta_{ech}=2\xi_e p_e=2\xi p\left[1+\varepsilon\left(\frac{63}{4}r_1\sigma^2_{\dot{x}}+\frac{2835}{16}r_3\sigma^4_{\dot{x}}\right)\right], \tag{19}$$

Because

$$E\{\dot{x}x\}=0,\quad E\{\dot{x}x^3\}=0,\quad \frac{E\{\dot{x}^6\}}{E\{\dot{x}^4\}}=\frac{63}{4}\sigma^2_{\dot{x}},\quad \frac{E\{\dot{x}^4\}}{E\{\dot{x}^2\}}=\frac{45}{4}\sigma^2_{\dot{x}}\,. \tag{20}$$

The elastic characteristic is given by

IOP Conf. Series: Materials Science and Engineering **659** (2019) 012013 doi:10.1088/1757-899X/659/1/012013

$$\gamma_{ech}=p_e^2=\frac{E\{xh\}}{E\{x^2\}}=2\xi p\left[\frac{E\{x\dot{x}\}}{E\{x^2\}}+\varepsilon\left(r_1\frac{E\{x\dot{x}^3\}}{E\{x^2\}}+r_3\frac{E\{x\dot{x}^5\}}{E\{x^2\}}\right)\right]+p^2\frac{E\{x^2\}}{E\{x^2\}}+\alpha p^2\frac{E\{x^4\}}{E\{x^2\}} \tag{21}$$

Because

$$E\{x\dot{x}\}=0,\quad \frac{E\{x^4\}}{E\{x^2\}}=3\sigma_x^2,\quad \frac{E\{x\dot{x}^3\}}{E\{x^2\}}\cong 0,\quad \frac{E\{x\dot{x}^5\}}{E\{x^2\}}\cong 0, \tag{22}$$

We can write

$$\gamma_{ech}=p_e^2=p^2(1+3\alpha\sigma_x^2). \tag{23}$$

Therefore, the equivalent linear equation is written in the form

$$\ddot{x}(t)+2\xi p\left[1+\varepsilon\left(\frac{63}{4}r_1\sigma_x^2+\frac{2835}{16}r_3\sigma_x^4\right)\right]\dot{x}(t)+p^2(1+3\alpha\sigma_x^2)x(t)=w(t) \tag{24}$$

If the Fourier transform is applied, the transfer function associated with the oscillator system is

$$H(\omega)=\frac{1}{m\left(-\omega^2+i\omega\frac{E\{\dot{x}h\}}{E\{\dot{x}^2\}}+\frac{E\{xh\}}{E\{x^2\}}\right)}, \tag{25}$$

where from

$$|H(\omega)|=\frac{1}{m\sqrt{\left(\frac{E\{xh\}}{E\{x^2\}}-\omega^2\right)^2+\omega^2\left(\frac{E\{\dot{x}h\}}{E\{\dot{x}^2\}}\right)^2}} \tag{26}$$

The frequency response function, in this cas, of the single degree of freedom system is

$$|H(\omega)|=\frac{1}{m\sqrt{\left(p^2+3\alpha p^2\sigma_x^2-\omega^2\right)^2+4\xi^2p^2\omega^2\left[1+\varepsilon\left(\frac{63}{4}r_3\sigma_x^2+\frac{2835}{16}r_5\sigma_x^4\right)\right]^2}} \tag{27}$$

The displacement variance [5] of the single-degree of freedom system under Gaussian white noise excitation can be expressed as,

$$\sigma_x^2=R_x(0)=\int_{-\infty}^{\infty}\left|H(\omega)\right|^2 S_F d\omega=\int_{-\infty}^{\infty}\left|H(\omega)\right|^2 mS_{\ddot{x}_g}\,d\omega \tag{28}$$

Using the transfer function, the response as a function of frequency has expression

$$\bar{X}(\omega)=H(\omega)\bar{F}(\omega), \tag{29}$$

where

$$\bar{x}(\omega)=\mathrm{F}\,(x(t)),\qquad \bar{F}(\omega)=\mathrm{F}\,(F(t)). \tag{30}$$

The power spectral densyti of the response is

$$S_x(\omega)=\frac{S_F(\omega)}{m^2\left[\left(\frac{E\{xh\}}{E\{x^2\}}-\omega^2\right)^2+\omega^2\left(\frac{E\{\dot{x}h\}}{E\{\dot{x}^2\}}\right)^2\right]}. \tag{31}$$

or

$$S_x(\omega)=\frac{S_F(\omega)}{m^2\left\{\left(p^2+3\alpha p^2\sigma^2_x-\omega^2\right)^2+4\xi^2p^2\omega^2\left[1+\varepsilon\left(\frac{63}{4}r_1\sigma^2_x+\frac{2835}{16}r_3\sigma^4_x\right)\right]^2\right\}} \tag{32}$$

The most used spectrum to describe the earthquake ground acceleration is earthquake spectrum which is expresses as

$$S_{\ddot{x}_g}=mS'_0,\ S'_0=(0,5...1,1)\frac{m^2}{s^3}. \tag{33}$$

We obtain:

$$\sigma^2_x=\int_{-\infty}^{\infty}\left|H(\omega)\right|^2 m^2S'_0d\omega=$$

$$S'_0m^2\int_{-\infty}^{\infty}\frac{1}{m^2}\frac{1}{\left(p^2+3\alpha p^2\sigma^2_x-\omega^2\right)^2+4\xi^2p^2\omega^2\left[1+\varepsilon\left(\frac{63}{4}r_1\sigma^2_x+\frac{2835}{16}r_3\sigma^4_x\right)\right]^2}d\omega= \tag{34}$$

$$=S'_0\int_{-\infty}^{\infty}\frac{1}{\left(p^2+3\alpha p^2\sigma^2_x-\omega^2\right)^2+4\xi^2p^2\omega^2\left[1+\varepsilon\left(\frac{63}{4}r_1\sigma^2_x+\frac{2835}{16}r_3\sigma^4_x\right)\right]^2}d\omega.$$

The displacement variance is given by:

$$\sigma^2_x=\frac{S'_0\pi m^2}{ck}-\alpha E\{xG(x)\}. \tag{35}$$

Similarly, the velocity variance of the system [5,6] can be expressed as

$$R_{\dot{x}}(\tau)=-\frac{d^2R_x(\tau)}{d\tau^2}=\int_{-\infty}^{\infty}\omega^2\bar{S}_x(\omega)e^{i\omega\tau}d\omega. \tag{36}$$

Obtain for the velocity variance

$$\sigma^2_{\dot{x}}=E\{\dot{x}^2\}=R_{\dot{x}}(0)=\int_{-\infty}^{\infty}\omega^2S_x(\omega)d\omega=S'_0m^2\int_{-\infty}^{\infty}\frac{\omega^2}{m[(p_e^2-\omega^2)^2+4\xi^2p^2\omega^2]}d\omega=\frac{S'_0\pi}{2\xi p}=\frac{\pi S'_0m}{c}. \tag{37}$$

The relative acceleration of this structure is unbounded. On the other hand, the absolute acceleration $\ddot{x}_{abs}$ is bounded and can be obtained as follows:

$$\ddot{x}_{abs}=\ddot{x}(t)+\ddot{x}_g(t)=2\xi_ep_e\dot{x}(t)+p_e^2x(t). \tag{38}$$

3. Numerical results:

In this example, $S'_0=0,52\frac{m^2}{s^3}$, $m=2,2\cdot10^5kg$, n=4 columns, $d=0,5m$, $h=2m$, $E_b=0,2\cdot10^{11}Pa$, $\xi=0,25$, $\alpha=7m^{-2}$, $k=29\cdot10^6\frac{N}{m}$; $p=12,04s^{-1}$; $r_1=3\cdot10^2s^2/m^2$, $r_3=12,142\cdot10^2s^4/m^4$, $c=1204\cdot10^3\frac{N\cdot s}{m}$, $\varepsilon=0,01$.

The power spectral density for excitation $S_F(\omega)$ is $S_F(\omega)=m^2S'_0=2,08\cdot10^{10}N^2\cdot s$.

Obtain in this case for the displacement variance:

$$\sigma^2_x=105\cdot10^{-5}m^2. \tag{39}$$

In this case, the coefficient p_e can be expressed

$$p_e^2=p^2(1+15\alpha\sigma_x^4)=15,21s^{-2}. \tag{40}$$

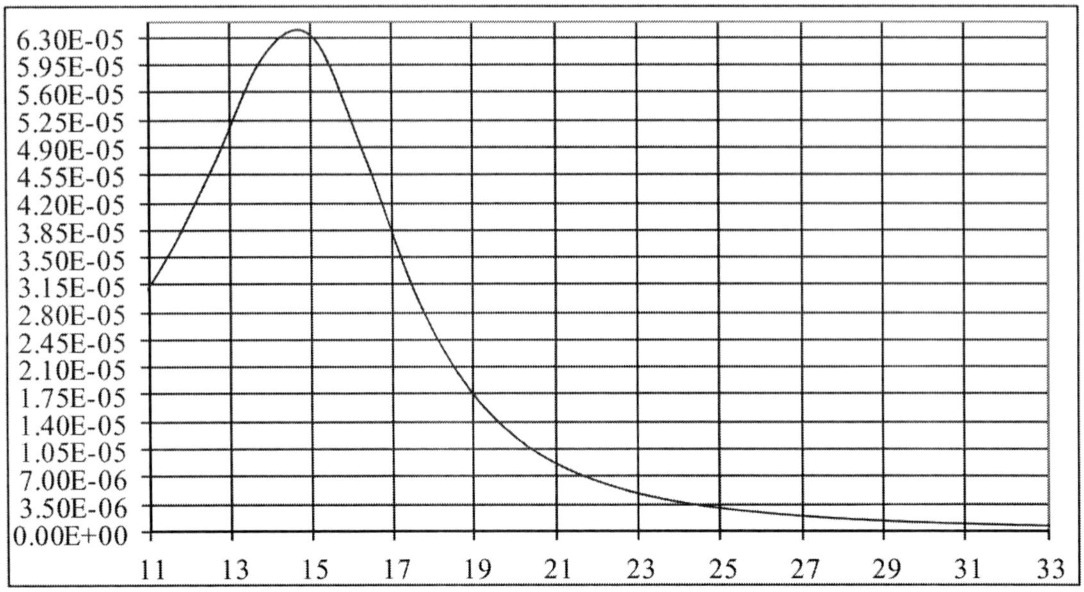

$\omega / 2\pi$

Figure 2. The power spectral density of response for $\alpha = 3m^{-4}$, in the frequency range (13,33) rad/sec.

4. Remarks

This suggests that the variance of absolute acceleration decreases as the damping increases when the damping ratio is smaller than $0,5$. The variance increases as the damping increases when the damping ratio is bigger than $0,5$.It can be seen that stiffening structure (increase stiffness) can reduce displacement but would result in the increase of absolute acceleration. A positive value of α represents a hardening system while a negative value represents a softening system behavior. On the other hand, increasing mass can reduce absolute acceleration but increase displacement. It seems that the only damping increase (when $\xi < 0,5$) can result in the simultaneous reduction of displacement and absolute acceleration. These conclusions are very useful when designing structure under seismic condition.

References

[1] Zhao L and Chen Q 1997 *An equivalent non-linearization method for analyzing response of nonlinear systems to random excitations*, Applied Mathematics and Mechanics

[2] Stan P 2009 *Analysis of single-degree of freedom non-linear structure under Gaussian white noise ground excitation,* The Sciencific Book, University of Pitesti

[3] Stan P and Stan M 2017 The study spectral analysis to random vibrations for nonlinear oscillators, *International Congress of Automotive and Transport Engineering CAR 2017, Pitesti, Automotive Series* **27**

[4] Stan P and Stan M 2015 *Nonlinear vibrations in automotive suspensions*, International Conference Computational Mechanics and Virtual Engineering, Brasov Romania

[5] Stan P and Stan M 2015 *Random vibration of Houdaille shock absorber*, International Conference Computational Mechanics and Virtual Engineering, Brasov Romania

[6] Stan M and Stan P 2017 Numerical solution of the proces of hydroplaning at wheelswith envelope, *International Congress of Automotive and Transport Engineering, Mobility Engineering and Environment, CAR 2017, 017 IOP Conf. Ser. Mater. Sci. Eng.* **252**

The response of a random vibrations of nonlinear structure under white-noise excitations

P Stan[1] and M Stan[1]

[1]University of Pitești, Faculty of Mechanical Engineering and Technology, Romania

E-mail: petre_stan_marian@yahoo.com

Abstract: Random vibration is the phenomenon wherein random excitation applied to a mechanical induces random response. Random vibration is the most practical of the areas discussed in this paper. In this paper, the sistem is excited by a force W which is a random process described. A method is to replace the governing set of non-linear defferential equations by an equivalent set of linear equations the defference betwen the sets being minimized in somme appropiate sense. In this way, the parameters of the linearised system are determined.

1. Introduction

We present a method for estimating the power spectral density of the stationary response of oscillator with a nonlinear restoring force under external stochastic wide-band excitation. An equivalent linear system is derived, from which the power spectral density is deduced. Recently, considerable attention has been paid towards approximate solutions for analytically solving nonlinear differential equation. A clasical representation in seismic response analizes its that of a single or multi-degree-of-nonlinear oscillator subjected to earthuake ground motion at its support. An approximate method is presented for estimating the power spectral density of the response of multi-degree-of nonlinear random structure under white noise excitations. The non-linearities are assumed to be single-valued functions of accelerations, velocities and displacements. Using a property of gaussian vector processes, the closed forms of the coefficients of the equivalent linear system are obtained by the direct application of partial differentiation and expectation operators to the non-linear terms. The efficiency of the method is checked by comparing results with those numerical simulations.

2. System model

Recently, considerable attention has been paid towards approximate solutions for analytically solving nonlinear differential equation. To illustrate the procedure of equivalent linearization theory, let us consider a three-story building. A three-story building is modeled by four identical columns of Young's modulus E and height h and two rigid floors of weight m. The damping can be approximated by an equivalent damping constant c_1 and c_2. The ground acceleration due to an earthquake is assumed to be a Gaussian white noise with a constant spectrum S_0.

The method of the stochastic equivalent linearization [1,2] is based on the idea that a nonlinear system may be replaced by a linear system by minimizing the mean square error of the two systems. The most commonly applied and convenient procedure uses Yingfang, L. Zhao, Q. Chen suggestion to estimate the linearization coefficients in context with Wen's introduction of an analytical expression for the restoring force. This method has seen the broadest application because of their ability to accurately capture the response statistics over a wide range of response levels while maintaining relatively light computational burden. The method will be briefly discussed in the following sections.

The columns have cilinder sections of diameter D. Note that as the ground acceleration is assumed to be a Gaussian white noise of constant spectral density S'_0, the spectral density of the earthquake force [3] that acts on the structure can be found to be $(0,5...1,1)m^2S'_0$. This can be readily seen from the definition of spectral density function which is the Fourier transform of the autocorrelation function.

The equations of motion, while neglecting very small terms we get

$$\begin{pmatrix} m & 0 & 0 \\ 0 & m & 0 \\ 0 & 0 & m \end{pmatrix}\begin{pmatrix} \ddot{x}_1 \\ \ddot{x}_2 \\ \ddot{x}_3 \end{pmatrix}+\begin{pmatrix} c_1+c_2 & -c_2 & 0 \\ 0 & c_2+c_3 & -c_3 \\ 0 & -c_3 & c_3 \end{pmatrix}\begin{pmatrix} \dot{x}_1 \\ \dot{x}_2 \\ \dot{x}_3 \end{pmatrix}+\begin{pmatrix} k_1+k_2 & -k_2 & 0 \\ -k_2 & k_2+k_3 & -k_3 \\ -k_3 & 0 & k_3 \end{pmatrix}\begin{pmatrix} x_1 \\ x_2 \\ x_3 \end{pmatrix}+$$
$$+\alpha\begin{pmatrix} k_1+k_2+k_3 & -k_2-k_3 & 0 \\ -k_2 & k_2+k_3 & 0 \\ -k_3 & -k_3 & k_3 \end{pmatrix}\begin{pmatrix} x_1^3 \\ x_2^3 \\ x_3^3 \end{pmatrix}=\begin{pmatrix} W(t) \\ 0 \\ 0 \end{pmatrix}. \tag{1}$$

where the component nonlinear [4] is of the form $[A]=\begin{pmatrix} x_1^3 \\ x_2^3 \\ x_3^3 \end{pmatrix}$, $W(t)$ is the external excitation signal with zero mean.

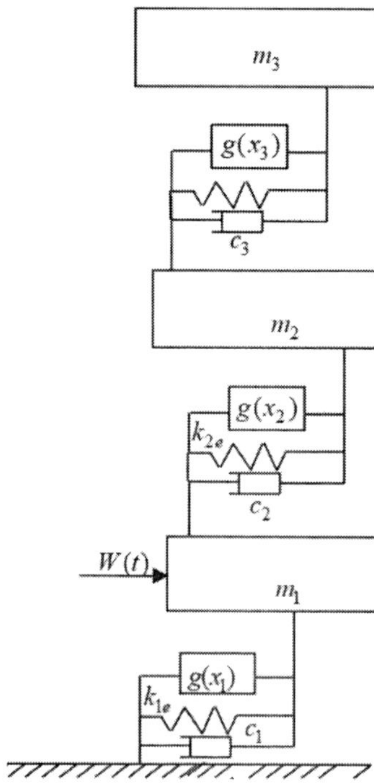

Figure 1. A three-story building.

The equations of motion, while neglecting very small terms we get

$$\begin{pmatrix} m & 0 & 0 \\ 0 & m & 0 \\ 0 & 0 & m \end{pmatrix}\begin{pmatrix} \ddot{x}_1 \\ \ddot{x}_2 \\ \ddot{x}_3 \end{pmatrix}+\begin{pmatrix} c_1+c_2 & -c_2 & 0 \\ 0 & c_2+c_3 & -c_3 \\ 0 & -c_3 & c_3 \end{pmatrix}\begin{pmatrix} \dot{x}_1 \\ \dot{x}_2 \\ \dot{x}_3 \end{pmatrix}+\begin{pmatrix} k_1+k_2 & -k_2 & 0 \\ -k_2 & k_2+k_3 & -k_3 \\ -k_3 & 0 & k_3 \end{pmatrix}\begin{pmatrix} x_1 \\ x_2 \\ x_3 \end{pmatrix}+$$

$$+\alpha\begin{pmatrix} k_1+k_2+k_3 & -k_2-k_3 & 0 \\ -k_2 & k_2+k_3 & 0 \\ -k_3 & -k_3 & k_3 \end{pmatrix}\begin{pmatrix} x_1^{\,3} \\ x_2^{\,3} \\ x_3^{\,3} \end{pmatrix}=\begin{pmatrix} W(t) \\ 0 \\ 0 \end{pmatrix}. \tag{2}$$

where the component nonlinear is of the form

$$[A]=\begin{pmatrix} x_1^{\,3} \\ x_2^{\,3} \\ x_3^{\,3} \end{pmatrix} \tag{3}$$

The linear equation can be write

$$\begin{pmatrix} m & 0 & 0 \\ 0 & m & 0 \\ 0 & 0 & m \end{pmatrix}\begin{pmatrix} \ddot{x}_1 \\ \ddot{x}_2 \\ \ddot{x}_3 \end{pmatrix}+\begin{pmatrix} c_{1e}+c_{2e} & -c_{2e} & 0 \\ 0 & c_{2e}+c_{3e} & -c_{3e} \\ 0 & -c_{3e} & c_{3e} \end{pmatrix}\begin{pmatrix} \dot{x}_1 \\ \dot{x}_2 \\ \dot{x}_3 \end{pmatrix}+\begin{pmatrix} k_{1e}+k_{2e} & -k_{2e} & 0 \\ -k_{2e} & k_{2e}+k_{3e} & -k_{3e} \\ -k_{3e} & 0 & k_{3e} \end{pmatrix}\begin{pmatrix} x_1 \\ x_2 \\ x_3 \end{pmatrix}=\begin{pmatrix} W(t) \\ 0 \\ 0 \end{pmatrix}. \tag{4}$$

The difference between the nonlinear stiffness and linear stiffness terms [4,5] is

$$e=\begin{pmatrix} e_1 \\ e_2 \\ e_3 \end{pmatrix}=\begin{pmatrix} k_1x_1+k_2(x_1-x_2)+\lambda(k_1+k_2+k_3)x^3{}_1-\alpha(k_2+k_3)x^3{}_2-(k_{1e}+k_{2e})x_1+(x_2+\alpha x^3{}_2)-2(k_{1e}+k_{2e})x_1+(x_2+\alpha x^3{}_2)+k_{2e}x_2 \\ k_2(x_2-x_1)+k_3(x_2-x_3)+\alpha(k_2+k_3)x^3{}_2-\alpha k_2x^3{}_1+k_{2e}x_1-(k_{2e}+k_{3e})x_2+k_{3e}x_3 \\ k_3(x_3-x_1)+\alpha k_3x^3{}_3-\alpha k_3x^3{}_1+k_{3e}x_1-k_{3e}x_3 \end{pmatrix} \tag{5}$$

The value of k_{1e} and k_{2e} can be obtained by minimizing the expectation of the square error

$$\frac{dE\{e_1^{\,2}\}}{dk_{1e}}=0\,, \tag{6}$$

and

$$\frac{dE\{e_2^{\,2}\}}{dk_{2e}}=0\,. \tag{7}$$

Because

$$\begin{aligned} E\{e_1^{\,2}\}=&k_1^{\,2}E\{x_1^{\,2}\}+k_2^{\,2}E\{x_1^{\,2}\}+k_2^{\,2}E\{x_2^{\,2}\}-2k_2^{\,2}E\{x_1x_2\}+\alpha^2(k_1+k_2+k_3)^2E\{x_1^{\,6}\} \\ &+\alpha^2(k_2+k_3)^2E\{x_2^{\,6}\}+(k_{1e}+k_{2e})^2E\{x_1^{\,2}\}+k_{2e}^{\,2}E\{x_2^{\,2}\}+2k_1k_2E\{x_1^{\,2}\}-2k_1k_2E\{x_1x_2\}+ \\ &+2\alpha k_1(k_1+k_2+k_3)E\{x_1^{\,4}\}-2\alpha k_1(k_2+k_3)E\{x_1x_2^{\,4}\}-2k_1(k_{1e}+k_{2e})E\{x_1^{\,2}\}+2k_1k_{2e} \\ &\cdot E\{x_1x_2\}+2\alpha k_2(k_1+k_2+k_3)E\{x_1^{\,4}\}-2\alpha k_2(k_1+k_2+k_3)E\{x_1^{\,4}x_2\}-2\alpha k_2(k_2+k_3) \\ &\cdot E\{x_2^{\,3}x_1\}+2\alpha k_2(k_2+k_3)E\{x_2^{\,4}\}-2k_2(k_{1e}+k_{2e})E\{x_1^{\,2}\}+2k_2(k_{1e}+k_{2e})E\{x_1x_2\}+ \\ &+2k_2k_{2e}E\{x_1x_2\}-2k_2k_{2e}E\{x_2^{\,2}\}-2\alpha^2(k_1+k_2+k_3)(k_2+k_3)E\{x_1^{\,3}x_2^{\,3}\}-2\alpha(k_1+k_2+ \\ &+k_3)(k_{1e}+k_{2e})E\{x_1^{\,4}\}+2\alpha(k_1+k_2+k_3)k_{2e}E\{x_1^{\,3}x_2\}+2\alpha(k_2+k_3)(k_{1e}+k_{2e})E\{x_2^{\,3}x_1\} \\ &-2\alpha(k_2+k_3)k_{2e}E\{x_2^{\,4}\}-2(k_{1e}+k_{2e})k_{2e}E\{x_1x_2\}, \end{aligned} \tag{8}$$

$$\begin{aligned}
&E\{e_2^{\,2}\}=k_2^{\,2}E\{x_1^{\,2}\}+k_2^{\,2}E\{x_2^{\,2}\}+k_2^{\,2}E\{x_1x_2\}+k_3^{\,2}E\{x_2^{\,2}\}+k_3^{\,2}E\{x_3^{\,2}\}-2k_3^{\,2}E\{x_3x_2\}\\
&+\alpha^2(k_2+k_3)^2E\{x_2^{\,6}\}+\alpha^2k_2^{\,2}E\{x_1^{\,6}\}+k_{2e}^{\,2}E\{x_1^{\,6}\}+(k_{2e}+k_{3e})^2E\{x_2^{\,2}\}+k_{3e}^{\,2}E\{x_3^{\,2}\}+\\
&+2k_2k_3E\{x_2^{\,2}\}-2k_2k_3E\{x_3x_2\}-2k_2k_3E\{x_1x_2\}+2k_2k_3E\{x_1x_3\}+2k_2\alpha(k_2+k_3)E\{x_2^{\,4}\}\\
&-2k_2\alpha(k_2+k_3)E\{x_1x_2^{\,3}\}-2k_2^{\,2}\alpha E\{x_1^{\,3}x_2\}+2k_2^{\,2}\alpha E\{x_1^{\,4}\}+2k_2k_{2e}E\{x_1x_2\}-\\
&-2k_2k_{2e}E\{x_1^{\,2}\}-2k_2(k_{2e}+k_{3e})E\{x_2^{\,2}\}+2k_2(k_{2e}+k_{3e})E\{x_1x_2\}+2k_2k_{3e}E\{x_3x_2\}-\\
&-2k_2k_{3e}E\{x_3x_1\}+2k_3\alpha(k_2+k_3)E\{x_2^{\,4}\}-2k_3\alpha(k_2+k_3)E\{x_3x_2^{\,2}\}-2k_2k_3\alpha E\{x_1^{\,3}x_2\}+\\
&+2k_2k_3\alpha E\{x_1^{\,3}x_3\}+2k_3k_{2e}E\{x_2x_1\}-2k_3k_{2e}E\{x_3x_1\}-2k_3(k_{2e}+k_{3e})E\{x_2^{\,2}\}+\\
&+2k_3(k_{2e}+k_{3e})E\{x_3x_2\}+2k_3k_{3e}E\{x_3x_2\}-2k_3k_{3e}E\{x_3^{\,2}\}-2\alpha^2(k_2+k_3)k_2E\{x_1^{\,3}x_2^{\,3}\}+\\
&+2\alpha(k_2+k_3)k_{2e}E\{x_2^{\,3}x_1\}-2\alpha(k_2+k_3)(k_{2e}+k_{3e})E\{x_2^{\,4}\}+2\alpha(k_2+k_3)E\{x_2^{\,3}x_3\}-\\
&-2\alpha k_2k_{2e}E\{x_1^{\,4}\}+2\alpha k_2(k_{2e}+k_{3e})E\{x_1^{\,3}x_2\}-2\alpha k_2k_{3e}E\{x_1^{\,3}x_3\}-\\
&-2k_2(k_{2e}+k_{3e})E\{x_1x_2\}+2k_{2e}E\{x_1x_3\}-2(k_{2e}+k_{3e})k_{3e}E\{x_3x_2\},
\end{aligned} \tag{9}$$

$$\begin{aligned}
&E\{e_3^{\,2}\}=k_3^{\,2}E\{x_3^{\,2}\}+k_3^{\,2}E\{x_1^{\,2}\}+\alpha^2k_3^{\,2}E\{x_3^{\,6}\}+\alpha^2k_3^{\,2}E\{x_1^{\,6}\}+\alpha^2k_3^{\,2}E\{x_2^{\,6}\}+\\
&+k_{3e}^{\,2}E\{x_1^{\,6}\}+k_{3e}^{\,2}E\{x_3^{\,2}\}-2k_3^{\,2}E\{x_1x_3\}+2\alpha k_3^{\,2}E\{x_3^{\,4}\}-2\alpha k_3^{\,2}E\{x_1^{\,3}x_3\}-\\
&-2\alpha k_3^{\,2}E\{x_3x_2^{\,3}\}+2k_3k_{3e}E\{x_1x_3\}-2k_3k_{3e}E\{x_3^{\,2}\}-2\alpha k_3^{\,2}E\{x_1x_3^{\,3}\}+\\
&+2\alpha k_3^{\,2}E\{x_1^{\,4}\}+2\alpha k_3^{\,2}E\{x_1x_2^{\,3}\}-2k_3k_{3e}E\{x_1^{\,2}\}+2k_3k_{3e}E\{x_1x_3\}-\\
&-2\alpha^2k_3^{\,2}E\{x_1^{\,3}x_3^{\,3}\}-2\alpha^2k_3^{\,2}E\{x_2^{\,3}x_3^{\,3}\}+2\alpha k_3k_{3e}E\{x_1x_3^{\,3}\}-\\
&-2\alpha k_3k_{3e}E\{x_3^{\,4}\}+2\alpha^2k_3^{\,2}E\{x_1^{\,3}x_2^{\,3}\}-2\alpha k_3k_{3e}E\{x_1^{\,4}\}+2\alpha k_3k_{3e}E\{x_1^{\,3}x_3\}-\\
&-2\alpha k_3k_{3e}E\{x_1x_2^{\,3}\}+2\alpha k_3k_{3e}E\{x_3x_2^{\,3}\}-2k_{3e}^{\,2}E\{x_1x_3\},
\end{aligned} \tag{10}$$

while neglecting very small terms we get

$$k_{1e}=\frac{3\alpha\left(k_1+k_2+k_3\right)\sigma_{x_1}^4+(k_2-k_1)\sigma_{x_1}^2}{\sigma_{x_1}^2}, \tag{11}$$

$$k_{2e}=\frac{3\alpha k_2\sigma_{x_1}^4+3\alpha(k_2+k_3)\sigma_{x_2}^4+(k_2+k_3)\sigma_{x_2}^2+k_2\sigma_{x_1}^2-k_3\sigma_{x_2}^2}{15\sigma_{x_1}^6+\sigma_{x_2}^2}, \tag{12}$$

$$k_{3e}=\frac{3\alpha k_3(\sigma_{x_1}^4+\sigma_{x_3}^4)+k_3(\sigma_{x_1}^2+\sigma_{x_3}^4)}{15\sigma_{x_1}^6+\sigma_{x_3}^2}. \tag{13}$$

Using the Fourier transform [2] and having the relations

$$\mathrm{F}\ (\dot{x}_1(t))=i\omega\bar{x}_1(\omega),\ \mathrm{F}\ (\dot{x}_2(t))=i\omega\bar{x}_2(\omega),\ \mathrm{F}\ (\dot{x}_3(t))=i\omega\bar{x}_3(\omega), \tag{14}$$

$$\mathrm{F}\ (\ddot{x}_1(t))=i\omega\,\mathrm{F}\ (\dot{x}_1(t))=-\omega^2\,\bar{x}_1(\omega), \tag{15}$$

$$\mathrm{F}\ (\ddot{x}_2(t))=i\omega\,\mathrm{F}\ (\dot{x}_2(t))=-\omega^2\,\bar{x}_2(\omega), \tag{16}$$

$$\mathrm{F}\ (\ddot{x}_3(t))=i\omega\,\mathrm{F}\ (\dot{x}_3(t))=-\omega^2\,\bar{x}_3(\omega), \tag{17}$$

$$\mathrm{F}\ ((W(t))=\bar{W}(\omega), \tag{18}$$

obtain for the response

$$\bar{x}_1(\omega)=\frac{A}{D}\bar{W}(\omega),$$
$$\bar{x}_2(\omega)=\frac{B}{D}\bar{W}(\omega), \qquad (19)$$
$$\bar{x}_3(\omega)=\frac{C}{D}\bar{W}(\omega),$$

with

$$D=[-m\omega^2+i\omega(c_{1e}+c_{2e})+k_{1e}+k_{2e}]\{[-m\omega^2+i\omega(c_{3e}+c_{2e})+k_{2e}+k_{3e}](-m\omega^2+ \\ +i\omega c_{3e}+k_{3e}-i\omega^2 c_{3e}^2+i\omega k_{3e}c_{3e})\}-k_{2e}(-m\omega^2+i\omega c_{3e}+k_{3e})+k_{3e}(i\omega c_{3e}-k_{3e}) \qquad (20)$$

$$A=\begin{vmatrix} \bar{W}(\omega) & i\omega c_{2e}-k_{2e} & 0 \\ 0 & -m\omega^2+i\omega(c_{3e}+c_{2e})+k_{2e}+k_{3e} & i\omega c_{3e}-k_{3e} \\ 0 & -i\omega c_{3e} & -m\omega^2+i\omega c_{3e}+k_{3e} \end{vmatrix}, \qquad (21)$$

$$B=\begin{vmatrix} -m\omega^2+i\omega(c_{1e}+c_{2e})+k_{1e}+k_{2e} & \bar{W}(\omega) & 0 \\ -k_{2e} & 0 & i\omega c_{3e}-k_{3e} \\ -k_{3e} & 0 & -m\omega^2+i\omega c_{3e}+k_{3e} \end{vmatrix}, \qquad (22)$$

$$C=\begin{vmatrix} -m\omega^2+i\omega(c_{1e}+c_{2e})+k_{1e}+k_{2e} & i\omega c_{2e}-k_{2e} & \bar{W}(\omega) \\ -k_{2e} & -m\omega^2+i\omega(c_{3e}+c_{2e})+k_{2e}+k_{3e} & 0 \\ -k_{3e} & -i\omega c_{3e} & 0 \end{vmatrix}. \qquad (23)$$

The frequency response function of the system is given by equation

$$\bar{H}_1(\omega)=\bar{W}(\omega)\frac{(-m\omega^2+i\omega c_{3e}+i\omega c_{2e}+k_{2e}+k_{3e})(-m\omega^2+i\omega c_{3e}+k_{3e})-i\omega^2 c_{3e}^2+i\omega k_{3e}c_{3e})}{D} \qquad (24)$$

$$\bar{H}_2(\omega)=\bar{W}(\omega)\frac{k_{2e}(-m\omega^2+i\omega c_{3e}+k_{3e})-k_{3e}(i\omega c_{3e}-k_{3e})}{D}, \qquad (25)$$

$$\bar{H}_3(\omega)=\bar{W}(\omega)\frac{ik_{2e}\omega c_{3e}+k_{3e}[-m\omega^2+i\omega(c_{3e}+c_{2e})+k_{2e}+k_{3e}]}{D}. \qquad (26)$$

The mean square value for the displacement of the system is given by equation

$$\sigma^2_{x_1}=R_{x_1}(0)=\int_{-\infty}^{\infty}\left|\bar{H}_1(\omega)\right|^2 m^2 S'_0 d\omega=\frac{1}{m^2}\int_{-\infty}^{\infty}\left|\bar{H}_1(\omega)\right|^2 m^2 S'_0 d\omega \qquad (27)$$

for the second structure and for the third structure

$$\sigma^2_{x_2}=R_{x_2}(0)=\int_{-\infty}^{\infty}\left|\bar{H}_2(\omega)\right|^2 S'_0 d\omega, \qquad (28)$$

$$\sigma^2_{x_3}=R_{x_3}(0)=\int_{-\infty}^{\infty}\left|\bar{H}_3(\omega)\right|^2 S'_0 d\omega. \qquad (29)$$

The power spectral density of response for the first structure (in $m^2\cdot s$) is given by equation

$$S_1(\omega)=\left|H_1(\omega)\right|^2 S_W=\left|H_1(\omega)\right|^2 m^2 S'_0(\omega)=\left|\frac{1}{m}\bar{H}_1(\omega)\right|^2 m^2 S'_0(\omega)=\left|H_1(\omega)\right|^2 S'_0(\omega) \qquad (30)$$

and the power spectral density for the second and the third structure (in $m^2\cdot s$) is given by equation

$$S_2(\omega)=\left|\bar{H}_2(\omega)\right|^2 S'_0(\omega), \qquad (31)$$

$$S_3(\omega)=\left|\bar{H}_3(\omega)\right|^2 S'_0(\omega). \tag{32}$$

3. Numerical results

Considered a third story building with weight $m=2\cdot10^5 kg$, by four identical columns of Young's modulus $E=0,2\cdot10^{11} N/m^2$ and height $h=2m$, with the diameter $d=0,5m$, the damping factor $\xi=0,25$, with the nonlinear factor to control the type and degree of nonlinearity $\alpha=20m^{-2}$, and $S'_0=0,52\dfrac{m^2}{s^3}$, which means that the power spectral density of excitation $S_F=m^2 S'_0=2,08\cdot10^{10} N^2\cdot s$.

Obtain $k=29\cdot10^6\dfrac{N}{m}$, $p=12,04s^{-1}, c=1204\cdot10^3\dfrac{N\cdot s}{m}$.

Because $c_1=c_2=c_3=c$, for $\alpha=20m^{-2}$, obtain

$$\sigma^2_{x_1}=193,5\cdot10^{-5}m^2,\ \sigma^2_{x_2}=201\cdot10^{-5}m^2\ \sigma^2_{x_3}=213\cdot10^{-5}m^2. \tag{33}$$

The values of k_{1e}, k_{2e} and k_{2e} is

$$k_{1e}=31,57\cdot10^6\frac{N}{m},\quad k_{2e}=33,1\cdot10^6\frac{N}{m},\ k_{3e}=33,1\cdot10^6\frac{N}{m}. \tag{34}$$

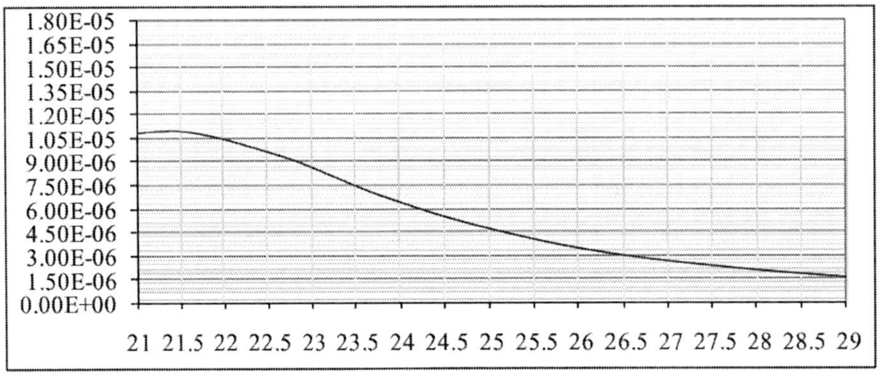

$\omega / 2\pi$

Figure 2. The power spectral density of the response $S_1(\omega)\ [m^2\cdot s]$ for the first structure in the freqency range (21,29) rad/sec.

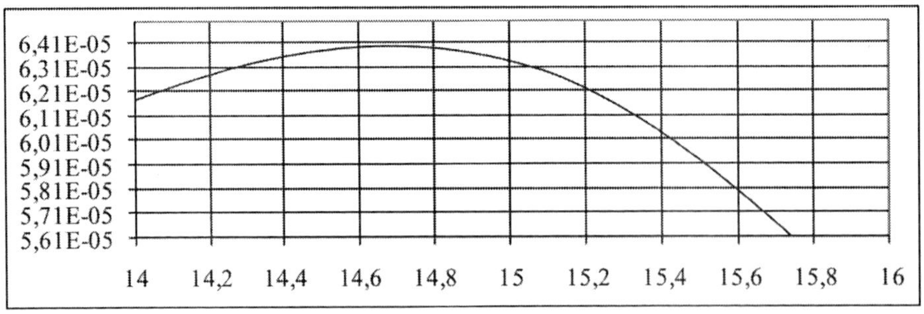

$\omega / 2\pi$

Figure 3. The power spectral density of the response $S_2(\omega)\ [m^2\cdot s]$ for the second structure in the freqency range (14,16) rad/sec.

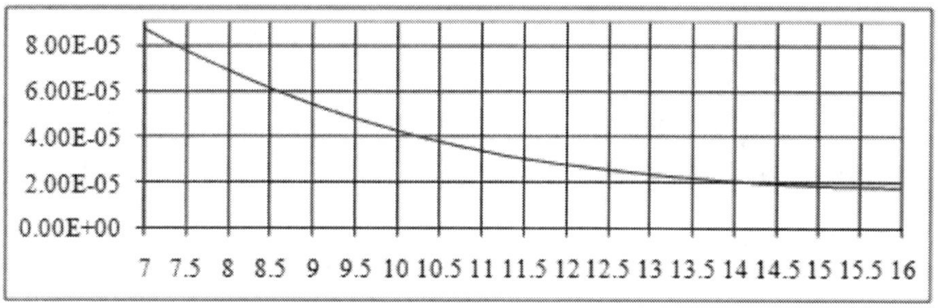

Figure 4. The power spectral density of the response $S_2(\omega)$ $[m^2 \cdot s]$ for the three structure in the freqency range (14,16) rad/sec.

4. Remarks

In the figures 2, 3 and 4, the power spectral density of the excitation is plotted for the differents parameters. The graphs show how each floor of the structure is affected differently by the effect of seismic movement. By acting on the parameters in the calculation formulas, there is the advantage of taking preventive measures on the effect of seismic movement. The statistical linearization technique can also tackle a wide variety of problems and also provides approximate information on the frequency domain characteristics of the stochastic response. Due to the fact that response statistics of such a model can be evaluated analytically, in general, statistical linearization is computationally very effcient.

References

[1] Elishakoff I and Cai G Q 1993 Approximate solution for nonlinear random vibration problems by partial stochastic linearization, *Probabilistic Engineering Mechanics*

[2] Stan P and Stan M 2015 *Random vibration of Houdaille shock absorber*, International Conference Computational Mechanics and Virtual Engineering, Brasov Romania

[3] Stan M and Stan P 2017 Numerical solution of the proces of hydroplaning at wheelswith envelope, *International Congress of Automotive and Transport Engineering, Mobility Engineering and Environment, CAR 2017, 017 IOP Conf. Ser. Mater. Sci. Eng.* **252**

[4] Stan P and Stan M 2015 *Nonlinear vibrations in automotive suspensions*, International Conference Computational Mechanics and Virtual Engineering, Brasov Romania

[5] Stan P 2009 *Analysis of single-degree of freedom non-linear structure under Gaussian white noise ground excitation*, The Sciencific Book, University of Pitesti

Automatic measurement of precision and accuracy from the hit pattern of small arms using electronic target system

A B Živković[1], N P Hristov[2], D D Jerković[2], B S Bogdanović[3] and J M Milutinović[1]

[1]University of Kragujevac, Faculty of Engineering, Sestre Janjić 6, 34000 Kragujevac, Serbia
[2]University of Defence in Belgrade, Military Academy, Pavla Jurišića Šturma 33, 11000 Belgrade, Serbia
[3]Hexagon Metrology Serbia, Sestre Janjić 6, 34000 Kragujevac, Serbia

E-mail: nebojsahristov@gmail.com

Abstract. The basic parameters of the combat training and weapon practicing are the determination of the accuracy and precision. The automatic systems for acquiring of the hit pattern are very expensive and inaccessible. Modern automatic systems are designed as non-material target system for accurate spatial positioning of the projectile trajectory in one or more planes. In the paper will be described the design of the non-material high-resolution target acquisition system developed on the optoelectronics IR (infra-red) sensors. The frame of the multiple IR sensors coupled with the microcomputer platform is the optical frame. The complete target acquisition system is consisted of the two frames for positioning in vertical and horizontal direction. Additionally, in the target system is incorporated the algorithm for the determination of the appropriate statistical values from the group of shots, as accuracy and precision parameters. The assembly hardware components are commercially accessible. The designed algorithm is applied through simple software code. The main aim of the developed target system is for the measurement of the small arms shooting. The presented target system is open for enhancement and can be used for different weapon system.

1. Introduction

The quality of the small arms and ammunition is represented with accuracy and precision. The determination of the accuracy and precision depends on the reliability of methods and equipment. Despite the numerical, theoretical or semi-empirical methods for the determination of the precision parameters, experimental test and measurements of the hit parameters still represents the standard procedure and it is required by the armament and ammunition factories and especially customers.

The conventional experimental methods are consisted of the test fire from the small arms on the material target, on the proving test facilities. The results are the hits of the bullets on the target. The test material target is made of special type of paper with adequate dimensions. The distance of the test target from the test weapon depends on the caliber, muzzle velocity, type of the weapon and required test conditions (different standards as CIP, SAAMI, STANAG, etc.).

The test procedure is namely consisted of the required number of fire groups (usually three groups). The test group depends on the time that is required to fulfil the test conditions, and have to be less then several minutes. Every fire group is consisted of the required number of the hits, according to standard (usually ten hits). The determination of the precision begins from the measurement of the every hit.

The group of hits is represented as hit pattern and can be visually described as „image of shots“ on the test target, [1].

The hit pattern enables the estimation of the precision parameters from the position of the every single hit in relation to the standard size and position of the test target, and in the specific relation to the other hits in the same fire group. The precision parameters depends on the standard and can be described through several types of measures as: hit pattern span („half-perimeter“), average radial derivation, best-half circle radius, probable derivation, core span, circular error probable, [1], etc. The specified precision measures are derived from the statistical parameters and probability distribution according to statistical methods.

The precision is defined as value of the precision parameters, i.e. better precision characteristics are represented as smaller size of the dimension of the hit pattern, and also as lower values of the precision measures. The one of the specific characteristic of the hit pattern is the average hit, and can be calculated or graphically determined, [1].

However, the term accuracy, in the paper and related references, represents the relation of the hit pattern characteristics to dimensions and characteristics of the specified target. In domestic military practices, the accuracy is described as the distance from the average hit to the centre of the specific target,(CT), Figure 1.

The goal of the research is the increasing of the reliability and the rate in the process of the hit detection, the decreasing of the processing time and the increasing of the safety and comfort during experimental research. The design of the device with performances of automatic hit detection and computing processing of detected results, according to implemented mathematical model, is the research proposition, in order to achieve the goal.

The performance of the hit detection coupled with programmable hardware interface, enables registration of the multiple results as data. The process of detection is automated and results are saved as array of data. Data of the group of detected hits is treated through mathematical, i.e. statistical methods due to determination of precision parameters.

2. Mathematical model

The detection of the single hit begins with registration of the particular sensors in frame, that are mounted in two opposite directions – vertical and horizontal. Every hit is saved as data from specific sensors, as following information: number (i.e. position) of the sensors in vertical and horizontal direction of mounted frames, in specific time. Different hits have different time tag.

In the research, the precision parameters have to be evaluated from the relation of the specific hit to the average hit. The average hit is the point, specific centre of the hit pattern, and doesn't represent any real hit, [1, 2].

The determination of the average hit can be done graphically or numerically, depends on the number of the hit. In the paper, as we have the information of the every hit, as numeric data, the determination of the average hit is numerical.

The average hit is the point determined as the cross of the vertical and horizontal axes of the average values of the coordinate of hits in the group. The vertical coordinate of the average hit is average value of the all hits, measured in the vertical direction – height. The horizontal coordinate of the average hit is average value of the all hits, measured in the horizontal direction – width. Every detected hit have the particular value of the derivation from the average hit.

The procedure of the determination of the average hit, on the material test target is:

- placement the horizontal line through the centre of the lowest hit on the test target (horizontal axe);
- placement the vertical line through the centre of the top left hit on the test target (vertical axe), as all hit positions can be perpendicular projected on the axe;
- measurement of the vertical coordinate of all hits from horizontal line to the centre of the hit;
- measurement of the horizontal coordinate of all hits from vertical line to the centre of the hit;

The coordinates of the average hit (AH) are calculated as average value as following, [1,2], and represent the values of horizontal and vertical coordinates in Cartesian coordinate frame of test target fram, Fig. 1,

$$\bar{x} = \frac{x_1 + x_2 + x_3 + \ldots + x_n}{n} = \frac{\sum_{i=1}^{n} x_i}{n}, \; i = 1,2,\ldots,\text{n} \tag{1}$$

$$\bar{y} = \frac{y_1 + y_2 + y_3 + \ldots + y_n}{n} = \frac{\sum_{i=1}^{n} y_i}{n}, \; i = 1,2,\ldots,\text{n} \tag{2}$$

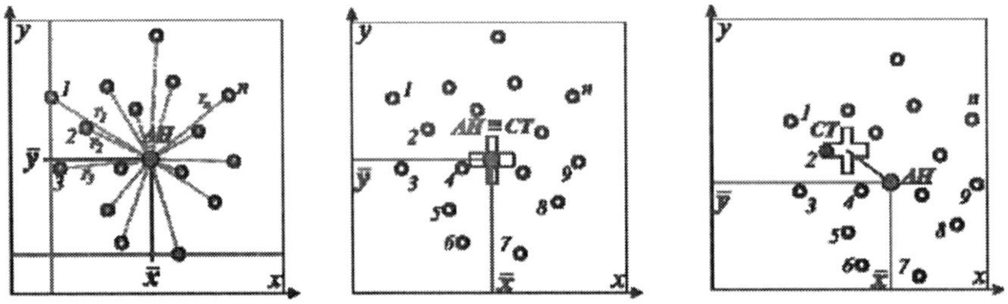

Figure 1. The hit pattern and precision parameters.

The precision parameters can be described, in relation to the average hit, with precision measure of average radial derivation. The average radial derivation (R_D), represents the arithmetic mean of the radial distances of the centre of every hit from the average hit, (r_i),

$$R_D = \frac{r_1 + r_2 + \ldots + r_n}{n} = \frac{\sum_{i=1}^{n} r_i}{n}, \; i = 1,2,\ldots,\text{n} \tag{3}$$

The single radial derivation is represented as,

$$r_i = \sqrt{(\bar{x} - x_i)^2 + (\bar{y} - y_i)^2}, \; i = 1,2,\ldots,\text{n} \tag{4}$$

In the Figure 1 is presented scheme of the hit pattern with ten hits with radial derivations (r_1, r_2, …, r_{10}) in the relation to the average hit ($\bar{x}$, $\bar{y}$) in Cartesian coordinate frame.

The estimation of the precision can be described, according to the presented relations (1) to (4), as better precision characteristics as lower value of the average radial derivation, i.e. as the hits are placed in lower radial distances.

The estimation of the accuracy is defined as matching, or lower distances between average hit (*AH*) and the center of the target (*CT*). In the Figure 1 are shown the differences between high and low accuracy.

The estimation of the precision and accuracy of the weapon can be described through four characteristic states, in relation to the value of the precision parameters (3) and the position of the average hit on the target:

- high accuracy and precision – average hit matches the target centre and low value of the average radial derivation;
- high accuracy and low precision – average hit matches the target centre and high value of the average radial derivation;
- low accuracy and high precision – average hit does not matches the target centre and low value of the average radial derivation;
- low accuracy and precision – average hit does not matches the target centre and high value of the average radial derivation;

IOP Conf. Series: Materials Science and Engineering **659** (2019) 012015 doi:10.1088/1757-899X/659/1/012015

As the quality of the small arms depends on the estimation of the presented characteristics of the accuracy and precision, the reliability and processing time are the key reason to made automated system. In the further part of the paper will be described the conceptual design of the virtual target system for automatic detection and determination of the precision and accuracy parameters. The proposed system is based on the signal communication of the infra-red sensor (IR diode) and photo transistor.

3. The design conception

The idea of the design of the device for automatic detection of the hits is based on the interruption of the established signal between transmitting sensor (IR diode) and receiving sensor (Phototransistor). The transmitter is IR sensor, source of the IR light, and the receiver is the detector of the transmitted light as phototransistor. The scheme of the conceptual design is presented in Figure 2.

Figure 2. The conceptual design of the target system.

The figure shows two groups of the coupled transmitter and receiver in two perpendicular directions. The receiver is placed opposite to transmitter, in one plane, for one direction. Second group of transmitter and receiver is the same, except that is mounted in perpendicular direction, in the plane slightly translated from first test plane. The group of transmitter is consisted of the number of the single IR diodes, that are mounted in equal, small and required distance. According to the distribution of the IR diodes, in the opposite side of the frame are mounted coupled pairs of the phototransistor, in direct communication with diode separated from other diodes. Two frames are represented as optical net of the high resolution (distances between sensors are 2,54 mm i.e. 0.1") for detection the motion of small rigid body, as bullet (diameter of pellet is 4,5 mm, rifle bullet diameter is 5,56 or 7,62 mm).

The result of the detection is the position (h_i, v_i) of the bullet in the moment of interrupting of the signal between each pair of transmitter and receiver in two frames, horizontal (h) and vertical (v) frame. The prototype of the receiving sensors and microcontroller board are shown in Figure 2.

The data of interrupting signals are detected and saved using hardware device as microcontroller board Arduino Due, [4], connected to PC. The software code processes the received data, and according to the presented mathematical model, determines the values of the precision parameters.

3.1. Transmitting Sensors

The transmitting sensor is infrared diode, and represents the light source. The purpose of the IR diode is transmission of the infrared light to the receiver. The receiving sensor detects the IR light until interruption, as IR diode is connected to the supply voltage. The voltage supply of the diode is provided by microcontroller board Arduino Due, [4], Figure 3.

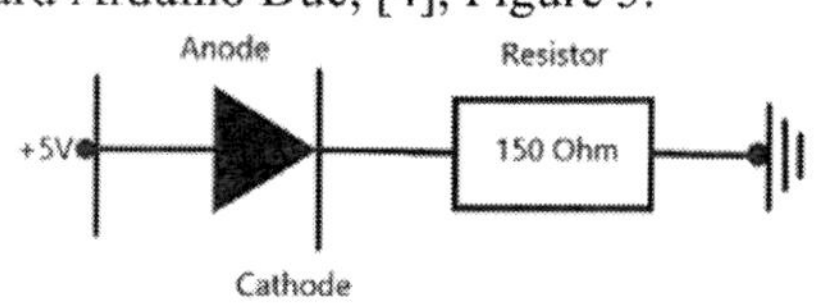

Figure 3. Scheme of IR diode.

The single IR diode transmits non-polarized light. Every phototransistor detects the light source from any direction. The light polarization is included in the research, in order to provide that one receiver sensor detects the light from just one transmitter. In the paper, polarizing filter is made, by the Malus's law, Figure 4 [3].

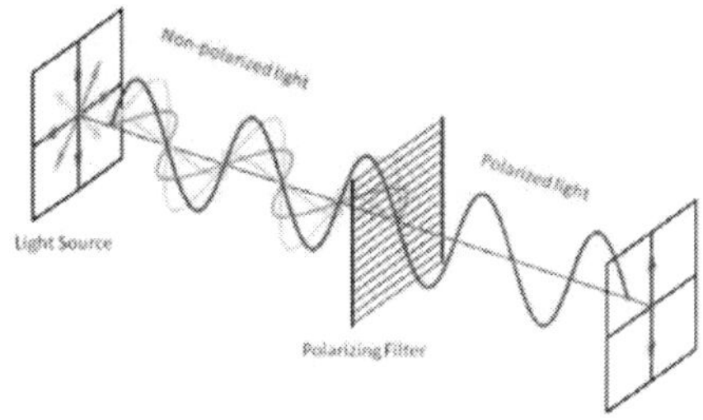

Figure 4. Scheme of polarizing filter.

3.2. Receiving Sensors

The receiving sensor is phototransistor. Phototransistor is an electronic switching and current amplification component, which relies on exposure to light to operate. It is capable of converting light energy into electric energy. Along the detection function, phototransistor has the function of amplifier. The rate of detection on the light is very high, as the reason for the research. Every phototransistor is connected to microcontroller board, in order to detected signal interruption be captured and saveed for further data acquisition. The scheme of phototransistor is presented in Figure 5.

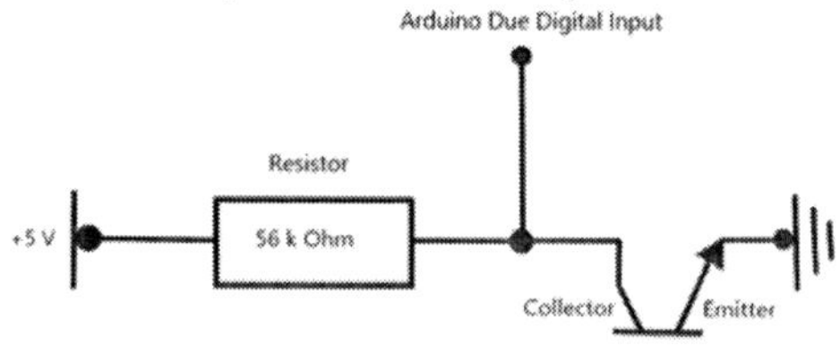

Figure 5. Scheme of phototransistor connection.

3.3. Microcontroller Board

In the research, four microcontroller board Arduino Due was implemented in the target system. The capabilities and availability of the board enables the design of the target system. The microcontroller is AT91SAM3X8E with technical specification in the Table 1, [4].

Table 1. Technical specifications of Arduino Due Microcontroller Board.

Operating Voltage	3.3V
Input Voltage (recommended / limits)	7-12V / 6-16 V
Digital I/O Pins	54 (of which 12 provide PWM output)
Analog Input Pins	12
Analog Output Pins	2 (DAC)
Total DC Output Current on all I/O lines	130 mA
DC Current for 3.3V Pin / 5 V Pin	800 mA / 800 mA
Flash Memory	512 KB all available for the user applications
SRAM	96 KB (two banks: 64KB and 32KB)
Clock Speed	84 MHz

The purpose of the microcontroller board in the research is to provide the voltage supply to the IR diode, to detect the signals from phototransistors, provide the detected signal to PC and enables data processing. The graphic user interface of the Arduino Due enables communication with connected sensors and components. The software code is made for the research, in Arduino Environment, [4], for monitoring of the signal interruption between IR diode and phototransistor, according to the specific frame and position of the pair of transmitter and receiver. The interruption signal as information of the hit position in target plane is displayed to user and recorded.

4. The Model of the Hit Pattern Determination

The model of the determination of the hit pattern is based on the automatic calculation of the detected signals of position of the hit on the test target plane. The detected values of the position and calculation of the precision parameters is provided by microprocessor acquisition system, as early described.

According to the described equipment, the design of the automatic electronic target system is consisted of two frames. The detection frame is initially designed for detection of sixty signals along the one direction. The detections of the every hit are results of interruption detection of both frames in the same time (i.e. predefined short period). The software receives signals, compiles and records as array of the data. According to the described mathematical model, automatic target system calculates the precision parameters and displays the results.

The determination process is consisted of the calculation the position of the every hit in the group, calculation of the average hit, and estimation of the precision parameter, as average radial derivation. The precision estimation software can be improved, by implementation of calculation of the other precision measures, [1].

The initial phase of the research, described in the paper, the accuracy parameter will not be estimated, because of the current technical limitations. The scheduled target frame is made of 60 connected sensors (30 sensors per 1 microcontroller board). The dimension of the active part of the frame is approximately 150 mm.

5. Conclusion

The paper deals with the research of the start-up design of the electronic target system for automatic detection of the bullet hits and determination of the precision and accuracy parameters from the group of the hits, as didactical resoure. The initial design of the system is made from available commercial electronic and IT components and custom made support mountings. The initial tests shows that designed system have improved performances in regard to the standard procedure with material test targets. Improved characteristics are the decreased time of test, automatic generation the results of the precision parameters, relatively simple and low-cost design and production, the possibility of perpetual usage. The automatic target system is designed and applicable for the different didactical purposes, fire practicing facilities, as sport fire practicing, military fire training, fire proving ground test for the research and development of the small arms and quality management of the ammunition.

References

[1] Regodić D 2007 *Exterior Ballistics Measurements*, In book: Exterior Ballistics (in Serbian) VIZ, Belgrade

[2] Regodić D and Jerković D 2015 *The Analysis of Systematic Disturbances and the Estimation of Probable Variation of Projectile Trajectory*, Sinergija University – Proceeding s International Scientific Conference 91-98, http://dx.doi.org/10.7251/ZRSNG1501091R

[3] Matioli E, Brinkley S, Kelchner F, Harada K, Hu Y L, Nakamura S, DenBaars S, Spreck J and Weisbuch C 2012 *High-brightness polarized light-emitting diodes*, Light: Science & Applications **1**(22), http://dx.doi.org/10.1038/lsa.2012.22

[4] https://www.arduino.cc/en/, 10.12.2018, 15.02.2019

IOP Conf. Series: Materials Science and Engineering **659** (2019) 012016 doi:10.1088/1757-899X/659/1/012016

The application of the ballistic pendulum for the bullets velocity measurements

J M Milutinović[1], N P Hristov[2], D D Jerković[2], S Z Marković[2] and A B Živković[1]

[1]University of Kragujevac, Faculty of Engineering, Sestre Janjić 6, 34000 Kragujevac, Serbia
[2]Military academy, University of Defence, Pavla Jurišica Šturma 33, 11000, Belgrade, Serbia

E-mail: nebojsahristov@gmail.com

Abstract. The accurate values of the initial velocity of bullets are significant for the quality and precision of small arms. The application of the simple measurement methods, during the tests of the weapon and ammunition, enables the determination of the velocity values from the group of shoots. The accuracy of bullet trajectory, appropriate trajectory correction parameters and terminal ballistics parameters are depended of the quality of initial conditions as initial velocity. The ballistic pendulum, as simple and old-design device, enhanced with optoelectronic encoder sensor and computer acquisition system, can be one of the good start-up device platform for measurement of velocity and observation of the terminal ballistics effects. The function principle of the considered device is based on the energy conservation. Initial data are mass of bullet, mass and dimension of pendulum, and result is velocity, according to the values of time and angle of pendulum. The output signals of measured angle in time are captured, and as required values for calculation the velocity on the microprocessor platform, for each shoot in the test group. The microprocessor platform saves measured and calculated values in memory and generates statistic report of results. The presented method can improve weapon and ammunition tests, by decreasing the time of measurement acquisition and increasing the quality and speed of results without errors. The method and system is simple and low-cost, and enables the design of small arms ammunition database of testing results.

1. Introduction

The range of projectiles directly depends on the muzzle velocity. The muzzle velocity of projectiles are significant for the quality and precision of classic weapon systems. This parameter is of utmost crucial for the proper function of ballistic systems. The quality of ammunition and weapons are determined by this essential quantitative parameter. It effects on flight stability of the projectiles and precision and accuracy parameters [1]. The muzzle velocity deviation values are defined by standards (MIL, STANAG, SORS, CIP, ..) according to which a qualitative assessment of ammunition and weapons. Depending on the type of weapons, the muzzle velocity ranges from 100 to 2000 meters per second. Measuring the projectile speed as a physical size is very complex. Generally it is performed by various indirect methods. The main methods are with chronographs, electromagnetic fields, sensors for measuring force, displacement, acceleration or other physical size. At first method, the law of energy conservation has been applied. The Benjamin Robins, a British mathematician and military engineer, was the first who invented device and practical method for measuring the velocities of projectiles. He described his work in New Principles of Gunnery, first published in 1742. By developing science and technology, this method has been improved.

A ballistic pendulum is a device for measuring a projectiles momentum, primarily for small arms. From which it is possible to calculate the velocity and kinetic energy. Ballistic pendulums have been largely rendered obsolete by modern chronographs, which allow direct measurement of the projectile velocity. It can be used to measure any transfer of momentum [2, 3, 4].

In a perfectly inelastic collision, a bullet is fired into the stationary pendulum, which captures the bullet and absorbs its energy. The stationary pendulum now moves with a new velocity just after the collision. While it isn't all of the energy from the bullet transformed into kinetic energy for the pendulum (some is used as heat and deformation energy), the momentum of the system is conserved. By measuring the height of the pendulum's swing, the potential energy of the pendulum, when it stops can be measured. In the case of a pendulum total mechanical energy is conserved. So kinetic energy of the pendulum (after firing) is fully converted to potential energy. Thus the pendulum's initial velocity can be calculated. Using the law of conservation of momentum, the velocity of the bullet can be computed [2, 3].

The improvement of the described measurement method is possible with application of sensors and microprocessor platforms. The data can be permanently stored in the digital records and automatic statistical processing can be performed for the fired group of bullets.

2. Mathematical method

There are two ways for the calculation of the velocity of the bullet. The first method (approximate method) assumes that the pendulum and bullet together act as a point mass located at their combined center of mass. This method does not take rotational inertia into account. It is somewhat quicker and easier than the second method, but not as accurate.

The second method (exact method) uses the actual rotational inertia of the pendulum in the calculations. The equations are slightly more complicated, and it is necessary to take more data in order to find the moment of inertia of the pendulum, but the results obtained are generally better [2,3].

2.1. Approximate Method

Begin with the potential energy of the pendulum at the top of its swing,

$$\Delta E_p = M \cdot g \cdot \Delta h \tag{1}$$

where M is the combined mass of pendulum and ball, g is the acceleration of gravity, and Δh is the change in height. Substitute for the height,

$$\Delta h = R(1-\cos\theta) \tag{2}$$

where R is the distance from the pivot point to the center of mass of the pendulum/ball system [2,3].

$$\Delta E_p = M \cdot g \cdot R(1-\cos\theta) \tag{3}$$

This potential energy is equal to the kinetic energy of the pendulum immediately after the collision,

$$E_k = \frac{1}{2} M v_p^2 \tag{4}$$

The momentum of the pendulum after the collision is just,

$$P_p = Mv_p \tag{5}$$

which we substitute into the previous equation to give,

$$E_k = \frac{P_p^2}{2M} \tag{6}$$

Solving this equation for the pendulum momentum gives,

$$P_p = \sqrt{2M \cdot E_k} \tag{7}$$

This momentum is equal to the momentum of the bullet before the collision,

$$P_b = mv_b \tag{8}$$

Setting these two equations equal to each other and replacing kinetic energy with our known potential energy gives us,

$$mv_b = \sqrt{2M^2 gR(1-\cos\theta)} \tag{9}$$

Solve this for the bullet velocity and simplify to get,

$$v_b = \frac{M}{m}\sqrt{2gR(1-\cos\theta)} \tag{10}$$

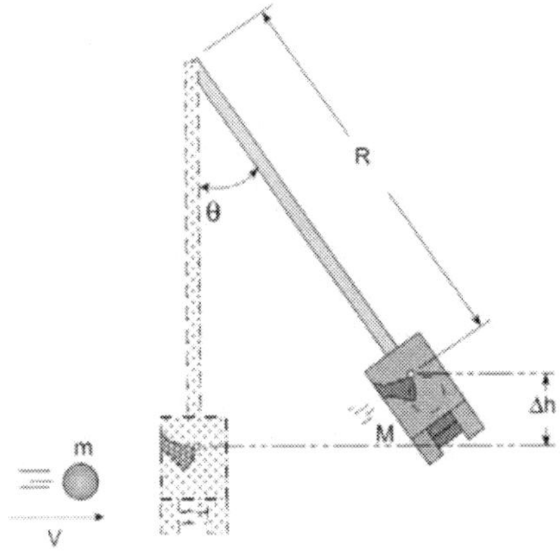

Figure 1. The motion of pendulum.

2.2. Exact Method

The potential energy is found in a way identical to the way shown previously,

$$\Delta E_p = M \cdot g \cdot R(1-\cos\theta) \tag{11}$$

For the kinetic energy, we use the equation for angular kinetic energy instead of linear, and substitute into it the equation for angular momentum [2, 3].

$$E_k = \frac{1}{2}I\omega^2 \tag{12}$$

Here I is the moment of inertia of the pendulum and bullet combination, and ω is the angular velocity immediately after the collision.

The moment of movement of the pendulum is,

$$L_P = I\omega \tag{13}$$

$$E_k = \frac{L_P^2}{2I} \tag{14}$$

$$L_P = \sqrt{2IE_K} \tag{15}$$

This angular momentum is equal to the angular momentum of the bullet before the collision, as measured from the pendulum pivot point.

$$L_b = mR_b^2\omega = mR_b v \tag{16}$$

where R_b is the distance from the pendulum pivot to the bullet. This radius is not in general equal to R, which is the distance from the pivot point to the center of mass for the pendulum and bullet system.
These two angular momenta are equal to each other, and,

$$mR_B v = \sqrt{2IMgR(1-\cos\theta)} \tag{17}$$

$$v = \frac{1}{mR_B}\sqrt{2IMgR(1-\cos\theta)} \tag{18}$$

Now we need to find I, the moment of inertia of the pendulum and bullet. To do this, we start with the rotational equivalent of Newton's second law,

$$\tau = I\alpha \tag{19}$$

where τ is torque, I is moment of inertia, and α is angular acceleration.

The force on the center of mass of the pendulum is just Mg, and the component of that force directed towards the center of the pendulum swing is, Figure 2,

$$F = -Mg\sin\theta \tag{20}$$

The torque on the pendulum is thus,

$$I\alpha = -RMg\sin\theta \tag{21}$$

For small angles θ, $sin\theta \approx \theta$, so if we make this substitution and solve for α we get,

$$\alpha \approx -\frac{MgR}{I}\theta \tag{22}$$

This angular equation is in the same form as the equation for linear simple harmonic motion,

$$\alpha = -\omega^2 x \tag{23}$$

So if we compare these two equations, linear and angular, we can see that the pendulum exhibits simple harmonic motion, and that the square of the angular frequency (ω^2) for this motion is just,

$$\omega^2 = \frac{MgR}{I} \tag{24}$$

Solving this for I gives us the desired result,

$$I = \frac{MgR}{\omega^2} = \frac{MgRT^2}{4\pi^2} \tag{25}$$

where T is the period of the pendulum.

IOP Conf. Series: Materials Science and Engineering **659** (2019) 012016 doi:10.1088/1757-899X/659/1/012016

The projectile velocity calculation is more accurate using the exact method, but the difference is minimal (less than 2 %). In this case, the approximate method was used for simpler calculation.

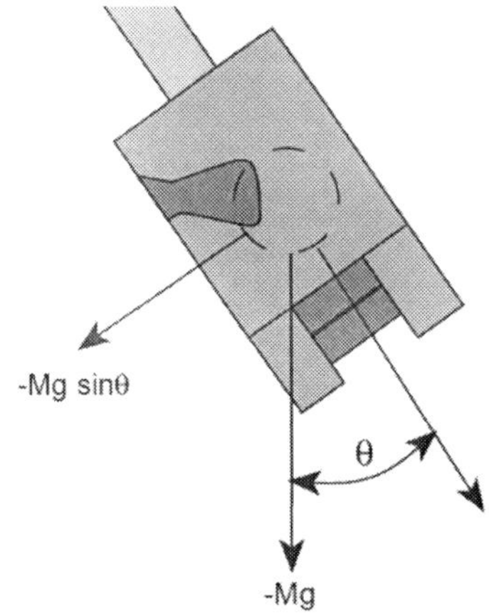

Figure 2. The forces direction.

3. Conceptual design

Conceptual design for model is presented in the Figure 3. The system consists of the pendulum, the encoder and microprocessor platforms connected to the computer. The encoder registers the angle of movement after the collision of the projectile and the pendulum. The mass of pendulum and bullet are known in advance. The microprocessor calculate the output signal of the encoder with the corresponding mathematical model. The characteristics of the encoder is presented in the Table 1.

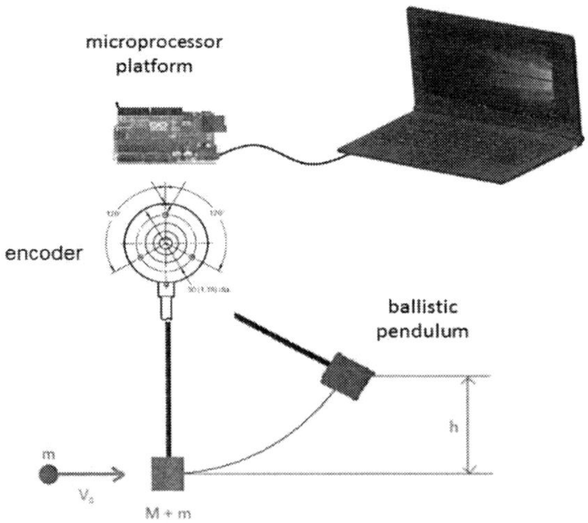

Figure 3. Principal scheme of system.

Table 1. Technical specifications of Encoder.

Model E6B2-CWZ6C	Model E6B2-CWZ6C
Operating Voltage	5 ÷24 V
Resolution (pulses/rotation)	2000
Output capacity	Applied voltage: 30 VDC max. Sink current: 35 mA max. Residual voltage: 0.4 V max. (at sink current of 35 mA)
Maximum response frequency	100 kHz
Rise and fall times of output	1 µs max. (Control output voltage: 5 V, Load resistance: 1 kΩ, Cable length: 2 m max.)
Starting torque	0.98 mN·m max.
Maximum permissible speed	6,000 r/min
Temperature range	Operating:-10 to 70°C (with no icing)
Shock resistance	Destruction: 1,000m/s^2 3 times each in X, Y, and Z directions
Weight	100 g

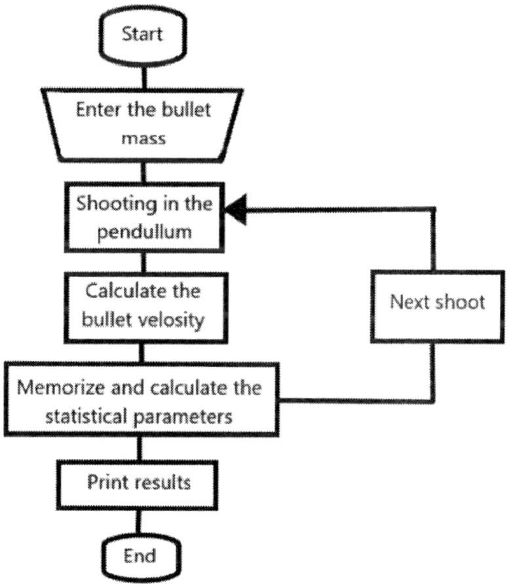

Figure 4. Algorithm of solution.

For this system is used Arduino Uno microprocessor board platform with USB interface. This platform has all functions for controls and can be directly connected to the computer through the USB connection. The software solution was made according to the mathematical model and algorithm, presented in the Figure

4. The software is designed to record velocities of each single shoot, then process the statistical parameters of the shooting group (10 to 100 shoots). The calculation outputs are the mean of projectiles velocity, maximum and minimum velocity value, standard deviation and mean square deviation as statistical parameters. The software solution was made in the open Arduino code.

4. Conclusion

In the paper is presented solution for the bullets velocity measurements using application of the ballistic pendulum. This is a practical application of theoretical settings in combination with modern sensors and microprocessors. The concept required measuring the physical measures indirectly and translating it into an electrical digital signal. Also, the research requires signal processing and application of statistical tools. This is very demanding challenge for engineering practice and requires knowledge of different technical disciplines. The next step is to check the measuring technique by a validated method, and this is in progress. The testing will be performed for different types of small arms, and then a final evaluation of measuring method will be given.

References

[1] Cerović P and Subotić Z 2007 *Zaustavna moć projektila kao element optimizacije izbora kalibra streljačkog oružja i municije u razvoju*, OTEH 2007
[2] Instruction Manual and Experiment Guide for the PASCO scientific Model ME-6830/ME-6831, *Roseville* 1993
[3] Marković S 2017 *Komparativna analiza metoda određivanja dejstva na cilju modela streljačkog naoružanja*, Master Thesis N-321, Vojna akademija, Belgrade
[4] Martinović V 2015 *Eksperimentalno ispitivanje i analiza dejstva projektila malih kalibara na cilju*, Faculty of Engineering, Kragujevac

Analysis of the wind effect on the mechanical structures in the case of the four-side sunshade

B Marković[1], M Krajišnik[2] and A Đurić[1]

[1]University of East Sarajevo, Faculty of Mechanical Engineering, Vuka Karadžoća 30, 71123 East Sarajevo, RS, B&H
[2]Orao a.d. Bijeljina, Šabačkih đaka bb, 76300 Bijeljina, , RS, B&H

E-mail: biljana.markovic@ues.rs.ba

Abstract. This paper presents the analysis of the wind effect of four-side sunshade construction. The aim of the paper is to demonstrate the possibility of applying analogy in research, with examples from civil constructions, in order to simplify and reduce the analysis. The paper presents the analytical calculation of the pressure on the sides of the four-sided sunshade, analogously to the calculation of pressure on the four-sided roofs of civil construction objects.

1. Introduction

Sunshade are used in different places and in different weather conditions, from high mountain areas with large amounts of snowfall, to sea beaches where tempest and storm can often do damage if the parasols are not adequately closed. As the sunshade has burdened by the wind (most cases), this paper presents the method of calculating the wind pressure on the sunshades, analogous to the method of calculating the influence of the wind on the roofs of building structures.

The wind load belongs to the climate effects of the environment. The wind force depends primarily on the actual characteristics of this natural phenomenon, from the speed and direction of the air movement, the mode of action (constant, variable or both). The effect of the wind on a particular construction depends on the dimension of the structure and its position in the area. Although the dynamic nature of the wind is, in most cases, an equivalent static load can replace it. The effect of wind on the stability of structures has defined by norms. The description of the wind action is based on the adoption of the reference wind velocity (v_{ref}) defined based on meteorological measurements [1]. The influence of wind on the structures has investigated considerably in the construction industry, and special attention has paid to the study of the effects of the wind on bridges to European norms [2, 3].

2. Analytical determination of wind pressure affecting the sunshade

Under special conditions - during weather conditions, the sunbathing is exposed to the wind effect and therefore the pressure of the wind (Wi) is operating on the surface of the dome, and the stated load is further transferred to the other elements of the sunshades, where they can break if they are not adequately designed. In order to ensure the carrying capacity of the sunshades at the time of strong winds, it is necessary to carry out the calculation of all of its elements; therefore, it is necessary to know the load, wind pressure on the sunshade. Figure 1 shows broken Collage 8x8 sunshade from the wind in Dubrovnik on August 13, 2017. In this paper, it has shown the calculation of wind pressure on the sunshade, just on the example given in the figure.

Figure 1. Deformation of sunshade Collage 8x8 – Dubrovnik 13.8.2017.

On First place, the published metrological data for Dubrovnik from the State Hydro meteorological Institute DHMZ has published in the year, for the day on August 13, 2017., and these data are given in Table 1.

Table 1. Metrological data.

Location	Date	min daily temperature	max daily temperature	average wind speed	max hit the wind
Dubrovnik	13.8.2017.	21°C	30°C	54 km/h	84 km/h

The pressure of the wind on the outer surfaces w_e, as well as the wind pressure on the inner surfaces w_i, is calculated by the terms:

$$w_e = q_{ref} \cdot c_e(z_e) \cdot c_{pe} \tag{1}$$

$$w_i = q_{ref} \cdot c_e(z_i) \cdot c_{pi} \tag{2}$$

where are:

q_{ref} - Medium pressure caused by an average wind speed

$c_e(z_e), c_e(z_i)$- Coefficient of exposure

c_{pe}, c_{pi} - Coefficient of external and internal pressure

2.1. Determination of the mean pressure caused by the average wind speed

Net pressure on the wall, roof or element is equal to the difference in pressure on the opposite surfaces, taking into account their signs. Pressure, directed at the surface has taken as positive. Negative pressure has considered as a pressure that is direct from the surface. Examples has shown in Figure 2.

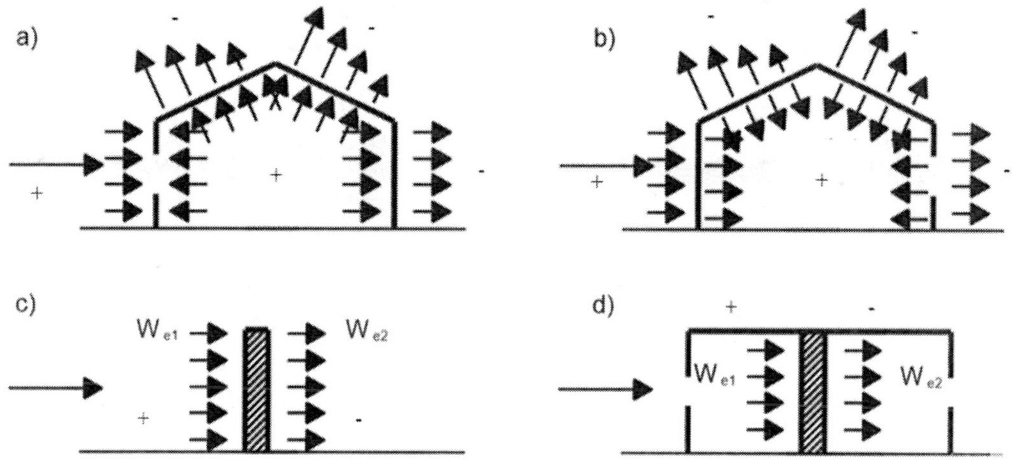

Figure 2. Surface pressure [4].

The average pressure caused by the average wind speed is calculated using the following form:

$$q_{ref} = \frac{1}{2} \cdot \rho \cdot v_{ref}^2 \tag{3}$$

where ρ is the air density, which is read from the diagram shown in Figure 3. at an average temperature of 25 ^{0}C and is ρ=1,18 kg/m^3.

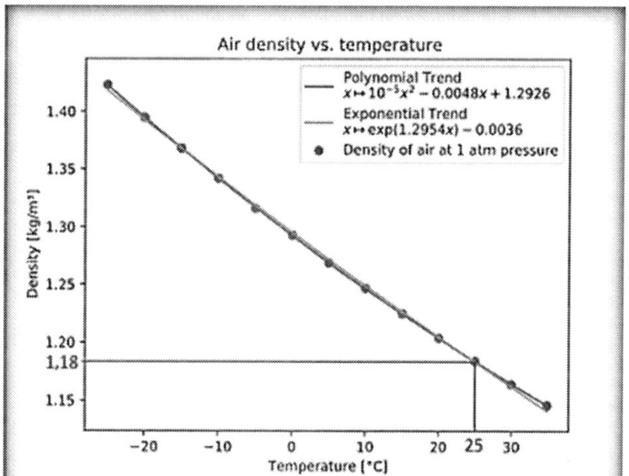

Figure 3. Diagram of air density and temperature relationship.

By adding known values to equality (3), wind pressure is obtained:

$$q_{ref} = \frac{1}{2} \cdot \rho \cdot v_{ref}^2 = \frac{1}{2} \cdot 1{,}184 \frac{\text{kg}}{\text{m}^3} \cdot \left(\frac{15\text{m}}{\text{s}}\right)^2 = 133 \frac{\text{N}}{\text{m}^2} = 0{,}133 \frac{\text{kN}}{\text{m}^2} \tag{4}$$

2.2. Determination of the coefficient of external and internal pressure

The dome is at a height of h=3,5 m, which is smaller than the width of the dome 3.88 m on which the wind blows and on this basis the reference height is z_e=h=3,5 m. The dome of sunshade has viewed as a four-conductor roof and, according to Figure 4; a division has made into eight surfaces in which the coefficients of the external pressure were determined c_{pe}.

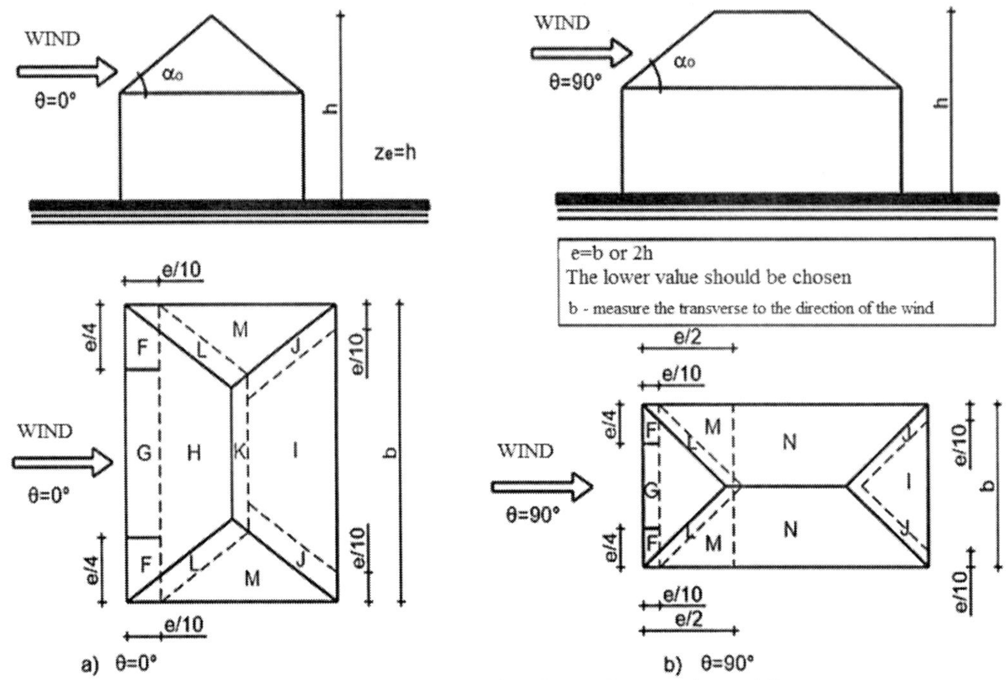

Figure 4. Division of quadruple roofs to surfaces [4].

Based on the known dimensions of the sunshade, the dimensions from figure 5 are adopted, e=3, 88 m, b=3, 88 m; $2h$=7 m. According to Figure 4 for the direction of wind 0, the four-stream roof of the dome of the sunshade is 19^0 and it has divided into eight surfaces, as shown in Figure 5.

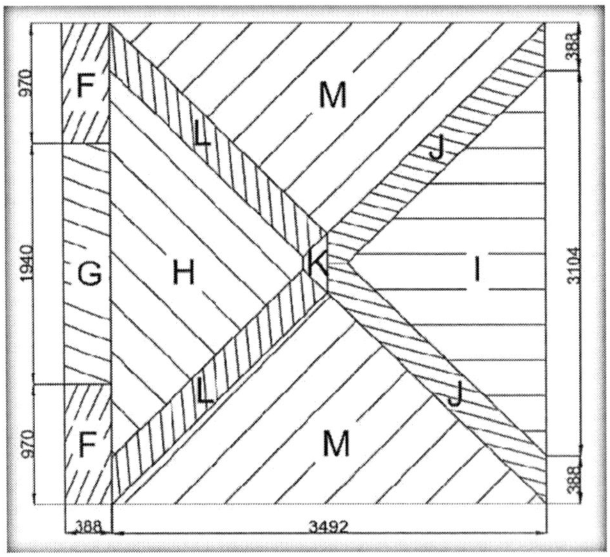

Figure 5. Division of the dome of sunshade to surfaces

The coefficients of the external pressure c_{pe}, depends on the dimensions of the loaded surfaces. The coefficient c_{pe1} is suitable for calculating small elements, with an area of up to $1m^2$ or less, while the coefficient c_{pe10} is suitable for calculating areas larger than 10 m^2. These codes are readable from Table 2.

Table 2. External pressure coefficients for four-conductor roofs [4].

α_0 $\Theta = 0°$ α_{90} $\Theta = 90°$	$\Theta = 0°$ $\Theta = 90°$									
	F		**G**		**H**		**I**		**J**	
	$c_{pe,10}$	$c_{pe,1}$	$c_{pe,10}$	$c_{pe,1}$	$c_{pe,10}$	$c_{pe,1}$	$c_{pe,10}$	$c_{pe,1}$	$c_{pe,10}$	$c_{pe,1}$
5°	-1,7	-2,5	-1,2	-2,0	-0,6	-1,2	-0,3		-0,6	
15°	-0,9	-2,0	-0,8	-1,5	-0,3		-0,5		-1,0	-1,5
	+0,2		+0,2		+0,2					
30°	-0,5	-1,5	-0,5	-1,5	-0,2		-0,4		-0,7	-1,2
	+0,5		+0,7		+0,4					
45°	+0,7		+0,7		+0,6		-0,3		-0,6	
60°	+0,7		+0,7		+0,7		-0,3		-0,6	
75°	+0,8		+0,8		+0,8		-0,3		-0,6	

α_0 $\Theta = 0°$ α_{90} $\Theta = 90°$	$\Theta = 0°$ $\Theta = 90°$							
	K		**L**		**M**		**N**	
	$c_{pe,10}$	$c_{pe,1}$	$c_{pe,10}$	$c_{pe,1}$	$c_{pe,10}$	$c_{pe,1}$	$c_{pe,10}$	$c_{pe,1}$
5°	-0,6		-1,2	-2,0	-0,6	-1,2	-0,4	
15°	-1,2	-2,0	-1,4	-2,0	-0,6	-1,2	-0,3	
30°	-0,5		-1,4	-2,0	-0,8	-1,2	-0,2	
45°	-0,3		-1,3	-2,0	-0,8	-1,2	-0,2	
60°	-0,3		-1,2	-2,0	-0,4		-0,2	
75°	-0,3		-1,2	-2,0	-0,4		-0,2	

For the calculation of the coefficient of external pressure, the area between $1 m^2$ and $10\ m^2$ was used:

$$c_{pe} = c_{pe1} - (c_{pe1} - c_{pe10}) \cdot \log_{10} A \tag{5}$$

A is the corresponding surface for which the coefficient is calculated.

Since in Table 2. value cannot be read for angle 19^0, it is necessary to use linear interpolation, and finally it is obtained:

$F = 0{,}376\ m^2 \rightarrow c_{pe} = -1{,}868$

$G = 0{,}752\ m^2 \rightarrow c_{pe} = -1{,}50$

$H = 2{,}500\ m^2 \rightarrow c_{pe} = -0{,}273$

$L = 0{,}620\ m^2 \rightarrow c_{pe} = -2$

$K = 0{,}055\ m^2 \rightarrow c_{pe} = -1{,}6$

$M = 2{,}970\ m^2 \rightarrow c_{pe} = -0{,}728$

$15° \rightarrow c_{p1} = -1{,}2;\ c_{p10} = -0{,}6; \rightarrow c_{pe} = -1{,}2 - (-1{,}2 + 0{,}6) \cdot \log_{10} 2{,}97 = -0{,}669$

$30° \rightarrow c_{p1} = -1{,}2;\ c_{p10} = -0{,}8; \rightarrow c_{pe} = -1{,}2 - (-1{,}2 + 0{,}8) \cdot \log_{10} 2{,}97 = -0{,}892$

$J = 0{,}665\ m^2 \rightarrow c_{pe} = -1{,}42$

$I = 2{,}471\ m^2 \rightarrow c_{pe} = -0{,}47$

The internal pressure coefficient can be readed from Figure 6 and is $c_{pi} = -0.5$, since it is clearly seen that this is an hole with a opening coefficient $\mu = 1$.

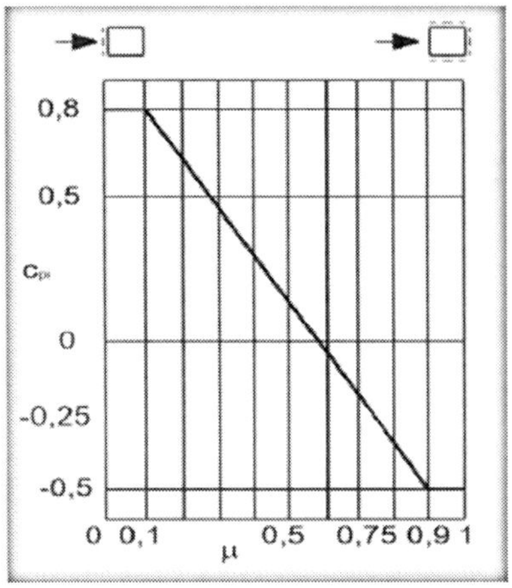

Figure 6. Diagram of the ratio of the coefficient of the opening and the coefficient of internal pressure [4].

The sunshade is located on the ground level III category, and for a height of z = 3.5 m above the ground, the exposure coefficient can be read from Figure 7 and it will be c_e (z) = 1.6.

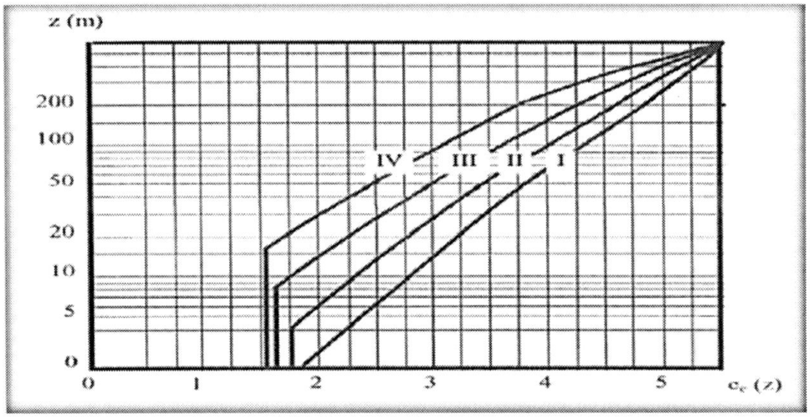

Figure 7. Exposure coefficient [4].

3. Results and discussion

Wind pressure on the external surface is:

$w_e(F) = 0{,}133 \cdot 1{,}6 \cdot (-1{,}868) = -0{,}3975\ kN/m^2$

$w_e(G) = 0{,}133 \cdot 1{,}6 \cdot (-1{,}50) = -0{,}3192\ kN/m^2$

$w_e(H) = 0{,}133 \cdot 1{,}6 \cdot (-0{,}273) = -0{,}0581\ kN/m^2$

$w_e(L) = 0{,}133 \cdot 1{,}6 \cdot (-2) = -0{,}4256\ kN/m^2$

$w_e(K) = 0{,}133 \cdot 1{,}6 \cdot (-1{,}6) = -0{,}3404\ kN/m^2$

$w_e(M) = 0{,}133 \cdot 1{,}6 \cdot (-0{,}7282) = -0{,}1549\ kN/m^2$

$w_e(J) = 0{,}133 \cdot 1{,}6 \cdot (-1{,}42) = -0{,}3021\ kN/m^2$

$w_e(I) = 0{,}133 \cdot 1{,}6 \cdot (-0{,}4736) = -0{,}1007\ kN/m^2$

Wind pressure on the inner surfaces is:

$w_i = 0{,}133 \cdot 1{,}6 \cdot (-0{,}5) = -0{,}1064\ kN/m^2$

The wind load has determined by algebraic addition of external and internal pressure:

$$w_1(F) = (-0{,}3975 + |-0{,}1064|) = -0{,}2911\frac{kN}{m^2} = -0{,}000291\frac{N}{mm^2}$$

$$w_1(G) = (-0{,}3192 + |-0{,}1064|) = -0{,}2128\frac{kN}{m^2} = -0{,}000212\frac{N}{mm^2}$$

$$w_1(H) = (-0{,}0581 + |-0{,}1064|) = 0{,}0483\frac{kN}{m^2} = 0{,}0000483\frac{N}{mm^2}$$

$$w_1(L) = (-0{,}4256 + |-0{,}1064|) = -0{,}3192\frac{kN}{m^2} = -0{,}000319\frac{N}{mm^2}$$

$$w_1(K) = (-0{,}34048 + |-0{,}1064|) = -0{,}23408\frac{kN}{m^2} = -0{,}000234\frac{N}{mm^2}$$

$$w_1(M) = (-0{,}1549 + |-0{,}1064|) = -0{,}0485\frac{kN}{m^2} = -0{,}0000485\frac{N}{mm^2}$$

$$w_1(J) = (-0{,}3021 + |-0{,}1064|) = -0{,}1957\frac{kN}{m^2} = -0{,}000195\frac{N}{mm^2}$$

$$w_1(I) = (-0{,}1007 + |-0{,}1064|) = 0{,}0057\frac{kN}{m^2} = 0{,}0000057\frac{N}{mm^2}$$

4. Conclusion

The method of reseraching the influence of wind on mechanical constructions based on the example of wind impact studies on construction structures, by the method of analogy, is shown in this paper. An example of wind pressure arrangement on structures that are not controlling the wind effect has presented. Also, an example of sunshadeds construction analysis has used in specific climatic conditions (temperature and air density) is, in fact, based on an example of four-leaf roofs, using known patterns for calculating the external and internal wind pressure in order to determine the load from the wind. The purpose of this research is simplification, because experimental research, besides being expensive, is often very time-consuming and technologically demanding. Therefore, it has been shown here that using the symmetric model simplified the construction analysis procedure, the time period of production is shortened, and the obtained results are reliable and can be used in the next research in the field of wind impact on machine contructions. In addition, the obtained analytical consistency of the wind pressure allows mechanical constructions to be calculated both analytically and numerically.

References

[1] Gjetvaj G 2006 *Rukopis predavanja iz predmeta Hidraulika-Djelovanje vjetra*, University of Zagreb, Faculty of Civil Engineering

[2] Čaušević M, Špalj I and Žic E 2008 Djelovanje vjetra na mostove prema europskoj normi, *Građevinar*, **60**(1), 21-35

[3] Kovačević D and Bjelanović A 2011 *Analiza lučnog drvenog mosta posebne konstrukcije*, In book: Zbornik radova Sveučilišta u Rijeci - Manualia Universitatis Studiorum Fluminensis, Građevinski fakultet Sveučilišta u Rijeci

[4] Rajčić,V, Mandić A and Čizmar D 2009, *Djelovanje snijega i vjetra na nosive konstrukcije*, Zavod za konstrukcije - Građevinski Fakultet Sveučilišta u Zagrebu, Zagreb

Failure diagnosis and prognosis of sliding bearings

R Antunović[1], N Vučetić[1] and A Halep[2]
[1]University of East Sarajevo, Faculty of Mechanical Engineering, Vuka Karadžića 30, 71123 East Sarajevo, Bosnia and Herzegovina
[2]Cement Factory, Selima efendije Merdanovića 146, 72240 Kakanj, Bosnia and Herzegovina

E-mail: rankoantunovicmf@gmail.com

Abstract. Sliding bearings are widely used machine elements from which the correct operation depends on the availability of many technical systems and great attention is paid to the study of the prediction method for sliding bearings. Problems that occur with sliding bearings lead to high levels of vibration and noise. One of the important parameters in the diagnosis and prediction of the failure of sliding bearings is the thermal analysis. In addition to these parameters, frequent vibration spectrum analysis, oil film thickness and other relevant parameters are often used. Using advanced technologies, it is possible to significantly reduce the duration of measurement, and on the other hand increase the accuracy of diagnostic methods. Phase logic is a mathematically formalized mode of representation modeling uncertainty. In this paper, a selection of damage in the operation of a sliding bearing has been made and the way of diagnosing these problems is shown. Methods that integrate several diagnostic parameters into one show reliable methods of early detection of slip bearing damage.

1. Introduction

The predictive concept of maintenance consists of the use of non-destructive techniques for the very early detection of potential failures, so appropriate corrective actions, that is, repairs are only planned for the time when there is a real need for it. Proactive maintenance involves taking corrective actions with the emergence of the basic causes of degradation and failure, and based on the appearance of symptoms of the resulting damage and upcoming failures or on the basis of the plan, regardless of the real state of the system. In Figure 1, a comparison of the concept of a predictive and proactive approach to maintenance is presented.

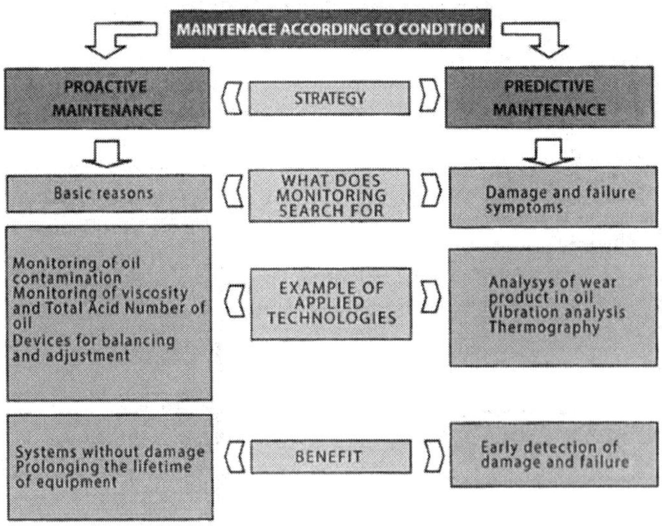

Figure 1: Comparison of predictive and proactive approach to maintenance.

Small problem in process oriented complex production system, as sliding bearing failure, can often cause long deadlocks during plant working, which results to huge financial costs in company business [1]. Monitoring and diagnostics sliding bearings failure are given in the literature [2]. Plain or sliding bearings are lubricated by the formation of a hydrodynamic film of lubricant, where the wedge formed lifts the shaft or journal off the bearing [3].

Basic causes that cause damage and failure of sliding bearings include many aspects of construction, material selection, materials mistakes, production and processing, assembly, control, testing, storage, transportation, maintenance, unforeseen exposure to overload, direct mechanical or chemical damage during operation [4]. Often, multiple causers contribute to the sliding bearing failure. The frequency of the individual causers given in Table 1 was obtained by monitoring 530 cases of sliding bearing leakage [5].

Table 1. Causers and frequency of sliding bearing failures.

Failure causer	Failure frequency, %
Errors during production:	23,4
• inadequate budgeting,	9,1
• material mistakes,	3,6
• irregularities during production and installation.	10,7
Errors during exploitation	39,1
Wear during long-term operation	30,5
Other	7,0

Failures and damages caused by said groups of causers are manifested, most often, as wear, breakage and plastic deformation of the material. These are, at the same time, the basic types of failures that can be divided into two categories, relative to material properties. One that is connected and dependent, above all, on the strength of the material and the other which is the function of tribological processes on the bearing-shaft ground surfaces. Fracture and plastic deformation are damages in the function of strength, while wear and tear with all their manifestations are related to tribological processes.

2. Selection of Sliding Bearings Failure

There is general agreement that wear is the most frequent manifestation of failure in sliding bearings, as confirmed by the authors [6, 7]. The results show that abrasive and adhesive wear is the most present, then wear due to surface fatigue, while other types of wear are considerably less represented. These are, also, indicators that need to be taken into account when defining the bearing condition monitoring program during exploitation. Table 2 shows few sliding bearings failure mechanisms.

Table 2. Sliding bearings failure mechanisms.

Failure type	Appearance	Cause	Manifest	Solution
Abrasion	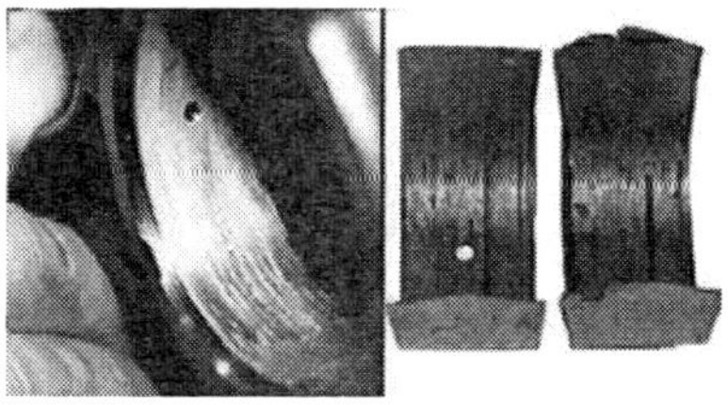	Presence of solid impurities	Visible rolling of the bearing surface	The bearing damaged by abrasive wear should be dismantled and a new one installed

IOP Conf. Series: Materials Science and Engineering **659** (2019) 012018 doi:10.1088/1757-899X/659/1/012018

Wear and tear		Inadequate gap Inadequate supply of lubricants or inadequate supplies characteristics of lubricants	Frequent color surface change due to heating	The bearing can be used again if the causes are removed
Surface fatigue	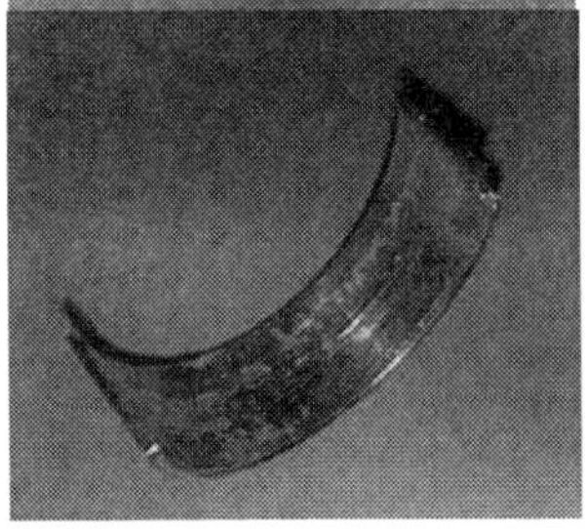	Dynamic loading of sliding bearings	Cracks occurrence	Eliminate the cause of surface fatigue
Cavitation erosion	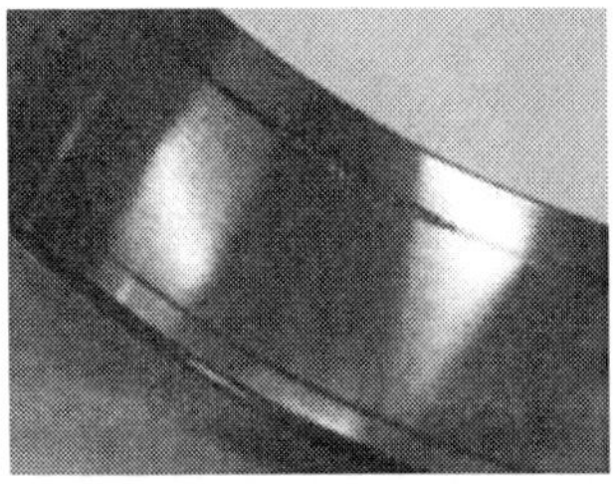	Fluctuation of pressure in the bearing, vibration of the shaft, inadequate flow lubricants through holes and channels	The occurrence of material destruction is always at the same or similar bearing location	Improving the bearing construction by increasing the oil pressure in the lubrication system
Corosion	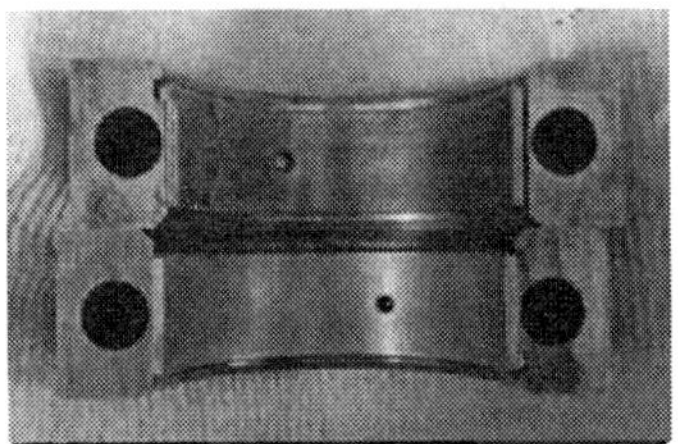	Corrosive action of the material	Visible corroded surfaces of materials	Changes in lubricant characteristics
Plastic deformation	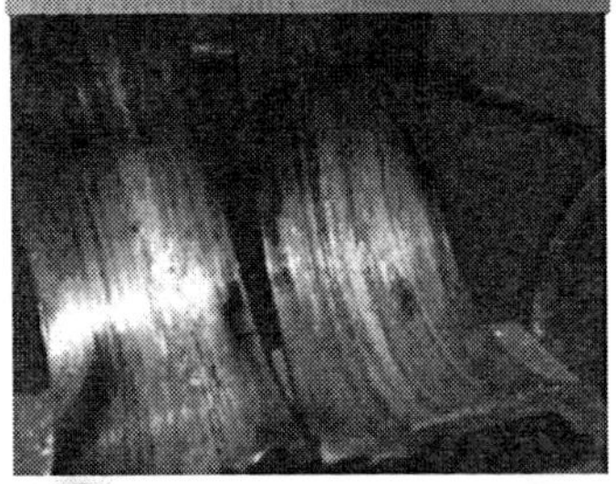	Irregularities in the lubrication system, inadequate lubricant layer	Roofing of the bearing material on the surface and plastic streaming	Changes in lubrication system or lubricant characteristics
Fracture		Overload, impact load	Loss of bearing material from larger surfaces	Removing possible overloads or expressive impact loads

3. Diagnostics of sliding bearing damage

Problems that occur in sliding bearings lead to high levels of vibration and noise. These problems mainly arise as a result of an inadequate gap in the bearing (oil gap) or the appearance of oil instability. An excessive gap in the bearing results in looseness and irregular lubrication.

A sliding bearing with excessive grip, which is mainly caused by the wearing of white metal, causes a relatively small imbalance, a disturbance of centricity or the appearance of some other disturbing force that will cause mechanical looseness, Figure 2.

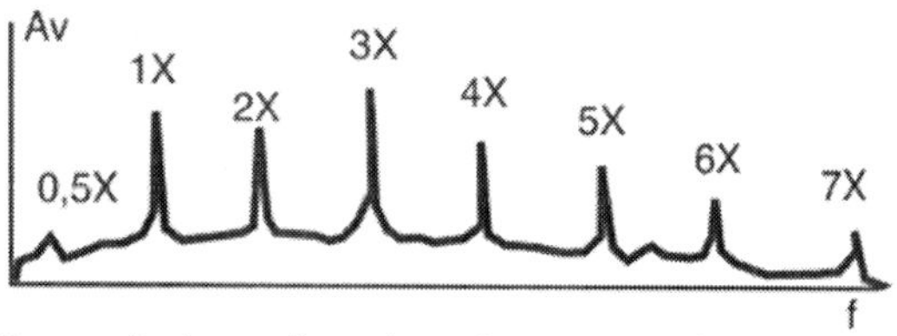

Figure 2. Spectral view when there is an excessive gap in the sliding bearing.

Oil instability is manifested in the appearance of oil vortex and oil whip. The oil vortex occurs as a result of the orbital motion of the rotor and occurs at a frequency that is proportional to the rotation frequency and amounts usually 0.4 to 0.5 from the main frequency, Figure 3.

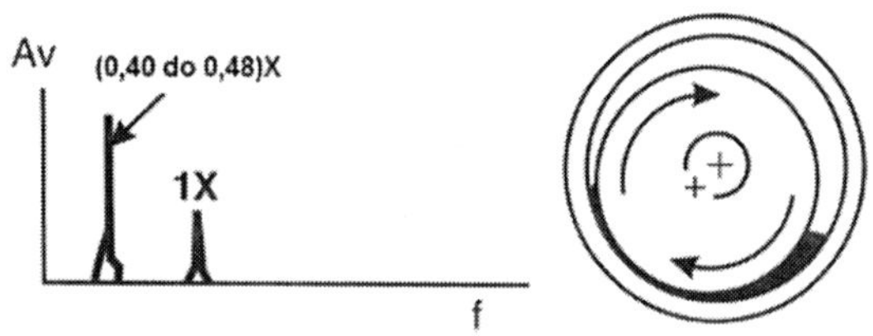

Figure 3. Spectral display due to oil instability.

In machines that operate above the first critical speed, the oil vortex usually passes into an oil whip which frequency is equal to the frequency in the formation of the same.

The smallest rotational speed in which fluid instability occurs, and this is theoretically in the case of zero-eccentricity, is called the threshold of stability. In actual rotary machines, the eccentricity is significantly greater than 0 (usually from 0.4 to 0.6). However, due to the action of one-way forces on the rotor (such as force due to the non-centering of the coupling, fluid forces due to the working fluid, the electric forces, etc.), the rotor shaft can be brought into the bearing axis and then the mean eccentricity approaches to zero. These problems are best seen by monitoring the cascaded spectral display during the machine slip as in Figure 4 [13].

Bearing temperature measurement is necessary to determine possible overheating of the bearing. The cause of overheating of the sliding bearing can be poor lubrication, insensibility, overload, and also the bearing can be overheated due to the presence of an external heat source.

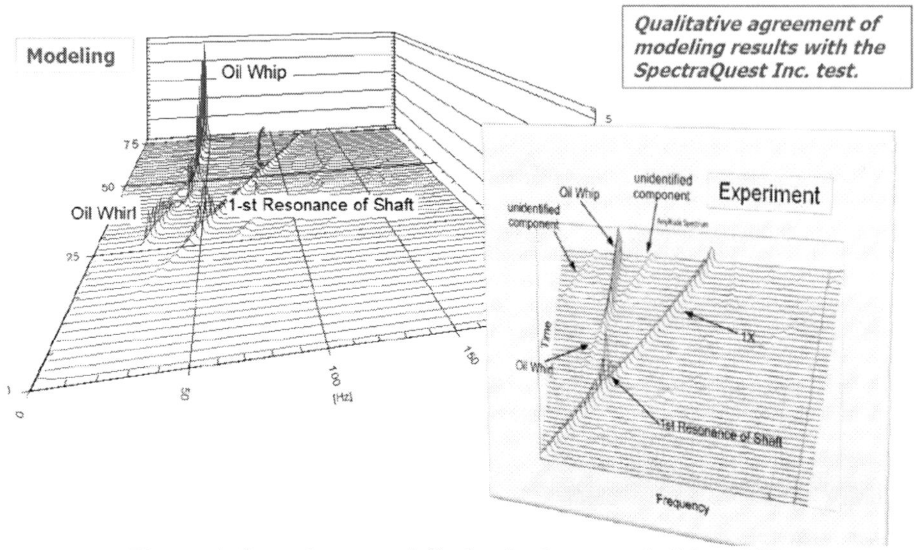

Figure 4. Cascade spectral display in the case of oil instability.

4. Predicting of sliding bearings failure

In order to reliably determine the condition and prediction of sliding bearing failure, it is necessary to observe and analyze simultaneously several parameters of the condition. One such method is based on the fuzzy logic and enables the diagnostics of the state of the element and determining the urgency of the need for intervention on the machine element. Thus, by applying the disjunctive probabilistic fuzzy operator, a vibration-thermal indicator of malfunction of the sliding bearings (defect factor DFJB) can be defined which contains information about both the temperature and the vibration of the bearing defined as [17]:

$$DFJB = (x(q) + w(v)) - x(q) \times w(v) \tag{1}$$

If the temperature and vibrations are within the allowed limits, i.e. below the warning limit, then the value of the DFJB indicator = 0, and if at least one of the parameters (temperature or vibration) is above the limit, then the value of the DFJB indicator = 1 in accordance with the rules for disjunction. If any of the parameters are in the fuzzy warning area, then the indicator of the malfunction in the fuzzy area. It should be noted that in this factor of malfunction, besides the level of absolute and rotor vibrations in the analysis, other parameters such as the vibration spectrum or the thickness of the oil film can be included, which depends on the need for diagnosis.

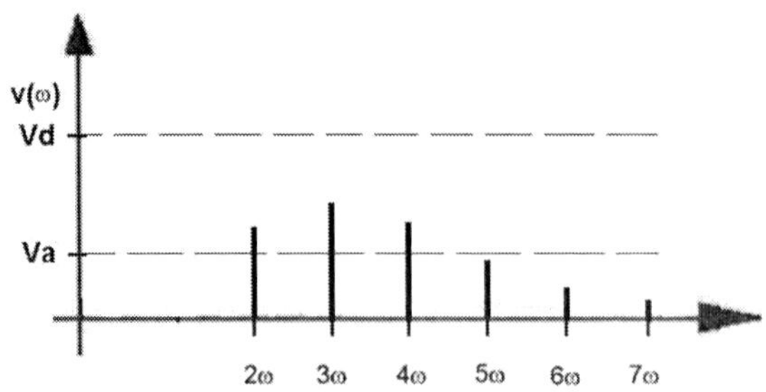

Figure 5. Vibration Limits Expressed for Harmonic Components [12].

When it comes to considering the frequency spectrum of vibrations then it is necessary to take into account the fact that the failure of the slide bearing in the frequency spectrum of vibrations is manifested in the form of higher harmonic components of the base frequency ω as illustrated in Fig.

Therefore, they take into consideration the more harmonic component of the frequency 2·ω, 3·ω, 4·ω, etc. and for the amplitudes of these harmonic components, the limits v_a and v_d are given in the same way as they are set for total vibrations.

The basic harmonic component of the frequency ω is not taken into account due to the fact that the imbalance of the rotor, which is always present, to a greater or lesser extent, leads to the appearance of this harmonic component in a spectrum with an amplitude that is significantly large. By processing harmonic components using the disjunctive probabilistic phase of the operator, a frequency spectrum is obtained on the basis of which a much more precise diagnosis can be made, since the effects of the interference signal are eliminated.

5. Case Study

The first example. The electric power mill of the raw material of 1.6 MW has sliding bearings with a diameter of the shaft 180 mm, Figure 6.

Figure 6. Electric motor of 1.6 MW.

By measuring the temperature and absolute vibration velocity in the vertical direction on the housing of said sliding bearings, absolute vibrations at the housing v = 9.3 mm / s and temperature Θ = 66 °C were measured on the first bearing. For concrete application, the vibration limits were va = 4 mm / s and vd = 12 mm / s, and the temperature limits were Θa = 60 °C and Θd = 90 °C. After the phasisation of the measured values of the vibrational speed and temperature is carried out, the following is obtained:

$$x(\Theta) = (66 - 60) / (90 - 60) = 0,20 \tag{2}$$

that is

$$w(v) = (9,3 - 4) / (12 - 4) = 0,66. \tag{3}$$

By calculating the vibration-thermal indicator of malfunction, we obtain:

$$DFJB = (0,20 + 0,66) - 0,20 \cdot 0,66 = 0,73. \tag{4}$$

It can be noted that the value of the vibration-thermal indicator is greater than the two-phase variables x and w. After the bearing was dismantled, significant damage was observed on the lower bearing hemisphere which is visible in Figure 7. Thus, this example demonstrates the applicability and reliability of the vibration-thermal indicator of the defect of the sliding bearing.

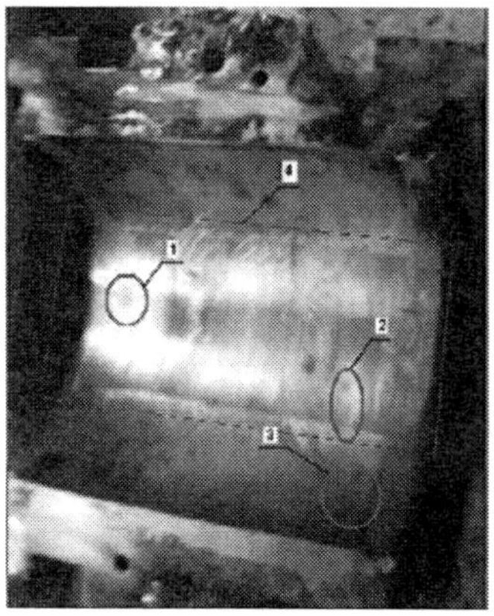

Figure 7. Damage to the bottom half of the sliding bearing.

The frequency spectrum of absolute vibrations on the housing of the bearing is given in Figure 8 and the shape of the frequency spectrum characteristic of the bearing wear can be seen. In other words, the form of the FFT spectrum of Figure 8 indicates that this is a worn-out bearing.

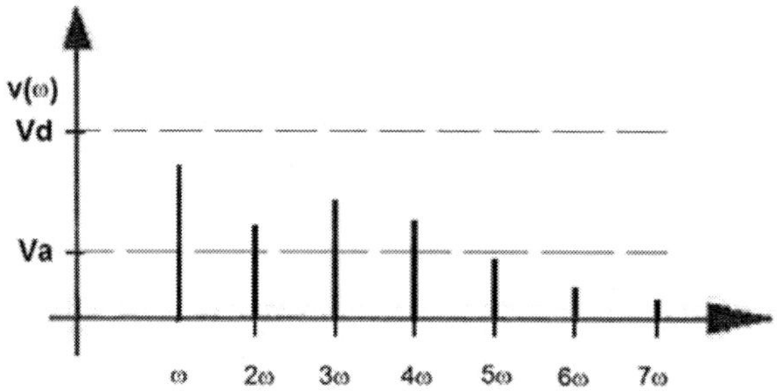

Figure 8. Frequency spectrum of absolute vibrations measured on the housing.

The frequency spectrum of the vibration-thermal indicator of DFJB defects obtained by processing harmonic vibration components using the disjunctive probabilistic phase of the operator is given in Figure 9 and it can be observed that this is a spectrum characteristic of the sliding bearing failure which, by its form, corresponds to the FFT spectrum of Figure 8, which means that the frequency spectrum of the vibration-thermal indicator of the defect of the DFJB can also be used to identify the type of malfunction of the bearing in the same manner as well as the frequency spectrum of absolute vibrations.

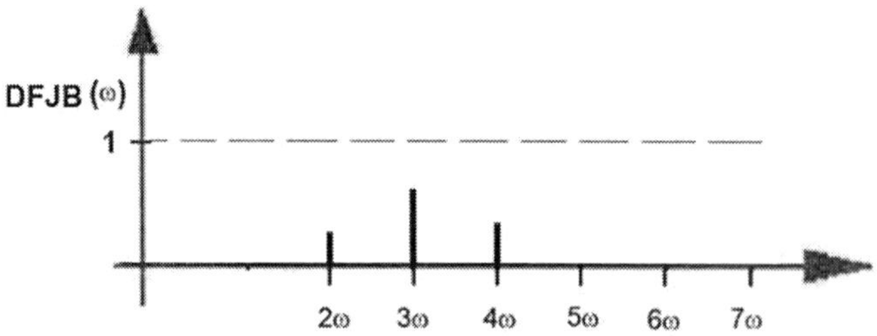

Figure 9. Frequency spectrum DFJB.

The second example. Electric motor of receptracer with power of 18.5 kW, sliding bearings are on 50 mm shaft diameter, Figure 10.

Figure 10. Electric motor of 18.5 kW.

The nominal speed of the electric motor is 1456 rpm. By measuring the temperature and absolute vibration velocity on the housing of said sliding bearings on the second bearing, the absolute vibrations measured on the housing v = 2.8 mm / s and the temperature Θ = 35 °C. For the concrete application of the bearing, the vibration limits were va = 1.5 mm / s and vd = 4 mm / s, and the temperature limits Θa = 40 °C and Θd = 60 °C. After the phasisation of the measured values of the vibrational speed and temperature is carried out, the following is obtained:

$$x(\Theta)=0 \text{ i } w(v)=(2{,}8-1{,}5)/(4-1{,}5)=0{,}52. \quad (5)$$

The value of the vibration-thermal indicator of malfunction is:

$$DFJB=(0+0{,}52)-0\cdot 0{,}52=0{,}52. \quad (6)$$

After the bearing was disassembled it was observed that the bed of the bearing (bearing shaft) was significantly worn out. Thus, this example demonstrates the applicability and reliability of the vibration-thermal indicator of the defect of the sliding bearing.

The third example. Cement mill, built-in sliding bearings with a diameter of 1554 mm. The nominal speed is 15.7 rpm and the mill diameter is 4.1 m. The mill is shown in Figure 11.

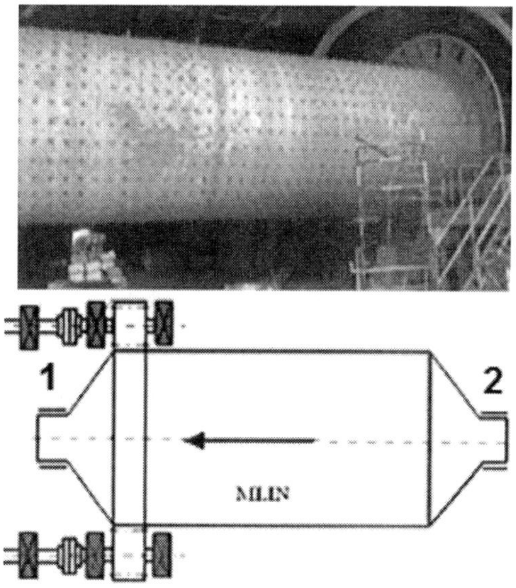

Figure 11. Cement mill.

The lubrication of the bearings during startup is performed hydrostatically, using a high pressure pump of 200 bar, and after starting, a hydrodynamic lubrication is carried out with a low pressure pump of 2 bar. By measuring the temperature and the absolute vibration velocity on the housings of said sliding bearings, the absolute vibration in the horizontal direction was measured on the second bearing, at the housing v= 16 mm/s and the temperature $\Theta=81^0$C. For the concrete application of the bearing, the vibration limits were va = 18 mm / s and vd = 30 mm / s, and the temperature limits were Θa = 50 °C and Θd = 90 °C. After the phasisation of the measured values of the vibration and temperature velocity is carried out

$$x(\Theta)=(81-50)/(90-50)=0{,}77 \text{ i } w(v)=0. \tag{7}$$

The value of the vibration-thermal indicator of malfunction is:

$$DFJB=(0{,}77+0)-0{,}77\cdot 0=0{,}77. \tag{8}$$

After analyzing the cause of abnormal warming, it was found that the pressure of the oil pump, due to the wear of the rotor, fell below the required level. It can be noted that in this example the vibration-thermal indicator of malfunction pointed to the present fault, which means that this example also proves the applicability and reliability of the vibration-thermal indicator of the defect of the sliding bearing.

6. Conclusion

Failures and damage of sliding bearings are manifested most often as wear, fracture and plastic deformation of materials. These are, at the same time, the basic forms of failure, which can be divided, in relation to the characteristics of the material, into two categories. One that is connected and depends primarily on the strength of the material and the other that is the function of tribological processes on the coupled surfaces of the housing-shaft.

Fracture and plastic deformation are damage in the function of strength, while wear with all its manifestations is related to tribological processes.

By applying the defined vibration-thermal indicator of the defects of sliding bearings (factors of malfunction), the procedure for diagnosing sliding bearings is considerably simplified, because the two most important technical indicators (temperature, vibrations) are integrated intelligently into a single technical indicator of malfunction. This is particularly important when it comes to complex systems with a large number of measurement points where there is often an overload of operators.

Vibration-thermal malfunction indicator (fault factor) DFJB contains processed temperature information and bearing vibration, but in some applications, other than temperature and vibration, other parameters such as, for example, the thickness of the oil bearing film or the strength of the ultrasonic bearing, etc., which in some cases would increase the level of reliability of the method shown for the prediction of the sliding bearing failure.

References

[1] Antunović R and Halep A 2014 *Defekt faktor kliznih ležaja*, Konferencija "Održavanje 2014", Zenica BiH, 97-101

[2] Rac A 2003 *Praćenje stanja i dijagnostika otkaza kliznih ležaja*, Osma internacionalna konferencija o tribologiji, Beograd 8-10, 271-275

[3] Swan P G 2006 Studies in plain bearing failures, Technical Analysys, *Tribology & Lubrication Technology*, 28-31

[4] Bartz W 1976 The Influence of Lubricants on Failures of Bearings and Gears, *Tribology*, **9**(5)

[5] Grupa autora 1978 *Handbook of Loss Prevetion*, Springer-Verlag

[6] Lamjua P 2006 *tribological behaviour study of sliding bearing of leaded red brass,* Bangkok (Thailand): King Mongkut's Institute of Technology North Bangkok, Doctoral Thesis

[7] www.neotehnika.co.rs (accessed 30.07.2018)

[8] Rac A 2003 *Praćenje stanja i dijagnostika otkaza kliznih ležaja*, Proceedings of the 8th International Tribology Conference, Belgrade, Serbia 8-10 October, 271-275

[9] Antunović R 2011 *Rotor dynamic models with sliding bearings,* Internacional Scientific-Professional Symposium Infoteh-Jahorina, March 2011, **10**(C-7) 268-272

[10] San Andrés L S, Rivadeneira L, Gjika K, Groves C and LaRue G 2006 A virtual tool for prediction of turbocharger nonlinear dynamic response, *ASME Paper* GT2006-90873

[11] Bartz W 1976 The Influence of Lubricants on Failures of Bearings and Gears, *Tribology International* **9**(5) 213-224

[12] Radil K and Zeszotek M 2004 An experimental investigation into the temperature profile of a compliant foil air bearing, *Tribology Transactions* **47**(4) 470-479

[13] Tran V T and Yang B S 2010 Machine fault diagnosis and condition prognosis using classification and regression trees and neuro-fuzzy inference systems, *Control and Cybernetics*, **39**(1) 25-54

[14] Dadouche A and Conlon M 2016 Operational performance of textured journal bearings lubricated with a contaminated fluid, *Tribology International* **93** 377–389

[15] Singla A, et al. 2014 experimental determination of temperature and pressure profile of oil film of elliptical journal bearing, *International Journal of Advanced Mechanical Engineering* **4**(5) 469-474

[16] Antunović R, et al. 2018 Mathematical model for temperature change of a journal bearing, *Thermal science*, **22**(1A) 323-333

[17] Antunović R, et al. 2018 vibration and temperature measurement based indicator of journal bearing malfunction, *Tehnički vjesnik-Technical Gazette*, **25**(4) 991-996

[18] Antunović R and Veg A 2014 Analysis of dynamic behavior of rotating machines, *Annals of Faculty Engineering Hunedoara- International Journal of Engineering Romania*, **9**(2) 205-210, Tome XII

IOP Conf. Series: Materials Science and Engineering **659** (2019) 012019 doi:10.1088/1757-899X/659/1/012019

Examples of kinematic analysis of complex mechanism using modern software applications

M Imamović[1,2], F Hadžikadunić[1], A Talić-Čikmiš[1] and A Bošnjak[1]

[1]University of Zenica, Faculty of Mechanical Engineering, Fakultetska 1, 72000 Zenica, Bosnia and Herzegovina
[2]ArcelorMittal Zenica, Bulevar Kralja Tvrtka 17, 72000 Zenica, Bosnia and Herzegovina

E-mail: hfuad@mf.unze.ba

Abstract. Analysis of planar or spatial mechanisms as segments of real complex systems and machines in the domain of kinematics, kinetostatic, dynamic, functional and other aspects of analysis can be done by analytical methods and by applying contemporary software applications. This paper shows some examples of complex mechanism analysis using SolidWorks software, which provides a special module for quick analysis with a visual representation of characteristic data mechanism for different members position as a diagrammatic representation of certain parameters throughout the observed motion cycle. Such representations are very useful in designing mechanisms in the manner of analysis or synthesis. Analysis of characteristic points of moving members of the mechanism with the presentation of its path during the motion cycle, the analysis of velocity and acceleration vectors of members' moving points, forces on the members of the mechanism, definition of the moment or force on the drive member of the mechanism, the path of the executive member of the mechanism, the possibility of collision in the movement of the members of the mechanism, the geometry of the members of the mechanism based on the required path of the movement of the executive member of the mechanism, etc., are only some of the possible areas of analysis.

1. Introduction

Realistic machine systems and devices for manufacturing, transportation, service and many other uses are mostly complex planar or spatial type mechanisms that incorporate a number of kinematic-bound elements that have a drive, transfer, or executive character in performing a complex task. These include, for example, lifting and lowering systems for aeroplane wheels and cargo platforms, machine mechanisms for sorting large quantities of small miscellaneous products, mechanisms for opening and closing the door of the lifts, mechanisms for movement and securing the force at the production machinery of various uses, mechanisms of robotic complex systems, operating and executive elements of construction machinery, etc.

In the process of analyzing existing solutions for optimization and improvement, or in the process of creating new conceptual solutions of complex mechanisms, the kinematic analysis of characteristic points of executive members of the mechanism, the analysis of the motion of moving points, the analysis of the driving forces or forces on the executive and other members are some of the important segments, where the use of an analytical approach to a vector calculations, the application of analysis by the relative velocity and acceleration method, the use of the current centres method, etc., are very significant, [1,2,3].

Certainly, contemporary software solutions for different purposes, including these, derive from the needs of the real industry for the purpose of realizing reliable and rapid analyzes, and in order to create optimal solutions, according to the default goal function.

This paper will present examples of application of software and its modules for kinematic analysis of mechanisms that provide a quick analysis with visual representation of characteristic data for the positions of one or more members of the mechanism or for the overall cycle of the movement of the mechanism.

When applying software solutions, the basis of the overall analysis is to create a 2D or 3D model of all the components of the mechanism, then connect the component kinematic links that limit the movement, or reduce the number of degrees of freedom of the mechanism. The degree of model approximation to the actual situation is higher in more demanding systems than in a simpler software, which increases hardware requirements. Certainly, the creation of members of the mechanism in terms of geometric characteristics, the type of connection between them, and the assignment of driving characteristics to one of the members of the mechanism are features that have a crucial effect on the overall performance of the overall mechanism and its functionality.

Certainly, after kinematic analysis, for realistic systems it is necessary to carry out kinetostatic and dynamic analysis, especially in those where speeds, accelerations and masses of system members are significant, as well as in a system that exerts significant forces on executive and transmitting members, e.g. in manufacturing machines for deformation technology, etc.

Benefits of using software packages, among others, include: visualization of the movement of the mechanism in the virtual environment with all the details, quick data processing and obtaining output data for various input parameter combinations, the ability to use output data for different applications, and the possibility of direct transfer to dynamic analysis.

2. Examples of some software solutions for complex mechanism analysis

There are various softwares that allow for the creation and analysis of complex mechanisms, as well as very detailed analysis of all member movements in characteristic points, individual positions of the mechanism members, or for total movement cycles. Some of them are: SolidWorks, MechDesigner, SAM - the ultimate mechanism designer, Linkage, Pro/Engineer, etc., [1, 4, 7].

2.1. Linkage - mechanism designer and simulator

It is a computer program that allows the user to create and edit planar-based mechanisms and simulate their movement. This program cannot perform detailed motion analyzes, such as determining velocity and acceleration of individual members of the mechanism, but is extremely useful in cases where it is necessary to see how each member of the mechanism is connected and what their mutual mobility is. By entering different input parameters and relationships between members, it is possible to determine the function of movement and make appropriate changes to improve the functioning of the mechanism itself. Also, in this way, it is easy to check whether the dimensions and path of the mechanism are obtained by analytical or graphical method properly defined. Figure 1 shows Whitworth's mechanism with drawn paths of selected moving points. If there is a fault in the design of the mechanism or dimensions, e.g. if the length of the sliding path is not long enough and the slider will slip out of it, the program will report an error, stop with the simulation of the motion, and display the location on the mechanism where irregularities occur.

Once the members of the mechanism have been formed, it is necessary to define the driving members, as well as the input parameters such as the speed of the drive member, and the dimension of individual members.

IOP Conf. Series: Materials Science and Engineering **659** (2019) 012019 doi:10.1088/1757-899X/659/1/012019

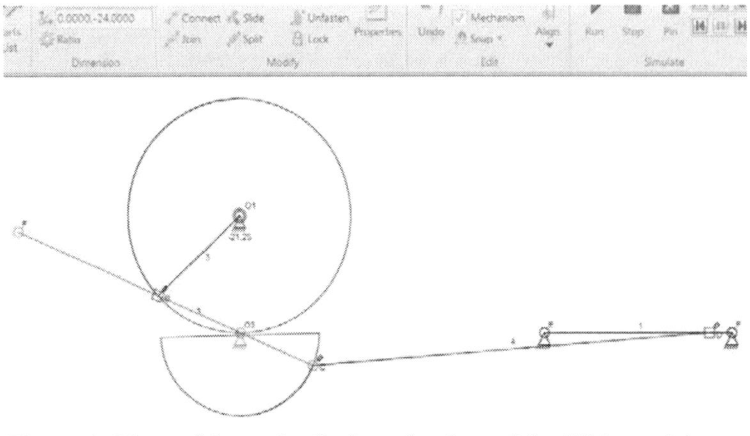

Figure 1. View of the path of selected points of the Whitworth mechanism.

2.2. SAM – the ultimate mechanism design software

The SAM software solution allows the design, analysis and optimization of planar mechanisms. The way of assembling the mechanism is very easy using basic components such as beams, sliders, gears, springs, absorbers and friction elements. This program integrates preprocessing, numerical analysis and postprocessing, which involves creating animations and xy-diagrams in a simple to use environment.

DXF import / export option allows the translation of the created concept of mechanism into any CAD program for further editing and analysis, as well as entering CAD data, for simple editing or animation of the final mechanism.

When the mechanism is created and input parameters defined, it is possible to calculate any kinematic size including the member position, shift, velocity, acceleration, angles, angles of velocity and acceleration, figure 2. Additionally, the SAM program also provides dynamic analysis.

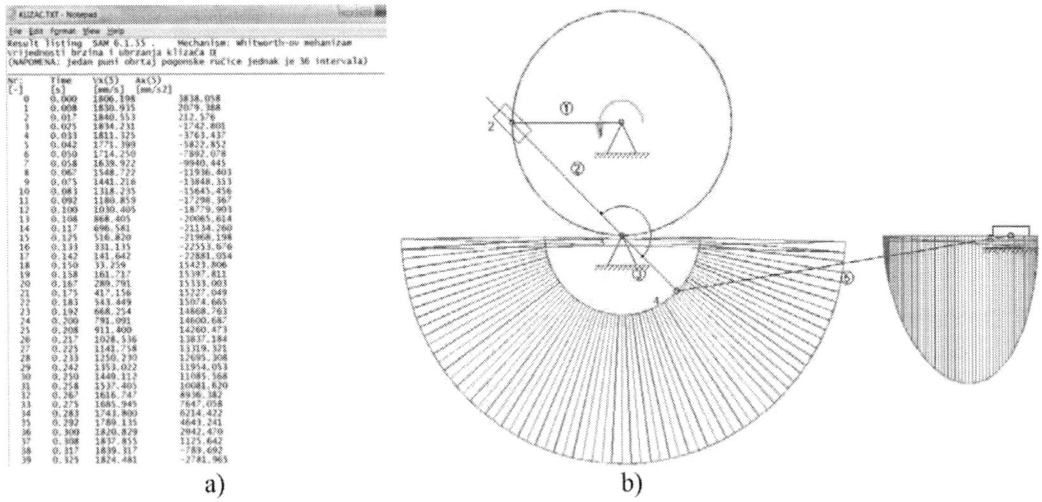

Result listing SAM 6.1.55 . Mechanism: Whitworth-ov mehanizam
Vrijednosti brzina i ubrzanja klizača [illegible]
(NAPOMENA: jedan puni obrtaj pogonske ručice jednak je 36 intervala)

Nr. [-]	Time [s]	Vx(5) [mm/s]	Ax(5) [mm/s2]
0	0.000	1806.198	3838.058
1	0.008	1830.935	2079.388
2	0.017	1840.553	212.576
3	0.025	1834.231	-1742.803
4	0.033	1811.325	-3763.437
5	0.042	1771.399	-5822.852
6	0.050	1714.250	-7892.078
7	0.058	1639.922	-9940.445
8	0.067	1548.722	-11936.403
9	0.075	1441.216	-13848.333
10	0.083	1318.235	-15645.456
11	0.092	1180.859	-17298.367
12	0.100	1030.405	-18779.903
13	0.108	868.405	-20065.614
14	0.117	696.581	-21134.260
15	0.125	516.820	-21968.198
16	0.133	331.135	-22553.676
17	0.142	141.642	-22881.054
18	0.150	33.259	15423.806
19	0.158	161.717	15397.811
20	0.167	289.791	15333.003
21	0.175	417.156	15227.049
22	0.183	543.449	15074.665
23	0.192	668.254	14868.763
24	0.200	791.091	14600.687
25	0.208	911.400	14260.473
26	0.217	1028.536	13837.184
27	0.225	1141.758	13319.321
28	0.233	1250.230	12695.308
29	0.242	1353.022	11954.053
30	0.250	1440.112	11085.568
31	0.258	1537.405	10081.620
32	0.267	1616.747	8936.382
33	0.275	1685.945	7647.058
34	0.283	1743.800	6214.422
35	0.292	1780.135	4643.241
36	0.300	1820.829	2942.470
37	0.308	1837.855	1125.642
38	0.317	1839.317	-789.692
39	0.325	1824.481	-2781.965

a) b)

Figure 2. Some of Whitworth's mechanism analysis results display: a) Tabular display of velocity and acceleration results; b) Display of the speed of the selected points.

2.3. SolidWorks

SolidWorks is 3D CAD (Computer Aided Design) and CAE (Computer Aided Engineering), a Dassault Systèmes program. According to statistics of companies that commercially offer this software, this program currently utilizes over 2 million engineers and designers, and more than 165.000 companies around the world. SolidWorks uses the principle of 'solid modelling' and parameter access to model design and 'assembly' options for complex systems. The software is written on the Parasolid kernel.

SolidWorks in options of SolidWorks Standard, Premium, and Professional configuration software contains a significant number of modules for the corresponding tasks:

- Engineering Data Management: SolidWorks Enterprise PDM,
- Engineering calculations: SolidWorks Simulation Professional, SolidWorks Simulation Premium, SolidWorks Flow Simulation,
- Electrical design: SolidWorks Electrical,
- Creating an internet documentation: SolidWorks Composer,
- Machining, CNC: CAMWorks,
- System Check: CAMWorks Virtual Machine,
- Quality control: SolidWorks Inspection,
- Technology Analysis: SolidWorks Plastics, DFM, etc.,
- E-drawing Technologies: SolidWorks MBD (Model-Based Definition).

For the purpose of creating and analyzing examples from domain of mechanisms, the subforms for the creation of 3D models and SolidWorks Motion are the most commonly used.

3. Example of complex systems analysis in the SolidWorks software solution

Through this section, over selected example of the planar complex mechanism, the application of kinematic analysis of the complex mechanisms in SolidWorks software will be shown.

Firstly a 3D model (Module Assembly) must be created. In addition to the use of 3D models, kinematic or dynamic analysis of planar mechanisms, it is important to note that in the case of the use of the Motion Analysis module, a 2D drawing of the mechanism can also be used. After the creation of the mechanism assembly with the geometrical characteristics of all members of the mechanism, the supports and the interconnections of individual members, a module for the analysis of movement of the mechanism during the multi-cycle motion processes or for certain parts of particular interest can be used. Solving tasks from the domain of kinematics mechanisms depends on the geometrical parameters of the members of the mechanism, the relation between the members and the boundary conditions of the supports and movements, as well as the initial driving parameters of one of the members of the mechanism.

SolidWorks Motion can also take into account other effects because, apart from the option to add driving motors, there are options for adding other modules such as:

- Spring (definition of spring together with stiffness coefficient),
- Damper,
- Force (definition of force acting together with intensity and place of action),
- Contact (definition of contact between two surfaces with definition of friction and type of material in contact),
- Gravity (adding gravity to the whole mechanism).

After establishing calculation of motion kinematic parameters of the mechanism, structural analysis can be performed due to the force induced by motion dynamics (acceleration and force in the bonds) or external acting forces on the members of the mechanism. After the calculation of the movement of the mechanism, it is possible to obtain time-based motion diagrams, trajectory of characteristic moving points of the members of the mechanism, speed vectors and accelerations for characteristic points, using the „Results and Plots" option.

IOP Conf. Series: Materials Science and Engineering **659** (2019) 012019 doi:10.1088/1757-899X/659/1/012019

3.1. An example of a complex planar mechanism analysis

Figure 3a) shows a 3D model of examples with all the geometry parameters of the mechanism members, and Figure 3b) shows an example of using the 'Trace Path' option to obtain the path of the required moving points of the mechanism, [5, 6, 7].

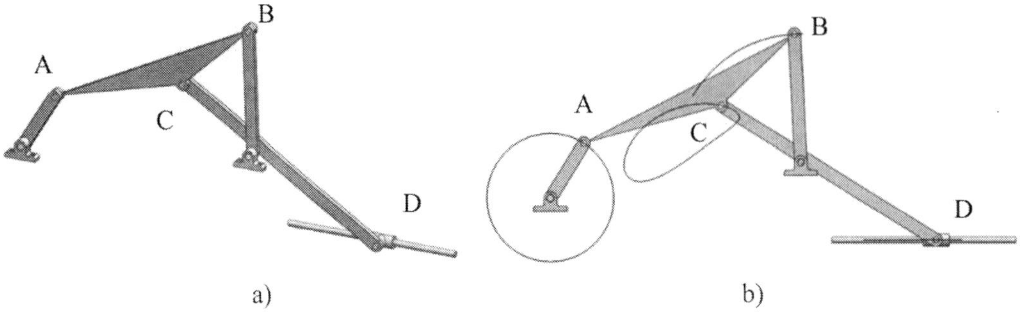

Figure 3. Planar mechanism with all members: a) 3D view of the model, b) Display of trajectory of characteristic mechanism points using the 'Trace Path' option.

Figure 4 shows the speed change diagrams for the B and C points of mechanism in the time domain.

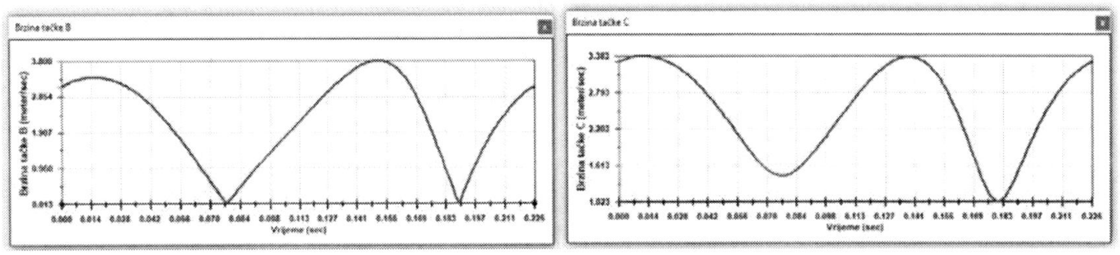

Figure 4. Speed change diagrams for individual points of mechanism in the time domain.

Figure 5 shows the acceleration diagrams for the B and C points with the ability to display the normal and tangential components of acceleration of the moving points of the mechanism.

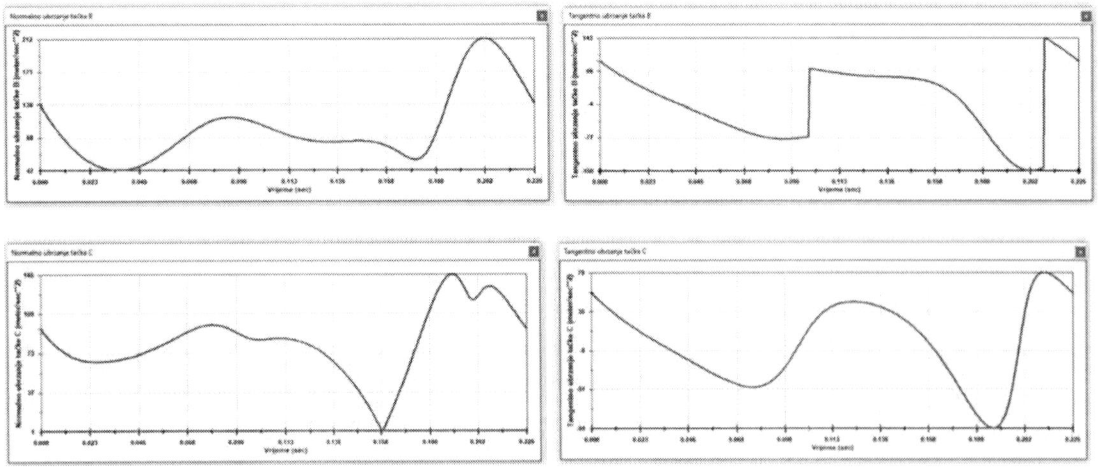

Figure 5. Change of acceleration for individual points of mechanism in the time domain with views of normal and tangential acceleration components.

Also, it is possible to create velocity and acceleration vectors that automatically change the size and direction according to the current position value in an appropriate simulation. Figure 6 shows two selected positions from the total cycle of movement of mechanism members.

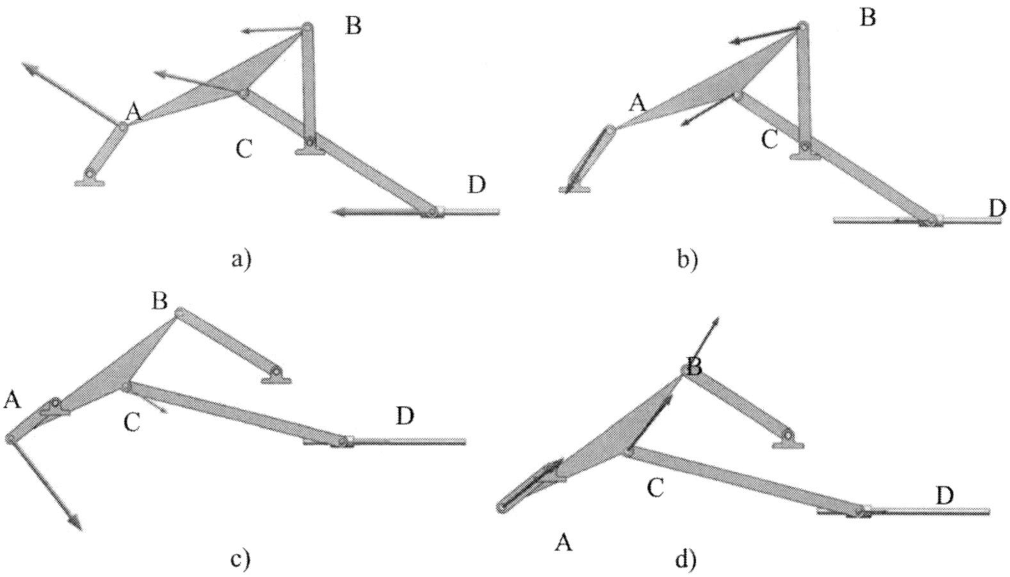

Figure 6. Two characteristic positions of the mechanism with speed and acceleration vectors.

4. Conclusions

The example presented shows the importance of possible analysis through the associated diagrams, the most important of which are the path diagrams with respect to the motion requirements of the individual points of the mechanism, and the speed and acceleration diagrams with special emphasis on the sudden changes in the speed values giving maximum acceleration, which are important for the dynamic mechanism stability. This reflects the scientific and practical application of the implemented software application methodology. In the case of analysis of mechanisms as parts of real devices or machines, such as: maximum mechanism member position, maximum speed or acceleration, motion point path, collision detection and real conduct of mechanism members motion, as well as creating an executive member of the mechanism according to the precisely defined motion curve for synthesis of mechanisms, etc. are just some essential segments that require a good understanding of vector calculation and application of software solutions.

References

[1] Imamović M 2016 *Teorija mehanizama i mašina*, IPI

[2] Zlokolica M Ž, Čavić M V and Kostić M D 2014 *Mehanika mašina*, FTN

[3] Chen Y 2003 *Design of Structural Mechanisms*, University of Oxford, Doctoral Thesis

[4] Fomin A and Ivanov W 2019 Development of a Mixing Mechanism with a Complex Motion of the End-effector, *Strojniški vestnik Journal of Mechanical Engineering* **65**(5) 319-325

[5] Vavro J, Vavro J Jr, Kovačikova P and Bezbedova R 2017 Kinematic and dynamic analysis of planar mechanisms by means of the SolidWorks software, *Procedia Engineering* **177** 476-481

[6] Vavro J, Vavro J Jr, Kovačikova P and Bezbedova R 2016 Kinematic and dynamic analysis of the manipulator for removal of rough tyres, *Procedia Engineering* **136** 120-124

[7] Leharika I A S and Bhanuprakash T V K 2018 Kinematic analysis of planar and spatial mechanisms using Mathpack, *International Research Journal of Engineering and Technology* **5**(11) 416-421

IOP Conf. Series: Materials Science and Engineering **659** (2019) 012020 doi:10.1088/1757-899X/659/1/012020

Case study on topology optimized design for additive manufacturing

A M Aliyi[1] and H G Lemu[2]

[1] Addis Ababa University of Science and Technology, Department of Mechanical Engineering, Addis Ababa, Ethiopia
[2] University of Stavanger, Faculty of Science and Technology, N-4036 Stavanger, Norway

E-mail: corresponding author: Hirpa.g.lemu@uis.no

Abstract. Additive Manufacturing (AM), also known as rapid prototyping, rapid manufacturing, layer manufacturing and three-dimensional printing, represents one of the most promising aspects of manufacturing for highly complex geometries. In particular, AM is nowadays seen as provider of possibilities to realize true design optimized manufacturing through topology optimization. Topology optimization is an approach that is considered powerful in design because it contributes to a design that can save energy, materials and time that are not economically achievable using other manufacturing processes. This paper is intended to explore the potentials of topology optimized design approach for AM in developing products that are lightweight and at the same time efficient and have load bearing capacity. A case study has been conducted to demonstrate the steps involved in topology optimization and its benefits in terms of weight reduction.

1. Introduction

Additive manufacturing (AM), also referred to as Rapid prototyping (RP), Additive layer manufacturing (ALM), 3D printing, and many other names is a new manufacturing technique that allows fast fabrication of computer models designed with three-dimension (3D) computer aided design (CAD) software. Nowadays, this technology is seen as a potential gamechanger in a wide variety of industries, from shoes to aircraft manufacturers. It employs a group of technologies capable of performing the layer by layer manufacturing processes almost completely under computer control, with little or no human intervention once the process has begun [1]. The initial concept of AM technology was to assist the design engineering in better visualization and smooth communication through rapid prototyping. In its latter developments, however, the technology transformed itself to a tool of manufacturing functional products as a result of the innovations in different printing machines, materials and printing processes. In deciding whether a process can be classified as a rapid prototyping process or not, five criteria, as specified by Burns [2], are used. These are:

1. Raw material intake is in some shapeless form such as blocks, sheets or a fluid, and the products are solid objects with a definite shape.
2. The manufacturing process takes place without a significant amount of human intervention.
3. The produced shapes have some degree of three dimensional geometrical complexity. This eliminates extrusion and cutting of tubes or rods drilling of holes.

4. The process must not involve manufacturing of part specific new tools. This eliminates all types of molding and casting, electro-discharge machining (EDM) die sinking and copy milling operations.
5. Each produced item must be a single object not an assembly of component parts. This eliminates joining operations such as gluing, welding and riveting.

Employing the above-listed criteria, some of the established fabrication techniques can be classified as illustrated in Figure 1. One of the expected benefits of AM technology is the realization of optimized design through topology optimization. This optimization technique is a mathematical tool used in conceptual design stage to reduce part weight by optimally distributing the material throughout the body. To implement topology optimization technique during the concept design stage, it is necessary to determine the goal function(s) as well as the desired constraints. Material properties, essential geometric features of the part and loading conditions are normally considered as the desired constraints in the optimization process. The optimization tool will then attempt to spread the material within the design boundaries while meeting the design requirements [3]. As a result, there exists no question about the feasibility of the design due to not taking into account the design for manufacturing requirements.

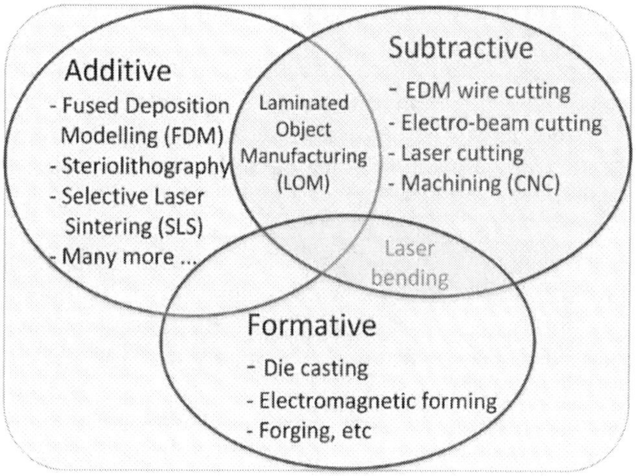

Figure 1. Classification of major component manufacturing methods.

The aim of this paper is to make a closer study of research and development works in application of topology optimization in design of parts for manufacturing using AM technology. In addition to highlighting the recent works in the area in Section 2, the paper explains the key formulation procedures involved in a topology optimization based design in Section 3, whose application has been illustrated using a case study in Section 4. Finally, the withdrawn conclusions are briefly presented in Section 5.

2. Literature review

As stated earlier, topology optimization provides the design freedom because it allows the designer to perform very flexible design optimization problems at early stage of the design process, thus, creating a design concept in terms of a general material distribution as either solid or void. The existing challenge of topology optimization is, however, to initiate design parameterization that leads to physically optimal designs. With the advances in finite element analysis (FEA), however, it has been possible to generate finite element meshes and define the meshed elements in design domain as a design variables. This allows a variation in density (homogenization, Solid Isotropic Material with Penalization (SIMP)) [4 - 7] or void-solid (bidirectional evolutionary structural optimization (BESO)) [8 - 9].

In recent years, in particular, topology optimized design has been an active topic of research from the perspective of using AM technology to manufacture functional parts with complex geometry. A study focusing on topology optimization based design approach for AM with a case study on lightweight design of jet engine bracket was also reported [10]. The angle bracket was redesigned considering four loading conditions and the work was intended to reduce its weight while satisfying all stated design requirements. Altair Hypemesh 14 Optistruct, which is a dedicated topology optimization tool, was used for the optimization purpose. The results of the study showed that topology optimization is a powerful design technique to enable reduction of the weight (65% reduction) of additive manufactured part while maintaining the design requirements.

Gardan and Schneide [11] reported optimization and fabrication of a hip joint implant interior part using AM technique. The implant has an interior pore that makes complex geometry and unachievable with conventional manufacturing techniques. Use of AM enabled manufacturing of the part with its complex geometry and allowed to find porous implants that allow the ingrowth of bone and cells, which could have been impossible if design optimization and AM were not used together in the design process. A topology Optimized Design, Manufacturability, and Performance Evaluation of Ti-6Al-4V Porous Structures Fabricated by Selective Laser Melting (SLM) was also reported by Xu et al. [12]. The porous structures were designed by using topology optimization under the condition of human skeletal stress, obtained a unit cell and constructed a series of porous structures with porosity from 40 to 80 % and unit cell size from 2 to 8 mm. The manufacturability of these parts was investigated for SLM technology. Among others, properties such as compression strength and dynamic elastic modulus were measured. The compressive behavior was explained and three failure models of porous structures were proposed.

Implementation of topology optimization approach for Fused Deposition Modeling (FDM) process is reported by Rezaie et al. [13], who investigated the issues and opportunities for the application of topology optimization methods for additive manufacturing. The topology optimization output files were converted to usable additive manufacturing input data for production and investigation of meso-scale structures for realizing intermediated density regions. Based on the implemented AM technology (i.e. FDM), a case study was redesigned, fabricated and evaluated.

Material usage for the supporting structure in AM is considered one of the major material wastages. Leary [14] introduced the idea of self-supporting designs, where the topology optimized design was altered to include features similar to support structures. Based on existing experience, support structures were introduced as design features. This alters the structural load path because the self-supporting design is introduced after topology optimization. Bracket et al. [15] have also recommended the integration of topology optimization and AM in to minimize the need for support structures. They suggested a penalization scheme on overhanging surfaces, and an edge analysis was carried out on a benchmark 2D example. The overhang constraint was suggested but not demonstrated. Based on the recommendation of Bracket et al. [15], Gaynor and Guest [16] employed a smooth heavy side approximation method to penalize overhanging surfaces within a SIMP based topology optimization. They demonstrated that, for 2D compliance minimization, this scheme changes the topology to be AM friendly. In particular, they demonstrated that it is possible to eliminate support structures by suitably changing the topology optimization process. Though the results seem encouraging, convergence issues when the overhanging penalization was imposed are reported as remaining issues. Another method proposed to reduce the need for support structure in topology optimized design is using the concept of support structure topological sensitivity [17]. This method combines performance sensitivity with the result in a topology optimization framework that maximizes performance, subject to support structure constraints. The robustness and efficiency of the proposed method was demonstrated through numerical experiments, and validated using FDM process.

The research on reduction of the need for support structure is part of the measures to reduce the material cost in AM process. Among recent efforts, the work of Wang et al. [18], who proposed a novel strategy to reduce the material cost by first extracting the frame structure of the design, can be mentioned. In this proposed design frame, solution of a multi-objective optimization problem (MOOP) was used to minimize the number of struts while accounting for stability and printability. The combined topology optimization and shape optimization is also considered [19] as a means of providing more self-supporting printing process by generating a volumetric tetrahedral mesh and mapping the overhang tetrahedra onto a Gauss sphere that is minimally rotated to a self-supported state. This method is also reported to be an effective method in finding optimal build direction.

3. Formulation of topology optimized design

3.1. Design optimization – backgrounds

Topology optimization for mechanical design is a mathematical approach implemented in a finite element mesh (the design space) and finds the tradeoffs in both minimizing material usage and maximizing part strength for a given design. The approach functions as a solver as well as a high-end analysis tool because it requires a pre-processor solver for meshing. The earlier optimization works extensively focused on size and shape optimization that are employed to manipulate design variables to improve performance of structures and/or mechanical systems. For a truss design, for instance, the design variable during a size optimization uses the cross sectional area of its members. The structure is optimized by finding the cross sectional areas that maximize its stiffness for a particular weight [5]. Shape optimization is also employed in performance enhancement of energy converters such as gas turbine blades [20] and hydropower converters [21]. The basics of shape optimization is often demonstrated when it is applied to alter the shape of holes to reduce the stress concentrations, resulting in a more structurally efficient part. The design variables in such cases are the parameters that control the shape of the holes in the original design.

Topology optimization is far more comprehensive than both size and shape optimization techniques. It is widely applied using the SIMP approach that involves modification of the stiffness matrix of the model so that it depends continuously on a function that is interpreted as a density of material. The optimal distribution of material is then obtained through making the material density as a design variable. This leads not only the optimum shapes of the model, but also the material distribution.

3.2. Topology optimized design process

Topology optimization in design may be applied to two different types of structures; namely, continuum and discrete. While the former refers to optimized design of single objects like an engine block or a turbine blade, the latter is optimization of truss-like constructions composed of many members. The topology optimization algorithms solve design problems by applying boundary conditions, design responses and constraints to a design domain. The design domain involves all possible configurations of the design. Using finite element meshes, the optimization algorithms find the optimal distribution of material and voids in the given design space, depending on the loading and boundary conditions such that the resulting structure meets the prescribed performance targets.

Figure 2 shows processes involved in a design process based on topology optimization. The process starts with the development of the original design using a 3D CAD modeling software and then the design and non-design spaces are identified (Figure 3), where the former represents the part of the model that is subjected to topology optimization. In other words, the topology optimization algorithm is "allowed" to remove materials from areas defined as the optimization design space. The non-design space, on the other hand, remains untouched by the optimizer, and all boundary conditions including loads and constraints are applied to this space. The next step involves meshing and conducting finite element analysis (FEA), which provides stress and strain distributions. The topology optimizer utilizes the stress and strain distributions for the optimization procedure. Based on the distribution of stress and strain/displacement, the topology optimizer removes materials from areas that have insignificant contribution in carrying the loads.

The material removal usually leaves an irregular geometry that needs smoothing using proper algorithms or remodeling using approximate surface of the exterior boundaries of the optimized shape. As illustrated in the figure (Figure 2), the topology optimization based design involves iterative procedures of changing the geometry of the model within a specified design domain constrained by the given boundary conditions. Provided that the design optimization is verified, i.e. it satisfies the design requirements, a physical verification model such as physical prototypes can be developed. If not, the part is remodeled until it is verified. The final design model is then developed for manufacturing using AM technology.

The key steps given in Figure 2 [10] were used to conduct a case study on a jet engine angle bracket, which was intended to demonstrate the potentials of topology optimization. With reduction of the weight of the angle bracket as an objective, the bracket was subjected to different loading conditions (horizontal, vertical and torsional loads). As reported earlier in Section 2, the topology optimization resulted in a 65% weight reduction, which is very significant for aircraft components. The step-by-step application of the procedure is illustrated in Figure 3.

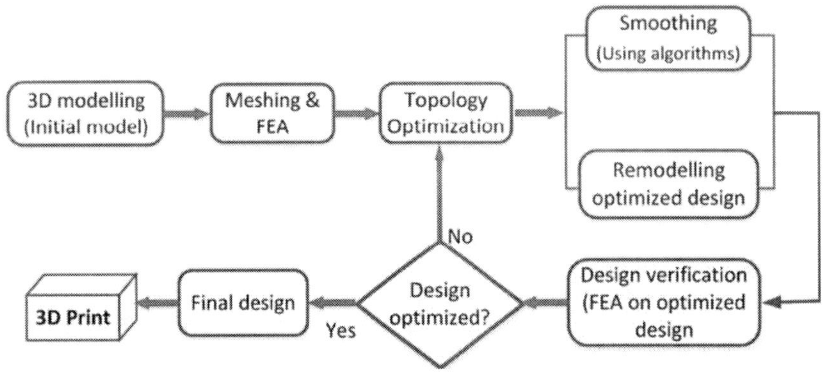

Figure 2. Topology optimization based design process.

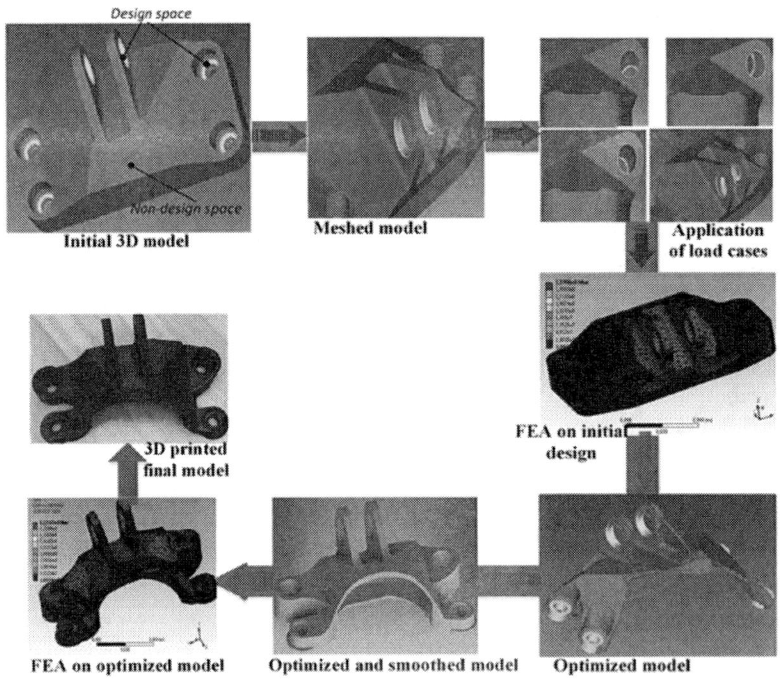

Figure 3. Sequential demonstration of topology optimization design process (Adapted from [10]).

3.3. General topology optimization problem statement

Topological optimization is sometimes referred to as layout optimization [22]. As stated earlier, its goal is to find the best use of material for a load carrying object such that an objective criterion (i.e., global stiffness, natural frequency and the like) is achieved subject to given constraints such as weight or volume reduction. In topology optimization, the material distribution function serves as optimization parameter.

To formulate a topology optimization based design optimization, five steps are commonly employed [23].

1. Developing the optimization problem statement: This involves definition of the design goals, the objective of the optimization and defined group of criteria.
2. Collecting data and information: This step involves collection of all the necessary information.
3. Identifying and defining design variables: At this step, the design variables describing the system are identified and defined.
4. Identifying the optimization criterion: This involves identification of the objective function and the stop criteria of the optimization process.
5. Identifying constraints: These represent the restrictions on the problem and can be extracted from the resources and stated performance requirements.

In general, a topology optimization problem can be formulated as:

$$\begin{aligned} &Minimize\text{: } F = \textstyle\int_{\Omega} f(u(x),x)dV \\ &Subject\ to\text{: } G_0 = \textstyle\int_{\Omega} x dV - V_0 \leq 0 \\ &\qquad G_j(u(x),x) \leq 0 \text{ with } j = 1, 2, \dots m \end{aligned} \quad (1)$$

Where $f(u(x), x)$ is the objective function, which represents the quantity that is being minimized for the required best performance. In structural optimization, for example, compliance function, which maximizes the stiffness of a structure, is often used as the objective function. $u(x)$ is the material distribution variable described by the density of the material at each location of the given volume dV. Using this variable, the optimization is defined by either 1 (presence of material) or 0 (absence of material). Ω is design space and it indicates the allowable volume within which the design value exist or valid. $G_j(u(x), x)$ define the m constraints that the solution must satisfy. Examples include maximum amount of material, which is volume constraint or the maximum stress values.

4. Case Study on a Triangular Bracket

In this section, a case study conducted on a triangular bracket or angle bracket with three holes is presented. The bracket (shown in Figure 4(a)) has arbitrarily selected dimensions of $D_1 = 2\ mm$, $H_1 = 13.9\ mm$, $H_2 = 18.4\ mm$, $L = 25.7\ mm$, $V_1 = 20.8\ mm$ and $V_2 = 25.3\ mm$. The stress distribution (von Mises) from FEA on the original design with contour plots is shown in Figure 4(b). As the contour plots indicate, the portion of the part shown in blue color represent the inefficient use of material. This indicates the probability that materials can be removed from this areas of the model because they have insignificant effect on the mechanical performance of the part in carrying loads.

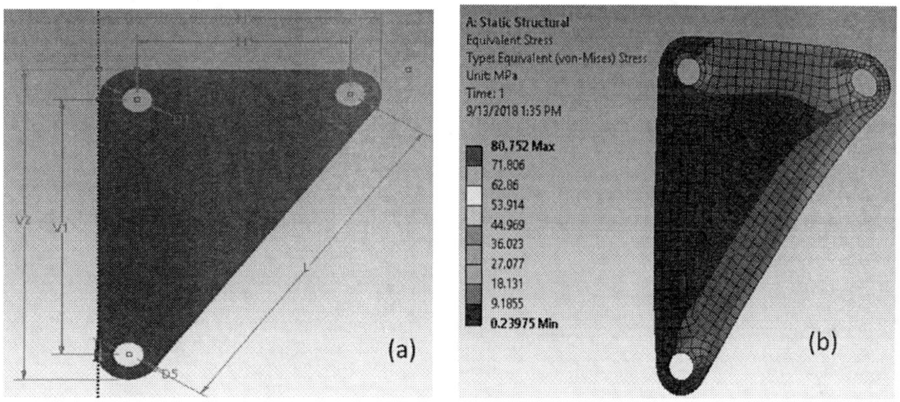

Figure 4. Original triangular bracket with three holes, (a) Dimensions and (b) FEA of the model.

In this study, the triangular bracket is redesigned using topology optimization based design approach using the three holes as the non-design spaces, where the two left side holes are fixed support and a bearing load is applied to the right hand side hole. The objective of this study is to reduce the weight of the bracket while satisfying all the design requirements related with its performance including:

- The optimized geometry must fit within the original part envelope.
- The material is structural steel with tensile yield and ultimate strength of 250 MPa and 460 MPa respectively at room temperature (20 °C).
- The wall thickness is 2.0 mm.

Figure 5(a) shows the model processed by the topology optimizer in ANSYS Workbench, where the portion that contributes to the load carrying (gray) and non-load carrying (red) are demarcated after the computation of the stress and displacement distribution on the part for load bearing. As optimized (Figure 5(b)), the geometry looks rough and very natural showing the layout of the optimized material within the design space.

To smooth the irregular surface of the optimized model to not only make it aesthetically good looking, but also to represent the geometry by solid modelling functions, i.e. curves and surfaces, for further FEA for verification, it is important to create a new model that roughly circumscribes the optimized shape. Then the new model is remeshed and followed by computation of the stress and displacement distribution under the same loading conditions as the first analysis. If required, the final part is prepared for the second round of topology optimization process as shown in Figure 6.

This final design has to be verified with the given design criteria so that the von Mises stress values for the load cases should not exceed the yield strength. For this triangular bracket, the final design satisfied the yielding condition with safety factor that varied from 1.067 to 50. The optimized design weighed only 1.188e-003 kg, compared to 4.378e-003 kg in the original design, which is a significant weight reduction of 72.86%. Though this case study is conducted on a simple geometry to demonstrate topology optimization steps and benefits with practical examples, the real benefit is the parallel use of AM technology to produce the resulting complex shape with no manufacturing constraints.

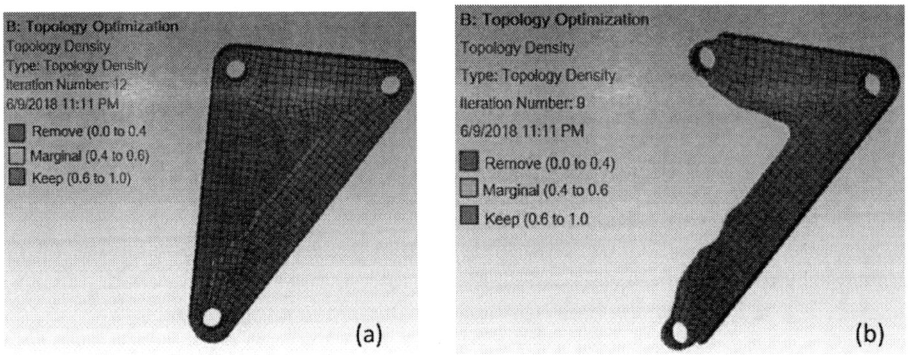

Figure 5. Optimization phase (a) Optimization region generation and (b) Design topology optimized geometry of the triangular bracket.

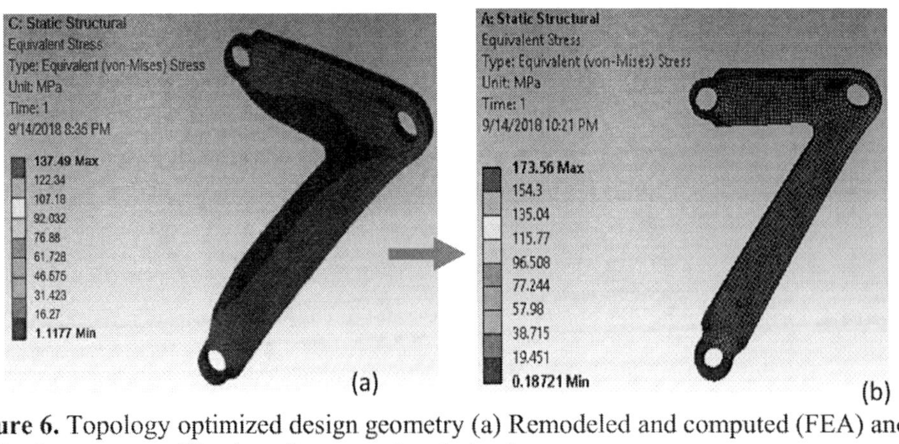

Figure 6. Topology optimized design geometry (a) Remodeled and computed (FEA) and (b) Design stress verification after second optimization.

5. Conclusions

Topology optimization based design may provide more freedom to change design than conventional design for structural products. In this era of additive manufacturing, in particular, the designers design freedom can better be realized through topology optimization that results in possibilities of fabricating parts at low cost and lightweight. However, there still exist remaining challenges in the topology optimization process to make it effective and user friendly. Among these, the existing optimization algorithms are not yet fully capable to smooth the geometry that remains after optimization, which appears natural than a model defined by modelling functions. Though the considered case for topology optimization is simple and intended only to apply the optimization procedures on a practical example, the case study reported in this paper has demonstrated that topology optimization is a powerful design concept that can reduce the weight of structural products. Among others, reduction of weight can save large amount of material, processing energy and hence cost of the product.

As demonstrated by the case study, the weight reduction of approx. 73% implies a significant achievement in terms of products that require lightweight design such as in aircraft component design. Because the finally optimized geometry can be of irregular and complex shape, it is possible to conclude that real optimized design can be realized only through application of AM technology.

References

[1] Onuh S O and Yahaya Y Y 1999 Rapid prototyping technology: Applications and benefits for rapid product development, *J. Intell. Manuf.* **10** 301–311

[2] Burns M 1993 *Automated fabrication improving productivity in manufacturing*, 1st Edition. PTR Prentice Hall New Jersey

[3] Sigmund O K and Maute K 2013 Topology optimization approaches, *Struct. Multidiscip. Optim.* **48** 1031–1055

[4] Bendsoe M P and Kikuchi N 1988 Generating optimal topologies in structural design using a homogenization method, *Comput. Methods Appl. Mech. Eng.* **71** 197 - 224

[5] Sigmund O K and Maute K 2004 *Topological optimization, theory, methods and application*, Springer Verlag, Berlin

[6] Rozvany G I N, Zhou M and Birker T 1992 Generalized shape optimization without homogenization, *Struct. Optim.* **4** 250-252

[7] Rozvany G I N 2001 A critical review of established methods of structural topology optimization, *Struct. Multidiscip. Optim.* **37** 217-237

[8] Querin O M, Steven G P and Xie Y M 1998 Evolutionary structural optimization using a bidirectional algorithm *Eng. Comput.* **15** 1031-1048

[9] Huang X and Xie Y M 2007 Convergent and mesh-independent solutions for the bi-directional evolutionary structural optimization method, *Finite Elem. Anal. Des.* **43** 1039-1040

[10] Gebisa A W and Lemu H G 2017, *A case study on topology optimized design for additive manufacturing*, In: Lemu H G, Pavlou D G, Jakobsen J B, Ong M C, Gudmestad O T and Siriwardane S C)(Eds,) Proc. of COTech Conf. Stavanger, NorwayNov. 30 – Dec 1, 2017, IOP Conf. Series: Mater. Sci. Eng. **276**

[11] Gardan N and Schneider A 2015 Topological optimization of internal patterns and support in additive manufacturing, *J. Manuf. Syst.* **37** 417-425

[12] Xu Y, Zhang D, Zhou Y, Wang W and Cao X 2017 Study on topology optimization design, manufacturability, and performance evaluation of Ti-6Al-4V porous structures fabricated by Selective Laser Melting (SLM), *Mater.* **10** 1048

[13] Rezaie R, Badrossamay M, Ghaie A and Moosavi H 2013, Topology optimization for fused deposition modeling process, In: The 17th CIRP Conf. on Electro Physical and Chemical Machining (ISEM), *Procedia CIRP*, **6** 521 – 526

[14] Leary M, Merli L, Torti F, Mazur M and Brandt M 2014 Optimal topology for additive manufacture: A method for enabling additive manufacture of support-free optimal structures, *Mater. Des.* **63** 678–690

[15] Brackett D, Ashcroft I and Hague R 2011 Topology optimization for additive manufacturing, In: *22nd Annual international solid freeform fabrication symposium*, 348–362, Austin, TX

[16] Gaynor A T and Guest J K 2014 *Topology optimization for additive manufacturing: Considering maximum overhang constraint*, 15th AIAA/ISSMO Multidisciplinary Analysis and Optimization Conf., June 16–20, Atlanta, GA

[17] Mirzendehdel A M and Suresh K 2016 Support structure constrained topology optimization for additive manufacturing, *Comput. Aided Des.* **81** 1 – 13

[18] Wang W, Wang T Y, Yang Z, Liu L, Tong X, Tong W, Deng J, Chen F and Liu X 2013 Cost effective printing of 3D objects with skin-frame structures, *ACM Trans Graph*, **32** 1–10

[19] Hu K, Jin S and Wang C C L 2015 Support slimming for single material based additive manufacturing, *Comput. Aided Des.* **65** 1–10

[20] Safari A, Lemu H G and Assadi M 2013 A novel combination of adaptive tools for turbomachinery airfoil Shape optimization using a real-coded genetic algorithm, In: *Proc. of ASME Turbo Expo 2013*, vol. 6B, San Antonio, Texas, USA, 3-7 June, 2013

[21] Woldemariam, E T, Lemu H G and Wang G G 2017 *Geometric parameters' effect characterization and design optimization of a micro scale cross-flow turbine for an improved performance*, The Int. Society of Offshore and Polar Engineers Conf. ISOPE 2017, San Francisco, CA, June 26 - 30

[22] Rozvany G I N, Bendsoøe M P and Kirsch U 1995 Layout optimization of structures, *Appl. Mech. Rev.* **48** 41-119

[23] Arora J 2004, *Introduction to optimum design,* San Diago- Elsevier Academic Press

Trends of using polymer composite materials in additive manufacturing

Y Regassa[1], H G Lemu[2] and B Sirabizuh[1]

[1]Addis Ababa University of Science and Technology, Department of Mechanical Engineering, Addis Ababa, Ethiopia
[2] University of Stavanger, Faculty of Science and Technology, N-4036 Stavanger, Norway

E-mail: corresponding.author: Hirpa.g.lemu@uis.no

Abstract. Additive manufacturing (AM) technology is nowadays one of the advanced manufacturing technologies used to convert three dimensional product data to physical objects without any special tooling. The emergence of the technology has made production of complex geometries having consumer demands for more customized products and services. Though the underlying technology was initially intended for rapid prototyping, the technology has recently getting wider attention in diverse industrial sectors using different material types commonly polymers, metals, ceramics, and other compatible researched materials as input. Because the technology processes the input materials with no or very low wastes, the use of polymer composite materials leads to a no waste manufacturing system. This article aims to review and analyse the opportunity and existing challenges for AM technology in processing polymer composites and provide an insight on the state-of -the art applications. The paper provides an overview of the materials and processes currently available for this novel production technology to fill the industrial needs for engineering application. In addition, closely reviewing published research works in the area, the existing research gaps are identified for possible future research in the field.

1. Introduction

Additive manufacturing (AM) is one of the emerging manufacturing technologies that is used to form three dimensional products by adding materials consequently layer by layer at a time through the bottom-up approach. Due to the way parts are produced, it is often referred to as 3D printing. Thus, in this article, the words additive manufacturing and 3D printing are used to imply the same technology. The concept Rapid Prototyping (RP) is also, sometimes, synonymously used though Rapid prototyping is the earlier development of the technology that allows engineers to understand and explain their design through computer-aided design to conceive as a technique for production of physical models or prototyped parts from virtual model that enhance the efficiency and reliability of production system. The improvement for 3D part printing process and advancement of material processing attracts market to prefer this technology that helps to lead the transition of old production system towards additive manufacturing system to be used as user interfaced parts production method are among 21st century technology to fulfill basic human engineering needs which is under alarming growth [1].

Emerging as a rapid prototyping tool, AM is timely mechanism for adoption flexible manufacturing system (FMS) that can be applied to diverse material types including thermoplastics, metals, concretes and ceramics. However, 3D printing predominantly targets the production of polymeric parts and models it prefers suitability associated with photo polymerization and availability of polymers. This is partly due to the fact that polymer composites are easy to reprocess the wastes and convert into new raw materials for AM.

The 3D printing technology covers a wide range of technologies and gives a clear advantage over the existing manufacturing systems, especially for complex geometric parts, precision resources, short manufacturing lead time and customized products. According to Attaran [2], only in the year 2014, the key players in the manufacturing industries spent nearly about $3 billion in additive manufactured components. These include industrial machines 18.5%, consumer products 18.0%, automotive industry 17.3%, medical/dental sector 13.7%, aerospace industry 12.3%, and others 19.1% such as academic institutions, the military sector, architecture, etc.. The global sales of AM machines, related materials and services have also shown an average annual growth of about 33% in the period 2013 – 2015 [3]. The same study indicates that AM technology is largely used to produce functional parts and prototypes for fits and assembly. In terms of the printing materials, polymer-based materials account for the largest share, particularly in the production of medical implants [4].

Though the technology emerged in the product design and manufacturing environment not more than three decades ago, it has already left its footprints with a potentially huge impact to transfer the manufacturing method to a real automated, flexible, fully customizable and digitally controlled process. However, there still exist a number of bottlenecks including reluctance of the manufacturing sector to adopt as a viable manufacturing method. In addition, several process specific characteristics, machine related constraints, materials and achievable qualities both as dimensional and geometric deviations and surface qualities influence the opportunities that this technology can provide [5].

The aim of this paper is to conduct a review of the opportunities and challenges associated with additive manufactured functional parts using polymer composite materials. Following this introduction section, the paper consists of 5 sections. The next section presents the key polymer materials including ceramics and cement based composites used in AM based part fabrication process. In addition to some of their mechanical properties, their application areas, benefits and main limitations are highlighted. Then the trends in industrial application of AM as functional part fabrication method is presented, and followed by the outlook for the technology roadmap and standardization works for AM application. Based on the conducted review and analysis, some observed research gaps are highlighted and then drawn conclusions of the study are summarized.

2. Polymer composite materials for AM

2.1. Mechanical and functional properties of additive fabricated polymers

As mentioned earlier in this article, polymers are not only the first material types used for rapid prototyping using the layer by layer fabrication approach; they are also preferred as commonly used materials for the AM industry. This is primarily because of the diversity of polymer-based composites and the simplicity to adopt them to different AM processes. Polymer-based materials for AM are available as thermoplastic filaments, reactive monomers, resin or powder. They are marketed as ABS, polyamide (PA), polycarbonate (PC) and polylactic acid (PLA), and thermosetting powders such as polystyrene, polyamides and photopolymer resins, are the most common types of polymers for additive fabrication [6] by many AM machines such as stereolithography apparatus (SLA), selective laser sintering (SLS), fused deposition modelling (FDM) and 3DP inkjet printers.

Some of the benefits of fabricating polymer composites using AM technology include the ability to customize complex geometries with high accuracy [7]. Contrary to the attractiveness of polymer composites with respect to their diversity and ease of adoption, their inferior mechanical properties have impact on their application for load carrying structures. As a result, a large size of research and development has focused on developing various methods and materials for prototyping purposes. In order to overcome the limitations of lack of mechanical properties of polymer composites, which are limited to prototyping purposes, the reinforcement of polymers with fibres and nano-materials has been introduced in recent years. These efforts are aimed to enhance the mechanical properties of the printed products to be used as load-bearing or functional components [6, 8]. In this regard, nylon based polymer composites i.e. PA is one of the widely used polymers for SLS operations because of its easy melting using lasers, compared with other forms of polymers [9 - 11]. Other polymers such as polystyrene, thermoplastic and elastomers are also well known as suitable materials for SLS process [12]. Among the polymer composite materials processed by FDM technology, ABS is most popular [13] for general applications, while a recent developed high performance polymer called ULTEM has attracted interests [14, 15] for additive fabrication of parts that require higher strength, durability, toughness, high tensile and flexural strength including resistance to heat and chemicals.

Due to its popularity as industrial class additive fabrication for polymer composites, FDM is recently under extensive investigation. Due to the inherent poor mechanical properties of the polymers, however, FDM process itself suffers critical limitations such as unpredictable dimensional accuracy and surface quality. As several fabrication parameters, such as raster angle, contour width, air gap, etc. (Figure 1 [15]), influence the accuracy and surface roughness, there exists no reliable mechanism to control the accuracy. For instance, a recent study conducted in our 3DP laboratory on 3D printed parts using FORTUS 450mc, with 45° and 90° orientation, show significant variation of the surface quality in terms of all measured parameters.

As can be observed from Table 1 [16], the surface roughness has better qualities in the 90° orientation while the 45° orientation provides better flatness accuracy. Though geometric complexity, lack of mechanical strength and functionality are often mentioned to be the forefront challenges limiting the wide applications of additive manufactured polymer products, the polymers in liquid state are the most widely used materials in AM industry due to their low melting point, low weight, low cost and processing flexibility. Combining various materials for achieving desired mechanical and functional properties is a promising way to solve the challenges in AM of polymer products. Therefore, development of composite materials that are compatible with the available additive fabrication technology has recently attracted tremendous attention. Short fiber-reinforced polymers, in particular, are most realistic for FDM based process and result in moderate improvement of mechanical properties [17, 18].

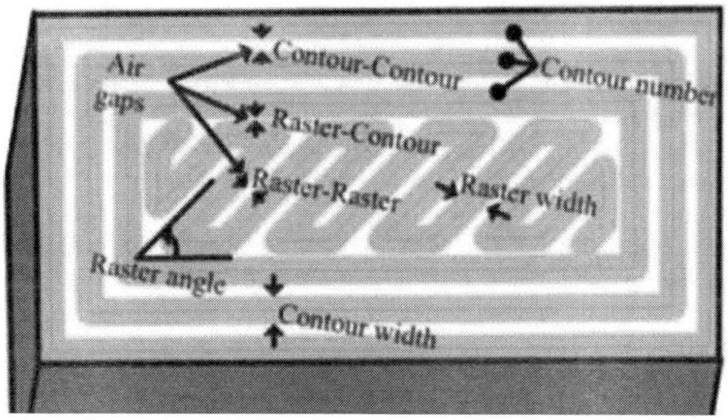

Figure 1. Process parameters for FDM technology.

Table 1. Values of surface quality measured in two different print orientations.

Print orientation		45° [μm]	90° [μm]
Surface finish parameters	S_a	15.389	14.113
	S_q	20.339	18.290
	S_v	76.964	86.888
	S_z	135.438	140.677
	S_{sk}	-0.370	-0.227
	S_{ku}	3.312	3.249
	FLT_t	135.438	140.677

Notations: S_a = Average height; S_q = Root-Mean-Square height; S_v = Max. valley depth; S_z = Max. height; S_{sk} = Skewness; S_{ku} = Kurtosis; FLT_t = Flatness using least squares reference of the selected area.

The progressive developments in AM technology has enabled processing of a wide range of materials from chocolate to advanced multifunctional materials. Materials can be processed in the forms of filaments, wire, powder, paste, sheets and inks. Among these, polymer composites are considered as the most common materials applied in aerospace, automotive, sports, medical, architectural and toy industries. While those used in FDM technology are in the form of filaments, those in powder form or auxiliary binders in the powder-bed processes and resins in liquid state are used in SLA machines [6].

2.2. Ceramics for additive fabrication

Additive manufactured ceramics have created a trend to tailor-design materials with a high strength to weight ratio and facilitated the creation of complex ceramic lattices for many applications [6]. In particular, strong and versatile ceramic scaffolds with complex shapes for tissue engineering is one of the main applications. The possibility to embed biosensors in medical devices or human organ has significantly advanced biomedical sector [19]. Ceramic materials for additive fabrication are in the form of powders that are sintered using laser or bond together via an auxiliary adhesive or in the form of ink for ink-jet printing through suspension of ceramic particles at high sintering temperature followed by post-treatment. The main challenge is the limited availability of ceramic materials for additive fabrication using currently existing methods, relatively poor dimensional accuracy and quality. On the other hand, additive fabrication of ceramics provides better control over the microstructure and composition of the part. Therefore, research and development in the optimisation of AM technologies, as well as the expansion of materials selection for fabrication of ceramics are opportunities that are yet to be exploited [20].

The main application area, pros and cons of ceramic and polymer materials for AM is presented in Table 2. Forther details are also available in the literature [21].

Table 2. Ceramic and polymer materials, Pros and cons for AM.

Material	Main applications	Benefits	Limitations
Polymer composites	Aerospace, sports, medical, toys and architecture	Allow fast prototyping, cost effective, fit for complex struct., cost effective and customized product.	Mechanically weak, limited availability incl. reinforcements, anisotropic
Ceramics	Biomedical, aerospace, automotive & chemical industries	Control of lattice porosity and printing complex structures and scaffolds for human organs, reduced fabrication time, etc.	Limited avail. of printable ceramics, inaccuracy and poor surface finish, and inaccuracy if not post processed.

2.3. Concrete based composites

Concrete is the most ancient and widely used man-made material (composite) particularily within construction and infrastructure sector. Employing the concept of the deposition process from other small size processes like FDM and 3DP technologies, the first cement-based additive fabrication [22] attempted to glue sand layers using Portland cement paste. The process can be considered as an expansion of 3D printing to large-scale application [23] in the domains such as design, construction and architecture, using cementitious materials [24 - 26].

Though the adoption within this industrial sector has been at poor progress, its benefits such as mass-customisation, building process automation without the need for formwork and the like has created interest in the field. The main techniques used to additively fabricate concrete is the extrusion process, though the powder-bed method has been also researched well. Many factors influence the suitability of the concrete type both in terms of its simplicity to extrude, shape stability and dimensional accuracy after fabrication. The formwork-free fabrication may not fit for self-compacting concrete as it may not hold its shape. Furthermore, other problems such as anisotropic mechanical properties and poor inter-layer adhesion are the main challenges that need to be addressed. Despite these challenges, freedom of design and opportunity to build complex and lightweight structures are promising [1, 6].

2.4. Multimaterial printing and process control

One of the key motivation for multimaterial printing is in order to get multifunctional properties. For instance, current AM technologies can provide a multifunctionality of additive manufactured parts by composite materials while providing their required composition property [6]. For such composites, though there are some existing limitations, high-pressure jetting systems can combine only polymers with good flow ability and similar curing temperatures and extrusion-based methods. While FDM process can couple only materials with similar melting temperatures, other AM technologies, such as direct energy deposition (DED), allow for the combination of metal alloys or ceramics, and as a result, more materials can be extruded at the same time. However, additional care must be taken to avoid the formation of unwanted phases between two materials in contact [6].

Multimaterial AM has been successfully achieved in some powder-based technologies using powders with different melting temperatures. In such cases, issues related with remelting or degradation of materials with lower melting points can be encountered adjacent to the powders with higher melting points. Though performance prediction is expected to be complex, the development for optimization through numerical modelling approaches can assist with the multimaterial manufacturing of 3D parts within their designed multifunctional purpose.

3. Development trends of AM as industrial process

The adoption of AM and other advanced manufacturing technologies is heralding the future manufacturing process in which value chains are shorter, smaller, more localized and customized, more collaborative, and provide sustainability benefits. Compared with traditional subtractive processes, the following potential benefits stand out [4]:

- Improved resource efficiency.
- Extended product life.
- Reconfigured value chains.
- Wastage minimization of costly metals like titanium.
- Advancing design for assembly method to improve properties and minimize cost.
- Highly costomized products to individual needs.

There are sufficient indications that make us believe that AM continues to be implemented into the modern industrial processes as well as human livelihood as this technology continues to develop, broaden its scope of applications and continue to be available at reasonable costs. Only within the last three decades, the AM related industry including the technological and material developments and related services have shown an exponential growth. As shown in Figure 2, almost a double growth is observed only within two years [3].

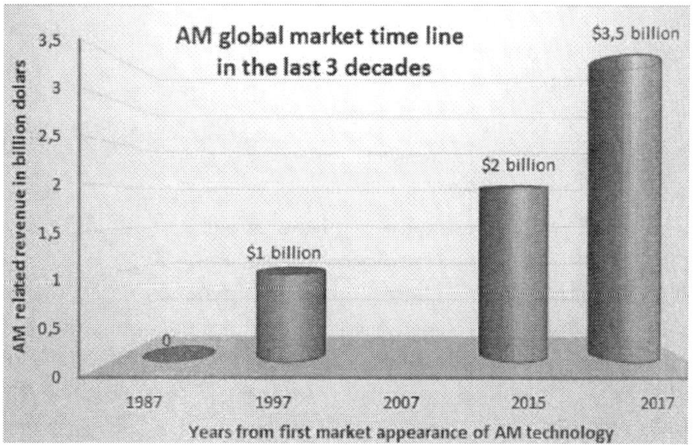

Figure 2. Plots of the growth trends of the additive manufacturing industry.

4. Technology road map and fabrication guideline standards for AM

Surplus availability of materials for AM and their compatability are among the chronic problems of the field to explore more of its application areas. Moreover, the challenges rise due to AM systems that are designed and developed for particular materials. Thus, innovations on new materials will represent next significant progress expected for the engagement of future researchs to tackle the challeneges through optimized and different processing conditions. For example, FDM is a plastic based technology, which is usually unfit for those products that require high strength, magnetic and electrical conductivity, application for human implants, etc. In addition, more focus is needed to develop commercial standards for testing of such materials and their resulting parts. Their consistencies in terms of mechanical, physical and metallurgical properties are not yet established for those based on alternative materials. In order to tackle such critical issues, it is important to define roadmaps [27] that facilitate, among others, adoption of industrial standards and effective means of promoting how new material developments and other research outcomes are shared.

There are many additive manufacturing methods and system operation guidelines provided by different standardization companies that depict the methods and specific constraints of machines and listed considerations that provide a set of 'design specifications'. For instance, a committee that develops AM-related standards has been established by the International Organization for Standardization (ISO/TC 261) and the American Society for Testing and Materials (ASTM F42) [28]. This committee has produced standards related to nomencluters, single processes, flow of processes including required software and hardware, test procedures, quality parameters, means of customer–supplier agreements, and basic elements. Recent additions address data processing and consider the relevance and specify variations to existing standards [20].

As additive manufacturing of metals, polymers, and other materials are entering into a widespread industrial use, it is critical that the infractures with additive manufactured parts do not reduce the safety or reliability of the systems. The potential roadmaps are thus expected to indicate the work that is needed to assure safe application of the technology and to identify the unique activities to be performed [29]. In the area of functionally graded materials, in particular, AM technology is considered to provide exclusive opportunities that other conventional methods cannot.

5. Observed research gaps

Despite the fact that AM systems are proved useful in hard-to-reach locations, such as military bases, the international space station, and also widely employed in the medical applications especially for patient-specific orthopedic implants, there is still a need to fill the gap for many consumer products. Furthermore, many more good parts of this technology are yet to come for the well-being of human needs.

Among others, an intensive materials research and development is needed in order to broaden the selection of suitable materials, prepare a database of the mechanical properties of parts fabricated by AM, and determine the interaction between materials and process parameters. The mechanical property of the fabricated parts is still not consistent and the process lacks repeatability of achieved accuracy and surface quality within each machine and across different machines operating under the same principle. Both the applicable design procedures and process modelling and control are considered as critical issues that require closer study in the future research [21].

After rigorous review of recent and on-going researches, the following research directions were identified as some of possible research gaps.

- Design of new alloys using mixed powder solutions to create novel control of microstructures.
- Fabrication of the so-called high entropy alloys, which constitute the latest discovery in design of metallic materials.
- Design of novel material combinations and metal powders to create new functional properties.
- Development of biomaterials that serve as cell delivery media or as biomolecules in the bioprinting field.
- Conducting tests on how mixed powder solutions using additives can be applied to tune solidification and hence microstructural development during the AM process.

- Research on correlation between process parameters, for instance, how the various process parameters affect the material and the resulting component.
- Reproducibility issues, i.e. how to produce the same part with maintained material properties repeatdly in different machines.
- Methodology to obtain minimum iteration of the process.

6. Conclusions

Nowadays, AM techniques are in a position to secure a key role in the modern manufacturing sector to provide 3D part production with different materials composition and this can be a unique turning point of the manufacturing industry. The possibility of producing three dimensional real objects directly from a 3D model data that can have controllable properties will for sure open opportunities for additional research as well as developments for upgrading and implementations of AM. Beyond all the unlimited opportunities to design and produce parts from composite materials as per the client interest and engineering flexibility for customization of products through AM system are the fundamental uniqueness of this technology.

In this article, the opportunities offered by additive fabrication of polymer composites and the accompanied challenges are reviewed and elaborated. Through the process of this study, it is observed and concluded that polymer composite based AM fabrication is at flourishing stages to pave the way to implement 3D/AM system from limited application of prototyping techniques to automated fabrication systems and such needs are not too far away to knock out the traditional manufacturing process soon. Time efficiency and customer satisfaction are the basic parameters fulfilled by AM/3D for demand and supply of products at reasonable costs. The unique characteristics of the technology such as providing excellent adaptation, fabrication with composite materials, i.e. multi-material fabrication of functional customer needs and fabrication of 3D part products with full satisfaction of intended material properties are significant steps in the manufacturing sector. Moreover, the need for complicated parts in diverse sectors such as the automotive, aerospace, biomedical, electronics, robotics industries, as well as fast fabrication of components for maintenance purposes has made high demand of polymer composite product through AM to achieve and provide timely service. Fiber reinforced polymer composite materials, in particular can maximize the near future needs of the upcoming aerospace and automotive industries.

In conclusion, parallel with the necessary advances in the fabrication process technology, broadening its application areas and availability of materials for diverse application, significant work is needed to get acceptance of the technology in the industry. This needs diverse forms of changing the existing mind-set to gain industrial acceptance.

References

[1] Gao W, Zhang Y, Ramanujan D, Ramani K, Chen Y, Williams C B, Wang C C L, Shin Y C, Zhang S and Zavattieri P 2015 The status, challenges, and future of additive manufacturing in engineering, *Comput. Des.* **69** 65–89

[2] Attaran M 2017 The rise of 3-D printing: The advantages of additive manufacturing over traditional manufacturing, *Bus. Horiz.* **60**(5) 677 – 688

[3] Wohlers T 2018 *Additive Manufacturing and 3D Printing State of the Industry*, Annual Worldwide Progress Report, Published by Wohlers Associates, USA, 2018

[4] Stansbury J W, and Idacavage M J 2016 3D printing with polymers: Challenges among expanding options and opportunities, *Dent. Mater.* **32**(1) 54 – 64

[5] Parandoush P and Lin D 2017 A review on additive manufacturing of polymer-fiber composites, *Compos. Struct.* **182** 36–53

[6] Ngo T D, Kashani A, Imbalzano G, Nguyen K T Q and Hui D 2018 Additive manufacturing (3D printing): A review of materials, methods, applications and challenges, *Compos. Part B: Eng.* **143**(15) 172-196

[7] Park S, Lih E, Park K, Ki Y and Keun D 2017 Progress in polymer science biopolymer-based functional composites for medical applications, *Prog. Polym. Sci.* **68** 77–105

[8] Ford S 2016 Additive manufacturing and sustainability: an exploratory study of the advantages and challenges, *J. Cleaner Prod.* **137** 1573 – 1587

[9] Kruth J P, Mercelis P, Vaerenbergh J V, Froyen L and Rombouts M 2005 Binding mechanisms in selective laser sintering and selective laser melting, *Rapid Prototyping J.* **11** 26–36

[10] Caulfield B, McHugh P E and Lohfeld S 2007 Dependence of mechanical properties of polyamide components on build parameters in the SLS process, *J. Mater. Process. Technol.* **182**(1–3) 477–488

[11] Zarringhalam Z, Majewski C and Hopkinson N 2009, Degree of particle melt in Nylon-12 selective laser-sintered parts, *Rapid Prototyping J.* **15**(2) 126–132

[12] Dziadek M, Stodolak-zych E, and Cholewa-kowalska K 2017 Biodegradable ceramic-polymer composites for biomedical applications: A review, *Mater. Sci. Eng. C.* **71** 1175–1191

[13] Ahn S H, Montero M, Odell D, Roundy S, and Wright P K 2002 Anisotropic material properties of fused deposition modeling ABS, *Rapid Prototyping J.* **8**(4) 248–257

[14] Gebisa A W and Lemu H G 2018 *Influence of 3D printing FDM process parameters on tensile property of ULTEM 9085*, Procedia Manuf., Proc. of the 14th Global Congress on Manufacturing and Management (GCMM2018), Brisbane, Australia, Dec. 5-7

[15] Gebisa A W and Lemu H G 2018 Investigating effects of Fused-Deposition Modeling (FDM) processing parameters on flexural properties of ULTEM 9085 using Designed Experiment, *Mater.,* **11**(4)

[16] Lemu H G 2018 *On opportunities and limitations of additive manufacturing technology for industry 4.0 era*, 8th Int. Workshop of Adv. Manuf. and Automation (IWAMA2018), Changzhou University, Jiangsu, China, Sept. 25 – 26.

[17] Zhong W, Li F, Zhang Z, Song L, Li Z 2001 Short fiber reinforced composites for fused deposition modeling, *Mater. Sci. Eng.* **301** 125-130

[18] Karsli N G and Aytac A 2013 Tensile and thermomechanical properties of short carbon fiber reinforced polyamide composites, *Compos.* Part B, **61** 270-275

[19] Leigh S J, Bradley R J, Purssell C P, Billson D R, and Hutchins D A 2001 A simple, low-cost conductive composite material for 3D printing of electronic sensors, *PLoS ONE* **7**(11)

[20] Hegab, H A 2016 Design for additive manufacturing of composite materials and potential alloys : a review, *Manuf. Rev.* **3**(11)

[21] Guo N and Leu M C 2013 Additive manufacturing: technology, applications and research needs, Front, *Mech. Eng.* **8**(3) 215–243

[22] Pegna J 1997 Exploratory investigation of solid freeform construction, *Autom. Constr.* **5** 427–437

[23] Gosselin C, Duballet R, Roux Ph, Gaudillière N, Dirrenberger J and Morel Ph 2016 Large-scale 3D printing of ultra-high performance concrete – a new processing route for architects and builders, *Mater. Des.* **100** 102–109

[24] Le T, Austin S, Lim S, Buswell R, Gibb A and Thorpe T 2012, Mix design and fresh properties for high-performance printing concrete, *Mater. Struct.* **45** 1221–1232

[25] Le T, Austin S, Lim S, Buswell R, Law R, Gibb A and Thorpe T 2012 Hardened properties of high performance printing concrete, *Cem. Concr. Res.* **42** 558–566

[26] Lim S, Buswell R, Le T, Austin S, Gibb A and Thorpe T 2012 Developments in construction-scale additive manufacturing processes, *Autom. Constr.* **21** 263–268

[27] Stansbury J W and Idacavage M J 2016 3D printing with polymers: Challenges among expanding options and opportunities, *Dental Mater.* **32**(1) 54 – 64

[28] ISO/ASTM52900-15, *Standard Terminology for Additive Manufacturing – General Principles – Terminology*, ASTM Int., West Conshohocken, PA, www.astm.org, 2015

[29] Kruth J P, Leu M C and Nakagawa T 1998, Progress in additive manufacturing and rapid prototyping, *CIRP Annals* **47** 525–540

IOP Conf. Series: Materials Science and Engineering **659** (2019) 012022 doi:10.1088/1757-899X/659/1/012022

Automotive rubber part design using machine learning

D Huri[1] and T Mankovits[2]

[1]University of Debrecen, Doctoral School of Informatics, Kassai út 26, 4028 Debrecen, Hungary

[1,2]University of Debrecen, Faculty of Engineering, Department of Mechanical Engineering, Ótemető utca 2-4, 4028 Debrecen, Hungary

E-mail: huri.david@eng.unideb.hu

Abstract. In rubber products design finite element analysis is a widely used technique. In many cases, the pre-defined operating conditions can be achieved by changing the geometric dimensions of the product which is the well-known iterative design method. Using more than one design parameter the number of possible combinations will increase significantly. The application of Support Vector Machine (SVM) can handle the large number of data in a special way and helps to find the optimal design parameters. In this paper an optimization process of a rubber jounce is presented using nonlinear finite element analysis and SVM.

1. Introduction

In rubber bumper design one of the most important technical characteristics of the product is the force-displacement curve under compression load. This behaviour is the most critical customer need, in many cases its fulfilment requires general iterative design method. Design engineers can handle this task with the modification of the product shape. This kind of shape optimization problem can be solved with several standard optimization methods, if the parametrization of the design process is determined. As it can be seen in the following chapter, the numerical method is a good way to evaluate the working characteristics of the rubber part. Automation of the whole process feasible with the use of Visual Basic for Applications (VBA) which allows to directly access Femap from Excel. Thereby the finite element model pre- and post-processing were controlled with macro running in excel.

The application of SVM in optimization process has the advantage that the transformation function between the input space ($\boldsymbol{D}$) and so-called feature space (E) can be hidden, and supervised machine learning procedures can be applied to find an appropriate regression function.

2. Finite element model of the rubber bumper under compression load set

Rubber behave as a nonlinear, elastic, isotropic and incompressible material, which can be described accurately with hyperelastic constitutive model. Within this several material models and material constants can be found. The material models for rubbers are generally given by the strain energy potential. A successful finite element simulation of rubber parts hinges on the selection of an appropriate strain energy function and on the accurate determination of material constants. Because of material incompressibility, the strain energy function can be divided [1]

$$W = W_D(\bar{I}_1, \bar{I}_2) + W_b(J) \quad (1)$$

where $W_b(J)$ denotes the volumetric terms of the strain energy function and J is for the Jacobian and $W_D(\bar{I}_1, \bar{I}_2)$ is for the deviatoric terms of the strain energy function. The polynomial form of the strain energy potential is based on the first $\bar{I}_1$ and second $\bar{I}_2$ strain invariants of the right Cauchy-Green tensor [2]

$$W = \sum_{i+j=1}^{N} c_{ij}(\bar{I}_1 - 3)^i(\bar{I}_2 - 3)^j + \sum_{k=1}^{N} \frac{1}{d_k}(J-1)^{2k} \tag{2}$$

where determination of c_{ij} and d_k material constants are required in material model. The κ bulk modulus can be calculated as

$$\kappa = \frac{2}{d} \tag{3}$$

where d is the material compressibility parameter.

Mooney-Rivlin, Yeoh and Neo-Hookean material models are available within the polynomial form of the strain energy potential. There are two-, three-, five- and nine-term Mooney Rivlin models. With N=1 substitution the expression of the polynomial form is equivalent with the two-term Mooney-Rivlin model.

$$W_{MR} = c_{10}(\bar{I}_1 - 3) + c_{01}(\bar{I}_2 - 3) + \frac{1}{d}(J-1)^2 \tag{4}$$

According to [3-4] two-term Mooney-Rivlin model with $c_{10} = 1{,}28801\,MPa$, $c_{01} = 1{,}1371\,MPa$ and $\kappa = 1000\,MPa$ values were selected for the finite element investigation of the rubber part.

The geometry of the he investigated rubber specimen is axisymmetric, furthermore the boundary conditions are symmetric as well, thereby the deformation of the shape is independent from the φ axis. In such a case it is worth choosing axisymmetric element (isoparametric quadrilateral elements) for meshing. The size of the element was $1mm$. The investigated geometry can be seen on 'Figure 1' and it was created with $H = 40\,mm, \alpha = 3°, D_1 = 108\,mm$ and $D_2 = 33\,mm$ dimensions. Under working conditions, the rubber jounce comes into contact on the bottom and on the top with flat steel parts. Therefore, frictional contact was defined between the surfaces of the product and the steel parts. The coefficient of static friction was selected $\mu_s = 0{,}6$ according to [5]. At boundary conditions it is given $12\,mm$ prescribed displacement for the top edge of the upper one steel plate, furthermore the bottom curve nodes on the lower one steel were constrained along z axis. Finally finite element analysis was run and as a result the 'Figure 1' shows the deformation state of the rubber jounce while the named 'Optimum characteristics' curve on 'Figure 2' shows the load displacement characteristics of the investigated product.

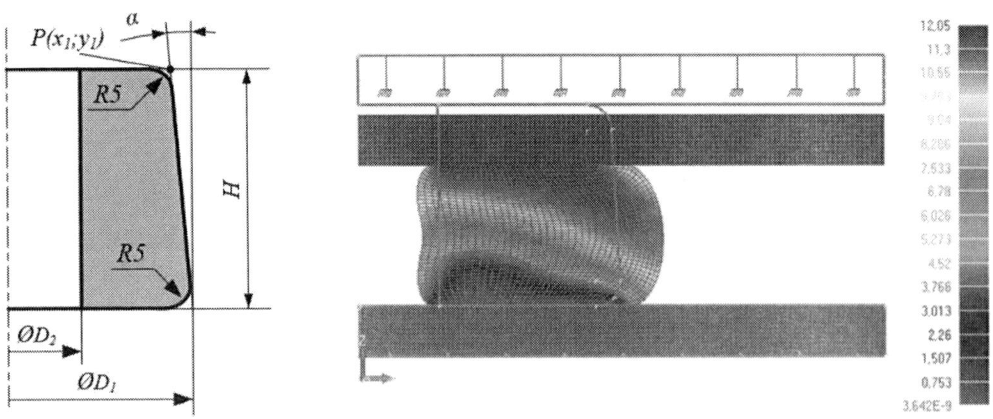

Figure 1. The investigated rubber jounce's geometry and the obtained deformation state from finite element analysis.

3. Two-Dimensional Shape Optimization Problem

In many cases the dimensions of the rubber bumper built in air spring are constrained by other parts. During the current investigation, the product's height ($H = 40\ mm$) and draft angle ($\alpha = 3°$) were fixed while D_1 outer diameter and D_2 hole dimeter are variables, see in 'Figure 1'. Thus in the shape optimization the design parameters are defined in mm, according to the following conditions:

$$\boldsymbol{D} = (D_1; D_2), where \begin{cases} D_1 \in [70,71, \dots 130] \\ D_2 \in [10,11, \dots 60] \end{cases} \text{and } x_1 - \frac{D_2}{2} \geq 15 \quad (5)$$

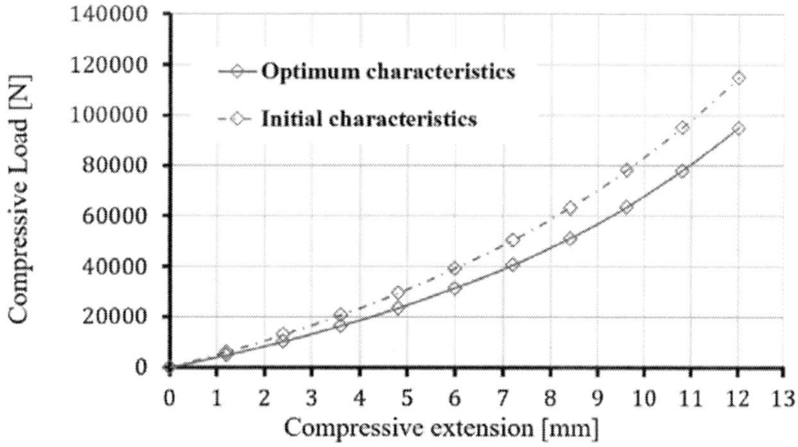

Figure 2. Working characteristics of rubber jounces with optimum shape and a possible initial shape.

The goal of the shape optimization process is to minimize the difference between the initial characteristics and the desired characteristics see on 'Figure 2'. The aim of this research is to investigate the applicability of support vector regression machine, therefore the desired characteristics determined from predefined optimum shape $\boldsymbol{D}_{opt} = (108; 33)\ mm$. The finite element analysis was solved in 100 step and every 10th step was created as output. The difference was calculated as sum of the square error (SSE):

$$E(\boldsymbol{D})_{FEA} = \sum_{i=1}^{10} \left(F_i(\boldsymbol{D}_{opt}) - F_i(\boldsymbol{D})\right)^2\ [kN^2] \quad (6)$$

where $E(\boldsymbol{D})$ is the error value in an investigated design point, $F_i(\boldsymbol{D}_{opt})$ is the ith results of required compression force value in the optimal design point while $F_i(\boldsymbol{D})$ is the ith results of required compression force value in the initial design point. Table 1 contains the calculated error value for an initial design point $\boldsymbol{D} = (130; 60)\ mm$. The objective function of the shape optimization process is to minimize this error value.

Table 1. Calculated Error values in different design points.

	$D_1[mm]$	$D_2[mm]$	E_{FEA}
Optimum Shape	108	33	0
Initial Shape	130	60	1275,84

4. Train support vector regression models

The objective of the machine learning is to discover a function $E = f(\boldsymbol{D})$ that best predict the value of E associated with each value of $\boldsymbol{D}$. At the first step 128,36,22 and 8 pieces of vertex pairs (Learning

Points) were selected from ***D*** according to 'Figure 3', than *E* values were determined by using finite element analysis. Thereby the training set was produced for the machine learning algorithm.

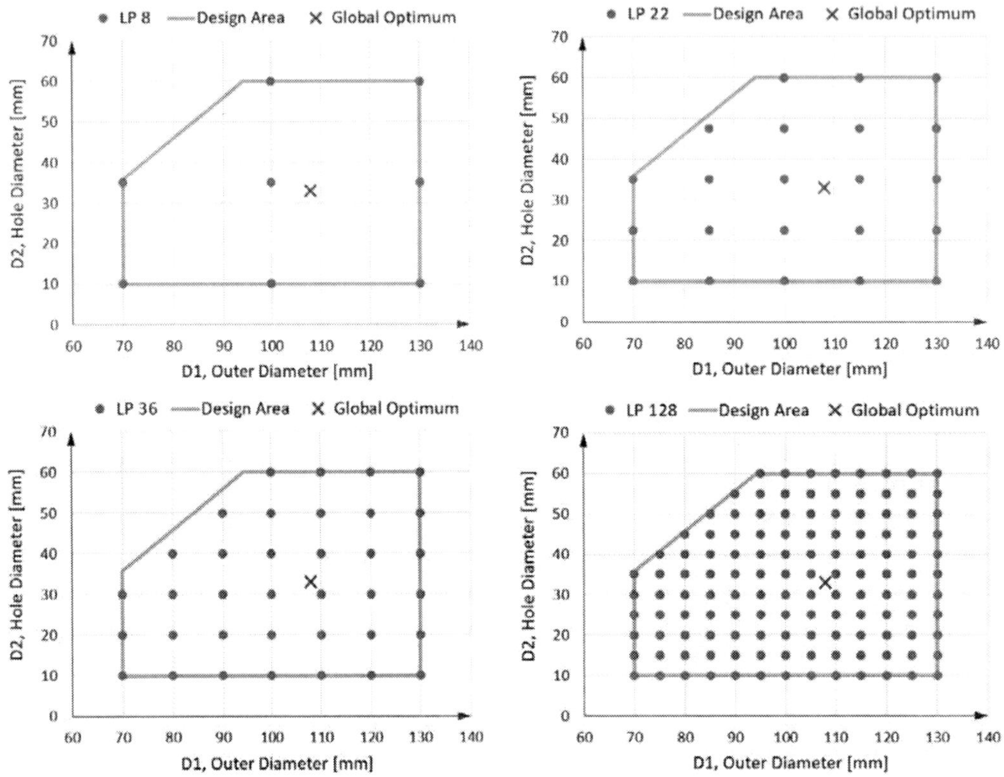

Figure 3. The selection of different learning points set from the design area.

Matlab has a Regression Learner application. With the use of this application we could perform automated training to search for the best regression model type, including linear regression models, regression trees, Gaussian process regression models, support vector machines, and ensembles of regression trees. Manual regression model training was ran without validation data set for all available SVM types.

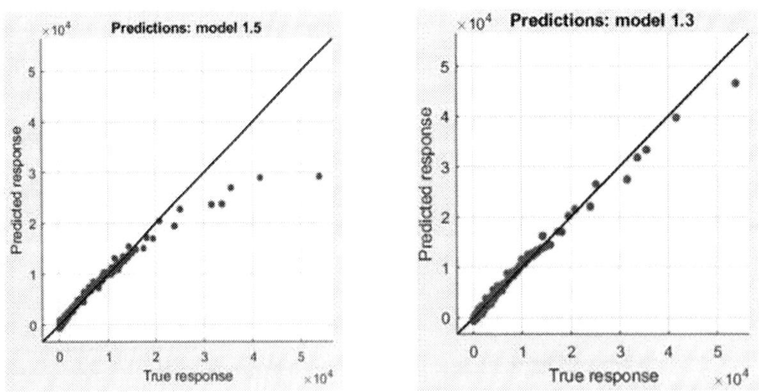

Figure 4. Predicted response versus true response by the use of different kernel function.

The predicted response of Medium Gaussian SVM model is plotted against the actual, true response, see the left one picture in 'Figure 4', while the right one shows the Cubic SVM model's predicted values. A perfect regression model has a predicted response equal to the true response, so all the points lie on diagonal line. The vertical distance from the line to any point is the error of the prediction for that point. The predictions are scattered more farther from the line with the use of Gaussian model like with Cubic model so this has smaller errors for the data set.

As it could see, the goodness of the prediction highly hinges on the kernel function type. To choose the best model, the root mean square error (RMSE) value was calculated on the predicted set. As it can be seen below in the Table 2, the best trained model was the Cubic SVM for each set of Learning Points.

Table 2. Comparison the goodness of the kernel functions with the use of different number of Learning Points and root mean square error calculation.

kernel function	RMSE			
	LP 8	LP 22	LP 36	LP 128
Linear SVM	15765	12972	11516	9526
Quadratic SVM	11968	9053	5598	3751
Cubic SVM	8645	5304	2619	1012
Fine Gaussian SVM	11695	9135	7643	3514
Medium Gaussian SVM	12202	8419	6382	2903
Coarse Gaussian SVM	16553	12710	11123	8520

5. Investigate the minimum number of required learning points

Using the different trained Cubic SVM models for different data sets, predictions were made for each combination of design parameters. Table 3 contains the smallest predicted object function value for each data set and the associated design parameters. Based on the values, the minimum error of the possible solution is $E = -660{,}84$ for which the optimum design parameters are $\boldsymbol{D}_{opt}^{SVM} = (108; 33)\ mm$. This object function value is not possible because of the SSE calculation method, but this cannot be eliminated. Thereby and due to the verification the error values were calculated by finite element method as well, see in Table 3. As it was expected these numbers are different, however the error values are close to zero except for that predictions which one was made by the data set containing 8 pieces of learning points. The rubber bumper working characteristics were also determined for each predicted optimum shape, see in 'Figure 5'. Compared the different possible optimum shape with the known optimum, it can be stated that each geometry shows nearly the desired characteristics except for the *LP 8*.

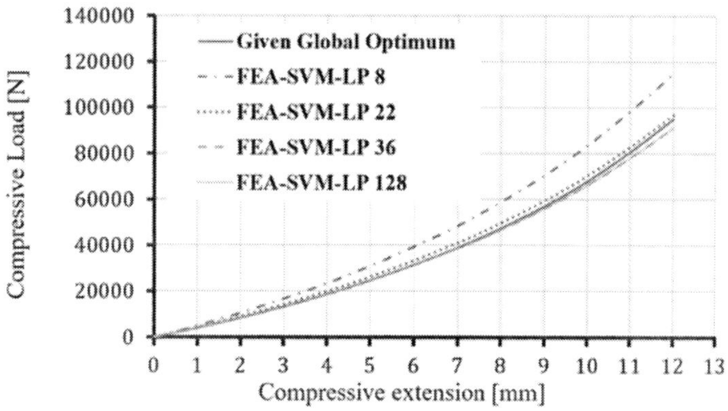

Figure 5. Comparison of the predicted different optimum shape's working characteristics.

Table 3. The smallest predicted and calculated object function value for each data set and the associated design parameters.

Method	$D_1[mm]$	$D_2[mm]$	E_{SVM}	E_{FEA}
Global Optimum	108	33	-	0
LP 8	130	60	2409,65	1275,84
LP 22	124	60	-282,65	30,02
LP 36	122	60	-625,44	23,54
LP 128	119	55	-660,84	23,28

6. Conclusion

Foremost the finite element model of the rubber jounce was built with the use of prescribed displacement as load set. The deformation and working characteristics of the product were determined from the analysis results. Next step was the introduction of two-Dimensional Shape Optimization Problem, where the design parameters and object functions has been defined also. Training data set with different number of learning points was created for supervised machine learning. Support vector regression method was selected to predict the optimum shape of the rubber jounce. With the use of calculated training data set several kernel functions were trained. Root mean square error (RMSE) value was calculated to choose the best model. For every data set the Cubic kernel function shows the best match, therefore it was chosen for the prediction of each combination of design parameters. From the predicted values the optimum shape was determined for each data set. The predicted values did not agree with the finite element calculations, however the determined optimum shape was sufficiently accurate if the number of the learning points are at least 22. As a result, it can be stated that support vector machine is a good way to predict the optimum shape of rubber products. It would be worth in the future to test the efficiency of the algorithm for more complex design problem.

Acknowledgment

The described work was carried out as part of a project supported by the EFOP-3.6.1-16-2016-00022 project. The project is cofinanced by the European Union and the European Social Fund. This research is partly supported by the ÚNKP-18-3 New National Excellence Program of the Ministry of Human Capacities.

References

[1] Bonet J and Wood R D 1997 *Nonlinear Continuum Mechanics for Finite Element Analysis*, Cambridge University Press
[2] Hossain M and Steinmann P 2013 More Hyperelastic Models for Rubber-Like Materials: Consistent Tangent Operators and Comparative Study, *Journal of the Mechanical Behavior of Materials* **22** 1-24
[3] Huri D 2016 Incompressibility and Mesh Sensitivity Analysis in Finite Element Simulation of Rubbers, *International Review of Applied Sciences and Engineering* **7** 7-12
[4] Huri D and Mankovits T 2018 Comparison of the Material Models in Rubber Rinite Element Analysis, *IOP Conference Series: Materials Science and Engineering* **393**
[5] Cruz Gómez M A, Gallardo-Hernández E A, Vite Torres M and Pena Bautista A 2013 Rubber steel friction in contaminated contacts, *International Journal on the Science and Technology of friction Lubrication and Wear* **302** 1421-1425

Assembly systems planning with use of databases and simulation

Š Vaclav[1], P Kostal[1], D Michal[1] and Š Lecky[1]

[1]Slovak University of Technology, Faculty of Materials Science and Technology, Jána Bottu č. 2781/25, 917 24 Trnava, Slovakia

E-mail: stefan.vaclav@stuba.sk

Abstract. The work deals with the assembly workplaces and systems planning. The aim is to create a literary overview and describe problems in this topic of assembly workplaces and systems. Another aim is to design a proposal for solving the assembly systems planning problems. The focus of this work is therefore to propose a methodology for the design of assembly systems and thus to create an integrated methodological procedure and tools. In the introductory chapter, the paper is devoted to the current state of assembly systems and their planning. Based on the theoretical knowledge, the analysis of the assembly systems and workplaces will be carried out according to the pre-selected criteria. Given the need to develop the classification of assembly systems and workplaces in terms of product or assembled subjects. Next chapters of the paper is devoted to the creation of databases for the needs of designing assembly systems. Created database need proper algorithm to design correct assembly system. Proposal for solution in this paper is describe in last chapters.

1. Introduction to assembly systems planning

In the area of designing assembly workplaces and systems, it is necessary to address a whole range of problems, respectively tasks that arise from the specifics of designing and assembly processes [1].

As part of the project design of assembly plants, there is an attempt to create methodological procedures for individual assembly processes, but most of the framework procedures developed to support the design of assembly workplaces and systems are designed for specific cases or do not address this issue as a whole.

Making a broader methodology for designing assembly workplaces or systems is a purposeful task for shortening design times and trying to create a form of general practice is evident on the basis of already published concepts. Comprehensive methodology can be a required solution to the problem at planning stages where it would save substantial time pools.

At present, the term project is differently defined and generally applicable definition has not been met. Madauss [2] defines the project as "a goal with a defined beginning and end, characterized by time constraints, uniqueness, complexity and novelty. Similarly, it defines the Frese project [3], but, unlike the previous characteristic, it speaks of relative novelty, and the definition is complemented by the statement that "each project is characterized by a significant risk for a unique character." These characteristics are also found in the Haberfellner definition [4], which complements them by stating that each project must have a defined goal.

Several authors apply the definition according to DIN 69 901, which describes the project as "an intention that is in its overall essence characterized by the uniqueness of the conditions, such as the given objective, time, financial, personnel or other constraints, boundaries in terms of other intentions and project specific organization".

2. Simulation in assembly system planning

At present, designing deals with specialized companies that provide customers with a complex design of manufacturing systems. Such a team is made up of experts in engineering, industrial engineering, electrical engineering and computer science who are engaged in the development of innovative solutions using the latest technologies. They are more specific than the individual tasks, as well as the whole complexes, from the design of the concept to the production of the first series of products, from designing workstations and production lines to balancing, verifying and optimizing based on simulation as well as planning and production management. Development is conditional on the maximum use of digital technologies. Digitization helps shorten delivery times for projects and allows flexible responses to customer requirements.

Simulation is an experimental method in which we are replacing a real system with a computer model. On this model, it is possible to perform a number of experiments, evaluate them or modify them and apply the results to the real system. There is no other method or theory that would allow experimentation with a complex system before it was put into operation. There is no other algorithm that would allow for a few minutes to play complicated processes on the computer that actually last for weeks or months. It is the ideal tool to support decision making at various levels in the company [5].

According to collective [6], simulation is: Simulation of systems is a specific form of cognitive process.

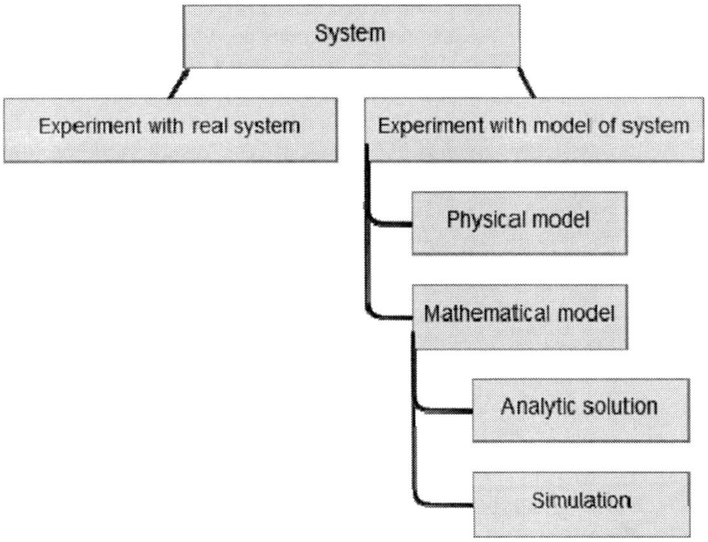

Figure 1. Model representations of systems.

3. Proposed solution based on databases

By defining the categories, the necessary first format of the database will be obtained, which will be assigned to the individual categories. Each entry must have its own attributes or properties, the properties on which the algorithm will work. Assignment of data properties is based on practice or after discreet simulation.

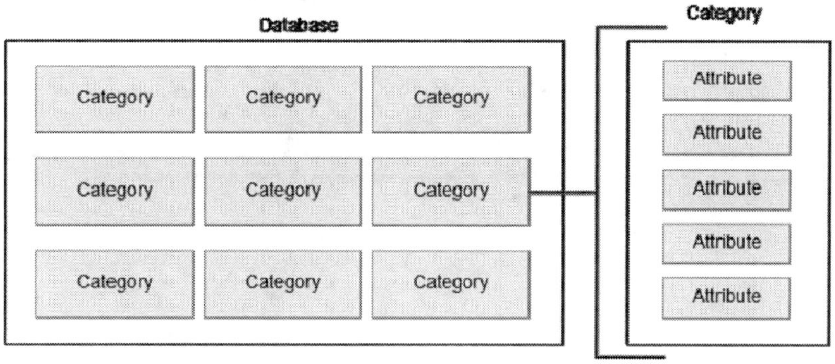

Figure 2. Scheme of categories in database.

3.1. Creation of database

Mounting workplaces and systems can be designed in two ways:

- from down to top,
- from top to down.

In case of down to top and top to down design, the same database should be used for designing. This can be ensured by assigning such features to data that will focus on component requirements as well as requirements for the assembly system / workplace.

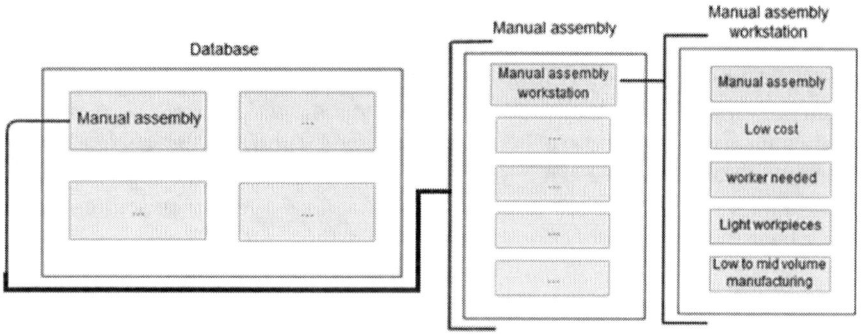

Figure 3. Scheme of attributes in database.

3.2. Approach to database handling

The algorithmic solution has the task of correctly identifying data based on the selected properties. After selecting the correct data, it is necessary to perform and evaluate the capacities. This step should be simplified by designing an interactive capacitive calculator that can work on the basis of the algorithm of the selected data.

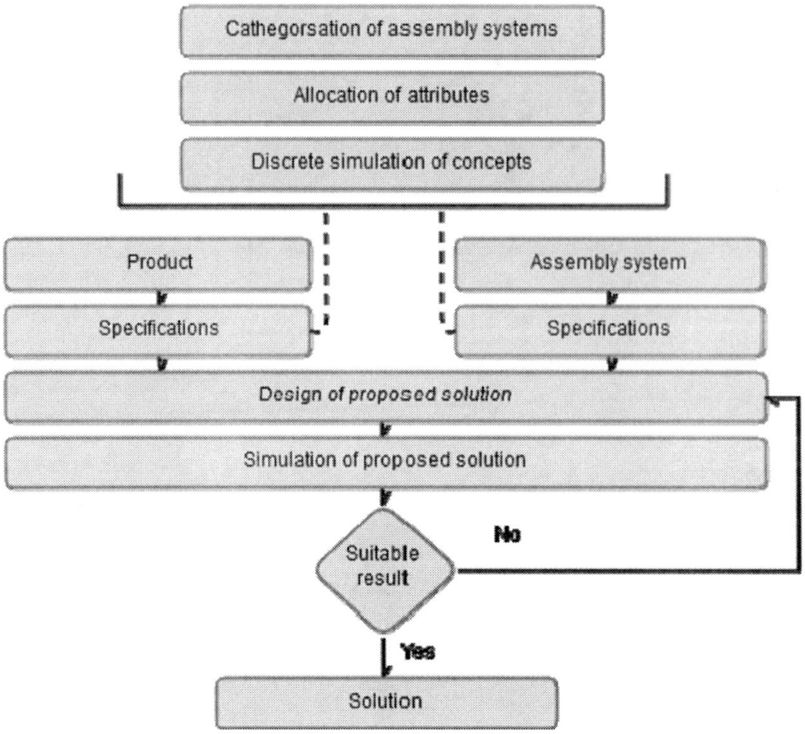

Figure 4. Proposed idea of solution.

If the simulation model does not confirm the assumptions, it is necessary to return to the design and change the simulation model and thus the assembly system or workstation.

If the simulation model proves to be appropriate and confirms the design assumptions of the assembly system or workplace, we can talk about a design that meets the requirements. Such a model is suitable for future planning of plant or system modifications where it saves time and can also serve as a test model for planning changes in production. In the case of incorporation of the developed methodology into the software solution, we can talk about computer support of design in the field of assembly workplaces and systems.

4. Conclusion

The article shows the ideological design of the assembly system design methodology. The aim of the work was to clarify and to create a proposal of solutions and partial solutions, which will enable and help the design of the assembly system design methodology.

Paper defines access to designing databases that can be used for productbased access or assembly system requirements. Consequently, use the databases in the field of assembly systems to reduce design times.

Creating the design methodology of the assembly systems allows for shortening the initial stages in the projects. Database based on the categories that will include simulation verified data as well as associated with the correct properties, in cooperation with the appropriate algorithmic solution to create a useful tool.

Use of this approach for the drafting and subsequent simulation model al-lows not only shorten the projection times but also to creating the first simulation model that could be used for further planning and verification of changes.

After verifying the functionality of the design idea, it is planned to create a software solution. Incorporating the entire proposed solution into the software can have a positive benefit for practice, especially in terms of simplicity and usability for the user.

Acknowledgement
This paper was created thanks to the national grant: KEGA 021STU-4/2018 - Development of a laboratory for the design and maintenance of production systems supported by the use of Virtual Reality

References

[1] Václav Š, Senderská K and Benovič M 2011 Assembly technology and CAA systems, *Trnava: AlumniPress* **249** ISBN 978-80-8096-141-1
[2] Madauss B J 2000 *Handbuch Projektmanagement*, 6., überarb. und erw. Aufl., Stuttgart, ISBN 3-7910-0694-0
[3] Frese E 1980 *Projektorganisation*, In book: Handwörterbuch der Organisation, 2. Aufl., hrsg. v. Erwin Grochla, Stuttgart, 1960-1974
[4] Haberfellner R 1992 *Projektmanagement*, In book: Handwörterbuch der Or-ganisation, 3. Aufl., hrsg. v. Erich Frese, Stuttgart, 2090-2102
[5] Debnár R, Košturiak J and Kuric I 1998 Simulation as a tool for increasing productivity and profit in a business, *Web electronic journal – Computer-aided systems in engineering, Žilina*, Web: http: PPfstroj.utc.sk/
[6] Authors' team: *Agreement on Understanding "Simulation of Systems"* (version 5/85). In book: Proceedings of the conference Simulation of systems in designing, planning, forecasting. Prague, DT SVTS 1985

Application of the multi-criteria decision making in the selection of materials of composite shaft

Z Djordjevic[1], S Jovanovic[1], S Kostic[2], M Blagojevic[1] and D Nikolic[1]

[1]University of Kragujevac, Faculty of Engineering, Sestre Janjić 6, 34000 Kragujevac, Serbia
[2]Technical College of Applied Studies in Kragujevac, Kosovska 8, 34000, Kragujevac, Serbia

E-mail: dviks@kg.ac.rs

Abstract. Thanks to the outstanding combination of construction and special features such as high resistance and stiffness, increased wear resistance to fatigue, higher values of critical speeds, lower mass, long lifetime, and so on, today a growing number of shafts, especially in the automotive industry, are made of composite materials. The optimal values of these parameters can be obtained by selecting the appropriate material.
The selection of materials in the process of product design requires efficient decision-making. Incorporating hybrid materials, in addition to providing the potential for improved performance, can make the decision-making process in the selection of material difficult.
In this sense, the use of multi-criteria decision making (MCDM) was used as an adequate tool and support for making optimum decisions. This method has the potential to improve practically all areas of decision making in engineering from design to product.

1. Introduction

The shafts belong to the group of elements for rotational motion. They carry on their machinery machine parts (gears, belts, sprockets, couplings, working elements, etc.), making with them a functional system, providing the conditions for rotation with the required speed and load transmission. The shafts are exposed to complex stress: torsion (due to transmit a torque from one rotating part to the other), bending (due to the force in the gears of the gear teeth, etc.) and axial stress (due to the existence of axial forces). Also, the shafts are loaded with their own weight as well as the weight of the parts on it, the inertial forces due to the unbalance of the mass, the forces of sudden shock and overload, etc. However, the stresses due to these loads are relatively low intensity, and most often during the calculation of the shafts, they are neglected.

For the construction of shafts most commonly used general structural steels, tempering steels or cementing steels, depending on the required strength, durability, cost of production, etc.
In recent years, due to its good characteristics, composite materials increasingly replace steel in shaft production. These advantages are reflected in the high accuracy of production, without the need for subsequent mechanical and heat treating, the possibility of combining different materials of the matrices and fiber, thus achieving the improvement of mechanical properties, energy savings, etc.

Despite the advantages of composite in comparison with conventional metal shafts, their application is limited for now. The main reason for this is still the high cost of raw materials and the production process, which is mostly individual. It is a challenge for constructors to reach a compromise between the tendency to obtain the minimum mass of the shaft with the maximum mechanical characteristics and the cost and difficulties in the organization of production.

The most commonly used shafts are carbon or glass fibers in combination with epoxy or polyester resins or hybrid constructions obtained by combining these fibers. Also, the laminates obtained by the combination of a metal matrix-composite are suitable for high resistance, load and specific stiffness.

When choosing metal materials for shaft production, more criteria must be considered (required load capacity, stiffness, stability, ...). When it comes to composite materials, the problem becomes even more complicated. Namely, it is known that the composite characteristics may vary depending on the type of material applied, the amount, the angle of orientation of the fibers, etc.

In order to include as many criteria as possible when deciding on material selection, the multi-criteria decision making process (MCDM) is recommended. Such a decision-making process can be encountered in a large number of papers.

In the paper [1], it is emphasized how much the proper choice of materials plays a major role in the product development process. The selection was carried out using two methods of multi-criteria decision-making (VIKOR and ELECTRA). The application of these methods has been demonstrated through concrete examples of analysis.

The problem of material selection is very complex, as explained in [2]. By using hybrid materials, such as composites, which seek to improve some of the characteristics, the problem becomes even more complicated. That's why more and more approach to multi-criteria decision-making in the selection of materials, using different software.

Paper [3] proposes a multi-criteria decision making approach (MCDM) when selecting materials for engineering needs. The paper emphasizes how this method is actually very simple to implement, and includes numerous qualitative and quantitative properties of the material in selection. The results obtained showed good agreement with other methods of analysis.

A multi-criteria decision making approach (MCDM) in the selection of materials has also been proposed in [4]. According to the authors, the number of selection criteria should be limited to seven plus/minus two. The paper emphasizes the advantages of the VIKOR method compared to other MCDM decision-making methods.

Paper [5] deals with analysis of multi-criteria decision-making (MCDM) in material selection, taking into account the ecological aspect. The model is based on engineering analysis of the entire life cycle of the product - from design, product development and product recycling opportunities after the end of the service life.

2. Multi-criteria decision making in material selection

Multi-criteria decision making is a relatively new discipline that, through its development, should provide support to decision-makers who, in most cases, face very numerous and often contrary influencing factors. Through the review of the historical development of this method [6], we discover the fact that the term Multi-Criteria Decision-Making (MCDM) was introduced for the first time in 1972 in the management sciences in the United States. The European "version" of this method is an analysis of multicriterial decision making (MCDA - Multi-Criteria Decision-Analysis). These methods aim to achieve an optimal and compromise solution within a transparent process and to maximize the quality of decisions that involve meeting multiple criteria.

In the usual MCDM approach, the final set of alternatives (options, potential solutions) $A_i \in A$. Any alternative solution can be evaluated from the aspect of multiple criteria (attributes), $C_j \in C$. Alternatives (A_i) and criteria (C_j) and their interrelations can be presented in the form of a matrix table or the so-called decision matrix (Table 1).

Table 1. Correlation between alternatives and criteria – the decision matrix.

Alternative (scenario)		Criteria					
		$C_1\,(W_1)$	$C_2\,(W_2)$	…	$C_j\,(W_j)$	…	$C_n\,(W_n)$
	A_1	x_{11}	x_{12}	…	x_{1j}	…	x_{1n}
	A_2	x_{21}	x_{22}	…	x_{2j}	…	x_{2n}
	⋮	⋮	⋮	…	⋮	…	⋮
	A_i	x_{i1}	x_{i2}	…	x_{ij}	…	x_{in}
	⋮	⋮	⋮	…	⋮	…	⋮
	A_m	x_{m1}	x_{m2}	…	x_{mj}	…	x_{mn}
	max/min	max	max	…	min	…	min

Each of the criteria (C_j) can be the maximization (max) or minimization (min) type according to the nature of the size it characterizes. In Table 1, the size x_{ij} represents the value of the i-th alternative in relation to the j-th criterion, and W_j is the weight coefficient of the j-th criterion. Each of the criteria has no equal importance for decision-makers, so their significance is a weight coefficient - W_j. These coefficients have their absolute (ratios) and relative values (weight coefficients). In this segment of the MCDM procedure, in determining the rating of certain criteria, there is a significant influence of the individual or group subjectivism of the decision-makers of certain decisions. It is necessary, first and foremost, to improve the quality of the analysis, to minimize this subjectivism.

In the final phase of each MCDM procedure, the stability of the solution (choosing one or more alternatives) on changes in input data, as well as changes in relative weighting coefficients of the criteria can be investigated.

Linear normalization of the value of x_{ij} in r_{ij} is carried out, depending on the type of criteria in the following way:

$$\text{for } \mathbf{max} \text{ criteria: } r_{ij} = \frac{x_{ij} - x_j^{min}}{x_j^{max} - x_j^{min}}, \tag{1}$$

$$\text{for } \mathbf{min} \text{ criteria: } r_{ij} = \frac{x_j^{max} - x_{ij}}{x_j^{max} - x_j^{min}}, \tag{2}$$

Both of these methods for normalizing data reduce the value of alternatives by criteria between zero and one .

SAW (Simple Additive Weighting Method) [7] is one of the most well-known and widely used methods, which provides similar results as well as significantly more complex methods of multi-criteria decision making. This method takes into account the weight coefficients of the criteria. By each criterion, it is necessary to join the weight factor assigned directly by the decision maker, or obtained by applying some of the known methods for determining the weight coefficients of the criteria. For each of the considered alternatives, the aggregate characteristic, or the value of the sum of multiplication of relative weight factors and normalized performance values, is calculated according to all criteria. Quotient an alternative with the highest value is the best of the offered solutions:

$$A^* = \left\{ A_i \middle| \max_i \sum_{j=1}^{n} W'_j r_{ij} \right\} \tag{3}$$

W_j` represents the normalized value of the weight coefficient W_j:

$$W'_j = \frac{W_j}{\sum_{j=1}^{n} W_j} \tag{4}$$

Values r_{ij} are obtained by the aforementioned linear normalization process.

2.1. Normalization the value of selected parameters and determining weight coefficients

In this case, four different materials (see Table 2) for the production of shafts were analysed the influence of six characteristic values (performance) that have the role of criteria in the MCDM process. In the Table 2. the values of the x_{ij} are given and in this way the Matrix of the decision is formed.

Table 2. Values of x_{ij}.

Material	Young's Modulus E_1 (MPa)	Young's Modulus E_2 (MPa)	Shear Modulus G_{12} (MPa)	Density ρ (kg/m^3)	Max. deflection f (mm)	Max bending stress σ (MPa)
Steel	210000	210000	83000	7830	0.087	7.589
Carbon fibers / epoxy resin	131600	8200	4500	1550	0.591	6.31
Fiberglass / epoxy resin	43300	14700	4400	2100	0.691	7.95
Aramid fibers / epoxy resin	81800	51000	1510	1380	1.15	9.7
Type of criteria	max	max	max	min	min	min

Linear data normalization from the Table 2. was made on the basis of the equation (1) and (2) depending on whether the maximization or minimization type is the criterion. Results of normalization are presented in the Table 3.

Table 3. Values of r_{ij}.

Material	Young's Modulus E_1 (MPa)	Young's Modulus E_2 (MPa)	Shear Modulus G_{12} (MPa)	Density ρ (kg/m^3)	Max. deflection f (mm)	Max bending stress σ (MPa)
Steel	1	1	1	0	1	0.62
Carbon fibers / epoxy resin	0.53	0	0.04	0.97	0.53	1
Fiberglass / epoxy resin	0	0.03	0.04	0.87	0.43	0.52
Aramid fibers / epoxy resin	0.23	0.21	0	1	0	0

In tables 4, 5 and 6 the values of weight coefficients (three variants), which are calculated on the basis of the expression (4) and applying the Seaty procedure. Additionally, for the sake of transparency, the diagram of comparative values of weight coefficients is given six criteria for all three considered variants (Figure 1).

Table 4. Determination of weight coefficients (Seaty scale – procedure), variant 1.

	k_1	k_2	k_3	k_4	k_5	k_6		
k_1	1	7	5	0	3	3	19	0.260274
k_2	0	1	0	0	0	0	1	0.013699
k_3	0	3	1	0	0	0	4	0.054795
k_4	3	9	7	1	4	5	29	0.39726
k_5	0	4	4	0	1	3	12	0.164384
k_6	0	4	3	0	0	1	8	0.109589
							73	1

Table 5. Determination of weight coefficients (Seaty scale – procedure), variant 2.

	k_1	k_2	k_3	k_4	k_5	k_6		
k_1	1	1	1	1	1	1	6	0.166667
k_2	1	1	1	1	1	1	6	0.166667
k_3	1	1	1	1	1	1	6	0.166667
k_4	1	1	1	1	1	1	6	0.166667
k_5	1	1	1	1	1	1	6	0.166667
k_6	1	1	1	1	1	1	6	0.166667
							36	1

Table 6. Determination of weight coefficients (Seaty scale – procedure), variant 3.

	k_1	k_2	k_3	k_4	k_5	k_6		
k_1	1	3	3	0	1	0	8	0.115942
k_2	0	1	1	5	3	5	15	0.217391
k_3	0	1	1	0	0	0	2	0.028986
k_4	3	5	5	1	3	1	18	0.26087
k_5	1	3	3	0	1	0	8	0.115942
k_6	3	5	5	1	3	1	18	0.26087
							69	1

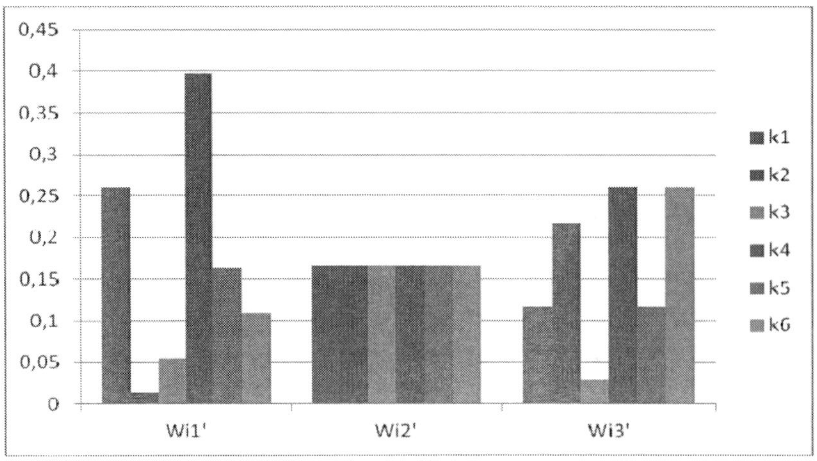

Figure 1. Weight coefficients by variants.

2.2. Results of ranking selected materials

As the final result of the analysis, using the expression (3), the values of the aggregate characteristics for the three variants of the weight coefficients of the performances - criteria were obtained. In Table 7 and diagram (Figure 2) it is noted that for the values of the first variant of weight coefficients, the carbon fiber composite and the epoxy resin are best chosen. For the second group of weight coefficients, steel was shown as the most favorable material, while in the third case, these two materials were practically identical. The remaining two materials received less or lesser grades in all three variants.

Table 7. Collective characteristics for different values of weight coefficients.

	S	C	F	A
A_{i1}	0.493151	0.722192	0.47589	0.46
A_{i2}	0.666667	0.511668	0.315006	0.24
A_{i3}	0.64	0.637972	0.420144	0.333189

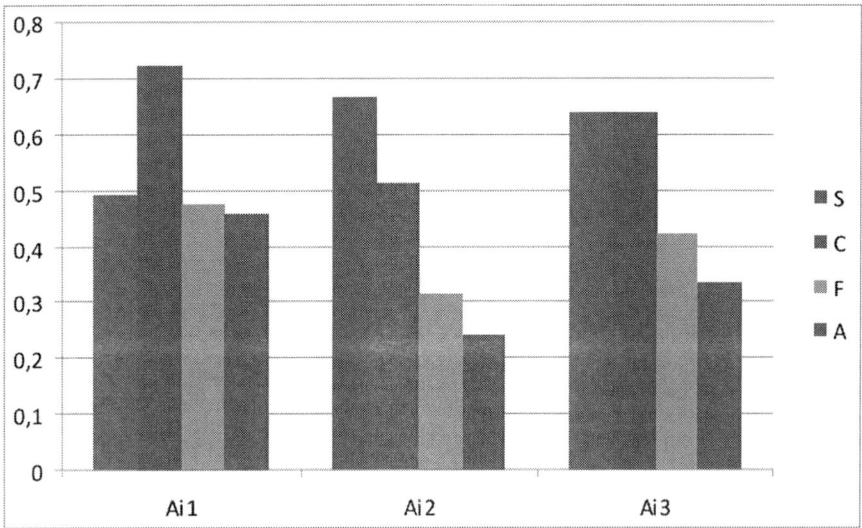

Figure 2. Diagram showing aggregate characteristics for three variants of weight coefficients of preformance - criteria.

3. Conclusions

The phase of selecting adequate material within the product design process is a task for the designer who has to make the appropriate decisions. Based on the characteristics of the various available materials and certain constructional requirements, it is often not easy to effectively optimize the choice between possible options only based on the subjective assessment of the designer. Solving such and similar problems is greatly facilitated by the use of one of the multi-criteria decision-making methods (MCDM). Based on the analysis presented for the four materials considered and the evaluations of the six selected characteristics, it can be concluded that the steel and carbon fiber composites with epoxy resins are the best estimate of the potential solution. The analysis was made for three variants of the weight coefficients of the criterion - characteristics. In each of the variants, to a greater or lesser extent, these two materials received the highest marks in the multi-criteria decision-making process.

Acknowledgment

This is result of the TR33015 project, which is investigation of the Technological Development of the Republic of Serbia. The project is titled "Research and development of a Serbian net-zero energy house". We would like to thank to the Ministry of Education, Science and Technological Development of the Republic of Serbia for their financial support during this investigation.

References

[1] Chatterjee P, Athawale V M and Chakraborty S A 2009 Selection of Materials Using Compromise Ranking and Outranking Methods, *Materials & Design* **30**(10) 4043-4053

[2] Jahan A, Edwards K L and Bahraminasab M 2016 *The Importance of Decision Support in Materials Selection*, In book: Multi-Criteria Decision Analysis for Supporting the Selection of Engineering Materials in Product Design 1-23

[3] Jahan A, Ismail M Y, Mustapha F and Sapuan S M 2010 Material Selection Based on Ordinal Data, *Materials & Design* **31**(7) 3180-3187

[4] Chakraborty S and Chatterjee P 2013 Selection of Materials Using Multi-Criteria Decision-Making Methods with Minimum Data, *Decision Science Letters* **2** 135-148

[5] Huang H, Zhang L and Liu Z 2011 Multi-Criteria Decision Making and Uncertainty Analysis for Materials Selection in Environmentally Conscious Design, *Int J Adv Manuf Technol* **52** 421-432

[6] Costa B and Pirlot M 1997 *Thoughts on the Future of Multicriteria Field: Basic Convictions and Outline for General Methodology*, Springer

[7] Fishburn P 1967 *Additive Utilities with Incomplete Product Set: Applications to Priorities and Assignments*, ORSA Publication

IOP Conf. Series: Materials Science and Engineering **659** (2019) 012025 doi:10.1088/1757-899X/659/1/012025

Design of algorithm for creation of modular lines of special-purpose machines in the automotive industry

J Galík[1], R Kohár[1] and T Capák[1]

[1] University of Žilina, Faculty of Mechanical Engineering, Univerzitná 8215/1, 010 26 Žilina, Slovakia

E-mail: jan.galik@fstroj.uniza.sk

Abstract. This article deals with modularity of special-purpose machines in the automotive industry. The Slovak economy, by its interconnection to engineering and most to the automotive industry, often solves the lowest costs, whether in the sphere of production or in the field of assembly. This question is associated with the question of special-purpose machines. The design and specificity of each such device results in high development and construction costs. It is important to mention the modularity that is very often used in the design of such devices. Due to the large variability of the automotive environment, it is a big prerequisite for the future rebuilding of equipment, which is usually expensive and time consuming. By looking for connections and bonds between elements, can be created some modules. By creating a modular construction of the device, it is possible, if a requirement comes, to reconfigure the device and prepare it for new production requirements, respectively. Certain algorithms and procedures are needed to design such devices and utilize the modular arrangement of the device. By creating such modular lines, whether in assembly or in production, we are able to save energy, work and money.

1. Introduction

The production of cars is the engine of the Slovak economy, while high demands are made and automation is supported, whether robotic automation or the placing of single-purpose machines in production and assembly. This is mainly due to requirements for production quality, greater flexibility and increased productivity. Automation is also the result of demographic change (the problem of finding enough qualified workers) and to ensure the competitiveness of companies.

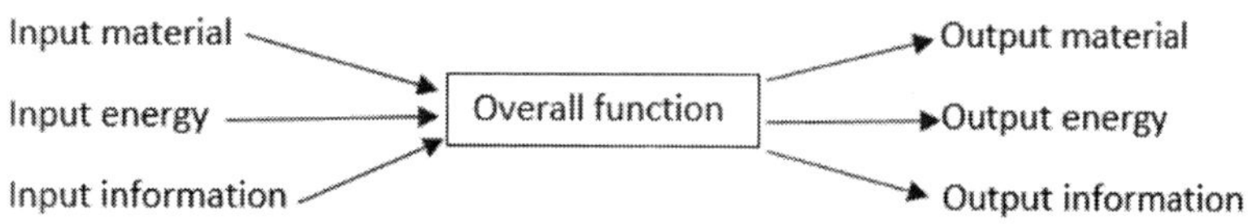

Figure 1. Example of a functional structure of a machine [1].

The trend is to use dedicated devices for assembly closely related to man, while facilitating the handling of loads or performing more demanding activities, respectively one complex activity at one workplace. Single-purpose machines that do not need (need small) human presence are included in the production. Establishments should perform their function (Figure 1) and by modular construction we are able to design devices capable of performing specified tasks. Possible simple modification or inclusion of other module (s) in a dedicated device to meet the requirement for change in the production or assembly.

2. Conceptual design

For any design simpler or more complicated, the designer(s) comes out of conceptual design. The role of conceptual design is the specification of basic machine solutions. The basic step is to identify the main problem, construct functional structures, find work principles and combine them to create a working diagram [2]. The conceptual design specifies a more fundamental solution while providing answers to a few basic questions that arose during the concept approval process.

- Is there a clear definition of the task for the next design process?
- Is there a conceptual design or a well-known solution for realization and design of detail?
- If the conceptual design is used, to what extent should it be used?

The conceptual design and phases they have their succession and must not walk to the end to meet the need for the product [3]. Emphasis should be placed on the creation of product families, respectively products of a very related and modular nature, thus creating the future concept of a reconfigurable production system (RMS). This mode of production is closely related to the product and to the generation of profit and plays a major role in the integration of new products into production [4].

3. Conditions in the automotive industry

Modularly designed single-purpose equipment has a large share in the automotive industry, whether in the manufacturing or automobile assembly. The presence of such equipment in production or assembly indicates the degree of maturity of production. On the basis of this claim, it can be said that car production is one of the most technologically advanced productions.

The design of single-purpose machines is subject to a modular solution for future possible rebuilding, while their development and design is the cheaper matter of different purchased modules compared to robotic solution (depending on the complexity of the solved problem). From the point of view of the degree of automation, the price of equipment is also increasing. Advantages of such devices are lower costs, increased production efficiency, possibly increased equipment automation. The trend of different production changes results in an increasing number of special single or multipurpose modular construction machines. It can be said that the main requirements are:

- Modular solutions, whether manufacturing or assembly machines and equipment
- Development of higher automated machines without human or with little human presence;

Designers should, when designing equipment, place emphasis on concept creation and advancement in the design process to modularity. Purchased parts are used to make the construction cost-effective so that the cost of the equipment is as low as possible. Parts from well-known manufacturers and suppliers are used, while simple complaints are also a requirement.

a. Production

Manufacturers of any single-purpose production machines are subject to a trend of increasing automation and modular machine design. Despite the high price, the equipment is needed because of the growing volume of production and this trend will continue in the future. The need for such machines pushes on their prices, performance, accuracy, environmental and cost-effectiveness.

The concept of manufacturing equipment is, based on a possible transformation, designed to be of a modular type and capable of so-called. "Flexible automation" (module replacement by another module). The trend is set to multi-professional machines, with high dynamics, lightweight construction, reduced need for cutting fluids and the least amount of machining needed (almost finished products from subcontractors). When designing such production machines, it is also necessary to consider the exchange of tools, the principles of automatic exchange, the presence of quick couplers, tracking gauges, magazine, feeders or other parts of the machine with regard to the degree of automation of the equipment [5].

Any design of a modular or modular system is an effective way to reduce production costs and produce a customer-friendly machine. It is advantageous for the customer if the machine is made of serial components. This assumes that spare parts and possible failures will be easily repairable and economically inexpensive. By using a modular structure and buying larger component units, it is possible to achieve lower prices for the constructed equipment as the purchased parts are produced in larger series and cheaper. With modular construction, it is possible to achieve a short time for a constructed device as well as quick commissioning, low operating costs. The user of such a single-purpose machine receives a machine that is built of tested and matching construction nodes [6].

The development of modular machining centers is directly linked to the development of cutting materials, which allow the work of so-called. High-Speed Cutting (HSC) technology. The HSC designation itself carries high cutting speeds (100 to 10,000 m.min-1) when machining. By using HSC tools, we are able to shorten production times, increase the surface quality of the product, reduce the workpiece temperature and reduce the cutting forces. There is a trend that demand for machines capable of manufacturing HSC technology is growing [7].

The production of parts for the automotive industry is dominated by entire modular lines, which are designed to form a system of interconnected devices, individual manufacturing operations and do not need an operator. Typically, such lines are split, either to current (production sites in the series) or to multi-flow (parallel to series workstations). With the modular characteristics of the line, it is between the individual workstations via conveyors and conveyor systems. All functional relationships are secure and operate in an automatic cycle [7].

b. Assembly

Assembly single-purpose machines increase the degree of automation on production lines that replace people or make it easier for him. These include a variety of fully automatic, semi-automatic, manipulators, feeders, or a vision-checker and similar type control function.

The trend for the future is the concept of cellular production [8], so assembly will take place in a given cell, with the product being assembled going from cell to cell and acquiring new product contours for the customer. At such workplaces, the presence of modular single-purpose, but rather multi-purpose, devices will be required to fill multiple functions at once. Modular construction of such devices will be a necessity, for a possible or a very likely production reconfiguration [9].

4. Modularity

Modularity and design of modular components combine the use of the same or similar components to achieve a particular modular solution. Parts and components used in modular designs must ultimately be connectable and assembled. The creation of modular products can be perceived as a design of units fulfilling their function, making it possible to connect between them and to function as one complex product. At design parts, the emphasis is on making the interaction between the components as low as possible and making the proposed components independent. Each modularity component should support one or more function. When assembling parts together, a whole is created that supports one or more functions. By structuring the overall product, the resulting product should be able to support a larger function or a general function, it means the basic analysis of functions, which can be supported by different modules, is very important by creating modular products.

A good example of a comprehensive modular device is a car that, with its modular structure, has as its main objective to move the driver from point A to point B, and can perform several functions, such as transportation of other persons, goods, etc. It consists of a number of modules such as steering, brakes, body, chassis, tires, engine, and transmission. A customer buying a car can upgrade the car and configure it with additional modules (air conditioning, various steering, starting, tracking…) [10].

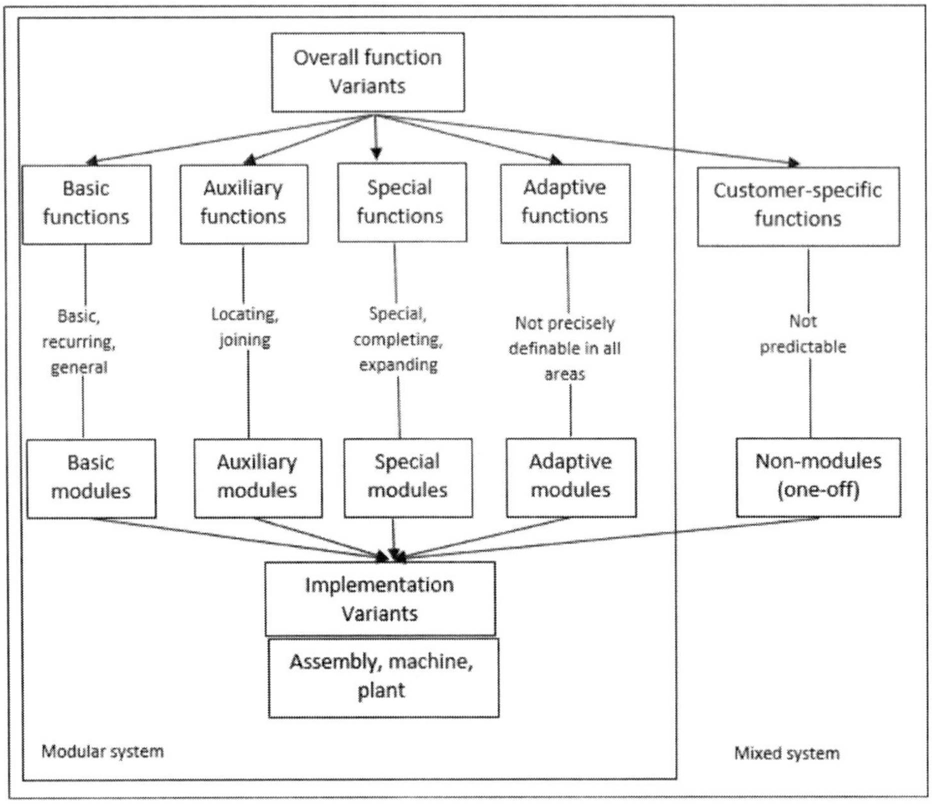

Figure 2. Module functions and modules and mixed production system [3].

a. Types of modularity

The modularity system can be applied in a number of sub-activities such as product design, process design, manufacturing system design, or combinations thereof. The best choice is to combine these three industries, their combinations and their simultaneous use, designing modular products, manufacturing them with modular processes and modular manufacturing processes.

- Product modularity- products that perform different functions through a combination of different building blocks or modules that perform the main function and this function can be subdivided into functions that can be filled by other components or modules. While the core should have sufficient capacity to handle all expected and unexpected variations.
- Modularity in product design- most design problems can be broken down into smaller, simpler sub-issues, where a small change in the problem can lead to a change in other higher-resolution solutions. An important part is the separation of individual problems and solutions so that they do not interact.
- Modularity in manufacturing systems- it should be said that it is primarily a focus on the construction of standardized modular machines. In practice, there is a large number of requirements that, together with the lack of agreement, has created a great deal of diversity in the design and building blocks of machines in this area - there are no standards in this area.

When creating any production system from dedicated devices, it is necessary that the production machines are defined in groups to form a modular system to meet the requirements of each group. The production machines can be divided into 4 groups and are basic or "primitive" parts, drives, modular basic elements, configurable control units. By using these parts, it is possible to build a diversifiable, efficient, automatic and integrated production system [10].

b. Category of modules

Any modular type device is built up of units with little or no dependency on each other. Two basic types of modules are function modules and production modules. For function modules, modules need to perform their function, either alone or in combination with other modules. Production modules are mainly subject to production requirements, so they need to be filled. The modules themselves can be based on different types of functions, and these functions are repeated throughout the system (basic, auxiliary, special, adaptive and customer) [10].

Modularity can e of a different character. If a modular series is created where the individual elements are similar and differ only in size, it is size modularity. When modularity elements are included in the system so that it is possible to use different types of modules per basic module, it is functional modularity.

c. Creating modular units

In modularity, it is important how individual sub-module assemblies are interconnected and can be based on different module combinations. The relationships between the modules are based on the interaction analysis with each other and with the final product. We have several variations of linking:

- Component interchangeability- type modularity that belongs to the same product family, combining two or more components with the same core component. Figure 3a shows the construction scheme of the aforementioned modularity and describes the interchangeability of the individual components with the ability to connect to the base module.
- Sharing components- this is a type of product linking when different products from different modular designs share the same core component. A very similar type of modular connection like the previous type (Figure 3b).
- Size variation- has several standardized components with one or more additional devices. Frequency is often associated with physical dimensions. (Figure 4a).
- Bus modularity- this is a type of modularity when one module can be exchanged for another module in the base assembly. This type of modularity allows for a large number of interchangeable variations (Fig. 4b).

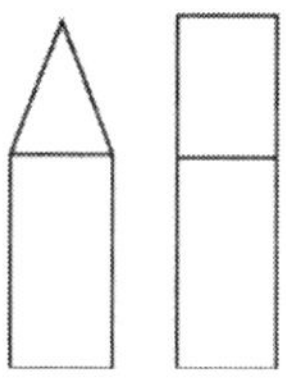

Figure 3.a Component interchangeability.

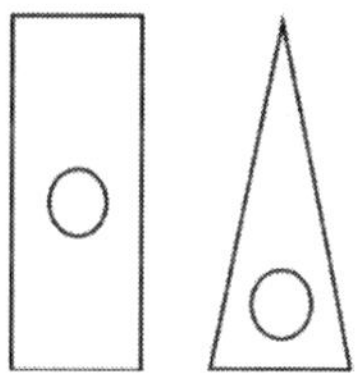

Figure 3.b Sharing components.

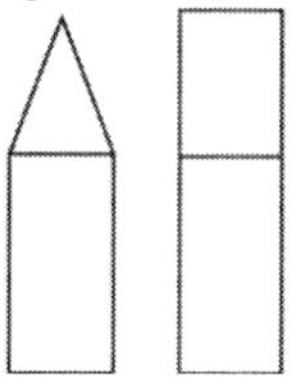

Figure 4.a Size variations.

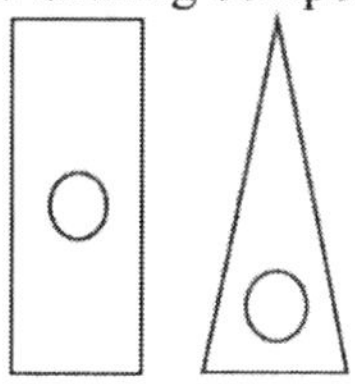

Figure 4.b Bus modularity.

d. Creating modular systems

The design of modular systems is possible only on the basis of a thorough analysis and division of the system into basic functional elements that characterize the basic physical principles and are integrated into the components of the modular system and capable of performing their function. The first key point in the creation of a modular system is the decomposition to basic problems. The second point is the integration of individual subsystems into an overall solution.

The structural problem decomposition system has at least two pros. The first is the simplification of the overall problem to smaller, solvable parts and the second is the possibility of a parallel solution of individual parts of the complex problem (to shorten the solution of the overall problem). Decomposition methods can be categorized according to where they were used, e.g. product decomposition, problem decomposition, process decomposition.

Product modularity is the ability to replace some component of the same or similar physical properties with other products of the same or similar properties. Modularity depends on physical properties between components. The system is divided into subsystems, sub-assemblies, assemblies, components leading to products and a particular design. The entire decomposition is captured in a hierarchical structure, along with dependencies between the subsystems.

The distribution of the whole system consists of constant simplification up to the basic ("primitive") element, which is independent of each other, respectively. Dependent as little as possible. The interrelationships between products should be taken into account as constraints on higher sub-issues. Problem decomposition is composed of decomposition of requirements, decomposition of constraint parameters and decomposition of basic design values. The requirements are constraints on solving a design problem. It starts with basic requirements and ends with specific (customer) requirements. The layout of the requests and the relationships between them are represented by a tree diagram.

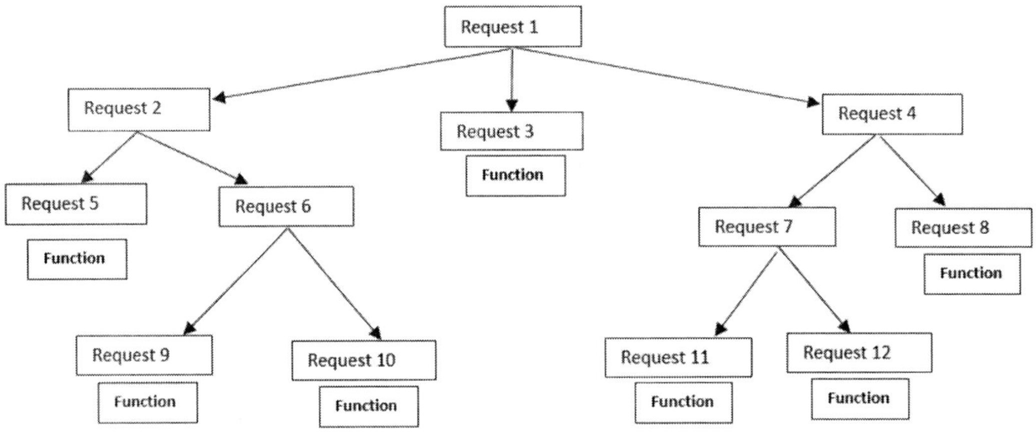

Figure 5. Decomposition based on requirements [10].

5. Design algorithm of modular lines

In this part of the article, which deals with the creation of an algorithm for the design of single-purpose machines and devices in the automotive industry, it is necessary to say that the proposal is based on three key parts:

- Device planning (5.1)
- Concept creation (5.2)
- System design (5.3)

Subsequently, the whole algorithm is discussed in chapter 5.4, taking into account the fact that it is an automotive environment and what is often an environment that is subject to change and the construction of machines working there is an expensive matter.

a. Device planning

Designing using conventional methods does not give designers as many solutions as possible using new methods. By incorporating new materials, technologies, processes and other factors into the design process, we can design new combinations that will serve better solutions. Designers should occupy design points that need to be improved, summarized in the following points:

- Improvement of technical, quality, safety functions
- Weight and space reduction
- Reducing economic intensity
- Reduce delivery time
- Improving production methods

The importance of these points is different from proposal to proposal. While the impetus for improvement may be the need to improve one of the aforementioned points and create a new working principle on the basis of which the device will operate. New inventions include a fundamental solution, modifiable in manufacturing methods, based on lower prices and shorter delivery times. Therefore, the plan needs to be well thought out and to determine precisely what the facility is going to perform and to determine the exact working conditions together with the continuation of other working procedures, possible future changes [1].

b. Concept creation

Creating concepts can be said to be one of the most important parts of the equipment design process (often underestimated), because it is precisely in the creation of concepts that we are able to influence future equipment, its shape, cost of production, selling price, performance, and so on. By design, it is necessary to formulate the working principle of our modular product at first. The principles must meet the functional requirements, most of which are based on the competitor's experience, patents, and catalogs [11]. The formulation of construction requirements plays a role in mapping relationships between individual working principles. All the principles are evaluated in the table using important criteria, thereby separating the non-essential ones for the designer. The matrix itself is later decomposed and the design modules are defined. By systematically combining working principles, we are able to create variations of construction modules and then build upon them to build a modular design family scheme of the desired device [2].

c. System design

The design of the device is progressively made up of modules that are suitable for use in equipment, both economically and technically. After analyzing and evaluating the modules based on the criteria, the product structure is written and the modular product family is assembled. The final product is subjected to a practical assessment. If the design of a modular construction does not meet the requirements, a return to the design process is needed, and more specifically to the point of combination design and selection of device module variants. Subsequently, the design process proceeds to the next points, where the design of the device is detailed [1].

d. Algorithm design

The resulting algorithm is based on the previous chapters; their task was to bring the process of creating a device to a modular character, and this can be true not only in the automotive industry but also in other industries. The automotive industry has its rules and time plays a major role, along with other important conditions (see Chapter 3).

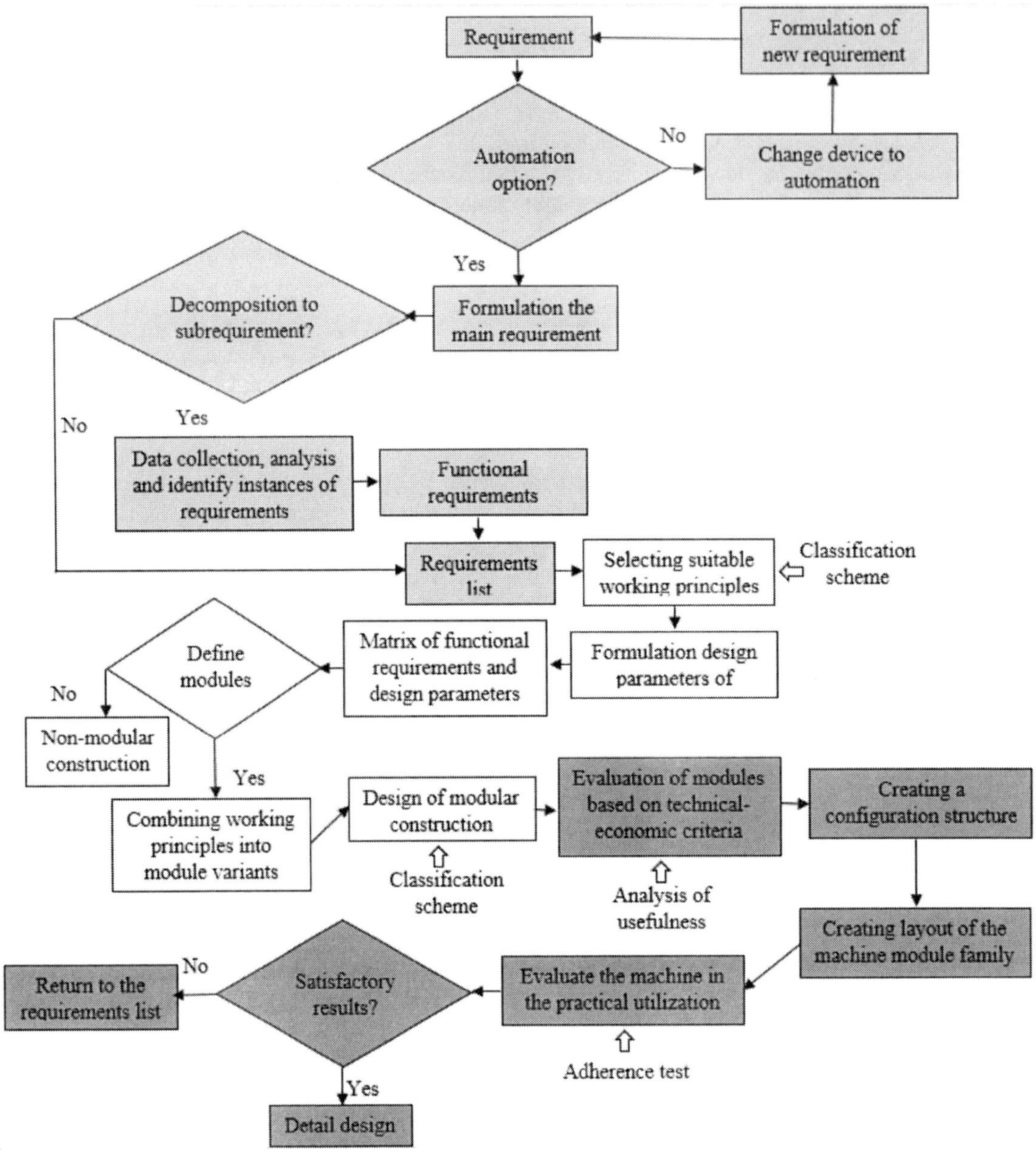

Figure 6. Modular design algorithm.

6. Conclusion

The article was intended to provide the reader with an insight into the design of machinery in the automotive industry, whether manufacturing or assembling. Mostly designed devices relieve people from labor, either work with it or replace it, thereby reducing costs. The proposed algorithm is the nature of a flowchart and gives the reader guidance on how to deal with the problem. Obviously, the design needs to take into account the requirements of the automotive industry and it depends on the designer/-s how to cope with the problem and what values (importance) they will give when creating classification schemes for individual requirements.

Acknowledgments
The research is supported by the Cultural and Educational Grant Agency of the Ministry of Education, Science, Research and Sport of the Slovak Republic under the project No. 046ŽU-4/2018.

References

[1] Gauss L, Lacerda D P and Selitto M A 2019 Module based machinery design: a method to support the design of modular machine families for reconfigurable manufacturing systems, *The international Journal of Advanced manufacturing Technology* **102**(9-12) 3911-3936
[2] Bašťovanský R, Bronček J and Žarnay M 2017 *Metodika konštruovania* (Methodology of construction) EDIS (in Slovak)
[3] Pahl G, Beitz W, Feldhusen J and Grote K-H 2007 *Engineering Design: a systematic approach-3rd ed,* Springer
[4] Andersen A-L, Rösiö C, Bruch J and Jackson M 2016 Reconfigurable Manufacturing – An Enabler for a Production System Portfolio Approach, *Procedia CIRP* **58** 139-144
[5] Múdry M 2016 *Predaj robotov s dvojciferným rastom* (Sales of robotos with double-digit growth), Engineering, MEDIA/ST (in Slovak)
[6] Bronček J, Kučera Ľ, Žarnay M, Poljak S and Čillík L 2015 *Konštruovanie 1* (Mechanical designing 1) EDIS (in Slovak)
[7] Pilc J, Stančeková D, Mičietová A and Salaj J 2001 *Jednoúčelové stroje a výrobné linky* (Single-purpose machines and production lines) EDIS (in Slovak)
[8] Gregor M 2016 Nový koncept budúcej výroby (the new concept of future production), *Engineering,* MEDIA/ST (in Slovak)
[9] Belorit M, Hrcek S, Gajdosik T, et al. 2017 *Description of the bearing check program for countershaft gearboxes*, 58th International Conference of Machine Design Departments (ICMD 2017), Prague, Czech Republic, Sep. 06 – 08, 32-35
[10] Kamrani A K and Salhieh S 2000 *Product Design for Modularity*, Springer
[11] Caco M, Kohar R, Hrcek S, et al. 2017 *Use the method of TRIZ in optimizing automated machine for ultrasonic welding,* 12th International Conference of Young Scientist on Sustainable, Modern and safe Transport, High Tatras, Slovak Republic, May 31 – Jun 02, Procedia Engineering vol. 192, 80-85

IOP Conf. Series: Materials Science and Engineering **659** (2019) 012026 doi:10.1088/1757-899X/659/1/012026

Transaction applications of enterprise information system

V Prajová[1], M Homokyová[1] and M Horvátová[1]

[1]Slovak University of Technology, Faculty of Materials Science and Technology, Institute of Industrial Engineering and Management, J. Bottu 25, 917 24 Trnava, Slovakia

E-mail: vanessa.prajova@stuba.sk

Abstract. The huge amount of data currently available to enterprises needs to be exploited to their advantage, as the lack of information can decide the future of the enterprise in the near future. On enterprises development in recent years has been greatly influenced by the rapid development of technologies for processing and organizing them. These technologies share the role of an information integrator to support decision-making at all levels of the manufacturing enterprise. Enterprise resource planning is a key transactional application of enterprise information systems. Its main features include the ability to automate and integrate key business processes, functions and data across the enterprise.

1. Introduction

In terms of defining enterprise data categories, transactions have been characterized as transforming a state that has atomicity, durability, consistency, and isolation characteristics. Business transactions then change business, usually during activities or processes [2].

In order to support business transactions, transactional applications. These applications integrate a variety of enterprise management tasks into one application sharing a common data base. This ensures the minimization of inconsistencies and errors in business data and eliminates inefficiencies in operations with these data. Enterprise Resource Planning is a key transactional application of enterprise information systems.

Enterprise Resource Planning (ERP) is an enterprise resource planning and management tool that integrates sub-agendas, comprehensively supports core business processes, and provides the necessary business management information. There are two basic views of ERP systems. In a narrower perspective, the ERP system represents a solution that integrates in-house areas, especially manufacturing, logistics, finance and human resources. In a broader perspective, ERP systems also include applications for managerial decision-making, MIS, and applications to support the environment. In the case of suppliers, it is SCM (Supply Chain Management), in the case of customers, CRM (Customer Relationship Management) [3].

2. ERP application features

The functions of ERP applications can be divided into the basic functions that are associated with the three basic circuits of transactions (economy, logistics and human resources), and the extensions. The individual functions are grouped into application modules. The range of functions and range of application modules can vary widely across different ERP applications.

ERP applications provide a wide range of features and are most comprehensive in relation to other types of applications. If we want to compare the functionality of ERP applications of different suppliers, then it is good to draw attention to these pitfalls [1]:

- ERP functionality is very extensive and therefore more detailed comparative analysis is quite complicated and time consuming;
- the structure of the functions, and thus the menu structure, is very different between different ERP applications;
- various ERP application vendors use partially different terminology, different names for similar functions.

The basic ERP modules or feature groups include [1]:

- ***Economic governance*** - must provide a comprehensive view of the economy of the entire organization and the effective implementation of financial operations. It usually includes features related to the general ledger and individual journals, bank relationship management functions, asset management, cost accounting. The module provides a comprehensive overview of financial operations in the company, evaluation of the economic performance of the company and its individual business units and continuous provision of compliance of the information system with legislation.
- ***Sales and marketing*** - includes, in particular, integrated support for customer management, sales activity management and marketing, displaying relationships between customers, buyers, suppliers, employees and competitors, managing business opportunities, sales management, support for creating and managing marketing campaigns, and evaluating their results.
- ***Purchasing and warehouse management*** - Provides support for processing purchase requirements, for stock assessment and for own supply operations, supply price analysis, for material requirements records of individual manufacturing and other centres, and communication of purchase and inventory management requirements [6].
- ***Human resources management*** - Ensures personal records and, in particular, supports the management of the company's staff development, its efficient use and recruitment.
- ***Production module*** - is focused mainly on production planning, or production orders, monitoring their status and performance with respect to deadlines, monitoring and evaluation of stock, production management at the level of operative and workshop management.

Most manufacturers supply specialized vertical solutions for individual areas, which differ from each other by the composition of modules, functionality and support for individual processes. The overall degree of process automation by implementing an ERP system is different in every industry. The common area occurring in all verticals is support for purchasing and accounting processes.

3. Linking ERP to Business Process Management

There are various explanations for the concept of process in the literature. However, it can simply be defined as "transforming inputs into outputs". The ISO 9001: 2008 standard characterizes the process as "a set of interrelated or interacting activities that change inputs into outputs." If we look at definitions by author [10], we understand the process - a set of activities that require inputs and generates value-added outputs for the customer.

If we look at this definition, a closer process is not only linked to the production process, as the word process might seem. The process concerns all areas of the business, administrative or non-administrative, and thus, of course, manufacturing. In addition, processes go far beyond the enterprise and thus create and shape relationships with other businesses, partners, customers, employees, government [7].

The process is triggered by an initial event. The event can be created after any activity, with the most important end goal being a customer-oriented outcome [1]. The process or process result is determined by a variety of parameters, such as process costs or process efficiency. In this case, the value for the endpoint - customer. The process owner and the process parameters are responsible for the process.

There are so many business processes in the company that there would be very little overview of them without their categorization. The information system implementer works with these categories. Therefore, three basic process categories have been created [4]:

- ***key processes***: are processes that lead to the company's core vision, to profit, and thus to customer satisfaction;
- ***supporting processes***: these processes are equally important as key processes, they are also customer-focused, but internal. Internal customers are employees of the enterprise and thus one of the inputs to the process;
- ***side processes***: are support processes that can be excluded from the organizational structure of the enterprise and thus can be "outsourced".

The link between the information system and the business process is confirmed by the implementation of enterprise information systems by the need to carry out an "inventory" and, of course, to change all processes in the company, whose activities are affected by the new information system. In today's companies, it is possible to say almost all areas from logistics, business and technical departments, through employee records and all related tasks - remuneration of merchants based on delivered goods, remuneration of technical workers based on repairs and installations, remuneration. BackOffice workers for their work, to quantify all work and results in the hands of management [8]. As a result, it is not only the achievement of data availability but also the method and thus the processes that obtain and handle the data. Finally, foreign surveys [3] demonstrate this information system symbiosis with processes, which show that the introduction of ERP systems has brought not only cost reductions and quality improvements, but also a space for decision-making in decision-making and customer-driven decision-making. Also improve processes throughout the enterprise.

Most ERP vendors offer present processes for each vertical. In the following section, we look at an overview of the basic processes delivered in the E-Business Suite for manufacturing.

The basic processes provided by the E-Business Suite ERP package are [1]:

- Order entry, planning and customer service
  - ✓ from forecasting to planning;
  - ✓ from order to delivery;
  - ✓ from order to delivery from manufacturer;
  - ✓ from claim to credit note;
  - ✓ from complaint to exchange.

- New product development, workshop management
- from design to release to production;
- from planning to production schedule;
- from schedule to production;
- from standard production costs to warehouse valuation.

- Shopping processes
- from request to income - indirect purchase;
- from request to receipt - direct purchase;
- from analysis to contract.

Business processes need to be mapped to maximize efficiency. Mapped processes are recorded in the business process model. The role of this model is to summarize all the activities of the company that generate high efficiency gains in one place. Such an overview then makes it easy to look at the entire process structure of the company, allows the processes to monitor, control and then continuously optimize, train new employees, each of which is easier to access, thereby saving time and thus the whole company [1].

The business process shown in the Figure 1. begins with the creation of an application, the request to send a quotation, the creation of a contract or a purchase order, the processing of goods receipt, the processing of the incoming invoice, the payment. In manufacturing plants, the primary source of demand is the purchase of direct materials planning - the most common run of MRP.

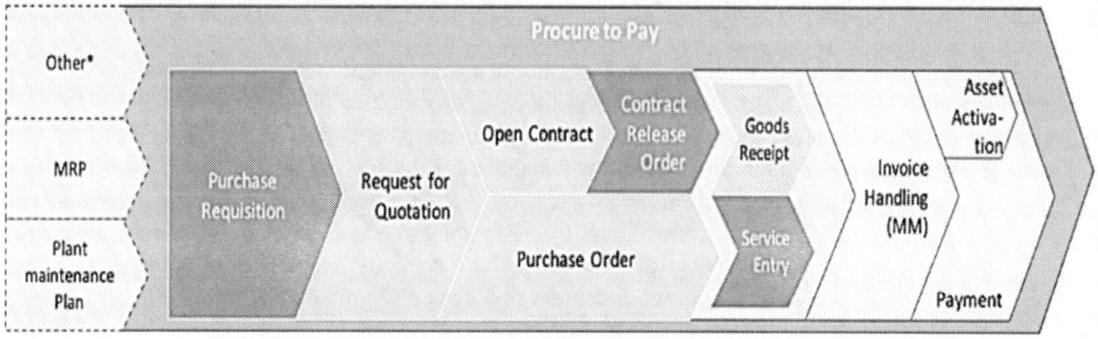

Figure 1. SAP Process Diagram, From Purchase to Payment [11].

4. Metrics as an indicator of ERP efficiency and performance

Many IT projects in the past and present in businesses are unsuccessful. The projects do not meet their objectives and do not bring the expected benefits of their introduction. IT support for business processes is so often sceptical. Solutions that were originally intended to bring productivity gains, time savings, cost reductions are often out of control and ultimately do not meet customer expectations. It is no exception if the situation becomes even worse after the introduction of the information system. Summary of IT Problems by [9]:

- the benefits of introducing new technologies from the IT sphere are not being met;
- implemented systems are not being used properly, failing their possible benefits to the organization;
- many projects are closed or cancelled due to budget or time frame overrun;
- users reject changes in information technology;
- there is a misunderstanding between the needs of business and IT.

For accessing enterprises to deploy ERP by [9], the recommended basic steps are as follows:

- ensuring support for a new project in the company;
- use of external consultant services - deployment partner;
- thorough mapping of key task holders;
- identification of business processes;
- project scope definition;
- written documentation of requirements;
- identifying potential solutions;
- solution presentations;
- reference projects;
- final decision.

One of the key issues encountered in introducing a methodology to measure the benefits of deploying an ERP system or other major change affecting an enterprise is to use an appropriate methodology to evaluate these benefits. As the comprehensive deployment of the new ERP will generally affect all of the company's departments with the desire to cover and optimize key in-house processes, it is appropriate to address a methodology that can quantify the benefits and improvements from a corporate perspective. According to [5], the metric is a system of parameters or pathways for the quantitative and periodic evaluation of processes that will be measured. Metrics are typically specialized by area of application and cannot be mechanically transferred to other applications.

5. ERP effects

ERP applications bring users the full range of effects among the main features include [1]:

- Labor productivity increases in normal business and administrative activities through the use of existing data (about customers, suppliers, goods...) and copying them into existing documents.
- It reduces the risk of errors and errors in management activities, business or financial transactions by using embedded control mechanisms in the software.
- It reduces the risk of errors and errors in management activities, business or financial transactions by using embedded control mechanisms in the software.
- The accuracy of decision-making operations is enhanced due to the interconnection of individual ERP modules, for example, by more objective assessment and planning of customer orders with immediate evaluation of their material capacity and economic efficiency.
- The overall level of business management is enhanced by using ERP embedded management methods, such as controlling, order planning, and so on, with many previous installations, often worldwide.

6. Conclusion

Nowadays, the automation of in-house processes and their optimal implementation is an important success factor. ERP systems provide efficient business process management at a low cost. They are a solid foundation on which to build and operate new types of enterprise software applications that can provide strategic benefits in the competitive world of today. Only efficiently executed processes are able to move the company to the competition, increase its competitiveness, flexibly respond to constantly changing customer requirements and strengthen the market position. Implementation of modern ERP systems is a means by which it is possible to cope with ever more demanding customer requirements, increase competition pressure, fast technological development and globalization of business. The ERP lifecycle is closely linked to the lifespan and support for the technologies in which it was developed. Already today, ERP systems have become the information ecosystem of any future-thinking company.

Acknowledgement

The paper has been written as a partial outcome of the project KEGA No.015 STU-4/2018: "Specialised laboratory suppoted by multimedia textbook for subject: "production systems design and operation" for STU Bratislava.

References

[1] Gála L, Pour J and Šedivá Z 2015 *Enterprise Informatics*, 3rd, updated edition, Praha, Grada Publishing, a.s., ISBN 978-80-247-5457-4
[2] Bruckner T, Voříšek J 2012 Buchalcevová A 2012 *Creation of information systems*, Praha, Grada Publishing, a.s., ISBN 978-80-247-4153-6
[3] Basl J and Blažíček R 2012 *Business information systems*, Praha, Grada Publishing a.s., ISBN 978-80-247-4307-3
[4] Řepa V 2006 *Business processes*. Praha, Grada Publishing a.s., ISBN 80-247-1281-4
[5] Učeň P 2001 *Metrics in Computer Science*, Praha, Grada Publishing a.s., ISBN 80-247-0080-8
[6] Basl, J 2011 *Innovation of enterprise information systems*, PBtisk Příbram, ISBN 978-80-7431-045-4
[7] Sodomka P 2016 *Information systems in business practice,* Computer Press, Brno, ISBN 80-251-1200-4

[8] Vrana I and Richta K 2017 *Principles and procedures of enterprise information systems implementation*, Grada Publishing, Praha, ISBN 80-247-1103-6
[9] Kováč M 2009 *Metrik ERP system applications,* In: Transfer inovácií 13/2009, Technical University of Košice, ISSN 1337-7094
[10] Hammer M 2002 *Process Management and the future of six sigma*, MIT Sloan Management Review
[11] https://erproof.com/mm/free-training/sap-procure-to-pay-process/

IOP Conf. Series: Materials Science and Engineering **659** (2019) 012027 doi:10.1088/1757-899X/659/1/012027

Tools of organizational-economic mechanism of internal control functioning

M K Aristarchova[1] and V S Fakhrutdinov[1]

[1]Ufa state aviation technical University, Institute of economy and management, department of taxes and taxation, K. Marx St., 12, 450008 Ufa, Republic of Bashkortostan, Russian Federation

E-mail: ninufa@mail.ru

Abstract. The article touches upon the topical issues of the internal control functioning at the enterprise. The tools developed by the organizational-economic mechanism of functioning of internal control, including the principles, functions, logicaly, time aspect, controlling, norms. The structural and functional components of internal control are identified, the technology of each of them is considered. The adaptive organizational structure of the risk management and internal control system is formed, which can be the basis for the formation of the model of internal control of the gas transportation enterprise.

1. Introduction

For almost a decade, since the introduction of article 19 of the accounting Act, there has been a continuing interest in internal control among the scientific community, consultants and practitioners. Innovation of the Law of 06.12.2011 № 402-FL obliged economic entities to organize and carry out internal control of committed facts of economic life. To date, the legislation on internal control in the Russian Federation is represented by numerous normative legal acts. At the same time, the analysis found that "in Russia at the moment there is no single normative document regulating the activities of internal control, the information is "blurred" among a large number of normative and subordinate acts [1]". Accordingly, in the absence of its own source of methodology, the construction of an internal control system always implies an element of creativity, i.e. each individual system of internal control is unique, which opens up the opportunity to design their own systems.

However, on the other hand, the organization and implementation of internal control – the way to the unknown: who and how should exercise internal control? What principles should correspond to the functioning of the organizational and economic mechanism of internal control? How to establish the process of obtaining information on the effectiveness and efficiency of activities, the reliability of all types of reporting and compliance with applicable law?

To answer the questions posed in the framework of the study, a review of the existing literature [1] was conducted and it was found that the function of "Internal control" came from a practical field: in the West, its legacy can be considered an external audit, in Russia – state financial control and audit.

It is also understood that internal control is inextricably linked to the management process and is an integral part of it, is present at all stages of management: planning, implementation, accounting and control, analysis.

The basis of the internal control task is to identify deviations of the current actual state from the desired result for the subsequent adjustment of management actions and, as a consequence, the

achievement of such a desired result. In a General global sense, control is follow-up, that is, after action has been taken. The disadvantage of this format is the loss of time and functioning after the fact.
Thus, a conscious approach to the function of "Internal control" is the ability in the current mode to fix deviations from the goals and adjust management decisions, which in turn always leads to a competitive advantage.

The structure of the conceptual approach of the organizational and economic mechanism of internal control is shown in figure 1 and includes the following steps:

1. Formation of the ideology of organizational and economic mechanism of internal control, containing: the basic values, the rationale for the interdependence of all subsystems of the enterprise with the mechanism of internal control, the basic laws of its organization, the requirements for the organization of the structural components of the internal control system.
2. Development of tools of organizational-economic mechanism of functioning of internal control, including: principles, functions, logicaly, time aspect, controlling, normal.
3. Creation of an adaptive organizational structure of the organizational and economic mechanism of internal control functioning through the use of the developed tools.

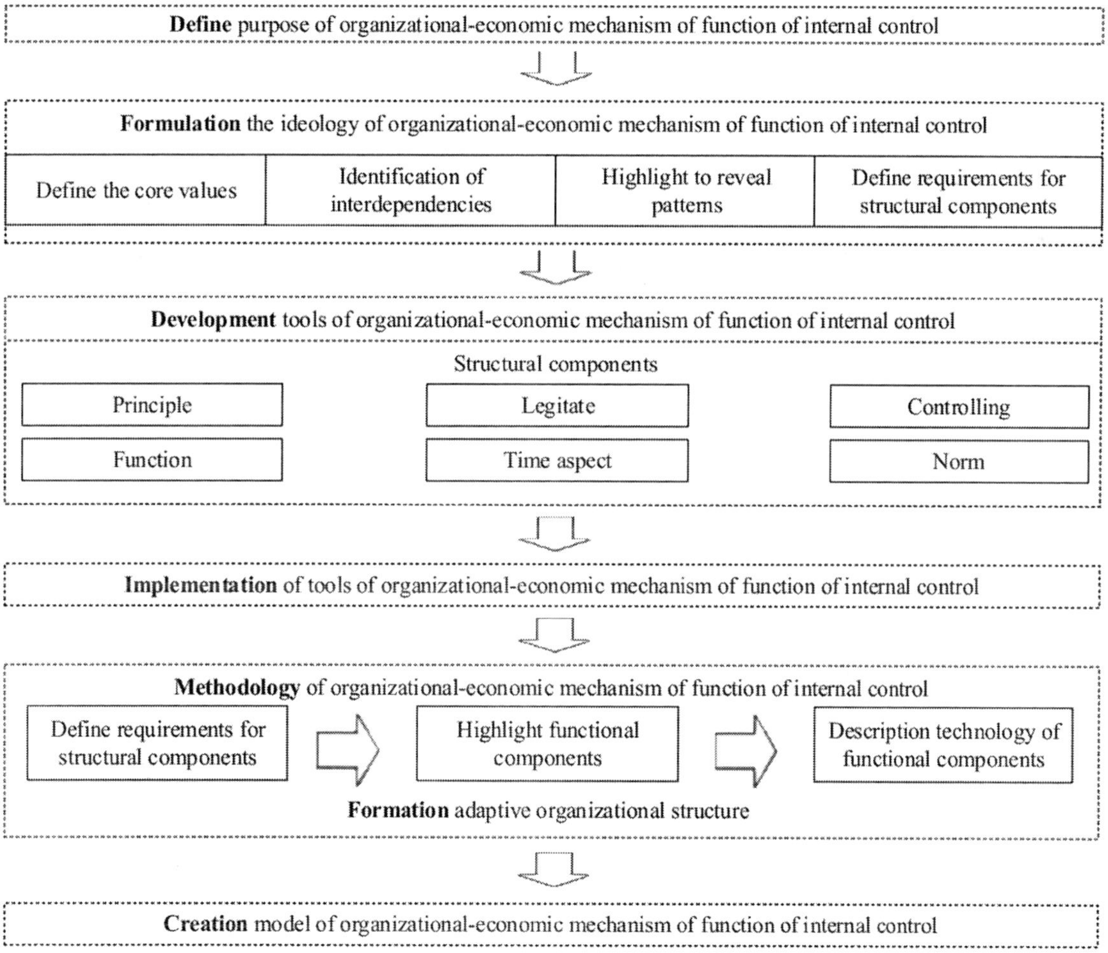

Figure 1. Concept of organizational-economic mechanism of function of internal control.

A. Formation of the ideology of organizational and economic mechanism of internal control

The initial stage of developing the concept of the organizational structure of the organizational and economic mechanism of internal control is to determine the purpose of its organization. According to the Ministry of Finance of the Russian Federation № PL-11/2013 internal control is designed to obtain sufficient confidence that the economic entity ensures the efficiency and effectiveness of its activities, including the achievement of financial and operational indicators, the safety of assets [2].

V. V. Burtsev defines the purpose of internal control as "ensuring information transparency of the management object for the possibility of making effective decisions. In this case, the concept of "information transparency" of the control object reflects the idea of the degree of controllability of the object, that is, to what extent as a result of management is provided to maintain the desired state or action of the control object at the appropriate time (period)" [3].

In other words, the objectives of the internal control mechanism is to ensure sufficient confidence in the achievement of the goals of the economic entity, as well as proper control over financial and economic activities.

To achieve this goal, the internal control system should address the following main tasks:

– effectively manage risks, improve decision-making processes to respond to risks;
– perform internal control procedures properly and in a timely manner;
– develop internal control procedures (including those aimed at preventing and combating corruption) for timely response to risks;
– timely identify deficiencies, deviations and violations of internal control procedures in the activities of the enterprise;
– create conditions for the timely preparation and submission of reports, as well as other information to be disclosed in accordance with the law;
– improve the efficiency of use and allocation of resources for risk management and internal control procedures;
– to ensure compliance with the legislation and local regulations of the enterprise.

Accordingly, the management of internal control should be aimed at providing sufficient guarantees to achieve the goals and objectives through a set of interrelated organizational measures and processes, organizational structure, local regulations, other documents, methods and procedures (regulations, regulations, standards and guidelines), norms of corporate culture and actions taken by employees of structural units.

All of the above proves that the organizational and economic mechanism of internal control is determined by the laws of formation of complex systems. Presentation of the organizational and economic mechanism of internal control functioning as an integral part of the enterprise management system determines the need to highlight the role, i.e. subordination to synergetic laws. In the process of building an internal control system should take into account a number of laws, the presence of which is due to the role of control in the economy of the enterprise.

Summarizing and summarizing the above provisions of the ideology of internal control, we note that they are in a strict logical relationship. The formulated basic values of internal control are aimed at creating awareness among managers and employees of the enterprise, the importance of monitoring in their daily activities. With this in mind, the second provision of the ideology of the organization of the internal control system is the allocation of the relationship with all the functional subsystems of the enterprise. The internal control system is an integral part of corporate governance, integrated with existing planning systems, project and program management, industrial safety management and other management systems. Internal control processes must be built into all business processes. Third position ideology contains selected and specified regularities of internal control. The fourth provision of the ideology is the definition of the requirements for the organization of structural components of internal control, one of the key requirements of which is strict compliance of the structural construction of the internal control system with the logic of all types of work related to it.

B. Development of tools of organizational-economic mechanism of functioning of internal control.
The presented ideology is the theoretical basis for the development of tools of organizational and economic mechanism of internal control, which include:

– principles of internal control organization;

– internal control functions;
– legitate internal control;
– time aspect;
– controlling;
– norm as a condition of goal-attainment.

The principles of functioning of internal control determine the basic requirements for its organizational structure, formulated on the basis of the above regularities of functioning of internal control. Management regularities are distinguished by the group of General management principles. The laws and principles of Cybernetics allow us to distinguish the General rules of the organization of internal control, considering it as a complex system. Synergetic laws are implemented by the principles of synergy. The regularities due to need to take into account the role of internal control in the economy of the enterprise are identified by a group of special principles. Thus, four groups of principles were identified, generalized and formed: General management, principles of Cybernetics, principles of synergy, and special principles. The classification of the principles of the organizational and economic mechanism of internal control is presented in figure 2.

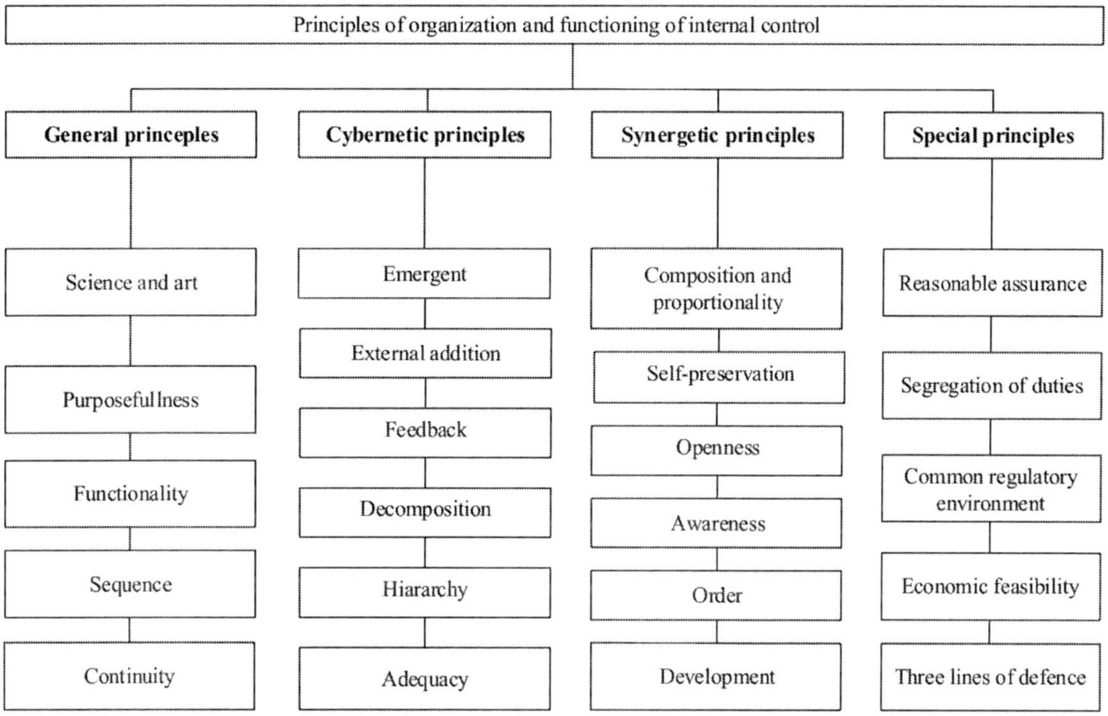

Figure 2. Principles of organization and functioning of internal control.

Internal control functions are the tool that determine the composition and content of the work of each structural element of the internal control system. Each structural component of internal control is assigned a corresponding function. Figure 3 shows the scheme of assigning the allocated functions to the relevant participants of the internal control system.

The functions presented in figure 3 are based on a risk-based approach, which is now considered to be the most popular way of organizing any activity, since its use provides the possibility of spot checks, saves time, allows you to adjust the intensity, duration and frequency of control actions. Also, the obvious advantage of using a risk-based approach in the planning and implementation of the "Internal control" function is seen as resource savings, since management is carried out exclusively at critical

IOP Conf. Series: Materials Science and Engineering **659** (2019) 012027 doi:10.1088/1757-899X/659/1/012027

moments, and preventive control and risk measures can be integrated into management and decision-making processes.

A similar conclusion in favor of risk-oriented technologies in internal control can be made on the basis of the study of the labor functions of the internal controller set out in the professional standard "Internal control specialist (internal controller)", approved by the Order of the Ministry of labor of Russia from 22.04.2015 № 236n [4], namely:

– identification and assessment of conditions that contribute to the emergence of significant risks and their transformation into risk events, development of recommendations for effective risk management;
– business process risk analysis;
– assessment of compliance of the performed economic operations and efficiency of use of assets to legal and regulatory framework;
– assessment of the correctness of financial and economic operations;
– evaluation of the effectiveness of the developed control procedures.

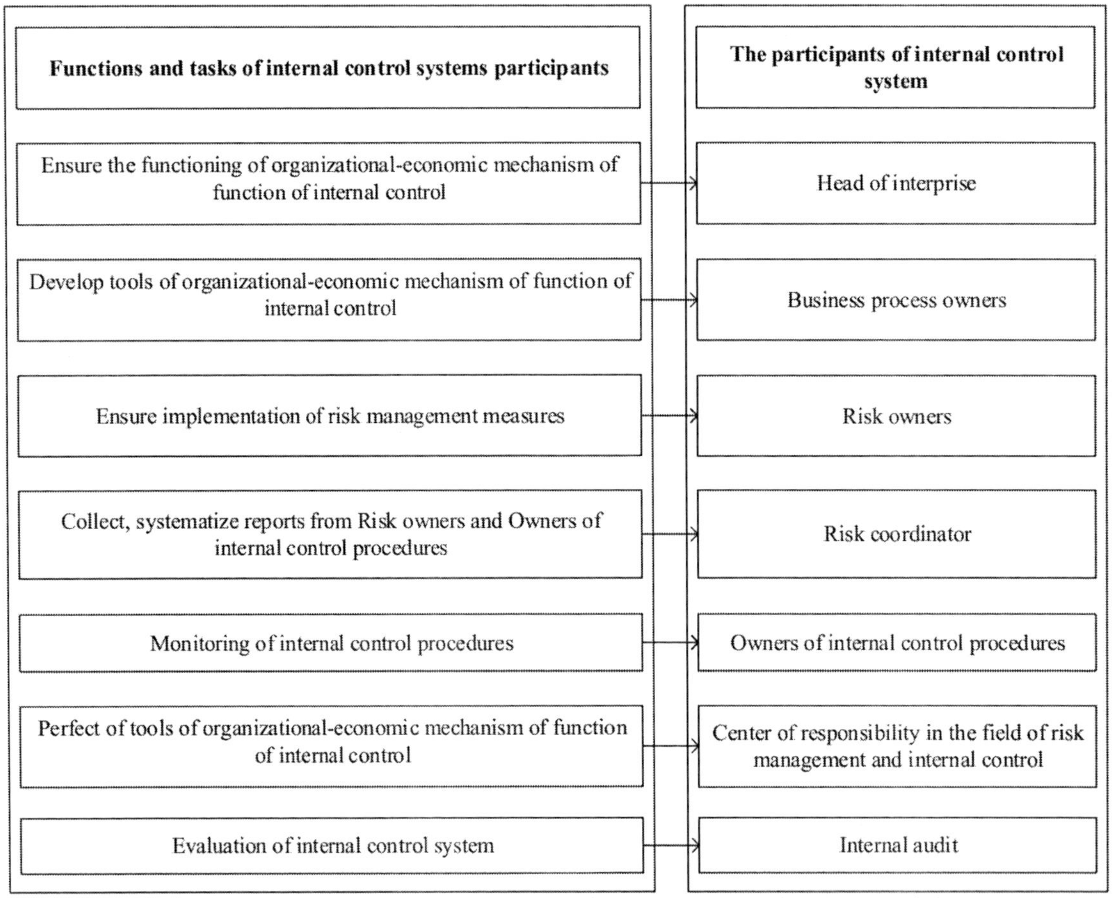

Figure 3. Consolidation of the functions of the internal control system participants.

There are also other variations in the way internal controls are organized and implemented:
– system approach based on Western concepts (SAS 55/78, SAC, COSO, COBIT);
– integrative approach of control subsystems based on ranking the degree of importance of losses;
– on the basis of accounting data and formation of organizational and instructional cards;
– analytical approach involving the creation and monitoring of key performance indicators, etc.

C. Creation of an adaptive organizational structure of the organizational and economic mechanism of internal control functioning.

Approaches can be highly individual, complement and even replace each other at certain stages. At the same time, it is important to build a system that will contribute to the achievement of the goal of internal control – to form an adequate, timely system of information support for all levels of management.

The Foundation of the future internal control system is laid at the stage of creation of the enterprise (company): when designing the organizational structure and building business processes. The organizational structure is a scheme of hierarchical subordination, demonstrates the way of interaction between departments and management bodies in relation to the implementation of authority, information flows and performance of functional duties. The organizational structure, answering questions about the number and composition of units, their subordination, however, does not disclose the mechanism of work. To understand the essence and purpose of what is happening, the persons responsible for the process (sub-processes) and the results of the persons allow properly structured business processes. A business process is a system of consistent, focused, and formalized activities in which inputs are converted through leadership and resources into outputs (process results) of value to the consumer. In other words, the organizational structure is a field of business processes, the main task of which is to promote their work as much as possible without creating "friction" and "obstacles"

In this study, it is proposed to identify and justify the role and importance of such a structural component as the Center of responsibility in the field of risk management and internal control. It is assumed that the center of responsibility in the field of risk management and internal control created on the basis of a matrix organizational structure will provide adaptability to changes in the external and internal environment, flexibility, the ability to send the most suitable employees to the right places, the speed of decision-making, complexity.

The formation of the staff of the Center of responsibility is planned to be based on identified and classified risks (General, operational, legal, tax, etc.) from among the employees of such structural units as the Engineering center, Financial and economic management, accounting, Corporate protection service, property management Department, legal Department, labor protection Department, internal audit Department, as well as delegated employees of production departments.

The main tasks of the Center of responsibility in the field of risk management and internal control will be:

– coordination of the activities of the structural units for risk management and internal control;

– definition of methodological approaches to risk management and internal control processes and internal control procedures;

– realization of generalization and analysis of reporting the company's structural subdivisions on the realized risk of the data on the contractors, on contracts with counterparties on the financial instruments of internal control;

– formation of consolidated reporting on risks and internal control procedures for provision to management bodies;

– monitoring the implementation of risk management processes and internal control procedures;

– formation of a portfolio of key risks, their analysis in order to provide results to the head of the enterprise;

– analysis of measures to improve internal control procedures developed by structural units;

– ensuring the improvement and development of approaches, methods and tools of risk management processes and internal control procedures of business processes.

The interrelation of internal control functions determines the interrelation of the specialists of the responsibility Center performing the corresponding functions, therefore, the works performed by all participants of the internal control system are also interrelated. Interaction of risk management and internal control system participants includes vertical and horizontal interaction. Vertical interaction includes interaction between the head of the enterprise, structural units and risk coordinator. Horizontal interaction is carried out between structural units, risk coordinator, risk owners, owners of internal control procedures, owners of business processes and the enterprise as a whole. Interaction of the

Company with external stakeholders, including state supervision bodies, external auditors, banks, insurance companies, shareholders, investors is carried out in compliance with the requirements of the current legislation, local regulations. Application circuit control logicaly when designing a responsibility Centre in the field of risk management and internal control is presented at figure 4.

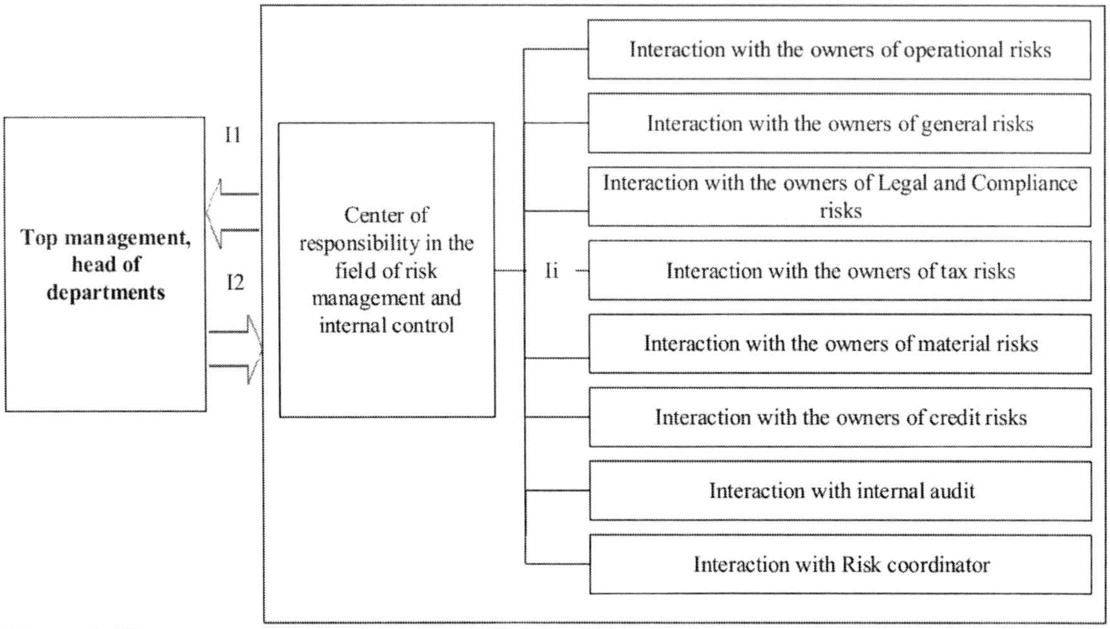

Figure 4. The content of the control of logicalization: ""I1" - information which Center's staff transfer in various structural subdivisions of the enterprise; "I2" - outside information, "Ii" - inside nformation.

The selected relationship within the Center of responsibility, as well as the developed technology of work performance, requires the integration of the internal control system into an integrated enterprise management system. It is proposed to apply logicaly and temporal aspects of internal control. With the help of logicaly coordination on business management and management control on the basis of organizational activities, systematic consultation of experts internal control unit with specialists from other departments, as well as on the basis of information transfer. The potential of the control logicaly allow us to consider it as a bridge through which there is integration and coordination of the internal control system and management of the enterprise at every level: strategic, tactical, operational.

Figure 5 shows a model of spatial representation of the organizational structure of the enterprise and the internal control system. The advantage of this model is the ability to clearly see that the business processes of the enterprise are combined with the levels of strategic, tactical, operational management. In addition, this model contains a vertical plane that permeates the architecture of the enterprise and demonstrates that internal control is present at each stage of enterprise management (strategic, tactical, operational), in its implementation involves all the staff of the economic entity (ordinary employees, line managers and enterprise management). Thus, the structural components of internal control-integrated enterprise management system by means of instrumentation: principles, functions, logicaly, time lines, controlling rules.

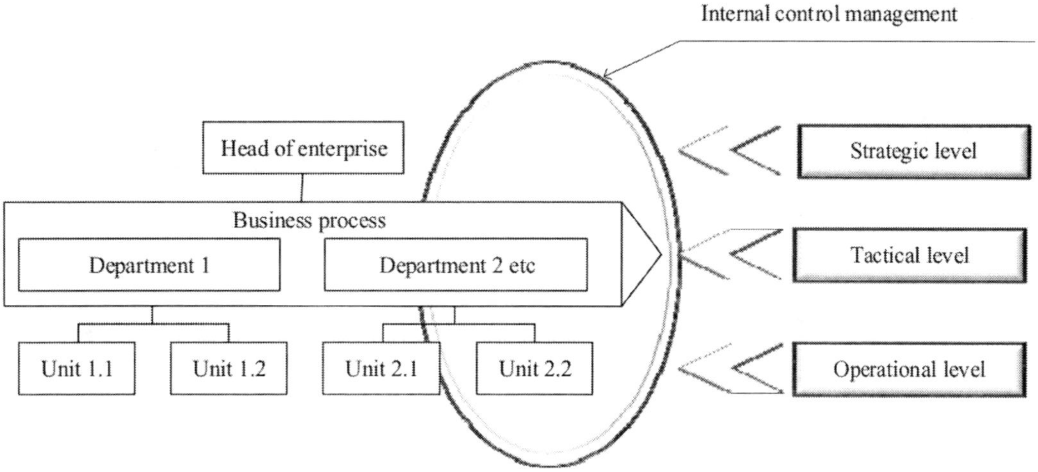

Figure 5. Model of spatial representation of the organizational structure of the enterprise and internal control management.

Summarizing all the above, we can conclude that the selected tools are closely interrelated and represent an integrated system, the application of which is determined by the developed methodology of internal control, within which three stages of the formation of an adaptive organizational structure are identified. At the first stage, the structural components of internal control are determined. At the second stage, the functional components of internal control are distinguished. At the third stage, the technologies of each functional component of internal control are developed. The composition and characteristics of the structural components of internal control, the type of its organizational structure, the form of organization of internal control will be individual for each enterprise, since all of the above must be consistent with the overall organizational structure, and in addition to take into account the wishes of customers and legal requirements.

2. Results

The main purpose of internal control is to identify inconsistencies in any time period in all areas of enterprise management. The functioning of internal control is aimed to provide assistance and support to management, being an additional channel of information for it, it provides a methodological basis for other control systems, coordinates all control functions within the organization.

In essence, internal control is formed by a set of subjects of internal control; internal control procedures, i.e. the tools by which the controller carries out its work; norms of behavior and actions of employees. Within the framework of the internal control system, the norm is designed to maintain its functioning in a stable state, providing a certain degree of freedom, avoiding negative trends.

Also, when building a system of internal control there is an excessive overregulation of economic activity. To solve the existing problems, we used a risk-oriented approach, which has provided an opportunity to choose the object of inspection, as well as the intensity of control measures.

In this case, as a result, we got no unwanted controls, adequacy of control procedures in form and content, transparency and control of business processes. All this, certainly, moves apart the horizons and opens a way to creativity.

References

[1] Aristarkhova M K and Fakhrutdinov V S 2019 *Prerequisites for the formation of internal control and some problems of its functioning*, All-Russian competition of scientific, methodological and creative works on Economics – 2019, economic Olympus

[2] Information of the Ministry of Finance of the Russian Federation № PZ-11/2013 *Organization and implementation by the economic entity of internal control of committed facts of economic*

life, accounting and preparation of accounting (financial) statements, [Electronic resource]. Access mode: http://www.minfin.ru ahhh! [Electronic resource]. Access from ATP consultant Plus

[3] Burtsev V V 2002 *Internal control in the organization: methodological and practical aspects*, Audit reports, 2002, (8)

[4] Order of the Ministry of labor of Russia dated 22.04.2015 № 236n *On approval of the professional standard* "Internal control specialist (internal controller)" (Registered in the Ministry of justice of Russia 13.05.2015 № 37271) [Official Internet portal of legal information]. URL: http://www.pravo.gov.ru (date accessed: 10.11.2018).

[5] Aristarkhova M K and Gunyakova A A 2010 Organizational aspects of corporate tax management, *Vestnik UGATU* **14**(4-39) 245-255

[6] Aristarkhova M K and Fakhrutdinov V S 2018 Tools of organizational and economic mechanism of internal control functioning at the enterprise, *Auditor* **2018**(10) 17 - 26

[7] Aristarkhova M K and Fakhrutdinov V S 2019 *The Mechanism of internal control at the enterprise*, Collection of articles on the results of XI all-Russian scientific and practical conference with international participation, dedicated to the 100th anniversary of the Republic of Bashkortostan "Innovative technologies of management of socio-economic development of regions of Russia", may 23-24, 2019

[8] Fakhrutdinov V S 2019 Features of formation of economic actors within the system of internal control, *Scientific review, Economics* **2019**(2) 26-32, URL: https://science-economy.ru/ru/article/view?id=1002 (date accessed: 23.07.2019)

[9] Fakhrutdinov V S 2018 The questions of preparation of qualified personnel for the management of internal control, *Scientific review, Economics* **2018**(4) 27-31, URL: https://science-economy.ru/ru/article/view?id=987 (date accessed: 27.01.2019)

[10] *Documents issued by the Committee of sponsoring organizations of the Treadway Commission (COSO)* [Electronic resource], URL: http://www.coso.org (date accessed: 01.09.2018)

IOP Conf. Series: Materials Science and Engineering **659** (2019) 012028 doi:10.1088/1757-899X/659/1/012028

Quality of employee education, training needs in small and medium enterprises in the Slovak Republic

O Poniščiaková[1], I Litvaj[2] and J Makarovič[2]

[1] Department of Economics, The Faculty of Operation and Economics of Transport and Communication, University of Žilina, Univerzitná 1, 010 26 Žilina, Slovakia
[2] Department of Power Electrical Systems and Electric Drives, The Faculty of Electrical Engineering and Information Technology, University of Žilina, Univerzitná 1, 010 26 Žilina, Slovakia

E-mail: litvaj@fel.uniza.sk

Abstract. The article analysis the quality of employee education, training needs in small and medium enterprises in Slovakia. The analysis is based on survey among educational institutions and enterprises. Based on the survey, educational needs and requirements on cooperation of institutions and recommendations are proposed and elaborated. These are further incorporated in a quality control system requirements and structure to ensure open and flexible educational environment able to fulfill current and future needs of the Slovak industry.

1. Introduction

Social systems are currently forced to operate in very demanding, variable, and turbulent conditions, characterized by increasing complexity and dynamic change. These changes are of a different nature. The democratic processes of our society were caused by changes in social and political conditions, while the scope of diapason of the changes affect whole society and all its social systems. The result is, on the one hand, a number of stimulating impulses for the development of related economic, social, cultural conditions, but on the other hand, the increasing pressure on people operating in social systems towards the necessary adaptation to changing conditions and thus the revaluation of traditional values, value systems and value orientation.

The people, directly concerned, are continually contemplating ways and solutions to the effects of a certain synergistic effect of turbulence [1, 2, 3, 4, 5]. Realizing that changes in traditional functioning affect both social systems and individuals, they strongly support the rise of the value of education and education itself. This process takes on a modified social function and becomes an integral part of social systems. From the traditional perception of education and development, social systems have shifted to the philosophy of lifelong learning and learning beyond the originally age and social groups.

This trend is not random, but it is based on some empirical experience. In particular, it supports the fact that the principle of lifelong learning leads, on the one hand, to the development of both the professional and the personal aspects of the individual. According to Charted Institute of Personnel and Development, the increased adaptability and employability of people are also beneficial [3]. On the other hand, education and training have a socially beneficial effect. It helps to improve economic competitiveness and employability and, according to the Memorandum on Lifelong Learning, it is the optimal way to combat social exclusion as well as enhance the efficiency and sustainability of the organization [5].

The philosophy of lifelong learning in Slovakia began to be implemented on the basis of the Lifelong Learning Strategy and Lifelong Guidance in Slovakia. The original version was created in 2007 and has been applied over time at the level of secondary schools and universities, employers as well as by personnel agencies and interested public [4, 5]. The current state of implementation of the Strategy's objectives and the level of success of its implementation has brought empirical surveys into these areas, supplemented by information from interviews with experts from lifelong education field. Specific objectives were also set within the objective of the empirical survey:

- to identify and analyze implementation of the Lifelong Learning (LLL) and Lifelong Consulting (LLC)in to the real life,
- to identify and analyze opinions and attitudes towards the LLL and LLC,
- to propose corrections and updates to fulfill the mission strategies in accordance with EU policies.

Survey questions focused on:

1. The current situation in LLL and LLC,
2. Monitoring and forecasting of educational needs,
3. Quality management system in formal and non-formal education,
4. Employability, readiness for the labour market,
5. Interconnection of subsystems and approval of results,
6. Advisory services,
7. Motivation to participate in further education,
8. Statistical data on LLL and LLC,
9. Information level about the LLL and LLC strategy.

In the following section we will discuss the results of the empirical survey. Findings of this survey are decisive information for improving the quality of LLL and LLC, particularly in the context of lifelong learning in small and medium-sized enterprises in Slovakia.

2. The current situation in LLL and LLC

In the first question of the survey, it was important to analyze the way of providing lifelong learning at the secondary schools and universities levels, as well as at the level of employees as well as interviewed individuals. The result was satisfactory concluding that educational activities were carried out to the required extent in all areas analyzed. In the school environment, the activities were focused on expanding, deepening professional knowledge and skills; in the corporate environment, educational activities were carried out to deepen and extend knowledge and skills in the relevant field (in the workplace or outside the workplace), on the level of individuals, the activities were focused mostly on their workplace, expanding and deepening their knowledge and skills in the field they work in. Individuals who stated they had not been educated, mentioned reasons as high costs of courses or the lack of courses with their specialization.

Educational activities were mostly carried out by the classical form of education (lectures, seminars), and modern methods such as e-learning or distance learning were used only minimally. In terms of educational content, the need for teaching soft skills in formal education prevailed, and schools in particular expressed a critical attitude to teaching new knowledge about technology in formal education, which lags behind the demands and ideas of employers. From the point of view of the labour market, all respondents expressed the same opinion - the need for better communication between schools and employers in order to create a better balance between workforce qualification and employers' real demands.

Both employers and interviewees agreed that positive attitudes to adult learning prevail in the public, as confirmed by nearly three quarters of employers and three-quarters of individuals (79.5 percent respondents). A dominant position was equally positive for consultancy services, and 41 percent of respondents considered this service absolutely necessary. In the area of LLL and LLC, the lack of interconnection of theory with practice has proven critical, often the education does not meet market requirements.

3. Monitoring and forecasting educational needs

In the area of monitoring of educational needs, it was important to recognize way how school's select and open their learning and educational courses and programs. The priority for the selection at secondary schools and universities is the interest of students and their parents and consequently the situation on the labour market. Furthermore, it was necessary to find out what factors affect particular educational institutions in planning of educational events. Employers are most affected by the prognosis of educational needs in relation to employers' needs. Further education institutions plan their educational programs, depending on requirements of individuals, and then follow the labour market situation, and personnel agencies plan mainly in the context of the labour market situation. The monitoring of educational needs is influenced by additional factors, such as those that dominate the choice of the field of study of individual's children and the choice of further education. In the case of secondary school, the choice of parents was influenced by the possibility of a good preparation of the child for university and the choice of the university was influenced mainly by the child's interest. In the selection of further education, the potential participant does not give a significant role to obtaining a certificate or the quality of the course, but primarily the potential participant considers as important the content of a course.

4. Quality management system in formal and non-formal education

The attitude of schools to the system of accreditation of study and study fields was in balanced with a higher relevance of the negative assessment of the accreditation system. Other educational institutions responded differently to the question of certification and accreditation experience, with accredited programs prevailing. However, the other educational institutions had a stronger attitude compared to schools, with positive views on the accreditation system. They accentuated the accreditation system more positively as a quality management system in the form of certification. Personnel agencies have largely assessed negatively the accreditation system and the same attitude has also been expressed by agencies providing mostly non-accredited and only non-accredited courses. The quality management system in the form of certification is considered by the personnel agencies to be more suitable than unsatisfactory.

5. Employability, readiness for the labour market

The market requirements were met by students of both secondary schools and universities at the level of theoretical knowledge, to a greater extent at universities. Schools ranked as second the positive assessment of graduates skilled in practical skills. Least positive was the readiness in soft skills.

According to employers, university graduates are ready for the labour market in some selected key competencies and soft skills more than secondary school graduates. Similar assessments were presented by personnel agencies.

With regard to the employment of secondary school graduates, the greatest obstacle to obtaining a job is the lack of experience and the required practical skills. In the case of university graduates, the low salary of the offered job is a drawback.

6. Interconnection of subsystems and approval of results

In the survey, most schools consider the linkage of formal, non-formal and informative learning systems to be functional. Other educational institutions were more critical in this context, and up to 65 percent rated linkage as malfunctioning. The link between systems by personnel agencies was also largely negative.

In the area of approval of the results of non-formal education and informative learning, schools and employers have agreed that formal education is most important in recruitment. The second important factor is the knowledge and skills acquired in an informal way. Employers recruiting graduates have given minimal importance to certificates and certificates obtained in non-formal education.

Interesting was the comparison with the attitudes of other educational institutions and personnel agencies, which considered work experience from previous jobs and the level of education acquired at school as a decisive reason for recruiting of new employees.

7. Advisory services

The interviewed schools have said they have the position of adviser used by pupils, students, but also parents. More than half of other educational institutions also provide counseling services, especially in the selection of another educational pathway. In particular, personnel agencies provide consultancy for employers.

Individuals are interested in counseling services in education, career counseling and career choices.

8. Motivation to participate in further education

In the survey, it was important to find out which employers' use as motivational tools to motivate their employees to participate in further education It has been found that companies prefer non-financial motivators such as job retention and career growth. Furthermore, it was examined how motivational tools are evaluated for their effectiveness. Most employers evaluate motivational tools as effective.

However, they would like to add tax reliefs. Surveys have found that other educational institutions consider further education and getting employed as most motivating factors. According to personnel agencies, adults are motivated to maintain a job by continuing their education. The interviewed individuals often considered as a motivator the professional growth, personal growth, a desire for education and gaining new information.

9. Statistical data on LLL and LLC

Statistical data refer to the school's awareness of the number of graduates who have found employment on the labour market. Most of these schools have information and are mainly obtained from the Office of Labour, Social Affairs and Family. However, they also use other sources, internal statistics or their own surveys from graduates. Other education institutions use to a much lesser extent access to statistical data, they obtain information from their own statistics, from the Office of Labour, Social Affairs and Family and also from Eurostat.

10. Information level about the LLL and LLC strategy

Assessing the information level about the LLL and LLC strategy was the same for all interviewed, because they responded consistently that they know the strategy document only partially, depending on which part is more interesting to them. The survey results provided information that could be evaluated in order to be useful in the form of certain recommendations. These recommendations for LLL an LLC have a common denominator that is:

- systematic approach to LLL and LLC,
- cooperation and communication of stakeholders,
- willingness of motivation,
- openness of the subjects to changes leading to the improvement of the system of LLL and LLC.

These general recommendations can be specified for the area of companies that favor LLL and LLC. Employers in companies need to increase the level of connectivity with schools and help them to use new technologies in education. Similarly, a greater connection between employers and schools is recommended for the need to better meet the educational needs of practice at education and transfer of new information.

However, based on the results of the survey, employers also found certain facts, namely that educational needs to be prepared in line with labour market requirements so that different courses would not produce graduates who do not have a chance in the labour market due to low demand. They also highlighted the need for better promotion of LLL, especially for the younger generation.

They recommend incorporating various innovative forms of education in formal education and devoting themselves to leadership along with the development of soft skills already at secondary school level. They also pointed out the need for financial support from the state (e.g. tax reliefs) that they would use to create their own academies or to intensify their cooperation with existing educational institutions.

10.1 Evaluation and conclusion

In the context of a set of inquiry areas that represented a set of 10 above-mentioned issues, it was necessary to define a more specific recommendation for some of them, as the analysis of the results brought some shortcomings in the monitoring and forecasting of educational needs, a consistency of opinions was necessary to reach in the area of quality management system in formal and non-formal education. It was also necessary to define recommendations for employability, interconnection of subsystems and recognition of results in more details, as well as the motivation to participate in further education and advisory services. In the field of monitoring and forecasting, it is necessary to create compact information and to ensure continuity in forecasting. Demand needs to be forecasted - job development by sector, professions and qualifications and supply side - education and training by age and gender, all at macroeconomic, sectoral level and at the level of individual schools and employers. In forecasting, it is recommended to link an important information center - the National System of Occupations with the resources for forecasting educational needs, with sectoral analysis and studies that monitor the state of development of individual sectors in terms of economic parameters and follow up the development of skills needs. The national system of professions must also be linked to representative research on the employers' qualification needs, carried out periodically on the basis of a detailed questionnaire.

Employers should be obliged to cooperate with the relevant organizations, institutions, public authorities in defining those needs. The clear dominant recommendation is communication and solutions "win-win", with inspiration from abroad and from good domestic practice. It can also be stated that an intensified focus on modern forms of in-house training is part of the need to improve the quality of educational needs. These include: openness and flexibility of the system, functional expert centers, incubators of ideas, internal seminars and lectures and the like.

11. Recommendations in the area of education quality management system

Quality management system recommendations in formal and non-formal education aim at quality management through investment in quality and assessment of teachers and trainers. In terms of employment and readiness for labour market, the recommendations concern the acquisition of a skilled and flexible retraining lecturers through better education, better teacher appraisal, greater focus on skills and technical disciplines, equalization of formal and non-formal education, and in particular, active school-employer cooperation. The interconnection of subsystems and the recognition of results could be optimized by increasing employers' confidence in certificates acquired through courses, but also by guaranteeing the quality of output of these training programs. Motivation to participate in further education could be unambiguously increased by financial motivations based on recommendations. Following the example of Ambition 2020 from the UK, individual training accounts could be introduced for those who are not educated at ISCED to complete it. Another option is to create vouchers, similar to Germany and Austria. Another option is to create funds by sector, to which employers contribute to education. It is also recommended to introduce grants, loan funds for vocational education and savings and education. In addition to proposals of a financial nature, it is recommended to increase the motivation to educate people with low education and the reasoning behind would be an increase of the financial value after successful completion of courses. It is also very important to increase the availability of training courses for low-income groups. Higher awareness through campaigns is important, by increasing the assurance of use and subsequently evaluating the acquired knowledge and skills, and the public awareness of the general premise to "educate". Answering a question about advisory services revealed a lack of functional advisory institutions and recommendations are aimed at developing advisory services in the choice of employment, and at introducing the position of so-called advisory services, career advisor in all kinds of schools with formal education.

Acknowledgment
The paper was prepared with the framework of the support of the project KEGA 19-003-00 – Implementation of the Integrated GPS System for Specification and Verification of Products for Teaching Engineering Programs and Practice.

References

[1] *Moderné formy vnútropodnikového vzdelávania,* Available on: http://www.efocus.sk/webcasty/kategoria/top-clanok/clanok/moderne-formy-vnutropodnikoveho-vzdelavania/
[2] https://www.cipd.co.uk
[3] https://www.minedu.sk/data/files/2606_akcny-plan-strategie-czv-2011_final.pdf
[4] https://www.minedu.sk/data/files/2607_2000_memorandum_o_celozivotnom_vzdelavani.pdf
[5] Storey J 2007 *Human Resource Management, A Critical Text,* London: Open University 292 ISBN 978-18-448-0615-7
[6] Toth D 2010 *Personální manažment,* Praha: Powerprint, 475 ISBN 978-80-87415-05-4
[7] Stýblo J 2003 *Personálni práce v malých a středních podnicích,* Praha: Management Press 146 ISBN 80-7261-097-X 28.
[8] Stýblo J, Urban J and Vysokajová M 2009 *Meritum personalistika 2009,* Praha: ASPI 912 ISBN 978-80-7357-429-1
[9] Steelman L A and Wolfeld L 2018 The Manager as Coach: The Role of Feedback Orientation, *Journal of business and psychology* **33**(1) 41-53
[10] Pousa C, Richards D and Trepanier C 2018 Managerial coaching of frontline employees: The moderating role of gender, *Human resource development quarterly* **29**(3) 219-241
[11] Ratiu L, David O and Baban A 2017 Developing Managerial Skills Through Coaching: Efficacy of a Cognitive-Behavioral Coaching Program, *Journal of rationalemotive and cognitive-behavior therapy* **35**(1) 88-110

IOP Conf. Series: Materials Science and Engineering **659** (2019) 012029 doi:10.1088/1757-899X/659/1/012029

Model of "short cycles" as innovative product development

E A Gromova[1]

[1]Peter the Great St. Petersburg Polytechnic University, Politechnicheskaya st., 29, St. Petersburg, 195251, Russia

E-mail: lizaveta-90@yandex.ru

Abstract. The fourth industrial revolution is a phenomenon that is actively becoming a reality and that initiates a change in the strategic guidelines of industrial enterprises. The previous strategic indicators such as scale of production, cost and quality have been replaced by a value system such as rapid response, flexibility, disruptive innovation and speed. Organization of production in today's reality has an important characteristic, reflecting the concentration at the stage of product development. Therefore, the purpose of the study is to offer effective product development concept in the context of reducing time to market. The concept of the model of "short cycles" is considered. The example of implementing this model in PJSC Severstal, which is the flagship of the Russian industry, is analyzed. The company has launched a global transformation, one of the points of which is the provision of excellent customer experience. Its implementation requires innovative and effective product development models, such as the model of "short cycles", which has already brought initial success to the company.

1. Introduction

In the early 2000s, the National Research Council of the United States identified the main tasks for industrial enterprises, which represented gaps in existing practice and determined the vision of production in 2020 [1]:

- ensure concurrency of operations;
- "instantly" transform information gathered from a wide variety of sources into useful knowledge for effective decision-making;
- integrate human and technical resources to improve workforce efficiency and satisfaction;
- rapidly restructure production facilities in response to changing needs and opportunities.

These tasks are extremely relevant for the current state of the industrial sector in general, and in particular, Russian industrial sector. In today's reality, an industrial enterprise faces a number of challenges. Along with those that have already become standard, which consists in achieving a high quality indicator and minimal costs, the following are becoming topical:

- reduction in time to market;
- wide variety of product range;
- increase in the rate of flow of production processes;
- use of the latest technological advances;
- accurate customer satisfaction.

Currently, great attention is paid to saving time in the course of production and sale of products. This is due to the onset of the Fourth industrial revolution – an increase in the degree of unpredictability and market instability, rapid changes in technology and increased competition.

Previously, R. Suri [2,3] proposed the concept of quick response manufacturing. It means responding to customer requests so that you can quickly develop and release products that are responsive to those requests. This concept focuses on the continuous reduction of the time required to perform all types of activities in the company, while ensuring quality improvement, cost reduction and faster response to changes in the situation. The roots of the concept go to the strategy " time-based competition", which was proposed by G. Stalk and T. M. Hout [4]. This model was based on the use of speed to gain a competitive advantage: a company that uses this strategy provides products or services to consumers faster than competitors. Also, many scientists [5-12] consider agile manufacturing as a concept of production organization based on the foresight of changes in the business environment and timely response to the rapidly changing needs of the market through the effective use of internal and external resources. Today A. Borovkov [13] rightly points out: «The key thesis is time to market. We are now witnessing the window of opportunity slamming in front of entire corporations and industries. The window existed for a year or two, during which time it was necessary to have time to enter the market. If this did not work out for some reason, you will have to wait for the next window».

So, the purpose of the study is to offer a product development concept, which will be effective in terms of reducing the time of product market entry in the realities of the Russian industry.

2. Materials and Methods

The model of "short cycles" is based on the Deming cycle (PDCA) [14], which is a model of continuous improvement. Each cycle of the development process is a mini-project and includes all the necessary range of operations: design, prototyping, testing, evaluation of results. This project is divided into the required number of "short cycles" depending on the identification of the most appropriate prototype (the standard procedure for product development often consists of one large cycle). It allows to visualize the scope of work and explain to the customer how the finished product will look. The duration of each cycle is from 1 to 4 weeks. The principal feature is the ability to influence each stage of the "short cycle" at any time (figure 1).

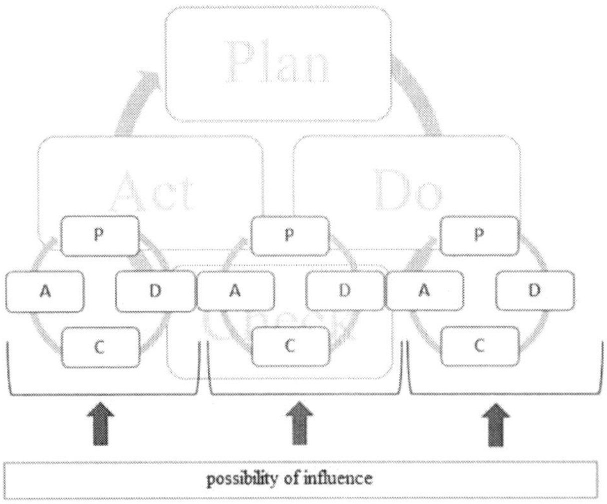

Figure 1. Visualization of the model of "short cycles" (developed by the author).

An interdisciplinary team in cooperation with the customer develops a prioritized list of tasks for the upcoming cycle (with the exception of the very first cycle, where there is virtually no feedback, the chosen concept prevails). The basic rule is that if the team has agreed on a certain number of tasks that need to be performed in one cycle, then it is impossible to add new ones. Small groups of end users have the opportunity to get acquainted with the new prototype and provide feedback. At the end of each "short cycle", a stakeholder assessment will be conducted, which will result in a list of tasks for the next cycle, taking into account the adjustments received through feedback and based on the discussion of the results by the team.

Fast and frequent releases are not a new idea, it goes back to the idea of evolutionary development proposed by T. Gilb [15]. In his words, "evolution is a technique designed to create the appearance of stability. The chances of successful creation of a complex system will be maximum if it is implemented in a series of small steps and if each step contains a clearly defined success, as well as the possibility of "rollback" to the previous successful stage in case of failure. Before putting into action all the resources intended for the creation of the system, the developer has the opportunity to receive feedback from the real world and correct possible errors in the project».

Among the main advantages of the model of "short cycles" can be distinguished:

- reduction of product's time to market;
- increase the speed of execution of a specific order;
- high speed of product development generates disruptive innovations [16];
- reduction of total costs;
- improving the quality of finished products;
- better satisfaction of consumer preferences;
- minimize risk by reducing development to a series of short cycles.

3. Results and Discussion

Since the beginning of 2018, PJSC Severstal, as one of the world's leading steel and mining companies, has sold 85.5 tons of new complex products. The project "Product innovations" started in 2017. The project includes the work of product teams in 5 categories: coated steel, cold rolled and hot rolled steel, thick sheet with large diameter pipes and long products. Product innovations brought $ 12 million in profits only in 2018. In 2019, product innovations are expected to yield about $ 20 million. During the period 2018-2023 Severstal plans to earn more than two billion dollars of additional income. Half will come from product, process and business model innovations.

In 2018, the company moved from local projects to a large-scale transformation of the organization. Severstal has started a major transformation that will require a lot of effort from all employees. According to A. Mordashov – chairman of the Board of Directors of the company, the creation of excellent customer experience is its most important element. "Excellent customer experience" is aimed at bringing the service to a fundamentally new level, quality customer service and finding new ways to bring them exceptional benefits [17]. Today Severstal analyzes new markets and develops further processing of products to strengthen relations with end users and joint development of products with major customers. The main goal is to further differentiate Severstal from its competitors and to bring products that have no analogues to the target markets. So, company need to find new opportunities in the market, produce new products and offer unique solutions. The model of "short cycles" helps to make the process of creating new innovative products more efficient. The model has proven itself well in Severstal in the process of development and market launch of a new product. The work is divided into two-week cycles. Before the beginning of each cycle, goals are set, tasks that need to be performed to achieve the goal are painted. The results are summarized at the end of each cycle, a retrospective is held and measures to improve the efficiency of the team are determined. There are also monthly demo days, where each cross-functional team, where each member is an expert in their field, has the opportunity to talk about their results and share best practices.

The result of each short cycle deserves special attention. A minimum viable product helps to get meaningful feedback from users: understand what they need and not create what they are not interested in and what they are willing to pay for. It's a hypothesis that needs to be confirmed or disproved. Procedure for identification of the minimum viable product:

1) clearly formulate the hypothesis;
2) determine the criteria by which its viability will be determined;

3) make the minimum viable product to confirm the hypothesis and run it;
4) measure the performance indicators on the basis of the obtained results;
5) draw conclusions, check the following hypothesis, if necessary.

The minimum viable product is schematically shown in figure 2.

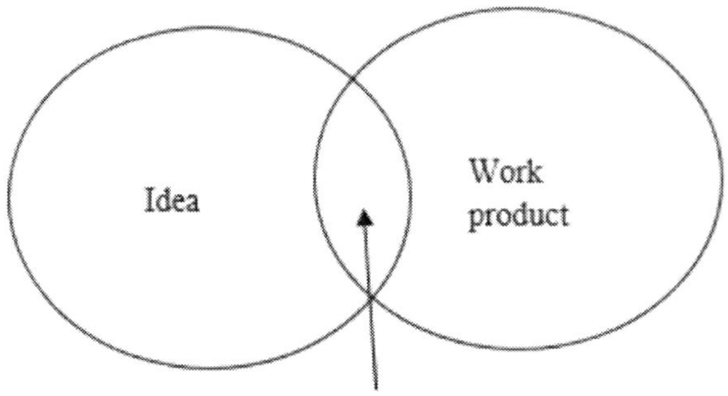

Figure 2. Visualization of the minimum viable product (developed by the author).

Minimum viable product does not mean raw product made in a hurry. Minimal time is spent on its development, and it contains only key characteristics, the relevance of which for real clients should be checked. Research shows that up to 60% of product properties are generally not important to customers. The concept of a minimum viable product allows to reduce the time to start a project by creating only the necessary functions and characteristics and begin to receive real feedback on your product. That's the trick. Minimum viable product will allow to include only the most necessary options and work out the necessary characteristics.

By creating a minimally viable product, it becomes possible:

- save money without investing them in a failed project;
- check if the product is of interest to potential users;
- choose the optimal direction of development due to iterations;
- create a potential customer base and find early product loyalists.

A pilot project in the development of a new type of product was the development of a new, more durable plastic packaging tape for the production of flat products. To release a new product is planned not only for internal customers, but also for external ones. One packaging tape is used in almost every redistribution of metallurgical production. The tape must have high strength, withstand heavy loads, as well as high ductility so that it is not torn during transportation. Thanks to the model of «short cycles», at the beginning the experts selected the optimal chemical composition, which should satisfy the customers' requirements for the packing tape. Next new product was tested. After the customer and the cross-functional team have reached a final agreement, they will be engaged in the release of a new cold-rolled tape in flat-rolled products. The team consists of employees of the directorate for sales, marketing, technical quality development, customer support services. The leader of the team is the Business System Development Center. Single criterion of success is to bring the finished product to the client, satisfy him, and bring the company profit. Director of Business System Development "Severgrupp" A. Kolobov argues that the use of model of «short cycles» in the work helps in the conditions of "very large uncertainties." He emphasizes constant contact with the client as an integral part of successful activity. Turning to the results of a pilot project for the production of a new type of product - high-strength packaging tape, the efficiency of the achieved indicators is reversed: the product creation period was 9 months, instead of 2-3 years, obtained using traditional methods, that is, acceleration of product launch to than 3 times. There were 28 short cycles for obtaining the final result. The product was distinguished by unique properties that gave reason to recognize this product as innovative. Another value is the increased demand among consumers. The pilot project was recognized as successful by the company's management.

One of the notable innovative solutions in 2018 is the production of damping steel. Severstal was the first company to produce such steel in Russia. After analyzing the market and listening to customers, the company has developed a new brand of steel, simultaneously absorbing noise and vibration, which is very important for architects of modern cities. In addition, the production of cryogenic steel for liquefied natural gas production and storage facilities was started within the framework of the program of import substitution of construction materials, implemented jointly with PJSC Gazprom. More one, "SeverFarm" [18] is a galvanized metal with a special coating that is resistant to the damaging effects of biologically and chemically aggressive environments. The characteristics of the new brand provide high resistance to corrosion even under the influence of acids, ammonia, disinfectants, organic solvents, alkalis and steam. The coating has maximum UV protection and increased resistance to mechanical damage. It became the first premium brand fully developed within the framework of the project "Product innovations". More innovative products are expected in the near future.

4. Conclusion

A key priority of the updated strategy of PJSC Severstal is to provide an excellent customer experience. The future belongs to those companies that can quickly provide consumers with unique solutions based on a deep understanding of their needs. The goals set for the company cannot be achieved without a number of changes in internal processes, including the organizational design of the company, which are necessary for the new format of cooperation with customers. The model of "short cycles" reduces the time to develop new products from several years to several months. Also, this concept of product development contributes to the genesis of innovative products due to the achieved speed. Although Severstal has started to actively introduce innovations relatively recently, today the results of work in this direction are already noticeable. Now there are more than 1400 ideas at different stages of implementation in the product portfolio of the project, which can potentially bring a huge effect.

Severstal is the first steel company in Russia to make such profound changes, transforming its structure to suit the client. As a result of these changes, customers of Severstal will have an even more sustainable and reliable partner that is able not only to offer high-quality steel, but also steel-based solutions that meet their business objectives. And the model of "short cycles" can be promising for enterprises of the industrial sector of the Russian economy.

References

[1] Camarinha-Matos L M, Afsarmanesh H, Galeano N and Molina A 2009 Collaborative networked organizations – Concepts and practice in manufacturing enterprises, *Computers & Industrial Engineering* **57**(1) 46-60

[2] Suri R 1998 *Quick response manufacturing: a companywide approach to reducing leadtimes*, Productivity Press

[3] Suri R 2010 *It's About Time: The Competitive Advantage of Quick Response Manufacturing*, Productivity Press

[4] Stalk G and Hout T M 1990 *Competing against time: how time-based competition is reshaping global markets*, Free Press

[5] Goldman S L, Nagel R N and Preiss K 1995 *Agile competitors and virtual organizations: strategies for enriching the customer*, Van Nostrand Reinhold

[6] Gunasekaran A 1998 Agile Manufacturing: Enablers and an Implementation Framework, *International Journal of Production Research* **36**(5) 1223-1247

[7] Gunasekaran A 2017 *International Journal of Production Research* 1-13

[8] Dubey R and Gunasekaran A 2014 Agile Manufacturing: framework and its empirical validation, *The International Journal of Advanced Manufacturing Technology* **76**(9-12) 2147–2157

[9] Kidd P T 1994 *Agile manufacturing: forging new frontiers*, Addison-Wesley

[10] Larman C 2004 *Agile and iterative development: A manager's guide*, Addison-Wesley
[11] Sharp J, Irani Z and Desai S 1999 Working towards agile manufacturing in the UK industry, *International Journal of Production Economics* **62** 155-169
[12] Sutherland J 2014 *SCRUM. The art of doing twice the work in half the time*, Crown Business
[13] *Do not oversleep the revolution 2019*, http://atomicexpert.com/page747353.html (accessed on Jun. 2019)
[14] Deming E 1982 *Out of the crisis*, MIT Press
[15] Gilb T 1988 *Principles of Software Engineering Management*, Addison-Wesley
[16] Christensen C M 1997 *The innovator's dilemma. When new technologies cause great firms to fail*, Harvard Business School Press
[17] *Creation of metallurgy of the future. Business development today 2019*, https://www.severstal.com/files/23850/Annual_report_2018_RUS.pdf (accessed on Jun. 2019)
[18] Gromova E 2019 Introduction of Flexible Manufacturing Systems as a Necessary Measure for the Russian Industrial Development, *Materials Science Forum* **957** 195-202

IOP Conf. Series: Materials Science and Engineering **659** (2019) 012030 doi:10.1088/1757-899X/659/1/012030

Indicators of tax reliability of innovative activity

M K Aristarkhova[1], O K Zueva[1] and M S Zueva[1]

[1]Ufa state aviation technical University, Institute of economy and management, department of taxes and taxation, K. Marks St., 12, 450008 Ufa, Republic of Bashkortostan, Russian Federation

E-mail: ninufa@mail.ru

Abstract. Corporate tax management – the organizing subsystem of the enterprise promoting creation of conditions for formation and the direction of stable financial flows in the state budget.
Management of this subsystem is based by means of formation of the indicators reflecting technologies of performance of the functions assigned to it, values of indicators of corporate tax management characterize internal work of the enterprise for lawful optimization of tax payments. At the same time quite often there is a need of assessment of tax opportunities of the enterprise which would predetermine its ability to implementation of tax payments during the concrete period of time.

1. Introduction

Now the concept – the corporate tax management (KNM) absolutely fairly takes root into practice. According to economic community KNM it is designed to form, optimize and redirect means of tax payments to the government budget.

Now, KNM in structure of business management is represented as the independent subsystem having specific characteristics. The set of the called characteristics can be presented four levels:
The I level – characteristic of corporate tax management;
The II level – characteristic of the entered being tax categories (TC);
The III level – characteristic of the structural elements entered by the Tax Code;
The IV level – characteristic of the conditions of achievement entered by the Tax Code.
Let's stop on representation of the called categories.
As the main estimated categories, I of level are offered – the tax stability (TS) and the tax reliability (TR), i.e.:

$$KNM = f\,(HY;\ PP) \tag{1}$$

The marked-out characteristics are borrowed from technical science.

Stability is defined as ability of a system to resist to process of destruction and to support the chosen functioning mode which can be characterized through separate indicators during certain time. Taking into account told it is possible to tell that tax stability – this property of a system to keep ability to performance of the tax obligations at the scheduled time and given sizes in the presence of external and internal influences.

Tax stability has the following characteristics: viability, flexibility, adaptability.

The viability for the organization is an opportunity not only to fulfill the tax obligations under the changing external and internal conditions, but also to provide at the expense of it sustainable development.

For the organization, it is important to have flexibility that represents variation by ability of involvement of external and internal factors for ensuring tax payments.

The adaptability represents ability of the organization to adapt under the changing environment, in due time and quickly to carry out necessary adjustments in various spheres of action.

Reliability – ability of a system to keep the most essential properties at the set level during the fixed period under certain conditions functioning. Social and economic systems belong to the class of the restored systems which operability at refusal is subject to restoration in the course of functioning.

The operability of a control system is understood as a condition of a control system at which she is capable to perform the set functions by the set parameters. In this research, the main place is taken by ability of the organization to pay taxes. Therefore, tax reliability for the organization it is logical to interpret as its ability in due time and in full to carry out the tax obligations. Respectively, non-failure operation, durability, ability to restoration – characteristic of the organization having tax reliability.

The non-failure operation is a property of the organization to repay the tax obligations in full. That is the organization has to have enough money supply for repayment of debt to the budget in time, it is also possible to say about such organization that it financially – is well founded [2].

The durability is an ability of the organization to pay taxes during any period irrespective of when it is required.

Ability to restoration – it is a possibility of the organization to quickly find sources of financial resources for repayment of debt to the budget, that is is an ability to mobilization of the required size of financial resources.

Tax reliability – it is a prerequisite and a condition of tax stability. Reliability gives the chance to estimate internal state of the organization, their ability for payment of the tax obligations. In addition, stability, in turn, considers these possibilities at the changing internal and external environment.

Tax stability and reliability help not only to estimate position of the enterprise, but in view of knowledge of indicators of the studied characteristics, an opportunity to improve its situation is represented. Due to understanding and judgment of the marked-out characteristics, the organization can ensure steady tax position in the system of the economic relations.

A. Definition of tax stability and tax reliability.

Tax stability characterizes the organization from a position of viability (Zh), flexibility (G), adaptability (A), that is the tax stability (TS) is their function:

$$TS = f\,(Zh;\ G;\ A) \qquad (2)$$

Each functional component can be presented by set of the characteristics promoting deeper and to detailed studying of stability.

The organizations allocated in a formula (2) an indicator of viability characterizes its opportunities to fulfill the tax obligations independently from the changing conditions. Taking into account told it is possible to assume that the viability (Zh) depends on internal (VU1) and external conditions (VU2) of work of the organization and on existence of the financial resources (FR) necessary for tax payment:

$$Zh = f\,(VU1,\ VU2,\ FR) \qquad (3)$$

Flexibility (F) – the possibility of payment of taxes by the enterprise is not dependent on changes of external (VF1) of internal (VF2) conditions of its work. The flexibility in fact means variation by internal and external opportunities of the enterprise, which is carried out by its personnel (SPVF):

$$F = f\,(VF1;\ VF2;\ SPVF). \qquad (4)$$

The ability and ability of personnel to vary external and internal conditions can be interpreted differently. In this case, acceptance for a basis of characteristic of the certain employee is advisable or the whole collective in such parameters as – professionalism (P), skill to communicate (K), vigor (AA):

$$SPVF = F\,(\Pi;\ K;\ AA) \qquad (5)$$

The adaptability (A) represents ability of the enterprise to adapt to changes of both the vernal and internal environment. This concept can be spread out on separate components – external (BC1) and the internal environment (BC2) and also property of adaptation of the organization to changes in the environment (POIS):

$$A = f(BC1, BC2, POIS) \tag{6}$$

Adaptation of the organization to changes in the environment in many respects depends from the term of its functioning and experience of personnel, owing to the fact that the organization can get into various situations.

This circumstance demands development of the qualifier of the standard changes (SC) by the organization and an algorithm of response to the standard changes happening in external and internal environment. Besides, it is necessary to systematize deviations (O) about standard changes in a section of the reasons and responsible. The functional dependence of adaptation of the organization to changes can be shown as follows:

$$POIS = f(SC, O) \tag{7}$$

The Tax Reliability (TR) provides ability in due time and in full to pay fiscal charges and characterizes the organization as the financially independent (FI) capable to pay taxes during any period of time (SPN) and capable to mobilization of the required size of financial resources (SMFR), what is presented by functional dependence:

$$TR = f(FI, SPN, SMFR) \tag{8}$$

Each of the elements making a concept of tax reliability has own characteristics which allow to realize features of tax reliability better.

The Financial Independence (FI) for the organization is a property to pay the tax obligations in full that is at the organization has to be rather financial resources (FR), effectively to carry out the duties before the budget. In this regard at the enterprise, the control system of a possibility of a proplachivaniye of taxes (UVUN) at the operational, tactical and strategic levels has to be created. Actually, for taxation planning the organization makes the tax calendar (TC), where the due dates for tax payment undersign, as provides timeliness of execution of the tax obligations. Thus, the financial independence represents function from all allocated parameters:

$$FI = f(FR, UVUN, TC) \tag{9}$$

The ability to pay taxes (SPNV) during any period is most connected with temporary streams so for its assessment it is necessary to consider rates of development of the enterprise (TRO). Besides, it is necessary to consider the sizes of financial resources of the enterprise (FR):

$$SPNV = f(FR, TRO) \tag{10}$$

Rates of development of the organization can be estimated in various ways and for this purpose, there is a set of techniques. For example, in this case it is expedient to speak about assessment of change of net profit, as on this indicator it is possible to track not only profit level, but also share of taxes.

The ability to mobilization of financial resources (SMFR) is an ability quickly and qualitatively to cooperate the financial resources (LFR) and professionalism of employees (DP) ready to the correct organization of work on transformation of resources into money supply:

$$SMFR = f(LFR, DP) \tag{11}$$

The personnel for such work have to have certain business qualities (DP), such as professionalism (P), speed of work (B), skill to communicate (J). That is activity of personnel can be present in a look:

$$DP = F(P, S, C), \tag{12}$$

where DP are business qualities of personnel;
P – professionalism;
S – speed;
C– skill to communicate.
Characteristics of corporate tax management as system.

B. The characteristics given above form a big system that in total defines concepts of tax stability and tax reliability.

The characteristics given above form a big system that in total defines concepts of tax stability and tax reliability. Their studying demands consideration of activity of the organization from various parties.

It is obvious that an independent indicator can present each structural element. All set of indicators, equally as well as each of them, affect value of tax stability and tax reliability. At the same time, there is a need of quantitative assessment of these indicators [1].

For this purpose, the allocated set of indicators needs to be systematize on levels:
The I level – characteristics of corporate tax management – tax stability and tax reliability;
The II level – components of characteristics of tax categories;
The III level – structural elements of tax characteristics;
The IV level – conditions of achievement of the marked-out characteristics.

The given indicators are separate and not unified. This circumstance complicates development of a technique of their quantitative assessment. It is obvious that this technique will integrate all complex of methods – expert assessment, questioning, the direct account, the determined analysis.

Technique of quantitative assessment of characteristics of corporate tax management. The logic of structure of a technique of quantitative assessment of characteristics of corporate tax management allows tracking visually interrelation of conditions, elements, components, characteristics on each of the allocated levels.

In this regard, it is expedient to begin reasonings on the called technique from the IV level reflecting conditions, which support by the manifestation functioning of separate elements of tax characteristics. Therefore, conditions – professionalism, skill to communicate, vigor – provide ability of personnel to variation with external and internal factors. For implementation of estimates of this type, it is possible to use expert methods, having subjected to inspection of the performer, his head and colleague applying a 10-mark scale [4].

Except called, at the IV level the group of the conditions causing ability of personnel to variation by external and internal factors is allocated. "Standard changes of activity of the enterprise under the influence of external and internal environment" are referred to those conditions. Certainly, before speaking about standard changes, it is necessary to carry out the retrospective analysis of changes of external and internal environment, having established frequency of concrete changes that will allow calling them standard. For example, treat changes of the external environment change of the legislation, contractors, and the general economic situation in the market. Change of the internal environment is accompany by change of strategy of the enterprise, its structure, production structure, development of new technologies.

Practice claims, that there are cases of emergence of deviations from standard changes that causes special complications and demands from personnel of masterly possession of a situation and not permissions to it to get out of hand. For this purpose necessary to estimate the arisen deviations from standard changes by means of introduction of conditional estimates of the importance of deviations, for example:

• less than 5 deviations – are appropriated assessment zero;
• from 5 to 10 cases of deviations – are appropriated assessment 5;
• more than 10 cases of deviations – are appropriated assessment 10.

The given scale can be detailed and adapted for assessment not only deviations, but also changes.
The following, III level, provides a research of structural elements of tax characteristics. Initial elements of this level are relate to such characteristic of corporate tax management as tax stability and are present by assessment of external and internal conditions, and financial resources of the enterprise. It is obvious that key the concept "condition" meaning a circumstance, the existing long time acts and the having impact on activity of the organization.

The current legislation, current situation in the market, the general state of the economy belong to external conditions. Are carried to internal conditions the current strategy of the enterprise; its organizational and production structure; condition of the production technology. Assessment of degree of usefulness of external and internal conditions of activity of the enterprise should be made, having involved the mechanism of expert estimates. Experts have to put down marks to each condition on a ten-mark scale where the greatest point (Bmax) will be got by optimum conditions for business, i.e.:

$$B_{max} = \sum^{m}_{i=1} 10N_i , \tag{13}$$

where N is an external or internal condition;
i – number of an external or internal condition;
m – number of external and internal conditions/

One more element of this level – financial resources of the enterprise. Experts have to carry out the general analysis of financial resources of the enterprise and for analogy with the above-stated technique to carry out assessment of degree of their sufficiency for implementation of tax payments.
The further research of structural elements of tax characteristics concerns assessment of external and internal factors, and also ability of personnel to variation by external and internal factors. Structure of external factors, unlike internal, it is rather heavy to predict. In this case, having involved retrospective data and statistical methods, it is possible to reveal probability their manifestations during the concrete periods of time. For example, the changes in the legislation capable to influence stable work of the organization or statistics on changes in labor market and dynamics of demand on certain goods and services.

Internal factors are shown in each of aspects of activity of the enterprise. Work has to predetermine their assessment on optimization of factors during each separate period.

C. Structural elements of tax characteristics.
We pass to consideration of such structural elements of tax characteristics as the external environment, the internal environment, adaptation of the organization to changes in the environment, which represent very important characteristic of corporate tax management – adaptability. Classical representation of the environment – it is a socioeconomic system which there is an enterprise. As each system, Wednesday has the structural components, which based on the available retrospective material, it is possible to reveal and estimate in the order given above (branch 13).

The further research of this level is accompanied by consideration of the characteristics of corporate tax management concerning to tax stability. The first of such characteristics – financial solvency as which elements it is necessary to recognize – financial resources vision of a possibility of tax payment at various levels, the tax calendar. The technology of assessment of financial resources is given above.

Vision of a possibility of tax payment at various levels and also the tax calendar it is offered to estimate in common as the second is reflection and logical continuation of the first. The quality of planning of payments in the budget is exposed to assessment: degree of correctness of drawing up schedules, existence of mistakes or inaccuracies [6].

In this case it is offered to involve the expert assessment given above with application of the 10th mark rating scale where 10 points have to be given to absolutely correct planning and the organizations of calculations with the budget at the operational, tactical and strategic levels.
The following elements, which are subject to a research - assessment of financial resources and rates of development of the enterprise that can be made by means of the algorithm, stated above, with use of a 10-mark scale.

Final group of elements of this level – liquid financial resources and active personnel are related to assessment of ability of the enterprise to mobilization of financial resources.
Liquid financial resources – these are resources, which for extremely short period can be translated in money supply. Can be them – receivables, the underpaid dividends from other organizations, etc.
It is possible to apply the algorithm stated above focused on involvement to assessment of liquid financial resources 10-mark scale.

The technology of assessment of the following element – "active personnel" is stated above.

IOP Conf. Series: Materials Science and Engineering **659** (2019) 012030 doi:10.1088/1757-899X/659/1/012030

Assessment of characteristics of tax management of the enterprise – viability, flexibility, adaptability, financial solvency of the enterprise, ability of the enterprise to pay taxes, abilities of the enterprise to mobilization of financial resources is made by summation of estimates of the structural elements and estimates of conditions of achievement of the marked-out characteristics making them.
The Tax Stability (TS) is defined by summation of estimates of viability (Zh), flexibility (G), adaptability (A):

$$TS = Zh + G + A \tag{14}$$

The Tax Reliability (TR) is defined by summation of estimates of financial solvency of the enterprise (FS), ability of the enterprise to pay taxes (SNP), abilities of the enterprise to mobilization of financial resources (SMR):

$$TR = FS + SNP + SMP \tag{15}$$

2. Results

In article, the attempt of introduction of new characteristics of corporate tax management is stated and approach is stated to creation of a technique of their assessment.
Thus, need and opportunity introductions of new characteristics of corporate tax management of the enterprise – tax stability and tax reliability is proved. Specific features at each of the entered characteristics are marked out. The possibility of their quantitative assessment is predetermined. Multilevel representation of structural elements of the marked-out characteristics is offered. The technique of quantitative assessment of the marked-out characteristics by means of involvement a complex of methods – expert assessment, questioning, the direct account, and the determined analysis is stated.

References

[1] Aristarkhova M K 2017 Conceptual approach to statement of accounting of results of introduction of innovations, *UGNTU* **November** 29-35
[2] Volosnikov A P 2011 Accounting of efficiency of innovations, *Accounting* **2011**(1) 31-34
[3] Borodkin A S 2005 *Accounting of costs of production in the system of cost accounting is Kiev*, the Higher school, - 232 pages
[4] Gradstein A 2015 Analysis of innovations on results of activity of the enterprise, *Accounting* **2015** 29
[5] Gallium V F 2005 Accounting in the system of economic information – M., *Finance and statistics*
[6] Grabina T K 2007 About reflection of the actual results of introduction of innovations in accounting in b., *Efficiency of innovations of M., Science,* -152 pages

Study accuracy of a transportation system positioning of a test rig for automated mounting of luster terminals

R K Dimitrova[1], V Zhmud[2], N N Petrov[3] and T A Vakarelska[4]

[1] Technical University, Faculty of Mechanical Engineering, Blvd Kl. Ohridski 8, 1000 Sofia, Bulgaria
[2] NGTU, Novosibirsk, Russia
[3] Technical University, Faculty of Mechanical Engineering, 8 Kl. Ohridski Blvd, 1000 Sofia, Bulgaria
[4] Technical University, KEE, 8 Kl. Ohridski Blvd, 1000 Sofia, Bulgaria

E-mail: rkd@tu-sofia.bg

Abstract. A test rig for study of parameters of a process of automated assembly of luster terminals is designed and developed. The aim of the developed test rig is to study the accuracy of positioning of the transport system during the automated assembly process of luster terminals. On the basis of data collected during the experiments, reliability analysis of the whole system has been made.

1. Introduction

Subject to automation is a plastic luster terminal with twelve inputs/outputs and mounting holes. It consists of a body in which brass details are placed with threads, each having two screws. An automated assembly system test rig had been designed and manufactured. The design is based on the test and assembly rigs presented in [1, 2, 3, 4, 5, 6, 8]. It has the following main modules (Figure1 and Figure 2):

- Magazine - collectors (MC) for each of the three different elements, which are assembled into a final product "luster terminal" - a total of 10 pieces;
- A linear conveyor (LC), which moves the elements between the different positions;
- Interceptors for Element B and Element B (element A is cut off by MC by LC);
- Module for carrying out technological operation (TO) - assembling by threaded connection of 6 pcs. Elements B to 3 pcs. Elements B in position 4.

After the production and purchase of all the necessary items, a test rig for the automated assembly of luster terminals is assembled. For programming of the stand the used program is "WinPISA 4.51" by "Festo" [7]. The input and output parameters are responsible for the information sent and received by all pickups, sensors and pneumatic cylinders, connected in the test rig. By using this information and depending on the value of each parameter, the blocks in the prepared block diagram are executed one after the other (Figure 5). Each individual block requires a specific value of one or several parameters in order to send the program signal for a subsequent operation to the test rig and to move to a next block where again the program will require specific values from the input parameters. After executing one program cycle, it starts again. Each pneumatic cylinder sends a signal of the position in which it is located and this signal is constantly checked by the program. The same applies to the sensors responsible for the presence of elements in the MC. The test rig enters pause mode in case there are no items available in any of the MCs. The program set input and output parameters shown in Figure 3 and Figure 4.

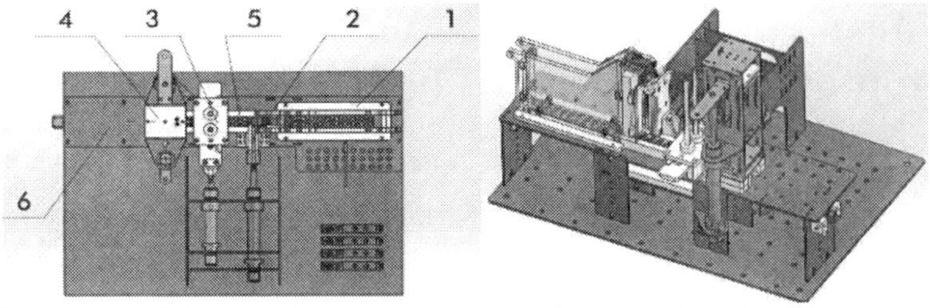

Figure 1. 3D model of a test rig of automated luster terminal mounting [3]: 1 - MC; 2 - MC with interceptor; 3 - Two MCs with interceptor; 4 - Operating position for mounting screw-type element; 5 - Linear conveyor.

Figure 2. General view of the test-rig [3].

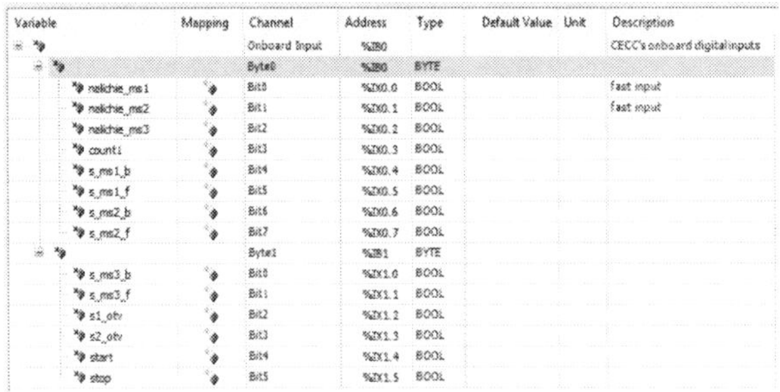

Variable	Mapping	Channel	Address	Type	Default Value	Unit	Description
		Onboard Input	%IB0				CECC's onboard digitalinputs
		Byte0	%IB0	BYTE			
nalichie_ms1		Bit0	%IX0.0	BOOL			fast input
nalichie_ms2		Bit1	%IX0.1	BOOL			fast input
nalichie_ms3		Bit2	%IX0.2	BOOL			
count1		Bit3	%IX0.3	BOOL			
s_ms1_b		Bit4	%IX0.4	BOOL			
s_ms1_f		Bit5	%IX0.5	BOOL			
s_ms2_b		Bit6	%IX0.6	BOOL			
s_ms2_f		Bit7	%IX0.7	BOOL			
		Byte1	%IB1	BYTE			
s_ms3_b		Bit0	%IX1.0	BOOL			
s_ms3_f		Bit1	%IX1.1	BOOL			
s1_otv		Bit2	%IX1.2	BOOL			
s2_otv		Bit3	%IX1.3	BOOL			
start		Bit4	%IX1.4	BOOL			
stop		Bit5	%IX1.5	BOOL			

Figure 3. Input parameters [3].

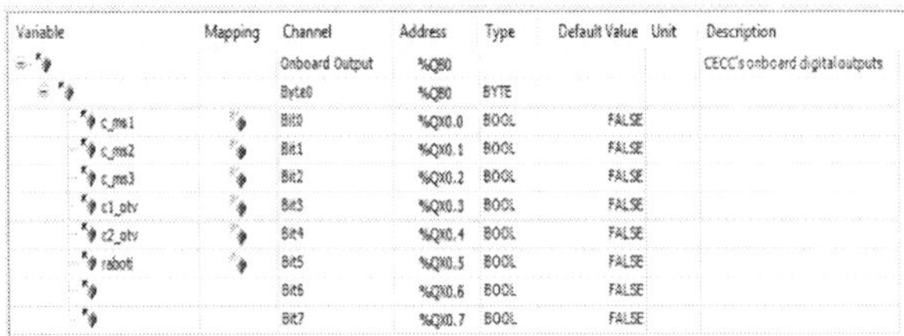

Variable	Mapping	Channel	Address	Type	Default Value	Unit	Description
		Onboard Output	%QB0				CECC's onboard digitaloutputs
		Byte0	%QB0	BYTE			
c_ms1		Bit0	%QX0.0	BOOL	FALSE		
c_ms2		Bit1	%QX0.1	BOOL	FALSE		
c_ms3		Bit2	%QX0.2	BOOL	FALSE		
c1_otv		Bit3	%QX0.3	BOOL	FALSE		
c2_otv		Bit4	%QX0.4	BOOL	FALSE		
raboti		Bit5	%QX0.5	BOOL	FALSE		
		Bit6	%QX0.6	BOOL	FALSE		
		Bit7	%QX0.7	BOOL	FALSE		

Figure 4. Output parameters [3].

A block diagram of the operations performed in accordance with the operating cycle of the test rig is shown in Figure 5.

2. Methodology for experimental studies

When performing a certain number of observations or measurements on the subject, the results are obtained $x_1, x_2, \ldots x_q$, which form an elementary statistical aggregate (elementary statistical order) of volume n. Observed or measured values are the final number of independent random variables that have the same distribution law as the random variable X. The values x_1, are called variations, and their ranking in ascending order - ranging order [1, 2, 3].

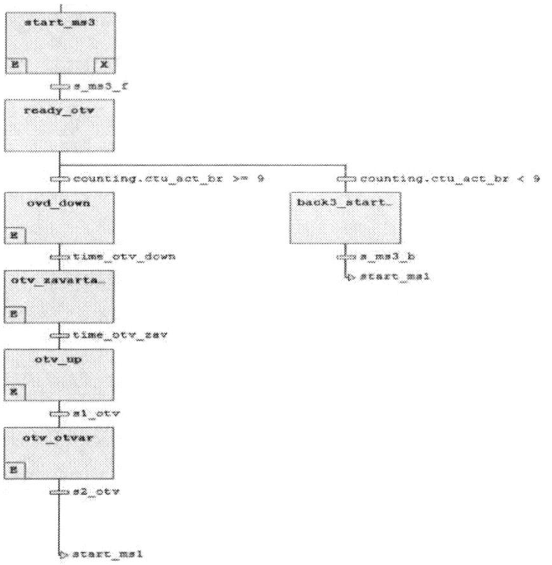

Figure 5. Block diagram of the program used [3].

The resulting statistical set of values x_i, (i = 1, 2,..., q) of the random value X together with the absolutes v_i and relative frequencies f_i form a statistical distribution law. The conversion count of X determines the absolute frequencies - v_i, and their relationship to volume determines the relative frequencies f_i. In that:

$$\sum_{i=1}^{q} v_i = n, \sum_{i=1}^{q} \frac{v_i}{n} = 1 \quad (1)$$

For a continuous random quantity, the statistic is broken down into classes (groups, orders) that are usually of equal lengths of intervals. In this way, an interval statistical sequence is obtained with k - class. The length of the interval $h = x_i - x_{i-1}$ (i = 1, 2,..., k) is determined by the dependence:

$$h = \frac{x_{max} - x_{min}}{k} = \frac{\hat{R}}{k} \quad (2)$$

Processing of experimental data x_i (i = 1, 2,..., n) from small samples is performed in the following sequence of calculation, analysis, comparison of the relevant parameters (parameters):

- Variation order is created.
- The values of the numerical characteristics are determined.
- Absolute and relative mean errors for individual numerical characteristics are determined.
- The maximum absolute and relative errors of the individual numerical characteristics are determined by replacing the theoretical with the empirical values.
- Trusted limits and confidence intervals are determined at a preset confidence probability y.
- The minimum sample size required is determined.

Statistical analysis and processing of experimental data x_i ($i = 1, 2,..., n$) from large samples is performed in the following order:

- An interval statistic line is compiled and the range R is determined, the length of the interval h and the number of classes k.
- The statistical set is represented graphically by an empirical distribution function, histogram and polygon.
- The values of the numerical characteristics are determined.
- The absolute and relative weighted errors for the individual numerical characteristics of disposition and scattering are determined.
- The values of the numerical characteristics of the moments (conditional start and center moments) are determined.
- Statistical estimates of asymmetry and excess (skewness and kurtosis) are determined.

By the type of the histogram and the polygon of the statistical law, the empirical values of asymmetry and excess and the physical essence and the physical sequence for obtaining and measuring the random quantity can be made the choice of the theoretical distribution law. For a more accurate assessment of the similarity of the experimental and theoretical distribution, the theoretical frequencies are calculated and a theoretical curve is constructed on the same scale [1, 2, 3].

2.1. Processing of results from engineering studies under "normal distribution"

The methodical sequence for constructing a normal distribution, which is most often subordinated to the distribution of random variables in all statistics, is subdivided into k classes with equal lengths of intervals h; apply the exact values of the lower and upper limits of the classes $x_{i-1} \div x_i$; the average of each class and the corresponding frequencies are determined v_i [1, 2, 3]:

- the arithmetic mean is calculated:

$$\bar{x} = \frac{\sum_{i=1}^{k} \bar{x}_i v_i}{\sum_{i=1}^{k} v_i} \tag{3}$$

- the corrected dispersion and the mean quadratic deviation are calculated:

$$s^2 = \frac{\sum_{i=1}^{k} \bar{x}_i^2 v_i}{\sum_{i=1}^{k} v_i} - \bar{x}_i^2 \tag{4}$$

- Differences are determined for each class $|\bar{x}_i - \bar{x}|$ and the values of the normalized deviations:

$$t_i = \frac{|\bar{x}_i - \bar{x}|}{s} \tag{5}$$

- The tabulated values of $f(t_i)$ for the found values of it are determined;
- the theoretical frequencies are determined and round up to an integer:

$$v_i = \frac{nh}{s} f(t) \tag{6}$$

The experimental (polygon) and the theoretical (equalized) distribution by normal law are constructed on the same scale: the differences between the experimental and the theoretical distribution curve are determined and the error is evaluated [1, 2, 3].

2.2. Result processing under the "exponential distribution" hypothesis

The methodology includes the following steps:

Step 1: Chronometry of fault-free time and time of failure.

Step 2: A statistical hypothesis on the exponential law of distribution of the probability of faultless work is adopted.
Step 3: Based on experimental data, the number of intervals for non-faultless operation n is determined and the limits for their modification t_i, i = n.

The time of faultless operation m_{mid} is calculated according to:

$$m_{mid} = \frac{\sum_{i=1}^{n}(\Delta N_i t_i)}{n} \tag{7}$$

- Calculation of the statistical probability curve for faultless operation $P_i(t_i)$

$$P_i(t_i) = \frac{n-\sum_{i=1}^{n}(\Delta N_i t_i)}{n} \tag{8}$$

- Calculation of the reliability function P(t)

$$P(t) = e^{\frac{t}{m_{mid}}} \tag{9}$$

- Determining the average withdrawal time T_v

$$T_v = \frac{\sum_{i=1}^{n}(\Delta N_i t_{mid})}{\sum_{i=1}^{n}(\Delta N_i)} \tag{10}$$

where: t_{mid} is the average of the time interval
- We define the failover flow parameter w

$$w = \frac{t_c}{m_{mid}} \tag{11}$$

where: t_c is cycling time
- Determination of actual productivity Qf

$$Q_f = \frac{1}{t_c + t_{mid}} = \frac{m_{mid}}{t_c(m_{mid}+T_v)} \tag{12}$$

where: t_{mid} are the stays of AT in the execution of one cycle.
- The results of the experimental studies are presented in a table.
- The experimental curve for faultless operation $P_i(t_i)$ and the theoretical curve P(t) are built.
- The error is determined and compared to the allowable one.

3. Conducting the experiments and processing the results of the experimental research

The following experimental studies are performed:
- Position accuracy of the transport system.
- Reliability of the automated mounting test rig.

3.1. Position accuracy of the transport system

Experimental studies have been conducted to determine the positioning accuracy at different transfer speeds of the linear conveyor. Figure 6 shows the schematic diagram of the test setup for measuring positioning accuracy. The results of the experiments are shown in Table 1.

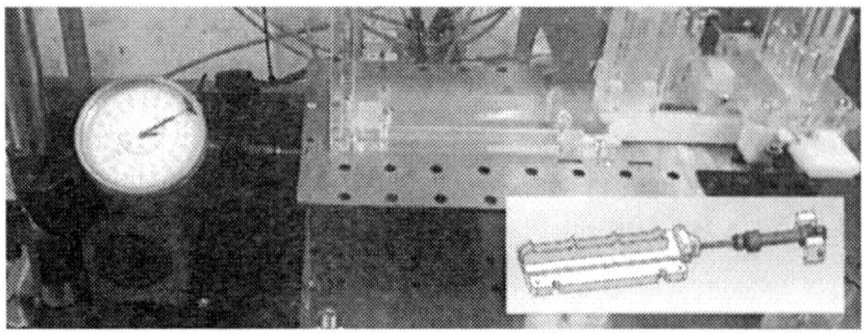

Figure 6. Positioning accuracy measurement.

$$\Delta X_{mid} = \bar{x} = 2{,}5 \qquad \sigma = \sqrt{\frac{21{,}25}{10}} = 1{,}458$$

The normal distribution function Y is calculated using the formula:

$$Y = \frac{1}{\sqrt{2\pi\sigma}} e^{-\frac{(\Delta x_{mid} - \Delta x_{imid})}{2\sigma^2}} \tag{13}$$

Table 1. Experimental research results.

№ of the interval	Interval [microns]	Frequency of occurrence n	$(\Delta x_{mid} - \Delta x_{imid})$	$(\Delta x_{mid} - \Delta x_{imid})^2$	Y
1	-2 ÷ -1,5	3	2.5	6.25	0.00036
2	-1,49 ÷ -1	5	2	4	0.00389
3	-0,09 ÷ -0,5	10	1.5	2.25	0.02503
4	-0,49 ÷ 0	14	1	1	0.09454
5	0,01 ÷ 0,5	28	0.5	0.25	0.20985
6	0,51 ÷ 1	16	0	0	0.27374
7	1,01 ÷ 1,5	12	-0.5	0.25	0.20985
8	1,51 ÷ 2	8	-1	1	0.09454
9	2,01 ÷ 2,5	3	-1.5	2.25	0.02503
10	2,51 ÷ 3	1	-2	4	0.00389

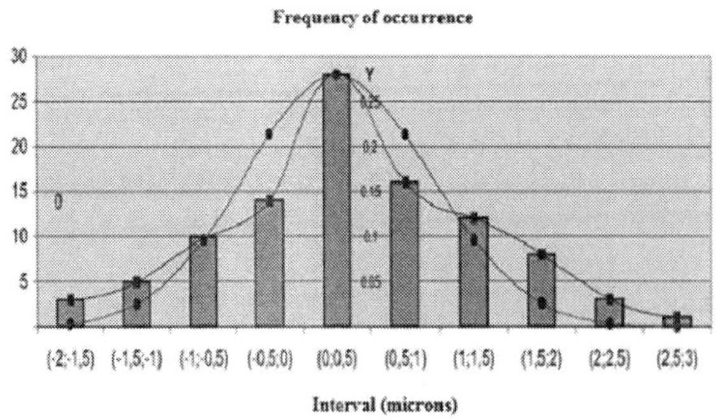

Figure 7. Experimental results for positioning accuracy.

Figure 7 shows the processed experimental results for positioning accuracy of the transport system of the test rig in graphical form. The graph shows that the accuracy of the positioning is about +2,5 [microns], which corresponds to the positioning accuracy for mounting operations.

3.2. Determination of the reliability of the test rig for the automated assembly of luster terminals.
Experimental tests have been carried out for the reliability of the automated assembly system and the results are shown in Table 2.

Table 2. Experimental results for the reliability of the test rig.

№ of the interval	Interval boundaries [min]	Average interval value(t_i)	ΔN_i	$\sum\Delta N_i$, i=1÷n	N - $\sum\Delta N_i$, i=1÷n	$P_i(t_i)$	t/m_{mid}	P(t)
1	0 - 2	1	3	3	23	0,88	0,10	0,92
2	2 - 4	3	2	5	21	0,81	0,30	0,72
3	4 - 6	5	4	9	17	0,65	0,51	0,58
4	6 - 8	7	1	10	16	0,62	0,71	0,47
5	8 - 10	9	2	12	14	0,54	0,91	0,36
6	10 - 12	11	3	15	11	0,42	1,12	0,25
7	12 - 14	13	5	20	6	0,23	1,32	0,18
8	14 - 16	15	2	22	4	0,15	1,52	0,12
9	16 - 18	17	1	23	3	0,12	1,73	0,05
10	18 - 20	19	3	26	0	0,00	1,93	0,02

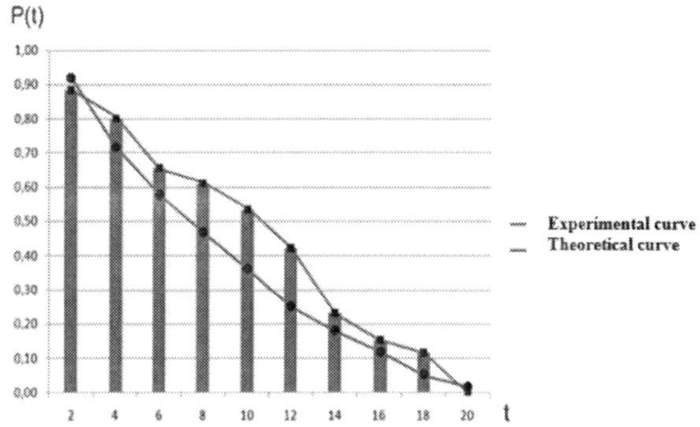

Figure 8. Results of experimental reliability studies.

On Figure8 a graphical interpretation of the experiment results is presented. It can be seen that the error between the statistical and theoretical reliability curves is small, i.e. the two curves almost coincide.

For the automated assembly test rig under consideration:

$$m_{mid} = \frac{256}{26} = 9{,}84\ [min]$$

$$K_G = \frac{T_w}{T_w + T_v} = 0{,}97$$

where T_w=m_{mid}; T_v – average recovery time after failure;

The coefficient K_G=0,97 (for K_{Gmin}=0,8), which fully satisfies the operational reliability requirements.

4. Conclusion

- A test rig for automated luster terminals mounting is developed.

- From the experimental tests for the positioning accuracy of the transport system can be seen that the positioning accuracy is around +2,5 [microns], which corresponds to the positioning accuracy for mounting operations.
- From the experimental tests carried out for the reliability of the developed test rig for automated mounting of luster terminals is seen that the coefficient KG = 0,97 (for KGmin = 0,8), which fully satisfies the operational reliability requirements.
- The results of experimental studies that can be used in engineering practice are processed.

References

[1] Dimitrov B S, Dimitrov L, Dimitrova R and Nikolov S 2019 Examination of the process of automated closure of containers with screw caps, *Studies in Systems, decisions and control,* **199** 502-514

[2] Petrov N N 2016, *Tipovi efektivni resheniya za avtomatizatziya na montazha na vyzli ot elektronnoto proizvodstvo (Efficient automation solutions for the assembly of electronic manufacturing units)*, Technical University of Sofia, Doctoral Thesis

[3] Jiang Z, Harrison D K and Cheng K 2002 An integrated concurrent Engineering approach to the design and manufacture of complex components, *International Journal of Advanced Manufacturing Technology* **20**(5) 319 – 325

[4] Sarkis J and Parsaei H 2011, *Advanced Manufacturing Systems - Strategic Management and Implementation,* Springer

[5] Wang Q, Sowden M and Mileham A R 2013 Modelling human performance within an automotive engine assembly line, *International Journal of Advanced Manufacturing Technology* **68**(1) 141-148

[6] Festo Didactic, *WinPisa for SPC 200,* https://www.festo-didactic.com/us-en/products/software-e-learning/programming-software/winpisa-for-spc-200.htm?fbid=dXMuZW4uNTc5LjE3LjE4LjU2Ni4zNzUy (accessed on Jan. 2018)

[7] Dimitrov V L, Michalopoulos D, Apostopoulos C and Neshkov T 2009 Investigation of contact fatigue of high strength steel gears subjected to surface treatment. *Journal of Materials Engineering and Performance* **18** (7) 939-946

[8] Tomov K P 2017 Increasing the efficiency of automation of production processes by reporting the parameters of the parts' flow, *TEM Journal* **6** (3) 484-487

[9] Guergov S 2005 *Mathematical Modelling of a Flexible Multifunctional Manufacturing System for Mechanical Treatment*, 2nd International Conference on Manufacturing Engineering (ICMEN), Kassandra-Khalkidhiki, Greece, October 5-7, 781-790

Fatigue behaviour of friction stir welded AA-2024 aluminium alloy sheets

T Vuherer[1], J Kramberger[1], D Milčić[2], M Milčić[2] and S Glodež[1]

[1]University of Maribor, Faculty of Mechanical Engineering, Smetanova 17, 2000 Maribor, Slovenia
[2]University of Niš, Faculty of Mechanical Engineering, Aleksandra Medvedeva 14, 18000 Niš, Serbia

E-mail: janez.kramberger@um.si

Abstract. The presented paper deals with the mechanical and fatigue properties of the alloyed aluminium alloy AA-2024 welded by the Friction Stir Welded (FSW) process. Using the optimized tool and welding procedure, 6 mm thick plates made of about quoted material were connected. In the first part of the paper, the influence of welding speed on the microstructure and mechanical properties of treated FSW welded joints is studied. In the continuation, the fatigue behaviour of treated welded joints is analysed based on the experimental testing on a servo-hydraulic fatigue testing machine at room-temperature under the stress ratio R=0.1 The experimental results have shown that the welding parameters have a significant influence on the mechanical and fatigue properties of AA-2024 welded sheets.

1. Introduction

The high-strength aluminium alloys such as 2xxx, 6xxx and 7xxx series are often used to produce the components for many important areas in automotive, aerospace and defence industries due to their reasonable mechanical properties for relatively low densities [1-3]. Their strength has been increased by the addition of elements such as Cu, Mg etc. which adversely affect the corrosion resistance of the alloy [4, 5].

Not long ago, riveting and fastening procedures were the main joining technologies for components made of high strength aluminium alloys because of their poor weldability [6, 7]. However, this problem has been overcome with implementation of the Friction Stir Welding (FSW) technique developed by the TWI-institute [8]. As presented by da Silva et al [9], Zimmer et al [10], Avettand-Fenoel et al [11] and Ahmed et al [12], there are three main magnitudes influencing the quality of FSW-joints: (i) tool geometry; (ii) welding parameters and (iii) joint construction. A peculiar feature of FSW-joints is the complex microstructure which significantly depends on the structural tool design, the rotation speed, the welding speed, the pressure of the tool on the plates along the vertical axis, the angle under which the tool acts on the material and the characteristics of the material being welded. As described by Ouyang et al [13] and Balokhonov et al [14], the following zones can be identified in the microstructure of the FSW-joints: the heat-affected zone (HAZ), the thermo-mechanical zone (TMZ), the stir zone (SZ) and the zone of base material. From that aspect it is very important to choose the right combination of welding parameters in order to obtain a welded joint of the corresponding quality without defects [15].

In last two decades, many authors have published comprehensive investigations where different influencing parameters on the mechanical, structural and corrosion properties of FSW-joints of aluminium alloys are studied in details. Vidal et al. [16] analysed the improvement of FSW-joints of the aerospace aluminum alloy AA2024-T351 using the Taguchi method and statistical analyses to obtain the optimal welding parameters. Perović et al. [17] analysed the influence of the rotation speed and welding speed on the impact strength, microstructure and cross-section micro-hardness of FSW-joints of Al-Zn-Mg-Cu high strength aluminum alloy. The combined effect of a small difference in pin geometry and rotation and welding speed on the weldability, mechanical and structural properties of FSW 2024-T351 Al plates was studied by Radisavljević et al [18]. Hussain [19] conducted a study to determine and evaluate the influence of main welding parameters on the Vickers hardness, tensile strength and microstructure of FSW-joints of AA6351 aluminium alloy. Gupta et al [20] analysed the process parameters (tool rotational speed between 300 to 1000 rpm and welding speed of 50 mm/min) and energy input on the microstructure, hardness, tensile strength and corrosion property of 7475 aluminum FSW- joints. The similar work has also been proposed by Su J Q et al [21] who investigated the grain structure, dislocation density and second phase particles in various regions (dynamically recrystallized zone- DXZ, thermo-mechanically affected zone- TMAZ and heat affected zone- HAZ) of FSW aluminum alloy 7050-T651.

As described above, numerous researches have been carried out in the past to study the mechanical properties of FSW-AA2xxx-joints and their dependence on the tool geometry and welding parameters. On the other hand, the comprehensive investigations of the fatigue behaviour of FSW-AA2xxx-joints dated in the last decade. Aydin et al [22] studied the influence of the rotating speed and translation welding velocity on the fatigue behaviour of FSW-AA2014-T6 alloy. Authors concluded that the highest fatigue strength corresponds to the rotating speed of 1520 rpm and translation velocity 80 mm/min. Lemmnt et al [23, 24] investigated the influence of welding orientation and residual stresses on the fatigue crack initiation. Authors concluded that the influence of residual stresses is significant when the welding orientation is parallel to the applied load. Ilman et al [25] and Milan et al [26] investigated the fatigue crack propagation of FSW-AA2024-T3-joint. Their experimental results have shown that the FSW-joint is more resistant to the longitudinal crack growth due to beneficial influence of compressive residual stresses in this direction. As shown by Hatamleh et al [27], the fatigue behaviour of FSW-AAxxxx-joints could be improved with additional shot or laser peening. In last period, the numerical modelling was sometimes performed (Nikfam et al [28], Kraedegh et al [29], Đurđević et al [30], Živojinović et al [31]) to simulate the fatigue crack initiation and growth in the critical areas of FSW-joints. In [32], Neto et al presented a literature review of the numerical modelling of friction stir welding process. They concluded that FSW-modelling helps to visualize the fundamental behaviour of the welded materials and allows to analyse the influence of different welding parameters (including tool design) and boundary conditions, without performing costly experiments. The comprehensive numerical simulation of material flow in FSW of AA7075-T6 aluminium butt joints has also been done by Fratini et al [33], who proposed the experimental verification of numerical results.

The main objective of these research is to determine the influence of welding speed on the mechanical properties and fatigue behaviour of FSW-joints made of aluminium alloy AA 2024 T351. When studying the fatigue behaviour of treated FSW-joints, the special attention is destined to both, low cycle fatigue (LCF) and high cycle fatigue (HCF). Namely, it is assumed that the fatigue behaviour may be significantly dependent on the expected fatigue life of treated FSW-joint.

2. Material and welded joint characterization

The experimental testing has been performed on the high strength aluminium alloy AA2024-T351 with the chemical composition and mechanical properties given in Table 1. Two plates 500 mm × 65 mm × 6 mm were welded using FSW welding machine (see Figure 1) where the final length of the weld was 400 mm.

There are a large number of input parameters required for the production of friction stir welded joints, such as tool rotation speed, welding speed, axial force, tool pitch, geometry of tool etc. The microstructure and the mechanical properties of the welded joint are most influenced by the tool rotation speed and welding speed. The microstructure of FSW welded joint is characterized by 3 zones: the heat-affected zone (HAZ), the thermo-mechanical zone (TMZ) and the stir zone (SZ) (Fig. 2).

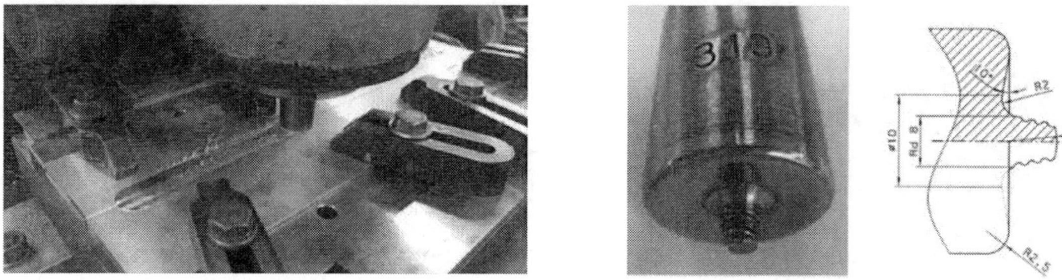

Figure 1. FSW welding machine (left) and welding tool (right).

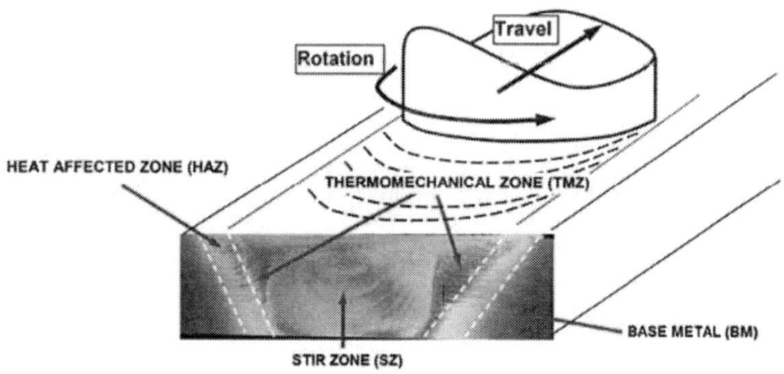

Figure 2. Schematic illustration of FSW.

In the conducted researches, the samples were welded with a constant tool rotation speed $n = 750$ rpm and three different welding speeds as presented in Table 2. After the welding process was completed, the visual control as well as the radiographic control of welded samples on the weld face and root of the seam were performed. No defects were detected (visually, touch or magnifier).

Table 1: Chemical composition and mechanical properties of AA2024-T351.

Chemical composition [%]							Mechanical properties			
Cu	Mg	Mn	Fe	Si	Zn	Ti	Yield strength R_e [MPa]	UTS R_m [MPa]	Elongation A_5 [%]	Hardness *HV*
4.70	1.56	0.65	0.17	0.046	0.11	0.032	370	481	17.9	137

Table 2. Friction stir welding parameters.

Sample designation	Rotation speed *n* [rpm]	Welding speed *v* [mm/min]	Ratio *n*/*v* [rev/mm]
A – I		73	10.27
B – II	750	116	6.47
C – III		150	5

Metallographic observation was carried out by optical microscopy using Leica M205A optical microscope. The specimen for optical microscopy was ground, polished and etched using Tucker's (45 ml HCl, 15 ml HNO_3, 5 ml HF and 25 ml H_2O) reagent. Figure 3 shows four microstructures which correspond to the base material (*BM*) and welded joints for different welding speeds (*A-I*, *B-II*, *C-III*) as already presented and appropriate designated in Table 2. In general, the microstructure is dependent on the amount of the heat input during the welding process. From that respect, the maximum heat input will appear by the lowest welding speed (the case *A-I*) and vice versa. Therefore, the fine microstructure corresponds to the welding regime *C-III* and coarse microstructure to the regime *A-I*.

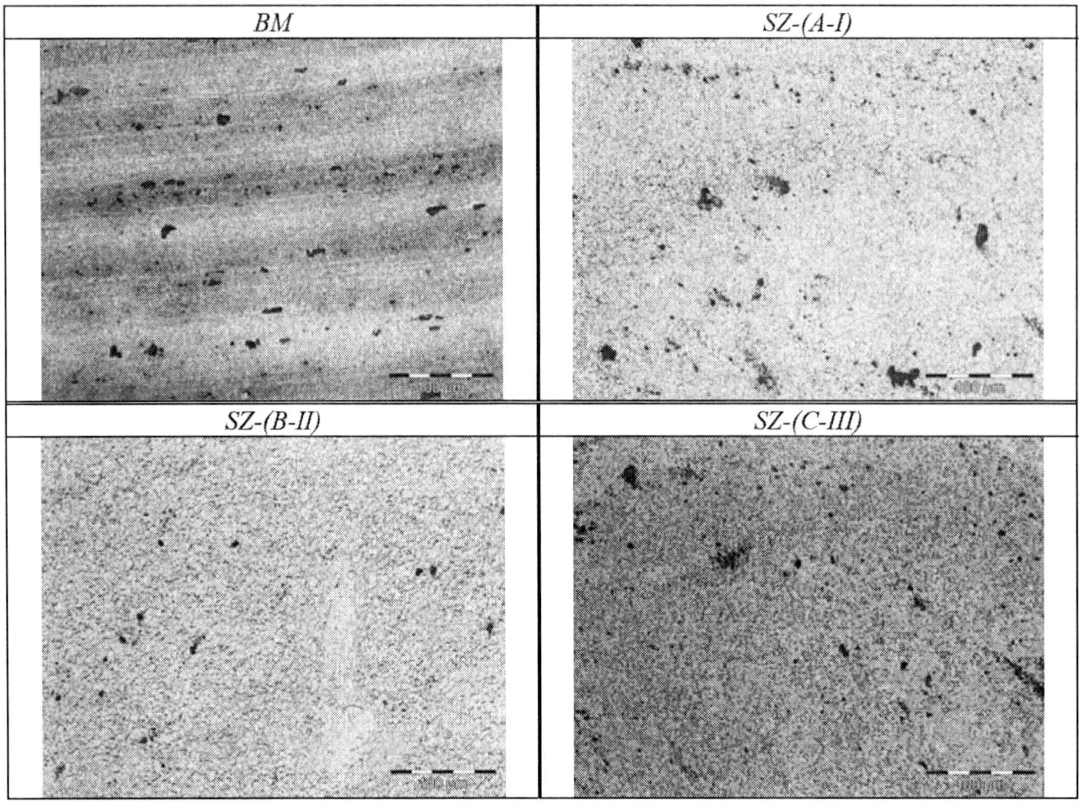

Figure 3. Microstructures of base material (*BM*) and welded joints for different welding speeds .

3. Experimental testing

Specimens for the further mechanical and fatigue testing were manufactured from FSW-samples. As shown on Figure 4, two types of specimens were provided: specimen for tensile test and specimen for fatigue test. Tensile tests were carried out at room-temperature under strain rate of 3.3×10^{-3} s^{-1} according to the ASTM E8M standard. Two tensile specimens were tested for each welding regime (*A-I*, *B-II* and *C-III*) and average values have then been considered when determining the mechanical properties of FSW-joints (see section 4). Fatigue tests were carried out on a servo-hydraulic fatigue testing machine (Shimadzu Servopulser E100kN, Shimadzu Co.) at room-temperature under the stress ratio R=0.1 and loading frequency of 35 Hz. For each welding regime, the fatigue tests have been performed at different stress levels. Based on the obtained experimental results, the S-N curves were constructed as presented in section 4.

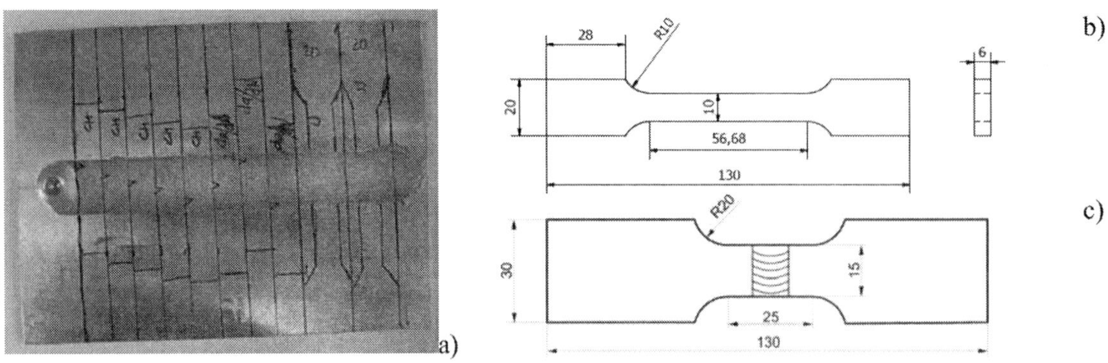

Figure 4. Specimens locations in FSW-welded joint (a), specimen for tensile test (b) and specimen for fatigue test (c).

4. Results and discussion

Figure 5 shows the mechanical properties (ultimate strength R_m, yield stress $R_{p0.2}$ and elongation *EL*) of welded joints for different welding regimes (*A-I*, *B-II* and *C-III*). It is evident, that the best mechanical properties correspond to the welding regime *B-II* with the middle welding speed $v = 116$ mm/min. This welded joint also shows a relative good ductility with elongation at breakage about 7.43 %. On the other hand, a very brittle behaviour (small difference between R_m and $R_{p0.2}$ and extremely short elongation at breakage) under quasy static loading can be observed for welding regime *C-III*, where the welding speed was the highest ($v = 150$ mm/min). For welding regime *A-I* with the lowest welding speed ($v = 73$ mm/min), the mechanical properties of welded joint can be found in the between of the properties for *B-II* and *C-III*.

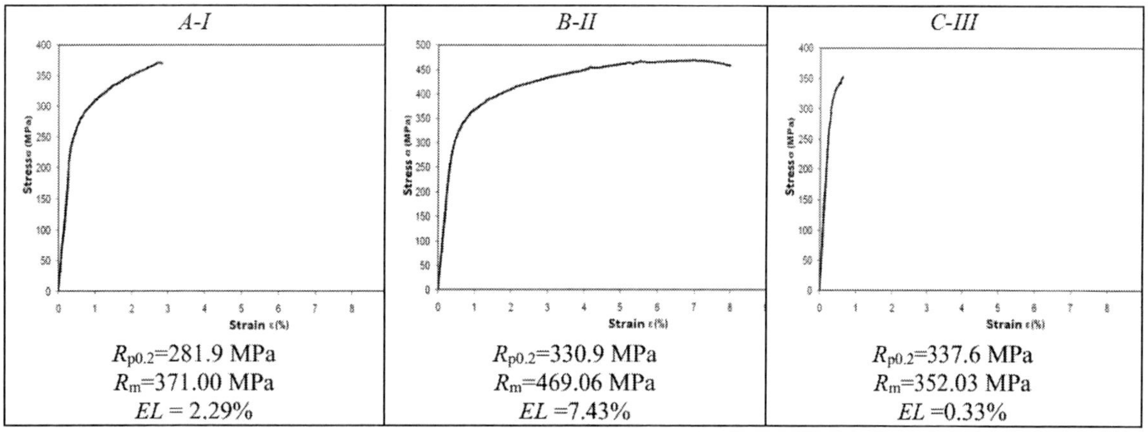

Figure 5. Mechanical properties of welded joints for different welding regimes.

Figure 6 shows the S-N curves (relations between amplitude stress σ_a and number of cycles until failure *N*) for different welding regimes. When creating the S-N curves the fatigue limit of about 10^7 was assumed. It is evident, that the highest fatigue strength corresponds to the welding regime *B-II* where the welding speed was $v = 116$ mm/min. For this case, the fatigue limit at 10^7 loading cycles is approximately 68 MPa. The S-N curve for welding regime *C-III* (the highest welding speed) shows a relative high fatigue strength in the area $N < 10^5$. It can be explained with the fact that the yield stress $R_{p0.2}$ is the highest for welding regime *C-III* which results in the high fatigue strength in this area. On the other hand, the fatigue strength is for the welding regime *C-III* significantly decreasing with increase of *N*. The latter can be explained on the assumption that the hard and brittle material which corresponds to the regime *C-III* is very sensitive to the high cycle fatigue regime.

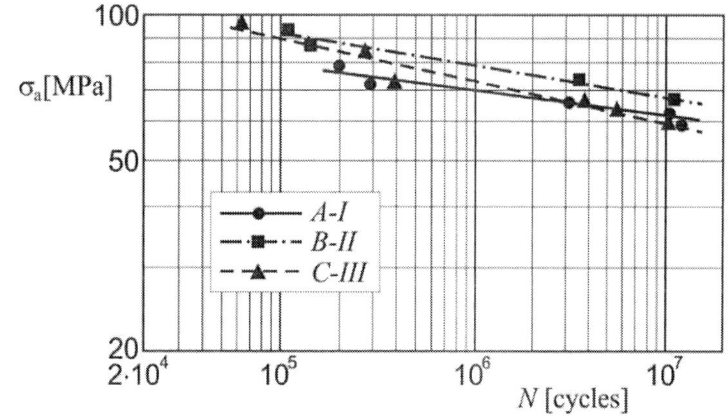

Figure 6. S-N curves for different welding regimes.

5. Conclusions

In the presented work, the mechanical and fatigue behaviour of the alloyed aluminium alloy AA-2024 welded by the Friction Stir Welded (FSW) process were studied. The FSW-process was performed by using a vertical milling machine for constant rotation speed of welding tool 750 rpm and three different welding speeds: 73, 116 and 150 mm/min. Based on the obtained experimental results the following conclusions can be made:

- As expected, the minimum heat input due to friction between welding tool and welding plates has been obtained for the highest welding speed 150 mm/min (welding regime *C-III*) and the maximum heat input has been obtained for the lowest welding speed 73 mm/min (welding regime *A-I*).
- The welding speed and consequently the heat input have a significant influence on the microstructure of welded joints. Because the maximum heat input appears by the lowest welding speed (the case *A-I*), the relative coarse microstructure has been detected in this case. The finest microstructure (the smallest grain size) corresponds to the welding regime *C-III* (the highest welding speed).
- The experimental results of the static tensile tests have shown that the best mechanical properties (ultimate strength, yield stress) correspond to the welding regime *B-II* with the middle welding speed 116 mm/min, which at the same time assure a relative good ductility (elongation at breakage is about 7.43 %). On the other hand, a very brittle behaviour (extremely short elongation at breakage) has been obtained for welding regime *C-III*, where the welding speed was the highest.
- The experimental results of the fatigue tests have shown that the highest fatigue strength corresponds to the welding regime *B-II* (welding speed 116 mm/min) for both, LCF and HCF area. The fatigue behavior for welding regime *C-III* (the highest welding speed) shows a relative high fatigue strength in LCF-area ($N < 10^5$) and low fatigue strength in HCF-area ($N > 10^5$). It can be explained with the fact that the hard and brittle structure which appears by high welding speed is very sensitive to the high cycle fatigue regime.

For future work the influence of the residual stresses on fatigue life of FSW-joint may be addressed. Namely, the higher heat input in the case of low welding speed could induces greater residual stresses which are in high cycle fatigue regime usually not relaxed and may significantly influence the fatigue life of treated FSW-components.

Acknowledgment

The authors acknowledge the financial support from the Slovenian Research Agency
(Bilateral cooperation No. BI-RS/14-15-029: Influence of friction stir welding parameters on fatigue strength of joints of alloy 2024T351)

References

[1] Gunes I, Erdogan M and Cengelci B 2016 Mechanical and electrical properties of artificially aged aluminium alloy AA-2024, *Kovove Mater* **54** 379-387

[2] Bertrand R, Robe H, Texier D, Zedan Y, Feulvarch E and Bocher P 2019 Analysis of aa2xxx/aa7xxx friction stir welds, *J Mater Process Technol* **271** 312-324

[3] Franchim A S, Fernandez F F and Travessa D N 2011 Microstructural aspects and mechanical properties of friction stir welded AA2024-T3 aluminium alloy sheet, *Materials & Design* **32** 4684-4688

[4] Ziemian C W, Sharma M M, Bouffard B D, Nissley T and Eden T J 2014 Effect of substrate surface roughening and cold spray coating on the fatigue life of AA2024 specimens, *Materials & Design* **54** 212-221

[5] Princ W, Geng S, O'Keefe M, Fahjrenholtz W and O'Keefe T 2009 Effect of acid and alkaline based surface preparation on spray deposited cerium based conversion coatings on Al2024-T3, *Appl Surf Sci* **255**(7) 4061-4065

[6] Kumar M, Sotirov N and Chimani C M 2014 Investigations on warm forming of AW-7020-T6 alloy sheet, *J Mater Process Technol.* **271** 1769-1776

[7] Al-Obaisi A M, El-Danaf E A, Ragab A E and Soliman M S 2016 Precipitation hardening and statistical modelling of the aging parameters and alloy compositions in Al-Cu-Mg-Ag alloys, *J. Mater. Eng. Perform* **25** 2432-2444
[8] Thomas W M, Nicholas E D, Needham J C, Murch M G, Temple-Smith P and Dawes C J 1991 *Friction stir butt welding*, GB Patent No. 9125978.8
[9] Da Silva A A M, Arruti E, Janeiro G, Aldanondo E, Alvarez P and Echeverria A 2011 Material flow and mechanical behaviour of dissimilar AA2024-T3 and AA7075-T6 aluminium alloys friction stir welds, *Mater. Des.* **32** 2021-2027
[10] Zimmer S, Langlois L, Laye J and Bigot R 2010 Experimental investigation of the influence of the FSW plunge processing parameters on the maximum generated force and torque, *Int J Adv Manuf Technol* **47** 201-215
[11] Avettand-Fenoel M N and Simar A 2016 A review about Friction strir welding of metal matrix composites, *Material Characterisation* **120** 1-17
[12] Ahmed M M Z, Wynne B P, Rainforth W M, Addison A, Martin J P and Threadgill P L 2019 Effect of tool geometry and heat input on the hardness, grain structure and crystallographic texture of thick-section friction stir-welded aluminium, *Mater. Trans. A Phys. Metal. Mater. Sci.* **50** 271-284
[13] Ouyang J, Yarrapareddy E and Kovacevic R 2006 Microstructural evolution in the friction stir welded 6061 aluminum alloy (T6-temper condition) to copper, *Journal of Materials Processing Technology* **172** 110–122
[14] Balokhonov R, Romanova V, Batukhtina E, Sergeev M and Emelianova E 2018 A numerical study of the microscale plastic strain localization in friction stir weld zones, *Facta Universitatis Series: Mechanical Engineering* **16**(1) 77-86
[15] Podržaj P, Jerman B and Klobčar D 2015. Welding defects at friction stir welding, *Metalurgija* **54**(2) 387-389
[16] Vidal C, Infantea V and Vilacaa P 2010 Assessment of Improvement Techniques Effect on FatigueBehaviour of Friction Stir Welded Aerospace Aluminium Alloys, *Procedia Engineering* **2** 1605-1616
[17] Perović M, Baloš S, Kozak D, Bajić D and Vuherer T 2017 Influence of kinematic factors of friction stir welding on the characteristics of welded joints of forged plates made of EN AW 7049 aluminium alloy, *Tehnical gazette* **24**(3) 723-728
[18] Radisavljević I, Živković A, Grabulov V and Radović N 2015 Influence of pin geometry on mechanical and structural properties of butt friction stir welded 2024-T351 aluminum alloy, Hem. Ind. **69**(3) 323-330
[19] Hussain A K 2010 Evaluation of parameters of friction stir welding for aluminium AA6351 alloy, *International journal of engineering science and technology* **2**(10) 5977-5984
[20] Gupta R K, Das H and Pal T K 2012 Influence of Processing Parameters on Induced Energy, Mechanical and Corrosion Properties of FSW Butt Joint of 7475 AA, *JMEPEG* **21** 1645-1654
[21] Su J Q, Nelson T W, Mishra R and Mahoney M 2003 Microstructural Investigation of Friction Stir Welded 7050-T651 alloy, *Acta Mater.* **51** 713-729
[22] Aydin H, Tutar M and Durmus A 2012 Effect of welding parameters on tensile properties and fatigue behaviour of friction stir welded 2014-T6 aluminum alloy, *Trans Indian Inst Met* **65** 21-30
[23] Lemmen H J K and Alderliesten R C 2010 Benedictus R, Fatigue initiation behaviour throughout friction stir welded joints in AA2014-T3 joints, *International journal of fatigue* **32** 1928-1936
[24] Lemmen H J K, Alderliesten R C and Benedictus R 2010 Evaluating the fatigue initiation location in friction stir welded AA2014-T3, *International journal of fatigue* **33** 466-4736
[25] Ilman M N and Iswanto P T 2013 Fatigue crack growth rate behaviour of friction-stir aluminium alloy AA2024-T3 welds under transient thermal tensioning, *Materials & Design* **50** 235-243
[26] Milan M T, Bose Filho W W, Ruckert C O F T and Tarpani J R 2008 Fatigue behaviour of

friction stir welded AA2024-T3 alloy: longitudinal and transverse crack growth, *Fatigue Fract Engng Mater Struc* **31** 526-538

[27] Hatamlech O, Lyons J and Forman R 2007 Laser and shot peening effects on fatigue crack growth in friction stir welded 7075-T7351 aluminum alloy joints, *International journal of fatigue* **29** 421-434

[28] Nikfam M R, Zeinoddini M, Aghebati F and Arghaei A A 2019 Experimental and XFEM modelling of high cycle fatigue crack growth in steel welded T-joints, *International journal of mechanical sciences* **153/154** 178-193

[29] Kraedegh A, Li W, Sedmak A, Grbovi A, Trisovic N and Mitrovic R 2017 Simulation of fatigue crack growth in A2024-T351 T-welded joint, *Struct Integr and Life* **17** 3-6

[30] Đurđević A, Živojinović D, Grbović A, Sedmak A, Rakin M, Dascao H and Kirin S 2015 Numerical simulation of fatigue crack propagation in friction stir welded joint made of Al2024-T351 alloy, *Engineering failure analysis* **58** 477-484

[31] Živojinović D, Sedmak A and Grbović A 2013 Crack growth analysis in friction stir welded joint zones using extended finite element method, *Structural integrity and life* **13** 179-188

[32] Neto D M and Neto P 2013 Numerical modelling of friction stir welding process: a literature review, *Int J Adv Manuf Technol* **65** 115-126

[33] Fratini L, Buffa G, Palmeri D, Hua J and Shivpurio R 2006 Material flow in FSW of AA7075-T6 butt joints: numerical simulations and experimental verification, *Science and Technology of Welding and Joining* **11**(4) 412-421

CAD modeling of the closing HIGH TIBIAL OSTEOTHOMY

N F Cofaru[1], I I Cofaru[1], V Marjanović[2], N Marjanović[2], M Blagojević [2] and R E Petruse[1]

[1]University "Lucian Blaga" of Sibiu, Faculty of Engineering, Emil Cioran 3, 550025 Sibiu, Romania
[2]University of Kragujevac, Faculty of Engineering, Sestre Janjić 6, 34000 Kragujevac, Serbia

E-mail: nicolae.cofaru@ulbsibiu.ro

Abstract. The article proposes a generalised CAD modeling for a surgical strategy applied to the axial deviations of the human leg. These deviations appear most of the cases from a most common disease nowadays, ghonartrosys characterised by usage of the knee's articular cartilage. The surgical strategy is HIGH TIBIAL OSTEOTOMY and it is very appropriate in order to correct the axial deviation due several reasons such as: repair with good results the axial deviations, good rate price/quality, strongly recommended method for young patients, not needed a maintenance activity as in case of the prosthesis. In the article it starts with the presentation of the mechanical axis of the lower limb and of the axial deviations that might appear and then using the CAD type modeling tools it is simulated and modelled the surgical procedure: Closing High tibial osteotomy. The approach is very important both from didactically consideration and possible evaluations presurgical, intrasurgical and postsurgical.

1. Introduction

The 3D modeling of the bones of the human lower limb, assembly of these structures in different ways for healthy subjects but especially for the subjects affected by different diseases in this part of the body, is an important problem that must be taken in consideration in order to establish the best way for treatment and for the correctitude of the simulation of the surgery procedures. CAD modeling of the anatomical structures is a very useful method in the biomedical field since the objects of the study, human body parts, cannot be removed and studied as experimental objects.

The utility of the method results in a clear manner from the previous approaches which are founded in the anterior researches [8, 9, 7]. In those articles CAD-CAE methods proved its efficiency.
The advantages of CAD modeling are: a better assessment of the concrete situation on the custom model, compatibility analysis with implants and transplants found in 3D databases, and the educational role in supporting the learning these techniques using virtual models.

The CAD modeling which we will achieve in this article is applied for simulation of the surgical procedures: Closing High Tibial Osteotomy. For the beginning in order to have a clear image of the subject, a presentation of the mechanical axis of the leg is made.

A correct alignment of the bones of the human lower member is defined by collinearity of the line segments which unite the centre of the hip joint, the centre of the knee articulation and the centre of the ankle (figure 1). Any swerve from this axis becomes an axial deviation and it has important consequences on the health and on the comfort of the patient affected by these deviations.

The main cause of this deviation is unicompartmental ghonartrosis malady that unfortunately is more and more common today. The effect of this malady is the wear of the articular cartilage which leads to an incorrect loading of the knee and thus to the appearance of the deviations. These deviations cause an abnormal strain of the knee & displacement of the mechanical axis from the centre to the medial or lateral (figure 2).

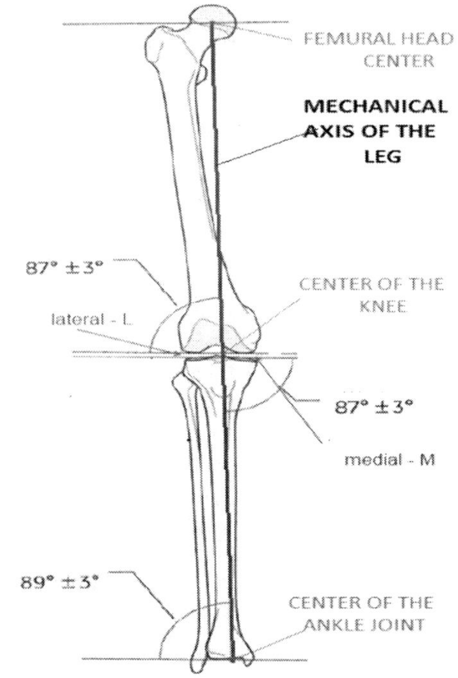

Figure 1. Mechanical axis of the leg.

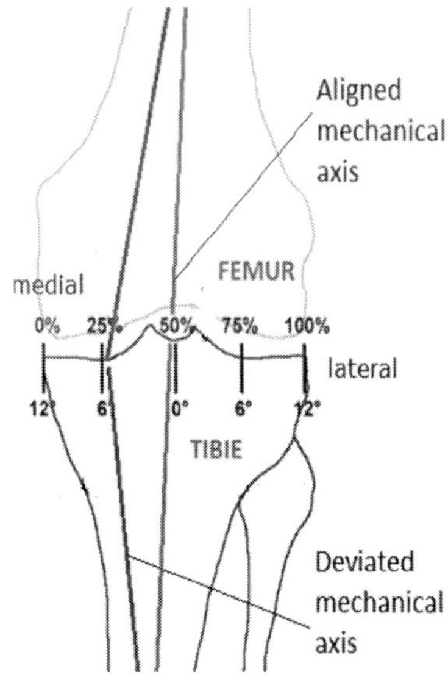

Figure 2. Aligned and Deviated mechanical axis.

The worn-out cartilage will be overstrained, which further causes its wear and therefore further increase the deviation. The two effects potentiate each other, a vicious circle which will lead to generalized arthrosis changes in the knee joint.

One of the most common surgical procedure used to correct these axial deviations is surgical technique called High Tibial Osteotomy.

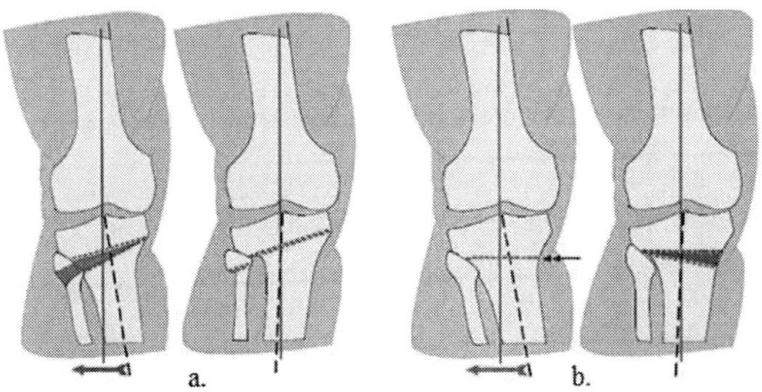

Figure 3. High tibial osteotomy (a – closing, b – opening).

Tibial osteotomy is a surgical technique that involves creating a bone wedge at the proximal tibia level to realign the bones of the lower limb. The osteotomy technique is fully recommended for younger patients because in their case it the recovery is relatively quick. In addition, this technique does not exclude the possibility of partial or total prosthetic intervention, if needed later on.

The purpose of the proximal tibial osteotomy is the realignment of the mechanical axis of the lower limb and the redistribution of the forces in the joint to unload the impacted area.

High tibial osteotomy can be one of a closing or of an opening (figure 3) [6].

For the closing osteotomy (figure 3.a.) the cuts are made, the bony wedge that results being removed, and the bony fragments are being realigned by closing the wedge. In the case of an opening only one cut is made, the fragments are realigned, resulting in an empty space by opening the wedge (figure 3.b.).

The both surgical techniques have advantages and disadvantages [2]. For opening high tibial osteotomy the advantages are: accurate corrections, no fibular osteotomy, correction in 2 plans, no limb shortening and disadvantages are: bone graft/spacer material, risk of delayed and non-union, risk to increase tibial slopes, patella height and patella-femoral pressure. In the case of lateral closing wedge osteotomy the most important advantages are the greater potential of correction, no need for bone grafting and faster healing. The disadvantages are risk to peroneal injury, muscle detachment, shortening of the leg and loss bone stock.

Taking in account the fact that closing wedge osteotomy has the important advantage of a more accurate correction with less morbidity in this paper we achieved the 3D CAD parameterized modeling of this surgical techniques.

2. Geometrical elements that are necessary for the planning of the intervention

As presented in the last chapter the mechanical axis of the lower member merging the centre of the femur's top with the centre of the ankle. We observe (figure 4) that in case of the knee affected by axial deviation this axis does not cross the middle of the knee and it is closer to the medial zone of the knee or is possible even to be less than 0% and it passes outside the joint, aspect which needs to be corrected.

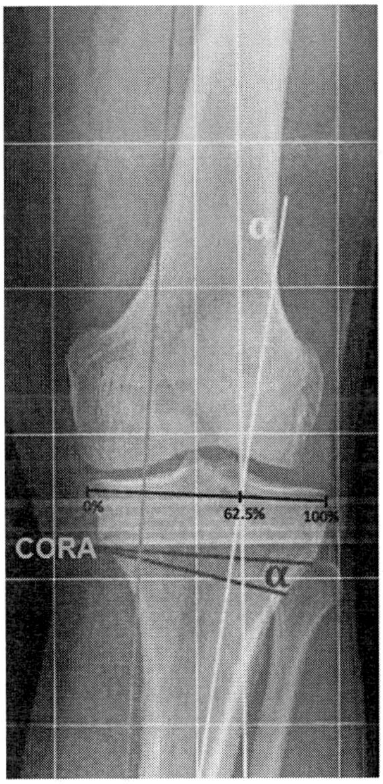

Figure 4. Determination of the correction angle α [1].

Generally patients with ghonarthrosis have a predisposition to a new axial deviations. From this reason the goal of the closing high tibial osteotomy is to lead the mechanical axis of the leg (the green line in figure 4) not in the neutral position (50%) but in the point corresponding to the 62.5% of the tibial plateau width. It is achieved an over-correction. In young patients with a minimal degeneration of the medial compartment, correction to a neutral mechanical axis is planned [1].

The angle of the correction α is formed by the yellow lines from figure 4. The yellow lines run from the point situated at the 62.5% of the tibial plateau width to the centre of the hip and centre from the ankle.

The establishment of the hinge point CORA (figure 4) is made close to the border of the tibia so that the angular rotation is realized, but while assuring that don't affect the medial cortex of the tibia. The first cutting plane is placed parallel to the tibial plateau, 2–2.5 cm below the articular joint line, taking in account the inclination of the tibial plateau. From the bone's lateral it is recommended to leave a distance of 5-10 (mm). The second cutting plane is a normal plane on frontal plane of the knee and make the correction angle with the first cutting plane. Obviously CORA is a line situated at the intersection of the cutting planes.

The presentation of the geometrical elements mentioned above highlights a part of the geometrical parameters which are important for achieving the intervention. These are: the correction angle α, the placement of the hinge point or CORA. Other parameter which can be studied is the diameter of the relieving stress hole realized in CORA.

So that these parameters are easy to control (even by the laymen in aided modeling) we propose a parameterized modeling, with wide possibilities of customization, of the tibia closing osteotomy.

3. The parameterized modeling of the tibia closing osteotomy

In order to achieve a parameterized modeling [4], [5] which, only by modifying the parameters on the model, creates all the possible situations, the creation of a system of points and planes, according to which the parameterization will be realized, has to be done.

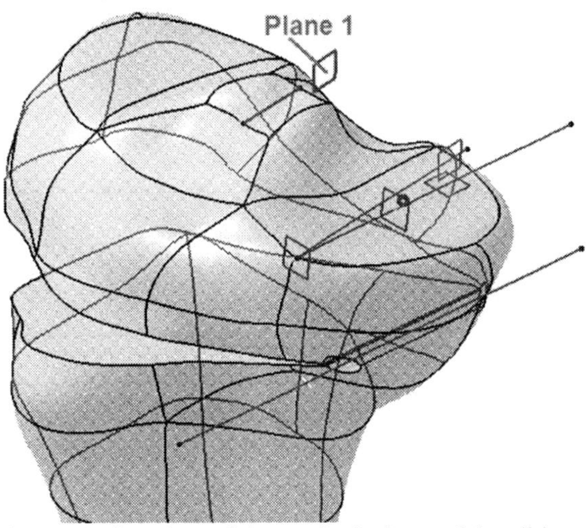

Figure 5. Creation of the sagittal plane of the tibia.

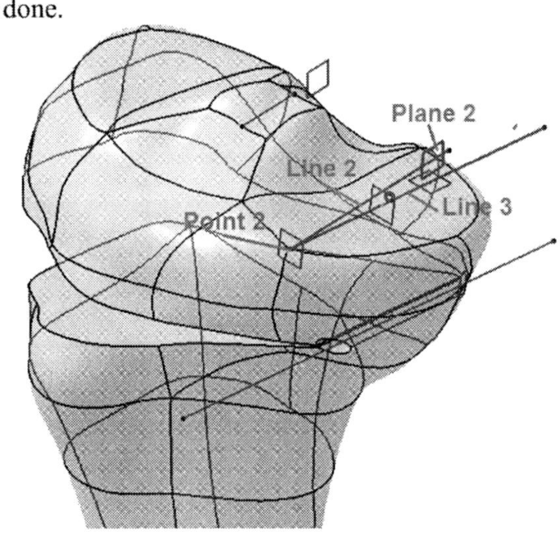

Figure 6. Modeling of the tibial plateau inclination.

First step is the creation of the sagittal plane of the tibia. To materialize this plane, a parallel line with the sagittal plane of the human body is created. This line crosses the middle of the tibia line and through this a plane is being constructed (Plane 1). The plane forms a 90° angle with the horizontal plane (figure 5).

In order to highlight the posterior tilt with a 7° angle of the tibial plateau, Point 2 has to be defined on the tibial plane, in the centre and boundary to the intersection with the anterior surface of the tibia. Through this point Line2 will be defined as parallel to the one defined at the previous step. Through this line, a Plane 2 rotated with 0° (parallel) towards Plane 1, and in his plane Line 3 at 7° towards Line 2 (figure 6.). For the modeling 3 planes, perpendicular on the inclination line of the tibia plane, are constructed: Plane 4 through a point situated at the middle of the tibia and normal to Line 3, Plane 5 through Point 2 and also normal of the line 3 and Plane 3 through Line 3 and normal of the Plane 4 and 5 (figure 8).

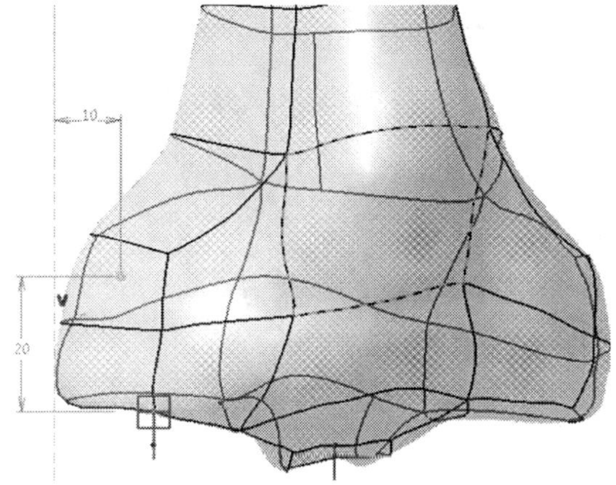

Figure 7. Determination of the CORA positions.

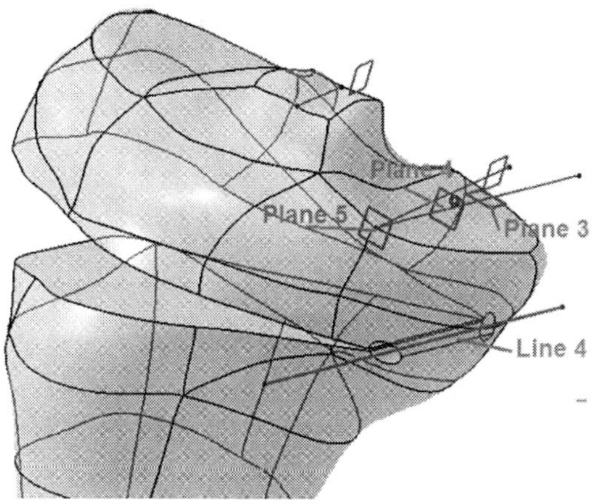

Figure 8. Determination of the hole axis in CORA.

The next step is the execution of the relieving stress hole in the hinge (CORA). This step consists in the making of a hole which has a double role: eliminating the stress concentrators from the bottom of the osteotomy wedge and limiting the lateral cutting plan. For this it was made a first sketcher (figure 7.) in the Plane 5 in order to determine the position of the CORA. This point is situated at 10 mm measured from the tangential plan of the medial cortical surface of the tibia and 20 mm measured from the Plane 3. These coordinates could be parameters that will be taken into account in further approaches. Through this point Line 4 will be defined normal to the Plane 5. Line 4 will be axis for the relieving stress hole. It is worth mentioning that the hole is not horizontal but parallel to the tibia planes. This method of modeling, apparently more complicated, was also created with the purpose of offering the model a high degree of generality [1].

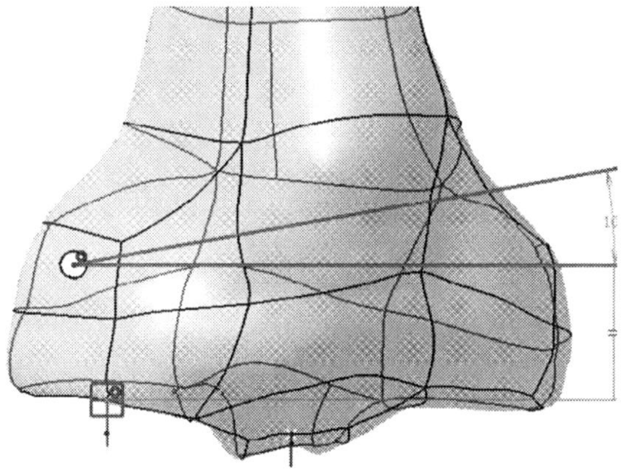

Figure 9. Sketcher for the closing wedge.

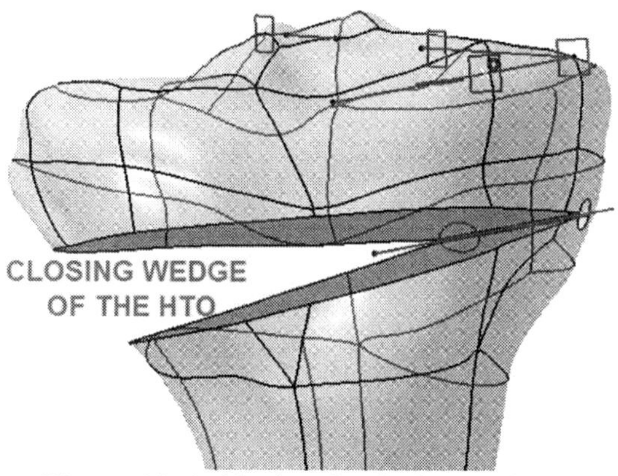

Figure 10. Achieving the closing wedge.

The last step is the execution of the closing wedge (figure 9). For this Plane 5 was chosen as sketcher plane and the first cutting is parallel with Plane 3 and passes through CORA. The second cutting occurs in the same plane and is positioned at the requested correction angle. Correction angle also could a variable parameter. In the figure 10 is shown the tibia prepared for closing high tibial osteotomy.

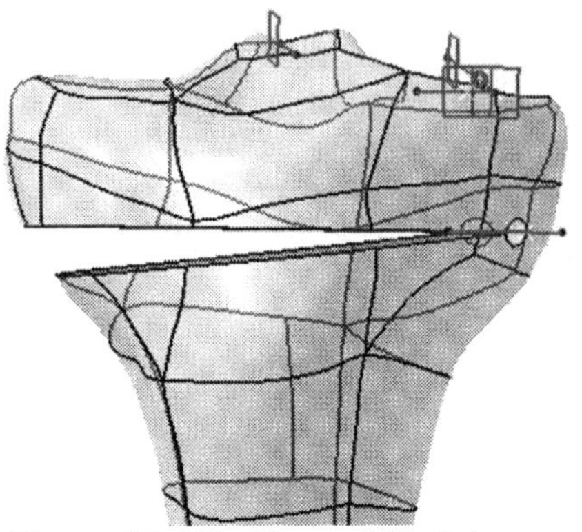

Figure 11. V1=10 mm, V2=6 degree.

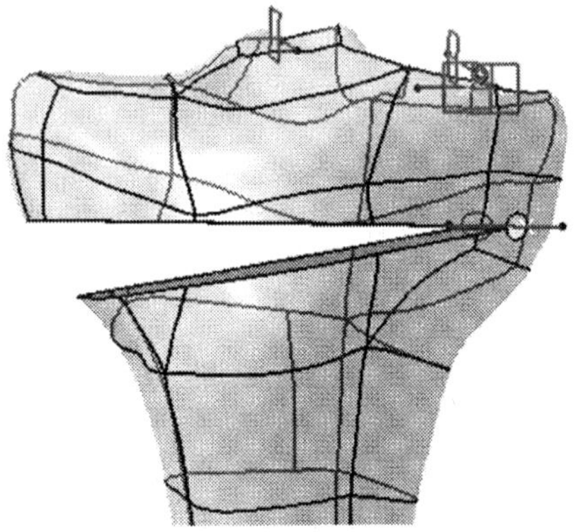

Figure 12. V1=10mm, V2=10 degree.

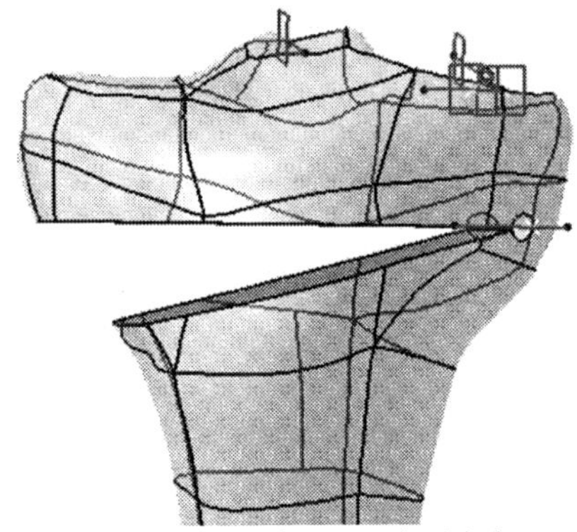

Figure 13. V1=10mm, V2=14 degree.

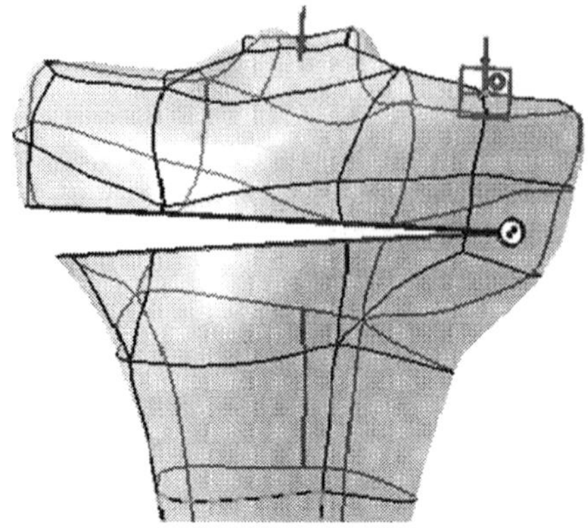

Figure 14. V1=8 mm, V2=6 degree.

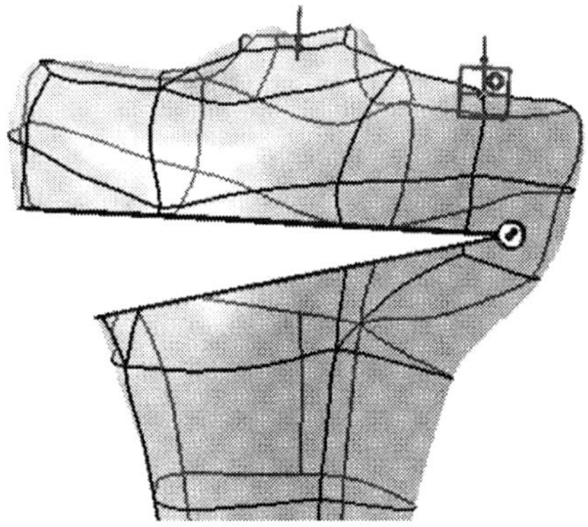

Figure 15. V1=8 mm, V2=14 degree.

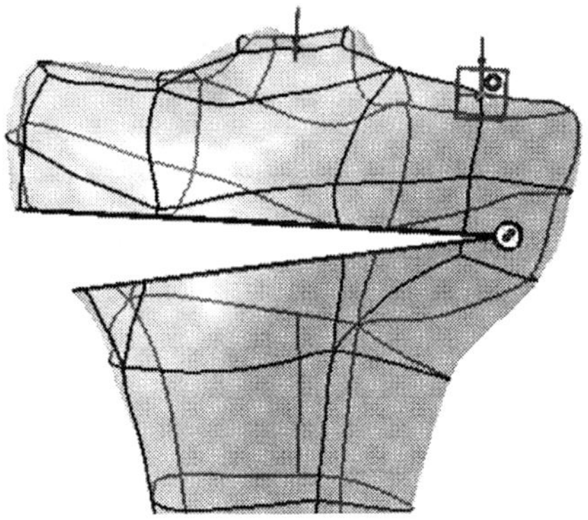

Figure 16. V1=12 mm, V2=10 degree.

The 3D generalized model allows us to obtain a multitude of cutting possibilities only by modifying the presented parameters. In this case for closing high tibial osteotomy the most important parameters could be the position of CORA related to the tangential plane of the lateral cortical surface of the tibia V1(with values 8, 10 and 12 mm) and the correction angle V2 (with values 6,10 and 14 degrees). In the figures 11-16 it is shown some combinations of these parameters.

4. Conclusions

The created model was designed so as it will present a high degree of flexibility and so as to ensure that, only by modifying parameters, a large variety of particularized models will result. With the model we can realize any configuration and we can easily prepare the geometric structures in order to undergo the analyses through the finite element method.

The analyses using the finite element method proves its full efficiency in the study of the osteoarticular surgery, being a method that we can use in order to simulate a high range of situations, in the following scenarios: pre-surgery, during the surgery and post-surgery.

Numerical simulation applied of the tibia's behaviour in osteotomy surgeries offers a pretty accurate image of the stress state and allows us to take the most appropriate decision regarding the surgery planning.

The generalised CAD models developed in this article will be used in the further FEM researches in order to establish the best way to performing the surgical procedure CLOSING HIGH TIBIAL OSTEOTOMY.

Acknowledgments
This work is supported by the DiFiCIL project (contract no. 69/08.09.2016, ID P_37_771, web: http://dificil.grants.ulbsibiu.ro), co-funded by ERDF through the Competitiveness Operational Programme 2014-2020.t

References

[1] Mattei L, Lea S, Nicolaci G, Ferrero G, Marmotti A and Castoldi F 2017 *Closing wedge tibial osteotomy: is it an actual procedure nowadays?,* Annals of Joint, *AOJ,* Ann Joint, doi: 10.21037/aoj.2017.06.04

[2] Landreau P 2019 on Twitter: *"High tibial osteotomy techniques"* ttps://twitter.com/plandreau/status/953283838054207489 (accessed on May. 2019)

[3] Bae D K, Song S J and Yoon K H 2009 Closed-wedge high tibial osteotomy using computer-assisted surgery compared to the conventional technique, *J Bone Joint Surgery Br. 2009 Sep* **91**(9) 1164-71, doi: 10.1302/0301-620X.91B9.22058

[4] Cofaru I I 2013 *Researches regarding the biomechanics of the axial deviations of the human lower member and the development of the correspondent surgical devices,* Sibiu, Doctoral Thesis

[5] Amis A 2013, Biomechanics of high tibial osteotomy, *Knee Surgery Sports Traumatology Arthroscopy Jan* **21**(1) 197-205, doi: 10.1007/s00167-012-2122-3

[6] Kyung D, Sang J, Hwan Kim J and Wan Seo J 2012 Change in limb length after high tibial osteotomy using computer-assisted surgery, *Knee Surgery Sports Traumatology Arthroscopy* **21**(1), DOI: 10.1007/s00167-012-1898-5

[7] Pena J, Lerner A, Yao J, Salo A and Totterman S 2003 *Evaluation of distance maps from fast GRE MRI as a tool to study the knee joint space,* Proceedings of SPIE - The International Society for Optical Engineering 5031:551-562, DOI: 10.1117/12.480435

[8] Müller-Karger C and Larrazabal A 2005 *Finite Element Bone Model Incorporating Heterogeneity And Anisotropy* From Ct, ISB XX-th Congress - ASB 29th Annual Meeting July 31 - August 5, Cleveland, Ohio

[9] Soncini M, Vandini L and Redaelli A 2004 Finite Element Analysis Of A Knee Joint Replacement During A Gait Cycle, *Journal Of Applied Biomaterials & Biomechanics* **2** 45-54

IOP Conf. Series: Materials Science and Engineering **659** (2019) 012034 doi:10.1088/1757-899X/659/1/012034

Modeling of 3D temperature field in butt welded joint of 6060 alloy sheets using the ANSYS program

M Matuszewski[1]

[1] Czestochowa University of Technology, Faculty of Mechanical Engineering and Computer Science, Armii Krajowej 21, 34-201 Częstochowa, Poland

E-mail: matuszewski@itm.pcz.pl

Abstract. In work, the modeling of a three-dimensional temperature field in a butt weld connection of two 6060 aluminum alloy sheets using Finite Element Method is presented. The calculations were performed for two welding methods: TIG (Tungsten Inert Gas) and MIG (Metal Inert Gas). The Goldak's double ellipsoidal heat source model has been used in modeling. The thermal-mechanical properties of the material were assumed to depend on the temperature. The Workbench, DesignModeler, Mechanical, Fluent and CFD-Post modules of the ANSYS program were used for numerical simulations. In the description of the geometry of joints, cube type elements were used, with density of grid in the heat affected zone. The parabolic shapes of face and root were assumed based on the literature and results of the experiment. The temperature distributions in cross-sections of welded joints as well as welding thermal cycles at selected points were analyzed. The results of numerical simulations were verified experimentally. Comparison of calculated and obtained in the experiment the characteristic limits of heat affected zones showed satisfactory compatibility. The directions of heat propagation determined by vectors of cooling rates coincide with the longitudinal axis of dendritic grains determined on the basis of metallographic tests.

1. Introduction

Thermal phenomena are important in the technological processes of metals and their alloys. In many processes, such as casting [1, 2], heat and laser treatment [3, 4], welding [5, 6], welding coatings [7, 8], they are the essence of the technological process, i.e. calling the right temperature is to achieve the desired state of the material leading to a change in the state of focus or its properties. In other processes, the increase of the temperature of the processed material is aimed at determining such properties that enable the final technological method, eg. FSW (Friction Stir Welding) [9], coating [10]. In some processes, thermal effects are a secondary effect, such as in machining [11]. Depending on the material, thermo-mechanical properties change as the temperature changes [12], and in the case of steel (cast steel, cast iron), there are structural changes (phase transformations in solid state).

In modelling the temperature field of welding processes, the selection of the heat source model and the calculation method is very important. The analysis of models of heat sources used in the description of thermal phenomena in welding processes is presented, among others in the works [13, 14]. The first proposals were point heat source models [15, 16]. Because the point models did not allow for describing the temperature field near the weld axis, Eagar [17] proposed a Gaussian distributed surface source. The breakthrough moment was the double ellipsoid model of the heat source proposed by Goldak [18]. Goldak's proposal was the first model taking into account the volumetric nature of the welding source. In work [19] a volumetric model with Gaussian power distribution in the horizontal plane and parabolic change in the vertical direction was proposed.

A further direction of the search was to consider the inclination of the electrode (welding head) in heat source models [20, 21] and to analyze the influence of this inclination on the temperature distribution in welded joints [22, 23].

Because single-distributed models of heat sources only took into account the heat of the welding arc, disregarding the heat transferred to the weld pool through the molten additional material (eg electrodes), in works [24-27] two-distributed models were proposed. It allowed, among others, to obtain an irregular line of fusion (often observed in welding practice), impossible to reproduce with the use of single-modal source.

Despite many models and numerous publications on temperature field modelling, intensive search for models and algorithms is still in progress to describe the temperature field enabling to obtain temperature distributions as close as possible to real values [28-33].

In the descriptions of the temperature field during welding, analytical [34, 35], semi-analytical [36] and numerical [33, 37-39] solutions for the differential heat conduction equation are used.

The proposed solutions concern the possibility of describing the temporary temperature field during particular welding methods and related processes: GMAW (Gas Metal Arc Welding) [40, 41], SAW (Submerged Arc Welding) [42-45], GTA (Gas Tungsten Arc) [25], hardfacing [46] and surfacing [47], PAW Plasma Arc Welding) [48], PAC (Plasma Arc Cutting) [49], LBM (Laser Beam Welding) [50], hybrid laser-arc welding [51], hybrid laser-waterjet micro machining [52,53] and metallization [54].

This paper presents modelling of 3D temperature field in butt weld joint of 6060 alloy sheets using the ANSYS program. Numerical simulations have been carried out for two welding methods (GTA and GMA) verified experimentally.

2. Experimental work

Welding tests of single pass butt welded joint of 6060 aluminum alloy sheets were carried out in the Welding Laboratory of Czestochowa University of Technology. Welded joints were made using two methods (in the argon shield):

- GTA method (141) – PN-EN ISO 4063:2011 [55],
- GMA method (131) – PN-EN ISO 4063:2011 [55].

The scheme of single-pass butt welding of aluminum alloy sheets is presented in Figure 1. The chemical composition and thermomechanical properties of has been summarized in Tables 1 and 2.

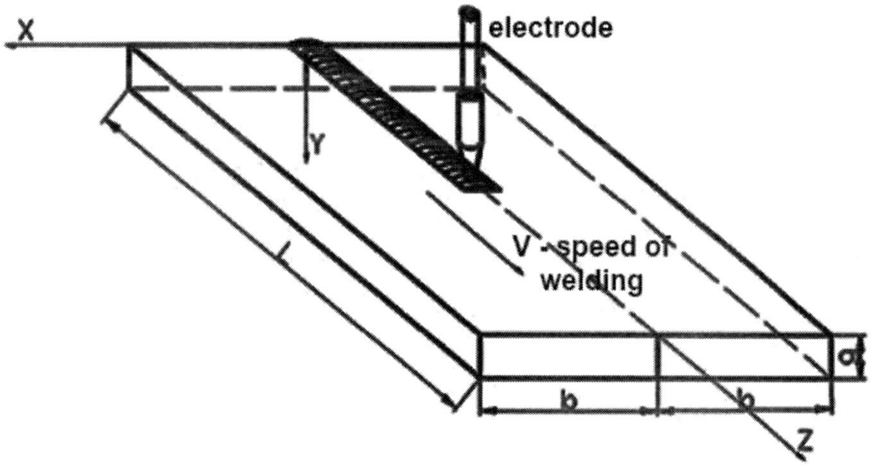

Figure 1. The scheme of single-pass butt welding process.

The welding process involved the manufacture of a butt joint of two 6060 aluminum alloy sheets with dimensions 244x110x4 mm by GTA (141) method was carried out on the Faltig 315 welding device (OZAS, Opole, Poland) [56] (figure 2).

Table 1. Thermomechanical properties of Alloy 6060.

Property	Value
Density	2700 (kg/m^3)
Specific heat	898 (J/kgK
Thermal conductivity	209 (W/mK)
Solidus temperature	610 (°C)
Liquidus temperature	655 (°C)

Table 2. The chemical composition of Alloy 6060 (with accordance with EN 573-1[57]).

Element	Participation (%)
Mg	0.35 – 0,60
Mn	≤0.10
Fe	0.10-0.30
Si	0.30 – 0.60
Cu	≤ 0.10
Zn	≤ 0.15
Cr	≤ 0.05
Ti	≤ 0.10
other	≤ 0.05
Al	rest

Faltig 315 allows welding with a non-clay electrode (141) and coated electrode (111) direct current and alternating current. The parameters of the 6060 aluminum alloy GTA welding process are shown in Table 3.

Figure 2. Faltig 315 AC/DC.

Figure 3. Synermig 400.

The welding process involved the manufacture of a butt joint of two 6060 alloy sheets with dimensions of 200x60x5 mm using the GMA (131) method. The welding process was carried out on the Synermig 400 welding device (OZAS, Opole, Poland) presented in Figure 3. The device is intended for joining construction steels, high-alloy steels of type 18 - 8 and alloys (AlMg5 and AlSi5) [56]. We can weld with direct current and alternating current. The welding parameters using GTA method is setting in Table 4.

Table 3. Welding parameters during GTA welding.

Parameter	Value
Current type	Alternating
Voltage	15.2 A
Current	130 A
Welding speed	8.6 cm/min
Diameter of electrode	2.4 mm
Composition of additional material	AlMg5
Shielding gas	Ar 4.5
Intensity of shielding gas	15 l/min

Table 4. Welding parameters during GMA welding.

Parameter	Value
Current type	Alternating
Voltage	15.2 (A)
Current	130 (A)
Welding speed	8.6 (cm/min)
Diameter of electrode	2.4 (mm)
Composition of additional material	AlMg5
Wire feeding speed	8.5 (m/min)
Shielding gas	Ar 4.5
Intensity of shielding gas	15 (l/min)

3. Examples of numerical simulations

The problem of modeling the temperature field in welding processes using the finite element method was made using Ansys packages (4 programs from Ansys software):
- Ansys DesignModeler used to make a solid geometry,
- Ansys Meshing used to divide the solid into finite elements,
- Ansys Fluent used to define the model and calculations,
- Ansys CFD-Post for the development and analysis of results,

Ansys Fluent uses the following equation to solve heat transfer problems for a solid:

$$\frac{\partial}{\partial t}(\rho h)+\nabla\cdot(\vec{v}\rho h)=\nabla(k\nabla T)+Q \tag{1}$$

where: ρ – density, h – enthalpy, k – conductivity, T – temperature, Q - volumetric heat source, $\vec{v}$ - speed field. When modeling the three-dimensional temperature field in welding processes, the dual ellipsoidal moving heat source proposed by Goldak presented in the figure 4 was used. The Goldak's model consists of two semi-ellipsoidal volumes that create a heat flux. For points (x, y, z) belonging to the semi-ellipse located in the front part of the source, the heat flux is described by the equation:

$$q_f(x,y,\xi)=\frac{6\sqrt{3}f_fQ}{abc_f\pi\sqrt{\pi}}exp\left(\frac{-3x^2}{a^2}\right)exp\left(\frac{-3y^2}{b^2}\right)exp\left(\frac{-3\xi^2}{(c_f)^2}\right) \tag{2}$$

However, for points (x, y, z) belonging to the back of the source:

$$q_r(x,y,\xi)=\frac{6\sqrt{3}f_rQ}{abc_r\pi\sqrt{\pi}}exp\left(\frac{-3x^2}{a^2}\right)exp\left(\frac{-3y^2}{b^2}\right)exp\left(\frac{-3\xi^2}{(c_r)^2}\right) \tag{3}$$

where: a, b, c_f, c_r are the parameters of the ellipsoidal heat source, f_f, f_r proportionality coefficients corresponding to heat in the front and rear parts of the heat source, where $f_f+f_r=2$, ξ distance of current source position to point(x,y,z) [18].

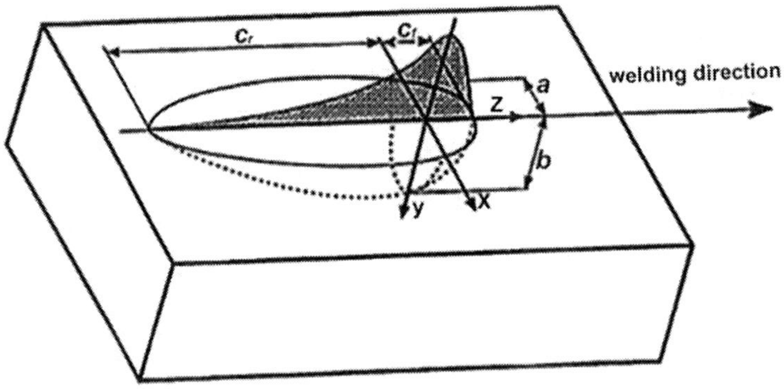

Figure 4. Double ellipsoidal Goldak's model of heat source.

Boundary Conditions:

- First-type boundary condition (Dirichlet):

$$T_s = f(x, y, z, t) \tag{4}$$

- Second-type boundary condition (Neumann):

$$q_s = -k\frac{dT(0,t)}{dx} \tag{5}$$

- Convection boundary condition:

$$q_{conv} = h(T_{free}\text{-}T_w) \tag{6}$$

- Radiative condition:

$$q_{rad} = \varepsilon\sigma(T_\infty^4\text{-}T_w^4) \tag{7}$$

IOP Conf. Series: Materials Science and Engineering **659** (2019) 012034 doi:10.1088/1757-899X/659/1/012034

3.1. *Modeling of the temperature field during butt welding of 6060 aluminum alloy sheets using the GTA method (141).*

Figure 5. Finite element mesh for temperature field modeling during GTA welding.

Table 5. Parameters used in the simulation.

	Arc power 1970 [W]	Efficiency	Heat transfer coefficient [W/m^{2}K]	Speed of welding [m/s]	External emissivity
GTA	1970	0.6	25	0.0014	0.1

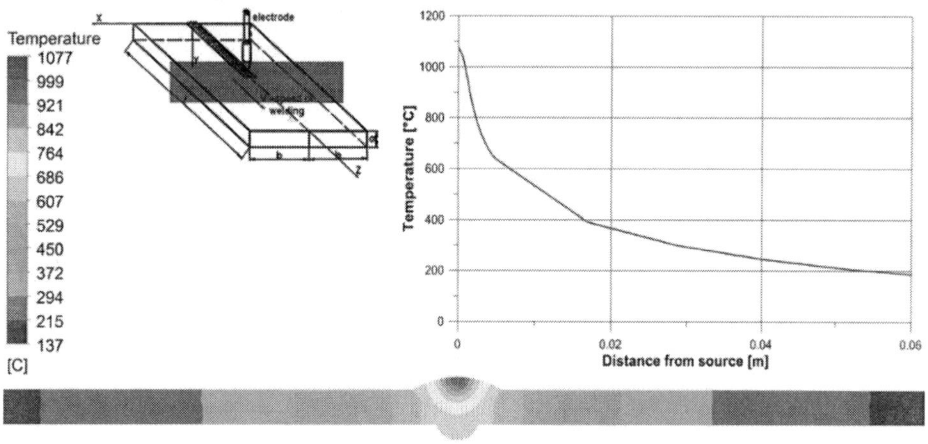

Figure 6. Temperature distribution during welding of Al6060 alloy sheet with the GTA method at time t = 69 s from the beginning of welding in the cross-section.

The cross-section shows the temperature distribution at time t = 69 s from the beginning of welding. The graph shows the temperature change with the change of distance from the heat source. It may observe a sharp drop in temperature in the initial phase of the graph, which presents densely arranged isotherms near the source. Figure 7 shows two points for which a diagram of thermal cycles was performed during the time of connecting the two butt-welded alloy plates using the GTA method.

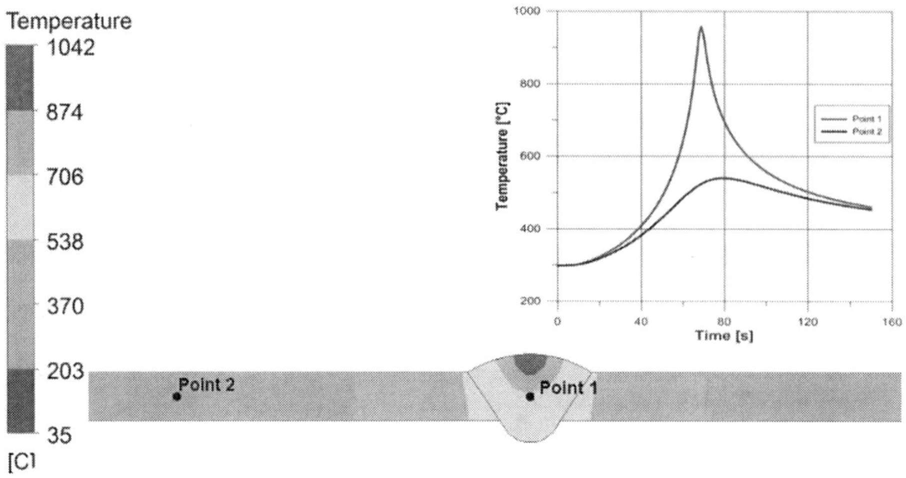

Figure 7. Heat cycle for two points during butt welding of GTA aluminum alloy.

The results obtained give satisfactory results. The red color on the left is the area in which we obtained a temperature above solidus where the material melts. The dimensions of the remelting zone obtained during modeling in relation to experimental studies show divergences less than 5%.

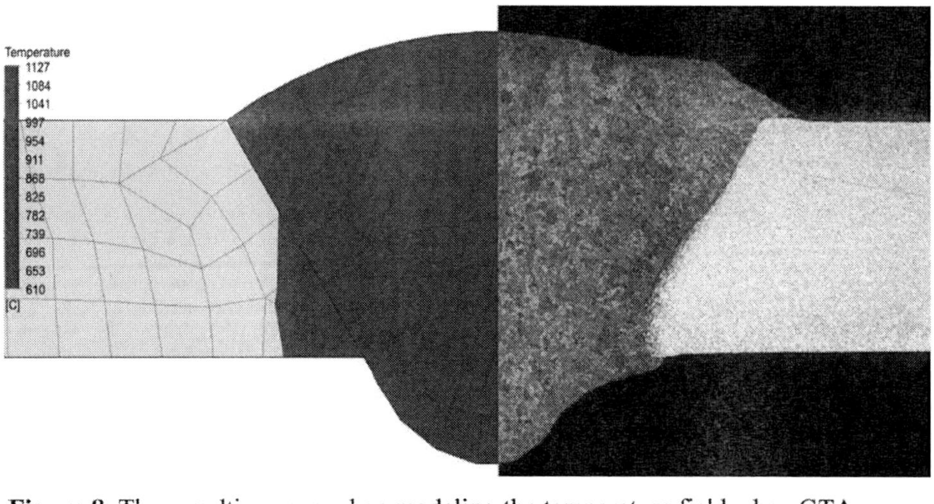

Figure 8. The remelting zone when modeling the temperature field when GTA welding (141) compared to the experimental study (metallographic fracture).

3.2. Modeling of the temperature field during butt welding of 6060 aluminum alloy sheets using the GMA method (131)

During the modelling of the temperature field during the butt welding using the GMA method, two sheets of 6060 alloy, the solid was divided into 444609 cubic elements and 487760 nodes with a densegrid in the joint area [57].

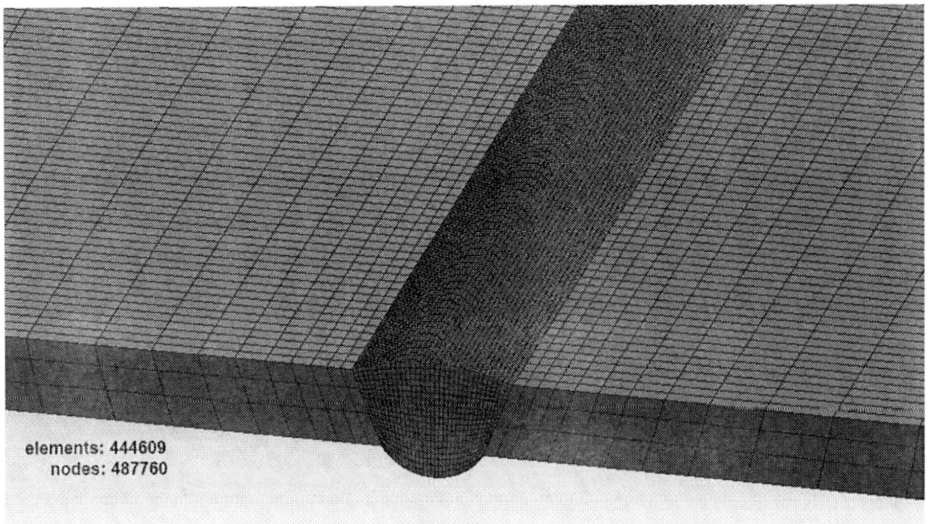

Figure 9. Finite element mesh for temperature field modeling during GTA welding.

Table 6. Parametres used in the simulation.

	Arc power 1970 [W]	Efficiency	Heat transfer coefficient [W/m^{2}K]	Speed of welding [m/s]	External emissivity
GTA	1970	0.6	25	0.0014	0.1

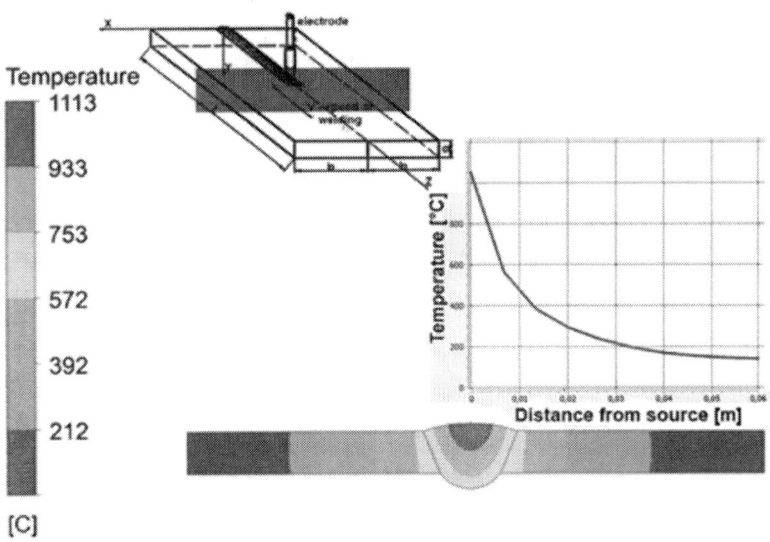

Figure 10. Temperature distribution during welding of Al6060 alloy sheet with the GMA method at time t = 24 s from the beginning of welding in the cross-section.

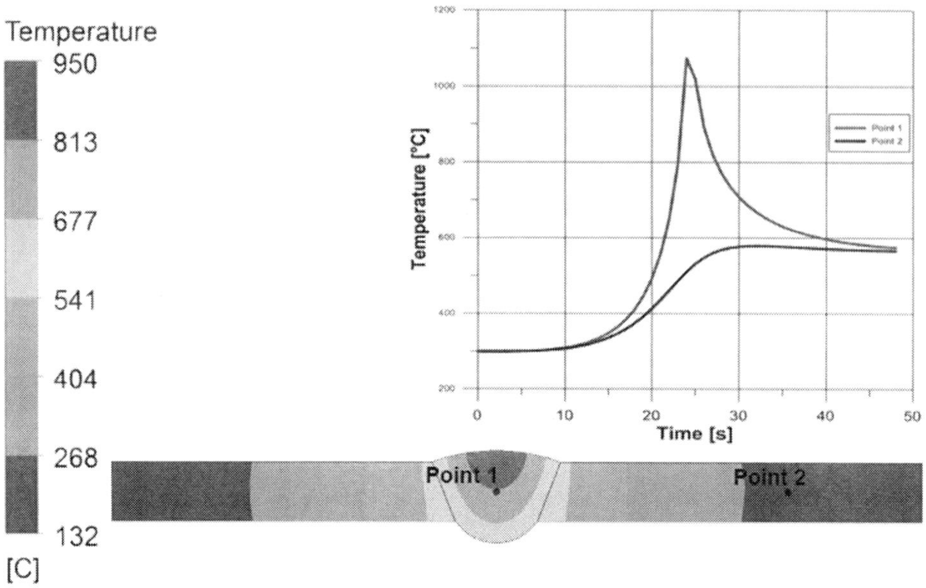

Figure 11. Heat cycle for two points during butt welding of GTA aluminum alloy.

Figure 11 shows two points for which a diagram of thermal cycles was made during the joining of two butt-welded alloy plates using the GTA method.

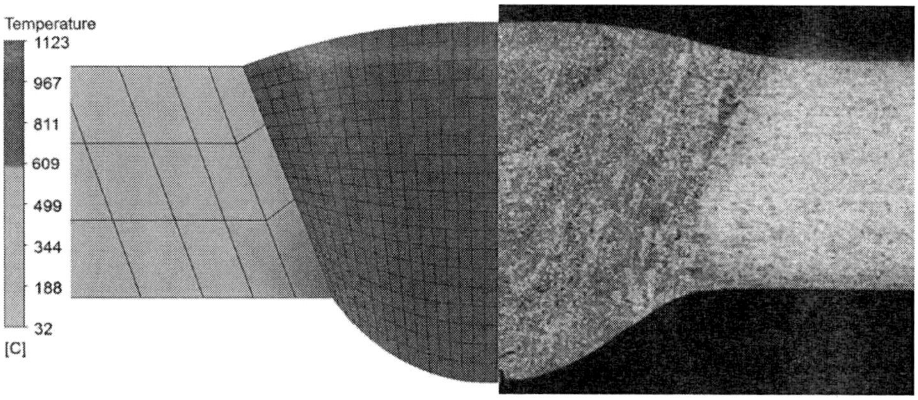

Figure 12. The remelting zone when modeling the temperature field when GMA welding (131) compared to the experimental study (metallographic specimen).

The result of numerical simulations gives satisfactory results. The red color on the left shows the area in which we reached the temperature above the solidus where the material melts. The difference in dimensions obtained in the simulation with respect to experimental tests is below 5%.

4. Conclusion

Numerical simulations of the temperature field in welding processes for sheets made of aluminum alloys:

- butt welded joint made with the GTA method (using a infusible (tungsten) electrode in the Argon shield with the addition of a deposited metal in the form of a wire),
- butt welded joint made with the GMA method (using a fusible electrode in the Argon shield),

allowed to determine the fusion zone of welded sheets in the mentioned processes.

The numerical simulation results were verified experimentally by comparing the shapes and dimensions of remelting zones obtained theoretically (computationally) and experimentally (on the basis of metallographic specimen).

The obtained results are the origin point for the calculation of strain and stress states in the welding processes considered in the article.

References

[1] Sowa L, Skrzypczak T and Kwiatoń P 2018 Analysis of Temperature and Velocity Fields During Filling of Continuous Casting Mould, *Archives of Foundry Engineering* **18** 115-118

[2] Skrzypczak T, Węgrzyn-Skrzypczak E and Winczek J 2015 Effect of Natural Convection on Directional Solidification of Pure Metal, *Archives of Metallurgy and Materials* **60** 835 – 841

[3] Kumar U, Gope D K, Srivastava J P, Chattopadhyaya S, Das A K and Krolczyk G 2018 Experimental and Numerical Assessment of Temperature Field and Analysis of Microstructure and Mechanical Properties of Low Power Laser Annealed Welded Joints, *Materials* **11** 1514, doi:10.3390/ma11091514

[4] Winczek J, Modrzycka A and Gawrońska E 2016 Analytical Description of the Temperature Field Induced by Laser Heat Source with any Trajectory *Procedia Engineering* **149** 553 – 558

[5] Jiang Z, Hua X, Huang L, Wu D and Li F 2017 Effect of Multiple Thermal Cycles on Metallurgical and Mechanical Properties during Multi-pass Gas Metal Arc Welding of Al5083 Alloy, *International Journal of Advanced Manufacturing Technology* **93**(9-12) 3799-3811, DOI: 10.1007/s00170-017-0771-6

[6] Górka J 2018 Assessment of Steel Subjected to the Thermomechanical Control Process with Respect to Weldability, *Metals* **8** 169, doi:10.3390/met8030169

[7] Berezshnaya O V, Gribkov E P and Kuznestov V D 2016 Investigation of Thermostressed State Coating Formation at Electric Contact Surfacing of Shaft Type Parts, *Advances in Materials Science and Engineering*, Article ID 6597317, 1-14, http://dx.doi.org/10.1155/2016/6597317

[8] Gucwa M, Winczek J, Bęczkowski R and Dośpiał M, 2016 Structure and Properties of Coatings Made with Self Shielded Cored Wire, *Archives of Foundry Engineering* **16**(3) 39-42

[9] Haghpanahi M, Salimi S, Bahemmat P and Sima S 2013 3D Transient Analytical Solution Based on Green's Function to Temperature Field in Friction Stir Welding, *Applied Mathematical Modelling* **37**(24) 9865-9884

[10] Golański D, Dymny G, Kujawińska M and Chmielewski T 2016 Experimental Investigation of Displacement/Strain Fields in Metal Coatings Deposited on Ceramic Substrates by Thermal Sprying, *Solid State Phenomena* **240** 174-182

[11] Murčinková Z and Vasilko K 2016 Thermo-physical Aspects of Chip Machining, *High Temperatures - High Pressures* **45**(4) 273-289

[12] Winczek J, Mičian M and Ivanov V 2018 The Modelling of Temperature-Dependent Stress-Strain Curves for Weldable Steels, *Journal of Applied Mathematics and Computational Mechanics* **17**(3) 111-117

[13] Winczek J 2016 The Influence of the Heat Source Model Selection on Mapping of Heat Affected Zones During Surfacing by Welding, *Journal of Applied Mathematics and Computational Mechanics* **15**(3) 167-178

[14] Rochalski D, Golański D and Chmielewski T 2017 Models of Welding Heat Sources in the Analysis of the Temperature Field, *Welding Review* **89** 109-116

[15] Rosenthal D 1941 Mathematical Theory of Heat Distribution during Welding and Cutting. *Welding Journal* **20** 220s-234s

[16] Rykalin N 1947 *Fundamentals of heat flow in welding*. AN SSSR, Moskva

[17] Eagar T W and Tsai N S 1983 Temperature Fields Produced by Travelling Distributed Heat Sources. *Welding Journal* **62** 346-355

[18] Goldak J, Chakravarti A and Bibby M 1985 *A Double Ellipsoidal Finite Element Model for Welding Heat Source*, II W Doc. No.212-603-85

[19] Winczek J 2010 Analytical Solution to Transient Temperature Field in a Half-Infinite Body Caused by Moving Volumetric Heat Source, *International Journal of Heat and Mass Transfer*

53 5774- 5781
[20] Hongyuan F, Qingguo M, Wenli X and Shude J 2005 New General Double Ellipsoid Heat Source Model, *Science and Technology of Welding and Joining* **10** 361-368
[21] Parkitny R and Winczek J 2013 Analytical Solution of Temporary Temperature Field in Half-Infinite Body Caused by Moving Tilted Volumetric Heat Source, *International Journal of Heat and Mass Transfer* **60** 469- 479
[22] Ghosh A, Yadav A and Kumar A 2017 Modelling and Experimental Validation of Moving Tilted Volumetric Heat Source in Gas Metal Arc Welding Process, *Journal of Materials Processing Technology* **239** 52-65
[23] Winczek J, Gucwa M, Mičian M and Makles K 2019 Numerical Analysis of the Influence of Electrode Inclination on Temperature Distribution during GMAW Overlaying, *Mathematical Problems in Engineering*, Article ID 9048025, https://doi.org/10.1155/2019/9048025
[24] Jeong S K and Cho H S 1997 An Analytical Solution for Transient Temperature Distribution in Fillet Arc Welding Including the Effect of Molten Metal, *Proceedings of the Institution of Mechanical Engineers* **211** 63-72
[25] Kang S H and Cho H S 1999 Analytical Solution for Transient Temperature Distribution in Gas Tungsten Arc Welding with Consideration of Filler Wire, *Proceedings of the Institution of Mechanical Engineers* **213B** 799- 811
[26] Winczek J 2011 New Approach to Modeling of Temperature Field in Surfaced Steel Elements, *Int. International Journal of Heat and Mass Transfer* **54** 4702-4709
[27] Winczek J 2017 Modeling of Temperature Field during Multi-Pass GMAW Surfacing or Rebuilding of Steel Elements Taking into Account the Heat of the Deposit Metal, *Applied Sciences* **7**(1) 6 1-19, doi:10.3390/app7010006
[28] Gu Y Y, Li Y D, Yong Y, Xu F L and Su L F 2019 Determination of Parameters of Double-Ellipsoidal Heat Source Model Based on Optimization Method, *Welding in the World* **63**(2) 365-376, https://doi.org/10.1007/s40194-018-00678-w
[29] Jia X, Xu J, Liu Z, Huang S, Fan Y and Sun Z 2014 A New Method to Estimate Heat Source Parameters in Gas Metal Arc Welding Simulation Process, *Fusion Engineering and Design* **89** 40-48
[30] Flint T F, Francis J A, Smith M C and Balakrishnan J 2017 Extension of the Double-Ellipsoidal Heat Source Model to Narrow-Groove and Keyhole Weld Configurations, *Journal of Materials Processing Technology* **246** 123-135
[31] Sun J, Klassen J, Nitschke-Pagel T and Dilger K 2018 Effects of Heat Source Geometric Parameters and Arc Efficiency on Welding Temperature Field, Residual Stress, and Distortion in Thin-Plate Full-Penetration Welds, *The International Journal of Advanced Manufacturing Technology* **99** 497–515, https://doi.org/10.1007/s00170-018-2516-6
[32] Lv C, Wang G, Chen H and Wan S 2019 Estimation of the Moving Heat Source Intensity Using the Multiple Model Adaptive Inverse Method, *International Journal of Thermal Sciences* **138** 576-585
[33] Wu C S, Hu Q X and Gao J Q 2009 An Adaptive Heat Source Model for Finite-Element Analysis of Keyhole Plasma Arc Welding, *Computational Materials Sciences* **46** 167-172
[34] Nasiri M and Enzinger N 2018 An Analytical Solution for Temperature Distribution in Fillet Arc Welding Based on an Adaptive Function, *Welding in the World*, **63**(2) 409- 419, https://doi.org/10.1007/s40194-018-0667-6
[35] Nasiri M B and Enzinger N 2019 Powerful Analytical Solution to Heat Flow Problem in Welding, *International Journal of Thermal Sciences* **135** 601-612
[36] Flint T F and Smith M C 2019 HEDSATS: High Energy Density Semi-Analytical Thermal Solutions, *SoftwareX* **10** 100243
[37] Sun Y L, Obasi G, Hamelin C J, Vasileiou A N, Flint T F , Francis J A and Smith M C 2019 Characterisation and Modelling of Tempering during Multi-pass Welding, *Journal of Materials Processing Technology* **270** 118-131
[38] Mahiskar G I, Chadge R B, Ambade S P and Patil A P 2014 Thermo-mechanical Analysis of Multi-Pass Bead-on-Plate Welding, *Procedia Materials Science* **5** 2522-2531

[39] Joshi S, Hildebrand J, Aloraier A S and Rabczuk T 2013 Characterization of Material Properties and Heat Source Parameters in Welding Simulation of Two Overlapping Beads on a Substrate Plate, *Computational Materials Science*, **69** 559-565
[40] Winczek J 2016 Modeling of Heat Affected Zone in Multipass GMAW Surfacing S235 Steel Element, *Procedia Engineering* **136** 108-113
[41] Frei J, Alexandrov B T and Rethmeier M 2019 Low Heat Input Gas Metal Arc Welding for Dissimilar Metal Weld Overlays, Part III: Hydrogen-Assisted Cracking Susceptibility, *Welding in the World* **63**(3) 591-598, https://doi.org/10.1007/s40194-018-0674-7
[42] Winczek J, Gawronska E, Gucwa M and Sczygiol N 2019 Theoretical and Experimental Investigation of Temperature and Phase Transformation during SAW Overlaying, *Applied Sciences*, **9** 1472 1 -17, doi:10.3390/app9071472
[43] Ghosh A, Krolczyk G, Saramdzic I and Mitra R K 2014 Thermal Transport Phenomenon of Submerged Arc Welding Process, *Tehnički vjesnik* **21**(6) 1303-1306
[44] Winczek J and Parkitny R 2017 Modelling of Heat Affected Zone in Submerged Arc Welding Butt Joint with Thorough Penetration, *Procedia Engineering* **177C** 241-246
[45] Winczek J, Makles K, Gucwa M, Gnatowska R and Hatala M 2017 Modelling of Strains during SAW Surfacing Taking into Heat of the Weld in Temperature Field Description and Phase Transformations, *IOP Conf. Series: Materials Science and Engineering* **225**, 012038 doi:10.1088/1757-899X/225/1/012038
[46] Gucwa M, Bęczkowski R, Winczek J and Wyleciał T 2017 The Effect of Type of Welding Sequence During Hardfacing Chromium Cast Iron for Erosion Resistance. Archives of Foundry Engineering, **17**(3) 51-54
[47] Winczek J 2012 Modelling of Heat Affected Zone in Cylindrical Steel Elements Surfaced by Welding, *Applied Mathematical Modeling* **36** 1514-1528
[48] Li Y, Feng Y, Zhang X and Wu C 2018 An Evolutionary Keyhole-Mode Heat Transfer Model in Continuous Plasma Arc Welding, *International Journal of Heat and Mass Transfer* **117** 1188-1198
[49] Moarrefzadeh A 2011 Numerical Simulation of Workpiece Thermal Profile in Plasma Arc Cutting (PAC) Process, *WSEAS Transactions on Applied and Theoretical Mechanics* **6**(4) 160-166
[50] Franco A, Romoli L and Musacchio A 2014 Modelling for predicting seam geometry in laser beam welding of stainless steel, *International Journal of Thermal Sciences* **79** 194-205
[51] Turichin G, Kuznetsov M, Pozdnakov A, Gook S, Gumenyuk A and Rethmeier M 2018 Influence of Heat Input and Preheating on the Cooling Rate, Microstructure and Mechanical Properties at the Hybrid Laser-Arc Welding of API 5L X80 steel, *Procedia CIRP* **74** 748-751
[52] Feng S, Huang C, Wang J, Zhu H and Yao P 2017 An Analytical Model for the Prediction of Temperature Distribution and Evolution in Hybrid Laser-Waterjet Micro Machining, *Precision Engineering* **47** 33-45
[53] Chmielewski T, Golański D, Włosiński W and Zimmerman J 2015 Utilizing the Energy of Kinetic Friction for the Metallization of Ceramics, *Bulletin of the Polish Academy of Sciences, Technical Sciences* **63**(1), *DOI: 10.1515/bpasts-2015-0023*
[54] Chmielewski T, Golański D and Włosiński W 2015 Metallization of Ceramic Materials Based on the Kinetic Energy of Detonation Waves, *Bulletin of the Polish Academy of Sciences, Technical Sciences* **63**(2), DOI: 10.1515/bpasts-2015-0051
[55] PN-EN ISO 4063:2011. *Arc welding process numbers*
[56] EN –731-1. *Aluminium and aluminium alloys - Chemical composition and form of wrought products*
[57] *Operation and maintenance manual documentation*, Device Faltig - 315 AC/DC
[58] *Dokumentacja Operation and maintenance manual documentation*, Device Synermig 400
[59] Kik T, Slavacek M, Wyględacz B 2016 Numerical Analysis of Multipass T-joint Welding and Post Welding Heat Treatment, *Welding Review* **88**(5) 101-106

IOP Conf. Series: Materials Science and Engineering **659** (2019) 012035 doi:10.1088/1757-899X/659/1/012035

Possibilities of intelligent flexible manufacturing systems

P Kostal[1], A Mudrikova[1] and D Michal[1]

[1] Slovak University of Technology, Faculty of Material Science and Technology, J. Bottu 25, 91724 Trnava, Slovakia

E-mail: peter.kostal@stuba.sk

Abstract: The basic requirement of production is to produce as many products as possible, at the lowest cost, with the highest possible quality. These requirements can be met with the new "Intelligent Production" paradigm. In particular, this paradigm involves the use of intelligent manufacturing systems and the introduction of intelligent production management. In this article, we'll look at intelligent manufacturing systems. The intelligent manufacturing system itself is a flexible manufacturing system that can flexibly respond to changes in production requirements as well as changes in its surroundings and interact with its surroundings. As a result of these flexible responses, there is less space, reduced production and investment costs, and increased productivity.

1. Introduction

The time analysis of the entire production process is influenced by the current trend of increasing product requirements. The basic requirement for production is to produce as many products as possible at minimum cost, with the required quality. For these reasons, a new paradigm called "intelligent production" is being put into practice.

Industry automation continues. Today, it's not just about using classic automation tools, but about a flexible manufacturing system. We can now talk about a new generation of intelligent manufacturing systems.

This new manufacturing paradigm involves a high degree of production automation as well as all processes related to production.

This means that on the one hand (the hardware part) is a highly automated production system on which its various parameters relating to the production itself are constantly monitored, as well as those related to the system's own operation (for example oil temperature, bearing vibration and others).

On the other hand, the software section is an intelligent control system that can handle all of these data and make decisions based on both the production and the manufacturing system itself (for example, schedule individual activities for regular maintenance, customize the production process to the current system status, and more).

These two parts together form an intelligent manufacturing system that can flexibly respond to and adapt to the current situation.

Complex component machined with zero defects is a top performance in mass production and it becomes a new challenge required for the new generation of intelligent machine-tools. Increasing the precision and accuracy of machines, products and processes offers substantial benefits to a wide range of applications from ultra-precision to mass production with higher quality and better reliability [1].

The next generation of manufacturing machines will be described as new intelligent reconfigurable manufacturing systems which realises a dynamic fusion of human and machine intelligence, manufacturing knowledge and state-of-the-art design techniques. This may lead to low-cost self-optimising integrated machines. It will encompass fault-tolerant advanced predictive maintenance facilities for producing high-quality error-free workpieces using conventional and advanced manufacturing processes.

Machining process monitoring and control is a core concept on which to build up the new generation of flexible self-optimising intelligent NC machines. In-process measurement and processing of the information provided by dedicated sensors installed in the machine, enables autonomous decision making based on the on-line diagnosis of the correct machine, workpiece, tool and machining process condition, leading to an increased machine reliability towards zero defects, together with higher productivity and efficiency [1].

This new generation of manufacturing systems is characterized by the use of a large number of different sensors that constantly gather information about the system itself and its surroundings. These sensors are located in individual production machines as well as in all production system accessories.

All the data collected by the sensors is sent to an intelligent control system that evaluates it based on various modern decision algorithms to take decisions that can affect not only production but also the production system itself.

1.1. Manufacturing initiatives in the different regions

In 2013, Germany unveiled its Industry 4.0 strategy, which directed a great deal of global attention to the advances in manufacturing systems technology [2]. In the United States, the government launched the Advanced Manufacturing Partnership (AMP) in 2011. Since then, many other initiatives have been rolled out, including the Advanced Manufacturing Partnership Steering Committee "2.0" in 2013; the National Network for Manufacturing Innovation (NNMI) in 2014; and the Revitalize American Manufacturing and Innovation Act, which was signed into law by the President of the United States in December 2014 [3]. Most recently, Manufacturing USA was officially launched by the US government in order to further "leverage existing resources... to nurture manufacturing innovation and accelerate commercialization" by fostering close collaboration between industry, academia, and government partners [4]. In 2015, the Chinese government officially published a 10-year plan and roadmap toward manufacturing: Made in China 2025 [5]. The largest international collaborative program, Intelligent Manufacturing Systems (IMS), which is led by Japan, is also rolling out a roadmap for its next step with its IMS2020 project.

2. Intelligence and intelligent systems

The conventional manufacturing control structure is very efficient under relatively steady operating conditions as it provides a single central control over a range of operations. However a centrally managed, hierarchical control strategy is:

- unresponsive in the face of disturbances such as rush orders, breakdowns, supply stock-outs,
- cumbersome to reconfigure for new products, new manufacturing technologies or different line configurations.

In simple terms, the decoupling of routing, scheduling and planning functions—in both hardware and software—from the changing state of physical shop floor operations means that there is often little or no correspondence between the decisions made and the resulting activities of the resources that execute them. [6]

Automated manufacturing systems are capable of ensuring the long-term required level of quality and production efficiency. This must be achieved without operator intervention. This means that all automated production system subsystems have an impact on the resulting quality and production efficiency. In order to realize the production process in automated production systems, it is necessary to create a proper sequence of individual functions and links between individual subsystems [7]. The exact time coordination of these functions is very important. It is very important to follow the system dynamics in this process.

Indeed, flexible monitoring systems are required under the actual market requirements and thus, reliable process diagnosis is necessary under different cutting conditions. Nowadays, a common problematic shared by conventional process monitoring approaches for part and tool condition monitoring is the lack of reliability under changing cutting conditions hence limiting the flexibility of such automation systems. As a characteristic example of this problematic for process condition based tool condition monitoring (TCM), the process condition is not only influenced by changes in tool condition, but it is also directly affected by cutting conditions. Furthermore, under different cutting conditions, different wear mechanisms can be activated on the tool, each one having its particular impact on process and part condition [1].

Due to the increased complexity of modern manufacturing systems—particularly after all the units/elements been integrated into a common system—process decisions have become much more difficult. There is a strong need to leverage vast amounts of manufacturing data and to utilize the power of computing intelligence to enhance the decision-making process in manufacturing.

Intelligent capability refers to three functions, which operate in an analogy of a human body: sensing, decision-making, and action. With today's rapid advances in sensing and control technologies, there is no lack of sensors or actuators in manufacturing systems.

The challenge is how to process information and knowledge so that the right decision can automatically be made by a computer at the right time and in the right location, with little or no human intervention. New technologies are emerging in this areas, such as big data analytics, machine learning (ML), and cloud computing, which provide great potential for enhanced intelligent capability in manufacturing [8].

2.1. Intelligence

The notion of intelligence is not unambiguously defined, but in general, the rapid and correct orientation in new situations, which is based on a quick insight, the correct assessment of the necessary information, and the correct conclusions, is considered to be positive manifestations of intelligence. Furthermore, it is mindfulness, understanding, logic, criticality, originality, ability to reveal subtle, hidden, difficult to perceive connections and relationships, similarities and differences, versatility and ability of interests, rich vocabulary and precise concise expression.

Intelligence expresses some general (intellectual) ability, which is the basis for appropriate response in situations where experience is not sufficient.

Intelligence can be called the ability to retrieve information from its surroundings, store it, process it, and respond to the current situation based on the results of the processing.

2.2. Artificial intelligence

Artificial intelligence is the intelligence realized by machines. It is a field of study in computer science that tries to reproduce human brain activity in solving various tasks. This means perceiving your surroundings, learning, planning, deciding and solving problems.

Like any computer program, artificial intelligence is based on the use of algorithms. An algorithm is a set of clear instructions executed on a computer that lead to the desired result.

In recent decades, great progress has been made in AI's artificial intelligence. The methods by which artificial intelligence solves problems and seeks solutions are taken from the established practices of people. The main tools for calculations are mainly searching and mathematical optimization, neural networks, methods based on statistics, probability.

Expert systems are the oldest and most advanced type of intelligent systems that incorporate the "know-how" of a human expert into a computer program. The representation of knowledge in these systems is given symbolically in the form of rules and frameworks.

Nowadays, the most advanced subset of machine learning is learning through neural networks. The artificial neural network is divided into multiple layers, the inlet passing through the inlet layer and then through the hidden inner layers, with each layer extracting more abstract information. Finally, the output is transmitted through the output layer. In the case of the presence of multiple inner layers, we call the network a deep neural network and the approach is called deep learning.

In recent years, considerable effort has been devoted to the development and use of distributed artificial intelligence systems.

The problems that today's artificial intelligence research has to cope with are, in addition to the limited computing power, the system itself. Although partial tasks (like speech recognition, answering, image recognition, and so on), are already solved at a certain level, the complex solution is still far away.

3. Intelligent production

Intelligent manufacturing is a progressive building of integrated production management that connects all technological aspects (sensor use, process control, IT systems, production planning, ...) with the addition of intelligence through modelling, advanced management including cognitive automation concepts, diagnostic tools. Optimization, simulation, and expert knowledge are in harmony and interaction with human intelligence [9].

The concept of intelligent manufacturing remodels the ability of decision support systems to form generative systems that can acquire knowledge, learn, and adapt to changing environments and the actual composition of system components. A characteristic of smart production is the ability of the system to learn as well as the ability to obtain the information needed to manage an integrated production system.

The main components of intelligent manufacturing are shown in Fig. 1

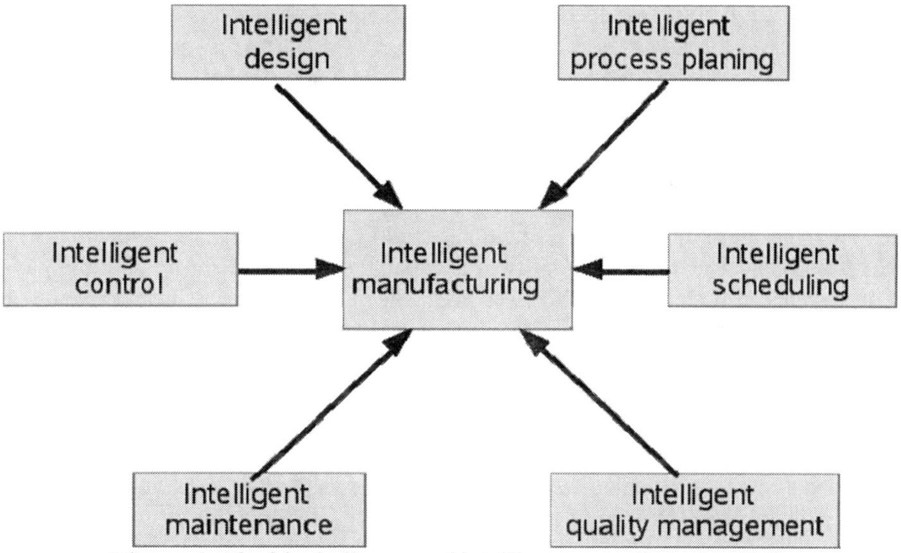

Figure 1. The block diagram of intelligent manufacturing [10].

Products and production are mainly related to the following parts:

- intelligent design - information is produced at the time of product design that can affect production both positively and negatively. (technology design, tolerance,…). This information carries the 3D model of the product and all its components as part of the input to all subsequent pre-production and production phases;
- the intelligent design of production process - a selection of suitable production technology based on product design, finding suitable technological parameters that are close to optimal, selection of suitable production process, design of material flow during production,…
- intelligent planning - scheduling production so that it can be completely realized at minimum time, including purchasing all the necessary material, availability of free capacity of all necessary equipment and their operation,…

- Intelligent Quality management - correct recognition of important (required) and unimportant parameter controls correctly determine the methodology and scope of quality control ...

The production system itself relates mainly to the following two parts:

- intelligent management - the intelligent manufacturing system can react to some unexpected situations that may occur and be monitored by the control system (blunted or broken tool, the control system is able to detect and respond to this problem, for example, temporarily replaces it with another if possible if not, report a fault, fixture detect improperly inserted workpiece,...). Of course, this manufacturing system can only work if its peripherals (eg fixture) have a certain level of intelligence, so they can monitor their functions while communicating with the master system's control system, which in turn communicates with its superior system.
- Intelligent maintenance - the control system monitors the status and performance of the individual components of the production system. Based on the evaluation of these data, it suggests smaller maintenance interventions flexibly and possibly plans. (for example, by monitoring the cutting forces it can assume the need for a tool change, or postpone the tool change term, the machine work time is measured and, based on the actual measured values, it can make a proposal for scheduled maintenance).

All of the above intelligent manufacturing components must work closely together.

In order to implement intelligent manufacturing technology, industries are preparing to develop cloud computing platforms based on IoT. It has been estimated that investment in IoT is expected to reach $60 trillion in the next 15 years. By 2020, it is predicted that more than 50 billion devices will be connected to the Internet. In order to unleash the benefits of IoT for industrial applications, many software platforms are being developed and deployed, such as the aforementioned Predix platform by GE.

GE's Predix is a comprehensive, purpose-build industrial platform for the implementation of intelligent systems to monitor and control physical devices or systems through the Industrial Internet. A significant feature of all such platforms is the capability to build "digital twins." A digital twin is a computerized model of a physical device or system that represents all functional features and links with the working elements. A digital twin is more than a virtual computer system for simulation study. It provides the operation status, insights, outcomes, and knowledge that are associated with the proper functions of the physical system. A digital twin is capable of communicating with the physical system it represents via real-time sensing devices, so as to keep it almost synchronized with the real-time status, working condition, position, and environment situation. Digital twins allow for the prediction of future conditions [8].

4. Intelligent manufacturing system

As industrial production is evolving, new generation generations of manufacturing systems are also being developed—intelligent manufacturing systems. These systems are equipped with "intelligent" control systems. Nowadays, the use of real artificial intelligence in these systems is a far future. First steps are being taken to increase the intelligence of these systems by implementing various modern decision algorithms. A basic part of such systems is also a subsystem that allows extended monitoring of the operating status of the system and its surroundings. The information obtained in this way constitutes the basic input for the said decision algorithms.

For a better understanding of the term "intelligent manufacturing system", it is best to compare its behaviour with the classic ("non-intelligent") production system.

Today, the automated manufacturing system is known as a production facility with various possible levels of automation of both operational and off-site operations and with different levels of subsystem integration (technological, transport, handling, control) [10]:

- technological (production equipment)
- Control (own control systems of all devices)
- transport and handling (using industrial robots, manipulators and conveyors)
- monitoring and monitoring (monitoring observes the status of the production system and its individual subsystems, can be connected to maintenance, inspection checks production quality)

• Energy (supplies the production system with the necessary energy and also monitors and evaluates its consumption)
• warehouse (used to store semi-finished products and finished products)

An intelligent manufacturing system is a system that has the ability to adapt to unexpected changes, such as product changes, market demands, technological change, social needs, etc.

He is able to collect and process information from his surroundings. Based on the results of the evaluation of the information obtained, it is able to make a separate decision affecting the production facility, the production process or its surroundings.

Intelligent manufacturing systems also consist of subsystems, as well as automated manufacturing systems (technological, surveillance, transport, handling). However, these subsystems are highly integrated and equipped with technical means that allow the subsystems concerned to have a certain level of intelligent response to external stimuli. These intelligent manufacturing systems can be seen as a higher level of flexible manufacturing systems [9].

The control system of manufacturing and assembly systems is possible to understand as the integration of combined elements of control system. The control system is realized as a mechanical, pneumatic, hydraulic, electro-pneumatic or electro-hydraulic control structure. That means the control system could be to accept as an individual system that is represented by a hierarchically ordered control system of manufacturing and assembly process [11].

In terms of automation, assembly is one of the most complex operations. Sequences, such as proper gripping, orientation, and placement of a component entering the assembly system in an unordered state (e.g., freely in a container) are very manageable. However, in terms of automation, these seemingly simple tasks are one of the most difficult issues. Usually, we try to solve this task in an automated assembly process so that individual components enter the assembly system already oriented and in a location specified by various feeders, containers or pallets [12].

In case individual components that are unpositioned and unoriented enter into the automated system, intensive cooperation between the subsystems is needed to determine the current position and orientation of the component and other various intelligent mechanical peripherals. Different types of sensors (contact, contactless, pressure sensors, force and torque sensors, cameras, etc.) are used depending on the specific application requirements. See e.g. [13], [14], [15] and others.

The simultaneous combined use of different types of sensors provides a solution to the possibility of a very complex observation task. The different types of sensors can also vary in their output signals. Some sensors have only simple binary output signals, others may have more complex output signals, which consist of several simple binary signals (eg a colour sensing sensor), and others may provide an analogue signal (eg a resistance thermometer). All these signals must be processed and evaluated correctly in the control system because it is only on the basis of this information that it is possible to correctly respond to the current state of the production system of its individual subsystems as well as to the current state of the technological process.

Equipping manufacturing systems with sensors and a control system that can send the data for evaluation and processing to its superior system is one of the fundamental steps towards increasing the intelligence of these systems. Sensory systems provide the ability to monitor the various functions of the manufacturing process, manufacturing technology, properties of the objects being processed, as well as environmental properties. These monitoring systems are always implemented depending on the specific purpose of use.

An extensive survey of manufacturing systems has made it possible to identify the main current trends for manufacturing systems, which can be summarized as follows:

- specialization that is characterized by a broad focus on key competencies;
- the transition from vertical to horizontal management systems, the change from highly centralized to decentralized (each individual element, adapted to decision-making ability / intelligence);

- Developing the system's own capabilities and capabilities (eg adaptation), which usually occur at lower levels of the system. Production systems with these characteristics have a high level of integration, are easily expandable and adaptable;
- development of technologies and applications to support all requirements of current distributed production systems;
- competitiveness, businesses must remain competitive in terms of costs and benefits; (corresponding equipment and machines enabling new production processes);
- sustainability (eg taking environmental protection into consideration);
- production technologies and equipment (eg to evaluate different system configurations with regard to quality, system reliability);
- integration of people, software and machines, functional features such as fault tolerance;
- openness, adaptation; each unit of the production system should take appropriate and justifiable decisions (eg in terms of resource utilization, including planning algorithms, planning and execution control) that have a common goal and work together to achieve it;
- performance analysis [12]

5. CONCLUSION

At present, we are often witnessing a disproportionate reduction in the product life cycle, market liberalization, and ever-changing customer demands. As a result of these phenomena, manufacturers are forced to gradually rebuild their production from mass production to small series with a wide variety of similar product variants.

The concept of all production facilities is subordinate to the nature of production. New generation production systems differ from a flexible production system not only in construction but especially in their properties and capabilities. For these reasons, the implementation of new generation production systems in the manufacturing process brings wide possibilities for increasing productivity and reducing production costs.

Acknowledgment:
This paper was written thanks to the national KEGA Grant: 021STU-4/2018 - Development of a laboratory for the design and maintenance of production systems supported by the use of Virtual Reality.

References

[1] Mekid S, Pruschek P and Hernandez J 2009 Beyond intelligent manufacturing: A new generation of flexible intelligent NC machines, *Mechanism and Machine Theory* **44**(2) 466–476
[2] Kagermann H, Wahlster W and Johannes H 2013 *Recommendations for implementing the strategic initiative INDUSTRIE 4.0.* National academy of science and engineering
[3] Reed T 2014 H.R.2996 - 113th Congress (2013-2014): Revitalize American Manufacturing and Innovation Act of 2014, 16-Sep-2014, https://www.congress.gov/bill/113th-congress/house-bill/2996, [Accessed: 09-Aug-2019]
[4] Obama B 2011 *Strategy for American innovation: Driving towards sustainable growth and quality jobs*. DIANE Publishing
[5] Lee X E 2015 *Made in China 2025: A new era for Chinese manufacturing*, CKGSB Knowledge
[6] McFarlane D, Sarma S, Chirn J L, Wong C Y and Ashton K 2003 Auto ID systems and intelligent manufacturing control, *Engineering Applications of Artificial Intelligence* **16**(4) 365–376
[7] Velíšek K and Košťál P 2007 Self Organizing Manufacturing Cell Philosophy, *Scientific Bulletin*. **XXI**(May)
[8] Chen Y 2017 Integrated and Intelligent Manufacturing: Perspectives and Enablers, *Engineering* **3**(5) 588–595

[9] Meziane F, Vadera S, Kobbacy K and Proudlove N 2000 Intelligent systems in manufacturing: current developments and future prospects, *Integrated Mfg Systems* **11**(4) 218–238
[10] Danišová N and Velíšek K 2007 Intelligent manufacturing and assembly system, *MD* 413–416
[11] Ruzarovsky R, Horvath S and Velisek K 2008 *Designing of automated manufacturing and assembly systems*, In book: *Annals of daaam for 2008 & proceedings of the 19th international daaam symposium*, Vienna Univ Technology, Karlsplatz 13, Wien, a-1040, Austria 1201–1202
[12] Holubek R and Kostal P 2013 The Intelligent Manufacturing Systems, *Advanced Science Letters* **19**(3) 972–975
[13] Danisova N, Ruzarovsky R and Velisek K 2011 Design alternatives of intelligent camera system for check parts at the intelligent manufacturing-assembly cell, *Information Technology for Manufacturing Systems II, PTS 1-3* **58**(60) 2262–2266
[14] Danišová N, Ružarovský R and Velíšek K 2013 Designing of Intelligent Manufacturing Assembly Cell by Moduls of System Catia and E-Learning Module Creation, *Advanced Materials Research* **628** 283–286
[15] Danišová K and Velíšek N 2010 Shelf storage system running at the intelligent manufacturing cell, *Proceedings of World Academy of Science, Engineering and Technology* **70** 529–533

Welding method as influential factor of mechanical properties at high-strength low-alloyed steels

A Ilić[1], L Ivanović[1], V Lazić[1] and D Josifović[1]

[1] University of Kragujevac, Faculty of Engineering, Sestre Janjić 6, 34000 Kragujevac, Serbia

E-mail: andreja.coka@gmail.com

Abstract. In this paper the analysis of welding technology to mechanical properties of welded joints at high strength low alloyed steel S690QL is presented. Experimental testing was done at models with V-groove butt joints that are done by MMA or MIG for root pass and MAG for other passes of multipass welded joints with related parameters and consumables. Yield and tensile strength, so as answer to load of models are determined experimentally in order to analyses the effect of welding technology to resulted mechanical properties. On the basis of the obtained results it can be concluded that welding parameters have significant influence to mechanical properties of welded joints at high-strength low-alloyed steel. Furthermore, it can be concluded that welded joints with root pass done by MIG and other passes by MAG provide better mechanical properties. Presented research point out that welding parameters at high strength low alloyed steels must be selected and controlled precisely to obtain welding joints with adequate mechanical properties. Welding, as dominant method of joining, is primary factor that provide beneficial application of high-strength low-alloyed steels. The paper highlight the influence of material degradation due to welding. This paper point out the importance of analysing the welded joints at different levels of dimensions, while further, more detailed, research can be continued through development of a numerical model of the welded joints which will complete the experimentally obtained results. As experimentally obtained results correlate to material degradation due to welding it is implicated that future development of high-strength low-alloyed steels must be followed with development and modification of welding processes.

1. Introduction

High-strength steels were developed in order to improve the mechanical characteristics and corrosion resistance of conventional carbon steels during seventh decade of the 20th century and linked to the beginnings of commercial production of iron alloys, especially ferroniobium. The typical microstructures of high-strength low-alloy steels are fine grain, consist of small size ferrite (α) grains with shape homogeneity. The small quantity of cementite is also present in microstructures, so as the fine dispersed particles of carbon nitride (can be identify only by using of electron microscope). During the final rolling process advantageous conditions for creation of the large number of referent locations within the discrete formation of α metal grains. The density of those locations is a consequence of a production processes and material deformation. The effects of regeneration of deformed microstructure are induces, as recuperation and recristalization. By those phenomena, the deformed microstructure goes into the undeformed form means decrease of dislocation density in the microstructure. The recristalization processes are blocked by decrease of speeds of grains' nucleus formation and by the reduction of movement of metal grains and sub grains boundaries. The recristalization processes are caused by the presence of alloying elements' in the solid solution and induced by continual rolling with small pause periods, when dominant is the effect of niobium. The recristalization processes are made as the consequence of precipitation during the reversible rolling with longer pause periods, when the separation of carbon nitride is dominant process.

Welding procedures are developing by the intensive applications of the obtained results in fundamental and applied scientific disciplines. Common welding process is based on localized heating and, after that, cooling, means inhomogeneous temperature fields at heat affected zones of welded joints. Present welding technology processes bring numerous and heterogeneous consequences by its nature, as inhomogeneity of material, alteration of chemical compositions, microstructural transformations, etc. Material degradation of high-strength low-alloyed steels due to welding can be related to increasing of hardness and transition temperature, a decrease of toughness, different material discontinuities and etc. Formation of brittle structures and initialization of cracks due to welding are related to the high cooling speed of weld zone in diapason of temperature at which austenite is highly unstable.

Material discontinues (inclusions, defects, cracks, sharp cuts, etc.) and imperfections often exist at welded joint zones means that they are critical from the aspect of structural integrity. Also, from the same aspect, safety and reliability analysis, construction answer to load is crucial. The welded construction answer to load and its ability to absorb energy with preserved integrity come in focus of many researches.

Reference [1] presents a comparative analysis between the mechanical properties of high strength steel weld metals obtained by shielded metal arc (SMAW) and gas metal arc (GMAW) welding processes. Multipass welding by SMAW and GMAW processes was performed with preheating of 200 °C in plates. After welding, a post weld heat treatment (PWHT) at 600 °C for 1 h was performed and this condition was compared to the as welded one. Mechanical tests and metallographic examination by optical microscopy (OM), scanning electron microscopy (SEM) and electron backscattered diffraction (EBSD) were performed for mechanical and microstructural characterization.
It is concluded in this reference that presenting impact toughness behaviour slightly different due to the chemical composition and carbide precipitates, as predicted by computational simulation, both processes showed a good relationship between mechanical strength and impact toughness for all analysed conditions, even after PWHT. It can be concluded that GMAW process can be applied as an interesting alternative for welding of high strength steels, once this process promotes a significant improvement in productivity with good quality.

Nathan with associates in reference [2] present that naval grade high-strength low-alloy (HSLA) steels can be easily welded by all types of fusion welding processes. Fusion welding of these steels leads to the problems such as cold cracking, residual stress, distortion and fatigue damage. It is concluded that these problems can be eliminated by solid state welding process such as friction stir welding (FSW). The reference [2] present comparative evaluation of mechanical (tensile, impact, hardness) properties and microstructural features of shielded metal arc (SMA), gas metal arc (GMA) and friction stir welded (FSW) naval grade HSLA steel joints. It is concluded that the use of FSW process eliminated the problems related to fusion welding processes and also resulted in the superior mechanical properties compared to GMA and SMA welded joints.

Reference [3] presents the analysis of the weldability of three different modern high strength steel plates, with a thickness of 8 mm. Two of the test materials were produced by a thermo-mechanically controlled process (TMCP) and one by a quenching and tempering method (Q&T). Two-passes MAG (metal active gas) welding was used with four different heat inputs. The tests implemented on all the materials included tensile, hardness profiles (HV5), Charpy-V impact toughness tests, and microstructure analysis using scanning electron microscope (SEM). For one of the TMCP steels, some extended tests were conducted to define how the tensile properties change along the weld line. These tests included tensile tests with digital image correlation (DIC), and 3-point bending tests. The most notable differences in mechanical properties of the welds between the materials were observed in Charpy-V impact toughness tests, mostly at the vicinity of the fusion line, with the Q&T steel more prone to embrittlement of the heat affected zone (HAZ) than the TMCP steels. Microstructural analysis revealed carbide concentration combined with coarse bainitic structures in HAZ of Q&T steel, explaining the more severe embrittlement. During the tensile tests, the DIC measurements have shown a strain localization in the softest region of the HAZ. Increasing the heat input resulted in earlier localization of the strain and less maximum strength. The tensile properties along the weld line were investigated in all welding conditions, and the results emphasize relevant and systematic differences of the yield strength at the transient zones near the start and end of the weld compared with the intermediate stationary domain.

Zgripcea with associates in reference [4] presents application domains of high strength weld, fabrication technologies and tests. High strength weld is used for assembly of high strength low alloyed steel (HSLA) where the same or greater mechanical properties must be obtained in weld line. The method presented in this paper is based on micro alloying inside the root of the weld, with minimum supplementary cost and keeping the same productivity for welding process. The applications of this method are realizations of storage pressure vessels, shields, crane arms and generally, all applications where high strength at a low weight are required.

The presented literature overview point out that properties and characteristics of welded joints are crucial element of structural integrity of the whole welded constructions. Welded constructions are complex systems that can be considered from many different aspects. Safety and reliability analysis of welded constructions also point out that welded joints are critical zones.

The aim of determining answer to load and mechanical properties of welded joints is the qualitative analysis of structural integrity due to the integrity of its welded joints. Capacity calculations analyse and prove mechanical response and integrity of welded structures for the anticipated loads and exploitative conditions. The most dominant factors to relevance of analytical models used for calculations are stress concentrations at zones of welded joints. Exploitation of welded constructions showed that properties of welded joints, due to theirs nature, were not adequately take in consideration in present analytical models, so as capacity calculations. It can be concluded that only results obtained by experimental testing can be taken as highly relevant.

2. Experimental testing of welded joints

The experimentally tested models are made of high-strength low-alloy steel that related to EN standard EN 10025-6, steel number 1.8928 classification 3.1/3.2 – EN 10204 / TCM, commercially Weldox 700 - S690QL. The analysed steel is produced by SSAB Oxelösund AB, 613 80 Oxelösund, Sweden. Chemical composition of alloying elements is presented in table 1.

Table 1. Chemical composition of steel Weldox 700 (S690QL).

Alloying element	Content, %	Alloying element	Content, %
C	max 0.20	V	max 0.09
Si	max 0.60	Cu	max 0.30
Mn	max 1.6	Ti	max 0.04
P	max 0.020	Al	total max 0.015
S	max 0.010	Mo	max 0.70
B	max 0.0005	Ni	max 2.0
Nb	max 0.04	N	max 0.015
Cr	max 0.70		

The considered high steel grade are developed to provide the favourable mechanical properties and higher corrosion resistance and not classified as alloyed steel grades in common manner. Those steels contain small amounts of nickel, molybdenum, copper, nitrogen, vanadium, niobium, titan, zirconium and boron. Combination of allying elements and their quantities provide the proper weldability and deformability are obtained. Also, this is achieved through highly controlled production processes and a high degree of purity with respect to non-metallic inclusions. The nature of microstructure of high-strength steels, so as theirs characteristics must be considered adequately in design process. The microstructure of considered high-strength low-alloyed steel is fine grained and consist of fine ferrite (α) grains with uniformity of shapes. The small amounts of cementite is present in microstructures and fine dispersed particles of carbon nitride (figure 1) [5, 6, 7].

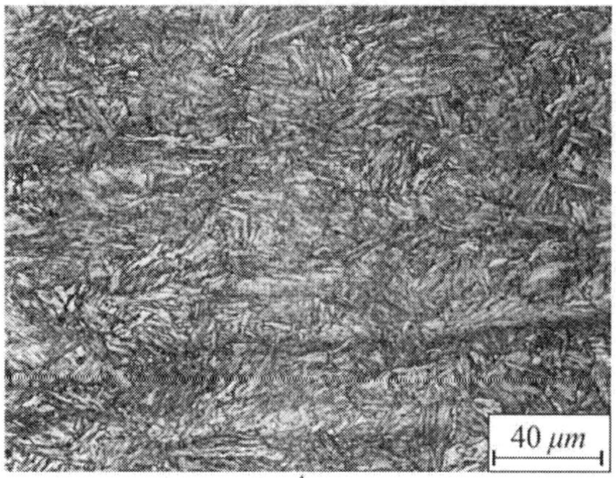

Figure 1. Typical microstructure of high-strength low-alloyed steels.

Two different groups of samples with butt V-joint are prepared by welding of 15 *mm* thick plates. For the first group of samples, root pass was done by MMA welding process with welding consumables of lower strength, while other passes were done by MAG welding process with welding consumables of higher strength. For the second group of samples, root pass was done by MIG welding process, while other passes were done by MIG welding process with related welding consumables [5, 6, 7]. Welding parameters and used filler material are presented in Table 2.

Table 2. Parameters and filler material used for welding.

Parameter	Root pass MMA	Root pass MIG	Other passes MAG
Welding consumables	INOX B 18/8/6; Ø 3.25 mm	MIG 18/8/6 Si; Ø 1.2 mm	MIG 75; Ø 1.2 mm
Current, Iz	≈ 120 *A*	≈ 110 *A*	≈ 250 *A*
Voltage, U	≈ 24 *V*	≈ 24 *V*	≈ 25 *V*
Welding speed, vz	≈ 0.2 *cm/s*	≈ 0.35 *cm/s*	≈ 0.35 *cm/s*
Heat input, ql	≈ 12 *kJ/cm*	≈ 13 *kJ/cm*	≈ 15 *kJ/cm*
Penetration, δ	≈ 1.8 *mm*	≈ 1.8 *mm*	≈ 1.9 *mm*
Protective atmosphere	-	100% Ar (M11)	82% Ar + 18% CO_2 (*M21*)

Welding of plates and reparation of samples was done in industrial environment. Presence of imperfections and defects in zones of welded joints is inevitable. Potential risks for formation and presence of welding defects are related to every specific welding process. Presents of defects and imperfections as result of those parameters come in focus of mechanical construction testing. Welding defects in welded joints are defined and classified in standard EN 26520:1992, while quality of welded joints are defined by EN 25817:1992. Used welding processes were verified and quality of welding joints were checked before experimental testing. Testing was done Experimental testing are done at ambient temperature by universal testing device type z100 producer Zwick Roell GmbH & Co. KG according to regulation ISO 6892–1:2009 Metallic materials. Tensile testing. Method of test at ambient temperature (figure 2).

Figure 2. Testing stand Zwick Roell GmbH & Co. KG z100.

Testing methodology involve loading to fracture in quasi-static conditions at ambient temperature with automatic registration of dependence tension force-elongation and determination of basic mechanical properties. Speed of force increase was suitable for quasi-static testing [5, 6, 7].

Specimens were tested in series, made of parent metal without welded joint, and other two with considered welded joint types. Steel plates are cut into pieces, with minimal heat input during cutting with intensive cooling to prepare specimens with welded joint is at the middle of referent zone. Appearance and geometrical characteristics of specimens are presented in figure 3.

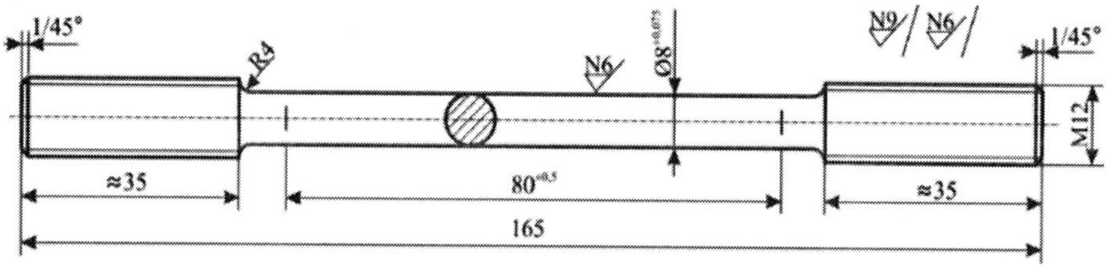

Figure 3. Geometrical characteristics of test specimens.

Location of the fractures surfaces at tested specimens are out of zones of welded joins and by that point out that applied welding processes are done adequately. Mechanical answer to load of the specimens with and without welded joint during tension to fracture have same related characteristics. Specific zones can be accepted during tension to fracture at specimens with welded joint MMA/MAG or MIG/MAG: zone of linear dependence during elastic deformation, zone of elastic deformation without of linear dependence, zone of plastic deformation, with material yielding at beginning, zone of force increase during plastic deformation to maximal force and fracture zone. Comparison of experimentally determined yield and tensile strength of the tested specimens is presented at figure 4.

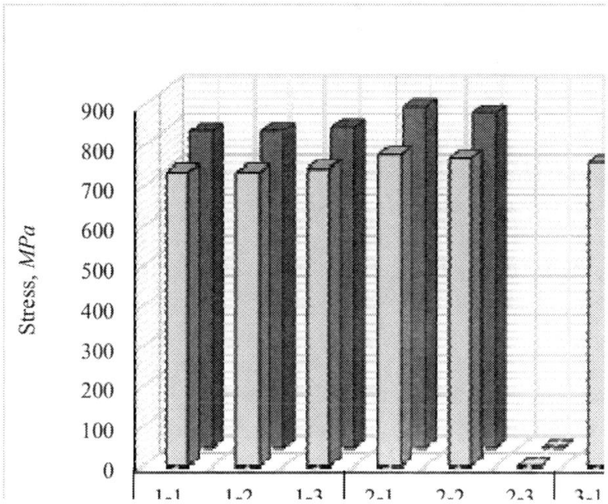

Figure 4. Histograms of yield and tensile strength of tested specimens.

Experimentally determined values of yield strength and tensile strength at tested specimens have insignificant deviations so results can be taken as relevant. Experimentally determined yield strength Rp0,2=739.59 MPa and tensile strength Rm=796.66 MPa at tested specimens made of considered material fulfil requirements defined by norm for considered high-strength low-alloy steel. Experimentally determined elongation to fracture is within range for considered material. At one specimen with MMA/MAG joint fracture happened at significantly low tensile force, as consequence of defect during welding, so this specimen is excluded from further evaluation. Presented facts verifies significance of welding defects. The experimentally obtained yield strength and tensile strength showed that MMA/MAG welding process provide better mechanical properties then MIG/MAG. But, MIG/MAG welding process provide higher productivity, possibility of automatization in present industrial environment. Those fact also point out that selection of welding process at high strength low alloy steels is multi criterion analysis [5, 6, 7].

3. Results evaluation and discussion

Characteristic and properties of welded joints depend on excessive number of factors. Presence of defects must be considered in details during forming and exploitation of welded constructions. Welding defects are defined and categorized by norm EN 26520:1992 Classification of imperfections in metallic fusion welds and quality of welding is defined by norm EN 25817:1992 Arc-welded joints in steel. Guidance on quality levels for imperfections.

Zones of welded joints by nature as zones of conditional material discontinuity caused multiple stress concentrations. Redistributions of stresses have local characters. Stress concentrations at welded joints, usually caused increase of stresses, but it can also relax stress-strain state [8, 9, 10]. Experimentally obtained results of mechanical properties showed that both considered MMA/MAG and MIG/MAG welding caused comparable stress concentration.

Residual stresses at zones of welded joints are consequence of interaction of different phenomena caused by welding. Characteristics of residual stresses at weld metal, heat affected zone and surrounding zone depend on number of complex factors. The basic mechanisms that caused residual stresses at welded mechanical constructions are identified, but estimate of values and distribution is very complex. Mechanical answer to load and values of yield and tensile strength of tested specimens showed that level of residual stresses at considered welding processes are equal [11, 12, 13]. Welding consumables for root pass done by MMA and MIG process with low strength and high plasticity provide relaxing of residual stresses.

Solidification processes, mean processes of microstructural transformations under thermal cycles due to welding are most important factor that influent to final microstructure of material at zone of welded joints. Factors, that influent to final microstructure of material at zones of welded joints are numerous with complex interactions. Those factors are very similar for both of considered welded processes MMA/MAG and MIG/MAG.

Joining by welding at high-strength low-alloy steels is complex process with large number of influential factors. Producers of high-strength low-alloy steels due to importance and complexity of welding and sensitivity of those steels to welding also recommend welding processes and parameters.

Weldability is complex characteristic of material that is related to influence of welding process and its parameters to characteristics and properties of welded joint. Presented facts lead to conclusion that overall characteristics and mechanical answer to load from stress-strain aspect are highly influenced by welding. Evaluation of weldability in details must be done as analysis of design solution, selection of material, applied welded process and their interactions [14, 15, 16, 17].

4. Conclusion

Present norms and standards of welding joints design at high-strength low-alloy steels are based on heterogeneous backgrounds. Limitations are established during design process from different reasons in order to provide welded joint with adequate mechanical properties. Improving resistance of welding joints to forming of defect and inclusions are based on relaxing residual stresses, reducing hydrogen content and obtaining preferred thermal cycles. Present standards and recommendations are not still fully developed and precise. Effects of specific welding parameters to microstructural state at different zones of welding joins are not analysed in details. Distribution of microhardness and mechanical properties of the joint as complex characteristic are caused by large number of factors. Those relations are in focus of research presented in this paper on qualitative level. From the practical aspect, mechanical characteristics and properties of welded joints at this steel grade is crucial.

High-strength low-alloy steels welding is far more complex than welding of other general purpose constructional steel grades. Development of steel grades with advanced properties such as high-strength low-alloy steel caused intensive research of its welding. High-strength low-alloyed steels due to importance and complexity of its welding usually came with recommend welding technologies and its parameters. Weldability is complex characteristic of material related to influence of welding and its parameters in relation to material sensitivity. Weldability analysis must be done as multi criterion analysis of design solution, selection of material, applied welded process and their mutual interactions.

Present demand related to welded constructions become far more complex. Simultaneously, exploitative and load conditions are more serious and much more significant to construction answer to load and its integrity. Welded constructions are complex systems of heterogeneous elements with mechanical properties that are highly influenced by its welded joints.

High-strength low-alloy steels due to specific microstructure and alloying design are sensitive to welding influences, but applications of those steels instead of conventional, general purpose structural steels provide many advantages. It can be concluded that full benefit of applications of those steels can be obtained only by suitable welding. Appropriate welding is the main condition for preserving material characteristics and properties means minimalizing of its degradation, that are the basis of proper mechanical properties, and condition of joining. Share of usage of high-strength low-alloyed steels continually increase and forming of joints is critical factor from the aspect of structural integrity and load capacity. Mechanical properties of high-strength low-alloy steels are highly influenced by its specific characteristics, as limited plasticity reserve due to high strength of steels, possible formation of local zones with lower plasticity in relation to rest of the construction and possibility of material discontinuities and initialization of cracks (primarily, hydrogen) during welding at weld metal and heat affected zone.

References

[1] Jorge C F J, Monteiro J L D, Gomes A J C, Bott I S, Souza L F G, Mendes M C and Araújo L S 2019 Influence of welding procedure and PWHT on HSLA steel weld metals, *Journal of Materials Research and Technology* **8**(1) 561-571, https://doi.org/10.1016/j.jmrt.2018.05.007

[2] Nathan S, Ragu Balasubramanian V, Malarvizhi S and Rao A 2015 Effect of welding processes on mechanical and microstructural characteristics of high strength low alloy naval grade

steel joints, *Defence Technology* **429**, https://doi.org/10.1016/j.dt.2015.06.001

[3] Lahtinen T, Vilaça P, Peura P and Mehtonen S 2019 Welding Tests of Modern High Strength Steels with Minimum Yield Strength of 700 MPa, *Applied Sciences* **9**, doi:10.3390/app9051031 www.mdpi.com/journal/applsci

[4] Zgripcea L, Lupu O, Putan V and Birtok Baneasa C 2019 Researches for improving mechanical characteristics of high strength weld, *International Conference on Applied Sciences IOP Conf. Series: Materials Science and Engineering*, doi:10.1088/1757-899X/477/1/012008

[5] Ilić A, Ivanović L, Rakić B, Josifović D and Lazić V 2014 Influence of welding technology to mechanical properties of welding joints at high strength low alloy steels, *Machine and industrial design in mechanical engineering-KOD* 2014, 255-260

[6] Ilić A 2015 *Influence of shape complexity, material, Stress concentration and temperature to designing of welded constructions*, University in Kragujevac, Faculty of engineering, Doctoral Thesis

[7] Ilić A, Ivanović L, Josifović D, Lazić V and Živković J 2018 Effects of welding on mechanical and microstructural characteristics of high-strength low-alloy steel joints, *IOP Conference Series: Materials Science and Engineering*, **393**, 012020. 10.1088/1757-899X/393/1/012020

[8] Lei X W, Dong S, Huang J H, Yang J, Chen S H and Zhao X K 2018 Phase evolution and mechanical properties of coarse-grained heat affected zone of a Cu-free high strength low alloy hull structure steel, *Mater Sci. Eng. A* **718** 437-448

[9] Lazić V, Aleksandrović S, Nikolić R, Prokić-Cvetković R, Popović O, Milosavljević D and Čukić R 2012 Estimates of weldability and selection of the optimal procedure and technology for welding of high strength steels, *Procedia Engineering* **40** 310-315

[10] Cho B Y, Kim H Y, Yang S C and Chae S U 2015 A Study on the Microstructure of Welded Structural Steel Members at High Temperature, *Applied Mechanics and Materials* **784** 397-402

[11] Frei J, Alexandrov B T and Rethmeier M 2016 Low heat input gas metal arc welding for dissimilar metal weld overlays part I: the heat-affected zone, *Welding in the World* **60**(3) 459-473

[12] Saedi A H, Hajjari E and Sadrossadat S M 2018 Microstructural Characterization and Mechanical Properties of TIG-Welded API 5L X60 HSLA Steel and AISI 310S Stainless Steel Dissimilar Joints, *Metallurgical and Materials Transactions A* **49**(11) 5497-5508

[13] Kimapong K and Triwanapong S 2018 Influence of Gas Metal Arc Welding Parameter on Lap Joint Properties of SS400 Carbon Steel and SUS304 Stainless Steel, *Key Engineering Materials* **789** 110-114

[14] Alipooramirabad H, Ghomashchi R, Paradowska A and Reid M 2016 Residual stress-microstructure-mechanical property interrelationships in multipass HSLA steel welds, *Journal of Materials Processing Technology* **231** 456-467

[15] Maurer W, Ernst W, Rauch R, Vallant R and Enzinger N 2015 Evaluation of the factors influencing the strength of HSLA steel weld joint with softened HAZ, *Welding in the World* **59**(6) 809-822

[16] Schneider C, Ernst W, Schnitzer R, Staufer H, Vallant R and Enzinger N 2016 Influence of different welding processes on the mechanical properties of structural steel S960MC *Thermec Graz* **1**

[17] Gosavi P D, Sarkar K K, Khunte S K, Pawar V R and Basu B 2019 Microstructure and mechanical properties correlation of weld joints of a high strength naval grade steel, *Procedia Structural Integrity* **14** 304-313

Quality control of closed-cell metal foam produced by direct foaming

A Gábora[1] and T Mankovits[1]

[1]University of Debrecen, Faculty of Engineering, Department of Mechanical Engineering, Ótemető Street 2-4, 4028 Debrecen, Hungary

E-mail: andrasgabora@eng.unideb.hu

Abstract. Metal foams have a lightweight cellular structure with excellent mechanical and physical properties. Although metal foams are popular, they are still not sufficiently characterized thanks to their extremely complex structure which is highly stochastic in nature. In this paper the influence of the technological parameters on the structure is analyzed. In laboratorial circumstances the production of closed cell aluminum foams depends on several factor. The purpose of the paper is to analyze the human factor on the quality of the metal foam specimens.

1. Introduction

Metal foams are relatively new and advanced materials with high stiffness to weight ratio, good thermal conductivity, good acoustic insulation and excellent energy absorption capability which make them ideal materials for a variety of applications [1–3]. Therefore, they have increasingly been employed for a wide range of applications, such as structural elements, automotive parts, sound and vibration absorbers or even biomedical implants [4–8]. There are different types of metal foam structures; open and closed metal foams [9-10], metal matrix syntactic foams [11-13] from different raw materials. Closed cell metal foams are produced by various methods, but the key step of their manufacture is the inclusion of air in the metal structure. The fact that gas pockets are present in their structure provides an obvious weight advantage and other favourable physical, mechanical, thermal, electrical and acoustic properties.

2. Materials and methods

Different metal alloys are foamed by mixing into them a so called foaming agent that releases gas when heated. The most common foaming agent is titanium hydride (TiH_2) which begins to decompose when heated above around 465°C into Ti and gaseous H_2. By adding titanium hydride to an aluminum melt hydrogen gas are rapidly produced. It creates bubbles that can lead to a closed cell foam. The process begins by melting aluminum and stabilizing the melt temperature between 670 and 690°C. The melt is then aggressively stirred and 1–2% of TiH_2 is added in the form of 5–20 μm diameter particles. As soon as these are dispersed in the melt, the stirring system is withdrawn, and a foam is allowed to form above the melt. When foaming is complete the melt is cooled to solidify the foam before the hydrogen escapes and the bubbles collapse.

In our experiments, for the raw material the F3S.20S (AA 359/SiC/20p) aluminum-based metal matrix composite with up to 20% silicon carbide (SiC) particles (Figure 1.) were used. The most useful features of Duralcan® composites are their high strength, stiffness, wear resistance, thermal conductivity, their improved elevated temperature tensile and fatigue strengths and their low density and coefficient of thermal expansion. The silicon carbide particles are sufficient for viscosity modification. The foaming agent was the titanium hydride (Figure 2.). It is commercially available as a stable grey/black powder, which is used as an additive in the production of sintering powdered metals, production of metal foam, and in pyrotechnics.

Figure 1. The F3S.20S composite.

Figure 2. The foaming agent.

For the experiments we used a Goldbrunn 3000 furnance (Maximum temperature: 1100 °C, Mass: 3 kg, Volume: 294 cm³, Power: 1900 Watt, 230V / 50Hz, Accuracy: ±0.5% temperature, Dimensions: 335 x 280 x 360 mm). The original graphite crucible is not applicable so we design and produce a new dismountable crucible. We used CAD systems to design the components and illustrate an assembly of them (Figure 3.). This crucible has two main parts: an outside closed form and an inner part which consist of two symmetric half parts.

Further used devices (Figure 4.): Kern EW400 Precision Scale (Range: Max=600g, Min=0,5g, Accuracy: 0,01 g), Struers Labotom-3 cutting machine, Metkom Forcipol 2V polishing machine, Metabo SBE 750 (0-3000 rpm) and Makita DDF458Z (0-2000 rpm) drilling machines, Turning machine, and for foaming process a mixing head. The used crucible was coated with a borone nitride suspension to be easier removable.

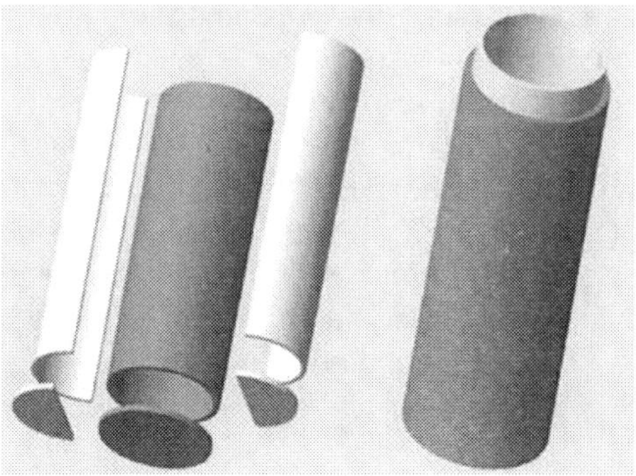

Figure 3. The design of new crucible.

Figure 4. Experiment devices.

The experiments took place at the Department of Mechanical Engineering, University of Debrecen (Figure 5. and Figure 6.). For the quality of the metal foam specimen (e.g. structure, specimen size, compressive properties) the human factor has great effect. At all experiment the temperature when the TiH_2 is added was 750°C, while the percentage of the TiH_2 to the raw material was 1,5%. The mixing was made manually and human precision was also a factor in the adjustment of cutting cube specimens. After the foaming and cooling processes the samples were cut in 30x30x30 mm^3 cubes. All the investigations were done according to the ISO 13314. To provide information from the cell distribution digital quantitative image analyses were performed using macroscopic records from the specimen's surfaces. The compression tests were performed on an INSTRON 8874 type universal testing machine at room temperature. The compression tests were carried out with the application of lubricant. The deformation rate was maintained in quasi-static condition at 8.7mm/min.

Figure 5. After cooling process.

Figure 6. Metal foam samples.

3. Result and discussion

For the investigation 4 experiment was analyzed and evaluated. Table 1 shows the result of the specimen cut using universal cutting machine Struers Labotom-3. The porosity of each specimen was also calculated. The porosity evaluation was based on weight and volume measurements.

Table 1. Geometrical and structural parameters of the specimen.

Properties	AF-01	AF-02	AF-03	AF-04
Edge length a1 (mm)	30.25	30.75	30.2	29.45
Edge length a2 (mm)	30.18	30.98	29.74	30.09
Edge length a3 (mm)	29.97	30.55	29.16	29.67
Porosity (%)	90.7	81	92.7	88.3

From Table 1 it can be seen that the edge length are not exactly the same and the porosity shows great difference. Dimensional inaccuracy affects subsequent investigations (e.g. compressive test) while the cube is not regular. The porosity difference is because of the mixing process which is difficult to control if it is done manually.

Applying surface analysis from macroscopic images and the ImageJ software the area percentage of the cells (Figure 7.) and the number of the cells (Figure 8.) for all faces of the specimen can be calculated and determined. The process of the surface analysis is detailed in [14]. From the figures it can be stated that the above geometrical properties highly depend on the mixing and from where the specimens are cut from the produced sample. All of those have effect on the compressive properties of the aluminum foams.

IOP Conf. Series: Materials Science and Engineering **659** (2019) 012037 doi:10.1088/1757-899X/659/1/012037

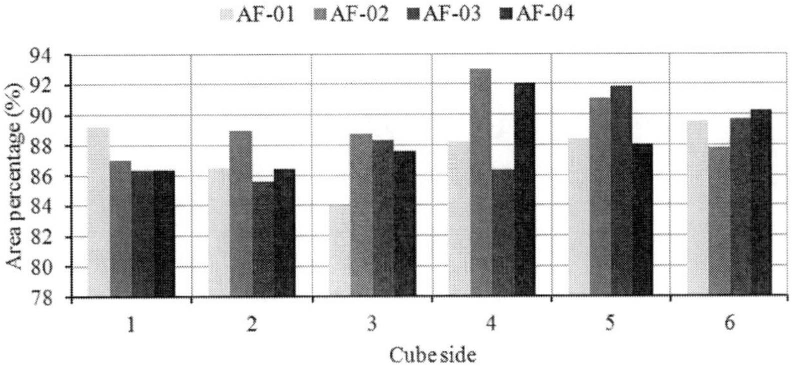

Figure 7. Results of the area percentage determination.

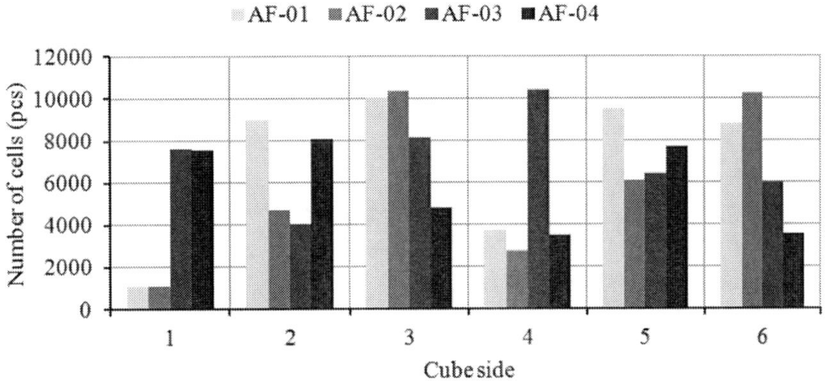

Figure 8. Results of the cell number calculation.

The compression tests were performed on an INSTRON 8874 type universal testing machine at room temperature (Figure 9.). The compression tests were carried out with the application of lubricant. The deformation rate was maintained in quasi-static condition at 8.7mm/min.

Figure 9. Compression test.

During the tests, the force-displacement curve (Figure10.) were registered and processed according to the ruling standard for the compression test for porous and cellular materials.

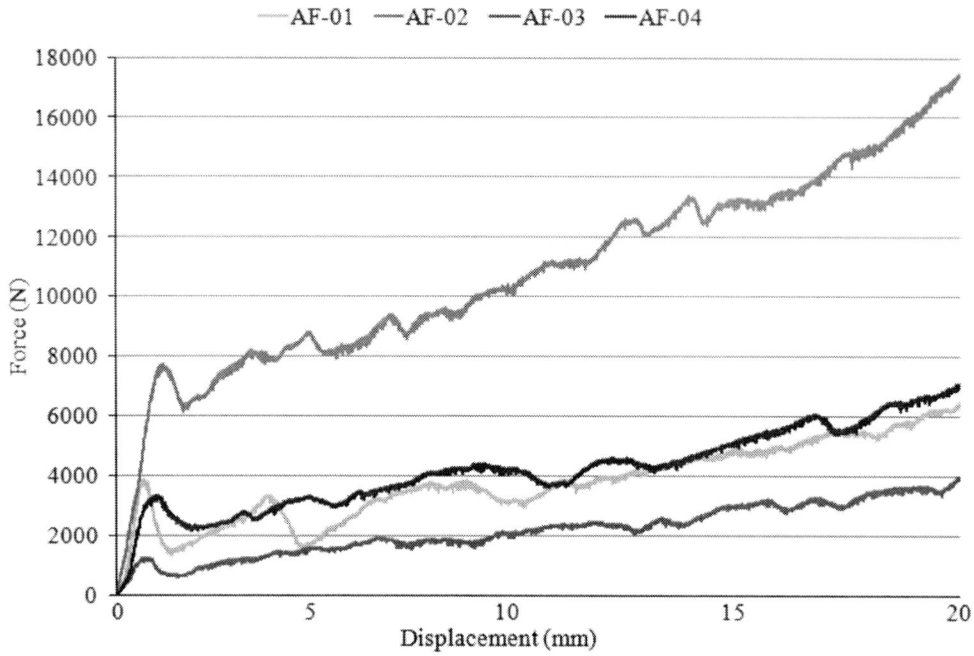

Figure 10. Force-displacement curves of the specimens

The force-displacement curves show great deviation thanks to the different geometrical and structural properties of the specimens. The deviation can be observed even in the elastic region of the force-displacement curve which is an essential part for application purposes.

4. Conclusions

In our experiments Duralcan F3S.20S aluminum based metal matrix composite as a raw material was used with direct foaming in the melt by titanium hydride addition. Applying same temperature and TiH_2 concentrate the effect of the human factor for the quality of the produced aluminum foam specimens were analyzed. As a result great deviation on physical and geometrical properties can be observed between certain specimens which can be explained with the several manual manufacturing steps. The cellular structure of our specimens is not enough homogeneous, so for the future we are planning to change the technological process to more automated, using machines, particularly for the mixing process.

References

[1] Ashby M F, Evan A G, Fleck N A, Gibson L J, Hutchinson J W and Wadley H N G 2000 *Metal Foams: A Design Guide*, Butterworth-Heinemann

[2] Czekanski A, Attia M S, Meguid S A and Elbestawi M A 2005 On the use of a new cell to model geometric asymmetry of metallic foams, *Finite Elements in Analysis and Design* **41**(13) 1327–1340

[3] Banhart J 2001 Manufacture, characterization and application of cellular metals and metal foams, *Progress in Materials Science* **46**(6) 559–632

[4] Vendra L J and Rabiei A 2007 Evaluation of modulus of elasticity of composite metal foams by experimental and numerical techniques, *Materials Science and Engineering: A* **527**(7–8) 1784–1790

[5] Tuncer N and Arslan G 2009 Designing compressive properties of titanium foams, *Journal of Materials Science* **44**(6) 1477–1484

[6] Kádár Cs, Chmelík F, Rajkovits Zs and Lendvai J 2004 Acoustic emission measurements on metal foams, *Journal of Alloys and Compounds* **378**(1–2) 145–150

[7] Djebbar N, Serier B, Bouiadjra B B, Benbarek S and Drai A 2010 Analysis of the effect of load direction on the stress distribution in dental implant, *Materials & Design* **31**(4) 2097–2101

[8] Kashef S, Asgari A, Hilditch T B, Yan W, Goel V K and Hodgson P D 2010 Fracture toughness of titanium foams for medical applications, *Materials Science and Engineering: A* **527**(29–30) 7689–7693

[9] Devivier C, Tagliaferri V, Trovalusci F and Ucciardello N 2015 Mechanical characterization of open cell aluminium foams reinforced by nickel electro-deposition, *Materials and Design* **86** 272-278

[10] Xiao L, Song W, Tang H, Zhu Z, Wang J and Wang H 2015 High temperature compression properties of open-cell Ni–20Cr foams produced by impregnation, *Materials and Design* **85** 47-53

[11] Orbulov I N and Májlinger K 2013 Description of the compressive response of metal matrix syntactic foams, *Materials and Design* **49** 1-9

[12] Orbulov I N 2012 Compressive properties of aluminium matrix syntactic foams, *Materials Science and Engineering: A* **555** 52-56

[13] Szlancsik A, Katona B, Bobor K, Májlinger K and Orbulov I N 2015 Compressive behaviour of aluminium matrix syntactic foams reinforced by iron hollow spheres, *Materials and Design* **83** 230-237

[14] Mankovits T, Budai I, Balogh G, Gábora A, Kozma I, Varga T A, Manó S and Kocsis I 2014 Structural analysis and its statistical evaluation of a closed-cell metal foam, *International Review of Applied Sciences and Engineering* **5**(2) 135-143

IOP Conf. Series: Materials Science and Engineering **659** (2019) 012038 doi:10.1088/1757-899X/659/1/012038

Designing and optimizing extractors for automated dispensers

A Hajdučík[1], L Smetanka[1], Š Medvecký[1] and J Škrabala[1]

[1]University of Žilina, Faculty of Mechanical Engineering, Univerzitná 1, 010 26 Žilina, Slovakia

E-mail: adrian.hajducik@fstroj.uniza.sk

Abstract. Currently, the process of manufacturing the goods and then distributing them to the customer is constantly optimized, accelerated and automated. For these aspects, goods handling and automated storage play a key role. This article focuses on designing and optimizing the mechanisms that are parts of vending machines. Both static and dynamic models of mechanisms are presented, also the thesis describes the configuration and operating principle of given mechanisms. The mechanisms have been designed to handle for instance boxes, reliably and precisely, they were subjected to dynamic analysis. Several principles of the various mechanism are presented. After this, the functionality of some mechanisms was verified by analysis. These models of extractors can be used in food dispensers, vending machines in libraries, pharmacies, and stores of specific goods. The presented mechanisms are particularly suitable for medicine dispensers and pharmacy environments.

1. Introduction

Currently, the process of manufacturing the goods and then distributing them to the customer is constantly optimized, accelerated and automated. For these aspects, goods handling and automated storage play a key role. This article focuses on designing and optimizing the mechanisms that are parts of vending machines. Both static and dynamic models of mechanisms are presented, also the thesis describes the configuration and operating principle of given mechanisms. The mechanisms have been designed to handle for instance boxes, reliably and precisely, they were subjected to dynamic analysis. Several principles of the various mechanism are presented. After this, the functionality of some mechanisms was verified by analysis. These models of extractors can be used in food dispensers, vending machines in libraries, pharmacies, and stores of specific goods. The presented mechanisms are particularly suitable for medicine dispensers and pharmacy environments [4, 5].

2.Methodology

As the proposed mechanisms are designed primarily for vending machines and pharmacies, specific requirements that are part of the environment need to be taken into account. Handling mechanisms are primarily intended for supplying and dispensing objects. When designing extractors, it is necessary to pay attention to the shape, weight and other associated operations. Figure 1 shows the division as well as the intended application of the mechanism.

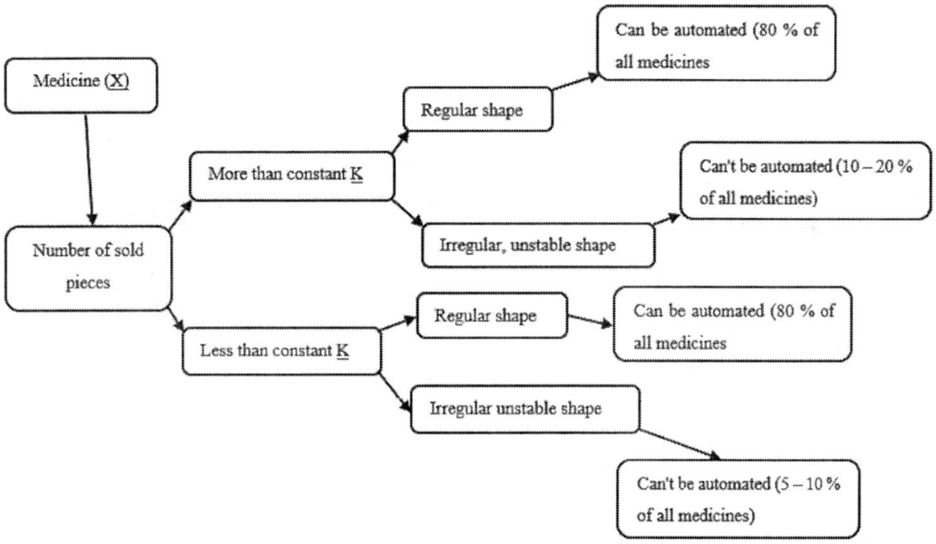

Figure 1. Possibilities of medicine storage.

When designing, it is also necessary to take into account the principle of the equipment to work. In most cases, these are mainly the principles: pneumatic, electromechanical and hydraulic. In our case, we exclude hydraulic operation. Disadvantages have far outweighed the advantages of the hydraulic system. Especially because of costly operation of the system, the presence of dangerously high pressures, pollution caused by leakage. When designing the mechanism for supplying the dispenser and for dispensing objects from the dispenser, it was mainly based on the shape diversity of the medicines. For this purpose, we have communicated with pharmacists and worked closely with pharmacies, conducted a survey, and developed an extensive list of the most widely sold drugs. Pharmacies have been involved in the research to provide us with their data [5].

It has also been shown that the marketing of drugs and, in particular, their species is different from the season. For these aspects, a consensus has been created that meets market requirements.

Partial goal of data collection was also input conditions for solving problematic packaging forms of some types of drugs. As can be seen from Figure 2, in particular the irregular shapes of the vials and bags.

It has also been found that it is not possible to automate the sale of all kinds of drugs in the present automated device. According to the drug sales list, a formula has been created according to which the design of extractors has started.

If a given type of drug has not exceeded a certain threshold of sell ability (constant K) and the shape of the packaging has been complicated, it has been classified as non-automated and must be stored manually, see Figure 1. In general, drug vending machines can automate the sale of 80% of the drugs offered on the market. However, this number may increase over time as mechanical principles are constantly innovated, and drug manufacturers will also consider the shape and packaging of the drug. However, we expect that within a few years we will see more than 90-95% of all drugs automated. Nevertheless, the processes mentioned above need to be carried out [5].

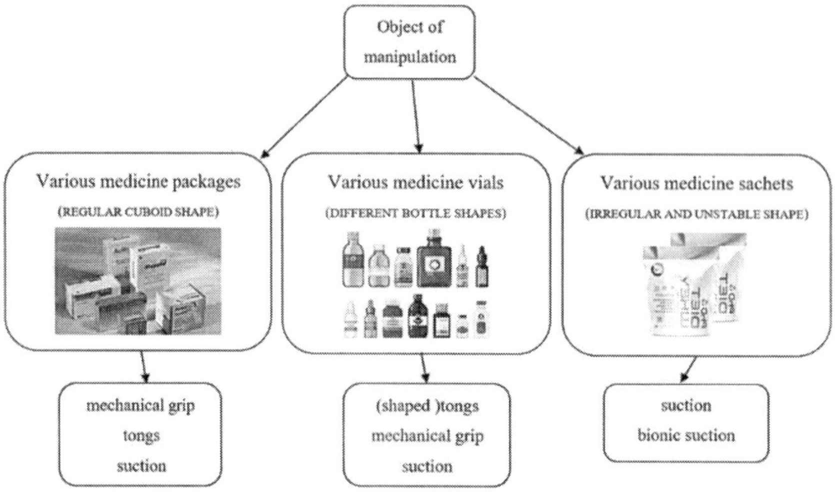

Figure 2. Medicine division by shape and methods of the gripping.

Following the analysis of the drug market, the development of mechanisms for the drug dispenser was started. The fully automated drug delivery system is characterized by the fact that it can automate the supply and delivery of drugs. This means that one supply extractor takes care of the supply and one extractor ensures the dispensing itself. The machine we developed is based on two parts. In the back, there are shelves in the horizontal direction. These shelves balance stocks and serve as a temporary store. In front of the dispenser, there are shelves that have the character of a gravity rack. The shelves are installed at an angle of 30 °. In the following chapters, we analyse the supply extractor and the dispensing extractor.

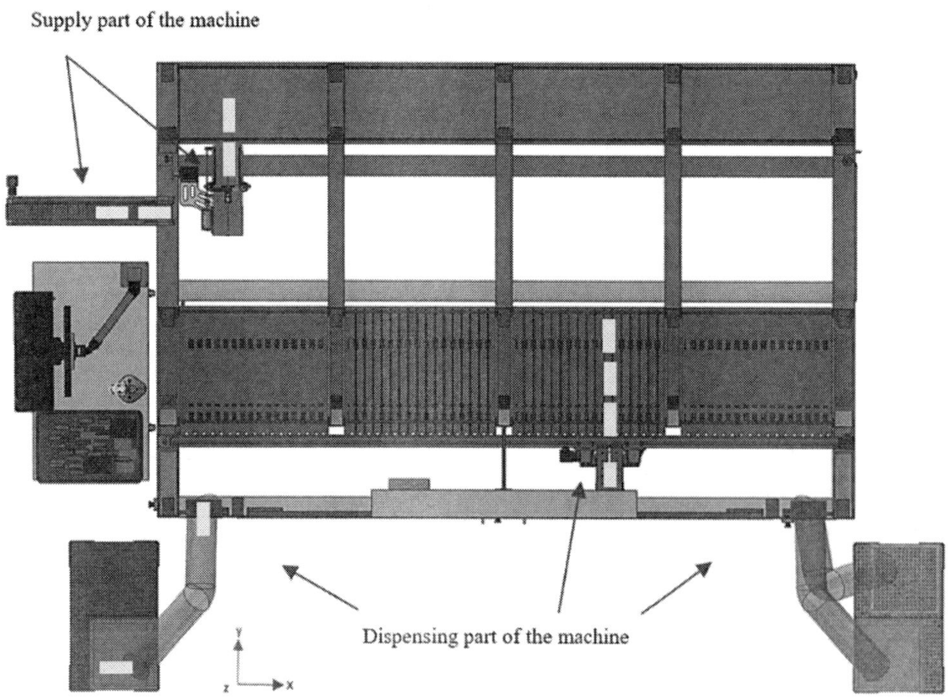

Figure 3. Schematic movement of the medicine across the machine.

Figure 3 schematically shows the movement of the medicine in the dispenser. The medicine comes through a conveyor belt to pick it up by supplying extractor and then the supplying extractor stores the medicine in the shelves. From the front shelves, the vending dispenser extracts the medicament, and through the dispensing site, through the pipeline, enters the pharmacist's work area.

3. Conceptual design of a supply extractor

We have selected a conveyor to deliver drugs to the dispenser. The conveyor can be made in various designs, depending on the customer's request. The drugs are deposited on a conveyor and passed through an optical gate, where their bar code is scanned, and the data is recorded into the system. The following systems have been proposed for the collection and subsequent incorporation of drugs [4].

3.1 Supply extractor with suction pad and scissors mechanism

One of the first proposals was to use a suction cup and a scissor mechanism. In another concept, we have replaced the scissor mechanism with a pneumatic piston. The advantage of this solution lies in its simplicity of design, predictable life. However, the big disadvantage is the small number of medicines per hour, because we are able to grab only one box. Another drawback is that the position of the suction cup must be oriented as low as possible to the conveyor because there is a risk that the suction cup will not catch the object, especially small boxes. Thus, the versatility of the device can be more difficult than with other prototypes. The forces that need to be overcome are very low since the drug box only slides over the pad, where friction is considered (k= 0.1, between pad and box). Thus, only the self-weight and weight of the suction cup acts on the scissor stroke. In addition, this mechanism is supported by auxiliary guide, see figure 4 [1].

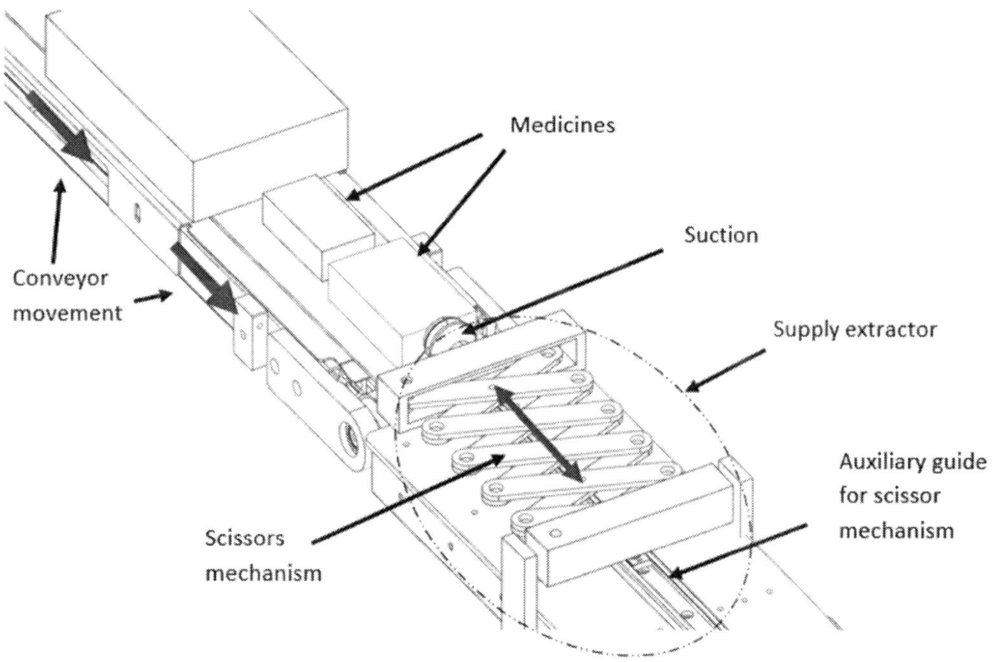

Figure 4. Schematic view of extractor with suction cup.

3.2 Extractor with forks and slider

We have tried to eliminate the shortcomings identified in the concept above by using other, in particular more universal mechanisms. We went to a full electromechanical drive. There are 4 electric motors on the extractor. One of them provides rotational movement of the extractor around itself axis. The gripping of the boxes is ensured by the so-called forks that move along the side guides. The motion is provided by the motor which produces the torque on the threaded rod. The principle of function and the individual elements of the mechanism can be understood from the following figures [4].

The advantages of using this system lie in the higher capacity of the stored medicines. 1 to 3 boxes at a time can be grasped. This system is much quieter. Plastic lines are used for several reasons. There is no need to lubricate them with an external lubricant, which means a cleaner and more hygienic environment. They are on the order of 50% lighter than steels, without stiffness and strength. They have a better predictable life [1, 4, 5].

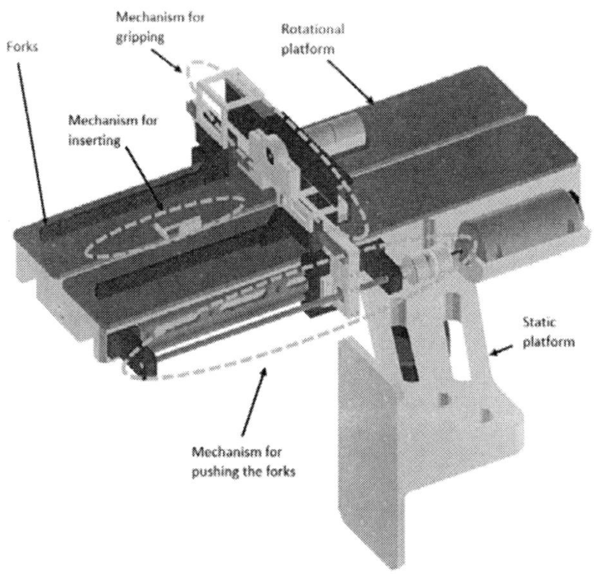

Figure 5. Main functional mechanisms of the fork extractor.

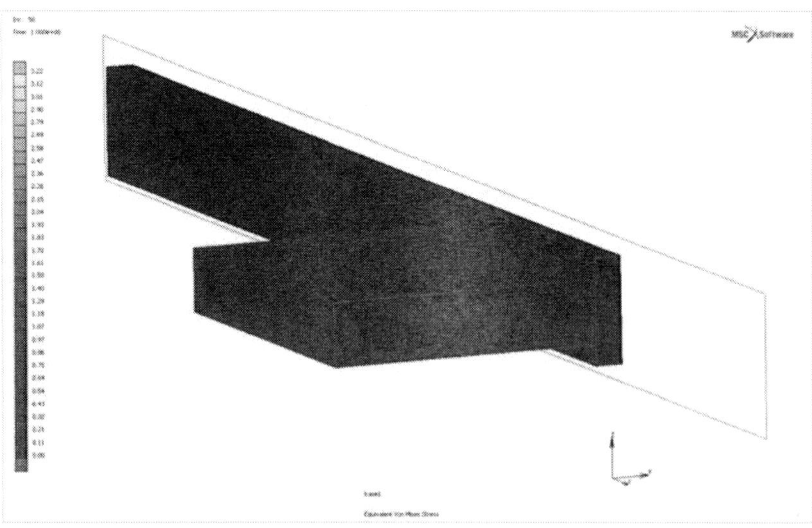

Figure 6. Von Misses strain at force 80 N.

IOP Conf. Series: Materials Science and Engineering **659** (2019) 012038 doi:10.1088/1757-899X/659/1/012038

We made a simulation calculation of the design of the fork clamp force. Based on the calculation we propose to use a force of F = 80 N per one fork, see figure 6. As the fork material, S235 was chosen with Young's modulus of E = 210,000 MPa and a Poison number of 0.3, and the paper was chosen as the homogeneous material with Young's modulus of E = 10,500 MPa and Poison number of 0.35. When a force of about 500N is applied to a fork, the box begins to deform, see figure 7 [2, 3].

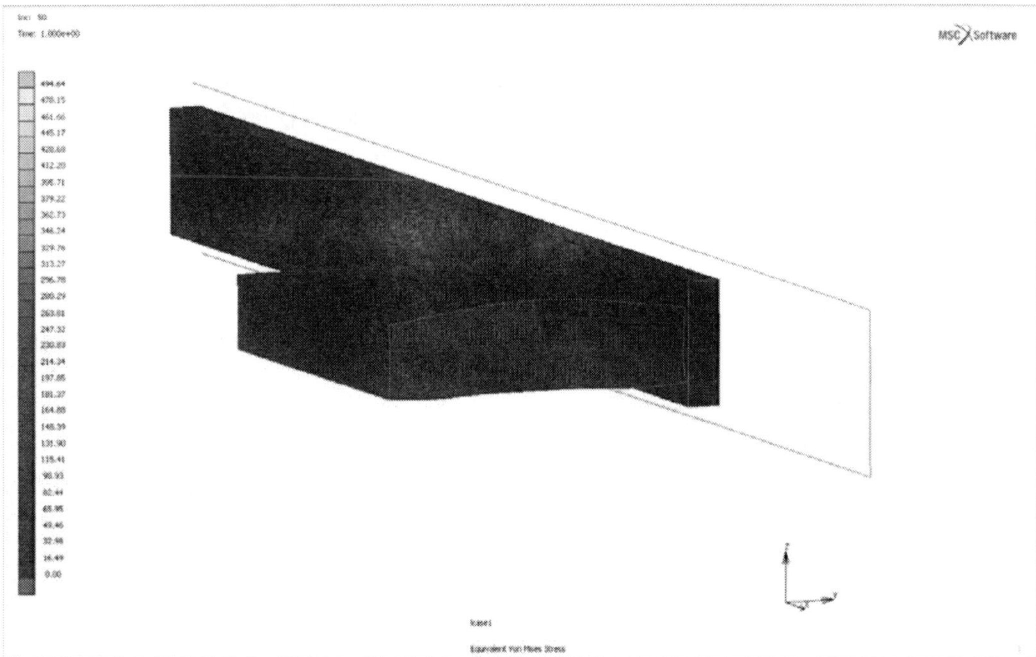

Figure 7. Von Misses strain at force 500 N.

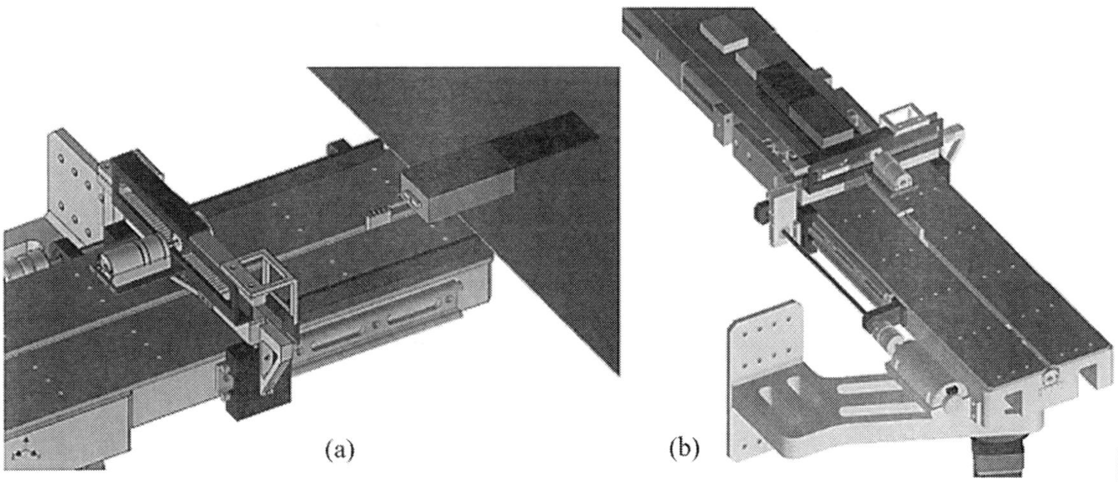

Figure 8. The process of inserting the medicine (a), the process of pull the medicine from the conveyor (b).

4. Conceptual design of dispensing mechanism

Various types of dispensing extractors are used to remove drugs from the front shelves. Medicines can be stored in horizontal shelves, but also in shelves that have the character of a gravity rack and they are stored at the certain angle. If the front shelves are horizontal, we have designed a suction and slip extractor. However, if the front drug shelves were tilted at an angle, we designed an extractor that includes a so-called thumb mechanism. In a particular position in front of the rack, with the ordered medicine at the complete stop of the extractor, the dispensing mechanism is then triggered.

The drug is released from the gravity shelf and lowered into the extractor storage area. If there is more than one medicine order, this cycle is repeated. After collecting medicine from the shelf (which, depending on their size, may be more than five pieces), the extractor moves to the dispensing place. At the same time, the extractor door is opened by reaching the position of the dispensing place and its content is displaced spontaneously through the tube to the selected point at the cash register. Both extractors are designed to weigh up to 10 kilograms [4, 5].

4.1 Design of extractor with thumb mechanism.

The main parts are the kinematic mechanism of the thumb, the engine securing the movement of the thumb mechanism, the actuator opening the door, the structure itself, the cover, supporting and connecting elements, see figure 9. Opening of the door ensures micro-linear actuator. This door opening principle was chosen for the following facts: the door weight is 200 g, on the hinges they perform a simple two-way movement of 90 °. Actuators provide quiet and precise motion control, are effective, have a long lifetime and virtually zero maintenance. The selected actuator consists of a motor, gear and trapezoidal screw [5].

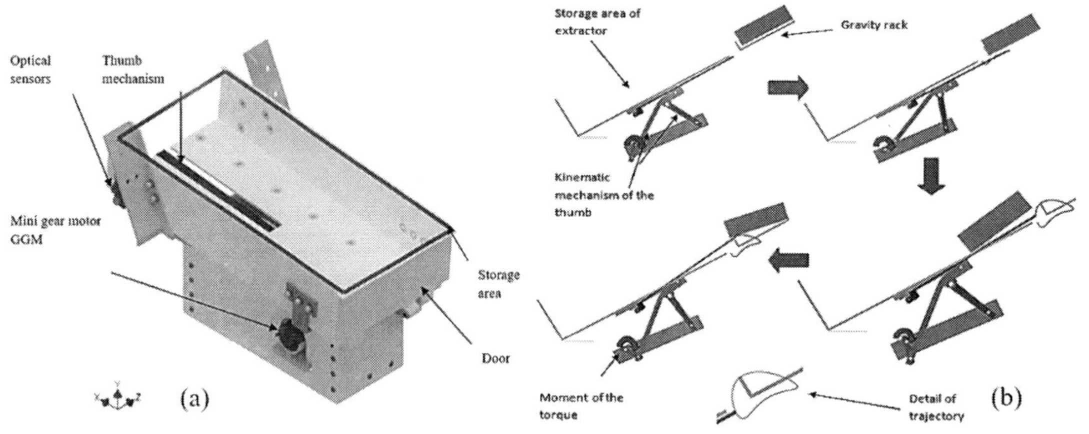

Figure 9. Dispensing extractor (a), process of releasing the medicine from the gravity rack (b).

The kinematic mechanism of the thumb was subjected to dynamic analysis in Inventor 2018, and one of the outputs was the endpoint trajectory of thumb, see fig. 9. With the help of dynamic simulation, it was possible to analyse the acting forces and the torque needed to ensure the rotational motion of the kinematic mechanism. One cycle lasts 2 seconds, the time it takes for the thumb to go through the entire trajectory from the starting point while removing the box from the gravity shelf. In the analysis, the mechanism was loaded with the heaviest box weighing 450 grams.

As can be seen from the figure 10, the maximum torque required was reached at 1.35 s, when the thumb lifts the box over the edge of the shelf. The value reached 101 Nmm (0.1 Nm), which was one of the data needed to select a suitable engine [5].

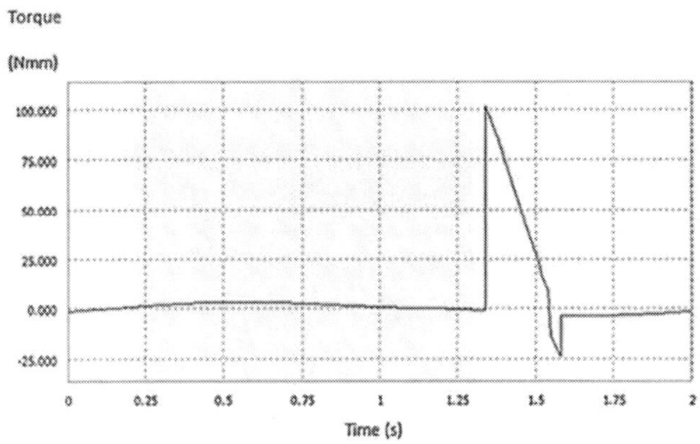

A miniature brush motor with a planetary gearbox was selected to move the thumb mechanism. The motor is powered at 12V. The gear ratio is 1/198.

Figure 10 . Graph of torque and cycle time with defined parameters for motor.

4.2 Extractor with a suction cup and slip surface

In addition to the clamp grip, a suction cup and scissor mechanism has been designed to move medicines from the horizontal shelves to the extractor. When the scissor mechanism is lifted, the suction cup grabs the medicine. The medicine is simply pulled by the suction cup. Thus, the scissor stroke only has its own weight, which allows it to be used in such an application. After reaching the maximum contraction point, the drug is released and the inclined plane self-releases to the door.

The storage surface on an inclined plane is sufficient for 2 to 4 packs of drugs. Similar to the previous case, after the storage space has been filled, the extractor is moved by a manipulator on which it is placed at the dispensing points. At the dispensing point, the rear door opens and the medicines spontaneously advance through the conduit.

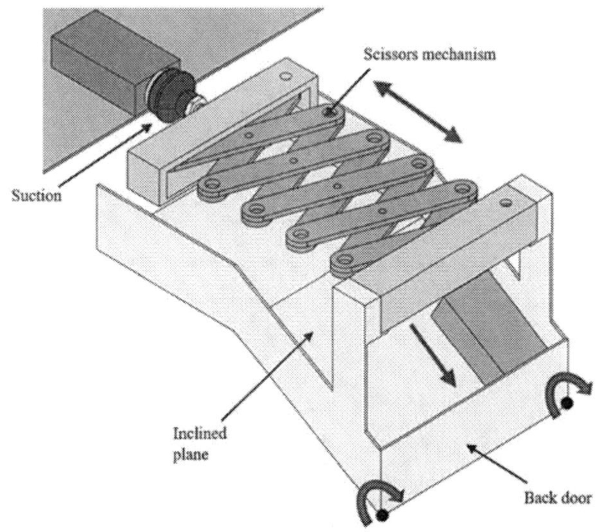

Figure 11. Schematic view on extractor with a suction and slip surface.

5. Visualization of the automat

In the figure 12, 13 we can see the design of the cladding, the access door and other supporting elements. The implementation of vending machine is individual and depends on the requirements.

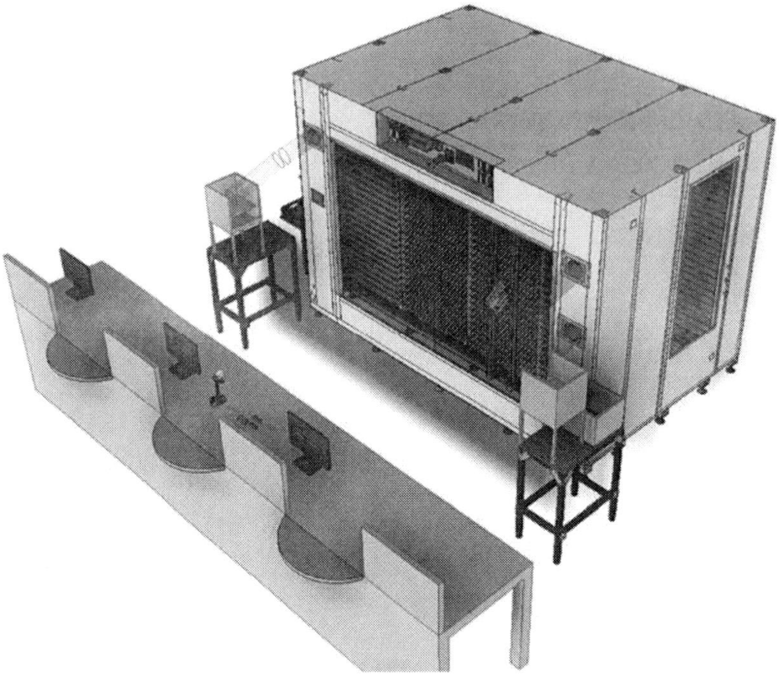

Figure 12. Visualization of the automat.

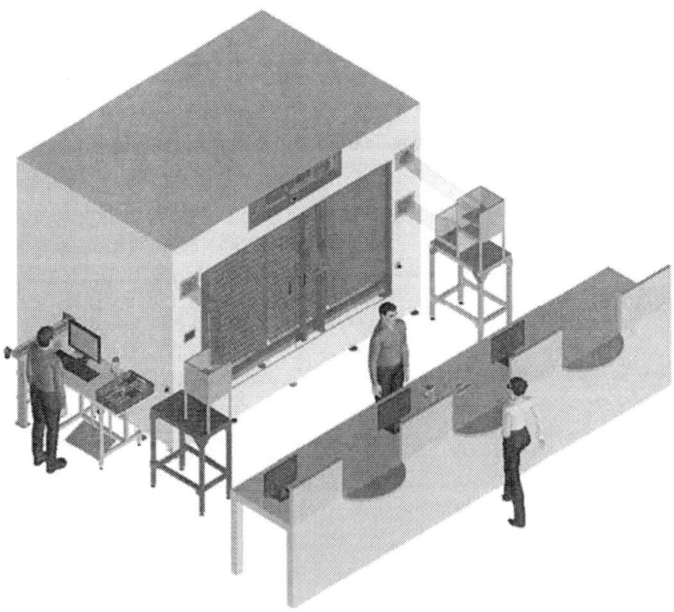

Figure 13. Visualization of the automat with staff and customer.

6. Conclusion

The article reflects the issue of medicine storage in automated devices. The results of a survey conducted in pharmacies have produced statistics on medicine storage options. It was clear from the results that so far it is not possible to automate the sale of all drugs on the market. In the next section, some of our extractor designs were presented, their advantages and disadvantages were evaluated. We have analysed some of them and discussed their technical principles in more detail. This text can serve as a review article as an inspiration for new solutions and ideas for developing new mechanisms.

Acknowledgments

This publication is the result of the Project implementation: Competency Centre for Knowledge technologies applied in Innovation of Production Systems in Industry and Services, ITMS: 26220220155, supported by the Research & Development Operational Programme funded by the ERDF.

References

[1] Martikan M, Brumercik F and Bastovansky R 2015 Development of mechatronic deformation system, *Applied Mechanics and Materials* **803** 173-178
[2] Kohar R and Hrcek S 2014 Dynamic analysis of a rolling bearing cage with respect to the elastic properties of the cage for the axial and radial load cases, *Communications* **3A**(16) 74-81
[3] Gramblička S, Kohár R and Stopka M 2017 *Dynamic analysis of mechanical coveyor drive system*, In book: Procedia Engineering [elektronický zdroj] **192** 259-264
[4] Graňák J 2018 *Koncepčný návrh zásobovacieho mechanizmu pre automat na lieky,* Žilinská univerzita v Žiline, Slovakia, Engineer Thesis
[5] Hajdučík A 2018 *Koncepčný návrh výdajneho mechanizmu pre automat na lieky*, Žilinská univerzita v Žiline, Slovakia, Engineer Thesis

Effect of reinforcement on mechanical characteristics of A356 alloy nanocomposites

S Veličković[1], B Stojanović[1], Z Djordjević[1], S Miladinović[1] and J Blagojević[1,2]

[1]University of Kragujevac, Faculty of Engineering, Sestre Janjić 6, 34000 Kragujevac, Serbia
[2]Metalac a.d., Metalac Inko d.o.o., Kneza Aleksandra 212, 32300 Gornji Milanovac, Serbia

E-mail: sandrav@kg.ac.rs

Abstract. Nanocomposites with aluminum base reinforced with nanoparticles, have wide application in many industries, for example, aerospace, marine, automobile, due to its lightweight, high tensile strength and high wear resistance, etc. The study of the microstructure, mechanical and tribological properties of the nanocomposite is very important in order to reliably and optimally carry out the design of new materials. In this study for the mechanical characterization of the A356 composites with a ceramic particle reinforcements nanoindentation was performed. Nanoindentaion is largely used for the analysis of mechanical properties. The samples of the nanocomposites were developed using 0, 0.2, 0.3 and 0.5 wt.% SiC, with an average size of particles of about 50 nm, produced by a compocasting process with pre-processing mechanical alloying. Nanoindentation was carried out based on Berkovich's test method with the indenter in shape of three sided diamond pyramid. The nanocomposite testing showed that the nanocomposites containing 0.5 wt.% of the reinforcement have slightly higher tensile strength and a lower value of the elastic modulus compared to the aluminum alloy base.

1. Introduction

In recent years, many composite materials have been developed that replace monolithic materials in many engineering applications. Composite materials consist of a base and an reinforcement that can be in the form of particles, fibers (long or short), laminates etc. Depending on the requirements of the industry, the properties of the base and the reinforcement are combined in order to respond to the set tasks that cannot be answered by conventional engineering materials. The composite can be with a polymeric, ceramic or metal base. Metal matrix composites (MMCs) are a special group of engineering materials due to the use of a light metal base to which different types of reinforcement are added, thereby improving the characteristics of the material. The most common materials used as base of composites are aluminum (Al) [1-3] magnesium (Mg) [4, 5], copper (Cu) [6], and titanium (Ti) [7] or their alloys.

Recent years great attention is paid to aluminum nanocomposites strengthen with particles due to application in the production of parts in the automotive and airplane industry [8, 9]. Metal matrix nanocomposites are a new class of materials that contain a nano-level reinforcement (1-100 nm), e.g. nanoparticles that are often used as an reinforcement. The most commonly used ceramic nanoparticles are Al_2O_3 and SiC, followed by TiC [10, 11], TiO_2 [12, 13], and TiB_2 [14, 15], which is confirmed by a large number of published researches in previous years [8].

Ceramic particles can be embedded in the base of aluminum alloy with different techniques. The most commonly used are mechanical alloying and casting. The method of making the nanocomposite with a metal base in liquid state is promising in terms of more economical production (shorter processing time and production cost) [16]. However, the efficient supply and dispersion of the nano-reinforcement is still a big challenge for many researchers [17]. Also, one of the useful techniques for achieving an equal distribution of nanoparticles is powder metallurgy technique [8].

In the last decades, testing of nanocomposite materials has been in expansion in order to find a nanocomposite that will meet certain requirements and replace monolithic materials in automotive, airplane, military, and many other industries.

The aim of this paper is the production of aluminum nanocomposites with different content of silicon carbide reinforcements by using an innovative process that represents a combination of mechanical alloying and compocasting process. Examination of the mechanical characteristics of the obtained nanocomposites with the base of the A356 alloy and with the low content of ceramic nanoparticles, is in order to find the dependence of the reinforcement content on hardness, elastic modulus and other feature of these materials. The tribological properties of these nanocomposites were tested and shown in the paper [18].

2. Overview of mechanical properties of nanocomposites with aluminum base

Today, great attention is paid to aluminum composites reinforced with particles in order to detect the best combination of base and reinforcement (type, size and content) and applied production techniques in order to achieve superior mechanical characteristics. Based on mechanical characteristics such as hardness, elasticity, elongation, density, porosity and tensile strength, a overall characteristics of the test material are obtained. A large number of scientists tested tribological characteristics [8], and the mechanical characteristics of nanocomposites with an aluminum base. Due to the diversity of the methods of obtaining and testing, combinations of the type, content and size of the reinforcement in the nanocomposite, as well as the way of displaying the results, the comparative results of the characteristics of the nanocomposite cannot be unified and diagrammatically illustrate. Some researchers carried out tests of the specific, i.e. the required characteristics, while other researchers carried out a complete analysis of the nanocomposite.

Rajmohan et al. carried out mechanical testing of the composite on hardness with the use of the Vickers method. The reinforcements used to form the composites are: CuO and SiC in mass ratio: CuO of 0 - 2 wt.% and SiC of 10 wt.%. The hardness results showed that the hardness increased with the increase of nanoparticles of the reinforcement in an aluminum base [19]. Also, the test of the hardness by the Vickers was carried out by Sharifi and Karimzadeh who then compared the results of aluminum alloys and hybrid nanocomposites with various reinforcement admixtures [20]. Based on the obtained hardness results, they concluded that the hardness increases with increase of the mass ratio of the reinforcement. The hardness test, depending on the mechanical grinding time, when obtaining the nanocomposite with the CNT and nSiC reinforcements, was considered by Kwon et al. [21]. They concluded that the hardness of the nanocomposite increases with the increase in the mechanical grinding time.

R. Taherzadeh Mousavian and others, in the production of an aluminum composite, tried to use micro particles instead of using nanoparticles to form SiC nanoparticles by increasing the grinding time. They concluded that with increase of grinding time the particle size was reduced, but not completely, because the SEM (Scanning Electron Microscopy) analysis revealed the presence of SiC particles of about 10-20 μm. They found that the particle size cannot be completely altered but they recorded an increase in the strength and hardness of the composite [22].

The combination of SiC particles of nano and micro sizes using stir casting technique for the production of composites was carried out by K. Amouri et al. The average particle size of nano and micro-size SiC powders was 50 nm and 5 μm. Testing the strength of the composite with the A356 base and the nano SiC particles of 0.5 and 1.5 wt.% and micro SiC with a content of 5 wt.% recorded a significant improvement in strength. More precisely, the highest strength was obtained in the composite containing 1.5 wt.% nano SiC. With the addition of micro/nano SiC particles the MMC has a significantly increased strength while ductility is reduced. Composite with a content of 1.5 wt.%

nano SiC possesses the highest tensile and compression strengths. The greatest strength of the nanocomposite is attributed to an optimized combination of sizes, equal distribution, and volume of particle fraction [23].

Hassan Tazari et al. carried out mechanical testing of aluminum nanocomposites with different SiC particle content of 1 wt.% to 5 wt.% produced by powder metallurgy [24]. They discovered that micro-hardness increases with an upward trend with the addition of particles up to 3 wt.% and then adding 5 wt.% this trend slightly decreases. In the analysis, they also found an increase in the porosity of the nanocomposite with the content of the reinforcement of 5 wt.%, which was also uncovered by the decrease in relative density. The improvement in micro-hardness values of 95%, 83% and 166% was recorded with an increase in reinforcement content of 1, 3 and 5 wt.%, respectively.

In addition to the hardness of the material, tensile strength monitoring is also important. Poovazhagan and others performed all mechanical tests on the nanocomposite [25]. For the reinforcements, they used SiC and B_4C with mass ratio for both reinforcements: 0.5 wt.%, 1 wt.% and 1.5 wt.%. The results obtained by tension test showed that the maximum tensile strength of the composite with 1 wt.% SiC + 0.5 wt.% B_4C was 265 MPa. While the composite with 1.5 wt.% SiC + 0.5wt.% B_4C had a tensile strength of 200 MPa. Mahmoud et al. [26] performed mechanical testing of nanocomposite samples with the Al_2O_3 and Ni reinforcements in the following ratio: Al_2O_3 of 2, 3 wt.%, and Ni of 5, 10, 15 wt.%. The highest value of the tensile strength of 380 MPa was obtained for the composite with 2 wt.% Al_2O_3 and 5 wt.% Ni, while the significantly lower strengths of 300 MPa were obtained with the composite ratio of the reinforcement from 3 wt.% Al_2O_3 and 5 wt.% Ni.

Comparison of experimental and theoretical results for density was carried out by Muley et al. They used a composite made of only one variation of the mass fraction of the reinforcement [27], while Sharifi and Karimzadeh made a comparison between the density of the composites obtained by experimental paths and between the composites with multiple variations of the reinforcement [20]. Generally, the density values obtained by the experiments and theoretically are very close.

On the basis of the obtained results for density [28, 29] it is noticed that the density of the nanocomposite is mostly influenced by the content of the mass fraction of the material, as well as the choice of the material of the reinforcement because different reinforcements result in different densities of the final nanocomposite. With the increase in the mass fraction of CuO in the composite, the density increases [28]. The stretch test of the nanocomposite and comparison with the base of the nanocomposite [25, 26] led to the conclusion that increasing the reinforcement content may result in more or less elongation of the nanocomposite than in the base alloy.

Based on previous research and literature review, it is noted that the properties of the nanocomposite with the metal base are greatly influenced by the size, shape, uniform distribution, the hardening mechanism, as well as the thermal stability of the nano reinforcement [30]. Also, the great influence on the mechanical behavior of the material depends on the porosity of the materials which is need to be controlled in order to avoid undesirable properties [31, 32]. Researchers tend to achieve an improvement in mechanical properties without affecting ductility by applying certain quantities of particles at the nano level [33].

For nanocomposites strengthened by nanoparticles, strength and ductility can be improved at the same time if the particles are correctly or evenly distributed on the base. Unlike micron particles, nanoparticles with a similar and even smaller particle volume in the base can improve the mechanical properties of the nanocomposite [2, 34]. Moreover, in the previous research, the optimization methods have been observed in order to speed up the research process and optimize the processing parameters and conditions, as well as the nanocomposite testing. Temel Varol et al. [35] created a model for predicting the effect of SiC reinforcement size, reinforcement content, and grinding time on the density and strength of the nanocomposite AA2024-SiC by using a neural network. They have confirmed that the results obtained by the multilayer neural model have consistency with experimental values that they found on the basis of error in results which was less than 1%. The results of this study show that the neuronal network model can be used to predict the density and hardness of the AA2024 nanocomposite with great precision. Also, a powerful means of achieving appropriate data for predicting and optimizing the desired parameters in various processing technologies of advanced materials is possible by using the combination of the artificial neural networks (ANN) and genetic algorithms (GA) method [36] and many other optimization methods.

3. Experiment

3.1. Materials

A356 alloy is selected as the base of the nanocomposites because it possesses high mechanical strength, ductility, hardness, fatigue strength, fluidity and the possibility of light machining and good weldability. The chemical composition of the used hypoeutectic Al-Si alloy A356 is as follows: Si (7.20 wt.%), Cu (0.02 wt.%), Mg (0.25 wt.%), Mn (0.01 wt.%), Fe (0.18 wt.%), Zn (0.01 wt.%), Ni (0.02 wt.%), Ti (0.11 wt.%) and rest is Al.

To obtain the nanocomposite, prior to the compocasting procedure, a pre-treatment was applied - the mechanical alloying of the metal chips of the base alloy with nanoparticles of the SiC reinforcement of about 50 nm. One of the most important ceramic reinforcements for metal matrix composites are SiC particles. The importance of use of SiC particles in MMC is due to its properties such as good thermal stability, low density, high wear resistance, high melting point and high hardness. The SiC nanoparticles were washed in ethyl alcohol before the alloying procedure, and heated to a temperature of up to 400°C in order to free the moisture and possible chemical impurities. The metal chips of the A356 alloy was washed in trichlorethylene and then in ethyl alcohol for degreasing. A more detailed procedure for obtaining the nanocomposite with and the used apparatus is described in [18].

The mechanical properties of the A356 alloy can be improved by thermal treatment. In this study, the applied thermal treatment T6 implied the first solvent irradiation for 8 hours at a temperature of 525°C, followed by tempering in water at 30°C to room temperature. And finally, natural aging for 5 hours in the air or artificial aging (precipitation) for 8 hours at a temperature of 165°C. The goal of thermal treatment is to increase hardness and reduce toughness, to obtain an optimal combination of strength and toughness, as well as removing residual stresses.

3.2. Specimen preparation

The quality of the treated surface greatly influences the exploitation properties of the machine parts, as well as their lifetime. It is generally known that the surface of any machine element after finishing is never absolutely smooth. So before testing the material, it is necessary to prepare the surface of the samples after cutting the castings to certain dimensions.

The application of the grinding and polishing method tended to achieve certain roughness of the surfaces after the milling process. The grinding process was carried out by treating the contour surfaces with abrasive paper grit of 1000, 2000 and 3000, while the polishing was performed with an emulsion with abrasive grains of 1 μm in order to obtain surfaces of the appropriate quality. The preparation of the samples was carried out on the MetaServ 250 polisher. This polisher can be found in the paper [37] in which the analysis of tribological characteristics of nanocomposites was carried out. The computer-assisted device for testing of the mechanical properties of the tested materials is Nanoindenter and Micro Scratch Tester, and this device is shown in figure 1. Technical specification of the used nanoindenter is: maximum load 500 mN, load resolution 0.04 μN, maximum depth 40 μm (optional 200 μm), depth resolution 0.04 nm and movement of measuring table XY, 150 x 80 mm.

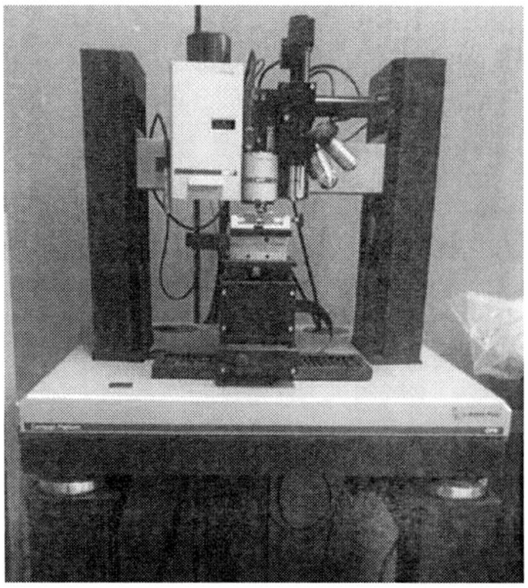

Figure 1. Nanoindenter and Micro Scratch Tester.

With this device is possible to continuously measure normal load and movement of the indenter during the hardness testing. Measurement of the elastic modulus and hardness using this device is possible based on the budget method developed by scientists Pharr and Oliver [38].

3.3. Nanoindentation

The mechanical properties of the nanocomposite and the A356 alloy are tested by penetrating the surface of the test sample. During these tests, Berkovich's three-sided diamond pyramid was used as an indenter. Experimental tests were carried out in the conditions shown in table 1. The chosen type of test is in the form of a matrix, and it is 3x3, which means tests were carried out in 9 points.

Table 1. Conditions of nanoindentation experiments.

Parameter	Unit	Value
Normal load	mN	50
Maximum load holding time	s	15
Loading speed	mN/min	100
Unloading speed	mN/min	100

In this study, the value of the normal load was 50 mN, because the previous research of these materials proved that force variation does not significantly affect the results of hardness. The value of Poison's coefficient in this test was taken as the value for aluminum and it is 0.33.

4. Results and discussion

As the results of nanoindentation a number of representative values can be obtained, in this paper the following will be shown: hardness (HV), elastic modulus (E) and value of maximum indentation depth (hmax). Also, based on the measurements carried out, in figure 2, the diagram of indentation is shown, which represents the dependence of the normal load of the indenter and the maximum depth of penetration. The illustrated diagram of indentation was selected according to the corresponding hardness values that were closest to the mean values of the hardness (figure 3).

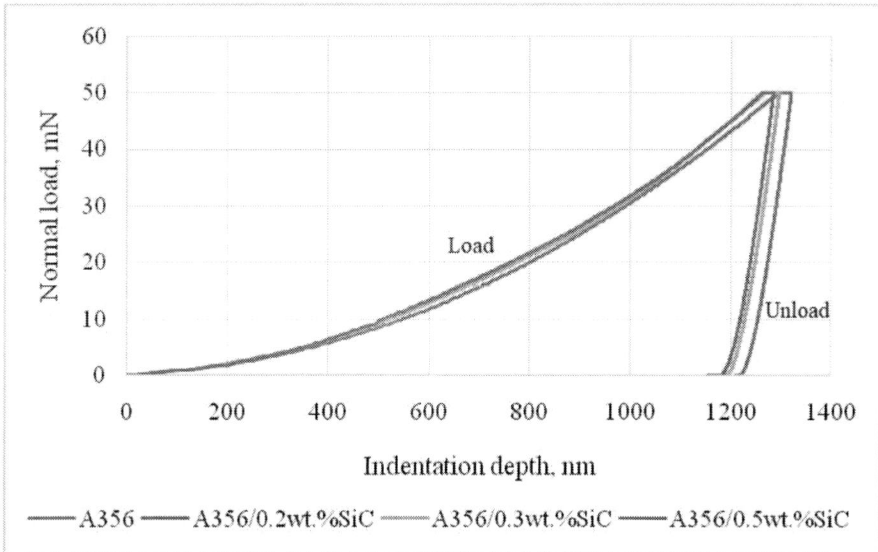

Figure 2. Nanoindentation curves for nanocomposites with SiC reinforcement and A356 alloy.

Based on the trend of indentation curves of all the examined materials, there is no large difference between the tested materials in the appearances of the curves. These curves show the elastic and plastic deformations during and after the process. The first is the result of the elastic deformation of the tested materials and it ends with the indentation depth about ~ 300 nm. Thereafter, there are no significant changes in the curve trends. Also, it is possible to notice two stages in the load process (load and unload), which are denoted in figure 2.

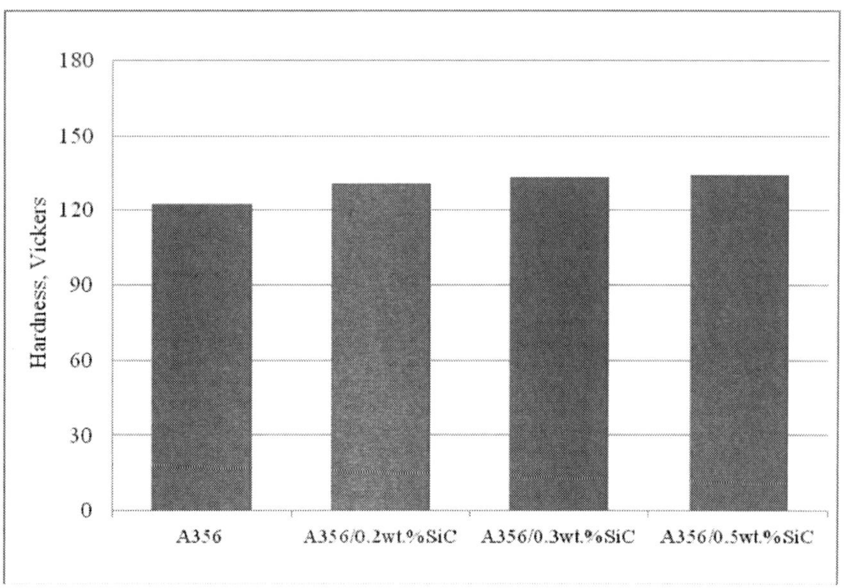

Figure 3. Experimental values of hardness.

Based on the curves obtained by nanoindentation of samples of A356 alloys and nanocomposites with reinforcement content of 0.2, 0.3 and 0.5 wt.% SiC, it can be noticed that the obtained values of the hardness of the nanocomposite are higher compared to the base alloy. The difference in the hardness of the material is evident from the difference in the indentation depths of the indenter into the test samples (figure 4), which means that the greater indentation depth is followed by a lower hardness value.

What has to be noted is that during the implementation of the experiment, it was taken into account that the matrix is positioned in such a way that it is approximately in the same position for each material, more precisely, the same number of imprints are in areas with and without the presence of eutectic silicon on the surface of the material. However, some of the results were not taken into consideration because there was much greater hardness, and this happened when the indenter hit the eutectic silicone. Previously mentioned was done to reduce the dispersion in the averaging of the test results.

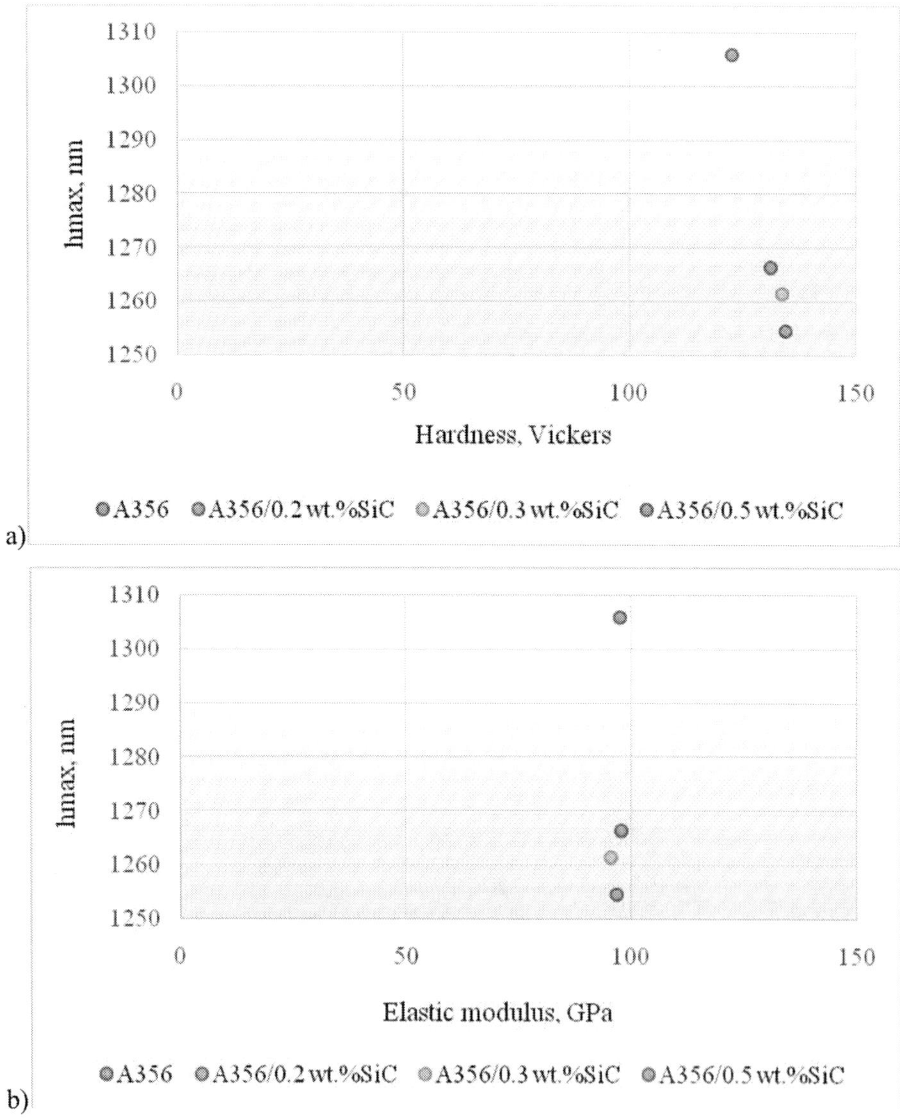

Figure 4. Mechanical properties as a function of the impression depth: (a) hardness; (b) elastic modulus.

Based on the results shown, for the elastic modulus, it can be said that the increase in the percentage content of the reinforcement does not greatly affect the value of the elastic modulus. It can also be concluded that the lower indentation depth indicates the strengthening of the base with silicon carbide, and Behzad Sadeghi and others have come to similar conclusions [39]. It is generally known that the presence of agglomeration and porosity has a negative impact on the mechanical properties of nanocomposite materials [40]. Analyzing the nanocomposite surfaces by SEM analysis in [37], no agglomeration of nanoparticles was observed, which confirms the increase in the hardness of these materials.

5. Conclusions

Hardness tests were conducted with the aim of analyzing the effect of nanoparticles of silicon carbide as a reinforcement in the A356 base. Based on the obtained values of the hardness in surface layer of the nanocomposites, a noticeable increase in hardness was observed in comparison to the base alloy of the nanocomposite. Increasing the reinforcement content of 0.2, 0.3 and 0.5 wt.% increases the hardness of the nanocomposites ~131, 133 and 134 Vickers, respectively. Also, it can be concluded that the elasticity of the material decreases with the increase of reinforcement content. When comparing the hardness and elastic modulus of the tested nanocomposites there is a small deviation in the results, which is due to the small difference in reinforcement content in the nanocomposites.

However, it has to be noted that the extreme values obtained for eutectic silicon have not been taken into consideration and this requires additional research because it has an impact on the hardness values of the surface layers of the material.

Acknowledgments

This paper presents the results obtained during research within the framework of the project TR 35021, supported by the Ministry of Education, Science and Technological Development of the Republic of Serbia.

References

[1] Shalaby E A, Churyumov A Y, Solonin A N and Lotfy A 2016 Preparation and characterization of hybrid A359/(SiC+ Si3N4) composites synthesized by stir/squeeze casting techniques, *Materials Science and Engineering: A* **674** 18-24

[2] Kandemir S 2017 Microstructure and mechanical properties of A357/SiC nanocomposites fabricated by ultrasonic cavitation-based dispersion of ball-milled nanoparticles, *Journal of Composite Materials* **51**(3) 395-404

[3] Sameezadeh M, Farhangi H, and Emamy M 2015 Nanocomposites of aluminum alloy-MoSi2: Synthesis and characterization, *Journal of Composite Materials* **49**(25) 3145-3155

[4] Shen M J, Wang X J, Li H, Zhang M F, Ying T, Nie K B and Wu K 2018 Effects of (micron+ submicron+ nano) multisized SiC particles on microstructure and mechanical properties of magnesium matrix composites, *Journal of Composite Materials* **52**(15) 2055-2064

[5] Penther D, Ghasemi A, Riedel R, Fleck C and Kamrani S 2018 Effect of SiC nanoparticles on manufacturing process, microstructure and hardness of Mg-SiC nanocomposites produced by mechanical milling and hot extrusion, *Materials Science and Engineering: A* **738** 264-272

[6] Uddin S M, Mahmud T, Wolf C, Glanz C, Kolaric I, Volkmer C, Höller H, Wienecke U, Roth S and Fecht H J 2010 Effect of size and shape of metal particles to improve hardness and electrical properties of carbon nanotube reinforced copper and copper alloy composites, *Composites Science and Technology* **70**(16) 2253-2257

[7] Prakash K S, Gopal P M, Anburose D and Kavimani V 2016 Mechanical, corrosion and wear characteristics of powder metallurgy processed Ti-6Al-4V/B4C metal matrix composites. *Ain Shams Engineering Journal* **9**(4) 1489– 1496

[8] Veličković S, Garić S, Stojanović B and Vencl A2016 Tribological properties of aluminium matrix nanocomposites, *Applied Engineering Letters* **1**(3) 72-79

[9] Veličković S, Stojanović B, Ivanović L, Miladinović S and Milojević S 2018 *Application of nanocomposites in the automotive industry*, International Congress Motor Vehicles &

Motors, Kragujevac, Serbia, Oct 4 - 5, pp. 323-332

[10] Reddy M P, Himyan M A, Ubaid F, Shakoor R A, Vyasaraj M, Gururaj P, Yusuf M, Mohamed A M A and Gupta M 2018 *Enhancing thermal and mechanical response of aluminum using nanolength scale TiC ceramic reinforcement,* Ceramics International **44**(8) 9247-9254

[11] Gao Y Y, Qiu F, Geng R, Chu J G, Zhao Q L and Jiang Q C 2019 Effects of nanosized TiCp dispersion on the high-temperature tensile strength and ductility of in situ TiCp/Al-Cu-Mg-Si nanocomposites, *Journal of Alloys and Compounds* **774** 425-433

[12] Mehrizi M Z, Mostaan H, Beygi R, Rafiei M and Abbasian A R 2018 Reaction Pathways of Nanocomposite Synthesized in-situ from Mechanical Activated Al–C–TiO_2 Powder Mixture, *Russian Journal of Non-Ferrous Metals* **59**(1) 117-122

[13] Qanavatian H, Mehrizi M Z and Raoufi M2018 Synthesis of Fe3Al–TiC-SiC Nanocomposite by Mechanical Alloying of FeSiTiO3–Al–C Powders, *Silicon* 1-7

[14] Javadi A, Pan S and Li X 2018 Fabrication of High Strength Al Nanocomposites with Populous TiB2 Nanoparticles *Procedia Manufacturing* **26** 629-632

[15] Singh H, Raina A and Haq M I U 2018 Effect of TiB_2 on Mechanical and Tribological Properties of Aluminium Alloys–A Review, *Materials Today: Proceedings* **5**(9) 17982-17988

[16] Bobić B, Vencl A, Ružić J, Bobić I and Damnjanović Z 2019 Microstructural and basic mechanical characteristics of ZA27 alloy-based nanocomposites synthesized by mechanical milling and compocasting, *Journal of Composite Materials* **53**(15) 2033-2046

[17] Javadi A, Cao C and Li X2017 Manufacturing of Al-TiB2 Nanocomposites by Flux-Assisted Liquid State Processing, *Procedia Manufacturing* **10** 531-535

[18] Veličković S, Stojanović B, Babić M, Vencl A, Bobić I, Bognár G V and Vučetić F 2019 Parametric optimization of the aluminium nanocomposites wear rate, *Journal of the Brazilian Society of Mechanical Sciences and Engineering* **41**(1) 19

[19] Rajmohan T, Palanikumar K and Arumugam S 2014 Synthesis and characterization of sintered hybrid aluminium matrix composites reinforced with nanocopper oxide particles and microsilicon carbide particles, *Composites Part B: Engineering* **59** 43-49

[20] Sharifi E M and Karimzadeh F 2011 Wear behavior of aluminum matrix hybrid nanocomposites fabricated by powder metallurgy, *Wear* **271**(7-8) 1072-1079

[21] Kwon H, Saarna M, Yoon S, Weidenkaff A and Leparoux M 2014 Effect of milling time on dual-nanoparticulate-reinforced aluminum alloy matrix composite materials, *Materials Science and Engineering: A* **590** 338-345

[22] Mousavian R T, Khosroshahi R A, Yazdani S, Brabazon D and Boostani A F 2016 Fabrication of aluminum matrix composites reinforced with nano-to micrometer-sized SiC particle, *Materials & Design* **89** 58-70

[23] Amouri K, Kazemi S, Momeni A and Kazazi M2016 Microstructure and mechanical properties of Al-nano/micro SiC composites produced by stir casting technique, *Materials Science and Engineering: A* **674** 569-578

[24] Tazari H and Siadati M H 2017 Synthesis and mechanical properties of aluminum alloy 5083/SiCnp nanocomposites,*Journal of Alloys and Compounds* **729** 960-969

[25] Poovazhagan L, Kalaichelvan K, Rajadurai A and Senthilvelan V 2013 Characterization of hybrid silicon carbide and boron carbide nanoparticles-reinforced aluminum alloy composites, *Procedia Engineering* **64** 681-689

[26] El-Labban H F, Abdelaziz M and Mahmoud E R 2016 Preparation and characterization of squeeze cast-Al–Si piston alloy reinforced by Ni and nano-Al2O3 particles, *Journal of King Saud University-Engineering Sciences* **28**(2) 230-239

[27] Muley A V, Aravindan S and Singh I P2015 Mechanical and tribological studies on nano particles reinforced hybrid aluminum based composite, *Manufacturing Review* **2** 26

[28] Baradeswaran A and Perumal A E 2014 Study on mechanical and wear properties of Al 7075/Al2O3/graphite hybrid composites, *Composites Part B: Engineering* **56** 464-471

[29] Yazdani B, Xu F, Ahmad I, Hou X, Xia Y and Zhu Y2015 Tribological performance of Graphene/Carbon nanotube hybrid reinforced Al2O3 composites, *Scientific reports* **5** 11579

[30] Kannan C and Ramanujam 2017 Comparative study on the mechanical and microstructural characterisation of AA 7075 nano and hybrid nanocomposites produced by stir and squeeze casting, *Journal of advanced research* **8**(4) 309-319

[31] Sadeghi B, Shamanian M, Ashrafizadeh F, Cavaliere P and Rizzo A 2018 Friction stir processing of spark plasma sintered aluminum matrix composites with bimodal micro-and nano-sized reinforcing Al2O3 particles, *Journal of Manufacturing Processes* **32** 412-424

[32] Mazahery A and Ostadshabani M 2011 Investigation on mechanical properties of nano-Al2O3-reinforced aluminum matrix composites, *Journal of composite materials* **45**(24) 2579-2586

[33] Ezatpour H R, Sajjadi S A, Sabzevar M H and Huang Y 2014 Investigation of microstructure and mechanical properties of Al6061-nanocomposite fabricated by stir casting, *Materials & Design* **55** 921-928

[34] Akbari M K, Baharvandi H R and Mirzaee O 2013 Fabrication of nano-sized Al2O3 reinforced casting aluminum composite focusing on preparation process of reinforcement powders and evaluation of its properties, *Composites Part B: Engineering* **55** 426-432

[35] Varol T, Canakci A and Ozsahin S 2018 Prediction of effect of reinforcement content, flake size and flake time on the density and hardness of flake AA2024-SiC nanocomposites using neural networks, *Journal of Alloys and Compounds* **739** 1005-1014

[36] Esmaeili R and Dashtbayazi M R 2014 Modeling and optimization for microstructural properties of Al/SiC nanocomposite by artificial neural network and genetic algorithm, *Expert Systems with Applications* **41**(13) 5817-5831

[37] Veličković S, Stojanović B, Vencl A, Babić M, Džunić D, Pantić M and Šljivić V 2019 *Tribological behaviour of A356/SIC nanocomposite*, 16th International Conference on Tribology, Kragujevac, Serbia, May 15 – 17, pp. 115- 123

[38] Oliver W C and Pharr G M 1992 An improved technique for determining hardness and elastic modulus using load and displacement sensing indentation experiments, *Journal of materials research* **7**(6) 1564-1583

[39] Sadeghi B, Shamanian M, Ashrafizadeh F, Cavaliere P and le Valerini D 2018 Nanoindentation characterization of Al-matrix nanocomposites produced via spark plasma sintering, *International Journal of Materials Research* **109**(1) 50-62

IOP Conf. Series: Materials Science and Engineering **659** (2019) 012040 doi:10.1088/1757-899X/659/1/012040

Extension of the steel sieve during the spherical gun-powder screening

N Bojić[1], R Nikolić[2], D Milčić[3], M Banić[3] and B Hadzima[2]

[1]FASIL A. D. Arilje, Svetolika Lazarevića 18, 31230 Arilje, Serbia
[2]University of Žilina, Univerzitna 8215/1, 010 26 Žilina, Slovakia
[3]University of Niš, Faculty of Mechnanical Engineering, Aleksandra Medvedeva 14, 18000 Niš, Serbia

E-mail: ruzicarnikolic@yahoo.com

Abstract. Influence of the steel sieve extension on screening of the spherical gun-powder was investigated. The objective of this research was to increase the screening productivity and reduce the production downtimes. The sieve's mechanical characteristics – elongation – test was performed on the tensile testing machine ZWICK ROEL 110. Through analysis of results for two sieve types the conclusion was drawn that the steel sieves with the smaller wire diameter are less extended. Their advantage lies in the fact that the cleaning ring, used to deblind the blocked mesh, can come closer to the sieve, due to the smaller clearance between the sieve's opening and the cleaning ring and thus clean it more efficiently. That additionally reduces the possibility for sieve the extension due to the residual gun-powder mass.

1. Introduction

The size distribution of particulate materials is very important in determining their properties (chemical, physical, mechanical etc.) in processes in various branches of industry and in numerous practical engineering operations - mining, metallurgy, dry mortar and artificial sand, production of food powders, chemicals, colorants, paints and pharmaceuticals. The sieves/screens are the oldest and most widely used working elements for separation of solid particles by size. The term screening is usually used for continuous sizing operation, while sieving usually means a batch process.

The aim of mechanical classification (screening) is to divide particulate material into groups of grains according to their sizes. For this purpose, sieves are equipped with either one or several screens. The selection of a correct sieve for a given particulate material determines the course of screening, [1].

A stack of sieves with decreasing mesh size is usually used for materials separation, which can be done in order from the coarse particles to fine, when the process is called the "sifter cascade". If the particles are sifted in a fine to coarse order, by the multiple sieving steps, with each step using a single sieve, such a process is called the reverse sieve method. Of all the elements of the sieving operation, the sieve blinding is considered as the most important and direct controlling factor. It occurs when particles block up and lodge in the sieving mesh, causing reduction of the effective transfer area on the sieve surface, resulting in reduction of the sieving rates (sieving performance or capacity), as well as of the degree of sharpness of particle separation (sieving efficiency).

Many factors have been identified that affect the screening, including the size and shape of particles relative to the aperture of the sieve, the mesh size of the sieve itself, the amount of material on the sieve surface, the direction of movement of the sieve, the rate of movement of the material relative to the sieve surface, [2].

Screening machines may be classified based on the direction of the particulate material flow through a screen into two main groups: screens with linear material flow through the machine and screens with either radial or spiral flow of the segregated material through the machine. The second class is based on the design configuration, i.e. circulating and rotary screens, screens with vibrating sieves, circulating screens with vibrating sieves, screens with spatial sieves motion and fluid-flow screening machines. The principal element of a screening machine is the vibrator, i.e. a device that provides the vibrations to the sieve, [3].

Vibrating screens of different types are the main sieving tool for large-scale separation and classification of solid particles by size. The sieving efficiency is an essential indicator for evaluation of the sieving performance. It is hard to be predicted due to the comprehensive effect of many parameters within the particle sieving process with multiple factors that influence the selection or determination of those parameters.

There are numerous researches on the sieving process by the DEM simulations (Discreet/Distinct Element Method) giving qualitative relation between the sieving efficiency and sieving parameters in a vibrating screen such as amplitude, vibration frequency, screen mesh size, particle size, and vibration direction angle has been analyzed, which provides references for in-depth study, [4-5].

The mesh aperture size is an important parameter in evaluation of the sieving efficiency. When the mesh aperture size is small, the sieving efficiency is low due to the lower penetration probability of the fine particles and a greater proportion of fine particles that remain in overflow particles and *vice versa*, when the mesh aperture size is larger, the sieving efficiency is higher due to the larger penetration probability of the fine particles. However, if the mesh aperture size is too large, the sieving efficiency becomes lower due to the higher penetration probability of the particles whose size is larger than the separation size and undersize particles are mixed with a greater proportion of larger particles, [2].

The vibro-sieve consists of the screen, which is fixed to the supporting frames by the cylindrical coil springs and the electro-mechanical vibratory system, which causes mechanical vibrations of the sieve that separates material to factions. The shaft on the rolling bearings is mounted at the screens COG axis and it has two eccentric flywheels at its ends. The following parameters are important for operation of a vibro-sieve: mass of material that is being screened, sieve's vibrations' amplitude, vibrations' frequency, the screens slope angle with respect to horizontal plane and the incoming angle of the excitation force, which acts on the screen. Change of the vibrations' amplitude influences the intensity of screening and this change is achieved by mutual turning of the eccentric flywheels, [6]. Change of the vibrations frequency, on the other hand, influences the quality of the gun-powder screening; it is realized directly by change of the rotational speed of the driving electric motor.

The sieves driving is realized through the electric motor and the belt transmission. Increase of the mass on the screen induces an increase of the electric motor power. Depending on the position along the technological production line, the sieves are divided into the primary and secondary ones.

To obtain the optimal vibro-sieve, a comparison is performed in [7] of the dynamic behavior of the vibro sieve's mathematical models. In the first model, an electric motor with imbalance was used, while in the other model such an electric motor was not used. It was noticed that the biggest deficiency of the system without the motor is just the inability of controlling the particles.

The spherical gun-powder is the nitro-cellulose gun-powder. The basic component is the nitro-cellulose to which the nitroglycerine and stabilizer diphenylamine are added. The whole production process is rigorously controlled, starting from the raw materials, via the semi-finished product up to the finished product, while all the tests are carried out on a daily basis, on the state-of-the-art devices, both in laboratories and on the testing grounds. The production process of the spherical gun-powder must be in accordance with requirements of the following standards: ISO 9001/2015, ISO 14001/2015 [8-9].

The gun-powder separation process, based on geometrical comparison of the shape and size of particles to shape and size of the openings on the sieve's surface is called screening. Machines at which the separation to fractions is done by mechanical vibrations, according to the granulation sizes, are the vibro-sieves. Number of fractions depends on the number of levels on the vibro-sieve. The screening can be done in the dry conditions or with washing with addition of the washing system.

2. Steel sieves

The steel sieve is presented in Figure 1. Due to the safety reasons, the sieve is shown in the still condition without water. The steel sieve is tightened on a specially made ring with a rim, which is attached to the metal drum using a shaft. There is a metal ring between the metal drum and the steel sieve, which has the role of a scrubber of the stale gun-powder. Technical documentation is not available due to manufacturer protection.

The average sieving area (screen) can be made in different ways and from different materials, depending on the material being sown. For the screening of spherical gun-powder, the most commonly used are steel screens with square or rectangular openings.

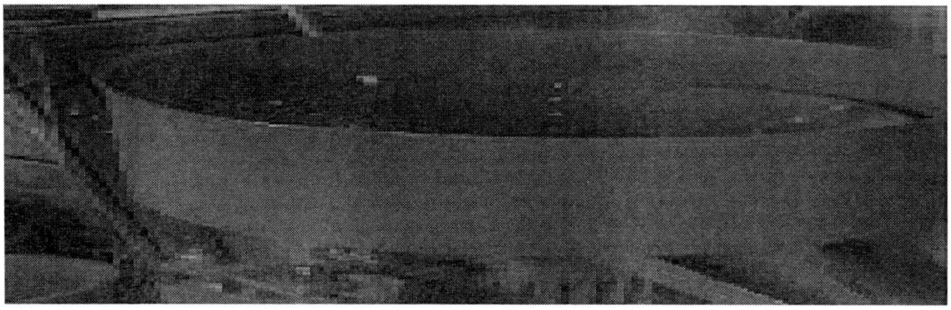

Figure 1. The vibro-sieve.

The sieve represents the wires interlace, which is formed by joining the two systems of wires by weaving, namely by alternate interlacing of the base and weft wires at the right angle. The longitudinal system of wires, which is parallel to the direction of the sieves motion during weaving, is called the *base*, while the *weft* is the lateral system of wires, placed at a right angle with respect to base. The interior dimensions of the sieve are combination of the weft wires diameters and their distances. The space formed between the two wires of base and two wires of the weft is called the sieve's eye or the opening, Figure 2.

The variables that define the sieve's dimensions, Figure 2, are: d – the wire diameter, w – the size of the opening, t – the step, N_0 – number of wires per cm, F_0 – the illuminated area. By knowing these variables, one can calculate the sieve's number, which represents the number of wires and openings, namely, the number of steps per unit length. Generally, there are two numbers: N_w – the weft number and N_0 – the base number; only in special cases are those two values equal. In the US and UK the sieve's number value is called the mesh. The sieve geometry is given in Figure 2.

Sieving of the gun-powder of the irregular form is a complex process and it influences the efficiency of the sieving if the powder particles size are approximately the same as the sieve's opening size, [10]. It should be said that a vibration sieve is an above resonant sieve with circular vibrations. Due to the increased impact oscillations, when spherical gun-powder is poured onto the vibration sieve, stretching, breaking, damage in some places and, in the worst case, breaking of the sieve occurs. There are several reasons for this extremely negative effect: low mechanical resistance of the sieve, inadequate fixing of the sieve's structure, poor construction and small sieve thickness, low stiffness of the screen, inadequate mounting on the construction, the high impact and vibration load and long-term operation under unfavorable conditions, [11-12]. Therefore, in this paper, the mechanical characteristics of the sieves, i.e. the influence of one of the parameters, is analyzed, which is the stretching (extension) of the steel sieve when the spherical gun-powder is being screened.

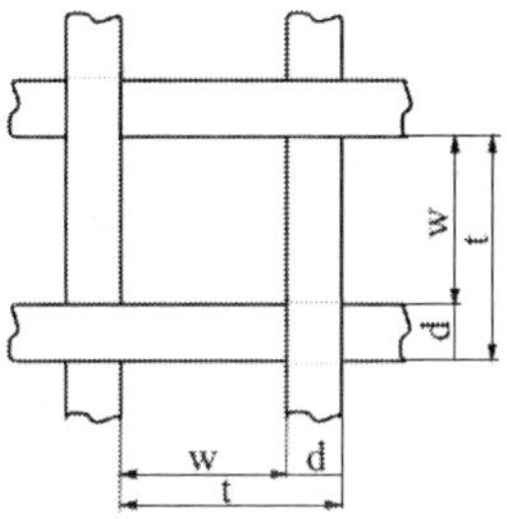

Figure 2. The sieve's geometry.

3. Experimental setup and results

The two metal single-layered sieves were used in this experiment of the same type - TYPE 110, but with different numeration. Data on the used wire types are given in Table 1. Material for all the wires was WNr. 1.4301. Technical characteristics of the tested sieves are given in Table 2.

Table 1. Basic data on two types of sieves.

	Sieve Type 1	Sieve Type 2
Base number N_o (wires per cm)	5	30
Base wire diameter (mm)	0.5	0.15
Weft number N_w (wires per cm)	5	30
Weft wire diameter (mm)	0.5	0.15

Table 2. Technical characteristics of the tested sieves.

Number of screens (steps)	2
Single screen area (m^2)	2.4
Electric engine power (kWh)	4
Sieve's maximum length (mm)	2590
Sieve's maximum width (mm)	1500
Sieve's maximum height (without a stand) (mm)	1570

The following data were recorded during the test:

- a_0 – sample thickness
- b_0 – sample width
- L_0 – sample length
- F_{max} – maximum force
- F_{break} – breaking force
- ε_{Fmax} – extension at maximum force
- ε_{break} – extension at break.

Test results for the sieve of type 1 are given in Table 3 and in Figure 3 and for the sieve of type 2 in Table 4 and Figure 4. Two samples were tested for each sieve type.

Table 3. Test results of the sieve TYPE 110 N_o 5.

Sample number	a_0	b_0	L_0	F_{max}	F_{break}	ε_{Break}	ε_{Fmax}
	mm	mm	mm	N	N	%	%
1	0.3	1.1	150.20	2064.08	1928.44	21.84	21.56
2	0.3	1.1	150.34	2218.06	2205.97	22.17	22.06

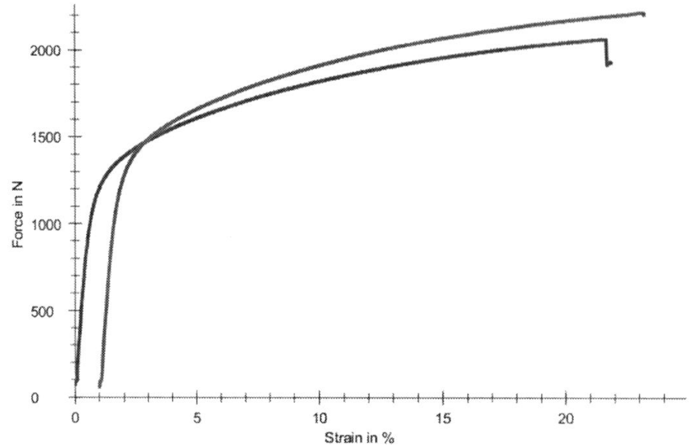

Figure 3. The stress-strain diagram for the sieve TYPE 110 N$_0$5.

Table 4. Test results of the sieve TYPE 110 N$_o$ 30.

Sample number	a_0	b_0	L_0	F_{max}	F_{break}	ε_{Break}	ε_{Fmax}
	mm	mm	mm	N	N	%	%
1	0.3	30	150.19	1077.95	969.28	6.32	6.08
2	0.3	30	150.12	1160.74	1151.64	9.19	9.12

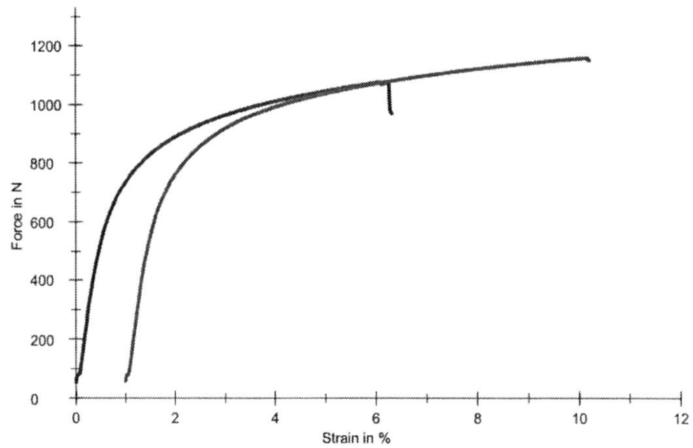

Figure 4. The stress-strain diagram for the sieve TYPE 110 N$_0$30.

4. Discussion of results and conclusions

From obtained results one can see that the sieve of type 1, i.e. TYPE 110 N$_0$5, has bigger extension for about 50 % and 50 % bigger breaking force that the sieve of type 2, i.e. TYPE 110 N$_0$30. This corresponds to the results from the technical practice that it is necessary to use the sieves with the smaller extension %. The reason for this requirement, namely for the sieve to have as small extension (stretching) as possible, lies in the fact that this guarantees better screening (sifting). The system with rinsing of the sieve with water under compression, which is usual in sieving the gun-powder, can cause the blinding (clogging) of the sieve's openings, creating a powder residue, what has as a consequence reducing of the sieving area, thus decreasing the sieving efficiency. In addition, the gun-powder residue increases the load on the sieve, thus causing its additional stretching. The stretched sieve does not lie in a single plane with the whole surface, i.e. a certain portion of its area is inclined with respect to the original surface, what additionally aggravates the screening. As the openings are reduced due to inclination, the percentage of the screened gun-powder is decreased. Thus, one can conclude that out of the two tested sieves, the better one to be used is the sieve with the smaller wire diameter and the higher number, i.e. the sieve of type 2 - TYPE 110 $N_0$30.

Acknowledgements
Research presented in this paper was partially financially supported by the project of Operational Program Research and Innovation: "Research and development activities of the University of Zilina in the Industry of 21st century in the field of materials and nanotechnologies", No. 313011T426, co-funded by European Regional Development Fund.

References

[1] Liu K S 2009 Some factors affecting sieving performance and efficiency, *Powder Technology* **193**(2) 208-213
[2] Zhang B, Gong J, Yuan W, Fu J and Yi H 2016 Intelligent Prediction of Sieving Efficiency in Vibrating Screens, *Shock and Vibration* **2016** Article ID 9175417, 7 pages
[3] Lawinska K, Modrzewsli R 2017 Analysis of sieve holes blocking in a vibrating screen and a rotary and drum screen, *Physicohemical Problems of Minerals Processing* **53**(2) 812-828
[4] Grozubinsky V, Sultanovitch E and Lin I J 1998 Efficiency of solid particle screening as a function of screen slot size, particle size, and duration of screening, *International Journal of Mineral Processing* **52**(4) 261-272
[5] Chen Y H and Tong X 2010 Modeling screening efficiency with vibrational parameters based on DEM 3D simulation, *Mining Science and Technology* **20**(4) 615-620
[6] Bojic N and Stojanovic N 2017 *Key performance indicators for production of the screens for paper and pulp industry*, XXII international symposium in the field of pulp, paper, packaging and graphics, Zlatibor, Serbia, June 13 - 16, 59-62
[7] Bojic N and Stojanovic N 2018 *Risk management of steel and plastic wires used for construction of formatted and dried sieves*, XXIII international symposium in the field of pulp, paper, packaging and graphics, Zlatibor, Serbia, June 19 - 22, 101-106
[8] Djoković J M, Tanikić D I, Nikolić R R and Kalinović S M 2017 Screening efficiency analysis of vibrosieves with the circular vibrations, *Civil and Environmental Engineering* **13**(1) 77-83
[9] Milisavljević J, Ćirić I, Petrović E and Djekić P 2011 *Mathematical models behavior of vibrating sieve with and without electric motor with imbalance*, 28th Danubia-Adria Symposium on Advances in Experimental Mechanics, Siofok, Hungary, Sept. 28 - Oct. 01, 239-241
[10] Trumić M S and Trumić M Ž 2012 *Analysis of Sieving Kinetics of Mineral and Secondary raw Materials*, 7th Symposium Recycling Technologies and Sustainable Development, Soko Banja, Serbia, Sept. 5 - 7, 50-55 (In Serbian)
[11] Bojić N V, Nikolić R, Jugović B Z, Jugović Z S and Gvozdenović M 2013 Uniaxial tension of drying sieves, *Chemical Industry* **67**(4) 655-662
[12] Bojic N, Nikolic R, Banic M and Hadzima B 2018 Evaluation of mechanical properties of the two PVC conveyor belts, *Communications* **20**(4) 47-51

Effect of nanosized particles on the bainitic transformation in austempered ductile irons

V P Mishev[1], J A Kaleicheva[1], V C Anchev[1] and Z K Karaguiozova[2]

[1] Faculty of Industrial Technology, Technical University of Sofia, 8 Kl. Ohridski Blvd, 1000 Sofia, Bulgaria
[2] Space Research and Technology Institute, BAS, Acad. Georgy Bonchev Str. bl.1, 1113 Sofia, Bulgaria

E-mail: jkaleich@tu-sofia.bg

Abstract: The austempered ductile iron was austenitized at 900°C for 1 h and quenched in an isothermal quenching furnace at 380 °C, 0,5 h, 1 h, 2 h, 4 h, 6 h. This paper aims to investigate the effects of nanosized particles (50nm) of titanium carbite+titanium nitride (TiCN+TiN), titanium nitride (TiN) and cubic boron nitride (cBN) on the bainitic transformation in austempered ductile irons (ADI). The samples upper bainitic microstructure is studied by metallographic microscope (OM), X-ray diffractometer (XRD) and scanning electron microscope (SEM). The influence of the nanosized additives on the kinetic of the bainitic transformation and on the morphology of the bainitic structure is investigated. The abrasive wear testing and hardness measurements are carried out. It is established that the presence of nanoadditives in the bainitic cast irons leads to the changes in their microstructure which increases their abrasive wear resistance. The studied nanocomposite materials expand the potential for new ADI applications in the industry.

1. Introduction

Bainitic cast iron is a structural material characterized by high strength, toughness and wear resistance. It is obtained by heat treatment involving isothermal quenching in the bainitic area [1]. Depending on the isothermal quenching mode, cast iron with a structure of lower bainite (220÷280°C) or upper bainite (350÷450°C), which have high strength and increased toughness, are produced. Graphite present in the cast iron provides resistance to mechanical wear and improves workability. Changes in the composition of the cast iron, changes in the heat treatment regime, or a combination of these two processes can influence the morphology of the graphite and the structure of the metal base in order to obtain castings with optimal properties [2].

Isothermal quenching in the bainitic area is widely used in the heat treatment of structural steels and ductile cast iron [1,3]. Incomplete isothermal quenching is also applied to the heat treatment of some hypereutectoid and ledeburite steels. The possibility of wider practical application of this type of heat treatment requires additional data on the bainitic conversion of iron-carbon alloys of different composition, including alloys with nanosized particles.

Nanosized particles added to the casting of cast iron in small quantities can alter the graphite morphology, also increase the amount of graphite and reduce the size of the graphite sphere, alter the structure of the metal matrix of the cast irons and austempered ductile irons (ADI), which increases the mechanical and tribological properties of the cast irons [4÷7].

The aim of this study is to investigate the kinetics of the bainitic transformation, the bainitic structure morphology in the upper temperature range of the bainitic area, the hardness and the wear resistance of the austempered ductile cast irons (ADI) containing additives of nanosized particles - titanium carbonitride + titanium nitride (TiCN + TiN), titanium nitride (TiN) and cubic boron nitride (cBN).

2. Materials and investigation methods

The samples tested are made of unalloyed ductile cast iron with the following composition: Fe-3,55C-2,67Si-0,31Mn-0,009S-0,027P-0,040Cu-0,025Cr-0,08Ni-0,06Mg wt.%. During the casting in the cast iron melt nanosized particles of *TiCN+TiN, TiN* and *cBN* (table 1) are added. The nanosized particles are coated by electroless nickel coating EFTTOM-NICKEL prior to the edition to the melt [9]. The nickel coating improves the particles wetting into the melt and their uniformity distribution into the casting volume.

Table 1. Retained austenite quantity *A* and hardness *HV10* of austempered ductile irons (ADI).

№ of sample	Nanosized additives	Time of austempering									
		0,5 h		*1 h*		*2 h*		*4 h*		*6 h*	
		A (%)	*HV10*	*A (%)*	*HV10*	*A (%)*	*HV10*	*A (%)*	*HV10*	*A (%)*	*HV10*
1.1÷1.5	-	25.1	297	29.3	306	40.4	320	23.1	313	20.0	321
2.1÷2.5	*TiCN+TiN*	26.7	314	26.8	304	27.1	322	12.0	309	7.9	318
3.1÷3.5	*TiN*	25.3	294	26.0	292	31.8	312	19.3	317	19.4	311
4.1÷4.5	*cBN*	26.2	298	27.9	305	30.2	308	19.3	314	13.6	301

Bainitic cast irons are obtained by isothermal quenching in the upper temperature range of the bainitic area, including heating at 900°C for an hour, after that isothermal retention at 380°C, 0,5 h, 1 h, 2 h, 4 h, 6 h. The austempered ductile iron samples' microstructure is observed by means of an optical metallographic microscope GX41 OLIMPUS. The samples surface is treated with 2 vol. % solution of HNO3 in ethanol. The microstructure of the austempered ductile irons (ADI) is tested by scanning electron microscopy (SEM). The scanning microscope EVO® MA10 „Carl Zeiss" is used. The hardness testing is performed by Vickers method.

The austempered ductile iron samples are tested by X-Ray diffraction analysis the retained austenite quantity in the structure to be defined. X-ray powder diffraction patterns for phase identification are recorded in the angle interval 22 ÷104° (2θ), on a Philips PW 1050 diffractometer, equipped with Cu–Kα tube and scintillation detector. Quantitative analysis is carried out by BRASS - Bremen Rietveld Analysis and Structure Suite [10].

For the samples with an upper bainitic structure obtained for 2 hours isothermal retention at 380°C (samples 1.3; 2.3; 3.3; 4.3;) an abrasion test is performed.

The experimental study of the wear is carried out using method and device for accelerated testing in kinematic scheme "thumb-disc" under friction over a fixed abrasive [6]. The impregnated material Corundum 220 is used for the austempered ductile cast iron samples. The test data are: nominal contact pressure, $Pa=0,4.10^6$ (Pa); average sliding speed, V=24,5(cm/s); nominal contact surface, Aa=50,24 (mm^2); density, $\rho=7,80.10^3$ (kg/m^3).

3. Experimental results and discussion

In the cast condition, the unalloyed ductile cast iron has a structure consists of ferrite, perlite and graphite. The quantitative metallographic analysis carried out show that nano-sized particles influence the amount and size of the graphite phase without altering the shape of the graphite.

They decrease the average diameter of the graphite sphere Dmid from 11,00 to 10,34 μm and increase the quantity of the graphite phase with 35÷94% [5].

SEM analysis of the fracture of the impact destructed cast iron sample with nanoparticles additives, show the nanoparticles presence in the graphite [6].The results from SEM analysis and quantitative metallographic analysis prove the modifying effect of the nanosized particles on the graphite phase in the cast iron [5, 6].

The spheroidal graphite cast irons with and without nanosized additives are subjected to austempering in order to obtain a bainitic structure of the metal base. The austempered ductile iron is austenitized at 900°C for 1 h and quenched in an isothermal quenching furnace at 380°C, 0,5 h, 1 h, 2 h, 4 h and 6 h. As a result of this heat treatment, the cast iron obtains an upper bainitic structure 'figure 1'.

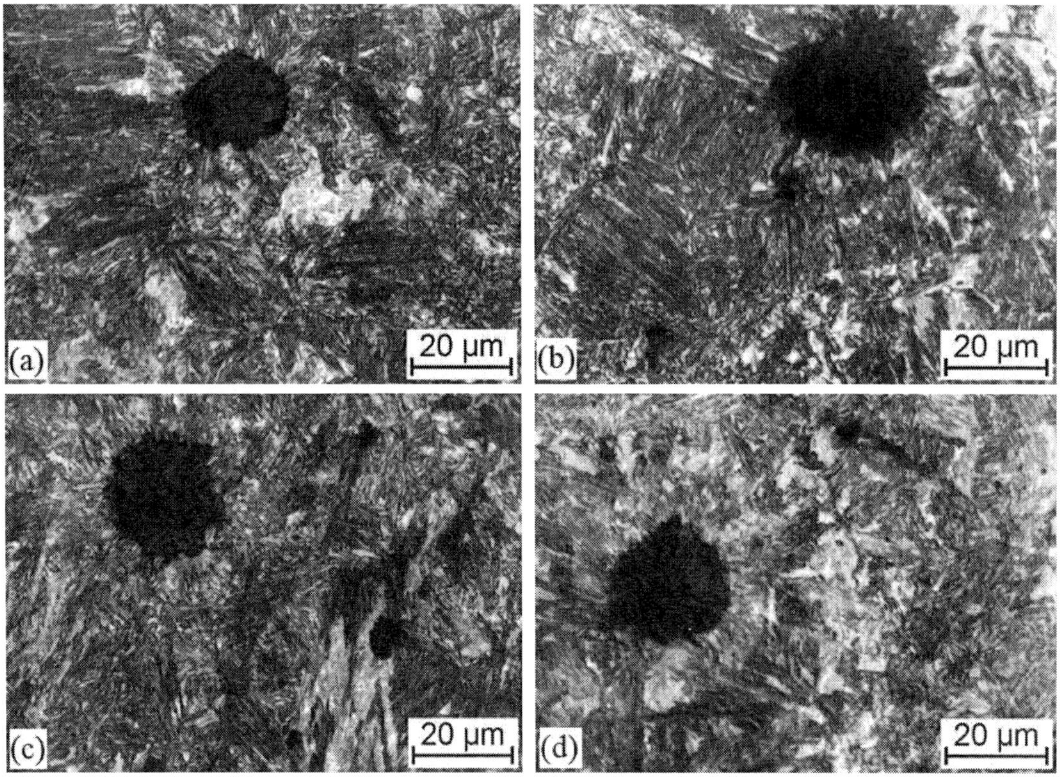

Figure 1. Upper bainitic structure in ADI samples without nanoadditives (a) and with nanoadditives of (TiCN+TiN) (b), TiN (c) and cBN (b) after isothermal retention for 4 hs at 380°C.

The bainite is an oriented structure consisting of α-phase needles (bainitic ferrite), carbides and non-converted austenite. The α-phase is formed by a martensitic mechanism from the austenitic regions of low carbon content [1,3]. During cooling from the temperature of the isotherm to a room temperature, part of the non-converted austenite undergoes martensitic transformation, and another part remains in the structure as retained austenite A. The bainitic transformation of the austenite begins with the formation of separate α-phase needles (bainitic ferrite) and develops with the formation of new oriented needles placed close to each other and forming a package of alternating α-phase plates and non-converted austenite enriched with carbon A (c) [1,3].

In an optical metallographic analysis, this package appears as a separate needle in the lower bainitic structure. In the upper bainitic structure α-phase and A(c) in the package are perceived as separate phases. The carbide phase is formed as a result of self-tempering of the α-phase or directly from A(c). Silicon in ductile cast iron (2 ÷ 3%) hinders the process of a carbide formation. Isothermal retention up to 2 ÷ 4 hours produces a formation of a bainitic ferrite structure and carbon-enriched non-converted austenite A(c). This structure is characterized by high mechanical properties. At austempering modes over 4 ÷ 6 hours, it is possible to observe carbides separation directly from A(c) or decomposition of the carbon-enriched austenite A (c) to the ferrite-carbide mixture (α + carbide), which reduces the mechanical properties of the cast iron and it is not actually carried out in practice [1,3].

Nanosized particles accelerate the transformation of the austenite to bainite. The amount of the retained austenite in the samples without nanoparticles additives for 2 hours isothermal retention at 380°C is 40.4%. In the samples with nanoadditives, the amount of the retained austenite decreases to 30.2÷27.1% for 2 hours of isothermal retention at 380°C 'figure 2'.

With the development of the bainitic conversion from 2 to 6 hours, the amount of the retained austenite in all samples tested decrease, (table 1, figure 2). The hardness of the tested ADI samples is in the range from 294 to 322 HV10 (table 1).

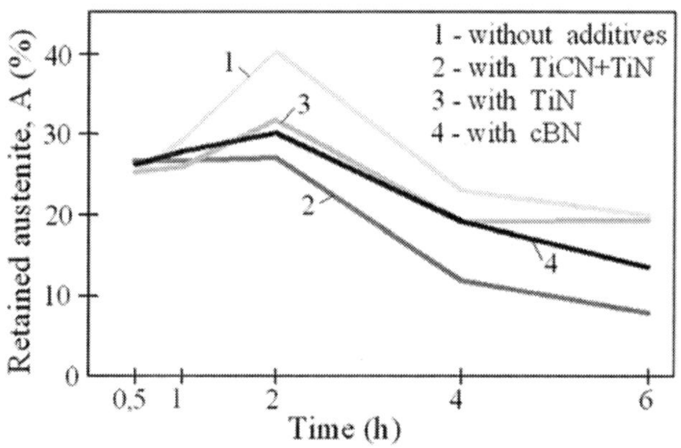

Figure 2. Dependence of the retained austenite quantity *A* on the time of austempering *h* at 380°C for austempered ductile irons (ADI) without and with nanoadditives of (TiCN+TiN), TiN and cBN.

After completion of the isothermal quenching and final cooling to ~20°C, a complex structure is formed consisting of bainitic ferrite (α-phase), carbides, retained austenite and martensite formed by the unreacted carbon-enriched austenite A (c), with cooling from 380°C to room temperature (~20°C). The properties of austempered ductile irons (ADI) depend on the type of phases and their quantity in the structure of ADI.

'Figure 3' shows a microstructure observation by SEM analysis of the fracture of the impact destructed sample 2.3 with nanoparticles (TiCN + TiN) and with upper bainitic structure obtained by isothermal quenching at 900°C, 1h + 380°C, 2h.

For the samples after an isothermal retention for 2 hours at 380 ° C, an abrasion resistance test is conducted (table 2). An increase in the wear resistance for ADI specimens with nanoadditives is found compared to those without nanoadditives 'figure 4'. The amount of the retained austenite in the cast iron samples is determined before and after the tribological test 'figure 4'.

The austempered ductile iron samples hardness with an upper bainitic structure varies from 308 to 322 HV10 for 2 hours of isothermal retention at 380°C (table 2).

The standard measured properties (hardness, etc.) not always are reliable criteria for the steel and cast iron wear resistance. The tribological properties of the metallic materials significantly depend on the structural condition, forming on the contact surface during the friction. When the material is exposed to an intensive plastic deformation the structural transformation goes off in the frictional contact area in the metastable structures (retained austenite, martensite, bainite). They greatly influence on the effective surface strength and respectively on the material tribological properties.

During friction the retained austenite transforms partly into strain-induced martensite with the same amount of carbon as in high carbon austenite. This strain-induced martensite is untempered martensite characterised with a high hardness and ability for intensive strengthening by wear [7,11].

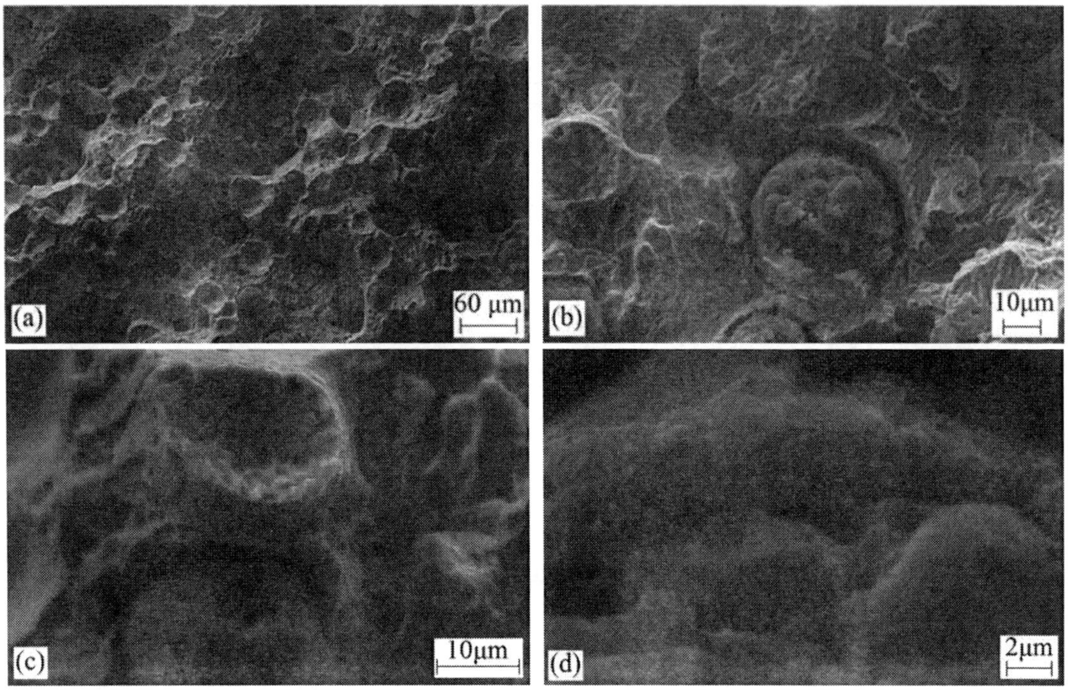

Figure 3. SEM analysis of the fracture of the impact destructed sample 2.3 from ADI with nanoadditives of (TiCN+TiN) at different magnification (a,b,c,d).

Table 2. Hardness HV10, wear resistance I and retained austenite quantity A of austempered ductile irons (ADI).

№ of sample	Austempering mode	Hardness *HV10*	Wear resistance *I*	Retained austenite *A* (%)	
				before wear test	*after wear test*
1.3	900°C, 1 h + 380°C, 2 h	320	$7.67.10^6$	40.4	31.3
2.3		322	$9.42.10^6$	27.1	11.6
3.3		312	$7.72.10^6$	31.8	31.2
4.3		308	$8.03.10^6$	30.2	25.3

The amount of the retained austenite decreases in all ADI samples tested after the tribological test. Mostly, the amount of the retained austenite decreases in the sample with (TiCN + TiN) nanoparticles - from 27.1% to 11.6% (table 2, figure 4). This specimen also has the highest abrasion resistance (I=$9.42.10^6$).

The formation of strain-induced martensite from the metastable retained austenite in the frictional area (microtrip-effect), probably is one of the reasons for the wear resistance increase of these materials.

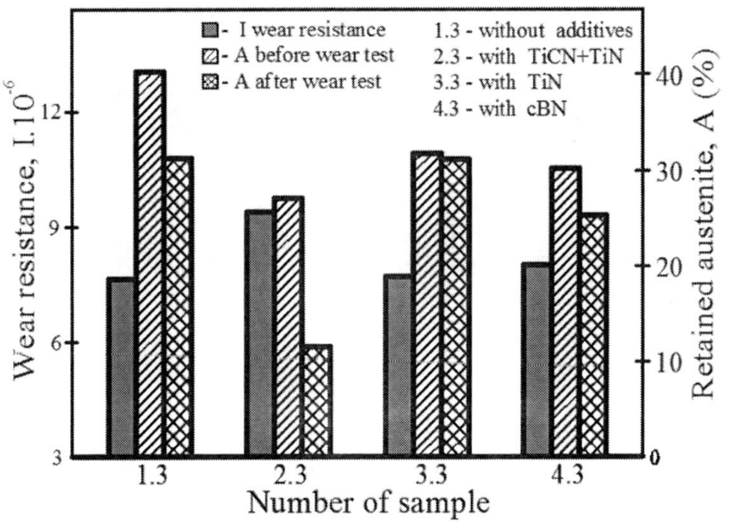

Figure 4. Wear resistance *I* and retained austenite *A* of austempered ductile iron samples without nanoadditives (1.3) and with nanoadditives of (TiCN+TiN) (2.3), TiN (3.3) and cBN (4.3) after isothermal retention for 2 h at 380°C.

4. Conclusions

Nanosized additives in the spheroidal graphite cast irons influence on the structure formation in the temperature range of the bainitic area. They alter the kinetics of the bainitic transformation and accelerate the transformation of the undercooled austenite into bainite.

Austempered ductile irons (ADI) with nanosized additives possess higher wear resistance (until 23%) in comparison to the samples without nanoadditives. The influence of the nanoadditives on the graphite phase characteristics and on the extent of the transformation of the austenite to bainite explain higher abrasion wear resistance of the tested austempered ductile irons with nanoadditives compared to the same without nanoadditives. Partially transformation of the metastable retained austenite to strain-induced martensite during friction which is observed in the austempered ductile irons also affects their wear resistance.

Acknowledgment

The authors would like to thank the Research and Development Sector at the Technical University of Sofia for the financial support.

References

[1] Dorasil E 1985 *High-strenght Bainitic Nodular Cast Iron*, Prague

[2] Kandeva M, Ivanova B, Karastoyanov D and Assenova E 2014 *Static and Dynamic Friction of Sphero-graphite Cast Iron with Sn Microalloy*, Proceedings: 2nd Austrian-Indian-Symposium on Materials Science and Tribology, Wiener Neustadt 35-39

[3] Bhadeshia H K D H 2001 *Bainite in Steels* 2st ed. London

[4] Wang Y, Pan Z, Wang Z, Sun X and Wang L 2011 Sliding wear behavior of Cr–Mo–Cu alloy cast irons with and without nanoadditives, *Wear* **271** 2953-2962

[5] Kaleicheva J, Mishev V, Lasarova R and Karaguiozova Z 2015 Study of Spheroidal Graphite Cast Irons with Nanosized Additives, *Nanoscience & Nanotechnology'15* **15**(1) 40-43

[6] Kalicheva J, Mishev V and Karaguiozova Z 2019 Tribological characteristics of austempered ductile irons with nanosized additives, *Journal of the Balkan Tribological Association* **25**(2) 280–290
[7] Kaleicheva J, Mishev V, Avdeev G, Karaguiozova Z and Dineva B. 2014 *Influence of Nanoadditives on the Structure and Properties of Austempered Ductile Irons*, Conference Proceedings: European Conference on Heat Treatment and 21st IFHTSE Congress, Munich 537-543
[8] Sun X, Wang Y, Li D Y and Wang G 2013 Modification of carbidic austempered ductile iron with nano ceria for improved mechanical properties and abrasive wear resistance, *Wear* **301** 116–121
[9] Gavrilov G and Nicolov C 1985 *Electroless Nickel and Composite Coatings*, Sofia
[10] Birkenstock J, Fischer R X and Messner T 2003 The Bremen Rietveld Analysis and Structure Suite, *BRASS 2003* **15**(1) 21
[11] Wen F, Zhao J, Zheng D, He Ke, Ye W, Qu S and Shangguan J 2019 The Role of Bainite in Wear and Friction Behavior of Austempered Ductile, *Materials* **12**(5) 767, DOI: 10.3390/ma12050767

Parameter optimisation and failure load prediction of resistance spot welding of aluminium alloy 57547

A Đurić[1], D Klobčar[2], D Milčić[3] and B Markovic[1]

[1]University of East Sarajevo, Faculty of Mechanical Engineering, Vuka Karadžoća 30, 71123 East Sarajevo, RS, B&H
[2]University of Ljubljana, Faculty of Mechanical Engineering, Aškerčeva cesta 6, 1000 Ljubljana, Slovenia
[3]University of Niš, Faculty of Mechanical Engineering, Aleksandra Medvedeva 14, 18106 Niš, Serbia

E-mail: aleksija.djuric@ues.rs.ba

Abstract. This paper will also present single objective optimization and Failure load prediction of Resistance spot welding of Aluminium alloy 57547. The experimental studies were conducted under varying welding currents I, electrode forces F, welding times T, pred preheating currents IA. The settings of welding parameters were determined by using the Taguchi experimental design of L9 Orthogonal array method. For optimization and prediction will be used analysis of Signal-to-Noise (S/N) ratio and Response surface modelling RSM.

1. Introduction

Resistance spot welding (RSW) is the most commonly used method of joining in the automotive industry, more than 60% of the total number of joining technology is by the RSW [1], so one car has, for example, over 5000 RSW points [2] and each car factory has more than 200 RSW welding machines [3]. Traditionally, steels have been the material of choice for the fabrication of automobile structures. However, in order to respond appropriately to economic and environmental requirements for lighter but faster vehicles, many automobile manufacturers are re-directing their research and development efforts towards advanced high strength steels (e.g. TRIP steels and dual phase steels) and aluminium alloys [4]. Aluminium alloy has already gained popularity among automakers, because it is lightweight and virtually non-corrosive. Aluminium car bodies have already reached the level of mass production, although only on expensive models. Many parts of the chassis have been mastered from aluminium instead of steel, as well as lighter components [5]. Recently several groups of aluminium alloys are applied in car manufacturing, among them for example AA5754 H22 and AA6082 T6. Concerning for example AA5754 (AlMg3) alloy, it is widely used in sports cars. This alloy has medium strength among aluminium alloys [6]. Therefor, a lot of author in different paper analysing the joining aluminium alloy using RSW, what is shown in review paper of S. M. Manlada and el [7].

Zhen Luo and el. [8] were analysing application of pre-heating to joining two sheet from aluminium alloy AA5052 and they conclude that all parameters need to be studied and optimized for specific application such as different aluminium alloys, thicknesses, or same aluminium alloy but with different aging time, which will have different oxide layers on the surface of the sheet. Harry Atherton and el. [4] conclude that mechanical properties of resistance spot welded structures from AA5754 are dependent on the geometry of the spot weld distribution.

The parameter settings of each welding machine have been difficult because there are many sensitive factors [3]. Therefore, it is necessary to analyse the parameters affecting the quality and mechanical properties of the RSW joint using optimization method. There are plentiful literature about optimisation and lot of papers who dealt with parameter optimisation for RSW. Most of them have been done on single objective optimization, what is showed in in review paper of Zoha Nasir and M. I. Khan [9]. Karthikeyan R. and el [10] applies response surface methodology (RSM) to explore the effects of process parameters of spot welded aluminium alloy AA2024-T3 on the Tensile shear strength. Arick. M. Lakhani and Darji P.H. [11] in them research use Taguchi to generate design matrix for experimental work.

This paper will also present single objective optimization and failure load prediction of Resistance spot welding of aluminium alloy 57547. The experimental studies were conducted under varying welding currents *I*, electrode forces *F*, welding times *T*, pred preheating currents I_A. The settings of welding parameters were determined by using the Taguchi experimental design of L9 Orthogonal array method. For optimization and prediction will be used analysis of Signal-to-Noise (S/N) ratio and Response surface modelling RSM.

2. Experimental set-up

Sheet metals of aluminium alloy AlMg3 - AW5754 H22 as the parent metal to be lap welded were used in this research. The dimensions of the specimen are defined according to the standard ISO 14273:2016 (figure 1).

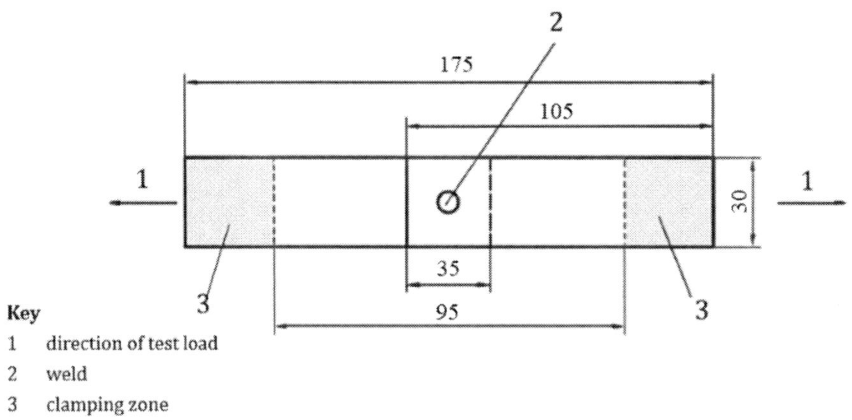

Figure 1. Dimensions of specimen.

The experiment involved joining of two sheet metals using RSW machine manufactured by Kocevar & sinovi which is managed using the BOSH 6000 software. Welding was carried out using electrode type A0 with radius 75 mm. The welding machine and dimension of electrode obtained from standard ISO 5821:2009, are showed on figure 2.

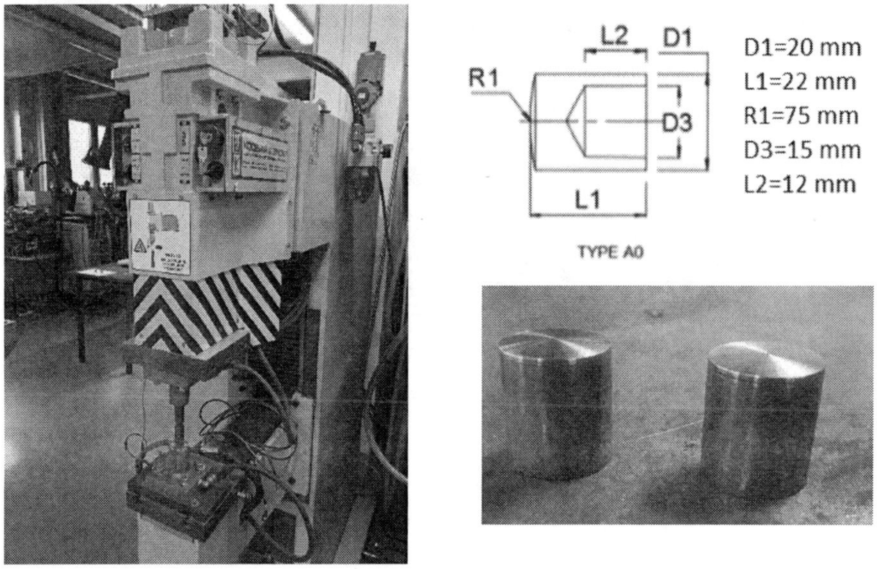

Figure 2. The welding machine and dimension of electrode.

Four welding parameters such as weld current I, electrode force F, weld time T and preheating current I_A were selected for experimentation for three levels of factors. The value of welding process parameter at different levels is tabulated in Table 1. Other welding parameters such as squeeze time (SQZ), hold time (HLD), pre-heating time (Pre-Weld), Cool Time (CT), Up Slope Time (UST) and Down Slope Time (DST) were constant during the experiment. Schedule obtained from software BOSH 6000, with values of this parameters is shown in figure 3.

Table 1. Selected levels for the parameters of the welding.

Factor/level	L1	L2	L3
Weld current I [kA]	20	28	35
Electrode force F [kN]	3,68	4,91	6,14
Welding time T [ms]	80	140	200
Pre-heating current I_A [kA]	6	8	10

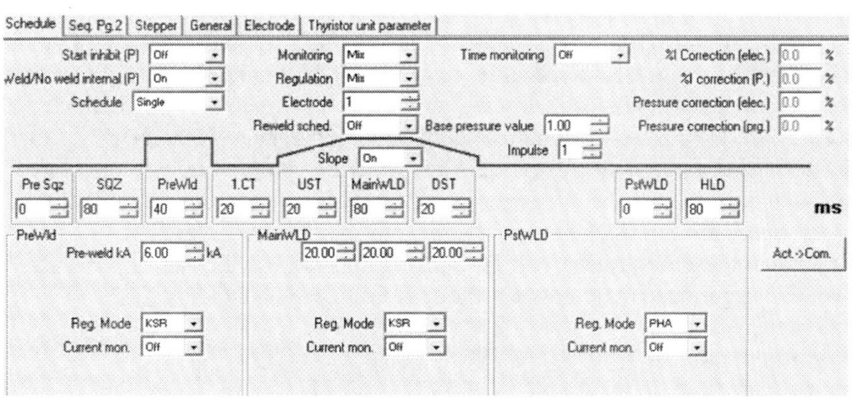

Figure 3. Schedule of RSW.

The tensile-shear tests were performed according to standard ISO 14273:2016 at cross-head speed of 2mm/min with a Beta 50-7 / 6x14 testing machine, as shown in Figure 4. As per the L9 orthogonal array for each combination of process parameters results obtained from the test are given in table 2.

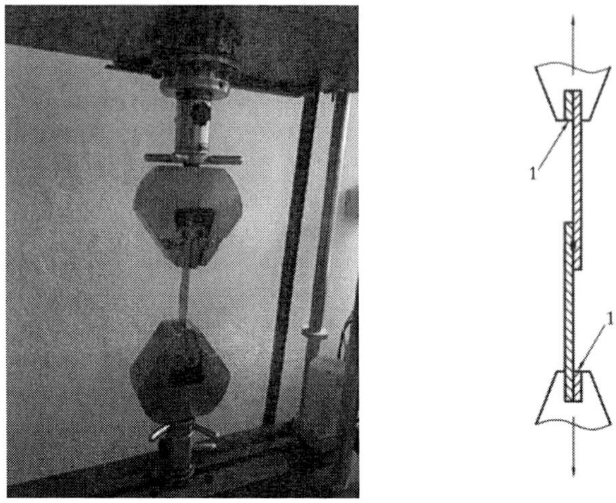

Figure 4. Specimen prepared for testing.

Table 2. Experimental layout using L9 OA and result from tensile-shear tests (Failure load).

Runs	Weld current *I* [kA]	Electrode force *F* [kN]	Welding time *T* [ms]	Pre-heating current I_A [kA]	Failure load *F* [N]
1	20	3,68	80	6	1504
2	20	4,91	140	8	1830
3	20	6,14	200	10	1638
4	28	3,68	140	10	2046
5	28	4,91	200	6	1581
6	28	6,14	80	8	1316
7	35	3,68	200	8	2719
8	35	4,91	80	10	1989
9	35	6,14	140	6	1862

2.1. Taguchi optimization and prediction

The total number of experiments can be substantially reduced with the help of a well-designed experimental plan without affecting the accuracy during the experimental study of any manufacturing process [12]. Taguchi suggested a specially designed method called the use of orthogonal array to study the entire parameter space with lesser number of experiments to be conducted. Taguchi thus, recommends the use of the loss function to measure the performance characteristics that are deviating from the desired target value. The value of this loss function is further transformed into signal-to-noise (S/N) ratio. Usually, there are three categories of the performance characteristics to analyse the S/N ratio. They are: nominal-the-best, larger-the-better, and smaller-the-better [13].

In this research, the optimization will be done to maximizing the failure load, so it will be used larger-the-better categories to obtaining S/N ratio, as following equation:

$$\eta(S/N) = -10\log_{10}\left(\frac{1}{n}\sum_{i=1}^{n}\frac{1}{y^2}\right) \tag{1}$$

2.2. Response surface modeling (RSM)

Response surface methodology (RSM) is a multivariation analysis technique, in RSM, all the input process parameters are as sum to be measurable. It is required to find a suitable approximation for the true function relationship between independent variables and the response surface. Usually, a second-order regression model as given below is utilized in RSM:

$$y = b_0 + \sum_{i=1}^{p} b_i x_i + \sum_{i=1}^{p} b_{ii} x_i^2 + \sum_i \sum_j b_{ij} x_i x_j \quad (2)$$

where all *b's* are the regressions coefficients, determined numerically by using least square fit method, x_i are the different control factors changing from $i=1$ to p, p is the total number of control factors, y_j are the different responses (j are the number of responses used) [12].

3. Results and discussion

Optimization of the parameters of the process was carried out using the Taguchi method by computational software that indicated an optimal configuration according to the equation (1) for S/N ratio. The effect of different control factors on S/N ration is shown in Table 3. The optimum levels of different control factors for higher failure load are weld current at level 3 (35 kA), electrode force at level 1(3,68 kN), weld time at level 3 (0.2 s), and pre-weld current at level 3 (10 kA). The plot of S/N ration is showed in figure 5. It can also be concluded from table 3 that the welding current has the greatest influence on the failure force, while the pre-weld current has the slightest influence on the meaning of this force.

Table 3. S/N response and the level of parameters.

Level	A	B	C	D
1	64.36	66.15*	63.97	64.31
2	64.19	65.07	65.62	65.44
3	66.69*	64.02	65.65*	65.49*
Delta	2.49	2.13	1.68	1.18
Rank	1	2	3	4

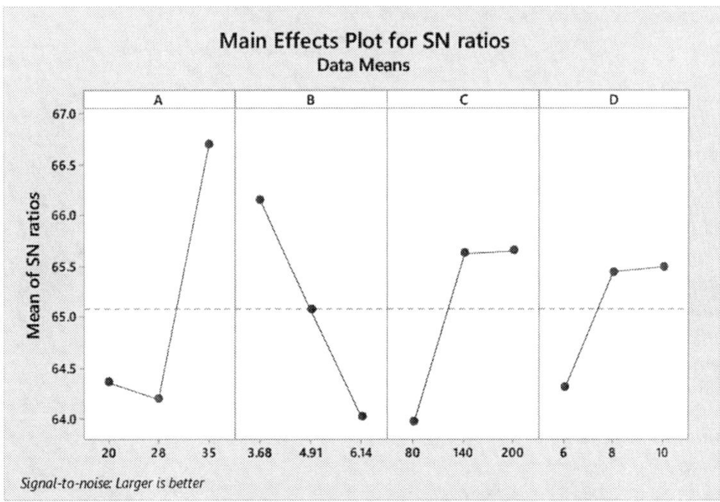

Figure 5. Plot of S/N ratio.

Using the analysis of variance (ANOVA), the previous assertion that the most influence on failure load has a weld current is confirmed. The results in table 4 show that the welding current is with 43.65% contribution to maximize failure load.

Table 4. Results of ANOVA for Failure load.

Factors	DoF	Sum of Squares	Mean of squares	F	P	Contribution (%)
A	2	11.656	5.828	-	-	43,65%
B	2	6.786	3,393	-	-	25,41
C	2	5.573	2,7865	-	-	20,87
D	2	2.689	1,345	-	-	10,07
Error	0	2.91E-11	-	-	-	-
Total	8	26.704	-	-	-	100%

The predicted S/N radio at the optimal level of the welding parameters can be calculated as:

$$\eta = x + \sum_{i=1}^{n} Y_i - x \quad (3)$$

where x is the mean of S/N ration and Y is the average S/N at the optimal level. Using this equation, it turns out that S / N ratio at an optimal level is 68.74, so that failure load at the optimum level is 2735.179 N.

The response surface model (RSM) for failure load F has been developed from the experimental response values obtained using L9 OA experimental matrix. The above-mentioned model is presented below as the equation:

$$F = 1971 - 253A + 505{,}2B + 12{,}59C + 800{,}5D + 5{,}246A^2 + 31{,}4B^2 - 0{,}03375C^2 - 46{,}25D^2$$

To obtain the influencing nature and optimized condition of the process on failure load F, the surface plots and contour plots which are the indications of possible independence of factors have been developed for the proposed empirical relation by considering two parameters in the middle level and two parameters in the x- and y-axes as shown in Fig. 6 and 7.

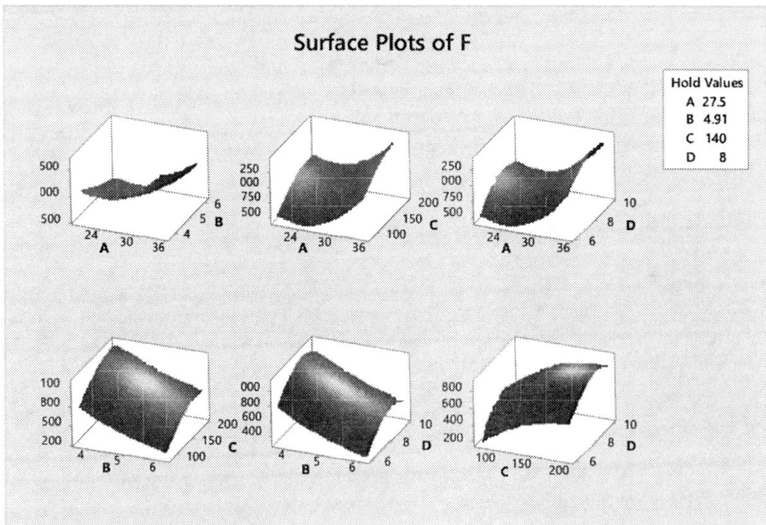

Figure 6. Surface plots of failure load (F).

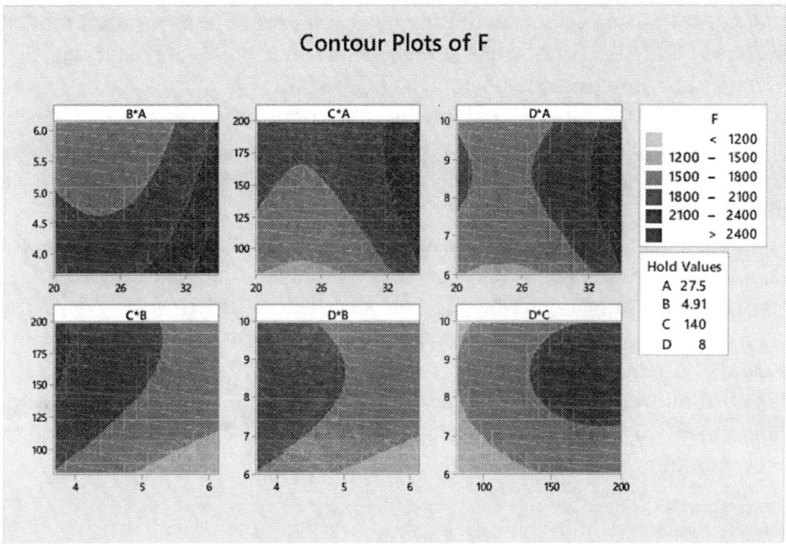

Figure 7. Contour plots of failure load (F).

By analyzing the response surfaces and contour plots (Fig. 6 and 7), the maximum achievable failure load *F* value is found to be 2744,9630 kN. The corresponding parameters that yielded this maximum value are weld current 35 kA, electrode force 3,68 kN, wled time 0.1866667 s, and pre-weld current 8,667 kA, what is howen on the figure 8.

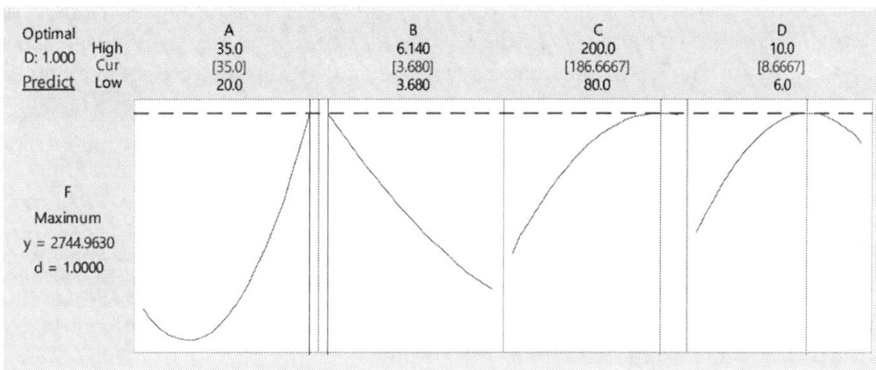

Figure 8. Response optimizer of RSW joint.

The percentage error between confirmation experiment and prediction is 10,3 % for the welding parameters I=20 kA, F=3,68 kN, T=120 ms, I_A=8 kA what is showed in table 5. The same table also shows the prediction error for parameters I=30 kA, F=3,68 kN, T=160 ms, IA=6 kA and its value is 15.5%.

Table 5. Results of the confirmation tests for failure load (F).

Verifications	Actual F (N)	Predicted F (N)	Error %
1 (A=20 kA, B=3,68 kN, C=120 ms, D=8 kA)	2279	2043	10,3
2 (A=30 kA, B=3,68 kN, C=160 ms, D=6 kA)	2317	1955,72	15,5

4. Conclusion

Taguchi method and response surface modelling has been used in this research for optimizing welding process parameters of RSW joint of AA5754 aluminium alloy. Based on the modelling and optimization results, the following conclusion can be drawn:

(I) Optimum conditions for four parameters (weld current, welding time, electrodes force and pre-heating current) affecting failure load of the joints. Based on the results, the effectiveness of the parameters as follows: 1-weld current, 2- electrodes force, 3-welding time, and 4- pre-heating current;

(II) The optimum conditions for the parameters based on Taguchi method as follows: current of 35kA, 3.68 kN as the electrode force, 0.2s as the welding time and 10 kA as the pre-heating current. The optimum conditions for the parameters based on RSM as follows: current of 35kA, 3.68 kN as the electrode force, 0.186667 s as the welding time and 8.667 kA as the pre-heating current;

(III) The confirmation test results demonstrate that the use of parameter design of the Taguchi method and RSM provides a simple, systematic, and efficient methodology for optimizing the process parameters.

Acknowledgment

The authors gratefully acknowledge partial support by CEEPUS network CIII-BG-1103-04-1920-M-133090 – Modelling, Simulation and Computer-aided Design in Engineering and Management.

References

[1] Smith B, Adela S, Shashank M and Terni F 2017 *Technology Roadmaps: Intelligent Mobility Technology, Materials and Manufacturing Processes, and Light Duty Vehicle Propulsion,* Center for Automotive Research, Ann Arbor, MI

[2] Bertin L 2014 *Tensile Strength of Automotive Aluminum Joints Using Resistance Spot Welding, Self-Piercing Riveting and Adhesive Hybrid Joining*, University of Windsor, Doctoral Theses

[3] Thongchai A, Kawin S, Phisut A and Kreangsak T 2014 Resistance Spot Welding Optimization Based on Artificial Neural Network, *International Journal of Manufacturing Engineering,* **2014** 1-6

[4] Harry A, Yuying X and John K 2017 The mechanical properties and geometrical configuration efficiency of multiple resistance spot welds, *Advances in Materials and Processing Technologies* **3**(3) 449–462

[5] Hovorun T P, Berladir K V, Pererva V I, Rudenko S G and Martynov A I 2017 Modern materials for automotive industry, *Journal of Engineering Sciences* **4**(2) 8-18

[6] Tisza M and Lukác Z 2018 High strength aluminum alloys in car manufacturing, International Deep Drawing Research Group 37th Annual Conference, *IOP Conf. Series: Materials Science and Engineering* **418** 012033

[7] Manladan S M, Yusof F, Ramesh S, Fadzil M, Luo Z and Ao S 2016 A review on resistance spot welding of aluminum alloys, *The International Journal of Advanced Manufacturing Technology* **90**(1–4) 605–634

[8] Zhen L, Sansan A, Yuh J C, Xuetuan C, Yang L and Ye L 2015 Application of Pre-heating to Improve the Consistency and Quality in AA5052 Resistance Spot Welding, *Journal of Materials Engineering and Performance*

[9] Zoha N and Khan M I 2016 Resistance spot welding and optimization techniques used to optimize its process parameters, *International Research Journal of Engineering and Technology (IRJET)* **3**(5) 887-893

[10] Karthikeyan R and Balasubramanian V 2017 *Optimization of Electrical Resistance Spot Welding and Comparison with Friction Stir Spot Welding of AA2024-T3 Aluminum Alloy,* Materials Today: Proceedings 5th International Conference of Materials Processing and Characterization (ICMPC 2016) **4** 1762–1771

[11] Arick M L and Darji P H 2018 Experimental Investigation on Resistance Spot Welding of Aluminium Alloy 6082t651 Using Interlayer of SS304, *International Journal of Mechanical*

Engineering and Technology (IJMET) **8**(7) 313-324
[12] Ghazalia F A, Berhanb M N, Manurungb H Y, Sallehb Z and Abdullahc S 2015 Tri-Objective Optimization of Carbon Steel Spot-Welded Joints, *Jurnal Teknologi (Sciences & Engineering)* **76**(11) 69–73
[13] Athreya S and Venkatesh Y D 2012 Application Of Taguchi Method For Optimization Of Process Parameters In Improving The Surface Roughness Of Lathe Facing Operation, *International Refereed Journal of Engineering and Science (IRJES)* **1**(3) 13-19

IOP Conf. Series: Materials Science and Engineering **659** (2019) 012043 doi:10.1088/1757-899X/659/1/012043

The influence of different lamina positions on buckling properties of composites plates under biaxial compression

D Jeremić[1] and N Radić[1]

[1] University of East Sarajevo, Faculty of Mechanical Engineering, Vuka Karadžića 30, 71123 East Sarajevo, RS, B&H

E-mail: dejan.jeremic@ues.rs.ba

Abstract. The present work focused on the buckling of composite plates subjected to biaxial compression using classical laminated plate theory (CLPT). The two composite plates are bonded by an elastic medium. Classical laminated plate theory (CLPT) is utilized for deriving the governing equations. Difference between 4-ply symmetric lamination scheme with different material properties and fiber orientation angle (0° and 30°) is shown. The study show that the mechanical properties of the composite materials depending on the fiber orientation angle and lamina positions.

1. Introduction

Composite materials are materials made from two or more constituent materials with specific different physical or chemical properties that when combined produce a material with characteristics different from the individual components. Generally, a composite material is composed of reinforcement (fibres, particles, flakes, and/or fillers) embedded in a matrix (polymers, metals, or ceramics). The matrix holds the reinforcement to form the desired shape while the reinforcement improves the overall mechanical properties of the matrix. Currently, the analysis of the behavior of the laminated plates is an active research area because of their complex behavior [1]. The majority of the research work carried out on the laminated plates, is devoted to the stress determination, deformations or displacements of flexional origin. The structural instability becomes an important concern in a reliable design of composite plates. Several studies on laminated plates stability were concentrated on rectangular plates [2-5]. It is known that buckling strength of the rectangular plates depends on the boundary conditions [3], plies orientation [3,4,6] and geometrical ratio [3,5-7]. The thin composites structures which are largely used become unstable when they are subjected to mechanical or thermal loadings which leads to buckling. Consequently, the buckling behavior is a significant factor of the sure and reliable design [8]. To predict buckling load and deformation mode of a structure, the linear analysis can be used as an evaluation technique [9]. Generally, the analysis of the laminated plates is more complicated than the analysis of an isotropic and homogeneous material [10]. Finite element method is used for the analysis of the buckling behavior of the notched antisymetrical fibers plates under compression [11].

The main contribution of this work is to perform a composite laminated plates analysis by using the Classical Laminated Plate Theory (CLPT) is described in [12-14]. Various geometries of the plates subjected to compressive load are studied.

2. Governing equations of biaxially compressed composites plates

Let us consider composite plate the length of a , width b and height h, as shown in figure 1.

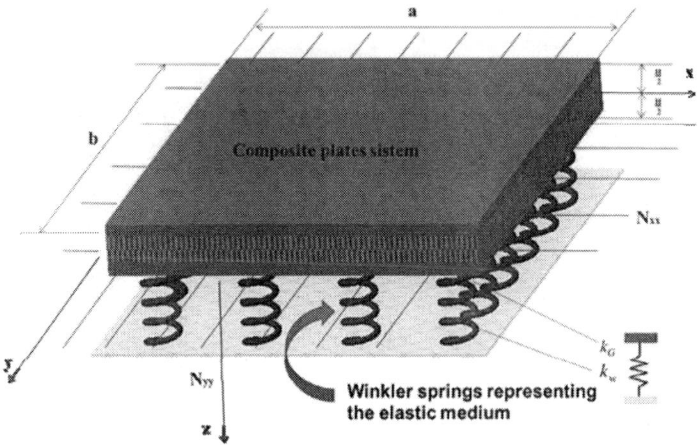

Figure 1. Schematic diagram of composite plates system subjected to biaxial compression.

Composite plates are assumed to be coupled by an elastic medium and subjected to biaxially compression. The composite plates are surrounded by external elastic medium. The external medium is modeled as Pasternak-type foundation which is equivalent to Winkler modulus parameter k_w and shear modulus parameter k_G of polymer matrix. Bending displacements of the plate-1 and plate-2 are $w_1(x, y, t)$ and $w_2(x, y, t)$, respectively. It was assumed that each composite plate had the length, a and width, b. We assume that composite plates are biaxially compressed by forces N_{xx} and N_{yy} in the directions of x and y axes, respectively (figure 1).

The governing equation for biaxially compressed orthotropic composite plate embedded in an elastic medium [15], which is based on Classical Laminated Plate Theory CLPT, have following form

$$D_{11}\frac{\partial^4 w}{\partial x^4}+2(D_{12}+2D_{66})\frac{\partial^4 w}{\partial x^2 \partial y^2}+D_{22}\frac{\partial^4 w}{\partial y^4}+N_x\frac{\partial^2 w}{\partial x^2}+N_y\frac{\partial^2 w}{\partial y^2}+k_w w-k_G\left(\frac{\partial^2 w}{\partial x^2}+\frac{\partial^2 w}{\partial y^2}\right)=0 \quad (1)$$

We assume that composite plate is biaxially compressed in the directions of x and y axes, N_x=N_y. Now we can define compression ratio which equals the ratio between the forces acting in y and x directions

$$\delta=\frac{N_{yy}}{N_{xx}} \rightarrow N_{yy}=\delta N_{xx} \quad (2)$$

Substitution of equation (2) in equation (1) we derive the general form of governing equation

$$D_{11}\frac{\partial^4 w}{\partial x^4}+2(D_{12}+2D_{66})\frac{\partial^4 w}{\partial x^2 \partial y^2}+D_{22}\frac{\partial^4 w}{\partial y^4}+N_x\left(\frac{\partial^2 w}{\partial x^2}+\delta\frac{\partial^2 w}{\partial y^2}\right)+k_w w-k_G\left(\frac{\partial^2 w}{\partial x^2}+\frac{\partial^2 w}{\partial y^2}\right)=0 \quad (3)$$

Before solving constituent equation (3), boundary conditions should be defined. In this study it is assumed that all edges on both nanoplates are simply supported. This means that both the displacements and moments at the composite plates edges are zero. This can be expressed by following equations

$$w_i(0,y,t)=0\,,\ w_i(a,y,t)=0\,,\ w_i(x,0,t)=0\,,\ w_i(x,b,t)=0\ \ i=1,2 \tag{4}$$

$$M_i(0,y,t)=0,\ \ M_i(a,y,t)=0,\ \ M_i(x,0,t)=0,\ M_i(x,b,t)=0 \tag{5}$$

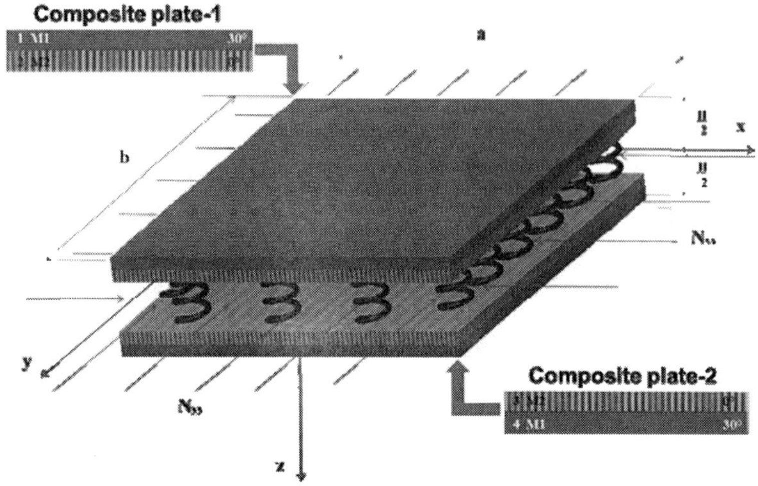

Figure 2. Composite plate system coupled by an elastic medium.

If in the middle of the four-layer laminate sistem we insert an elastic medium that separate the laminate into two symmetric parts, we will have two composite plates with two laminae (figure 2) whose main equations are:

Plate 1

$$D_{11}\frac{\partial^4 w_1}{\partial x^4}+2(D_{12}+2D_{66})\frac{\partial^4 w_1}{\partial x^2\partial y^2}+D_{22}\frac{\partial^4 w_1}{\partial y^4}+N_x\frac{\partial^2 w_1}{\partial x^2}+N_y\frac{\partial^2 w_1}{\partial y^2}+k_w\left(w_1-w_2\right)-k_G\nabla^2\left(w_1-w_2\right)=0 \tag{6}$$

Plate 2

$$D_{11}\frac{\partial^4 w_2}{\partial x^4}+2(D_{12}+2D_{66})\frac{\partial^4 w_2}{\partial x^2\partial y^2}+D_{22}\frac{\partial^4 w_2}{\partial y^4}+N_x\frac{\partial^2 w_2}{\partial x^2}+N_y\frac{\partial^2 w_2}{\partial y^2}+k_w\left(w_2-w_1\right)-k_G\nabla^2\left(w_2-w_1\right)=0 \tag{7}$$

$$\nabla^2=\frac{\partial^2}{\partial x^2}+\frac{\partial^2}{\partial y^2} \tag{8}$$

3. Buckling loads of biaxially compressed composite plates

In this section different explicit cases of biaxial buckling will be considered. The composite plate system is subjected to both biaxial as well as biaxial compressive forces. The cases studied will be composite plates buckling with out-of-phase (asynchronous); in-phase (synchronous); and when one of the composite plates is considered to be fixed.

3.1. Asynchronous-type bi-axial buckling (out-of- phase)

Composite plates system is assumed to be bi-axially buckled. Figure 3 shows the three-dimensional configuration of double composite plates system with the asynchronous (out-of-phase) sequence of buckling:

$$w_1(x,y,t) - w_2(x,y,t) \neq 0$$

In out-of-phase, sequence of buckling the nanoplates is buckled in opposite directions. We evaluate the buckling load for the out-of-phase (asynchronous) type buckling and use equations (6,7) for the biaxial buckling solution of double composite plates system.

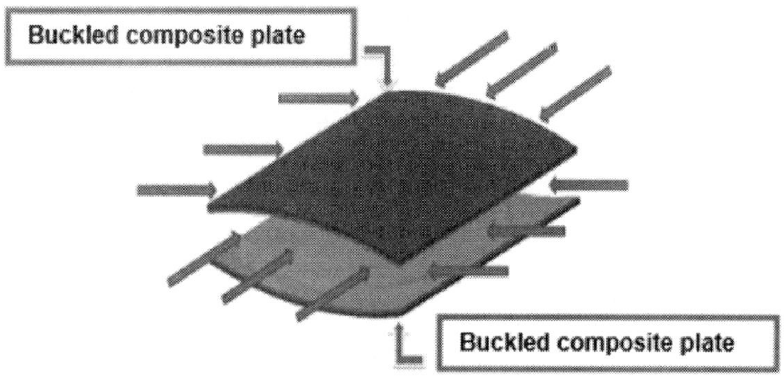

Figure 3. Asynchronous-type bi-axial buckling.

Subtracting equation (6) from equation (7) we get:

$$D_{11}\frac{\partial^4 w}{\partial x^4} + 2(D_{12} + 2D_{66})\frac{\partial^4 w}{\partial x^2 \partial y^2} + D_{22}\frac{\partial^4 w}{\partial y^4} + N_x\frac{\partial^2 w}{\partial x^2} + N_y\frac{\partial^2 w}{\partial y^2} + 2k_w w - 2k_G\left(\frac{\partial^2 w}{\partial x^2} + \frac{\partial^2 w}{\partial y^2}\right) = 0 \quad (9)$$

$$w = w_1 - w_2 - w_2 + w_1 = 2w_1 - 2w_2 = 2(w_1 - w_2) = 2w \quad (10)$$

We assume that the buckling mode of the double-nanoplate system as

$$w = \sum_{m=1}^{\infty}\sum_{n=1}^{\infty} W_{mn}\sin(\alpha x)\sin(\beta y) \quad (11)$$

In the upper equation:

$$\alpha = \frac{m\pi}{a}, \quad \beta = \frac{n\pi}{b} \quad (12)$$

where m and n are the half wave numbers.

Substituting equation (11) into equation (9), we get critical buckling load for asynchronous type of buckling

$$N_{cr} = \frac{D_{11}\alpha^4 + 2(D_{12} + 2D_{66})\alpha^2\beta^2 + D_{22}\beta^4 + 2k_w + 2k_G(\alpha^2 + \beta^2)}{(\alpha^2 + \delta\beta^2)} \quad (13)$$

3.2. Synchronous-type bi-axial buckling (in-phase)

In-phase (synchronous) sequence type of buckling of the double composite plates under bi-axial compression is also being considered here. The schematic illustration buckling of the orthotropic composite plates is shown in figure 4, which is the first mode synchronous type buckling. In synchronous buckling both the composite plates buckle in the same direction. For the present system, the relative displacements between the two composite plates are

$$w_1(x,y,t) - w_2(x,y,t) = 0$$

In synchronous buckling state, the double composite plates system can be considered to be as one of the composite plates.

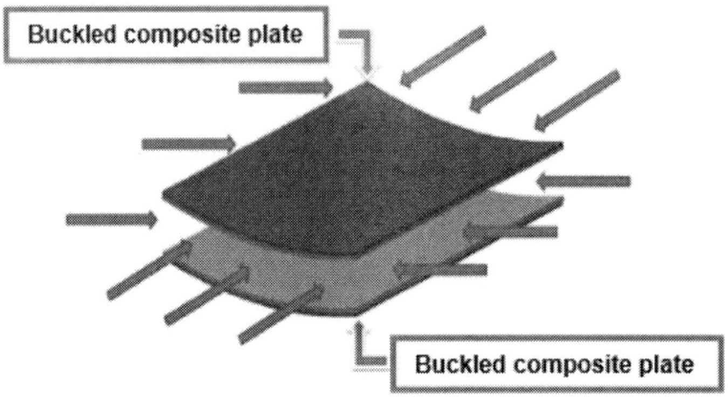

Figure 4. Synchronous-type bi-axial buckling.

We apply the same procedure as earlier for solving equations (6) and (7). For this case of buckling governing equation can be written as

$$D_{11}\frac{\partial^4 w_2}{\partial x^4} + 2(D_{12} + 2D_{66})\frac{\partial^4 w_2}{\partial x^2 \partial y^2} + D_{22}\frac{\partial^4 w_2}{\partial y^4} + N_x\frac{\partial^2 w_2}{\partial x^2} + N_y\frac{\partial^2 w_2}{\partial y^2} = 0 \tag{14}$$

$$\begin{aligned} w_1 &= w_2 \\ w = w_1 - w_2 &= 0 \end{aligned} \tag{15}$$

Following procedure similar to that of out-of-phase buckling, critical buckling load for synchronous type of buckling can be written as

$$N_{cr} = \frac{D_{11}\alpha^4 + 2(D_{12} + 2D_{66})\alpha^2\beta^2 + D_{22}\beta^4}{(\alpha^2 + \delta\beta^2)} \tag{16}$$

For this case the critical buckling load is independent of the stiffness of the coupling springs and biaxial compression of double composite plates sistem can be effectively treated as a single composite plate.

3.3. Bi-axial buckling with one composite plate fixed

Consider the case of composite plates system when one of the two composite plates is stationary

$$w_2 = 0$$

which is shown in figure 5. For this case of buckling governing equation can be written as

$$D_{11}\frac{\partial^4 w_1}{\partial x^4}+2(D_{12}+2D_{66})\frac{\partial^4 w_1}{\partial x^2\partial y^2}+D_{22}\frac{\partial^4 w_1}{\partial y^4}+N_x\frac{\partial^2 w_1}{\partial x^2}+N_y\frac{\partial^2 w_1}{\partial y^2}+k_w w_1-k_G\left(\frac{\partial^2 w_1}{\partial x^2}+\frac{\partial^2 w_1}{\partial y^2}\right)=0 \quad (17)$$

and critical buckling load for this type of buckling can be written as

$$N_{cr}=\frac{D_{11}\alpha^4+2(D_{12}+2D_{66})\alpha^2\beta^2+D_{22}\beta^4+k_w+k_G\left(\alpha^2+\beta^2\right)}{(\alpha^2+\delta\beta^2)} \quad (18)$$

In fact, when one of the composite plates in composite system is fixed ($w_2 = 0$), the composite system behaves as composite plate on an elastic medium.

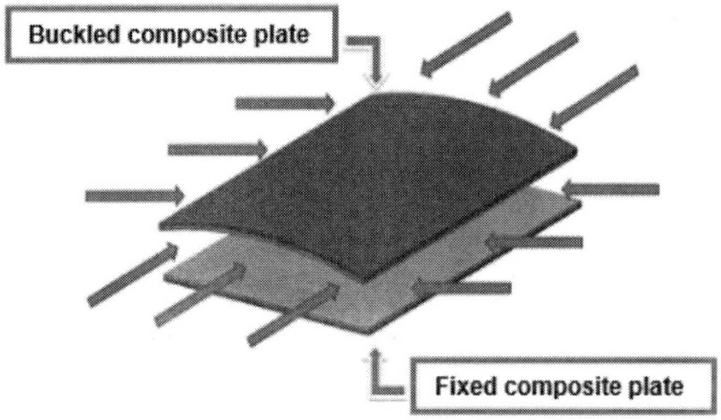

Figure 5. Bi-axial buckling with one composite plate fixed.

4. Numerical results and discussion

In this section follows analysis of four-layer symmetric laminate made of two types of material with different orientation of the fibers. Two types of materials are combined:
-Kevlar 49/CE 3305 (material M1)
-Graphite-Epoxy AS-1/3501-5A (material M2)

For laminates of total thickness of 1mm with four sheets of individual thickness of 0.25mm, bending stiffness matrix D has the following form [15]:

$$D_{ij}=\frac{1}{3}\sum_{k=1}^{N}\left(\overline{Q}_{ij}\right)^k\left(h_k^3-h_{k-1}^3\right)$$

$$=\frac{1}{3}\left(\overline{Q}_{ij}\right)_\theta\left(h_1^3-h_0^3\right)+\frac{1}{3}\left(\overline{Q}_{ij}\right)_\theta\left(h_2^3-h_1^3\right)+\frac{1}{3}\left(\overline{Q}_{ij}\right)_\theta\left(h_3^3-h_2^3\right)+\frac{1}{3}\left(\overline{Q}_{ij}\right)_\theta\left(h_4^3-h_3^3\right)$$

$$=\frac{1}{3}\left(\overline{Q}_{ij}\right)_\theta\left[(-0,25)^3-(-0,5)^3\right]+\frac{1}{3}\left(\overline{Q}_{ij}\right)_\theta\left[(0)^3-(-0,25)^3\right]+\frac{1}{3}\left(\overline{Q}_{ij}\right)_\theta\left[(0,25)^3-(0)^3\right]+$$

$$\frac{1}{3}\left(\overline{Q}_{ij}\right)_\theta\left[(0,5)^3-(0,25)^3\right]$$

Based on the above expression and using the MATLAB software package, bending stiffness matrix for selected laminate schemes $\theta = 30^\circ$ and $\theta = 0^\circ$ are obtained.

$$[D] = \begin{bmatrix} 6,7954 & 0,3167 & 0,3990 \\ 0,3167 & 0,4800 & 0,1298 \\ 0,3990 & 0,1298 & 0,4841 \end{bmatrix} \quad [D] = \begin{bmatrix} 9,8649 & 0,3548 & 0,2588 \\ 0,3548 & 0,9093 & 0,0941 \\ 0,2588 & 0,0941 & 0,5000 \end{bmatrix}$$

Substituting the values of bending stiffness matrix in the previously set equations we obtain values of non-dimensional critical force for three types of buckling.

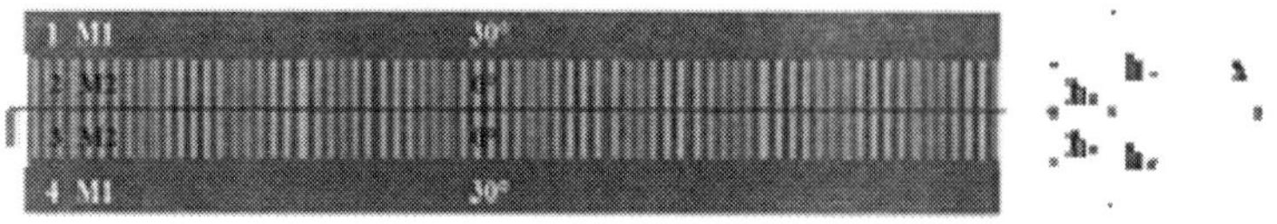

laminate a)

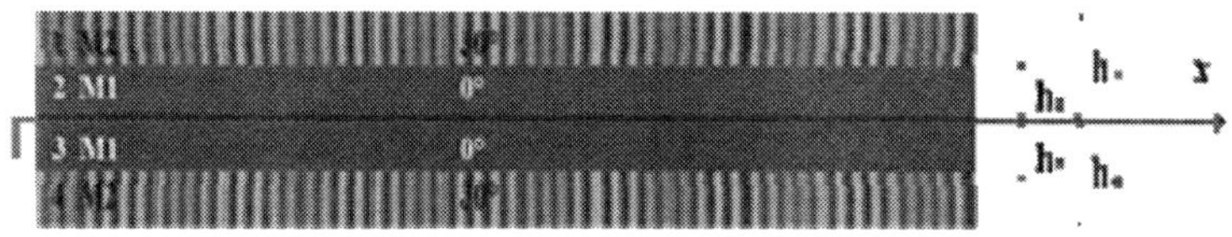

laminate b)

Figure 6. Schematic layout of composite laminate.

Based on equations (13), (16) and (18), in this section follows analysis of impact of aspect ratio a/b or lamina positions on the non-dimensional buckling load. Nondimensional buckling load is calculated for the value of Winkler modulus (k_w) which took following values: 10N/m^3, 100N/m^3 and 1000N/m^3, while the shear modulus parameter (k_G) took following values: 1N/m, 10N/m and 100N/m. The number of half waves was m = 1, n = 1, while the compression ratio was δ=0,5. The thickness of one composite plate is h = 0,25 mm, while the length takes values within a=0–0,6m range and width takes value b=0,3.

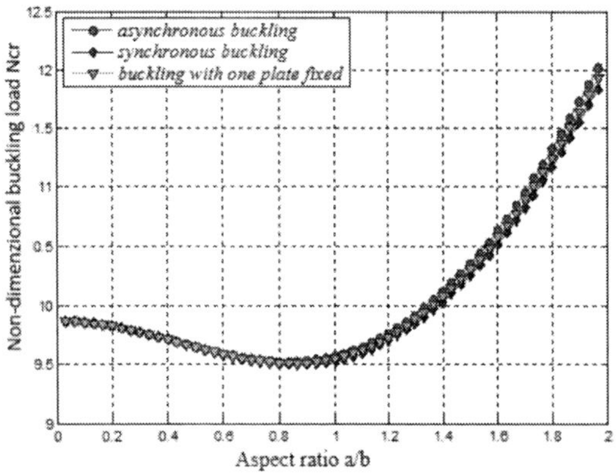

laminate a)

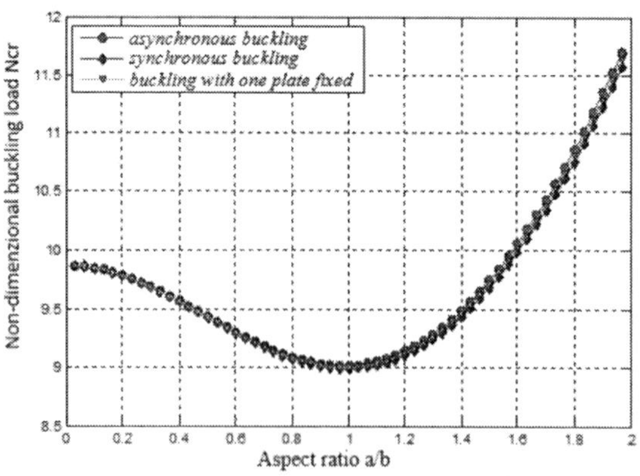

laminate b)

Figure 7. Effect of aspect ratio a/b on non-dimensional buckling load k_w=10N/m^3, k_G=1N/m.

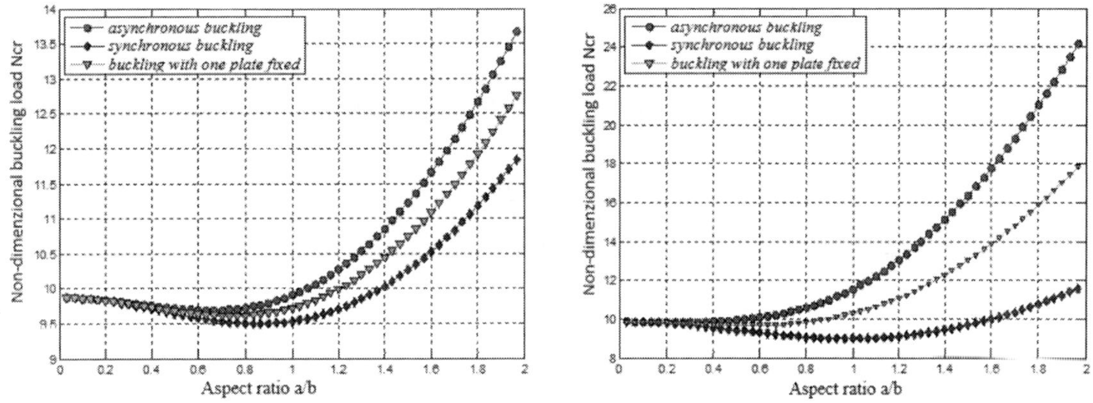

laminate a) laminate b)

Figure 8. Effect of aspect ratio a/b on non-dimensional buckling load k_w=100N/m^3, k_G=10N/m.

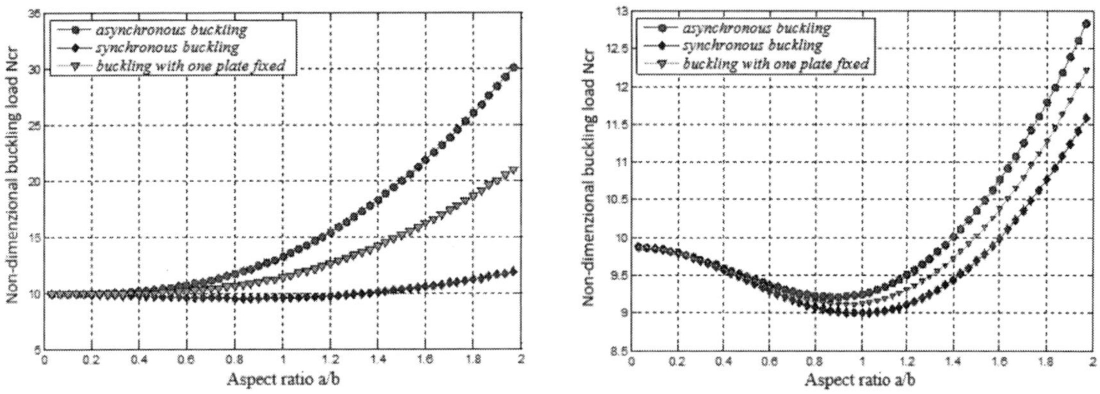

laminate a) laminate b)

Figure 9. Effect of aspect ratio a/b on non-dimensional buckling load k_w=1000N/m^3, k_G=100N/m.

The influence of Winkler modulus parameter k_w, which represents the influence of polymer matrix on the two composite plates, can be seen in figures 7-9. It can be concluded that higher values of Winkler modulus parameter increase the nondimensional buckling load the same for all three buckling types. By analyzing all the previous diagrams it can be concluded that the laminate types a) and b) should be avoided in the required boundary conditions and parameter values. For aspect ratio a/b=0-0.8, the curves are very close together for all three buckling types. For value of aspect ratio a/b=0.8-1 a very small value of the non-dimensional critical force leads to the deformation of the composite plates and the occurrence of instability of the system.

By increasing the relationship of dimensions a/b=0.8-2, the non-dimenzional critical force value is constantly increasing. It is obvious that the type of material of individual laminates has a great influence on the value of critical buckling force and should be taken into account. It can be concluded that in the case of laminate a) and b) we have behavior similar to the behavior of the isotropic and orthotropic plates.

IOP Conf. Series: Materials Science and Engineering **659** (2019) 012043 doi:10.1088/1757-899X/659/1/012043

5. Conclusion

Based on CLPT, in this paper was analyzed influence of aspect ratio a/b, fiber orientation angle and lamina positions on the non-dimensional buckling load on biaxial compressed composite plates embedded in elastic medium. In this paper, there are analytical expressions for non-dimensional buckling load for three characteristic cases of buckling of simply supported composite plates. It was shown that the increase of Winkler modulus parameter and shear modulus parameter also increases the nondimensional buckling load for all three characteristic buckling cases. The layout of the lamina inside the laminate has a significant effect as well as fiber orientation angle on the buckling and can not be ignored. It has been shown that with the change of lamina positions in the laminate, the value of the non-dimensioning critical load is changed for all three characteristic buckling cases. Laminates have different minimum and maximum values of non-dimensional critical force at the same value of aspect ratio. For asynchronous type of buckling non-dimensional critical buckling force has the highest value in both schemas of laminate plates.

References

[1] Leissa A W 1985 *Buckling of laminated composite plates and shell panels, AFWAL-TR-85-3069*, Flight Dynamics Laboratory, Air Force Wright Aeronautical Laboratories, Wright–Patterson Air Force Base, Ohio

[2] Muc A 1988 Optimal fibre orientation for simply-supported angle-ply plates under biaxial compression, *Composite Structures* **9**(2) 161–172

[3] Hu H-T and Lin B-H 1995 Buckling optimization of symmetrically laminated rectangular plates with various geometry and end conditions, *Composite Science and Technology* **55**(3) 277–285

[4] Nemeth M P 1988 Buckling behavior of compression-loaded symmetrically laminated angle-ply plates with holes, *AIAA Journal* **26**(3) 330–336

[5] Vellaichamy S, Prakash B G and Brun S 1990 Optimum design of cutouts in laminated composite structures, *Computers and Structures* **37**(3) 241–246

[6] Zor M, Sen F and Toygar M E 2005 An investigation of square delamination effects on the buckling behavior of laminated composite plates with a square hole by using three-dimensional FEM analysis, *J Reinf Plast Compos* **24**(11) 1119–1130

[7] Shufrin I, Rabinovitch O and Eisenberger M 2008 Buckling of laminated plates with general boundary conditions under combined compression, tension, and shear-A semi-analytical solution, *Thin-Wall Struct* **46**(7-9) 925–938

[8] Akbulut H and Sayman O 2001 An investigation on buckling of laminated plates with central square hole, *J Reinf Plast Compos* **20**(13) 1112–1124

[9] Kong C W, Hong C S and Kim C G 2001 Postbuckling strength of composite plate with ahole, *J Reinf Plast Compos* **20**(6) 466–481

[10] Ghannadpour S A M, Najafi A and Mohammadi B 2006 On the buckling behavior of crossply laminated composite plates due to circular/elliptical cutouts, *Compos. Struct.* **75**(1-4) 3–6

[11] Hamani N, Ouinas D, Benderdouche N and Sahnoun M 2012 Buckling analyses of the antisymetrical composite laminate plate with a crack from circular notch, *Advanced Materials Research* **365** 56-61

[12] Reddy J N 2004 *Mechanics of Laminated Composite Plates and Shells: Theory and Analysis*, CRC Press LLC, London and New York

[13] Spencer A J M 1984 *Constitutive Theory for Strongly Anisotropic Solids in Continuum Theory of the Mechanics of Fibre Reinforced Composites*, Springer-Verlag, Wien

[14] Powell P C 1994 *Engineering with Fibre-Polymers Laminates*, Chapman & Hall, England

[15] Jeremić D 2013 *Analiza stabilnosti laminantnih kompozitnih ploča*, East Sarajevo, Master Thesis

IOP Conf. Series: Materials Science and Engineering **659** (2019) 012044 doi:10.1088/1757-899X/659/1/012044

Non-woven composites intensification properties for air filters by plasma pre-treatment

M P Neznakomova[1], M-L Klotz[2] and D N Gospodinova[3]

[1] Technical university – Sofia, Department of Material Science and Technology of Materials, 8 Kliment Ohridski Blvd., 1000 Sofia, Bulgaria

[2] Rhine-Waal University of Applied Sciences, Department Werkstoffe, Qualität und Ökologie, Marie-Curie-Straße 1, 47533 Kleve, Germany

[3] Technical university – Sofia, Faculty of Electrical engineering, 8 Kliment Ohridski Blvd., 1000 Sofia, Bulgaria

E-mail: dilianang@abv.bg

Abstract. The nonwovens use in the automotive industry has increased significantly. Nonwovens for air filters can be changed in a desired direction to have specific properties. The use of non-woven textiles in the automotive industry has the following advantages: it reduces the weight of the vehicle, increases comfort and provides improved insulation, fire retardancy and resistance to water, fuels, extreme temperatures and abrasion. The electrospinning process is one of the nanofibres, two-layer and multilayer composites with their participation obtaining methods. The low frequency plasma surface modification of needle-punched polyethilenthereftalat (PET) nonwoven fibres is good possibility to intensification the process and the rate of increase of the mass of nanolayer regarding to its' hydrophobicity. The plasma physicochemical treatment type allows adhesion change and change of electro physical properties at the expense of new active centres formation that do not exist in the native fibre. In this study, different gases (Ar, O_2 and air) and electrical conditions were employed. The surface changes of PET fibres were investigated by cation exchange capacity (CEC) determination and its influence on the mass nanofibres increase rate. Observation revealed the plasma treatment significantly change surface wettability of PET fibres. The effect of caused change on the electrospinning process is analysed.

1. Introduction

The use of nonwovens in the automotive market has increased significantly in recent years. Frequently, nonwovens can be engineered to have specific properties. By building on the essential properties that are necessary for good performance and safety, nonwovens help reduce the weight of the car, enhance comfort and aesthetics and provide advanced insulation, fire retardancy and resistance to water, fuels, extremes of temperature and abrasion.

A promising method of obtaining fibrous materials with changed porosity is the process of electrospinning (ESP-process), which provides the possibility of applying to the surface substrate – classic non-woven filter materials over a layer (mat) consisting of nanofibres. Elctrospinning is a widely used technique for constructing special high-performance filters. With this technique, a wide variety of polymers can be spun. One of these is also polyvinyl alcohol (PVA), which has very good properties in this field. Its water solubility can be adjusted with crosslinking techniques – chemical or physicochemical crosslinking [1].

The principle of electrostatic spinning is based on applying a strong electromagnetic field to the polymer solution to create an electrically charged stream of polymer solution or melt. The tension created between the two electrodes then act on the polymer located on the spinning electrode. At the final stage of electrospinning, phase transformations take place, including molecular diffusion of the solvent to the filament surface, evaporation of the solvent, the formation of a solid phase of the fibre-forming polymer.

The properties of the fibres and their mats depend on the process parameters, the chosen polymer and nature of the collector on which the layer is applied. By changing these parameters it is possible to modify and to control the appearance and quality of the deposited fibres as well as the morphology of the deposited mat. The fibres are paving and oriented on the collector, and the appearance of the surface of the overlay, when the purpose is to obtain double layer or multilayer composite also is affected. To increase the mass of matte and the efficiency of the process, it is of particular importance to choose the couple – a polymer solution and a suitable surface of the collector (organic/inorganic material, smoothness / fluffiness, hydrophilicity / hydrophobicity, morphology, density, etc.). The reason for this is the interaction of the solution with the electric field, because simultaneously with the drawing of the fibres, evaporation of the solvent is carried out [2].

The collectors are used to deposit the resulting fibres. They are mostly made of metal. Changing the surface and the type of collector leads to a change in the morphology of the native nanomat. Fabric collector porosity or evenness is the result of the constructional parameters combination - aluminium foil, fabric or nonwoven material. This is because the final fibres still not completely stable because of the presence of residual solvent therein. Thus, the hydrophilicity / hydrophobicity of the pad on which the fibres are applied affects the adhesion of the mat of nanofibres and the layer formed by them to the pad [3, 4].

Surface modification via plasma treatment not only eliminates the need of wet processing, but several modifications are carried out: hydrophilicity/ hydrophobicity alterations, surface roughening, grafting, flame retardant, antimicrobial, insect repellent, stain resistant, and single or multiple surface functionalization [5]. Many novel changes using plasmas have been thoroughly analysed and well documented.

The atmospheric plasma treatment allows specific modifications of the surface properties to be carried out without causing a change in the fibre or bulk properties. Plasma provides an excellent way to study surface modifications, and is reasonable for batch processing of small product quantities. It is not, however, feasible for on-line processing in a textile mill. Thus, the hydrophobicity/hydrophility of the fibre-matrix adhesion, for example to perform finishing processes with synthetic fibres increased, which is important for the amount of nanofibres mat on the textile collector. The increase in adhesion strength of the boundary surface depends on the growth of roughness and the emergence of new active groups.

Most of the studies in the field of textile modification to increase adhesion relate to two-layer film materials or duplication of these materials. For inert polymers, as polyethylene terephthalate (PET), formation of the reaction centre is the only way for increased strength at the boundary between the two layers.

Polyester fibre (PET), as a very important textile fibre, has been widely used in production of nonwovens for some industrial areas in particular with different application filters due to its superiorities of high strength, mechanical stability and low cost [6, 7]. Bat it is known that PET fibres are characterized by a dense, fine-grained surface structure which in many cases impedes adhesion and diffusion in their volume. PET fibres has high hydrophobic behaviour and at 100 % air relative humidity absorb 1% humidity only. Therefore, it is desirable to change their surface properties because, it is very difficult for hydrophobic surface of polyester fibres to be wetted.

In that paper have been investigated the effect of pad pre-treatment in cold plasma environment. Through the electrical discharge and radicals creation, we are managed to hydrocarbons deposition on the PVA layer and increase surface hydrophobicity in a rapid way. The surface chemistry change had a relatively small effect on the fibre structure sample and individual fibre morphology. The surface properties of PET fabrics were then characterized using ion exchange capacity determination.

In our other studies it has been found that the hydrophilicity/hydrophobicity of the material upon which the fibres are deposited in order to obtain a two-layer structure is of importance at the beginning of the process [8]. The amount of the deposited fibres depends on collector surface (hydrophobicity/ hydrophility) properties with the same other electrospinning process parameters. The nanofibres laying process is stabilized by increased PET pad hydrophilicity and subsequent more uniform and rapid moisture diffusion in the volume of nonwoven base. When the pad has high hydrophobicity and lower thickness, the solvent evaporation in the chamber depends on the ambient humidity and as a result there is a greater unevenness of the fibre deposition process. A mat of higher density nanofibres is obtained in the case of hydrophilic pad (cotton tissue).

So, through different type of physic-chemical treatment, adhesive and electrophysical properties can be changed at the expense of formation of boundary molecular layers and thus change the nature of the polyester fibres surface that do not exist in the native fibre. It is known that after processing in cold plasma, the hydrophilic properties of the polymers change. The picture is considerably more complicated in the case of non-woven needle-punched textile materials with varying area and volume widely used in the automotive industry. As a result of the plasma treatment, the ionic groups increased on the PET fibres surface, as a result of which the interaction with the deposited PVA mat increased also. Furthermore, depending on the processing mode the amount of accelerated ions and their energy are different. At the same time there are two areas where ultraviolet radiation in the chamber has different intensity and therefore possible a photooxidative reaction occurred with variety of chemical products. This affects the amount and type of new active centres created on the polyester fibres surface.

In general, plasma treatment is associated with several effects, each of which influences on the adhesion between the two layers of the composite: the non-woven pad and the nanofibres mat.

Surface plasma activating of fibres results in:

- Plasma etching, increasing the relative and active surface as a result of bombardment with accelerated ions leading to micro cracks and craters formation on the fibres surface;
- New active sites formation as a result of beat out electrons and radicals formation on the fibres surface; Other polar moieties that are formed during treatment include carbonyls (–CO) and carboxyls (–COO). This change in surface polarity has been widely studied for adhesion increase. Several excellent reviews are referenced here that detail the specific and mechanisms of the treatment [9-11].
- Cross-linking due to new functional groups formation with oxygen or water from the air and cross-linking with hydrogen bridges; Cross-linking is also associated with a reduction in the number of hydroxyl groups already introduced into the surface and they become inaccessible to interact with water [12].

It can be seen that plasma exposure leads to changes in the polymer composition surface. This includes free radicals formation. These radicals enable reactions such as cross-linking by activated species of inert gas surface graft polymerization, as well as the incorporation of functional groups.

The present work attempts to elucidate these complex and simultaneous changes in the materials as well as their influence on the surface properties of the non-woven fabric. It is necessary to evaluate these changes so that a suitable pretreatment regime can be selected, ensuring the desired good or poor adhesion between the two layers of the composite. Depending on the gas used in the chamber there is a difference in radical intensity for constant exposure time and duration of exposure 180 s [13]. There was is difference in radical intensity or the various substrates and gases, as specified in next order: linen > cotton > polynosic rayon > standard rayon > nylon6 > PET and CF_4 > CO > Ar > N_2 > O_2. Due to relatively low radical intensities measured for O_2 in comparison to CF_4, it is concluded that the O_2 plasma generated free radicals are unstable with a short lifetime. Oxygen and air plasma treatment leads to surface wetting improvement of synthetic polymers (PA, PE, PP, PET, PTFE). By oxygen or air plasma treatment, it can be controlled the hydrophilicity of the PET polymer in the desired direction.

For plasma-exposed PET nonwoven collector, the hydrophilicity/hydrophobicity of the substrate surface can be altered by the gas selection. In other words, oxygen can create –OH groups, which increase hydrophilicity.

Good adhesion property can be achieved if between the two layers, in our case needlefelt non-woven fabric and a nanofibre mat, can initiate a boundary contact area, figure 1. This happens when the contact distance is less than the cut off distance of the van der Waals attractive force [14].

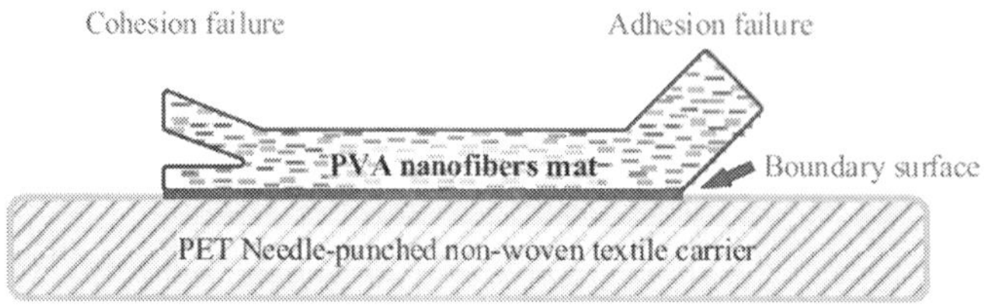

Figure 1. Structure of double-layer gas filter media consist of PET Needle-punched non-woven textile carrier, boundary surface and PVA nanofibres mat.

For inert polymers such as PET, reactive centres formation is the only possibility to create a strong bond between the two layers [15].

Electrospun mat has a different adhesion to the collector, depending on its surface smoothness. Extraordinary adhesion properties are demonstrated by Wong and co-workers [16, 17].

The aim of the work is to define empirically the changes in nonwoven fabric designed for pad of the nanofibres layer, related to the surface modification of the fibres, their nature and their impact. Durable bilayer composites on the basis of needle punch PET non-woven fabric with a layer of nanofibres are designed for gas filters. The influence of low temperature plasma on the hydrophilic properties of a non-woven from polyester fibre was examined.

2. Full ion exchange capacity Determination

In the present work are presented only results related to the change of cation-exchange properties after plasma chemical modification of polyester carrier of the PVA nanofibres layer. These properties determine the adhesion strength between the two used layers due to the increase of the negatively charged group's quantity on the surface.

The nonwoven fabric used for the carrier of the layer of nanofibres is: needle felted filters produced from polyester fibres with 3.4 dtex density and 57 mm staple length, 400 g/m^2 surface density. They are treated by plasma. As plasma- forming gas was used argon and oxygen gases, applied plasma treatment duration is 1 min and 2 min.

The prepared samples are washed at 60°C in an ultrasonic bath for 2 min treatment time. The drying process is accomplished at 105°C.

The input parameters of the plasma unit varied within the following limits: discharge power Wp = 0.5-2.0 kW, working pressure in the discharge chamber P = 80 Pa; plasma gas consumption G = 0.04-1.2 g/s; generator frequency f = 1.76 MHz, treatment time t = 1 and 2 min. The indicated times are selected because they are quite sufficient for the creation of new active centres or so called plasma-chemical modification.

Argon and oxygen are used as plasma-forming gas; arc current $I_d = 65 \pm 5$ A, sample rotation speed 1.5-2 revolutions per minute.

Ion exchange capacity, is the measure of the number of replaceable H+ /–OH ions per unit mass of the exchange. Determination of full cation exchange capacity is defined by titration method with 0.01N sodium hydroxide solution at room temperature. In parallel, a blank sample is also run. The complete cation/ anion exchange capacity in mgeqv/g is calculated according to well-known formula.

3. Results and discussion

As has already been indicated in the treatment of textile materials with plasma their properties are changed – mechanical, physical, chemical, and electrical. This happens simultaneously.

The measurement of the thickness of the plasma-treated non-woven fabric under the above conditions was performed at the follow conditions: climatic condition $T = 20°C$, humidity = 62 %; work pressure $P = 1$ kPa. The results obtained are presented in table 1.

Table 1. Change in thickness depending on the type of gas used.

No	Untreated	Plasma treatment in oxygen (O_2) for 1 min	Plasma treatment in oxygen (O_2) for 2 min	Plasma treatment in argon (Ar) for 1 min	Plasma treatment in argon (Ar) for 2 min
Thickness average value $\overline{x}$ (mm)	1.948	1.929	2.043	1.853	1.910
Dispersion s (mm)	0.042	0.027	0.057	0.024	0.032

The thickness change is greater when O_2 is used as the plasma-forming gas at 2 min treatment time. This is due to the introduction of bigger quantity oxygen-containing groups. This leads to an increase in surface free energy and an increase in fabric's wettability. These oxygen-containing groups may also cause additional surface modifications, which may contribute to increasing adhesion between the two layers of the filter.

Depending on the number of hydrophilic groups capable of attracting and retaining water near themselves textile fibres have more or less hydrophilicity therefore. At the same relative humidity and air temperature, studied samples have different moisture content, figure 2. That fact can be explained by the material polarization.

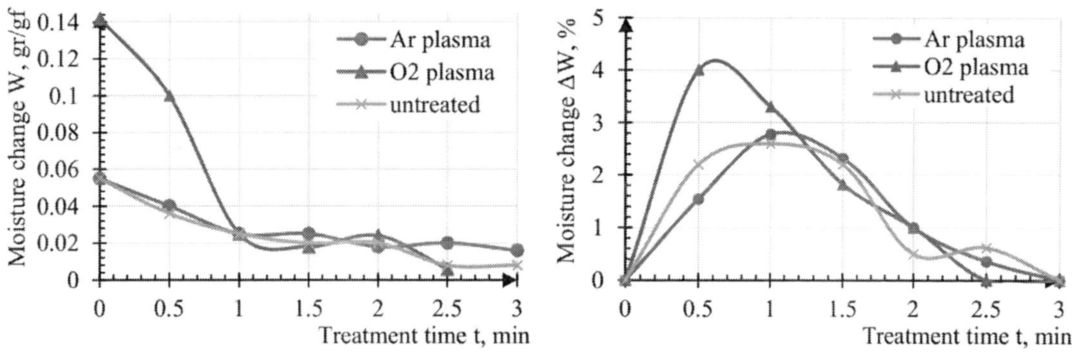

Figure 2. Summary experimental data on the change of moisture in the samples at the working pressure $P = 80$ Pa and O_2 and Ar as gas in the chamber.

To determine the plasma treatment influence on the change of the quantity of active groups on the surface and hence the change of adhesion with the applied second layer of the nanofibres composite the study of the change of moisture content is not sufficient. The electrophysical properties of the material changes as a result of the change in the amount of negatively charged active centres on the surface, the thickness of the nonwoven fabric increases and the size of the capillaries is changed at the same time. At the same time, the influence of the radius on the capillary and the amount of the negative/positive side of the active site is opposite. Therefore, it is necessary to use a parallel method by which the influence of these simultaneous changes in the material can be identified. In that study differential scanning calorimetry (DSC) analysis of treated samples is performed and the change in PET fibres shows the difference in the amount of heat required to increase the temperature, table 2.

Plasma treatment leads to additional oxygen-containing functional groups such as –OH, –C=O, –COOH to the samples surface that are evaluated by DSC analyses to detect possible changes in thermal stability and melting behaviour.

Table 2. DSC evaluation of plasma treated PET non-woven samples.

No	Type of plasma-forming gas	Plasma treatment duration *t*, min	Entalpy Δ H, J/g	Peak Area, mJ	Max. peak value °C	Peak height, mW
1	Ar	1	60.4969	120.994	252.11	6.8683
2	Ar	2	62.2919	124.584	251.25	7.3700
3	O_2	2	70.3840	126.691	251.66	6.7632
4	O_2	1	63.0338	125.461	251.21	6.0325
5	Untreated	0	80.1161	208.302	240.31	8.7330

According to the results presented in table 2 is observed enthalpy decrease of the system after treatment in oxygen and Ar plasma compared to untreated material, which means changing the crystallinity of the surface of the PET fibres. The melting temperature of the polymer also changes. Free radicals formation on the surface due to the oxygen-containing and argon-containing gas plasma treatment leads to the formation of a chemically active surface layer and increasing the adhesion of the nonwoven fabric to the nanofibrous electrospin layer on it.

F. Rombaldoni et al. [18] used low-temperature oxygen plasma treatment to enhance the adhesion between polypropylene nonwoven and poly(ethylene oxide) and polyamide-6 nanofibrous mats deposited onto polypropylene nonwoven. Improving the adhesion between the nonwoven and the nanofibrous mat, which was attributed to the increased wettability of the supporting polypropylene fabric and polar functional groups introduced by plasma treatment, allowed a stronger interaction with the treated fabric. It should be expected that after increasing the amount of negatively charged groups on the surface it becomes negatively loaded due to the increase in weakly acidic groups as carboxylate group (COO-). This was attributed to the increased surface roughness of the PET fibres leading to enhanced mechanical adhesion between the two layers of non-woven PET fabric and the PVA nanofibres layer. With the IEC determination, it is observed that the amount of oxygen-associated groups increases after the plasma treatment, making the fabric surface more wettable, Figure .

After the plasma treatment the PET fibres surface are activated and the CEC increases. This is evident from the obtained results presented in table 3.

Table 3. CEC (cation exchange capacity) evaluation of plasma treated PET non-woven samples.

No	Type of plasma-forming gas	Treatment time *t*, min	CEC, mEq/100g
1	Ar	1	10.9
2	Ar	2	11.2
3	O_2	2	19.3
4	O_2	1	18.6
5	Untreated	0	7.8

4. Conclusion

Cold plasma activation in the O_2 and Ar gas media was used as a good possibility for polyester non-woven modification used as a carrier for PVA nanofibres, designed to produce high-grade double-layer dust filters for automobile industry.

Plasma activation is an environmentally friendly (green) technology for filter functionalization which consist in improving hydrophilicity, wettability, wicking, and ion exchanging interaction.

Polyester plasma activation produce different new functional groups on the surface of the fibres as – C–OH, C–O–OH, C–OH, C=O, O=C–O–OH, C–O–, O=C–O–, C–H+, C–H_2O+ etc.

Experimentally, as a result of pre-treatment in plasma, the adhesion between the two layers of the double layer gas filter consists of needle-punched nonwoven PET fibre and PVA nanofibres mat, has been improve.

The most significant change in the values of moisture in the samples is in O_2 medium. This fact can be explained by that this textile non-woven PET material is significantly thicker and characterized by a much larger surface and the greater activity of O_2 vs. Ar gases.

Acknowledgments

The authors would like to thank the Research and Development Sector at the Technical University of Sofia for the financial support.

References

[1] Geoffrey I T and van Dyke MD 1969 Electrically driven jets, *Proc. R. Soc. Lond. A.* **313**(1515) 453–75

[2] Deitzel J M, Kleinmeyer J, Harris D and Beck Tan N C 2001 The effect of processing variables on the morphology of electrospun nanofibers and textiles, *Polymer* **42** 261–72

[3] Neznakomova M and Geninska E 2011 Analysis of thermoplastic non-woven textile materials designed for nanofibres, *Textile and garment magazine* **12** 326–31

[4] Geninska E and Neznakomova M 2013 Producing bilayer materials containing nanofibers, *Sci. conf. with international participation EMF'2013 (Sozopol, Bulgaria)*

[5] Wakida T and Tokino S 1996 Surface modification of fiber and polymeric materials by discharge treatment and its application to textile processing, *Ind. J. Fibre and Textile Res.* **21**(69) 69–78

[6] Varesano A et al 2016 A systematic study on the effects of doping agents on polypyrrole coating of fabrics, *J. Applied Polymer Science* **133**(1) 133

[7] Yu T Z, Pang Z and Wei Q 2015 Flame retardancy and conductive properties of polyester fabrics coated with polyaniline, *Textile Research J.* **86**(11) 1171–9

[8] Gospodinova D, Neznakomova M and Dineff P 2017 Influence of cold air plasma treatment on thermobonded PET nonwoven textile, *Proc. XV-th Int. Conf. Electrical Machines, Drives and Power Systems ELMA 2017 (Sofia)*

[9] Podhajny R M 1988 Corona treatment of polymeric films *J. Plastic Film & Sheet.* **4**(3) 177–88

[10] Zhang D, Sun Q and Wadsworth L C 1998 Mechanism of corona treatment on polyolefin films, *Polymer Engineering & Science* **38**(6) 965–70

[11] Nitschke M 2008 Plasma modification of polymer surfaces and plasma polymerization, *Polymer surfaces and interfaces: characterization, modification and applications (Springer)* 324

[12] Inagaki N 1996 *Plasma Surface Modification and Plasma Polymerization 1-st ed* (CRC Press) p 280

[13] Chen J R 1991 Free-radicals of fibers treated with low temperature plasma, *J. Applied Polymer Science* **42**(7) 2035–7

[14] Lee H J et al 2007 Adhesion test of nanostructured materials by a Novel AFM probe, *Key Engineering Materials* **353-358** 2253–6

[15] Gerasimova H A, Kuzmichev and V E, Veselov V E 1966 *Improving the physicochemical methods of increasing the strength of adhesive compounds*, Ivanovo: Ivanovo Textile Institute.

[16] Huber G, Mantz H, Spolenak R, Mecke K, Jacobs K, Gorb SN and Arzt E 2005 Evidence for capillarity contributions to gecko adhesion from single spatula nanomechanical measurements, *PNAS* **102**(45) 16293–6

[17] Shi Q., Wan K T, Wong S C, Chen P and Blackledge T A 2010 Do electrospun polymer fibers stick?, *Biology Letters* **26**(17) 14188–93

[18] Rombaldoni F, Mahmood K, Varesano A, Bianchetto Songi M, Aluigi A, Vineis C and Mazzuchetti G 2013 Adhesion enhancement of electrospun nanofiber mats to polypropylene nonwoven fabric by low-temperature oxygen plasma treatment, *Surface and Coatings Technology* **216** 178–84

Estimation of laser cutting process efficiency

C C Girdu[1], L A Mihail[1] and M V Dragoi[1]

[1]Transilvania University of Braşov, Manufacturing Engineering Department, Bdul Eroilor 29, 500036 Brasov Romania

E-mail: dragoi.m@unitbv.ro

Abstract. The efficiency of laser cutting processes is generally treated in technical literature in qualitative terms, referring to ways to increase it. The present paper is focussed on metal cutting by laser and proposes some quantitative means to estimate the process efficiency. For certain working conditions – machine-tool, material to be processed, specific costs and other – the effectiveness and the specific power consumption are computed based on the main cutting parameters: laser power and cutting speed. The proposed mathematical relationship can be successfully used when the criterion of process optimization is the environment friendliness. A relevant case study is presented, as well. When significant different samples are to be compared, the criterion used to evaluate laser cutting efficiency becomes very important.

1. Introduction

Generally speaking, efficiency of machining processes is a relative vague concept. It can be interpreted in terms of economical or technical aspects, it can refer to different specific processes, or to a certain material. In the technical literature, the efficiency of machining processes is approached mainly from the perspective of the ways to increase it.

A survey of literature reveals different kinds of the concept's approach. First of all, different materials processed by laser are studied in research works: metal, polymers, glass, textile, human tissue, ceramic and others. Thinking about the specific of laser processing, different types were found: cutting, remote cutting, laser-assisted machining. In terms of areas of applying laser processing, the main ones were industry and medicine. Indifferently the material, type of laser processing, or area of applying, the most frequent approach of efficiency's study is dedicated to ways of improving it.

So that, in [1], are presented researches regarding the efficiency of the remote laser metal cutting. In context of the cited paper, authors mean by '*efficiency*' the capability of an installation to cut an as possible thick steel plate. It was proved that a steel wall having the thickness of 60 mm can be cut with power of 7.5 kW of the laser radiation. In the paper [2] an optimization method of laser cutting parameters is presented. Rate of removal of the melted material by assistant gas was used as basis to define a model to estimate the cutting process. Though the title of [3] refers to the efficiency of laser cutting of electrodes for Li-Ion batteries, a main aspect treated is focused on physical phenomena encountered at this kind of processing. This paper presents the advantages of laser cutting of electrodes compared to classical machining, and the efficiency of the laser cutting is estimated in relationship with cutting parameters. The articles [4] and [5] discuss laser-assisted cutting, for that the laser is an auxiliary process that only assists mechanical cutting, making it more effective and accurate. The reference [4] presents the results of a combined process in which laser heats selectively the zone of glass to be cut, thus becoming easier machinable. The type of chip is modified, and the main consequence is a better geometrical accuracy and roughness of the part. The reference [5] presents a hybrid method to process materials: a classical machining preceded by a preheating by means of a laser beam. Neither the type of cutting (milling, turning, other), nor the material are specified. The cutting parameters are selected in accordance with the modified properties of the material. It was proved that cutting combined with laser pre-heating was more effective and the dimensional accuracy and roughness were better compared to classical machining. Power consumption was reduced, as well. In [6] the efficiency of cutting textile materials is appreciated by means of percentage of laser power lost by the translluminate phenomenon. This loss is directly influenced by the angle of fibers. The paper [7] provides an overview on the impact on environment of laser cutting technologies. It is a discussion mostly in qualitative terms, with regard to the way to generate the beam. The recent advances in laser technologies provide industry with equipment more efficient in terms of power consumption. The main drawback of this equipment is that the quality of the machined surface quality is lower, especially for sheets with bigger thickness. In [8] a study on the efficiency of laser cutting of ceramics is presented. Ceramics machining has some specific features related to the need of high laser power, because of high temperature needed to cut such a material. In these conditions, the problem of efficiency becomes more relevant and cutting parameters have to be carefully selected to get a good quality of the product. Thermodynamic and efficiency analysis are combined in the research. The paper [9] presents the investigations which had as a goal finding the optimum in terms of requirement of specific energy to cut holes in Kevlar plates. The needs of specific energy is computed related to different scanning speed, laser power, and diameter of machined holes. The papers [10, 11] refer to using laser to work on human tissue, and present the problem from the perspective of cutting efficiency. In [10] efficiency is appreciated mainly in terms of the quality of the surface obtained. Talking about human tissue, an important aspect was the measure it was affected by processing, for instance if the dentine tubules were occluded or remained opened.

As it can be seen in the approaches presented above, none of the researches have put the efficiency of laser cut processing in a direct relationship with the cutting parameters. Furthermore, in some cited papers, efficiency expresses rather the capability of the process to produce some certain output, than its effectiveness. The present paper proposes a discussion on different ways to interpret efficiency as effectiveness, and how can it be directly, or indirectly influenced by the cutting parameters.

2. Efficiency indicators of laser cutting processes

To narrow down the area of research, and to be more specific, the present paper focusses on laser cutting of metal. For this particular category of materials, the process consist mainly of melting (and even vapouring) and removing the melt. The volume previously filled by the melt becomes the so called kerf that separates the part and the metal sheet. In this context, efficiency can be interpreted in different ways, or expressed by different indicators, as follows:

- Effectiveness, that is how fast a part is produced;
- Linear specific energy (E_l). This expresses one way to appreciate the measure the process is friendly to environment. This can be estimated by the quantity of energy spent to process a unit of length of part edge. Of course, the lower the linear specific energy is, the more environmentally friendly the process is considered to be;
- Surface specific energy (E_s). This indicator is intended to take into account the thickness of metal sheet. It expresses the energy spent to process a unit surface on the side of cut part;
- Volume specific energy (E_v). This is a third mean to appreciate the measure the process environmentally friendly. This can be estimated by the quantity of energy spent to melt a unit of volume, and is interpreted in the same way as the previously specified indicator.

These indicators are influenced by some input parameters as: laser power, cutting speed, pressure of the assistant gas, the processed material itself (characterized by different properties as specific heat, conductivity and other), the metal sheet thickness. It is important to note that these parameters influence directly by their values the efficiency indicators, but also indirectly, by means of their combined effect/interaction (especially the couple power-speed). The cutting parameters can be directly adjusted in the process, to influence in a direction or in the opposite one any efficiency indicators. Some other data are given, and cannot be changed for a specific batch, e.g. the part material itself or the thickness of the metal sheet. This is the reason for that the given data will not be involved in computing efficiency indicators. Discussions on each of the mentioned indicators are provided in the following paragraphs.

2.1. Effectiveness

Effectiveness is already used as an indicator to express efficiency. It is easy to be computed, based only on cutting speed and the length of the part contour. Obviously, the higher the cutting speed is, the efficiency is better. It has to be mentioned, that this is a rough indicator, used mainly when the target is minimizing cutting time. Furthermore, changing cutting speed should be accompanied by an adjustment of laser power that produces some side consequences that cannot be appreciated exclusively by this indicator.

2.2. Linear specific energy

Linear specific energy (E_l) measures the quantity of energy spent to process a unit length [mm] of part edge. This depends on the power of laser, and the time spent to process 1 mm of part edge (3).

$$E_l = (P * t)/L \quad (1)$$

$$t = L/v \quad (2)$$

$$E_l = \frac{P}{v} \quad (3)$$

In (1), (2), (3) we have:
E_l – Linear specific energy [W*s/mm], [J/mm];
P – Laser power [W];
t – Processing time [s];
L – Length of part side edge (contour) [mm];
v – Cutting speed [mm/s]. **Note** that in context of NC (Numerically Controlled) machine-tools the usual unit measure for speed is [mm/min], so the appropriate transformation to [mm/s] should be performed before using the value in formula.

Replacing (2) in (1), easily results (3), the linear specific energy E_l.
This is a synthetic indicator, since it does not take into account thickness of the part, or other input data. It can be used to appreciate the efficiency of the cut within a certain batch, where thickness is a given data and cannot be modified, hence, has to be not involved in any computing relationship. According to (3), a process is more efficient if ratio P/v has lower value. Though, specific technological restrictions do not allow decreasing laser power below a certain limit, on one hand, and equipment limitations restrict increasing cutting speed over a limit, on another hand. Furthermore, the combined action of power and speed force user to adjust the mentioned ratio in a reasonable range. With all these, E_l still remains a useful indicator to appreciate cutting efficiency in terms of eco-friendly technology criteria, and guides user to adjust cutting parameters towards the goal of saving energy.

2.3. Surface specific energy

Surface specific energy (E_s) is already used as an efficiency indicator by some authors [11,12]. Such an approach takes into account the thickness of the part, to calculate the energy spent to process a unit of side surface of the part (6) [11,12]. Similar relationships are often used to describe the specific energy consumption expressed by how large side surface can be processed spending a unit of energy.

$$E_s = (P * t)/A \quad (4)$$

$$t = L/v;\ A = L * h \quad (5)$$

$$E_s = \frac{P}{v * h} \quad (6)$$

where:
E_s – Surface specific energy [W/mm^2];
A – Side surface area of the part [mm^2];
h – Metal sheet thickness [mm].

Anyway, such a comparison should be performed with a grain of salt, especially when the difference of thickness is big. In such a situation the physical phenomena that is developed during cutting might differ significantly (at least in their magnitude) and alter the result of comparison. When it comes to apply it for practical purposes, it allows to compare processing parts having different thicknesses in terms of efficiency in respect to environment protection.

2.4. Volume specific energy

Volume specific energy (E_v) is an indicator that shows how much energy has to be spent to melt a unit volume of material. It takes into account the average width of kerf (9). In a derivate shape, some authors name it melting efficiency [11].

$$E_l = (P * t)/V \tag{7}$$

$$t = L/v; \quad V = L * A_k; \quad A_k = W_{avg} * h \tag{8}$$

$$E_V = \frac{P}{v * W_{avg} * h} \tag{9}$$

where:
E_v – Volume specific energy [W/mm^3];
V – Volume of kerf (of melt material [mm^3];
S_k – Area of cross section of kerf [mm^2];
W_{avg} – Average kerf width [mm] (according to figure 1).

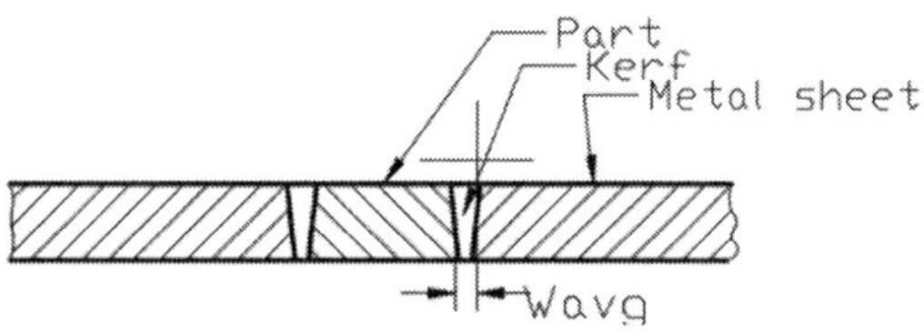

Figure 1. Cross section through metal sheet and part. Average width of kerf.

We consider that the relationship (9) is important because depending on input variable parameters: laser power and cutting speed, on one hand, pressure of assistant gas and even focus point and given properties of the material (specific heat, conductivity, reflexivity) on other hand, the shape and width of kerf (gap between the part and raw material) varies. This variation, can be sometimes relatively significant, but it does not influence the part itself if trajectory corrections are correctly applied. The wider the gap is, a bigger volume of material have to be melted. In these terms, the less quantity of material to be melt (that is the narrower kerf) is advantageous in terms of energy consumption.

3. Case study

3.1. Case study setup

A case study was conducted to illustrate using the indicators of laser cutting processes efficiency. Hardox 400 and Hardox 450 steels have been cut with different combinations of input data. Hardox is a difficult to be machined alloy and it is often processed by means of non-conventional methods, as abrasive waterjet [13] or laser cutting. Combinations of laser power (*P*) and cutting speed (*v*) levels were selected from data sets, according to design of experiments (DOE) procedures. For the case study presented here, the trial values of *P* and *v* are displayed in table 1, along with other variable input data as the steel sheet's thickness and the average width of the kerf. Note that the average width of the kerf is considered an input data for this case study. In fact, it was measured for each part; its value is influenced by the value of *P* and *v*, and their combined action on the metal sheet, as well. The combinations used limit values for *P* and *v*, which were determined by means of initial test screening trials, as to be sure that cutting process develops properly, that means that the part is completely cut (separated from the sheet material) and the machined surface's roughness is appropriate. Even if the combinations where the extreme low, or extreme high values for both the input parameters were coupled, the part output quality was acceptable.

All the tests were performed for the same shape of part, formed of three segments and an arc. The length of contour was L= 91.4 mm.

The specific energy in its different forms was computed for each combination and the output data were centralized in table 2. The link between table 1 and table 2 is done by means of number of test (column Test#).

Table 1. Input data for case study.

Test#	Material	Power P [w]	Cutting speed v [mm/min]	Sheet thickness h [mm]	Average width of kerf, W_{avg} [mm]
1	Hardox 400	4100	1200	10	1.285
2	Hardox 400	4100	1600	10	1.105
3	Hardox 400	4200	1600	10	1.470
4	Hardox 400	4200	1200	10	1.090
5	Hardox 400	4300	1600	10	1.190
6	Hardox 400	4300	1200	10	1.135
7	Hardox 450	3700	1450	12	0.610
8	Hardox 450	3700	1650	12	0.870
9	Hardox 450	3800	1450	12	0.890
10	Hardox 450	3800	1650	12	0.610
11	Hardox 450	3900	1450	12	1.060
12	Hardox 450	3900	1650	12	0.610

Table 2. Output data.

Test#	El [J/mm]	Es [J/mm^2]	Ev [J/mm^3]
1	205.000	20.500	15.953
2	153.750	15.375	13.914
3	157.500	15.750	10.714
4	210.000	21.000	19.266
5	161.250	16.125	13.550
6	215.000	21.500	18.943
7	153.103	12.759	20.916
8	134.545	11.212	12.887
9	157.241	13.103	14.723
10	138.182	11.515	18.877
11	161.379	13.448	12.687
12	141.818	11.818	19.374

Table 3. Ranking the samples according to different criteria.

Rankng	Test# Criterion E_l	Criterion E_s	Criterion E_v
1	8	8	3
2	10	10	11
3	12	12	8
4	7	7	5
5	2	9	2
6	9	11	9
7	3	2	1
8	5	3	10
9	11	5	6
10	1	1	4
11	4	4	12
12	6	6	7

3.2. Discussions

As it can be observed relatively easily, the values of *P* and *v* for the two materials differ. As was previously stated, they were selected based on test processing performed to find the limit values for which cutting still develops properly. Values lower than the minimum ones, or higher than the maximum ones cause improper running of the process, with non-conform output parts. The different values for the two materials are imposed by the different thermal properties of the materials.

Observing the output data displayed in table 2, can be drawn several conclusions:

1. It is obvious that for both the materials, the linear specific energy E_l, that depends exclusive on *P* and *v*, according to (3) the highest value of E_l is recorded for the combination with lowest *P* and highest *v*; the difference in terms of E_l value between the two materials is given by their different properties.
2. The influence of *h* on surface specific energy E_s, shows that this indicator becomes more useful when it is to compare in terms of cutting efficiency related to environment protection parts made of different materials and having different thickness. In these conditions, parts made of Hardox 450, that is those having a bigger thickness are placed at the top of the list, the first four having been ranked the same as according to criterion E_l.
3. Criterion based on volume specific energy E_v. comes with some apparently random results. Being involved more input data, it is almost impossible to appreciate without this indicator which is the least and most energy consuming sample. According to this criterion, the best combination is #3: Hardox 400, thickness of 10 mm, laser power of 4200 (medium), and cutting speed of 1600 mm/min (the highest). The worst sample (the most energy consumer per melted volume unit) is Hardox 450, thickness of 12 mm, laser power of 3700 (the lowest), and cutting speed of 1450 (the lowest). In this context, referring to lowest and highest value of a parameter we mean the extreme value used for a certain material. This is an important remark, because from technological reasons, different minimum-maximum *P* and *v* values must be set up for different materials, depending on their physical properties.
4. Table 3 provides a good imagine of ranking the samples according to different criterion, all of these being related to eco-friendly technologies. Values in columns 2,3, and 4 indicate the number of sample placed on rank stated in column *Ranking*
5. It is important to select correctly the indicator used to appreciate a certain case (set of part samples), according to the input variable data that can be controlled. We have to keep in mind that in certain cases, some input data could be fixed, they may not be changed (e. g. material, or thickness of the part).
6. Depending on the appreciation criterion ranking may differ, sometimes drastically, as can be seen in table 3 and depicted in figure 2.

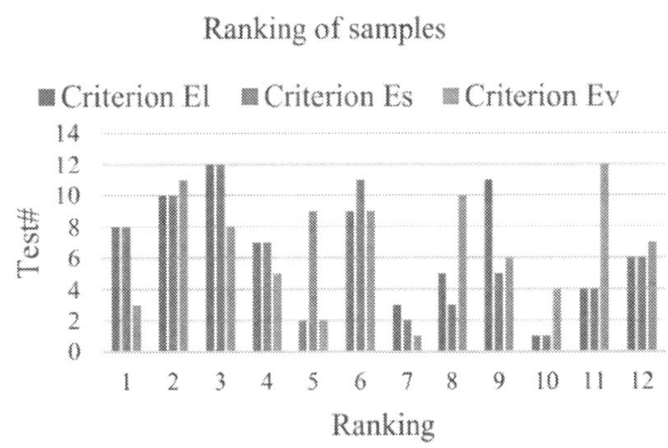

Figure 2. Ranking of probes differ depending on appreciation criterion

4. Conclusions

Nowadays, when environment protection becomes a stringent problem, taking care about energy consumption is an important task. In this context, the energy consumption becomes an important criterion of optimization in manufacturing. Of course, the technical/technological criterion remains the most important, but the environment protection cannot be neglected anymore. The research presented here has proved that even respecting the technological restrictions, that is obtaining parts conform to specifications, still optimization in terms of energy consumption can be performed. The task is not easy, because of the many input data involved and of the particularities specific to each application apart. A mean to appreciate the efficiency of laser cutting of metals are the efficiency indicators, which have to be selected correctly, according to the specific available input data.

Acknowledgments

The researches presented in this paper were possible due the support provided by Bystronic Laser SRL to cut the parts by laser.

References

[1] Gyozdev SV, Glova AF, et al. 2018 *Attenuation of Laser Radiation by the Flame of Burning Hydrocarbons and Efficiency of Remote Cutting of Metals,* III International conference on laser and plasma researches and technologies, Journal of Physics Conference Series, **941**, Natl Res Nucl Univ MEPhI, Moscow, Russia, Jan 24-27, 2017

[2] Kazarev V A, Galushkin M G, Golubev V S, Panchenko V Y and Filipova E O 2005 *Influence of Basic Parameters of Metal Cutting with Gas Laser on Process Energy Efficiency*, XV International symposium on gas flow, chemical lasers, and high-power lasers Pt 1 and 2, Prague, Czech Republic Aug 30-Sep 03, 2004. Book series: Proceedings of the society of photo-optical instrumentation engineers (SPIE) **5777** 864-73 Part: 1&2

[3] Lee D 2018 Investigation of Physical Phenomena and Cutting Efficiency of Laser Cutting of Anode for Li-Ion Batteries, *APPL SCI-BASEL* **8**(2) 266

[4] Ito Y, Kizaki T, Shinomoto R, Ueki M, Sugita N and Mitsuishi M 2017 High-efficiency and Precision Cutting of Glass by Selective Laser-assisted Milling, *PREC ENG-JISPEN* **47** 498-507

[5] Ahn J W, Woo W S and Lee C M 2016 A Study On The Energy Efficiency Of Specific Cutting Energy In Laser-Assisted Machining, *APPL THERM ENG* (94) 748-53

[6] Fuchs A and Zaeh M F 2016 *Efficiency of laser cutting of carbon fiber textiles,* Laser assisted net shape engineering 9 international conference on photonic technologies proceedings of the lane, Furth, Germany, Sep 19-22, 201

[7] Kellens K, Rodrifues G C, Dewulf W and Duflou J R 2014 *Energy and resource efficiency of laser cutting processes,* 8th International Conference on Laser Assisted Net Shape Engineering (LANE), Edited by: Schmidt, M; Vollertsen, F; Merklein, M, Book Series: Physics Procedia **56** 854-864, Furth, Germany, Sep 08-11

[8] Sahin A Z, Yilbas B S and Ayar T 2012 *Laser cutting of ceramic and Thermal Efficiency Analysis*, 14th International Conference on Advances in Materials and Processing Technologies, Edited by: Yigit F, Hashmi M S J, Book Series: Advanced Materials Research, **445** 400-+, Istanbul, Turkey, Jul 13-16, 2011

[9] Sahin A Z, Ayar T and Yibas B S 2009 Efficiency analysis of laser hole cutting, *INT J EXERGY* **6**(4) 592-604

[10] Youssef M, Quinelato A, Youssef F, et al. 2008 Dentinal surface-cutting efficiency using a high-speed diamond bur, ultrasound and laser, *LASER PHYS* **18**(4) 472-477

[11] Pocorni J, Petring D, Powell J, et al. 2016 The Effect of Laser Type and Power on the Efficiency of Industrial Cutting of Mild and Stainless Steels, *J MANUF SCI E-T ASME* **138**(3)

[12] Adelmanna B, Hellmanna R 2011 *Fast Laser Cutting Optimization Algorithm*, Lasers in manufacturing 2011: Proceedings of the SIXTH International WLT conference on lasers in manufacturing, Munich Germany May 23-26, 201, Book series Physics Procedia, **12** 591-598

[13] Filip A C, Morariu C O, Mihail L A and Oancea G 2019 Research on the Surface Roughness of Hardox Steel Parts Machined with an Abrasive Waterjet, *STROJ VESTN-J MECH E* **65**(4) 230-237

IOP Conf. Series: Materials Science and Engineering **659** (2019) 012046 doi:10.1088/1757-899X/659/1/012046

Determination of operating parameters of turbines for micro hydroelectric power plants for optimal use of hydropower

M Prodanović[1], A Miltenović[2] and M Nikodijević[1]

[1]JP EPS Beograd, Ogranak Održivi razvoj, Belgrade, Serbia
[2]University of Niš, Faculty of Mechanical Engineering, A. Medvedeva 14, Niš, Serbia

E-mail: aleksandar.miltenovic@masfak.ni.ac.rs

Abstract. Energy trends have been mostly focused on renewable energy sources in recent years. In the world, about 20% of electricity is produced using energy of water flows, ie. hydropower with a total installed power of 720 GW. The production of hydroelectric power has a number of advantages from fossil fuels or nuclear energy and the greatest is that it does not pollute the environment. In mountainous areas, small and micro hydro power plants are often used, which can be installed on small rivers or streams with little or negligible impact on the environment. The choice of the type of turbine is influenced by a number of parameters, which should be determined before the complete project of the micro hydroelectric power plant. For turbines that are most commonly used in micro hydroelectric power plants, the process of determining operating parameters for optimal utilization of available hydro power is presented in the paper.

1. Introduction

The basic function of the turbine is the transformation of the kinetic energy of the water into the mechanical energy of the rotating parts of the turbine. The operating conditions of the turbine depend on a large number of influencing parameters such as field configuration, possible flow and fall combinations, etc. Many different types of turbines have been developed to cover a wide range of working conditions. It can be said that turbines used in small hydropower plants are a reduced version of turbines used in conventional large hydroelectric power plants.

According to the qualification of the turbines according to the delivered electricity of the micro hydro power plants, they include a power range of 5 to 100 kW.

Considering that in the case of micro-flow hydroelectric power plants, the flow of water varies widely, which is why water turbines are chosen that work efficiently at wide flow limits. Based on the diagram in Fig. 1 in micro hydroelectric applications Pelton, Francis, Kaplan, and crossflow turbines are applied.

2. Parameters of micro hydro power plants

2.1. Operational parameter

Today, technically defined and available to engineers a larger number of turbines, which can be applied to mini and micro hydro power plants. For this reason it is first necessary to show the most important components of a mini hydro power plant with their operating parameters. Figure 2 shows the layout of the most important components and devices as well as the parameters of small and micro hydro power plants.

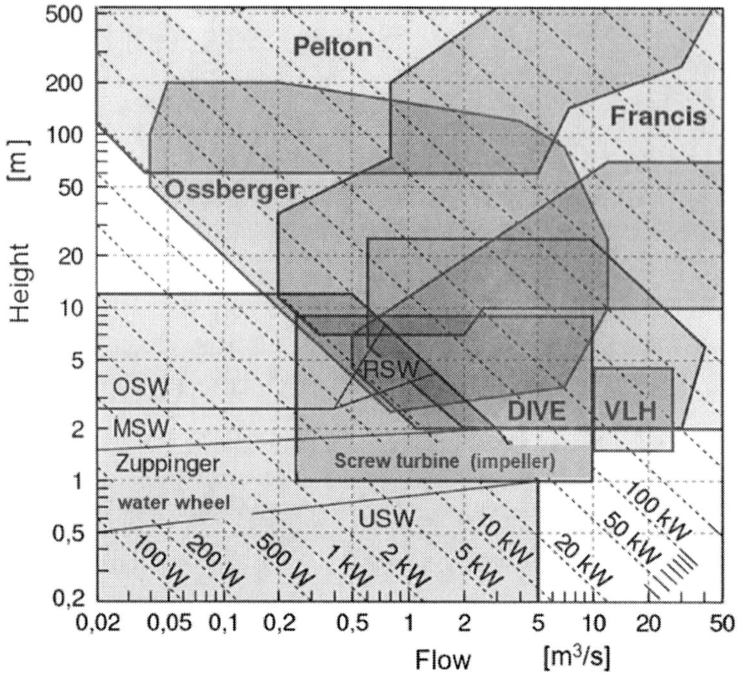

Figure 1. Selection of turbine from the aspect of power, flow and height of the fall.

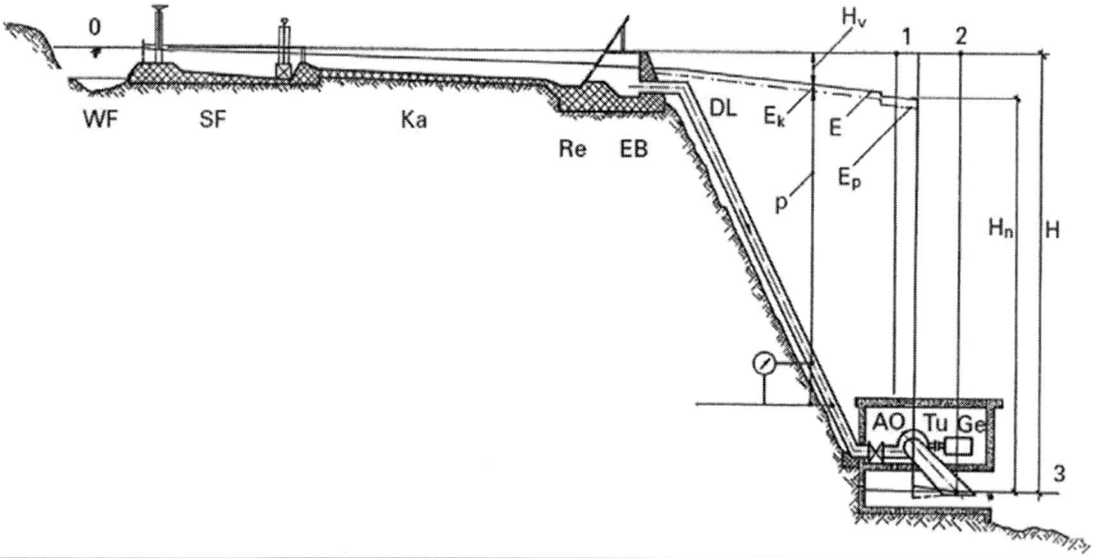

WF – water accumulation	Tu – turbine	E_k – kinetic energy
SF – crossing	Ge- generator	E_p – potential energy
Ka – channel	0 – ref. point of accumulation	p – pressure
Re – grille	1 - ref. point at the entrance to the turbine	H – gross height drop
EB – input workspace	2 - ref. point at the exit from the turbine	H_n – net height drop
DL – input line	3 - ref. point of the drainage channel	H_v – hydraulic energy losses
AO – input body of the turbine	E - energy line	

Figure 2. Profile of micro hydro power plant with components and devices.

The most important parameters of the hydroelectric power plant are:

- Flow - the amount of water that passes through the turbine in the Q unit [m3 / s].

• Gross height drop - height difference between the water level in the accumulation (point 0) and the level of water at the exit from the turbine (point 3) H [m].
• Net height drop - which represents the available potential energy of the turbine H_n [m]
• Hydraulic energy losses expressed through altitude difference represent losses in flow of water through grilles, channels, pipelines and turbine H_v [m]. Determined according to:

$$H_v = A \cdot Q^2 \quad (1)$$

- Hydraulic power of the turbine is determined according to [W]:

$$P_{hyd} = \rho \cdot Q \cdot g \cdot H_n \quad (2)$$

- Mechanical power on the shaft of the turbine [W]:

$$P_{mech} = T \cdot \omega = T \cdot \frac{\pi \cdot n}{30} = \frac{T \cdot n}{9.549} \quad (3)$$

- Efficiency of the turbine is determined as the ratio between the mechanical and hydraulic power of the turbine

$$\eta_T = \frac{P_{mech}}{P_{hyd}} \quad (4)$$

2.2. Specific parameters

Different types of turbines can be classified according to characteristic sizes that include the most important operating parameters. The most commonly used are the following sizes.

1. The specific speed from the aspect of power n_s

It is defined as the number of rotations of the turbine n_s in min^{-1}, where 1 kW power level is obtained for a height drop of 1 m.

$$n_s = n \cdot \frac{\sqrt{P}}{H_n^{5/4}} \quad (5)$$

This term has the advantage that the utilization rate is encompassed by the power P.

2. The specific speed from the aspect of flow n_q

It is defined as the rotational speed of the turbine n_q in min^{-1} at a height drop of 1m and a flow of 1 m^3/s.

$$n_q = n \cdot \frac{\sqrt{Q}}{H_n^{3/4}} \quad (6)$$

Corelation between n_s and n_q:

$$n_s = 3.65 \cdot \sqrt{\eta} \cdot n_q \quad (7)$$

D - nominal diameter of the turbine circuits [m]
A - coefficient
g - acceleration due to gravity: g=9,81 [m/s^2]
ρ - water density ρ=1000 [kg/m^3].
n - working speed of the impeller [min^{-1}]
P - turbine power [kW]
ω - angular velocity [rad/s]

3. Operating parameters of turbines for application to micro power plants

3.1. Pelton turbine

The basic geometric dimensions of the Pelton turbines are (Fig. 3) the mean diameter of the impeller D_0 and the diameter of the jet D_2. The flow of water per nozzle is:

$$Q_1 = 0{,}97 \cdot \frac{\pi}{4} \cdot D_2^2 \cdot \sqrt{2g \cdot H_n} \quad \left[m^3/s\right] \tag{8}$$

For the number of nozzles z_i, the total flow of water is

$$Q = z_i \cdot Q_1 \quad \left[m^3/s\right] \tag{9}$$

The diameter of the water jet

$$D_2 = 0{,}545 \cdot \frac{\sqrt{Q_1}}{\sqrt[4]{H_n}} \quad [m] \tag{10}$$

The middle diameter of the impeller D_1 where the resulting force of the jet of water is operating is determined according to:

$$D_1 = (37...41) \cdot \frac{\sqrt{H_n}}{n} \quad [m] \tag{11}$$

H_n - net height drop [m]
n - speed of the impeller [min^{-1}]

Depending on the specific speed, empirically it can be taken

$$\frac{D_2}{D_1} = 1/7 \,....\, 1/30 \tag{12}$$

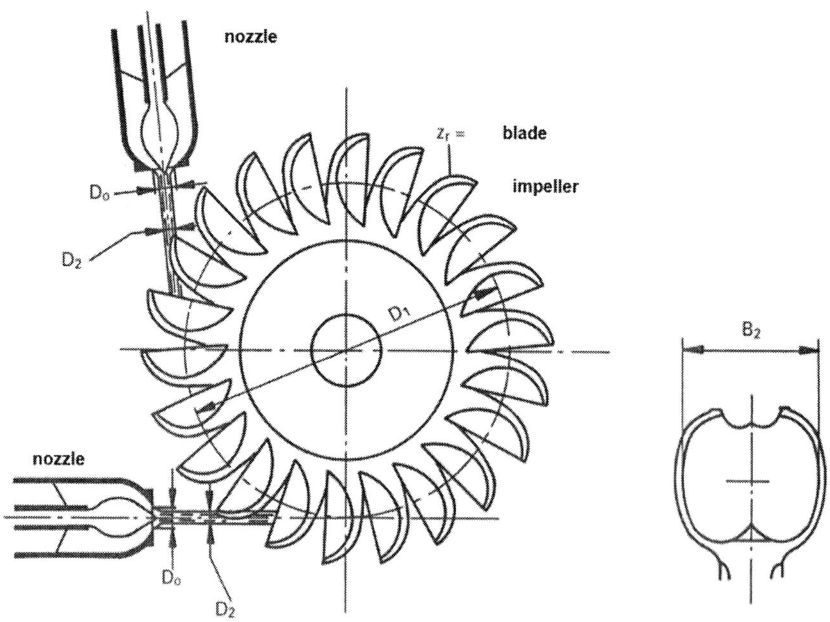

Figure 3. Principal scheme of the Pelton turbine.

3.2. Crossflow turbine

Crossflow turbines are pulse turbines in which water passes twice over the blades of the impeller. It has a simple construction with three main components (fig.4):

• Rectangular introductory device, which directs the flow of water over two profiled blades placed side by side.

• Impeller drum with arched profiled shovels.

• Enclosures that, in accordance with the flow of water, surround the impeller and on which the impeller of the impeller is clamped.

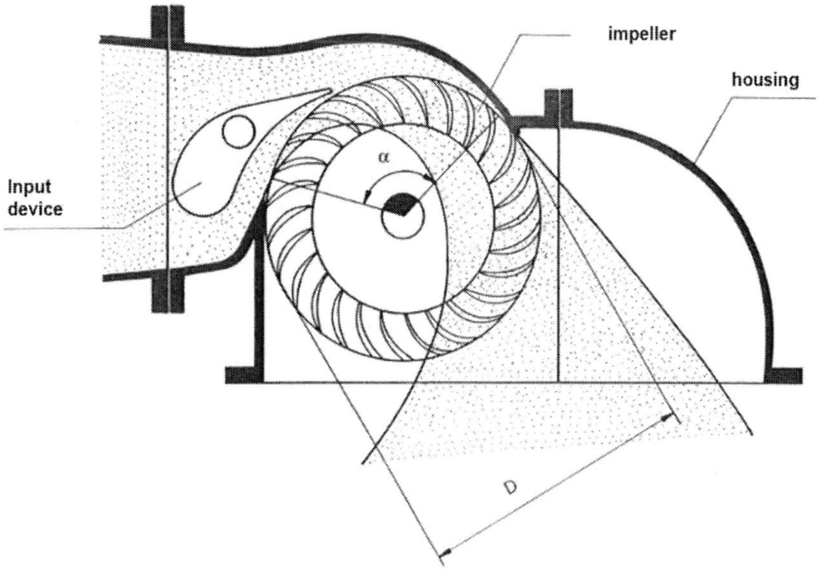

Figure 4. Principal diagram of the crossflow turbine.

The flow of water through the flow turbines is determined according to:

$$Q = 0,25 \cdot \alpha \cdot \frac{D \cdot B}{2} \cdot \sqrt{2g \cdot H_n} = (0,2 .... 0.3) \cdot D \cdot B \cdot \sqrt{2g \cdot H_n} \qquad \left[m^3/s\right] \tag{13}$$

and

$$B \cdot D = (1,13 .... 0,75) \cdot \frac{Q}{\sqrt{H_n}} \tag{14}$$

D - turbine wheel diameter [m]
B - width of the turbine circuits [m]
H_n - net height drop [m]
α - angle of water flow [rad]

The angular velocity, or the speed of the turbine impeller, is determined according to:

$$\omega = 0,45 \cdot \frac{2}{D} \cdot \sqrt{2g \cdot H_n} = 0,9 \cdot \frac{\sqrt{2g \cdot H_n}}{D} \qquad [rad/s]; \quad n = \frac{30}{\pi} \cdot \omega \quad \left[min^{-1}\right] \tag{15}$$

For α=120….90°:

$$D = 38 \cdot \frac{\sqrt{H_n}}{n}; \qquad B = (0{,}02 \ldots . 0{,}03) \cdot \frac{Q \cdot n}{H_n} \tag{16}$$

Depending on the height of the drop H_n and the load of the material of the impeller, the ratio of width and diameter moves within the limits B/D = 0.3 ... 4. The number of blades of the impeller moves within the limits z_r = 24 ... 32.

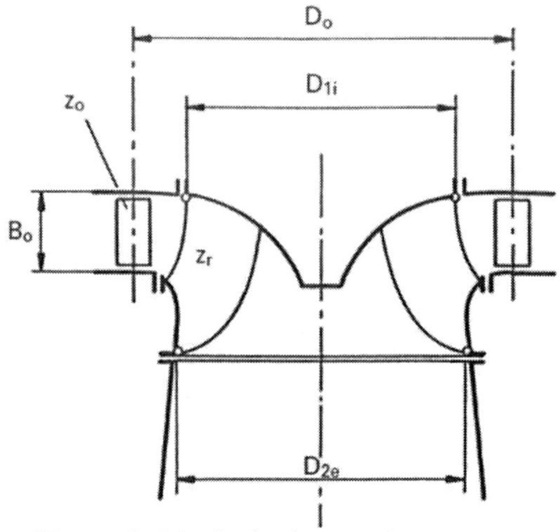

Figure 5. Principal scheme of Francis turbine.

3.3. Francis turbine

With the Francis turbine, the introducer device is a fixed wheel, where the introductory blades are located. The blades are articulated so that the turbine operation can be controlled. The water is fed by a circular channel and directs it to the impeller via the opening blades. The impeller is connected to a shaft that is clamped over the main bearing of the turbine and connected to the gear and generator. Selection of the turbine is made according to the specific speed n_q (Fig.5).

Depending on the flow Q and the speed rpm, the outer diameter of the impeller is determined according to:

$$D_{2e} = 4{,}44 \cdot \left(\frac{Q}{n}\right)^{1/3} \tag{17}$$

The internal diameter depends on the height of the drop Hn and the speed rpm of the impeller, and can be roughly determined according to:

$$D_{1i} = 64{,}4 \cdot \frac{\sqrt{H_n}}{n}; \tag{18}$$

3.4. Kaplan turbine

The Kaplan turbine impeller is similar to a propeller. It is primarily intended for low altitude falls, where it has a good degree of efficiency. The blades of the impeller at Kaplan's turbine are attached to the hub so that their position can be adjusted. The impeller is most often placed in the spiral housing so that the water flows radially through the input device and radially through the impeller. A complete turbine can be integrated into the pipeline, which enables a compact construction.

The oscillatory dimensions of the Kaplan turbine are shown in Fig.6. The outer diameter of the impeller D_e can be approximately determined by

IOP Conf. Series: Materials Science and Engineering **659** (2019) 012046 doi:10.1088/1757-899X/659/1/012046

$$D_e \cong 84,46 \cdot c_{ue} \cdot \frac{\sqrt{H_n}}{n} \quad [m] \tag{19}$$

c_{ue} - velocity coefficient
n - number of turbine circuits [min^{-1}]
H_n - net height drop [m]

The velocity coefficient depends on the specific speed n_q

c_{ue}	1.2	1.4	1.51	1.65	1.75	1.85	1.95
n_q	100	125	150	175	200	225	250

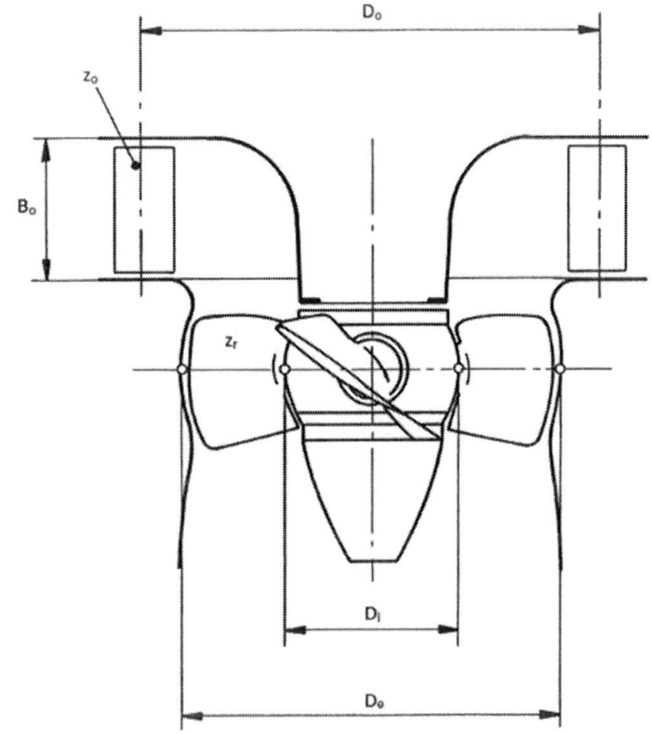

D_e - external (revolving) diameter of the impeller [m]
D_i - hub diameter [m]
D_0 - diameter of the introductory device [m]
B_0 - height of the introductory blades [m]
z_0 - number of introductory blades
z_r - number of wheel blades

Figure 6. Principal diagram and basic dimensions of Kaplan turbine.

4. Determination of operating parameters for selecting the turbine of the micro power plant

Based on the expressions given in items 2 and 3, as well as the nomograms in Fig. 1, turbines for micro hydroelectric power plants can be selected. For the purpose of further analysis, the operating parameters for power of 50 kW (4 variants) and 100 kW (4 variants) were determined for Pelton, throughput, Fransis and Kaplan turbines. The results are shown in Tables 1 to 4.

Table 1. Pelton turbine parameters with 1 nozzle.

H [m]	Q [m^3/s]	η [–]	P [kW]	n_q [min^{-1}]	n_s [min^{-1}]	n [min^{-1}]	D_1 [m]
100	0.06	0.9	50.00	8.00	27.70	1238.85	0.31
90	0.06	0.9	50.00	6.00	20.78	814.49	0.44
80	0.07	0.9	50.00	4.00	13.85	468.65	0.73
70	0.08	0.9	50.00	2.00	6.93	198.30	1.60
100	0.11	0.9	100.00	8.00	27.70	876.00	0.43
90	0.13	0.9	100.00	6.00	20.78	575.93	0.63
80	0.14	0.9	100.00	4.00	13.85	331.39	1.03
70	0.16	0.9	100.00	2.00	6.93	140.22	2.27

Table 2. Parameters of crossflow turbines.

H [m]	Q [m^3/s]	η [–]	P [kW]	n_q [min^{-1}]	D [m]	B [m]	n [min^{-1}]	n_s [min^{-1}]
24	0.25	0.84	50.00	65	0.11	0.23	1639.69	218.26
20	0.30	0.84	50.00	57	0.15	0.30	1138.67	190.37
12	0.51	0.84	50.00	39	0.32	0.64	409.92	129.78
10	0.61	0.84	50.00	34	0.42	0.84	284.67	113.19
24	0.51	0.84	100.00	46	0.23	0.45	819.84	154.34
20	0.61	0.84	100.00	40	0.30	0.60	569.34	134.61
12	1.01	0.84	100.00	27	0.64	1.28	204.96	91.77
10	1.21	0.84	100.00	24	0.84	1.69	142.33	80.04

Table 3. Parameters of the Francis turbine.

H [m]	Q [m^3/s]	η [–]	P [kW]	n_q [min^{-1}]	n_s [min^{-1}]	n [min^{-1}]	D_{1i} [m]	D_{2i} [m]
30	0.19	0.9	50.00	35.00	121.19	1203.37	0.29	0.24
20	0.28	0.9	50.00	30.00	103.88	621.35	0.46	0.34
14	0.40	0.9	50.00	25.00	86.57	331.53	0.73	0.47
10	0.57	0.9	50.00	20.00	69.25	174.16	1.17	0.66
30	0.38	0.9	100.00	55.00	190.45	1337.14	0.26	0.29
20	0.57	0.9	100.00	40.00	138.51	585.82	0.49	0.44
14	0.81	0.9	100.00	30.00	103.88	281.32	0.86	0.63
10	1.13	0.9	100.00	24.00	83.10	147.78	1.38	0.88

Selection of turbine is significantly influenced by the available flow and its change during the year. In this regard, the following cases should be distinguished:

- Constant flow - a turbine with a fixed hole is selected at the input of the propeller turbine (Kaplan's with fixed inlet blades) and Pelton to a constant diameter of the nozzles.
- Small change of flow - turbine is relatively little exposed to variable flow. A Fransis or Kaplan turbine is selected with a fixed fixed hole at the inlet for maximum efficiency.
- A large change in flow - turbine very often does not have enough water. Despite the reduced efficiency, it is recommended to use a flow turbine or a Fransis turbine. Pelton turbines with multiple nozzles, as well as Kaplan turbines with 2 degrees, can be used as they can be adjusted according to the flow.

Table 4. Kaplan turbine parameters.

H [m]	Q [m^3/s]	η [–]	P [kW]	n_q [min^{-1}]	n_s [min^{-1}]	n [min^{-1}]	D_0 [m]
9	0.63	0.9	50.00	120.00	415.52	916.04	0.39
7	0.81	0.9	50.00	105.00	363.58	585.45	0.54
5	1.13	0.9	50.00	90.00	311.64	329.52	0.80
3	1.89	0.9	50.00	75.00	259.70	145.01	1.41
9	1.26	0.9	100.00	120.00	415.52	647.74	0.55
7	1.62	0.9	100.00	105.00	363.58	413.98	0.76
5	2.27	0.9	100.00	90.00	311.64	233.01	1.14
3	3.78	0.9	100.00	75.00	259.70	102.54	2.00

From the aspect of optimal exploitation of the available hydroelectric power plant, it is necessary, in accordance with hydrological conditions, to harmonize the choice of the turbine itself, its construction parameters with the parameters of the machine plant or the generator itself. It should be noted that the working speeds of the turbine circuits are not fixed and can be changed within certain limits by changing the design parameters of the turbine.

The current generator is connected to a network having a constant network frequency of 50 Hz. Therefore, it is necessary that the number of revolutions of the generator does not change and that it is harmonized with the other parameters of the hydroelectric power plant. The following should be taken into account:

- With the increase in the speed, the centrifugal force increases significantly, and therefore the mechanical mapping of the machine group, and therefore the nominal speed is limited to 1500 min^{-1}.
- When applying a generator with a rotational speed below 600 min^{-1}, the overall dimensions and therefore the specific price of the genarator are growing, while at the same time reducing the energy efficiency due to magnetic losses.
- It is therefore recommended that, with turbine speeds below 750 min^{-1}, using a multipixer power transmission, the rpm rises to 1500 min^{-1} or 1000 min^{-1} and optimally exploits available hydropower.

5. Conclusion

The energy efficiency of small hydropower plants depends on a number of influencing parameters: flow, altitude, selected turbine plant, variation of flow and height drop during the year, terrain configurations, etc. The problem that arises when choosing a type of micro-hydroelectric turbine is to determine the influencing parameters for the correct turbine precipitation with the optimum use of available hydro power. For the most commonly used turbines of micro hydroelectric, Pelton, flow, Fransis and Kaplan, the process of determining operating parameters, based on which it is possible to make the choice of the type of turbine, is presented in the paper.

Changing hydrological conditions mostly reflect the available flow and its change during the year. It is easiest to choose the type of turbine for ideal conditions when the flow rate is constant. It is then recommended for a smaller altitude drop of Kaplan turbine with fixed inlet blades, and for Pelton turbine higher altitude drops to constant nozzle diameter. In case of a small change in flow, it is recommended in accordance with the altitude drop of the Fransis or Kaplan turbine with the tuned opening at the inlet for the maximum degree of efficency. In case of large flow changes, one of the flow turbines or Fransis turbines is recommended. Pelton turbines with multiple nozzles, as well as Kaplan turbines with 2 degrees, can be used as they can be adjusted according to the flow.

From the aspect of optimal use of the available hydroelectric power plant, it is necessary, in accordance with hydrological conditions, to harmonize the choice of the turbine itself, its construction parameters with the parameters of the machine plant or the electricity generator itself. In this regard, the application of the generator with the numbers of revolutions of 1500 min^{-1} or 1000 min^{-1} is purposeful, with adequate harmonization with the speeds of the turbine circuits with the use of a power transmission - multipixer.

References

[1] Chapallaz J-M 1995 *Kleinwasserkraftwerke Wasserturbinen*, Bundesamt für Konjunkturfragen, 3003 Bern

[2] Ivanović D 2014 *Hidroenergija*, Inženjerska komora Crne Gore

[3] Miltenović V, Marković B and Tica M 2018 Konstrukcioni elementi u mašinogradnji 1 – oblici, proračun, primena, Univerzitet u Istočnom Sarajevu, Mašinski fakultet

[4] Nuernberg D 2006 *Wasserräder zur Elektroenergiegewinnung*, DOI: 10.13140/RG.2.1.4738.2805.

[5] Project of Innovation Fund, 2019, *Development and technical design of planet multiplicator of special purpose for micro hydropower plants*. Innovation center of University of Niš and Technos Profi Niš

[6] Detering M and Roger S 2013 *Ökologischer Forschung(sbedarf) rund um die Wasserkraft*, Wasser als Energieträger, Internationales Wasserbau-Symposium, Aachen

[7] Spasić Ž, Jovanović M, Bogdanović-Jovanović J and Milanović S 2019 Numerical Investigation of the Influence of the Doubly Curved Blade Profiles on the Reversible Axial Fan Characteristics, *Facta Universitatis: Mechanical Engineering* **17**(2)

[8] Benad J 2019 Numerical Methods for the Simulation of Deformations and Stresses in Turbine Blade Fir-Tree Connections, *Facta Universitatis: Mechanical Engineering* **17**(1) 1-15

Energy performance of the Serbian and Estonian family house with a selective absorption facade

N Lukić[1], A Nešović[1], N Nikolić[1], A Siirde[2], A Volkova[2] and E Latosov[2]

[1]University of Kragujevac, Faculty of Engineering, Sestre Janjić 6, 34000 Kragujevac, Serbia
[2]Tallinna Tehnikaülikool, Energiatehnoloogia instituut, Ehitajate tee 5, 19086 Tallinn, Estonia

E-mail: lukic@kg.ac.rs

Abstract. In this paper the influence of selective absorption facade on heating energy consumption of the family house is analyzed. Two houses built according to Serbian and Estonian building energy performance regulation (minimum requirements) are simulated by EnergyPlus software. For simulation process the appropriate weather data are used (Serbia – Kragujevac and Estonia – Tallinn). Heat transfer by radiation, gains and losses can be a significant part of the energy consumption, during heating season. Using selective absorption facade those gains and losses could be increased and reduced, respectively. Several simulation scenarios of Serbian and Estonian house are carried out. Different weather conditions in case of the defined houses point on specific impact of the applied improvement on family house energy performance. Maximum annual percentage savings of heating energy compared to house without selective facade, in case of Serbian house are obtained as 8.32%. In case of Estonian house maximum savings are obtained as 1.92%.

1. Introduction

Certain analyses have shown that in developed countries around the world, the consumption of final energy in buildings in 2008 was higher than consumption in the industrial and transport sector [1]. In order to preserve the environment, renewable energy sources (primarily solar and geothermal energy) in the residential and public sector should be more use. The application of solar technologies in heating, cooling and ventilation systems in order to reduce the consumption of final energy was considered in [2]. In the literature, a large number of papers can be found relating to the application of solar facades for the same purpose [3-6]. An integral component of solar installations is an absorber. In order to reduce the amount of heat emitted from the surface of the absorber, the use of various selective coatings and paints, whose performance was tested in [7-14]. However, the absorbers of solar facades can be external walls, and therefore [15-18] tested the use of low-emission coatings and paints in order to reach energy-efficient buildings. Similar effects can be achieved by coating external walls on the inside [19-21]. Cozza E. S. and others used coatings in [22] to increase the IR reflection from the facade walls in order to better cool the object in the summer period. A facade with grooved cavities was proposed in [23]. The grooved cavities are configured so that solar flares are reflected from their superficial, while absorption in the winter accelerates. The model was developed for a house that is 41 degrees north latitude, but can be adapted for any climatic area. Investigation of the influence of colored facades with selective properties on the energy balance of buildings located in Freiburg, Germany was done by comparing real data with data calculated using the ESP-r simulation program [24]. The results showed that the influence of solar radiation and heat losses caused by convection are the most dominant factors affecting the energy bill of the building. Another conclusion is made: in weakly isolated houses, greater selectivity can be achieved by using selective facades than in better insulated houses.

The aim of this paper is to examine the possibilities of using selective absorption facades in the construction of Serbian and Estonian houses in order to reduce the energy consumption during the heating season. Defined simulations will be carried out by EnergyPlus. This software is a collection of many program modules that work together to calculate the energy required for building heating and cooling, using a variety of systems and energy sources. It does this by simulating the building and associated energy systems when they are exposed to different environmental and operating conditions. The core of the simulation is a model of the building that is based on fundamental heat balance principles.

2. Model of the family house

The subject of research is the family house shown in Figures 1 and 2. The net area of the house is 180.75 m^2. The net conditioned building area is 171 m^2 (storage room and closet are unconditioned rooms). The total surface of the thermal shell is 432 m^2 and window-wall ratio is 14%. The family house has two floors (ground floor and first floor). Arrangement of the rooms on the ground floor shown in Figure 3. Arrangement of the rooms on the first floor shown in Figure 4. The house is intended for the permanent stay of a six-member family.

Figure 1. Northwest façade.

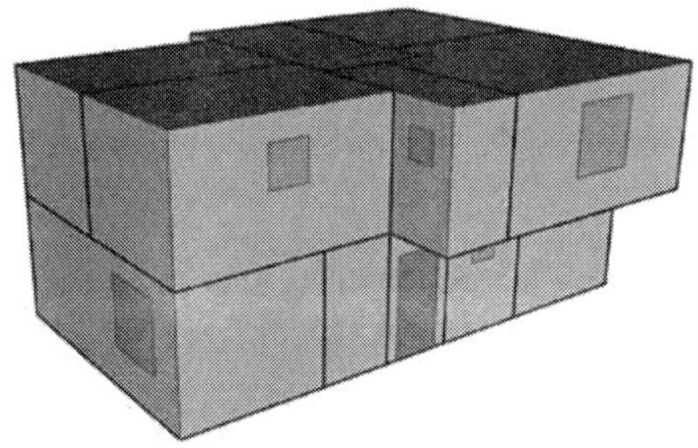

Figure 2. Southwest façade.

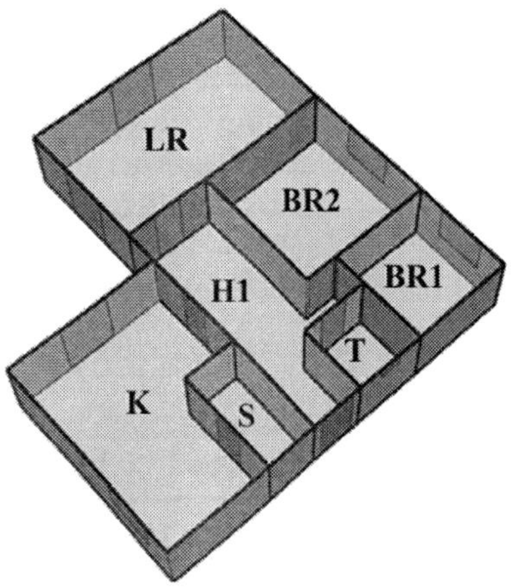

Figure 3. Arrangement of the rooms on the ground floor (K – kitchen, LR – living room, H1 – hall 1, T – toilet, SR – storage room, BR1 – bedroom 1, BR2 – bedroom 2).

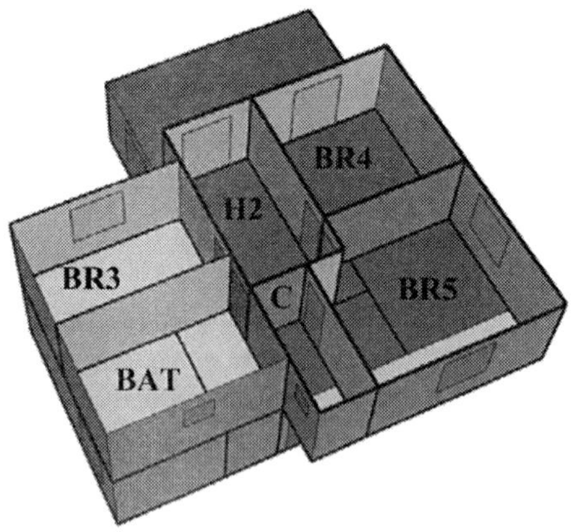

Figure 4. Arrangement of the rooms on the first floor (H2 – hall 2, C – closet, BAT – bathroom, BR3 – bedroom 3, BR4 – bedroom 4, BR5 – bedroom 5).

2.1. Weather files for Kragujevac (Serbia) and Tallinn (Estonia)

To simulate weather conditions for the city of Kragujevac (latitude of 44.02°N, longitude of 20.92°E, the average altitude of 209 m) and the city of Tallinn (latitude of 59.41°N, longitude of 24.83°E, the average altitude of 39.9 m) the EnergyPlus weather files were used [25]. Kragujevac (with the time zone of GMT+1h) is characterized by moderate continental climate with pronounced seasons. Tallinn (with the time zone of GMT+2h) has a humid continental climate with warm, mild summers and cold, snowy winters. Some meteorological data for Tallinn and Kragujevac are shown in the Figures 5-10.

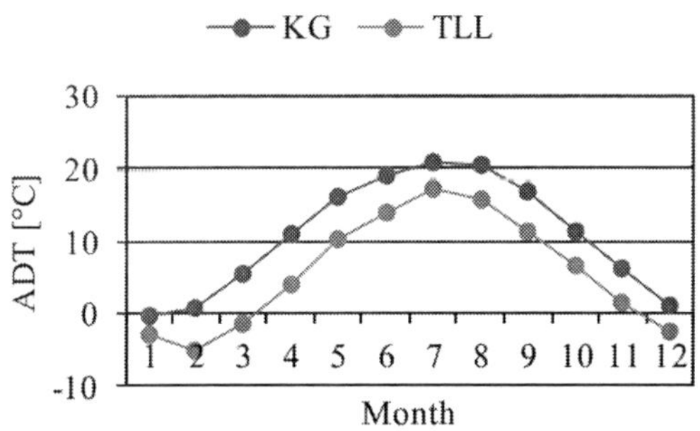

Figure 5. Air drybulb temperature.

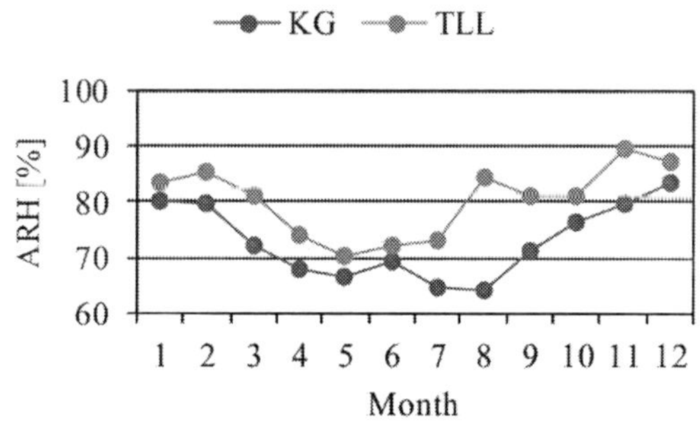

Figure 6. Air relative humidity.

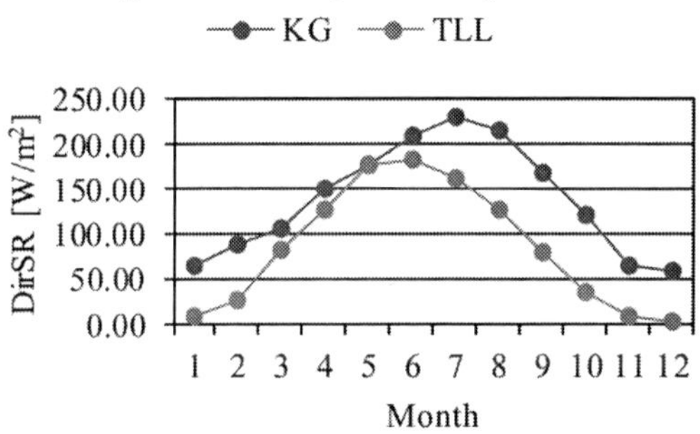

Figure 7. Direct solar radiation.

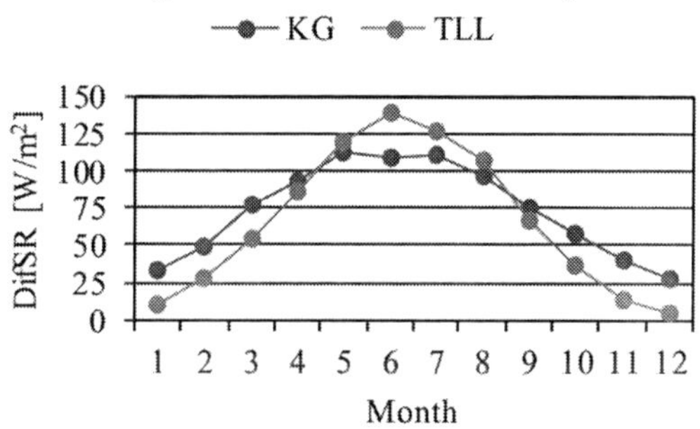

Figure 8. Diffuse solar radiation.

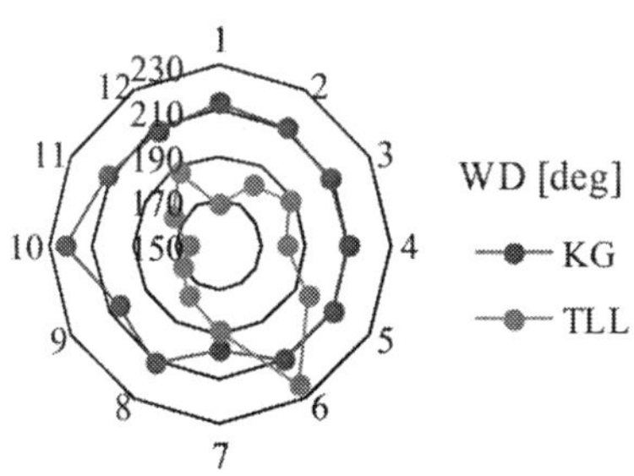

Figure 9. Wind direction.

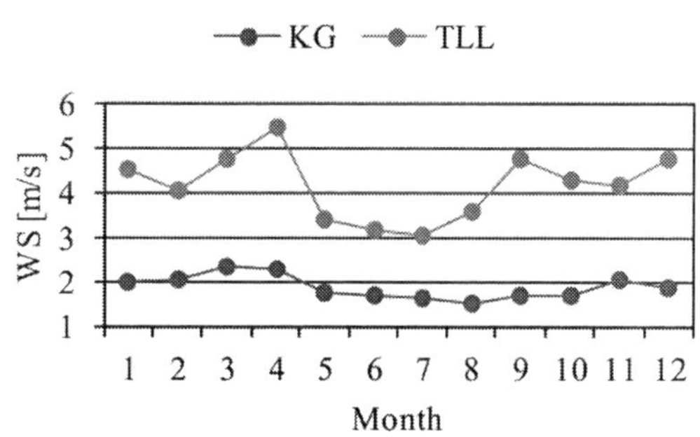

Figure 10. Wind speed.

2.2. Serbian and Estonian building energy performance regulations

Family house is modelled according to the Serbian and Estonian regulations on energy efficiency in the residential sector. In Table 1 (minimum requirements). Some allowed values (heat transfer coefficients for external and internal building elements, internal project temperatures, required air changes depending on the purpose of the room, as well as energy consumption) are shown in these regulations for easier comparison.

Table 1. Serbian and Estonian building energy performance regulations.

<table>
<tr><th>Parameter</th><th>Description</th><th>Serbia</th><th>Estonia</th></tr>
<tr><td rowspan="2">Energy consumption [kWh/(m^{2}y)]</td><td>Total</td><td>-</td><td>160</td></tr>
<tr><td>Only for heating</td><td>65</td><td>-</td></tr>
<tr><td rowspan="13">Heat transfer coefficient [W/m^{2}K]</td><td>Exterior wall</td><td>0.3</td><td>0.22</td></tr>
<tr><td>Flat roof (above heated room)</td><td>0.15</td><td rowspan="3">0.15</td></tr>
<tr><td>Flat roof (above unheated room)</td><td rowspan="2">0.3</td></tr>
<tr><td>Floor</td></tr>
<tr><td>Interconnected construction (between heated rooms)</td><td>0.9</td><td rowspan="6">-</td></tr>
<tr><td>Interconnected construction (above unheated room)</td><td rowspan="2">0.3</td></tr>
<tr><td>Interconnected construction (under unheated room)</td></tr>
<tr><td>Interconnected construction (to outside)</td><td>0.2</td></tr>
<tr><td>Interior wall (between heated rooms)</td><td>0.9</td></tr>
<tr><td>Interior wall (to unheated room)</td><td>0.4</td></tr>
<tr><td>Window</td><td rowspan="2">1.5</td><td rowspan="3">1.1</td></tr>
<tr><td>Exterior door (glass)</td></tr>
<tr><td>Exterior door (normal)</td><td>1.6</td></tr>
<tr><td rowspan="2">Heating set-point [°C]</td><td>Bathroom</td><td>22</td><td rowspan="2">21</td></tr>
<tr><td>Other rooms</td><td>20</td></tr>
<tr><td rowspan="4">Outdoor air flow rate
For Serbian [h^{-1}]
For Estonian [l/(sm^2)]</td><td>Kitchen</td><td rowspan="3">1.5</td><td rowspan="4">0.42</td></tr>
<tr><td>Toilet</td></tr>
<tr><td>Bathroom</td></tr>
<tr><td>Other rooms</td><td>0.5</td></tr>
<tr><td>References</td><td></td><td>[26]</td><td>[27]</td></tr>
</table>

2.3. Simulation scenarios

According to Table 2, four simulation scenarios are performed. Since, the minimum regulation requirements of thermal characteristics are different for Serbian and Estonian house, all simulations are carried out for two houses, separately. Initial scenario (DHM) has been implied two houses (Serbian and Estonian), built according to current regulations of building thermal characteristics as minimum requirements (the highest allowed heat transfer coefficients). Thermal and solar absorptance (and emittance) are adopted as usual for a building façade, 0.9 and 0.7, respectively. This means that building façade of DHM absorbs solar radiation and emits thermal (infrared longwave) radiation very well. Other three simulation scenarios has been implied two houses with changed absorptance characteristics of the south (S), south and west (S, W) and south, west and east oriented wall. The changed absorptance characteristics means a good absorptance of solar radiation and relatively small emittance of thermal radiation. Defined façade has similar characteristics as selective absorber surface in a solar collector. Depending on intensity of solar radiation, northern latitude, wall orientation, insulation thickness, air temperature, the selective façade effects on heating energy consumption could be positive or even negative. Generally, during winter on northern latitude, vertical surface has better position for solar radiation absorption than horizontal one.

Table 2. Simulation scenarios for Serbian and Estonian house models.

Model	Absorptance and emittance [-]		Thickness of insulation [cm]	
	Thermal	Solar	Serbian house	Estonian house
Default house model (DHM)	0.9	0.7	9.69	13.9
South wall changed (S)	0.2	0.9	9.69; 5; 0	13.9; 5; 0
South and West walls changed (S, W)	0.2	0.9	9.69; 5; 0	13.9; 5; 0
South, West and East walls changed (S, W, E)	0.2	0.9	9.69; 5; 0	13.9; 5; 0

3. Results and discussion

In Figure 11 the percentage savings of the annual heating energy consumption of Serbian house using different insulation thickness and selective façade is given. The maximum insulation thickness (9.69 cm) corresponds with the minimum regulation requirements in Serbia. Thicker insulation layer inhibits heat losses but in the defined case also solar gains on façade surface. The maximum energy savings (8.32%) Serbian house achieves in scenario of S, W and E selective facade and maximum insulation thickness. In case of 5 cm insulation thickness, the selective facades give still positive energy savings, similar for all three scenarios (S; S, W; S, W, E). Serbian house without insulation layer on south wall and selective south façade spends almost the same amount of heating energy as (DHM) with ordinary façade and maximum insulation thickness on south wall (difference is only 1.3%). This means that additional solar gains are approximately equal with additional convective losses in scenario S. Other scenarios give distinctively negative energy savings.

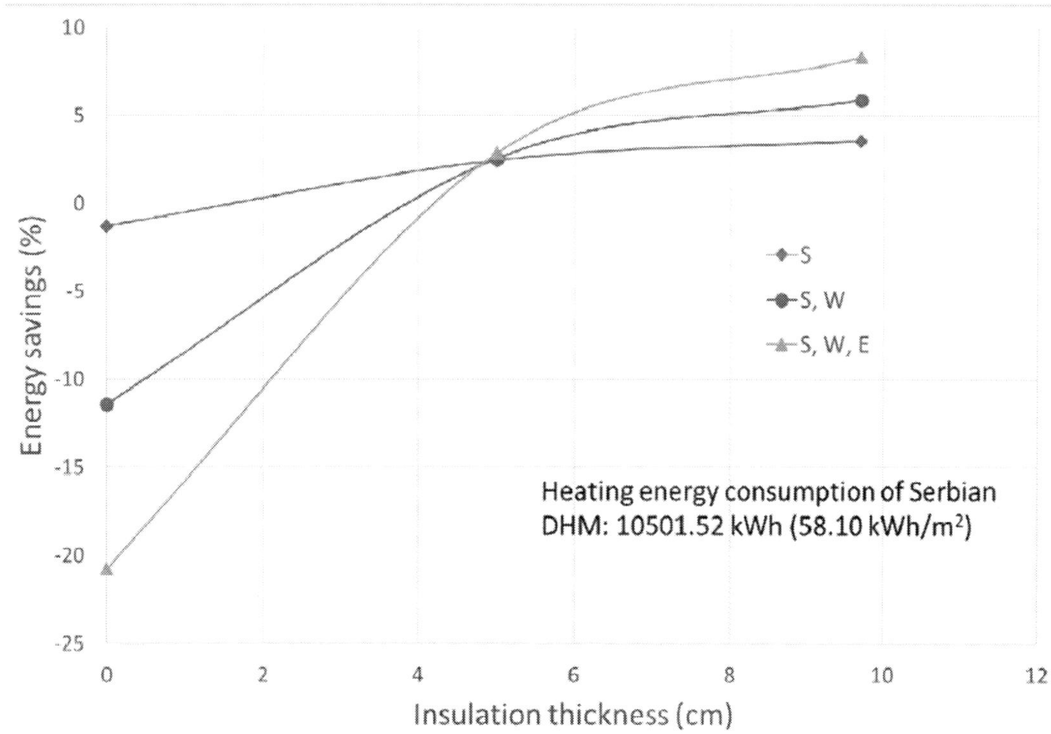

Figure 11. Percentage savings of the annual heating energy consumption of Serbian house using different insulation thickness and selective façade.

In Figure 12 the percentage savings of the annual heating energy consumption of Estonian house using different insulation thickness and selective façade is given. The maximum insulation thickness (13.9 cm) corresponds with the minimum regulation requirements in Estonia. The maximum and insignificant energy savings (1.92%) Estonian house achieves in scenario of S, W and E selective facade and maximum insulation thickness. Decreasing of insulation thickness in any scenario of selective façade leads to significantly increasing of heating energy consumption. In case of Estonian house, during heating season, solar radiation is not a significant factor that reduces heating demands.

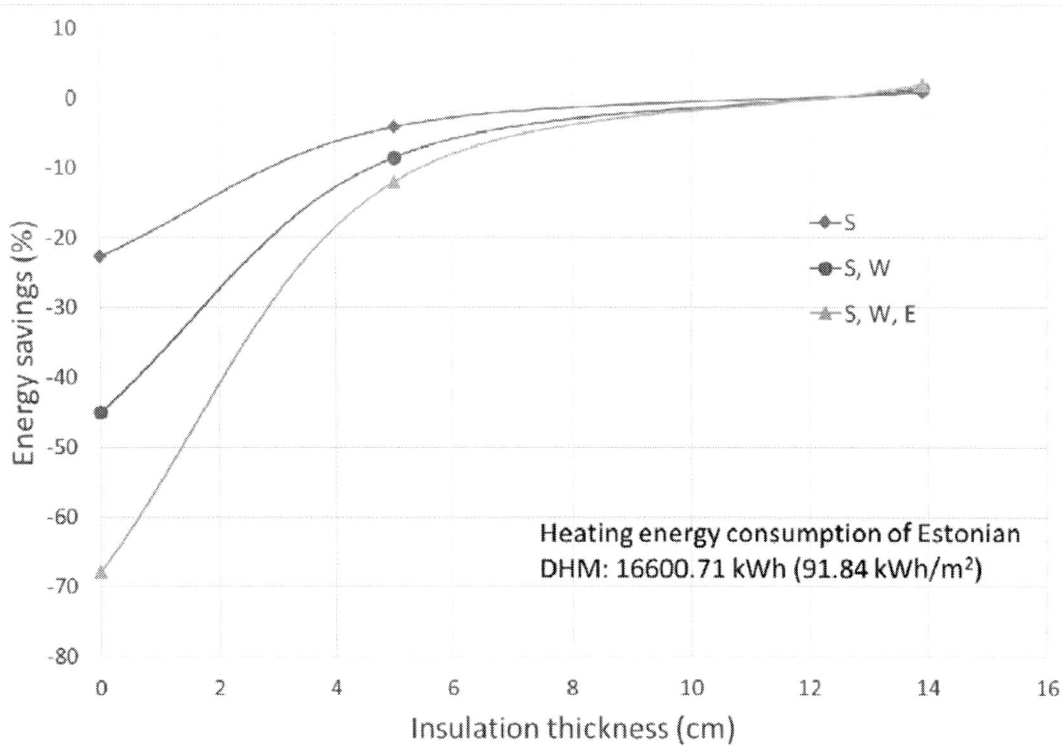

Figure 12. Percentage savings of the annual heating energy consumption of Estonian house using different insulation thickness and selective façade.

In Figure 13 the percentage savings of the monthly heating energy consumption of Serbian and Estonian house using different insulation thickness and selective façade is given. For this review the warmest and coldest heating season months in Kragujevac, Serbia (April and January, respectively) and Tallinn, Estonia (October and January, respectively) are considered. For all presented cases the simulation scenario is S (selective south façade). Figure 13 shows that in case of Serbian house for the warmest heating month (April) the percentage savings are positive even for south wall without insulation layer. The maximum energy savings (8.51%) Serbian house achieves with maximum insulation thickness (south selective façade). For the coldest month (January) only for south wall without insulation layer savings are negative (-4.00%). Also, figure 13 shows that in case of Estonian house for the both considered heating months (October and January) the percentage savings are positive only for maximum insulation thickness. In case of the coldest month (January) and maximum insulation thickness the energy savings effects of south selective façade are insignificant (0.41%).

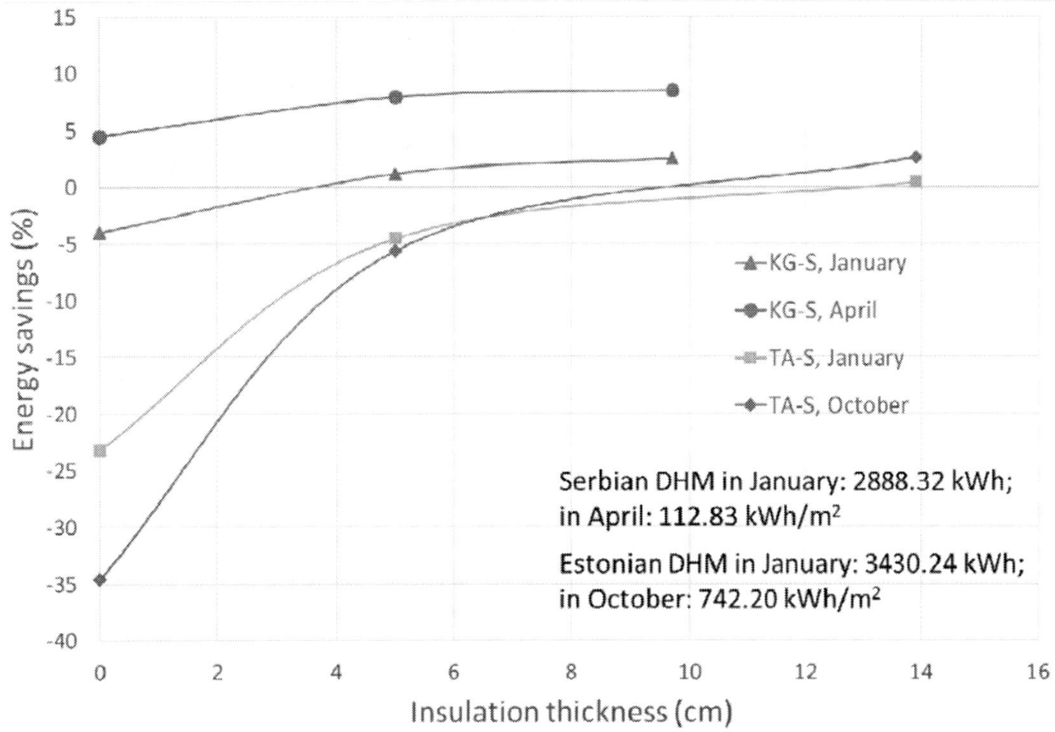

Figure 13. Percentage savings of the monthly heating energy consumption of Serbian and Estonian house using different insulation thickness and selective façade for selected months.

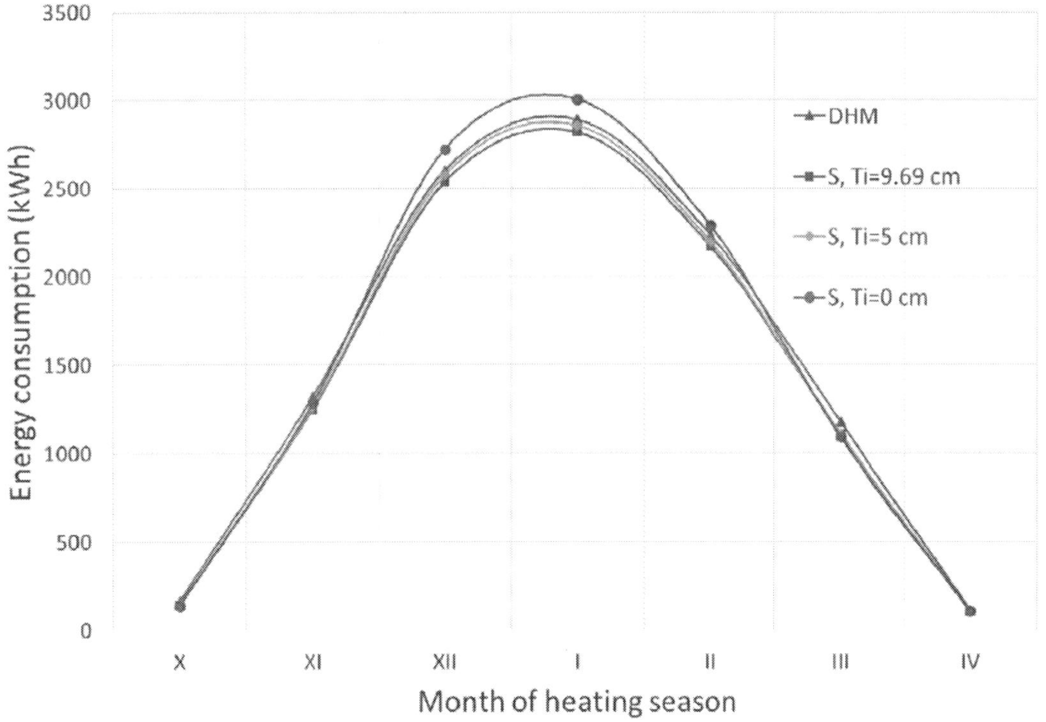

Figure 14. Heating energy consumption per month of Serbian house during heating season (October-April) for DHM and S simulation scenarios.

In Figure 14 the heating energy consumption per month of Serbian house during heating season (October-April) for DHM and S simulation scenarios is given. Figure shows that the south selective façade always gives positive energy savings compered to DHM, except three coldest heating months (December, January and February) in case of south wall without insulation layer.

In Figure 15 the heating energy consumption per month of Estonian house during heating season (October-April) for DHM and S simulation scenarios is given. Figure shows that the south selective façade gives positive energy savings compered to DHM, only in case of maximum thickness of the south wall insulation layer. In cases of insulation layer thickness decreasing the energy savings are negative for all heating months.

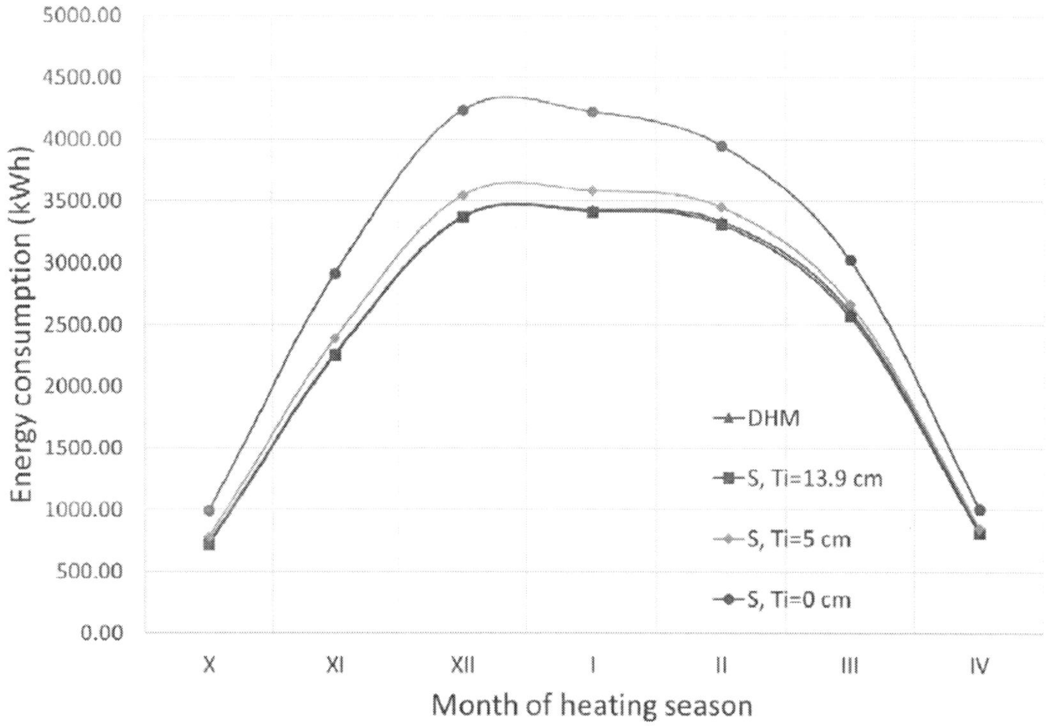

Figure 15. Heating energy consumption per month of Estonian house during heating season (October-April) for DHM and S simulation scenarios.

4. Conclusion

A Changeable (intelligent) building envelope is logical development of the heating energy efficient consumption system. Radiation selective surfaces could be applied as final building façade layer and on that way use solar gains more efficiently in order to decrease heating energy consumption in buildings. On the example of Serbian (location Kragujevac) and Estonian (location Tallinn) family house modelling according to the actual (Serbian and Estonian) regulation of the minimum building thermal envelope requirements, using simulation software EnergyPlus, is shown how selective façade could decrease energy consumption during heating season. Under weather conditions of the enhanced solar radiation and higher outside temperature during heating season, October-April (case of Serbian house) a selective façade could be helpful to decrease heating energy consumption compared to the same house without selective façade. Annual percentage savings of heating energy are done as 8.32%. Simulated saving effects could be also considered through reduction of insulation thickness and investment costs. Otherwise, under weather conditions of the poor solar radiation and lower outside temperature during heating season (case of Estonian house) a selective façade shows the weak effects on energy consumption decreasing. The maximum annual reduction of energy consumption is done as 1.92%. In this simulation case a reduction of insulation layer is not an option. Problem with overheating of a selective façade during summer conditions could be solve with using of reflective curtain.

Acknowledgment
This investigation is a part of the project TR 33015 of the Technological Development of the Republic of Serbia. We would like to thank the Ministry of Education, Science and Technological Development of the Republic of Serbia for their financial support during this investigation.

References

[1] Pérez-Lombard L, Ortiz J and Pout C 2008 A review on buildings energy consumption information, *Energy and Buildings* **40**(3) 394-398
[2] Chan H Y, Riffat S B and Zhu J 2010 Review of passive solar heating and cooling technologies, *Renewable and Sustainable Energy Reviews* **14**(2) 781-789
[3] Quesada G, Rousse D, Dutil Y, Badache M and Hallé S 2012 A comprehensive review of solar facades. Opaque solar facades, *Renewable and Sustainable Energy Reviews* **16**(5) 2820-2832
[4] Lai C M and Hokoi S 2015 Solar façades: A review, *Building and Environment* **91** 152-165
[5] Visa I, Moldovan M, Comsit M, Neagoe M and Duta A 2017 Facades integrated solar-thermal collectors – challenges and solutions, *Energy Procedia* **112** 176-185
[6] Zhang X *at al.* 2015 Active Solar Thermal Facades (ASTFs): From concept, application to research questions, *Renewable and Sustainable Energy Reviews* **50** 32-63
[7] Sarkar A, Sinha S, Palai D, Dey A and Mallick A B 2018 Delineating the role of surface characteristics on the solar selectivity of colored chromium oxide coating on 304 stainless steel substrate, *Solar Energy Materials and Solar Cells* **182** 354-361
[8] Koželj M, Vuk A Š, Jerman I and Orel B 2009 Corrosion protection of Sunselect, a spectrally selective solar absorber coating, by (3-mercaptopropyl) trimethoxysilane, *Solar Energy Materials & Solar Cells* **93**(10) 1733-1742
[9] Vince J, Vuk A Š, Krašovec U O, Orel B, Köhl M and Heck M 2003 Solar absorber coatings based on CoCuMnOx spinels prepared via the sol-gel process: structural and optical properties, *Solar Energy Materials & Solar Cells* **79**(3) 313-330
[10] Jerman I, Koželj M and Orel B 2010 The effect of polyhedral oligomeric silsesquioxane dispersant and low surface energy additives on spectrally selective paint coatings with self-cleaning properties, *Solar Energy Materials & Solar Cells* **94**(2) 232-245
[11] Orel B, Spreizer H, Perše L S, Fir M, Vuk A Š, Merlini D, Vodlan M and Köhl M 2007 Silicone-based thickness insensitive spectrally selective (TISS) paints as selective paint coatings for coloured solar absorbers (Part I), *Solar Energy Materials & Solar Cells* **91**(2-3) 93-107
[12] Orel B, Spreizer H, Vuk A Š, Fir M, Merlini D, Vodlan M and Köhl M 2007 Selective paint coatings for coloured solar absorbers: Polyurethane thickness insensitive spectrally selective (TISS) paints (Part II), *Solar Energy Materials & Solar Cells* **91**(2-3) 108-119
[13] Japelj B, Vuk A Š, Orel B, Perše L S, Jerman I and Kovač J 2008 Preparation of a TiMEMO nanocomposite by the sol-gel method and its application in coloured thickness insensitive spectrally selective (TISS) coatings, *Solar Energy Materials & Solar Cells* **92**(9) 1149-1161
[14] Duta A, Isac L, Milea A, Ienei E and Perniu D 2014 Coloured solar-thermal absorbers – a comparative analysis of cermet structures, *Energy Procedia* **48** 543-553
[15] Gordon J M 1982 Selective coatings in passive solar heating, *Solar Energy* **29**(1) 13-17
[16] Azemati A A, Hadavand B S, Hosseini H and Tajarrod A S 2013 Thermal modeling of mineral insulator in paints for energy saving, *Energy and Buildings* **56** 109-114
[17] Čekon A and Čurpek J 2019 A transparent insulation façade enhanced with a selective absorber: A cooling energy load and validated building energy performance prediction model, *Energy and Buildings* **183** 266-282
[18] Guan T H, Wu C Q, Liu H M, Shen Y and Xu Y B 2014 Influence of Reflective Insulation Coating on Heat Transfer Characteristics of Composite Thermal Insulation Wall, *Applied Mechanics and Materials* **633-634** 909-912

IOP Conf. Series: Materials Science and Engineering **659** (2019) 012047 doi:10.1088/1757-899X/659/1/012047

[19] Weinlaeder H, Pottler K, Burger T, Kuhn J and Fricke J 1999 Energy savings with low-emission wall paint, TAB, *Technik am Bau* **30** 63-69

[20] Simpson A, Fitton R, Rattigan I. G, Marshall A, Parr G and Swan W 2019 Thermal performance of thermal paint and surface coatings in buildings in heating dominated climates, *Energy and Buildings* **197** 196-213

[21] Hugo G 2001 Effects of low emissive wall coatings on thermal comfort and energy consumption, *High Temperatures - High Pressures* **33** 1-8

[22] Cozza E S, Alloisio M, Comite A, Tanna G D and Vicini S 2015 NIR-reflecting properties of new paints for energy-efficient buildings, *Solar Energy* **116** 108-116

[23] Naraghi M H and Harant A 2012 Configuration of Building Façade Surface for Seasonal Selectiveness of Solar Irradiation-Absorption and Reflection, *Journal of Solar Energy Engineering* **135**(1) 1-9

[24] Prager C, Köhl M, Heck M and Herkel S 2006 The influence of the IR reflection of painted facades on the energy balance of a building, *Energy and Buildings* **38**(12) 1369-1379

[25] EnergyPlus, Energy Simulation Software (Weather File)

[26] *Правилник о енергетској ефикасности зграда*, Закон о планирању и изградњи Републике Србије 2011,
http://www.ingkomora.org.rs/strucniispiti/download/ee/PRAVILNIK_O_EEZ_za%20obuku.pdf (accessed on Jun. 2019)

[27] *Minimum requirements for energy performance*, Building Act of the Republic of Estonia, https://www.riigiteataja.ee/en/eli/520102014001 (accessed on Jun. 2019)

IOP Conf. Series: Materials Science and Engineering **659** (2019) 012048 doi:10.1088/1757-899X/659/1/012048

Approximation of the bridge deck diffusion coefficient and surface chloride concentration from field data

P Konecny[1], P Lehner[1], D Vorechovska[2], M Somodikova[2], M Hornakova[1] and P Rovnanikova[2]

[1]VSB – Technical University of Ostrava, Faculty of Civil Engineering, Ostrava-Poruba, Czech Republic
[2]Brno University of Technology, Faculty of Civil Engineering, Brno, Czech Republic

E-mail: petr.konecny@vsb.cz

Abstract. The aim of this work is to analyse the data from in-situ measurements performed on 5 reinforced concrete bridges built between 1937 and 1986, that are a part of the road network in the Czech Republic. The bridges were subjected to the field measurements in order to evaluate the ability of the concrete to protect the steel reinforcement from corrosion, namely the pH and the amount of water-soluble chlorides were evaluated on drilled core samples of concrete. Based on the actual chloride content, the diffusion coefficient and surface chloride concentration were determined. All the data were statistically summarized and the correlations among them were provided. The main focus of this study is to evaluate the non-proportional effect of the amount of chlorides per mass of concrete on the risk of corrosion initiation and to determine the unknown parameters of concrete – diffusion coefficient and surface chloride concentration, based on the numerical model of diffusion.

1. Introduction

Bridges in the Czech Republic are the typical constructions exposed to the combined effect of carbonation, chloride ingress and mechanical load. Such structures are prone to corrosion that threatens their durability and reduces the service life and bearing capacity.

The current design of typical structural elements with respect to exposure conditions is executed by the deemed-to-satisfy rules prescribed by EN 1992-1 and EN 206. Currently, the discussion and benchmarking on durability calculations based on these rules have been undertaken by the European technical committees and fib Commissions. It is intended to replace them with a performance-based design approach (PBD) in future standards [1] –[4]. An important part of preparation for PBD is analysis of the durability of existing structures under service in order to improve the authenticity of available numerical tools. Some of them are described for instance in [5] – [12].

Even though the physical law for the analysis of the penetration of agents inducing steel reinforcement corrosion (such as chlorides or carbon dioxide) are known, the modelling still contains a lot of unknowns, especially in case of heterogeneous material (such as concrete). It is worth to mention that even more questions arise when consider the selection of input parameters for models. Of course, there are recommendations in codes [2], [4], [13], [14]; however, the fidelity of such recommendations to every structure are of question. Thus, interpretation of the results of such durability analysis needs to be taken with care. The simulation results, such as length of service life, might be used in case of comparison of different scenarios and selection of the most durable one. The actual time shall not be extracted from the analysis by itself [13].

Numerical modelling of chloride penetration might be simplified using deterministic one dimensional numerical diffusion model called Crank's solution applied by Collepardi [15]:

$$C_{z,t} = C_0 \left[1 - erf\left(\frac{z}{\sqrt{4D_c t}}\right)\right], \quad (1)$$

where:

- $C_{z,t}$ – concentration of chlorides [% by mass of concrete],
- C_0 – concentration of chlorides [% by mass of concrete] at the surface layer of concrete,
- D_c – apparent diffusion coefficient of chloride ions in concrete [$m^2 \cdot s^{-1}$],
- t – chloride exposition time [s],
- z – depth of the reinforcement (concrete cover) [m].

The more elaborated model that considers random scatter of input parameters was used in [8], [14] and punctual model with respect to cracking of steel reinforcement protection, such as 2D model with crack consideration, in [9]. Some of the models imitate heterogeneity, example of such mesoscopic model is presented in [16]. The more detailed model is used, the more demanding the calculation in terms of the computer hardware and simulation time is, and the total cost of analysis increases. Thus, the balance between the fidelity and quality of the model and input parameters is the challenge for researchers specialized in areas related to modelling of the steel corrosion in concrete.

With respect to the matters mentioned above, the attention shall be focussed, beside of the improvement of the numerical models, also on the improvement of recommendations for the input parameters used in the models in order to provide effective, yet reliable models to the engineering community.

In this paper, the data from regular inspections of highway bridges in the Czech Republic are analysed with respect to the corrosion related durability aspects and modelling of reinforced concrete. Some of the bridges were already analysed in the previous works focusing on the critical value of chlorides and pH to initiate the corrosion and on estimation of surface chloride concentration, however, the diffusion coefficient was not determined yet, see e.g. [17].

The attention is concentrated to a simple 1D model of the chloride penetration, where the diffusion coefficient of chloride penetration D_c, time and the surface chloride concentration C_0 are of interest – eq. (1).

Since the limit state definition is an important part of the evaluation, the effect of carbonation on the chloride threshold to initiate the corrosion is of high importance [17], [18]. It needs to be noted that the effect of the pH change related to the carbonation of concrete is usually not taken into account in the publications from this area. Thus, further evaluation of the common effect of carbonation and chloride presence is necessary.

Special attention is paid to the chloride concentration profiles and to determination of the unknown parameters of concrete – diffusion coefficient and surface chloride concentration, based on the numerical model of diffusion.

2. Scope

The subjects of the inspection and subsequent analysis are 5 highway bridges in the Czech Republic. The bridges are marked by numbers: no. 54-040 over a stream that was built in 1937 and inspected in 2014, no. 55l-026a over a local way built in 1976 and inspected in 2014, no. 55l-030 over a river built in 1953 and inspected in 2015, no. 55-033 over a river built in 1963 and inspected in 2016 and no. 57-039 also over a river built in 1985 and inspected in 2013.

The procedure of the selection of corrosion related data and their processing are discussed in the section Experimental methodology. The process of evaluation of data is shown in the section Numerical analysis of the bridge 55l-030 and overall results are discussed in the section Discussion and conclusions. The comprehensive results for the other selected bridges are given in the Appendices.

3. Experimental methodology

The corrosion related data were collected during the regular inspections of the highway bridges under service that are in the portfolio of Directorate of Highways and Motorways of the Czech Republic. The representative locations of girders, abutments, bearing seats or joints were selected and concrete cores were sampled. In each location, the samples were taken from three different depths of 0–10, 10–20 and 20–30 mm from the concrete surface.

Subsequently, the value of pH and the amount of water-soluble chlorides were measured in water leaches in laboratory [20]. Based on these parameters, the diffusion coefficient and surface chloride content were determined. The ratio between the concentrations of Cl^- and OH^-, that indicates the ability of concrete to protect the reinforcement, was determined in previous work [17]. In the other words, the higher the ratio is, the higher is the potential risk of corrosion initiation and propagation. As a critical value of the ratio $c(Cl^-)/c(OH^-) = 0.6$ can be considered [21].

3.1. Calculation of the diffusion coefficient and surface chloride concentration

The lacking parameters, that are, in this case, the diffusion coefficient D_c and the chloride concentration on the surface C_0, were determined iteratively by the analogy to the least square method. Regression analysis served for minimizing the sum given by equation (2) [22]:

$$S_j = \sum_i^N \Delta C_{(i)}^2 = \sum_i^N \Delta (C_{m(i)} - C_{c(j,i)})^2, \tag{2}$$

where:

- S_j – sum of the squares in the j-th iteration [(% by mass)^2],
- N – number of layers in profile,
- $\Delta C_{(i)}$ – difference between the measured and calculated concentration of chlorides in the N-th layer [% by mass of total cementitious materials],
- $C_{m(i)}$ – measured concentration of chlorides [% by mass of total cementitious materials],
- $C_{c(j,i)}$ – calculated concentration of chlorides in the j-th step iteration and the i-th layer [% by mass of total cementitious materials].

Chloride concentration can be calculated using the equation (3) also [19], [23]. For the first iteration, the diffusion coefficient D_c and surface concentration of chlorides C_0 are selected. The time t corresponds to the exposure time to the chloride mode.

$$C_{z,t} = C_0 \left\{ 1 - \frac{2}{\sqrt{\pi}} \sum_{n=0}^{14} \frac{(-1)^n \left(\frac{z}{\sqrt{4 D_C t}} \right)^{2n+1}}{n!(2n+1)} \right\}, \tag{3}$$

where the symbols have been already explained in relation to equation (1) and n is the number of the polynomial set.

However, due to the approximation of the chloride profile by the equation (3), that does not take into account the aging of the concrete over the time, the value of the diffusion coefficient can be overestimated.

4. Numerical analysis of the bridge 55I-030

All the investigated bridges were subjected to the numerical analysis leading to the estimation of diffusion coefficient and surface chloride concentration from the measured data. For the purpose of the paper, the numerical procedure is described in detail in case of one of the bridges. The subject of the example is a single pole bridge no. 55I-030, which transfers the road I/55I across the Dřevnice river in the centre of Otrokovice [20]. The bridge was built in 1953 and diagnostic drilling was executed in 2015, so the exposure time is 62 years.

The layout of the bridge is available in the Figure 1. The lower structure is composed of two monolithic concrete supports with short parallel wings installed in the upper part of the abutments. There is stone paving in concrete bed that creates relatively steep slope in front of the abutments along the entire length of the retaining walls. The horizontal load bearing structure is made of 16 prefabricated prestressed "T" beams with the span of 28 m. The beams are based on steel bearings.

The selected points of the drilled cores for aforementioned chemical analysis are also shown in the Figure 1 and described in Table 1. It should be noted that for bridge no. 551-030 in Figure 1, there are 11 sample cores (points from 1 to 11). There are also 4 sites where the repairing concrete overlay has been applied (points 12 to 15).

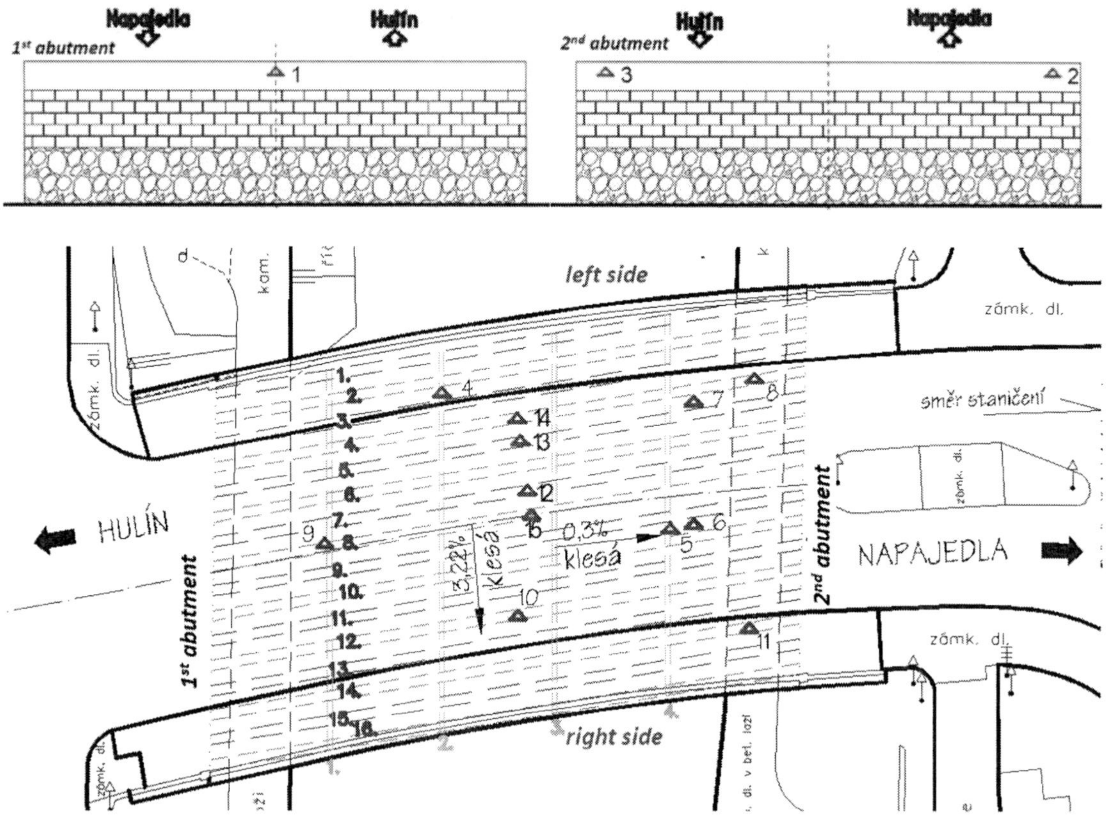

Figure 1. Front views of the abutments and the scheme of the bridge including sample core spots. Red triangles with black numbering represent core positions and red numbers represent the number of girders.

From Table 1 suitable chloride profiles, that are characterized by decreasing value of chloride concentration as shown in Figure 2, were selected. The chosen samples are given in Table 2. Approximations of the chloride profile are shown in Figure 3.

Table 1. Chloride profiles and concrete strengths sampled at bridge no. 55I-030 (built in 1953, inspected in 2015, over a river).

ID	No.	Location	Chloride profile $[\%Cl^-]$ 0-10 *[mm]*	10-20 *[mm]*	20-30 *[mm]*		Concrete	Strength *[MPa]*
55I-030-1	1	1st abutment, middle part under bearing seat	0.02	0.02	0.02	Unsuitable	C45/55	55.3
55I-030-2	2	2nd abutment, right under bearing seat	0.50	0.55	0.38	Unsuitable	C45/55	55.3
55I-030-3	3	2nd. abutment, left under bearing seat	0.23	0.16	0.16	Suitable	C45/55	55.3
55I-030-4	4	2nd cross girder, between 2nd and 3rd girder	0.03	0.01	0.01	Suitable	C40/50	55.5
55I-030-5	5	4th cross girder, between 9th and 10th girder	0.01	0.01	0.07	Unsuitable	C40/50	55.5
55I-030-6	6	Longitudinal joint between 9th and 10th girder, 1st abutment	0.02	0.02	0.01	Suitable	C40/50	55.5
55I-030-7	7	Longitudinal joint between 4th and 5th girder, 1st abutment	0.03	0.01	0.02	Unsuitable	C40/50	55.5
55I-030-8	8	Girder no. 4, 1st cross girder	0.04	0.03	0.02	Suitable	C45/55	55.2
55I-030-9	9	Girder no. 8, 2nd abutment	0.01	0.02	0.02	Unsuitable	C45/55	55.2
55I-030-10	10	Girder no. 12, half of span of the girder	0.03	0.02	0.02	Suitable	C45/55	55.2
55I-030-11	11	Girder no. 14, 2nd abutment	0.07	0.05	0.02	Suitable	C45/55	55.2

Note: Selected sample locations are marked in the bridge scheme in Figure 1.

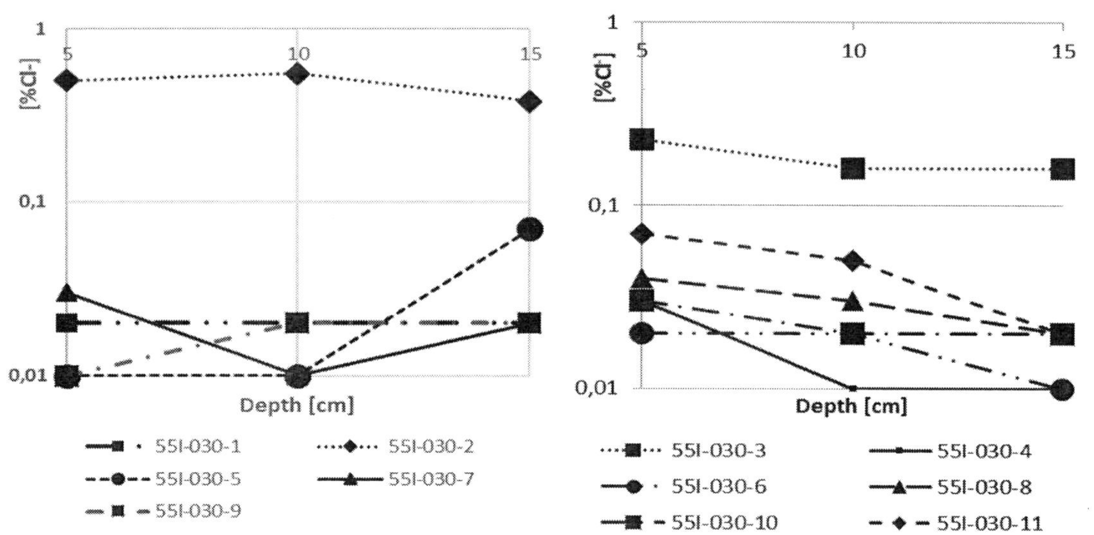

Figure 2. Graphical representation of chloride profiles from Table 1. Left – unsuitable profile, right – suitable profile.

Table 2: Calculated surface concentration of chlorides C_0 and diffusion coefficient D_c for selected suitable samples.

ID	pH			$c(Cl^-)/c(OH^-)$			Surface concentration [%Cl^-]	Diffusion coefficient [$m.s^{-2}$]
	0-10 [mm]	0-10 [mm]	10-20 [mm]	20-30 [mm]	10-20 [mm]	20-30 [mm]		
551-030-3	11.64	11.65	11.70	0.10	0.16	0.16	0.238	6.501E-13
551-030-4	11.11	11.08	11.70	0.08	0.01	0.01	0.037	8.025E-14
551-030-6	11.51	11.55	11.69	0.14	0.02	0.01	0.024	3.443E-13
551-030-8	11.69	11.54	11.76	0.06	0.03	0.02	0.046	2.748E-13
551-030-10	11.50	11.57	11.74	0.04	0.02	0.02	0.031	5.330E-13
551-030-11	11.75	11.76	11.79	0.06	0.05	0.02	0.087	1.347E-13

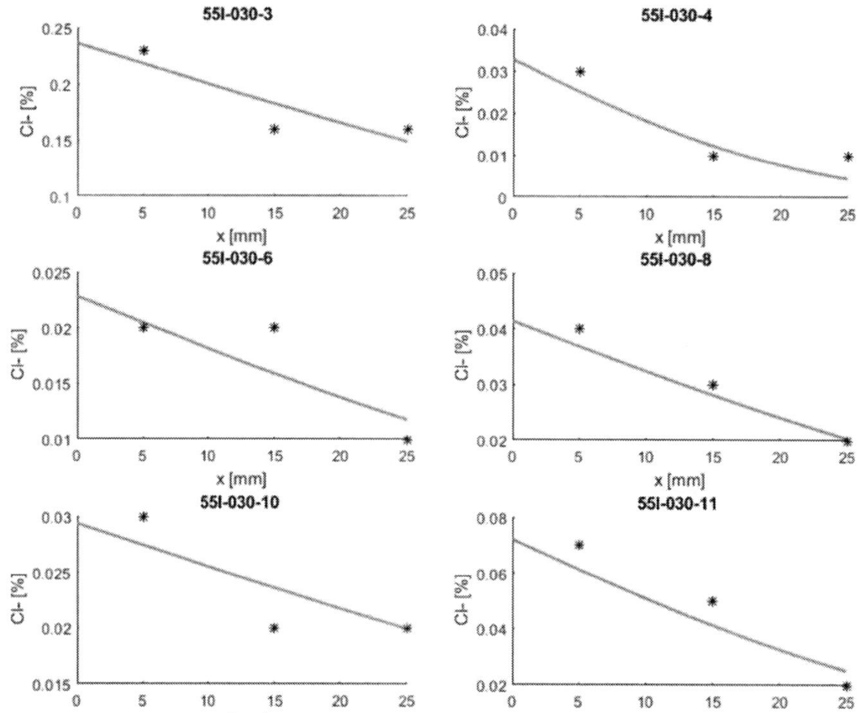

Figure 3. Approximation of the chloride profiles used for calculation of the surface concentration of chlorides C_0 and diffusion coefficient D_c.

The values of chloride concentration of every suitable sample were approximated as shown in Figure 3 and diffusion coefficient and surface chloride content were computed. The approximated curve is based on the determined diffusion coefficient. The next graph (given in Figure 4) is constructed in order to show the connection between individual data points of chloride profiles and the curve fitted among the data points. Thus, the concentration of chloride ions in three evaluated profiles and diffusion coefficient are displayed in Figure 4. It is worth mentioning that the source data are given in two tables (Table 1 and 2) including approximated surface chloride concentration values.

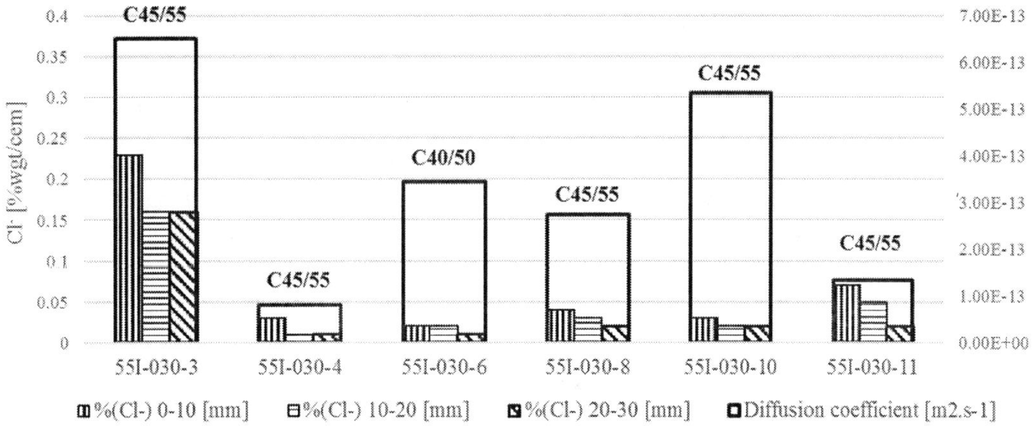

Figure 4. The chloride profile and calculated diffusion coefficient D_c for suitable samples of the bridge no. 551-030.

It can be seen that the trend of the ratio between chloride concentration and the diffusion coefficient is not proportional. It might appear, that if there is a higher concentration of chloride ions, the diffusion coefficient value should be higher. The assumption might be that if more chlorides got into the construction over the time then the concrete might be less resistant to chloride ingress and the diffusion coefficient is higher. Since the diffusion process is more complex phenomenon and the chloride ingress depends also on the surface concentration, temperature, moisture, type properties of the concrete, age of the construction etc., it is not that simple [7]. Because of this information, the results were divided into categories based on the class of the concrete represented by compressive strength. As a satisfactory result would be considered the values of diffusion coefficient in the same decimal level. It is shown in Figure 4 that the diffusion coefficient is not in the order of magnitude in one category. Values of the diffusion coefficient for the class C45/55 range between 0.8×10^{-13} and 6.5×10^{-13} [m^2s^{-1}]. The similar trend is visible also in the bridges 54-040 and 57-039. The bridge 551-026a has variation of the diffusion coefficient within one class up to 40 percent. Unfortunately, authors do not have access to the strength class of the bridge 55-033. Thus, results from four bridges might be compared. The resulting values and respective graphs are given for the four bridges in the Appendices.

5. Discussion and conclusions

The article is focused on the analysis of data related to corrosion measured on 5 existing constructions in the Czech Republic. The numerical analysis is shown on one specific bridge (no. 551-030) in detail and the results of the remaining bridges are given in the appendices A – D. Based on the provided data of concentration of chlorides, suitable samples were selected and the initial value of surface concentration and diffusion coefficient were estimated.

The diffusion coefficient was determined for every suitable sample from the approximation of three values given by measurements. Difference in the estimation of the diffusion coefficient in specific concrete classes in the level of decimal place, that have been found in three bridges from four, seems to be unacceptable. The difference in case of one bridge (551-026a) have the difference up to 40 percent which might be acceptable due to the effects of natural variation of concrete properties and environmental conditions affecting the measurements.

Based on the visual results, in this case, it cannot be proved that the diffusion coefficient is correlated to the concentration of chloride ions, nor the type of the concrete. Thus, it can be concluded that for the successful determination of the diffusion coefficient, three measured values are not sufficient; hence, the approximation and the subsequent analysis is burdened with a numerical error. The above mentioned applies also to the estimation of surface chloride concentration values. The recommended value for more precise determination of diffusion coefficient is 5 or 6 measurements for every drilled core.

The visual comparison of the relationship between the concrete class and diffusion coefficient was used to evaluate the quality of approximation. Proposed procedure with quantified error analysis deserves further evaluation in the future research in order to be used as the tool for the evaluation of the quality of chloride profile data.

Acknowledgment

This contribution has been developed as a part of the research project GACR 18-07949S "Probabilistic Modeling of the Durability of Reinforced Concrete Structures Considering Synergic Effect of Carbonation, Chlorides and Mechanical Action" supported by the Czech Science Foundation.
We are grateful to Mr. Igor Suza from Mostní a silniční, s.r.o. company that enabled the processing of presented data from in-situ measurements carried out by him and his colleagues.

References

[1] Helland S, Norge S and Norway O 2016 *Performance-Based Service Life Design in the 2021 Version of the European Concrete Standards-Ambitions and Challenges*, In book: Hans Beushausen, editor, Symposium 2016 Performance-Based Approaches for Concrete Structures, Cape Town, South Africa

[2] Special Activity Group 5 „Model Code 2010 - Final draft, Volume 2", fib bulletin No. 66, Lausanne, Switzerland, 2012. [online] Available at: https://www.fib-international.org/publications/fib-bulletins/model-code-2010-final-draft,-volume-2-143-detail.html [Accessed: 13 June 2019]

[3] Andrade C 2016 fib T.8.3: *Operational document to support Service Life Design*

[4] JCSS Probabilistic Model Code, http://www.jcss.byg.dtu.dk/Publications/Probabilistic_Model_Code.aspx, (2016)

[5] Novák D, Vořechovský M and Teplý B 2014 FReET: Software for the statistical and reliability analysis of engineering problems and FReET-D: Degradation module, *Advances in Engineering Software*, doi:10.1016/j.advengsoft.2013.06.011

[6] Matesová D, Teplý B, Chromá M and Rovnaník P 2007 *Stochastic Degradation Models for Durability Limit State Evaluation*, Sara – Part VI 1–10

[7] Boddy A, Bentz E, Thomas M D A and Hooton R D 1999 Overview and sensitivity study of a multimechanistic chloride transport model, *Cement and Concrete Research*, doi:10.1016/S0008-8846(99)00045-9

[8] Stewart M G and Rosowsky D V 1998 Time-dependent reliability of deteriorating reinforced concrete bridge decks, *Structural Safety* **20** 91–109, doi:10.1016/S0167-4730(97)00021-0

[9] Ghosh P, Konečný P and Tikalsky P J 2011 SBRA model for corrosion initiation of concrete structures, *RILEM Bookseries* **5** 85–100, doi:10.1007/978-94-007-0677-4_5

[10] Lehner P, Konečný P, Ghosh P and Tran Q 2014 Numerical analysis of chloride diffusion considering time-dependent diffusion coefficient, *International Journal of Mathematics and Computers in Simulation* **8** 103–6

[11] Vořechovská D, Podroužek J, Chromá M, Rovnaníková P and Teplý B 2009 Modeling of chloride concentration effect on reinforcement corrosion, *Computer-Aided Civil and Infrastructure Engineering* **24** 446–58, doi:10.1111/j.1467-8667.2009.00602.x

[12] Bentz E C and Thomas M D A 2013 *Life-365 Service Life Prediction Model: and Computer Program for Predicting the Service Life and Life-Cycle Cost of Reinforced Concrete Exposed to Chlorides*, Life-365 User Manual 1–87

[13] Konecny P and Lehner P 2017 *Frat. Ed Integrita Strutt.* **11** 29–37

[14] Tikalsky P J, Pustka D and Marek P 2005 *ACI Struct. J.* **102** 481–486

[15] Collepardi M, Marcialis A and Turrizuani A R 1972 Penetration of Chloride Ions into Cement Pastes and Concretes, *Journal of American Ceramic Research Society* **55**(10) 534-535

[16] Jiang W, Shen X, Xia J, Mao L, Yang J and Liu Q 2018 A Numerical Study on Chloride Diffusion in Freeze-Thaw Affected Concrete, *Construction and Building Materials* **179** 553–65
[17] Vořechovská D, Konečný P, Šomodíková M, Rovnaníková P 2018 Preliminary analysis of durability related field inspection of highway bridge, *IASTEM 2018* **57-039** 55–58
[18] Raharinaivo A and Genin J M R 1986 Mater, *Construcción* **36** 5–16
[19] ACI 2008, ACI 201.2R-01 Guide to Durable Concrete reported by ACI Committee 201, http://ccl.worldcat.org.ccl.idm.oclc.org/oclc/244388069
[20] Diagnostic exploration of bridge 55I-026a over the local roadway nearby town Otrokovice. Brno, Czech Republic: 2014
[21] Raharinaivo A and R Genin J M 1986 Sobre la corrosión de armaduras de hormigón en presencia de cloruros, *Materiales de Construcción* **36** 5–16, doi:10.3989/mc.1986.v36.i204.881
[22] NT Build 443 1995 Concrete hardened: accelerated chloride penetration. *NT Build 443*. https://doi.org/10.1176/ps.47.9.985
[23] Tikalsky P 2003 *Chapter 20 Durability and Performance-Based design using SBRA*, In book: [24]
[24] Marek P, Brozzetti J and Guštar M 2003 ed. Probabilistic assessment of structures using Monte Carlo simulation: background, exercises and software. Prague: Institute of Theoretical and Applied Mechanics Academy of Sciences of the Czech Republic, *TeReCo*, 2ND edition, ISBN 80-86246-19-1

Appendix A **Bridge 54-040 (built in 1937, diagnostics in 2014, over a stream).**

ID	No.	Location	Chloride profile [%Cl-] 0-10 [mm]	10-20 [mm]	20-30 [mm]		Concrete	Strength [MPa]
54-040-1	1	1st abutment, left, bottom part	0.04	0.04	0.03	Suitable	C8/10	10.2
54-040-2	2	1st abutment, right, bottom part	0.11	0.06	0.04	Suitable	C8/10	10.2
54-040-3	3	1st abutment, left under bearing seat	0.22	0.28	0.26	Unsuitable	C8/10	10.2
54-040-4	4	1st abutment, right, last cross girder	0.23	0.37	0.32	Unsuitable	C8/10	10.2
54-040-5	5	2nd abutment, left, bottom part	0.04	0.03	0.03	Suitable	C8/10	10.2
54-040-6	6	2nd abutment, middle part, bottom	0.07	0.06	0.06	Suitable	C8/10	10.2
54-040-7	7	2nd abutment, left under bearing seat	0.11	0.14	0.18	Unsuitable	C8/10	10.2
54-040-8	8	2nd abutment, middle part under bearing seat	0.14	0.11	0.17	Unsuitable	C8/10	10.2
54-040-9	9	girder no. 1 (2nd cross girder)	0.16	0.04	0.05	Unsuitable	C16/20	22.4
54-040-10	10	girder no. 2 (2nd cross girder)	0.04	0.04	0.03	Suitable	C16/20	22.4
54-040-11	11	girder no. 3 (2nd cross girder)	0.03	0.03	0.02	Suitable	C16/20	22.4
54-040-12	12	girder no. 4 (3rd cross girder)	0.02	0.02	0.03	Unsuitable	C16/20	22.4
54-040-13	13	girder no. 5 (3rd cross girder)	0.03	0.02	0.01	Suitable	C16/20	22.4

ID	pH 0-10 [mm]	pH 10-20 [mm]	pH 20-30 [mm]	c(Cl-)/c(OH-) 0-10 [mm]	c(Cl-)/c(OH-) 10-20 [mm]	c(Cl-)/c(OH-) 20-30 [mm]	Surface conc. [%Cl$^-$]	Diffusion coefficient [m.s^{-2}]
54-040-1	10.35	10.34	9.81	2.64	2.51	6.72	0.044	9.94E-13
54-040-2	10.47	10.12	9.55	5.23	6.14	15.55	0.131	1.05E-13
54-040-5	10.48	10.27	10.56	1.8	2.57	1.27	0.041	8.03E-13
54-040-6	10.03	9.86	9.72	8.81	12.43	17.15	0.071	2.56E-12
54-040-10	8.68	8.47	8.60	116.8	183.89	98.02	0.044	1.00E-12
54-040-11	8.82	8.62	8.66	59.06	100.57	61.94	0.034	5.78E-13
54-040-13	8.79	8.64	8.61	74.47	74.84	27.18	0.036	1.17E-13

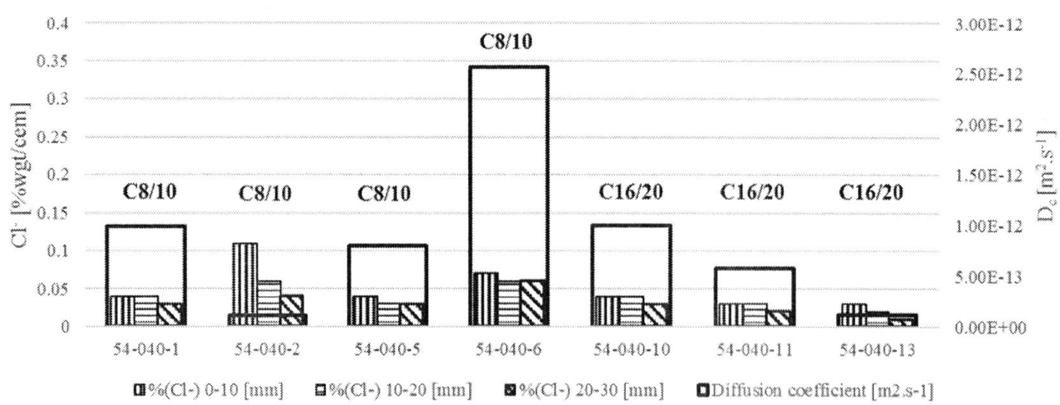

Appendix B **Bridge 55I-026a (built in 1976, diagnostics in 2014, over a local way).**

ID	No.	Location	Chloride profile [%Cl-]				Concrete	Strength [MPa]
			0-10 [mm]	10-20 [mm]	20-30 [mm]			
55I-026a-1	1	1st abutment, left, bottom part	0.06	0.05	0.04	Suitable	C12/15	16.3
55I-026a-2	2	1st abutment, right, bottom part	0.1	0.08	0.06	Suitable	C12/15	16.3
55I-026a-3	3	1st abutment, center part under bearing seat	0.22	0.27	0.3	Unsuitable	C12/15	16.3
55I-026a-4	4	girder no. 4, 1st abutment	0.05	0.07	0.07	Unsuitable	C40/50	54.7
55I-026a-5	5	2nd abutment, left, bottom part	0.02	0.02	0.02	Unsuitable	C12/15	16.3
55I-026a-6	6	2nd abutment, right, bottom part	0.03	0.03	0.02	Suitable	C12/15	16.3
55I-026a-7	7	2nd abutment, middle part under bearing seat	0.02	0.02	0.02	Unsuitable	C12/15	16.3
55I-026a-8	8	2nd abutment, right, bearing seat	0.02	0.02	0.02	Unsuitable	C12/15	16.3
55I-026a-9	9	girder no. 8, 2nd abutment	0.03	0.02	0.02	Suitable	C40/50	54.7
55I-026a-10	10	girder no. 12, 2nd abutment	0.04	0.03	0.02	Suitable	C40/50	54.7
55I-026a-11	11	concrete patch, left above 1st abutment	0.19	0.13	0.13	Suitable	C30/37	38.4
55I-026a-12	12	concrete patch, right above 2nd abutment	0.07	0.05	0.04	Suitable	C30/37	38.4

ID	pH			c(Cl-)/c(OH-)			Surface conc. [%Cl⁻]	Diffusion coefficient [m.s⁻²]
	0-10 [mm]	10-20 [mm]	20-30 [mm]	0-10 [mm]	10-20 [mm]	20-30 [mm]		
55I-026a-1	9.06	9.31	9.15	78.88	37.51	42.4	0.065	1.02E-12
55I-026a-2	9.44	9.51	10.11	49.82	32.87	7.04	0.111	7.07E-13
55I-026a-6	8.84	8.75	8.73	51.42	63.26	62.86	0.034	1.20E-12
55I-026a-9	11.53	11.88	12.01	0.12	0.04	0.03	0.031	8.62E-13
55I-026a-10	11.63	11.95	12.07	0.35	0.19	0.16	0.046	4.46E-13
55I-026a-11	11.88	11.97	12.04	0.12	0.04	0.02	0.197	9.78E-13
55I-026a-12	11.17	11.28	11.85	0.67	0.35	0.07	0.077	5.67E-13

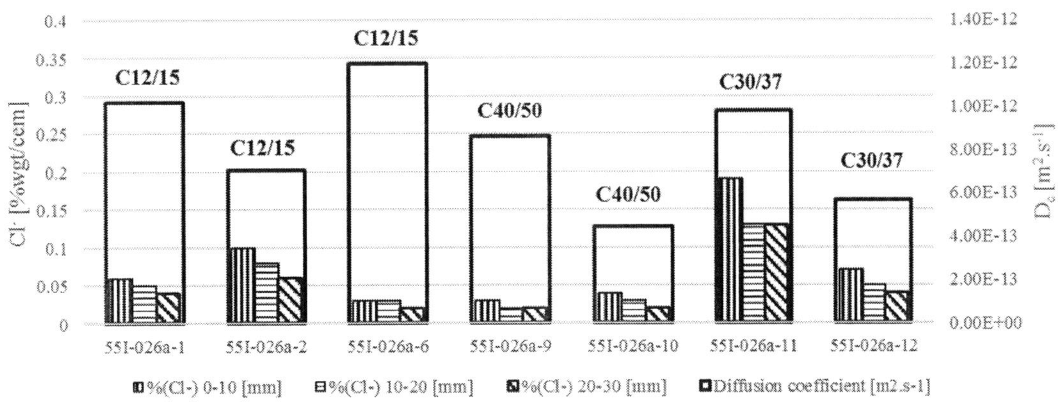

Appendix C: **Bridge 55-033 (built in 1963, diagnostics in 2016, over a river).**

ID	No.	Location	Chloride profile [%Cl-]			
			0-10 [mm]	10-20 [mm]	20-30 [mm]	
55-033-1	1	1st abutment, bearing seat, left	0.21	0.24	0.23	Unsuitable
55-033-2	2	1st abutment, bearing seat, middle part	0.11	0.18	0.09	Unsuitable
55-033-3	3	1st abutment, bearing seat, right	0.08	0.11	0.09	Unsuitable
55-033-4	4	4th support, bearing seat, left	0.06	0.1	0.08	Unsuitable
55-033-5	5	4th support, bearing seat, middle part	0.02	0.02	0.02	Unsuitable
55-033-6	6	4th support, bearing seat, right	0.06	0.03	0.03	Suitable
55-033-7	7	2nd support, bearing seat, left	0.02	0.01	0.01	Suitable
55-033-8	8	2nd support, bearing seat, middle part	0.02	0.02	0.01	Suitable
55-033-9	9	2nd support, bearing seat, right	0.01	0.01	0.01	Unsuitable
55-033-10	10	3rd support, bearing seat, left	0.04	0.03	0.02	Suitable
55-033-11	11	3rd support, bearing seat, middle part	0.02	0.01	0.01	Suitable
55-033-12	12	3rd support, bearing seat, right	0.01	0.01	0.01	Unsuitable
55-033-13	13	Girder, 2nd span, half of the span, left	0.02	0.01	0.01	Suitable
55-033-14	14	Girder 1st span, 3 m next to 1st support	0.03	0.02	0.01	Suitable
55-033-15	15	Girder, 1st span, middle part, front of 3rd abutment	0.02	0.02	0.02	Unsuitable
55-033-16	16	Girder, 2nd span, middle part, 2nd abutment	0.02	0.01	0.01	Suitable
55-033-17	17	Girder, 2nd span, half of the span, middle part	0.02	0.01	0.01	Suitable
55-033-18	18	Girder, 2nd span, middle part, 3rd abutment	0.02	0.02	0.02	Unsuitable
55-033-19	19	Girder, 3rd span, left side, next to 4th abutment	0.22	0.09	0.06	Suitable
55-033-20	20	Girder, 3rd span, left side, half of the span	0.11	0.08	0.03	Suitable

ID	pH			c(Cl-)/c(OH-)			Surface conc. [%Cl-]	Diffusion coefficient [m.s-2]
	0-10 [mm]	10-20 [mm]	20-30 [mm]	0-10 [mm]	10-20 [mm]	20-30 [mm]		
55-033-6	11.89	11.86	11.83	0.1	0.07	0.05	0.066	2.32E-13
55-033-7	11.89	11.93	11.91	0.03	0.02	0.02	0.022	2.29E-13
55-033-8	11.51	11.61	11.75	0.07	0.05	0.03	0.024	4.05E-13
55-033-10	11.83	11.85	11.86	0.09	0.05	0.03	0.046	3.22E-13
55-033-11	11.73	11.83	11.65	0.05	0.03	0.04	0.022	2.32E-13
55-033-13	10.62	10.76	11.17	0.55	0.35	0.13	0.022	2.35E-13
55-033-14	11.70	11.89	11.89	0.1	0.04	0.03	0.036	1.66E-13
55-033-16	11.33	11.84	11.95	0.12	0.03	0.02	0.022	2.32E-13
55-033-17	10.24	10.63	11.51	1.33	0.48	0.04	0.022	2.28E-13
55-033-19	11.58	11.90	11.93	0.83	0.16	0.1	0.278	9.03E-14
55-033-20	11.47	11.52	11.82	0.51	0.32	0.07	0.137	1.55E-13

Note: *The type of the concrete was not measured.*

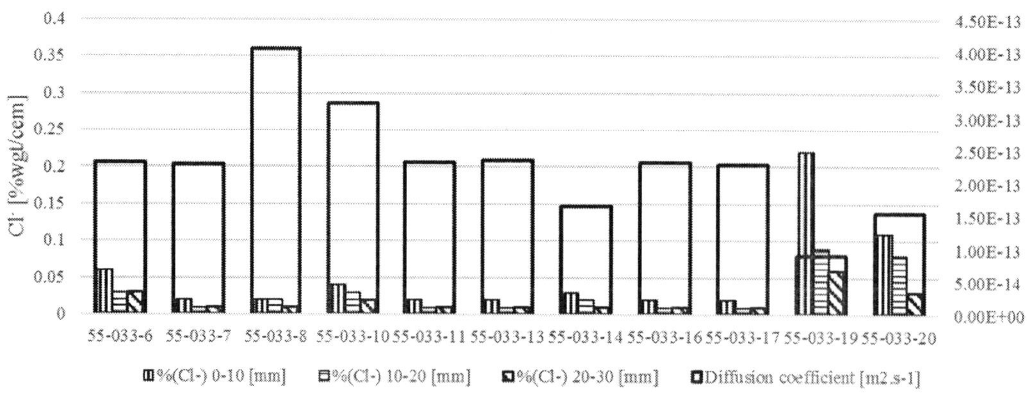

Appendix D: **Bridge 57-039 (built in 1985, diagnostics in 2013, over a river).**

ID	No.	Location	Chloride profile [%Cl-]				Concrete	Strength [MPa]
			0-10 [mm]	10-20 [mm]	20-30 [mm]			
57-039-1	1	2nd abutment, left, wing wall	0.04	0.02	0.01	Suitable	C16/20	20.4
57-039-2	2	2nd abutment, left, concrete patch	0.04	0.03	0.05	Unsuitable	C12/15	15.1
57-039-3	3	2nd abutment, right, wing wall	0.03	0.03	0.02	Suitable	C16/20	20.4
57-039-4	4	2nd abutment, right, concrete patch	0.22	0.11	0.08	Suitable	C12/15	15.1
57-039-5	5	1st abutment, middle bottom part	0.1	0.06	0.04	Suitable	C12/15	15.1
57-039-6	6	1st abutment, middle part under bearing seat	0.13	0.18	0.19	Unsuitable	C12/15	15.1
57-039-7	7	2nd abutment, middle bottom part	0.09	0.09	0.08	Suitable	C12/15	15.1
57-039-8	8	2nd abutment, middle bottom part	0.19	0.22	0.19	Unsuitable	C12/15	15.1
57-039-9	9	2nd abutment, middle part under bearing seat	0.26	0.29	0.3	Unsuitable	C12/15	15.1
57-039-10	10	Girder no. 4, section 3	0.03	0.03	0.03	Suitable	C40/50	53.9
57-039-11	11	Girder no. 5, section 3, 2nd abutment	0.05	0.04	0.03	Suitable	C40/50	53.9
57-039-12	12	Longitudinal joint between 3rd and 4th girder	0.05	0.03	0.02	Suitable	C25/30	34.7
57-039-13	13	Longitudinal joint between 4th and 5th girder	0.04	0.03	0.03	Unsuitable	C25/30	34.7

ID	pH			c(Cl-)/c(OH-)			Surface conc. [%Cl⁻]	Diffusion coefficient [m.s⁻²]
	0-10 [mm]	10-20 [mm]	20-30 [mm]	0-10 [mm]	10-20 [mm]	20-30 [mm]		
57-039-1	8.66	8.06	8.27	126.27	218.56	105.89	0.05	1.97E-13
57-039-3	9.11	8.58	8.14	33.74	92.88	207.76	0.034	1.63E-12
57-039-4	8.48	8.53	11.06	1022.66	449.11	1.04	0.261	2.70E-13
57-039-5	7.50	7.30	7.43	4296.5	4536.66	1931.27	0.117	3.56E-13
57-039-7	8.23	8.01	7.86	750.44	1213.89	1608.27	0.095	1.00E-11
57-039-10	10.71	11.60	11.89	1.42	0.13	0.05	0.056	9.54E-13
57-039-11	10.77	11.75	11.84	1.11	0.08	0.04	0.058	3.55E-13
57-039-12	10.43	9.78	11.34	2.16	6.34	0.16	0.041	2.20E-12

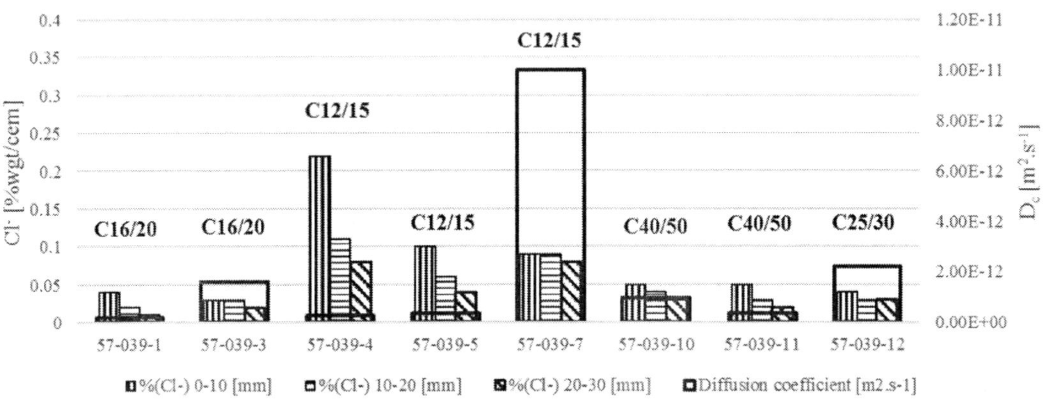

Innovative solution of fine horizontal trash rack for small hydroelectric power stations

A Miltenović[1], M Prodanović[2], L Beju[3], V Miltenović[4] and N Velimirović[5]

[1]University of Niš, Faculty of Mechanical Engineering, A. Medvedeva 14, Niš, Serbia
[2]JP EPS Beograd, Ogranak Održivi razvoj, Belgrade, Serbia
[3]Lucian Blaga University of Sibiu, Faculty of Engineering, Str.Emil Cioran 4, Sibiu, Romania
[4]University of Niš, Innovation Center, Univerzitetski Trg 2, Niš, Serbia
[5]State University of Novi Pazar, Department for Technical Science, Vuka Karadžića bb, Novi Pazar, Serbia

E-mail: aleksandar.miltenovic@masfak.ni.ac.rs

Abstract. Alternative energy sources have a number of advantages over energy derived from fossil and nuclear power plants. In this way, the environment is protected, and many countries are encouraging the use of alternative energy sources. Large potential from this aspect has small hydropower plants, which are not expensive, they can be installed on small rivers, and in this way the received electricity is much cheaper. One of the problems that need to be solved is to prevent the penetration of impurities into turbine plants. This is solved by placing protective trash rack on water intakes. The paper presents an innovative solution for horizontal fine trash-rack with cleaning function for small hydro-electric power stations.

1. Introduction

Energy trends have been mostly focused on renewable energy sources in recent years. In the world, about 20% of electricity is produced using energy of water flows - hydropower with a total installed power of 720 GW. The energy obtained from hydroelectric power plants is significantly cheaper than the energy generated from fossil fuels or nuclear energy.

As an alternative energy source, especially in remote areas where other energy sources are not available, small hydropower plants are used. They can be installed on small rivers or streams with little or negligible impact on the environment. When installing a small hydroelectric power plant, it is necessary to take into account the amount of water that is permanently available, since insufficient rain can jeopardize the operation of the power plant, as well as the height drop, ie the height difference between the operation and the outlet of the water.

Water for hydroelectric power plants comes from mountainous areas and passes through its various paths through various forested, uninhabited and populated areas, carrying large amounts of impurities or waste. For the correct and undisturbed operation of turbines, water intakes, and water drainage systems, it is necessary to prevent the penetration of impurities into turbine plants. This is achieved by the installation of appropriate trash-racks. Due to the long distance of the water intake, there is a problem of cleaning these trash-racks. Various automatic cleaners are installed for cleaning, but electric power is required for their operation. Given that these sites are also several kilometers away and in an inaccessible terrain, the application of this solution is unrealistic, increases the production of kilowatts in hydropower plants and requires additional labor. For this reason, the idea was to develop an innovative solution of a completely autonomous trash-rack with a unified function and a cleaner as well as a reliable performance of the function.

2. Solutions of trash-rack cleaner

Most often, trash-racks made of longitudinally placed rods at the appropriate distance and transversely positioned rods for securing the position of longitudinally placed rods are most often used for the protection of turbine plants. Depending on the dimension of the trash-rack, transverse support brackets are also installed. The design shape and dimensions are shown in Fig.1 [4].

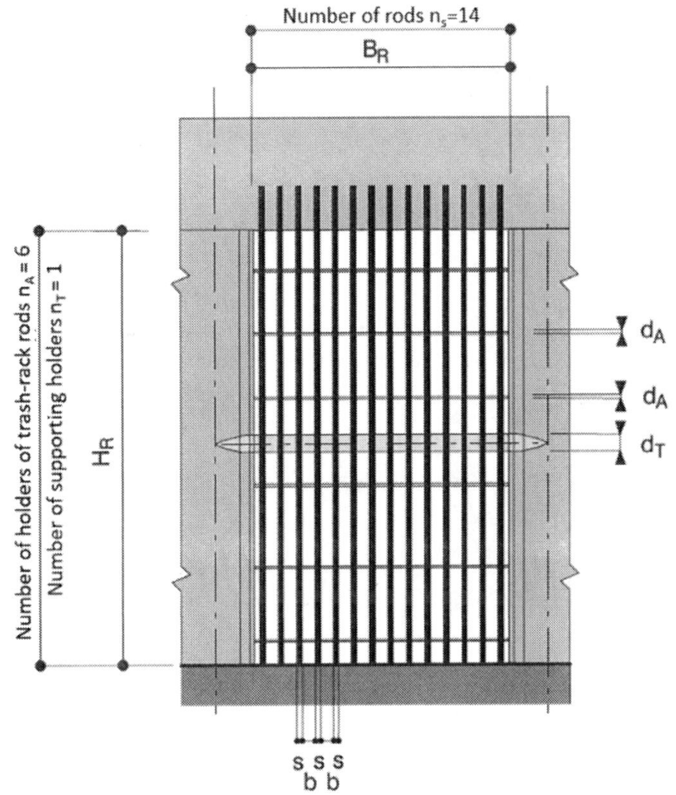

Figure 1. Design shape of trash-rack.

From the point of view of energy losses, an important parameter of the trash-rack is the degree of resistance to the passage of water - the degree of openness of the trash-rack (Verbauungsgrades) P, which is determined as the ratio of the surface of the A_P resistance and the total area of the arc A_{RF}:

$$P = \frac{A_{RS} + A_{AH}}{A_{RF}} = \frac{A_P}{A_{RF}} \tag{1}$$

$$A_{RS} = n_s \cdot s \cdot (H_R - n_A \cdot d_A - n_T \cdot d_T);\quad A_{AH} = B_R \cdot (n_A \cdot d_A + n_T \cdot d_T);\quad A_{RF} = B_R \cdot H_R \tag{2}$$

In practice is degree P in the limits between 0,1 and 0,4.

There are different types of trash-rack cleaner (ger. Rechenreiniger) hydropower plants. The principles on which they work can be manual, electrical, electromechanical, automatic, semi-automatic, etc. The shape and principle of the work largely depend on the type of trash-rack.

It can be concluded that there are no completely autonomous trash-rack cleaner, without the need for direct connection to the power grid. This is especially true for small and mini hydro power plants, whose water intakes are far away from the machine building and inaccessible terrain. Consequently, the idea was to develop an innovative solution of a fine horizontal trash-rack with a unified grid function and a cleaner that will be completely autonomous and use water energy for its work and with automatic monitoring of the working condition.

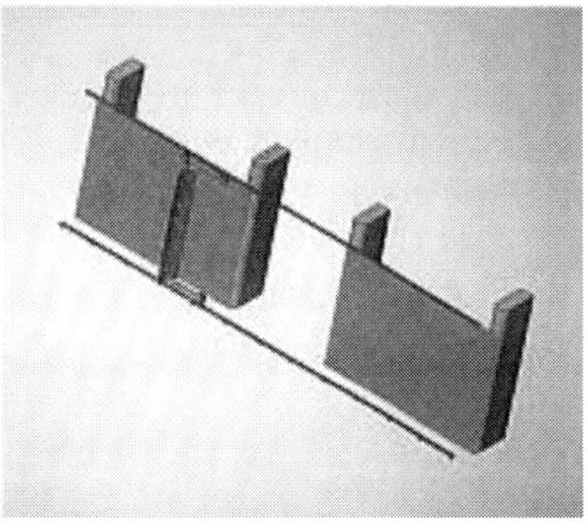

Company Fincos (http://www.fincos.eu/rechenreiniger.html) has also developed and manufactured a horizontal trash-rack cleaner. The cleaner of the appropriate construction is placed in the guides and at the appropriate time intervals it moves horizontally along the trash-rack, cleaning it from impurities. It belongs to a group of fully automatic cleaners, but it requires electricity for its operation.
Fincos manufactures and grates (rough and fine) for the protection of turbine plants from impurities. The distance of the rods in the fine trash-rack is 15.mm, which enables reliable protection of turbine plants.

Company Lukas (Fa. Lukas Anlagenbau GmbH) has developed a cleaner for small hydropower plants, which, during cleaning, covers a working width of up to 2 m and a cleaning length of up to 1.5 m. The cleaning brush is made of plastic with smooth and serrated work. This is an articulated sweeper intended for handling and cleaning on-site. It can also clean horizontally and vertically the trash-rack.

Company Wild Metal (www.wild-metal.com) has developed a family of trash-rack cleaners with horizontal movement of the cleaning part. The cleaners are fully automated with electro-mechanical drive. The device for movement is protected by a hood from the influence of the outside environment. Cleaning brushes have plastic teeth.

Figure 2. Overview of trash-rack cleaners for hydropower plants.

3. Designing the conceptual solutions

Ideas of variant solutions describing the mode of operation are shown in Figures 3, 4 and 5.

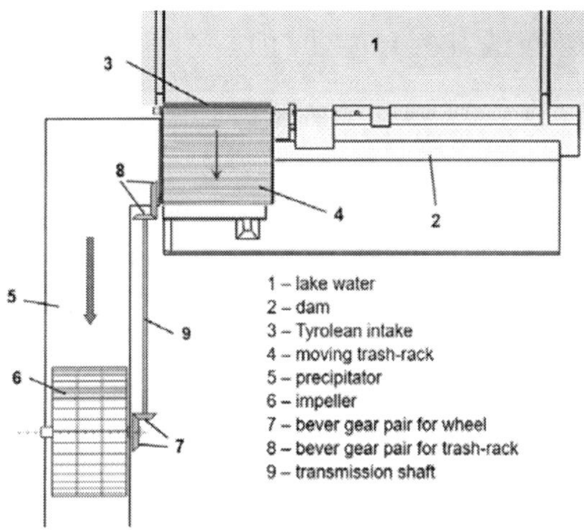

The trash-rack is mobile and is made according to the model of the conveyor belt with chain drive. Perforated sheet metal is used to prevent the penetration of impurities into the inlet pipeline or turbine plant. Water from the lake over the Tyrolean intake 3 comes to the trash-rack 4 and through the holes on the perforated sheet flows further into the precipitator 5 and the pipeline. In the waterfall of the precipitator, a working wheel 6 is installed, which receives a drive from the flow of water into the precipitator bed. A conical gearing pair 7 is mounted on the output shaft of the impeller. The shaft 9 moves through the shaft 9 and the cone pair 8 via the chain conveyor. With its motion of the trash-rack, it immediately expels the possible penetration of impurities from the lake, or the surrounding environment, which ensures the reliable operation of the turbine of the hydropower plant.

Figure 3. Trash-rack – variant 1.

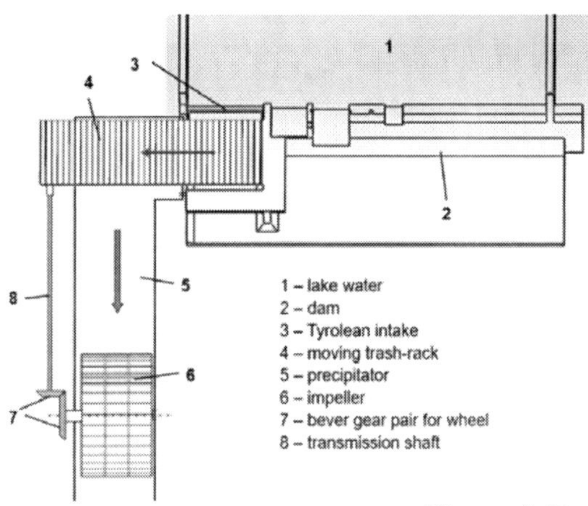

The conceptual solution of the 2 fine trash-rack variant is similar to the solution of variant 1. The difference is that the trash-rack is set transversely in relation to the direction of flow of water over the Tyrolean intake. The trash-rack is driven by a drive from the conical pair 7 and the shaft 8, which is directly connected to the drive chain of the sprocket. This variant is more favorable than the variant 1, because the drive is simpler, and the water is easier to pass through the grate. In addition, the collected impurities go away, making it easier to remove it.

Figure 4. Trash-rack – variant 2.

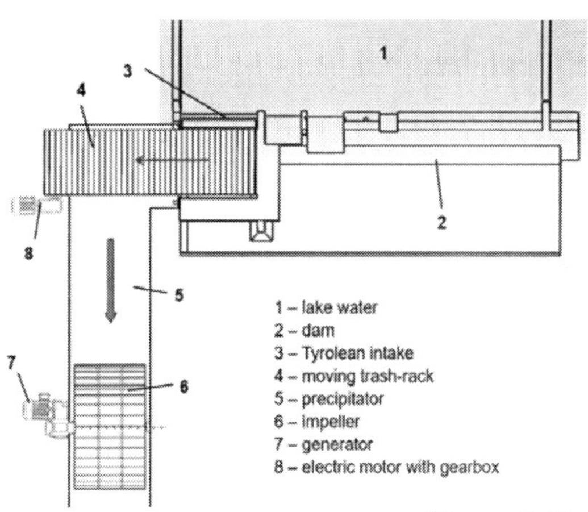

In contrast to variant 2, here the transmission of the drive from the working point to the trash-rack is not carried out directly by mechanical connection, but indirectly by electrical means. A current generator 7 is installed on the impeller of the impeller through the corresponding multiplier. The drive unit is now running via an electric motor with gearbox 8, which is connected to the chain drive chain. This variant is more expensive compared to variants 1 and 2, but provides a number of advantages.

Figure 5. Trash-rack – variant 3.

4. Evaluation of variant solutions

The analysis and evaluation of the variant solutions is given in Table 1. Based on the analysis, it was adopted for the further development of the construction of the variant 3. Regardless of the fact that this is more expensive, the variant enables reliable execution of the operating function and has a number of advantages in relation to variant 1 and 2.

Table 1. The analysis and evaluation of the variant solutions.

Variant	Advantages	Disadvantages
1	• Combining the trash-rack function (preventing the penetration of impurities) and cleaner (unhindered passage of water); • A simple design solution; • Low price	• Direction of water flow and movement of grids coincide, which may interfere with operation; • Vital drive parts (shaft, gears) are exposed to the environmental impact, which can affect the performance of the function. • Filling the operating function depends on the flow of water. • It is not possible to automatically monitor the state of operational correctness.
2	• Integrated trash-rack functions (preventing impurity penetration) and cleaners (unhindered passage of water); • A simple design solution; • Low price; • The direction of flow of water and the movement of the grid does not match, which positively influences the fulfillment of the function;	• Vital drive parts (shaft, gears) are exposed to the environmental impact, which can affect the performance of the function. • Filling the operating function depends on the flow of water. • It is not possible to automatically monitor the state of operational correctness.
3	• Integrated trash-rack functions (preventing impurity penetration) and cleaners (unhindered passage of water); • Fully autonomous cleaner; • The direction of flow of water and the movement of the trash-rack does not match, which positively influences the fulfillment of the function; • Filling the operating function does not depend on the flow of water. • Automatic control of operating status.	• More expensive than variants 1 and 2

5. Further development of an innovative solution

5.1. Design solution of drive system

Small hydropower plants are characterized by flow-through operation with very small accumulation. The low overflow threshold or the Tyrol intake is intended to slow the water stream before entering the feed channel. Considering the working conditions in which the flowing mini hydroelectric power plant works, it is advisable to choose a water wheel as a solution for the operation of the trash-rack cleaner.

Guide wheels have a number of advantages for autonomous operation. They are simple constructions, they have a low price, are reliable in operation and do not require special working conditions. They have adequate inertia in operation and stability of the speeds. They can work for normal flow rates. There are several types of these wheels [5]. For the conditions on the water intakes of the power plants for construction reasons as the driving wheel, the Poncelet water wheel was selected (Fig. 6).

For the flow of water in small hydroelectric power plants, the following starting data can be taken:

$Q_{min} = 0,027\ m^3/s$ - the minimum amount of water flowing through the trash-rack to enable the hydroelectric plant turbine to work;

Q_{max} = 0,24 m^3/s – the maximal amount of water flowing through the trash-rack to enable the hydroelectric plant turbine to work;

Q_{BM} = 0,011 m^3/s – biological minimum – minimal amount of water for the drive of wheel.

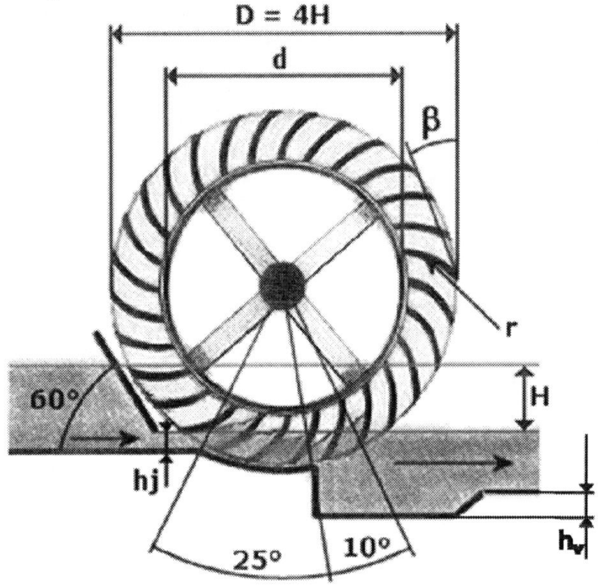

Figure 6. Design measurements of Poncelet water wheel [2].

The power of the Poncelet water wheel for water fall values in the range of 0.2 m to 4 m is:

$$P_t = \eta_t \cdot \rho \cdot g \cdot Q \cdot H \tag{3}$$

P_t - Power of water wheel [kW]
g - gravitational acceleration (g = 9.81 m/s^2)
η_t - Level of efficiency of water wheel
ρ - Water dencity (ρ = 1000 kg/m^3)
Q - flow [m^3/s]
H - Netto fall [m]

For slow moving water as the value of drop H, the difference between the input and the output water level is taken, which in our case is H = 400 mm. The usefulness of the Poncelet wheel is η_t = 0.65. Accordingly, the maximum power is:

$$P_t = \eta_t \cdot \rho \cdot g \cdot Q \cdot H = 0.65 \cdot 9.81 \cdot 1 \cdot 0.24 \cdot 0.4 = 0.612 \text{ kW} = 612 \text{ W} \tag{4}$$

The diameter of the Poncelet wheel (Fig.6) is: DRK = 4·H = 4·400 = 1600 mm. Other technical data are given in Table 2.

According to the data given in Table 2, the constructive solution of the Poncelet wheel shown in Fig. 7, b. The supporting structure of the wheel is made of structural steel by welding from the steam-rod rods. The hub point is connected with a wreath with 6 paws symmetrically arranged on both sides of the wheel. The supporting rings are longitudinally also connected by rectangular cross-reels. Through this construction, a drum is placed with blades. Due to the large width of the impeller (Bk = 930 mm), two circles of circular cross-section are installed as the blade holder.

The width of the blade is b_l = 89mm, length l = 930mm, and the radius of curvature R_l = 440mm. The fixation was carried out evenly at 4 places along the length of the blade. The blade is inserted into specially designed arched holes, and in length it is additionally attached to 2 hoops so that it has a stable operating position.

Table 2. Technical data for Poncelet water wheel.

Size		Unit	Value
Power for Q_{max} = 0.24 m^3/s	P_t	kW	0.612
Speed at maximum power	n_t	min^{-1}	16.24
No load speed	n_{t0}	min^{-1}	32.5
Torque	T_t	Nm	360
Efficiency	η_t		0.65
External diameter	D_{RK}	mm	1600
Internal diameter	d_t	mm	1080
Maximal width	B	mm	1000
Impeller width	B_k	mm	930
Depth of the water jet	h_j	mm	100
Depth of water at the exit (Fig.6)	h_v	mm	240
Speed of the jet	v_m	m/s	2.72
Blade width	b_l	mm	89
Blade radius	R_l	mm	440
Blade angle	β_l	°	28
Number of blades	Zl	-	35

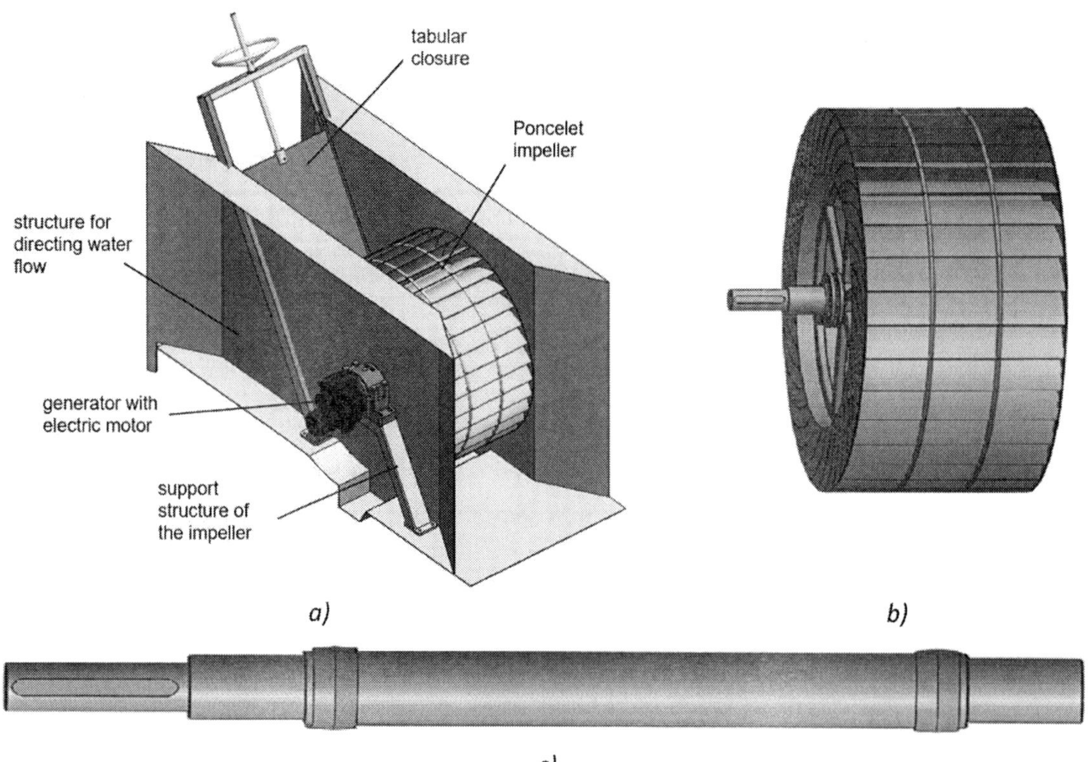

Figure 7. Mini turbine plant (a), Poncelet impeller (b) and shaft (c).

The complete plant for autonomous electrical energy production is shown in Fig.7, a. It consists of side panels and sheet metal, the function of which is to direct the flow of water properly. The shape of the sheet metal is harmonized with the construction of the Poncelet wheel (Fig.6). A water flow control mechanism is also installed with the appropriate guides of the stop plate (plate closure). The flow can be regulated via a threaded spindle. On the side are the mounts on which the impellers of the impeller are fixed. For bearings are selected 22324 CCK / W33 with complete SKF sleeve SAF 22624. The torque transfer from the impeller to the generator is carried out over the stepped shaft (Fig. 7, c). The link of the shaft with the impeller is made via conical clamping rings and with the generator via the coupling.

5.2. Design solution of horizontal trash-rack

The horizontal fine trash-rack with cleaning function is constructed according to the model of conveyor belts with movable perforated plates (Fig.8). The drive is operated via a chain conveyor.

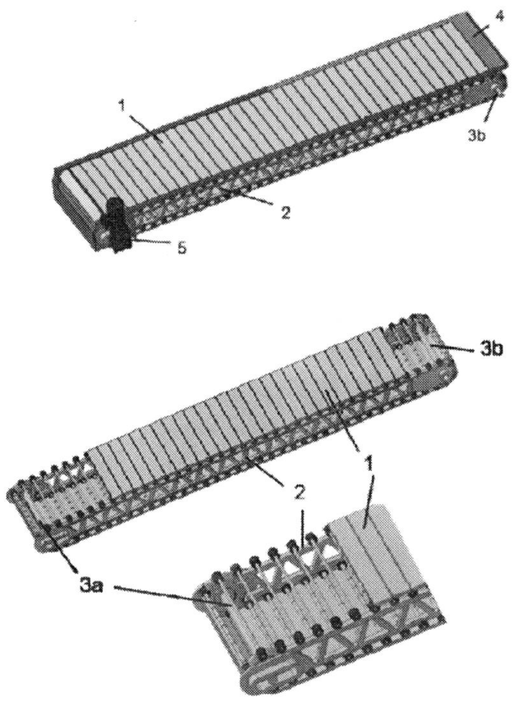

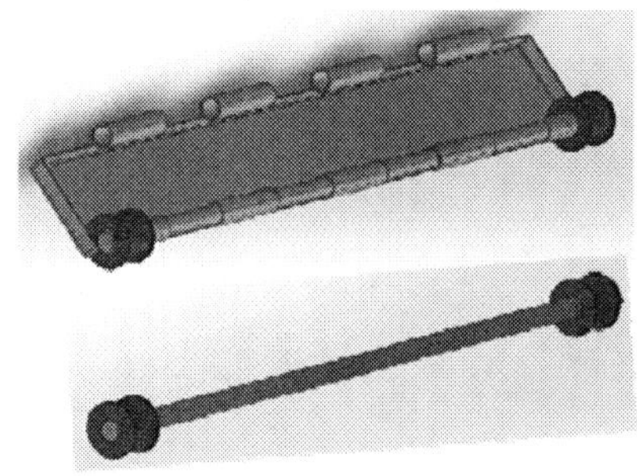

b) Articulated perforated plates with rollers

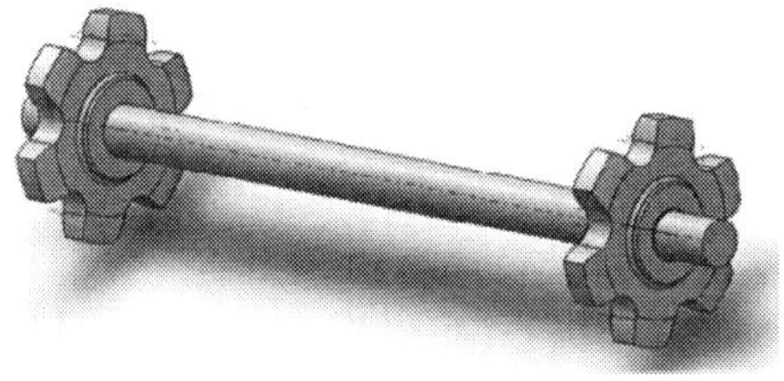

a) Solution - assembly *c) Set of chain sprocket with shaft*

Figure 8. Design solution for horizontal trash-rack.

The basic components (Fig. 8, a) or components of the trash-rack are:

1. Perforated plates are mutually articulated with rollers for operation and guiding (Fig.8, b);
2. A supporting structure of the trash-rack, which has a guide function;
3. Set the sprocket with the drive (3a), ie the driven (3b) shaft (Fig.8, c);
4. Protective sheet for trash-rack, with a brush for removal of impurities and waste from the grid;
5. Electric motors for the peripheral motor (24V / 600W DC motors), connected to the set of drive sprockets 3a.

The innovativeness of such a construction solution is reflected in the integration of the trash-rack function (preventing the impurity from entering the turbine plant) and the cleaning function (preventing the trash-rack congestion, ie, removing impurities from the trash-rack itself). This was achieved by installing the trash-rack (Fig.8, a). The trash-rack itself is made of perforated plates, which are articulated jointly (Fig.8, b). At the ends of the sticks (shaft) are placed plastic rollers, which move along the tracks of the supporting structure of the trash-rack (Fig. 8, b). The rollers come in the middle of the chain of the sprockets, which drives the movement through the electric motor and the drive shaft of the grate. In this way, the impurities that come to the horizontal trash-rack are immediately removed and thrown off the outside of the Tyrolean intake, with additional brush cleaning from the bottom. The complete construction of the trash-rack is made by welding with adequate protection against the influence of the surrounding environment.

Perforated panels are made of perforated stainless steel sheet 1 mm thick. The diameter of the opening is w = 4mm, while the distance between the holes is t = 6mm (Fig.9). In this way, the conditions for reliable operation of turbine plants are fulfilled, as it prevents the penetration of all objects larger than 4mm. The trash-rack openness is 40%, which allows water resistance to water to be minimal.

Figure 9. Perforated sheet metal for trash-rack panels (W=4 mm; t=6 mm; openness 40%).

5.3. Selection of generator and battery

Due to the power generated by the Poncelet impeller (Pt = 612W), the following type of 24V / 600W DC motors, PM296 of the Ningbo Jirun Electric Machine Co. Ltd. The generator is supplied with the multiplier to increase the speed with $n_t = 16.2\ min^{-1}$ at $n_G = 1450\ min^{-1}$.

The generator current is charged with AKU batteries and is powered by an electric motor to drive the trash-rack. The electric motor 24V / 600W DC motors, PM296 (High torque permanent magnetic brush DC motor) was selected for the trash-rack drive, with a gearbox so that the velocity of the reservoir is 0.1 m/s.

We selected 2 AKU batteries of NARADA 6-GFM-50F, power of 300W, capacity 50Ah, so that they can work for 3h without charging with full load. Since the trash-rack drive is not permanent, but, as needed, the batteries provide enough electricity for reliable protection against impurities and automatic monitoring of the working condition.

6. Conclusion

The horizontal fine trash-rack with cleaning function for small hydro power plants is designed to reliably function as a function, because its drive is via an electric motor, which is powered by over 2 battery cells. Consequently, in the event of minimal water flow, reliable cleaning of the impurities is ensured. On the other hand, the trash-rack openness is 40% with the diameter of the opening 4mm, which ensures an uninterrupted flow of water and a reliable prevention of impurity penetration into the turbine plant.

Since the energy source - the Poncelet impeller and the grating with the electric motor are physically completely separated, as well as from the aspect of the design of the trash-rack itself, universal application on water intakes with a horizontal trash-rack is enabled.

It is envisaged that the vital components and components of the cleaner are made of quality stainless material, which should enable long-term operation with minimal maintenance costs. The cleaner prevents the penetration of impurities not only in the turbine plant, but also in its own energy source via the Poncelet wheel, which is also an important parameter of reducing maintenance costs. Since an autonomous source of electric power is provided through the generator and battery cells, it is also possible to install automatic control of the state of operational safety.

An innovative construction solution is installed on the trash-rack, which is reflected in the trash-rack integration (preventing the impurity penetration into the turbine plant) and the cleaning function (preventing the trash-rack congestion, ie removing impurities from the trash-rack itself). This was achieved by installing a trash-rack of perforated plates, which are mutually connected jointly.

References

[1] Abgottspon A and Staubli T 2010 *Verschmutzung von Kleinwasserkraftwerken*, Eidgenössisches Departement für Umwelt, Verkehr, Energie und Kommunikation UVEK, Bern, Switzerland

[2] Hesham H, Ahmed M and Kassem Medhat E 2016 Hydro Matrix Power Wheels Generate More than 5 GW/h from Main Branch Canals (River Nile) in Egypt, *Journal of Power and Energy Engineering* **4**(3) 71-78

[3] Project of Innovation Fund 2018 *Development and technical design of automatic cleaner for fine trash-rack for small hydropower plants*, Innovation center of University of Niš and Koni Sim Zaječar

[4] Meusburger H 2002 *Energieverluste an Einlaufrechen von Flusskraftwerken*, Technischen Hochschule, Zürich, Switzerland

[5] Nuerneberg D 2006 *Wasserräder zur Elektroenergiegewinnung - Water wheels generating electrical energy,* Tagung des ATW Arbeitsgemeinschaft Thüringer Wasserkraftwerke

[6] Detering M and Roger S 2013 *Ökologischer Forschung(sbedarf) rund um die Wasserkraft, Wasser als Energieträger*, Internationales Wasserbau-Symposium, Aachen

[7] Spasić Ž, Jovanović M, Bogdanović-Jovanović J and Milanović S 2019 Numerical Investigation of the Influence of the Doubly Curved Blade Profiles on the Reversible Axial Fan Characteristics, *Facta Universitatis: Mechanical Engineering* **17**(2)

[8] Benad J 2019 Numerical Methods for the Simulation of Deformations and Stresses in Turbine Blade Fir-Tree Connections, *Facta Universitatis: Mechanical Engineering* **17**(1) 1-15

IOP Conf. Series: Materials Science and Engineering **659** (2019) 012050 doi:10.1088/1757-899X/659/1/012050

Insulation impact on appliance acoustic characteristics

P Zvolensky[1], L Lestinsky[1], J Dungel[2] and J Grencik[1]

[1]University of Zilina, Faculty of Mechanical Engineering, Univerzitna 1, 01026 Zilina, Slovakia
[2]CEIT, a.s., Univerzitna 6A, 01008 Zilina, Slovakia

E-mail: peter.zvolensky@fstroj.uniza.sk

Abstract. The paper is aimed on types of insulation actually used in washing machines and raising requirements on its characteristics. Next part is design of unconventional insulations, which will be applied directly on appliance. Then, series of tests will pass (according to regulations), so it will be possible to determine acoustic power. After that, we will use simulation to determine impact of insulation rigidity on washing machine shell (with identification of acoustic benefit). Relatively important factor is rigidity of washing machine shell and forces affecting the shell. Force affecting on the shell depends on inner and outer tub weight (and all equipment attached), load, unbalance in a drum and revolutions during spin cycle. Usually more rigid washing machine shell produces less noise than weaker shell. That's why is relatively important to stiffen shell. Doing this by changing type of insulation is advantageous because it is possible to do without any significant side effects.

1. Introduction

This article is concerned with impact of insulation rigidity on washing machine acoustic characteristics. Washing machine body is affected by forces during washing cycles which originate from mass of water and laundry. The forces also originate from an unbalance inside the washing machine tub. The unbalance is changing after every inner tub stoppage during a spin cycle. Unbalance inside the washing machine tub is big acoustic issue, not only for increased noise, but also for increased load of components.

Nowadays, comfort requirements are raising so premium washing machines are equipped with noise reduction elements such as insulations and other damping items. Also, rigidity of the washing machine body has big impact on overall noise. Rigid washing machine bodies does not deform in excessive way. If there is no excessive relative movement of panels and other components there is no excessive noise also [1].

2. Strength analysis of the washing machine body

Washing machine body is usually well designed in terms of deformation resistance. Panels are equipped with many extrusions which improves their rigidity. Panels of different producers differ in thickness also. Thickness of panels has also big impact on the washing machine rigidity. In ideal conditions, washing machine body should be as robust as possible, to prevent deformations. If producers want to achieve higher level of a robustness, they have to invest into production of the washing machine body. But there is an issue of increased production costs. Efficient way is to use current panels in production and improve their rigidity without big intervention into its structure. The solution is to use noise insulation as a panel stiffener. Production costs increase would be significantly smaller than production of the thicker panels. Nowadays, washing machines have got installed noise insulation on the panels, but the insulation has not got any measurable impact on the panel rigidity. It is because sealing tape, which is used to secure insulation on panels is not strong enough to transfer forces from panel to insulation. New solution relies on a new way of gluing insulation on the panels. To secure solid connection between an insulation and a panel, the insulation is equipped with a crosshatch. The crosshatch, made of glass-fabric, is part of the insulation (located between insulation and panel). Adhesive foil on the crosshatch secures solid connection between panel and insulation and allows to transfer forces from panel to insulation, so the insulation can serve as the stiffener. For determination of the insulation impact on the panel stiffness, 3D models of the panels were made. For this simulation only panels without washing machine body was used in simulation. The main reason for this was, that we did not want to affect simulation with other variables such as washing machine body stiffness. With this type of simulation, we determined the sole effect of the insulation stiffness on panels only. In real life conditions, with panels mounted on washing machine, deformation of panels would be smaller and it would be harder to investigate insulation stiffness effect on panels. The 3D models were exposed to strength analysis in the FEM software Ansys.

To determine amount of forces used in simulation, we applied one of our internal tests used multiple times in the past with washing machine experiments. Test consists of full load of wet laundry in the washing machine, plus 50 kg ballast located on the top of the washing machine. During this test, side panels are affected by the downward force of 255 N.

The biggest impact on a rigidity of a panels was achieved on the side panels, which on a top loader washing machine transfer the most of the forces [2], [3].

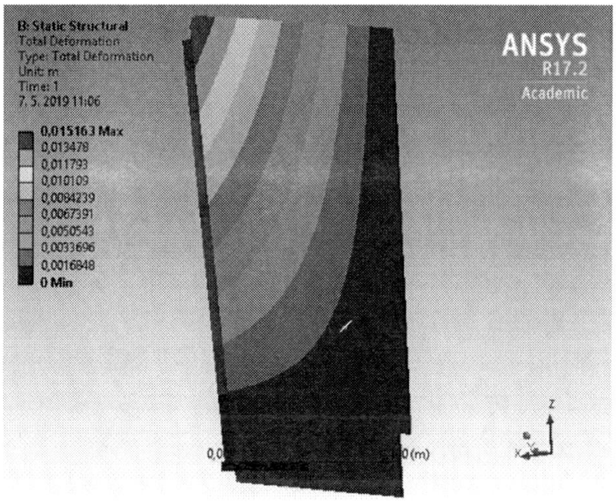

Figure 1. Side panel with OEM insulation.

In the figure 1 is side panel of the top loader washing machine. The panel is equipped with OEM insulation and is loaded with downward force. The biggest deformation was achieved in the upper corner (because of ballast location). Deformation amount is 15 mm.

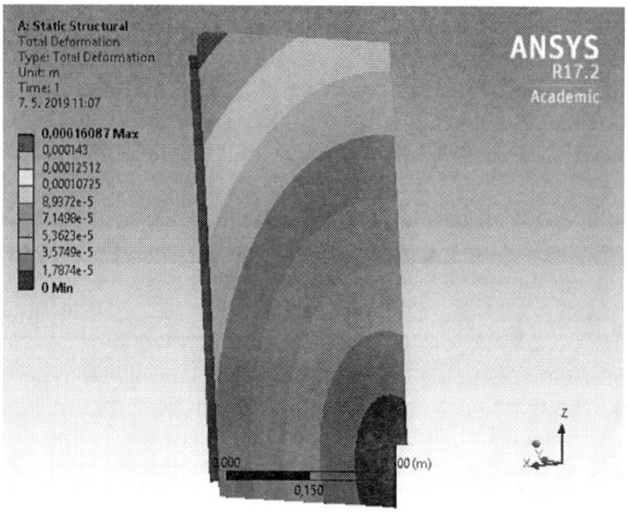

Figure 2. Side panel with upgraded insulation.

In the figure 2 is the same panel loaded with same force but in this case, the panel is equipped with upgraded sticking of the insulation. In this case, smoother deformation was achieved with amount of 0.16 mm. Deformation is evenly distributed on the whole panel surface. In comparison with OEM insulation, deformation was present only in upper part of the panel. The reason is the upgraded insulation sticking transfers forces to insulation, which can serve as the stiffener [4].

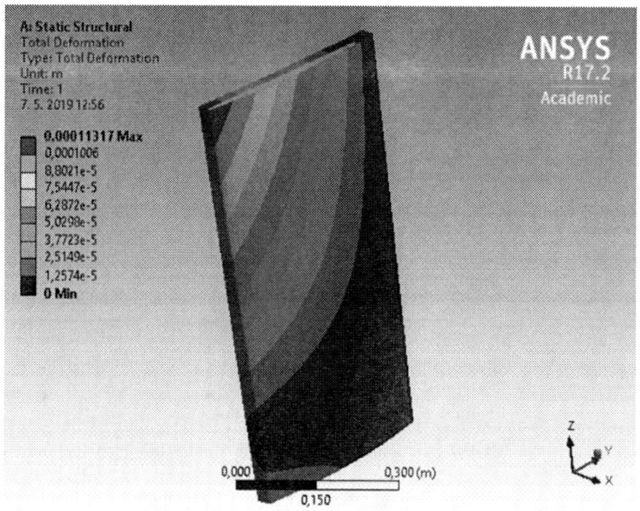

Figure 3. Front panel with upgraded insulation.

In the figure 3 is front panel equipped with upgraded sticking of the insulation. This panel achieved 5 times better deformation resistance than panel with OEM insulation. Front panel is affected by smaller amount of forces than side panels. Because front panel is naturally stiffer than side panels (because of curved shape), we decided to use same amount of forces than on the side panels, to highlight insulation stiffness benefit [5].

3. Insulation impact on overall noise

To make final decision, if upgraded insulation has impact on overall acoustic power, a real scale washing machine equipped with OEM and upgraded insulation was exposed to noise tests. With upgraded insulation was achieved improvement of overall noise. We measured acoustic power, which was slightly reduced [6]. Experiments were performed at maximum rpm (1200 rpm) and 600rpm. The reason, that we have chosen 600 rpm, this model of the washing machine during its washing cycle mostly works at or around 600 rpm.

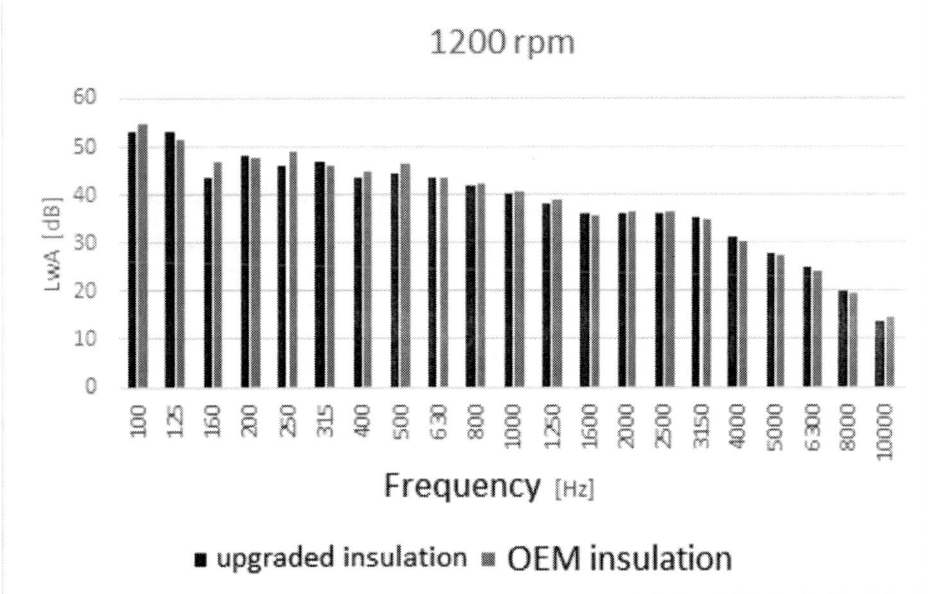

Figure 4. Acoustic power at 1200 rpm.

In the figure 4 and 5 is graph of 1/3 octave spectrum, measured on washing machines equipped with OEM and upgraded insulation.

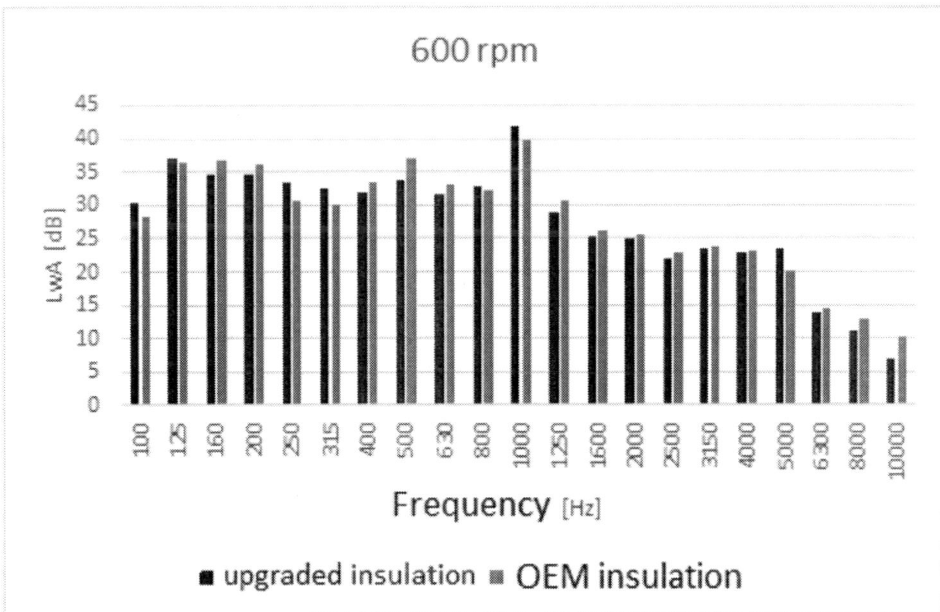

Figure 5. Acoustic power at 600 rpm.

IOP Conf. Series: Materials Science and Engineering **659** (2019) 012050 doi:10.1088/1757-899X/659/1/012050

4. Conclusion

The upgraded insulation has significant impact on overall rigidity and noise of the washing machine. The biggest benefit of upgraded insulation was on side panels of a top loader washing machine, because those panels are usually more affected by the forces occurred during washing cycle. Front panel achieved minor benefit only. It is because of its curved shape, so the front panel achieved better results than side panel even without any insulation installed. Rear panel was not tested, because on this particular washing machine was not possible to install such a type of insulation. We expect that rear panel would not achieve such an improvement as side panels, because rear panel on top loader washing machines is affected only by minor forces.

Acknowledgment

This work was created by the implementation of the project "Low Cost Logistic System Based on Mobile Robotic Platforms for Industrial Use", ITMS: 26220220205 supported by the Operational Programme Research and Development and by the European Fund for Regional Development.

References

[1] Michalczyk J 1993 Noise reduction problems of vibratory machines, *Archives of Acoustics* **18** 2300-262X
[2] Braut S, Pavlovic A, Beno P and Babic M 2019 Measuring Strain in Sheet Metals, *FME Transactions* **47** 2406-128X
[3] Pierchala M D 2018 Testing the Acoustic Silencers, *Archives of Acoustics* **43** 2300-262X
[4] Ibrahim M A and Melik R W 2003 Optimized sound absorption of a rigid polyurethane foam, *Archives of Acoustics* **28** 2300-262X
[5] Mu R, Toyoda M and Takahashi D 2016 Effect of Urethane Foam Cushioning on Structure-Borne Sound Transfer by a Slab with Panel Flooring, *Archives of Acoustics* **41** 2300-262X
[6] Wszołek T 2007 Uncertainty of sound insulation measurement in laboratory, *Archives of Acoustics,* 2300-262X

IOP Conf. Series: Materials Science and Engineering **659** (2019) 012051 doi:10.1088/1757-899X/659/1/012051

Vehicles optimization regarding to requirements of recycling *Example: Bus dashboard*

S Milojević[1], R Pešić[1], J Lukić[1], D Taranović[1], T Skrucany[2] and B Stojanović[3]

[1] University of Kragujevac, Faculty of Engineering, Department for Motor Vehicles and IC Engines, Sestre Janjić 6, 34000 Kragujevac, Serbia
[2] Department of Urban Transport, University of Žilina, Slovakia
[3] University of Kragujevac, Faculty of Engineering, Department for Mechanical Construction and Mechanization, Sestre Janjić 6, 34000 Kragujevac, Serbia

E-mail: sasa.milojevic@kg.ac.rs.

Abstract. It is very important during production to take into account the requirements regarding to recycling of vehicles. On this manner it is possible reuse of consisting materials and optimization of End-of-Life Vehicles. In European Union has defined targets to be achieved in the future for the recycling rate of vehicles at the end of their service cycle. Directive of the European Parliament and of the Council 2000/53/EC regulate waste management in the vehicle sector. This manuscript presents in global the international legislative in this field, regarding to the situation in Republic of Serbia. As contribution, during production of domestic city bus running on natural gas, performed the replace of material for bus dashboard on the basis of plastic and polyurethane foam, with ecological material with carbon inserts. The requirements for dismantling, reuse and recycling are integrated in the design of new bus.

1. Introduction

In 2016, more than 22.5 million of trucks and busses (approx. 7.5 million in U.S.) and approx. 72.1 million of personal cars (approx. 4.2 million in U.S.) were produced globally. According to prognosis, a total of around 107 million vehicles will be manufactured worldwide in 2020 [1]. Worldwide, some 79 million automobiles are expected to be sold by the end 2019. From the second side, in 2017, some 26 million vehicles were sold globally, and U.S. is the largest market for commercial vehicles.

Parallel the number of end-of-life vehicles (ELV) arising constantly, and there are gaps between the numbers of deregistered and ELV in many countries. There is no detailed information available because of a certain number of de-registered vehicles are commercially exported to the underdeveloped countries as second hand. Also, it is important to reduce illegal shipments of ELV; because inspections of transports within and out of the EU should be intensified.

If current trends continue and no relevant legislation from design to disposal vehicles, The Globe can become overloaded vehicles at the end of the service cycle. Vehicle recycling and reuse of materials incorporated is very important from that point of view and presents a solution the problem.

1.1. Problems with the dismantling of the vehicle at end-of-life

Regarding dismantling (consisting of collection, depollution and crushing) of ELV in Europe, two main aspects have to be considered.

First, there is evidence suggesting that ELV are treated illegally in some cases. However measures which make the disposal of an old vehicle in the country of de-registration more attractive, such as refunds obtained in a deposit-refund system, would help reduce illegal shipments of ELV.

Second, there is evidence suggesting that even in authorized treatment facilities dismantling (specifically depollution) is not in full compliance with the relevant requirements of the existing ELV Directive.

Liquids seem to be removed to a certain extent. Certain types of fluids or components such as brake fluids, windscreen washer fluid, oil filters or shock absorbers, however, are not always removed or de-polluted.

Usually little effort is put into the removal of components containing heavy metals, such as Hg, containing display backlights or switches.

Lead-Acid Batteries are generally removed from ELV because lead may be used as a source of income and because of constraints for the shredder process if not removed. Gas tanks and air bags are usually removed because of the well-known risks for shredder plants.

Shredder means any device used for tearing into pieces or fragmenting ELV, including for the purpose of obtaining directly reusable metal scrap.

In order to improve depollution of ELV, measures should be taken against illegal waste car dismantlers and unauthorized treatment facilities.

Depollution of ELVs should lead to shredder residues with low contents of hazardous substances. According to some available information on the composition of shredder residues, it is in particular the content of hydrocarbons that indicates that the depollution of ELV is not always sufficient.

1.2. Legislation in domain of recycling and recovery of the vehicle

Every year ELV in the Community generate between 8 and 9 million tons of waste, which must be managed correctly. EU regulation for the automotive sector regarding to ELV is Directive (2000/53/EC) [2].

The requirements for dismantling reuse and recycling of ELV and their components should be integrated in the design and production of new vehicles.

According to ELV Directive demands, the reuse/recyclable target for components of a vehicle must constitute a minimum of 85 wt. % (weight percentage) and the recoverable parts a minimum of 95 wt. % per vehicle.

ELV Directive requires the number of improvements on the vehicle and equipment as it is: engine efficiency, the use of biofuels and other alternative fuels, weight reduction by application of new materials etc. Responsibility and awareness of vehicle manufacturers is very important here.

However, recycling and recovery rates are different and overestimated due to differ between States, and different national interpretations. There exist the need for improvement regarding the recycling and recovery of ELV materials.

Dismantling and subsequent material recycling of glass and plastics for instance takes place in minor quantities in several States. Recovery of glass after shredding, however, prevents glass recycling because of the bad quality of the glass. Recovery of plastics from shredder residues is still limited. Whereas in some States post shredder technology has been installed and landfilling of shredder residues prohibited or made very expensive, several States still deposit shredder residues on landfills. Overall, in this area is very important and necessary to establish harmonization.

The job is complex and due to the intensive application of new propulsion technologies. Current issues are solving the problem for recycling of electric vehicles with recovery their parts, such as permanent magnets etc. [3].

The recycling of all plastics from ELV should be continuously improved. The Commission is currently examining the environmental impacts of PVC.

There are several other legislations dealing indirectly with plastics and plastic waste, be it safety or emissions related. The second most important problem is in domain of recycling and recovery of polyurethane (PU) foams from vehicles [4].

Polyurethanes are dominant also in transport sector, in passenger cars, as well as in trucks and buses, not only in construction sector, inside buildings and for furniture.

Average plastic content is approx. 150 to 120 (kg/car), of which PU constitutes a rather large, but widely varying share. On avg. polyurethane represent around 10 to 15%, depending on manufacturing technology.

PU materials reduce the fuel consumption directly through lightweight design, either in conventional models or in new hybrid metal or plastics designs. PU bonding capabilities enable new manufacturing technologies, such as replacing welding, and improved design, lightweight metals technology etc.

There are many different types of PU used in vehicles given the broad performance standards requiring reduced weight, increased fuel economy and comfort and improved resistance, insulation and sound absorption. The variety of applications is equally broad, including items such as seating cushions, steering wheels, sound and energy absorbing elements, ceilings and dashboards.

Dashboards protect the mechanics of the vehicle against moisture, impact, abrasion, excessive heat, and harmful UV rays. For several decades, plastics have been playing an increasingly vital role in the performance and design of these panels. Acrylonitrile-Butadiene-Styrene (ABS), polycarbonate alloys, polycarbonates, polyester, and polypropylene are the most common materials associated with these important applications.

The numerous and varied PU automotive applications leads to a rather broad product mix being utilized in vehicles. The product properties vary from light foam, as used for seating, to very rigid and tough thermoset plastic materials, as in steering wheels, instead of thermoplastics.

In this area there is also a problem because that dismantling of seat cushions or larger parts containing PU, such as the instrument panel, is not eco-efficiency.

Shredder treatment technology and the further refinement by companies such as Salyp or Sicon, and new post shredder technologies have shown the bigger cost and limits [4].

2. Proper experience in the field of buses designing regarding to ELV recycling requirements

Domestic recycling industry is in this moment in the stage of development. The problem is the design of the vehicles, which does not allow an adequate and rapid implementation of the process of dismantling, and the lack of unified system.

Due to existence of several domestic bus manufacturer in the market solution to the problem of recycling these vehicles is very important. Designing regarding to recycling demands inside ELV Directive allow not only to meet domestic ecological requirements, but also to contribute to the increase in the export potential of the vehicle trademark.

As an example, Figure 1 shows the structure of redesigned local city bus running on natural gas. Dismantling information for the correct and environmentally sound treatment of ELV formed, too.

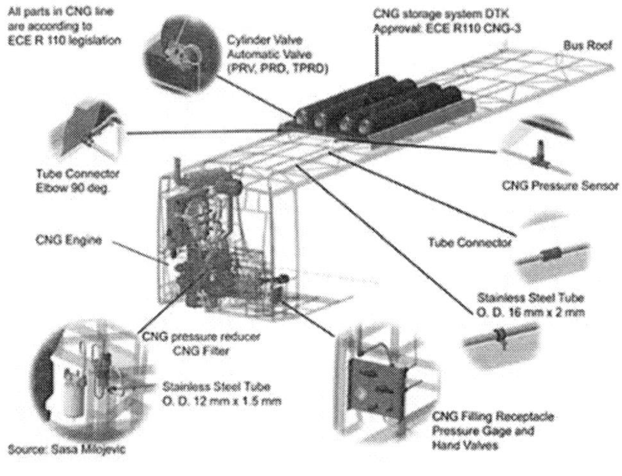

Figure 1. Redesigned domestic city bus, running on natural gas.

Figure 2. Module with aluminum gas tanks for alternative fuels.

Of light composite materials are also produced and tanks on vehicles running on conventional and alternative fuels, such as hydrogen and natural gas. Under such conditions of application, the material of the tank must sustain the high pressures and other impacts such as corrosive effects of fuel etc. Figure 2 shows a module with aluminum tanks for alternative fuels [5-8].

Applying the light materials for tanks, is achieving a weight reduction of vehicles to alternative drive which contributes to the reducing fuel consumption and preserving the environment.

Based on the results of the analysis, the buses are very complex system regarding to their compatibility for recycling and reuse.

Inside recycling factories, is very difficult to organize recycling process, starting from complex process for dismantling of parts, sorting of materials, etc.

To increase bus recycling rates, the complex interconnected material cycles originating from recycled buses have to be optimized from a technological as well as economical point of view. The optimization of the resource cycle of buses involves the interaction of various disciplines, as shown in Figure 3 [9].

To links the recycling of buses from a material and energy perspective as depicted by that figure's golden circle; recycling technology and design as represented by the grey circle; and cognizance of economy, environmental impact, and legislation as shown by the blue circle.

Fundamental knowledge of recycling processes, such as shredding, mechanical separation processes and metallurgy, and material characteristics of recycling (intermediate) products, such as material type and liberation, have to be combined with that of the design of the product (material combinations and connections). In order to optimize the resource cycle and maximize the recycling rate of future buses, the parameters determining the recovery rate for each of the materials present in the multilateral ultralight bus of the future, as well as the dynamics and statistically distributed nature of the resource cycle system have to be fully understood.

Figure 3. The diverse and distributed properties of recycling materials (clockwise: steel approximately 650 t. from 1,153 ELVs; copper fraction; Al/Mg/Cu/Zn fraction; plastic fraction; Al/Mg/Cu/Zn fraction; steel fraction).

Generally, during buses production according to international and domestics' requirements in domain of ELV recycling adopted is the following concept:

- Bus concept according to characteristics requested by the market and customer exceptions. Optimization of the bus design with minimization of energy and material consumption during service cycle and design with easy disassembly;
- Use of ecological materials suitable for reuse, with minimal percentage of included hazardous materials;

- Mass reduction contributes to lowering fuel consumption in the traffic, as well as logistics costs during production process;
- Optimization of production technology (technological process with minimal energy consumption and toxic emission); and
- Design according to ELV Directive to allow disassembly and separation of materials and possibility to reuse components. Materials must be labeled (specifically elastomers and polymers).

3. Recyclability and recoverability; Calculation method

The International Standard (ISO 22628:2002) specifies a method for calculating the recyclability rate and the recoverability rate of a new road vehicle, each expressed as a mass fraction of the vehicle. Under this procedure, performed by the vehicle manufacturer when a new road vehicle is put on the market, potentially, the vehicle can be recycled, reused or both (recyclability rate), or recovered, reused or both (recoverability rate) [10].

Recyclability rate of the bus is determined by the recovery by each of the materials present in the bus. Indicators of recyclability and recoverability rates are determined by the calculation of the masses of materials of bus components identified in the technical documentation of the bus maker with subsequent monitoring of actual deviations in production.

Recyclability rate (R_{cyc}), or (recyclability indicator for reuse) of the bus, is calculated as a percentage by mass (mass fraction in percent), using the following equation (1):

$$R_{cyc} = \frac{m_P + m_D + m_M + m_{Tr}}{m_V} \cdot 100 \geq 85\% \tag{1}$$

Recoverability rate (R_{cov}), or (recovery indicator) is calculated using the equation (2):

$$R_{cov} = \frac{m_P + m_D + m_M + m_{Tr} + m_{Te}}{m_V} \cdot 100 \geq 95\% \tag{2}$$

where is:

- m_P - Mass of materials taken into account at the preliminary process (pretreatment); (all liquids, battery, oil filters, tires, catalytic converters, gas tanks etc.);
- m_D - Mass of materials taken into account at the dismantling (large, easily removable parts from polymers, elastomers, as it is bumper, lining, dashboard etc.);
- m_M - Mass of metals taken into account during Shredder processing at the metal separation process;
- m_{Tr} - Mass of nonmetallic materials for recycling taken into account during Shredder processing etc.; and
- m_{Te} - Mass of materials taken into account at the nonmetallic residue treatment step and which can be considered for energy recovery (polymers and elastomers difficult to dismantle or less than 100 and 200 g, respectively) and other combustible materials (leather, wood, cardboard, etc.).

Time for dismantling the bus is determinate by the features of its design, the elements of fastening the units and parts, the availability of technical instructions for dismantling etc. The bus recycling costs at ELV is very dependent upon dismantling time. Many manufacturers are investing time and resources to shorten the time required for dismantling the bus at ELV.

4. New design and development of buses podium dashboard

In order to minimize the time and financial costs for the dismantling of buses, special solutions and limitations are being put at the design stage, simplifying disassembly. Using the computer-aided design (CAD) to aid in the creation, modification, analysis, or optimization of a design system, helps to simplifying dismantling.

IOP Conf. Series: Materials Science and Engineering **659** (2019) 012051 doi:10.1088/1757-899X/659/1/012051

An example of modeling the dismantling of plastic parts of the city bus type (203 CNG-S) podium dashboard is shown in Figure 4 and 5.

The result is the catalog for the dismantling of city bus, which meets the requirements of the EU regulations on the suitability of vehicles for disposal.

Podium is an integrated platform, a generic dashboard solution (ISO16121) compliant, including mainly:

- A main console with adjustable steering column, steering wheel (with optional integrated commands) and stalk switches;
- MultiBUS modules like MultIC cluster in the center and Switch Packs in the left and right part can be adapted: number of CAN buses (1 to 3) and number of gauges (2, 4 or 6 gauges);
- A slot for Public transport system, a gearbox selector, up to 4 push buttons for doors and a rotary switch for exterior lighting (different number and type of switches; low current switches, door buttons…); and
- A harness interface for the chassis.

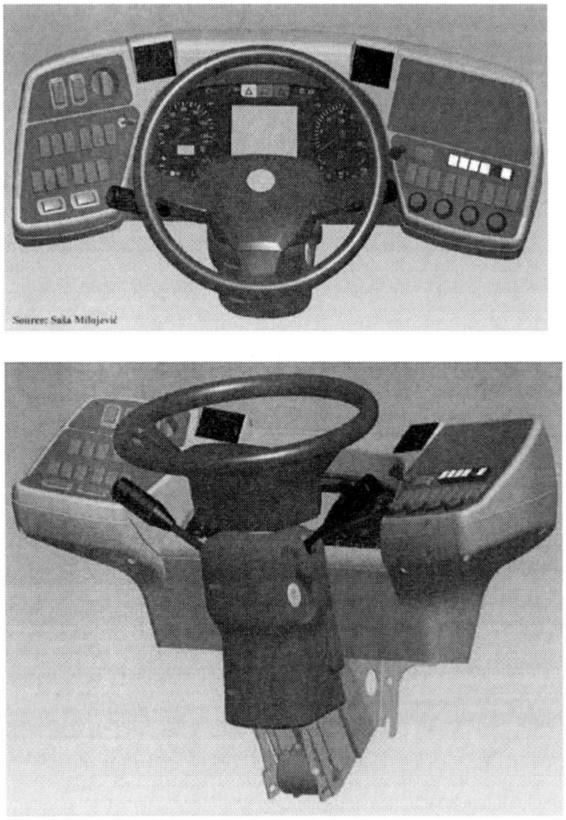

Figure 4. Integrated platform with new design of podium dashboard for city bus.

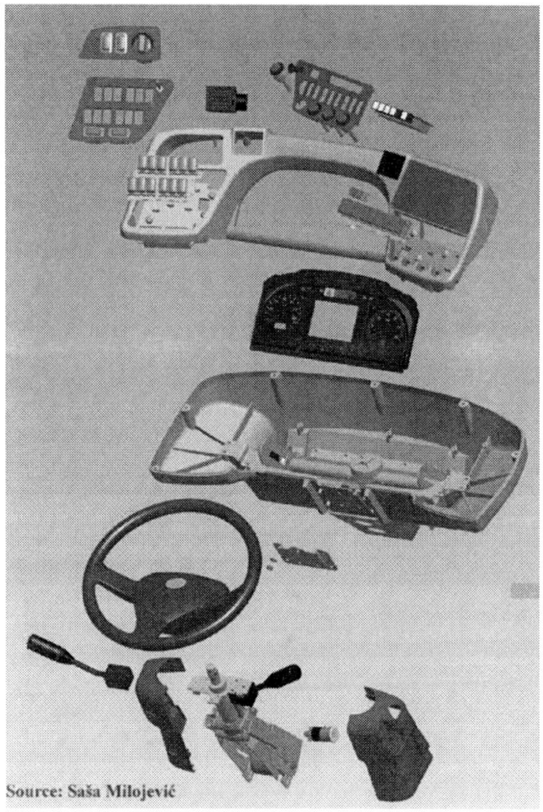

Figure 5. Internal design of podium dashboard for easy dismantling.

In order to be as close as possible to Bus and Coach Market requirements, podium is a flexible and modular unit which can integrate several options. Dedicated catalogues centralize the available options.

The main Podium advantages compared to competitors' solutions are:

- A modern style and design compatible with the latest Bus and Coach developments;
- Better Man Machine Interface (cluster with large color TFT display, backlighted switches…);
- An empty space dedicated to public transport system on the right part of the console; and
- Improved ergonomics (more space for driver legs, centralized commands etc.).

5. Conclusions

If current trends continue and no relevant legislation from design to disposal vehicles, The Globe can become overloaded vehicles at the end of the service cycle. Vehicle recycling and reuse of materials incorporated is very important from that point of view and presents a solution the problem.

EU regulation for the automotive sector regarding to ELV is Directive (2000/53/EC). The requirements for dismantling reuse and recycling of ELV and their components should be integrated in the design and production of new vehicles.

Domestic recycling industry is in this moment in the stage of development. The problem is the design of the vehicles, which does not allow an adequate and rapid implementation of the process of dismantling, and the lack of unified system.

The recycling of all plastics from ELV should be continuously improved. The Commission is currently examining the environmental impacts of PVC. The variety of applications is equally broad, including items such as seating cushions, steering wheels, sound/energy absorbing elements, ceilings and dashboards.

As contribution, during production of domestic city bus running on natural gas, performed the replace of material for bus dashboard on the basis of plastic and polyurethane foam, with ecological material with carbon inserts. The result is also the catalog for the dismantling of city bus, which meets the requirements of the EU regulations on the suitability of vehicles for disposal.

Acknowledgment

The paper is a result of the research within the project TR 35041 financed by the Ministry of Science and Technological Development of the Republic of Serbia.

References

[1] Motor vehicle production - Statistics & Facts, https://www.statista.com/topics/975/motor-vehicle-production/ (accessed on May. 2019)

[2] Directive 2000/53/EC of the European Parliament and of the Council of 18 September 2000 on end-of life vehicles, 2008, Council of the European Union and European Parliament

[3] Bukvić M, Stojanović B, Ivanović L and Milojević S 2017 Recycling of the Hybrid and Electric Vehicles, *Acta Technica Corviniensis - Bulletin Of Engineering* **10**(3) 107-114

[4] Recycling & recovery of polyurethane foams from cars, http://www.isopa.org/media/1122/may-2012_fact-sheet_recycling-recovery-pu-foam-from-cars.pdf/ (accessed on May. 2019)

[5] Milojević S and Pešić R 2011 CNG buses for clean and economical city transport, *Int. J. Mobility and Vehicle Mechanics (MVM)* **37**(4) 57-71

[6] Milojevic S and Pesic R 2012 Theoretical and experimental analysis of a CNG cylinder rack connection to a bus roof, *International Journal of Automotive Technology* **13**(3) 497-503

[7] Milojević S 2016 Reconstruction of existing city buses on diesel fuel for drive on Hydrogen, *Applied Engineering Letters* **1**(1) 16-23

[8] Milojevic S 2017 Sustainable application of natural gas as engine fuel in city buses - Benefit and restrictions, *Journal of Applied Engineering Science* **15**(1) 81-88

[9] Schaik van A and Reuter M A 2004 The Optimization of End-of-life Vehicles Recycling in the European Union, *JOM: the journal of the Minerals, Metals & Materials Society (JOM-US)* **56**(8) 39-43

[10] ISO 22628:2002 Road vehicles - Recyclability and recoverability - Calculation method, *International Organization for Standardization* ed.1 1-8

Assessment of dynamic properties of a carriage using multibody simulation considering rigid and flexible tracks

J Dižo[1] and M Blatnický[1]

[1]University of Žilina, Faculty of Mechanical Engineering, Univerzitná 1, 010 26 Žilina, Slovakia

E-mail: jan.dizo@fstroj.uniza.sk

Abstract. Recently, we cannot imagine investigation and analysing mechanical properties of railway vehicles without using state-of-art tools. They are implemented in whole process of design, analysing, verification, testing and optimising of railway vehicles. Such an approach saves time for development, production costs and finds application across whole spectrum of produced types of railway vehicles. In order that the approach based on simulation computations can reliably simulate the reality, a virtual model of a railway vehicle has to take into account as many factors as important. When we investigate behaviour of a railway vehicle in terms of dynamics, we used the multibody approach for creating its computational model. The goal of this article is evaluation of output quantities of a passenger car from the dynamics point of view. The main objective is assessment of selected values for rigid and flexible models of a track. There are evaluated quantities, which belong to main assessed outputs for investigation of dynamic properties of a passenger car. There are values of vertical forces in the wheel/rail contact as the indicator of dynamic load of a railway track as well as the bodies of a railway vehicle.

1. Introduction

Nowadays, passeneger cars design has to meet the quite strict requirements before commissioning for regular operation. Phases of production and operation of passenger cars closely associated with economic factors. Modern approaches of computational simulations significantly help determine mechanical properties of passenger cars and verify they dynamical behaviour [1], [2]. Correct dynamical analyses require setting-up a representative multibody model of a passenger car, which correspond to actual operational conditions [3]. There are several commercial software working based on multibody system principle and inbuilt functionalities allow to create such a virtual model of a railway vehicle/railway track multibody model, which simulate actual running conditions [4], [5].

When we set-up a multibody model, we usually consider just a rigid track model. But, such an approach to the process of modelling does not always correspond to the real conditions, which are reached in a railway vehicle/a track couple reached [6], [7]. Therefore, in many studies the flexibility of a track is defined in the multibody model in order to evaluate the dynamic response of the entire mechanical system of a passenger car/a track [8-10].

2. Multibody model of the railway vehicle/railway track interface

This contribution introduces a study of influence of flexible track on dynamic response, i. e. selected output quantities. The multibody models of a passenger car and a track have been created in Simpack program package. There is a commercial program, which enables to an user building up many kind of virtual models, which can be quite simple subsystem of any transport mean, e. g. individual parts of suspension system, partial systems of a drive mechanism, such as an engine, gearbox, etc. [11-14], or there can be a bogie of a railway vehicle, even an entire train sets [15-17].

In our work we have submitted a passenger car to simulation computations. Its model created in the Simpack software package (figure 1).

Figure 1. A virtual model of the analysed passenger car created in Simpack software package.

There is a standard model passenger car, which consists of two bogies and one body of a wagon. Individual elements of the passenger car are represented by rigid bodies. Mechanical system of a passenger car composes of rigid bodies (15 totally), which are connected by so-called force elements of various types. There are force elements of "spring-damper" type as steel coil springs of primary and secondary suspension and as hydraulic dampers installed also in primary and secondary suspension system. Depending on positioning of hydraulic dampers they serve as dampers of oscillations of bogie or body of wagon on bogies in vertical direction, lateral direction and as anti-yaw dampers.

Each wheelset is guided in an axlebox, which functions as a swinging arm. As usually, the analysed passenger car uses primary as well as secondary suspension system. The primary suspension system connects unsprung masses of wheelset and axleboxes with a bogie frame and the secondary suspension system ensures desired conform level by joining the body of the passenger car to a bogie. Mass and inertia parameters of the analysed passenger car rigid bodies are listed in table 1.

Table 1. Parameters of the analysed railway vehicle rigid bodies.

Rigid body of the passenger car	Mass m [kg]	Moment of inertia I_{xx} [kg/m^2]	I_{yy} [kg/m^2]	I_{zz} [kg/m^2]	Number of bodies
Body of wagon	38,000	58,000	23,000	23,500	1
Frame of the bogie	2,850	1,950	1,665	3,330	2
Axlebox	215	4,55	7,1	7,1	8
Wheelset	1,690	880	190	880	4

Among other specific elements, which arise in a railway vehicle model, the wheel/rail contact represents one of the most important element, which significantly influences credibility, reliability as well as time-consuming of simulation calculation [18], [19]. We have defined the FASTSIM model of the wheel/rail contact in our multibody model. Its use is most widely defined in case of simulation of a railway vehicle dynamic analyses, because it is robust enough, calculation time is acceptable and it calculates all required quantities in the contact patch (namely values of tangential forces) with a sufficient accuracy [6], [20], [21].

If we consider flexible track foundation, a track model is no longer modelled by its layout in a space (the position coordinates (x, y, z) in respect of considered coordinate system and the orientation angles (φ, ψ, θ) representing roll, pitch and yaw angles), but, there is necessary to add into a model additional bodies representing track sleepers, ballast etc. [22], [23]. It means, the entire multibody model becomes more complicated (figure 2).

In the used software package, a user can set-up such a complex multibody model system consisting of a railway vehicle and a flexible track, what can be achieved more detailed model and there is possible to assume, that even more detailed results of simulation computations.

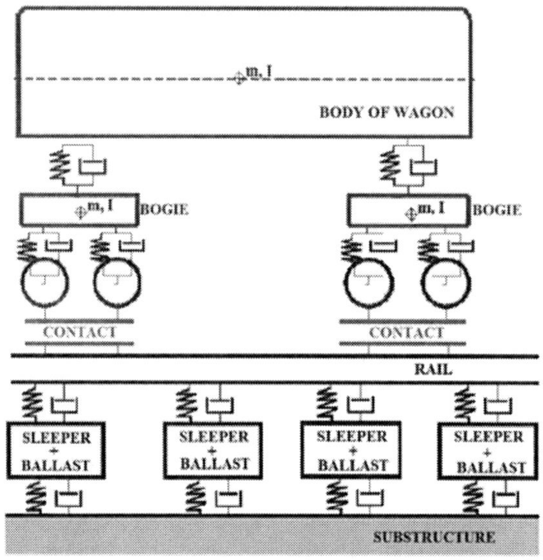

Figure 2. A dynamical model of a passenger car and a flexible track foundation.

In our research, we have defined the flexibility of a track by two manners. Dynamic response of the passenger car was evaluated for track flexibility, when coefficients of stiffness and damping were set-up on the constant value and for other set of simulations, coefficients of stiffness and damping were prescribed by the sinusoidal function. In the case of the sinusoidal function, the parameters of stiffness and damping are no longer constant during calculation, but they vary according to formulation (1) in certain bounds, which represents different stiffness-damping parameters of a track in compliance with sleepers' layout. Hence, for sinusoidal stiffness-damping coefficients the following formulation is considered:

$$c_i(s) = c_{i0} + c_{i1} \cdot \sin\left\{F_i \cdot \left(s(t) - s_0\right)\right\} \tag{1}$$

where c_{i0} is constant stiffness (or damping), c_{i1} is amplitude of the track stiffness (or damping), F_i is the nominal force resulting from the gravity of the passenger car, $s(t)$ is moved distance, which depending on the integration time and s_0 represents an offset of the start position. As there is obvious from eq. (1), in simulation calculation the stiffness-damping coefficients vary depending on the moved distance (not depending on time).

For purposes of dynamical analyses, we have performed relatively large number simulation computations for various running speeds, and several track foundation. In case of the flexible track foundation, we have chosen two levels of values of stiffness-damping coefficients. In order to differentiate them to each other we have marked them as "soft" flexibility for lower values of stiffness and damping coefficients and "stiff" for their higher values. Thus, in following section contains results of simulation computations for four different track flexibility foundation, namely "constant soft", "constant stiff", "sinusoidal soft" and "sinusoidal stiff".

In our multibody model we have defined the straight track model, because we have wanted to eliminate influence of centrifugal forces or other additional dynamic effects resulting from the running in curves and we have aimed mainly at evaluation of the dynamic response due track flexibility. In the track model, track irregularities were defined in discrete form.

3. Results from simulation computations

In this section, results from numerical calculations of the passenger car running on a track with various track flexibility foundation are presented. They are introduced in the form of graphs.

As there was mentioned above, from the large number of simulation for various running speeds, we have selected only several speeds, which can concrete show influence of different running speed on waveforms of output quantities. We have decided that the representative examples of results are waveforms for 60 km/h, 100 km/h and 140 km/h.

In our research we have focused on dynamic response of a passenger car running on a track represented by the dominant forces, which act in the railway wheel and railway track during operation of a railway vehicle, namely the vertical wheel force. This force affects significantly the dynamic loads of a track as well as dynamic loads of main structural units of a railway vehicle [18], [20], [23].

Figure 1 contains waveforms of vertical wheel force for "constant soft" and "sinusoidal soft" track foundation. We can see, when constant values of stiffness-damping coefficients are defined in the track model, dynamic vertical force is for every evaluated running speed almost the same and it varies only negligible.

The different situation can be observed, when the "sinusoidal soft" track foundation is defined in the track model. As we can see in figure 3 (lower), in this case the track flexibility foundation influences dynamic response of the passenger car during its running on the track. Here we can observe that a value of the vertical wheel force oscillates regularly with the noticeable amplitude and frequency. Further we see, the amplitude of the vertical wheel force is maximal for lower running speed, namely for 60 km/h and with increasing running speed the amplitude of this force is being changed indirectly, i. e. the dynamic response of the vertical force is being decreased 100 km/h (green) and 140 km/h (red).

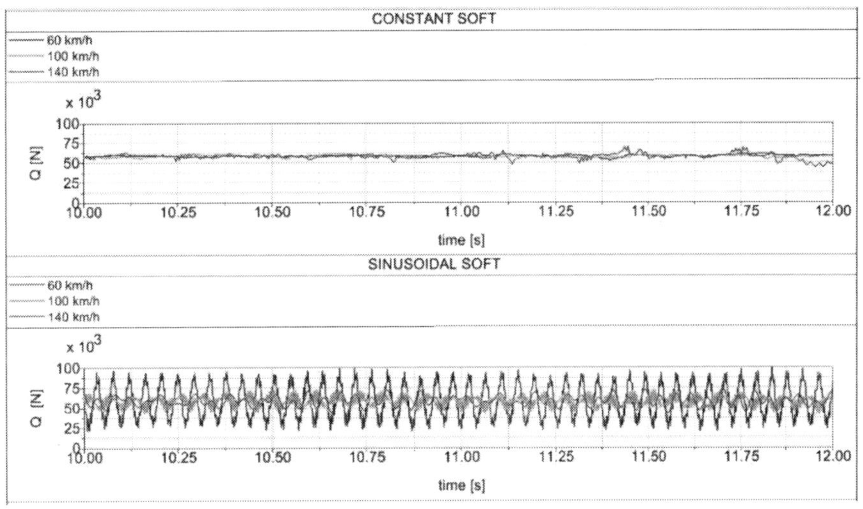

Figure 3. Dynamic vertical force for "constant soft" and "sinusoidal soft" track flexibility.

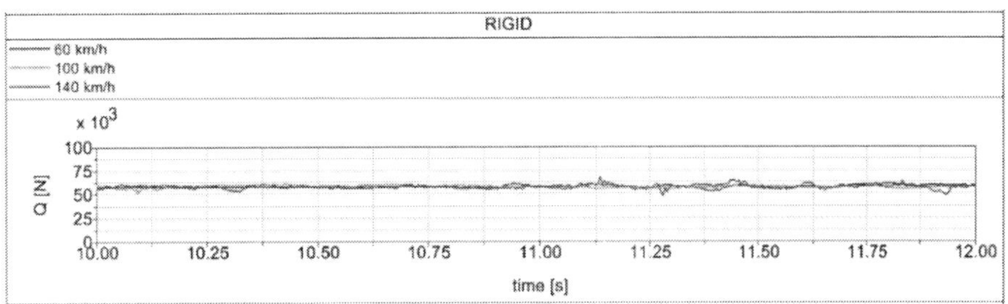

Figure 4. Dynamic vertical force for "rigid" track flexibility.

Figure 4 contains waveform of the observed quantity for "rigid" track foundation, i. e. without any stiffness or damping coefficients defined between rails and a ground. As we can see, waveform of forces are in this case similar to those, which can be observed for "constant soft" track foundation (figure 3 upper).

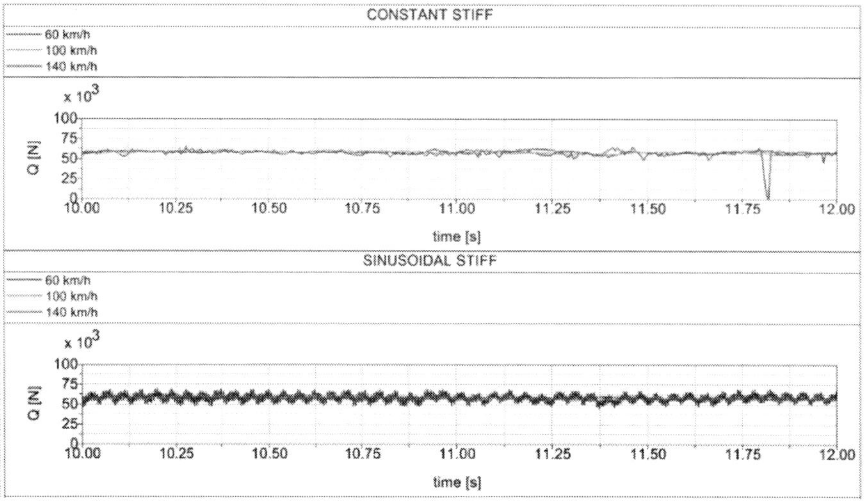

Figure 5. Dynamic vertical force for "constant stiff" and "sinusoidal stiff" track flexibility.

Waveforms of dynamic vertical wheel forces for higher values of stiffness-damping coefficients are shown in figure 5. Upper part of this figure contains results for constant stiffness-damping values and lower part for sinusoidal values. From reached results we can see, that in case of definition of track flexibility foundation using constant stiffness-damping coefficients, the vertical wheel forces do not vary such significantly as in comparison with sinusoidal stiffness-damping coefficients. Among other findings we can include, that higher values of stiffness and damping coefficients of the sinusoidal track foundation reduce amplitude of the observed result vertical wheel forces.

Following figures (figures 6, 7 and 8) show waveform of vertical wheel for individual running speed depending on various track flexibility foundation.

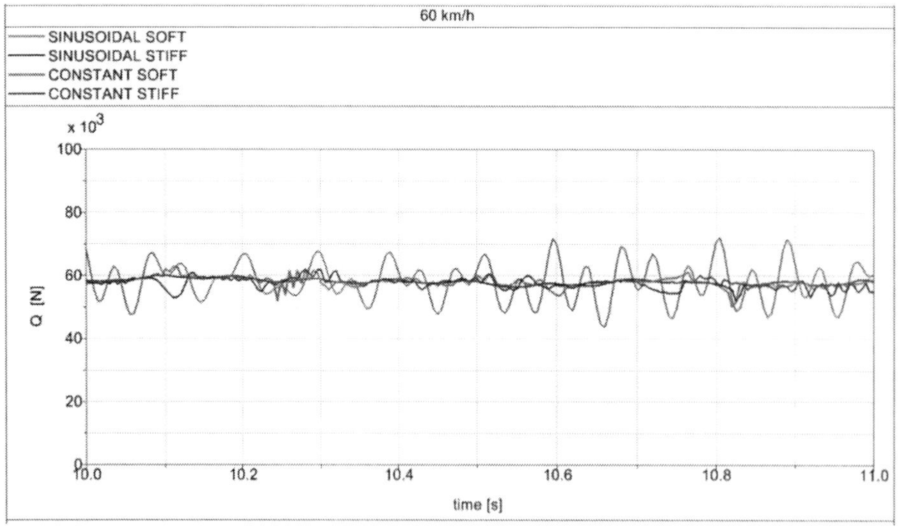

Figure 6. Waveform of the dynamic vertical force for various track flexibility foundation, 60 km/h.

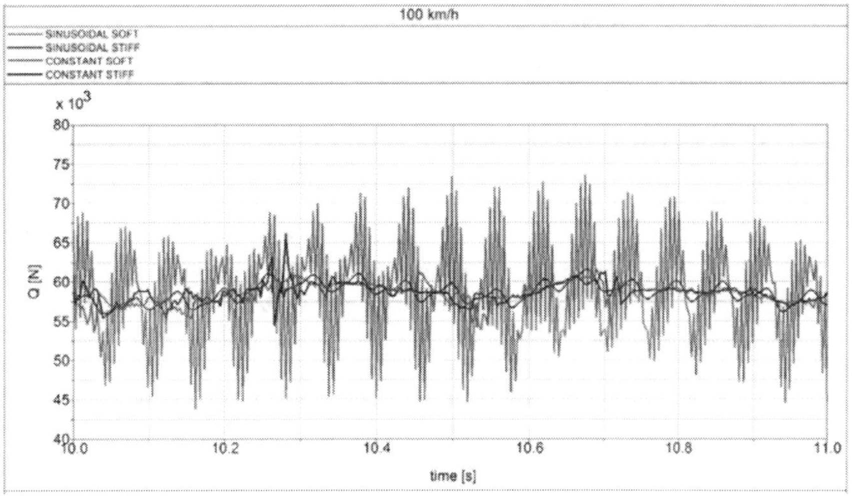

Figure 7. Waveform of the dynamic vertical force for various track flexibility foundation, 100 km/h.

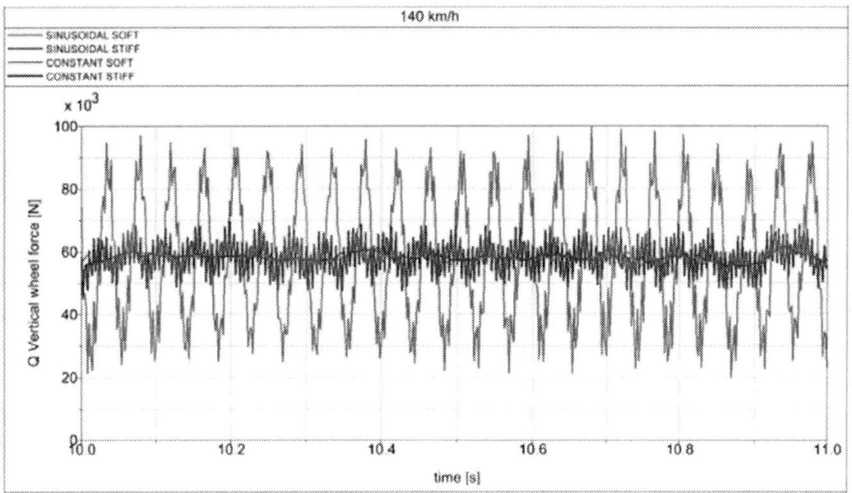

Figure 8. Waveform of the dynamic vertical force for various track flexibility foundation, 140 km/h.

These figures confirm the fact, that resulting dynamic response of a passenger car running on significantly depends on values of stiffness-damping coefficient. Lower values lead to higher amplitudes of resulting vertical wheel forces and as we would expect, also higher running speed results to higher amplitudes of the vertical wheel forces. The greatest amplitudes of vertical wheel forces were reached for "sinusoidal soft" track flexibility foundation. The "sinusoidal stiff" track foundation have led to greater increasing amplitudes of the vertical wheel forces in case of running at the speed of 140 km/h (figure 8). In other observed cases the waveform of vertical wheel forces oscillate about so-called quasi-static value of the vertical wheel forces and peaks corresponds to the dynamic effects related with the fact, that track irregularities are defined in the track model and they excite the mechanical system of the passenger car. The quasi-static value of the vertical wheel force corresponds to the part of the gravitational force of the passenger car, which accrues to the observed wheel.

Finally, based on observed results of our research we can formulate some findings.

When one wants to analyse dynamic properties of a railway vehicle using multibody simulation, complexity of a multibody model depends on available parameters. The flexible track foundation is closer to the reality, but it requires the credibility of the stiffness-damping parameters of the track foundation. As we have found out from our results, various values quite strongly influence accuracy of the results and it can lead to relatively mendacious output values. In our work we have decided to evaluate the dynamic vertical wheel force. Based on results we have observed that the credibility of the input data of the track will affect results of simulation computations. The evaluated vertical wheel force is the indicator of the track loading and also loads the mechanical system of a railway vehicle. Results have shown that the track is the most loaded, when the track substructure is soft and the increasing running speed leads to worse loading of the track.

The part of the future research in this field, other simulation computations will be performed, at which the track model will be improved by definition of layout in the space. In such a manner we will assess change of dynamic response when a passenger car will run in curve, evaluate the safety for various track flexibility foundation and also other dynamic parameters, such accelerations in different location of the passenger car body. Based on values of acceleration we can evaluate comfort for passenger [24], [25] under variable loading conditions and also other vibration properties of structural units of a railway vehicle (e. g. vibration of wheelsets, axleboxes, frames etc.)

If one wants to investigate only phenomenon related to wagon body, e. g. ride properties for passengers, there would be sufficient to use the rigid track foundation. As ride comfort for passengers is evaluated based on accelerations [22], [24], [25] resultant ride comfort indices would not be very different. But, if it would be necessary to assess affects a rail vehicle running on a track, even in case of any wheel untruenesses formed on a wheel tread surface, then definition of track flexibility influences results more markedly. Therefore, future research in thus field will be focused on investigation in more detail, how flexibility formulation in a multibody model effects more outputs parameters.

4. Conclusion

The article brought knowledge about the problem of modelling the flexible track in multibody system of a railway vehicle. There were compared results of simulation computations for the chosen stiffness-damping parameters of a track in the straight section. The analysed passenger car has run on the track at various speeds, from which the most interesting have been chosen for presentation. Based on reached results we have found out that the dynamic response of the passenger car, which indicate the track loading as well as the railway vehicle loading depends on the credibility of input data for track flexibility foundation.

Acknowledgment

This work was supported by the Cultural and Educational Grant Agency of the Ministry of Education, Science, Research and Sport of the Slovak Republic of the Slovak Republic in the project No. KEGA 077ŽU-4/2017: Modernization of the Vehicles and engines study program. The work was supported by the Slovak Research and Development Agency of the Ministry of Education, Science, Research and Sport of the Slovak Republic in the project No. VEGA 1/5058/18: Research of the interaction of a braked railway wheelset and track in simulated operational conditions of a vehicle running in a track on the test bench.

References

[1] Gerlici J, Gorbunov M, Nozhenko O, Pistek V, Kara S, Lack T and Kravchenko K 2017 About creation of bogie of the freight car, *Communications – Scientific Letters of the University of Zilina* **19**(2) 29-35

[2] Tartakovsky E, Falendysh A, Zinkivskyi A and Mikheev S 2015 Refining the models of performing service tests of upgraded locomotives, *Eastern-European Journal of Enterprise Technologies* **2**(3) 26-31

[3] Gerlici J, Lack T and Harušinec J 2014 Realistic simulation of railway operation on the RAILBCOT Test Stand, *Applied Mechanics and Materials* **486** 387-395

[4] Nozhenko O, Hauser V, Kravchenko K and Loulová M 2018 Reducing tram car's curve-pasing resistance by double treated wheel profile, *Manufacturing Technology* **18**(1) 79-83

[5] Hauser V, Nozhenko O S, Kravchenko K O, Loulová M, Gerlici J and Lack T 2017 Proposal of a mechanism for setting bogie wheelsets to radial position while riding along track curve, *Manufacturing Technology* **17**(2) 186-192

[6] Lack T and Gerlici J 2014 A modified strip method to speed up the calculation of normal stress between wheel and rail, *Applied Mechanics and Materials* **486** 359-370

[7] Golubenko A, Sapronova S and Tkachenko V 2007 Kinematics of point-to-point contact of wheels with a rails, *Transport Problems* **2**(3) 55-59

[8] Leitner B and Uricek J 2010 *A method for adaptive identification of stochastically loaded parts of mechanical systems*, 3rd International Multi-Conference on Engineering and Technological Innovation IMETI 2010 Vol II, Orlando,Florida, USA, Jun 29 - Jul 02, 174-179

[9] Leitner B, Mocova L and Hromada M 2017 A new approach to identification of critical elements in Railway infrastructure, *Procedia Engineering* **187** 143-149

[10] Stadnyk I, Vtenko T, Drozdziel P and Derkach A 2016 Simulation of components mixing in order to determine rational parameters of working bodies, *Advances in Science and Technology – Research Journal* **10**(31) 130-138

[11] Gerlici J, Sakhno V, Yefymenko A, Verbitskii V, Kravchenko A and Kravchenko K 2018 The stability analysis of two-wheeled vehicle model, *MATEC Web of Conferences* **157**

[12] Baran P, Štastniak P, Kukuča P and Brezáni M 2018 Investigation of kinematic parameters of two nonconventional piston machines with wobble board, *MATEC Web of Conferences* **157**

[13] Baran P, Kukuca P, Barta D, Labuda R, Drozdziel P and Pukalskas S 2017 The issue of balancing internal combustion engines with non-conventional crank mechanism, *Communications – Scientific Letters of the University of Zilina* **19**(2) 36-41

[14] Kašpárek J, Škopán M, Jonák M and Pokorný P 2016 *Experimental verification of FEM analysis of the forestry forwarder*, 20th International Scientific on Conference Transport Means 2016, Juodkrante, Lithuania, Oct. 5 - 7, 751-755

[15] Maňurová M and Suchánek A 2016 The analysis of a rail vehicle with a tilting bogie. *Manufacturing Technology* **16**(5) 1020-1027

[16] Svoboda M, Schmid V, Soukup J and Sapieta M 2018 Modal analysis of the vehicle model, *Springer Proceedings in Mathematics and Statistics* **249** 351-362

[17] Falendysh A, Volodarets M, Hatchenko V and Vykhopen I 2017 Software analysis for modeling the parameters of shunting locomotives chassis, *MATEC Web of Conferences* **116**

[18] Vaiciunas G, Bureika G and Steisunas S 2018 Research on metal fatigue of rail vehicle wheel considering the wear intensity of rolling surface, *Eksploatacja I Niezawodnosc – Maintenance and Reliability* **20**(1) 24-29

[19] Gerlici J and Lack T 2014 Modified HHT method for vehicle vibration analysis in time domain utilization, *Applied Mechanics and Materials* **486** 396-405

[20] Stastniak P and Moravcik M 2017 *Development of two types of freight wagons with bogies for non-standard wheelbase or track wheelset, complying with the criteria for interoperability, environmental issue, safety and reliability*, XXIII Konference s mezinarodni ucasti: Soucasne problemy v kolejovych vozidlech/23rd Conference: Current problems in rail vehicles 2017, Ceska Trebova, Czech Republic, Sep. 20 - 22, 401-408

[21] Sapronova S, Tkachenko V, Kramar N and Voron'ko A 2008 Regularities of shaping of a wheel profile as a results of deterioration of the rolling surface in exploitation, *Transport Problems* **3**(4) 47-54

[22] Kostrzewski M 2018 Analysis of selected acceleration signals measurements obtained during supervised service conditions – study of hitherto approach, *Journal of Vibroengineering* **20**(4) 1850-1866

[23] Melnik R and Koziak S 2017 Rail vehicle suspension condition monitoring – approach and implementation. *Journal of Vibroengineering* **19**(1) 487-501

[24] Gerlici J, Lack T and Ondrova Z 2007 Evaluation of comfort for passengers of railway vehicles, *Komunikacie* **9**(4) 44-49

[25] Loulová M, Suchánek A and Harušinec J 2017 Evaluation of the parameters affecting passenger riding comfort of a rail vehicle, *Manufacturing Technology* **17**(2) 224-231

Determining of the drive power of a transport machine for disabled persons using a computational model

M Blatnický[1] and J Dižo[1]

[1]University of Žilina, Faculty of Mechanical Engineering, Univerzitná 1, 010 26 Žilina, Slovakia

E-mail: jan.dizo@fstroj.uniza.sk

Abstract. The article deals with calculation of a stair lift drive and its design. The aim is the determining the necessary power of the used electric motor depending on the variable input values such as a passenger weight, a gradient of the seat lift etc. Calculating the required power for operation of the equipment is essential for selecting the optimal drive motor. In practice, the mathematical-physical principle of solving a problem such as drive calculation is often neglected. This results in the use of excessively powerful propulsion equipment, and thus in excessive financial cost in mass production. Consequently, by designing a drive-train, it will be possible to design the entire drive mechanism, i.e. motor, gearbox, gear module, etc. A significant part of the overall solution will be the design of the shaft for the drive pinion and its analytical dimensional calculation in comparison with the numerical solution. This will fulfil the prerequisites for optimizing the current shaft design by reducing its mass from the current value of 1.185 kg while the safety and reliability conditions of the construction are still met.

1. Introduction

One of the social aspects of a modern advanced society is its ability to help those in a difficult situation. One subset affliction of people is a restriction of a person's mobility. Several devices have been contrived for people with disabilities to facilitate surmounting various obstacles. A stair seat can also be found between these devices, which is also the subject of the issue solution (Figures 1 and 2). Since there are stairways of different shapes and dimensions, stair seats cannot be uniformly specified. In production of each stair lift, its design modifications must be done according to the specific requirements of each customer [1], [2].

Such a handling machine is usually made of more kind of material to reach suitable mechanical properties of particular components in combination with required lifetime and acceptable price. There are most often used combination of standard steel, high strength steel, light metals, mainly such as aluminium alloys, plastics etc. [3-5]. The latest types of stair lifts have the automatic process of tilting and they dispose of a remote control [6]. The drive-train is composed of an electric motor, a gearbox and a gear wheel, which is in the mash with a rack.

Previous steps of solving this issue [7], [8] cover: boundary and load conditions of the seat, structure material properties and force effects calculations related to the seat track.

Subsequently, the acquired results will be used to design the necessary power of the electric motor used as a seat drive, since determining the power of the electric motor for driving is an inevitable part of the design of each mechanism. Thanks to the calculation, it is possible to equip the device with an optimum engine that is powerful enough to handle the operation, but at the same time its power is not unnecessarily oversized. From the previous calculations [7], [8], it is clear that quantities of resistance can be expressed in the form of force or moment. The same principle applies at calculating the engine power when the desired speed of the moving device and the diameters of the contact circles of the wheel and the track are known. Calculation with considering resistances as moments of forces is appropriate in the case of known shaft speeds, the starting point being the approximate speed of the device and the pinion geometry. The speed calculated based on the approximate speed should be corrected in terms of a suitable gear ratio between speed of the engine and the pinion.

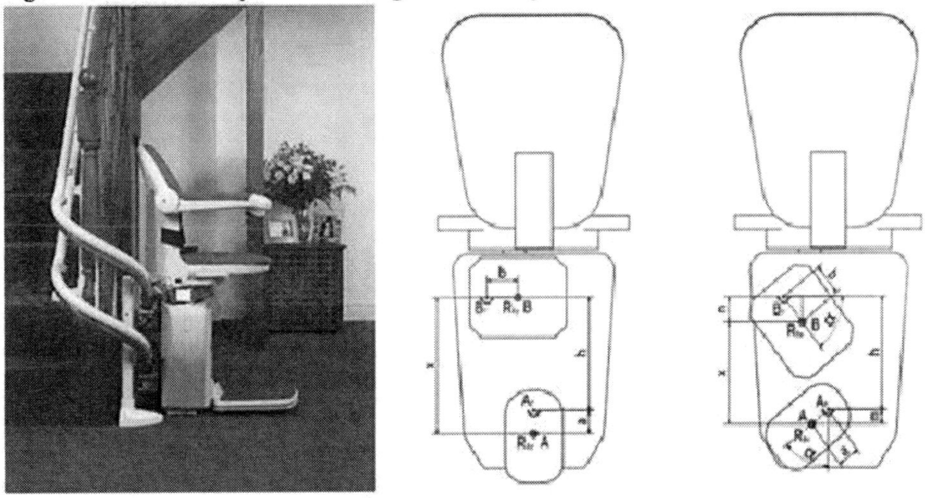

Figure 1. A real model of SA Alfa stair lift (left), Schematic change of geometry when the carriage is tilted due to the ascent of the track (right).

2. Theoretical aspects of electric motor power calculation

The calculation of the engine power of a stair seat consists of multiple variables and constant values. Therefore, it is necessary to state the boundary conditions when the resistive forces acting against the movement of the seat will be maximum. There has been found out, that from all resistances influencing the needed power of the used motor just the transported person weight is the most variable quantity. From this point of view there is necessary to consider different conditions in the contact of the driving wheel and the track [9-13].

From a structural point of view, a track with a maximum inclination of 52 ° is possible, which is the first boundary condition for the calculation. The second is the maximum load capacity that is in this case 130 kg. The third condition is the speed of the device, taking 0.1 ms^{-1}. From the three stated boundary conditions, the track inclination (or pitch) angle α and the weight of a person are the variables. Hence, it is suitable to illustrate the dependence of the required power on these variables. Resistance forces are dependent on the type of material and the machining contact areas in relative motion at mechanism operation.

Value of this resistance is determined by the normal force transmitted by the wheel, friction coefficient between the pin surface and material of the wheel, and the friction circle of radius [14], [15].

The staircase track (Figure 2) is made of steel S235JRG1 (STN 11 373) with powder coating comaxit. This ensures a surface resistance to mechanical damage and excellent durability and chemical resistance. Since the applied coating is hardened and its thickness is negligible, it can be assumed that it will not have a significant impact on the rolling resistance. A part of the track is also a gear rack made of steel C16E (STN 12 020) with modulus $m = 3.5$ (mm).

The components that come into contact with the track are made of polyamide ertalon 6 PLA. Its usage in industry ranges from the production of simple parts to gears and plain bearings. Its basic mechanical properties include high strength, stiffness, hardness and toughness, high fatigue strength, high mechanical damping ability, good sliding properties, abrasion resistance and good machinability. The wheels of the upper carriage are pressed on a composite plain bearing placed on a steel pin. The lower carriage pulley carrying the vertical load moves axially on the shaft. There is also located the pinion gear that transmits the power from the output shaft on the gear rack. The horizontal load bearing polyamide rollers of the lower carriage are mounted on steel quenched and polished pins. All connections between the rotating guide member and the pin are lubricated to reduce the coefficient of friction between the components. Output shaft of the gearbox is made of steel 17MnCr5 (STN 14 220). Because of the location of the rolling bearings on the shaft for the lower carriage, the shaft part is cemented and hardened to 60 HRC. The power flow is directed from the motor to the worm gear on the output shaft, from where it is transmitted to the track by means of a pinion on the track gear rack. The pinion is made of zinc galvanized steel E335 (STN 11 600). Since the pinion teeth engage more often than the teeth on the gear rack, the hardness of the pinion teeth needs to be higher due to wear reduction.

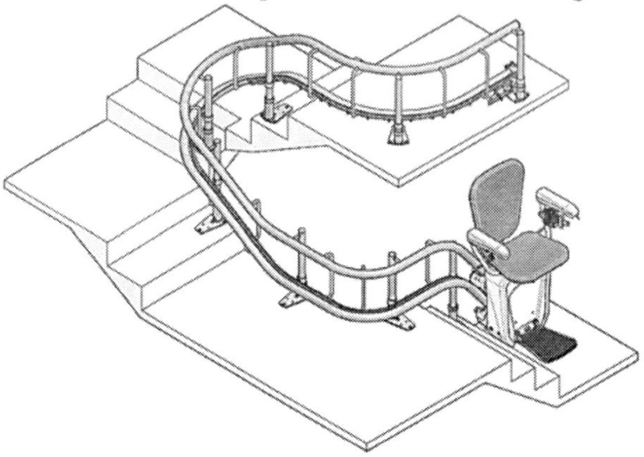

Figure 2. A curved path of the track for the designed stair lift.

3. Design of a drive motor

To calculate the necessary power of the used electric motor, a large number of input parameters are needed, the design of which was solved in [8]. The values of parameters obtained in previous calculations are stated in Table 1.

Table 1. Calculation results of input variables for calculation of electric motor performance.

no.	Input quantity	Value of quantity
1	G - weight of the seat	$G = 559.17$ N
2	Q - weight of a person	$Q_{max} = 1\,275.3$ N
3	α - slope of the staircase	$\alpha_{max} = 52\ °$
4	v - seat speed	$v = 0.11\ \mathrm{ms^{-1}}$
5	m – gear rack and pinion module	$m = 3.5$ mm
6	$r_{čav}$ - radius of pin friction of lower guide rollers	$r_{čav} = 4$ mm
7	$r_{čak}$ - radius of pin friction of the pulley carrying the vertical load	$r_{čak} = 17.5$ mm
8	$r_{čb}$ - radius of pin friction of upper guide rollers	$r_{čb} = 5.95$ mm
9	$f_{čb}$ - friction coefficient of upper guide wheels	$f_{čb} = 0.04$ (-)
10	$f_{ča}$ - friction coefficient of lower guide wheels	$f_{ča} = 0.1$ (-)
11	e_{av} - the coefficient of rolling resistance of the lower guide rollers	$e_{av} = 0.021$ mm
12	e_{ak} - the coefficient of rolling resistance of the lower guide pulley	$e_{ak} = 0.07$ mm
13	e_b - the coefficient of rolling resistance of the upper guide rollers	$e_b = 0.03$ mm
14	a - distance (figure 1)	$a = 49$ mm

Table 1. Calculation results of input variables for calculation of electric motor performance (continuation).

no.	Input quantity	Value of quantity
15	b - distance (figure 1)	$b = 67.5$ mm
16	h - distance (figure 1)	$h = 236$ mm
17	η_1 - worm gear efficiency	$\eta_1 = 0.66$ (-)
18	η_2 - rolling bearing efficiency	$\eta_2 = 0.98$ (-)
19	g - gravity acceleration	$g = 9.81$ m·s^{-2}
20	β - the angle between the normal reaction of the upper guide wheels and the vertical axis	$\beta = 32.7$ °
21	d - pitch diameter of the drive gear pinion	$d = 66.5$ mm
22	Fs - the force from the seat ascent	$F_{smax} = 2{,}570$ N
23	x - the distance between the reactions R_{ay} and R_{by}	$x = 212.977$ mm
24	n - shaft speed of the electric motor	$n = 28.72$ min^{-1}
25	η_c - the overall efficiency of the propulsion mechanism	$\eta_c = 0.634$ (-)

According to the relation (1), the power P (W) of the used engine depending on the resistive forces F_o (N), the speed of the seat movement (m·s^{-1}) and the overall efficiency of the mechanism η_c (-) is:

$$P = \frac{\sum F_o \cdot v}{\eta_c} \tag{1}$$

By adding the appropriate relations of all the considered resistive forces to equation no. 1 for power P we get:

$$P = \frac{\sum\left(F_{ob} + F_{oav} + F_{oak} + F_S\right)\cdot v}{\eta_c} \tag{2a}$$

$$P = \frac{\dfrac{2\cdot(G\cdot 119 + Q\cdot 180)}{x}\cdot\left(\dfrac{e_b + r_{čb}\cdot f_{čb}}{d_b\cdot\sin\beta} + \dfrac{e_{av} + r_{čav}\cdot f_{ča}}{d_{av}}\right)}{\eta_c} + \frac{(G+Q)\left[\left(\dfrac{2\cdot(e_{ak} + r_{čak}\cdot f_{ča})}{d_{ak}}\right)\cdot\cos\alpha + \sin\alpha\right]\cdot v}{\eta_c} \tag{2b}$$

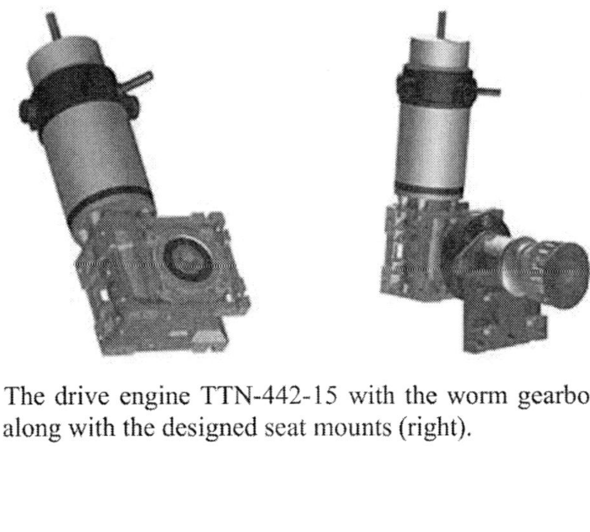

Figure 3. The drive engine TTN-442-15 with the worm gearbox CM 040 (left), along with the designed seat mounts (right).

According to the relation (2), the power P (W) of the used motor depending on the resistive moments M_o (N·mm), angular speed of the engine rotor ω (rad·s $^{-1}$) and the overall efficiency of the mechanism η_c (-) is:

$$P = \frac{\sum M_o \cdot \omega}{1000 \cdot \eta_c} \tag{3}$$

By adding the appropriate relations of all the considered resistance moments to equation no. 2 for power P we get:

$$P = \frac{\sum (M_{ob} + M_{oav} + M_{oak} + M_S) \cdot \omega}{1000 \cdot \eta_c} \tag{4a}$$

$$P = \frac{\frac{2 \cdot (G \cdot 119 + Q \cdot 180)}{x} \cdot \left(\frac{e_b + r_{čb} \cdot f_{čb}}{\sin \beta} + e_{av} + r_{čav} \cdot f_{ča} \right)}{1000 \cdot \eta_c} + \frac{(G+Q)\left[(e_{ak} + r_{čak} \cdot f_{ča}) \cdot \cos \alpha + \frac{d}{2} \sin \alpha \right] \cdot \omega}{1000 \cdot \eta_c} \tag{4b}$$

By substituting the appropriate values in (1) and (2) it is possible to obtain the value of the necessary power of the propulsion engine to meet the above requirements. The resulting power of the propulsion engine proposed for nominal equipment parameters is calculated as $P = 256.677$ W. The engine that will drive this equipment must therefore have the closest higher power in order to ensure the operation of the equipment without the risk of permanent overload. Therefore, the TTN-442-15 engine is proposed for drive. Its nominal power is $P_n = 270$ W, supply voltage $U = 24$ V, current $I = 16$ A, torque $M_n = 1.4$ Nm, speed $n = 1{,}800$ min^{-1} and weight $m = 5.5$ kg. The worm gearbox of gear ratio $i = 60$ (-) marked CM040 will be used. The entire drive unit assembly is shown in figure 3.

4. Design of drive mechanism

The transmission of torque from the motor into the system is ensured by two keys DIN 6885 A. The keys are located one behind the other on the shaft (figure 4 on the left). In this part of the shaft, the M8 thread is concentrically drilled. A bold with special washer is screwed there, when the shaft is mounted to the gearbox. This ensures the shaft against axial movement. In the middle of the shaft there is a space for the flange, seat cover, holder, washer, needle bearings and a movable plastic roll. A toothed pinion is mounted on the shaft from the other side. Between the pinion and the shaft there is a DIN 6885 A key, which transmits the torque. To achieve the minimum possible length of the key, the keyseat is located specifically extending into the side of the shaft cylinder. Similarly to the front of the shaft, an M8 thread is drilled into the shaft end, into which the screw is placed. Its purpose is to cover the tooth system by means of a washer and to axially secure the toothed pinion.

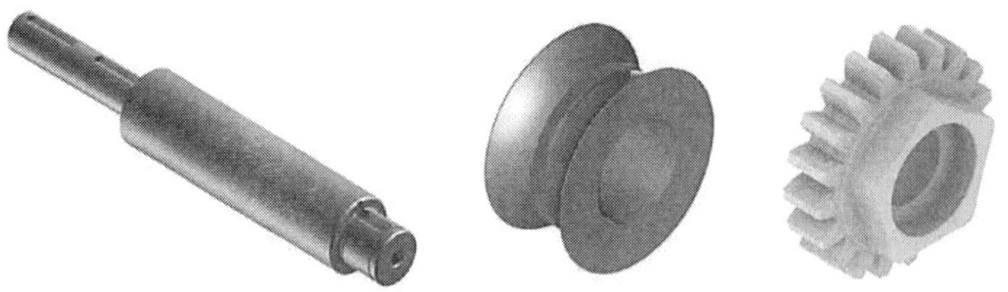

Figure 4. Shaft (left), plastic roll with groove (in the centre) and sprocket with pentagon flange (right).

Plastic roll (figure 4 in the centre) is clearance fit on the shaft so that it can move axially and copy the path shape mainly in curves. A special groove is formed on the roll, which is part of the safety mechanism. When the permitted seat speed is exceeded, a mandrel is lowered to the groove to secure the seat in place. In addition to the plastic roll, the gear also has a special adjustment (figure 4, right). The pentagonal flange is a part of the safety system. The bearing, which is placed on the stirrup moves on the flange. The mechanism controls the speed of the unit as well as the release of the safety mandrel.

Figure 5. 3D model of the drive mechanism with the back of the skeleton and safety mechanism (left) and mechanism with safety covers (right).

The drive mechanism is firmly attached to the seat frame to carry all reaction forces and moments. A flange is screwed onto the gearbox and the bolts are secured with spacers. The flange thus fastened is further screwed by means of bolts, washers and tabs to the back of the seat frame (figure 5). Pins with rollers are placed in the bearing housing. These help to move, stabilize the seat and transfer a portion of the forces acting on the lower carriage of the seat.

5. Expressing the change in the required engine power with respect to human weight

Of all the factors that affect the size of the resistance, and hence the performance of the drive motor, the weight of a man is the most variable value. The coefficients associated with passive resistances depend on the choice of material and the geometry of the components used, so their value is unchanged until the design of the device changes. Certain parameters are subjected to regulations and standards e.g. STN EN 81-40, therefore, the speed of a stair seat of various manufacturers is mostly 0.1 ms $^{-1}$.

Table 2. Values of resistance torque and power at a variable human weight at α = 52 °.

$m_č$ [kg]	Q [N]	M_{ob} [N·mm]	M_{oav} [N·mm]	M_{oak} [N·mm]	M_s [N·mm]	ΣM_o [N·mm]	P [W]
52	510.1	368.9	313.0	1,198.1	28,016.8	29,896.9	148.2
58	569.0	393.5	334.0	1,264.1	29,559.1	31,550.7	156.4
64	627.8	418.2	354.9	1,330.0	31,101.3	33,204.5	164.6
70	686.7	442.9	375.9	1,396.0	32,643.5	34,858,3	172.8
76	745.6	467.6	396.8	1,462.0	34,185.7	36,512.0	181.0
82	804.4	492.3	417.8	1,527.9	35,727.9	38,165.8	189.2
88	863.3	516.9	438.7	1,593.9	37,270.1	39,819.6	197.4
94	922.1	541.6	459.6	1,659.8	38,812.3	41,473.4	205.6
100	981.0	566.3	480.6	1,725.8	40,354.5	43,127.2	213.7
106	1,039.9	591.0	501.5	1,791.7	41,896.8	44,781.0	221.9
112	1,098.7	615.6	522.5	1,857.7	43,439.0	46,434.8	230.1
118	1,157.6	640.3	543.4	1,923.6	44,981.2	48,088.5	238.3
124	1,216.4	665.0	564.4	1,989.6	46,523.4	49,742.3	246.5
130	1,275.3	689.7	585.3	2,055.5	48,065.6	51,396.1	254.7

It follows that, while maintaining the design of the stair seat, the necessary power changes only in relation to the change in human weight and the slope angle. In the calculation for the power of the drive motor, two forces, namely force G and force Q, are exerted. The force G is formed by the weight of the stair seat and its size is fixed, while the force Q is the result of the weight of the person sitting on the seat.

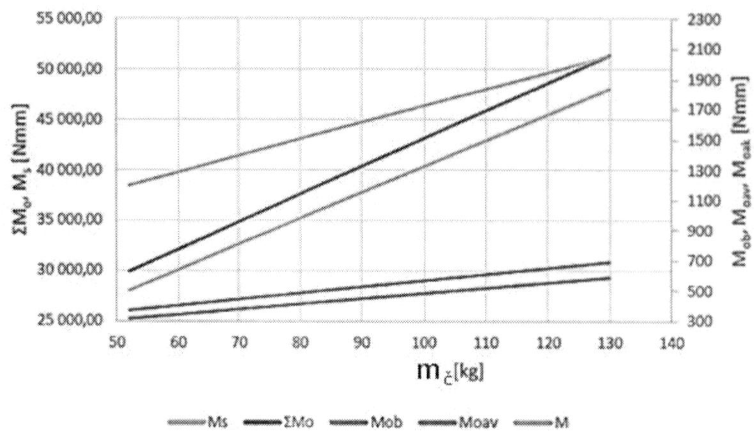

Figure 6. Dependence of resistive moments on passenger weight at constant inclination angle $\alpha = 52$ °.

Given that the weight of the occupants of the stair seat differs, the value of the necessary power will also be different. The starting weight of a human is 52 kg because it does not make sense to think of a human weight of zero or very low weight, since the primary purpose of the device is to transport a person. The change in the magnitude of the resistive moments is quantified in table 2 and graphically represented in figure 6 and figure 7.

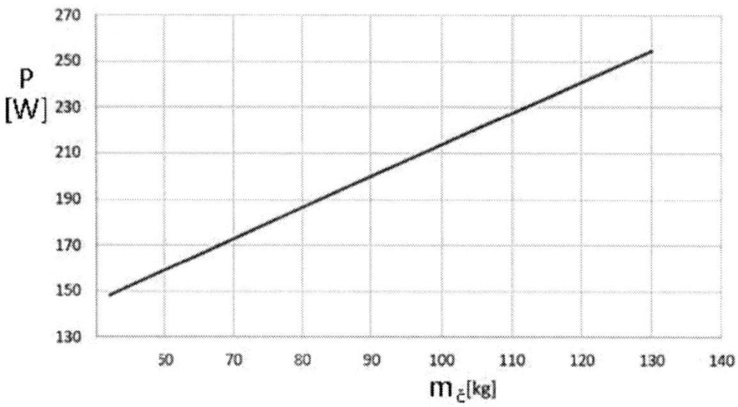

Figure 7. Dependence of the required power of the drive motor depending on the weight of the passenger at a constant inclination angle of the stair seat path $\alpha = 52$ °.

6. Numerical verification of shaft safety

Numerical calculation of stress in shaft was performed by means of the finite element method, which is common used method for standard analyses of structures of technical parts, assemblies and also whole constructions [16], [17]. In our research we have used Ansys Workbench. After starting the program, input material data was set in the EngineeringData section. The shaft model in format STEP was imported from Autodesk Inventor, and the shaft surfaces were split to simulate the surfaces under the pinion, plastic roll, bearings and gearbox. In addition, a surface has been created by dividing the front cylindrical surface resting on the gearbox in two halves. The larger radius of this area has been reduced to 15 mm, which corresponds to the radius of the contact portion of the gearbox. On the prepared surface it was possible to simulate the shaft-to-gear contact. All these actions were performed in the Geometry section. The model mesh was then defined in the Model section. The mesh was created with 2 mm long elements and was formed by 53 684 nodes and 31 234 elements (figure 8).

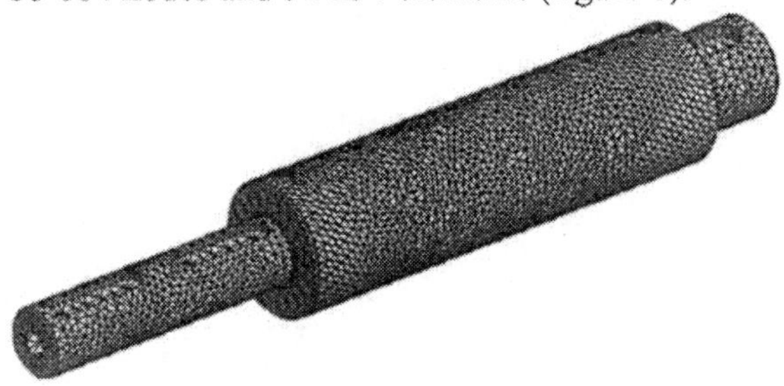

Figure 8. FEM mesh of the shaft.

After meshing the shaft, the necessary degrees of freedom have been provided to the prepared surfaces under the gearbox and the front cylindrical surface of the shaft. The forces and moments have been assigned to the area under the bearings, the area under the plastic roll, and the area on which the pinion is located. Under these conditions, the equivalent (von Mises) stress was selected as output and then the calculation was started. The calculation results are shown in figure 9.

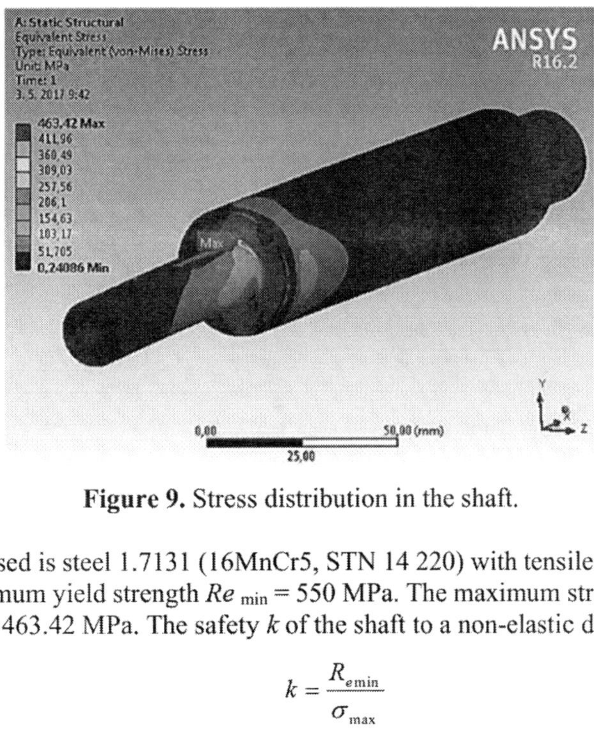

Figure 9. Stress distribution in the shaft.

The shaft material used is steel 1.7131 (16MnCr5, STN 14 220) with tensile strength R_m = 780 - 850 MPa and with the minimum yield strength $Re_{\min}$ = 550 MPa. The maximum stress value determined by the simulation is $\sigma_{\max}$ = 463.42 MPa. The safety k of the shaft to a non-elastic deformation is then (5):

$$k = \frac{R_{e\min}}{\sigma_{\max}} \tag{5}$$

By adding values to the relationship (5) we get:

$$k = \frac{R_{e\min}}{\sigma_{\max}} = \frac{550}{463.42} = 1.187 \tag{6}$$

The future research will be focused to analyse effects resulting from the structure dynamics. For these purposes, advanced dynamical model has to be developed in MBS software [18]. Solutionists will make an effort for creating a computational model of the solved lifting seat as a multibody model [19], [20] even including flexible model, which will aid to identify the most loaded elements of the mechanical system and further optimise them. By means of further simulations and analyses it will be possible to find out values of accelerations in individual components of the lifting seat. Accelerations are main dynamical outputs [21], based on which we are able to analyse other dynamical effects influencing comfort [22], [23] during operation of the seat (acceleration, deceleration, climbing, kneeling, moving along a curved track on a sharped staircase).

7. Conclusion

The article brings a summary of the state of the art of handling technology design for disabled people. This is the stair lift of the type with high availability for the civil sector. Along with previous calculations, article solves the design of the required power of the used electric motor, which was set to $P = 256.667$ W. From this point on, when a dimensional power calculation was made, it was possible to select a particular type of electric motor used along with gearbox.

Whole engine set TTN-442-15 with gearbox CM 040 will be delivered by Italian company Transtecno. To determine the instantaneous value of engine power depending on the weight of the passenger at the maximum inclination angle of the seat, a clear graph illustrating the given linear dependence was made. This seems logical, since human weight was considered as the only change in the parameters entering the calculation. Consequently, after determining the geometry of the engine, it was possible to design the seat frame into which the force effects are transferred from the operation of the device. Pinion drive shaft design and its numerical dimensional calculation were performed.

The analysis proved the suitability of the used shaft for the specified operation, but also revealed the space for its optimization because of the minimum stress values in the bulk of the shaft volume. Also, the safety factor $k = 1.187$ (-) does not correspond to the generally accepted fatigue damage prevention value to $k_{min} = 1.2$ (-). Therefore, it will be appropriate to optimize the shaft, which will be the main goal of further addressing this issue. It will also be necessary to perform security-oriented calculations for other parts and components used to complete the mechanism (figure 10).

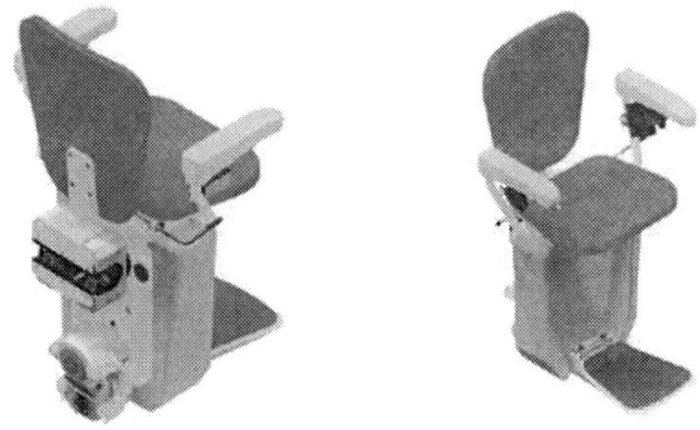

Figure 10. 3D visualization of the stair lift.

Acknowledgment

The work was supported by the Cultural and Educational Grant Agency of the Ministry of Education of the Slovak Republic in project No. KEGA 077UU-4/2017: Modernization of Vehicles and Engineers Study Program. The paper was prepared within the framework of the grant project of the Ministry of Education of the Slovak Republic KEGA, n. 018ŽU-4/2018:
„Innovation of didactic approaches and content of subjects of technical diagnostic as a tool for enhancing the quality of professional knowledge for practice needs."

References

[1] Behera P K and Gupta A 2018. Novel design of stair climbing wheelchair, *Journal of Mechanical Science and Technology* **32**(10) 4903-4908.
[2] Boyce M W, Al-Awar Smither J, Fisher D O and Hancock P A 2017 Design and instructions for evacuations disabled adults, *Applied Ergonomics* **58** 48-58.
[3] Moravec J 2018. *Unconventional method of forming circumferential rings on sheet metal parts*, 27th International Conference on Metallurgy and Materials METAL 2018, Brno, Czech Republic, May 23 - 25, 249-254
[4] Neslušan M, Trško L, Minárik P, Čapek J, Brončck J, Pastorek F, Čížek J and Moravec J 2018 Non-destructive evaluation of steel surfaces after severe plastic deformation via the

Barkhausen noise technique, *Metals* **8**(12)
[5] Zrak A, Meško J, Moravec J, Nigrovic R and Kadáš D 2016 Contactless thermal bending of steel sheets, *Manufacturing Technology* **16**(1) 309-313
[6] Kuligowski E, Peacock R, Wiess E and Hoskins B 2015 Stair evacuation of people with mobility impairments, *Fire and Materials* **39**(4) 371-384
[7] Blatnický M, Dižo J and Kurčík P 2018 Design of the handling device for disable people: Part I. (In Slovak), *Technológ* **10**(1) 36-41
[8] Kurčík P, Blatnický M and Dižo J 2018 *Functional calculation of the drive-train of a handling machine for disabled persons, part II* (In Slovak), 16th International Conference Dynamics of Rigid and Deformable Bodies DRDB 2018, Ústí nad Labem, Czech Republic, Oct. 17 - 19, 14 pages
[9] Gerlici J and Lack T 2009 Iterative method for railway wheel profile design, *Komunikacie* **11**(2) 49-56
[10] Gerlici J and Lack T 2003 Rail geometry analysis (from the point of view of wearing in the operation), *Komunikacie* **5**(1) 43-51
[11] Lack T and Gerlici J 2014 A modified strip method to speed up the calculation of normal stress between wheel and rail, *Applied Mechanics and Material* **486** 359-370
[12] Lack T and Gerlici J 2014 A modified strip method to speed up the tangential stress between wheel and rail calculation, *Applied Mechanics and Materials* **486** 371-378
[13] Sapietova A, Bukovan J, Sapieta M and Jakubovicova L 2017 Analysis and implementation of input lead effects on an air compressor piston in MSC.ADAMS, *Procedia Engineering* **177** 554-561
[14] Gerlici J, Gorbunov M, Kravchenko K, Domin R, Kovtanets M and Lack T 2017 Slipping and Skidding occurrence probability decreasing by means of the friction controlling in the wheel-breaking pad and wheel-rail contacts, *Manufacturing Technology* **17**(2) 179-186
[15] Hauser V, Nozhenko O S, Kravchenko K O, Loulová M, Gerlici J and Lack T 2018 Car body and bogie connection modification for track curves passability improvement, *MATEC Web* of Conferences **157**
[16] Falendysh A, Volodarets M, Hatchenko V and Vykhopen I 2017 Software analysis for modeling the parameters of shunting locomotives chassis, *MATEC Web of Conferences* **116**
[17] Soukup J, Skočilasová B and Skočilas J 2017 Vibration of mechanical system with higher degrees of freedom: solution of the frequency equations, *Procedia Engineering* **177** 17-24
[18] Kostrzewski M and Melnik M 2017 Numerical study of a rail vehicle with differential gears, *Procedia Engineering* **192** 439-444
[19] Gerlici J, Sakhno V, Yefymenko A, Verbitskii V, Kravchenko A and Kravchenko K 2018 The stability analysis of two-wheeled vehicle model, *MATEC Web of Conferences* **157**
[20] Rao P K V 2018 Design of stair-climbing wheelchair using tri-wheel mechanism, *International Journal of Mechanical and Production Engineering Research and Development* **8**(4) 685-694
[21] Smetanka L, Šťastniak P and Harušinec J 2018 Wear research of railway wheelset profile by using computer simulation, *MATEC Web of Conferences* **157**
[22] Loulová M, Suchánek A and Harušinec J 2017 Evaluation of the parameters affecting passenger riding comfort of a rail, *Manufacturing Technology* **17**(2) 224-231
[23] Vatulia G, Falendysh A, Orel Y and Pavliuchenkov M 2017 Structural improvements in a tank wagon with modern software packages, *Procedia Engineering* **187** 301-307

IOP Conf. Series: Materials Science and Engineering **659** (2019) 012054 doi:10.1088/1757-899X/659/1/012054

Technical solution of the under locomotives visual inspection system

A Miltenović[1], D Stamenković[1], M Banić[1] and M Simonović[1]

[1]University of Niš, Faculty of Mechanical Engineering, A. Medvedeva 14, 18000 Niš, Serbia

E-mail: aleksandar.miltenovic@masfak.ni.ac.rs

Abstract. In recent years, liberalisation of rail cargo operators market brought to the position of a number of small rail operators that are operating on rail infrastructure. Most of them are relatively small with specific cargo transport products, limited number of locomotives and wagons and with limited service resources. Independently of the cargo operator size, they are also obligated to have regular inspections according to the railway regulations. Inspections are frequent and in normal cases they need special depot with inspection pit that small operators do not possess. In the paper are given technical solutions that can be found in the market as well as developed and proposed equipment solution with different concepts for the under locomotives visual inspection without using inspection pit. In the paper is also carried out evaluation and choice of proposed solutions as part of regular product development process.

1. Introduction

Product development implies the whole process from the idea to the launch of the product on the market. It consists of number of steps.

In reality, product development depends from an investor while investor is defining starting and final point in the process of product development. Important steps in product development process are generating idea, developing concepts and their evaluation, development in details and manufacturing of prototype. Product development engineer has to follow these steps in order to solve task without losing possible partial solution or basic principle some into final solution [1].

Private rail cargo operators are operating in the market but they have limited inspection capacities. Independently of the cargo operator size, they are obligated to have regular inspections according to the railways regulations. Inspections are frequent and in normal cases they need special depot with inspection pit that small operators do not possess.

In the paper are given technical solutions that can be found in the market as well as developed and proposed equipment solution with different concepts for the under locomotives visual inspection without using inspection pit. In the paper is also carried out evaluation and choice of proposed solutions as part of regular product development process.

2. Product development task

According to railway regulations, railway vehicles, primarily locomotives, should be inspected relatively frequently, and inspections differ from those in which the visual inspection is sufficient to those for which it is necessary to measure something or open a part of the vehicle [2, 3, 4, 5].

Level of inspections depends from locomotive series and time or passed miles. Railway regulations prescribe "service inspections", P0, P1, P3, P6 and P12. Service inspection has to be done ones a week and others depending from series:

- P0 - 5000 - 7500 km
- P1 - 10000 - 15000 km
- P3 - 30000 - 45000 km
- P6 - 60000 - 90000 km
- P12 - 120000 - 180000 km

Smaller inspections like "service inspection" of electric locomotives includes:

- Visual control (from inspection pit and exterior) of boxes and equipment on it
- Visual inspection of handrails, axle assemblies, wheel rims, traction motor, primary and secondary suspension, winding springs and hydraulic shock absorbers.

Based on the described list of inspections, visual inspection from the pit, ie, the bottom view prescribes that an overview is performed:

- Wheelsets
- Traction motors
- Transverse coupling
- Nuts and bolts of transverse coupling
- Brake rigging
- Brake cylinders and
- Brake rigging regulator

Figure 1. Visual inspection of rail vehicle in the pit by rail inspector.

Normally, this overview is done from a pit by a rail inspector (Figure 1) who uses a hand lamp for lighting. However, very often a workshop (depot) equipped with a pit is distant, so it is advisable that the check is carried out on the ground, ie, on a secondary or industrial track. In order to visualize the vehicle from the bottom of the pit without a pit, a request was made to develop a visual control system that should be placed at the level of the thresholds between the rails and to allow the overview of the vehicle's running gear while the vehicle is standing or moving at a low speed (up to 5 km / h).

On the basis of the requirements set out by inspections regulations and the assessment of the potential positions of the camera from which the visual control of the partial (individual) parts of the running gear, in the Figure 1, a motion is proposed to position the camera in order to perform visual control of the locomotive.

Within the framework of visual inspection of the craft stand, as part of the service inspection, it is necessary to carry out the inspection and pay attention to the following:

- Wheelsets (see if there are any damage / drawings on the axles, wheels, wheel bars, gearboxes, and whether there is a leak from the gear unit)
- Traction motors (see if there are damage to the traction motor trains, coupling elements, gearbox, traction motor)
- Transverse coupling (see if there are damage to the transverse coupling, whether there are screws and nuts of the transverse coupling, see if there is damage / crack on the frame of bogy)
- Brake rigging (see if there are damage on three-way lever, brake controls, brake lugs-screw connections)
- Brake cylinders (see if there is damage to the brake cylinders and on the piston and brake links)

3. Existing and possible solutions on the market

In order to develop the appropriate system for visual inspection of the railway vehicle from the bottom on a track without inspection pit, research of technical solutions has been carried out, which can be applied with specific requirements on the railway.

By examining visual control in various areas of the technique, especially for parts that are less accessible, the following conceptual technical solutions that could (partially or completely) have been observed, will be applied in the conditions that we have explored:

- Manually operated inspection camera
- Under vehicle inspection system (UVIS)
- Industrial camera for inspection and mapping of pipes

3.1. Manually operated inspection camera

Figure 2 shows a hand-held inspection camera for inspecting road vehicles from the bottom that consists of a camera, which is fixed to a special holder, and a monitor.

Figure 2. Parking Security Undercarriage Inspection Camera SA920.

Solutions like this are widly used in inspections of cars and trucks on the borders or security points. It is easy to use and it needs flat surface.

3.2. Under vehicle inspection system (UVIS)

Under vehicle inspection system is a system used to inspect primarily road vehicles from below.

Figure 3 shows the under vehicle inspection system that is used to inspect the road vehicles from below (without the inspection pit). There are a number of producers for this solution that can be made to be fix in the ground or mobile so that can be easily moved from one place to another. It consist mostly with at least two cameras that are creating one picture. Pictures generated by cameras are processed by software and they are generating one picture from more pictures. Mostly these cameras are linear and they make it possible to create as long picture as the vehicle is long.

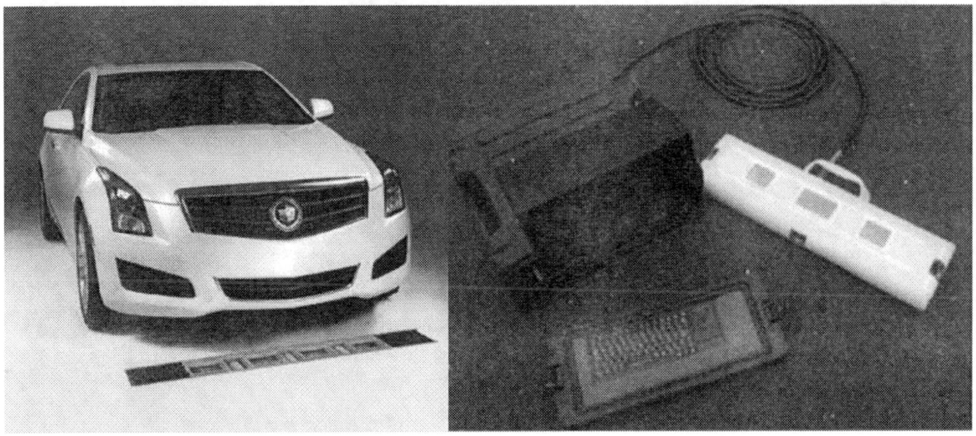

Figure 3. Visual inspecion system for road vechicles [Westminster International Ltd].

Figure 4 represents one picture made by this system. It is a single picture that contain whole undercarriage of vehicle.

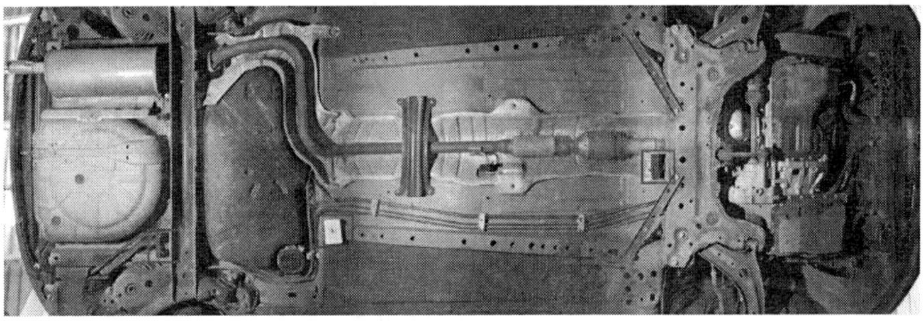

Figure 4. Picture made by visual inspection system for road vehicles.

Company Westminster International made UVIS system for railway vehicles (Fig. 5). Device with cameras has to be put between rail trucks and visual inspection is carried out when train goes over the device. System is mobile with two cameras and it can make inspection possible when train is running up to 40 km/h.

Figure 5. System UVIS for rail vehicles [Westminster International Ltd].

Figure 6 represents pictures made by such a system. It is a single picture that contain whole undercarriage of vehicle.

Figure 6. Picture of railway vehicle made by UVIS.

3.3. Industrial camera for inspection and mapping of pipes

In different areas, robots are often used to view hidden ones, hardly accessible spaces / parts. Figure 7 show examples of such robots. They are connected to the operator by wire and they are create to explore space up to 200 meters.

Figure 7. Robot for visual inspection.

4. Development of possible solutions - concepts

The conceptual solutions of the system are elaborated on the basis of two initial conditions:
• The system is mobile (mobile) and the locomotive is static (without moving) or vice versa
• At the location where the visual control is carried out, it is possible to set the appropriate surface
Four conceptual technical solutions have been developed in total:

- „Rod“
- „Tray“
- „Runway“
- „Trolley“

4.1. „Rod“

The technical solution "rod" is based on similar products used for inspection and control of passenger vehicles and trucks at border crossings, where the solution is adapted to the needs of inspection of railway vehicles and on-line inspections. The transverse and longitudinal movement, reliance on a single line around which the rod can be rotated, and all manipulation possible by one operator, is

IOP Conf. Series: Materials Science and Engineering **659** (2019) 012054 doi:10.1088/1757-899X/659/1/012054

designed. The power supply of the camera can be solved through a separate power supply, and a laptop can be viewed by another operator.

Figure 8. Concept of technical solution „rod".

Advantages:
- Low price
- Solution for which no additional installation of the structure and laying of the substrate is required
- Easily portable from vehicle to vehicle
- Low weight
- Quick setup

Disadvantages:
- Large operator manipulation is required
- Danger of damage to the camera
- Limited autonomy in work

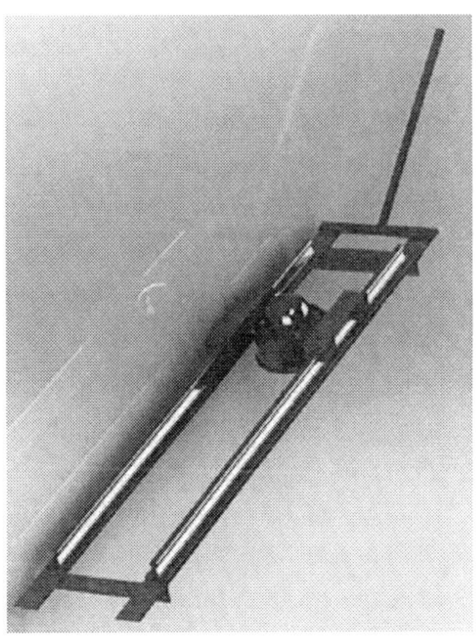

Figure 9. Concept of technical solution „tray".

4.2. „Tray"

Technical solution "tray" is an improved technical solution "rod" based on the installation of the entire device on the rails. Transverse and longitudinal movement is accomplished manually by one operator and the movement of the device is much easier than compared to the "rod". Longitudinal movement is achieved through the bearings, which greatly facilitates and accelerates the manipulation, and a special leverage would be the launch of cameras inside the device.

Advantages:
• A solution that does not require additional laying of the structure and handling of the surface
• Increased reliability and speed of longitudinal and transverse movement

Disadvantages:
• Great operator manipulation
• It takes more time to set up and more time to transfer the system between the rails
• Greater mass in relation to the "rod"

4.3. „Runway"

The technical solution of the "runway" is designed as a solution in which it is possible to easily position the camera to the desired location below the locomotive. For this purpose, a frame was constructed that would be placed between the rails and would be placed on the surface. The frame is projected at 3.6 meters which is limited by individual construction elements and 18 meters as estimated length is needed to suddenly perform visual control of the whole locomotive.

- Runway 3.6

Longitudinal and transverse suspension, guidance and drive system of the camera at a length of 3.6 m with automated visual control system; The locomotive need to be moved 5 times for complete visual control.

- Runway 18

Longitudinal and transverse suspension, guidance and drive system of the camera at a length of 18 meters with an automated visual control system (without movement of the locomotive).

Figure 10. Concept of technical solution „runway".

IOP Conf. Series: Materials Science and Engineering **659** (2019) 012054 doi:10.1088/1757-899X/659/1/012054

Advantages:
- Solution for which no additional installation of the structure and laying of the surface is required
- Increased reliability and speed of longitudinal and sub-movements (possibility of automation of transverse motion)

Disadvantages:
- Complex construction - complete assembly of parts, it is necessary to foresee the possibility of permanent construction of the structure
- A lot of work and time spent before the start of the inspection - the total inspection time is extended

4.4. „Trolley"

The technical solution of the "trolley" is designed as a solution where the camera with the device would move on the profiles back and forth and thus allow for longitudinal movement, and for the transverse movement on the device there are motors that allow the remote control to position the camera to the desired place. For this solution it is necessary to mount the path, ie. profiles (pipes) before inspection and it would be designed to quickly mount two tracks.

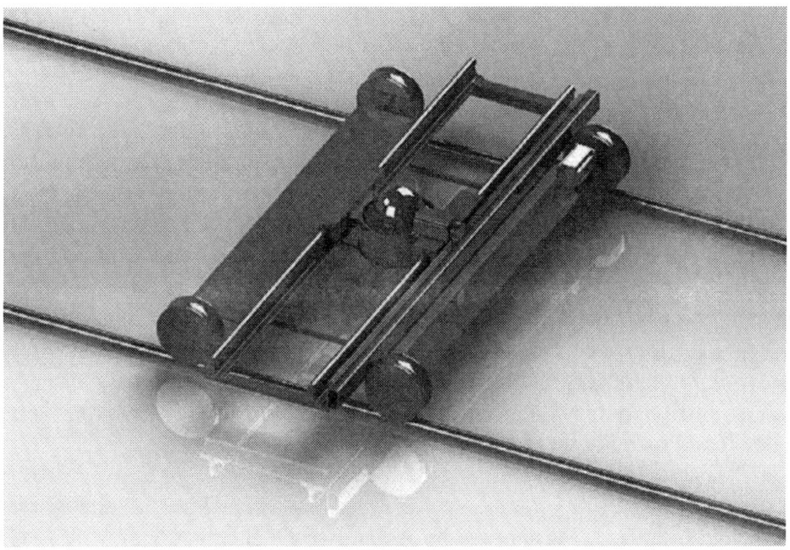

Figure 11. Concept of of technical solution „trolley".

Advantages:
- Full automation and ease of operation, the operator is completely free to focus on the inspection

Disadvantages:
- It is necessary to mount the profile, which increases the time of preparation of the inspection

5. Evaluation

Evaluation of the concepts has been done according to following criteria:
- Price – this is the most important factor for investor
- User friendliness. It is also important since the operators of this device can be people that has low education and awareness.
- Degree of automatization. It is important since higher degree of automatization will make whole inspection process easier.
- Inspection speed. It is important that the inspection is going to be done in some normal period.
- Mass. This should be more movable device and lower mass makes the whole process easier.

Weight coefficients are chosen in agreement with investor and experts in field of railway vehicles inspection.

Table 1. Evaluation according to criteria and weight coefficient.

Criteria	weight coefficient	Rod	Tray	Runway 3.6	Runway 18	Trolley
Price	0.4	10	8	3	1	6
User-frendliness	0.3	4	5	2	3	9
Degree of automatization	0.1	1	6	7	10	8
Inspection speed	0.1	2	7	6	10	8
Mass	0.1	9	6	4	3	6
Sum	1	6.4	6.6	3.5	3.6	7.3

According to Table 1. the best grade has concept Trolley. The lowest grades have concepts Runway due to their price. Concepts Rod and Tray are having a little lower Sum of grades than Trolley.

6. Conclusion

In the paper is shown case study of product development for under rail vehicle visual inspection system. For the development of under rail vehicle visual inspection system are presented steps of idea generation, concept development and evaluation.

First step in cases like this is to define product profile and to generate ideas. Product profile is given by investor. In the paper are presented three products that can be found on the market and they are solving similar problems which basic idea could be use in case of under rail vehicle visual inspection system. Basic ideas are generated on the base of existing products as well as authors give their own concepts for this specific problem. In the case of under rail vehicle inspection system are developed four different concepts and for every case are given advantages and disadvantages.

On the end is carried out evaluation which gives conclusion which concept has to be further developed in detail and manufactured.

References

[1] Miltenović V, Anišič, Z, Marjanović N, Adamović D, Banić M and Miltenović A 2015 *Razvoj proizvoda – osnove razvoja proizvoda; metode razvoja proizvoda; integralni razvoj proizvoda*, Univerzitet u Nišu - Mašinski fakultet, Niš

[2] Stamenković D 2014 Improve of railway vehicle maintenance by wayside detection, *Mechanics Transport Comunications* **12**(3) 64-70

[3] Vinberg E M, Martin M, Firdaus A H, Tang Y and Qazizadeh A 2018 *Railway application of condition monitoring*, Technical report, Stokholm

[4] Asplund M 2016 *Wayside Condition Monitoring System for Railway Wheel Profiles*, Lulea University of Technology, Doctoral Thesis

[5] Palo M 2012 *Condition Monitoring of Railway Vehicles A Study on Wheel Condition for Heavy Haul Rolling Stock*, Lulea University of Technology, Doctoral Thesis

[6] Bardenhagen A and Rakov D 2019 Advanced Morphological Approach in Aerospace Design During Conceptual Stage, *Facta Universitatis: Mechanical Engineering* **17**(2)

[7] Alkamachi A and Ercelebi E 2017 Modelling and Control of H-Shaped Racing Quadcopter with Tilting Propellers, *Facta Universitatis: Mechanical Engineering* **15**(2) 201-215

Features of tank car testing for dangerous cargoes transportation

M Kelrykh[1], O Fomin[1], J Gerlici[2], P Prokopenko[3], K Kravchenko[2] and T Lack[2]

[1]State University of Infrastructure and Technology, Faculty of Infrastructure and Railway Rolling Stock, St. Kyrylivska, 9, 040071, Kyiv, Ukraine
[2]University of Zilina, Faculty of Mechanical Engineering, St. Univerzitná 8215/1, 010 26, Zilina, Slovak Republic
[3]Branch "Research and design and technological Institute of Railway Transport "PJSC"Ukrzaliznytsya", St. Ivana Fedorova, 39, 030 38 Kyiv, Ukraine

E-mail: kkatherina@ukr.net

Abstract. The experience of the operation of rail transport shows that a significant part of emergency situations with significant negative environmental consequences are related to tank cars carrying liquid goods of a wide range, including liquefied gas, petroleum products, concentrated acids, toxic and explosive products of chemical origin. The article presents the peculiarities and result of carrying out of the corresponding tests. In order to address the question of the possibility of further safe operation of tank cars in the over-term period, their technical diagnostics and control tests are carried out. In this case, the control of the technical condition includes visual inspection of the tank wagon, measurements of the thickness of the metal at the control points of the boiler and the frame of the car, conducting of the magnetopowder, ultrasonic and acoustic emission control of the bearing metal constructions. Control tests include static vertical load tests, test for a low cyclic load by boiler pressure, typical, resource shock tests, and possible emergencies testing.

1. Introduction

In recent years, there has been a significant aging of the Ukrainian fleet of freight cars, including tank wagons for dangerous cargoes transportation. In conditions of Latvian railway the main damages of the barrels of oil tank wagons and their quantity are: cracks of welding seams of shaped pads – 50%, barrel damages in area of supports – 30%, damages of welding seams of the barrel – 15%, unfastening of shaped pads – 5% [1].

Currently, on the Ukrainian railways network, tank wagons with a service life that exceeds the deadline set by the manufacturer and even extended service life are in operation. Therefore, the urgent issue is either the replacement of rolling stock or the extension of the service life of the operated rolling stock.

For example, new railway tank wagon type Zans, which was developed by University of Žilina and Tatravagónka Inc. Poprad Results of calculations and prototype tests prove, that new designed construction of the tank wagon satisfies strength conditions [2].

Analysis of the technical condition of tank cars after the planned repairs shows that a significant part of them is in a satisfactory condition. Due to lack of financing, the purchase of new cars, to ensure the uninterrupted performance of freight transports by rail, remains the task of conducting research work on determining the residual resource and the possibility of continuing the safe operation of tank-wagons for particularly dangerous cargoes transportation.

In order to address the question of the possibility of further safe operation of tank cars in the over-term period, their technical diagnostics and control tests are carried out [3, 4]. In this case, the control of the technical condition includes visual inspection of the tank wagon, measurements of the thickness of the metal at the control points of the boiler and the frame of the car, conducting of the magnetopowder, ultrasonic and acoustic emission control of the bearing metal constructions. Control tests include static vertical load tests, test for a low cyclic load by boiler pressure, typical, resource shock tests, and possible emergencies testing.

In research of Voropai V [5] proposed the procedure of valuation of tank-cars' residual operation time for liquid gas transportation. The calculation allows to calculate equivalent stress using test results that rely on a change of internal pressure activity during drainage and filling operations, and the most unfavorable combination of simulation of active loads on a tank-wagon.

In the paper [6] presents numerically predict the number of cycles (or kilometers) to fracture of tank wagon railway axles in various theoretical conditions. It presented mechanism of railway axles due to the fatigue of material.

2. Formulation of the problem

It is necessary to form the necessity of conducting data of controlling tests of flat wagons.

After analyzing the park and the technical condition of tank cars for the particularly dangerous goods transportation, it was established that the main part of the park consists of tank cars of model 15-1407 - for transportation of propane (Fig. 1), 15-1408, 15-1408-01, 15-1408-02, 15 -1440, 15-1597, 15-1619 - for transportation of ammonia, 15-1409, 15-1556 - for transportation of chlorine (Fig. 2), 15-1519, 15-1780 - for transportation of propane-butane, and they are in good technical condition. The above-mentioned tank cars are structurally different from each other by the following features: the material from which the tank boiler is made, the diameter of the boiler, the number of shells in the boiler, the protective equipment of the bottom of the boiler from damage during accidents and protection of the boiler armature.

The purpose of the article. Illustration of theoretical and practical features of conducting technical diagnostics, control tests and tests with simulation of emergency situations of tank cars for the carriage of especially dangerous cargoes.

3. Analysis of literary data

Modern studies devoted to the reduction of the viability of freight wagons, and in particular, article [7-9] is devoted to the coverage of the proposed innovations for the construction of open wagon "railroad space 1520 mm" and the peculiarities of their design, but it limited the possibilities of applying such innovations for cars. The authors [10] reflect certain perspective directions of construction of railroad car bodies in order to improve the techno-economic indicators, but do not reveal the economic potential of the business use of special expensive carriage rolling of unlimited length. Thus, in [11] the characteristic features and results of dynamic characteristics of platform cars are described. And work [12] is devoted to the presentation of the proposed methods for determining the dynamic characteristics for different performances of bearing car designs. In paper [13] presents some of the options Ansys contact setting defines the real tank wagon contact.

The paper [14] presents a theoretical and experimental study on the resistance of the tank-chassis connection (seat) to the dynamical strains caused by the shock due to collision. The theoretical resistance computations using the Finite Element Method, for the compression on the buffers constituted fundamental information for the choice of the verified dangerous sections. The paper presents the collision testing conducted for this purpose with the conclusions imposed by the experimental study.

A number of modern publications are devoted to the new generation freight wagon designs, which are designed with the use of advanced materials and technologies. For example, in [15] the authors reflect the results of computer modeling of the prototype of the freight wagon with the main carrier elements, executed without excessive connections. In [16] the features of the innovations proposed by the authors in the module of the chassis are presented, and their influence on the body module is also reflected, but without taking into account the variation of its performances. In the article [11] published results of work on the generation of perspective configuration profiles, which can be used in the production of various types of rolling stock. But the authors do not offer data on the possibility of creating promising profiles in tight-fitted designs. Work [17, 18] highlights the proposed methods of analyzing railway structures of the future and ways to expand their functionality, but in it, as well as in article [15], there are no data on increasing the business functionality of intangible special profiles. Also, an important role in modern car-building is played by appropriate approaches in designing, for example, the article [2, 19] presents the developed methodological apparatus for the adoption of optimal solutions. But it also does not provide exhaustive opportunities for the formation of optimal length and configuration of articulated beams. In the works [7, 8, 20-23] features and results of conducted researches on definition of constructive components of freight wagons for creation of directional stress-strain state (based on the principle of previous stresses) were presented. The works [24] reflect the results of the production of round pipes in the construction of a freight wagon and the study of the strength of the carriage structure of the gondola body during transportation on the railway ferry. Taking into account the foregoing, it can be concluded that the results of the analysis of information sources on the issue under investigation indicate lack of sufficient methodological and practical materials on testing and technical diagnostics of tank cars for the transportation of dangerous goods, especially in emergency situations.

Emergency situation (special situation) - a situation that arises in the process of transportation due to a technical malfunction (failure) of parts of the railway transport system or the emergence of extreme external factors or their combinations and which is characterized by a violation of the system capacity, or the creation of hazardous operating conditions. Examples of emergency situations include: impact of a car in the bottom of the tank of the boiler, self-lifting of the wagons, increase of pressure in the boiler of the tank above the working and other.

4. Research results

Please follow these instructions as carefully as possible so all articles within a conference have the same style to the title page. The main part of the fleet of tank cars for the carriage of particularly dangerous cargoes is made up of tank cars of model 15-1407 - for transportation of propane, 15-1408, 15-1408-01, 15-1408-02, 15-1440, 15-1597- for carriage Ammonia, 15-1409, 15-1556 - for the re-introduction of chlorine, 15-1519, 15-1780 - for the transportation of propane-butane, their technical characteristics are given in Table 1. The experience of the operation of rail transport shows that a significant part of the emergency situations on the railways is connected with tank cars carrying liquid goods of a wide range, including liquefied gas, petroleum products, concentrated acids, toxic and explosive products of chemical origin. Most often such accidents are accompanied by the arrival of the carriage on the carriage and the transfer of tanks, which may result in a breach of the integrity of the boiler (breakdown of the bottom, damage to the neck for loading of the cargo, breaking the shell in the zone of connection of the boiler with a frame, etc.) and the outflow of environmentally hazardous cargo. For most emergencies, which arise when the carriages off the rails, hitting a car on a car or an obstacle, characteristic is the self- separating of cars and a clutch hit by a nearby wagon or its long load at the bottom of the boiler (Fig. 3).

At present, as additional means of protecting the bottom of tank-wagons in emergency situations, additional metal pads on the bottom (false bottom) or end shields are used. Overhead metal elements, reinforcing the lower part of the bottom and repeating its shape, are used on tanks of models 15-1619, 15-1408. These protective elements have low energy intensity, as they adhere tightly to the bottom of the tank. Taking into account the operational experience and the results of the inspection of the technical condition, it is possible to determine the probability of failure of the elements of the frame of the tank car, P_i ($i = 1, 2, 3 \dots N$, where N is the number of elements) by the formula:

$$P_i = \frac{\sum_{1}^{k} \frac{R_H}{R}}{k}, \tag{1}$$

where k - the number of surveyed wagons;

IOP Conf. Series: Materials Science and Engineering **659** (2019) 012055 doi:10.1088/1757-899X/659/1/012055

R_H - number of defective elements of the same type in the wagon.

For tank cars carrying ammonia, include: the zone of welding of the hatch, the shaped paws of a boiler, the zone of boiler obstruction on wooden bars, centre sill and a bolster beam of a frame.

Tests were carried out on the four-axle carriage of the model 15-1408 (Fig. 4) for transportation of ammonia, with the expired service life (20 years) and a prototype tank model 15-1619 (Fig. 5) with a boiler diameter of 3200 mm.

Figure 1. Tank wagon model 15-1407.

Figure 2. Tank wagon model 15-1556.

The platform consists of a frame of welded construction, which is mounted on two-axle bogies model 18-100. The frame is a welded system of beams of a box-shaped section: centre sill and two bolster. A boiler of the welded construction consisting of a cylinder and two elliptic bottoms.

Before the start of the tests, the technical diagnosis of the prototype, the gluing of the strain gauges on the frame of the tank wagon and the boiler, measurements of the thickness of the boiler and frame sheets, and the installation of measuring equipment were carried out.

The inspection of the technical condition included a visual inspection of the tank wagon, measurements of the thickness of the metal at the control points of the boiler and the frame of the wagon, conducting of the magnetopowder, ultrasonic and acoustic emission control of the bearing metal constructions.

The control tests included static vertical loading tests, low-cycle load tests with boiler pressure, typical and resource shock tests, and possible emergency situations testing.

Table 1. The main technical characteristics of the tank wagons.

Model	15-1407	15-1408	15-1408-02	15-1440	15-1597	15-1409	15-1556	15-1519
Function	For Propane	For Ammonia	For Ammonia	For Ammonia	For Ammonia	For Chlorine	For Chlorine	For Propane
Body material	09G2C, 09G2D, 09G2, 09G2CD-12	09G2C, 09G2D, 09G2, 09G2CD-12	09G2C, 09G2D, 09G2, 09G2CD-12	09G2C, 09G2D, 09G2, 09G2CD-12	09G2C, 09G2D, 09G2, 09G2CD-12	09G2C, 09G2D, 09G2, 09G2CD-12	09G2C, 09G2D, 09G2, 09G2CD-12	09G2C, 09G2D, 09G2, 09G2CD-12
Tare, t,	34.6							
min	36.7	32,3	32,4	32,0	35,7	28,9	27,2	34,8
max		36,7	32,4	33,5	38,8	30,7	30,0	38,8
Cargo capacity, t	22.9	30,7	31,2	30,7	43,0	47,6	57,5	46,0
Volume, m^3	54.0	54,0	54,8	54,0	76,0	38,7	46,0	75,7
Diameter of boiler inside, mm	2600	2600	2600	2600	3000	2200	2400	3000
Production year	1962	1963	1995	1978	1989	1964	1974	1990
Service life	40	20	40	20	40	24	24	40

Figure 3. Damage to the boiler of the tank.

Vertical static tests were performed by filling the boiler of the tank wagon with water to full filling with the subsequent drain, with the registration of the experimental data.

The test for a low cycle load of the boiler with excess pressure was performed by filling the boiler with water to full filling, followed by the supply of water to the boiler by the pumping station, to create an excess pressure of 2 MPa. After every 75 cycles of loading by working pressure, a load of test pressure of 3 MPa was performed.

The impact tests were carried out with the speed of rolling a percussive wagon (102 t. mass) on the experimental wagon tank from 1 to 12 km / h, while the wagon tank was in a stalled condition, and was supported by the retarded 4 wagons with a total mass of about 300 t. Longitudinal forces that acted on the tank-car through an auto-coupling device up to 3.0 MN and with several beats up to 4.2 MN. In the case of shock tests deformations were recorded in the design elements of the experimental wagon-tank in the most stressed places obtained during static tests.

Figure 4. Wagon tank 15-1408 for ammonia.

Figure 5. Wagon tank model 15-1619.

5. Simulation of emergency situations

Before and after the end of the test the boiler's tightness was checked by creating an excess pressure of 0.5 to 3 MPa. Pressure lasted for 5 minutes.

The boiler was tested for the following excessive loads:

- creation of excess pressure until the boiler is destroyed or its tightness is lost;
- a one-time stroke of a wagon with a weight of 90 t in an automatic coupler of a tank wagon at a speed of 22 km/h (fig. 6).
- strikes by the head of the automatic coupler of the shock wagon weighing 102 t in the bottom of the tank at a speed of 10,2 km / h (Fig. 7, 8).

Figure 6. Tank wagon after impact.

Estimation of the tensile condition and durability of the design of the boiler by the results of static tests was performed according to the calculated modes by comparing the total stresses from the action of the test loads with allowable stresses.

Mode I:

$$\sigma_{ver} + \sigma_{pI} \leq [\sigma]_I , \tag{2}$$

Mode III:

$$\sigma_{ver} + \sigma_{pIII} + \sigma_{dyn} + \sigma_{lat} \leq [\sigma]_{III} , \tag{3}$$

Test mode:

$$\sigma_{ver.test} + \sigma_{p.test} \leq [\sigma]_{test} , \tag{4}$$

where: σ_{ver} – stress from the vertical load, gross; σ_{pI}, σ_{pIII} – stress from the action of the calculated pressure, equal to the sum of working pressure and pressure from the hydraulic shock, respectively, for the I and III modes; σ_{dyn} – stress from the action of a dynamic additive; σ_{lat} – stress from the effect of lateral force; $\sigma_{ver.test}$ – stress from the vertical gross load when the boiler is fully filled with water in the tank; $\sigma_{p.test}$ – stress from the test pressure of 0.6 MPa.

Figure 7. Experimental sample after impact on the boiler by automatic coupler.

Figure 8. Test sample after emergency tests.

In the course of resource tests an analysis of the stressed state at the control points of the boiler of the tank-car was conducted. As can be seen in Fig. 9 the stresses at the control points did not exceed the permissible.

The criterion for assessing the strength of the tank against the effects of unusual loads is the preservation of the load and tightness of the boiler after the test, by controlling the pressure in the boiler.

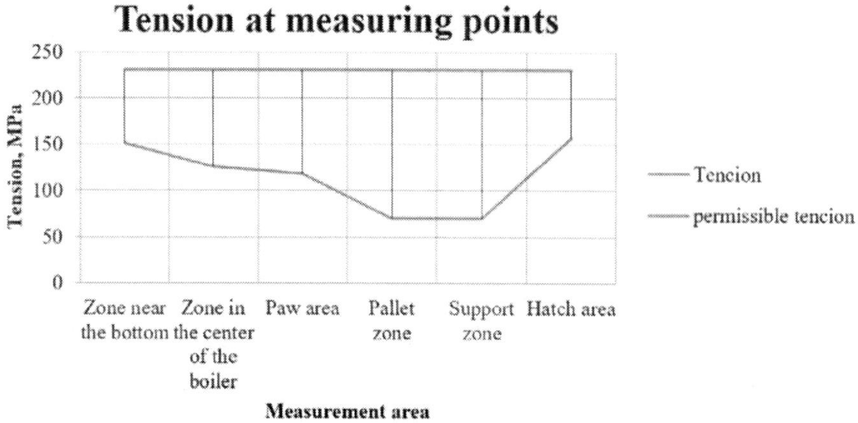

Figure 9. Test sample after emergency tests.

6. Conclusion

On the basis of technical diagnostics and control tests of the tank-wagon it was established that the tank-wagon is technically a typical representative of the operational fleet of Ukrainian railways, which have completed their intended service life. The tensile condition of the boiler of the tank-wagon from the action of normative static and shock loads did not exceed the permissible 236 MPa. When testing on a low-cycle load on a boiler with pressure from 0 to 2 MPa, 225 cycles were carried out, which corresponds to 5 years of operation. In tests with simulation of emergency situations: the impact on the bottom of the tank, the creation of a limit pressure in the tank, up to 6 MPa, an impact on the auto-coupling device of a tank carriage at a speed of 22 km/h, damage and depressurization of the boiler tank-tank was not detected.

References

[1] Boyko A 2012 *Influence of barrel damages on life time of tank wagon*, Proceedings of the International Conference of DAAAM Baltic "Industrial Engineering", 21-26.

[2] Šťastniak P, Smetanka L nad Moravčík M 2018 Structural Analysis of a Main Construction Assemblies of the New Wagon Prototype Type Zans, *Manufacturing technology* **18**(3) 510-517

[3] Gerlici J, Nozhenko O, Cherniak G, Gorbunov M, Domin R and Lack T 2018 The development of diagnostics methodological principles of the railway rolling stock on the basis of the analysis of dynamic vibration processes of the rail, *MATEC Web of Conferences* **157** 03007

[4] Blatnický M, Dižo J and Blatnická M 2017 Side plate strength analysis of the mechanism for vehicle axle scale calibration, *Manufacturing Technology* **17**(2) 147-151

[5] Voropai V 2017 Development of a design-experimental methodology for the prediction of reliable exploitation of freight railway cars, *Transport Problems* **12**(3) 21-34

[6] Locovei C, Răduţă A, Nicoară M and Cucuruz L R 2010 *Analysis of fatigue fracture of tank wagon railway axles*, 27th Danubia-Adria Symposium on Advances in Experimental Mechanics, DAS 2010, 219-223

[7] EN 12663–2. 2010 *Railway applications - structural requirements of railway vehicle bodies.* Part 2: Freight wagons 54 p

[8] Kelrykh M and Fomin O 2014 Perspective directions of planning of carrying systems of gondola, *Scientific and technical journal «Metallurgical and Mining Industry»* **6** 64-67

[9] Fomin O V 2015 Increase of the freight wagons ideality degree and prognostication of their evolution stages [In Ukranian] *Scientific Bulletin of National Mining University* **3** 68-76

[10] European Standard 14363 2005. *Railway applications – Testing for the acceptance of running characteristics of railway vehicles. Testing of running behaviour and stationary tests*, 113 p

[11] Myamlin S, Lunys O, Neduzha L and Kyryl'chuk O 2017 *Mathematical modeling of dynamic loading of cassette bearings for freight cars transport means*, Proc. of 21st Intern. Scientific Conference, 973-976

[12] Radu S M and Tănăsoiu B 2013 Researches in the field of the bearing structure of tank wagon, *WSEAS Transactions on Applied and Theoretical Mechanics* **8**(3) 222-231

[13] MăNescu T S, Zaharia N L, Avram D S, ȚIgăNașu N M and Stroia M D 2011 *Nonlinear finite element analysis used at tank wagons*, Proceedings of the 4th WSEAS International Conference on Finite Differences - Finite Elements - Finite Volumes - Boundary Elements, F-and-B '11, 110-113

[14] Copaci I, Tănăsoiu A and Tănăsoiu B 2010 *On the resistance of tank wagons to the shock caused by collision*, Proceedings of the 3rd WSEAS International Conference on Finite Differences - Finite Elements - Finite Volumes - Boundary Elements, F-and-B '10, 23-27

[15] Fomin O 2014 Modern requirements to carrying systems of railway general-purpose gondola cars, *Scientific and technical journal «Metallurgical and Mining Industry»* **5** 31-43

[16] Myamlin S, Neduzha L, Ten O and Shvets A 2010 *Spatial vibration of cargo cars in computer modeling with the account of their inertia properties*, Mechanics, Proc. of the 15th Internet. Conference, 325-328

[17] Fomin O 2015 Improvement of upper bundling of side wall of gondola cars of 12-9745 model, *Scientific and technical journal «Metallurgical and Mining Industry»* **1** 45-48

[18] Lovska A O 2015 Computer simulation of wagon body bearing structure dynamics during transportation by train ferry, *Eastern-European Journal of Enterprise Technologies* **3** 7(75) 9-14

[19] UIC Code 518 2009. *Testing and approval of railway vehicles from the point of view of their dynamic behaviour.* Safety – Track fatigue – Ride quality. International Union of Railways. September, 119 p

[20] Lunys O, Gorbunov M, Steišūnas S and Nozhenko O 2018 *Investigation on the reliability of axle-box failure detectors in Lithuanian railways,* Transport Means - Proceedings of the International Conference, 406-411

[21] Gevorkyan E, Lavrynenko S, Rucki M, Siemiatkowski Z and Kislitsa M 2017 Ceramic cutting tools out of nanostructured refractory compounds, *International Journal of Refractory Metals & Hard Materials* **68** 142 – 144

[22] Kondratiev A 2019 Weight-based optimization of sandwich shelled composite structures with a honeycomb filler, *Eastern-European Journal of Enterprise Technologies* **1**/**1**(97) 24-33

[23] Bychkov A S and Kondratiev A V 2019 Criterion-based assessment of performance improvement for aircraft structural parts with thermal spray coatings, *Journal of Superhard Materials* **41**(1) 53–59

[24] Lovska A 2015 Peculiarities of computer modeling of strength of body bearing construction of gondola car during transportation by ferry-bridge, *Metallurgical and mining industry* **1** 49-54

Determination of the strength of the containers fittings of a flat wagon loaded with containers during shunting

O Fomin[1], J Gerlici[2], A Lovska[3], K Kravchenko, Yu Fomina[2] and T Lack[2]

[1] State University of Infrastructure and Technology, Faculty of Infrastructure and Railway Rolling Stock, St. Kyrylivska, 9, 040071, Kyiv, Ukraine
[2] University of Zilina, Faculty of Mechanical Engineering, St. Univerzitná 8215/1, 010 26, Zilina, Slovak Republic
[3] Ukrainian State University of Railway Transport, Department of Wagons, 61050, Feuerbach Square 7, Kharkov, Ukraine

E-mail: kkatherina@ukr.net

Abstract. The article presents the results of the dynamic loading and strength study of the fittings for containers of a flat wagon loaded with containers during shunting. It has been found that with impact interaction of corner castings with fittings, there are increased stresses, both in structural components of a flat wagon and in containers. To provide the strength of the fittings of a flat wagon and corner castings of containers, it has been proposed to install elastic-viscous linkage between them. This solution is implemented by installing elements having appropriate properties in the corner castings. Mathematical modeling has been performed to determine the dynamic loading of the flat wagon and containers. The obtained values of dynamic loads were taken into account when determining of the strength indexes. It has been established that, the maximum equivalent stresses in the fittings and corner castings are within acceptable parameters taking into account the proposed solutions. The studies will contribute to provide the strength of the structural components of flat cars and containers during shunting, to reduce the cost of their unscheduled repairs, as well as to create recommendations for design of rolling stock with improved technical and economic indexes.

1. Introduction

The creation of foreign economic relations between the Eurasian states determined putting into operation of intermodal transport systems. At present container transportation is one of the most common transportations, owing to containers intermodality as vehicles.

Containers are transported by flat wagons equipped with fittings for container with respect by main tracks. The study of regulatory documents standardizing the containers dynamic loading in operation allows for the conclusion that one of the most loaded modes is a shunting collision of a flat wagon with containers placed on it. In this case, both the corner castings damage and the flat wagon container fittings damage may occur (fig. 1).

To improve the efficiency of container transportation and reduce the cost of unscheduled flat wagons and containers repair types, at the current stage of the transport industry development it is important to introduce new alternative solutions for the design features of vehicles, their interaction patterns, and so on [1-3].

2. Analysis of recent research

Requirements for the supporting structures of modern rolling stock, in particular freight wagons, are described in a paper [4]. These requirements are proposed to be applied at the stage of new wagons designs manufacturing, as well as those ones that are untitled to modernization.

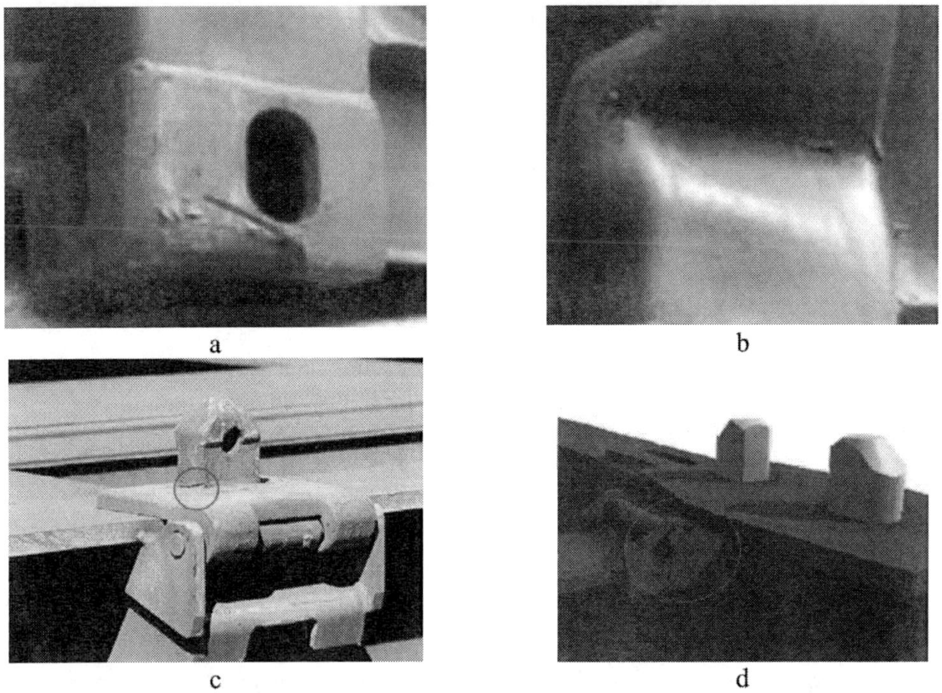

Figure 1. Damage to corner castings and container fittings: a) corner casting misalignment; b) container post break; c) cracks in the container fitting; d) damage to the fastening element of the container fitting to the frame.

However, attention to the issue of the wagons dynamic loading studying during shunting is not paid.

Features of pressure sintering of materials with alternating current are given in a paper [5]. It is found that the materials after this procedure have increased hardness and some elasticity.

Prospects for the use of this nanomaterial in wagons designs to reduce their dynamic loading in operation are not described in a paper.

Determination of the dynamic loading of flat wagons loaded with containers during shunting is carried out in an article [6]. The results obtained by mathematical modeling are proved by computer simulation. It has been found that there are increased values of dynamic loads of supporting structures when moving the casting corners relative to the container fittings of flat wagons.

It is important to note that measures to reduce the dynamic loading of flat wagons and containers during a shunting collision have been not proposed in the paper.

Features of mathematical modeling of spatial oscillations of the subframe – track system are described in an article [7]. The calculation is carried out using the finite element method in the Ansys software. Measures to improve the rolling stock design to reduce the dynamic load in operation are not given in the paper.

The dynamic loading study of the wagon body bearing structure during its transportation by a railway ferry is carried out in papers [8, 9]. The results of mathematical modeling are confirmed by computer simulation.

The issues of wagons supporting structures improving to reduce the dynamic loading during shunting are not considered in the articles.

Possible versions of improving the rolling stock technical and economic indexes are considered in a paper [10]. In this case, the paper does not describe possible measures to reduce wagons dynamic loading.

The analysis of innovative rolling stock designs constructed by "Uralvagonzavod" (Russia) for 1520 mm gauge is described in a paper [11]. Issues for the rolling stock and its components modernization to improve operating efficiency are considered. Measures to reduce the dynamic loading of the wagons bearing structures during a shunting collision are not considered.

The analysis of the BCNHL freight wagons designs is carried out in a paper [12]. Possible ways of improving the technical and economic indicators of wagons to improve the efficiency of their operation are determined.

The results of the freight wagon structural-elemental analysis are described in a paper [13]. The calculation has been carried out using the finite element method. Studies have been conducted on the example of an open wagon type "BOXN25", which is operated on the Indian railway.

Determination of wagons dynamic loading and measures of its reduce during a shunting collision is not carried out in these works.

3. The purpose and tasks of the study

The purpose of the article is to determine the strength of the container fittings of a flat wagon loaded with containers during shunting. To achieve this goal, the following tasks are defined:

1 To investigate the flat wagon dynamic loading taking into account the improved interaction scheme of container fittings with corner castings by means of mathematical modeling;

2 To investigate the flat wagon dynamic loading taking into account the improved interaction scheme of container fittings with corner castings by using computer simulation;

3 To calculate the strength of the flat wagon, taking into account the improved interaction scheme of container fittings with corner castings.

4. The main part of the study

To reduce the dynamic loading of the flat wagon loaded with containers during a shunting collision, in the case when there is a shock interaction of fittings with corner castings, it is proposed to place elastic, viscous or elastic-viscous couplings in the corner castings. Modeling of the container dynamic loading, taking into account the presence of elastic couplings in the corner castings, made it possible to conclude that this solution does not fully compensate for the container dynamic loading.

With a viscous and elastic-viscous coupling, the maximum accelerations acting on the container during flat wagon shunting do not exceed the permissible ones.

To determine the flat wagon dynamic loading during a shunting collision, taking into account the presence of a viscous linkage between the fittings and corner castings, a mathematical model (1) has been compiled.

$$\begin{cases} M_{FW}^{full} \cdot \ddot{q}_1 = P_{imp} - \sum_{i=1}^{n} \left(F_{FR} \cdot sign(\dot{q}_1 - \dot{q}_2) + \beta_f (\dot{q}_1 - \dot{q}_2) \right), \\ M_c \cdot \ddot{q}_2 = \left(F_{FR} \cdot sign(\dot{q}_1 - \dot{q}_2) + \beta_f (\dot{q}_1 - \dot{q}_2) \right), \end{cases} \tag{1}$$

where M_{FW}^{full} – is the flat wagon gross mass; P_{imp} – is the value of the longitudinal force acting on the automatic coupling; n – is the number of containers placed on the flat wagon; F_{FR} – is a friction force between container fittings and corner castings; M_c – is a container weight; β_f – is a coefficient of viscous resistance in the corner castings; q_1, q_2 – are coordinates that determine displacements, respectively, of the flat wagon and the container relative to the longitudinal axis.

The flat wagon model 13-401M has been chosen as the prototype wagon. Studies have been conducted in relation to the container size 1CC.

The forces scheme acting between the flat wagon and the container during a shunting collision taking into account improvement measures is shown in fig. 2.

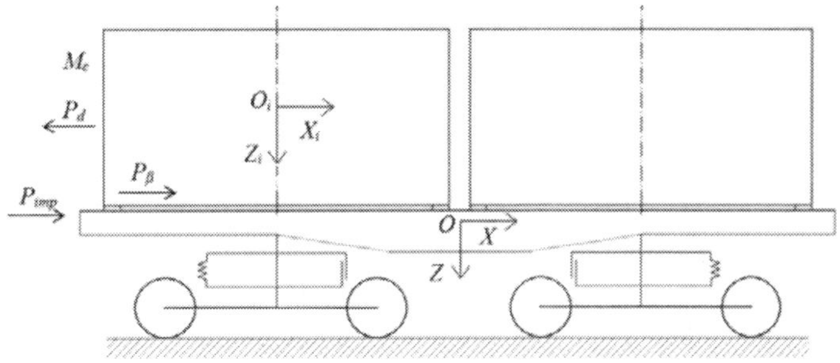

Figure 2. Scheme of the longitudinal for cesaction on the flat wagon with containers placed on it
P_{imp} – impact force acting on the flat wagon; P_d– dynamic force acting on the container; P_β – viscous resistance force to the container movement.

When compiling the model, the dry friction force arising when moving the corner castings relative to the horizontal surfaces of the container fittings [14] is taken into account. The study was conducted using plane coordinates.

When compiling the model, it was assumed that the container was loaded with a conditional load using the maximum allowable load capacity.

The solution of the mathematical model (1) has been implemented in the MathCad software [15, 16]. At the same time, it was reduced to the normal form of Cauchy $\dot{y}(t) = Q(t, y)$ [17, 18]. Where $q_1 = y_1; q_2 = y_2; \dot{q}_1 = y_3; \dot{q}_2 = y_4$.

$$Q(t,y) = \left| \begin{array}{c} y_3 \\ y_4 \\ \dfrac{P_{imp} - \sum_{i=1}^{n} \left(F_{FR} \cdot sign(\dot{q}_1 - \dot{q}_2) + \beta_f (\dot{q}_1 - \dot{q}_2) \right)}{M_{FW}^{full}} \\ \dfrac{\left(F_{FR} \cdot sign(\dot{q}_1 - \dot{q}_2) + \beta_f (\dot{q}_1 - \dot{q}_2) \right)}{M_c} \end{array} \right| \tag{2}$$

$Z = rkfixed\,(Y0, tn, tk, n', Q)$,

where *Y0* – is a vector containing the initial conditions, *tn, tk* – are the quantities that define the initial and final integration variables, *n'* – is a fixed number of steps, *Q* – is a symbol vector.

In this case, the viscous resistance to the movement of the container was taken in the range of 10 – 50 kN·s/m. On the basis of the calculations, accelerations, acting on the flat wagon taking into account the viscous coupling between corner castings and container fittings during a shunting collision, have been obtained (fig. 3). This acceleration value was about 40 m/s^2 (4g), that is, it does not exceed the standard value [19, 20].

When elastic-viscous interaction of corner castings with container fittings are used, the mathematical model for determining of the flat wagon dynamic loading will be:

$$\begin{cases} M_{FW}^{full} \cdot \ddot{q}_1 = P_{imp} - \sum_{i=1}^{n} \left(F_{FR} \cdot sign(\dot{q}_1 - \dot{q}_2) + C_{\phi}(q_1 - q_2) + \beta_f(\dot{q}_1 - \dot{q}_2) \right), \\ M_c \cdot \ddot{q}_2 = \left(F_{FR} \cdot sign(\dot{q}_1 - \dot{q}_2) + C_c \cdot (q_1 - q_2) + \beta_f(\dot{q}_1 - \dot{q}_2) \right). \end{cases} \quad (3)$$

where C_c – is the stiffness of the elastic elements in the corner castings.

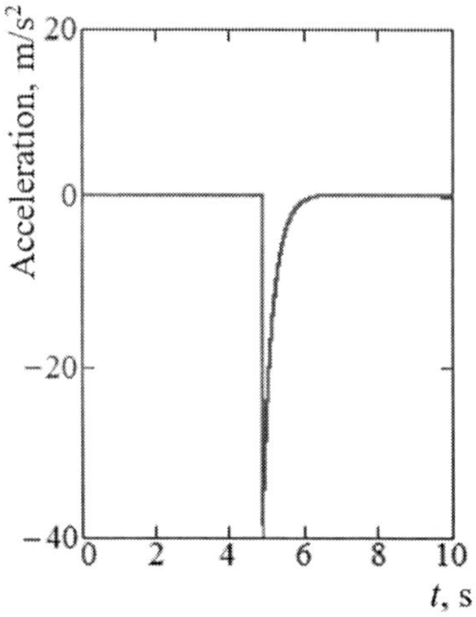

Figure 3. Acceleration acting on a flat wagon with viscous interaction of corner castings with fittings.

Then,

$$Q(t, y) = \left| \begin{array}{c} y_3 \\ y_4 \\ \dfrac{P_{imp} - \sum_{i=1}^{n} \left(F_{FR} \cdot sign(\dot{q}_1 - \dot{q}_2) + C_c(q_1 - q_2) + \beta_f(\dot{q}_1 - \dot{q}_2) \right)}{M_{FW}^{full}} \\ \dfrac{\left(F_{FR} \cdot sign(\dot{q}_1 - \dot{q}_2) + C_c \cdot (q_1 - q_2) + \beta_f(\dot{q}_1 - \dot{q}_2) \right)}{M_c} \end{array} \right| \quad (4)$$

$Z = rkfixed\,(Y0, tn, tk, n', Q)$,

The calculations results showed that when the elastic element stiffness is 20 kN/m and the viscous resistance coefficient is 30 kN·s/m, the accelerations acting on the flat wagon are about 40 m/s^2 (4g) (fig. 4), and are in limits of allowed values [19, 20].

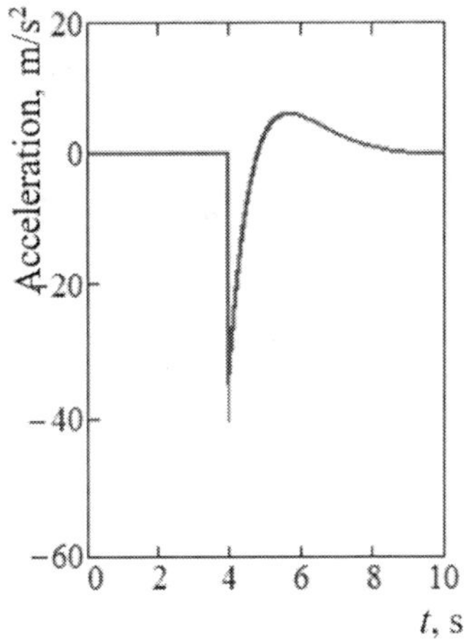

Figure 4. Accelerations acting on the flat wagon with the elastic-viscous interaction of corner castings with container fittings.

To study the flat wagon dynamic loading with the elastic-viscous interaction of corner castings with container fittings, computer simulation has been carried out using the finite element method, implemented in the CosmosWorks software [21, 22].

Spatial isoparametric tetrahedras were used as finite elements, the optimal elements number is determined by using the graphic analytical method. The number of mesh nodes is 285189 and the number of elements is 853256. The maximum element size is 100 mm, and the minimum one is 20 mm. The minimum number of elements in a circle is 9, the ratio of increasing the elements size in the mesh is 1.7. The maximum aspect ratio is 306.67, the percentage of elements with an aspect ratio of less than 3 is 29.6 and the percentage of elements with an aspect ratio of more than 10 is 21.2.

The design scheme of a flat wagon with containers during a shunting collision is shown in fig. 5. It is taken into account that a horizontal load P_h is acting on the corner castings, caused by the impact force P_{imp} action on the vertical surface of the rear draft lugs, as well as vertical load P_v is acting in the bearing areas of the corner castings on the container fitting. At the same time, the load effect on container walls is not taken into account. The container was fixed in the areas of its bearing on the flat wagon. It is taken into account that when a horizontal load P_h is applied to corner castings, it is moved relative to the initial position by 30 mm.

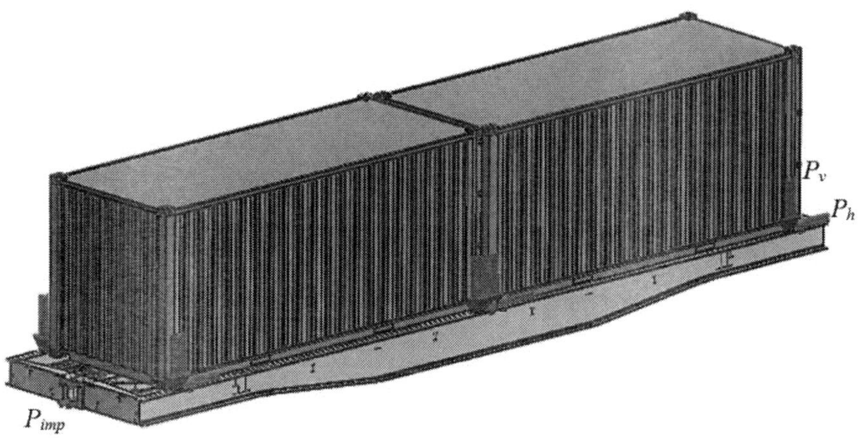

Figure 5. The design scheme of the flat wagon with containers during a shunting collision.

When modeling the container dynamic loading, taking into account the presence of viscous coupling in corner castings, the total viscous resistance to displacement of one container was taken to be 50 kN·s/m. To simulate the viscous coupling in the CosmosWorks software, the spring-damper option has been used. In this case, the stiffness value was set close to zero.

The material of the construction is steel 09G2S with the corresponding values of strength and yield point [19, 20]. The calculation results are shown in fig. 6.

The maximum accelerations acting on the container, taking into account the viscous interaction of corner castings with container fittings, arise in the end walls on the cantilever parts of the flat wagon frame and are about 20 m/s^2.

In the middle part of the container, the accelerations amounted to almost 15 m/s^2. The smallest acceleration occurs in the end walls of the container over the center of the flat wagon frame and is about 8 m/s^2.

The maximum accelerations acting on the flat wagon supporting structure occur in the cantilever parts and are about 38 m/s^2, and are about 30 m/s^2 over the bolster cross-sections of the frame. The accelerations are 21.4 m/s^2 in the middle part of the center girder. The smallest accelerations occur in the middle parts of the main longitudinal beams of the flat wagon frame and are 7.2 m/s^2.

When modeling the flat wagon with containers dynamic loading, taking into account the presence of elastic-viscous coupling in the corner castings, the stiffness of the elastic element is assumed to be 20 kN/m and the viscous resistance coefficient is assumed to be 30 kN·s/m.

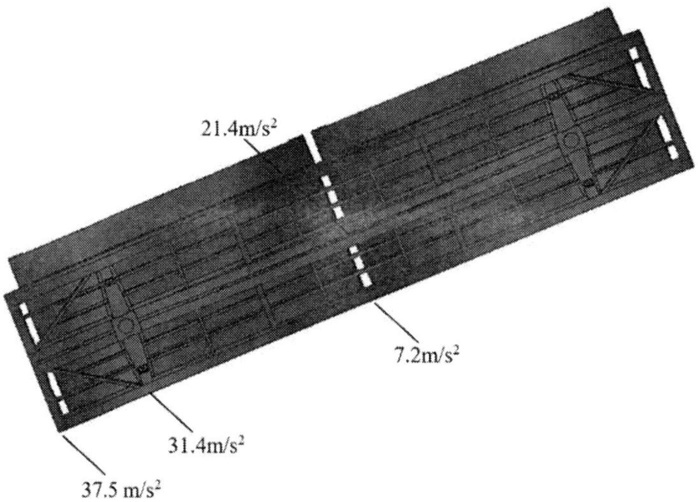

Figure 6. The design scheme of the flat wagon with containers during a shunting collision.

The calculation results made it possible to conclude that the maximum accelerations acting on the supporting structure of the container located on the flat wagon were 19.7 m/s^2 and 38.4 m/s^2, respectively.

Studies made it possible to conclude that the maximum accelerations acting on the container, taking into account the elastic-viscous coupling between the corner castings and the container fittings, do not exceed the allowable ones [19, 20].

In order to verify the developed models, Fisher's criterion has been applied [23]. The calculations showed that the adequacy hypothesis is not rejected.

The obtained accelerations are taken into account in the strength analysis of flat wagon container fittings (fig. 7). It is taken into account that the impact force P_{imp} acts on the flat wagon bearing structure, and the vertical load P_v from containers and the horizontal load P_h caused by the displacement of the corner castings relative to the container fittings act on the container fittings (fig. 8). The strength analysis has been carried out by using the finite element method.

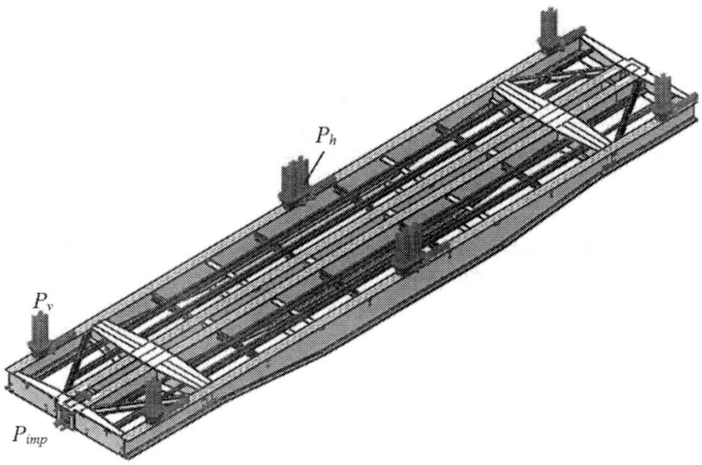

Figure 7. The design scheme of the flat wagon supporting structure.

Spatial isoparametric tetrahedras have been used as finite elements. The number of mesh nodes is 311601 and the number of elements is 932166. The maximum element size is 80 mm and the minimum one is 16 mm. The minimum number of elements in a circle is 9, the ratio of the increase in the size of mesh elements is 1.7. The maximum aspect ratio is 470.8, the percentage of elements with an aspect ratio of less than 3 is 41.5, and the percentage of elements with an aspect ratio of more than 10 is 8.17.

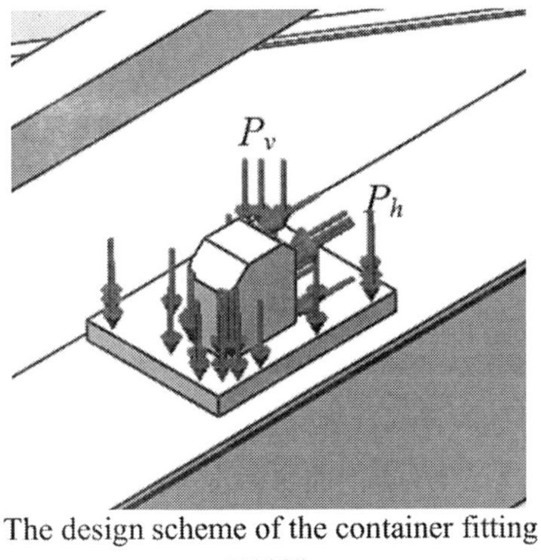

Figure 8. The design scheme of the container fitting of the flat wagon.

The calculation results are shown in fig. 9-13. It has been found that with the viscous interaction of corner castings with fittings during a shunting collision, the maximum equivalent stresses are about 270 MPa (fig. 9, 10) and are concentrated in the zone of interaction between the center girder and the bolster beam. Maximum displacements occur in the middle parts of the main longitudinal beams of the flat wagon frame and are 12.1 mm (fig. 11).

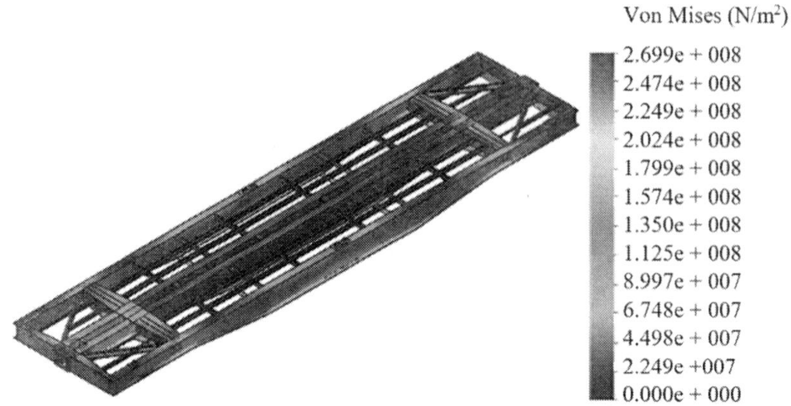

Figure 9. The stress state of the flat wagon bearing structure with viscous interaction of corner castings with container fittings (side view).

Using elastic-viscous interaction of corner castings with container fittings during flat wagon shunting, the maximum equivalent stresses are about 260 MPa (fig. 12, 13) and are concentrated in the zone of interaction between the center girder and the bolster.

The maximum displacements also occur in the middle parts of the main longitudinal beams of the flat wagon frame and are about 11,0 mm (fig. 14).

The proposed measures to improve the interaction scheme of the flat wagon with containers allow to reduce the maximum equivalent stresses acting in containers fittings almost three times (fig. 15), and in corner castings almost seven times (fig. 16).

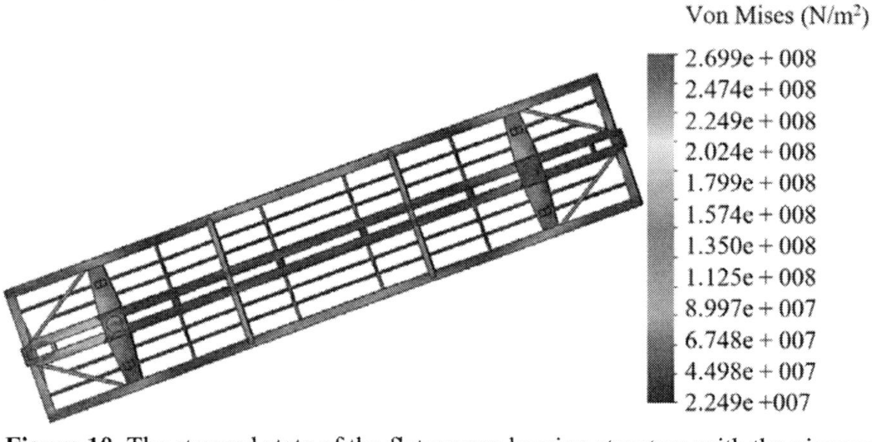

Figure 10. The stressed state of the flat wagon bearing structure with the viscous interaction of corner castings with container fittings (bottom view).

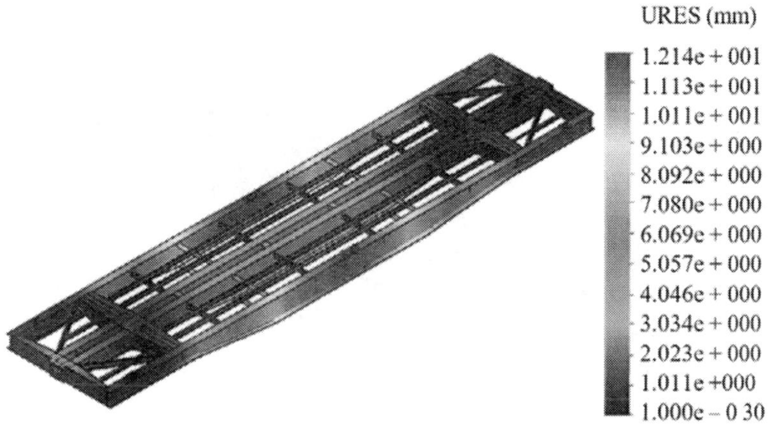

Figure 11. Moving in the nodes of the flat wagon bearing structure with the viscous interaction of corner castings with container fittings.

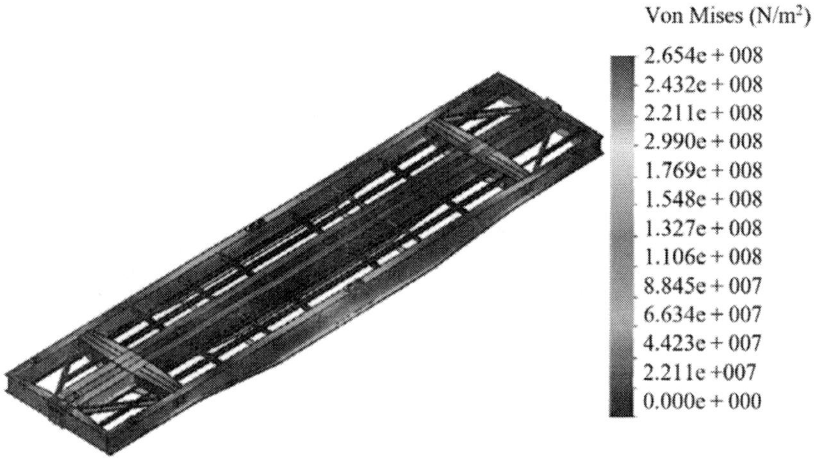

Figure 12. The stressed state of the flat wagon bearing structure with the elastic-viscous interaction of corner castings with container fittings (side view).

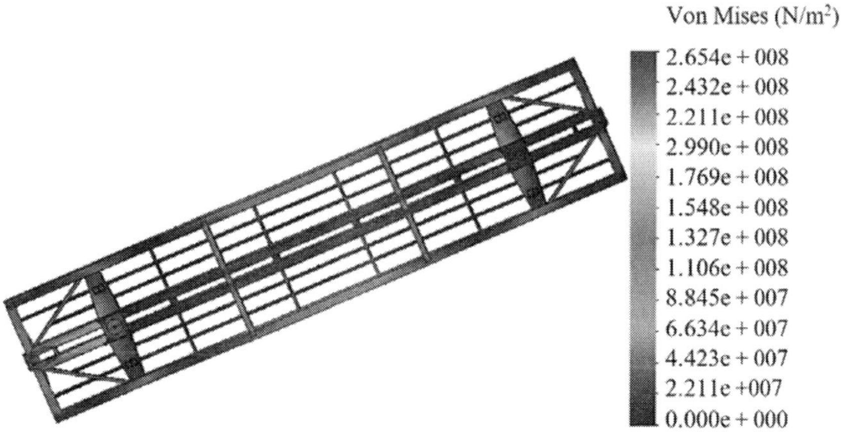

Figure 13. The stressed state of the flat wagon bearing structure with the elastic-viscous interaction of corner castings with container fittings (bottom view).

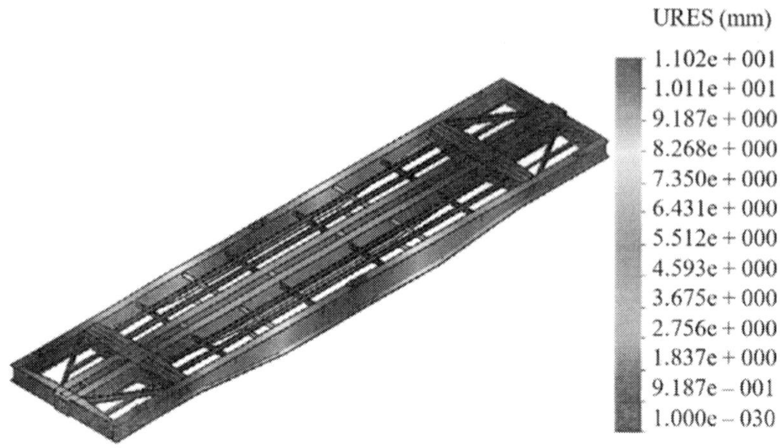

Figure 14. Moving in the nodes of the flat wagon bearing structure with the elastic-viscous interaction of corner castings with container fittings.

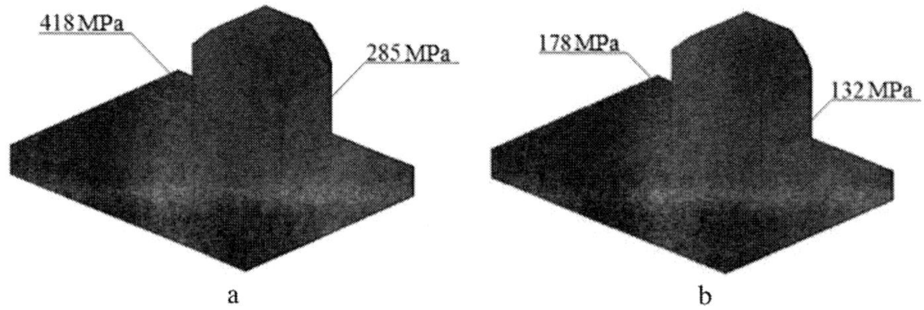

Figure 15. The stressed state of the container fittings of the flat wagon a) a typical scheme of interaction; b) an improved scheme of interaction.

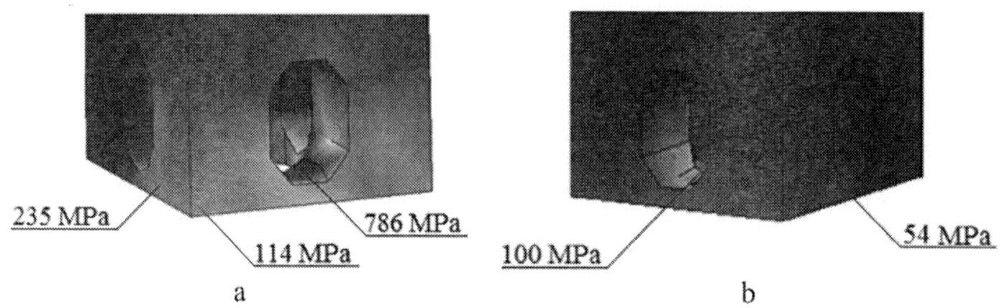

Figure 16. The stressed state of corner castings a) a typical scheme of interaction; b) an improved scheme of interaction.

5. Conclusions

Based on the research we can draw the following conclusions:

1 The dynamic loading of the flat wagon has been investigated taking into account the improved scheme of interaction of container fittings with corner castings using mathematical modeling. Mathematical models have been designed taking into account the presence of viscous and elastic-viscous couplings between fittings and corner castings. It has been established that the maximum

accelerations acting on the flat wagon are about 40 m/s^2 (4g), that is, they are within the allowable range;

2 The dynamic loading of the flat wagon has been investigated taking into account the improved scheme of interaction of container fittings with corner castings by using computer simulation. The finite element method, implemented in the CosmosWorks software, has been used in a calculation. The dislocation areas of accelerations acting on the flat wagon supporting structure with containers placed on it are determined. Verification of the designed models is made by the Fisher method;

3 The flat wagon strength has been calculated taking into account the improved scheme of interaction between container fittings and corner castings. It has been established that in case of viscous interaction of container fittings with corner castings during flat wagon shunting, the maximum equivalent stresses are about 270 MPa and are concentrated in the zone of interaction between the center girder and the bolster beam. Maximum displacements occur in the middle parts of the main longitudinal beams of the flat wagon frame and are 12.1 mm.

When using elastic-viscous interaction of fittings with corner castings during flat wagon shunting, the maximum equivalent stresses are about 260 MPa, the maximum displacements are about 11.0 mm. The research will contribute to improving the efficiency of container transportation, reducing the cost of unscheduled flat wagons and containers repair types, as well as creating recommendations for designing of modern rolling stock.

References

[1] Dižo J, Harušinec J and Blatnický M 2018 Computation of modal properties of two types of freight wagon bogie frames using the finite element method, *Manufacturing Technology* **18**(2) 208-214

[2] Šťastniak P and Moravčík M 2017 Freight bogie prototype properties analysis by means of simulation computations, *Manufacturing Technology* **17**(3) 381-388

[3] Hauser V, Nozhenko O S, Kravchenko K O, Loulová M, Gerlici J and Lack T Car body and bogie connection modification for track curves passability improvement, *MATEC Web of Conferences* **157** Article number 03009

[4] Fomin A 2014 *Modern requirements to carrying systems of railway general-purpose gondola cars*, Scientific and technical journal «Metallurgical and Mining Industry» **5** 31-43

[5] Gevorkyan E, Lavrynenko S, Rucki M, Siemiatkowski Z and Kislitsa M 2017 Ceramic cutting tools out of nanostructured refractory compounds, *International Journal of Refractory Metals & Hard Materials* **68** 142-144

[6] Lovska A and Rybin A 2016 The study of dynamic load on a wagon-platform at a shunting collision, *Eastern-European Journal of Enterprise Technologies* **3** 4-8

[7] Myamlin S, Lunys O, Neduzha L and Kyryl'chuk O 2017 *Mathematical Modeling of Dynamic Loading of Cassette Bearings for Freight Cars*, Proc. of 21st Intern.Scientific Conf. «Transport Means. 2017» 973-976

[8] Fomin O, Lovska A, Kulbovskyi I, Holub H, Kozarchuk I and Kharuta V 2019 Determining the dynamic loading on a semi-wagon when fixing it with a viscous coupling to a ferry deck, *Eastern-European Journal of Enterprise Technologies* **2/7**(98) 6-12

[9] Fomin O, Lovska A, Masliyev V, Tsymbaliuk A and Burlutski O 2019 Determining strength indicators for the bearing structure of a covered wagon's body made from round pipes when transported by a railroad ferry, *Eastern-European Journal of Enterprise Technologies* **1/7**(97) 33-40

[10] Boronenko Yu P 2013 Strategicheskie zadachi vagonostroitelej v razvitii tyazhelovesnogo dvizheniya, *Transport Rossijskoj Federaczii* **5**(48) 68–73

[11] Innovative rolling stock of Uralvagonzavod for railways in the area of 1520 mm 2010 *Transport of the Russian Federation* **3**(28) 20–21

[12] Shukla C P and Bharti P K 2015 Study and Analysis of Doors of BCNHL Wagons, *International Journal of Engineering Research & Technology (IJERT)* **4**(04) 1195-1200

IOP Conf. Series: Materials Science and Engineering **659** (2019) 012056 doi:10.1088/1757-899X/659/1/012056

[13] Harak S S, Sharma S C and Harsha S P 2014 Structural Dynamic Analysis of Freight Railway Wagon Using Finite Element Method, *Procedia Materials Science* **6** 1891-1898
[14] Bohomaz G I, Mechov D D, Pylypchenko O P and Chernomashencev yu G 1992 *Loading of tank-containers located on the railway platform, with shocks in automatic coupling,* [In Russian] Collection of scientific works "Dynamics and control of movement of mechanical systems" 87-95
[15] Djakonov V 2000 MATHCAD 8/2000 special reference book, [In Russian] St. Petersburg 592
[16] Kirjanov D V 2006 Mathcad 13, [In Russian] St. Petersburg 608
[17] Domin Yu V and Chernak A Yu 2003 *Basics of wagon dynamics*, Tutorial, [In Ukrainian] Kyev 270
[18] Okorokov A M, Fomin O V, Lovska A O, Vernigora R V, Zhuravel I L and Fomin V V 2018 Research into a possibility to prolong the time of operation of universal semi-wagon bodies that have exhausted their standard resource, *Eastern-European journal of enterprise technologies* **3/7**(93) 20-26
[19] DIN EN 12663-2-*2010 Railway applications – Structural requirements of railway vehicle bodies*, Part 2: Freight wagons. BSI, 54
[20] EN 13775-1-2-3:2003 Railway applications - Measuring of new and modified freight wagons - Part 1: Measuring principles
[21] Alamovsky A A 2007 SolidWorks/COSMOSWorks 2006 – 2007. *Engineering Finite Element Analysis* [In Russian] Moskow 784
[22] Gallager P 1984 *Finite element method.* [In Russian] Based. Moskow: World 428
[23] Rudenko V M 2012 *Mathematical statistics.* [In Ukrainian] Kyev textbook center 304

IOP Conf. Series: Materials Science and Engineering **659** (2019) 012057 doi:10.1088/1757-899X/659/1/012057

Estimation of the influence of the interaction of factors pairs on the coefficient of route execution possibility

V G Puzyr[1], O S Krasheninin[1], D S Zhalkin[1], Y M Datsun[1] and O M Obozny[1]

[1]Ukrainian State University of Railway Transport, Faculty of Mechanics and Energy, Feierbakh Square 7, 61050 Kharkiv, Ukraine

E-mail: sasha.obozny@gmail.com

Abstract. Successful execution of a route by a locomotive can depend on the combination of various operational factors. As an indicator that characterizes the reliable operation of locomotives in exploitation, the coefficient of route execution possibility is proposed to use. The task is to get a formula by which it would be possible to obtain the value of the coefficient of route execution possibility with different combinations of factors. As a tool for obtaining an interpolation formula for determining the coefficient of route execution possibility, a full factor experiment was chosen. The calculations of the model coefficients were carried out with a pairwise combination of four factors: the profile of the area, experience of the driver, the length of the shoulder, the weight of the train. The obtained models were checked for adequacy by F-test. On the received models in three-dimensional coordinates surfaces, reflecting the value of the coefficient of route execution possibility in the intervals of factorization, were constructed. An analysis of the obtained surfaces and the distribution of the selected factors by the degree of impact on the coefficient of route execution possibility were carried out.

1. Introduction

The analysis of the reliability of the work of locomotives in the operation indicates a rather high level of their failures in the route. It indicates that before the departure of the locomotive to the route, there is no assessment of its ability to successfully complete the route under the influence of various factors. Denials on the route may result in significant material losses that could have been avoided even at the stage of route preparation.

First, before sending the locomotive to the route, it is necessary to analyze its actual technical condition and to ensure that the state of its nodes allows executing the route without fail [1]. For this purpose an electronic passport of the locomotive can be used [2].

Secondly, it is necessary to make sure of sufficient qualification of the locomotive crew in order to avoid the failure of the locomotive due to improper actions of the locomotive crew.

Thirdly, it is necessary to take into account the influence of operational parameters of the future route to change the technical condition of the locomotive and its knots.

As a tool that will allow simulating and analyzing the interaction of various factors, the method of a full factor experiment can be used [3, 4].

Methods of the theory of experiment are used to find optimal conditions and obtain formulas reflecting the interaction of factors. Experiment planning applies for searching for optimal conditions, building interpolation formulas, choosing significant factors [5-8].

2. Purpose

The purpose of this article is to analyze the influence of the interaction of a pair of factors on the coefficient of route execution possibility and ranking of selected factors by degree of influence.

3. Main body

As factors that influence on the value of the coefficient of route execution possibility were selected: x_1 – the profile of the area, x_2 – experience of the driver, x_3 – the length of the shoulder, x_4 – the weight of the train.

Accepted values of the zero level, the interval of variation, the upper and lower levels of factors are given in Table 1.

Table 1. Values of variables and variables range.

Levels and interval of variation of factor	Factors			
	x_1	x_2	x_3	x_4
	i, ‰	T, years	L, km	Q, t
Zero level	0	10	300	4000
Interval of variation	10	10	200	2000
Lower level	-10	0	100	2000
Upper level	10	20	500	6000

For the transition from real values to values of ± 1 encoding of the factors is performed. Encoding is determined by the ratio

$$x_i = \frac{\tilde{x}_i - \tilde{x}_{i0}}{J_i}, \tag{1}$$

where x_i is the coded current value of the factor, $\tilde{x}_i$ is the natural current value of the factor, $\tilde{x}_{i0}$ is the natural value of the zero level and J_i is the natural value of the interval of variation.
The values of coded variables are

$$x_1 = \frac{i-0}{10},\ x_2 = \frac{T-10}{10},\ x_3 = \frac{L-300}{200},\ x_4 = \frac{Q-4000}{2000}$$

The matrix of a full factor experiment for interaction of profile of area and experience of the driver is given in Table 2.

Table 2. Matrix of a full factor experiment.

Experiment	Factors and their interaction				Coefficient				
	x_0	x_1	x_2	$x_1 x_2$	y_1	y_2	y_3	y_{cep}	$\hat{y}$
1	1	-1	-1	1	0.91	0.89	0.87	0.89	0.9044
2	1	1	-1	-1	0.89	0.88	0.87	0.88	0.8945
3	1	-1	1	-1	0.92	0.9	0.87	0.90	0.9112
4	1	1	1	1	0.88	0.94	0.93	0.92	0.8988

In this table, y_1, y_2, y_3 are the results of parallel experiments (obtained by simulation). Using them dispersions for each of the series of experiments were calculated by the formula

$$\sigma_i^2 = \frac{\sum_{j=1}^{m}\left(y_{ji} - \overline{y_i}\right)^2}{m-1} \tag{2}$$

where σ_i^2 is the dispersion at the i-th point, m is the number of parallel experiments, y_{ji} is the value of the optimization parameter in the j-th parallel experiment and $\overline{y_i}$ is the average value of the optimization parameter in this series of parallel experiments.

$$\sigma_1^2 = \frac{(0.91-0.89)^2 + (0.89-0.89)^2 + (0.87-0.89)^2}{3-1} = 0.0004$$

Analogically determined $\sigma_2^2 = 0.0001$, $\sigma_3^2 = 0.0006$ and $\sigma_4^2 = 0.001$.

To determine the homogeneity of dispersions, the Cochran's G-test is calculated by the formula

$$G = \frac{\sigma_{i\max}^2}{\sum_{i=1}^{n}\sigma_i^2} \tag{3}$$

where $\sigma_{i\max}^2$ – the largest value of the dispersion.

$$G = \frac{0.001}{0.0004 + 0.0001 + 0.0006 + 0.001} = 0.47$$

Table value is $G_{table} = 0.68$.

Since $G < G_{table}$, the dispersions are homogeneous.

The experimental error is determined by the formula

$$\sigma_y^2 = \frac{\sum_{i=1}^{n}\sigma_i^2}{n} \tag{4}$$

$$\sigma_y^2 = \frac{0.0004 + 0.0001 + 0.0006 + 0.001}{4} = 0.0005.$$

The coefficients of the model are determined by the formula

$$b_j = \frac{\sum_{i=1}^{n} x_{ij} y_i}{n} \tag{5}$$

$$b_0 = \frac{1 \cdot 0.89 + 1 \cdot 0.88 + 1 \cdot 0.9 + 1 \cdot 0.92}{4} = 0.8958$$

Analogically determined $b_1 = 0.0025$, $b_2 = 0.0108$ and $b_{12} = 0.0075$.

Knowing the coefficients of the model, it is possible to calculate the value of the coefficient of route execution possibility and to examine the adequacy of the model.

The Fisher's F-test is used to check the adequacy of the model

$$F = \frac{\sigma_{ad}^2}{\sigma_y^2} \tag{6}$$

where σ_{ad}^2 – the dispersion of adequacy.

$$\sigma_{ad}^2 = \frac{\sum_{i=1}^{n}(y_i - \hat{y}_i)^2}{n-k-1} \tag{7}$$

where $\hat{y}_i$ – the value of the optimization criterion provided by the equation for the i-th experiment.

If for the selected level of significance $\alpha = 0.05$ the calculated value of Fisher's F-test does not exceed the table value, then the model is considered adequate.

$$\sigma_{ad}^2 = \frac{(0.89-0.9044)^2 + (0.88-0.8945)^2 + (0.9-0.9112)^2 + (0.92-0.8988)^2}{4-2-1} = 0,00094$$

$$F = \frac{0.00094}{0,0005} = 1.75$$

Table value is $F_{table}(k_1 = 4, k_2 = 4) = 6.39$.

Since $F < F_{table}$, the model is adequate.

The formula for determining the coefficient of route execution possibility with the interaction of factors x_1 and x_2 will have the following form

$$k = 0.8958 + 0.0025x_1 + 0.0108x_2 + 0.0075x_1x_2 \tag{8}$$

On the received formula in three-dimensional coordinates it is possible to construct a surface that will reflect the value of the coefficient of route execution possibility in the intervals of variation the factors x_1 and x_2 (Figure 1).

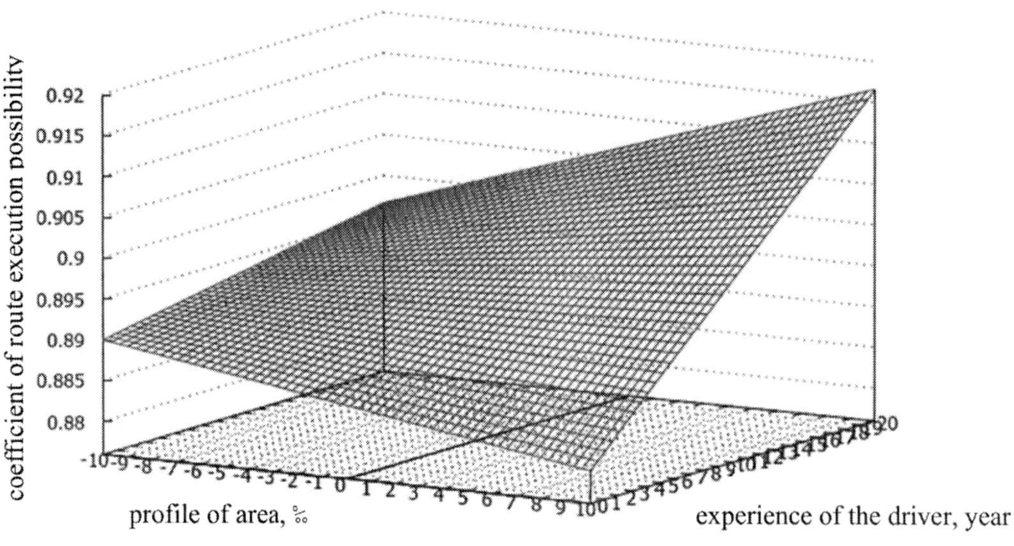

Figure 1. Dependence of the coefficient of route execution possibility on the profile of the area and experience of the driver.

The analysis of the graph shows that in the interaction of factors x_1 – the profile of the area and x_2 – experience of the driver the coefficient of route execution possibility varies from 0.88 to 0.916. The minimum value of the coefficient corresponds to the interaction of the maximum value x_1 and the minimum value of x_2, and the maximum value of the coefficient corresponds the interaction of the maximum values x_1 and x_2.

The next was a full factor experiment with the interaction of factors x_3 – the length of the shoulder, x_4 – the weight of the train. In this case, the matrix of the complete factor experiment was compiled, the dispersions, the coefficients of the model, the experimental error were calculated and the verification of the model for adequacy were made according to the methodology given above.

The formula for determining the coefficient of route execution possibility with the interaction of factors x_3 and x_4 will have the following form

$$k = 0.8715 - 0.0057x_3 - 0.0014x_4 - 0.0033x_3x_4 \quad (9)$$

The surface, which reflects the value of the coefficient of route execution possibility in the intervals of variation of factors x_3 and x_4 is shown on Figure 2.

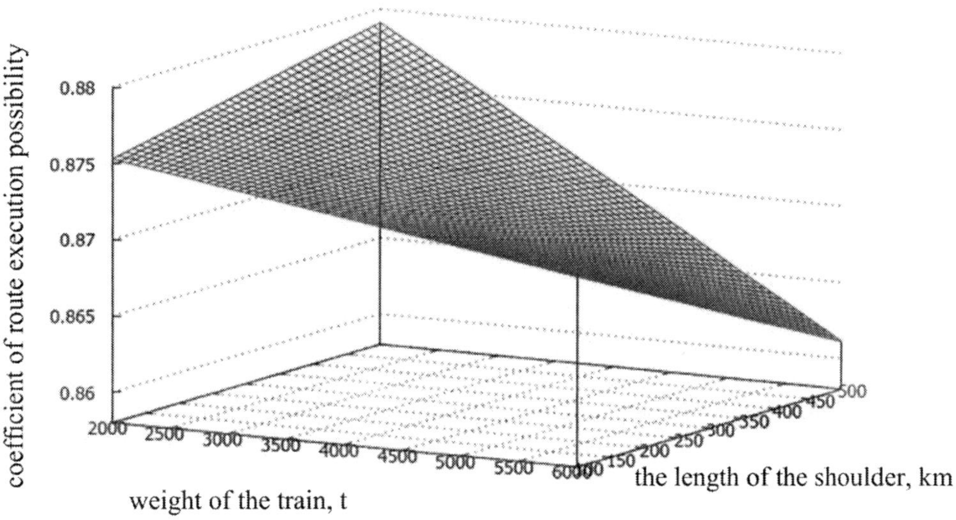

Figure 2. Dependence of the coefficient of route execution possibility on the weight of the train and the length of the shoulder.

In the interaction of factors x_3 – the length of the shoulder, x_4 – the weight of the train coefficient of route execution possibility varies from 0.862 to 0.878. The minimum value of the coefficient corresponds to the interaction of the maximum values x_1 and x_2, and the maximum value of the coefficient is the interaction of the maximum value of x_3 and the minimum value of x_4.

Then there was a full factor experiment with the interaction of factors x_4 – the weight of the train, x_2 – experience of the driver. In this case, the matrix of the complete factor experiment was compiled, the dispersions, the coefficients of the model, the experimental error were calculated and the verification of the model for adequacy were made according to the methodology given above.

The formula for determining the coefficient of route execution possibility with the interaction of the factors x_4 and x_2 will have the following form

$$k = 0.8837 - 0.0218x_4 + 0.0315x_2 - 0.0089x_4x_2 \quad (10)$$

The surface, which reflects the value of the coefficient of route execution possibility in the intervals of variation of factors x_4 and x_2 is shown in Figure 3.

In the interaction of factors x_4 – the weight of the train, x_2 – experience of the driver coefficient of route execution possibility varies from 0.866 to 0.926. The minimum value of the coefficient corresponds to the interaction of the maximum values of x_4 and x_2, and the maximum value of the coefficient is the interaction of the maximum value of x_2 and the minimum value of x_4.

The next pair of factors was chosen x_2 – experience of the driver and x_3 – the length of the shoulder. After carrying out a full factor experiment, the following formula was obtained for determining the coefficient of route execution possibility with the interaction of factors x_2 and x_3.

$$k = 0.8948 - 0.0253x_2 - 0.0071x_3 - 0.0105x_2x_3 \tag{11}$$

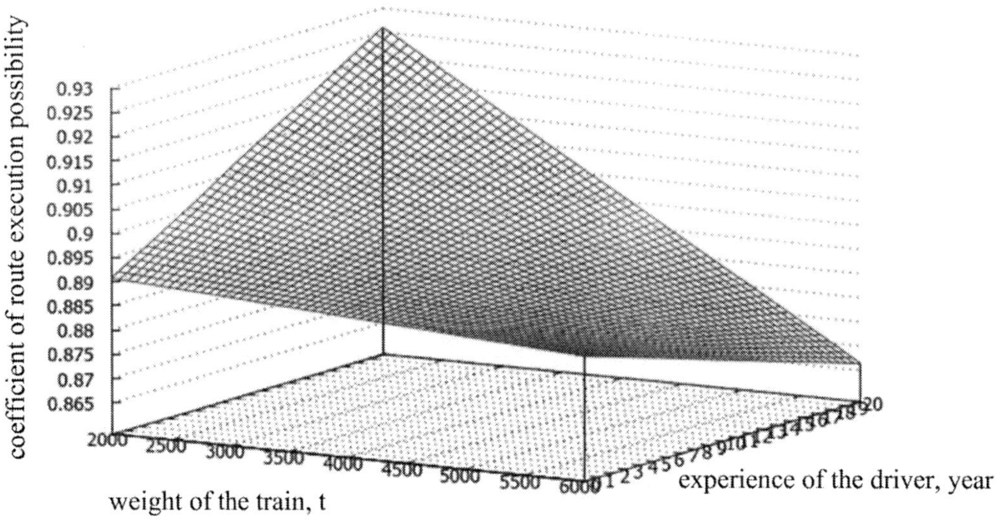

Figure 3. Dependence of the coefficient of route execution possibility on the weight of the train and experience of the driver.

According to the obtained formula, a surface that reflects the value of the coefficient of flight fulfillment in the intervals of the coagulation of the factors x_2 and x_3 (Figure 4).

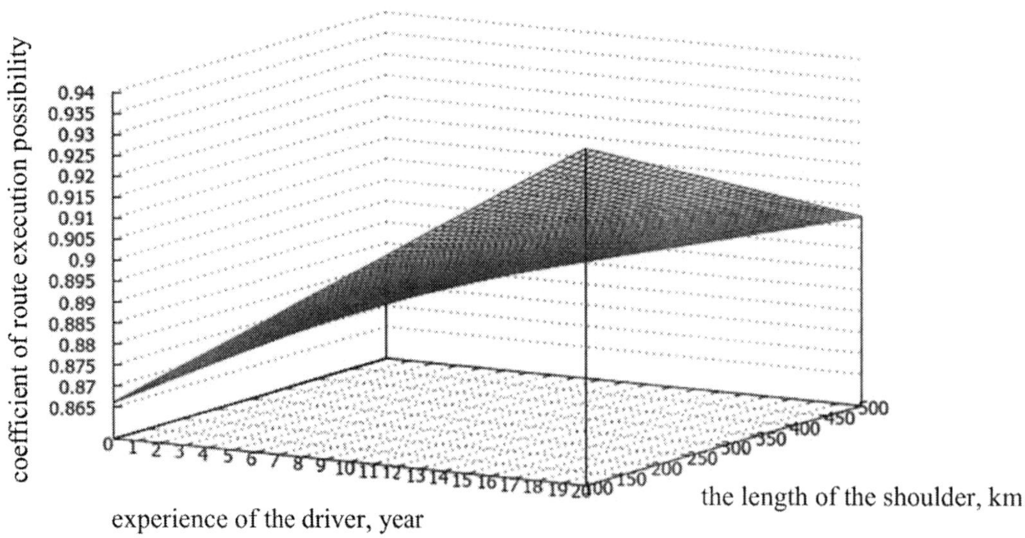

Figure 4. Dependence of the coefficient of route execution possibility on experience of the driver and the length of the shoulder.

In the interaction of factors x_2 – experience of the driver and x_3 – the length of the shoulder coefficient of route execution possibility varies from 0.866 to 0.913. The minimum value of the coefficient corresponds to the interaction of the minimum values x_1 and x_2, and the maximum value of the coefficient corresponds the interaction of the maximum value of x_2 and the minimum value of x_3.

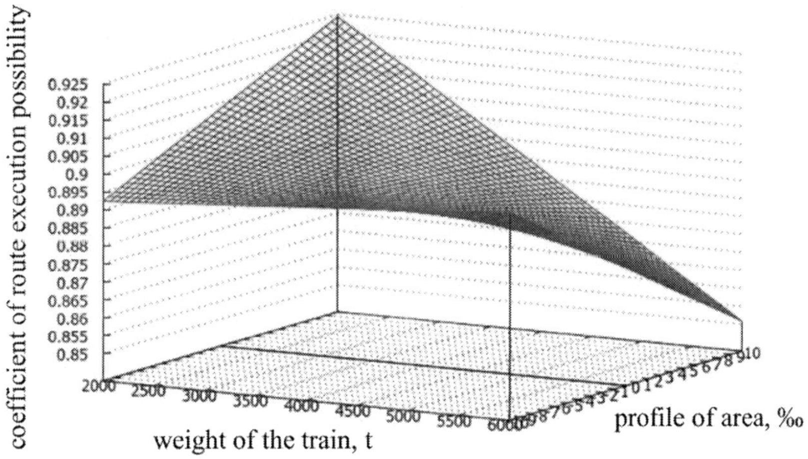

Figure 5. Dependence of the coefficient of route execution possibility on the weight of the train and the profile of the area.

The last group of factors on which the calculations were made were x_4 – the weight of the train and x_1 – the profile of the area. The made calculations showed that the formula for determining the coefficient of route execution possibility with the interaction of the factors x_4 and x_1 will have the following form

$$k = 0.8919 - 0.0165x_4 + 0.0043x_1 - 0.0203x_4x_1 \quad (12)$$

The surface, which reflects the value of the coefficient of route execution possibility in the intervals of the variation of factors x_4 and x_1 is shown in Figure 5.

The analysis of the graph shows that in the interaction of the factors x_4 – the weight of the train and x_1 – the profile of the area the coefficient of route execution possibility varies from 0.85 to 0.925. The minimum value of the coefficient corresponds to the interaction of the maximum values of x_4 and x_1, and the maximum value is the interaction of the minimum value of x_4 and the maximum value of x_1.

4. Conclusion

The obtained formulas are valid for the variables in the intervals of variation given in Table 1. According to them, it is possible to calculate the value of the coefficient of route execution possibility at different combinations of a pair of factors.

In this case, the coefficient of route execution possibility takes the value from 0.85 to 0.926. The minimum value of the coefficient corresponds to the interaction of the maximum values of x_4 and x_1, the maximum value corresponds the interaction of the maximum value of x_2 and the minimum value of x_4.

By the force of influence on the coefficient of route execution possibility, the factors are arranged in the following order: experience of the driver, the length of the shoulder, the weight of the train, the profile of the area.

References

[1] Tartakovskyi E, Ustenko O, Puzyr V and Datsun Y 2017 *Systems Approach to the Organization of Locomotive Maintenance on Ukraine Railways*, In book: Rail Transport - Systems Approach Sladkowski A (Ed.) 217-239, Springer

IOP Conf. Series: Materials Science and Engineering **659** (2019) 012057 doi:10.1088/1757-899X/659/1/012057

[2] Obozny O M 2015 Method of using electronic passport of locomotive in management system of preroute preparation, *EEJET* **3**(62) 56-58
[3] Jiju A 2003 *Design of Experiments for Engineers and Scientists*, Elsevier
[4] Adler Y P 1976 *Design of experiment for searching an optimal conditions*, Science
[5] Rasch D, Pilz J, Verdooren L R and Gebhardt A 2011 *Optimal Experimental Design with R* CRC Press
[6] Ajit C Tamhane 2012 *Statistical Analysis of Designed Experiments: Theory and Applications* John Wiley & Sons
[7] Pronzato L and Pázman A 2013 *Design of Experiments in Nonlinear Models: Asymptotic Normality, Optimality Criteria and Small-Sample Properties* Springer
[8] Davim J P 2015 *Design of Experiments in Production Engineering* Springer

IOP Conf. Series: Materials Science and Engineering **659** (2019) 012058 doi:10.1088/1757-899X/659/1/012058

Application of digital human models in determination of the pedal force while driving

S Mačužić[1] and J Lukić[1]

[1]University of Kragujevac, Faculty of Engineering, Sestre Janjić 6, 34000 Kragujevac, Serbia

E-mail: s.macuzic@kg.ac.rs

Abstract. One of the most important tasks of the car manufacturer is the design of the interior space of the vehicle. Anthropometric characteristics of drivers are also important for designing cars. In this paper the determination of the pedal force has been performed. The force determination was made on digital human models for different populations, a total of 11 different populations of men and women. Analysing was done in the software package Ramsis and subjects were presented using "mannequin". The obtained force values were different for different populations. The correlation between the anthropometric measures of the subjects and the force on the pedals was observed. By calculating Pearson R coefficient in both populations, female 5 % and male 95 %, it can be concluded that there is a good correlation between active foot pedal force and driver's height (R^2 = 0.84 and R^2 = 0.72, respectively). In the case of maximum values of passive forces R coefficient in both populations also have a good correlation between passive pedal force and driver's height (R^2 = 0.89 for female 5% population and R^2 = 0.69 for male 95 % population). The obtained results are important for designing the interior of the vehicle.

1. Introduction

The interior of the car can have a significant impact on both the mental and the physical health of drivers and passengers. Car manufacturers design the interior of the vehicle in order to better meet the needs of the driver. Ergonomics is the scientific discipline concerned with the understanding of interaction between human and other elements of a system, in this case - a vehicle [1]. The role of ergonomics is to increase the efficiency and productivity of production, improvement of health, safety and comfort man in his work environment. This paper explores the necessary force to activate the pedal while driving. Inadequate pedal design often leads to fatigue and discomfort of drivers [2]. Many studies have dealt with the determination of discomfort during driving. This is mostly done by subjective assessment methods. Studies have shown that the realization of the maximum force depends on the position of the seat, the type of seat and pedal [3]. Mortimer [4] suggested that the maximum pedal force for 5% female population should not exceed 400 N. According to [5] changes in the angles of the knees influenced the maximum force on the pedals. The smaller the knee flexion angle (from 20° to 80°) is, the greater the foot strength is. Mehta [6] concluded that the maximum force was achieved when the leg was in an almost elongated position, that is, the angle of bending the knees was between 35° and 45°.

In this research, all digital human models were placed in the virtual environment of Volkswagen UP! [9]. The anthropometric data of the male and female population are useful for the examination. The software packages Catia V5 R18 and Ramsis (Rechnergestutztes Anthropometerisches Mathematisches System zur Insassen-Simulation) were used [7]. With the Ramis software package, models can be set up and manipulated with their movements. This approach is the most important stage in the process of developing a new vehicle.

2. Methods

In order to examine how the design of the car and the position of the vehicle controls affect the inconvenience of drivers, certain digital human models are used. In this paper, the work environment of the driver in the vehicle was first created. The Catia V5 R18 software package is used to model the interior of the vehicle. Setting the model in the seat and defining the boundary conditions was done using the Ramsis software package. The vehicle seat is fixed and adjusted to the highest population.
Two types of human models can be used in Ramsis, kinematic and geometric. The kinematic model takes into account the human skeleton and the model is a wicker frame. The geometric model represents the surface of the model. In this paper, a mannequin (geometric model) with 5 fingers and shoes on the feet was used.

This study included the analysis of eleven different populations, men and women [11]. Their anthropometric characteristics are in Table 1. These populations belong to the age group 18-70 years for 5% and 95% of women and men. When designing a vehicle, manufacturers define a vehicle for a population of 5% to 95%, where 5% is defined in relation to female population, and 95% in relation to male population, with the aim of achieving a larger universality and placement of passenger cars on the market regardless of the geographical climate.

Good design of foot controls has an impact on the better efficiency of the braking system. The activating force on the pedals can be active and passive. The active force is only triggered by muscular activity, while the passive force is generated by muscular activity and additionally relying on the floor. The recommended values for pedal-force activation range from 25 to 400 N [10]. The change in anthropometric dimensions has an effect on the values of the activated pedal force.

Table 1. The anthropometric data of 5% female and 95% male population [11].

	Height (mm)		Sitting height (mm)		Foot length (mm)	
Population	*5%*	*95%*	*5%*	*95%*	*5%*	*95%*
West Africa	1440	1785	741	881	212	278
North India	1450	1765	771	919	205	270
Eastern Europe	1540	1845	831	959	225	285
North Europe	1585	1910	846	999	227	280
Australia	1565	1885	831	979	220	280
South East Europe	1525	1825	811	949	220	285
Central Europe	1560	1805	826	989	220	288
South East Africa	1480	1775	771	909	210	280
Middle East	1529	1800	801	944	222	280
South India	1395	1715	746	874	200	265
North Asia	1504	1820	801	961	207	275

3. Results

The Ramsis software package is based on a statistical analysis of the results obtained experimentally [7], [12]. On the basis of the collected anthropometric data of the subjects, digital models of man with boundary conditions were generated. With this software, the analysis of foot - pedal interaction was performed. Based on this, the pedal activation force was determined. The lowest value of active and passive force of the female population have the populations from Central Europe and North Europe, 138.1 N and 156.9 N, respectively for active force, and 149.2 N and 165.6 for passive force. The populations of West Africa, South India and North India have the highest values of active and passive forces (Figure 1). Maximum value of active force was 219.7 N (West Africa female 5% population), while the maximal passive force was 223.1 N.

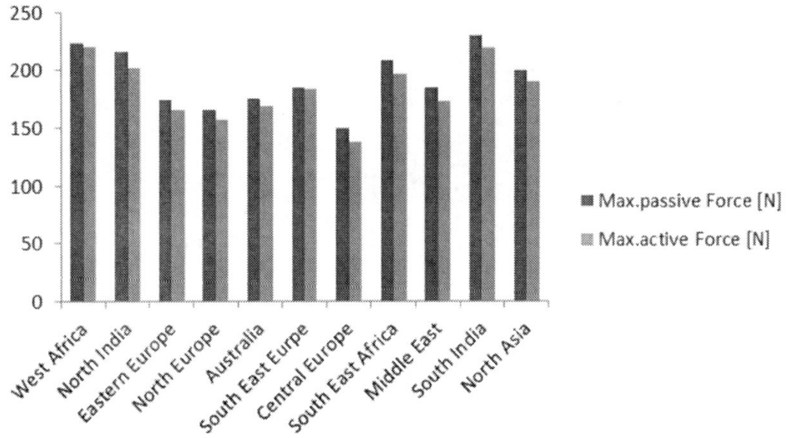

Figure 1. The maximum foot pedal force (active and passive) of eleven female 5% populations.

Figure 2 shows the values of active and passive forces of the male 95% population.

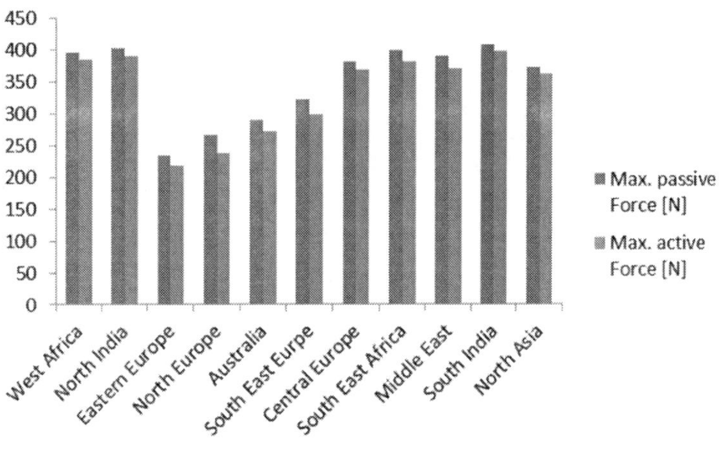

Figure 2. The maximum foot pedal force (active and passive) of eleven male 95% populations.

The smallest value of the active and passive force have population from Eastern Europe and North Europe 218.1 N and 238.5 N, respectively for active force, and 233.8 N and 266.2 N for passive force. The South India, North India and West Africa populations have the highest values of active and passive forces. Maximum value of active force was 397.1 N (South India male 95% population), while the maximum value of passive force was 408.7 N. With this value exceeding 400 N, the male 95% population of South India exceeds the permitted value of pedal passive force, so this position is uncomfortable.

By increasing the height of the male/female drivers, it can be noticed that there is a decrease in the active force values required to activate the pedal (Figure 3 and Figure 5), as well as passive force values (Figure 4 and Figure 6). By calculating Pearson R coefficient in both populations, female 5 % ($R^2 = 0.84$) and male 95 % ($R^2 = 0.72$), it can be concluded that there is a good correlation between active foot pedal force and driver's height. Also, in case of maximum values of passive forces R coefficient in both populations also have a good correlation between passive pedal force and driver's height ($R^2 = 0.89$ for female 5% population and $R^2 = 0.69$ for male 95 % population).

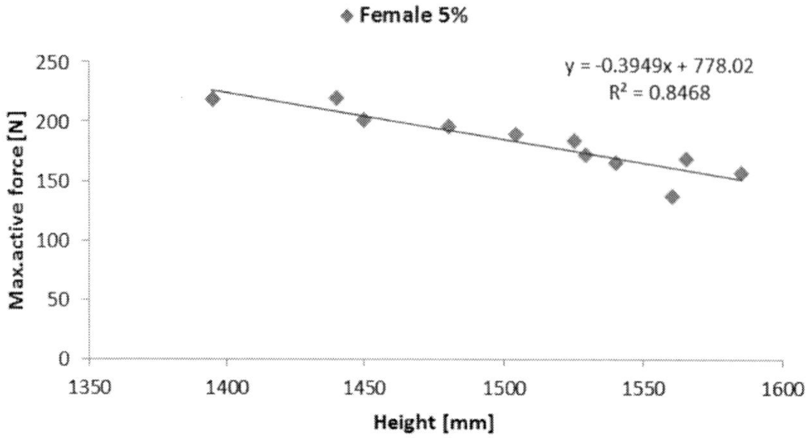

Figure 3. Maximum active foot pedal force of the female driver's vs. height.

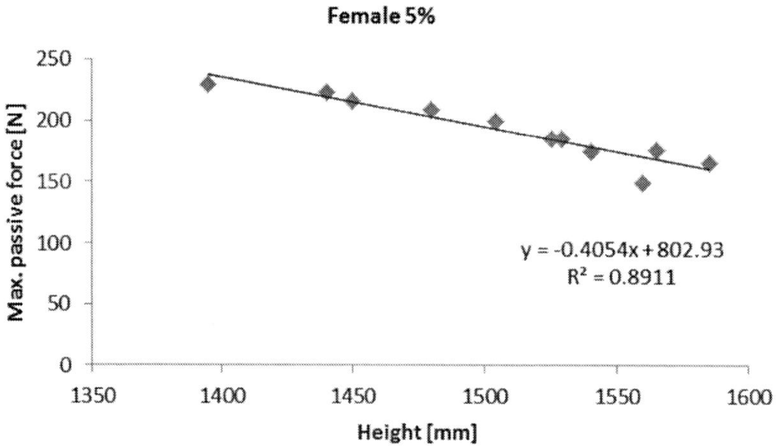

Figure 4. Maximum passive foot pedal force of the female driver's vs. height.

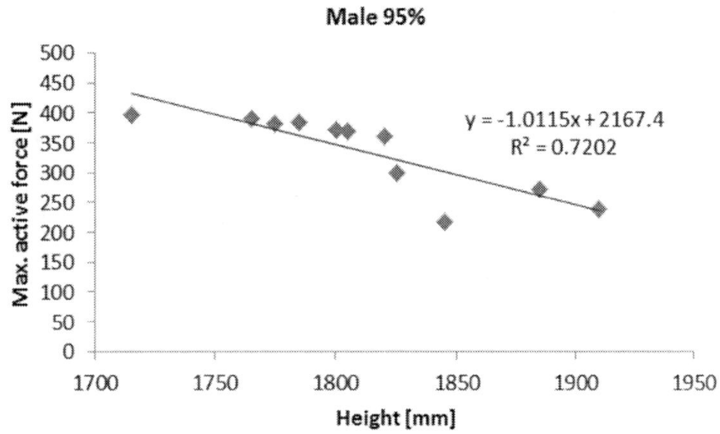

Figure 5. Maximum active foot pedal force of the male drivers vs. height.

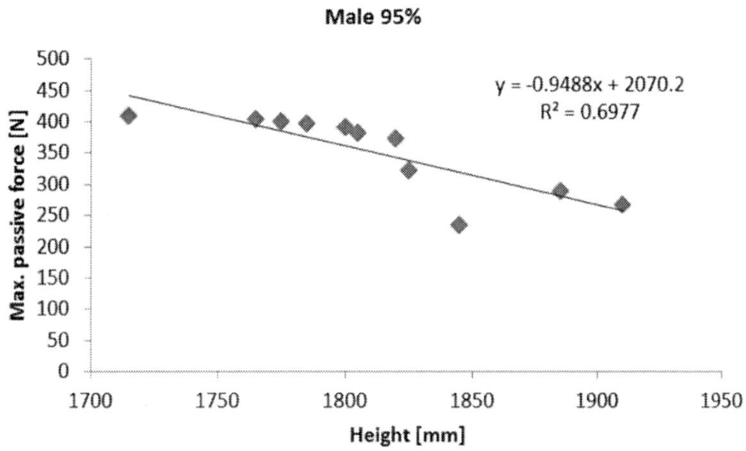

Figure 6. Maximum passive foot pedal force of the male drivers vs. height.

4. Conclusion

In this paper, the force required to activate the pedal from the driver's seat is determined. The study was conducted on eleven different populations, men and women. Ramsis software package is used for analysis and determination of force. Using this software package, car makers have the opportunity, before the first prototype, to graphically display how the model interacts with its environment inside the car. Results show that the smallest value of the active and passive force have population from Eastern Europe and North Europe 218.1 N and 238.5 N, respectively for active force, and 233.8 N and 266.2 N for passive force. The South India, North India and West Africa populations have the highest values of active and passive forces. Maximum value of active force was 397.1 N (South India male 95% population), while the maximum value of passive force was 408.7 N. By increasing the height of the male or female drivers, it can be concluded that there is a decrease in the active and passive force values required to activate the pedal. By calculating Pearson R coefficient in both populations, female 5 % (R^2 = 0.84) and male 95 % (R^2 = 0.72), it can be concluded that there is a good correlation between active foot pedal force and driver's height. Also, in case of maximum values of passive forces R coefficient in both populations also has a good correlation between passive pedal force and driver's height (over 0.69).

The significance of this research during the development and production of vehicles can save a money and time, and also lead to the final product driver friendly designed. The influence of position and indication angle of pedal activation force will be analysed in further research.

Acknowledgment
This research supported by the Ministry of Education, Science and Technological Development of Republic of Serbia through Grant TR 35041.

References

[1] Bridger R S 2009 *Introduction to Ergonomics*, Boca Raton: Taylor & Francis Group
[2] Sanders M S, McCormick J 1993 *Human Factors in Engineering and Design*, 7th Edition. McGrav-Hill, New York, 355
[3] Kroemer K H E 1971 *Foot operation of controls Ergonomics* **14**(3) 333–361
[4] Mortimer R G, Segel L, Dugoff H, Campbell J D, Jorgeson C M and Murphy R W 1974 *Brake force requirement study: driver-vehicle braking performance as a function of brake system design variables*, National Highway Safety Bureau U.S Department of transportation
[5] Lara-Lopez A, Aguilera-Cortes L and Barbosa-Castillo F 1999 Measurement of forces applied to handgrips and pedals for a sample population of Mexican males, *Applied Ergonomics* **30** 173-176
[6] Mehta C R, Tiwari P S, Rokade S, Pandey M M, Pharade S C, Gite L G and Yadav S B 2007 Leg strength of Indian operators in the operation of tractor pedals, *International journal of industrial Ergonomics* **37** 283-289
[7] Human Solutions 2010 RAMSIS models, available on April 7, 2017, http: //vv.human-solutions. com / mobility / front_content.php? idcat = 252
[8] https://www.volkswagen.rs/up-2, Accessed 10.06.2019
[9] Mačužić S, Lukić J, Glisovic J and Miloradović D 2017 Pedal force determination respect to ride comfort, *IOP Conference Series: Materials Science and Engineering* **252** 012010, 10.1088/1757-899X/252/1/012010
[10] https://www.unece.org/trans/main/wp29/wp29wgs/wp29gen/wp29glob_registry.html, ECE / TRANS / 180 / Add.3 : Agreement on the establishment of global technical regulations for wheeled vehicles, equipment and parts that can be fitted and / or used on wheeled vehicles, 21 December 2006, International Organization for Standardization, Geneva, Switzerland, accessed 06/07/2017
[11] http://dined.io.tudelft.nl/en, accessed 10.07.2016
[12] Heiner Bubb H, Bengler K, Grünen R and Vollrath M 2015 *Automobilergonomia*, Springer Vieweg

IOP Conf. Series: Materials Science and Engineering **659** (2019) 012059 doi:10.1088/1757-899X/659/1/012059

Indoor positioning and navigation system for autonomous vehicles based on RFID technology

M Regus[1], R Talar[1] and R Labudzki[1]

[1]Poznań University of Technology, Piotrowo 3, 60-965 Poznań, Poland

E-mail: michal.j.regus@doctorate.put.poznan.pl

Abstract. The localization and navigation systems play a very significant role in today's world. They are commonly used in different fields of industry as well as in our daily life. Location of objects is crucial in logistics and transport to provide the real-time monitoring and management of process chain. There are numerous methods which allows to follow people and objects within defined area. One of them is Global Positioning System (GPS), which provides geolocation of objects on the earth as long as an unobstructed line of sight to four or more satellites is assured. As a consequence of this limitation, GPS cannot be applied for indoor environments. Recently, a couple of Indoor Positioning Systems (IPS) based on different technologies have been developed. Despite the drawbacks and limitations, these systems have been successfully applied for different fields of industry. Notwithstanding the above, reliable, accurate, cost-effective and simple indoor positioning methods are still in the area of scientists interest. Present paper provides a general overview of existing indoor positioning systems as well as authors concept of 2D wireless indoor positioning and navigation system dedicated for autonomous vehicles based on Radio Frequency Identification (RFID) technology.

1. Introduction

The last few decades have brought a great technological progress. At present time, a strong trend of automation and data exchange process within the production area can be observed. This tendency is considered as a 4th Industrial Revolution (called also Industry 4.0). The concept of Industry 4.0 is based on the production environment where people and machines create the cyber-physical system, within which they communicate, cooperate and self-organize their job [1]. It is worth to mention that one of the major direction of Industry 4.0 is Human Robot Collaboration (HRC). According to this concept, collaborative robots (called also *cobots* or *co-robots*) - equipped with safety and collision avoidance systems – assist the human co-workers in the most physically difficult and monotonic tasks within one workspace. No additional safety fence is required. In the face of above mentioned idea, reliable and accurate Indoor Positioning System (IPS) seems to be indispensable component of modern production environment. Tracking of goods during the production process, navigation and localization of people, robots and autonomous vehicles are only a few potential applications of IPS in modern production plants.

It is worth to mention, that not only the industrial environments have changed over last few decades. We are the witnesses of changing lifestyle and living conditions. Nowadays, our society lead sedentary lifestyle. A huge percentage of people spend our time sitting at home in front of their computers, TVs and smartphones. They do the shopping through the internet and even build the relationships. It is predicted that in the near future people will spend more than 70% of their lifetime within indoor environments [2]. Consequently, reliable and accurate indoor positioning systems are highly demanded to complement widely used satellite-based technologies and to ensure tracking

IOP Conf. Series: Materials Science and Engineering **659** (2019) 012059 doi:10.1088/1757-899X/659/1/012059

within outdoor environments. Such hybrid system would provide real-time monitoring of objects and people position regardless they are within in- or out-door environment.

The Global Positioning System (GPS) is the most commonly used localization system based on the satellite technology. Nowadays, GPS receivers are built-in most of everyday use devices, like smartphones, watches or vehicles. The GPS system contains mainly 31 satellites and receivers. The satellites are evenly deployed in the Earth orbit and constantly broadcast the radio signal about their position and current time. The signals from multiple satellites, which arrive to the receiver are processed in order to calculate its current position. The object position can be calculated as long as an unobstructed line of sight to four or more satellites is assured. As a consequence, GPS technology is not sufficient for indoor applications [3, 4]. Furthermore, indoor environments are usually more complex and susceptible for noises which decrease the reliability and accuracy of positioning systems. All the obstacles like, walls, furniture, electrical devices or human being influence the propagation of electromagnetic field. In reference to above mentioned issues, it can be stated that appropriate solution for positioning within indoor environments became a new challenge for the present day.

Recently, the real time indoor positioning systems have become an object of interest for many scientist centers. Despite the huge progress in regards to IPS that has been done over last decades, a reliable, accurate, cost-effective, simple and flexible indoor location system is still searched for [3]. One of the biggest challenge that must be faced nowadays is not only to develop reliable and accurate positioning system containing advanced and highly specialized sensors but also to implement hybrid system which would be based on currently existing technologies and commonly used devices [2]. WLAN-based (Wireless Local Area Network) indoor positioning system can be mentioned as an example of such approach. WLAN technology reuses the devices and networks which are widely used almost everywhere. Consequently, any additional equipment or infrastructure are not required to implement IPS utilizing WLAN. Another approach which comes from an idea of reusing already deployed technologies are smartphone-based IPSs [5]. These are only two examples of many different indoor positioning systems (based on IR, RFID, WLAN, Bluetooth and other technologies) which have been already designed and implemented. More examples of locating systems are given in the further part of this article.

The following paper contains a general overview of currently existing indoor positioning systems as well as the authors concept of 2D wireless indoor positioning and orientation system dedicated for autonomous vehicles based on Radio Frequency Identification (RFID) technology.

2. General characteristic of indoor positioning systems

The purpose of indoor positioning system (IPS) is to determine continuously in real time the position of followed objects and people within defined workspace [3]. There are many different technologies and methods which can be used to create the indoor positioning system. Each of them has some advantages and limitations which are mentioned in further part of this section.

2.1. Location methods

There are a couple of methods which can be applied to determine the position of object within indoor environment (figure 1). They have been widely described in papers [2, 3, 4]. In this sub-section only the general principles of existing techniques are given.

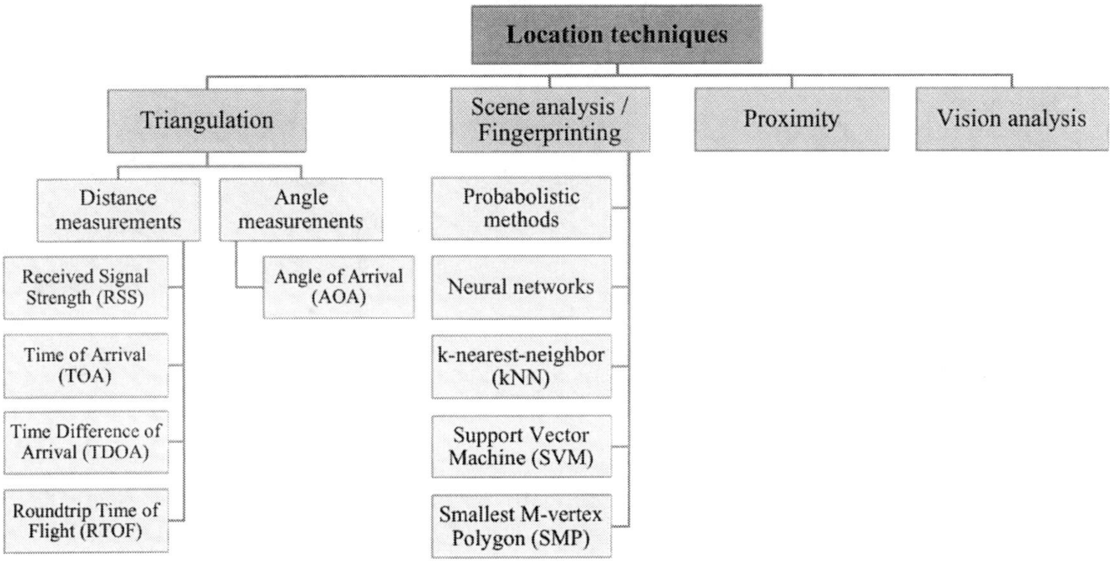

Figure 1. Location techniques and methods used for indoor positioning systems [3, 4].

2.1.1. Triangulation

The position of object is calculated based on the geometrical properties of triangle. The principle of method is shown in figure 2. If the geometrical positions (x_i,y_i) of at least three reference points are known (A, B, C), then the coordinates of object (O) can be calculated. The location can be determined either by measuring of distance between the object and reference points or by computing angles in relation to reference points [4]. Time of Arrival is considered as the most accurate methods, since it allows to filter out unwanted multi-path effects.

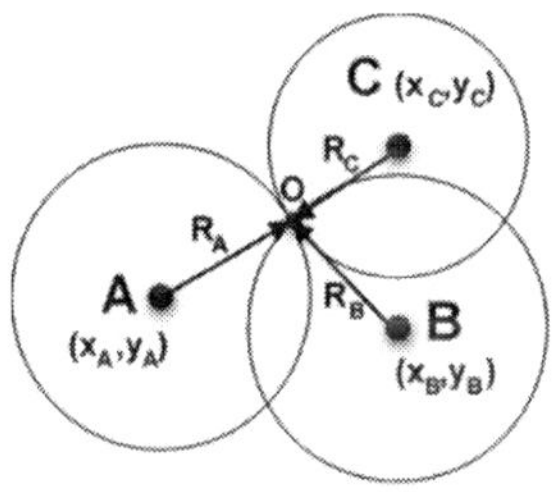

Figure 2. The principle of triangulation methods to determine the object position.

2.1.2. Fingerprinting

Another technique which has been applied for tracking systems is fingerprinting (called also mapping or scene analysis) [2]. The implementation process of fingerprinting method into indoor positioning system consists of two main stages: offline and online stage. During the offline stage the database of features (fingerprints) at known locations in relation to reference points is built. An example might be the measurements of Received Signal Strength Indication (RSSI) in different locations within the indoor environment in relation to the access points (figure 3). Other example of location-related feature might be the strength of magnetic field in relation to reference points where the coils are installed. In the online stage the same features are measured and then compared with those collected during the offline stage. Based on that, the location estimation is made. The fingerprinting technique is commonly used for localization systems, however it can be also deployed for tracking purpose [2]. The main disadvantage of described method is time-consuming offline phase as well as necessity of regular database up-dates which are essential to compensate the changes that have been done in the indoor environment.

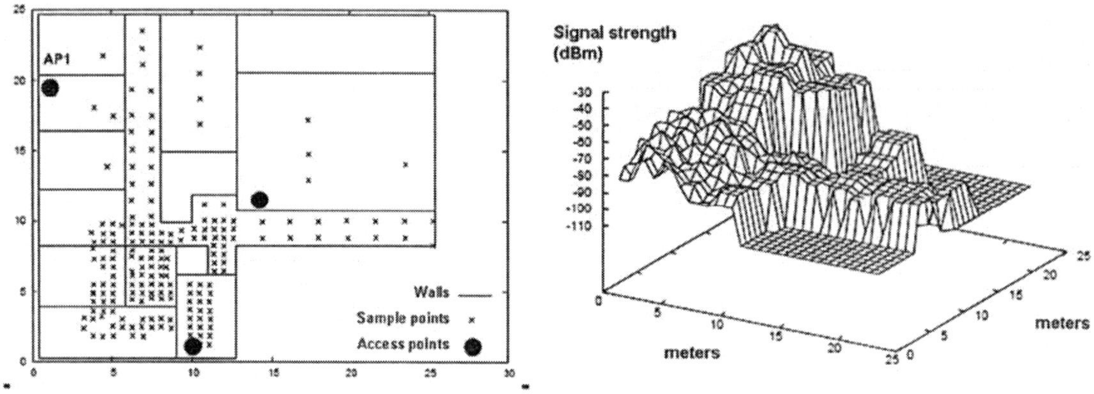

Figure 3. The principle of fingerprinting method, a) an example of indoor environment, b) the results of RSSI measurements in relation to access point 1 (AP1) [3].

2.1.3. Proximity
The proximity-based localization systems are usually built-up from grid of antennas which cover the indoor environment [4]. The positions of antennas/detectors are known. When the followed object gets into the range of particular antenna, then its position can be detected, since it is related with the well-known position of antenna. The accuracy of the method is strictly related with the density of detectors grid. The principle of proximity method is presented in the figure 4. The followed object (O) is placed in the range of detector 2 (D_2). Consequently its position can be collocated with the known position of detector 2.

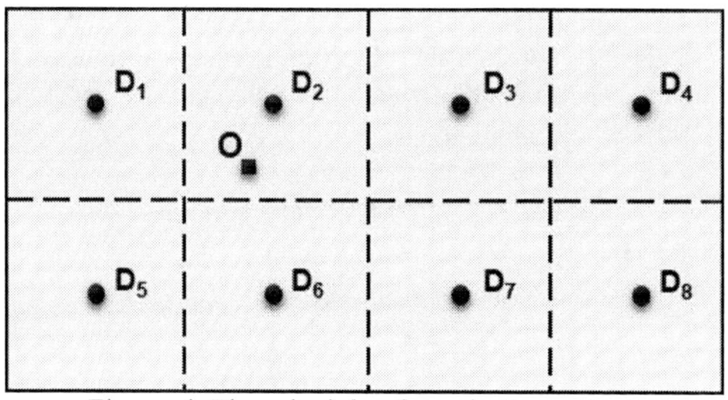

Figure 4. The principle of proximity method.

2.1.4. Vision analysis
This technique is based on the image processing of view which is usually captured by cameras which cover the whole indoor environment [3]. The cameras positions are known. The followed objects must be pre-defined in the database. When the tracked object gets into the range of particular camera, it can be identified and its position can be estimated.

2.2. Types of information provided by Indoor Positioning Systems
Depends on the application and the structure of IPS, different location information can be provided by the system. According to the [3], mainly three types of IPS information can be distinguished (figure 5). To provide the absolute information by IPS the map of indoor environment must be prepared and stored in the IPS. The map consists of reference points, according to which the object location can be estimated. The absolute location information is required for tracking and navigation systems, where exact position of the followed object must be determined. The relative location information is determined according to its own reference points [4]. An example of relative location might be the tracking of car door to check whether it is closed or open [3]. In such case, the relative position

between the tracked point (placed on the car door) and reference point (situated on the car body) is given. The exact position of the object is not always required. For some applications only the general information about the presence of the object within defined area is needed. This kind of location information is defined as a proximity. It provides only the information about the place where tracked object is [3]. As an example the tracking of medical equipment and patients within the hospital can be given.

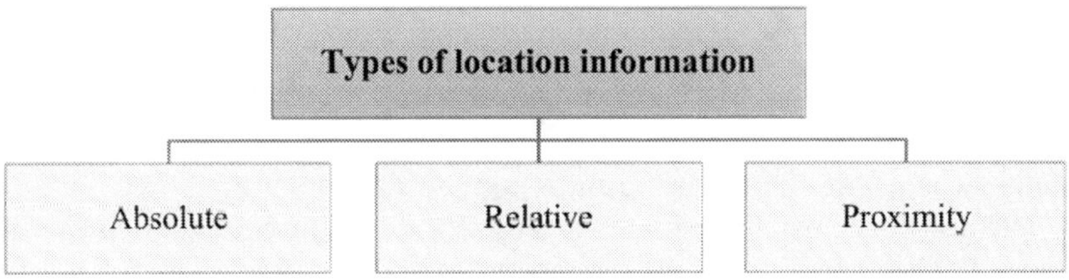

Figure 5. The types of information which can be provided by the indoor positioning systems [3].

2.3. Technologies used for Indoor Positioning Systems

There are different criteria according to which the IPSs can be distinguished. One of them might be the type of location information provided by the positioning system. Furthermore, indoor location systems can be divided into network-based and non-network-based [4]. However, the most natural and intuitive classification refers to the technology based on which the object location is determined (see figure 6). Like it was already stated in the present paper, different technologies have been already used for development of indoor positioning systems. A short characteristic of each technology is given accordingly.

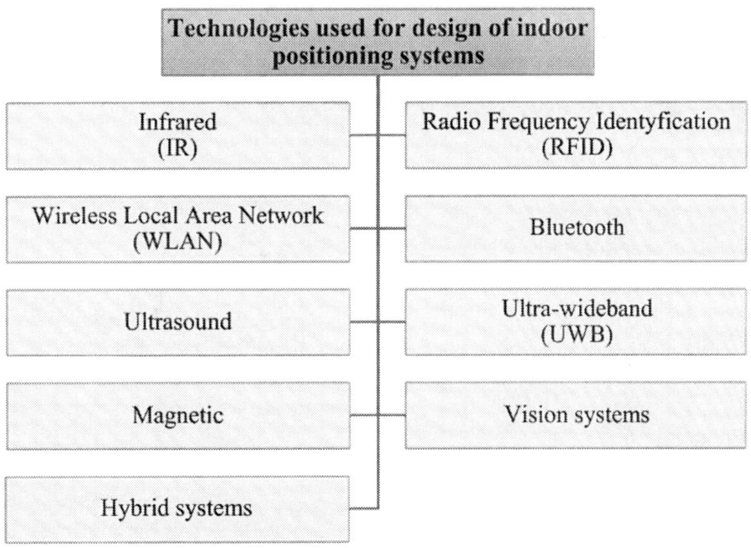

Figure 6. Classification of indoor positioning systems in relation to the technology based on which the object/human location is determined [3, 4].

2.3.1. RFID

The Radio-frequency Identification (RFID) technology ensures a wireless communication between reader and tags (called also transponders). These are two main components of RFID systems. The reading range depends on the RFID technique and operating frequency [6]. Strictly defined radio frequency and protocol are used to transmit and receive the data both, by reader and tags [4]. There are two types of tags: active and passive. The passive transponders are powered only by the electromagnetic field transmitted by reader. No additional battery is needed. This is the main advantage of passive tags, as well as small size and low cost. The passive RFID technology has been replacing gradually barcodes [3]. In opposite to the passive tags, the active ones have built-in battery. The tags works as transceivers which transmit their ID or further data (saved in chip) in reply to signal emitted by reader. One of the characteristic features of active RFID tag is wide reading range, even up to 100 m for Super High Frequency technology [6]. Furthermore, active transponders have usually larger memory and smaller antenna than passive tags. The batteries of active RFID tags have to be replaced periodically, what can be considered as main of their disadvantage.

2.3.2. IR
Infrared technology provides the wireless data transmission by the means of infrared radiation. The IR-based positioning systems requires unobstructed line-of-sight between transmitter and receiver. Moreover, the interference from strong source of light is not allowed [3]. For that reason the infrared indoor positioning systems are limited to transmission within single room. It is worth to mention that hybrid indoor location systems combining IR and RF technologies have been developed. Such combination takes the advantage of both mentioned techniques which increases the accuracy and applicability of the location system.

2.3.3. WLAN (IEEE 802.11)
Wireless local area networks operate usually in 2,4 GHz band. Over last years the WLANs have become very popular for domestic as well as commercial applications. Nowadays, WLAN technology has been implemented in almost all possible public areas, like department stores, hospitals, schools, universities, train stations, trains, offices, etc. The main advantage of WLAN-based indoor positioning systems is low cost of implementation, since they reuse already existing infrastructure. In the most of WLAN-based positioning systems the object location is determined by the measurements of received signal strength from a device (which is installed in the followed object) to a few access points [7]. The position of an object is calculated in relation to the known locations of access points. It must be noted, that the accuracy of WLAN-based indoor positioning systems is significantly affected by the environmental conditions. All the elements of indoor environments like walls, doors, furniture, electrical devices, human body and overlapping access points affect negatively the accuracy and reliability of RSS-based WLAN indoor positioning systems [3]. According to [4] the accuracy of positioning for typical WLAN-based location system using RSS measurements is approximately 3 – 30 m with refreshing of position information in about a few seconds.

2.3.4. Bluetooth (IEEE 802.15.1)
Bluetooth is one of the technologies which are used for wireless personal area network (WPAN). Bluetooth operates in a 2,4 GHz ISM band and depend on the signal power, provides wireless communication up to 100 m [2, 3]. Bluetooth standard is commonly used in mobile phones and has successively replaced the IR communication. Moreover, Bluetooth modules have been implemented into other devices, like laptops, TVs, printers, tablets, etc. Each Bluetooth tag (which works as a transceiver) has its own unique identification number and consequently it can be used to track the device in which it is installed.

2.3.5. Ultrasound
The design of ultrasound-based positioning systems is inspired by the bats which use the ultrasound signals for navigation purpose over the night. Ultrasounds are defined as waves with frequency above 20 kHz – which cannot be heard by human [8]. The systems based on ultrasound technology consists of tags which emit the ultrasound signals and receivers with known-position [3]. The location of object is calculated based on triangulation technique. The main disadvantage of ultrasound technology is the range limitation. The ultrasound signals are reflected by the walls [9].

2.3.6. Ultra-wideband (UWB)
The RF-based indoor positioning systems have limited scope of applications due to the multi-path distortion of radio signals reflected by walls - which affect greatly the accuracy of location. In ultra-

wideband technology the short-term pulses, less than 1 ns, are generated which allows to achieve wideband emission [3]. Consequently, the reflected signals can be filtered which increases the reliability and accuracy of the positioning system. Furthermore, ultra-wideband technology is characterized by low power consumption and fast data transfer.

2.3.7. Magnetic positioning systems
There are a few approaches where the geomagnetic field is used to find the device orientation [10]. Unfortunately, within the indoor environments many sources of disturbances can be found, what affect significantly the accuracy of the location system. Among others, the electrical devices, wires, piping or steel shell of buildings can be pointed out [10]. It can be stated, that geomagnetic field is useful for indoor positioning system only if it is stable over the long period of time. Another solution to determine the position of an object might be the measurement of artificially generated magnetic field [11]. This kind of positioning system contains electric coils with known positions and sensor of magnetic field attached to the followed object. If the strength of magnetic field from at least three coils is measured, then the triangulation principle can be applied to determine the position of object.

2.3.8. Vision systems
The vision systems provide the positioning and tracking of people and objects, even in the complex indoor environment. Furthermore, vision systems can recognize human or particular objects by their characteristic features. Such systems contain mainly the high resolution cameras which capture the view in the real time and CPU which is needed for image processing [12]. One of the advantages from vision system is lack of any additional tags or sensors which usually have to be carried by the objects which are followed [3]. Notwithstanding the above, it must be noted that vision systems have also a couple of limitations. Firstly, high computing power is required for in real-time image processing. Secondly, the vision systems are susceptible for noises which might be caused by reflecting light or changing the lighting level. What is more, the vision-based positioning systems are not suitable for dynamically changing environments and to track a few persons simultaneously [3].

2.3.9. Hybrid systems
As it was already highlighted, each technology has its advantages and limitations. Usually, the indoor positioning system implemented to a particular environment cannot be applied to different one. To overcome above mentioned problems, the hybrid systems combining different technologies are developed. An example might be IR-based positioning system with UHF-based positioning system or combination of RF with IR signals [4].

2.4. The application examples of indoor positioning systems
There are many fields and potential applications where the indoor positioning system could be implemented. Among others, the following examples can be pointed out:

- tracking of medical equipment within the hospital to prevent them from being stolen,
- monitoring of unauthorized access to restricted areas, for example in the hospitals [13],
- tracking the patients in the hospital to improve their flow and reduce transfer time [13],
- tracking of goods in the warehouse,
- tracking of goods and assets within the production plant,
- tracking of firefighters in the buildings where the rescue operation is conducted,
- navigation of cars in the parking to assist the drivers in finding the free place to park,
- navigation of customers within department stores to assist them in finding the shops they are searching for,
- navigation of readers in the library to assist them in finding the book they are searching for,
- navigation of people in the offices,
- navigation of collaborative robots which are dedicated to assist the human,
- tracking and navigation of elderly people,

- tracking and navigation of forklift and automated guided vehicles (AGV) within the production plant.

2.5. *The examples of indoor positioning systems*
Over last few decades, quite many indoor positioning systems - based on different technologies described in previous subsection – have been developed. Some of them have found a commercial application, others can be considered only as a scientific approaches. A couple of researches containing comprehensive overview of existing indoor positioning systems have been prepared recently [2, 3, 4, 9, 14, 15]. In these papers different indoor positioning systems based on variety of available technologies are described. The authors of above mentioned researches present the pros and cons of existing indoor location systems as well as try to evaluate their applicability in order to find the potential fields of implementation. In further part on this section only a few examples of indoor positioning systems are given with main focus on RFID-based IPSs.

2.5.1. *COMPASS*
COMPASS is an example of WLAN-based indoor positioning system which reuses existing WLAN infrastructure [16]. The position estimation of user is done by measurement of received signal strength in relation to different access points (APs) located in the indoor environment. The blocking effect caused by human body has been handled by digital compasses which detect the user orientation. Application of digital compasses has improved significantly the accuracy of the system [16]. The COMPASS system uses the fingerprinting method to determine the user position. The main principle of the COMPASS system is to sample the signal strength for preselected orientations at each reference point during the offline phase [16]. All collected data is stored as histograms in the database. In the online phase the measured values are compared with those collected in the offline stage. Consequently, the specific signal strength distributions for each positions can be calculated to improve the accuracy of position determination. The system has been tested in the hallway of an office building with the area of 312 square meters. There were six access points deployed within testing area. During the offline phase the measurements were conducted at 166 reference points (with approximately 1 m spacing). According to the authors results, the average positioning error less than 1.65 m was achieved. It has been proven by the authors that training set of 20 measurements at each reference is sufficient [16]. Further increasing of training set doesn't improve significantly the accuracy of positioning.

2.5.2. *Active Bat*
Active Bat is an ultrasound-based indoor positioning system which use triangulation method to determine the object or human position [17]. The system consists of tags (transmitters of ultrasound pulses) and receivers. The tag (called *Bat*) with size 7,5 x 3,5 x 1,5 cm is carried by the followed object [17]. The *Bats* are powered by 3.6V batteries which has to be replaced every 1,5 year. Each *Bat* has its own unique 48-bit code. The signal emitted by the *Bats* is captured by the receivers which are fixed to the ceiling. The receivers are situated in a grid with spacing of 1,2 m [17]. If the signal generated by the *Bat* arrives to three or more receivers, then the position of object can be determined. The orientation of an object can be calculated, if the relative positions of at least two *Bats* (attached to the followed object) are found. The tests of the system were conducted in the indoor environment of around 1000 square meters. The area was covered by 720 receivers. According to the authors the system can determine simultaneously the 3D position of 75 objects with accuracy around 3 cm [17].

2.5.3. *ViNav*
ViNav is a vision-based indoor navigation systems dedicated for smartphones [18]. The general idea of the system is to build 3D model of an indoor environment (in a form of 3D point clouds). According to this concept the 3D models are created and updated using crowdsourced visual and sensor data which is collected from smartphones [18]. Based on collected data an indoor space of interest is created by using SfM technique. Afterwards, the path meshes are created for path planning considering the obstacles that have been detected on 3D models. The representation of an indoor

environment created by this method may be imperfect and characterized by uneven points density distribution. To improve the system functionality, tracking of users (by motion sensors built-in the smartphones) and integration of registered paths to the navigation meshes have been proposed by authors [18]. System can be also implemented in the multi-stage buildings thanks to the barometric pressure sensor. With the 3D models made in SfM technique it is possible to determine the user location and face direction by using the picture taken in situ. Fast location of ViNav system is provided by the combination of image-based localization and fingerprinting method [18]. The 3D model is split into multiple smaller models based on the density of 3D points collected during an offline stage. The user send a request for location which contains picture and Wi-Fi fingerprints captured on site. These fingerprints are compared with fingerprints stored in the database. When the match is found, the location of user can be calculated. Once the destination is defined by user, ViNav can calculate the route based on the navigation meshes using path finding algorithms.

2.5.4. Non-satellite positioning system for vehicles

In paper [19] a non-satellite vehicle positioning system is presented. Despite the fact is not dedicated for indoor environment the authors of present article decided to mention it since it is based on RFID technology. The idea of described system is to determine the position of vehicle (equipped with RFID reader and antenna) on the road by reading the data saved in RFID tags which are placed on the road signs, traffic lights, trees, buildings etc. [19]. According to this concept the coordinates of tags are known and predetermined by any existing satellite positioning systems (like for example GPS). The map of tags with its coordinates must be pre-defined and stored in the database. The present RFID-based system can navigate the vehicle by determining the route from the nearest point (tag) where vehicle is located to the destination point through all the tags located between and saved in the database. The tags can also store in their memory information about permissible direction of driving, speed limitations, traffic signs, etc. To increase the reliability of the system there may be a few tags assigned to particular area coordinates. That would minimize the risk of missing the information saved in only one tag, in case the radio waves are reflected by any randomly occurred obstacle.

2.5.5. A RFID-based indoor positioning system using passive tags

A different approach of RFID-based indoor positioning system is described in paper [20]. In this concept the localization system takes the advantage of RSSI. The tags can be located on the floor or ceiling. In the experimental set-up the authors created a path of passive tags with spacing of 1,2 m. The vehicle with installed reader move forward parallel to this path at a distance of 2 m. A Kalman filter was implemented to decrease the reader location error. The average absolute position error achieved by the authors is about 0,1 m and the average absolute tag-reader distance error is 0,055 m [20]. The accuracy of the system can be improved by reducing the distance between adjacent tags or by decreasing the distance between the reader and path created from tags.

2.5.6. RFID-based navigation method

In the patent [21] a RFID-based positioning system was proposed. The navigation apparatus consists of RFID reader connected with three directional antennas and control unit. First antenna is mounted in the front of the object. Remaining two antennas are placed on the left and right side at angle of 45 ÷ 90 degrees to the front antenna. The tags with unique identification numbers and saved-in coordinates are placed along the roads/paths where the followed object is moving. The current object location and moving direction can be determined if at least one tag is read by at least one of the directional antenna connected with reader. Received signal strength is measured in order to estimate the object position. Based on this method also the navigation service can be realized.

2.5.7. RFID-based indoor positioning system combining tag detection probability and statistical distribution of measured received signal strength indication

An interesting approach for localization within indoor environments is presented in patent [22]. The room, in which the location was provided, had been covered by reference passive RFID tags with known positions. A couple of antennas (at least three) connected to the RFID reader are placed in selected points in order to cover the entire room by their range. The reader is connected with the central computing unit. The system is based on the learning method (fingerprinting). The central unit provides the location determination by measuring the received signal strength indication (RSSI) and considering the detection rate fingerprinting of reference tags. The location process consists of offline and online phase. At the offline stage (learning phase) the received signal strength of reference tags is measured. Then, the map with the positions of reference tags is created and saved in the database of central unit. In the online phase the real time localization of followed object is ensured by the system. The tracked object must carry two UHF RFID tags with different reading range. The tag with lower reading range is used to find the "zone" within the room in which the object is located. It can be stated that it is used for rough position estimation, since most likely it would be detected only by one antenna. The second tag has the same reading range like reference tags and it is used for exact position determination of the followed object. The RSSI of this tag is measured and then compared to the fingerprints saved in the database. Based on that, the exact position of an object can be easily determined. One of the advantages of presented system is shortened learning phase. The RSSI of tag attached to the followed object doesn't have to be measured at different positions within the indoor environment to create the database of fingerprints. The RSSI of reference tags is measured instead, what significantly speed up the learning process. Furthermore, thanks to the fixed reference tags the map can be easily "refreshed" and adopted to the indoor environments which is usually very complex and changes dynamically.

3. The concept of RFID-based indoor positioning system dedicated for autonomous vehicles

In the present paper the authors concept of RFID-based indoor positioning system dedicated for autonomous vehicles and robots is presented. The operation principle of the system is shown in the figure 7 and figure 8.

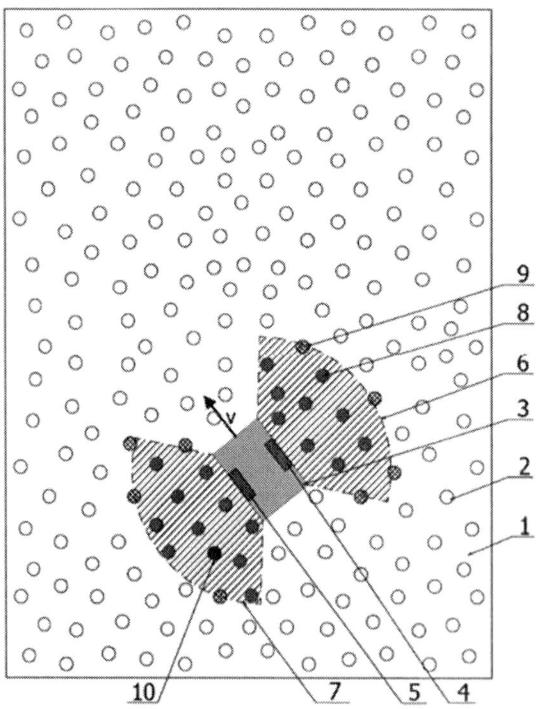

Figure 7. The operation principle of RFID-based indoor positioning system according to the authors concept, 1) indoor environment; 2) landmarks (tags) which cover the operational area; 3) autonomous vehicle; 4) right-side reader antenna; 5) left-side reader antenna; 6) range of right-side reader antenna; 7) range of left-side reader antenna; 8) landmarks (tags) detected by the reader antennas; 9) landmarks (tags) which are at the border of the antenna range; 10) a defected landmarks [22].

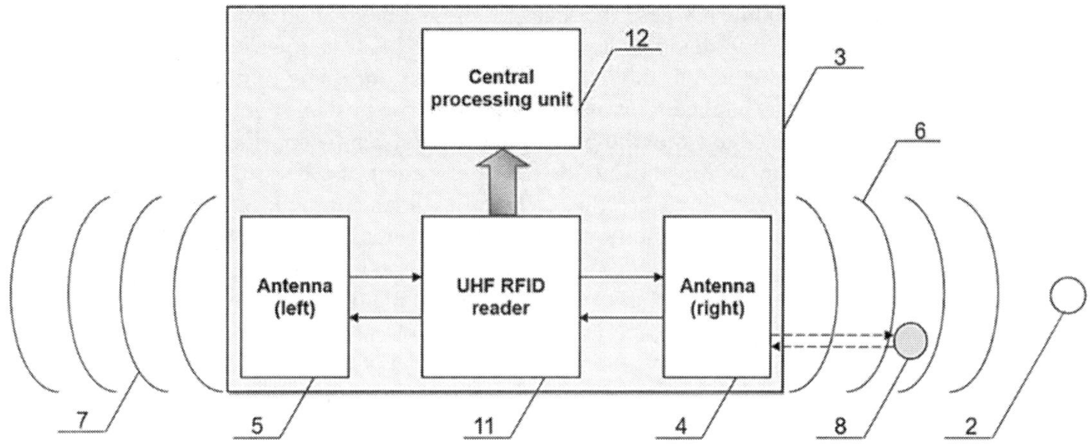

Figure 8. The scheme of control unit of present RFID-based indoor positioning system, 2) landmark (tag) which is not detected by the reader antennas - situated outside the operating range; 3) autonomous vehicle; 4) right-side reader antenna; 5) left-side reader antenna; 6) range of right-side reader antenna; 7) range of left-side reader antenna; 8) landmarks (tags) detected by the reader antennas; 11) UHF RFID reader; 12) Central processing unit [23].

The presented system is based on the fingerprinting technique. An autonomous vehicle (3) is equipped with UHF RFID reader (11) connected with two directional antennas, on the right (4) and left (5) side. The range of reader antennas is marked by broken lines and hatched (6, 7). The vehicle moves within the indoor environment (1) which is covered with "landmarks" (2). Each "landmark" consists of two RFID tags with different reading range: one with short reading range and second with long reading range. The tags can be attached on the floor randomly. There is no need to create any regular pattern or grid. During the deployment process of tags their positions (coordinates) are not known. The entire implementation process of described system consists of two phases: offline and online stage. Once the tags are situated in the indoor environment the offline learning process (mapping) can be started. In this process the coordinates of landmarks are found. The vehicle controlled remotely starts the mapping process from the known position (origin of the room coordinate system). It moves through the indoor environment and scan the whole area searching for the tags with short reading range. If the moving direction of vehicle and its distance from the origin are known, the position of each landmark can be determined. The angle of moving direction and distance can be measured by separate sensors built-in the autonomous vehicle. Once the tag with short reading range is detected, its position is calculated and saved in the database of central processing unit (12) which is connected to the UHF RFID reader. These coordinates are also saved in the memory of associated RFID tag with long reading range. When the map of landmarks is prepared and saved in the database of central processing unit, then the online navigation phase of vehicle can be started. In this process the vehicle drives and scans the area in search of RFID tags with long reading range. If at least two landmarks are detected by each antenna, the localization of vehicle can be determined.

Nevertheless, it is worth to mention about the disruptions which can occur and affect negatively the functioning of described localization system. One of the potential risks is a failure of tag with long reading range. Then, during the localization or navigation process the landmark with defected tag (10) would be not detected by reader antenna. Secondly, the landmark which is at the border of the antenna range (9) at a specific moment can be detected randomly. It also brings some uncertainty to the navigation process. Finally, in RFID-based systems the false negative reading (FNR) phenomena can occur [22]. It means that a particular tag might be not detected despite the fact it is covered by the antenna range. To face above mentioned problems, it is recommended that landmarks should be situated in the way that in every possible position of vehicle more than two of them can be detected by each antenna. The higher density of landmarks located in the indoor environment, the higher system accuracy and reliability.

The presented RFID-based indoor positioning system allows also to detect pre-defined obstacles situated within indoor environment where the localization and navigation are provided. The obstacles must be covered with the RFID tags with long reading range. The unique identification numbers of these tags must be saved and stored in the database of central processing unit. If the vehicle meets an obstacle, its trajectory can be changed to avoid the collision. The obstacles don't have to be situated in fixed locations. As an obstacle, also moving objects and human can be considered. Even if an obstacle is moved to another position within the indoor environment, it can be still detected by its unique identification number. However, to ensure this collision avoidance functionality the third antenna in front of the vehicle should be installed to cover its whole surrounding. In the same way the border of room can be marked to avoid the collisions with walls or another building structures. Within one indoor environments a navigation function can be provided to many vehicles what might be especially useful in the production plants. Then, each autonomous vehicle must be marked by tags to avoid the collision between them. It must be noted, that the vehicle does not detect randomly occurred obstacles which had not been pre-defined in the database of central processing unit.

4. Conclusion

In this paper the general overview of indoor positioning systems was presented as well as technologies and methods used to deploy of IPSs. Due to the quick development of new technologies the localization of human and objects has been getting more and more relevant. The application examples and potential areas where IPS could be deployed, were given in former sections of this paper. It is predicted that the global market value of indoor positioning systems can reach 10$ billion a year in 2024 [2]. That shows the relevance of this field of science and brings the attention of many scientific centers and researchers. The main development direction are IPSs based on the existing technologies and reusing already deployed infrastructures, if it is possible, like for example WLAN-based localization systems. Notwithstanding the above, simple, inexpensive, reliable and battery-free (energy efficient) solutions like RFID-based IPS still play a significant role.

The authors concept of RFID-based indoor positioning and navigation system dedicated for autonomous vehicles and robots was described. The following advantages of above mentioned indoor positioning system can be pointed out:

- low cost of the entire indoor positioning system,
- simplicity,
- wireless communication between the reader and landmarks (tags),
- battery-free design of tags based on RFID technology (the RFID passive tag, which constitute the main part of the sensor, is powered by electromagnetic field emitted by the reader),
- simultaneous communication between the reader and dozens of RFID tags,
- quick data processing,
- low computing power demand for signal processing,
- flexibility and adaptability of the system,
- reliability which increases with the density of deployed landmarks,
- once the map of indoor environment is done, it can be reused by other vehicles, robots and human for navigation purpose.

In further steps the experimental set-up of presented concept should be developed in order to define the technical characteristics like accuracy and repeatability of positioning etc. The reliability and resistance for disturbances must be investigated as well.

References

[1] TUV Austria, *Safety in Human-Robot Collaboration 2016*, www.tuv.at/i40 (accessed on May 2019)
[2] Dardari D, Closas P and Djurić P M 2015 Indoor Tracking: Theory, Methods, and Technologies, *IEEE Transactions on Vehicular Technology* **64**(4) 1263 – 1278
[3] Gu Y, Lo A and Niemegeers I 2009 A Survey of Indoor Positioning Systems for Wireless

Personal Networks, *IEEE Communications Surveys & Tutorials* **11**(1) 12 – 32

[4] Liu H, Darabi H, Banerjee P and Liu J 2007 Survey of Wireless Indoor Positioning Techniques and Systems, *IEEE Transactions on Systems, Man, and Cybernetics, Part C (Applications and Reviews)* **37**(6) 1067 – 1080

[5] Mulloni A, Wagner D, Barakonyi I and Schmalstieg D 2009 Indoor Positioning and Navigation with Camera Phones, *IEEE Pervasive Computing* **8**(2) 22 – 31

[6] Meng Z and Li Z 2016 RFID Tag as a Sensor - A Review on the Innovative Designs and Applications, *Measurement Science Review* **16**(6) 305 – 315

[7] Kaemarungsia K and Krishnamurthy P 2012 Analysis of WLAN's received signal strength indication for indoor location fingerprinting, *Pervasive and Mobile Computing* **8**(2) 292 – 316

[8] Hazas M and Hopper A 2006 Broadband Ultrasonic Location Systems for Improved Indoor Positioning, *IEEE Transactions On Mobile Computing* **5**(5) 536 – 547

[9] Harle R 2013 A Survey of Indoor Inertial Positioning Systems for Pedestrians, *IEEE Communications Surveys & Tutorials* **15**(3) 1281 – 1293

[10] Li B, Gallagher T, Dempster A G and Rizos C 2012 *How feasible is the use of magnetic field alone for indoor positioning?*, International Conference on Indoor Positioning and Indoor Navigation, Sydney, Australia, Now. 13 – 15, 1 – 11

[11] Blankenbach J and Norrdine A 2010 *Position Estimation Using Artificial Generated Magnetic Field*, International Conference on Indoor Positioning and Indoor Navigation (IPIN), Zürich, Switzerland, Sep. 15 – 17, 1 – 5

[12] Cooper A and Hegde P 2016 *An Indoor Positioning System Facilitated by Computer Vision*, IEEE MIT Undergraduate Research Technology Conference (URTC), Cambridge, USA, Nov. 4 – 6, 1 – 5

[13] Kamel Boulos M N and Berry G 2012 Real-time locating systems (RTLS) in healthcare: a condensed primer, *International Journal of Health Geographics* **11**(25) 1 – 8

[14] He S and Chan S-H G 2016 Wi-Fi Fingerprint-based Indoor Positioning: Recent Advances and Comparisons, *IEEE Communications Surveys & Tutorials* **18**(1) 466 – 490

[15] Fallah N, Apostolopoulos I, Bekris K and Folmer E 2013 Indoor Human Navigation Systems: A Survey, *Interacting with Computers* **25**(1) 21 – 33

[16] King T, Kopf S, Haenselmann T, Lubberger C and Effelsberg W 2006 *COMPASS: A Probabilistic Indoor Positioning System Based on 802.11 and Digital Compasses*, Proc. of the First ACM International Workshop on Wireless Network Testbeds, Los Angeles, USA, Sep., 34 – 40

[17] www.cl.cam.ac.uk/research/dtg/attarchive/bat/ (accessed on May 2019)

[18] Dong J, Noreikis M, Xiao Y and Ylä-Jääski A 2019 ViNav: A Vision-Based Indoor Navigation System for Smartphones, *IEEE Transactions on Mobile Computing* **18**(6) 1461 – 1475

[19] Simdiankin A, Byshov N and Uspensky I 2018 *A method of vehicle positioning using a non-satelite navigation system*, Thirteenth International Conference on Organization and Traffic Safety Management in Large Cities, Saint Petersburg, Russia, Sep. 27 – 29, 732 – 740

[20] Saab S S and Nakad Z S 2011 A Standalone RFID Indoor Positioning System Using Passive Tags, *IEEE Transactions on Industrial Electronics* **58**(5) 1961 – 1970

[21] Hyuk L, Jae-Hyung J and Young-Kwan K 2011 *Navigation Apparatus And Method Using RFID*, United States Patent, Patent No. US 7,978,066 B2

[22] Bekkali A and Kadri A 2016 *System and Method for RFID Indoor Localization*, United States Patent, Patent No. US 2016/0139238 A1

[23] Talar R and Regus M 2018 *Navigation System for Autonomous Vehicles in a Specific Workspace*, Poznan University of Technology, Poland, Patent Application

Simulation of vehicle's lateral dynamics using nonlinear model with real inputs

D Miloradović[1], J Glišović[1], N Stojanović[1] and I Grujić[1]

[1] University of Kragujevac, Faculty of Engineering, 6 Sestre Janjić Str., 34000 Kragujevac, Serbia

E-mail: neja@kg.ac.rs

Abstract. Basic understanding of vehicle's lateral dynamics may be achieved by using relatively simple mathematical-mechanical models. However, these models are often linearized and do not take into account nonlinear phenomena in vehicle's lateral dynamics and highly nonstationary nature of vehicle's vibrations in real driving conditions. One form of nonlinear nonstationary model of vehicle's lateral dynamics is developed in this paper. Simulations were conducted using experimentally recorded values of the steering wheel angle for generation of the front wheels steer angle, and vehicle longitudinal velocity as inputs to the model. Results of simulations of vehicle angular vibrations around the vertical axis, angular body side slip vibrations, tire slip angles, vehicle lateral acceleration wheel forces and inertial force are presented and discussed. Finally, conclusions regarding possible further use of the developed model were drown.

1. Introduction

Simple vehicle dynamics models that specifically represent its lateral, longitudinal and vertical dynamics have been established for basic understanding of its dynamic behaviour. These types of models are commonly used with linearized form of mechanical-mathematical equations, which is desirable in design of stability controllers, active suspension systems and other driver's assistance systems.

Required complexity of the vehicle model depends on its concrete application. For example, for research of the vehicle handling characteristics during its planar motion, simple models that include only degrees of freedom in the plane are sufficient. Two-axle models with one or two tracks are used in the analysis of vehicle stability. Single-track model is used for the analysis of vehicle road holding ability or the analysis of vehicle roll. In addition to single-track models, there are many other models modelling the vehicle lateral motion with different levels of complexity. These models are usually integrated dynamical models that observe vehicle longitudinal and lateral velocity simultaneously, by observing the tire slip angle.

The usual models for vehicle lateral dynamics are [1]:

- two-axle models with single track,
- two-axle models with two tracks - planar models,
- single-axle models with two tracks and
- spatial models.

Lateral dynamics of vehicles has been studied since 1950's [2]. However, Riekert & Schunk proposed a very often used single-track model, frequently referred to as "Riekert-Schunk single-track model", back in 1940. It involved some simplifications which did not significantly influence the basic analysis of vehicle behaviour for lateral accelerations up to 0.4g [3], but which notably reduced a number of degrees of freedom of the system. Years later, this model was used as an ESP input in practice [4].

Analysis of simple single-track models can explain a large part of dynamic behaviour during handling of the vehicle, especially during stationary ride [5-8]. This enables faster testing and analysis of vehicle behaviour during ride and easy translation of the models to the language of simulation programs.

Although they contain certain reductions, the single-track models of vehicle lateral dynamics are nowadays used in the design of steering system control units. However, due to their nature, these models do not take into account the redistribution of weight in the lateral direction, nor the angular oscillations of the vehicle around the longitudinal axis. That is why the spatial two-track models of the vehicle are used instead, which observe the behaviour of each tire separately and which may be used in the design of the electronic stability control systems of the vehicle. The two-track models of vehicle's lateral dynamics may have a complex structure and they demand modelling using modern computer programs.

The linear single-track models contain simplifications regarding the assumption of small angles and linear dependence between wheel forces and slip angle. In addition, they do not take into account highly nonstationary nature of vehicle vibrations induced by variable vehicle velocity or road roughness [9]. Considering the previous limitations, a nonlinear nonstationary single-track model of the vehicle was used for simulation of vehicle's lateral dynamics in this paper. The used model considered larger values of body slip angles and tire slip angles, in which case the linearized equations of motion were no longer valid. In addition, the selected model considered longitudinal, lateral and yow motion of the vehicle, under the assumption that the vehicle travels over even road and that there was negligible lateral weight shift and roll of the vehicle.

2. Nonlinear single-track model of the vehicle

Linear equations of motion of the vehicle single-track model do not adequately describe the vehicle's motion when large steering angles or body slip angles larger than 10° occur. That is why nonlinear equations describing the vehicle's lateral dynamics should be used. The nonlinear model of the vehicle lateral dynamics adopted in this research is shown in figure 1, where:

- m is vehicle mass,
- j_x, j_y are longitudinal and lateral acceleration of the vehicle, respectively,
- F_{ix}, F_{iy} are longitudinal and lateral components of the inertial force, respectively,
- Y_1, Y_2 are lateral forces acting on the front and the rear wheels, respectively,
- J_z is moment of inertia of mass m about vertical gravity axis,
- a, b are distances between the front and the rear axle and the centre of gravity, T, respectively,
- l is vehicle's wheel base,
- r is turn radius,
- v is vehicle velocity,
- v_1, v_2 are the front and the rear wheel velocities, respectively,
- φ is ride angle (angle between vehicle velocity direction and reference axis direction),
- ε is angle od rotation of vehicle body about vertical axis (angle between longitudinal vehicle axis direction and reference axis direction, yaw angle),
- θ - body side slip angle (angle between velocity direction and longitudinal vehicle axis direction),
- λ_1, λ_2 are steering angles of the front and the rear wheels, respectively,

- δ_1, δ_2 are tire slip angles of the front and the rear wheels, respectively.

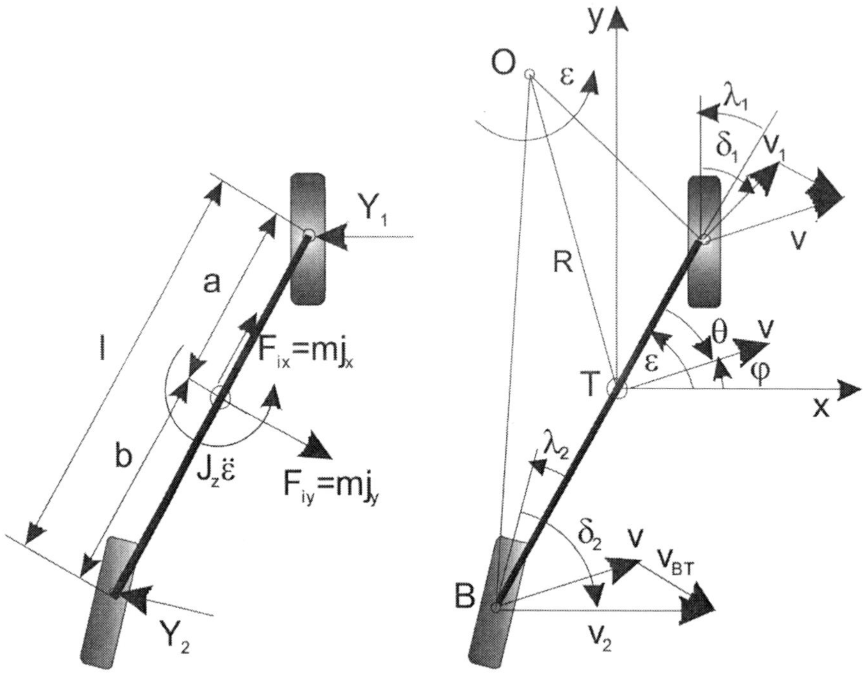

(a) (b)

Figure 1. Nonlinear single-track model of the vehicle: (a) forces acting on vehicle, (b) definition of the observed angles and velocity directions.

In simplified models, tire lateral forces, Y_1, Y_2, are perpendicular to vehicle's longitudinal axis. In reality, these forces act perpendicular to wheel longitudinal axis, figure 1 (a), and this must be taken into account when writing the equations of motion. Unlike the linearized equations of motion of a single-track vehicle model, the equations of motion of nonlinear model contain projections of forces in the direction of longitudinal axis of the vehicle, which, in this case, appear as deceleration forces acting on the centre of gravity in longitudinal vehicle axis, x_v:

$$m \cdot j_{x_v} = -Y_1 \cdot \sin \lambda_1 - Y_2 \cdot \sin \lambda_2 \tag{1}$$

and in lateral vehicle axis, y_v:

$$m \cdot j_{y_v} = Y_1 \cdot \cos \lambda_1 + Y_2 \cdot \cos \lambda_2 \ . \tag{2}$$

Acceleration vector of some point has its tangential and normal component:

$$\vec{a} = \dot{v} \cdot \vec{e}_t + \frac{v^2}{R} \cdot \vec{e}_n \tag{3}$$

According to figure 1, tangential ($\vec{e}_t$) and normal ($\vec{e}_t$) component of the unity vector are obtained in the following manner:

$$\vec{e}_t = \begin{pmatrix} \cos\theta \\ -\sin\theta \end{pmatrix}, \tag{4}$$

$$\vec{e}_n = \begin{pmatrix} \sin\theta \\ \cos\theta \end{pmatrix}. \tag{5}$$

Now, acceleration components of vehicle mass in direction of x_v- axis and y_v-axis are:

$$j_{x_v} = \dot{v}\cdot\cos\theta + \frac{v^2}{R}\cdot\sin\theta\,, \tag{6}$$

$$j_{y_v} = -\dot{v}\cdot\sin\theta + \frac{v^2}{R}\cdot\cos\theta\,. \tag{7}$$

Considering that:

$$\frac{v^2}{R} = \frac{v}{R}\cdot\dot{\varphi}\cdot R = v\cdot\left(\dot{\varepsilon}-\dot{\theta}\right), \tag{8}$$

the following relations are obtained:

$$j_{x_v} = \dot{v}\cdot\cos\theta + v\cdot\left(\dot{\varepsilon}-\dot{\theta}\right)\cdot\sin\theta\,, \tag{9}$$

$$j_{y_v} = -\dot{v}\cdot\sin\theta + v\cdot\left(\dot{\varepsilon}-\dot{\theta}\right)\cdot\cos\theta\,. \tag{10}$$

By substituting equations (9) and (10) into equations (1) and (2), respectively, the following equations are obtained:

$$m\cdot\left[\dot{v}\cdot\cos\theta + v\cdot\left(\dot{\varepsilon}-\dot{\theta}\right)\cdot\sin\theta\right] = -Y_1\cdot\sin\lambda_1 - Y_2\cdot\sin\lambda_2\,, \tag{11}$$

$$m\cdot\left[-\dot{v}\cdot\sin\theta + v\cdot\left(\dot{\varepsilon}-\dot{\theta}\right)\cdot\cos\theta\right] = Y_1\cdot\cos\lambda_1 + Y_2\cdot\cos\lambda_2 \tag{12}$$

By expressing the acceleration $\dot{v}$ from equations (11) and (12) and equating the two obtained relations, the following relations are gained:

$$\dot{v} = -\frac{Y_1\cdot\sin\lambda_1 + Y_2\cdot\sin\lambda_2}{m\cdot\cos\theta} - v\cdot\left(\dot{\varepsilon}-\dot{\theta}\right)\cdot\frac{\sin\theta}{\cos\theta} = -\frac{Y_1\cdot\cos\lambda_1 + Y_2\cdot\cos\lambda_2}{m\cdot\sin\theta} + v\cdot\left(\dot{\varepsilon}-\dot{\theta}\right)\cdot\frac{\cos\theta}{\sin\theta}\,, \tag{13}$$

or

$$\frac{Y_1\cdot\cos\lambda_1 + Y_2\cdot\cos\lambda_2}{m\cdot\sin\theta} - \frac{Y_1\cdot\sin\lambda_1 + Y_2\cdot\sin\lambda_2}{m\cdot\cos\theta} = v\cdot\left(\dot{\varepsilon}-\dot{\theta}\right)\cdot\left(\frac{\cos\theta}{\sin\theta}+\frac{\sin\theta}{\cos\theta}\right). \tag{14}$$

By multiplying the first and the second term of the left side of equation (14) with $\frac{\cos\theta}{\cos\theta}$ and $\frac{\sin\theta}{\sin\theta}$, respectively, and applying the sine theorem to the right side of the same equation, the following is obtained:

$$\frac{\left(Y_1\cdot\cos\lambda_1 + Y_2\cdot\cos\lambda_2\right)\cdot\cos\theta}{m\cdot\sin\theta\cdot\cos\theta} - \frac{\left(Y_1\cdot\sin\lambda_1 + Y_2\cdot\sin\lambda_2\right)\cdot\sin\theta}{m\cdot\cos\theta\cdot\sin\theta} = v\cdot\left(\dot{\varepsilon}-\dot{\theta}\right)\cdot\frac{1}{\sin\theta\cdot\cos\theta}\,, \tag{15}$$

or

$$\frac{Y_1\cdot\left(\cos\lambda_1\cdot\cos\theta - \sin\lambda_1\cdot\sin\theta\right)}{m\cdot\sin\theta\cdot\cos\theta} + \frac{Y_2\cdot\left(\cos\lambda_2\cdot\cos\theta - \sin\lambda_2\cdot\sin\theta\right)}{m\cdot\sin\theta\cdot\cos\theta} = \frac{v\cdot\left(\dot{\varepsilon}-\dot{\theta}\right)}{\sin\theta\cdot\cos\theta}. \tag{16}$$

By applying the angle addition formula $\cos(\alpha+\beta) = \cos\alpha\cdot\cos\beta - \sin\alpha\cdot\sin\beta$ to equation (16) and then expressing $\dot{\theta}$ from the same equation, a differential equation for body slip angle is formed:

$$\dot{\theta} = \dot{\varepsilon} - \frac{1}{m\cdot v}\cdot\left[Y_1\cdot\cos\left(\lambda_1+\theta\right) + Y_2\cdot\cos\left(\lambda_2+\theta\right)\right]. \tag{17}$$

Equation of equilibrium of moments for the adopted vehicle model is:

$$J_z\cdot\ddot{\varepsilon} = Y_1\cdot \mathrm{a}\cdot\cos\lambda_1 - Y_2\cdot \mathrm{b}\cdot\cos\lambda_2\,. \tag{18}$$

Substituting known relations between lateral forces and tire slip angles [1, 3]:

$$\begin{aligned} Y_1 &= k_1\cdot\delta_1\,, \\ Y_2 &= k_2\cdot\delta_2\,, \end{aligned} \tag{19}$$

(where k_1, k_2 are the front and the rear lateral force coefficients, respectively) and definitions of tire side slip angles, according to figure 1:

$$\delta_1 = \lambda_1 - \operatorname{arctg}\left(\frac{a \cdot \dot{\varepsilon} - v \cdot \sin\theta}{v \cdot \cos\theta}\right),$$
$$\delta_2 = \lambda_2 + \operatorname{arctg}\left(\frac{b \cdot \dot{\varepsilon} + v \cdot \sin\theta}{v \cdot \cos\theta}\right). \tag{20}$$

into equations (17) and (18), the following differential equations of motion are obtained:

$$\dot{\theta} = \dot{\varepsilon} - \frac{1}{m \cdot v}\left\{k_1 \cdot \left[\lambda_1 - \operatorname{arctg}\left(\frac{a \cdot \dot{\varepsilon} - v \cdot \sin\theta}{v \cdot \cos\theta}\right)\right] \cdot \cos(\lambda_1 + \theta) + k_2 \cdot \left[\lambda_2 + \operatorname{arctg}\left(\frac{b \cdot \dot{\varepsilon} + v \cdot \sin\theta}{v \cdot \cos\theta}\right)\right] \cdot \cos(\lambda_2 + \theta)\right\}, \tag{21}$$

$$\ddot{\varepsilon} = \frac{1}{J_z} \cdot \left\{k_1 \cdot \left[\lambda_1 - \operatorname{arctg}\left(\frac{a \cdot \dot{\varepsilon} - v \cdot \sin\theta}{v \cdot \cos\theta}\right)\right] \cdot \mathrm{a} \cdot \cos\lambda_1 - k_2 \cdot \left[\lambda_2 + \operatorname{arctg}\left(\frac{b \cdot \dot{\varepsilon} + v \cdot \sin\theta}{v \cdot \cos\theta}\right)\right] \cdot \mathrm{b} \cdot \cos\lambda_2\right\}. \tag{22}$$

Equations (21) and (22) describe mutual relations between different parameters of motion in nonlinear single-track model of lateral dynamics and parameters of vehicle and tires.

When only front wheels are steered ($\lambda_2 = 0$), equations (21) and (22) become:

$$\dot{\theta} = \dot{\varepsilon} - \frac{1}{m \cdot v}\left\{k_1 \cdot \left[\lambda_1 - \operatorname{arctg}\left(\frac{a \cdot \dot{\varepsilon} - v \cdot \sin\theta}{v \cdot \cos\theta}\right)\right] \cdot \cos(\lambda_1 + \theta) + k_2 \cdot \operatorname{arctg}\left(\frac{b \cdot \dot{\varepsilon} + v \cdot \sin\theta}{v \cdot \cos\theta}\right) \cdot \cos\theta\right\}, \tag{23}$$

$$\ddot{\varepsilon} = \frac{1}{J_z}\left\{k_1 \cdot \left[\lambda_1 - \operatorname{arctg}\left(\frac{a \cdot \dot{\varepsilon} - v \cdot \sin\theta}{v \cdot \cos\theta}\right)\right] \cdot \mathrm{a} \cdot \cos\lambda_1 - k_2 \cdot \operatorname{arctg}\left(\frac{b \cdot \dot{\varepsilon} + v \cdot \sin\theta}{v \cdot \cos\theta}\right) \cdot \mathrm{b}\right\}. \tag{24}$$

Equations (23) and (24) were the basis for development of MATLAB/Simulink program for simulation of lateral vehicle dynamics in case of nonlinear nonstationary vehicle model with real inputs, figure 2.

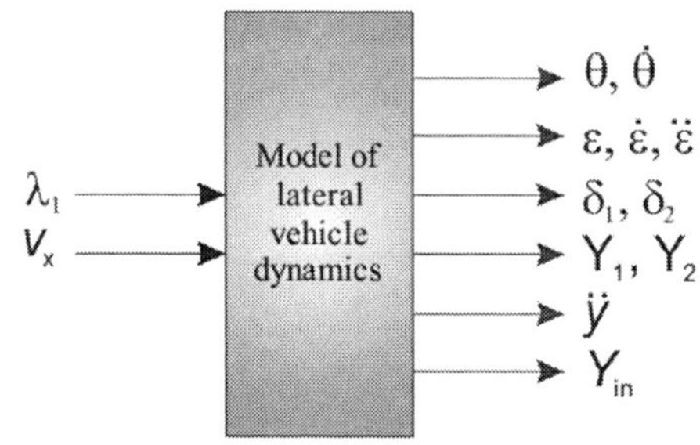

Figure 2. Inputs and outputs of the model of lateral vehicle dynamics.

The used mathematical model introduces the coupling between lateral and longitudinal vehicle dynamics through measured longitudinal vehicle velocity component, v_x. Namely, directly measured longitudinal component of vehicle velocity, v_x, and the front wheel steering angle, λ_1, determined from the measured steering wheel angle, β_v , were used as real inputs of the mathematical model.

Measured longitudinal component of vehicle velocity, v_x, is related to vehicle velocity, v, and body side slip angle, θ, in the following manner:

$$v_x = v \cdot \cos\theta . \tag{25}$$

The relation between the steering angle of the front wheels, λ_1, and the measured steering wheel angle, β_v, and steering system ratio, i_u, is as follows:

$$\lambda_1 = \frac{\beta_v}{i_u} \tag{26}$$

Developed mathematical model in figure 2 enables simulation of different output values:

- $\ddot{\varepsilon}, \dot{\varepsilon}, \varepsilon$ - vehicle angular vibrations around the vertical gravity axis,

- $\dot{\theta}$, θ - body slip angular vibrations,
- δ_1, δ_2 - side slip angles of the front and the rear tires, respectively,
- Y_1, Y_2 - lateral forces on the front and the rear wheels, respectively,
- $\ddot{y}$ - lateral acceleration at vehicle suspended mass centre of gravity and
- Y_{in} - lateral component of inertial force at the vehicle's centre of gravity.

3. Results of simulation of lateral vehicle dynamics

Simulations were conducted for a passenger vehicle with parameters given in table 1.

Table 1. Vehicle parameters [10].

Parameter, unit	Value
m, kg	1090
J_z, $\text{kg}\cdot\text{m}^2$	2000
a, m	1.4
b, m	1.1
k_1, $\text{N}\cdot\text{rad}^{-1}$	44500
k_2, $\text{N}\cdot\text{rad}^{-1}$	56500
i_u, –	17.4

Figures 3 to 11 present simulation inputs and outputs for vehicle motion with variable speed over asphalt in good condition. During experiment, the driver was asked to turn the steering wheel according to harmonic low.

Figure 3 (a) shows the recorded values of the steering wheel angle, β_v. The driver has, basically, achieved a quasiharmonic function of the steering wheel angle, with dominant frequency of 0.293 Hz, that is visible in diagram of spectrum intensity of the steering wheel angle shown in figure 3 (b).

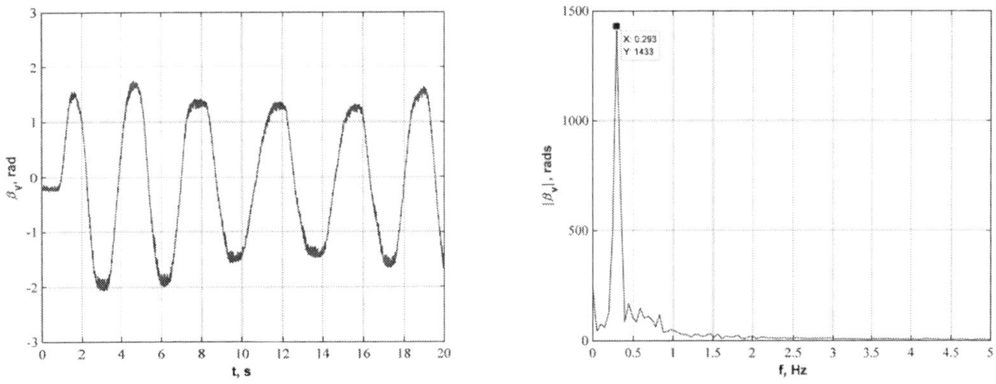

(a) (b)

Figure 3. Steering wheel angle: (a) in time domain, (b) in spectral domain.

Steer angle of the front wheels, λ_1, was calculated as an input, according to equation (26) and presented in figure 4. Longitudinal velocity of the vehicle, v_x, recorded during experiment is presented in figure 5. Vehicle's velocity was variable, with periods of vehicle accelerating and decelerating, inducing nonstationary behavior.

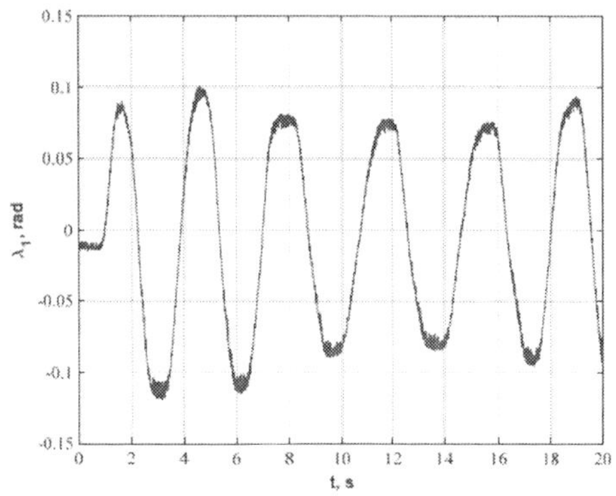

Figure 4. Steer angle of the front wheels, λ_1.

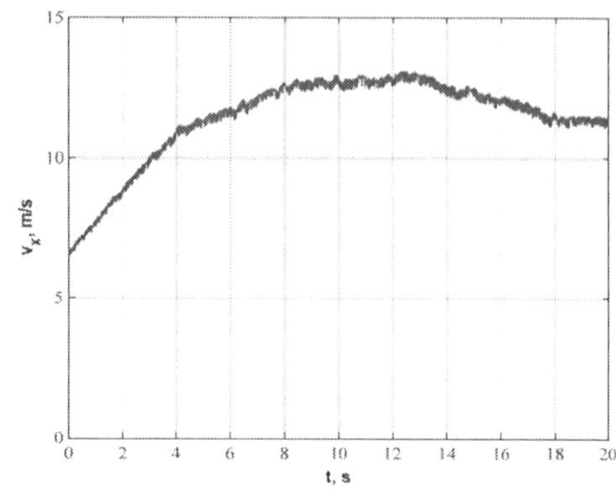

Figure 5. Longitudinal vehicle velocity, v_x.

As results of vehicle's lateral dynamics simulation using the adopted model, angular vibrations of the vehicle around the vertical axis (yaw angle, ε) were presented in figure 6. It may be seen that the curve presented in figure 6 has a wave character which corresponds to the model's basic excitation - steer angle of the front wheels, λ_1.

Figure 7 presents vehicle body slip angle, θ. Angular body slip vibration has also the wave shape that corresponds to model excitation in the form of quasiharmonic function of λ_1. Maximal values of body slip angle θ are found in the area of $\pm 1.4^{\circ}$.

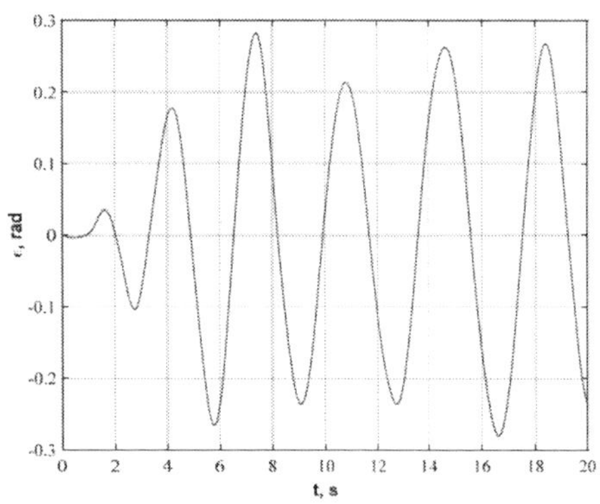

Figure 6. Angular vibrations of the vehicle around the vertical vehicle axis, ε.

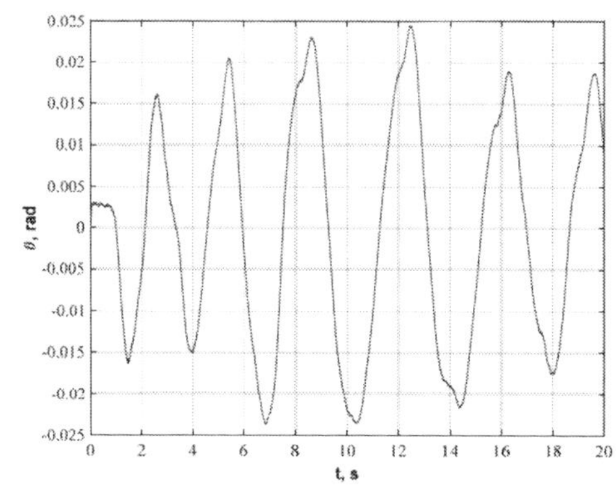

Figure 7 Body side slip angular vibrations, θ.

Values of tire side slip angles obtained by simulation of vehicles lateral dynamic are shown in figure 8. The front tire side slip angle is presented in figure 8 (a), while the rear tire side slip angle is presented in figure 8 (b). It may be seen that values of side slip angles are less than 4°, so the use of linear relationship between the tire side slip angles and corresponding lateral dynamic reactions on the wheels is justified.

Results of simulation of lateral dynamic reaction on the front wheels, Y_1, are shown in figure 9 (a), while those of lateral dynamic reaction on the rear wheels, Y_2, are shown in figure 9 (b). Values of the lateral dynamic reaction on the rear wheels are greater than corresponding values on the front wheels. This is due to the position of the vehicle's center of gravity, which is more distant from the front axle than from the rear axle$(a > b)$. For this reason, the portion of the lateral inertial force that is distributed to the rear axle is bigger than that distributed to the front axle.

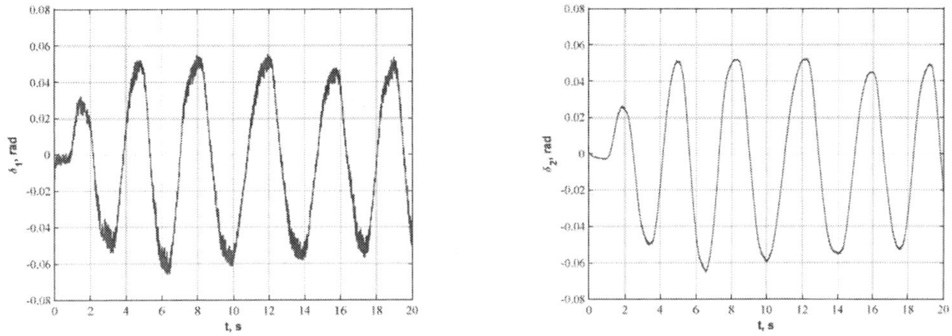

(a) (b)

Figure 8. Side slip angles of: (a) the front wheels, (b) the rear wheels.

Angular rotation of the wheels induces the vehicle lateral acceleration, $\ddot{y}$, shown in figure 10. This acceleration has a trend that corresponds to given input - steer angle of the front wheels, λ_1.

The resulting lateral inertial force is shown in figure 11. It may be seen that this force is approximately equal to the sum of lateral reactions on the front and the rear wheels, because, due to small values, the influence of the front wheels steer angle, λ_1, is practically negligible.

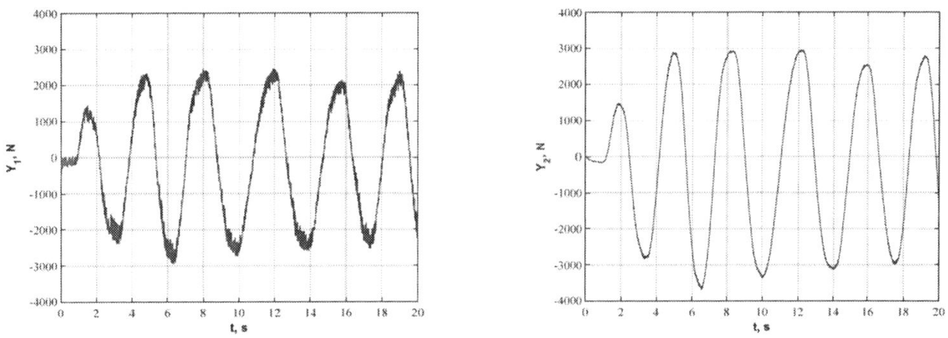

(a) (b)

Figure 9. Lateral dynamic reactions: (a) on the front wheels, (b) on the rear wheels.

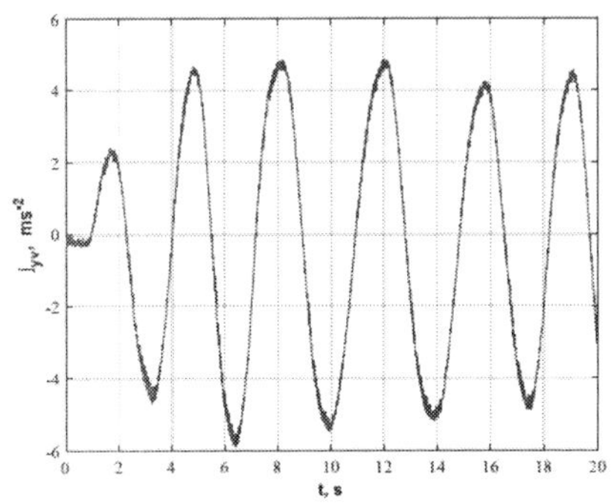

Figure 10. Vehicle's lateral acceleration.

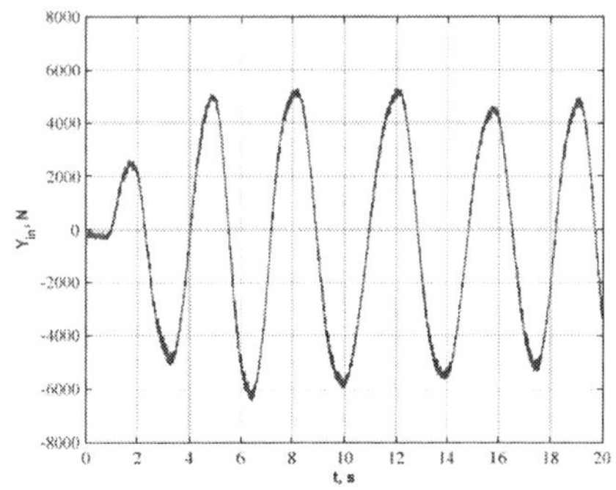

Figure 11. Vehicle's lateral inertial force.

4. Conclusions

Linear single-track models of vehicle lateral dynamics can help in understanding the vehicle handling behavior, especially during stationary ride. However, in order to analyze the nonlinear phenomena in lateral vehicle dynamics during nonstationary ride, nonlinear nonstationary single-track models ought to be used.

The developed nonlinear model of vehicle lateral dynamics was used for simulation with real, experimentally obtained inputs – variable: vehicle longitudinal velocity and steering wheel angle in the form of quasiharmonic curve. Model offers simulation results for several output variables that are important for understanding the vehicle lateral dynamics. The obtained results show that all output curves have a wave character which corresponds to the model's basic excitation - steer angle of the front wheels. In addition, all simulated curves have the same dominant frequency as the input curve of the front wheel steer angle. The influence of variable velocity and system nonlinearities can not be clearly seen in the time domain, so further analysis in spectral domain is necessary.

Acquired experimental data and the developed single-track model of the vehicle lateral dynamics may serve as a basis for application of estimation techniques in system parameter identification and for solving the problem of optimal control with the aspect of estimation of lateral vehicle dynamics.

Acknowledgment

Research presented in this paper was financially supported by the Ministry of education, science and technological development of the Republic of Serbia, grant number TR35041.

References

[1] Jankovic A 2008 *Car dynamics* (in Serbian), Faculty of Mechanical Engineering from Kragujevac
[2] Li L and Wang, F Y 2007 *Advanced Motion Control and Sensing for Intelligent Vehicles,* Springer Science+Business Media, LLC
[3] Schramm D, Hiller M and Bardini R 2014 *Vehicle Dynamics – Modeling and Simulation*, Springer-Verlag
[4] Mitschke, M 2005 Das Einspurmodell von Riekert-Schunck, *ATZ - Automobiltech Zeitschrift* **107**(11) 1030-31
[5] Jazar, R 2008 *Vehicle Dynamics: Theory and Applications,* Springer Science+Business Media, LLC, 2008
[6] Wong, J Y 2001 *Theory of ground vehicles,* John Wiley & Sons, Inc.
[7] Borner, M 2006 *Modellierung, Analyse und Simulation der Fahrzeugquerdynamik*, In book: Isermann R (Ed.) Fahrdynamik-Regelung, 3 47-79, Friedr. Vieweg & Sohn Verlag | GWV Fachverlage GmbH
[8] Pacejka, H B 2006 *Tire and Vehicle Dynamics,* Butterworth-Heinemann
[9] Rouillard, V and Sek, M A 2001 Simulation of non-stationary vehicle vibrations, *Proceedings of the Institution of Mechanical Engineers, Part D: Journal of Automobile Engineering* **215**(10), 1069–75
[10] Miloradović, D 2012 *Research of mechanical and functional couplings of vehicle dynamics system from the aspect of required performance*, Faculty of Engineering of the University of Kragujevac, Serbia, Doctoral Thesis

IOP Conf. Series: Materials Science and Engineering **659** (2019) 012061 doi:10.1088/1757-899X/659/1/012061

Railway carriage mass impact on retarder noise

P Zvolensky[1], M Kollar[1], L Lestinsky[1] and J Dungel[2]

[1] University of Zilina, Faculty of Mechanical Engineering, Univerzitna 1, 01026 Zilina, Slovakia

[2] CEIT, a.s., Univerzitna 6A, 01008 Zilina, Slovakia

E-mail: peter.zvolensky@fstroj.uniza.sk

Abstract. The paper is aimed on railway brake noise on set up track in a Teplicka nad Vahom arranging station. On the basis of noise measurements, it is possible to determine the presumption of impact of rail brake operation on municipalities in the vicinity of the set up rail area and to adjust its technology in the time frame so that the statutory noise limits are not exceeded. Also is possible to determine, if it is possible to use the technology in praxis to pass requirements for statutory noise limits. For operators of the arranging station and railway brakes is very important to set time schedule properly for every freight train set. It is precisely for this reason that technology can be designed according to the catchment density and the duration of the individual wagon alignment so that the freight trains concerned are not delayed. It is also possible to take into account that a shifted group of wagons with lower weight per axle will cause less braking noise.

1. Introduction

In the ŽSR railway network are many marshalling yards, but only few of them are classified as a hump yards. Equipment of those hump yard differs and in some occasions does not fulfil requirements for efficient train marshalling. Those marshalling yards can be equipped with various types of retarders. In general, equipment of ŽSR marshalling yards does not fulfil required standard every time [1]. The paper is focused on the weight impact on a retarder noise. The measurements were performed on the retarder PHB 04-S0 (RO), which is used in the Teplička nad Váhom marshalling yard [2].

2. Principe of a retarder

The braking effect of a single retarder unit is formed by two brake pads with brake rails pivotably mounted along one rail. The pressure is transferred by a hydraulic system to the brake pads which clamp the wagon wheels. For the reliable operation of the unit in the working pressure range, a hydraulic power unit located in a separate building near the retarder together with a four-stage hydraulic control. Hydraulic distributions are made by pressure hoses placed in plastic tubes with branched shafts. The output speed is evaluated by an electronic control unit with a radar speed meter. When the set output speed is reached, the retarder is activated (automatic mode). The retarder can be placed on the left or right rail, straight track or arc track with radius up to 160 m. The rail shape can be S 49 or R 65 for both flat and inclined mounting. According to the STN 34 2600 standard, the retarder is designed for work in an environment with a temperature range of -30 ° C to + 40 ° C [3].

Figure 1. Operation of retarder.

3. PHB 04-S0 (RO) retarder

PHB 04-SO (RO) retarder is single rail type of the retarder with spring-hydraulic mechanism, consisting of a set of four or more pads. The device is designed to regulate the speed of marshalling carriages in a gravity-type hump yard of marshalling station. It is designed as a modular element of mechanization and automation of the marshalling yards. To achieve a desired braking effect, the retarders consist of different numbers of pads. The smoothness of the carriage speed is ensured by their appropriate location. The brake is controlled by a hydraulic power unit that provides proportional contact force to the wheels of the carriage and thus the braking power. Wheelset mass is measured by tensometric sensor, and then appropriate braking force is calculated. Retarder consists of microprocessor control also. Control of the retarder can be performed manually or semi-automatically. Automatic control of the retarder can be done with microprocessor control unit. The retarder can be part of higher grade of automation [4].

Figure 2. PHB 04-S0 (RO).

4. Noise level of the retarder in the Teplička nad Váhom marshalling yard

On the basis of noise measurements performed at the Teplička nad Váhom marshalling yard, it is possible to determine the assumption of the impact of rail brake operation on municipalities in the vicinity of the marshalling yard. The PHB 04-S0 (RO) retarder is used in the marshalling yard. Various types of freight carriages were measured to determine how the carriage weight affects the retarder noise. Measurements were performed by STN EN ISO 3095:2013-12 standards, 7.5 m from track, in 1.2 m height, located in the middle of retarder.

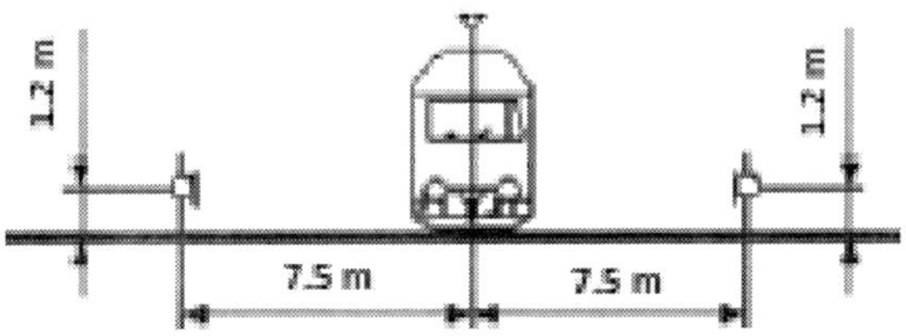

Figure 3. Measurement configuration.

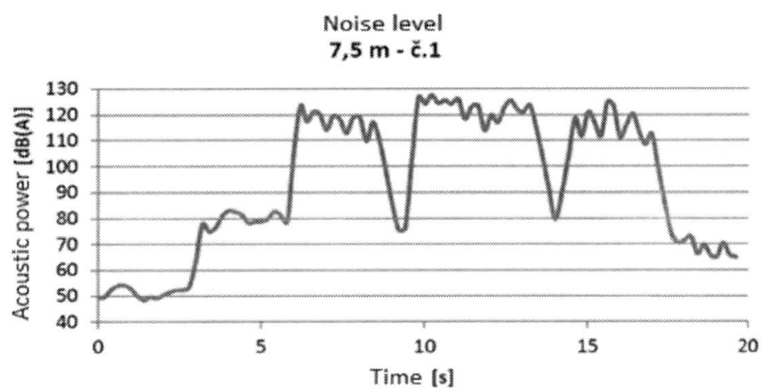

Figure 4. Noise level of 89 t carriage.

Weight of the carriage has got significant impact on overall noise of the retarder, because increased brake force has to occur.

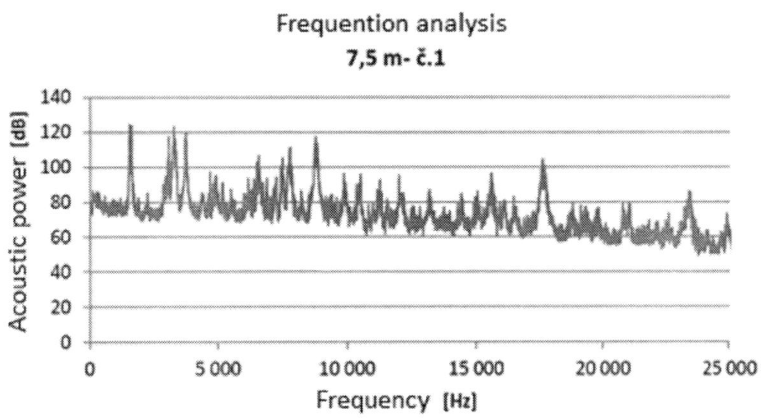

Figure 5. Frequention analysis of 89 t carriage.

Weights of measured carriages were in range of 21 t up to 89 t. When frequentional alnalysis was performed, we found out that weight of the carriage affects frequency also (Figure 5).

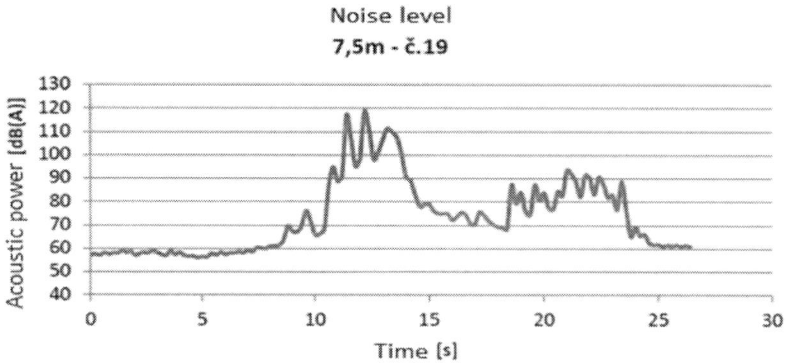

Figure 6. Noise level of 21 t carriage.

In comparisons of noise level, huge increase of noise was measured with heavier carriage (Figure 4 and Figure 6).

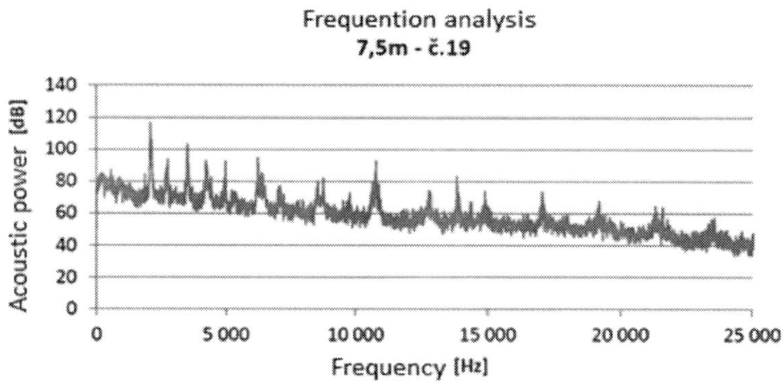

Figure 7. Frequention analysis of 21 t carriage.

Dominant frequency of the noise is reducing with increasing weight of the carriage. Lighter carriages achieved frequencies in range of 3200, 5400, 6200 Hz. Heavier carriages achieved frequencies in range of 1600, 2100, 3500 Hz (Figure 5 and Figure 7) [5].

5. Impact of a retarder operation in the marshalling yard on a surrounding municipalities

Unfortunately, the operation of the marshalling yard with retarders has a negative impact on the population in the surrounding municipalities [6]. From this point of view, it would be possible to adjust the marshalling technology and schedule the activity so that the marshalling yard has the least impact on the inhabitants of the municipalities in terms of the time effect of noise during the day. However, organizing the arrangement of heavier wagons that generate more noise out of the evening and night time of the day is unrealistic, so other solutions are possible [7].

6. Conclusion

Other solutions to reduce the impact of the marshalling yard on the surrounding municipalities are to build sound proof barriers or other objects such as buildings, vegetation [8]. With combination of those solutions is possible to reduce impact of the marshalling yard on surrounding municipalities [9]. Marshalling yards can have negative impact if build near municipalities. More effort is needed to reduce impact on the municipalities. Nowadays is possible to reduce impact in multiple ways [10].

IOP Conf. Series: Materials Science and Engineering **659** (2019) 012061 doi:10.1088/1757-899X/659/1/012061

Acknowledgment
This work was created by the implementation of the project "Low Cost Logistic System Based on Mobile Robotic Platforms for Industrial Use", ITMS: 26220220205 supported by the Operational Programme Research and Development and by the European Fund for Regional Development.

References
[1] Zvolensky P, Poliak J and Valent O 2005 *Technická diagnostika*, Tutoriál pre celoživotné vzdelávanie, *EDIS Žilina*
[2] STN ISO 1996-1 2006 Akustika. Opis, meranie a posudzovanie hluku vo vonkajšom prostredí. Časť 1: *Základné veličiny a postupy posudzovania, June 2006*
[3] Kukulski B and Wszolek T 2017 The Research on Impulsive Events in Railway Noise Generated During Passage Through a Railroad Switch, *Archives of Acoustic* **42** 0137-5075
[4] Tuma J 1997 Zpracování signálů získaných z mechanických systémů užitím FFT, *Príručka, Sdělovací technika s.r.o.*
[5] STN EN ISO 3095:2005 Zeleznice – Akustika – Meranie hluku emitovaneho kolajovymi vozidlami
[6] Zvolensky P 1998 Infrazvuk v prevadzke zeleznicnych vozidiel *Horizonty dopravy*
[7] Deak V 2002 Zdroje hluku a metody experimentalneho vyskumu hlukovych parametrov
[8] Zvolensky P 1997 Moderne metody sledovania hluku a kmitania v zeleznicnej doprave
[9] Batko W and Pawlik P 2012 New Approach to the Uncertainty Assesment of Acoustic Effects in the Enviroment, *Archives of Acoustic* **37** 0137-5075
[10] Salaiova B, Mandula J and Kovalakova M 2001 Vybrané kapitoly z cestných a železničných stavieb. *Hluk z dopravy, Košice*

Optimizing the braking system for handling equipment

D Varecha[1], R Kohár[1] and T Gajdošík[1]

[1]University of Žilina, Faculty of Mechanical Engineering, Department of Design and Machine Parts, Univerzitná 8215/1, 010 26, Žilina, Slovak republic

E-mail: daniel.varecha@fstroj.uniza.sk

Abstract. The article is focused on optimizing the friction disc plates of the braking system integrated into the drive wheel of an automated guided vehicle. The introduction is focused on the application of these types of braking systems. In the next step, the individual mathematical parameters and the acting forces in the braking system are expressed. Using the equation of motion that is written in Matlab, the braking distance and the stop time are graphically expressed. During the Matlab simulation, different diameters of the braking plates are monitored to determine the braking distance. The permitted pressure in the braking plates during the test must not exceed the specified pressure of 2 MPa. The functionality of the braking plates is ensured by hydraulic power. Part of the wheel is a sensor that monitors the speed during braking. Ultimately, a continuous reduction in kinetic energy is achieved without vigorous blocking of the slats during braking. The acquisition of the paper is the output of numerical calculations, which are necessary for the correct design of the wheel with the integrated braking system.

1. Introduction

The new technologies are coming to our market with industrial development. This is also the case in the sphere of logistics. Automated handling transport equipment is implemented in manufacturing processes to increase the productivity of work. These equipment are also referred to as AGVs (automated guided vehicles). The market offers transport equipment (see figure 1) that is braked by the engine brake (electric engine) in the present time [6]. This type of braking is only suitable in certain situations. This way of braking is successful at small speeds. Once the vehicle is driven at a speed of more than 2 $m.s^{-1}$, it should be equipped with a friction brake system [7].

Figure 1. Automated guided vehicle in present [6].

The reason why the braking system should be implemented is clear. To increase the protection of property and life. This fact is very important today. There is a danger for a persons or property damage in certain situations [12]. This isn't acceptable. This project will design a wheel with integrated braking system which maximum speed is 5 $m.s^{-1}$ (see figure 2). The handling devices have great applications in the automotive industry. These devices are found in three car factories in Slovakia. They allow transporting hundreds of kilograms. In this case, the device is designed to carry a load of 1100 kg. The weight of the reserve is set on 100 kg, which is thinking on the calculations. The reason is a safety at overload [11]. The handling equipment uses the energy of the battery, which powers an electric engine. Electric vehicles and automated guided vehicles have a significantly shorter range compared to conventional vehicles with an internal combustion engine [8]. These devices need to charge the battery more often.

The transport handling equipment is oriented in the space using sensors. These sensors monitor the space (warning zones) in front of the vehicle (see figure 3). The software evaluates the warning zones according to the distance to the vehicle. If there is an obstacle in the yellow or orange zone, the vehicle will slow down. If the obstacle is in the red zone, the vehicle stops. In crisis situations, the vehicle must be stopped immediately. For this purpose, the friction disc brake will be used as engine braking is ineffective in these cases [7].

Figure 2. Positioning the brake system in the wheel. Source: [3].

Figure 3. The safety zones for vehicle. Source: [7].

2 Technical knowledge of the topic

The design of the wheel will be created to base on the boundary conditions listed in table 1. It is intended to equip the wheel with a friction disc plate in the original wheel space. Therefore, it is necessary to optimize the size of the friction disc plate. The space in the wheel is between 80 and 220 mm. It should be noted that the entire brake assembly must be located inside the space of the wheel. It mustn't exceed pressure 2 MPa between the friction disc plates during braking. If the braking pressure between the friction disc plates exceeds 2 MPa during braking, the brake system will be damaged [10]. These boundary conditions must be borne in mind.

Table 1. Boundary conditions.

Maximum speed of the handling equipment	$5\ m.s^{-1}$
Internal space of the drive wheel	$80 - 220$ mm
Maximum allowed pressure at braking (between the friction disc plate)	2 MPa

Figure 4 represents the drive wheel in the graphic view with multiple dimensions of friction disc plates. As it can be seen from the figure, the wheel space inside (80-220 mm) is very limited.

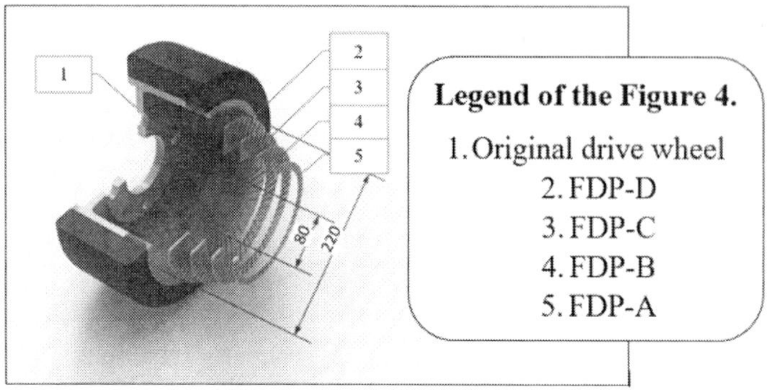

Figure 4. Original wheel with considered friction disc plates [9].

3. The theory on the theme of the braking system

Tangential and normal force act in the brake system with friction disc plates (see figure 5). The normal force imagines the pressure force in the friction disc plate. The tangential force characterizes the braking force, the size of which depends on the friction coefficient of the friction disc plate.

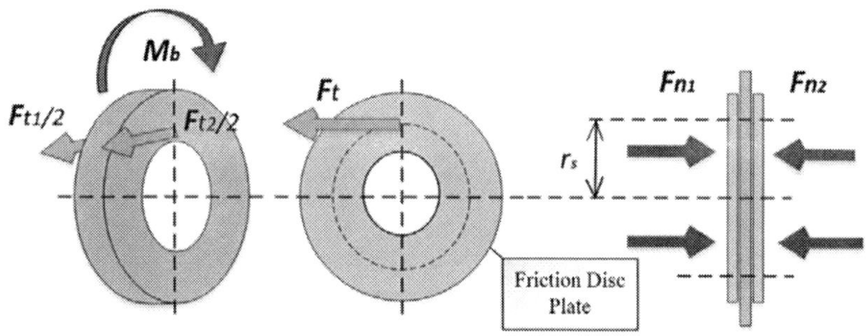

Figure 5. The acting forces in the friction disc plate [10].

The braking torque is generated during braking, the size is calculated according to the following equation (see equation 1). The value braking torque is very important in designing the brake body (static part of the brake).

$$M_b = \left(\frac{F_{t1}}{2} + \frac{F_{t2}}{2}\right).r_s \rightarrow \; M_b = F_t.r_s \; ; [\mathrm{Nm}] \tag{1}$$

where: M_b is braking torque [Nm]; F_{t1} a F_{t2} is tangential force [N]; r_s is the median radius of the friction disc plate [mm].

The construction of brake is designed with push springs, which return the brake to its original condition after releasing the brake. To calculate the real normal force, it is necessary to subtract the force from the compression springs (see equation 3). The normal force acting on both sides of the friction disc plates is the same (see equation 2) [10].

$$F_{n1} = F_{n2} \; ; \; [\mathrm{N}] \tag{2}$$

where: F_{n1}, F_{n2} is normal force [N].

$$F_{zb} = \frac{G}{8}.\frac{d^4}{D^3}.\frac{s}{n} \; ; [\mathrm{N}] \tag{3}$$

where: F_{zb} is the force exerted by one spring [N]; d is the diameter of the spring [mm]; D is the outside diameter of the spring [mm]; s is length after push [mm]; n is the number of spring thread active [-]; G is Modulus of Rigidity [GPa].

The sum of the two tangential forces acting on both sides' friction disc plate represents the result of the total tangential force. And also the multiplication of the normal force by the coefficient of friction for the brake disc plate (see equation 4) leads to tangential force.

$$\frac{F_{t1}}{2}+\frac{F_{t2}}{2}=F_t \rightarrow \quad F_t=F_n.f\,; \quad [\mathrm{N}] \tag{4}$$

where: f is the coefficient of friction [-]

Figure 6 shown the drive axle the handling equipment [14]. Individual arrows show the acting forces in the braking system and driving direction of the vehicle. The right part of the figure shows detail of the wheel with braking system [13].

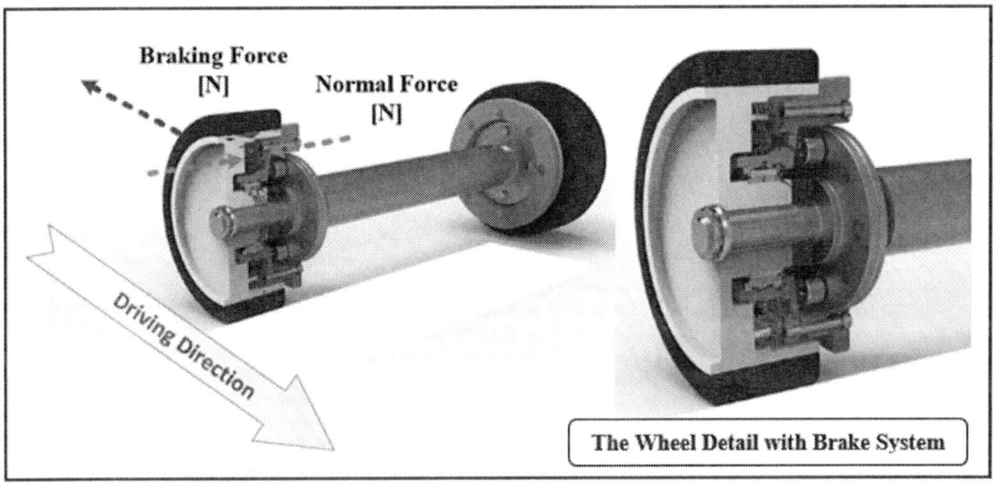

Figure 6. Drive axle in the graphical view [5, 14].

3.1. Mathematical analysis of forces during braking

Calculations for the friction disc plates will be listed below. They are needed to express the individual forces and parameters acting (occurring it) during braking [14]. The first step is calculating the middle radius of the friction disc plate (see equation 5), then next to calculate the effective area of the hydraulic piston (see equation 6) [10].

$$r_S=\frac{2}{3}\cdot\frac{R_2^3-R_1^3}{R_2^2-R_1^2}\,; \quad [\mathrm{mm}] \tag{5}$$

where: R_1 is the inside radius of the friction disc plate [mm]; R_2 is outside radius of the friction disc plate [mm]; r_S is middle radius of the friction disc plate [mm].

$$S_p=3{,}14.\left({R_2}^2-{R_1}^2\right);[\mathrm{mm}^2] \tag{6}$$

where: S_p is effective area of the hydraulic piston [mm^2].

As said, the design of brake is created with push springs. The reason is clear. The brake system must be released after braking. The acting force of springs (see equation 7) must be subtracted from the result of normal force [10]. In short, it is the force that pushes the direction out piston the friction disc plates, when the brake is released.

$$F_{zb} = \frac{G}{8} \cdot \frac{d^4}{D^3} \cdot \frac{s}{n} ; [\mathrm{N}] \tag{7}$$

For proper operation, the brake design must be equipped with three or more springs. This always depends on the design of the brake. In this case, three springs are used (see equation 8) [10].

$$F_{prz} = F_{zb} . P_p ; [\mathrm{N}] \tag{8}$$

where: F_{pr} is a force at compressed brake [N]; P_p is the number of push springs [-].

In the hydraulic brake system, the pressure is 1 Mpa, which acts on the friction disc plates [10]. Using the equation (see equation 9), the compressive force on friction disc plates (normal force) is expressed [10].

$$F_N = (S_p . p_{pr}) - F_{prz} ; [\mathrm{N}] \tag{9}$$

where: p_{pr} is the pressure of the hydraulic oil [MPa].

In the next step, the effective area of the friction disc plate is expressed (see equation 10).

$$S_L = \pi . (R_2^2 - R_1^2) ; [\mathrm{mm}^2] \tag{10}$$

The value of the tangential force is expressed as a multiple of the normal force (see equation 7) with a coefficient of friction (see equation 11) [10].

$$F_T = F_N . \mu ; [\mathrm{N}] \tag{11}$$

Braking torque is calculated with safety in mind at this point. This step is very important for the next calculation (see equation 12) [10].

$$M_b = F_t . r_s . k ; [\mathrm{Nm}] \tag{12}$$

where: k is coefficient of safety [-], M_b is the braking torque including the safety factor [Nm].

In the penultimate step, it is necessary to determine the number of friction areas needed (see equation 13) [10].

$$i \geq \frac{M_B}{F_N . f . r_s} ; [-] \tag{13}$$

where: i is count of the friction disc plate [-].

The result from equation 13 is very important for the next calculation of a number of friction disc plates (see equation 14). The one plate is added because of design reasons.

$$n = i + 1 ; [-] \tag{14}$$

The calculation for the braking torque has already been shown above. In this calculation, the coefficient of friction and the total number of friction disc plates are taken to account (see equation 15). This is a real braking torque for the existing situation at the braking. [10]

$$M_B = f . i . F_N . r_s ; [\mathrm{Nm}] \tag{15}$$

The above calculations must be applied to every friction disc plate. The aim will be to compare the individual calculated values (see table 2). The suitability of using the friction disc plate is determined by its size and maximum permitted braking pressure between the friction disc plates. These requirements were identified at the beginning [10]. The size of the friction plate must be between 80 and 220 mm considering the complete braking system.

Table 2. Calculated values of the friction disc plate.

Friction disc plate (next only FDP)	FDP-A	FDP-B	FDP-C	FDP-D
The median radius of the FDP (r_S) [mm]	72.6	77.9	83.4	89.0
Effective area of the hydraulic piston (S_P) [mm^2]	10 597.5	13 188.0	18 840.0	21 901.5
Normal force (F_N) [N]	9855.9	12 446.4	18 098.4	21 159.9
Effective area of friction disc plate (S_L) [mm^2]	4553	9734	15 543	21 980
Tangential force (F_T) [N]	1576.9	1991.4	2895.7	3385.6
Braking torque (M_B) [N]	114.5	155.2	241.5	301.4
Required number of friction areas (i) [-]	2	2	2	2
Selected number of friction disc plates (n) [-]	3	3	3	3

The calculation below expresses maximum pressure (see equation 16) between the friction disc plates. Between the friction disc plates mustn't exceed pressure 2 MPa. The results are in table 3.

$$P_{max} = \frac{F_N}{S_L} \quad \rightarrow \quad P_{max} < P_{Allow}\,, [\mathrm{MPa}] \tag{16}$$

Table 3. Pressure between the friction disc plates.

The friction disc plate	FDP-A	FDP-B	FDP-C	FDP-D
Pressure between friction disc plates [MPa]	2.165	1.279	1.164	0.963

As mentioned above, it is important that pressure greater than 2 MPa is no generated in the friction disc plates. For this fact, a graph is created that will present the braking pressure. It can be seen in figure 7, that the value of pressure for FDP-A isn't satisfactory [10].

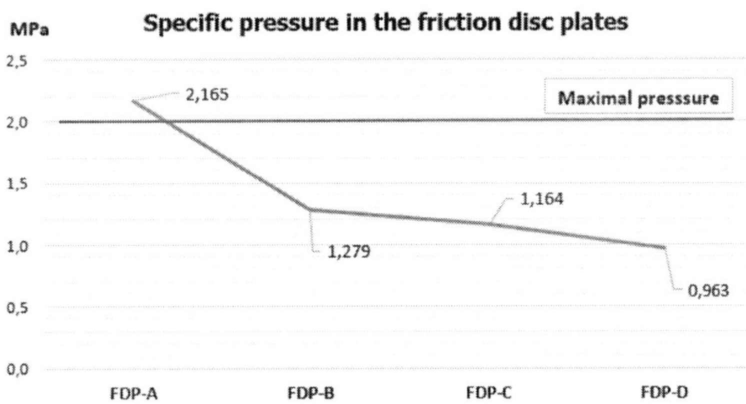

Figure 7. Specific pressure in the friction disc plates.

The size of the friction disc plate FDP-D cannot be applied to the intended inside size of the drive wheel. This is not possible due to size. This dimension only suits the permissible pressure. The most suitable friction disc plate is selected from FDP-C and FDP-B types.

3.2. Simulation the braking system with the friction disc plates

In the previous subsection 3.1 the two most suitable friction disc plates were selected. Subsequently, a braking simulation will be created in Matlab for both friction disc plates.

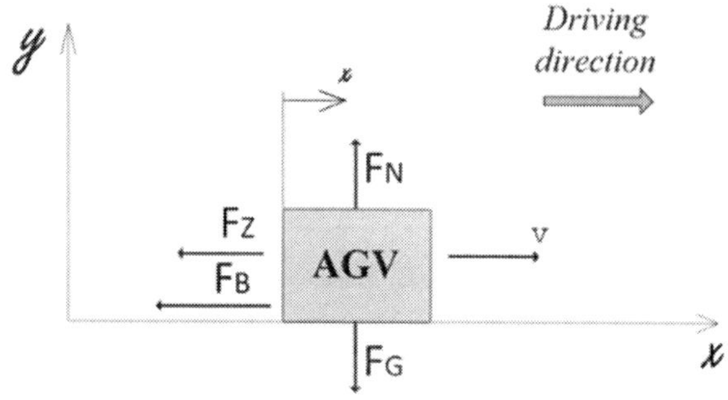

Figure 8. The forces to act on the handling equipment [1-2].

where: F_z is a inertia force [N]; F_N is a normal force [N]; F_B is a braking force [N]; F_G is a gravitational force [N]; v is speed [m.s^{-2}].

The values of braking forces are calculated and shown in table 2. Since the braking system is always designed in pairs, simulation in the Matlab is set for two wheels. Figure 8 shows the forces acting on the vehicle when braking [1]. The equation of motion (see equation 17) in the vector shape is created according to figure 8 [4]. The forces F_G and F_N act in the y-axis direction. However, it is not important for the equation of motion.

$$\overrightarrow{F_N} - \overrightarrow{F_B} - \overrightarrow{F_G} = \overrightarrow{F_Z} \tag{17}$$

Subsequently, the equation (see equation 17) is rewritten to a scalar form (see equation 18, 19).

$$x: -F_B = m.\vec{a} \tag{18}$$

$$y: F_N - F_G = 0 \tag{19}$$

where: x, y is coordinate system [-]; $\vec{a}$ is vector of acceleration [m.s^{-2}].

From the equation in the x-axis direction (see equation 18) is expressed relationship for acceleration (see equation 20) [1]

$$a = -\frac{F_B}{m} \; ; [\mathrm{m.s^{-2}}] \tag{20}$$

where: m is load [kg].

The acceleration (deceleration) of the handling equipment is constant. In this case, the device slows down using the braking force F_B, so the acceleration is negative. By integrating the acceleration of acceleration (see equation 20), the dependence of velocity in time is expressed (see equation 21). [1]

$$v(t) = \int a \; dt = \int -\frac{F_B}{m} dt = -\frac{F_B}{m}.t + c_1 \; ; \; [\mathrm{m.s^{-1}}] \tag{21}$$

where: c_1 is constant that is determined from the initial conditions; a is acceleration [m.s^{-2}] ; t is time [s].

In the next step, the equation above (see equation 21) is rewritten to the following shape (see equation 22).

$$v(t) = -\frac{F_B}{m}.t + v_0 \; ; [\mathrm{m.s^{-1}}] \tag{22}$$

where: v_0 is initial speed [m.s^{-1}].

By integrating the velocity from the relationship above (see equation 22), the dependence of the path from time is expressed (see equation 23).

$$s(t) = \int v(t)\,dt = \int \left(-\frac{F_B}{m}.t + v_0\right) dt = -\frac{F_B}{2.m}.t^2 + v_0.t + c_2\,;\ [\mathrm{m}] \qquad (23)$$

where: c_2 is constant that is determined from initial conditions; s is route [m].

Braking distance (see equation 25) and stopping time (see equation 24) are check calculation and listed in table 4 below.

$$t_b = \frac{m.v}{F_b}\,;\ [\mathrm{s}] \qquad (24)$$

$$s_b = \frac{m.v^2}{2.F_b}\,;\ [\mathrm{m}] \qquad (25)$$

Table 4. Parameters of the braking distance, and the braking time.

Type of the friction disc plate	FDP-B	FDP-C
Time braking [s]	3	2.1
Braking distance [m]	7.53	5.18

On the graph (see figure 9), it can be seen that from speed 5 m.s^{-1} the vehicle is stopped in 2.1 seconds (see figure 9). The device passes the distance of 5.18 meters (see figure 9). This graph belongs to friction disc plate FDP-C. The red point in the graph shows the moment when the device stops.

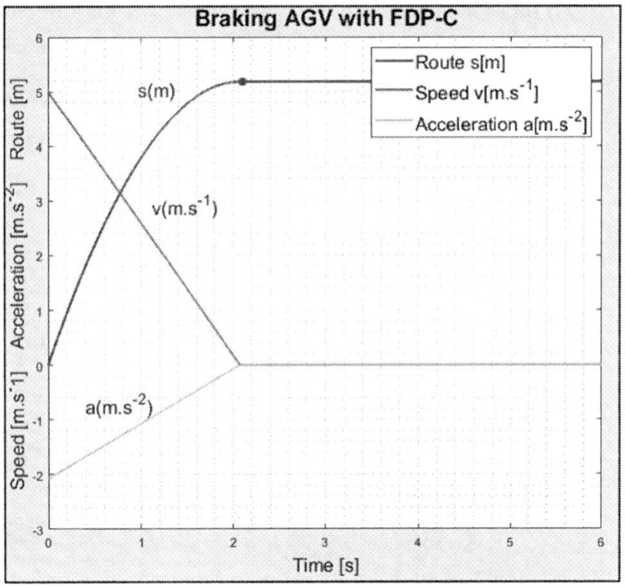

Figure 9. Simulation for friction disc plate type FDP-C.

Similarly to the previous example, a simulation for FDP-B is created. The braking time is 3.01 seconds (see figure 10). In this case, the device will pass 7.53 meters (see figure 10). The red point in the graph shows the moment when the device stops. The results in Matlab were verified by calculations (see equation 24 – 26) and are listed in table 4.

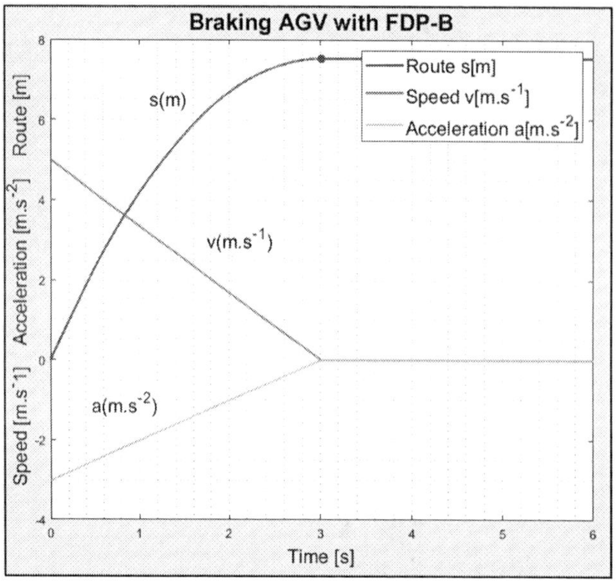

Figure 10. Simulation for friction disc plate type FDP-B.

The braking distance values have been satisfactory for the type of friction disc plate C (FDP-C). However, additional calculations are required for the final design. The brake design is a complex problem that involves many technical and operating factors. Such as heat dissipation during braking, etc. It is also necessary that the wheel is equipped with a sensor to monitor the speed during braking. The braking principle is similar to a car with an automatic braking system (ABS).

4. Discussion

By changing the mean radius of the friction plate surface, the size of the normal force changes. When the normal force changes, the tangential force and moment of the brake also change. As a result, the change in the size of the friction disc plate affects the distance of the braking distance of the AGV device [13]. Figure 4 above, shows the drive wheel and more types of friction discs plates (figure 11 is illustrative only).

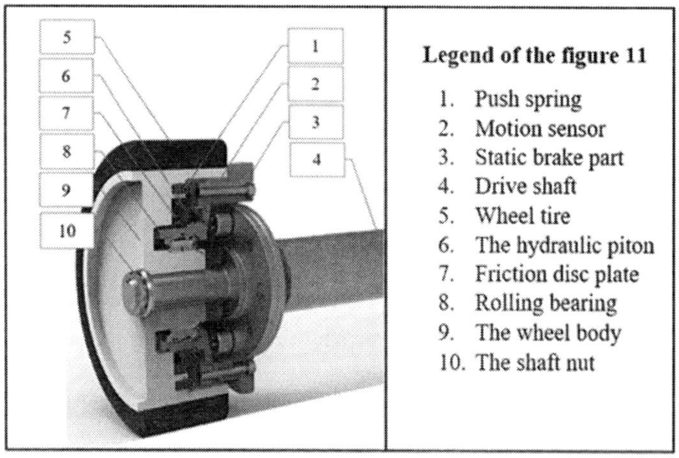

Figure 11. Optimize the drive wheel with friction disc plate braking system.

5. Conclusion

The aim of the work was to optimize the size of the friction disc plate for the drive wheel of the handling equipment. While the size of the brake system had to insert to the inside space of the original wheel of existing handling equipment. Of the four friction disc plates considered, two sizes (FDP-B and FDP-C) were recommended. The other two (FDP-A and FDP-D) did not fit the size and allowable pressure during braking. The simulation in the software Matlab is setting for two friction disc plates. The results were compared, and the most suitable friction disc plates were chosen. In this case, the friction disc plate FDP-C is accepted. At this setting, the equipment handling stops from the maximum speed of 5 $m.s^{-1}$ to zero in 2.1 seconds. The distance is 5.18 meters. The weight considered was 1200 kg. It must be said, that more effort is needed for the final design (heat dissipation during braking etc.) The braking system is a complex problem that involves many technical and operational factors.

Acknowledgment

The research is supported by the Cultural and Educational Grant Agency of the Ministry of Education, Science, Research and Sport of the Slovak Republic under the project No. 046ŽU-4/2018.

References

[1] Filo M and Lukáč M 2005 *Modelovanie a simulácia mechanizmov s počítačovou podporou (Modeling and Simulation of Mechanisms with Computer Support)*, Žilina, EDIS ZU, 42-62 (in Slovak)

[2] Rievaj V, Šulgan M and Hudák A 2013 *Automobil a jeho dynamika (Car and his Dynamics)*, Žilina, EDIS ZU, 22-35 (in Slovak)

[3] SPS Magazin, *Sicherheits-Laserscanner Erfahrung zählt 2019*, http://www.sps-magazin.de/ (accessed on Jan. 2017)

[4] MATLAB, *Modeling an Anti-Lock Braking System 2019*, http://www.mathworks.com/ (accessed on Jan. 2018)

[5] Brončekk J, Kučera Ľ, Žarnay M, Poljak S and Čilík L 2015 *Konštruovanie 1* (The design of mechanical parts), Žilina: EDIS ZU (in Slovak)

[6] Ceit, *Průmyslová automatizace 2019 (The industrial automation)*, http://www.ceit-cz.cz/ (accessed on Jan. 2017)

[7] Sick – Sensor intelligence, *Safe processes without interruptions 2019*, http://www.sick.com/ (accessed on Jan. 2016)

[8] Mruzek M, Gajdáč I, Kučera Ľ and Gajdošík T 2017 *The possibilities of increasing the electric vehicle range,* Transcom 2017: International scientific conference on sustainable, modern and safe transport 621-625, ISSN 1877-7058

[9] Schaeffler Technologies, *Schaeffler Technical Pocket Guide*, 2018 736 p. number 084713119-0000, Pocket book (STT).

[10] Málik L, Medvecký Š, et al. 2003 *Časti a mechanizmy strojov (The machine parts and mechanisms),* Žilina, EDIS, 25-80 (in Slovak)

[11] Lukáč M, Brumerčík F, Krzywonos L and Krzysiak Z 2017 Transmission system power flow model, *Communications: scientific letters of the University of Žilina*, ISSN 1335-4205 **19**(2) 27-31

[12] Medvecká I, Biňasová V and Kubinec L 2017 Planning and performance evaluation of the manufacturing organizations, *Procedia Engineering* **192**(2017) 46-51

[13] Belorit M, Hrček S and Gajdošík T 2017 *Description of the bearing check program for countershaft gearboxes,* Proceedings of 58th International Conference of Machine Design Departments (ICDM), Prague, Czech republic, Sep. 6 – 8, 32-35

[14] Bisták M, Medvecký Š and Hrček S 2017 *Axle weighing system.* Proceedings of 58th International Conference of Machine Design Departments (ICDM), Prague, Czech republic, Sep. 6 – 8, 36-39

Examination of vehicle impact against stationary roadside objects

S Karapetkov[1], L Dimitrov[2], Hr Uzuniv[1] and S Dechkova[1]

[1] Technical University of Sofia, Faculty and College – Sliven,
[2] Technical University of Sofia, 8 Kliment Ohridski bulv., Sofia 1000, Bulgaria

E-mail:lubomir_dimitrov@tu-sofia.bg

Abstract. In this paper, a model of a car impact against a fixed roadside objects/safety barrier is presented and discussed to solve a mixed task: to determine the pre-impact velocity of the vehicle center of mass taking into consideration the recorded post-impact final rest position. The principle of conservation of mechanical energy was applied, whereas the energy of the rigid barrier deformation was estimated by the finite element method/FEM for an elasto-plastic body. FEM analysis was performed by integrating SolidWorks - SW Simulation software, the output data being the deformation of the safety barrier in the three dimensional space subjected to limited boundary conditions. The post-impact car motion was modelled by a multiscale spatial model, well established in modern accident reconstruction. The findings of the study point out the precision and accuracy of the crash pattern revealed in the accident reconstruction and essential in clarifying the situation with appropriate real facts. Two cases involving real-world head-on accident collisions against a fixed flexible barrier and a tank body are considered, which exemplify the adopted method.

1. Introduction

In statistical analysis reports of traffic accidents, automobiles hitting against semi-rigid barriers and flexible barriers get a particularly large share [1-3, 7-9, 11, 14]. The analysis involves several different stages of computational procedures. On the whole, the accident can be divided into three stages. The first phase is prior to the moment of impact, when, generally, the vehicle moves at a loss of longitudinal stability or the automobile diverges to the metal crash barrier due to other reasons, such as individual differences associated with driving behavior. The second phase is at the point of impact, when the vehicle contacts the crash barrier, resulting in deformation of the car body and deformation of the barrier. At this stage part of the kinetic energy inflicted upon the vehicle is transformed into deformation energy of the car body and deformation energy of the metal crash barrier. The third phase includes the post-impact vehicle motion, which is the result of the residual kinetic energy of the vehicle, continuing lateral motion with relative sliding to its final rest position.

The following study proposes a solution to a traffic road: to define part of the initial data / pre-impact velocity of the center of mass / under certain final conditions / the post-impact final rest position of the vehicle/.

IOP Conf. Series: Materials Science and Engineering **659** (2019) 012063 doi:10.1088/1757-899X/659/1/012063

2. Design of a vehicle crash model and dynamic impact against a metal crash barrier and a rigid barrier

The impact phase between the vehicle and the metal flexible barrier is considered to be a deformation phase between two bodies where the transformed kinetic energy before the impact turns into deformation energy and post-impact vehicle kinetic energy. The pre-impact vehicle motion is considered to be a translational motion. According to the impulse-momentum change theorem and the kinetic energy theorem the impact impulse and its momentum to the center of mass imparted to the body make the vehicle lose its longitudinal stability and continue its motion to a final rest position. This motion is assumed to be a motion of a car on one or several inclined planes with different adhesion characteristics. The kinetic energy after impact is determined on the basis of a multiscale mechanics and mathematical model.

$$\frac{m_1.V_1^2}{2} = \frac{m_1.u_1^2}{2} + \frac{I_1.\omega_1^2}{2} + E_1 + E_2 \quad (1)$$

where:

m_1 – mass of vehicle

I_1 – mass inertia for the vehicle around the vertical axis Oz

V_1 – vehicle velocity of the center of mass at the point of impact

u_1 – post-impact vehicle velocity of the center of mass

ω_1 – post-impact angular velocity of the vehicle body around the vertical axis Oz

E_1 – deformation energy of car body

E_2 – deformation energy of metal crash barrier

Analysing crash tests with a car against a permanently fixed flexible barrier is a complex dynamic task, where the deformation energy applied to the barrier is usually taken into account. To estimate the necessary kinetic energy, too many components need to be known. The results obtained in the study are based on the implementation of a strategy based on the strain energy method.

3. Post-impact dynamic analysis of the vehicle motion

Undoubtedly, the proposed formula (2) is too general and does not cover the full dynamic analysis of the vehicle post-impact due to the fact that it does not count for the complexity of friction between tires and road surface, elasticity of suspension, damping and so on. Equation (1) clearly shows that not only the deformation energy of the two bodies, but also the post-impact vehicle velocity of the center of mass and the angular speed of the vehicle are exhibited. If we apply the center-of-mass motion theorem, it would be a rather inaccurate approach, for it would not be possible to estimate the kinetic energy for the rotation of the vehicle after the impact [13, 17, 20, 21].

A considerably more precise approach is to use a system of twelve differential equations that take into account friction forces, elasticity and suspension, damping forces, gravity force, air mass resistance, etc. [4, 10, 12, 22].

The automobile is considered to be a multi-mass system (Figure 1) and the initial conditions that meet the differential equations for vehicle motion determine the whole post-impact kinetic energy.

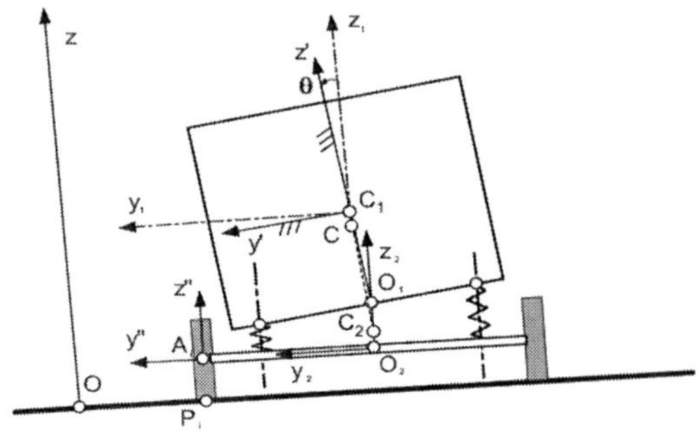

Figure 1. Rear view of the automobile.

The analysis shows post-impact vehicle velocity of the center of mass "u" and the angular velocity of the vehicle after the impact ω_z.

The differential equations of motion for each vehicle after impact, considered as a multi-mass spatial mechanical system, have the type

$$m.\ddot{x}_C = \sum_{i=1}^{4}[F_{ix}] + m.g.\sin\alpha - W_x.\sqrt{\dot{x}_C^2+\dot{y}_C^2}.\dot{x}_C$$

$$m.\ddot{y}_C = \sum_{i=1}^{4}[F_{iy}] + m.g.\sin\beta - W_y.\sqrt{\dot{x}_C^2+\dot{y}_C^2}.\dot{y}_C$$

$$m.\ddot{z}_C = \sum_{i=1}^{4} N_i - \frac{mg}{\sqrt{1+tg^2\alpha+tg^2\beta}} \tag{2}$$

$$\{[J_{C_1}]+[J_{C_3}]\}.[\omega] = [M_{C_\omega}]+[M_{C_3\omega}]+[M_{C_{a1}}]+[M_{C_{a2}}]+[M_{C_{a3}}]+[M_{C_k}]+[M_{C_m}]$$

$$[J_y].[\ddot{y}] = [M_y];\ M_{yi} = F_{i\tau}.r_i + sign(\dot{y}_i).[M_{di} - f_i.N_i - M_{si}]$$

$$F_{i\tau} = \mu.N_i\frac{v_{P_{ix}}}{v_{P_i}}\cos(\varphi_{z1}+\theta_k) + \mu.N_i\frac{v_{P_{iy}}}{v_{P_i}}\sin(\varphi_{z1}+\theta_k).$$

where the corresponding matrices are determined by the expressions

$$[M_{C_\omega}]\begin{bmatrix} M_{F_{x'}} + M_{N_{x'}} + J_{x'z'}\omega_{x'}\omega_{y'} - J_{x'y'}\omega_{x'}\omega_{z'} + (J_{y'}-J_{z'})\omega_{y'}\omega_{z'} \\ +J_{y'z'}\left(\omega_{y'}^2-\omega_{z'}^2\right) \\ M_{F_{y'}} + M_{N_{y'}} + J_{x'y'}\omega_{y'}\omega_{z'} - J_{y'z'}\omega_{x'}\omega_{y'} + (J_{z'}-J_{x'})\omega_{x'}\omega_{z'} \\ +J_{x'z'}\left(\omega_{z'}^2-\omega_{x'}^2\right) \\ M_{F_{z'}} + M_{N_{z'}} + J_{y'z'}\omega_{x'}\omega_{z'} - J_{x'z'}\omega_{y'}\omega_{z'} + (J_{x'}-J_{y'})\omega_{x'}\omega_{y'} \\ +J_{x'y'}\left(\omega_{x'}^2-\omega_{y'}^2\right) \end{bmatrix} \tag{3}$$

$$[M_{C_\omega}]\begin{bmatrix} J_{3x'z'}\omega_{3x'}\omega_{3y'} - J_{3x'y'}\omega_{3x'}\omega_{3z'} + (J_{3y'} - J_{3z'}).\omega_{3y'}\omega_{3z'} + \\ +J_{3y'z'}\left(\omega_{3y'}^2 - \omega_{3z'}^2\right) \\ J_{3x'y'}\omega_{3y'}\omega_{3z'} - J_{3y'z'}\omega_{3x'}\omega_{3y'} + (J_{3z'} - J_{3x'}).\omega_{3x'}\omega_{3z'} \\ +J_{3x'z'}\left(\omega_{3z'}^2 - \omega_{3x'}^2\right) \\ J_{3y'z'}\omega_{3x'}\omega_{3z'} - J_{3x'z'}\omega_{3y'}\omega_{3z'} + (J_{3x'} - J_{3y'}).\omega_{3x'}\omega_{3y'} \\ +J_{3x'y'}\left(\omega_{3x'}^2 - \omega_{3y'}^2\right) \end{bmatrix}$$

$$M_{Ca1} = \begin{bmatrix} \left[(\vec{r}_C - \vec{r}_{C_1}) \times (\vec{a}_{C_1} - \vec{a}_C)\right]_{x'} \\ \left[(\vec{r}_C - \vec{r}_{C_1}) \times (\vec{a}_{C_1} - \vec{a}_C)\right]_{y'} \\ \left[(\vec{r}_C - \vec{r}_{C_1}) \times (\vec{a}_{C_1} - \vec{a}_C)\right]_{z'} \end{bmatrix}$$

$$M_{Ca2} = \begin{bmatrix} \left[(\vec{r}_C - \vec{r}_{C_2}) \times (\vec{a}_{C_2} - \vec{a}_C)\right]_{x'} \\ \left[(\vec{r}_C - \vec{r}_{C_2}) \times (\vec{a}_{C_2} - \vec{a}_C)\right]_{y'} \\ \left[(\vec{r}_C - \vec{r}_{C_2}) \times (\vec{a}_{C_2} - \vec{a}_C)\right]_{z'} \end{bmatrix}$$

$$M_{Ca3} = \begin{bmatrix} \left[(\vec{r}_C - \vec{r}_{C_3}) \times (\vec{a}_{C_3} - \vec{a}_C)\right]_{x'} \\ \left[(\vec{r}_C - \vec{r}_{C_3}) \times (\vec{a}_{C_3} - \vec{a}_C)\right]_{y'} \\ \left[(\vec{r}_C - \vec{r}_{C_3}) \times (\vec{a}_{C_3} - \vec{a}_C)\right]_{z'} \end{bmatrix}$$

$$[M_{Ck}] = -\begin{bmatrix} \left[\sum_{i=1}^{4} J_{ky''i}.\ddot{\gamma}_i.\vec{J}_{2i}'' + \vec{\omega}_2 \times \sum_{i=1}^{4} J_{k2i}.\dot{\gamma}_i.\vec{J}_{2i}'' + \sum_{i=1}^{4} [J_{kz'i}.(\dot{\omega}_2 + \ddot{\vartheta}_{ki}) + \\ +m_{ki}.\left|C_2\vec{A}_i\right|^2.\dot{\omega}_2].\vec{k}_2\right]_{x'} \\ \left[\sum_{i=1}^{4} J_{ky''i}.\ddot{\gamma}_i.\vec{J}_{2i}'' + \vec{\omega}_2 \times \sum_{i=1}^{4} J_{k2i}.\dot{\gamma}_i.\vec{J}_{2i}'' + \sum_{i=1}^{4} [J_{kz'i}.(\dot{\omega}_2 + \ddot{\vartheta}_{ki}) + \\ +m_{ki}.\left|C_2\vec{A}_i\right|^2.\dot{\omega}_2].\vec{k}_2\right]_{y'} \\ \left[\sum_{i=1}^{4} J_{ky''i}.\ddot{\gamma}_i.\vec{J}_{2i}'' + \vec{\omega}_2 \times \sum_{i=1}^{4} J_{k2i}.\dot{\gamma}_i.\vec{J}_{2i}'' + \sum_{i=1}^{4} [J_{kz'i}.(\dot{\omega}_2 + \ddot{\vartheta}_{ki}) + \\ +m_{ki}.\left|C_2\vec{A}_i\right|^2.\dot{\omega}_2].\vec{k}_2\right]_{z'} \end{bmatrix}$$

$$[M_{Ck}] = -\begin{bmatrix} \sum_{i=1}^{2} \left[J_{mi} + m_{mi}.\left|C_2\vec{C}_{mi}\right|^2\right]\dot{\omega}_2.\vec{k}_{2x'} \\ \sum_{i=1}^{2} \left[J_{mi} + m_{mi}.\left|C_2\vec{C}_{mi}\right|^2\right]\dot{\omega}_2.\vec{k}_{2y'} \\ \sum_{i=1}^{2} \left[J_{mi} + m_{mi}.\left|C_2\vec{C}_{mi}\right|^2\right]\dot{\omega}_2.\vec{k}_{2z'} \end{bmatrix}$$

Here, m is the total mass of the vehicle; m_{ki} $/i = 1\div4/$ – mass of each of the wheels; m_{mi} $/i = 1\div2/$ - mass of each of the drives; x_c, y_c, z_c - coordinates of the vehicle center of mass in relation to the fixed coordinate system; ψ, θ, φ - Euler angles of body's corners (unsprung mass); θ_k - average angle of rotation of the steering wheels around their axles; γ_i $/i = 1\div4/$ - angles of rotation of the wheels on their own rotary axis; $\vec{F}_i$, $/i = 1\div4/$ - friction forces in the wheels; α, β - road surface angles of the longitudinal and transverse slope; N_i $/i = 1\div4/$ - normal reactions in the wheels; w –

drag coefficient; $\vec{\omega}$ - angular velocity of the movable coordinate system $C_1x'y'z'$ permanently connected to the unsprung mass; $[\dot{\omega}] = [\dot{\omega}_{x'}\ \dot{\omega}_{y'}\ \dot{\omega}_{z'}]^T$ - matrix column of the derivatives of the projections of angular velocity on the permanently connected to the unsprung mass coordinate axes; $\vec{\omega}_2$ - angular velocity of the sprung mass and the constantly connected to it movable coordinate system $O_2x_2y_2z_2$; ω_{kae} - angular velocity of the flywheel (if available); $M_{F.N_{x'}}, M_{F.N_{y'}}, M_{F.N_{z'}}$ - moments of the frictional forces on the wheels and the normal reactions to the permanently connected to the vehicle coordinate axes; $[J_{C_1}]$ - matrix of the mass inertia of the bodywork related to the coordinate axes, permanently connected to it; $[\omega]$ - matrix column of the projections of angular velocity on the same axes determined by Euler's formula; $J_{ky''i}, J_{kz''i}$ $/i = 1{\div}4/$ - mass inertia of each wheel relative to its own axis of rotation and its radial axis; J_{m_i} - $/i = 1{\div}2/$ - intrinsic mass moment of inertia of each of the drives relative to its central axis parallel to z_2; $\vec{F}_{i\tau}$ - tangential component of the friction force on the wheel; μ - coefficient of friction depending on the sliding speed of the contact spot, which is introduced graphically or analytically; r_i – radius of the wheel; f_i – coefficient of rolling friction; $[I_y]$ - square matrix of coefficients in front of the actual angular accelerations of the drive wheels, depending on the wheel and engine inertia moments; $[\ddot{\gamma}]$ - matrix column of its own angular wheel accelerations, of which two or four are propulsive; M_{d_i}, M_{s_i} - corresponding engine and brake torque applied to each wheel.

Coordinates of the center of mass of the unsprung mass are defined by the expressions

$$\begin{gathered}
x_{C_1} = x_C - \frac{m_3}{m_1+m_2+m_3}.\left(a_{11}.x'_{C_1} + a_{12}.y'_{C_1} + a_{13}.z'_{C_1}\right) - \frac{m_2}{m_1+m_2+m_3}. \\
.\left[a_{11}.x'_{O_1} + a_{12}.y'_{O_1} + a_{13}.z'_{O_1} + \left(\cos\varphi_{z1}.x_{2_{c2}} - \sin\varphi_{z1}.y_{2_{c2}}\right)\right] \\
y_{C_1} = y_C - \frac{m_3}{m_1+m_2+m_3}.\left(a_{21}.x'_{C_3} + a_{22}.y'_{C_3} + a_{23}.z'_{C_3}\right) - \frac{m_2}{m_1+m_2+m_3}. \\
.\left[a_{21}.x'_{O_1} + a_{22}.y'_{O_1} + a_{23}.z'_{O_1} + \left(\sin\varphi_{z1}.x_{2_{c2}} - \cos\varphi_{z1}.y_{2_{c2}}\right)\right] \\
z_{C_1} = \frac{m_2}{m_1+m_2+m_3}.z_C - \frac{m_3}{m_1+m_3}.\left(a_{31}.x'_{C_3} + a_{32}.y'_{C_3} + a_{33}.z'_{C_3}\right) - \frac{m_2}{m_1+m_3}.z_{C_2}
\end{gathered} \tag{4}$$

where m_1 is the unsprung mass; m_2 – sprung mass; m_{kae} – mass of flywheel (if available); a_{ij} $/i = 1 \div 3;\ j = 1 \div 3/$ - elements of the matrix formed by the mentioned cosines between the axes of the permanently connected to the sprung mass coordinate system and the fixed coordinate system.

The angle of rotation of the coordinate system $O_2x_2y_2z_2$ relative to the fixed coordinate system is determined by the integral

$$\varphi_{z1} = \int_{\varphi_{z10}}^{\varphi_{z1}} \varphi_z dt = \int_{\varphi_{z10}}^{\varphi_{z1}} \left[\dot{\psi}(t) + \cos\vartheta(t).\dot{\varphi}(t)\right]dt \tag{5}$$

Coordinates of the center of mass of the sprung mass have the notation

$$\begin{gathered}
x_{C_2} = x_{C_1} + a_{11}.x'_{O_1} + a_{12}.y'_{O_1} + a_{13}.z'_{O_1} + \left(\cos\varphi_{z1}.x_{2_{c2}} - \sin\varphi_{z1}.y_{2_{c2}}\right) \\
y_{C_2} = y_{C_1} + a_{21}.x'_{O_1} + a_{22}.y'_{O_1} + a_{33}.z'_{O_1} + \left(\sin\varphi_{z1}.x_{2_{c2}} - \cos\varphi_{z1}.y_{2_{c2}}\right) \\
z_{C_2} = const
\end{gathered} \tag{6}$$

where

m - total mass of the vehicle;

$\vec{a}_C$ – acceleration of the vehicle center of mass in relation to the fixed coordinate system;

$\vec{F}_i$ – forces acting on the vehicle, including friction, suspension elasticity, damping, air resistance.

4. Examining deformation energy of the car body

To determine the deformation energy of the car body E1, indicated in equation (1), it is assumed that the intensity of impact forces results in a linear relationship to plastic deformation [6, 5, 18]. It is determined by the formula:

$$\frac{dF}{dl} = q = A + B.c \tag{7}$$

where F is impact force; c – indentation depth of the plastic deformation; A, B - energy coefficients taking into account the stiffness characteristics of the frontal crumpledne.

The deformation energy is obtained, as follows:

$$dE = \left[\int_0^{\delta} q_{ел}.dx + \int_0^{c} q.dc\right].dl \tag{8}$$

Equation (8) can be expressed to obtain the following equation:

$$E = (1 + tg^2\theta_s)\cdot\frac{L}{5}\left[\frac{A}{2}\cdot\begin{pmatrix} c_1 + 2.c_2 + \\ +2.c_3 + 2.c_4 \\ +2.c_5 + c_6 \end{pmatrix} + +\frac{B}{6}\cdot\begin{pmatrix} c_1^2 + 2.c_2^2 + \\ +2.c_3^2 + \\ +2.c_4^2 + \\ +2.c_5^2 + c_6^2 + \\ +c_1.c_2 + \\ +c_2.c_3 + \\ +c_3.c_4 + \\ +c_4.c_5 \\ +c_5.c_6 \end{pmatrix} + 5\cdot\frac{A^2}{2.B}\right] \tag{9}$$

5. Investigation of the energy deformation of the metal crash barrier

The deformation energy of the metal crash barrier (Fig. 2) is determined from equation (1) by integrating the finite element method. This method is used as a universal tool for calculating and analyzing the behavior of mechanical structures in power and heat loads as well as for solving problems in fluid mechanics, heat engineering, theory of electric and magnetic fields, acoustics, nuclear physics, etc.

Figure 2. Models for identification of impact.

6. Design of 3d models in solidworks environment

The object of investigation was a car impact against a metal crash barrier with post-impact manifested residual energy as well as with lack of residual energy. Initial data was the known deformation value of the metal crash barrier and the final rest position of the vehicle after the impact.

The initial conditions of the analysed crash were the deformations of the given metal crash barrier, the shape and the geometrical dimensions of the deformation. Three dimensional patterns were created using SolidWorks software [16, 19] (Fig. 3 and Fig. 4).

The results showed that deformation energy could be obtained, in case of iterative setting of impact crash (distributed load). Thus, a car impact process could be simulated against a given barrier at maximum load.

The exact analogy between the scientific model and the real one can be found in each detail, referring to exact geometric dimensions and material that correspond to real-life characteristics according to the load.

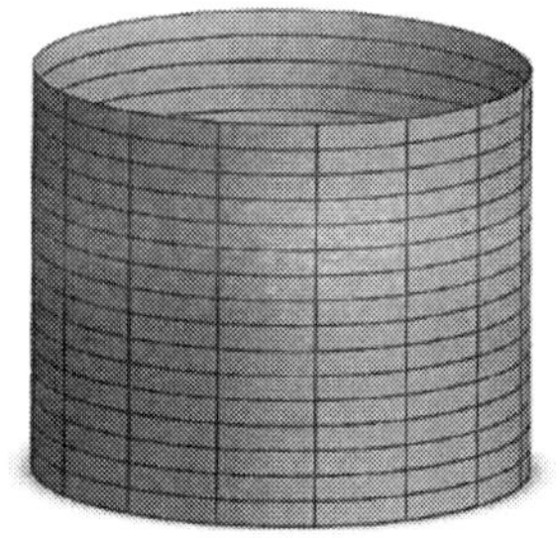

Figure 3. Three-dimensional tank model.

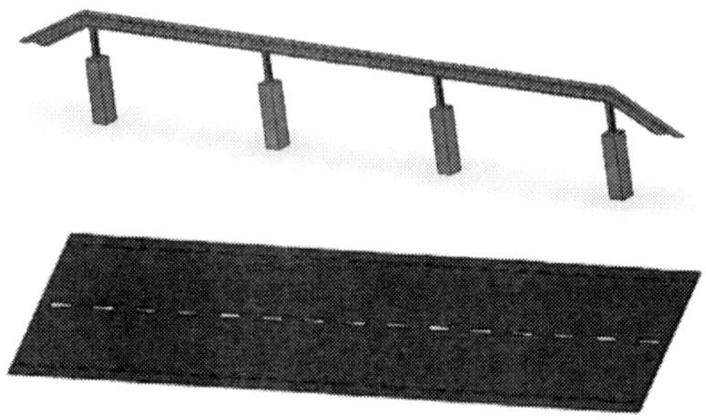

Figure 4. Three-dimensional patterns of metal beam crash barrier and road.

The road was modelled in the Oxy plane defining the stationary coordinate system. The longitudinal axis Ox coincides with the lane separation line. The length of the road section is selected from the scope of the accident.

7. Finite elements model

The finite element method provides a good justification for the magnitude of the resultant impact force, its magnitude and direction. In modelling, boundary conditions are introduced, which are known from the real depth deformation measurements. Thus, precision could be achieved by ensuring that the resulting deformation from the numerical experiment was estimated by the real magnitude of the impact force.

Spatial barrier models were analyzed applying Finite Element analysis in SolidWorks - SW Simulation [15] package. Each tetrahedron element of the second order has ten nodes (four angles and six inner ones) and each has three degrees of freedom. The edges and the planes of the second order elements take curvilinear forms after deformation. Therefore, these elements are usually modelled in curvilinear coordinates. In the model, second order elements were analysed, as they have a lower mesh density compared to first order elements.

For each barrier model, boundary conditions were set in accordance with the dynamics of the impact process. The objective was to achieve the most accurate modelling of the shape of the deformation from the preliminary data about its indentation depth.

In a complex of computational procedures, a solution to equation (1) on the vehicle pre-impact kinetic energy was obtained.

8. Analysis of a Solidworks 3D model of metal crash barrier

In the analysis, the elements of the metal crash barrier are modelled as flexible surface-mounted posts. These are concrete bases, which are not subject to complete destruction. The modelling of the dynamics of the impact process includes taking into account incomplete damage of the bollards, as well as their unremoval from the concrete slabs, i.e. there is no separation between post and base. This means that the impact process is in intensity up to a limited boundary condition - without complete damage of posts and bases or metal barrier elements. The deformation on the metal barrier does not include its damage or destruction, the same condition being valid for the posts as well, up to the separation phase from the concrete base. In the initial conditions, it is assumed that the limit value of the impact force should not be above the critical destructive force of the foundation.

Clamped boundary conditions were defined on the concrete slabs (Figure 5).

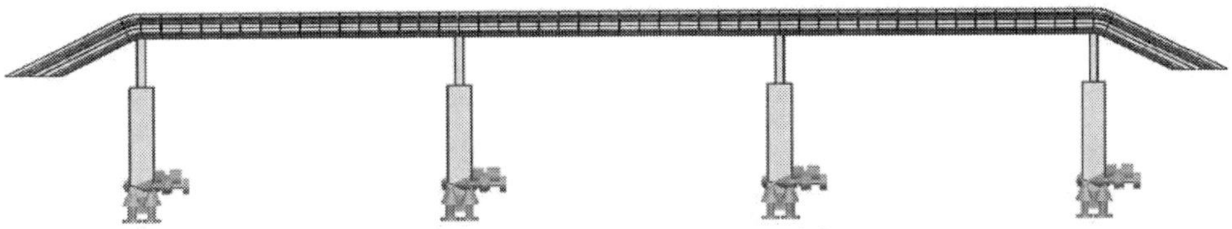

Figure 5. Clamped concrete slabs.

The metal crash barrier model was loaded with a distributed load in a normal and tangential direction over a defined area of the metal beam crash barrier, shown in Figure 6. Specified loads were distributed as pressure applied on the top.

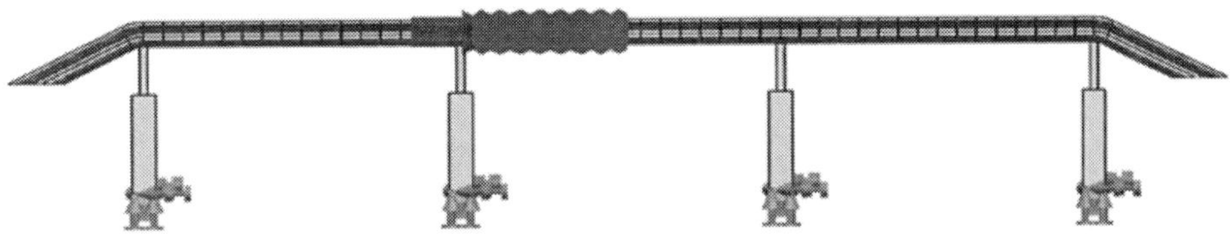

Figure 6. Applied distributed load on metal crash barrier.

9. Analysis of a solidworks 3d model of tank

Limit conditions were defined for the tightened cylindrical body with threaded fasteners on top and on both sides of the (Figure 7).

The tank model was loaded with a distributed load on the normal of the surface of the tank body (Figure 8). In the context of finite element modelling, the condition to use uniformly distributed loads is to be set by their counterparts and on the surface on which they are applied. These are loads of the so-called nodal pressure type.

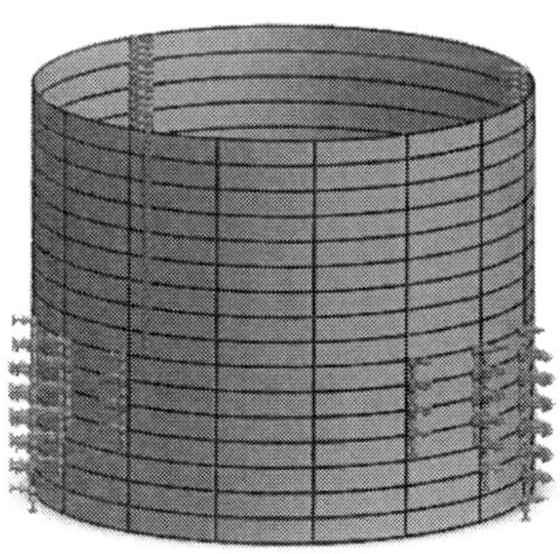

Figure 7. Cylindrical body tightened with threaded joints at the top and on both sides.

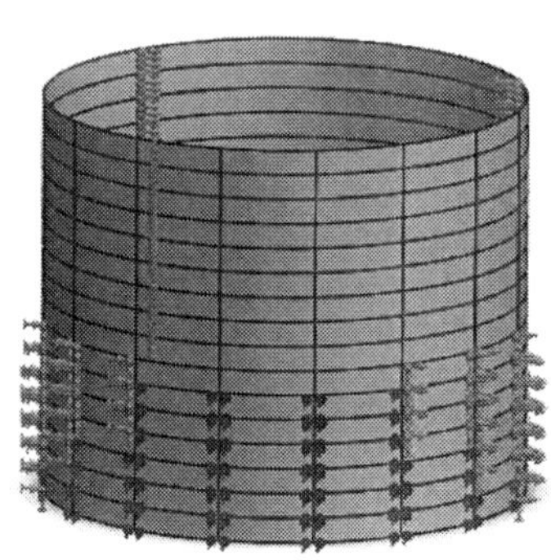

Figure 8. Applied distributed load on the tank body.

10. Impact results of a car crash against a metal crash barrier with post-impact manifested residual energy

After considering the mass inertial parameters

$$m = 1680\,kg;\; J_z = 3190\,kg.m^2 \tag{10}$$

the solution to the post-impact vehicle motion is obtained, shown in Fig. 9. Initial conditions for the velocity of the center of mass and the angular velocity of the vehicle, satisfying the differential equations for vehicle motion, determine the residual energy of post-impact vehicle motion. The graphs in Fig. 14 show the variation of the velocity parameters over time.

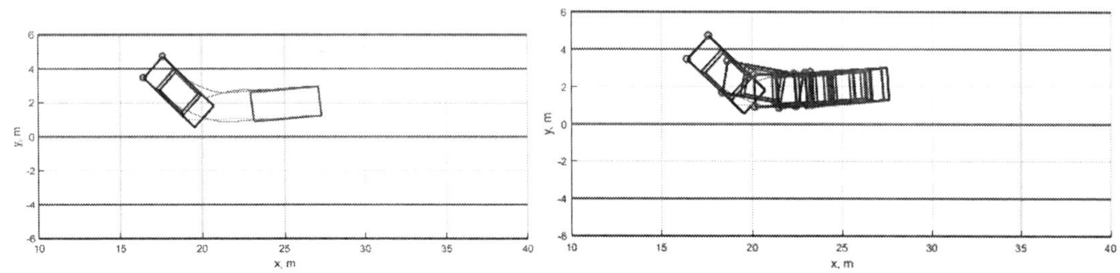

Figure 9. Post-impact vehicle motion against a metal crash barrier and manifestation of residual energy.

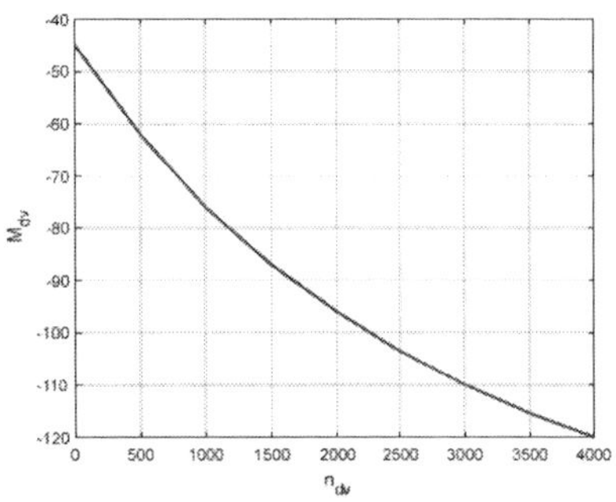

Figure 10. Motion graph in idle mode.

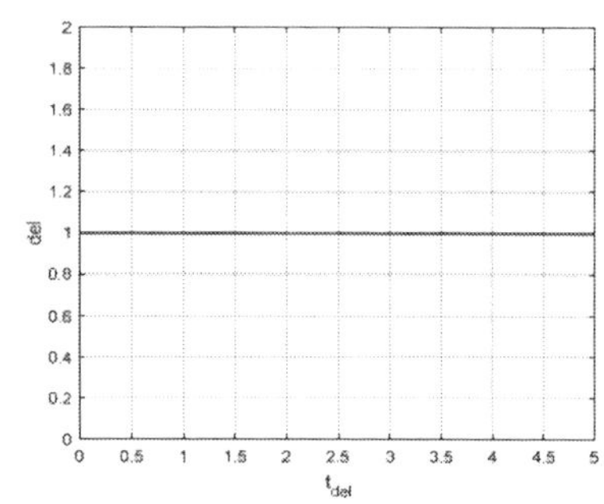

Figure 11. Rotation angle of front wheels.

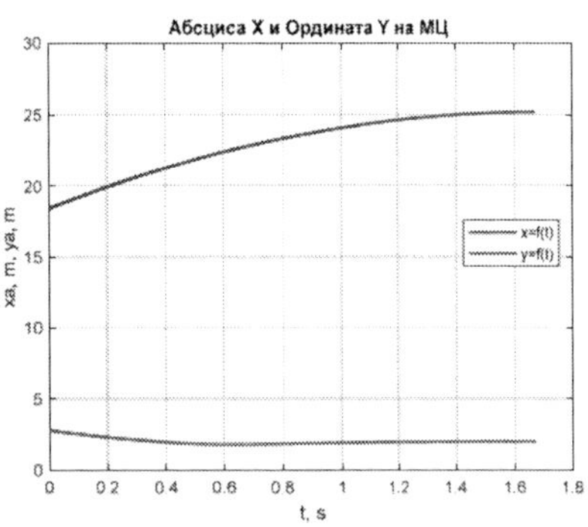

Figure 12. Change of coordinates of the center of mass.

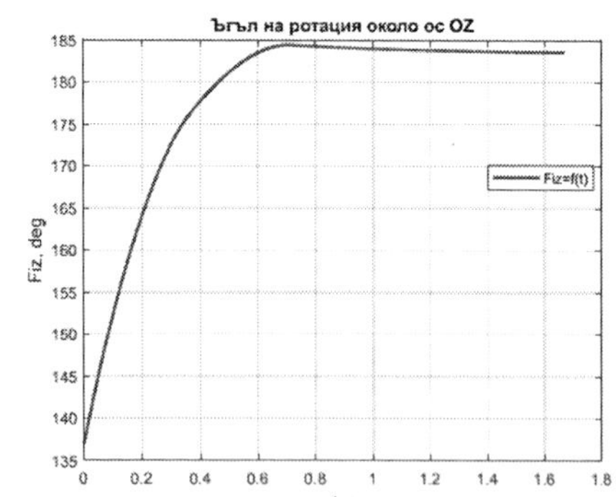

Figure 13. Rotation angle of the vehicle around the Oz axis.

IOP Conf. Series: Materials Science and Engineering **659** (2019) 012063 doi:10.1088/1757-899X/659/1/012063

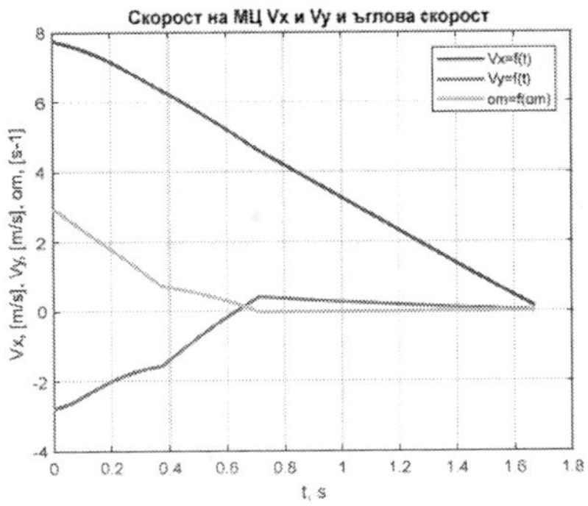

Figure 14. Velocity projections of center of mass.

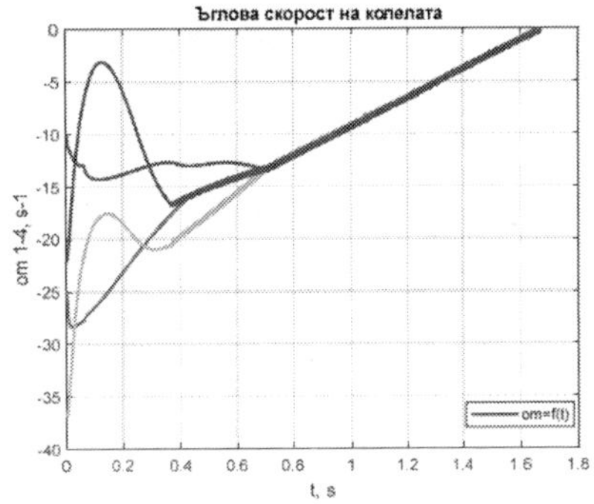

Figure 15.. Angular speed of each wheel.

Car body deformation from the impact against a metal crash barrier determines the impact energy by formula (8).

Table 1. Vehicle collision in a rigid barrier.

Vehicle total mass reported	1680,00
Depth of deformation zone reported C1 in[cm]	0,00
Depth of deformation zone reported C2 in [cm]	25,00
Depth of deformation zone reported C3 in [cm]	48,00
Depth of deformation zone reported C4 in [cm]	62,00
Depth of deformation zone reported C5 in [cm]	75,00
Depth of deformation zone reported C6 in [cm]	86,00
Reported angle between plane normal and impact force	6,00
Reported indentation width L in [m]	1,12
Coefficient - A energy in [N/cm]	362,00
Coefficient - B energy in [N/cm^2]	48,30
Deformation energy obtained [N.m]	108817

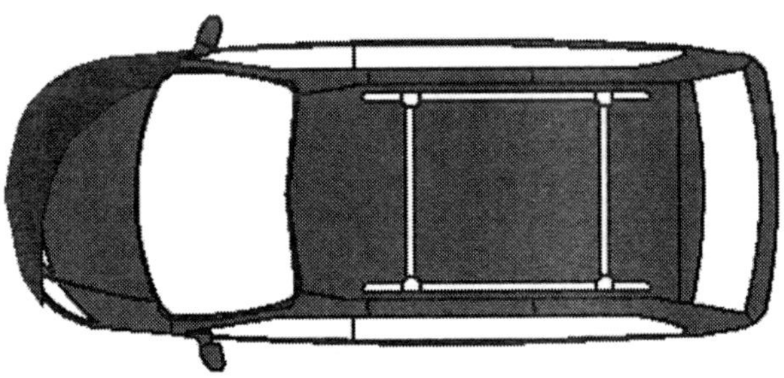

Figure 16. Reported depth of deformation zone of the vehicle crash against metal crash barrier.

Fig. 17 and 18 show the results of the analysis. At given loading with uniformly distributed load, the following deformation of the metal crash barrier was achieved.

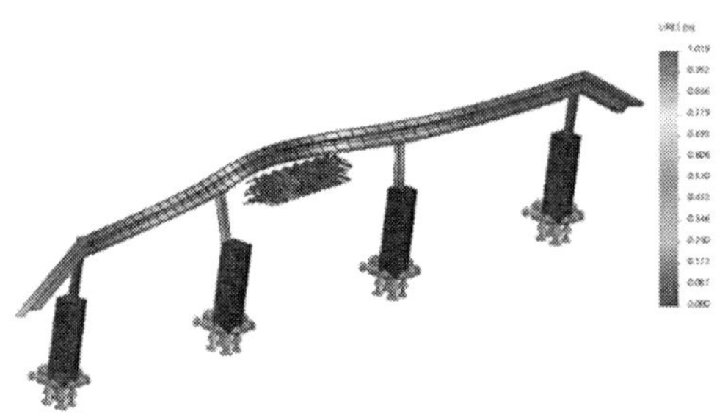

Figure 17. Deformation of the metal bend crash barrier.

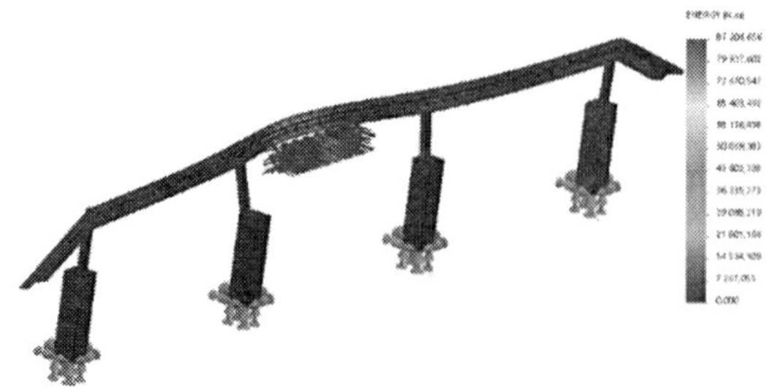

Figure 18. Deformation energy of the metal bend crash barrier.

The pre-impact vehicle velocity of the center of mass against the metal barrier is derived from the dependence following the law of conservation of mechanical energy:

$$V = \sqrt{\frac{m.u^2 + I_z.\omega_z^2 + 2.E_1 + 2.E_2}{m}} = \sqrt{\frac{1680.8{,}3^2 + 3190.2{,}95^2 + 2.108817 + 2.87024}{1680}} = \\ = 17{,}8\ m/s \qquad (11)$$

The value obtained gives a quantitative estimate of the energy needed to deform the car body, the barrier and to reach final rest position.

11. Impact results in a car crash against a tank body in a lack of residual energy after impact - perfectly plastic collision

Deformation energy of the car body is determined according to (6), shown in Table 2.

Table 2. Vehicle collision in a rigid barrier.

Vehicle total mass reported	1900,00
Depth of deformation zone reported C1 in[cm]	10,00
Depth of deformation zone reported C2 in [cm]	7,00
Depth of deformation zone reported C3 in [cm]	5,00
Depth of deformation zone reported C4 in [cm]	5,00
Depth of deformation zone reported C5 in [cm]	5,00
Depth of deformation zone reported C6 in [cm]	7,00
Reported angle between plane normal and impact force	0,00
Reported indentation width L in [m]	1,695
Coefficient - A energy in [N/cm]	541,00
Coefficient - B energy in [N/cm^2]	93,10
Deformation energy obtained [N.m]	11346

At the given loading with uniformly distributed load, the depth of deformation zone of the tank body was achieved, as follows (Fig. 19 and Fig. 20).

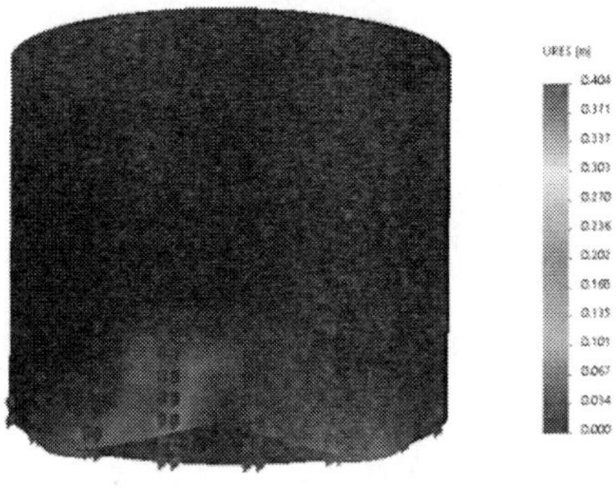

Figure 19. Deformation of the tank body.

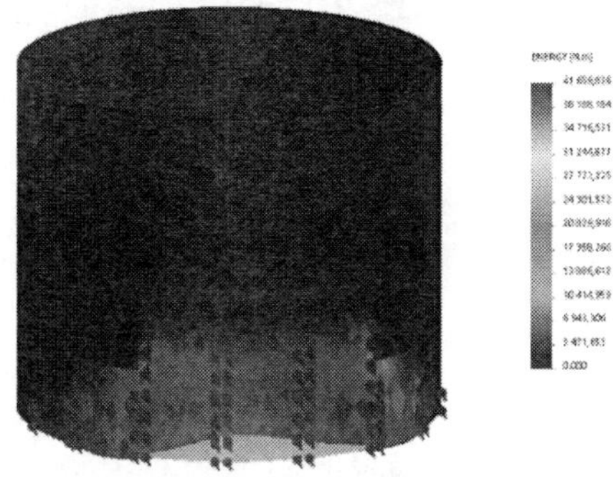

Figure 20. Deformation energy of the tank body.

Maximum reported deformation energy of the tank body as a result of impact and according to the adopted approach of finite element analysis is

$$E_1 = 41659{,}8\ N.m \tag{12}$$

The velocity of the mass center of the vehicle at the point of impact is determined by the formula

$$V = \sqrt{\frac{2.(E_1 + E_2)}{m}} = \sqrt{\frac{2.(41659{,}8\ + 11346)}{1900}} = 7{,}47\ m/s = 26{,}9\, km/h \tag{13}$$

The obtained value gives a quantitative estimate of the energy needed to deform the car body, the barrier in a perfectly plastic collision.

12. Conclusions

1. Solution to a common engineering task necessary for legal analysis and widely used in case-law is provided. It determines the pre-impact speed of a vehicle following roadside objects/barrier impact and the loss of longitudinal stability after the impact. Traditional methods analysing such crashes, which are generally applied to modern impact investigation and accident reconstruction, do not meet accuracy and precision requirements.
2. The proposed FEM approach is a modern innovative means of solving similar engineering tasks in order to identify traffic road accidents. It is based on credibility and justification of facts and evidence.
3. This method for investigating and analysing head-on vehicle impact with a rigid barrier can also be applied to other vehicles and trucking fleets - trailers and semitrailers. The solution to the task requires precise post-impact mechanical and mathematical model for vehicle crashes, which represent the motion of the composition of the vehicle fleet.

References

[1] Schmidt B, Haight W, Szabo T and Welcher J 1998 *System-based energy and momentum analysis of collisions, in Accident Reconstruction*: Technology and Animation VIII SP-1319, no. 980026, Society of Automotive Engineers, Warrendale, PA
[2] Sharma D, Stern S and Brophy J 2007 *An Overview of NHTSA's Crash Reconstruction Software WinSMASH.* In Proceedings of the Twentieth International Conference on Enhanced Safety of Vehicles, Lyon, France, Paper No. 07-0211
[3] Daily J, Shigemura N and Daily J 2006 *Fundamentals of Traffic Crash Reconstruction.* Jacksonville, Florida: Institute of Police Technololgy and Managment, University of North Florida

[4] Neptune J, Blair G and Flynn J 1992 *Amethod for quantifying vehicle crush stiffness coefficients*," SAE Technical Paper Series, no. 920607

[5] Wicher J 2006 *Energetyczne metody analizy zderzeń samochodów* (*Energy methods of the analysis of motor vehicle collisions*). The 3rd Scientific and Educational Conference "Development of automotive engineering versus motor insurance", Radom

[6] Campbell K 1974 *Energy basis for collision severity*, SAE Technical Paper Series, no. 740565.

[7] Prasad K 1990 *CRASH3 Damage Algorithm Reformulation for Front and Rear Collisions*, SAE 900098

[8] Prochowski L and Żuchowski A 2011 Comparative Analysis of Frontal Zone of Deformation in Vehicles with SelfSupporting and Framed Bodies, *Journal of KONES Powertrain and Transport, Warszawa* **18**(4)

[9] Stucki L and Fessahaie O 1998 *Comparison of Measured Velocity Change in Frontal Crash Tests to NASS Computed Velocity Change*, SAE 980649, Society of Automotive Engineers

[10] Nolan M, Preuss A, Jones L and O'Neill B 1998 *An Update on Relationships between Computed Delta-Vs and Impact Speeds for Offset Crashes*, In the Proceedings of the Sixteenth International Technical Conference on the Enhanced Safety of Vehicles (Windsor, Canada, June 1998), Paper No. 98-S6-O-07

[11] Hight P, Kent-Koop D and Hight R 1985 *Barrier equivalent velocity, delta v and CRASH3 stiffness in automobile collisions*, SAE Technical Paper Series, no. 850437

[12] Niehoff P and Gabler C 2006 *The Accuracy of WinSMASH Delta-V Estimates: The Influence of Vehicle Type, Stiffness, and Impact Mode,* In the 50th Annual Proceedings of the Association for the Advancement of Automotive Medicine, Chicago, IL

[13] Hibbeler R 2004 *Engineering Mechanics*: Dynamics. Upper Saddle River, NJ: Prentice Hall, 10th ed

[14] Owsiański R 2007 *Szacowanie energii deformacji nadwozi kompaktowych samochodów osobowych* (*Estimation of the bodywork deformation energy of compact passenger cars*). Paragraf na Drodze No. 4

[15] *SolidWorks Simulation 2009 – Training Manual*, Dassault Systemes SolidWorksCorporation, Massachusetts, USA

[16] *SolidWorks 2019 Users Manual*, Dassault Systemes SolidWorksCorporation, Massachusetts, USA

[17] Stronge W 2000 *Impact Mechanics. Cambridge*: Cambridge University Press

[18] Wach W 2003 *Analiza deformacji samochodu według standardu CRASH3.* Część 2: Pomiar głębokości odkształcenia (Analysis of motor vehicle deformation according to the CRASH3 standard. Part 2: Measurement of deformation depth), Paragraf na Drodze No. 12

[19] Dechkova S 2018 Creation of multi-mass models in the SolidWorks and Matlab environment for crash identification, *Machine Mechanics* **119** 28-32

[20] Karapetkov S 2005 *Auto Technical Expertise*, Technical University - Sofia, Bulgaria

[21] Karapetkov S 2010 *Investigation of road traffic accident, technical commentary on the lawyer*, Technical University of Sofia, Bulgaria

[22] Karapetkov S and Uzunov H 2016 *Dynamics of transverse resistance of a car*, Didakta Consult – Sofia, Bulgaria

IOP Conf. Series: Materials Science and Engineering **659** (2019) 012064 doi:10.1088/1757-899X/659/1/012064

Optimization of traction unit for low-cost automated guided vehicle

T Capák[1], R Kohár[1], J Škrabala[1] and J Galík[1]

[1]University of Žilina, Univerzitná 8215/1, 010 26 Žilina, Slovakia

E-mail: tomas.capak@fstroj.uniza.sk

Abstract. AGV system already known today is commonplace in any large-scale production, warehousing or supply process logistics. System is not only increasing process flexibility but also the quality of products and services. They are unattended vehicles designed for towing, guiding and carrying various types of cargo using peripheral devices designed for these vehicles. The paper describes the design of the traction unit of the automated guided vehicle (AGV). The main goal of this design is optimization of traction unit for a low-cost version of commercial AGV and thereby reducing overall costs of equipment. The traction unit must be designed so that the equipment that uses such a type of traction unit meets customer requirements.

1. Introduction

Industry 4.0 is a fundamental change not only in the industrial but also in the private sphere of each of us. Industry 4.0 is a technological (digital) revolution of the present, bringing many changes in work, communication and the way we live. This revolution also concerns automated systems, where it is raising awareness among, for example, individual members of the production process (device-to-device communication). Therefore, many companies are transforming more and more into so-called 'Smart Factory'. It is a wide range of production, which main goal is optimizing the production concept and thereby creating a flexible and configurable company.

2. Smart factory with cellular manufacturing

Changes in production and also in production logistics depend mainly on customer requirements. This type of production is driven by the customer and is necessary to adapt production to the product. Cellular production is a form of standard serial production that distributes work into separate teams known as cells that can be divided into three categories [1, 2]:

Production cells - cells (workstations) that contain production facilities (machining, casting, forging, welding…).

Assembly Cells - standard cells for mounting separate units into a unit.

Process cells - in other words, technological cells determined by technological processes of production (surface treatment, painting, heat treatment).

In general, cellular production has several advantages, such as:

- shortening production time and consequently shortening the time to market;
- reducing the cost of transforming the production line and the like.

There are also several drawbacks to cellular production, for example:
- high technical requirements for production preparation;
- high input costs;
- need for higher operator qualifications.

Many automated logistics systems of various kinds have been implemented in production processes at present. These systems are providing internal logistics within the manufacturing processes of companies. Such a combination of an automated logistics system and cellular production brings many savings not only within material flows, warehouse or handling operations, but also within the organization of opaque manufacturing processes, resulting in several major cellular production parameters associated with automated logistics and that [1, 2, 10]:
- increased flexibility;
- reduction of production times;
- shortening transport distances;
- direct material flow…

Figure 1. Example AGV cobot cooperating on workplace.

In the preceding figure 1, a simplified example of a cellular production connection (CNC machine and sorting position) and an automatic logistic system (cobot handling) is shown, which as a whole form a solution for performing multiple handling operations and logistics tasks throughout the manufacturing process [8].

3. Modular platform

Different levels of automation, resulting from costumer needs can be covered by AGV modular solution. At present, it is a requirement from customers, create more complex solutions by combining cellular production and an AGV system. These solutions providing the possibility of replacing physical operator with a fully automated (autonomous) platform of production cell. In many cases it will be able create more friendly working conditions for people [5, 6, 13].

CEIT, a.s., implements the development of equipment achieving the required parameters. One of these devices is the prototype platform being developed for handling and inter-operational purposes within individual cell manufacturing sections. This type of AGV platform will represent the middle class of autonomous vehicles and is designed to implement for inter-cell operations. It is a full-fledged autonomous guided vehicle with mark **AGV 400LC-F**, whose design has been developed as the basis of a modular platform, to which many peripherals (modular superstructures) can be attached. The shape and function of the peripherals depends on the cell production sections (cells). This modular platform can provide sections exchange during operation or directly supply or serve individual production cells.

4. AGV 400LC-F

This type of AGV can be described as a competitive modular platform, where is important to maintain the compatibility of the chassis with individual peripheral devices. Typically, when designing an autonomous logistics tractor, the first problem is to design a base frame (chassis) and determine the appropriate number and type of wheels for the traction unit and wheels for the carrier part. The device is based on a 6-wheel chassis with omnidirectional wheels oriented on the front and the rear section of the chassis and a traction unit (the axle) in the middle of the chassis so that the turning radius of the device is zero. The following figure 2 shows principle of differential steering of wheels.

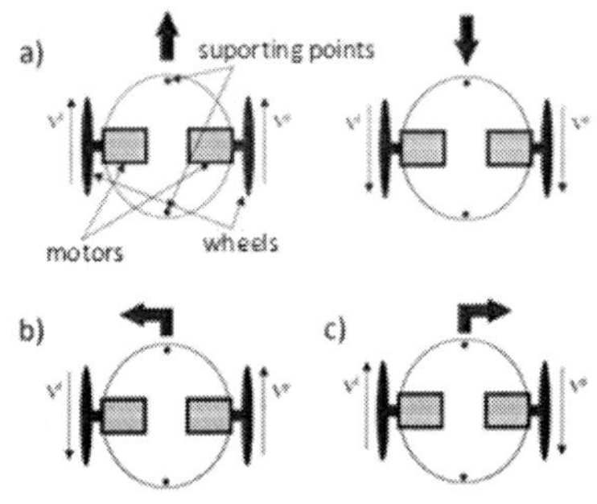

Figure 2. Principle of differential steering of wheels.

This orientation of the traction unit also serves as the equivalent of a mechanical differential. Differential type of drive is the easiest and most used type of mobile robot chassis used mainly for small and low-cost robots working indoor. It is independent control of traction wheels driven most often by DC electromotors with encoders or stepper motors.

The supporting points of chassis are usually the omnidirectional wheels with one or two degrees of freedom. The principle of operation of differential control lies in different speeds of traction wheels when the AGV changes the direction of movement. By turning the wheels at the same speed in the same direction, the AGV moves forward or backward. By rotating the wheels at the same speed in opposite directions, the AGV is rotated about its own axis. This bringing great benefits to the AGV, for example, a zero turning radius of the tractor. The disadvantage of this construction is the inability to overcome higher obstacles such as door thresholds and the like [7, 8, 9].

When designing the AGV concept, it is important that this concept satisfies the basic requirements, which represent a compromise between customer requirements and equipment manufacturer's capabilities.

As essential requirements, we define:

- maximum AGV height: 300mm;
- maximum load capacity: 400kg;
- maximum speed: 0,7 m/s;
- power supply: 24V (maintenance-free batteries)

4.1. Traction unit design

When designing a traction unit, it is necessary to identify the conditions in which the equipment will be operate. The weight of the load, weight of the AGV without load, lifting speed, AGV speed with and without load, the surface and the ascent of the route and many other factors, whether directly or indirectly, affect the traction of the equipment. The weight of the AGV is determined from a 3D model, which contains all the necessary elements such as base frame, lift mechanism, wheels, accumulators and other purchased and manufactured parts which scale up the total AGV weight. The weight of the AGV electronics was estimated. The total estimated AGV weight was 195 kg. The weight of the AGV is also increased by using peripheral devices. The total weight of the AGV is therefore set at 220kg [3, 4].

To calculate the power needed to run-up the AGV and determine the appropriate DC electric motor, gearbox and drive wheels, it is necessary to determine the set of resistances acting on the AGV during operation in the hall (indoor - no wind). When calculating, we consider an evenly accelerated movement to a speed of 0,7 - 1 m / s.

4.1.1. Driving resistances

To determinate AGV driving resistances is sufficient a basic calculation of driving resistances according to the following formula [3, 4].

$$F_{Total} = F_{AD} + F_{RR} + F_{ACC} + F_{RG} \quad (1)$$

The Aerodynamic drag (F_{AD}) is determined by the aerodynamic shape of the vehicle body and is described by parameters such as drag coefficient (C_d), projected frontal area of the vehicle (A_f), velocity (v) and air density (ρ_{Air}) [3, 4].

$$F_{AD} = C_d \cdot A_f \cdot \frac{\rho_{Air}}{2} \cdot v^2 \quad (2)$$

At speed AGV of 1 m/s and the specified operating conditions is Aerodynamic drag neglected for the calculation.

The rolling resistance force (F_{RR}) is mostly determined by the wheel tread, but also by parts of the traction unit. They are characterized by the rolling resistance coefficient (f_{RR}), which is dependent on the vehicle's velocity. The mass of the AGV (m_{AGV}) and mass of the load (m_L) also has a linear influence perpendicular to the road and the climb angle (α) is determined by certification of AGV with max value 2° [3, 4].

$$F_{RR} = (m_{AGV} + m_L) \cdot g \cdot f_{RR} \cdot cos(\alpha) \quad (3)$$

The Acceleration resistance (F_{ACC}) depends on the weight of the AGV with load. The inertia of the rotating parts (I_{RP}) is negligible at a speed of 1 m/s. [3, 4]

$$F_{ACC} = (m_{AGV} + m_L + I_{RP}) \cdot a \quad (4)$$

The climb resistance (F_{RG}) depends on the road gradient and the AGV mass with load. [3, 4]

$$F_{RG} = (m_{AGV} + m_L) \cdot g \cdot sin(\alpha) \quad (5)$$

Traction force (F_{Total})

F_{Total} = 520 N

Power (P_{Total})

P_{Total} = 520 W

Torque (T_{Total})

T_{Total} = 57,2 Nm

For the operation of differential chassis type equipment (two-wheel drive) is the total power distributed between the left and right traction units. Each of these parts includes an electric motor with a gearbox (or servo motor) and a traction wheel. Servomotors are the most suitable solution for AGV control, but they are very expensive compared to conventional DC electric motors. The proposed AGV prototype will be categorized as a low-cost type of AGV and it is preferable to select a system of an DC electric motor with a gearbox for traction unit. Transtecno s.r.o. offers, in its products, DC (electric) motors with worm gears. The advantage of such assemblies is their simple installation and operation with minimal maintenance [11, 12].

4.1.2. Selected assembly

The following tables 1 and 2 describes the parameters of the selected DC electric motor type and the selected worm gearbox [11, 12].

Table 1. Type of DC electric motor.

Type	NP [W]	NV [V]	NC [A]	NT [Nm]	NR [rpm]
EC 350.240	350	24	21	1,12	3000

Table 2. Type of worm gearbox.

Type	i_G	NR [rpm]	η_G	OT [Nm]	OP [W]
		900	0,61		
CM 063	50	1400	0,66	136	0,6
		2800	0,73		

Acronyms: NP – nominal power, NV – nominal voltage, NC – nominal current, NT – nominal torque, NR – nominal revolutions, i_G – gear ratio, η_G – gear efficiency, OT – output torque, OP – output power.

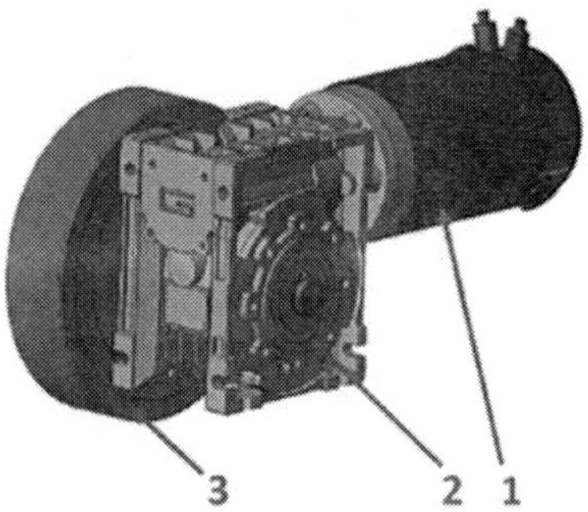

Figure 3. Assembly of 1 - DC electric motor EC 350.240, 2 - worm gearbox CM 063 and 3 - traction wheel.

The torque of the selected electric motor when the AGV acceleration is 1,12 Nm. It is the nominal torque value at the nominal motor speed indicated by the electric motor manufacturer. Check of the total output torque (T_{TO}) of traction unit is performed as follows [3, 4]:

$$T_{TO} = 2 \cdot (T_{Total} \cdot i_G \cdot \eta_G) \tag{6}$$

$$T_{TO} = 68{,}32\ Nm > T_{Total} = 57{,}2\ Nm$$

The resulting total torque of the traction unit is higher than the calculated torque required, and it can, therefore, be argued that the traction unit drive is dimensioned correctly and will be able to operate in the event of a small overload. This claim must be verified by the physical measurement of the AGV traction force under load. The equipment test conditions result from the required operating parameters, load: 400kg and max. speed: 0.7 m/s. The following figure 4 shows the position of the traction unit on the AGV frame.

Figure 4. Position of the traction unit
1 – traction unit, 2 – AGV frame.

Torque verification was done by measuring the AGV traction force. During the test were measured two physical quantities. The current between DC motor and traction converter using Fluke 345 clamp multimeter and the second variable was ductility AGV measured by the industrial electronic crane scale with load capacity 5000kg and sensitivity 0,5kg. The following figure 5 shows the way how to measure the ductility of a device. An industrial scale (3) is placed between the AGV (1) and the concrete column. The AGVs, together with the industrial scale, are firmly fixed by means of clamping straps. The AGV is loaded with 8 batteries (2). Each of these batteries weights 54kg, resulting in a total load of 432kg.

Figure 5. AGV with load
1 – AGV, 2 – load (batteries), 3 – industrial scale, 4 – clamp multimeter.

14 measurements were made after setting the measuring technique. Seven measurements at an electric motor temperature of 21 ° C and seven measurements at an electric motor temperature of 53 ° C.

At each measurement, the AGV acceleration was 0,05 m/s^2 and the constant speed was 0,1 – 0,7 m/s. After reaching a constant speed, the tensioning straps were tightened, and the electric motor stopped. The measured traction force at the stopped electric motor is equivalent to the maximum traction force needed to start the AGV. The following graphs show the torque curve of the electric motor, which is composed of the measured and calculated values given in the tables [14, 15].

Table 3. Results of first measurement.

CR [A]	PF [kg]	PF [N]	i_G	WR [m]	TR [Nm]	v [m/s]	WS [rps]	MS [rpm]
32,00	104	1020,24			1,12	0,1	0,145	434
32,50	105	1030,05			1,13	0,2	0,289	869
32,6	105	1030,05			1,13	0,3	0,434	1303
32,6	105	1030,05	50	0,11	1,13	0,4	0,579	1737
32,6	109	1069,29			1,18	0,5	0,724	2171
32,6	113	1108,53			1,22	0,6	0,868	2606
32,6	111	1088,91			1,20	0,7	1,013	3040

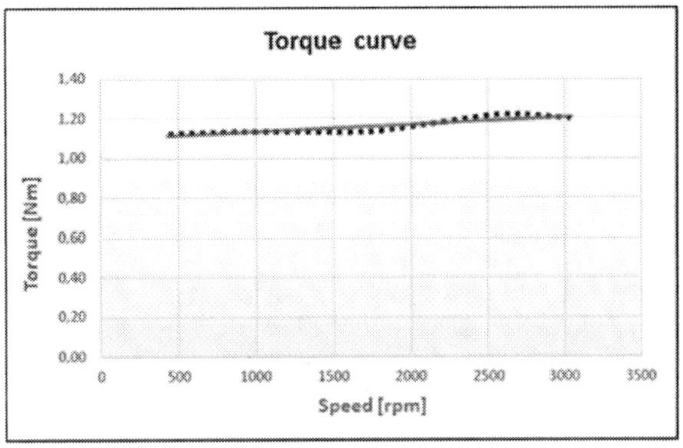

Figure 6. Torque curve of EC 350.240 – first measurement.

Table 4. Results of second measurement.

CR [A]	PF [kg]	PF [N]	i_G	WR [m]	TR [Nm]	v [m/s]	WS [rps]	MS [rpm]
32,7	107	1049,67			1,15	0,1	0,145	434
32,6	110	1079,1			1,19	0,2	0,289	869
32,6	110	1079,1			1,19	0,3	0,434	1303
32.6	113	1108,53	50	0,11	1,22	0,4	0,579	1737
32.6	114	1118,34			1,23	0,5	0,724	2171
32.6	118	1157,58			1,27	0,6	0,868	2606
33	120	1177,2			1,29	0,7	1,013	3040

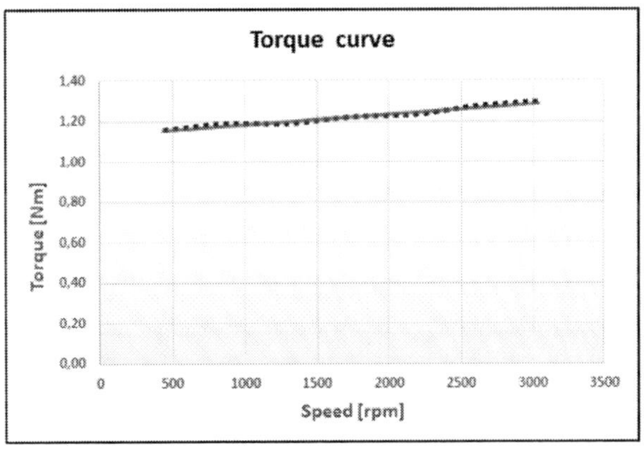

Figure 7. Torque curve of EC 350.240 – second measurement.

Acronyms: CR – current, PF – pulling force, i_G – gear ratio, WR – wheel radius, TR – motor torque, v – velocity, WS – wheel speed, MS – motor speed.

The measured traction force values of the AGV confirm that the designed parts of the traction unit (DC electric motor and worm gearbox) are suitable for a low-cost version of medium-class AGV equipment. From the resulting values of both measurements, it can be argued that the traction force of the AGV is ***1000N*** (see table 3 and 4) and the electric motor torque of ***1.12Nm will be sufficient*** for the start acceleration AGV (see figure 6 and 7) because it is greater than the calculated maximum torque. The cost of this type of drive is much lower than that of servo motors, which guarantees a lower price for the AGV and more competitiveness device. The figure below shows the newly developed AGV prototype for smart internal logistics, in which the proposed type of traction unit is used. This type of traction unit will be further tested in laboratory conditions and later directly in the warehouse operation for which this prototype is designed. The AGV type also includes the latest hardware components and software packages for autonomous location, navigation, and mapping of space.

Figure 8. Developed type of AGV 400LC-F during testing of the designed low-cost traction unit.

5. Conclusion and future work

This paper has described the optimization of traction unit for developed AGV. It is a low-cost version of the device for which it is also necessary to design a low-cost solution for each part of the AGV. In low-cost solutions, economic point is a major factor affecting those solutions. The main aim of the paper is to design a simple solution of parts of the traction unit and verify the proposed solution by physical measurement for the required parameters of the equipment. Testing of the equipment takes place in the laboratory of CEIT, a.s., which developed this equipment. The measurement results showed that the designed type of traction unit is a suitable equivalent of servomotors and it can be usable as AGV's drive.

In case of autonomous logistics tractors, it is necessary to use the parking brake when the AGV is stopped. This brake is usually located on the traction wheels or is part of the electric motors. The designed type of low-cost traction unit does not contain any braking system and therefore it is necessary to verify that the selected type of worm gear will meet the required braking parameters. On the developed prototype of AGV we will be testing brake force over the next few months of this year.

Acknowledgement

This publication is the result of the Project implementation: Competency Centre for Knowledge technologies applied in Innovation of Production Systems in Industry and Services, ITMS: 26220220155, supported by the Research & Development Operational Programme funded by the ERDF.

References

[1] Knoflíček R 2004 Roboty a pružné výrobní systémy, (Robots and flexible manufacturing systems) FSIVUT v Brně. (in Czech)

[2] Kárník L, Knoflíček R and Novák-Marcinčin J 2000 *Mobilní roboty,* První vydání, (Mobile robots. First edition.) (in Czech)

[3] Jasaň V, Košábek J and Szuttor N 1989 *Teória dopravných a manipulačných zariadení,* Prvé vydanie, (Transport and handling equipment theory. First edition.) Alfa STNL, (in Slovak)

[4] Košábek J a kol. 1990 *Teória dopravných a manipulačných zariadení,* Tretie vydanie, (Transport and handling equipment theory, Third edition) Bratislava: Alfa STNL, (in Slovak)

[5] Chironis N P 1991 *Mechanisms and Mechanical devices sourcebook,* First edition

[6] Ullrich G 2014 *Automated guided vehicle systems: A Primer with Practical Applications*, Second edition, Springer

IOP Conf. Series: Materials Science and Engineering **659** (2019) 012064 doi:10.1088/1757-899X/659/1/012064

[7] VUTBR, *Simulace činnosti mobilní robotické soustavy se všesmerovým podvozkem* 2006 (Simulation of the mobile system operation with omni-directional wheels) https://www.vutbr.cz/vav/vysledky/detail?vav_id=44180&aid_redir=1#vysledek-44180 (accessed on May 2019) (in Czech)
[8] VUTBR, *Mobilní roboty pro průmyslové využití* 2005 (Mobile robots for industrial use) https://www.vutbr.cz/vav/vysledky/detail?vav_id=66512&aid_redir=1#vysledek-66512 (accessed on May 2019) (in Czech)
[9] ROBOT PLATFORM, *Wheel control Theo*ry, http://www.robotplatform.com/knowledge/Classification_of_Robots/wheel_control_theory.html (accessed on April 2019)
[10] YASKAWA, Solutions/Application/Handling-Assembly, https://www.yaskawa.eu.com/en/solutions/application/handling-assembly/ (accessed on May 2019)
[11] RAVEO, *Šnekové prevodovky* (Worm gearboxes), http://www.raveo.cz/snekove-prevodovky (accessed on Oct. 2018) (in Czech)
[12] RAVEO, *Stejnosměrné motory* (DC electric motor), http://www.raveo.cz/stejnosmerny-motor-EC (accessed on Oct. 2018) (in Czech)
[13] CEIT, a.s, *Automation of internal logistics,* https://ceitgroup.eu/en/solutions/automation-of-internal-logistics (accessed on May 2019)
[14] Kučera Ľ and Gajdošík T 2014 *The vibrodiagnostics of gear* In book: Modern methods of construction design – 54th ICMD, Springer
[15] Lukáč M, Brumerčík F, Krzywonos L and Krzysiak Z 2017 *Transmission system power flow model* In: Communications: scientific letters of the University of Žilina

Optimal design of self-retaining full complement cylindrical roller bearings

C Ursache[1], A Barili[1,2], L Tudose[1,2] and C Tudose[2]

[1]Technical University of Cluj-Napoca, Faculty of Machine Building, Bd. Muncii 103-105, 400641 Cluj-Napoca, Romania
[2]RKB Europe SA, Via Primo Agosto 1, 6828 Balerna, Switzerland

E-mail: lucian.tudose@omt.utcluj.ro

Abstract: Due to the presence of the ribs on both sides of the outer ring, the self-retaining full complement (cageless) cylindrical roller bearings have a very interesting particularity: the outer ring must be heated to create enough room between the already mounted rollers to allow the last roller to be inserted in the row. In addition, after cooling down, the rollers must support each other with a certain small clearance. For this reason, the internal geometry of the bearing as well as the precision of execution are more than critical. On the other hand, one of the most important problem in rolling bearing design is to establish the optimal set of internal dimensions in order to obtain the maximum value of the basic radial dynamic load rating. Pursuing this goal, in this paper, the optimal design of certain self-retaining cylindrical roller bearing is performed by means of the PSO (Particle Swarm Optimization) algorithm. In this approach two design variables were considered (the roller diameter and the diameter of the inner ring raceway) and both geometrical and thermal constraints were taken into account. A study referring to the sensitivity of the design to the small variation of the geometrical parameters is also presented.

1. Introduction

Rolling bearings are one of the most interesting and, at the same time, complex mechanical component. There are many types of rolling bearings, starting from the simplest one, the single row ball bearings till most complex ones, like multi-row tapered roller bearings. Every bearing type has an unique internal geometry related to the shape of the rolling elements and to the afferent raceways. Apart the bearing internal geometry, there are also many other important features and variants. For example, for cylindrical roller bearings there are two main variants, with or without cage (full complement). Most of the rolling bearings are provided with a cage that guides and evenly angularly divide the rolling elements but, on the contrary, full complement bearings do not feature a cage, meaning that each rolling element is touching the next and the previous one, during rotation. The main benefit of the full complement bearing type is related to an increased number of rolling elements that, compared to an equivalent design with cage, leads to increased dynamic [1] and static [2] load ratings. Anyway, considering the applications where this types of bearings are used (e.g. planetary speed reducers), they have mandatory to reach a high value of fatigue life. As per [1], the maximization of the bearing fatigue life can be obtained only by maximizing its basic dynamic load rating due to an optimal internal bearing geometry. To achieve this goal, in the next paraghaphs, it has been decribed geometrically a self-retaining full complement cylindrical roller bearing, in order to obtain the necessary equations to permit the mathematical design of the optimization problem. The main innovation of this work is represented by the optimization of the internal bearing geometry of a self-retaining full complement roller bearing.

2. Literature review

The study of cylindrical roller bearing has been managed by many researchers. Dragoni [3] developed a design procedure to maximize the basic load ratings values of cylindrical roller bearings mounted inside a gearbox. Tudose et al. [4] considered two objective functions (the basic dynamic radial load rating and the minimum thickness of the elastohydrodynamic lubrication film between rollers and raceways) and performed the multi-objective optimization using Cambrian, a very powerful optimization platform. In order to describe completely and unique the bearing geometry, three design variables were considered: the geometric factor γ, roller diameter and roller length, respectively.

Considering full complement roller bearing, it is not easy at all to find similar work in the field. In fact, as it is in the real world, the focus of all the authors is to study the caged bearings that clearly represents the biggest part of the market. Anyway, about single objective optimization, it has been found some studies that consider the bearing mounting, see for example [5], where, in order to maximize the basic dynamic load rating, it has been imposed a constraint on the bearing diameter and number of balls. Then a constraint regarding the assembly angle was imposed to find a feasible set of designing variables that are compatible with the assembling method of the bearing.

Some works see for example [6], were focused on the maximization of the dynamic load rating of cylindrical roller bearing considering the influence of the interference between the rings and the afferent seats. Regarding other works, related to different bearing types (see, for example, [7]), it becomes clear that the maximization of the bearing fatigue life, i.e. of the basic dynamic load rating, can be defined as the most common and diffused objective function in the optimal design of rolling bearings. Nevertheless, considering the particular geometry and assembling method of the bearing in discussion here, the authors of this paper did not find anything in the open literature regarding the optimal design of the self-retaining full complement cylindrical roller bearings.

3. Bearing geometry

Since the focus of the present work is to study the maximization of the basic dynamic radial load rating considering the mounting of rollers, in Figure 1 it is presented only a bearing cross section perpendicular to the bearing axis. In this way, the objective function will not consider the roller length, which has nothing to do with the mounting procedure. As one can see, the rollers centres are staying on the bearing pitch circumference of diameter D_{pw} and δ is the space to insert the last roller. If the last roller is somehow mounted, δ is enough small that the all rollers support each other. But to mount the last roller the outer ring has to be heated and the roller is inserted radially. Note that the outer ring has ribs on both sides and the rollers cannot be mounted from lateral.

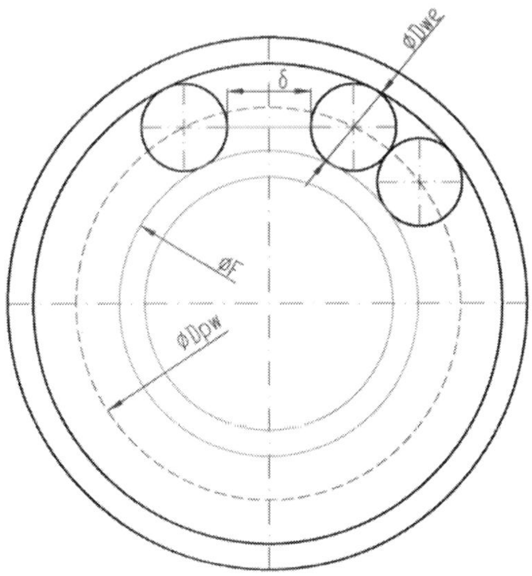

Figure 1. Bearing scheme.

4. Optimization problem

For radial cylindrical roller bearings, the basic dynamic radial load rating is given by the following equation:

$$C_r = b_m\, f_c (i\, L_{we} \cos\alpha)^{\frac{7}{9}}\, Z^{\frac{3}{4}}\, D_{we}^{\frac{29}{27}} \tag{1}$$

where b_m = 1.1 is the rating factor for contemporary, commonly used, high quality hardened bearing steel in accordance with good manufacturing practices, i represents the number of rows, L_{we} is the effective roller length, α is the bearing contact angle, Z is the number of rollers per row, and D_{we} is the roller diameter. The remaining f_c is a factor which depends on the geometry of the bearing components, the accuracy to which the various components are made, and the material. An accurate equation of this factor is given in [8]:

$$f_c = 172.455 \frac{\gamma^{\frac{2}{9}}(1-\gamma)^{\frac{29}{27}}}{(1+\gamma)^{\frac{1}{4}}} \cdot \left[1 + \left[1.04\left(\frac{1-\gamma}{1+\gamma}\right)^{\frac{143}{108}}\right]^{\frac{9}{2}}\right]^{-\frac{2}{9}} \tag{2}$$

where

$$\gamma = \frac{D_{we}}{D_{pw}} \tag{3}$$

4.1. *Design variables*

In order to describe completely and unique the bearing geometry, it has been found that only two design variables are necessary. Therefore, the design variable vector has the following form:

$$X = (D_{we}, F) \tag{4}$$

where F is the diameter of the inner ring raceway, and obviously

$$D_{pw} = F + D_{we} \tag{5}$$

It was preferred F instead of the pitch diameter D_{pw}, because the outer diameter of the inner ring raceway is controlled easier during the manufacturing process. As one will see in the followings, for both design variables (roller diameter and inner ring raceway diameter) constraints connected to the available manufacturing precision are taken into account.

The problem here is how to consider the length of the roller. In the clear majority of the cases the application where the bearing will be used is unknown and therefore the thickness of the ring ribs can be established only based on the manufacturer experience and feedback from the field. For this reason, the authors did not consider the roller length as a design variable of the optimization problem.

4.2. *Optimization function*

The main goal of an optimal bearing design is to achieve the maximum basic dynamic load rating. Obviously, in the present case i = 1 and cos α = 1, and since the roller length can be considered as a given value, the objective function can be expressed as:

$$f(X) = \frac{C_r}{b_m (L_{we})^{\frac{7}{9}}} = f_c\, Z^{\frac{3}{4}}\, D_{we}{}^{\frac{29}{27}} \tag{6}$$

where

$$Z = \mathrm{Int}\left[\frac{\pi}{\mathrm{asin}(\gamma)}\right] \tag{7}$$

4.3. *Constraints*

As it has been discussed in section 2, the usual constraints that shall be provided to the optimization problems are trivial, since the last roller can be inserted only after the heating of the outer ring. At the same time, once the last roller has been inserted, when the bearing is at environmental temperature, all the rolling elements must remain together, in contact with the outer ring raceway, and without falling (when the inner ring is not inserted). So, special constraints must be developed for this uncommon optimization problem.

Constraint g_1

As it was mentioned above, before the last roller is inserted, the initial distance δ (Figure 1) must be lower than the roller diameter:

$$g_1 = \frac{1}{2\gamma}\sin[(Z-2)\,\mathrm{asin}(\gamma)] - 1 < 0 \tag{8}$$

Constraint g_2

The second constraint has been developed considering the mounting method for the rollers: the bearing outer ring temperature is increased in order to mount the entire number of rollers. Considering that, when the outer ring is heated, the bearing pitch diameter D_{pw} increases to D'_{pw} and the factor γ decreases to γ' (as per the thermal linear expansion law) and taking into account that the rollers are not heated, it yields:

$$\gamma' = \frac{D'_{we}}{D'_{pw}} = \frac{D_{we}}{D_{pw}(1+\alpha_T\Delta T)} = \frac{\gamma}{1+\alpha_T\Delta T} < \gamma \tag{9}$$

where α_T the linear expansion coefficient of the bearing steel [9]. The new factor γ' must be the root of the following equation (the new distance δ' between the mounted rollers is equal to the roller diameter):

$$2\,\gamma' - \sin[(Z-2)\,\mathrm{asin}\,\gamma'] = 0 \tag{10}$$

The necessary difference of temperature to achieve this thermal expansion is given by:

$$\Delta T = \frac{1}{\alpha_T}\cdot\left(\frac{\gamma}{\gamma'} - 1\right) \tag{11}$$

Now it is possible to define the constraint g_2, considering that the maximum difference of temperature that can be applied to the bearing outer ring is equal to 90°C. In fact, considering an environmental temperature of 20°C, ΔT = 90°C leads to a heating temperature 110°C, perfectly compatible with the standard bearing steel [10]. Therefore:

$$g_2 = \frac{\Delta T}{90} - 1 < 0 \tag{12}$$

Constraints g_3 and g_4

To avoid unfeasible designs and after studying many existing bearing designs, other two simple constraints were used to complete the optimization problem and make it reliable and close to the real-world applications. The first one stipulates that the thickness of the inner ring

$$\delta_i = \frac{F-d}{2} \tag{13}$$

and the thickness of the outer ring

$$\delta_e = \frac{D-(F+D_{we})}{2} \tag{14}$$

cannot be very dissimilar:

$$g_3 = \frac{|\delta_e - \delta_i|}{0.05 \min(\delta_e, \delta_i)} < 0 \tag{15}$$

The last constraint refers to the minimum allowable thickness of the bearing rings:

$$g_4 = 1 - \frac{\min(\delta_e, \delta_i)}{\delta_{min}} < 0 \tag{16}$$

Again, the minimum allowable thickness of the bearing rings (as the thickness of ring ribs) is a very delicate issue and is rather connected to bearing manufacturer experience and feedback from the field than to rigorous mechanical calculations (which are possible only in concrete loading conditions of the bearings). Based on author experience, the minimum thickness was set to 18% of the bearing section height, i.e 0.09 $(D - d)$.

5. Optimization method

To solve the presented optimization problem and maximize (6), it was used a Particle Swarm Optimization algorithm that, in literature, has been widely applied and proved to be a solid, simple and reliable optimization tool. The scope of the present work was not a study of the performance of the PSO algorithm itself and, because the problem requests nothing special, a basic PSO with global best position update has been used (gbest algorithm). The main steps of this algorithm are the following ones:

1. Define the objective function $f(X)$ and the space of the possible solutions;
2. Set PSO parameters: population number, maximum number of iterations, inertia weight and acceleration coefficients;
3. Create, randomly, the initial population;
4. Calculate the fitness value $f(X_i(t))$ of each particle X_i at time t;
5. Calculate pbest for each particle;
6. Calculate gbest;
7. Update the velocity v_i of each particle;
8. Update the position X_i of each particle;
9. If the stopping condition is met, return gbest, otherwise restart from step 4.

Within the *gbest* algorithm, it is possible to define the velocity update of each particle as:

$$v_i(t+1) = wv_i(t) + C_1 r_1 \big(pbest_i(t) - X_i(t)\big) + C_2 r_2 \big(gbest(t) - X_i(t)\big) \tag{17}$$

where the particle velocity is composed by many contributes:

- $wv_i(t)$ provides the exploration ability of the optimization algorithm
- $C_1 r_1 \big(pbest_i(t) - X_i(t)\big)$ represents the cognitive part of PSO, where the change of the speed of the single particle is due to the relationship with its own best position
- $C_2 r_2 \big(gbest(t) - X_i(t)\big)$ represents the social part of the PSO, where the particles change their own velocity considering the information that are coming from other particles.

To ehnance the exploration ability in the initial stages of the optimization process and to increase the local search in the last phases, the inertia weight w was modified according to the folowing rule:

$$w(t) = w_{min} + \frac{(t_{max} - t)(w_{max} - w_{min})}{t_{max} - 1} \tag{18}$$

The constraint handling has been done by means of the penalty method: if one of the introduced constraints is violated, the value of the objective function is greatly penalized so that the new position of the particle will not be chosen in the future.

6. Case study

The optimization problem has been applied for the bearing NJG 2340. The main boundary dimensions of this bearing follow the standard prescriptions [11] and the roller chamfer form and dimensions, and

IOP Conf. Series: Materials Science and Engineering **659** (2019) 012065 doi:10.1088/1757-899X/659/1/012065

the roller length were kept as in the already existing design. The bearing geometry is summarized in Table 1.

Table 1. NJG 2340.

Dimension	Value [mm]
Bore diameter, d	200
Outer diameter, D	420
Width, B	138
Roller length, L_w	97
Roller chamfer, r_{ch}	2

Obviously, the useful roller length L_{we} (necessary for the final C_r calculation) is.

$$L_{we} = L_w - 2r_{ch} = 93 \text{ mm} \tag{19}$$

The range of the design variables, together with their requested precision are given in Table 2.

Table 2. Design variables range.

Variable	Range, mm	Precision, mm
D_{we}	60.0 … 71.5	0.001
F	207.5 … 291.5	0.010

To solve the above defined optimization problem, the parameters of the PSO algorithm are enlisted in Table 3 and the optimization results are presented in Table 4.

Table 3. PSO parameters.

Parameter	Value
Population	200
Max iteration	50
w_{max}	0.6
w_{min}	0.3
C_1	1.2
C_2	1.2

Table 4. Optimization results.

Design variable/Parameter	Value
D_{we}	69 mm
F	242 mm
Z	14
δ_i	21 mm
δ_e	20 mm
ΔT	75.2 °C
C_r	2249.87 kN

The obtained geometrical set of data was represented with a CAD software in both conditions, before (Figure 2a) and after (Figure 2b) heating of the outer ring.

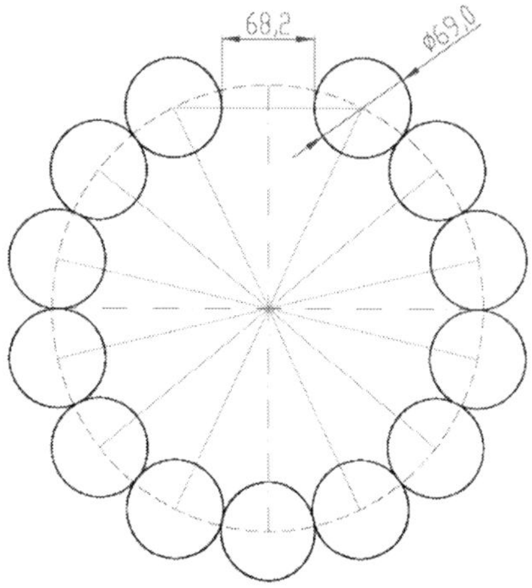

Figure 2a. Distance δ before the heating of the outer ring ($\delta < D_{we}$).

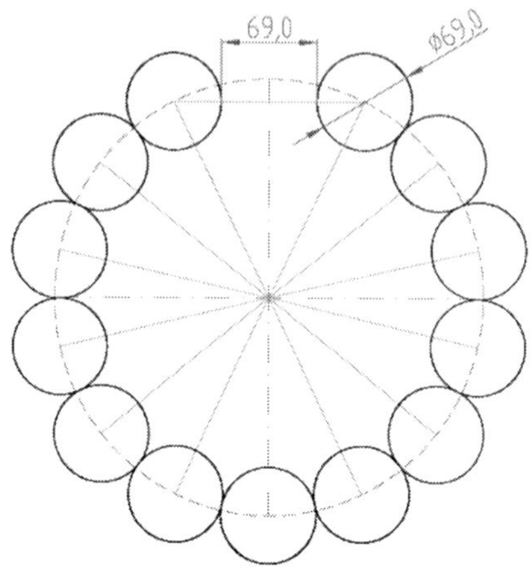

Figure 2b. Distance δ' after the heating of the outer ring ($\delta' = D_{we}$).

Despite being a single objective optimization process with only 2 designing variables, the constraints handling of this problem was not easy to deal with. In fact, the equation (10) is not so easy to be solved numerically, because it has been found that it is very sensitive to the provided inputs. In fact, a small perturbation of one of the particle components, for example D_{we}, can greatly modify ΔT value. In Table 5 one can find an example of the numerical criticality related with the temperature variation calculation.

Table 5. Example of numerical criticality.

Parameter	Set 1	Set 2
D_{we}	69 mm	**69.01 mm**
F	242 mm	242 mm
ΔT	75.2°C	**84.7 °C**

As it can be seen, with a slight variation of D_{we}, it resulted an important variation of ΔT. It is worth to notice that such behaviour is not at all strange in mechanics, especially when it is necessary to slightly modify some dimensions to obtain a specific bearing clearance or to adjust the roller size as per some manufacturing requirements.

7. Conclusion and future improvements

In this article, it has been described a novel approach in the optimization of the design of a self-retaining full complement cylindrical roller bearing, considering its peculiar assembly method. The maximization of the objective function, the basic dynamic load rating, has been obtained by means of a basic PSO, but modified to implement the innovative set of constraints.

Considering the mounting process and the self-retaining ability of the bearing geometry, the heating of the outer ring to fit the last roller also, lead to a feasible result in terms of internal geometry. The idea to embed this new type of constraint it can be considered solid and of extreme interest in many other cases.

The only criticality found, it is related to the sensitivity of the equation (10): it has been demonstrated that a small variation (acceptable if considering a normal mechanical manufacturing process) of only one of the two components of the best individual has a great influence on the heating temperature of the outer ring.

Another interesting and mandatory future development of the present work consists in taking into account of the inerent tolerances of the F dimension, and then to optimize the internal geometry considering the required range of the bearing radial internal clearance.

Acknowledgment
The authors would like to thank the RKB Group, the Swiss bearing manufacturer, for the permission to publish these results and the RKB staff for their great interest and support during the development of this project.

References

[1] ISO 281:2007 Rolling bearings — Dynamic load ratings
[2] ISO 76:2006 Rolling bearings — Static load ratings
[3] Dragoni E 2013 Optimal design of radial cylindrical roller bearings for maximum load-carrying capacity, *Proc IMechE Part C: J Mechanical Engineering Science* **227**(11) 2393-2401
[4] Tudose L, Kulcsar G and Stănescu C 2013 *Pareto Approach in Multi-Objective Optimal Design of Single-Row Cylindrical Rolling Bearings,* In book: Dobre G, Vladu M R (Eds.) Power Transmissions: Proceedings of the 4th International Conference, Sianaia, Romania, June 20-23, **13** 519-528, Springer
[5] Panda S, Panda S N, Nanda P and Mishra D 2015 Comparative study on optimum design of rolling element bearing, *TribologyInternational* **92**(2015) 595-604
[6] Ye Z-h and Wang L-q 2015 Optimization analysis on assembly interference of cylindrical roller bearings, *Advances in Mechanical Engineer* **7**(7) 1-13
[7] Waghole V and Tiwari R 2014 Optimization of needle roller bearing design using novel hybrid methods, *Mechanism and Machine Theory* **72**(2014) 71-85
[8] ISO/TR 1281:2008 Rolling bearings — Explanatory notes on ISO 281 — Part 1: Basic dynamic load rating and basic arting life
[9] ASTM, *52100 Bearing Steel,* http://www.astmsteel.com/product/52100-bearing-steel-aisi/ (accessed on July 2019)
[10] Radu C 2011 *It's All About the Steel: the Importance of Bearing Steel Cleanliness for Long-Life Applications,* https://www.rkbbearings.com/en/publications.php (accessed on July 2019)
[11] ISO 15:2017 Rolling bearings — Radial bearings — Boundary dimensions, general plan

IOP Conf. Series: Materials Science and Engineering **659** (2019) 012066 doi:10.1088/1757-899X/659/1/012066

Experimental determination of plastic gear durability

G Hlebanja[1], M Hriberšek[2] M Erjavec[2] and S Kulovec[2]

[1]Podkrižnik d.o.o., Lesarska cesta 10, 3331 Nazarje, Slovenia
University of Novo mesto, FME, Na Loko 2, 8000 Novo mesto, Slovenia
[2]Podkrižnik d.o.o., Lesarska cesta 10, 3331 Nazarje, Slovenia

E-mail: gorazd.hlebanja@podkriznik.si

Abstract. There are many plastic materials (POM, PA66, PEEK, ...) and material combinations available. The question is which combination is the correct one. It depends on many factors, like expected durability and price, means of production, e.g. injection moulding or cutting, lubrication, microstructure, fillers, etc. The available data bases and data sheets do not present conclusive results. So, it is of the utmost importance for a company to produce effective tests which could present correct results for each individual case.
The gear tooth flank profile could be also optimized. S-gear shape is characterized by a convex concave contact in meshing start and end areas. This implies improved properties in comparison to involute gears, like lower contact pressure, less sliding and lower frictional losses, stronger root and improved oil film when lubricated.
The paper presents some material properties for the combinations which were tested on the fatigue test benches, presents testing rigs, refers to experiments and discusses results.

1. Introduction

Plastic gears have become important alternatives to traditional metal gears in a wide variety of applications and their usage has expanded from simple low power motion transmission into demanding applications ranging from automotive, precision and medical equipment. Since designers push the limits of acceptable plastic gear applications, more is learned about the behaviour of plastics in gearing and how to take advantage of their unique characteristics. Plastic gears have lower weight and consequently lower inertia and run quieter than metal gears. They can run without external lubrication or their material can contain internal lubricants, e.g. silicone or PTFE. Their unit cost, based on manufacturing means, is lower than that of metal gears. They are corrosion resistant, depending on gear material and media. The most common materials are polyamide (PA) and polyacetal (POM). Due to harsher requirements, many new, high performance materials were developed. The use of fillers, reinforcements, internal lubricants, etc., became necessary, focusing on higher strength, wear and heat resistance. However, metallic materials data were systematically collected over decades, whereas there is apparent a lack of data (e.g. load carrying, wear performance, etc.) in the case of plastic materials, which coincides with rapid development of new materials. Various companies produce similar materials with similar properties as well. Proper combination of materials can result in long and reliable operation of such mechanisms and a cost-effective solution.

Despite the same basic material and similar fillers, materials disclose dissimilar mechanical and thermal behaviour. So, it becomes important for the gear producing company to have efficient means of testing and an effective database to use optimal solutions for each design task. Beside two old testing benches a new one was developed to facilitate testing according to the VDI 2736-4 [1], which is also important in the context of data inclusion in the KissSoft [2], which is a gear design software.

Some actual material combinations of gears, which can be used in complex parts are presented in the following chapter. It was discovered that the tooth flank shape has influence on gear efficiency. Some tests with small plastic gears conducted to assess Wöhler curves for gears with various material combinations and long-term tests with smaller loads to run up to 10 or 20 million cycles in the controlled environment have been conducted recently [3-6]. Results comparing S- and E-gear shape indicate better performance of S-gears. S-gear geometry feature lower contact pressure, less sliding and frictional losses, stronger root and improved oil film when lubricated.

Some accomplished experimental results (i.e. fatigue tests) with gears made of various material combinations are presented and discussed as well.

2. Material properties

Two series of tests were conducted: POM – PA66 and Steel C45 – PA66. POM and C45 gears were used as driving gears and PA66 as driven gear. Regarding PA66, four materials with different fillers were used. Basic properties of these materials are collected in Table 1. Some data are missing since they are not available on the corresponding data sheets.

Table 1. PA66 data of various producers derived from their datasheets.

Property		generic PA66[1]	LUBRICOMP RFP36[2]	Zytel 103HSL NC010[3]	Promyde B30 P2 GB30[4]	Luvocom 1/GF/30/TF/15/SI/2-2[5]
Additional description			30% glass fibres+ PTFE+silicone	lubricated	30% glass spheres	30% glass fibres + 15% PTFE+silicone
			wear resistant	heat stabilized	heat stabilized	heat stabilized
Density	ρ [g/cm^3]	1,13 … 1,16	1,46	1,14	1,36	1,48
Modulus of elasticity (tensile)	E[MPa]	3000		3100	4400 (dry)	9500
Tensile strength	R_m [MPa]	85	148		60	180
Tensile strengt at yield	R_e [MPa]			85		
Elongation at yield / break	δ [%]	5		4,5	-8	3,5
Poisson's ratio	ν [-]	0,4				
Service temperature (short term)	T_m [°C]	140 … 170				160
Service temperature (long term)	T_{op} [°C]	80 … 100				110
Thermal epansion coeff. (CLTE)	α [K^{-1}]	4 x 10^{-5}	2,5 x 10^{-5} / 11,3 x 10^{-5}	11 x 10^{-5} / 10 x 10^{-5}	2 x 10^{-5} / 6 x 10^{-5}	2,5 x 10^{-5}
Friction coefficient	μ [-]		0,56 / 0,48			0,12 / 0,14
Thermal conductivity	λ [W/(mK)]	0,23		0,25		0,24
Specific heat	c [J/(gK)]					
Thermal diffusivity	χ [mm^2/s]			0,05		

1 - generic PA66 - VDI 2736, Blatt 1 [7]
2 - Lubricomp RFP36 - Sabic [8]
3 - Zytel 103HSL NC010 - DuPont [9]
4 - Promyde B30 P2 GB30 - Nurel [10]
5 - Luvocom 1/GF/30/TF/15/SI/2-2 - Lehman&Voss&Co. [11]

According to the datasheet [8] is Lubricomp RFP36 distinguished by improved strength due to glass fibres and better frictional properties. DuPont's Zytel 103HSL NC010 is a basic PA66 with added lubricant and features high mechanical strength, excellent balance of stiffness and toughness, good high temperature performance, good electrical and flammability properties, good abrasion and chemical resistance [9]. Promyde B30 P2 GB30 is distinguished by high mechanical strength, hardness, rigidity, thermo stability (melting point 220°C), and resistance to hot lubricants and water. Parts made of this material have particularly high dimensional stability and creep strength [10]. Luvocom 1/GF/30/TF/15/SI/2-2 is characterized by improved sliding and wear behaviour and is self-lubricating.

It is meant for precision parts with high dimensional stability, high heat resistance and with tight tolerances [11]. POM used in these experiments is Lubricomp KA000M, which is a copolymer containing aramide fibres and a proprietary lubricant. Available data contain density of 1,36 g/cm^3, E=2600 MPa, R_m=41 MPa and friction coefficient of 0,21 [12].

Properties of various fillers (fibres, stabilizers, lubricants, etc.) are collected in Table 2. These fillers influence mechanical, tribological, electrical, technological, thermal and chemical properties of basic materials.

Table 2. Properties of fillers and reinforcing materials [13].

Filler	Positive influence	Negative influence
glass, carbon or aramid fibres	increase in stiffness, tensile strength, flexural fatigue strength and heat distortion resistance	reduction in impact strength
glass spheres	Improved processability, low, uniform shrinkage, low warpage, close tolerances, better surface finish, isotropic distribution of stress and deformation, etc.	
PTFE	reduction in friction and wear	reduction in impact strength and flexural fatigue strength
PE	reduction in friction	reduction in impact strength
graphite, boron nitride	reduction in friction and wear, increase in thermal conductivity	reduction in impact strength
silicone oil	reduction in friction and wear, increase in toughness	/
mineral fillers	increase in heat distortion resistance	/

3. Testing appliances

3.1. Small testing rigs for durability testing

Testing results presented in this paper are based on two small testing rigs were used to acquire gear durability data. Small gears (m = 1 mm, $z_1 = z_2 = 20$, b = 6 mm, with involute or S tooth flank shape) of varying material combinations (thermoplastic/thermoplastic or steel/thermoplastic) were exposed to a constant torque and driving speed and run until failure. In this way Wöhler curves are obtained. The testing rig is illustrated in Fig. 1. It consists of a framework on which two asynchronous squirrel cage electro-motors with 4 poles are mounted. On the motor rear side there are two fans to remove redundant heat. Two small gears are mounted in such a way to adapt centre distance. Power transmission from the motors to the shafts is by belts. Motor speed is governed by a frequency inverter. In this way load is implemented by a speed difference. It is important to have possibility of measuring contact temperature by a thermal camera, so the testing device was designed in such a way to enable frontal view. Motor power is 0,37 kW with synchronous speed of 1500 rpm. Maximal torque is 1,8 Nm.

3.2. Testing rig for tests according to VDI

While small gears are not demanding regarding production cost and are modest in operating power requirements, the new VDI Richtlinie 2736, Blatt 4 [1], prescribed some other gear sizes, depending on production means. Therefore, it was decided to produce a new testing rig. Two goals are 1) to produce tests comparable to a broader professional community and 2) to discover proper correction factors for tests with small testing rigs.

Special purpose testbench is developed and designed for testing the cylindrical spur or helical gear sets in accordance with [1], both »Size 01« or »Size 02«, or customer specific for a company's special purposes. The testbench is designed for either steel-polymer or polymer-polymer gear set. The device is represented in Fig. 2 schematically (left), the user interface of the control unit (right top) and size 01 meshing gears (right bottom).

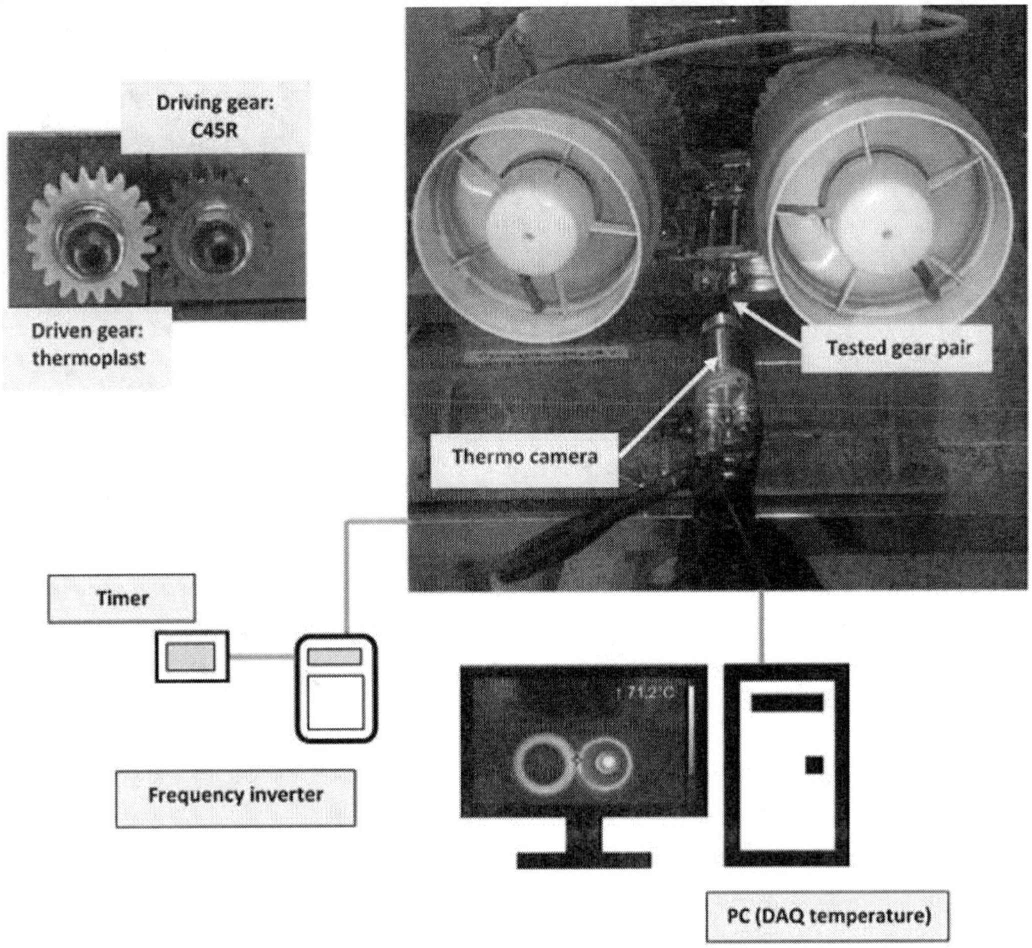

Figure 1. Illustration of the small testing rig for gear durability testing.

The testbench consist of two squirrel cage asynchronous motors (M1, M2) driven by frequency converters (FC1, FC2) with an internal PLC controlling the testbench operating parameters (load and speed). Further, the testbench is equipped with the thermo-camera providing the relevant information of the temperature distribution of the gear set being tested. The camera may be installed in a perpendicular direction to the gear meshing plane (shaft axis), with an observation point directly on the tooth flanks or it may be positioned to observe the gear face with a small inclination angle regarding the shaft axes orientation. The camera acquires spot temperature to the PLC, while the thermal images are sent to the PC periodically. The Optris Xi 400 device is used with automatic zooming, hot spot finder, IR sensor resolution of 382x288 points and frequency of 80 Hz.

The operating principle of the testbench is based on the operating characteristics of the squirrel cage asynchronous motors, driven by the frequency converter. The target speed of the primary gear is set, from which the excitation frequencies of both motors, considering gear ratios, are calculated (not necessary having the same value) by the PLC itself. After gear speed is reached the load is applied by altering the excitation frequency of the driven motor, hence the slip and consequently the load on both motor changes. The actual load transmitted through the gear set is evaluated with the pre-calibrated lookup tables. Once the failure of the gear set is detected, the testbench stops automatically. Since the counting of the load cycle is, beside the load level, an essential information, it is provided by the testbench PLC.

All relevant data i.e. the evaluated load, speed, temperature and the load cycle are stored in an internal PC in a digital form throughout the test-run.

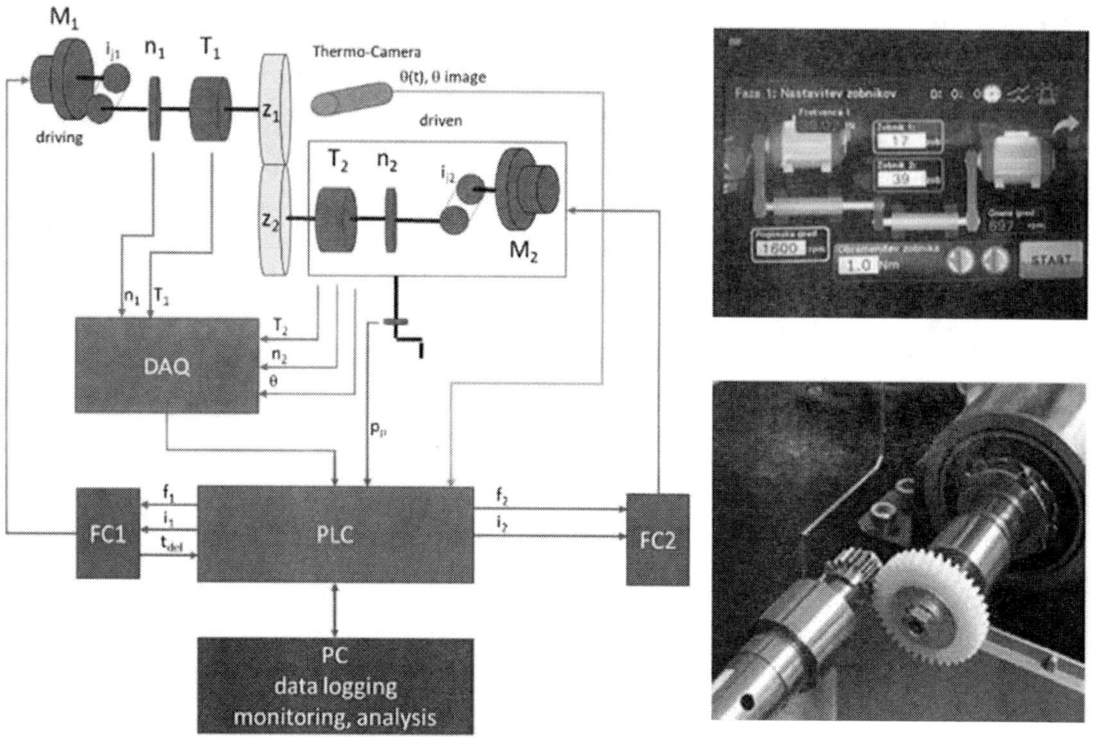

Figure 2. Scheme of the testing bench for durability testing (left); GUI of the PLC control unit (right top) Meshing gears, size 01 - z_1= 17 (driving), z_2= 39 (driven), (right bottom).

3.3. Gears

Manufacturing means of plastic gears can be injection moulding or cutting. Cut gears are of industrial interest for smaller lots and when a more precise tooth flank shape is prescribed or if alternative gear tooth flanks are of interest. Comparing cut and moulded gears of the same module and number of teeth by measurements of the quality Q according to DIN 3961 or DIN 3962 discloses at least for two grades higher Q of cut gears. Measuring gears on Wenzel GearTec coordinate measurement machine in our case showed Q=9 and Q=11 for cut and moulded gears, respectively [3].

Regarding our experience with moulded and cut gears it was decided that basic disks of plastic materials of interest are injection moulded, using an axial sprue enabling higher material flow to gain quality uniform physical characteristics and evenly distributed fillers (when applicable). Disks are then cut by a HSS cutting hobb on the CNC Koepfer machine tool as to enable effective manufacturing.
As already stated, gears with m=1.0 mm, z_1=z_2=20, b= 6.0 mm are in use in small testing rigs. The initial pressure angle of the S-gears α_{w0}=18° and a_p=1.5, whereas standard pressure angle α_w=20° is used for E-gears.

Gears planned for the new testing rig are (according to VDI 2736, Blatt 4):

a) size 1: m_n= 1 mm, z_1= 17, z_2=39, b_1=8 mm, b_2= 6 mm;
b) size 2: m_n= 2 mm, z_1= 30, z_2=30, b_1=13 mm, b_2= 12 mm;
c) custom gear combinations.

4. Important properties of S-gears

Several papers have been written clarifying the idea of S-gears in detail [13-16]. Some essential properties are summarized in this chapter. S gears feature a convex-concave contact in the meshing start and end areas, which can be achieved by a progressively curved path of contact, which is half symmetric. So, the generating rack profile (cutting tool) is defined in a following way:

$$y_p = a_p\left(1 - \left(1 - x_p\right)^n\right) \quad (1)$$

This parabolic type function defines the upper part of the rack profile whereas its counterpart is derived as a half-symmetric function. The unit of measurement is the module m. And a_p as a height factor and n as curvature exponent define a rack flank shape. y_p and x_p are the rack flank coordinates with the origin in the kinematic pole C. Such a rack profile defines a unique path of contact and out of there gears with any number of teeth can be derived with a bijective transformation.

The most important features of S-gears regarding plastic materials are summarized below:

- A tooth flank shape is optimized by a height factor a_p and exponent n, which enable gear tooth design according to prescribed properties. So, one can define a stronger tooth root, a longer convex-concave area, a curvature of the path of contact, a lower starting pressure angle in C, etc.
- Spur S-gears can operate with a low number of teeth, even as small as 6 or 4.
- S-gears feature convex-concave contact near meshing start and meshing end. This infers higher reduced radii of curvature and results in lower contact pressure.
- S-gears feature relatively longer dedendum part of a tooth flank (comparing to the involute gears). This implies sliding of a contacting pinion dedendum and gear addendum which results in less frictional work and lower temperature [15], which is illustrated in Fig. 3.

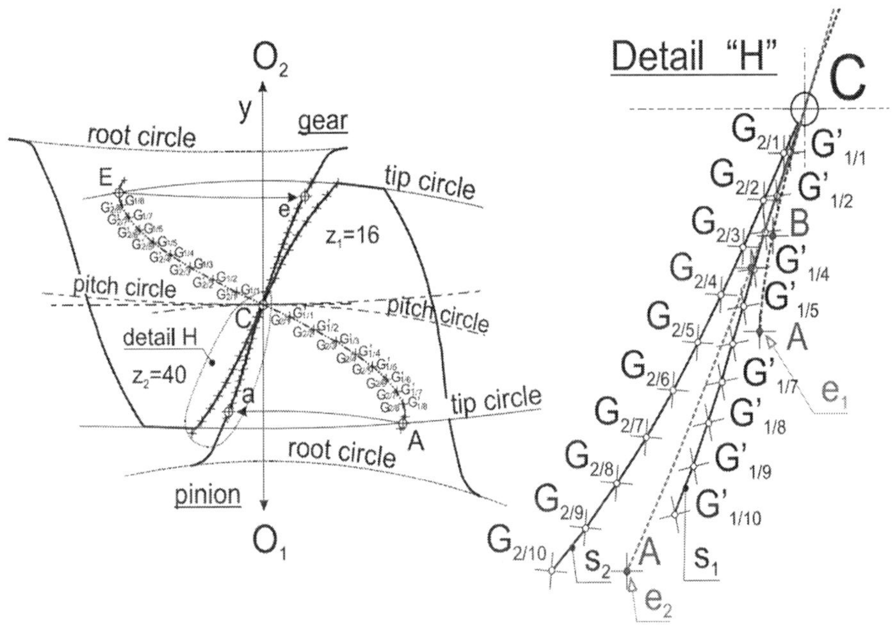

Figure 3. Contact density – the length of sliding of the gear addendum on the pinion dedendum.

The dedendum-addendum size difference depends on module, number of teeth, pressure angle. For S-gears the said difference also depends on forming factors – the height and the exponent. The size difference in the case of S-gears is comparatively more convenient, so less sliding is produced along the contact propagation compared to the involute case. Figure 1 (detail H) also shows the dedendum and the corresponding addendum of the E-gear. The starting pressure angle for the S-gear pair is 22° and that of the E-gear pair 20°. The contacting addendum flank length of the latter (for the case in Fig.3) is about twice as large as that of the pinion. And the obvious difference between E- and S-gears means more sliding and thermal impact in the case of E-gears.

5. Experimental results and discussion

Past research showed a clear distinction between S- and E-gear geometry in favour of former, which can be attributed to the mentioned features of the S-gears [3-6]. Two types of experiments were conducted:

a) durability tests where gears made of POM (driving) and PA66 (driven) run up to failure and tests where both gears were made of POM;
b) long term tests up to about 20 million cycles.

The rotational frequency (app. 1350 min^{-1}) and torque (from 1 up to 1,6 Nm) were set at experiment start. Tests were run at room climate circumstances. Tests confirmed theoretical assumptions. However, long term tests (1350 min-1 and 0,6 – 0,7 Nm) were time consuming and conducted in circumstances where ambient temperature varied to high extent, so additional tests in more stable environment should be arranged.

Current experimental work with small E-gears (m = 1 mm, z = 20, b = 6 mm, α_w = 20°) was conducted at room temperature 22 ± 0,5 °C. Input speed was 1400 min^{-1} and chosen torques were 1,1; 1,3 and 1,5 Nm. Each experiment was repeated 3 times to acquire statistical evaluation. The basic goal was to explore and evaluate the influence of the chosen material combination on attained number of cycles and heat flow and temperature rise in the gear meshing zone. S-gear experiments are already planned.

1. a C45R steel gear was driving and gears of various PA polyamides, basic and with reinforcements in the first experimental cycle.
2. The same PA driven gears and a POM with with fillers were used in the second experimental cycle. The corresponding material data are collected in Table 1.

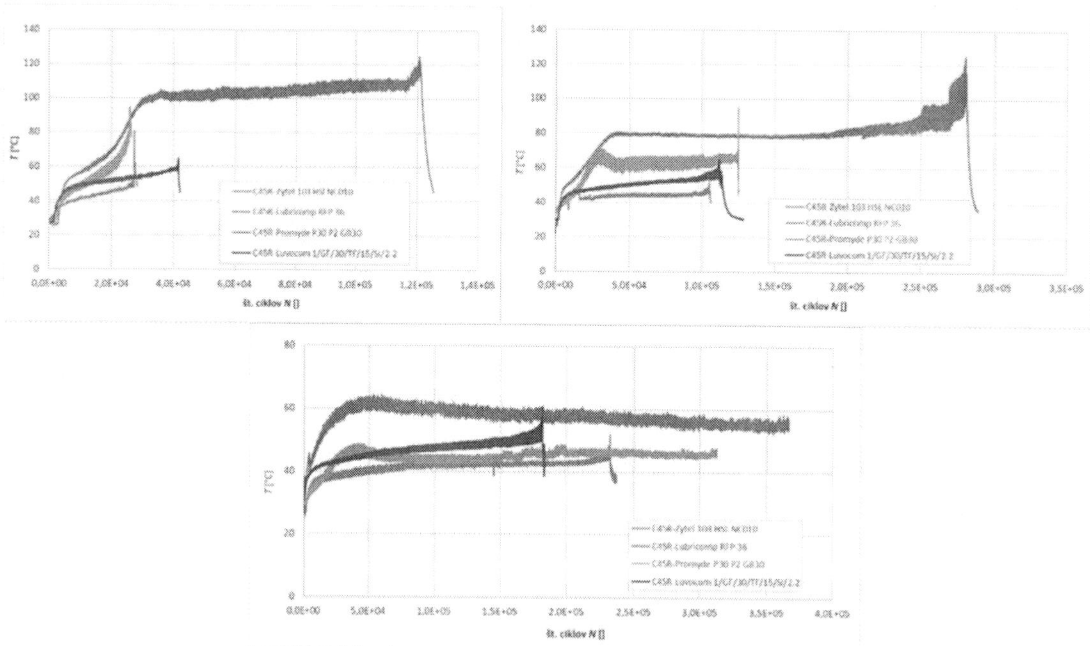

Figure 4. Temperature rise during durability testing of material combinations C45R – PA (n = 1400 min^{-1}, T_{ok} = 22°C) T = 1,5 Nm (top left), T = 1,3 Nm (top right),. T = 1,1 Nm (bottom) Colours: Zytel – green, Luvocom – blue, Lubricomp – orange, Promyde – grey)

Figure 4 shows temperature time diagrams for the gear pairs C45R/PA. Following conclusions can be derived based on these results:

- The best performance, the highest number of cycles N, was accomplished by Zytel 103 HSL NC010 made of basic polyamide 66 (PA 66), heat stabilized.
- The highest accomplished spot temperatures T were developed during meshing of gear combination C45R / Zytel 103 HSL NC010.

Table 3 collects average spot temperatures in a stable range for each polyamide, whereas Fig. 5 shows thermal camera snap-shots for the meshing C45R/ Zytel 103 HSL NC010 gear pairs loaded with torques 1,5, 1,3 and 1,1 Nm, respectively in a temperature stable region.

The highest temperatures in the meshing zone appear with gears made of Zytel 103 HSL NC010, which is expected, since this polyamide is non-reinforced, but heat stabilized, so it contains additives to improve properties at higher working temperatures. The combinations of working gear pairs C45R/Lubricomp RFP36 and C45R/Luvocom 1/GF/30/TF/15/SI/2-2 show the lowest temperature, which is also stable almost until gear breakage. This can be attributed to the presence of PTFE in both polymers. PTFE lowers friction in the contact and consequently lower temperatures are developed in the meshing zone. A more significant temperature drop with lower load was detected with gears of Zytel 103 HSL NC010 and Promyde-a B30 P2 GB30. These materials do not contain Teflon, which confirms conclusions for materials containing PTFE.

Table 3. Average spot temperatures in a stable range during testing (steel/PA).

Torque *M* [Nm]	**1,5**	**1,3**	**1,1**
Thermoplast	**Average spot temperature *T* [°C]**		
Zytel 103 HSI NC010	100	83	60
Lubricomp RFP 36	48	45	44
Promyde B30 P2 GB30	80	68	47
Luvocom 1/GF/30/TF/15/SI/2-2	50	53	48

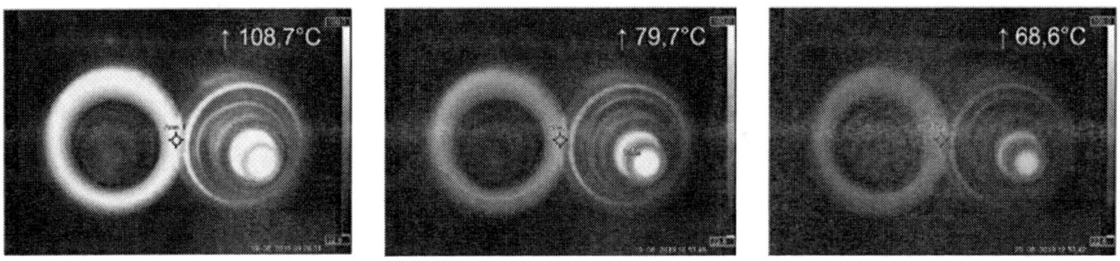

Figure 5. Thermal camera snap-shots for the meshing C45R/Zytel gear pairs in a stable region *T*=1,5 Nm (left); *T*=1,3 Nm (centre); *T*=1,1 Nm (right).

Table 4. Time/No. of cycles and range for various loads and material combinations (steel/PA).

Torque *M* [Nm]	**1,5**		**1,3**		**1,1**	
Thermoplast	**N**	δ [%]	**N**	δ [%]	**N**	δ [%]
Zytel 103 HSI NC010	115966	100	248085	100	354968	100
Lubricomp RFP 36	28225	24	105027	42	231828	65
Promyde B30 P2 GB30	26341	23	124527	50	308844	87
Luvocom 1/GF/30/TF/15/SI/2-2	42171	36	111165	45	179873	51

Table 4 shows how many cycles *N* were accomplished in tests with steel/polyamide gear pairs until breakage for loads 1,5, 1,3 and 1,1 Nm. Gears made of Zytel were the most durable, so their *N* values were put to 100% for comparison with other materials.

Table 4 reveals that the gear pair combination C45R/Zytel 103 HSL NC010 is the most durable, which can be attributed to the improved heat stabilized basic material and its deformability regarding stress state developed during meshing with the steel driving gear. Material combinations steel/PA with glass fibres have a negative influence on the thermoplastic material structure. One can say the glass fibres increase brittleness and, in this way increase a probability of breakage in dynamic fatigue of the used thermoplastic. Materials with glass spheres exhibit the same effect to a lower degree.

Fig. 6 shows temperature time diagrams of the second cycle of experiments. So, POM was used for a driving gear and PA for driven gear. POM in this case was Sabic's Lubricomp KA000M, which contains aramid fibres. PA materials are the same as in the first cycle, with the exception of Promyde.

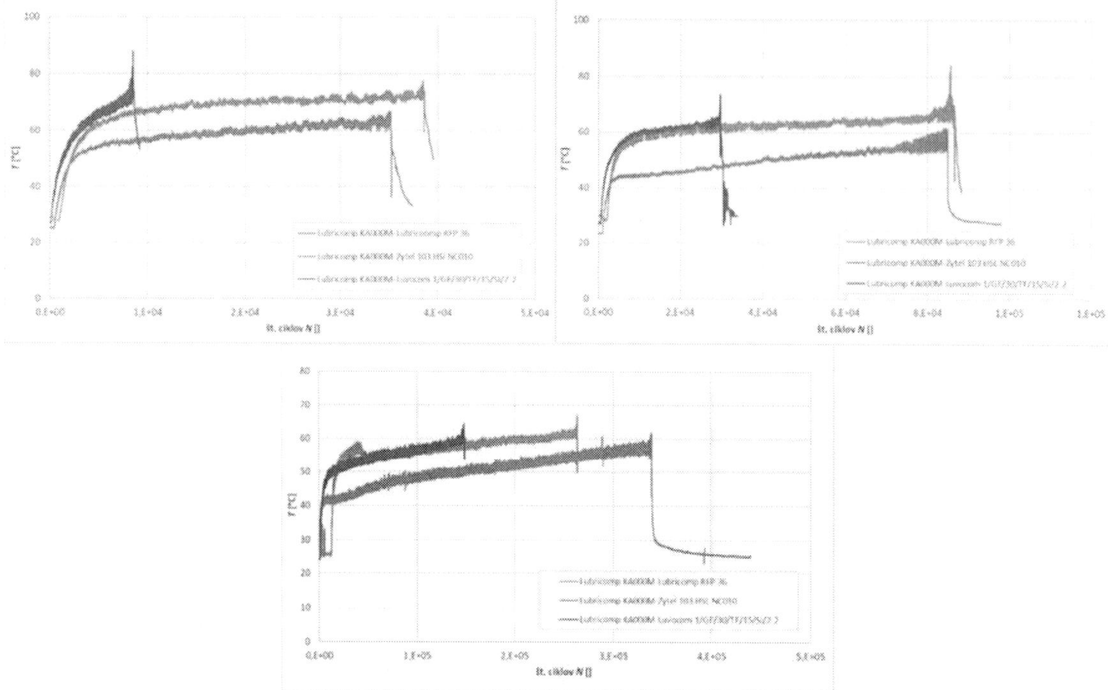

Figure 6. Temperature rise during durability testing of material combinations POM – PA (n = 1400 min^{-1}, T_{ok} = 22°C) T = 1,5 Nm (top left), T = 1,3 Nm (top right),. T = 1,1 Nm (bottom) Colours: Zytel – green, Luvocom – blue, Lubricomp – orange).

Table 5. Average spot temperatures in a stable range during testing (POM/PA).

Torque *M* [Nm]	1,5	1,3	1,1
Thermoplast	**Average spot temperature *T* [°C]**		
Lubricomp RFP 36	70	61	58
Zytel 103 HSI NC010	59	48	48
Luvocom 1/GF/30/TF/15/SI/2-2	66	62	55

Table 6. Time/No. of cycles and range for various loads and material combinations (POM/PA).

Torque *M* [Nm]	1,5		1,3		1,1	
Thermoplast	**N**	δ [%]	**N**	δ [%]	**N**	δ [%]
Lubricomp RFP 36	38093	100	81067	100	252939	100
Zytel 103 HSI NC010	34911	92	73892	91	326300	129
Luvocom 1/GF/30/TF/15/SI/2-2	8053	21	29022	36	142004	56

The best performance in this case can be attributed to combinations Lubricomp KA000M (POM)/ Lubricomp RFP 36 (PA66) and Lubricomp KA000M (POM)/ Zytel 103 HSL NC010. Lubricomp performs slightly better at higher loads and Zytel with lower loads. So, one could expect better performance of Zytel gears with loads at fatigue strength level, which is due to stable mechanical properties of this material at increased temperatures.

Table 5 collects average contact spot temperatures in a stable region for mating POM/PA gears. Zytel 103 HSL NC010 develops the lowest temperatures in the stable range among all materials. Both other materials contain glass fibres and PTFE and it appears that the temperatures in these cases are higher. Regarding temperature range (heat development) and durability (No. of cycles, Table 6) the basic, heat stabilized PA66, Zytel 103 HSL NC010, appears to be the most appropriate.

6. Conclusion

New plastic materials emerge all the time. Improved POM and PA66 were used in this paper. So, their mechanical and thermal properties were presented here. Regarding manufacturing raw parts were injected and gear cut subsequently. Despite net shape injected parts are preferred due to their price in mass production, cutting with professional hobs is necessary when gears of higher quality are required. Hob cutting is also more productive method.

Despite small testing rigs are quite effective for small gears, the new VDI Richtlinie, Blatt 4 [1] requires some other gear sizes, so a new testing rig was manufactured. The basic idea is, to produce tests, which can be directly compared to other. Some test cycles on the new testing are already in progress. Tests in this paper were conducted entirely on the small devices with small involute gears. A hob cutter for S-gears was already manufactured and tests are in plan later this year. S-gears are of interest due to lower frictional power comparing to E-gears.

In this paper presented test results show some interesting differences between different mating gears material combinations. It appeared that Zytel 103 HSL NC010, basic, heat stabilized PA66 serves the best as a driven gear with both driving gears, namely steel C45R and Lubricomp KA000M, aramid reinforced POM. This is true for material selection presented here.

Tasks which are in focus are S-gears experiments and adaptation of the results based on small testing rigs to the newer VDI based results.

Basic aim of performing durability tests of thermoplastic gears according to the VDI 2736, Blatt 4 is exportation of acquired temperature data (tooth flank and root) and number of cycles of tested gear pair at gear failure into the program system KISSsoft. Based on the input gear geometry, material properties, working conditions and data acquired from tests Wöhler curves are produced program aided for a chosen (tested) material and gear combination for prescribed operating temperatures.

Acknowledgment

The investment is co-financed by the Republic of Slovenia and the European Union under the European Regional Development Fund, no. SME 2/17-3/2017 and C3330-18-952014

EUROPEAN UNION
EUROPEAN REGIONAL DEVELOPMENT FUND

REPUBLIC OF SLOVENIA
MINISTRY OF ECONOMIC DEVELOPMENT AND TECHNOLOGY

References

[1] VDI 2736: Part 4. 2016 *Thermoplastic gear Wheels: Determination of Strength Parameters on Gears.* VDI - Richtlinien

[2] Pogačnik A 2018 *Web demo 27.09.2018 – Plastic gear calculation.* KissSoft, https://www.kisssoft.ch/english/webdemos/downloads/plastic-gears/WebDemo_EN.pdf

[3] Hlebanja G, Kulovec S, Hlebanja J and Duhovnik J 2016 *Experimental determination of S-gear characteristics made of plastics.* Polytrib 2016, 2nd International Conference on Polymer Tribology, 15th-16th September 2016, Ljubljana, Slovenia

[4] Hlebanja G, Kulovec S, Zorko D, Hlebanja J and Duhovnik J 2017 *Influence of the tooth flank shape on thermal load of the gear,* Europe invites the world, International Conference on Gears, Technische Universität München, Garching, Sept. 13th to 15th, 2017, (VDI-Berichte, ISSN 0083-5560, 2294.2). Düsseldorf: VDI 1583-1592

[5] Zorko D, Kulovec S, Tavčar J and Duhovnik J 2017 *Different teeth profile shapes of polymer gears and comparison of their performance.* Proceedings of MPT2017-Kyoto, The JSME International Conference on Motion and Power Transmissions February 28 –March 3, 2017, Kyoto, Japan

[6] Trobenar B, Kulovec S and Glodež S 2016 *The influence of gear flank shape on the service life of polymer gears,* Polytrib 2016, 2nd International Conference on Polymer Tribology, 15th-16th September 2016, Ljubljana, Slovenia
[7] VDI 2736: Part 1. 2016 *Thermoplastic gear Wheels: Materials, material selection, production methods, production tolerances, form design.* VDI - Richtlinien (2016)
[8] Sabic 2019 Lubricomp Compound RFP36, Technical Data Sheet
[9] DuPont 2005 nylon resin, Zytel® 103HSL NC010, Technical Data Sheet
[10] Nurel – Engineering Polymers 2012 Promyde B30 P2 GB30, Technical Data Sheet
[11] Lehman&Voss&Co. 2006 Luvocom 1/GF/30/TF/15/SI/2-2, Technical Data Sheet
[12] Sabic 2018 Lubricomp Compound KA000M, Technical data Sheet
[13] Hlebanja G and Hlebanja J 2012 *Contact circumstances of the highly loaded, low speed gears.* Proceedings KOD 2012, 24-26 May, 2012, Balatonfüred, Hungary, 21-26
[14] Hlebanja G 2012 *Characteristics of non-involute gears*, Proceedings, University of East Sarajevo, Faculty of Mechanical Engineering, 9-20
[15] Hlebanja G 2011 *Specially shaped spur gears - a step towards use in miniature mechatronic applications*, Proceedings, The 7th International Scientific Conference Research and Development of Mechanical Elements and Systems - IRMES 2011, Zlatibor, 27-28. April 2011, Serbia, 475-480
[16] Hlebanja, G and Hlebanja, J 2010 *Spur gears with a curved path of contact for small gearing dimensions*, International Conference on Gears, Garching near Munich, Germany, 4-6. 10. 2010 (VDI-Berichte, ISSN 0083-5560, 2108), VDI-Verlag, 1281-1294

IOP Conf. Series: Materials Science and Engineering **659** (2019) 012067 doi:10.1088/1757-899X/659/1/012067

Experimental investigation of conveyor idlers operational characteristics

R M Mitrović[1], Ž Z Mišković[1], Z V Stamenić[1], N D Soldat[1], N S Matić[1], M R Ristivojević[1] and A R Dimić[1]

[1] University of Belgrade, Faculty of Mechanical Engineering, Kraljice Marije 16, 11120 Belgrade, Serbia

E-mail: zmiskovic@mas.bg.ac.rs

Abstract. The efficiency of thermal power plants significantly depends on the efficiency of belt conveyor systems. Conveyor idlers, as key parts of belt conveyors, are often the main cause of bulk material transportation systems failures. Testing the operational characteristics of conveyor idlers in laboratory and exploitation conditions is a prerequisite for any progress in the field of increasing energy efficiency of conveyor systems. This paper describes an analysis and discussion of testing results for different types of conveyor idlers, in terms of their performances under the act of predefined radial load. Tests were conducted on a new conveyor idler's testing machine developed at the University of Belgrade - Faculty of Mechanical Engineering. The presented tests are performed in order to determine the quality of carrying and return conveyor idlers. The obtained results will help in solving the key problems in the critical areas of conveyor systems and will increase their energy efficiency. Finally, the expected results can potentially reduce the financial and energy losses which conveyor idlers cause in belt conveyor systems, as well as in the whole system of thermal power plants.

1. Introduction

Due to increasing global energy crisis and lack of natural resources, increasing the reliability and energy efficiency of mechanical systems in energetics is constantly in the focus of scientific research. Energy efficiency is one of the key elements in energy policy today, because it contributes to the improvement of economy globally and extends the lifetime of conventional energy sources, since the greatest part of total electrical energy is generated by fossil fuels, [1, 2]. Despite the increasing percentage of utilization of alternative energy sources, mostly still use the energy of fossil fuels, especially coal. Today's coal mining on open pit mine requires use of complex machinery with high productivity which allows continuous process of exploitation without unexpected malfunction with unplanned costs. For this reason, the tendency for daily analysis of existing and development of new, improved constructions of conveyor idlers is growing. An indicator of the actuality of the problem in our country is the fact that open pit mine in Kostolac has about 100,000 conveyor idlers in operation of which 10% need to be repaired or replaced. Existing level of quality production of conveyor idlers made by domestic manufactures don't satisfy the requirements of working conditions on open pit mine. Conveyor idlers made by reputable worldwide manufacturers have up to three times longer service life than domestic. Development of domestic industry, keeping of existing or creation the new jobs requires new approach to the design and manufacturing of conveyor idlers [2, 3]. A prerequisite for progress in field of improving energy efficiency of technical systems is introducing the testing of behaviour of conveyor idlers in laboratory and exploitation conditions.

Exploitation tests take a very long time and it can't be fast enough to provide information helped user to decide which of the offered product should use in transport system. Therefore, special attention is paid to the development of laboratory methods and test rigs for accelerated testing of conveyor idler quality. The subject of this paper is analysis and discussion of testing results for different types of new and repaired conveyor idlers, in terms of their performances under the act of radial load.

2. Testing of conveyor idlers operational characteristics

Because of the large influence of conveyor idlers on the performance of the overall belt conveyor system, there are many national standards dealing with this issues [4]. For the purpose of the quality and reliability inspection of conveyor idlers which are produced in our country, a methodology for accelerated testing and comparison of conveyor idlers performance under the action of the radial load of constant intensity was developed. This methodology was used to provide estimation for qualitative comparison between different types of conveyor idlers – new and repaired, domestic and imported, conveyor idlers with different greases applied etc. [5, 6]. According to this methodology, the basic parameters for determining the transport idlers quality are:

- Temperatures in idler's rolling bearings,
- Deceleration during rotation stopping,
- Reaction on rotation movement,
- Vibrations of idler's rolling bearings,
- Noise.

The analysis of existing constructions of conveyor idlers produced by the company PRIM - Kostolac is performed in order to obtain a full assessment of their quality. Beside results and appropriate conclusions of holding analysis, conveyor idlers' operation should be monitored in exploitation on the surface pit mine or through specific tests under laboratory conditions if more complete objectivity in assessment of the quality of existing and new conveyor idlers solutions is necessary.

3. Conveyor idler's testing machine and methodology

The procedure for determination of load capacity of conveyor idlers have been realized on a test rig SIV 300, which was developed in cooperation with thermal power plant and pit mines Kostolac. Its basic look and operation principle are shown in Figure 1. Laboratory installation is designed for comparative testing of conveyor idlers (quality assessment), and for measuring changes of their operational characteristics over time (temperatures and vibrations). All tests have been performed under the action of radial load of different intensity. Two pneumatic cylinders are used for radial force generation. They are controlled by system for data acquisition specifically developed for this purpose [7]. Therefore, a load distribution which is in accordance with exploitational one could be simulated by changing of pneumatic pressure. During the testing the idler shell is deformed because of its thin walls, so the loading value should be carefully selected. It is possible to test a large spectrum of conveyor idlers dimensions, with a diameter up to 200 mm and lengths up to 1000 mm. During accelerated tests, the data acquisition of micro-vibrations, temperatures and noise of rolling bearing of tested samples are performed – and results of analysis of obtained results are presented and discussed in the next section of this paper. Based on the result of experimental tests of conveyor idlers it is possible to estimate their service life in different load conditions. Due to time constraints, these tests are carried out with increased intensities of radial load. After the testing and processing of test results in one of the common software packages, it is possible to determine:

- sensitivity of specific types of conveyor idlers to radial loads,
- sensitivity of conveyor idlers from different manufacturers to radial loads,
- influence of different greases to conveyor idlers' service life,
- correlation between measured operational characteristics of conveyor idlers,
- temperature change in conveyor idlers during the examination.

Figure 1. Test frame for performance testing of conveyor idlers under the action of radial load of constant intensity.

4. Test samples and experimental setup

The conveyor idler test samples are provided by thermal power plant and pit mines Kostolac. The type of conveyor idlers with outer diameter of 159 mm and labyrinth seal are the most widely used on open pit mines systems, both in our country and abroad. Delivered test samples of different types of conveyor idlers are given in table 1.

Table 1. Conveyor idlers test samples.

Type of conveyor idler	Dimensions	Designation
Carrying idler	Ø159x670	NVPM
Carrying idler	Ø194x750	DVPM
Return idler	Ø215x1150	PV

All delivered test samples were produced by company PRIM Kostolac, except return idlers of type Ø215x1150 mm, which are manufactured by company ANSAL STEEL with rubber coating (cover) in the shape of concentric rings, non-uniformly distributed along idlers' length.

Experimental conditions for conveyor idlers testing were:

- Carrying idler Ø159x670 mm, radial load: ~9 kN, rotational speed: ~1116 RPM – testing acceleration: ~21,2 times;
- Carrying idler Ø194x750 mm, radial load: ~9 kN, rotational speed: ~905 RPM – testing acceleration: ~18 times;
- Return idlers Ø215x1150 mm, radial loading: ~2 kN, rotational speed: ~824 RPM

Evident signs of corrosion in all carrying idlers were observed by visual analysis during the preparation of test samples. This is a result of inadequate storage conditions, and has influence to accuracy of measurements of radial run-out.

5. Results review and discussion

During the conveyor idlers experimental testing, all listed operational characteristics were recorded in time domain – using specific data acquisition system. The examples of obtained diagrams are presented at Figure 2.

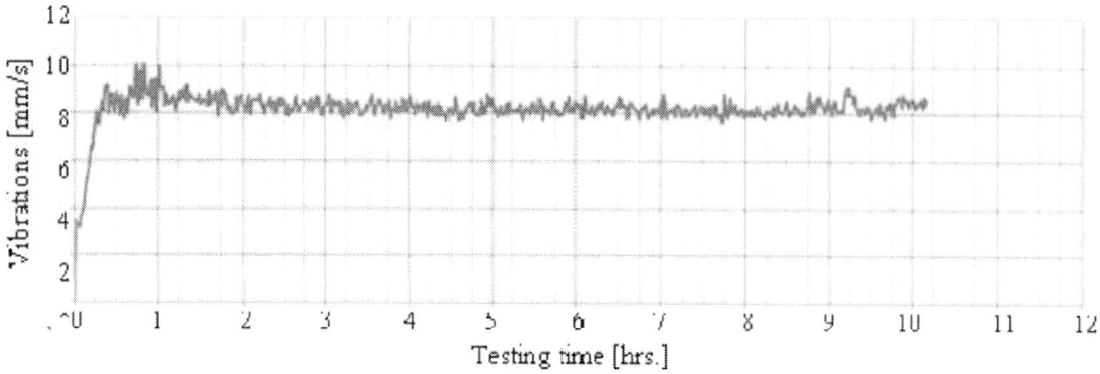

Figure 2. Example of obtained results (time domain diagram) for experimental testing of conveyor idlers vibrations – test sample NVPM3.

The processed results of experimental testing of carrying and return idlers operational characteristics in described loading conditions after 5 and 10 hrs. of testing (respectfully) are shown in table 2 and table 3.

Table 2. Results of accelerated experimental testing of carrying conveyor idlers.

Sample	The cumulative testing time [h]	V [m/s]	T1 [°C]	T2 [°C]	Noise level [dB]
NVPM1	5	3.571	52.538	54.736	4.404
	10	8.359	47.654	54.464	85.133
NVPM2	10	7.930	63.693	54.769	87.194
NVPM3	10	8.486	68.787	55.442	87.795
DVPM1	10	6.396	52.039	54.920	79.185
DVPM2	10	5.772	63.694	55.900	82.071
DVPM3	10	5.628	65.131	44.372	80.164

Table 3. Results of accelerated experimental testing of return conveyor idlers.

Sample	The cumulative testing time [h]	V [mm/s]	T1 [°C]	T2 [°C]	Noise level [dB]
PV1	10	3.079	54.076	48.834	69.656
PV2	10	2.823	62.260	50.771	73.755
PV3	10	4.585	57.900	44.958	70.273

Performance testing of conveyor idlers under the acting of radial load of constant intensity gave the following results:

- All tested conveyor idlers (carrying and return idlers) operate without failure after 10 hours of the accelerated test conditions, so it could be conclude that prescribed manufacturing technology was respected during conveyor idlers production,

- During the tests, the observed parameters of all tested conveyor idlers had approximately the same trends, representing a high level of uniformity of their production quality,
- The measurement variables of all tested conveyor idlers (carrying and return idlers) entered to the stationary state after 3-4 hours of accelerated tests,
- For all samples of tested conveyor idlers, it is observed that there is a significant difference between temperature measured at places of the first bearing and the second bearing (for each sample individually), and this applies for the measured macro-vibrations. This can be explained by manufacturing conditions, as the first and second ball bearings do not assembled in the same way,
- The results analysis leads to the conclusion that in initial stage of failure the macro vibration or noise levels does not come to change, but only the temperature of bearing registered changes. Bearing temperature of conveyor idlers was in the range of 40 to 70 °C after 10 hours of testing.
- Allowed vibration value (10 mm/s) of carrying idlers type Ø159 / 670 mm is exceeded only in the sample NVPM2 - after 8.5 hours of testing,
- Allowed noise value (85 dB) is exceeded in all samples of idlers bearing type Ø159 / 670 mm for ~4 dB, which fits into measurement uncertainty of test device (± 5 dB) - it can be concluded that noise of tested samples is at the permissible value limits,
- Allowed noise value (85 dB) during test of carrying idlers type Ø194 / 750 mm is not exceeded in any of the sample,
- Allowed vibration value (10 mm/s) during test of return idlers is not exceeded in any of the sample - it was observed that vibration value has a significant amount of oscillates during test, which can be explained by significant radial run-out of tested samples,
- Allowed temperature value of bearing of return idlers samples (70°C) is exceeded only in the sample PV2 - after 3 hours of testing,
- Allowed noise value (85 dB) during test of return reverse idlers is not exceeded in any of the sample - as in the case of vibration measurements, it was observed that this value has a significant amount of oscillates during test, which can be explained by significant radial run-out of tested samples,

Results of comparative analysis of test results show the following:

- Repaired conveyor idlers made by domestic manufacturers have higher values of operating temperatures for 7°C, in comparison with new conveyor idlers,
- New conveyor idlers made by domestic manufacturers have higher values of operating temperatures for 5°C, in comparison with conveyor idlers made by foreign manufacturers.

6. Conclusion

Implemented tests provide important information about the quality level of a large number of different types of conveyor idlers commonly used in open pit coal mines serving the thermal power plant Kostolac. The described conveyor idler testing methodology and equipment were developed in order to achieve potential increasement of conveyor idlers service life and reliability, reducing unnecessary and unplanned failures of belt conveyors as a system. The information on specific conveyor idler operational characteristics will allow the improvement of the regular maintenance of the entire transportation system planning. Obtained results will also contribute to improvement of the quality as well as energy and financial loses reductions which are caused by conveyor idlers – both in belt conveyors and in the entire system of thermal power plants.

Acknowledgment
Results presented in this paper were realized within the Project TR35029 so authors would like to express their sincere gratitude to the Ministry of Education, Science and Technological Development of the Republic of Serbia, as well as to the company thermal power plant and pit mines Kostolac (Serbia).

References

[1] Ristić L 2012 *Development of the algorithm for energy efficiency improvement of belt conveyor system on open pit mines*, University of Belgrade, Serbia, Doctoral Thesis.
[2] Mišković Ž 2017 *Exploitational contamination particles concentration influence on rolling bearing's operational characteristics*, University of Belgrade, Serbia, Doctoral Thesis.
[3] Mitrović R, Mišković Ž, Tasić M and Stamenić Z 2014 Conveyor idler's turning resistance testing methodology, *Machine design* **6**(4) 107-112
[4] Frittella A and Cohen G 2008 *Conveyor idler standards*, Melco Mining
[5] Mitrović R, Mišković Ž, Tasić M., Stamenić Z, Soldat N and Matić N 2012 *Conveyor idlers testing machine*, 29th Danubia-Adria-Symposium on Advances of Experimental Mechanics, Belgrade, Serbia, Sep. 26 – Sep. 29, pp. 278-281
[6] Mitrović R, Tasić M, Ivanović G, Mišković Ž and Stamenić Z 2009 *Test frame for radialy loaded conveyor idler accelerated testing*, Industrial prototype, Faculty of Mechanical engineering at the University of Belgrade and company 'Termoelektrane i kopovi Kostolac', Belgrade, Serbia
[7] Mitrović R, Tasić M, Mišković Ž, Stamenić Z and Jovanović D 2013 *Data acquisition and automatisation of a conveyor idler test stand*, Advanced Materials Research: Advances in Engineering Materials, Product and Systems Design (Special topic volume with invited peer reviewed papers only), Trans Tech Publications

IOP Conf. Series: Materials Science and Engineering **659** (2019) 012068 doi:10.1088/1757-899X/659/1/012068

Analysis of harmonic gearbox tooth contact pressure

M Majchrak[1], R Kohar[1], M Lukac[1] and R Skyba[1]

[1]University of Žilina, Faculty of Mechanical Engineering, Department of Design and Mechanical Elements, Univerzitná 1, Žilina 010 26, Slovakia

E-mail: Maros.Majchrak@fstroj.uniza.sk

Abstract. An analysis of a conventional gear wheel shot with MPK methodology has been processed by many authors before. Harmonic gearbox systems are composed of rigid and flexible parts through which the torque is transmitted, thus generating power. The contact pressure is specifically distributed through more teeth whilst against conventional gearbox systems. Firstly, a design description of a dynamic calculation model of a harmonic gearbox system in the planar strain is analysed in this article, as well as the definition of each contact constraint between the rigid and flexible parts of these gear systems in respect of "GLUE" and "TOUCHING" contact types. The last part of this article deals with the force distribution analysis in the teeth with respect to the frictional forces in them. The graphical conclusions of this pressure on each tooth were made. This article mainly focuses on the analysis of the stress in a gear tooth of the harmonic gearbox systems, as well as the research of a special trajectory of the rigid and flexible entity in a solid circular spline of such gear systems.

1. Introduction

The year 1955 is considered a milestone in the development of harmonic gears. In this year, Clarence Walton Musser, an American inventor, first published a new concept of the gear mechanism. He called this mechanism "Strain Wave Gearing". Later, in 1957, he managed to patent the invention, calling it "Harmonic Drive Gearing, Strain Wave Gearing". This mechanism was based on a theory which opposed the gear rigid-body mechanics that had been used many years until then. This new concept worked on the principle of a mechanical gear with a flexible spline. In fact, this invention was based on a theory of dynamics of an elastic body. This idea of the gear systems was significantly different from a globally accepted opinion on rigid gear systems. The first prototypes of this mechanism were successfully made in 1958 in America, in the United Machinery Group Corp — the Harmonic Drive Division. Nowadays, harmonic gears have been made in several countries. The greatest companies producing these gear systems include Harmonic Drive LLC (USA), Harmonic Drive Systems (JAP), Harmonic Drive AG (GER). Harmonic drive gearing systems have been widely used in the area of precise control. Precise reduction and positioning have been mainly used in robotics. Its high rate of precision and weight has been used in the aviation industry. The system is also used in the engineering industry for the transmission of high torque, conveyor and turbine drives, and in many other machine applications.

2. The Principle of the Harmonic Gear Functioning

Firstly, it is important to realise that harmonic gears (hereinafter referred to as "HG") were derived from traditional epicyclic gearing systems. In that case, a sun gear is the main input member and output speed and torque are obtained from planets that diverge from a central axis. HG is, in fact, a differential gear with a cylindrical gear pair in which gearing is achieved by elastic deformation of the flexible spline. This spline is characterised by its unique structure called a "Strain cam wave generator". The whole gear system is composed of four basic components, see Figure 1.

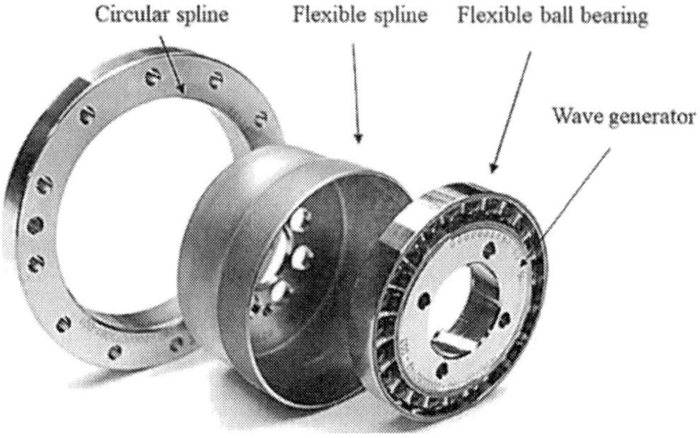

Figure 1. The basic components of a harmonic drive gear.

Input torque is applied to a wave generator. This component is specific for its ellipse-shaped geometry. However, this ellipse is not regular — to secure clearance between the generating teeth of a circular spline and a flexible spline when rolling. The shape of the ellipse is flattened in line with the prescribed geometric function characteristic for stretching or flattening the shape. On this component, a flexible ball bearing is pressed and flexibly deformed according to the ellipse shape of the wave generator. When gearing this system, waves are generated, which is the basic driving component of this gear system. Finally, a flexible spline is pressed on the wave generator system and the flexible bearing. The assembled internal part of this drive gear as a whole is then inserted into a circular spline. The circular spline is attached to a gearbox, thus ensuring a rigid embedding of the HG. Output torque and speed are obtained from the flexible spline. In the back part of this component, there are mostly two openings made for a pin and 6–12 openings, depending on the rate of the torque transferred to the screws. We need to realise that the flexible spline rotates counter to the rotation of the wave generator. The advantage of this mechanism is almost a zero-tooth clearance and its ability to transfer a rather huge gear ratio, mostly 1:50, 1:100, 1:200 [1].

3. Development of a Mathematical Calculation Model

To develop a calculation HG model, it is necessary to consider a number of requirements. First, an analysis of the geometric shape of the wave generator and then the dimensions of the flexible ball bearing is required. In the next stage, it is necessary to correctly calculate the tooth geometry of the circular spline and the flexible spline. These teeth must have the same module and pitch circles. It is important to acknowledge that the bodies are flexibly deformed after pressing. For this reason, their geometry must first be calculated for a non-deformed shape, and then for a deformed shape, too. This analysis is vital for the functioning of the whole HG and its fundamental operating characteristics such as backslash; rigidity, the gear ratio and a total vexillary size of the HG. Then, in the second stage, it is necessary to process the geometry to develop a 2D and 3D form in CAD software. This geometry must be correctly meshed in a 2D and 3D meshing appliance. Subsequently, a calculation model with boundary conditions and correct parameters must be created. In the last stage, the analysis results and assessments must be analysed correctly.

3.1. The Analysis of the Wave Generator Geometry in Matlab (software)

Matlab was used to develop a mathematical model. In this programme, equations for calculating the body geometries were mathematically defined. This geometry was exported from Matlab to a text document as a cloud of points with x and y coordinates. These points were later imported to Creo Parametric, a 3D CAD software, where a 2D and 3D model was created [2-3].

$$\mathrm{R} = A + B.cos\left\{2.\left[\frac{\pi}{2}.\left(\frac{\varphi}{2\pi}\right)^{n}\right]\right\} \tag{1}$$

$$R = \int_0^{\frac{\pi}{2}} ds = L_k = \frac{\pi D}{4} \tag{2}$$

$$ds = \left[\left(\frac{dx}{d\varphi}\right)^2 + \left(\frac{dy}{d\varphi}\right)^2\right]^{1/2} \tag{3}$$

Based on the equations (1), (2) and (3), forming parameters were calculated to describe the resulting ellipse shape of Figure 3. The coefficients A and B describe the length of the main and the minor semi-axis of the ellipse. The coefficient n impacts the flattening or stretching of the ellipse shape. The ellipse was described as a quarter, which was reflected in Creo Parametric 3D software.

3.2. Geometry Analysis of the Flexible Spline and the Circular Spline in Matlab

A mathematical calculation model was designed in Matlab to generate the geometry of these teeth bodies, as well as for the wave generator. This programme generated the geometry of both bodies simultaneously; see Figure 2. The generating was primarily for the shape geometry of the teeth. The following parameters were defined as programme inputs: teeth module, the number of teeth of both bodies, working addendum and the size of the meshing circles. For this gearing, we considered 200 and 202 teeth.

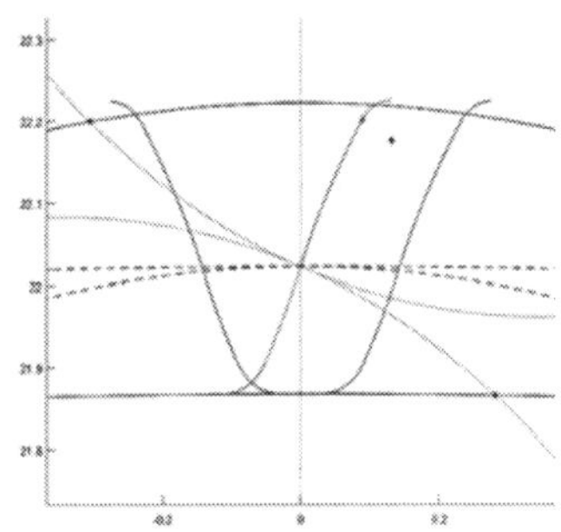

Figure 2. Detail of the shape of the flexible spline gearing (in red) and lines of action (in yellow).

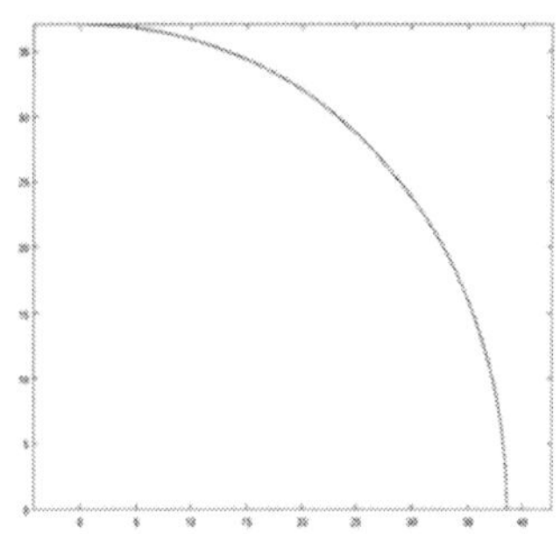

Figure 3. The ellipse shape in Matlab.

Established patterns and relations were used to calculate the gearing geometry. This programme calculated gearing pressures according to Hertzian contact stress and a material coefficient in line with the equation [4-5].

4. Designing the 2D and 3D Model for Simulation

The point clouds generated this way were then imported into a Creo Parametric sketchbook and they were used for developing volumetric and planar bodies, see Figure 4 and 5.

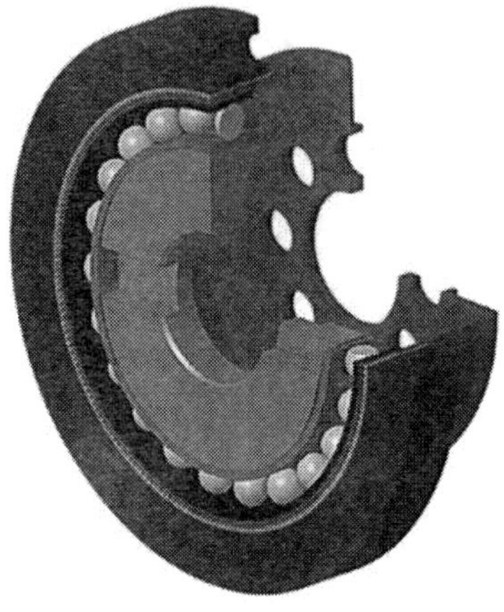

Figure 4. 3D model of the gear.

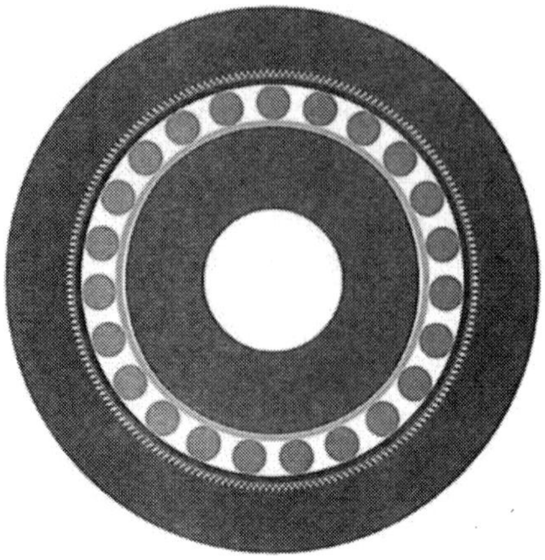

Figure 5. 2D model of the gear.

4.1. Developing the Mesh for 3D Simulation

A 2D and 3D mesh was developed for 2D and 3D models in MSC Apex Iberian Lynx. For the task regarding 2D, the Creo Parametric model was created by surface modelling through surfaces. It was exported in step format as a surface. It was meshed in Apex using linear quadrilateral elements. The 3D model was also exported in step format as a solid body. It was meshed using a linear hexagonal mesh. For a symmetrical mesh, only a section of the gear system was meshed. These bodies were reflected in the Marc Mentant and joined using a "sweep" function. The number of elements for 2D mesh amounted to 392,706 (Figure 7), and for 3D mesh it was 543,336 (Figure 6). The 2D task was meshed with a rather fine mesh - due to the lower number of elements. It is recommended to work with a model of up to 600,000 elements. If the number is exceeded, huge calculation parameters and a long calculation time are needed. At the same time, it leads to a problem with post-processing.

Figure 6. 3D mesh preview.

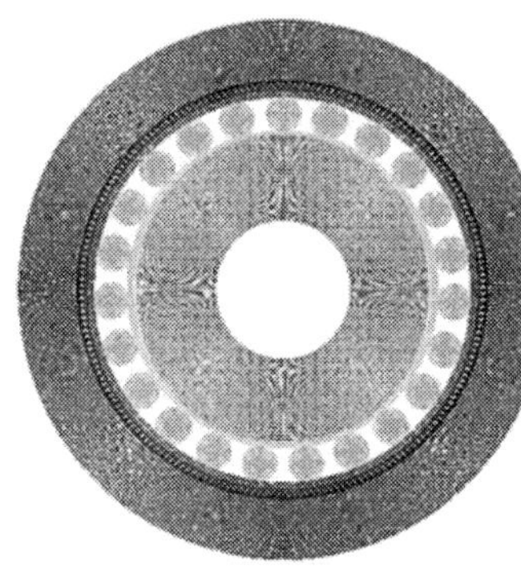

Figure 7. 3D mesh preview.

4.2. Developing a Simulation Calculation Model

The calculation model was created for the 2D and 3D tasks in Marc Mentant 2018. The task was broken down into several steps. Firstly, the gear bodies were assigned material characteristics and contact dependence was determined; see Figure 8 and 9. In Marc Mentant, it is possible to set a contact GLUE analysis for a solid bonding of bodies, without considering friction, and TOUCHING analysis for the contact of the bodies with prescribed friction. It is the option of defining the contact between the bodies that allows the whole gear to be set into rotating motion rather easily, thus allowing examination of individual characteristics directly "on the fly" of the harmonic drive gear. It is also possible to let the gear rotate as long as necessary and examine slips and wear on the teeth. This, however, is not a subject of this article [6].

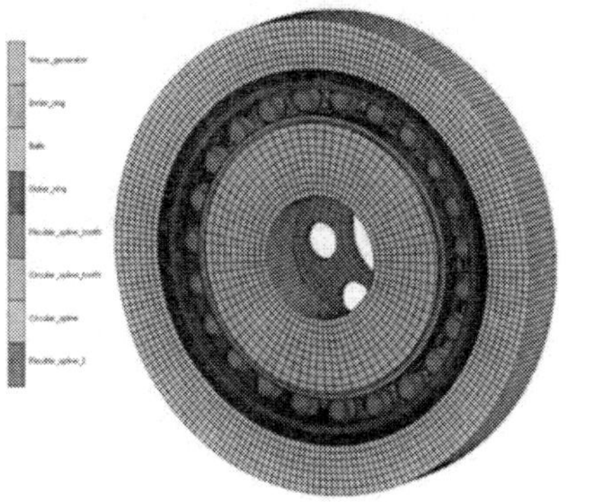

Figure 8. Preview of the 3D calculation model.

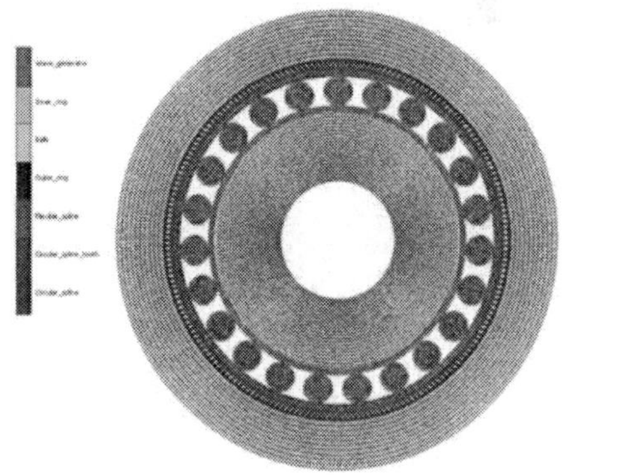

Figure 9. Preview of the 2D calculation model.

To make the simulation as realistic as possible, the calculation model was broken down into individual minor steps. The first step was to let the entire gear be pressed. After the pressing, the tooth contact was switched from GLUE to TOUCHING as the second step. After having switched the gearing contact dependence, the gear was set into rotating motion using the wave generator by 180°. After having finished the rolling, the wave generator was stopped and load torque M_k = 100 Nm was applied to the flexible spline in the opposite direction of the rotation. As a last step, the direction of the torque was switched to the direction of the rotation. Through the last two steps, the gear clearance (stressing the backslash) was discovered. The whole task had 261 increments. The results were analysed in each of them [7-8].

5. Evaluation of the Results

The evaluation of the results was post-processed in Marc Mentant. Contact stresses for 2D and 3D models engaged/meshed with the gear system were evaluated, see Figure 10 - 13.

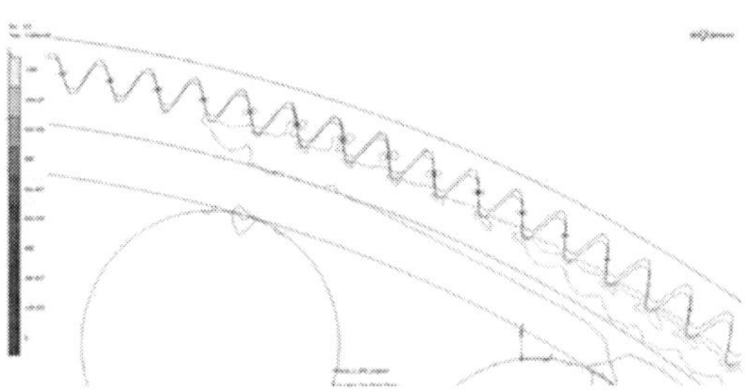

Figure 10. Image of the distribution of the contact stress with regard to the 2D model.

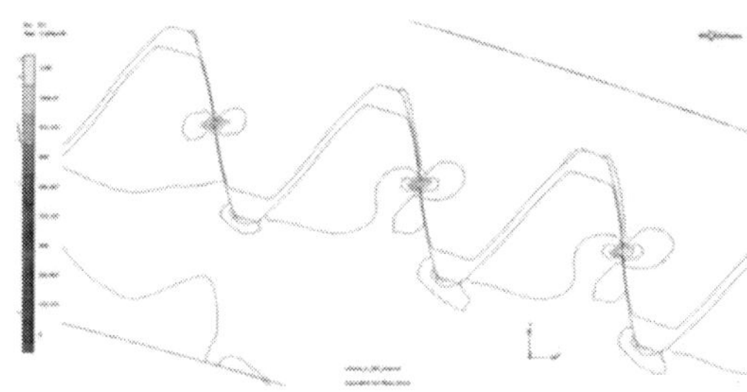

Figure 11. Details of the contact stress with regard to the 2D model.

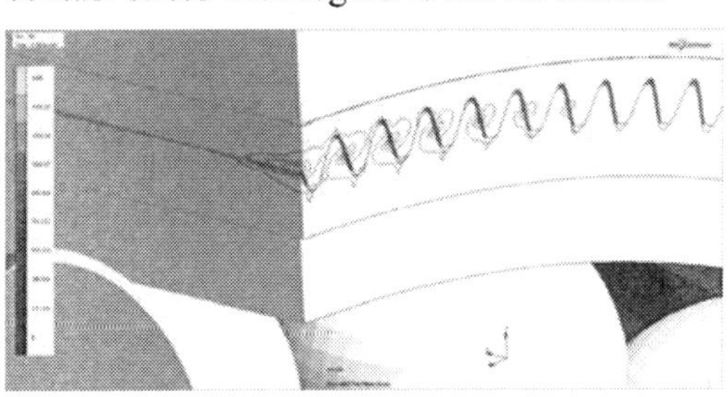

Figure 12. Image of the distribution of the contact stress with regard to the 3D model.

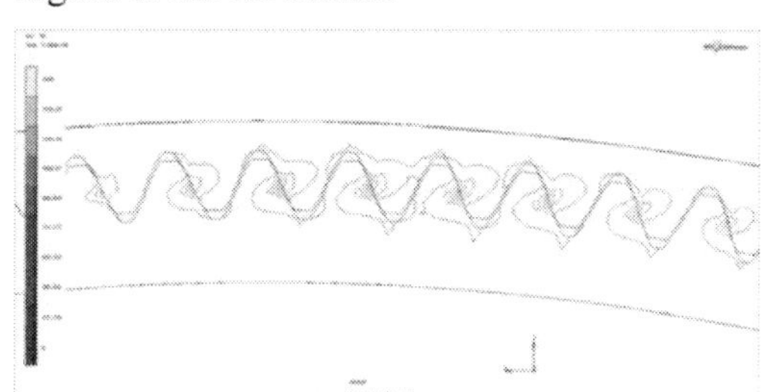

Figure 13. Details of the contact stress with regard to the 3D model.

The distribution of the stress is caused by "attaching" the flexible spline in the crown of the circular spline; see Figure 14 and 15. The stress on gearing is distributed evenly everywhere else.

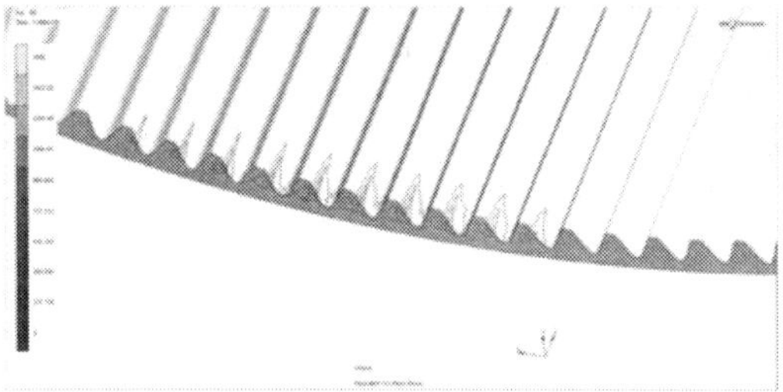

Figure 14. Displaying the distribution of the contact stress with regard to 2D.

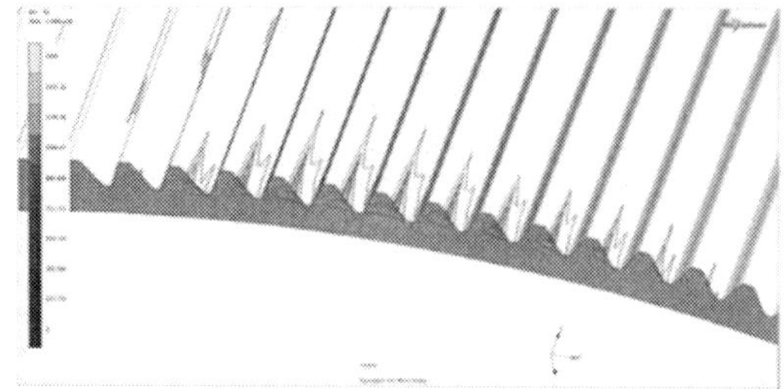

Figure 15. Details of the contact stress with regard to 2D.

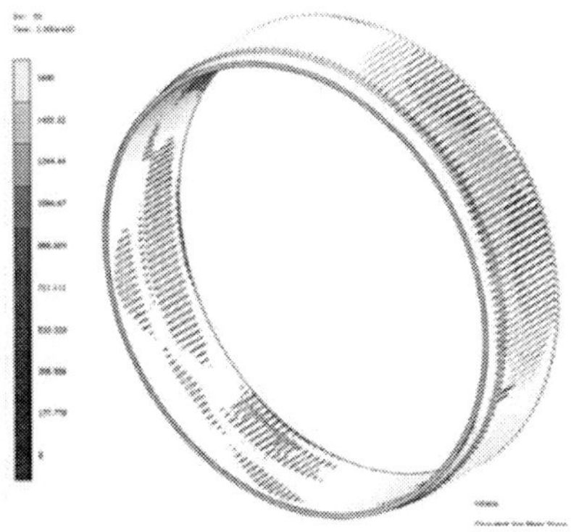

Figure 16. Displaying the distribution of the contact stress on flexible spline gearing.

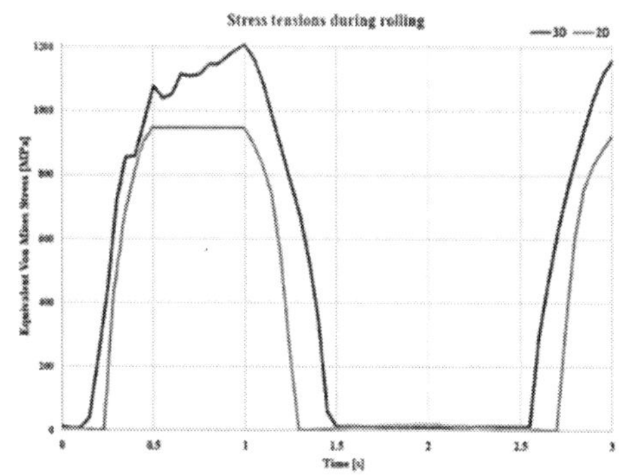

Figure 17. Comparison of the trajectory of maximum stress in the gearing of the harmonic gear system for 2D and 3D analyses.

6. Appendices

The simulation revealed that the flexible spline gearing is deformed in two directions simultaneously. The first depending on the pressing on the wave generator, and the second one depending on the impact of the torque. When transferring the performance from the gear, the flexible spline has been twisted in the circular spline; see Figure 16. This causes the greatest stress in the front part of the gearing. Figure 17 shows the course of the stress when the gear is rolling. The figures on the graph were acquired from the point with the greatest contact stress on a flexible spline tooth. The maximum stress with the 3D model was 1206 MPa and 995 MPa for the 2D model. The measured maximum stress with the 2D and 3D models differed by 17.49%, which was caused by different mesh densities. From the viewpoint of possible calculations, it is not possible to use as fine mesh for a 3D model as for a 3D analysis. Furthermore, this research proved that in the case of harmonic drive gears, several teeth mate simultaneously. With 200 teeth for a circular spline in a 2D task, 12 teeth mate/are engaged at the same time on the upper part of the gear and 12 teeth at the same time on the opposite part of the gear. It means that 24 teeth mate/are engaged at once. With the 3D model, the number amounted to 10 teeth on the upper part and 10 teeth on the opposite part of the gear. Consequently, 20 teeth mated/were engaged at once. This also implies that the rigidity and accuracy of these harmonic gearbox systems with a flexible component is huge in comparison to traditional rigid gear systems.

The main purpose of this article was to describe the basic topic concerning these special harmonic drive gears that have been increasingly popular, mainly in the area of robotics and precise control. Also, the article dealt with the main differences between conventional toothed gear systems. Given the recent exponential growth of the engineering industry, automation and robotization, we will surely come across the term of Harmonic Gearbox more often due to its great advantages and a really small size. The most important features of Harmonic Gearbox definitely include very big rigidity and high accuracy. The high stiffness is due to multiple teeth engaging on both sides of the gearbox. The geometry of the gearbox and its effect on the stiffness and durability will be the subject of the following research to address this issue.

Acknowledgement
This study was supported by Slovak Research and Development Agency under the contract no. APVV-18-0450 - Research of the influence design parameters of special transmissions with a high gear ratio with respect to kinematic properties.

References
[1] Lizhong Xu and Zhenglong Fu 2018, Applied Mathematical Modelling: Multifield coupled analysis for a harmonic movable tooth drive system integrated with shape memory alloys, *TECHNO-PRESS* **23**(4) 319-327
[2] Zou C, Tao, T, Jiang, GD, Mei, XS and Wu, JH 2017 MECHANICAL ENGINEERING SCIENCE: A Harmonic Drive Model Considering Geometry and Internal Interaction, *SAGE PUBLICATIONS LTD* **231**(4) 728-743
[3] Sahoo V and Maiti, R 2018 MECHANICAL ENGINEERING SCIENCE: Evidence of Secondary Tooth Contact in Harmonic Drive, with Involute Toothed Gear Pair, Through Experimental and Finite Element Analyses of Stresses in Flex-Gear Cup, *SAGE PUBLICATIONS LTD* **232**(2) 341-357
[4] Brumercik F, Lukac M, Majchrak M, Krzysiak M and Krzywonos L 2018 Scientific Journal of Silesian University of Technology: Teeth geometry and contact pressure calculation of external cycloidal gears, *Series Transport* **101**(1) 27-35
[5] Kohar R and Slavomir H 2014 Scientific Letters of the University of Zilina: „Dynamic Analysis of a Rolling Bearing Cage with Respect to the Elastic Properties of the Cage for the Axial and Radial Load Cases, *Communications* **16**(3A) 47-81
[6] Belorit M, Hrcek S and Gajdosik T 2017 *Description of the bearing check program for countershaft gearboxes*, ICMD 2017 58th International Conference of Machine Design Departments, Prague, Czech Republic, Sep. 06-08, 32-35
[7] Hrcek S and Kohar R 2014 *Roller Bearing Fatigue Life Calculation with Inclusion of the Fits Parameters*, ICMD 2014 55th International Conference of Machine Design Departments, Prague, Czech Republic, Sep. 09-12, 207-213
[8] Kohar R and Hrcek S 2014 *The Effect of Fits Stiffness on Distribution of Internal Loading in Rolling Bearings*, ICMD 2016 55th International Conference of Machine Design Departments, Prague, Czech Republic, Sep. 09-12, 235-241

An application of multicriteria optimization in selection of the two-speed two-carrier planetary gear trains

S Troha[1], J Stefanović-Marinović[2], B Rončević[3], B Anđelković[2], M Milovančević[2] and K Marković[1]

[1]University of Rijeka, Faculty of Engineering, Vukovarska 58, 51000 Rijeka, Croatia
[2]University of Niš, Faculty of Mechanical Engineering, Aleksandra Medvedeva 14, 18000 Niš, Serbia
[3]Vulkan Nova, Factory of Cranes & Ship Equipment, Rijeka, Croatia

E-mail: sanjin.troha@riteh.hr

Abstract. This paper outlines the optimum selection procedure for a two-speed planetary gear train with four external shafts based on two design and operational criteria: radial dimensions and efficiency. The paper outlines how to quickly determine the structure and important basic parameters of two-speed planetary gear trains that meet predefined transmission requirements. The procedure is followed by a numerical example in which the optimal two-speed planetary gear train is selected, defined by the teeth number of the central gears, modules and transmission ratios. The relevance of the presented results is reflected in the fact that these transmissions offer significant advantages in systems which require speed change under load.

1. Introduction

Compared to conventional gear trains, planetary gear trains (PGTs) offer numerous benefits, and over the past several decades their use has proliferated significantly in a variety of branches in mechanical engineering. Multi-stage planetary gear trains are built by linking one or two shafts between two different PGT stages from single-stage gear trains. A special multi-stage PGT type is a two-speed two-carrier PGT consisting of two coupling shafts and four external shafts. There are many important characteristics of this type of compound gear train, the most notable being the possibility of speed changes under load from the operator or control unit (machine tools, cranes...).The basic part of the paper is the review of possible choice of the transmission for any defined application by using a specially developed computer program. The choice between the computer obtained variants is then made by comparative analysis of the solutions.

Until recently, two-speed planetary change-gears with four external shafts and two coupled shafts were not investigated systematically. In general, some sporadic reviews of selected schemes were carried out in [1-4].

The review of 15 schemes of reversible transmissions of this type with orientation values for transmission ratio and achievable efficiencies is presented in [5]. The characteristics of some schemes are presented in [6] and [7], however the methodology of optimal scheme choice was not included. Symbolic review of structural schemes without detailed analysis is given in [7] and the authors describe the torque method, which can be employed for their analysis. The torque method was developed by Arnaudov as the principal tool of systematic analysis of multi-speed PGTs and the reader is referred to [8], where the method is explained and all source references are cited. The detailed systematic research carried out between 2006 and 2011 laid the foundation of the methodology for the selection of optimal PGT structures which was implemented in a computer program – all possible schemes of these transmissions and their working regimes are laid out by Troha in [9] and [10]. The choice of the optimal variant of a two-speed PGT intended for use in the driveline of fishing boat propulsion is given in [11].

The main objective of this paper is to demonstrate the capabilities of the computer program *DVOBRZ* that has been developed for the purpose of analysis and optimization of two-speed PGTs. This is also followed with an example in which the optimal solution determined by design parameters is obtained.

After the introductory section this paper is structured as follows. Section 1 introduces the synthesis methodology for two-speed planetary gear trains. Section 2 presents an analysis of numerical results obtained from the developed computer program, which is also complemented with a brief discussion. Section 3 presents the paper's conclusion.

2. Methodology of synthesis of two-speed planetary gear trains

2.1. Two-carrier planetary gear train structures and labelling method

In cases where two-speed transmissions are required, a mechanism obtained by connecting two simple planetary gear trains of type 2k-h, variant A [5] is one of the best suited design solutions. A symbolic representation of a simple planetary gear train with a Wolf-Arnaudov symbol, with the relations of torques acting on the shaft is outlined in [7]. Joining two shafts of one planetary gear train with two shafts of another planetary gear train forms a mechanism with four external shafts in total, Fig. 1.

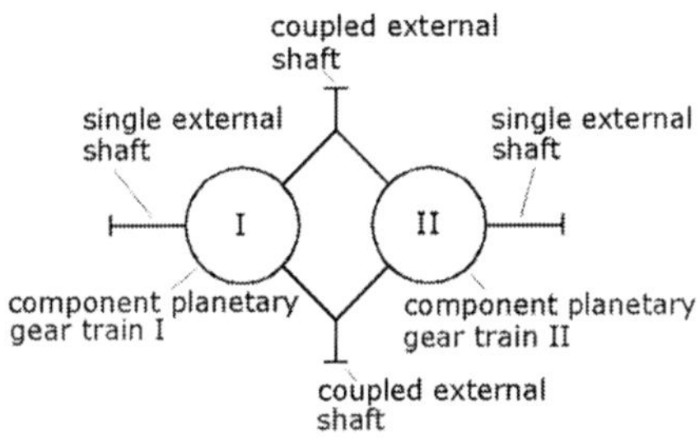

Figure 1. Symbolic representation of a compound planetary gear train with four external shafts).

We will refer to such an assembly hereafter as a *compound gear train*, and we will also refer to the two connected PGTs as *gear train I* (*stage I*) and *gear train II* (*stage II*).

In research published in [7] it was shown that there are a total of 120 different compound trains of this type, namely: 24 variants with brakes on single external shafts, 24 variants with brakes on the coupling shafts and 72 variants with brakes on both coupling and single shafts. Each of the 120 variants of two-speed planetary gearboxes has specific shifting capabilities.

By placing brakes on two shafts, a braking system is established in which activation of one brake (i.e. prevention of any shaft rotation) shifts the power flow through the planetary gear train, which also causes a change in the transmission ratio.

Each variant has its own features that determine the conversion possibilities. It could be assumed that some variants work either as reducers or as multipliers in both speeds, while other variants work in one speed as reducers and in the other as multipliers. Additionally, when the speed is changed, some variants change the direction of rotation, while others keep the direction of rotation during speed changing. Only the basic transmission ratios (i.e. ideal torque ratios) depend on the transmission ratio of the compound planetary gear train. It is important that some of these transmissions are capable of changing the transmission ratio under load. An example of transmission choice will be presented in the next section for a particular application.

This label of a compound two-speed two-carrier train transmits information unambiguously about the structural scheme (the way two PGTs are connected) and the layout variant (the role of each shaft in the transmission) and is determined from the program *DVOBRZ*. The label may also contain brake engagement information; for example, S16V2Br2 indicates the S16V2 compound train's operating regime when brake 2 is engaged [9]. An alphanumerical label (S11...S56) is assigned to structural schemes, which indicates the way in which the shafts of the first and the second PGT are connected, while the term *layout variant* refers to the particular placement of the input and output shaft and of the brakes. Consequently, any designation assigned to a particular PGT conveys information about both the structural scheme and the layout variant, and represents in essence a combination of the two.

2.2. The principle of synthesis of a two-speed planetary gear train

To ensure that the required transmission ratios i_{k1} and i_{k2} are realized, all possible variants that meet the imposed constraints are searched and this creates a set of feasible solutions. Fig. 2 shows a graphical explanation of this principle.

The intervals of ideal torque ratios need to be defined at the beginning as: $t_{Imin}...t_{Imax}$ and $t_{IImin}...t_{IImax}$. The intervals of the required transmission ratios should also be known: I_1 and I_2.

The graphic representation of the transmission ratio dependence on the ideal torque ratios (thus forming surface areas) is shown in Fig. 2a. Corresponding transmission ratio intervals I_1 and I_2 ($i_{k1} \in I_1$ and $i_{k2} \in I_2$ must be fulfilled) are shown on the vertical axis in figure 2b. Also, an arbitrary two-speed PGT as an example is presented in figure 2.

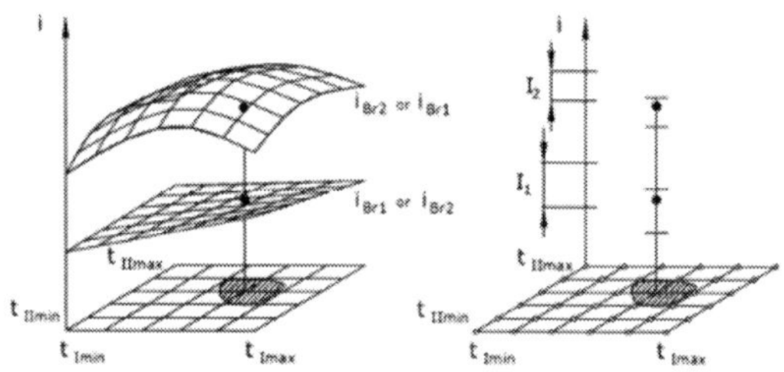

Figure 2. The principle of synthesis of a two-speed planetary gear train.

For every possible combination of ideal torque ratios, the computer program *DVOBRZ*, specifically developed for this purpose, calculates the values of the transmission ratio functions for each design structure of a two-speed PGT. The program then checks if the calculated transmission ratios fall within the user-defined I_1 and I_2 intervals, and if such torque ratio combinations exist, the program extracts them as possible solutions. The transmission ratio effected by the activation of brake Br1 is marked as i_{Br1} and the transmission ratio effected by the activation of brake Br2 is marked as i_{Br2}.

After all the obtained solutions are stored, the program is able to compare them according to the defined relevant criteria (e.g. minimal radial dimensions, maximum equivalent efficiency etc.) [9, 11-13].

3. Numerical example and discussion

The procedure is shown on an example problem to better illustrate how this program operates. We shall assume a two-speed PGT needs to be synthesized, subject to the following kinematic, design and operational requirements which present input data:

- the transmission ratio intervals: $5 \le i_{k1} \le 5.2$ and $3.5 \le i_{k2} \le 3.6$, or $-5.2 \le i_{k1} \le -5$ and
- $-3.6 \le i_{k2} \le -3.5$,
- no reversal of rotation between the two output speeds: sign i_{Br1} = sign i_{Br2},
- the number of teeth of the sun gears: $z_{1I} = z_{1II} = 18$,
- the number of planet gears in a single gear train: $k = 3$,
- torque on the input shaft: $T_A = 50$ Nm,
- input number of revolution $n_A = 500$ min^{-1}
- frequency of operation at nominal power: $\alpha_1 = 0.5$ and $\alpha_2 = 0.5$, respectively for the required transmission ratios,
- material of all gears 16MnCr5,
- the main optimality criterion is minimisation of radial dimensions.

Based on the requirements and assumptions listed above, the *DVOBRZ* program lists six possible two-speed PGT design structures as acceptable solutions, with the main parameters and kinematic capabilities summarized in table 1 (n_A – rotational speed of the input shaft, n_{2r} – relative rotational speed of the faster satellites, comparing satellites in two stages). In the next step (also by using the *DVOBRZ* software) it is possible to determine whether the required transmission ratio is achieved with the activation of brake Br1 or brake Br2. Transmission ratios i_{Br1} can be either upper or lower surface, as well as i_{Br2}.

With the obtained data of ideal torque ratios t_I and t_{II}, transmission ratios i_{Br1} and i_{Br2}, the teeth number of the ring gears z_{3I} and z_{3II}, as well as of the pitch diameters of the ring gears d_{3I} and d_{3II}, the program then determines the corresponding designation of the PGT and the required order of break engagements that produces the calculated transmission ratios.

Figure 3 presents the pitch diameter of the larger ring gear with the corresponding equivalent efficiencies and, in a separate diagram, the pitch diameter ratios of each design structure as shown in Table 1. We shall define here the pitch diameter ratio of a two-speed PGT as

$$D_{3,\mathrm{rel}} = \frac{\max(d_{3\mathrm{I}}, d_{3\mathrm{II}})}{\min(d_{3\mathrm{I}}, d_{3\mathrm{II}})} \tag{1}$$

The value of this parameter by definition always exceeds 1.

The value of the pitch diameter ratio is important as it indicates the housing shape. The housing can be of a cylindrical shape if it is close to one. In other cases, the form of housing must be used gradually.

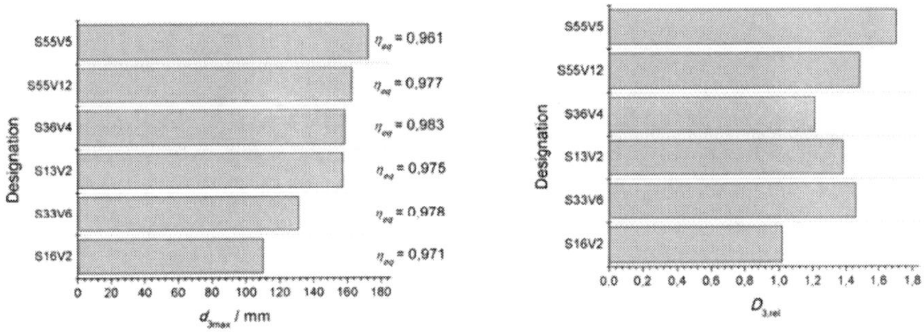

a) Absolute values given for the larger of the two pitch diameters. b) Pitch diameter ratios $D_{3,rel}$.

Figure 3. Radial dimensions of all acceptable design structures from Table 1.

The equivalent efficiency η_{eq} is calculated by using equation (2) as

$$\eta_{\mathrm{eq}} = \alpha_{\mathrm{Br1}}\,\eta_{\mathrm{Br1}} + \alpha_{\mathrm{Br2}}\,\eta_{\mathrm{Br2}} \tag{2}$$

where η_{Br1} is the efficiency when brake Br1 is activated, η_{Br2} is the efficiency when brake Br2 is activated, and α_{Br1} and α_{Br2} are the frequencies of operation at nominal power for their respective transmission ratios.

The equivalent efficiency can be precisely defined only when the transmission operates at nominal power with both speeds.

By closer examination of figure 3 and table 1, it is clear that the PGT S16V2 is the optimal solution according to the criterion of minimal radial dimensions of planetary gear sets.

On the other hand, when maximization of the equivalent efficiency is taken as the guiding criterion, the optimal solution is the S36V4 PGT.

Since minimal radial dimensions are the main criterion in this example, the optimal solution is S16V2.

By activating brake Br1 the input of the system is the sun gear of the first stage, the output is the sun gear of the second stage. The first stage carrier and the second stage ring gear are motionless, figure 4a.

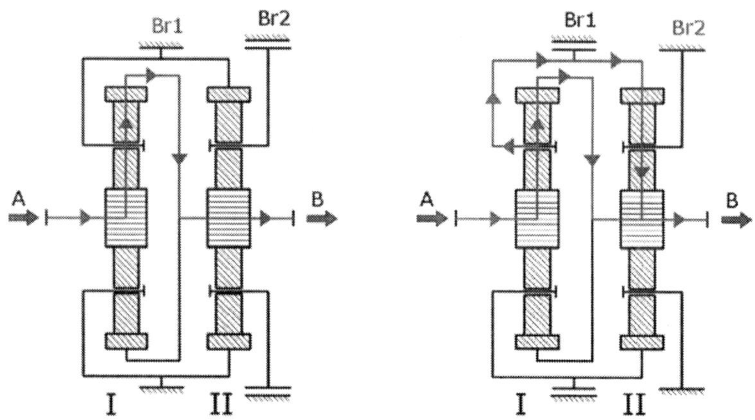

a) Brake Br1 activated b) Brake Br2 activated

Figure 4. Power flow through the transmission

(A–input shaft; B–output shaft).

The second stage operates in three-shaft mode, but without any resistance, which means it is idling. Because the carrier is reactive in the first stage, the power is transmitted to the first stage ring gear and then to the second stage sun gear, i.e. the power output B. There is no power branching, so the losses occur in just one PGT stage and only one power sink is present.

The power flow when Br2 is activated and brake Br1 disengaged is presented in figure 4b. The second stage carrier is motionless and the system's input element is the sun gear of the first stage. The power is divided into two components: one component is transmitted to the ring gear and the other to the carrier. The output element is the sun gear of the second stage, where both branches of the power flow finally meet and sum up to the total of the power output. Both planetary stages are activated in this case.

This example clearly outlines the procedure needed in the selection of a kinematic structure designed to operate optimally within a defined mechanical system. In the next step it is necessary to apply multicriteria optimization methods in order to define the PGT with design parameters [11].

IOP Conf. Series: Materials Science and Engineering **659** (2019) 012069 doi:10.1088/1757-899X/659/1/012069

Table 1. Possible two-speed PGT design structures as acceptable solutions.

Designation	Kinematic diagram	t_I [-]	t_{II} [-]	z_{3I} [-]	z_{3II} [-]	i_{Br1} [-]	i_{Br2} [-]	n_{2r}/n_A [-]	m_I [mm]	m_{II} [mm]
S16V2	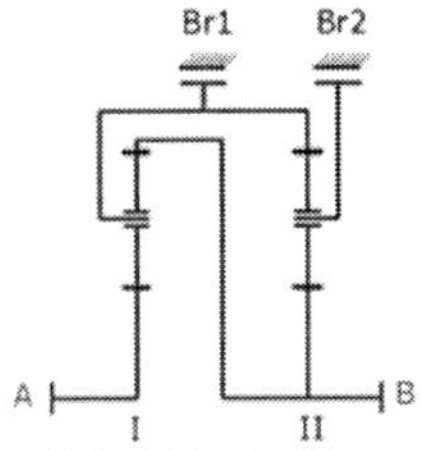	3.5	3	63	54	−3.5	−5	0.8	1.75	2
S33V6	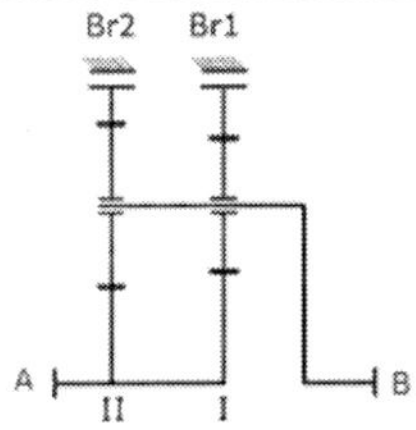	2.5	4.17	45	75	5.167	3.5	1.075	2	1.75
S13V2	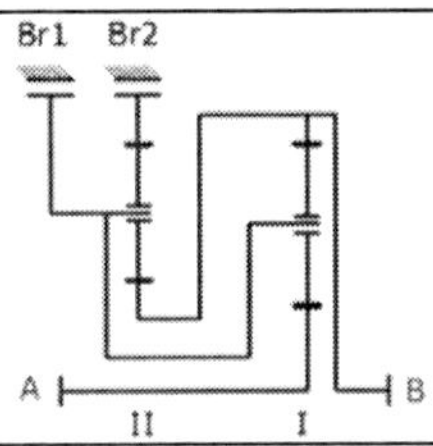	5	3.17	90	57	−5	−3.56	0.534	1.75	2
S36V4	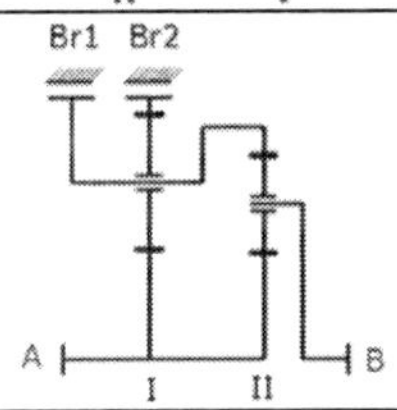	7.83	4.17	141	75	5.167	3.511	0.509	1.125	1.75
S55V12	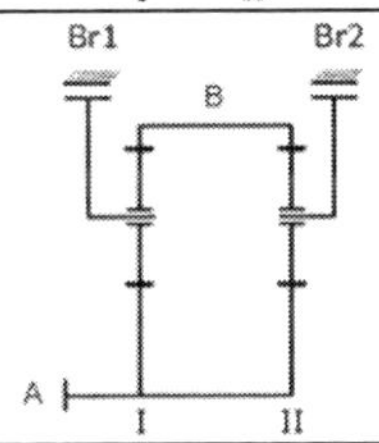	3.5	5.17	63	93	−3.5	−5.167	0.8	1.75	1.75
S55V5	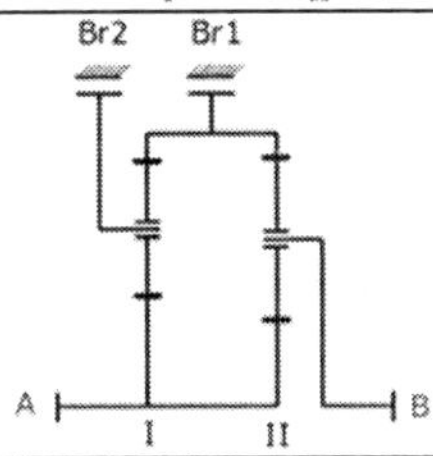	7.67	2.5	138	45	3.5	5.194	1.077	1.25	2.25

4. Conclusions

This paper presents the possibilities of a computer program designed to optimize the choice of two-speed planetary two-carrier gear trains. Optimization is possible only if the desired transmission ratios are defined in both operating regimes, along with the torque and number of input shaft revolutions, as well as some other input data required by the computer program. The final selection of the transmission's kinematic structure and brake layout depends primarily on the main governing criterion chosen by the designer among a potentially large number of feasible solutions. A numerical example demonstrates this principle, showing how conflicting criteria yield different solutions as optimality priorities change.

The procedure presented is a vital part of the optimal design of planetary gear transmissions, but other components such as shafts and clutches also need to be studied to fully investigate and optimize the functionality of the entire assembly.

References

[1] Merritt H E 1947 *Gear Trains*, London, Sir Isaac Pitman & Sons, LTD
[2] Lechner G and Naunheimer H 1999 *Automotive Transmissions*, Springer-Verlag, Heidelberg.
[3] Arnaudov K, Genova P and Dimitrov L 2005 For an unified and correct IFToMM terminology in the area of gearing, *Mechanism and Machine Theory* **40**(9) 993-1001
[4] Jelaska D T 2012 *Gears and Gear Drives*, Wiley
[5] Kudryavtsev V N and Kirdyashev Iu N 1977 *Planetary gears – Handbook*, Mashinostroenie, Leningrad
[6] Karaivanov D and Troha S 2006 Examining the possibilities for using coupled two-carrier planetary gears in two-speed mechanical transmissions, *Machinebuilding and electrical engineering* **5-6** 124-127
[7] Arnaudov K and Karaivanov D 2005 Systematik, Eigenschaften und Möglichkeiten von zusammengesetzten Mehrsteg-Planetengetrieben, *Antriebstechnik* **5** 58-65
[8] Arnaudov K and Karaivanov D 2013 The torque method used for studying coupled two-carrier planetary gear trains, *Transactions of FAMENA* **37**(1) 49-61
[9] Troha S 2011 *Analysis of a Planetary Change Gear Train's Variants*, Faculty of Engineering, University of Rijeka, Croatia, Doctoral Thesis
[10] Troha S, Žigulić R and Karaivanov D 2014 Kinematic Operating Modes of Two-Speed Two-Carrier Planetary Gear Trains with four External Shafts, *Transactions of FAMENA* **38**(1) 63-76
[11] Stefanović-Marinović J, Troha S and Milovančević M 2017 An Application of Multicriteria Optimization to the Two-Carrier Two-Speed Planetary Gear Trains, *FACTA UNIVERSITATIS Series: Mechanical Engineering* **15**(1) 85-95
[12] Stefanović-Marinović J and Milovančević M 2012 The Optimization Possibilities at the Planetary Gear Trains, *Journal of Mechanics Engineering and Automation* **2** 365-373
[13] Stefanović-Marinović J, Troha S, Anđelković B and Milovančević M Efficiency of Planetary Gear Trains as Criterion for Optimal Solution Selection, *Machine Design* **10** (3) 93-98

Analysis of the influence of direction of helical teeth in the universal helical gear reducer on service life of the bearings that support the reducer shaft

M Rackov[1], S Kuzmanović[1], I Knežević[1], M Čavić[1], M Penčić[1], D Stefanović[1] and M V Dragoi[2]

[1]University of Novi Sad, Faculty of Technical Sciences, Trg Dositeja Obradovića 6, 21000 Novi Sad, Serbia
[2]Transilvania University of Brasov, Technological Engineering and Industrial Management Faculty, 1, University str., Brasov, Romania

E-mail: racmil@uns.ac.rs

Abstract. The paper analyses the problems of the influence of direction of helical teeth in the universal helical gear reducer on service life of the bearings that support the reducer shaft, i.e. on the overall load carrying capacity of universal gear reducer. Directions of helical teeth for all gear units is adopted in that way that the overall axial force at the reducer shaft is lowest as possible, i.e. axial forces of a shaft are in counterbalance. It is a practice the same output gear pair is used within a helical gear reducer made in universal casing for two- and three-stage gear reducer. However, these gears are produced with different directions of helical teeth in order to obtain the lowest axial force on the fifth geared shaft. However, some gear reducer manufacturers, trying to simplify the production and reduce the number of manufactured components, produce the gears with the same helical teeth direction for both two- and three-stage gear reducers. In that way, they do not reduce the value of axial force on the fifth geared shaft in three-stage gear unit, but they make addition of axial forces that certainly affects the service life of the bearings and thus the overall load carrying capacity of the gearbox. Analysing the concrete case, this paper researches the justification of implementation this procedure.

1. Introduction

Universal helical gear reducers are usually produced as single-, two-, three- and multi-stage gear reducers. Although, there are many manufacturers who do not produce conventional single-stage gear drives [1, 2, 3]. Some manufacturers produce two-stage gear reducers in special housings only for two-stage drives. However, their three-stage gear drives are built by connecting two-stage and single-stage gearbox, or four-stage gear drives are made by connecting two pieces of two-stage gearboxes. Most of the manufacturers produce two-stage gear drives in universal housings for two- and three-stage gear drives. Therefore, their two-stage gear reducers are a little bit more expensive than two-stage gearboxes mounted in special housings for two-stage gear drive. Nevertheless, their three-stage gear reducers assembled in the same universal housings are much cheaper than three-stage gearboxes built by connecting two-stage and single-stage gearbox [2, 4]. Which concept of manufacturing will be adopted depends on the market orientation of the producer, i.e. which segment of the market they want to pay special attention. Namely, the most required gear ratios are in the area covered by high gear ratios of two-stage gear reducer and low gear ratios of three-stage gear reducer. Major manufacturers have generally chosen this second variant, i.e. the gear drives assembled in universal housings for two- and three-stage gear reducers. While, smaller manufacturers usually use the first variant, i.e. the gear drives built in special housings in order to ensure the better placement on the market [4].

As already mentioned, in the case of multi-stage helical gear reducers, but not only helical, it is practiced that axial forces (occurred on gears and transmitted from shaft to bearings) have to act against each other in order to reduce the overall axial value. In this case, axial loading of the bearing will be less, so their service life will be longer [5]. The directions of gear helical angles for single-stage gear reducers are usually adopted randomly. For example, pinion has right direction of helical angle and driven gear has left, but it can be reverse. However, for two-stage gear units, the direction of helical angle of the third gear (or fifth gear for the three-stage gear unit) is selected so that the axial force of the third (fifth) gear is reversed with the axial force of the second gear. Since it is used the same output gear pair for three-stage gear reducer, the third output gear pair should be made with opposite direction of helical angle in order to ensure the reduction of overall axial force on the fifth geared shaft. The production of two sets of output gear pairs (the largest and most expensive) is a major problem for most of smaller manufacturers of gear reducers. Therefore, they abandon this concept and produce only one set of output gear pair, so they produce only one direction of helical angles of the third gear pair. In this way, the axial forces of the fifth geared shaft are not reversed, so it is interested to consider the real impact of producing only one direction of output gear pair to the service life of the bearings on the fifth geared shaft of three-stage gear reducer [5, 6].

2. Problem Discussion

If single-stage gear reducer is observed, the output gear shaft rotates in the right rotation direction, looking toward the reducer (Fig.1). In that case, the axial force on the pinion acts from the motor and the force is held by the bearing from the electric motor since the gearbox is connected directly with special reducer motor [5]. Reducer electric motors are characterized by smaller flanges (in order to achieve smaller dimensions of motor gear units), they have smaller thickness of free end of shaft or have a hole at the front surface of the shaft (in order to mount small pinions), they have stronger bearings (in order to accept large force from the pinion) and they have better sealing with the motor shaft (in order to prevent oil leakage from the gearbox housing to the electric motor housing) [6].

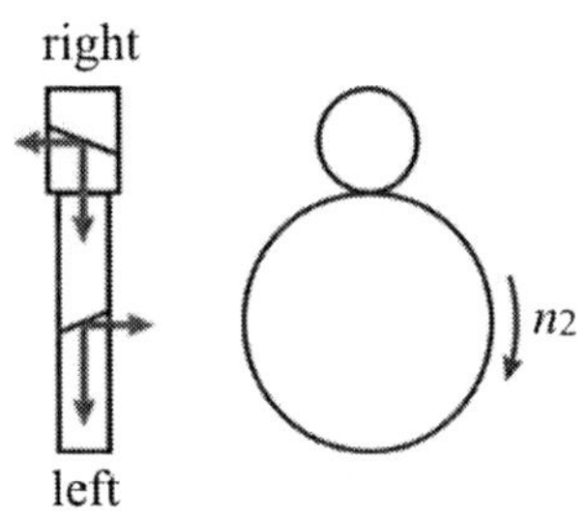

Figure 1. Graphic presentation of the axial forces action on the gear pair of a single-stage gear reducer.

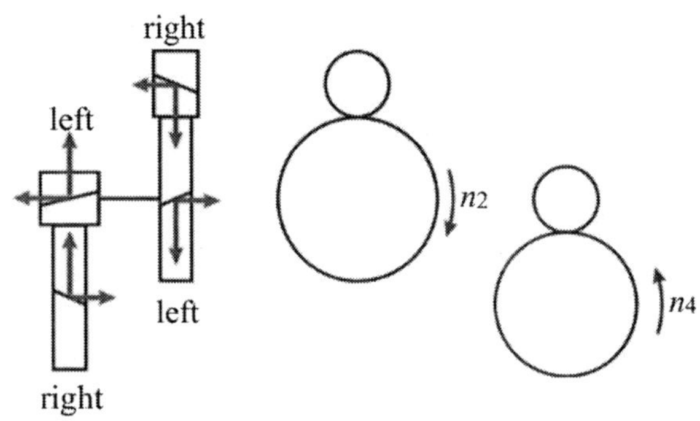

Figure 2. Graphic presentation of the axial forces action on the gears of a two-stage gear reducer for adopted rotation direction.

In the case of two-stage gear units, all the first gear pairs are kept the same as in single-stage gear unit because there are a large number of them (in order to connect different pinions with different second gear, so to provide gear ratio according the standard progression R20). The direction of helical angles of the second gear pair (or the third gear pair, since the same gears are used in three-stage gear unit, just with opposite helical angle direction) is adopted in that way so the axial forces on the third (or fifth) geared shaft are in a reduction as much as possible (Fig.2).

In the case of three-stage gear units, the first gear pair is also kept the same as in single-stage gear unit, but the second and third gear pairs are adopted with such direction of gear helical angle that the axial forces on the third and fifth geared shaft are in reduction at the shaft as much as possible (Fig.3).

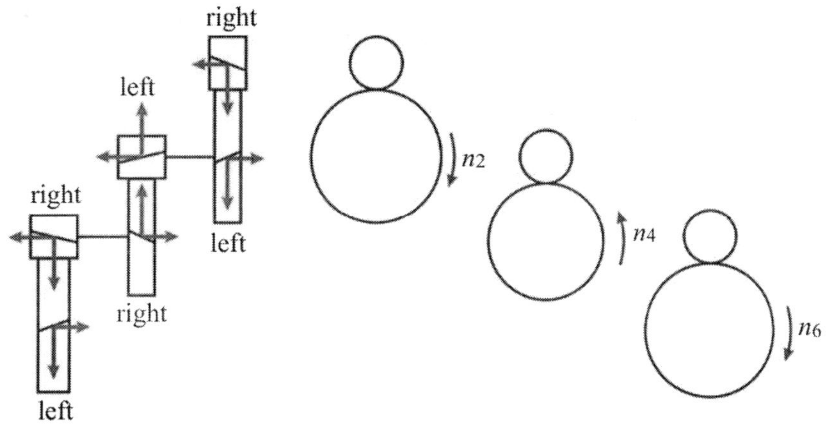

Figure 3. Graphic presentation of the axial forces action on the gears of a three-stage gear reducer where the output gears are with different directions of helical angles for adopted rotation direction.

In order to simplify the production and to reduce the number of necessary components, smaller manufacturers practice to use the third gear pair with only one direction of helical angle (Fig.4). In that case they could not provide the reduction of axial forces on the fifth geared shaft. Hence, the axial forces do not reverse each other on the fifth geared shaft of three-stage gear reducer. However, these forces are not large due to the small helical angle, but they are also not negligible due to the large torques on the fifth geared shaft. At the end, it is assumed that the sum of these axial forces will not be so large. The approval for such procedure is also reflected in the fact that the fifth geared shaft rotates significantly slower than the third geared shaft, which it certainly has more favourable impact on service life of the bearings [4, 5].

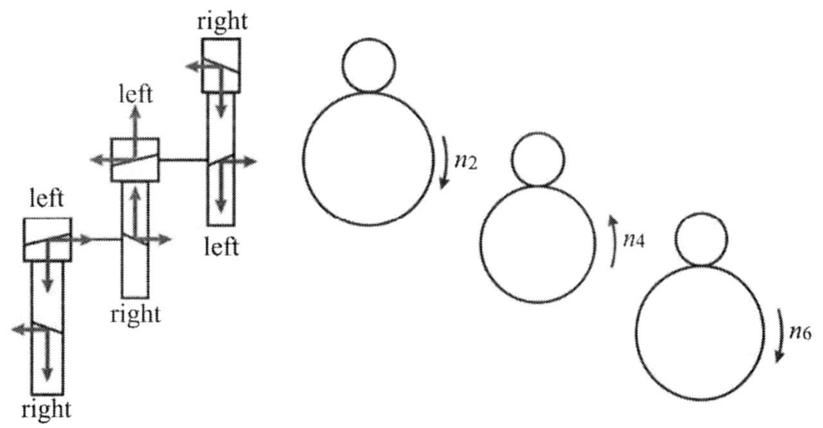

Figure 4. Graphic presentation of the axial forces action on the gears of a three-stage gear reducer where the output gears are with only one direction of helical angles (the axial forces are summarized) for adopted rotation direction.

Nowadays, the first gear pair usually provides gear ratios per standard progression R20, where the highest gear ratio value of some manufacturers reaches up to 15 [7]. Therefore, it is necessary to have two gears pairs, as the second gear pair in two-stage variant, which are the third gear pair for three-stage gear unit. In that case it is possible to have another second (third) gear pair with lower gear ratio for two-stage gear units, where the overall gear ratio of two-stage gear reducers is in the range of 3-4 till usually above 50. The problem of this approach can obtain for small gear ratio of the second gear pair when the large torque "enters" to the gear reducer which requires stronger components. Therefore, it is often case to see only one output gear pair, but some first gear pairs are produced as multipliers, which allows having small gear ratios for two-stage gear unit. For three-stage version, it is necessary to have only one pair of gears as the second gear pair of reducer, where the highest gear ratio usually exceeds 300 [7, 8].

3. Solving the Identified Problem

If three-stage gear reducer is observed with the third output gear pair with both directions of helical angles or with only one (Fig.3 and Fig.4), it is evident that in the first case axial forces acts against each other and reduce the overall axial force on the shaft, while in the second case the axial forces are summarized and additionally support the bearings, depending on the rotation direction of the output shaft. Of course, the axial forces could be also summarized on the third geared shaft, but here they are larger due to the larger helical angle. The rotation of the bearings is considerably faster, so for these reasons summarizing of axial forces is done at the fifth geared shaft [8, 9]. The fourth and the fifth gear are mounted on this shaft. These gears have much smaller helical angle, but the torque is larger, so the forces on the gears have also large values. In the case of three-stage gear reducer with axis height of 115 mm, with output nominal torque T_{2N} = 450 Nm and gear ratios per each stage $i_{1/2\max}$ = 6.36, $i_{3/4}$ = 7.09, $i_{5/6\max}$ = 5.6 and efficiency of each gear pair η = 0.98, it follows that the torque of the fifth geared shaft for the most favourable loading case is calculated:

$$T_4 = T_5 = \frac{T_{2N}}{i_{5/6}\,\eta_{5/6}} = \frac{450}{5.6 \cdot 0.98} = 81.997 \text{ Nm} \tag{1}$$

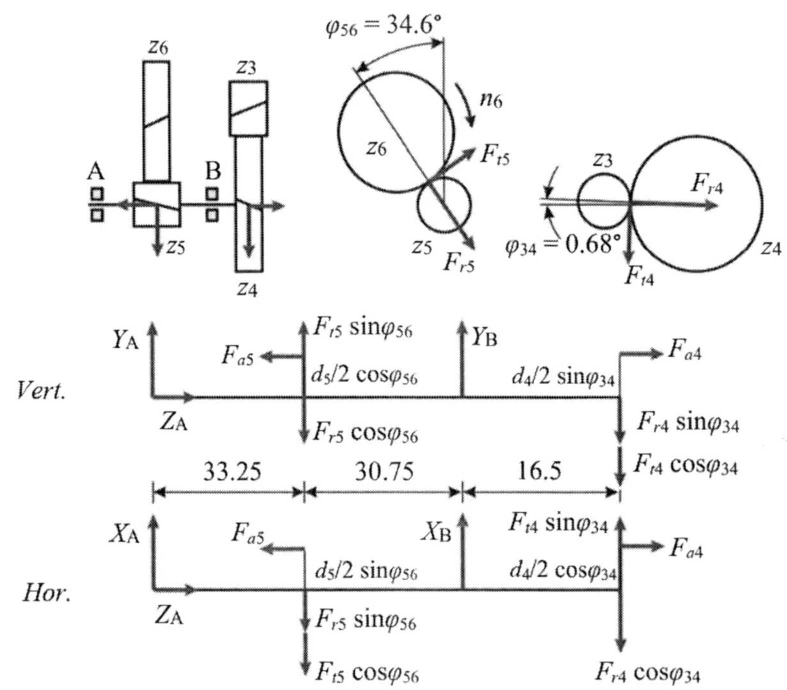

Figure 5. Analysis of forces on the gears of the fifth geared shaft for adopted rotation direction and right direction of helical angle of the fifth gear, where the axial forces have partially reduction.

All necessary data about the gears are known: z_4 = 78, $m_{n3/4}$ = 1.25 mm, $\beta_{3/4}$ = 20°, d_4 = 103.757 mm, F_{t4} = 1580.5 N, F_{r4} = 612.2 N, F_{a4} = 575.3 N, z_5 = 10, $m_{n5/6}$ = 2 mm, $\beta_{5/6}$ = 15°, d_5 = 20.705 mm, F_{t5} = 7920.3 N, F_{r5} = 2984.5 N, F_{a5} = 2122.2 N [10, 11].

If analysis of forces in vertical and horizontal plane is observed for the case that axial forces partially reverse each other and that left bearing (A) accepts the overall axial force (Fig.5), it follows that the reaction of the supports are F_A = 3694 N, Z_A = 1547 N, F_B = 5326.3 N, so the calculated service life of the bearing A designated by 6303Z is L_{hA} = 20200 h and the bearing B designated by NUP 2203 is L_{hB} = 20200 h for the lowest speed [12]:

$$n_4 = n_5 = \frac{n_{em}}{i_{1/2\max} \ i_{3/4}} = \frac{1450}{6.36 \cdot 7.09} = 32.2 \text{ rpm} \tag{2}$$

Considering the forces on the fifth geared shaft in the case the axial forces are summarized (Fig.6), it follows that the reaction of the supports are F_A = 3553.6 N, Z_A = 2697.5 N, F_B = 5805.8 N, so the calculated service life of the bearing A is less due to larger axial force L_{hA} = 10850 h and the bearing B is L_{hB} = 16500 h for the same rotation direction and the same lowest speed as in previous case [10, 11].

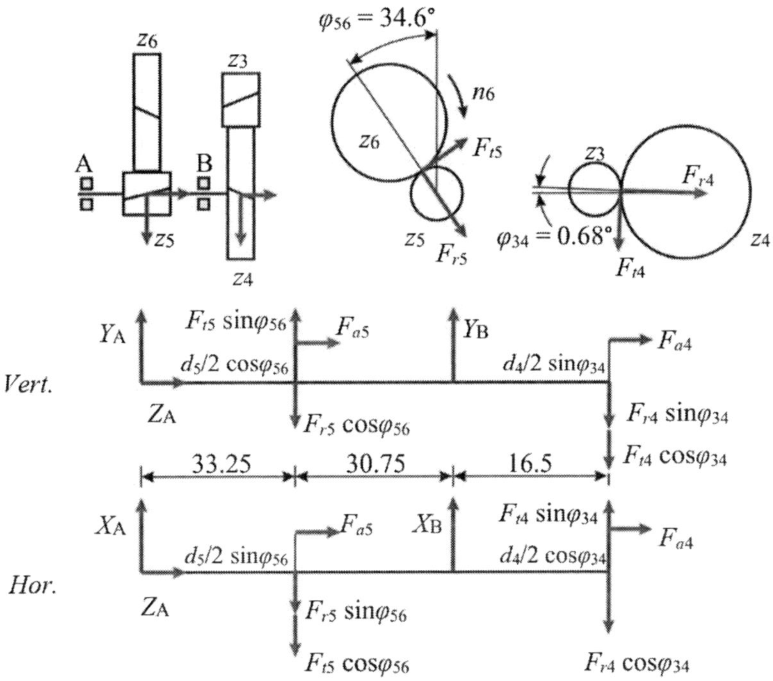

Figure 6. Analysis of forces on the gears of the fifth geared shaft for adopted rotation direction and left direction of helical angle of the fifth gear, where the axial forces are summarized.

If it is observed the case with opposite direction of rotation, the analysis of forces for the case that axial forces partially reverse each other is shown in Fig.7. The reaction of the supports are F_A = 4113.5 N, Z_A = 1547 N, F_B = 2755 N, so the calculated service life of the bearing A is L_{hA} = 17000 h for the lowest (the most favourable) speed. Since the support force in the bearing B is much lower for this rotation direction, its service life is not problematic and it is higher than 10^5 h [10, 11, 12].

Observing the analysis of forces on the fifth geared shaft for the opposite direction of rotation and the case the axial forces are summarized (Fig.8), it follows that the reaction of the supports are F_A = 4426.2 N, Z_A = 2697.5 N, F_B = 2729.3 N, so the calculated service life of the left bearing is less than required L_{hA} = 8130 h. Service life of the right roller bearing is also higher than required, so it is satisfied for this way of shaft loading.

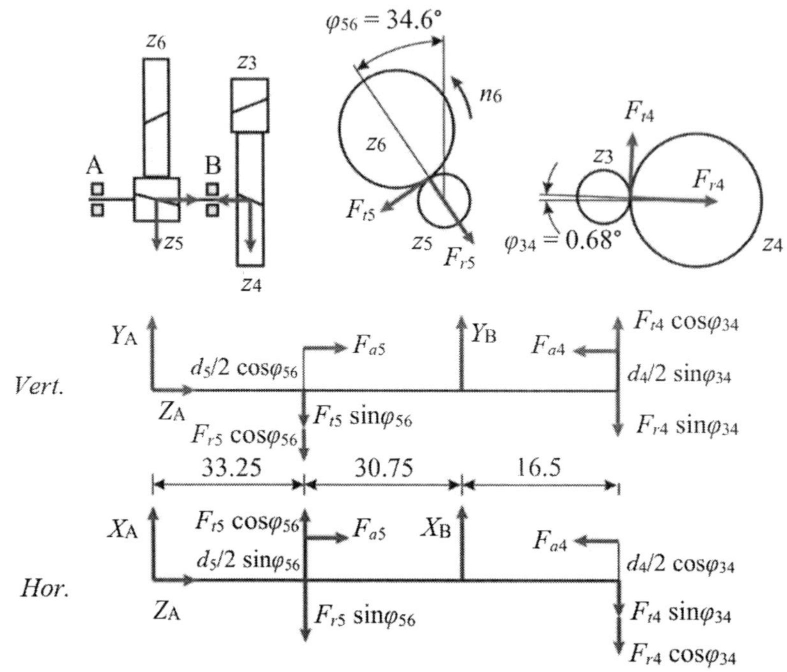

Figure 7. Analysis of forces on the gears of the fifth geared shaft for opposite direction of rotation and right direction of helical angle of the fifth gear, where the axial forces have partially reduction.

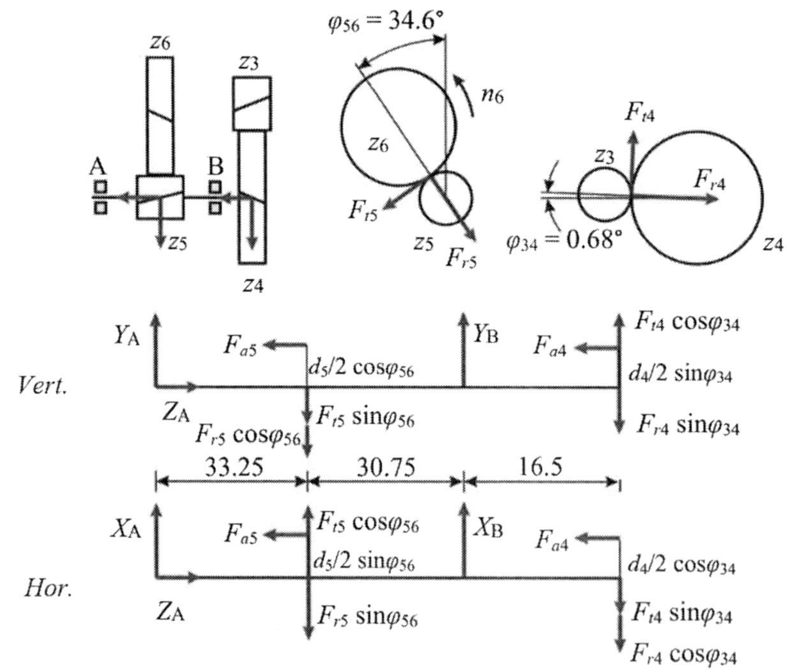

Figure 8. Analysis of forces on the gears of the fifth geared shaft for opposite direction of rotation and left direction of helical angle of the fifth gear, where the axial forces are summarized.

According to this analysis, it can be noticed that the service life of the bearings is not affected too much with the direction of axial forces. Service life of the left bearing is lower when the axial forces are summarized, but it is not great difference when the axial forces reduce each other. Decreasing of helical angle value of output gear pair could be interesting for reducing the axial forces on the fifth geared shaft. Since the helical angle of the output gears is less, the axial force on the fifth gear will be lower: z_4 = 78, $m_{n3/4}$ = 1.25 mm, $\beta_{3/4}$ = 20°, d_4 = 103.757 mm, F_{t4} = 1580.5 N, F_{r4} = 612.2 N, F_{a4} = 575.3 N, z_5 = 10, $m_{n5/6}$ = 2 mm, $\beta_{5/6}$ = 10°, d_5 = 20.308 mm, F_{t5} = 8075.1 N, F_{r5} = 2984.5 N, F_{a5} = 1423.9 N. If this analysis is calculated for both rotation directions and both helical teeth directions of the fifth gear, the service life of the ball bearing will be satisfied in all loading ways of the shaft [10, 11, 12].

4. Results of the Calculation

On the basis of implemented analysis and calculations, it can be concluded that service life of the bearings is pretty similar for different loading ways. Service life of the right roller bearing is not threatened by any rotation direction or direction of helical teeth, so it is always higher than required. The left roller bearing of the fifth geared shaft is problematic since it is mounted in the housing wall where much larger bearing of the output shaft is situated, so there is no place for adopting stronger bearing.

For the present situation of output gear pair and helical angle $\beta_{5/6}$ = 15°, there is a case when due to helical angle direction and rotation direction, left roller bearing can not reach required service life. However, if helical angle value is reduced $\beta_{5/6}$ = 15°, axial loading of the bearing will be lower and its service life will be satisfied for any loading way of the fifth geared shaft (Table 1).

Table 1. Support forces and service life of the left bearing of the fifth geared shaft for both rotation directions and both helical teeth direction of the fifth gear.

The way of loading	Forces in the left support and service life of ball bearing if helical angle is $\beta_{5/6}$ = 15°	Forces in the left support and service life of ball bearing if helical angle is $\beta_{5/6}$ = 10°
Right direction of rotation of output gear, where the axial forces have partially reduction	F_{rA} = 3694 N F_{aA} = 1547 N L_{hA} = 20200 h	F_{rA} = 3733 N F_{aA} = 848.6 N L_{hA} = 23700 h
Right direction of rotation of output gear, where the axial forces are summarized	F_{rA} = 3553.6 N F_{aA} = 2697.5 N L_{hA} = 10840 h	F_{rA} = 3642.3 N F_{aA} = 1999.1 N L_{hA} = 15200 h
Left direction of rotation of output gear, where the axial forces have partially reduction	F_{rA} = 4113.5 N F_{aA} = 1547 N L_{hA} = 17000 h	F_{rA} = 4228.5 N F_{aA} = 848.6 N L_{hA} = 16500 h
Left direction of rotation of output gear, where the axial forces are summarized	F_{rA} = 4426.2 N F_{aA} = 2697.5 N L_{hA} = 8130 h	F_{rA} = 4431.4 N F_{aA} = 1999.1 N L_{hA} = 11400 h

5. Conclusion

Based on the carried out analysis, it follows that the adopted directions of helical teeth on the fifth geared shaft do not affect much the service life of bearings. Therefore, it is completely justified to manufacture only one output gear pair with only one helical teeth direction. First of all this procedure is justified for small manufacturers of gear reducers, but also some large manufacturers use this way of producing.

By producing the output gears with smaller value of helical angle, it is possible to reduce axial forces and to reduce the axial loading of the fifth geared shaft and the bearings, which means longer service life of the bearings. It is very important since the possible space in housing wall is very limited and it is not possible to mount stronger bearing.

Using a simple calculation, that all manufacturers of gear reducer have, it can be easily determined the maximal torque at the output shaft of reducer, and that all components can transmit this load, where the most loaded bearings can reach the projected service life of 10000 h (industrial bearings).

Acknowledgment

This paper is part of a research on project "Research and Development of a New Generation of Wind Generators of High Energy Efficiency" TR 35005, supported by the Ministry of Education and Science, Republic of Serbia.

References

[1] Kudreavtev V N, Gierzaves I A and Glukharev E G 1971 *Design and calculus of gearboxes (in Russian)*, Mashinostroenie Publishing, Sankt Petersburg

[2] Aberšek B, Flasker J and Balic J 1996 Expert system for designing and manufacturing of a gear box, *Expert Systems with Applications* **11**(3) 397-405

[3] Kuzmanović S, Rackov M and Knežević I 2016 *Mechanical Gear Transmissions Nowdays*, The 5th International Conference on Power Transmission - BAPT 2016, Ohrid, Macedonia, Oct. 5-7, pp. 39 -46

[4] Rackov M 2013 *Conceptions of Development of Universal Gear Reducers (in Serbian)*, University of Novi Sad, Faculty of Technical Sciences, Serbia, Doctoral Thesis

[5] Kuzmanović S 2009 *Universal gear reducers (in Serbian)*, Faculty of Technical Sciences, Novi Sad

[6] Rackov M, Knežević I, Kuzmanović S, Čavić M, Penčić M and Ianici S 2016 *Proposal of Assessment Method of Different Quality Solutions of Universal Gear Reducers*, The Third International Scientific Conference "Conference on Mechanical Engineering Technologies and Applications" COMETa 2016, East Sarajevo, Bosnia & Herzegovina, Dec. 7-9, pp. 301-308

[7] Getriebebau NORD Drivesystems, Constant Speeds, G1000_IE3_50Hz_EN_2317, 2018, User Manual, www.nord.com (accessed on May 2019)

[8] Rossi, E04 – Coaxial Gear Reducers and Gearmotors (standard and for traverse movements), Edition December 2016, 4002BRO.ECO-en 1210 HQM, User Manual, www.rossi.com (accessed on April 2017)

[9] NORDBLOC.1® Helical In-Line Gear Units, 100200108 / 0517, User Manual, www.nord.com (accessed on May 2019)

[10] Miltenović V 2009 *Machine Elements*, University of Niš, Faculty of Mechanical Engineering

[11] Ognjanović M 2006 *Machine Elements*, University of Belgrade, Faculty of Mechanical Engineering

[12] SKF, Engineering Tools and CAD, https://www.skfbearingselect.com/#/type-arrangement/single-bearing (accessed on June 2019)

IOP Conf. Series: Materials Science and Engineering **659** (2019) 012071 doi:10.1088/1757-899X/659/1/012071

Transmission characteristics of simple cycloid drive with stepped planets

T Mačkić[1], M Tica[1] and R Šuba[2]

[1]University of Banja Luka, Faculty of Mechanical Engineering, Vojvode Stepe Stepanovića 71, 78000 Banja Luka, Bosnia and Herzegovina
[2]Slovak University of Technology, Faculty of Materials Science and Technology in Trnava, J. Bottu 25, 917 01 Trnava, Slovakia

E-mail: tihomir.mackic@mf.unibl.org

Abstract. Simple cycloid drive is a type of high-sensitive gear train, by which it is possible to realize high transmission ratios in single-stage. Their advantage is also reflected in the compactness and simplicity of production, as well as achieving greater efficiency compared to conventional planetary drive train. In this paper, the basic equations for obtaining the transmission ratios, torque and efficiency of simple cycloid drive with stepped planets are shown. The dependence between basic transmission ratio and basic efficiency was also examined to find the optimum value of the transmission ratio and determine the point at which the self-locking occurs.

1. Introduction

In many industries, there is a need for using more compact and cheaper mechanical power transmissions. These targets are very difficult to achieve with conventional drive trains with fixed axes, so the need to replace these drive trains with suitable planetary drive trains is imposed.
Very small dimensions of the gearbox and large transmission ratio can be achieved using planetary drive trains with cycloid gear. These gears are now mostly used in the case of single-stage cycloid reducers.

Planetary drive trains with cycloid gears have many advantages over classical planetary transmissions, but their usage is quite limited, mostly in the field of robotics. Due to the complicated and costly construction, in the past, the use of gearboxes with cycloid gears was avoided. However, with the development of the modern CNC machining centers, which enable the simple production of cycloid gears, there is the possibility of cheaper manufacturing and greater usage in commercial power transmissions.

2. Simple Cycloid Drive

In the analysis of a simple cycloid drive, it can be started from an elementary planetary mechanism with internal coupling, replacing, for example, classic involute gears with cycloid gears (figure 1). The cycloid gear (*1p*) meshes with a central ring gear (*1*), which on its periphery has mounted pins and rotates around the central axis. In this case, the cycloid gear rotates around its axis and, together with the member (*S*) - the eccentric shaft, rotates around the axis of the central gear (*1*). The axis of rotation of the planet is at a distance from the central axis for the value *e* (eccentricity). Thus, the cycloid gear (*1p*) is a planet, while the member (*S*) is the carrier of the planet. Based on this, a complete similarity with elementary planetary mechanisms is observed, in which involute gears are used.

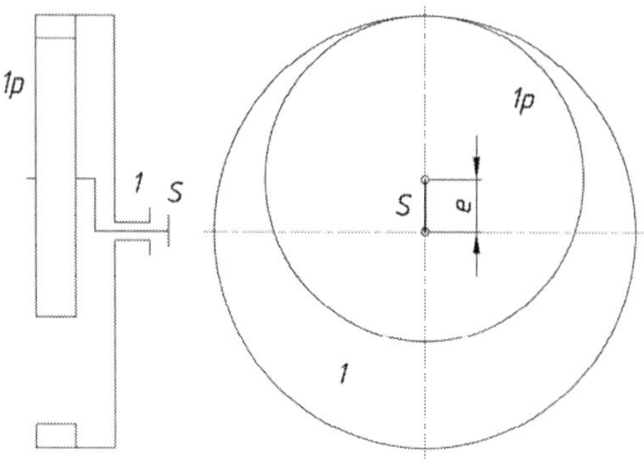

Figure 1. Elementary planetary mechanism.

Members of the planetary mechanism whose axis coincides with the central axis and receive the external torques are called the *basic members* [1]. The members of the elementary mechanism, the central gear (*1*) and the carrier (*S*), whose axes of rotation coincide with the base axis, can't be used in this case for the transfer of energy. In order to achieve this, one elementary member must be added to the elementary planetary mechanism. This gives a single-stage cycloid drive or *simple cycloid drive*. This can be accomplished in two ways:

- Adding a disc (*2*) with pins, this will accept the epicycle movement of the cycloid gear and transfer it to the shaft that coincides with the central axis (figure 2.a). This transfer method is used in today's cycloid reducers, and the disc (*2*) is called the output mechanism [2]. There are the holes on the cylinder, which allow the transfer of the rotation movement from the cycloid gear to the mechanism;
- Adding another central ring gear (*2*) with the pins placed on the periphery, which meshes with the second cycloid gear (*2p*) (figure 2.b). The cycloid gears (*1p*) and (*2p*) are tightly connected in this case. In literature, such planets are called stepped planets [3, 4].

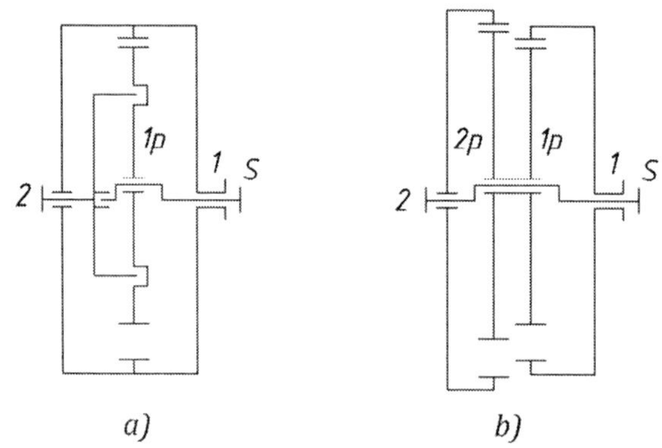

Figure 2. Simple cycloid drives.

In this paper, the transmission characteristics of the second method of obtaining a simple cycloid drive will be displayed. The equations and relations of the basic parameters defining the transmission performance will be presented.

In addition, it will be analyzed only drives with the numbers of pins of ring gear by one greater than the numbers of teeth of cycloid planet gear. For example, the number of pins of the *i*-th ring gear is:

$$z_i = z_{ip} + 1 , \tag{1}$$

where is: z_{ip} – numbers of teeth of *i*-th cycloid planet gear.

The gear drives shown in figure 2 have two degrees of freedom (DOF). By blocking one of the basic members, a *two-shaft cycloid drive* is provided which has only one DOF.

3. Transmission Ratios

If the eccentric shaft is stopped, one can be said that then the cycloid drive works as classical gearbox with a fixed axle, that is, with one degree of freedom. A simple work regime, when a carrier or eccentric shaft is stopped, can be termed as *the basic mode* and before that, define the transmission (speed) ratio, which is called the *basic transmission (speed) ratio*:

$$i_o = i_{12} = \left(\frac{n_1}{n_2}\right)_{n_S=0}, \tag{2}$$

where is: n_1- speed of ring gear shaft *1*,
n_2- speed of ring gear shaft *2*,
n_S- speed of eccentric shaft *S*.

Basic transmission ratio of simple cycloid drive with stepped planets is:

$$i_o = i_{11p} i_{2p2} = \frac{n_1}{n_{1p}}\frac{n_{2p}}{n_2} = \frac{n_1}{n_2} = \frac{z_{1p}}{z_1}\frac{z_2}{z_{2p}} = \frac{z_2(z_1-1)}{z_1(z_2-1)}, \tag{3}$$

where is: $n_{1p} = n_{2p}$ – speed of stepped planets,
$z_{1p} = (z_1 - 1)$ – the number of teeth on a cycloid gear *1*,
$z_{2p} = (z_2 - 1)$ – the number of teeth on a cycloid gear *2*,
i_{11p} – transmission ratio between ring gear *1* and cycloid gear *1*,
i_{2p2} – transmission ratio between cycloid gear *2* and ring gear *2*.

With the aid of the generalized Willis equation [5]:

$$n_1 - i_o n_2 + (i_o - 1) n_S = 0, \tag{4}$$

it is possible to derive the equations for all transmission ratios of two-shaft cycloid drive with stepped gear, as shown in Table 1.

Table 1. Equations of transmission ratios of two-shaft cycloid transmission.

	Transmission ratio $f(i_o)$	Transmission ratio $f(z_1, z_2)$
Minimum multiplication	$i_{12} = i_o$	$i_{12} = \frac{z_2(z_1-1)}{z_1(z_2-1)}$
Minimum reduction	$i_{21} = \frac{1}{i_o}$	$i_{21} = \frac{z_1(z_2-1)}{z_2(z_1-1)}$
Maximum multiplication	$i_{1S} = 1 - i_o$	$i_{1S} = \frac{z_2 - z_1}{z_1(z_2-1)}$
Maximum reduction	$i_{S1} = \frac{1}{1-i_o}$	$i_{S1} = \frac{z_1(z_2-1)}{z_2 - z_1}$
Reversible multiplication	$i_{2S} = 1 - \frac{1}{i_o}$	$i_{2S} = \frac{z_1 - z_2}{z_2(z_1-1)}$

Reversible reduction	$i_{S2} = \dfrac{i_o}{i_o - 1}$	$i_{S2} = \dfrac{z_2(z_1 - 1)}{z_1 - z_2}$

4. Torques

From the equilibrium conditions, during stationary state, the sum of all external torques on basic members shafts is equal to zero, ie: [2, 4]:

$$T_1 + T_2 + T_S = 0, \tag{5}$$

where is: T_1 - torque of ring gear shaft *1*,
T_2 - torque of ring gear shaft *2*,
T_S - torque of eccentric shaft *S*.

The ratio between the input and the output power is expressed through the efficiency of cycloid gear drive. The efficiency of a simple cycloid drive, in the basic mode, is called *the basic efficiency*:

$$\eta_o = \eta_{12} = -\left(\frac{P_2}{P_1}\right)_{n_S=0}, \tag{6}$$

where is: P_1 – power on ring gear shaft *1*,
P_2 – power on ring gear shaft *2*.

Equation (6) can also be written as [4, 6]:

$$\frac{T_2}{T_1} = -i_o \eta_o^w, \tag{7}$$

$$w = \frac{P_{R1}}{|P_{R1}|} = \frac{T_1(n_1 - n_S)}{|T_1(n_1 - n_S)|} = \pm 1, \tag{8}$$

where is: P_{R1} – rolling power which is transmitted by the ring gear shaft *1*.

The torque ratios are referred to as coefficients of transformation of torques or *energy transmission ratios* [71]. Accordingly, in order to simplified analysis, a *basic energy transmission ratio* or *basic energy ratio* can be defined, such as:

$$\bar{i}_o = i_o \eta_o^w. \tag{9}$$

If there are losses in the transmission, then on the basis of expressions (5), (7) and (9), the ratios of the torques can be obtained as:

$$T_1 : T_2 : T_S = 1 : -\bar{i}_o : \bar{i}_o - 1. \tag{10}$$

The simple cycloid drive has a possitive transmission ratio $i_o > 0$, so that is $\bar{i}_o > 0$, and accordingly to the equations (7) and (9) it follows that the torque ratio $(T_2/T_1) < 0$. This means that the torque of shaft *1* and shaft 2 have different signs. Therefore, one of them, which has a higher absolute value, must be a summation shaft.

Basic transmission ratio, given in equation (3), shows that $i_o < 1$ if the condition $z_1 < z_2$ is satisfied, so that: $0 < \bar{i}_o < 1$ if $w = +1$. Since $|T_2/T_1| < 1$, it can be concluded that the shaft *1* is a summation shaft. In case $w = -1$, (the input shaft *2*) it is possible to assume that the following three cases can occur, given the relationship between i_o and η_o:

- $i_o < \eta_o$, shaft *1* is summation shaft;
- $i_o = \eta_o$, locking of the gearbox, because $\bar{i}_o = 0$ and it follows from (10) that $T_1 = T_2$, $T_S = 0$;
- $i_o > \eta_o$, shaft *2* become summation shaft.

Therefore, when calculating losses inside the gearbox, theoretically a locking of the gearbox and change of the summation shaft can occur, but only if the shaft *2* is the input shaft.

In order to understand the nature of the change in transmission ratios, the theoretical model of a simple cycloid drive with stepped planets will be considered, where $z_2 = 20$, while the number of teeth of ring gear *1* changes from $z_1 = 4 \div 40$. In figure 3, the function of the change of the transmission ratios i_{S1} and i_{S2} is shown. If the teeth numbers are equal $z_1 = z_2 = 20$, self-locking can occur (the gearbox acts as a coupling). When $z_1 > z_2$, shaft *2* becomes a summation shaft.

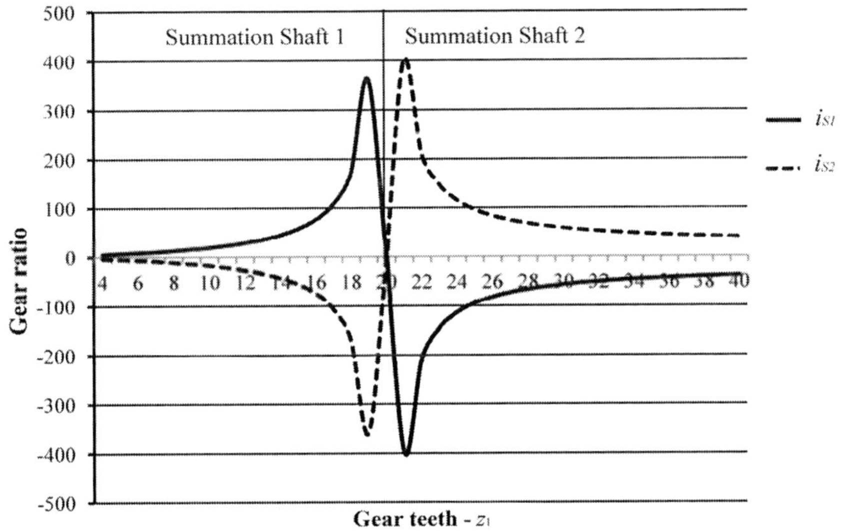

Figure 3. Change of transmission ratios i_{S1} and i_{S2}.

In figure 4, a change in the transmission ratios i_{12}, i_{21}, i_{1S} i i_{2S} is shown. It is obvious that there is no need for gears with a larger number of teeth z_1 (when the shaft 2 becomes a summation shaft), since similar transmission ratios can be achieved in the first period, or when the shaft *1* is a summation shaft. Therefore, in order to analyze losses and self-locking, it is enough to observe the first period of changing in the number of teeth of ring gear *1*, in order to define the relationship between the efficiency and the transmission ratio.

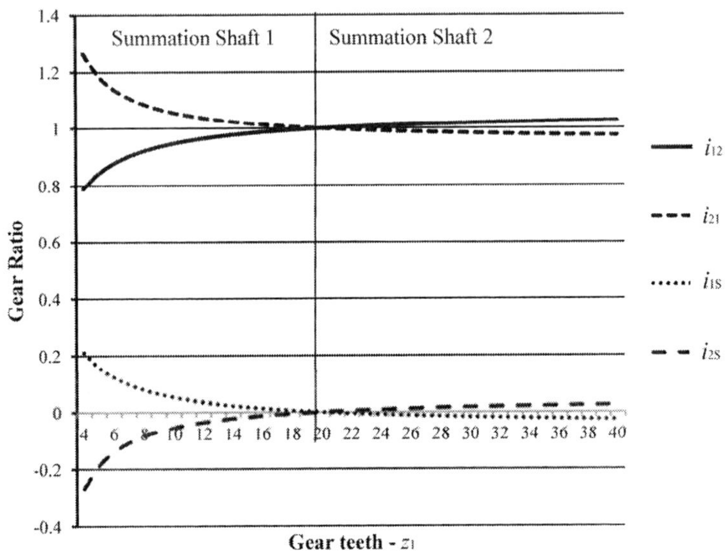

Figure 4. Change of transmission ratios i_{12}, i_{21}, i_{1S} and i_{2S}.

5. Efficiency

The efficiency of two-shaft cycloid drive can be presented as a function of energy ratio and speed ratio. In general case:

$$\eta_{xy} = -\frac{P_y}{P_x} = \frac{T_y}{T_x}\frac{\omega_y}{\omega_x} = \frac{\overline{i_{xy}}}{i_{xy}}, \tag{11}$$

where is: η_{xy} - the efficiency when shaft x is the input and shaft y the output shaft,

$\overline{i_{xy}} = \left(-\frac{T_y}{T_x}\right)$ - energy ratio between shaft x and y,

i_{xy} - speed ratio between shaft x and y.

Thus, in the case that the shaft *1* is the input shaft, and the shaft *S* is the output shaft, the efficiency can be written as:

$$\eta_{1S} = -\frac{P_S}{P_1} = -\frac{T_S}{T_1}\frac{\omega_S}{\omega_1} = \frac{\overline{i_{1S}}}{i_{1S}}, \tag{12}$$

and after introducing the equations from Table 1, as well as (9) and (10), the following is obtained:

$$\eta_{1S} = \frac{\overline{i_o}-1}{i_o-1}. \tag{13}$$

Since the input power is positive, then $T_1 > 0$ and $n_1 > 0$. Shaft *1* is summation shaft and therefore $T_S < 0$. Because the output power is negative, so the speed $n_S > 0$. Accordingly, in order to determine the coefficient „w" from the equation (8), it is necessary to calculate the speed ratio $i_{1S} = n_1/n_S$. From the equation in Table 1, it can be concluded that $i_{1S} < 1$, for basic transmission ratio $0 < i_o < 1$, so that $n_S > n_1$. Therefore, it is clear that the coefficient $w = -1$, because $P_{R1} = T_1(n_1 - n_S) < 0$. Equation (13) can be written as:

$$\eta_{1S} = \frac{i_o\eta_o^{-1}-1}{i_o-1}. \tag{14}$$

Similar analysis can be used to obtain equations for the efficiency of other variants of the two-shaft cycloid drive, as shown in Table 2.

Table 2. The efficiency equations of two-shaft cycloid drive.

	η_{12}	η_{21}	η_{1S}	η_{S1}	η_{2S}	η_{S2}
w	+1	-1	-1	+1	-1	+1
$f(\overline{i_o}, i_o)$	$\frac{\overline{i_o}}{i_o}$	$\frac{i_o}{\overline{i_o}}$	$\frac{\overline{i_o}-1}{i_o-1}$	$\frac{i_o-1}{\overline{i_o}-1}$	$\frac{i_o(1-\overline{i_o})}{\overline{i_o}(1-i_o)}$	$\frac{\overline{i_o}(1-i_o)}{i_o(1-\overline{i_o})}$
$f(i_o, \eta_o)$	η_o	η_o	$\frac{i_o\eta_o^{-1}-1}{i_o-1}$	$\frac{i_o-1}{i_o\eta_o-1}$	$\frac{i_o-\eta_o}{i_o-1}$	$\frac{i_o-1}{i_o-\eta_o^{-1}}$

From Table 2, it can be concluded that in cases where the eccentric shaft is output shaft (η_{1S}, η_{2S}) and for $i_o \geq \eta_o$, the two-shaft cycloid drive is blocked (self-locking occurs), because $\eta_{1S} \leq 0$ and

$\eta_{2S} \leq 0$. In addition, this shows that it is impossible for shaft 2 to become a summation shaft. This is possible only in the case of the basic mode, when shaft 2 is the input shaft.

If a constant value of the basic efficiency of $\eta_o = 99,6\%$ is assumed, a change in the efficiency of the other two-shaft cycloid drive modes can be displayed. In Figure 5, a change in the efficiency for the modes where the shaft 2 is locked can be observed, in the function of changing the number of gear teeth (z_1). The value of the efficiency, in case the shaft *1* is stopped, is slightly different from the change of the efficiency shown in Figure 5. Therefore, only cases when the shaft *2* is locked will be considered. It is obvious that the cycloid drive achieves a higher efficiency when working as a reducer.

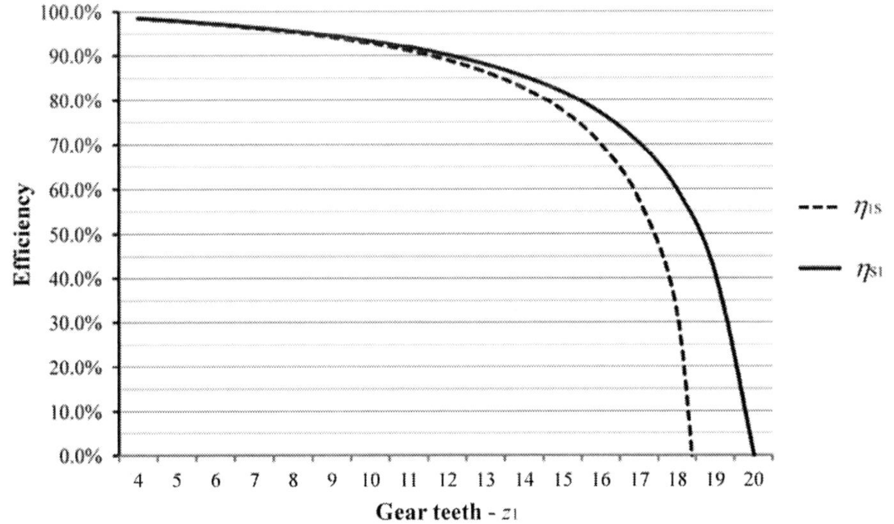

Figure 5. Change in the efficiency for the mode where the shaft 2 is locked.

It is very important to examine the dependence between the transmission ratio and the efficiency, in order to find the optimum value of the transmission ratio and determine the point at which the self-locking occurs. In figure 6, the relationship between the transmission ratio and the efficiency is shown when the shaft *1* is input and the shaft *S* is output shaft. The self-locking occurs when the number of teeth is $z_1 = 19$, since the value of the efficiency is $\eta_{1S} = -44.6\%$.

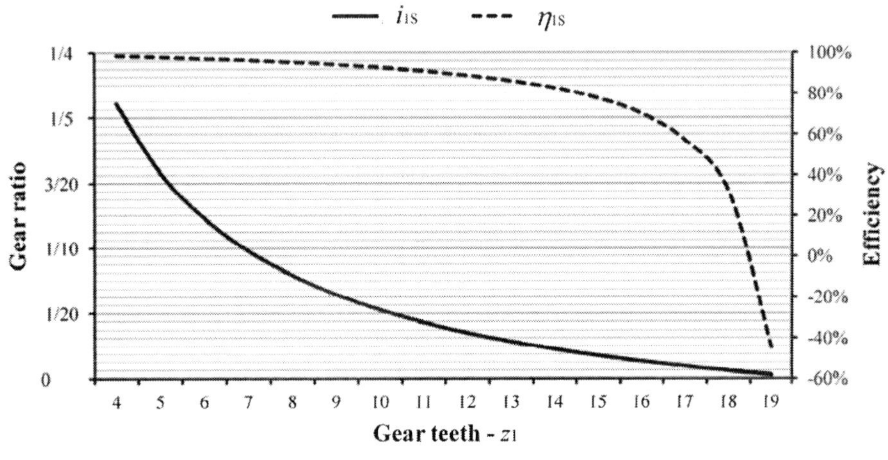

Figure 6. Relationship between the transmission ratio i_{1S} and the efficiency η_{1S}.

In figure 7, the relationship between the transmission ratio and the efficiency is shown, when the shaft *S* is the input and shaft *1* is the output shaft. It is obvious that the increase in the transmission ratio is proportional to the decrease in the efficiency. But even so, it can be concluded that in the particular area of transmission ratios, this kind of simple cycloid drive provides a high degree of efficiency than they are for conventional planetary drives.

For example, Wolfrom two-stage planetary drive, with transmission ratio 97:1, has the efficiency η=45% [8]. A simple cycloid drive, with a slightly higher transmission ratio (z_1=16, z_2=20, i_{S1}=108, η_o=99.6%), has an efficiency of η=70.1%.

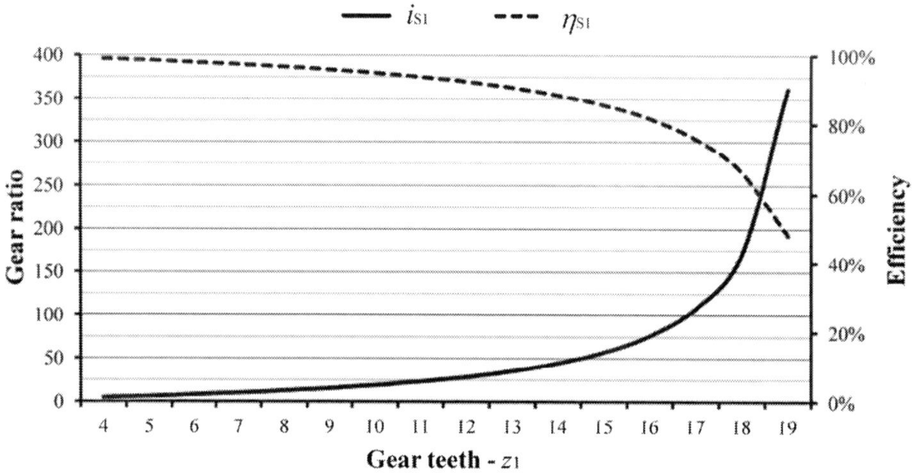

Figure 7. Relationship between the transmission ratio i_{S1} and the efficiency η_{S1}.

6. Conclusion

By theoretical analysis, it can be noticed that a simple cycloid drive with stepped gear can achieve very high transmission ratio, with great efficiency. In addition, their advantage is reflected in the compactness and simplicity of production, as well as achieving greater efficiency compared to conventional planetary drive train. It is recommended to design these gearboxes with a smaller number of teeth, as this results in smaller dimensions of the gearbox while preserving a large transmission ratio.

References

[1] Tanasijević S and Vulić A 1994 *Mehanički prenosnici*, Jugoslovensko društvo za tribologiju, Kragujevac

[2] Blagojević M 2008 *Naponski i deformaciono stanje elemenata cikloreduktora pri dinamičkim opterećenjima*, Mašinski fakultet Kragujevac, Doctoral Thesis

[3] Looman J 1996 *Zahnradgetriebe: Grundlagen, Konstruktion, Anwendung in Fahrzeugen;* Berlin: Springer-Verlag

[4] Müller H W 2001 *Die Umlaufgetriebe;* Berlin: Spinger-Verlag

[5] Willis R *Principles of Mechanism*, London: Parker 1841.

[6] Brandenberger H *Wirkungsgrad und Aufbau einfacher und zusammengesetzter Umlaufradergetriebe*, Maschinenbau 8 (1929), H. 8 u 9

[7] Pantić M 1997 *Gubici snage u ozubljenju kao parametar za formiranje optimalne kinematske šeme planetarnih prenosnika primenjenih u sistemu za prenos snage motornih vozila*, Mašinski fakultet Univerziteta u Beogradu, Doctoral Thesis

[8] Sensinger J 2013 Efficiency of high-sensitivity gear trains, such as cycloid drives, *ASME J Mech Des* **135** 071006

The use of the gas flow model to improve the design of the piston-rings-cylinder system of a diesel engine

G Koszalka[1]

[1] Lublin University of Technology, Faculty of Mechanical Engineering, Nadbystrzycka 38D, 20-618 Lublin, Poland

E-mail: g.koszalka@pollub.pl

Abstract. The article presents the results of simulation research aimed at assessing the impact of selected dimensions of the rings and piston on the ring-pack operation and developing proposals of changes that will improve its performance. The object of the research was a diesel engine for delivery vehicles. A mathematical model of the ring-pack developed by the author of this work was used for the research. The results have shown that with restrictions imposed by the engine manufacturer regarding the scope of modernization, it is possible to significantly reduce or even eliminate the reverse gas flow from the inter-ring space to the combustion chamber, but this must be done at the expense of increased blow-by. The effect of proposed changes on the axial displacement of the rings in the grooves has also been discussed.

1. Introduction

The main task of a ring-pack is to prevent leakage of the working gas from the combustion chamber to the crankcase. These leaks, called blow-by, worsen the performance and efficiency of the engine. However, minimizing the blow-by is not the only function of the piston-rings-cylinder system. In addition to the good sealing of the combustion chamber, this system should provide low resistance to movement (friction in the PRC system could amount up to 50% of mechanical losses of the entire engine [1]) and high durability, and low lubricating oil consumption at the same time. Increased engine oil consumption is undesirable because it requires refilling and increases service costs, but above all increases the emissions of hydrocarbons and particulate matters [2-6].

Improving the design of the system is not easy. Firstly because of the few functions that it performs. It turns out that changes in design that reduce blow-by often increase oil consumption. Measures to reduce oil consumption can lead to increased resistance to movement and wear intensity [4]. This forces the designers to look for compromise solutions. Secondly, because a large number of design parameters decide about the operation of the ring-pack. Due to the large number of these parameters, there are a huge number of possible design variants. Preparing many variants and their empirical testing means a long time and enormous costs of research. These costs are so high because this system can practically be tested only when mounted in an engine and when the engine is running on a dynamometer.

To reduce the time and cost of testing, simulations with the use of mathematical ring-pack models are carried out. Advanced models allow to study the effect of various design parameters on the operation of the ring-pack [7-19].

This paper discusses such simulation research carried out using the mathematical model of the ring-pack developed by the author of this work. These research was part of the wider work aimed at

IOP Conf. Series: Materials Science and Engineering **659** (2019) 012072 doi:10.1088/1757-899X/659/1/012072

meeting the Euro 6 standard by the engine. The engine was produced by an independent manufacturer and was used in delivery vehicles.

2. Objective and assumptions

The purpose of this work was to assess the possibility of improving the performance of the ring-pack of the engine by modifying the selected dimensions of the compression rings and piston.

In setting the assumptions to the simulations, the manufacturer was guided by changes introduced by other large manufacturers of similar engines. There was a tendency to increase some clearances in the piston-rings-cylinder assembly, including a large increase in the joint gap of the second ring, and thus reduce the blow-back (blow-back is a reverse gas flow from inter-ring spaces in direction of the combustion chamber, what happens when pressure in the inter-ring space is higher than in the combustion chamber) and axial displacements of the rings in the grooves at the expense of increased blow-by. Such modifications were done because results of many research indicated that blow-back causes increased oil consumption and emissions of hydrocarbons and particulates [20-21]. The axial displacements of the rings are unfavorable, because in addition to increasing the gas flow and pumping lubrication oil, increase the wear of the side surfaces of the rings and grooves, thus reducing the durability of the engine.

Dimensions, which the manufacturer allowed to change, are the joint gaps of the top and second compression ring $G1$ and $G2$ and the diameters of the piston between the top and second ring $D2u$ and $D2d$ and below the second ring $D3u$ (figure 1). However, these dimensions could only be changed in the direction of increasing the clearances, i.e. the joint gaps could be increased and the piston diameters reduced. Differentiation of diameters in the upper and lower part of the second land was allowed – in the initial piston these diameters were equal, i.e. $D2u = D2d$. The engine manufacturer did not allow the possibility of reducing the clearances for fear of seizing and damaging the engine. Particularly, as it was also planned to increase the maximum power of the engine.

The changes in the above dimensions that would reduce the blow-back and lower the pressure in the inter-ring spaces, and minimize the number of axial displacements of the rings in the grooves were to be sought. Since the modifications could only consist in increasing the clearances, the increase in the blow-by was counted. However, it was expected that this increase would be possibly small.

3. Object and method

3.1. *Engine specification*

The engine was a 4-cylinder, direct injection, turbocharged diesel engine with a displacement of 2.6 dm^3 and compression ratio 17.5, with intercooler and exhaust gas recirculation. The engine had a single overhead camshaft and 2 valves per cylinder and a Common Rail injection system. The engine block was made of cast iron and nominal diameter of cylinders was 94 mm. The stroke of the piston was 95 mm. The aluminum piston had a cast iron insert under the top compression ring. According to technical specification, the maximum power of the engine was 85 kW obtained at 3600 rpm and the maximum torque was 250 Nm in the range of 1800-2200 rpm.

The ring-pack of the engine consists of three rings (figure 1): a keystone, barrel faced compression ring made of heat treated spheroidal graphite cast iron with plasma plating, taper faced compression ring made of heat treated spheroidal graphite cast iron, and a double-beveled, chrome plated oil control ring made of standard grey cast iron equipped with a spiral spring. The nominal dimensions of the rings are presented in table 1.

3.2. *Model of the ring pack*

An integrated model of gas flow through the crevices between the piston, rings and cylinder and ring displacements in the ring grooves was used in this study. In the gas flow model the piston-rings-cylinder seal was treated as a labyrinth seal composed of a series of stages connected by throttling passages. The labyrinth stages are formed by the inter-ring and behind ring spaces, and the throttling passages correspond to the ring-end gaps and the clearances between the side surfaces of the rings and grooves (figure 1). Thermal deformations and positions of the rings in the grooves are taken into account when determining the instantaneous volumes of the stages and cross-sections of the throttling passages. The axial positions of the rings in the grooves are determined allowing for the following forces acting on them: gas pressure, inertia and friction against the cylinder. The gas pressure in individual labyrinth stages are determined applying the principles of preservation of mass and energy and the ideal gas law. The gas flow rates through the individual throttling passages are calculated assuming that the flow is isentropic, including cases of subcritical and critical flows, and taking into consideration the empirical flow coefficients. A detailed description of the model has been presented in previous works [22-23].

Table 1. Nominal dimensions of the rings.

	Height [mm]	Thickness [mm]	Joint gap [mm]
Top ring	3.00	3.90	0.35
Second ring	1.75	3.90	0.35
Oil ring	3.00	4.05	0.35

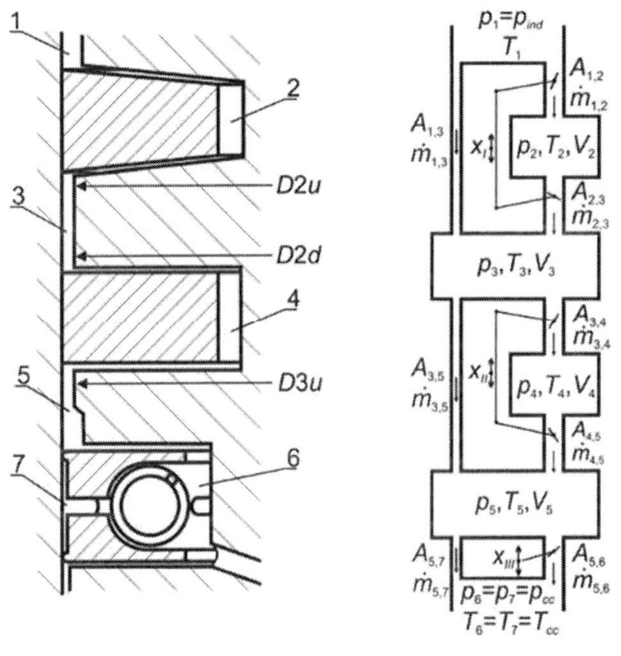

Figure 1. Drawing of the engine ring pack and corresponding orifice-volume representation used in the model (1 – top land, 2 – region behind the top ring, 3 – second land, 4 – region behind the second ring, 5 – third land, 6 and 7 – regions behind and below the oil ring, p_i, T_i and V_i – pressure, temperature and volume of gas in the i-th region, respectively, p_{ind} – pressure in the combustion chamber, p_{cc} and T_{cc} – pressure and temperature in the crankcase, $m_{i,j}$ and $A_{i,j}$ – mass flow rate and area of flow from region i to region j, x_I, x_{II}, x_{III} – axial position of the top, second and oil ring in the groove; *D2u*, *D2d* and *D3u* – examined diameters of the piston: second land under the top ring and over the second ring and third land under the second ring, respectively).

IOP Conf. Series: Materials Science and Engineering **659** (2019) 012072 doi:10.1088/1757-899X/659/1/012072

3.3. Simulations

In the first stage, calculations for the initial, existing version of the piston were made. The dimensions, temperatures and thermal deformations of the parts, as well as the course of pressure in the combustion chamber necessary for the calculations were provided by the engine manufacturer. The pressure in the combustion chamber came from measurements made while the engine was running at rated conditions on the engine dynamometer. The blow-by determined in the calculations was similar to the actual one measured on the engine test stand, so it was assumed that the results of simulations are reliable.

In the next stage, the influence of particular dimensions (*G1*, *G2*, *D2u*, *D2d* and *D3u*) on the operation of the ring-pack, ie: the blow-by and blow-back, pressure in the inter-ring spaces and axial displacement of the rings in the grooves, was examined. At this stage, only one dimension was altered, and the others remain unchanged and equal to the initial ones.

In the last stage, all examined dimensions were simultaneously changed and the best from the point of view of the research goals combination of dimensions of was sought.

4. Results and discussion

4.1. The initial version

The results of the simulations for the initial, not modified version of the ring pack are presented in figures 2-4. Pressure in the second land achieves relatively high value and drops very slowly. This pressure is higher than pressure in the top land for almost half of the working stroke, the entire exhaust and intake strokes and half of the compression stroke (figure 2). This means that gas flows from the second land to the top land for over two thirds of the engine cycle (figure 3). The top ring changes its positon in the groove only once, what is good. However, the disadvantage is that it lifts from the bottom flank of the groove already in the working stroke. The second ring does not change its axial position in the groove at all. The third ring moves up and downward three times during the engine cycle.

Summarizing, the advantages of the existing version of the ring pack is small blow-by and no axial movements of the second ring in the groove. The main drawbacks are high pressure in the second land and very high reverse gas flowrate equal to half of the blow-by (figures 5-6).

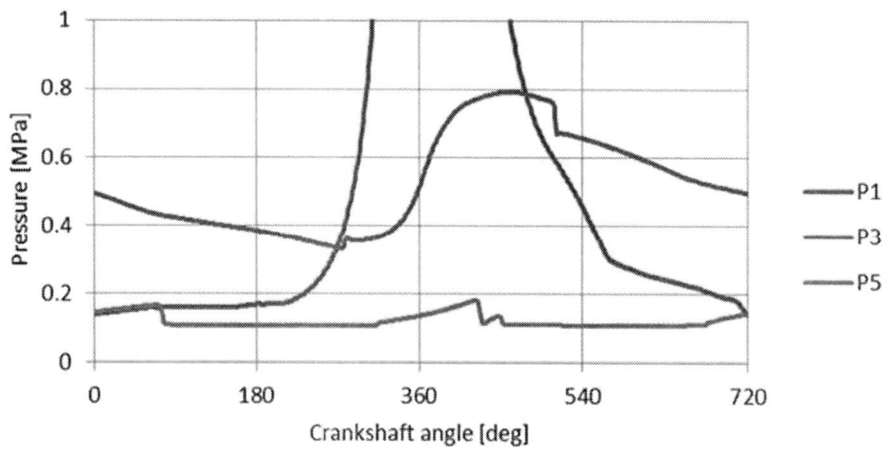

Figure 2. Pressure in inter-ring spaces for the ring pack before modification (denotation as in figure 1).

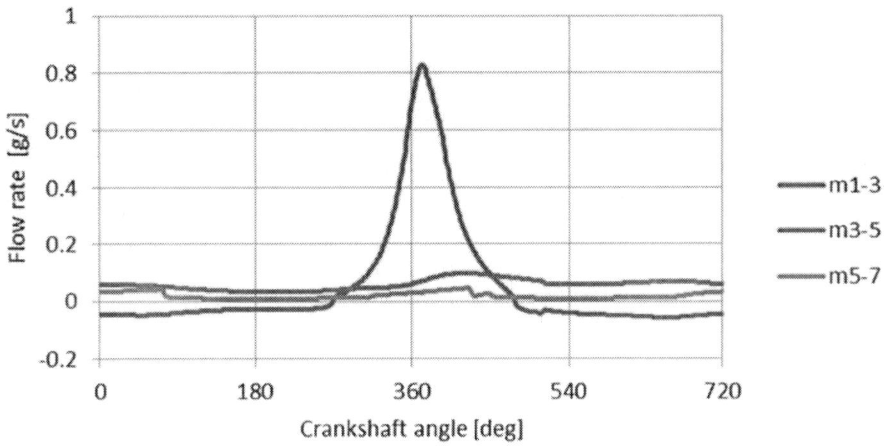

Figure 3. Flowrates of the gas in the ring end gaps for the ring-pack before modification (denotation as in figure 1).

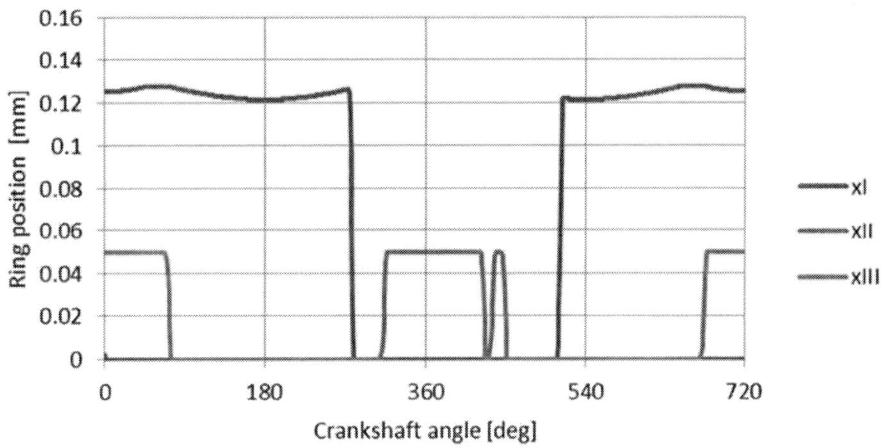

Figure 4. Axial positions of the rings in the grooves for the ring-pack before modification (denotation as in figure 1).

4.2. The influence of individual dimensions

The results of simulations show that the increase in the top ring end gap causes an increase in the blow-by and a particularly big increase in the blow-back (figure 5). That is because the bigger area of flow in the top ring gap results in a bigger pressure rise in the second land during the second half of the compression stroke and the first half of the working stroke (figure 7). This pressure drops much slower than the pressure in the combustion chamber and remains higher than the pressure over the top ring for the rest of the engine cycle forcing more intense gas flow in direction of the combustion chamber. The higher pressure in the second land also causes the bigger gas flow to the crankcase – higher blow-by.

The increase in the second ring end gap results in an increase in the blow-by and decrease in the blow-back (figure 6). The reason of this is lower pressure in the second land. For a big increase in the second ring end gap pressure in this land does not exceed pressure in the combustion chamber and there are no reverse gas flow (figure 7). However, the lower pressure in the second land also affects the displacements of the rings. The second ring changes its axial position in the groove twice and an additional oil ring displacements appear (figure 8).

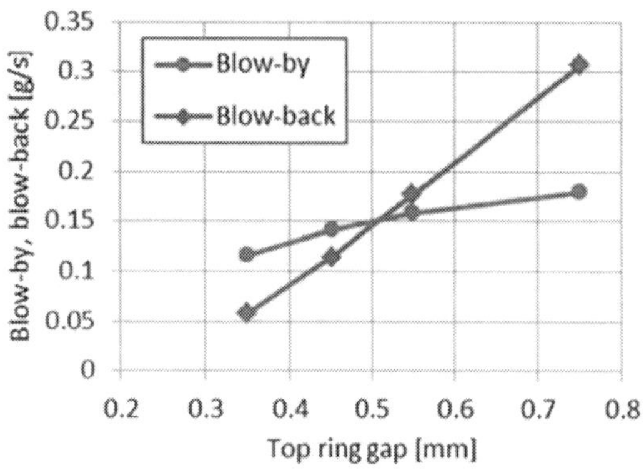

Figure 5. Influence of the top ring gap on the blow-by and blow-back.

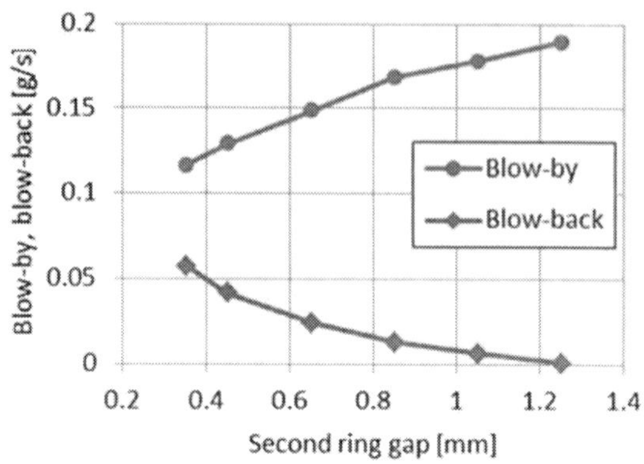

Figure 6. Influence of the second ring gap on the blow-by and blow-back.

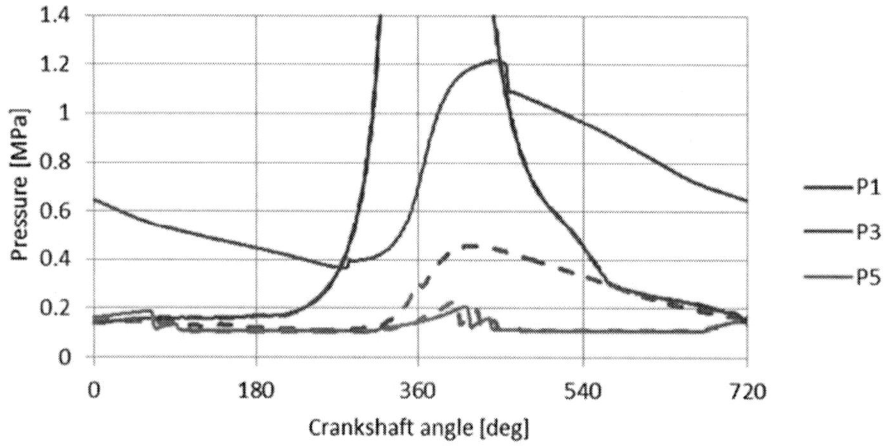

Figure 7. Pressure in inter-ring spaces for the top ring end gap enlarged to 0.55 mm (solid line) and for the second ring end gap enlarged to 1.25 mm (dashed line).

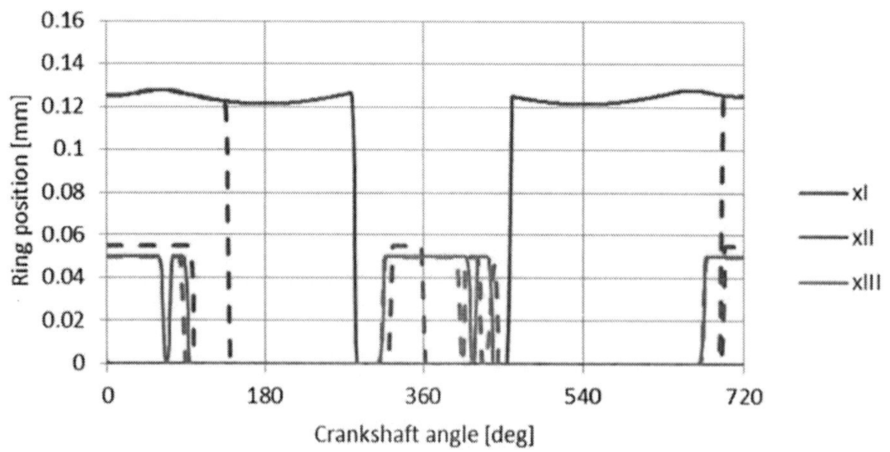

Figure 8. Axial positions of the rings in the grooves for the top ring end gap enlarged to 0.55 mm (solid line) and for the second ring end gap enlarged to 1.25 mm (dashed line).

A reduction of the diameter of the piston below the top ring *D2u* adversely affects both the blow-by and blow-back (figure 9). The results of increasing clearance in this region is similar to the effect of increasing of the top ring end gap. However, the influence on the blow-back is lower, in comparison to increase in the top ring gap, because the bigger clearance means also the bigger volume of the land and smaller increase in the second land pressure.

The influence of the change in diameter of the second land just above the second ring on the blow-by and blow-back is positive, but relatively small (figure 10). The outcome of reducing of the diameter of the piston below the second ring is favorable to blow-back, however unfavorable to blow-by (figure 11).

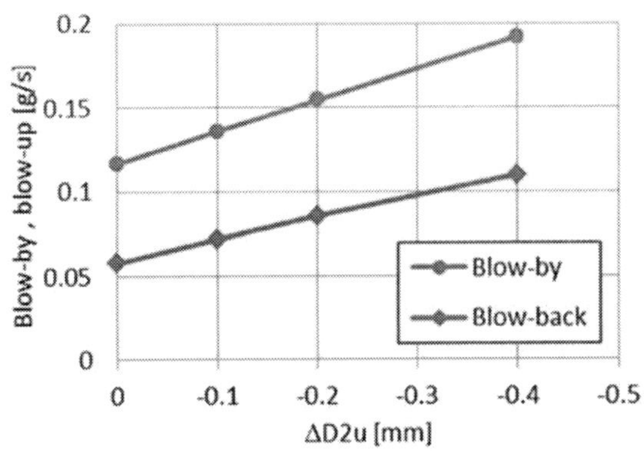

Figure 9. Influence of the change in the diameter of the upper part of the second land on the blow-by and blow-back.

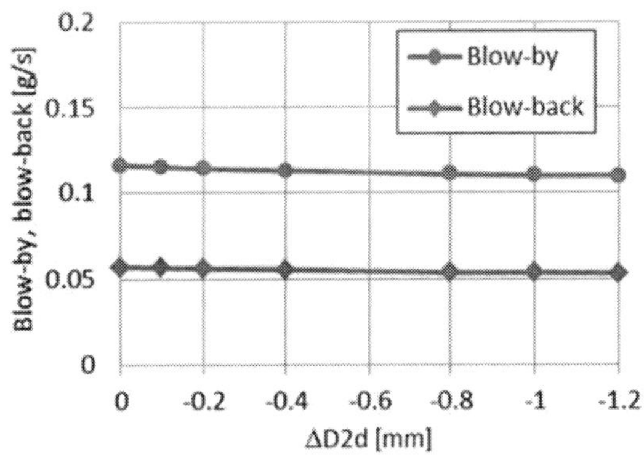

Figure 10. Influence of the change in the diameter of the lower part of the second land on the blow-by and blow-back.

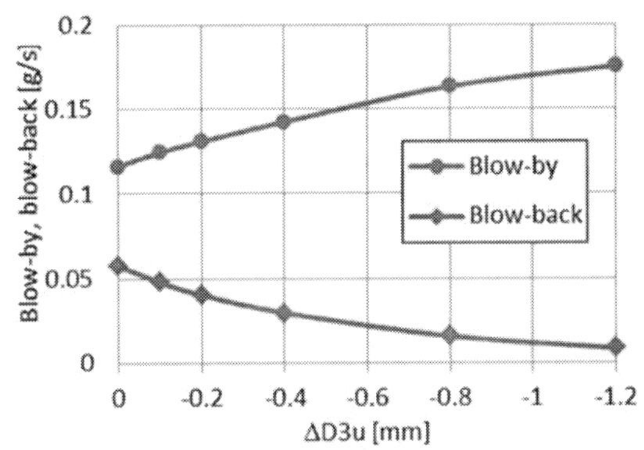

Figure 11. Influence of the change in the diameter of the upper part of the third land on the blow-by and blow-back.

4.3. The influence of simultaneous changes of several dimensions

The results of simulations revealed that the increase in the top ring gap *G*1 and the reduction of the diameter of the upper part of the second land *D2u* regardless of the other examined dimensions always increase the blow-by and blow-back. Therefore, it was decided that *G*1 and *D2u* should not be modified and only three other dimensions were changed in further studies. The impact of different combinations of the three dimensions on the blow-by, blow-back and displacements of the rings were studied. Selected results are shown in figure 12. Generally it can be stated that modifications of the examined dimensions in the assumed range do not allow for significant reduction of the blow-back without the increase in the blow-by. What is more, a given reduction in the blow-back is associated with a similar increase in the blow-by regardless of the combination of dimensions used. However, despite similar blow-by and blow-back, differences in pressure courses and ring displacements can be significant for different combinations of dimensions.

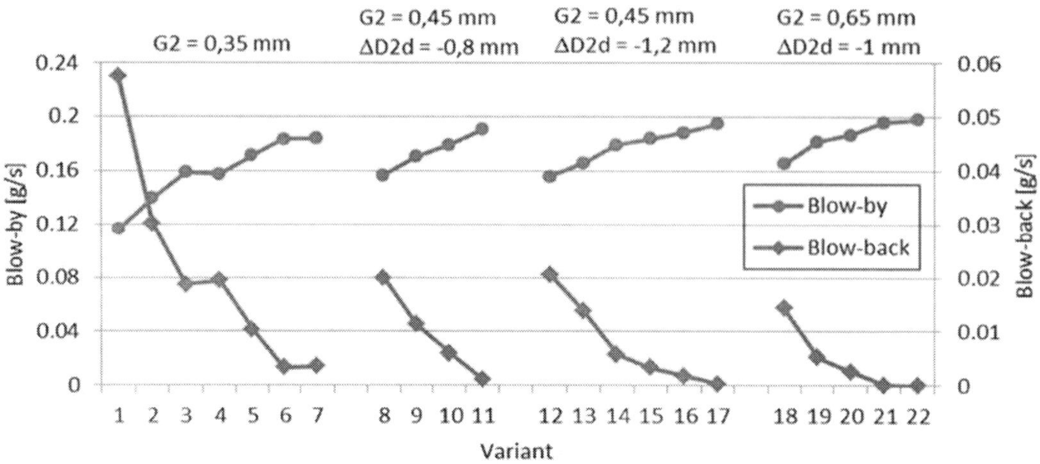

Figure 12. Influence of different combination of dimensions on the blow-by and blow-back (detailed dimensions for particular variant are listed in table 2).

Table 2. Second ring gaps *G2*, changes in second land diameter over the second groove $\Delta D2d$ and changes in third land diameter under the second ring groove $\Delta D3u$ used in simulations presented in figure 12.

Variant	1	2	3	4	5	6	7	8	9	10	11
G2 [mm]	0.35	0.35	0.35	0.35	0.35	0.35	0.35	0.45	0.45	0.45	0.45
DD2d [mm]	0	-0.8	-0.8	-1.2	-1.2	-1.2	-1.6	-0.8	-0.8	-0.8	-0.8
DD3u [mm]	0	-0.4	-0.8	-0.8	-1.2	-1.6	-1.6	-0.4	-0.6	-0.8	-1.2

Variant	12	13	14	15	16	17	18	19	20	21	22
G2 [mm]	0.45	0.45	0.45	0.45	0.45	0.45	0.65	0.65	0.65	0.65	0.65
DD2d [mm]	-1.2	-1.2	-1.2	-1.2	-1.2	-1.2	-1.0	-1.0	-1.0	-1.0	-1.0
DD3u [mm]	-0.4	-0.6	-0.8	-1.0	-1.2	-1.6	-0.2	-0.3	-0.4	-0.8	-1.2

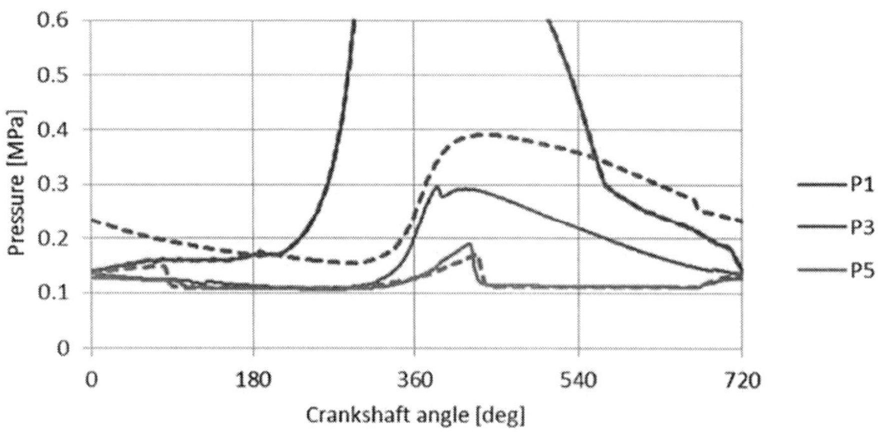

Figure 13. Pressure in inter-ring spaces for the selected variant ensuring elimination of the blow-back: $G2 = 0.6$ mm, $\Delta D2d = -1.0$ mm and $\Delta D3u = -1.0$ mm (solid line) and for the selected variant preventing movements of the second ring: $G2 = 0.45$ mm, $\Delta D2d = -1.0$ mm and $\Delta D3u = -1.0$ mm (dashed line).

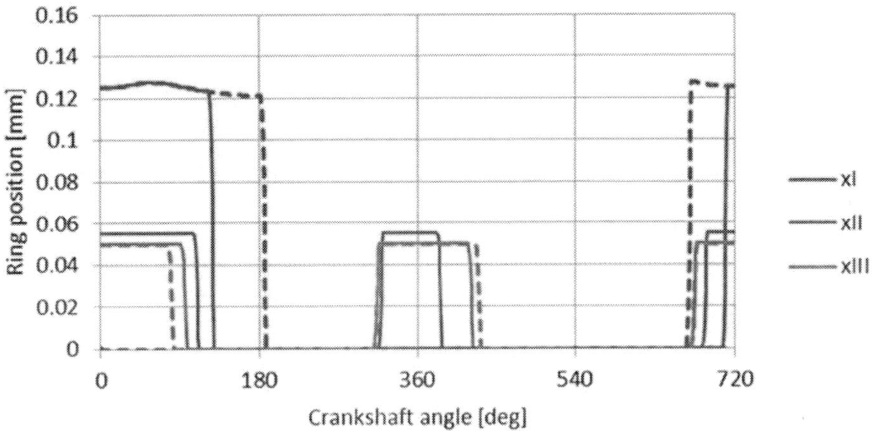

Figure 14. Axial position of the rings in the grooves for the selected variant ensuring elimination of the blow-back: $G2 = 0.6$ mm, $\Delta D2d = -1.0$ mm and $\Delta D3u = -1.0$ mm (solid line) and for the selected variant preventing movements of the second ring: $G2 = 0.45$ mm, $\Delta D2d = -1.0$ mm and $\Delta D3u = -1.0$ mm (dashed line).

The following changes: $G2 = 0.6$ mm, $\Delta D2d = -1.0$ mm and $\Delta D3u = -1.0$ mm were considered the most advantageous variant allowing for the complete elimination of the reverse flow (figure 13). Unfortunately, elimination of the blow-back is not possible without two displacements of the second ring (figure 14). For other variants, which eliminate the blow-back either the blow-by is higher or there are more displacements of the rings.

A combination of dimensions that would allow a possibly large reduction in the reverse flow and a small increase in blow-by, but at the same time would not cause the second ring to move in the groove was also sought. The best solution that fulfills this condition was considered to be: $G2 = 0.45$ mm, $\Delta D2d = -1.0$ mm and $\Delta D3u = -1.0$ mm (figures 13 and 14).

5. Conclusions

The paper presents the results of research aimed at assessing the possibility of reducing the reverse flow of gases from the inter-ring spaces towards the combustion chamber by increasing the compression rings end gaps and reducing the diameters of the ring part of the piston. An advanced mathematical model of the ring-pack was used for this assessment. This work was a part of a wider work aimed at reducing the emission of toxic components from the diesel engine used in delivery vehicles.

The results of the simulation showed that the top ring end gap should not be increased and the diameter of the shelf under this ring should not be reduced because it leads to a very big increase in the blow-by and blow-back.

Increasing the second ring end gap or reducing the diameter of the piston under this ring leads to increased blow-by, but at the same time causes a large reduction in the blow-back.

Increasing the diameter of the piston over the second ring slightly reduces the blow-by and blow-back, however in combination with changes in other dimensions it can significantly affect the pressure in inter-ring spaces and axial displacement of the rings in the grooves.

As a result of the work, two solutions were proposed: the first one, which allows to completely eliminate the reverse gas flow from the inter-ring space towards the combustion chamber. In this solution, the top ring moves axially in the groove once and the second and oil ring two times during one engine cycle. And the second solution, which ensures no displacement of the second ring and only two displacements of the oil ring (the top ring moves once in all examined versions), and a very large reduction of blow-back – 3.5 times in relation to the initial version. In both cases, the benefits in blow-back come at the cost of increased blow-by, respectively by 65% and 40%.

Summarizing, the application of the model allows to analyze the impact of many design variants on the operation of the ring-pack and to select one or more variants that will be manufactured and tested. This allows for a very large shortening and a huge reduction in the costs of an engine tests.

Acknowledgment

The research was financed in the framework of the project Lublin University of Technology – Regional Excellence Initiative, funded by the Polish Ministry of Science and Higher Education (contract no. 030/RID/2018/19).

References

[1] Richardson D E 2000 Review of power cylinder friction for diesel engines, *Journal of Engineering for Gas Turbines and Power* **122**(4) 506–519
[2] Thirouard B and Tian T 2003 Oil transport in the piston ring pack, *SAE Technical Paper* 2003-01-1952
[3] Yilmaz E, Tian T, Wong V and Heywood J 2004 The contribution of different oil consumption sources to total oil consumption in a spark ignition engine, *SAE Technical Paper* 2004-01-2909
[4] Ito A, Tsuchihashi K and Nakamura M 2011 A study on the mechanism of engine oil consumption, *SAE Technical Paper* 2011-01-1402
[5] Oliva A and Held S 2016 Numerical multiphase simulation and validation of the flow in the piston ring pack of an internal combustion engine, *Tribology International* **101** 98–109
[6] Duda K, Wierzbicki S, Smieja M and Mikulski M 2018 Comparison of performance and emissions of a CRDI diesel engine fuelled with biodiesel of different origin Fuel 212 202–222
[7] Tian T, Noordzij L B, Wong V W and Heywood J B 1998 Modeling piston-ring dynamics, blowby and ring-twist effects, *Journal of Engineering for Gas Turbines and Power* **120**(4) 843–854
[8] Koszalka G and Niewczas A 2006 The influence of compression ring clearances on the blowby in a diesel engine, *SAE Technical Paper* 2006-01-3356
[9] Koszalka G 2011 Predicting the durability of the piston-rings-cylinder assembly of a diesel engine using a piston ring pack model, *Eksploatacja i Niezawodnosc – Maintenance and Reliability* **51**(3) 40–44
[10] Rahmani R, Theodossiades S, Rahnejat H and Fitzsimons B 2012 Transient elastohydrodynamic lubrication of rough new or worn piston compression ring conjunction with an out-of-round cylinder bore, *Proc. IMechE Part J: Journal of Engineering Tribology* **226**(4) 284–305
[11] Wolff A 2014 Simulation Based Study of the System Piston-Ring-Cylinder of a Marine Two-Stroke Engine, *Tribology Transactions* **57**(4) 653-667
[12] Cheng C, Schock H and Richardson D 2015 The dynamics of second ring flutter and collapse in modern diesel engines, *Journal of Engineering for Gas Turbines and Power* **137**(11) 111504
[13] Lyubarskyy P and Bartel D 2016 2D CFD-model of the piston assembly in a diesel engine for the analysis of piston ring dynamics, mass transport and friction, *Tribology International* **104** 352–368
[14] Usman A and Park C W 2017 Numerical investigation of tribological performance in mixed lubrication of textured piston ring-liner conjunction with a non-circular cylinder bore, *Tribology International* **105** 148–157
[15] Koszalka G and Suchecki A 2017 Analysis of design parameters of pistons and piston rings of a combustion engine, *MATEC Web of Conferences* **118**,00013
[16] Wolff A 2018 Numerical analysis of gas flow through the labyrinth seal of piston rings of an automotive IC engine, *IOP Conf. Series: Materials Science and Engineering* **421**(4) 042082
[17] Selmani E, Delprete C and Bisha A 2019 Cylinder liner deformation orders and efficiency of a piston ring-pack, *E3S Web of Conferences* **95** 04001

[18] Delprete C, Bisha A and Selmani E 2019 Gas escape from combustion chamber to crankcase, analysis of a set of parameters affecting the blow by, *Lecture Notes in Networks and Systems* **42** 413–422
[19] Wolff A 2019 Influence of piston ring profiles and oil temperature distribution on cylinder liner lubrication of a marine two-stroke engine, *Combustion Engines* **178**(3) 257–263
[20] Burnett P J 1992 Relationship between oil consumption, deposit formation and piston ring motion for single-cylinder diesel engines, *SAE Technical Paper* 920089
[21] Petris De C, Giglio V and Police G 1996 A mathematical model for the calculation of blow-by flow and oil consumption depending on ring pack dynamic, *SAE Technical Paper* 961216
[22] Koszalka G 2010 Application of the piston-rings-cylinder kit model in the evaluation of operational changes in blow-by flow rate, *Eksploatacja i Niezawodnosc – Maintenance and Reliability* **48**(4) 72–81
[23] Koszalka G and Guzik M 2014 Mathematical model of piston ring sealing in combustion engine, *Polish Maritime Research* **21**(4) 66-78

Impact of a driving belt length on a device noise

P Zvolensky[1], M Benko[2], L Lestinsky[1] and J Dungel[2]

[1]University of Zilina, Faculty of Mechanical Engineering, Univerzitna 1, 01026 Zilina, Slovakia
[2]CEIT, a.s., Univerzitna 6A, 01008 Zilina, Slovakia

E-mail: peter.zvolensky@fstroj.uniza.sk

Abstract. The paper is concerned with design of high-rev planetary transmission and its overall noise. The transmission is able to reach high gear ratio, and provide very high revolutions per minute. This attributes can be achieved because of unconventionally mounted planet gears in two axis. Planet gears on the outside axis are in contact with ring gear and transfer torque to planet gears in the inner axis. Inner axis planet gears transfer torque to a sun gear. The paper defines an experiment of transmission design with possibility of changing number of planetary gears. Number of planetary gears has an impact on overall noise of transmission. So it is possible to spec the transmission to fulfil of producer requirements. Aim for the experiment was to find trade-off between number of planetary gears, maximum torque, gear design and an impact on overall noise. Next, the paper presents test stand for the transmission. The test stand design allows definition of transmission construction units, without affecting noise of the transmission. Experiment also includes drive belt testing, and possibility of BLDC electric motor usage.

1. Introduction

This article concerns about noise of a high revving planetary transmission and its correct setup in a test bench.

Producers requirements for the transmission are to cope with high gear ratio and high rpm. Its satellite gears are installed in two rows, which insures high gear ratio.

In an experiment, we were searching for compromise between parameters like acoustic power, number of satellite gears and maximum torque. When number of satellite gears was reduced, acoustic power was reduced also, along with torque limit. A lot of combinations was made with those changes. Very important part of the experiment was to test those combinations. There was need to find way, how to test those combinations, without affecting transmission acoustic power negatively [1, 2, 3].

Figure 1. Planetary transmission.

2. Test bench

For the experiment was necessary to design a test bench. The designed test bench consists of frame with a BLDC electric motor and the planetary transmission installed. Torque from the engine was delivered to the transmission via driving belt. Next, we were investigating the impact of the driving belt length on acoustic power level. Two lengths of the driving belt were used. We used two levels of rpms and the predefined place to measure acoustic power change.

Figure 2. Test bench.

3. Impact of the driving belt on the device noise

Before experiment, there was an anticipation, that shorter driving belt will be quieter. On the longer driving belt more vibrations have occurred. Amplitude of vibration may be source for increased acoustic power.

After all measurements were done, we compared results. No significant differences were measured between short and long driving belt, which could lead significant increase of overall noise.

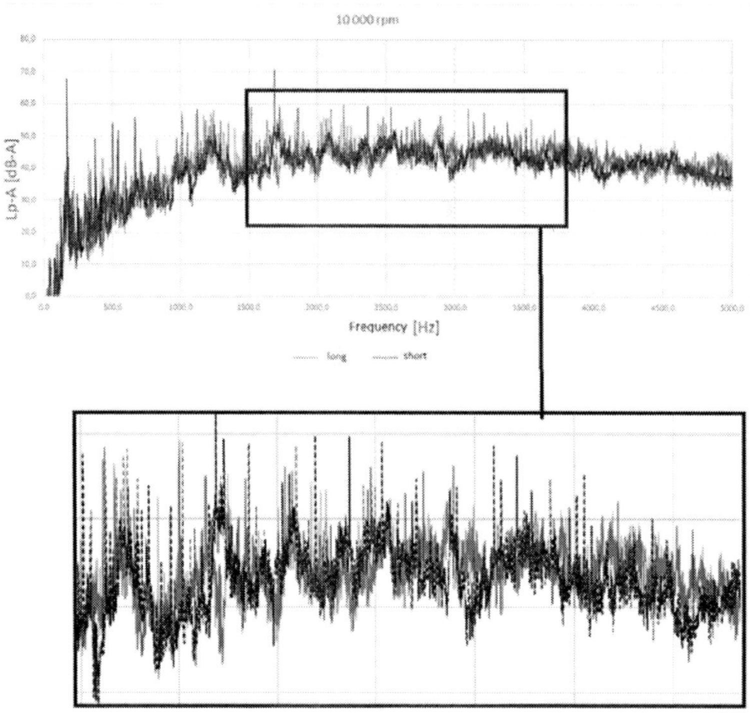

Figure 3. Comparison of long and short driving belt at 10 000 rpm.

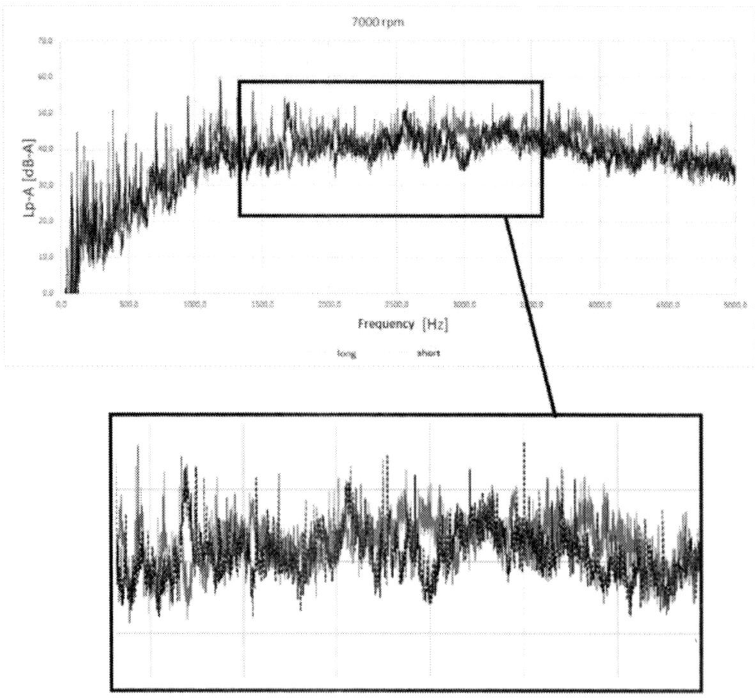

Figure 4. Comparison of long and short driving belt at 7000 rpm.

Experiments were made at 7000 and 10 000 rpm. With using longer driving belt was recorder lower peak figures, but it did not have impact on overall noise because of difference just 0.17 dB.

4. Conclusion

Length of the driving does not have any significant impact on overall noise of the device. Anticipation of the experiment was not confirmed. So this device is prepared for further noise testing of gear sets and other components.

Acknowledgment

This work was created by the implementation of the project "Low Cost Logistic System Based on Mobile Robotic Platforms for Industrial Use", ITMS: 26220220205 supported by the Operational Programme Research and Development and by the European Fund for Regional Development.

References

[1] Ciesielka W 2007 Active noise reduction system, *Archives of Acoustics* **32** 0137-5075
[2] Ciesielka W Gołaś 2006 An adaptive, active noise reduction system in closed space, *Archives of Acoustics* **31** 0137-5075
[3] Fiebig W 2014 Influence of the Inter Teeth Volumes on the Noise Generation in External Gear Pumps, *Archives of Acoustics* **39** 0137-5075

IOP Conf. Series: Materials Science and Engineering **659** (2019) 012074 doi:10.1088/1757-899X/659/1/012074

Price impact on acoustic comfort of a washing machine

P Zvolensky[1], L Lestinsky[1] and J Dungel[2]

[1]University of Zilina, Faculty of Mechanical Engineering, Univerzitna 1, 01026 Zilina, Slovakia
[2]CEIT, a.s., Univerzitna 6A, 01008 Zilina, Slovakia

E-mail: peter.zvolensky@fstroj.uniza.sk

Abstract. The article is focused on price impact on acoustic comfort of a washing machine. In the article is comparison of various washing machines price maximum rpm and acoustic power. Next article is focused on experiments how unbalance in a washing machine tub affect maximum rpm and how maximum rpm affect overall noise of various washing machines.

1. Introduction

Producers and customers are watching whole range of technical specs, when comparing appliances. Some of the technical specs are on energy label. This article is focused on price impact on the washing machine parameters like maximum rpm, noise during spin cycle. A comparison created in this paper is useful for producers to choose price range of their products and customers to decide, if they are willing to pay for more premium products. For comparison was chosen parameters that are most obvious for washing machines, and customers are mostly interested in it (price, maximum rpm, noise).
Maximum rpm of spin cycle is relatively important spec, because with higher rpm more water from laundry is drained in shorter time. With shorter spin cycle, less amount of energy is used. Another important spec is a washing machine acoustic power. The washing machine acoustic power depends on maximum rpm. With raising rpm raises acoustic power also. However, washing machines which are able to achieve high rpm with reduced noise, are improving acoustic comfort for customer. Those washing machines have some price-increasing technological solutions. Customer is searching for price-quality compromise constantly [3].

2. Price impact on acoustic comfort of a washing machine

Washing machines can be divided into two biggest categories: top loader and front loader type washing machines. In this comparison was used 28 washing machines from 21 producers in each category. In the comparison was used washing machines with capacity between 4 and 9 kg. For reference was used average market price of washing machines.

IOP Conf. Series: Materials Science and Engineering **659** (2019) 012074 doi:10.1088/1757-899X/659/1/012074

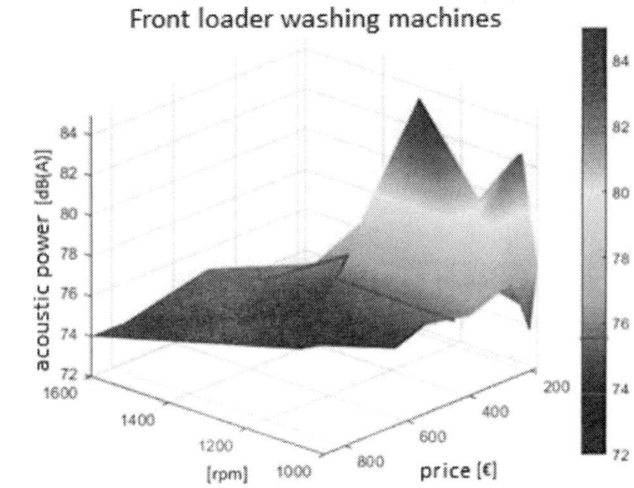

Figure 1. Front loader washing machines.

In the figure 1 is 3D graph of front loader washing machines. The graph is showing dependence between price, maximum rpm and spin cycle noise. The most expensive washing machines are achieving the highest rpm and the lowest amount of noise. In the figure 1, we can see that washing machine in the price range around 400 € are also achieving good results with acoustic power less than 75 dB(A). Washing machines in the price range around 800 € are achieving acoustic power less by 1.5 dB(A).

Optimal ratio between noise, price and maximum rpm was chosen from acoustic power median of front loader washing machines used in comparison (75,5 dB(A)). Optimal ratio is marked with red line in the 3D graph. With using this optimum ratio is possible to choose appliance requirements like price and other specs (in this case acoustic power).

Such as comparisons are very important for producers, because they can determine if they are competitive with an appliance in its price category.

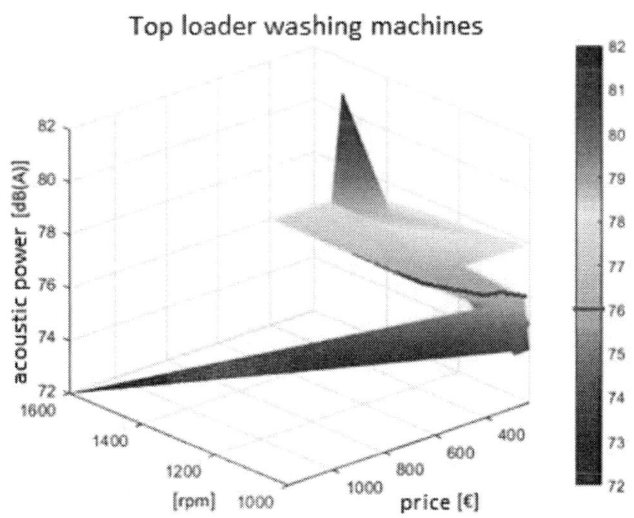

Figure 2. Top loader washing machines.

In the figure 2 is 3D graph of top loader washing machines. Again, the graph is showing comparison between price, noise and maximum rpm. In this case, situation differs little bit than with front loader washing machines. Price difference is significantly bigger. Median of the acoustic power during spin cycle is 76 dB(A). Optimal ratio is in range between 300 and 400 €. The best washing machines in the optimal range are able to achieve 1300 rpm. In this case the most expensive washing machine is able to achieve 1600 rpm. Undoubtedly, the most expensive washing machine is achieving the best parameters. The noisiest washing machine is able to achieve 1400 rpm which is good figure in the 350 € range. For customers that silent washing machine is not a must, this washing machine could be good choice.

Optimal range of the front loader washing machine has got wider choice than the top loader optimal range. However, optimal range of the top loader washing machine is in lower price range.

3. Maximum rpm during spin cycle

Maximum rpm during spin cycle has got big impact on noise increase. After wash cycle, in some occasions washing machine won't be able to achieve maximum rpm because of unbalance in the washing machine tub. It depends on quality of a washing machine, how it can handle unbalance. Premium washing machines attempt to balance the tub at first. If the attempt is not successful, they reduce maximum rpm during spin cycle. Cheaper washing machines does not try to balance the tub; they just reduce maximum rpm. Detection of unbalance is very important, if washing machines would not reduce maximum rpm, noise increase and damage may occur. During experiment with the washing machines, a different spin cycles were recorded which consisted of multiple phases. The laundry distribution in the tub is random and different unbalance values occur at each stop and re-spin.

Table 1. Unbalance vs maximum rpm comparison.

Maximum rpm		**1440 rpm**	
Load	**0 kg**	**3 kg**	**6 kg**
Unbalance 0 kg	1360	1362	1360
Unbalance 0.3 kg	1360	1362	1364
Unbalance 0.5 kg	1050	1050	1364
Unbalance 0.8 kg	820	820	1050
Unbalance 1 kg	0	0	820

Tab 1 shows how is unbalance affecting the maximum spinning speed of a particular washing machine. Unbalance was created with rubber ballast of different mass. An unbalance of up to 0.3 kg has a minimal effect on maximum speed because this unbalance is low compared to the gross weight of the washing machine tub and does not alter the spin cycle. The difference appears from an unbalanced weight of about 0.5 kg, in which the washing machine attempts to distribute laundry in the tub. If the attempt is not successful, reduction of rpm occurs. The washing machine was able to trigger spin cycle up to an unbalance of 1 kg, when the tub was unbalanced completely and did not start the spin cycle because of safety reasons. With an evenly loaded drum weighing 6 kg, the machine began to reduce the speed up to 0.5 kg. when the drum weighed 6 kg, unbalance of 1 kg had a smaller effect on the speed reduction, the machine was able to start spinning up to 820 rpm.

4. Unbalance impact on noise

With increasing unbalance is raising noise also. Unbalance in the tub is created with uneven position of laundry during washing cycles. The biggest impact of the unbalance is achieved during final spin cycle. In this cycle Washing machines must deal with unbalance in the tub. Depending on the price range, washing machines have different capability to handle unbalance in the tub. Regardless to its ability to handle unbalance, usually it's not possible to complete elimination of noise increase with increased unbalance.

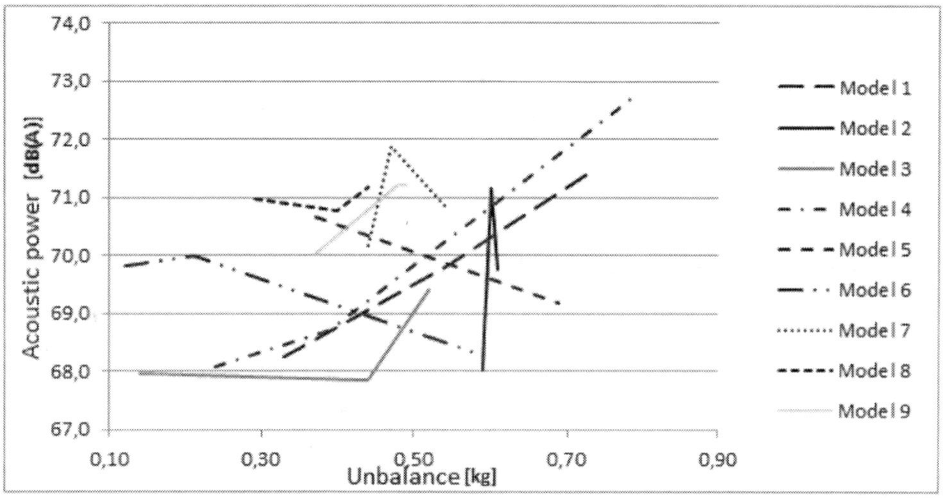

Figure 3. Unbalance impact on overall noise.

In figure 3 is graph, which shows how unbalance in the tub influences acoustic power. Results were achieved with experiments performed on 9 different models of washing machines. As figure shows, with increasing unbalance is increasing acoustic noise. However, some of the experiments unexpectedly resulted in different way (with increase unbalance was noise reduced). Potential reason could be, that position of the unbalance creating ballast located in the tub could change vibration frequency. Statistically, most common location of unbalance is on a front side of the tub. If unbalance location (ballast during experiments) is in the middle or rear side of the tub, it can change results and acoustic power can be lower.

5. Conclusion

Price of washing machines has significant impact on their acoustic comfort. However, some of the average priced washing machines can achieve decent noise reduction, but they are usually missing other functions or equipment in comparison with premium models.

In comparison of washing machines was determined that with unbalance above 0.3 kg reduction of maximum rpm during spin cycle may occur. With reduction of maximum rpm, noise and load is also reduced. Even different models from same producer have got different unbalance detection if different electric motor is used.

Acknowledgment

This work was created by the implementation of the project "Low Cost Logistic System Based on Mobile Robotic Platforms for Industrial Use", ITMS: 26220220205 supported by the Operational Programme Research and Development and by the European Fund for Regional Development.

References

[1] Ciesielka W 2007 Active noise reduction system, *Archives of Acoustics* **32** 0137-5075
[2] Ciesielka W Gołaś 2006 An adaptive, active noise reduction system in closed space *Archives of Acoustics* **31** 0137-5075
[3] STN EN 60704-1 2010: Acoustic. Electric appliances for home and similar usage. Skúšobný predpis na stanovenie hluku prenášaného vzduchom. Part 1: General requirements

Toward a sustainable mobility: A solar vehicle for a new quality of life

G Minak[1], M Lukovic[2], S Maglio[3] and S Kojic[4]

[1] InterDept. Centre for Industrial Research on Advanced Mechanics and Materials, Alma Mater Studiorum University of Bologna, Forlì, Italy
[2] Industrial Design Department, University of Arts Belgrade, Belgrade, Serbia
[3] Onda Solare Association, Castel San Pietro (BO), Italy
[4] Polytechnic School Kragujevac, Kragujevac, Serbia

E-mail: marko-design@hotmail.com

Abstract. The vehicular mobility causes 15% of greenhouse gases emission: one million tons of carbon anhydrite per hour. In addition, it produces CO, NOx, fine powders, carcinogenic and mutagenic elements: these substances will disappear in the presence of solar vehicles. And solar mobility would also mitigate indirect effects: fuel used to transport fuel, energy for the distillation of hydrocarbons, gas leaks, even fracking, explosions, rivers and oceans. In contrast, electric and hybrid vehicles do not allow this improvement in the quality of life. In almost all modern countries, the energy mix is strongly unbalanced towards fossil fuels: massive electrification would not make mobility sustainable, but rather risks worsening its effect on the environment by shifting the problem of emissions from cities to power plants. The Sun, indeed, can guarantee long-term sustainable mobility: for every circulating solar vehicle CO_2 production is really zero. From July to today, our solar racing car has travelled 3000 km, avoiding to emit half a ton of CO_2: reporting these data to a conventional use, each solar vehicle would avoid the release of 1.5 tons of CO_2 per year: like planting 10 large trees for each month in our garden. This study describes how to transform a solar super-car into an ordinary vehicle for urban and everyday mobility.

1. Introduction

The expression "quality of life" has been used since 1980 with reference to signs indicative of urban decay and criminality, and therefore of diminishing quality of life.

In the years after 2000 the concept of quality of life was often compared to the concept of sustainable development, especially in relation to the growing awareness of the limited nature of energy reserves related to oil and coal and to the negative effects of many modern technologies on the natural environment and resources.

Public opinion is beginning to wonder if all technological advances are always convenient, or if, on the other hand, long-term damage will nullify its immediate advantages, and even threaten the quality of life of future generations.

In these terms, another interesting expression more and more related to the modern idea of quality of life is the "Seventh Generation Standard". It proposes that the government should take into consideration the effect its decisions will have in the long term, up to the seventh generation, that is about a century later.

In the United States, for instance, the Seventh Generation Amendment was proposed to be included in the United States Constitution. In Canada, a similar amendment was proposed, called the Canada

Well-Being Measurement Act. This strategy seems very difficult to put into practice because predicting the future is never easy. At the same, often these laws have significant impacts in social, economic and industrial terms. But the essence of these legislative proposals lies in the importance they give to sustainable development and to forecasts of damage to the environment, an irreplaceable heritage of all peoples.

A strategic sector in which an intervention in such direction would be necessary for sure and, maybe, not-postponable, is represented by the personal mobility.

According to the European Environmental Agency (EEA), the vehicular mobility causes 15% of greenhouse gases emission in Europe, roughly equivalent, in proportion, to 106 t/h of CO2 worldwide. In addition, the fuel combustion produces further elements as CO, NOx, fine powders and other carcinogenic or mutagenic elements.

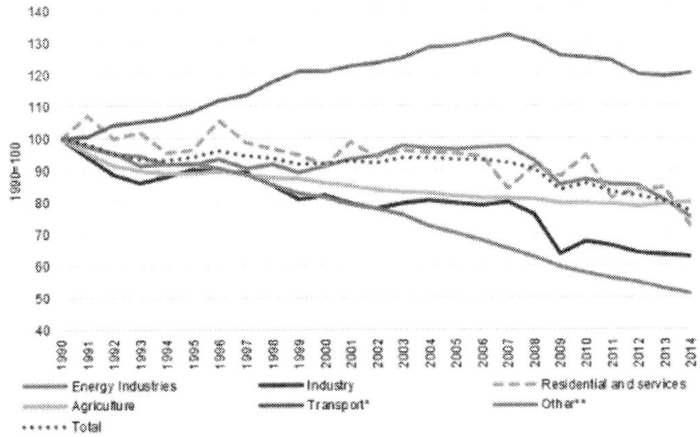

Figure 1. Greenhouse gases emission trends in Europe since 1990 (sources: EEA).

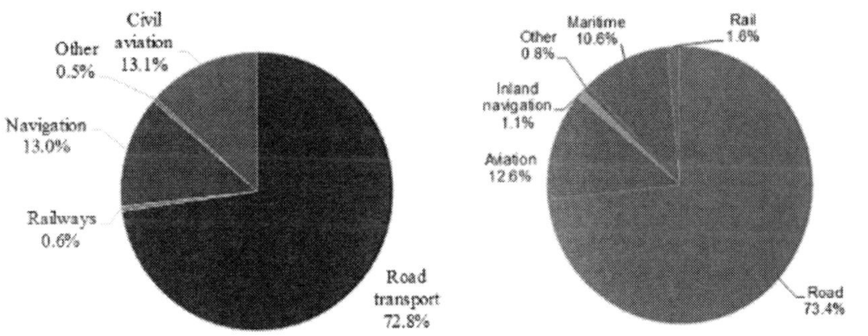

Figure 2. Greenhouse gas emissions (left) from transport by mode, compared with the transport energy demand (right) (sources: EEA, year 2014).

Finally, the vehicular transport is the main cause of air pollution in cities and it has not seen the same gradual decline in emissions as other sectors (Figs. 1 and 2).

At the same time, referring to the most common indicators used by national and international agencies in assessing the quality of life, the mobility surely represents a distinctive element, beyond the air pollution and environmental protection themselves, able to influence fundamental aspects as:

- Car ownership rate
- Urban congestion rate
- Accessibility to a parking lot
- Noise pollution
- Road accidents

- Subjective perception of security
- Injury rate outside work
- Quality of urban cleaning

together with a wide range of economic factors, since the relevance of the automotive industrial sectors and market segments in wealth creation and technology development.

In accordance with the EU strategy for a low-emission mobility, by mid-century greenhouse gas emissions from transport will need to be at least 60% lower than in1990 and be firmly on the path towards zero.

This is only a part of the overall strategy aiming to assure long terms benefits for European citizens and consumers by delivering improvements in air quality, reductions in noise levels, lower congestion levels and improved safety. Practically speaking, both national and supranational authorities are now looking for solutions able to guarantee less energy consumption and/or the use of alternative fuels, as a way for a long-term sustainable mobility.

It also means that the general interests are deeply focused on the developing technical answers aiming at:

- Increasing the efficiency of the transport system;
- Encouraging the shift to lower emission transport modes;
- Speeding up the deployment of low-emission alternative energy for transport;
- Investigating the potentialities of advanced biofuels, electricity, hydrogen and renewable synthetic fuels;
- Removing obstacles to the electrification of transport.

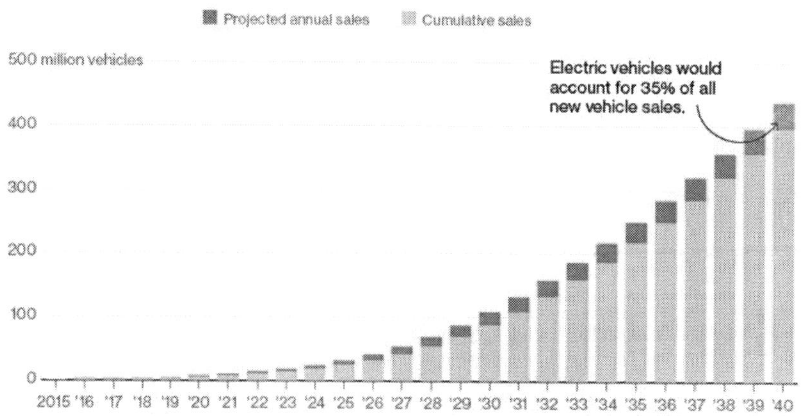

Figure 3. Expansion trend of electric vehicles (sources: Bloomber New Energy Finance).

But, a strategy moving towards zero-emission vehicles may turn out to be a chimera, if not properly defined, as repeatedly emphasized.

Hydrogen used as a fuel represents, i.e., an interesting solution, but still technologically not ready and full of risks. Furthermore, the production of hydrogen, usually through water electrolysis, is a process that requires a lot of water and a lot of energy, representing an option only combined with nuclear energy. Otherwise, planned benefits will miss on the environment or the economy.

The same contest can be assumed for the electrical vehicles, a market segment in quick expansion. Electric powertrain is an effective and modern mobility solution and all the major car manufacturers are directed in that direction. Toyota, for instance, has made hybrid engines the main development line, while Tesla is all about powerful electric motors. Compared to traditional vehicles, electric vehicles offer greater energy efficiency with increasingly competitive performances (as speed and acceleration).

On the contrary, many people think through the electric mobility nothing more than an alibi respect to a real sustainable mobility.

IOP Conf. Series: Materials Science and Engineering **659** (2019) 012075 doi:10.1088/1757-899X/659/1/012075

Undoubtedly this technology reduces to zero the gas emissions into the environment in our cities. But it also acts moving the emissions to our suburbs, where the thermal and coal power plants are usually located. Then, the powertrain electrification can certainly characterize the future mobility in our planet, but it is far from demonstrating, globally, a zero impact at the moment.

2. Solar vehicles

A restricted, but emerging technology is represented by the solar vehicles. A solar car consists of an electrical powertrain supplied by photovoltaic panels: the vehicle is able to extract all the energy it needs from the Sun and this energy is directly used through electric motors that transform it into motion, without passing by a combustion.

Therefore, all emissions characterizing combustion engine vehicles will disappear in the presence of solar vehicles. And this exclusivity remains true regardless of the particular fuel with which the comparison is proposed: petrol, diesel, methane, alcohol and so on, but only solar vehicles are able to guarantee a real zero-emission policy, for real, both at local and global levels.

Thus, the solar mobility seems to represent the only strategy able to mitigate the indirect effects: fuel used to transport fuel, energy for the distillation of hydrocarbons, gas leaks, even fracking, explosions, rivers and oceans. It also acts reducing the need of a massive electrification and the risk related to the fact that currently all countries present an energy mix strongly unbalanced towards fossil fuels.

Unfortunately the attention of carmakers for solar mobility seems quite far away in terms of large investment and production. Thus, at the moment, the largest part of solar vehicles crossing the roads were or are prototypes. Several hundreds of them are represented by racing single-seaters, designed with the sole purpose of racing at maximum speed possible in typical solar racing environments.

Figure 4. Single-seat solar car, made by the University of Bochum.

Made inside the top universities of the world (as Delft, Michigan, Tokyo, Sydney between others), these advanced prototypes represent an important step in demonstrating the future perspectives of the solar mobility. The real leap in their quality is offered by the recent development of another category of solar vehicles: the multi-occupants. In short, these vehicles intend to represent a middle ground from the light racing single-seaters and electric minicars, already available on the market.
At the same time, they also present technical features, such as energy efficiency and battery autonomy, capable of making them particularly attractive for a future potential commercialization.

Figure 5. Multi-occupant solar car, made by the University of Bologna.

For instance, the solar car, recently designed and developed by the University of Bologna for the American Solar Challenge 2018, won this 2700 km competition travelling with four occupants at an average speed over 60 km/h. It is characterized by advanced design, light materials and efficient energy solutions. This development has been largely reported in several investigations [1-13].

3. Sustainable vehicles

Next step in the direction of a sustainable mobility would be the transformation of a performing prototype like this in an attractive car, suitable for the modern transport needs.

This goal must involve the design and implementation of a different and new:

- vehicle profile, able to reduce the overall dimensions;
- location for batteries increasing space and habitability;
- mechanics, more suited for a conventional driving;
- luggage compartment for the transport of goods;
- cabin interior for a better driving experience.

With very light carbon fiber reinforced structures, high-performance monocrystalline silicon panels, ultra-modern lithium batteries, motors incorporated directly in the wheels, this new vehicle will have all the design and construction features to ensure the highest performance in terms of efficiency and use of energy (as for the race prototype).

In addition, an aluminum roll cage, anti-intrusion systems, numerical crash tests and many other expedients will ensure the best target in terms of personal protection, a very critical aspect for the city-cars.

At the same time, the design changes will permit to largely improve the vehicle features toward a better drivability, usability and livability.

4. Conclusion

Only the Sun is able to guarantee a long-term sustainable mobility: for every circulating solar vehicle CO_2 production is really zero. Never benefits are not limited to this, involving multiple aspects of the quality of life: elimination of carcinogenic substances, noise reduction, economic savings and so on.

Consequently, the development of solutions capable of making solar mobility increasingly valid and efficient should represent a goal of primary interest. At this time, solar vehicles are mainly seen as a curious attraction, but their performances should attract much more than this. Our solar racing car, for instance, has travelled 3000 km with without emission. Reporting these data to a conventional mobility, each solar vehicle would avoid the release of 1.5 tons of CO_2 per year.

This results can be obtained, according to the authors' opinion, transforming the solar extra-efficient prototypes, currently used inside solar races, in something more similar to the ordinary light vehicles, suitable for the urban and everyday mobility.

Acknowledgment

This research is funded by the Italian Ministry of Foreign Affairs and International Cooperation thought the Joint Research Projects as Particular Relevance, with a project named 'Two Seats for a Solar Car' within the Executive Programme of Cooperation between Italy and Serbia in the field of Science and Technology.

References

[1] Betancur E, Fragassa C Coy J, Hincapie S and Osorio G 2017 Aerodynamic effects of manufacturing tolerances on a solar car, *Smart Innovation, Systems and Technologies*, **68** 868-876

[2] Minak G, Fragassa C and De Carmago F V 2017 A Brief Review on Determinant Aspects in Energy Efficient Solar Car Design and Manufacturing, *Smart Innovation, Systems and Technologies* **68** 847-856

[3] De Camargo FV, Pavlovic A and Fragassa C 2018 *Crash Safety and the Principles for an*

Impact Numerical Simulation in Composite Materials, AIP Conference Proceedings **1981**, no. 020032, doi:10.1063/1.5045894

[4] Boria S, Pavlovic A, Fragassa C and Santulli C 2016 Modeling of Falling Weight Impact Behavior of Hybrid Basalt/Flax Vinylester Composites, *Procedia Engineering* **167**, 223–230

[5] De Camargo FV, Fragassa C, Pavlovic A and Martignani M 2017 Analysis of the Suspension Design Evolution in Solar Cars, *FME Transactions* **45**(3) 394-404

[6] Fragassa C, de Camargo, FV and Giorgini L 2018 *Quality Assessment of Carbon Fiber Automotive Parts when Using Polyurethane Foam as Pattern Material,* AIP Conference Proceedings, no. 020088; doi:10.1063/1.5045950

[7] Fragassa C, de Camargo, FV, Pavlovic A and Minak G 2019 Explicit Numerical Modeling Assessment of a Basalt Reinforced Composite For Low-Velocity Impact, *Composites Part B: Engineering* **163** 522-535. DOI: 10.1016/j.compositesb.2019.01.013

[8] Fragassa C, Minak G and Sassatelli M 2019 Reducing Defects in Composite Monocoque Frames, *FME Transactions* **47**(1) 48-53, DOI: 10.5937/fmet1901048F

[9] Fragassa C, Pavlovic A and Santulli C 2018 Mechanical and impact characterisation of flax and basalt fibre bio-vinylester composites and their hybrids, *Composites Part B: Engineering* **137** 247-259. doi: 10.1016/j.compositesb.2017.01.004

[10] Fragassa C, Pavlovic A and Massimo S 2014 Using a Total Quality Strategy in a new Practical Approach for Improving the Product Reliability in Automotive Industry, *International Journal for Quality Research* **8**(3) 297–310

[11] Martini A, Bellani G and Fragassa C 2018 Numerical assessment of a new hydro-pneumatic suspension system for motorcycles, *International Journal of Automotive and Mechanical Engineering*, **15**(2) 5308-5325

[12] Minak G, Brugo T M, Fragassa C, Pavlovic A, Zavatta, N and De Camargo F V 2019 Structural Design and Manufacturing of a Cruiser Class Solar Vehicle, *Journal of Visual Experiments* **143** no. e58525, doi:10.3791/58525

[13] Pavlovic A and Fragassa C 2015 General considerations on regulations and safety requirements for quadricycles, *International Journal for Quality Research* **9**(4) 657–674

Passenger car steering pull and drift reduction considering suspension tolerances

M De Rosa[1], A De Felice[2], C Fragassa[3] and S Sorrentino[2]

[1] Product Development, Maserati SpA, Via Emilia Ovest 911, 41123 Modena, Italy
[2] University of Modena and Reggio Emilia, Department of Engineering Enzo Ferrari, Via Vivarelli 10, 41125 Modena, Italy
[3] University of Bologna, Department of Industrial Engineering, Viale Risorgimento 2, Bologna, Italy

E-mail: silvio.sorrentino@unimore.it

Abstract. Pull to side, Steering wheel misalignment and Drift leeward are handling anomalies affecting passenger cars travelling on straight paths. They represent a cost factor in the automotive industry, occurring in small but not negligible percentage of the overall car production. The present contribution is aimed at understanding which are their causes in terms of suspension tolerances, and at reducing their extent, in the specific case of vehicles with front Double wishbone suspension and rear Five arms suspension. The most influential parameters on handling irregularities have been identified through a design of experiment analysis. Then, after performing a sensitivity analysis by multibody virtual modelling, a correlation with experimental data from presetting and wheel aligner benches has provided sufficient information for setting tolerance thresholds, able to keep the handling anomalies under study within acceptable bounds. Pull to side and Steering wheel misalignment have been found to be mainly related to the set up phase of wheel angles (front camber angles influencing Pull to side, front and rear toe angles influencing Steering wheel misalignment), while the main cause of Drift leeward has been identified with (rear) ride steer.

1. Introduction

Among handling anomalies affecting passenger cars travelling on a straight path, the Pull to side (PTS, lateral deviations from straight trajectory with steering wheel correctly aligned), the Steering wheel misalignment (SWM, straight trajectory with steering wheel not aligned) and the Drift leeward (DL, lateral deviations due to the effects of uneven ground surface on the rear suspensions) are particularly relevant. Even though they do not constitute safety problems, however they represent a cost factor in the automotive industry, occurring in small but not negligible percentage of the overall car production.

Several studies are devoted to investigate the contribution of tyre characteristics on Pull to side and Steering wheel misalignment (also referred to as drift). Among them, [1] presents steering pull as an effect of asymmetrical tyre cornering; [2] describes a method for measuring residual aligning torque, as a predictor of tyre contribution to steering pull; tyre–induced pull and drift problems are respectively discussed also in [3–5], with special emphasis on plysteer, and [6–7]; while in [8] a specific analytical model is developed, aimed at studying the effects of tyre asymmetric characteristics during straight–ahead running.

On the other hand, the literature aimed at investigating the contribution of tolerances of suspension systems on Pull to side and Steering wheel misalignment is less extensive. The analysis presented in [9] is focused on a simulation–based dimensional tolerance optimization process, to minimize vehicle pulls by reduction of dimensional variation in a front suspension system. A computational method is proposed in [10] to evaluate the impact of dimensional variations on vehicle steering pull, in the specific case of front Mac Pherson suspension and rear Twist bar suspension. In [11] an analytical vehicle model and a full–vehicle model are considered, for studying the effects of suspension geometry on steering pull and drift (i.e. Steering wheel misalignment). A robust design optimization method to reduce steering pull phenomenon is presented in [12], considering tolerance of suspension system elements, such as hard points, spring, damper and bushings. Clearly, these issues also have an impact on the design of control systems, such as active front steering control [13], as well as on steering performance of particular classes of vehicles, like autonomous drive vehicles [14] and ultra–light cars [15-17].

The present contribution is focused on understanding which are the causes and on reducing the extents of PTS, SWM and DL as handling anomalies due to suspension component and assembly tolerances, in the specific case of vehicles with front Double wishbone suspension and rear Five arms suspension [18]. The vehicle assembly process is first analyzed in terms of errors and tolerances by means of a design of experiment analysis (DOE), aimed at identifying which parameters generate PTS, SWM and DL. A sensitivity analysis by multibody virtual modelling is then performed, aimed at understanding how and at which level the relevant suspension parameters can affect the straight path handling of the vehicles under study. Finally, the results of simulations are correlated with experimental data from presetting and wheel aligner benches, for setting tolerance thresholds able to keep the anomalies under study within acceptable bounds.

2. Design of experiment analysis of the assembly process

Two vehicle models have been considered, with suspension systems of the same typology. A design of experiment (DOE) analysis has been performed aimed at understanding the role of tolerances on specific parameters that can affect handling behaviour. More in detail, displacements of the suspension hardpoints with respect to their ideal position (in a chassis reference system) have been estimated as functions of combined tolerances due to both component tolerances and coupling tolerances.

A two–wheel steering vehicle travels on a straight path if the average of slip angles at the front axle is equal to that at the rear axle. Since slip angles differ from toe angles because of tyre deformation [19], in this study the crab angle (δ_r) of an axle is considered, representing the contribution to the axle slip angle given by the toe angles of the two wheels (say, δ_1 and δ_2):

$$\delta_r = (\delta_1 - \delta_2)/2 \quad (1)$$

where δ_r takes the sign of δ_1. While on the rear axle the toe angles are imposed with respect to the mechanical symmetry axis (i.e. the axis passing by the front axle centre and the rear axle centre, which in general is not contained in the vehicle symmetry plane), the front toe angles are imposed with respect to the (rear) crab axis (defined by the rear toe angles), in order to reduce the differences between the front and rear axle slip angles.

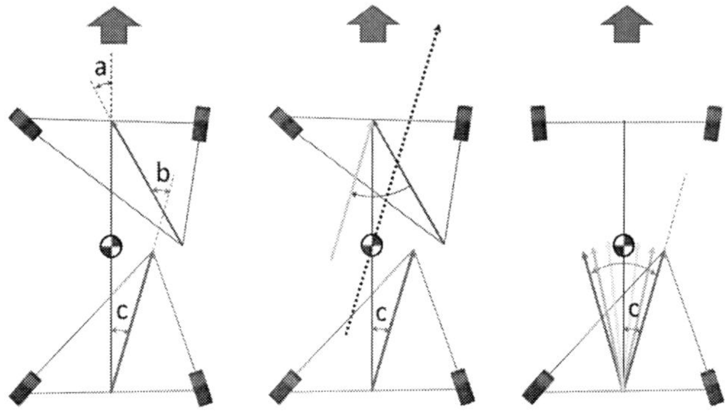

Figure 1. Crab axis (left: a = front crab angle; b = total crab angle; c = rear crab angle), SWM effect (centre) and DL effect (right).

Clearly, PTS, SWM and DL are strongly related to the axle slip angles. PTS can occur only if the front and rear axle slip angles are not equal. SWM depends by different front and rear axle slip angles as well; in this case, however, the vehicle reacts to lateral forces due to the tyres by modifying the position of the front wheels until the front axle slip angle coincides with the rear one; consequently, the steering wheel changes its position, producing misalignment. DL can be evaluated as a dynamic variation of the rear axle slip angle, due to a vertical parallel excitation of the axle (figure 1).

The front and rear axle slip angles are also influenced by axle positioning on the vehicle's body during the mounting phase, and by different loads acting on the wheels of each axle. Therefore, vehicle's handling is affected by non–symmetric values (on the same axle) of toe angles, camber angles, lateral displacement of wheel centre, vertical loads acting on wheel, ride steer behaviour in vertical travel.

At the front axle the vehicles mount Double wishbone suspensions, as shown in figure 2 (left). During the assembly process, the anomalies that can be generated are: non–symmetric king pin axes; non–symmetric caster, camber and toe angles (the last two anomalies are generally adjusted in the final phase of the assembly process); non–symmetric lateral and/or angular (around a vertical axis) positioning of subframe with respect to chassis (compromising symmetry of the suspension mechanisms at the two sides of the same axle); different ratios between spring travel and wheel travel.

A DOE analysis has been performed on front suspension. The influence on suspension parameters of hardpoint displacements due to worst combination of tolerances are represented by histograms as those displayed in figure 2 (including numbered hardpoints, with displacements expressed in a chassis XYZ reference system; in the histograms, the percentages are defined with respect to the total variation of the parameter under study: ride steer or ride camber). Among the most influential parameters, ride steer (Fig. 3, top) is affected mainly by the Z (vertical) displacements of hardpoints 3 (lower control arm) and 2 (steering box), while the largest variation on ride camber is given by the Z displacements of hardpoints 7 and 8 (upper control arm).

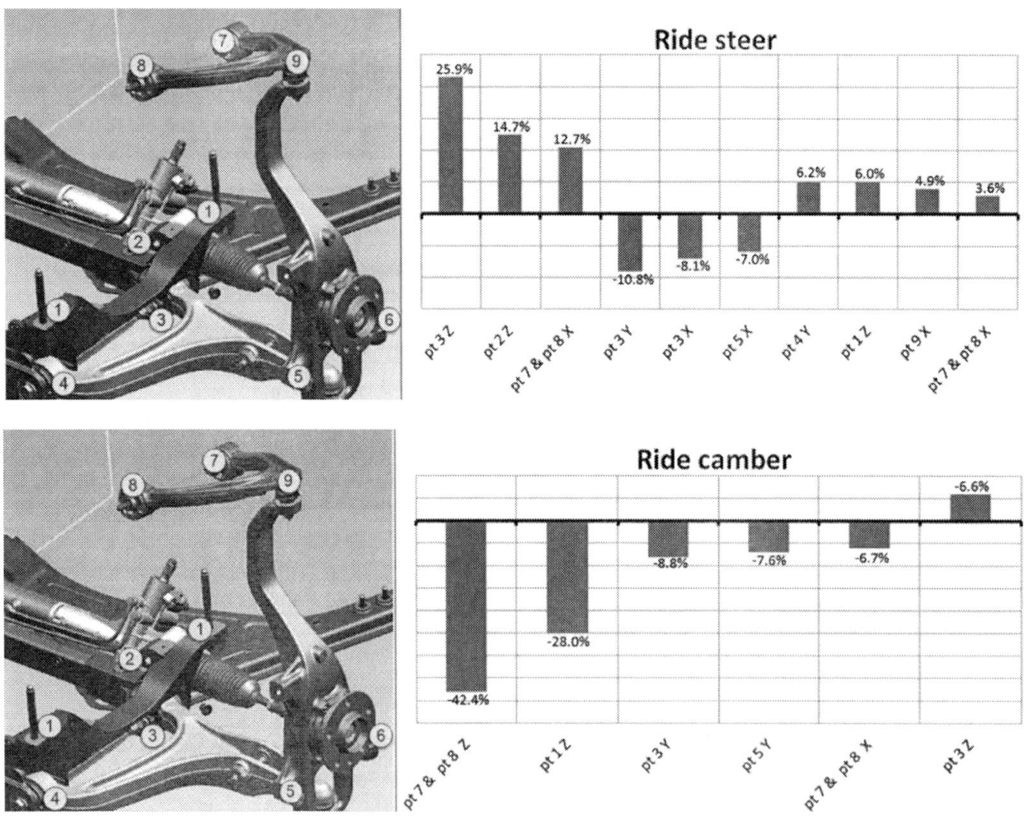

Figure 2. Front suspension: ride steer (top) and ride camber (bottom).

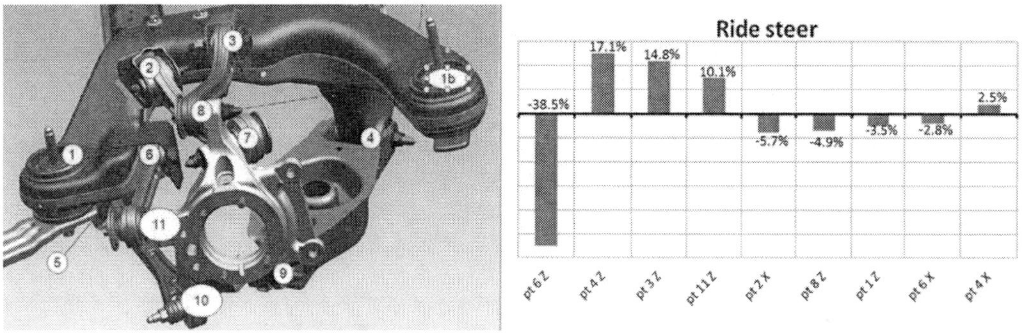

Figure 3. Rear suspension: ride steer.

At the rear axle the vehicles considered in this analysis mount multilink suspensions (Five arms), as shown in figure 3 (left). The rear suspension assembly process is a delicate matter: regulation button-holes are necessary to correctly complete the marriage procedure. A more complex DOE than that performed for the front suspension lead to histograms like the one represented in figure 3, showing the contribution of rear suspension tolerances on the ride steer (percentages defined with respect to the total variation of ride steer). The latter parameter is affected mainly by Z (vertical) displacement between hardpoints 6 and 11 (tie rod hardpoints).

Chassis, front and rear subframes are the main subsystems composing a full vehicle: they are connected together during the marriage, or coupling phase. Chassis can be affected by mass distribution anomalies (generating non–symmetric downforces on the wheels on opposite sides of the vehicle) and by geometric anomalies (twisting and bending; only the former is relevant within the present analysis, since it can cause different preloads on the opposite wheels of the same axle, and consequently different travel positions of the two suspensions). During subframe assembly process, there are two critical phases, potentially generating the anomalies under study: marriage and wheel alignment (last phase of assembly process, in which a vehicle complete in all its parts is positioned on the wheel aligner bench for setting the toe and camber angles of the front and rear axles). If the chassis is symmetric and not twisted, possible errors during marriage can lead to axle centre out of the chassis symmetry plane and non–parallel axles in lateral direction. On the other hand, during wheel alignment, possible errors can originate from incorrect wheel scanning on the aligner bench (the laser scanning process identify the wheel centres, along with toe and camber angles; from wheel centres, the positions of the two axles are determined, and consequently the vehicle mechanical symmetry plane, which, depending on the marriage phase, not necessarily coincides with the chassis symmetry plane). During the final setting of toe and camber angles, attention must be paid in avoiding any possible contribution yielding preloads on the elastic components, which after regulation would affect the setup.

3. Virtual simulation

After DOE analysis, a virtual sensitivity analysis has been performed using a multibody code (MSC Adams Car), aimed at understanding how and at which level the relevant suspension parameters can affect the handling on a straight path of the vehicles under study. Straight line manoeuvers have been simulated at an imposed vehicle speed of 100 km/h. Both a perfectly flat ground surface and a standard highway surface profile have been considered, the latter for taking into account vertical excitation due to irregularities of the ground. For PTS contribution, the steering wheel has been locked to 0°; while for SWM contribution, the steering wheel has been set free to rotate.

To evaluate the effects of asymmetry involving toe angles (front toe with respect to crab axis), camber angles, and lateral displacement of wheel centres, different hard–point configurations have been considered, in each case imposing the maximum tolerance differences found by DOE analysis; for vertical load asymmetry, different preloads have been imposed on the springs of the same axle; for different ride steer on the two wheels of the rear axle, the worst case evaluated with DOE analysis has been considered (largest error due to component tolerances), yielding different hard–point positions, and consequently different elasto–kinematic behaviours of the two rear suspensions.

Vertical load asymmetry on the front axle suspensions has been simulated by imposing a 6% difference in nominal preload on the two front springs, causing a displacement of centre of gravity from the symmetry plane, and consequently a lateral deviation of the vehicle (less than 5 mm over a 300 m travel) and a rotation of the steering wheel (lower than 2%). Asymmetry due to front subframe marriage errors has been simulated by imposing lateral displacements of lower control arms, and consequently of wheel centres (notice that after creating this asymmetry, camber and toe angles have been perfectly regulated), resulting in very small effects on PTS and SWM.

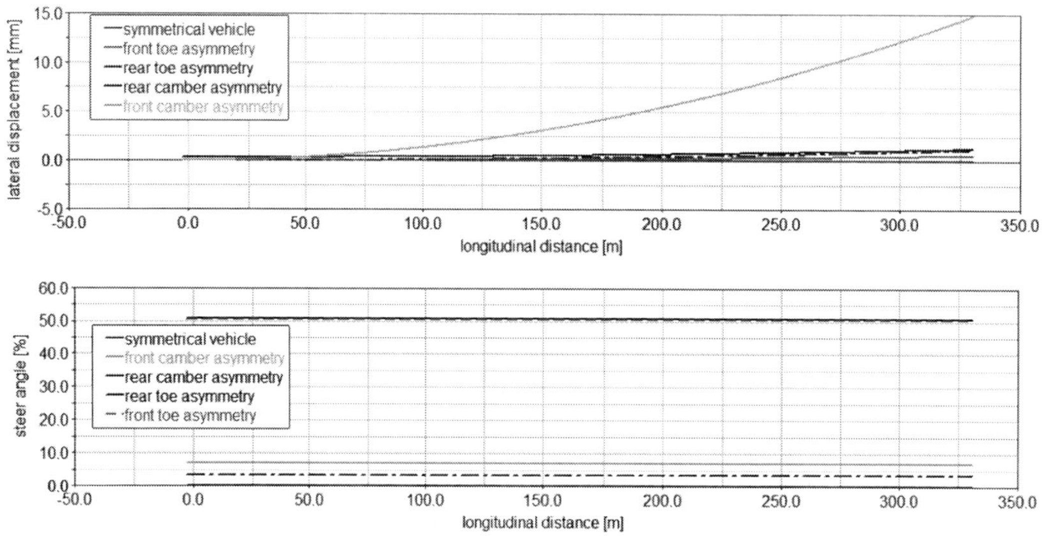

Figure 4. Effect of asymmetric camber and toe angles on lateral deviation (top) and steering wheel rotation (bottom).

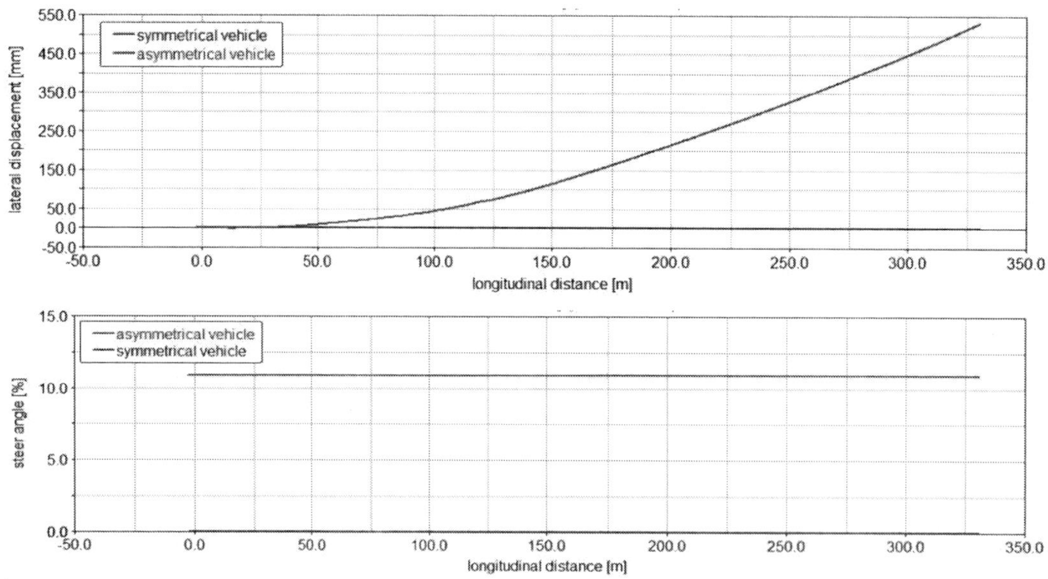

Figure 5. Effect of asymmetric rear axle ride steers on lateral deviation (top) and steering wheel rotation (bottom).

Asymmetry in wheel angles setup phase (camber and toe) is the most important one for investigating the anomalies under study. As shown in figure 4 (top), an asymmetrical regulation of the front camber angle produces the highest PTS effect on the vehicle (in fact, at low slip angles, camber variations are particularly effective on tyre lateral forces). As regards SWM, the front and rear toe angles have been recognised as the most influential parameters, as shown in figure 4 (bottom).
Different ride steer on the two wheels of the rear axle has been simulated by imposing the largest errors on hardpoint positions due to component tolerances (presetting process), while keeping camber and toe angles perfectly regulated (perfect static wheel setup). In this case anomalies occur only in presence of vertical oscillations of the suspensions (figure 5).

4. Experimental tests on presetting bench and wheel aligner bench

The vehicle assembly process has been improved by considering separately presetting bench regulation for reducing rear ride steer, and wheel aligner bench for reducing camber and toe static anomalies.

Scope of a presetting bench is imposing the correct position between rear subframe and uprights; if the components are all in tolerances, once that their relative positions are correct, also the arms will be correctly positioned. On the presetting bench only the position of part of the mechanism members is fixed, for allowing the regulation of camber and toe angles at the wheel aligner bench.

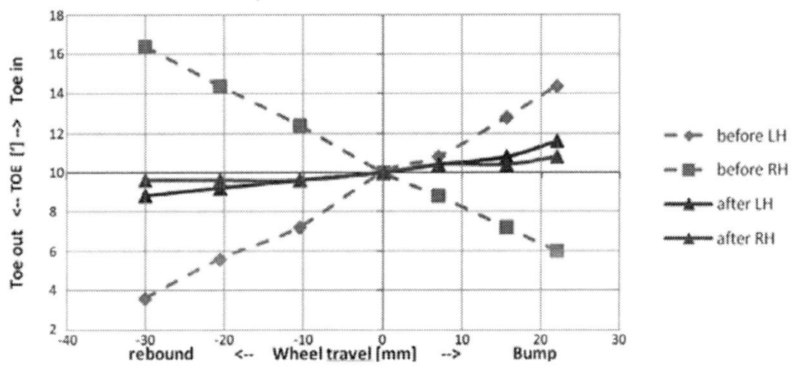

Figure 6. Toe angles ['] comparison before/after presetting regulation.

By testing some vehicles affected by DL, strong asymmetries in toe angles were detected, as shown in figure 6 (dashed lines). To evaluate whether this anomaly was related to the mounting phase or to the component tolerances, the components were disassembled to make again the mounting phase on the presetting bench. As a result, ride steer differences were strongly reduced. On some vehicles, presetting analysis was sufficient for reducing DL within acceptable bounds. On other vehicles, however, presetting analysis was not sufficient to correct this anomaly, which was caused by structural components. Consequently, a tolerance analysis has been performed for identifying the components out of tolerance.

The regulation of camber and toe angles at the wheel aligner bench is the most delicate phase during a vehicle set up. During this study, two critical aspects have been identified: correct positioning of laser scan columns on the bench, and correct locking of wheels on their supports. Regarding the first issue, it is important that the laser can scan a great surface of the tyre sidewall, in order to identify the closer point and to define correctly the centreline of the tyre. By improving sensor positioning, the reading errors of the lasers have been significantly reduced. The last phase which has been analyzed is locking of wheel supports: ideally the vehicle should be completely free, for reducing to zero bushing preloads; in practice the vehicle is locked for improving safety and reducing set up time. The improved procedure consists of locking the rear wheels while regulating the front wheels, and then locking one side of the vehicle for regulating the opposite rear wheel (in the existing procedure adopted for the production line under analysis, the front left plate was always locked, with the disadvantage of increasing preload on front bushings).

5. Results and discussion

Virtual analysis put in evidence that PTS and SWM are mainly related to the set up phase of wheel angles. In particular, both anomalies can be explained by the contribution that vehicle geometry gives to the sideslip angle. Front camber angles influence PTS, while front and rear toe angles influence SWM. Main cause of DL has been identified with (rear) ride steer, which in turn can be generated by components out of tolerances or/and presetting variation.

By improving all the set up process (presetting bench and wheel aligner bench) on a whole production line, the number of vehicles affected by PTS, SWM and DL was strongly reduced: from 3% down to about 1% of the production. Clearly, further improvements are still needed for correcting the anomalies of the residual 1% of vehicles.

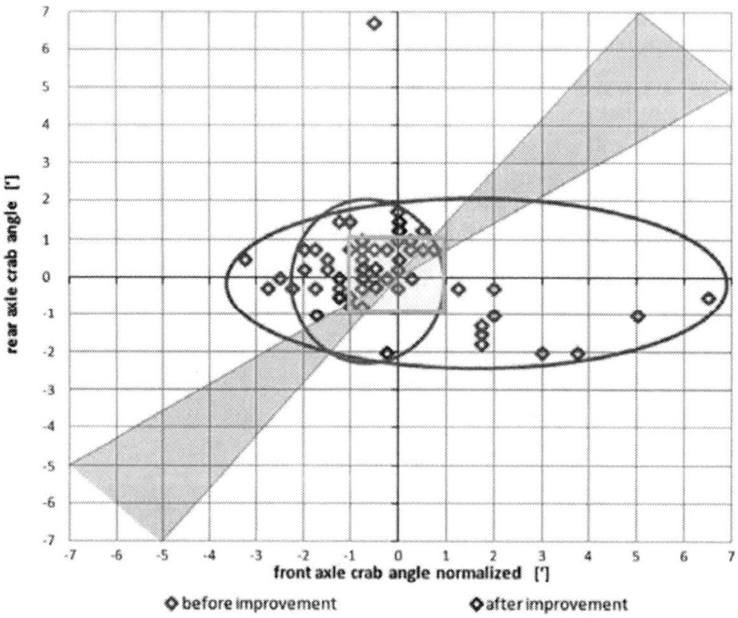

Figure 7. Crab angle ['] variation analysis.

The plot in figure 7 synthesizes the results of the present analysis, displaying a comparison of front and rear crab angles before and after process improvement, where especially the front crab angle variation has been strongly reduced. The yellow central rectangular area identifies the bounds for acceptable values (in terms of handling behaviour) of front and rear crab angles. In the diagram, the points on the first and third quadrant diagonal (equal front and rear crab angles) are acceptable in terms of handling behaviour (the vehicle travels on a straight trajectory); actually, also points close to these diagonals are still acceptable (green triangular areas).

6. Conclusions

In this study three relevant handling anomalies affecting passenger cars travelling on a straight path have been studied: Pull to side, Steering wheel misalignment and Drift leeward, in the specific case of vehicles with front Double wishbone suspension and rear Five arms suspension.

After analyzing the vehicle assembly process in terms of errors and tolerances to identify which parameters can actually generate these handling irregularities, sensitivity analysis by multibody virtual modelling, and subsequent correlation with experimental data from presetting and wheel aligner benches, has provided sufficient information for setting tolerance thresholds, able to keep the handling anomalies under study within acceptable bounds.

In conclusion, it can be stated that Pull to side and Steering wheel misalignment are mainly related to the set up phase of wheel angles. Front camber angles influence Pull to side, while front and rear toe angles influence Steering wheel misalignment. Main cause of Drift leeward has been identified with (rear) ride steer, which in turn can be generated by components out of tolerances and presetting variation.

Acknowledgment
The present paper was presented inside the '*Toward a Sustainable Mobility*' special session as part of the '*Two Seats for a Solar Car*' research project, an action funded by the Italian Ministry of Foreign Affairs and International Cooperation within the Executive Programme of Cooperation in the field of Science and Technology between the Italian Republic and the Republic of Serbia.

References

[1] Topping R W 1974 *Tire Induced Steering Pull*, SAE Technical Paper 750406
[2] Matyja F E 1987 Steering pull and residual aligning torque, *Tire Science and Technology* **15**(3) 207–240
[3] Lindenmuth B E 1974 *Tire conicity and ply steer effects on vehicle performance*, SAE Technical Paper 740074
[4] Pottinger M 1990 Tire/Vehicle Pull: An introduction emphasizing plysteer effects, *Tire Science and Technology* **18**(3) 170–190
[5] Mundl R et al. 2005 Simulation and validation of the ply steer residual aligning torque induced by the tyre tread pattern, *Vehicle System Dynamics* **43**(1) 434–443
[6] Yamazaki S, Fujikawa T, Suzuki T and Yamaguchi I 1998 Influence of wheel alignment and tire characteristics on vehicle drift, *Tire Science and Technology* **26**(3) 186–205
[7] Lee J H 2000 Analysis of tyre effect on the simulation of vehicle straight line motion, *Vehicle System Dynamics* **33**(6) 373–390
[8] Mastinu G, Lattuada A and Matrascia G 2018 Straight–ahead running of road vehicles – analytical formulae including full tyre characteristics, *Vehicle System Dynamics*, online December 2018
[9] Kim Y S and Jang D Y 2009 Optimization of geometric dimension & tolerance parameters of front suspension system for vehicle pulls improvement, *Transactions of the Korean Society of Mechanical Engineers A*, **33**(9) 903–912
[10] Murari T B et al. 2016 Vehicle steering pull: from product development to manufacturing, *Product Management & Development* **14**(1) 22–31
[11] Cho Y G 2010 Steering pull and drift considering road wheel alignment tolerance during high–speed driving, *International Journal of Vehicle Design* **54**(1) 73–91
[12] Park K S, Heo S J and Kang D O 2013 Robust design optimization of suspension system considering steering pull reduction, *International Journal of Automotive Technology* **14**(6) 927–933
[13] Aalizadeh B and Asnati A 2016 Integrated bees algorithm and artificial neural network to propose an efficient controller for active front steering, *International Journal of Automotive and Mechanical Engineering* **13**(2) 3476–3491
[14] Kajiwara S 2018 Improvement in steering performance by push–pull operation in car driving, *International Journal of Automotive and Mechanical Engineering* **15**(1) 4919–4934
[15] de Camargo F V, Fragassa C, Pavlovic A and Martignani M 2017 Analysis of the Suspension Design Evolution in Solar Cars, *FME Transactions* **45**(3) 394–404
[16] Minak G, Fragassa C and de Camargo F V 2017 *A brief review on determinant aspects in energy efficient solar car design and manufacturing*, International Conference on Sustainable Design and Manufacturing, Bologna, Italy, April 26 – 28, pp. 847-856
[17] de Camargo F V, Giacometti M and Pavlovic A 2017 *Increasing the energy efficiency in solar vehicles by using composite materials in the front suspension*, International Conference on Sustainable Design and Manufacturing, Bologna, Italy, April 26 – 28, pp. 801-811
[18] De Rosa M, De Felice A, Grosso P and Sorrentino S 2019 Straight path handling anomalies of passenger cars induced by suspension component and assembly tolerances, *International Journal of Automotive and Mechanical Engineering*, in press
[19] Pacejka H B 2006 *Tyre and Vehicle Dynamics*, Butterworth–Heinemann

IOP Conf. Series: Materials Science and Engineering **659** (2019) 012077 doi:10.1088/1757-899X/659/1/012077

Experimental analysis of jet slurry erosion on martensitic stainless steel

G Santacruz[1], A S Takimi[1], F Vannucchi de Camargo[1,2], C P Bergmann[1]

[1] Post-Graduation Program in Mining, Metallurgical and Materials Engineering, Federal University of Rio Grande do Sul - UFRGS, Osvaldo Aranha 99, Porto Alegre, RS 90035-190, Brazil

[2] Department of Industrial Engineering, University of Bologna, Viale del Risorgimento 2, 40136, Bologna, Italy

E-mail: galileo.santacruz@ufrgs.br

Abstract. Due to their enhanced tribological properties that contribute to an increased useful life of components, martensitic stainless steels are an excellent option for industrial applications such as hydroelectric, petrochemical, civil construction and mineral processing plants. In the present investigation, the erosive wear of AISI 410 martensitic stainless steel is evaluated after thermal treatment by quenching and tempering by mass loss, under erosive attack at 30° and 90° incidence angles, using a self-made jet slurry erosion equipment controlling parameters such as speed, volume of fluid, temperature and concentration of erosive particles of erodent. The characterization of the eroded samples was carried out in terms of the microstructure (SEM) and microhardness as well as the particle size distribution (LG) and morphology of the erodent. It was possible to establish the relationship between the slurry erosive wear and the physical properties inherent of stainless steel for this particular experimental configuration, concluding that the steel presents better resistance to jet slurry erosion wear when compared to austenitic steel commonly used in the industry.

1. Introduction

In different industrial applications, such as hydraulic turbines in hydroelectric plants, petrochemical industry, construction, and in the processing of minerals, the loss of metal due to slurry erosion caused by the particulate material in the liquid represents a major industrial problem affecting the life of the components and reducing their performance. Thus, the ideal choice of engineering materials is important in order to decrease the wear rate and to improve their tribological behavior. Among the different alternatives for such applications, the most current being studied is the austenitic and martensitic stainless steels. Austenitic stainless steels are used in many components where corrosion resistance is crucial. However, under the mechanical action of hard particles, they present a high plastic deformation and wear [1]. On the other hand, martensitic stainless steel presents better mechanical resistance to erosive particles than austenitic steel, with the compromise of a lower corrosion resistance [2-5].

In the present investigation, it was possible to evaluate the erosive wear of the martensitic stainless steel AISI 410 thermally treated with quenching and tempering, evaluating the volume loss under jet slurry erosion conditions at incidence angles of 30° and 90° between the axis of symmetry of the fluid flow and the surface of the samples, via the control of parameters such as angle and speed of impact, test temperature and concentration of erosive particles in the suspension. In this case, electrofused alumina was used as erodent.

The materials were characterized with regard to their microstructure by scanning electron microscopy (SEM), roughness, microhardness, and the particle size distribution analyzed with laser granulometry (LG) and morphology of the erodent.

2. Materials and Methods

2.1. Martensitic stainless steel AISI 410

AISI 410 martensitic stainless steel cylindric coupons with 30 mm in diameter and 10 mm in height were used. The chemical composition of the steel is presented in Table 1. The steel samples were tempered at 1263 K for 35 min then oil-quenched at 793 K for another 35 min. The microstructure obtained after this procedure was composed of martensite with its grain contours and some precipitated carbides, typical of quenching and tempering thermal treatments presented in Figure 1.

Table 1. Chemical composition of the AISI 410 martensitic stainless steel used in this investigation (wt.%).

Elements	Composition (%)
C	0.020
Cr	11.200
Mn	0.590
Si	0.720
Ni	0.350
P	0.022
S	0.002
N	0.021

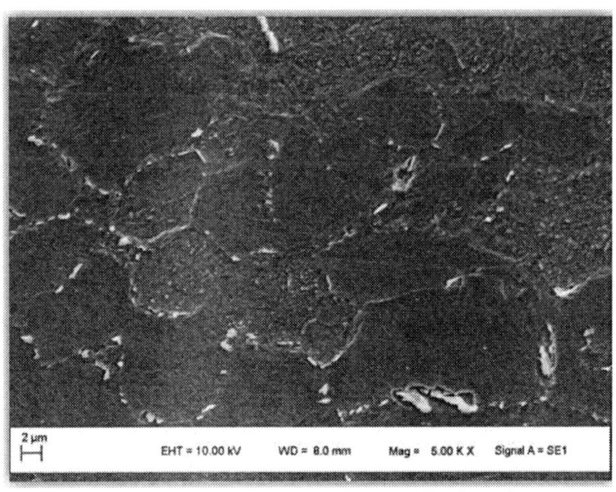

Figure 1. Microstructure of AISI 410 stainless steel quenched and tempered. (SEM 5000x, Vilella's acid etching).

2.2. Erodent particles

Cilas 1180 laser diffraction particle size analyzer was used to measure the particle size distribution of the alumina erodent (Al_2O_3). The technique follows the ISO 13320:2009 standard [6]. The abrasive particle size distribution is shown in Figure 2, yielding an average abrasive particle size of about 98.55 μm. SEM micrograph shows angular and irregularly shaped particles in Figure 3.

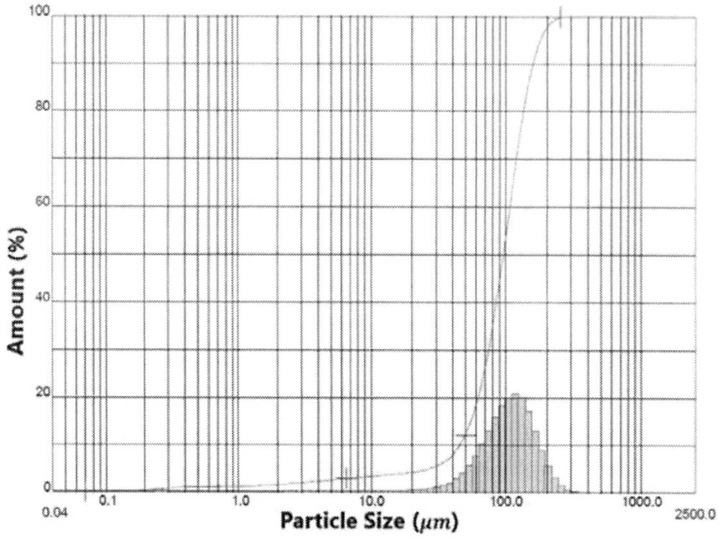

Figure 2. Particle size distribution of aluminum oxide.

Figure 3. SEM micrograph showing the abrasive particles shape and size. (56x).

2.3. *Microstructure characterization*

The microstructure characterization was done in an EVO MA10 SEM and an optical microscopy LEICA DM2700 M. Surface roughness of the steel samples was measured before the jet slurry erosion and after it using a surface roughness tester Mitutoyo SJ-400. Then, cross-sectional Vickers microhardness measurements were performed by using a Buehler Micromet 2001 microhardness tester (HV300 g, 30 s) following ASTM E-384-11 standard [7].

2.4. *Jet Slurry erosion tests*
The jet slurry erosion tests were carried out in a modified commercial high pressure washer which is based on ASTM G-76 standard [8-12] and adapted in the nozzle of the gun a system of feeding of erodent particles using an inner Venturi accelerator of particles inside a test chamber. This equipment allows controlling the angle of impact, the speed of impact, the concentration of erosive particles in the suspension and test temperature, all important parameters to determine the distribution of frictional energy along a surface [13]. Figure 4 shows the configuration of the testing device.

The tests were performed using water between 25 and 28°C of temperature with 960 g of (Al_2O_3) erodent. The samples were placed at the nozzle output to guarantee the incidence of impact fluid and mean jet velocity of erosive material in the suspension of 77 m/s. This velocity was calculated using flow rate, time and nozzle area measurements of flow output. The incidence angles studied were 30° and 90° between the axis of symmetry of the fluid flow and the surface of the samples. In all cases, the concentration of particles in the slurry was 7 wt% [14-16]. The erosion resistance was determined from the volume loss results per unit of time from the difference of mass loss considering the relation of the apparent density of the studied material (7.73 g/cm^3 for AISI 410 stainless steel [17-19]). Mass losses were measured every 1 min by using a scale with 0.01 mg resolution. The total duration of each test was 4 min. The samples were cleaned in an ultrasonic bath with deionized water before and after each test and dried and weighted afterwards.

3. Results and Discussion

3.1. *Mechanical Properties ofMaterials*
Microhardness reflects the microstructure and the physical and mechanical properties ofboth substrate and coating, which in turn are dependent on the materials and processes employed in their manufacture. Hardness is a property that can be considered variable throughout the material in certain zones due to eventual heterogeneities of the material. For this reason, when measuring this property, one must consider the preparation of the surface, the section analyzed, as well as the number of indentations performed. The microhardness of AISI 410 is presented in Table 2.

Figure 5 shows the indentation performed on the thermally treated martensitic AISI 410 stainless steel. The measured microhardness value was 219 HV0.3, which is very similar to that reported by [16] of 199 HV0.3.

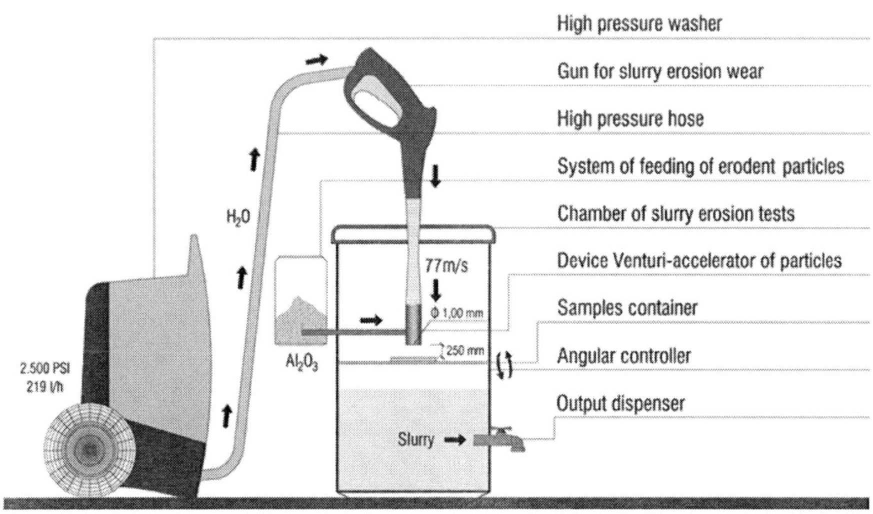

Figure 4. Schematic diagram of the jet slurry erosion tester.

Table 2. Results of Vickers microhardness measurements - cross section (HV0.3 - 2.94N)

Microhardness (HV)	
Average	219.30
Standard Deviation	±31.07
Coefficient of Variation (%)	14.16

Table 3 shows the accumulated erosion rate by mass of erodent impacted in loss of volume of the samples tested, as a function of the impact angle, in a total time of 4 minutes, for the samples used in the research of stainless steel AISI 410 martensitic.

Table 3. Accumulated erosion rate after 4 minutes.

Material	Impact Angle (°)	Volumetric Erosion Rate $(cm^3_{target}/g_{erodent})*10^{-5}$	Standard Deviation $(cm^3_{target}/g_{erodent})*10^{-5}$
Steel AISI 410	30	0.8449	±0.0114
	90	0.7025	±0.0495

Figure 5. Vickers indentation (219 HV0.3) performed for thermally treated AISI 410 martensitic stainless steel. (OM, 50x).

Figure 6 shows the average mass losses measured in 3 samples every 1 min until a total erosion time of 4 minutes for the impact angles of 30° and 90°. Steel exhibits a much higher erosive wear at 30° impact angle than at 90°, demonstrating a ductile behavior with a ploughing-predominant material removal mechanism, as reported in the literature [20,21]. Furthermore, many authors confirm that metals present an enhanced erosion rate at lower angles [22-25]. Figure 6 also shows that, for the impact angle of 90°, the samples presented an accumulated volumetric erosion rate slightly lower than the 30° impact angle.

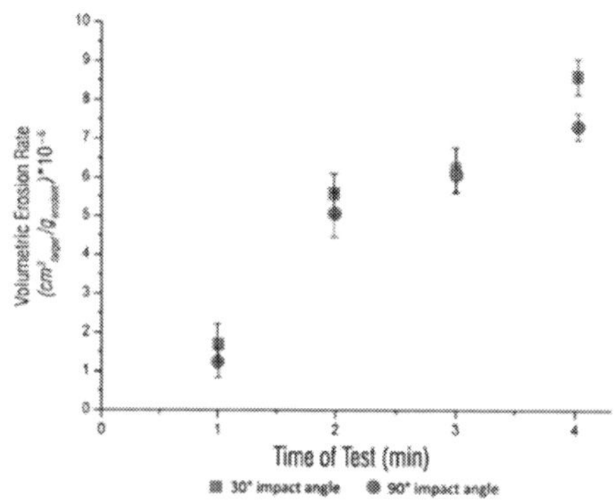

Figure 6. Variation of the accumulated volumetric erosion rate as a function of the impact angle of for AISI 410.

Figure 7 shows the areas of the erosion regions of the samples subjected to the jet slurry erosion test for the impact angles of 30° and 90°. It is observed that the eroded region of the pieces (highlighted in red) tested with the angle of 30° have an elongated shape and greater erosion area. The erosion of the pieces tested in the angle of 90° is more localized, has a smaller area and a greater depth. In the samples, it is possible to notice the formation of a concentric ring, described as "halo erosion" in the literature, which is a mechanical characteristic of the damage due to the development of craters that appear as macro pits on the surface for this angle [26-28]. The eroded region roughness in the samples, for the two impact angles, as shown in Table 4. Table 5 shows the areas of eroded regions according to the impact angle. It is observed that the eroded area is greater for the impact angle of 30°.

Table 4. Result of the average roughness measurement before and after jet slurry erosion.

Roughness					
Material	Before Erosion (μm)	Standard Deviation (μm)	Angle (°)	After Erosion (μm)	Standard Deviation (μm)
Steel AISI 410	0.07	±0.03	30	2.97	±0.47
			90	2.56	±0.11

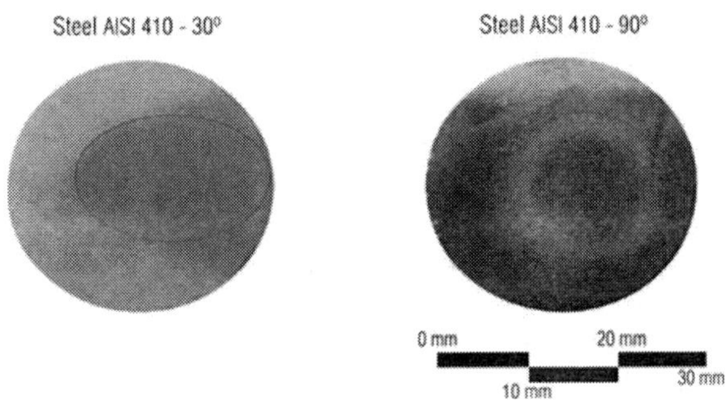

Figure 7. Areas of the eroded regions of the materials submitted to the jet slurry erosion test.

Table 5. Eroded area as a function of the impact angle.

Material	Impact angle (°)	Eroded area (mm^2)
Steel AISI 410	30	277.7
	90	147.1

Figure 8 shows the three-dimensional scanning of the samples submitted to the jet slurry erosion test obtained through the Geomagic Studio software, for impact angles of 30° and 90°, with which it was possible to estimate the average of the erosion depth (Table 6). As result, the average erosion depth of the samples tested at 90° is higher and more localized in relation to the samples tested at 30°.

Figure 8. Three-dimensional scanning of the samples submitted to the jet slurry erosion test.

Table 6. Average depth of the eroded regions.

Material	Impact angle (°)	Average depth (μm)	Standard Deviation (μm)
Steel AISI 410	30	93	±2.88
	90	102	± 15.87

The erosion study [20] conducted in steel tubes, varying impact angle and velocity, also found out that at a 90° angle, deeper eroded areas are found in comparison to samples tested at 30°, agreeing with the results presented in Table 6. Similarly, studies [28] have investigated the performance of the AISI 420 stainless steel subjected to particle erosion tests using two different abrasives at the angles of incidence of 30°, 45°, 60° and 90°. The results also showed that the samples evaluated at 90° presented a greater depth profile than those at 30° even when two different abrasives are adopted, obtaining 450 and 929 *nm* for the angle of 90°, and 180 and 400 *nm* for the angle of 30°, respectively.

Furthermore, it is possible to verify that large erosion rates at smaller impact angles are an intrinsic feature of metals, whereas in the case of brittle ceramics, for instance, higher erosion rates are identified in angles closer to 90° [29]. For this reason, ceramic coatings are an alternative to improve the erosion resistance of metallic materials [30, 31].

4. Conclusion

In the present investigation, completing provisional results from similar researches [32], the jet slurry erosion behavior of martensitic stainless steel AISI 410 was investigated. From the results obtained in the accomplishment of the experimental work, it is possible to infer the following conclusions:

- An experimental erosion wear simulation equipment of the jet slurry type was developed, considering conditions that allowed the control of the test parameters such as impact angle, impact velocity, erosive particle concentration in the suspension and test temperature, all significant variables in the determination of the behavior of the materials.
- The AISI 410 martensitic stainless steel presented a higher rate of accumulated volumetric erosion in the impact angle of 30°, presenting a predominantly ductile behavior, and a higher resistance to jet slurry erosion at the impact angle of 90° with the compromise of deeper eroded area.

Acknowledgment

The authors express their gratitude to the following laboratories: Laboratory of Ceramic Materials (LACER), Laboratory of Physical Metallurgy (LAMEF), Laboratory of Design and Selection of Materials (LdSM) and the Laboratory of Conventional Machining (LUC) of the Federal University of Rio Grande do Sul (UFRGS) for the investigation support and to the Research and Technology Center of Rijeza Metallurgy. G. Santacruz and F.V. de Camargo thank the Coordination of Improvement of Higher Education Personnel of the Brazilian Ministry of Education (CAPES/MEC) for the doctorate scholarship grants 88882.345871/2019-01 and 88882.345831/2019-01, respectively.

References

[1] Toro A, Sinatora A, Tanaka D K and Tschiptschin A P 2001 Corrosion–erosion of nitrogen bearing martensitic stainless steels in seawater–quartz slurry, *Wear* **251**(1) 1257–1264

[2] Levy A V and Yau P 1984 Erosion of steels in liquid slurries, *Wear* **98** 163–182

[3] Guaglianoni W C, Cunha M A, Bergmann C P, Fragassa C and Pavlovic A 2018 Synthesis, Characterization and Application by HVOF of a WCCoCr/NiCr Nanocomposite as Protective Coating Against Erosive Wear, *Tribology in Industry* **40**(3) 477-487

[4] Mesa D H, Toro A, Sinatora A and Tschiptschin A P 2003 The effect of testing temperature on corrosion–erosion resistance of martensitic stainless steels, *Wear* **255**(1-6) 139-145

[5] Babic M, Calì M, Nazarenko I et al. 2018 Surface Roughness Evaluation in Hardened Materials by Pattern Recognition Using Network Theory, *International Journal on Interactive Design and Manufacturing* **13**(1) 211-219
[6] BS ISO 13320 2009 *Particle Size Analysis-Laser Diffraction Methods.*
[7] ASTM E384 2011 *Standard Test Method for Knoop and Vickers Hardness of Materials.*
[8] ASTM G76 2013 *Standard Test Method for Conducting Erosion Tests by Solid Particle Impingement using Gas Jets.*
[9] Zu J B, Hutchings I M and Burstein G T 1990 Design of a slurry erosion test rig, *Wear* **140**(2) 331–344
[10] Wentzel E J and Allen C 1997 The erosion-corrosion resistance of tungsten-carbide hard metals, *Int J Refractory Met Hard Mat* **15**(1-3) 81–87
[11] Santa J F, Baena J C and Toro A 2007 Slurry erosion of thermal spray coatings and stainless steels for hydraulic machinery, *Wear* **263**(1) 258–264
[12] Santa J F, Espitia L A, Blanco J A, Romo S A and Toro A 2009 Slurry and cavitation erosion resistance of thermal spray coatings, *Wear* **267**(1) 160–167
[13] Willert E 2019 Energy Loss and Wear in Spherical Oblique Elastic Impacts, *Facta Universitatis, Series: Mech Eng* **17**(1) 75-85
[14] Hutchings I M 1981 A model for the erosion of metals by spherical particles at normal incidence, *Wear* **70**(3) 269– 281
[15] Finnie I 1995 Some reflections on the past and future of erosion, *Wear* **186** 1–10
[16] Maiti A K, Mukhopadhyay N and Raman R 2009 Improving the wear behavior of wc-cocr-based hvof coating by surface grinding, *J Mat Eng Performance* **18**(8) 1060
[17] Madsen B W 1988 Measurement of erosion-corrosion synergism with a slurry wear test apparatus, *Wear* **123**(2) 127–142
[18] Murthy J K N, Rao D S and Venkataraman B 2001 Effect of grinding on the erosion behaviour of a wc–co–cr coating deposited by hvof and detonation gun spray processes, *Wear* **249**(7) 592–600
[19] Liu X and Zhang B 2002 Effects of grinding process on residual stresses in nanostructured ceramic coatings, *J Mat Sci* **37**(15) 3229–3239
[20] Islam M D A and Farhat Z N 2014 Effect of impact angle and velocity on erosion of api x42 pipeline steel under high abrasive feed rate, *Wear* **311**(1-2) 180–190
[21] Okonkwo P C, Shakoor R A, Zagho M M and Mohamed A M A 2016 Erosion behaviour of api x100 pipeline steel at various impact angles and particle speeds. *Metals* **6**(10) 232
[22] Bayer R G 1994 Mechanical wear prediction and prevention, *Marcel! Dekker, Inc, P. O. Box 5005, Monticello, NY 12701-5185, USA, 1994. 657*
[23] Papini M and Spelt J K 1998 The plowing erosion of organic coatings by spherical particles, *Wear* **222**(1) 38–48
[24] O'Flynn D J, Bingley M S, Bradley M S A and Burnett A J 2001 A model to predict the solid particle erosion rate of metals and its assessment using heat-treated steels, *Wear* **248**(1-2) 162–177
[25] Biswas A, Satapathy A and Patnaik A 2010 Erosion wear behavior of polymer composites: a review, *J Reinf Plastics Composites* **29**(19) 2898–2924
[26] Wood F W 1986 Erosion by solid-particle impacts: a testing update, *J Testing and Evaluation* **14**(1) 23–27
[27] Souza V A and Neville A 2003 Corrosion and erosion damage mechanisms during erosion–corrosion of wc–co–cr cermet coatings, *Wear* **255**(1-6) 146–156
[28] Vite-Torres M, Laguna-Camacho J R, Baldenebro-Castillo R E, Gallardo-Hernandez E A, Vera-Cárdenas E E and Vite-Torres J 2013 Study of solid particle erosion on aisi 420 stainless steel using angular silicon carbide and steel round grit particles, *Wear* **301**(1-

2) 383–389

[29] Dauber C, de Camargo F V, Alves A K, Pavlovic A, Fragassa C and Bergmann C P 2019 Erosion Resistance of Engineering Ceramics and Comparative Assessment through Wiederhorn and Evans Equations, *Wear* **432-433**:202938. DOI: 10.1016/j.wear.2019.202938

[30] Cunha M A, Guaglianoni W C, Bezerra B F A, de Camargo F V and Bergmann C P 2019 Resistance to Abrasive and Erosive Wear of WCCo/NiCr HVOF Coatings: Comparative Evaluation of Commercial Nanostructured Spray Powders. *Tribology in Ind*, In press

[31] Campione I., Fragassa C, Martini A 2019 Kinematics optimization of the polishing process of large-sized ceramic slabs, *International Journal of Advanced Manufacturing Technology* **103** (1-4), 1325-1336. doi: 10.1007/s00170-019-03623-3.

[32] Santacruz G, Takimi A, Vannucchi de Camargo F, Bergmann CP, Fragassa C. 2019 Comparative Study of Jet Slurry Erosion of Martensitic Stainless Steel with Tungsten Carbide HVOF Coating. *Metals* **9**(5), 600; doi:10.3390/met9050600.

Solution combustion synthesis of Mo-Fe/MgO: Influence of the fuel composition on the production of doped catalyst nanopowder

R Y S Zampiva[1], C P Bergmann[1], A K Alves[1] and L Giorgini[2]

[1] Post-Graduation Program in Mining, Metallurgy and Materials Engineering (PPGE3M), Federal University of Rio Grande do Sul, Osvaldo Aranha 99, 90035-190, Porto Alegre, Brazil
[2] Department of Industrial Chemistry "Toso Montanari", University of Bologna, Viale del Risorgimento 4, 40136, Bologna, Italy

E-mail: rubiayoungsun@gmail.com

Abstract. Among the techniques for producing oxide catalysts, the solution combustion synthesis (SCS) has been widely used to produce high-quality nanostructured powders at low-cost. Through SCS, due to the high exothermic energy liberated by the reaction between the transition metal nitrate and the fuel, it is possible to homogeneously incorporate dopant ions into the catalyst. Besides, SCS allows the production of nanopowders with sizes in the decimal scale, which is extremely important for powdered catalysts. The smaller the particle is, the larger the surface area is. Based on these possibilities, the aim of this work is to produce Fe-Mo/MgO nanopowders by SCS. The influence of concentration and composition of the fuels glycine and polyethylene glycol (PEG) on the final product were investigated. The morphology and the physicochemical properties were characterized by X-ray diffraction (XRD), electron microscopy (SEM), granulometry and surface area analysis. The results indicated high crystallinity for the samples produced with PEG and a wide variation on the nanoparticles sizes depending on the fuel properties.

1. Introduction

Along the past few years, nanoscience has become a focus of research in several areas of development due to the different material properties evidenced in the nanometer scale [1-7]. For example, some materials when nanostructured are exceptionally strong, hard and ductile at high temperatures [8-9]. These characteristics are controlled by the size, composition and morphology of the material [10]. Among these nanomaterials a group that is attracting great interest is the nanostructured catalysts, sized between 1 and 100 nm. Such catalysts have high efficiency, because the surface of the particles have greater availability to carry out catalysis [11]. The smaller the particle diameter, the greater the number of atoms exposed on the surface. As the diameter, purity and homogeneity of the catalyst are of fundamental importance for the yield of the reactions in which it is employed [10]. Mo-Fe/MgO spinel is an excellent catalyst for multiwall carbon nanotubes production [11].

SCS technique has been increasingly applied in the production of oxides catalysts due to the possibility to produce, at low-cost, highly pure and homogeneous nanostructured powders. SCS consists basically of the combination of its reactants in an aqueous medium using a complexing agent (fuel) [12] such as citric acid [13,14] ; oxalic acid; tetraformol trisazine [12], glycine [4], urea [15], among others, and oxidizing agents (usually metal nitrates), responsible for oxidizing the fuel. The mixture is heated between 150 °C and 500 °C in order to carry out a self-ignition that leads to a fast combustion reaction that can reach more than 1700 °C [16] forming at the end of the process a solid product usually crystalline and well dispersed [17].

The high exotherm resulting from the reaction between the transition metal nitrate and fuel allow the incorporation of dopant ions in the catalyst [18]. The distribution of dopants throughout the host material is uniform because of the atomic mixing of the reactants in the initial solution. The dopants may drastically alter the properties of the catalyst even if the concentration in the crystalline structure is very low (1-3% atom) [12]. The easily incorporation of dopant ions is important for the production of structures such as Fe-Mo/MgO, where MgO is the host for Fe and Mo catalyst ions. Fe-Mo/MgO is an important catalyst for the production of single and multiwall carbon nanotubes. The synthesis parameters strongly affect the produced catalyst activity [19].

The stoichiometry of a combustion reaction is calculated through the chemistry of propellants [20] and it is generally expressed in terms of the equivalence ratio (Ø). An equivalence ratio of 1 signifies a stoichiometrically balanced combustion reaction, where the oxidizing power of the metal nitrates is equal to the reducing valence of the fuel. Positive departure (Ø>1) of the equivalence ratio signifies fuel-lean compositions, whereas negative departure (Ø<1) denotes fuel-rich compositions. Different products can be formed by varying the reducing potential of the precursor solution [21]. Furthermore, properties such as particle size, shape, morphology and agglomeration are directly influenced by the combustion conditions [22. 23].

Therefore, in this paper, in order to formulate the optimal methodology for preparing nanostructured oxides for catalysis, we studied the relationship of fuel-oxidant concentration, and the use of glycine and polyethylene glycol with molecular weight 200 (PEG) as fuel in the SCS of Fe-Mo/MgO nanopowder catalyst.

2. Materials and Methods

2.1. Synthesis of Fe-Mo/MgO

In order to produce Fe-Mo/MgO by SCS, iron nitrate nonahydrate ($Fe(NO_3)_3.9H_2O$), ammonium heptamolybdate tetrahydrate ($(NH_4)_6Mo_7O_{24}.4H_2O$) and magnesium nitrate hexahydrate ($Mg(NO_3)_2.6H_2O$) were used as ions precursors and oxidizing agents. The syntheses efficiency was studied varying the concentration of glycine ($C_2H_5NO_2$) and PEG 200 ($H(HOCH_2CH_2)nOH$) as fuel. All precursors were acquired from Sigma-Aldrich. The precursors concentration and the stoichiometric correlation for both fuels are indicated by Equations 1 and 2:

$$0.96\ Fe(NO_3)_3.9H_2O + 0.004\ (NH_4)_6Mo_7O_{24}.4H_2O + Mg(NO_3)_2.6H_2O + 2.615\ C_2H_5NO_2 \quad (1)$$

$$0.96\ Fe(NO_3)_3.9H_2O + 0.004\ (NH_4)_6Mo_7O_{24}.4H_2O + Mg(NO_3)_2.6H_2O + 1.76\ H(HOCH_2CH_2)_nOH \quad (2)$$

Briefly explaining, through the propellants chemistry, the stoichiometric correlation between fuel-oxidizer is reached with the valences neutralization. N is considered neutral, while for the equations above, Fe presents a valence of +3, Mg of +2, Mo of +6, H of +1, O of -2 and C of -4. The fuel-oxidizer correlation is given by the equivalence ratio Ø, equation 3:

$$Ø = Oxidizer / Fuel \quad (3)$$

When the sum of the oxidizer valences is equal to the sum of the fuel valences, the combustion reaction is considered stoichiometric (Ø = 1). Ø > 1 signifies fuel-lean compositions while (Ø<1) means fuel-rich compositions. Based on the stoichiometric combustions, rich and lean compositions were formulated in order to verify the influence of the fuel composition and concentration on the product. The samples were divided in glycine and PEG 200 groups and are described in Table 1:

Table 1. Composition and concentration of the fuels used to produce Fe-Mo/MgO by SCS.

Sample	Fuel	Ø	Description	Concentration (mol)
G1		0.66	Rich	3.975
G2	Glycine	1	Stoichiometric	2.615
G3		2	Lean	1.3
P1		0.665	Rich	0.855
P2	PEG	1	Stoichiometric	0.57
P3		1.995	Lean	0.285

The salts and fuels were individually dissolved in distilled water and then mixed (first the salts and then the specific fuel was added) in order to avoid precipitation during the solubilization processes. The solution was heated at 60°C during 5 min under constant stirring. This methodology was applied for all samples. After complete mixing, the samples were inserted in a 400 °C pre-heated electric muffle furnace during 15 minutes in order to undergo self-ignition. The samples were then calcined at 1100 °C in order to burn the carbonic residues and to organize the crystalline structure of the powdered product.

2.2. Characterization

The crystallinity of the samples was evaluated by X-ray diffraction (XRD) using a PHILIPS diffractometer (model X'Pert MPD) at 40 kV and 40 mA using a Cu anode. The particles surface area was obtained by the nitrogen adsorption method (BET - Quanta Chrome, Nova- 1000 model). As well the catalyst granulometric distribution was measured using a laser diffraction method in a Cilas-180L. the morphology of the Fe-Mo/MgO nanoparticles was characterized by scanning electron microscopy (SEM) using a JEOL microscope (JSM 6060) with a maximum operating voltage of 30 kV and a nominal resolution of 3.5 nm. The applied voltage was 10 to 20 kV. Transmission electron microscopy (TEM) was also carried out using a JEOL microscope (JEM 1200 EXII model) operating between 80 and 100 kV with a point resolution of 0.45 nm and line resolution of 0.20 nm. The powder sample was placed on a Cu grid and coated with a carbon film, to avoid magnetic interactions with the TEM microscope.

3. Results and discussion

According to diffractogram (figure 1), all samples produced using PEG present the characteristic peaks for periclase (MgO JCPDF # 00-001-1235, cubic Fm-3m) indicating the formation of MgO as host for Fe and Mo in the structure of Fe-Mo/MgO. It is possible to also observe a second phase that might be related to the formation of iron oxide. Interestingly, the fuel-lean combustion presented the purest composition among the PEG produced samples. The fuel concentration has a direct influence on the second phase formation; higher concentration leads to an ignition that reaches elevated temperature in really short time. The fast precipitation (crystallization) of the sample tends to produce nonstoichiometric or also heterogeneous samples, composed by more than one phase [24, 25]. The powders produced with glycine, did not presented the formation of the desired phases, indicating mostly the presence of amorphous phase. Since glycine did not show favourable results to produce Fe-Mo/MgO, the following analyses were performed only for PEG samples.

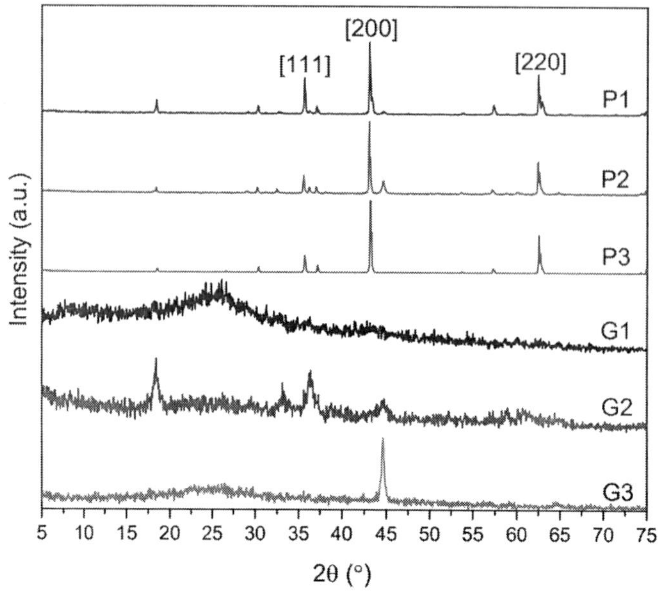

Figure 1. Samples diffractogram correlating the crystalline structure and SCS fuel concentration and composition. The peaks [111], [200] and [220] belong to periclase (MgO JCPDF # 00-001-1235, cubic Fm-3m).

Through the PEG micrograph is possible to observe that the particles morphology and diameter are very similar to all samples corroborating with the XRD analyses that also indicated a similar degree of crystallinity between samples. The elevated grain size (2 μm) might me related to the nanoparticles agglomeration tendency, leading to the formation of large grains, as observed in figure 2a, b and c.

Figure 2. PEG samples SEM micrograph: a) P1, b) P2 e c) P3.

The granulometric results (figure 3) indicated different grain sizes for each composition. However, all samples are manly composed by large grains. The elevated diameters are in accordance with the micrograph analysis, indicating also here the agglomeration of nanoparticles. This phenomenon occurs guided by forces such as Van der Waals in direction to a lower energy state, reducing the interfaces with the environment [4].

Sample P2 presented the major concentration of nanoparticles (under 1 μm). Both samples P2 and P3 showed particles with a maximum size under 100 μm, while 30% of P1 is formed of particles around 100 and 1000 μm. P1 has the most agglomerated particles while P2 presented the higher dispersion, also indicated by the 20% of nanoparticles in the sample. To correctly analyze these results is important to considerate that no process of ultrasonification, sifting or milling was applied to the final product.

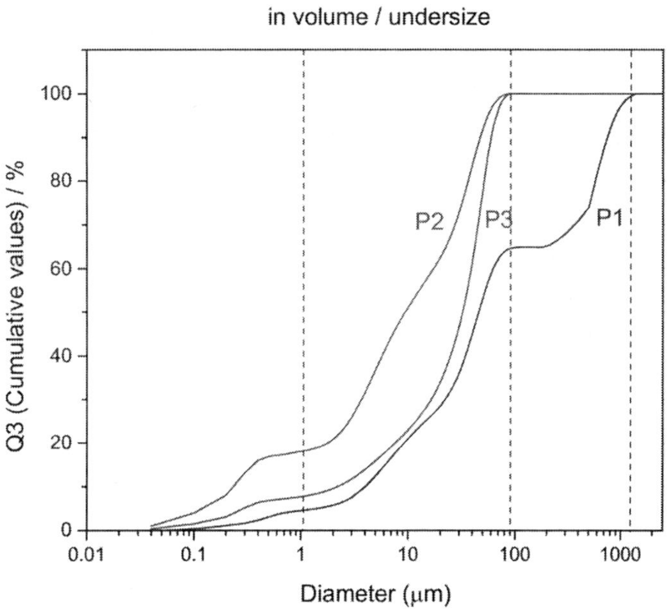

Figure 3. PEG samples granulometry.

Table 2 shows the surface area of the samples P1, P2 and P3. P3 presented the highest surface area among the samples: 29,712 m^2/g followed by the stoichiometric sample P2 (27,757 m^2/g). Indicating a higher degree of dispersion and also corroborating with the agglomeration theory discussed above. The smallest surface area was found for sample P1(10,3 m^2/g) and is in accordance to the granulometry results that indicated the largest grains for this sample. The smaller the grains, the higher the surface area.

Table 2. Surface area of the samples P1, P2 and P3

Sample	Superficial Area (m^2/g)
P1	10.292
P2	27.757
P3	29.712

Analyzing the table 2 is possible to infer that for the stoichiometric composition s not always the best choice for the production of nanostructured catalysts by SCS. The surface area is strongly dependent of the synthesis parameters, and the fuel concentration and composition directly alter the temperature, velocity and also the chemical composition during the crystal formation [4]. The influence of the fuel was also well observed through the XRD, granulometry and SEM analyses.

4. Conclusions

Nanostructured oxide catalysts were successfully obtained through the SCS method using PEG as fuel. The XRD analysis indicated that glycine was not efficient for the production of Fe-Me/MgO, presenting mostly an amorphous phase. All samples prepared with PEG showed elevated crystallinity and through the BET, granulometry and SEM analyses was possible to observe that the concentration of fuel strongly influences the properties of the final product. We conclude that not necessarily the stoichiometric formulation will be the best path to produce nanomaterials. For each final product, the precursors and fuels composition should be individually studied in order to choose the best combination of oxidizer/fuel composition and concentration. In the case of Fe-Mo/MgO catalyst, the best formulation was the PEG fuel-lean that presented the largest surface area among the samples.

Acknowledgement
The authors would like to thank the Coordination for the Improvement of Higher Education Personnel - CAPES for the financial support. The present paper was presented inside the 'Toward a Sustainable Mobility' special session as part of the 'Two Seats for a Solar Car' research project, an action funded by the Italian Ministry of Foreign Affairs and International Cooperation within the Executive Programme of Cooperation in the field of Science and Technology between the Italian Republic and the Republic of Serbia.

References
[1] Arcaro S, Wermuth T B, Zampiva R Y S, Venturini J, Caten C S, Bergmann C P, Alves A K, Oliveira A P N De and Moreno R 2019 Li2O-ZrO2-SiO2/Al2O3 nanostructured composites for microelectronics applications, *J. Eur. Ceram. Soc.* **39**(2-3) 491-498
[2] Zampiva R Y S, Acauan L H, Venturini J, Garcia J A M, da Silva D S, Han Z, Kassab L R P, Wetter N U, Agarwal A, Alves A K and Bergmann C P 2018 Tunable green/red luminescence by infrared upconversion in biocompatible forsterite nanoparticles with high erbium doping uptake, *Opt. Mater. (Amst).* **76** 407-415
[3] Venturini J, Zampiva R Y S, Piva D H, Piva R H, Cunha J B M da and Bergmann C P 2018 Conductivity dynamics of metallic-to-insulator transition near room temperature in normal spinel, *J. Mater. Chem. C* **6**(17) 4720–6
[4] Zampiva R Y S, Kaufmann Junior C G, Pinto J S, Panta P C, Alves A K and Bergmann C P 2017 3D CNT macrostructure synthesis catalyzed by MgFe2O4 nanoparticles—A study of surface area and spinel inversion influence, *Appl. Surf. Sci.* **422** 321-330
[5] Fotouhi M, Saghafi H, Brugo T, Minak G, Fragassa C, Zucchelli A and Ahmadi M 2017 Effect of PVDF nanofibers on the fracture behavior of composite laminates for high-speed woodworking machines, *P I Mech Eng C-J Mec* **231**(1) 31-43
[6] Kaufmann C G, Zampiva R Y S, Bergmann C P, Alves A K, Mortari S R and Pavlovic A 2018 Production of multi-wall carbon nanotubes starting from a commercial graphite Pencil using an electric arc discharge in aqueous medium, *FME Trans.* **46**(2) 151-156
[7] Zampiva R, Venturini J, Acauan L, Kaufmann C, dos Santos L, Alves A, Bergmann C and Caten C S ten 2018 The impact of the reaction atmosphere on the additive-free growth of $Mg_2B_2O_5$ nanorods *Ceram. Int.* **45**(5) 6228-35
[8] Guaglianoni W C, Cunha M A, Bergmann C P, Fragassa C, Pavlovic A 2018 Synthesis, Characterization and Application by HVOF of a WCCoCr/NiCr Nanocomposite as Protective Coating Against Erosive Wear, *Tribology in Industry* **40**(3) 477-487
[9] Fotouhi M, Fragassa C, Fotouhi S, Saghafi M, Minak G 2019 Damage Characterization of Nano-interleaved Woven CFRP Under Fatigue Loading, *Fibers* **7**(2) no. 13, doi:10.3390/fib7020013.
[10] Rao C N R, Müller A and Cheetham A K 2005 *The Chemistry of Nanomaterials: Synthesis, Properties and Applications*
[11] Dubey P, Choi S-K, Kim B and Lee C-J 2012 Synthesis of thin-multiwalled carbon nanotubes by Fe-Mo/MgO catalyst using sol-gel method, *Carbon Lett.* **13** 99–108
[12] Patil KC, Hegde M S and Aruna S T 2008 Chemistry of Nanocrystalline Oxide Materials, *Chemistry of Nanocrystalline Oxide Materials* (India: WORLD SCIENTIFIC) 362
[13] Liu Z, Bian J, Huang Y, Xu T, Wang X and Dai S 2017 Fabrication and characterization of mid-infrared emission of Pr 3+ doped selenide chalcogenide glasses and fibres, *RSC Adv.* **7** 41520–6
[14] Huang Y, Tang Y, Wang J and Chen Q 2006 Synthesis of MgFe2O4 nanocrystallites under mild conditions, *Mater. Chem. Phys.* **97** 394–7
[15] Satyanarayana L, Reddy K M and Manorama S V 2003 Nanosized spinel NiFe2O4: A novel material for the detection of liquefied petroleum gas in air, *Mater. Chem. Phys.* **82** 21–6
[16] Dinka P and Mukasyan A S 2005 In Situ Preparation of Oxide-Based Supported Catalysts by

IOP Conf. Series: Materials Science and Engineering **659** (2019) 012078 doi:10.1088/1757-899X/659/1/012078

Solution Combustion Synthesis, *J. Phys. Chem. B* **109** 21627–33
[17] Zampiva R Y S, Kaufmann C G, Alves A K and Bergmann C P 2018 Influence of the fuel composition and the fuel/oxidizer ratio on the combustion solution synthesis of MgFe2O4 catalyst nanoparticles, *FME Trans.* **46**(2) 157-164
[18] Dubey P, Choi S K, Choi J H, Shin D H and Lee C J 2010 High-Quality Thin-Multiwalled Carbon Nanotubes Synthesized by Fe-Mo/MgO Catalyst Based on a Sol–Gel Technique: Synthesis, Characterization, and Field Emission, *J. Nanosci. Nanotechnol.* **10** 3998–4006
[19] Panic S, Bajac B, Rakić S, Kukovecz, Kónya Z, Srdić V and Boskovic G 2017 Molybdenum anchoring effect in Fe–Mo/MgO catalyst for multiwalled carbon nanotube synthesis, *React. Kinet. Mech. Catal.* **122** 775–91
[20] Jain S R, Adiga K C and Pai Verneker V R 1981 A new approach to thermochemical calculations of condensed fuel-oxidizer mixtures, *Combust. Flame* **40** 71–9
[21] Hadke S, Kalimila M T, Rathkanthiwar S, Gour S, Sonkusare R and Ballal A 2015 Role of fuel and fuel-to-oxidizer ratio in combustion synthesis of nano-crystalline nickel oxide powders, *Ceram. Int.* **41** 14949–57
[22] Nasiri H, Bahrami Motlagh E, Vahdati Khaki J and Zebarjad S M 2012 Role of fuel/oxidizer ratio on the synthesis conditions of Cu-Al2O3nanocomposite prepared through solution combustion synthesis, *Mater. Res. Bull.* **47** 3676–80
[23] Sral O, Ianos R, Istratie R and Pa C 2016 *Solution combustion synthesis of strontium* 1150–7
[24] Jovanović R, Rašuo B, Stefanović P, Cvetinović D and Swiatkowski B 2013 Numerical investigation of pulverized coal jet flame characteristics under different oxy-fuel conditions, *Int. J. Heat Mass Transf.* **58** 654–62
[25] Effenberg, G., Ilyenko S . 2008 *Ternary Alloy Systems Phase Diagrams, Crystallographic and Thermo– dynamic Data* (Springer Berlin Heidelberg)

IOP Conf. Series: Materials Science and Engineering **659** (2019) 012079 doi:10.1088/1757-899X/659/1/012079

Synthesis of cobalt ferrite ($CoFe_2O_4$) by combustion with different concentrations of glycine

C G Kaufmann Junior[1, 2], R Y S Zampiva[1], A K Alves[1], C P Bergmann[1, 2] and L Giorgini[3]

[1] Post-Graduation Program in Mining, Metallurgy and Materials Engineering (PPGE3M), Federal University of Rio Grande do Sul, Osvaldo Aranha 99, 90035-190, Porto Alegre, Brazil
[2] Post-Graduation Program in Production and Transport Engineering, (PPGEP), Federal University of Rio Grande do Sul, Osvaldo Aranha 99, 90035-190, Porto Alegre, Brazil
[3] Department of Industrial Chemistry "Toso Montanari", University of Bologna, Viale del Risorgimento 4, 40136, Bologna, Italy

E-mail: gabriel.kaufmann@ufrgs.br

Abstract. Among the wide range of applications of Cobalt Ferrite ($CoFe_2O_4$), its usage for the manufacturing of modern electronic devices such as solar panels, capacitors and batteries has been studied' lately. Hence, solar-powered vehicles, which strongly rely their performance on energy efficient electronics, is a sector that could particularly benefit from enhanced applications of such material. In this work, the synthesis of Cobalt Ferrite was studied by means of Solution Combustion Synthesis (SCS) using iron nitrate nonahydrate and cobalt nitrate hexahydrate as precursors, and glycine as fuel. The nitrates were dissolved in distilled water and placed under stirring and heating for 5 minutes; when the temperature reached 60°C, glycine was added; after complete homogenization, the solution was placed in an electric oven (400°C) until complete combustion (approximately 15 minutes). Three syntheses were carried out using different concentrations of fuel: lean, stoichiometric and rich. The product crystallite size and composition were investigated in order to determine the influence of the fuel concentration on the structure of the produced cobalt ferrite.

1. Introduction

Ferrite is a ceramic material produced from burning large proportions of iron (III) oxide (Fe_2O_3) blended with small concentrations of one or more metallic elements, such as cobalt, manganese, nickel, magnesium, and zinc [1]. Ferrites usually present a spinel structure with the formula AB_2O_4, where A and B represent various metal cations, including iron (Fe) [2].

Many studies on ferrites nanoparticles including $CoFe_2O_4$ spinel have been recently presented due to their distinct magnetic and electronic properties, which open up numerous possibilities including cancer treatment [3], biosensors [4], sensors [5], memory shape alloys [6] and water purification [7]. In addition, ferrite cobalt has been emerging as a promising material for the development of modern electronic devices, such as capacitors, sensors or high capacity batteries [8–10]. Thus this material can be particularly useful in applications related to the sustainable mobility in general and to the solar vehicles in particular. Optimizing the synthesis and usage of such material, for instance, has become a topic of interest for the design of solar-powered electric vehicles, given that power management and energy efficiency are the key aspects of such mobility branch [11-13].

Solution combustion synthesis (SCS) has been increasingly applied in the production of oxide catalysts due to the possibility of producing low-cost, highly pure and homogeneous nanostructured powders. SCS mainly consists of combining the reactants in an aqueous medium using a complexing agent (usually glycine and urea), as well as oxidizing agents (usually metal nitrates), to oxidize the fuel [14,15]. The mixture is heated to 150°C and 500°C to carry out self-ignition in a rapid combustion reaction that can reach more than 1700°C [16]. A solid product that is typically well dispersed and crystalline is formed at the end of the process [16].

The concentration ratio between the complexing agent and oxidizing agent is directly related to the final properties of the powder produced [15]. Thus, this study aims to formulate a methodology to use SCS to produce, at a low cost, homogeneous $CoFe_2O_4$ catalyst nanoparticles. For this purpose, we studied the influence of different concentrations of fuel on the product composition and morphology. With this scope, we used an experimental approach that, coupling diffractometry and microscopy, permitted relevant and unexpected results in similar situations [17-19].

2. Materials and methods

2.1 *Synthesis of cobalt ferrite*

$CoFe_2O_4$ nanostructured particles were produced by SCS with various concentrations of fuel. $Co(NO_3)_2.6H_2O$ and $Fe(NO_3)_3.9H_2O$ were employed as precursors and oxidizers, while glycine ($C_2H_5NO_2$) is used as fuel. All of the reactants were obtained from Merck. The fuel stoichiometric, rich and lean ratios were established by the chemistry of propellants method [20] and applied according to Equations 1, 2 and 3, respectively.

$$2Fe(NO_3)_3.9H_2O + Co(NO_3)_2.6H_2O + 4.5C_2H_5NO_2 + 0.25O_2 \rightarrow CoFe_2O_4 + 9CO_2 + 35.25H_2O + 6.25N_2 \quad (1)$$

$$2Fe(NO_3)_3.9H_2O + Co(NO_3)_2.6H_2O + 5.5C_2H_5NO_2 + 4.75O_2 \rightarrow CoFe_2O_4 + 11CO_2 + 37.75H_2O + 6.75N_2 \quad (2)$$

$$2Fe(NO_3)_3.9H_2O + Co(NO_3)_2.6H_2O + 3.5C_2H_5NO_2 \rightarrow CoFe_2O_4 + 7CO_2 + 32.75H_2O + 5.75N_2 + 4.25O_2 \quad (3)$$

The nitrates were individually dissolved in water and then mixed. The solution was stirred and heated for approximately 5 min. When reaching 60°C, the fuel that had previously been dissolved in water was added to the solution. The solution was maintained under stirring and temperature for a few minutes to promote perfect homogenization. The solution was placed in an electric muffle furnace preheated to 400°C until complete combustion (around 15 min).

2.2 *Chemical and structural characterization*

The crystallinity of the samples was evaluated by X-ray diffraction (XRD) using a PHILIPS diffractometer (model X'Pert MPD) at 40 kV and 40 mA using a Cu anode. The powder crystallite sizes were calculated using Scherrer's equation 4:

$$D = \frac{K\lambda}{\beta \cos\theta} \quad (4)$$

where D represents the crystallite size, K is a constant that depends on the particle shape (we assume the particles to be spherical, meaning K=0.94), λ is the wavelength of the electromagnetic radiation used (1.5406 Å, a value related to the main characteristic radiation emitted by copper), θ is Bragg`s angle, and β is the contribution of the crystallite size to the full width at half maximum (FWHM) of the corresponding diffraction peak in radians.

The morphology of the $CoFe_2O_4$ was characterized by scanning electron microscopy (SEM) using a JEOL microscope (JSM 6060) with a maximum operating voltage of 30 kV and a nominal resolution of 3.5 nm. The applied voltage was 10 to 20 k.

3. Results and Discussions

The diffractograms in Figure1 shows the characteristic crystal peaks corresponding to the cobalt ferrite phase for all samples according to JCPDS 73-1960. The sample lean presented a high degree of purity and showed only the $COFe_2O_4$ phase. The stoichiometric and rich samples showed the formation of a second phase of cobalt oxide (JCPDS 73-1701).

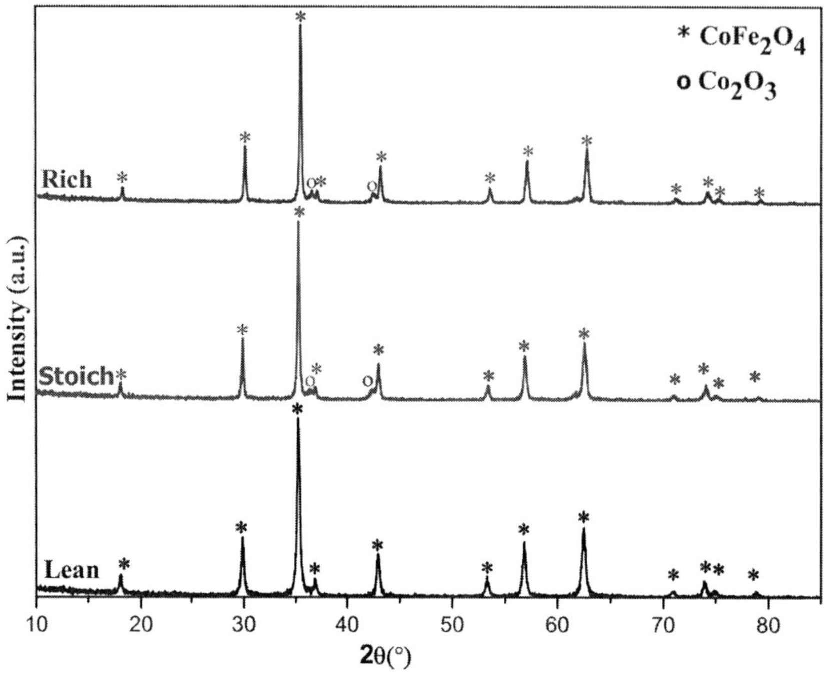

Figure 1. Diffractograms of the $CoFe_2O_4$ catalysts for different fuel concentrations.

SCS synthesis with elevated concentration of fuel leads to fast ignition and intense burn during reaction. With the excess of fuel, oxygen from the atmosphere is required in order to complete the reaction. This oxygen excess may cause the formation of secondary oxide phases besides the primary cobalt ferrite phase [21]. The crystallite sizes calculated by Scherrer's for showed for the different concentrations: 23.58 nm for lean (the smallest size among all samples), 31.14 nm for stoichiometric, and 33.16 nm for Rich. These results indicate that powder has a nanostructured crystallite. Figure 2 shows SEM micrographs of the $CoFe_2O_4$ samples just after combustion with no grinding process involved.

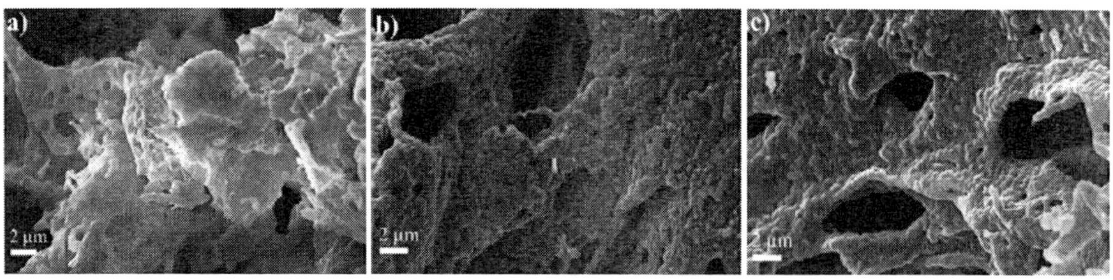

Figure 2. SEM images of samples: a) Lean, b) Stoich, c) Rich.

Figure 2 it is possible to observe the production of porous structures for all the compositions. The formation of the spongy features is attributed to the evolution of a large amount of gas during combustion [22].

The amount of oxygen involved in the reaction directly affects the velocity of the process and the product composition as discussed above. Glycine has a small carbonic chain, and the presence of N and OH in the structure provides an intense and fast high-temperature combustion during ignition, which releases a high volume of gases [21]. This release of gases leads to the production of more porous structures as can be seen when comparing the fuel-lean sample (Figure 2a) with the rich (Figure 2c) and stoichiometric sample (Figure 2b). It is possible to observe the increasing of structural porosity with the increasing of fuel concentration.

4. Conclusion

The synthetic route presented in this paper shown to be effective for production of nanostructured $CoFe_2O_4$. In this study was possible to conclude that in the SCS of $CoFe_2O_4$, higher concentrations of glycine lead to the formation of secondary phases of cobalt oxide. The fuel-lean sample was the only one that produced a nanometer pure phase of $CoFe_2O_4$. This sample is promising for applications in the electronics field, such as on solar-powered electric vehicles, since pure particles and good crystallinity are crucial in this area.

Acknowledgment

The authors would like to thank the Coordination for the Improvement of Higher Education Personnel - CAPES for the financial support. The present paper was presented inside the '*Toward a Sustainable Mobility*' special session as part of the '*Two Seats for a Solar Car*' research project, an action funded by the Italian Ministry of Foreign Affairs and International Cooperation within the Executive Programme of Cooperation in the field of Science and Technology between the Italian Republic and the Republic of Serbia.

References

[1] Carter, C. Barry, Norton M G 2007 *Ceramic Materials-Science and Engineering* (pringer Science & Business)

[2] Houshiar M, Zebhi F, Razi Z J, Alidoust A and Askari Z 2014 Synthesis of cobalt ferrite (CoFe2O4) nanoparticles using combustion, coprecipitation, and precipitation methods: A comparison study of size, structural, and magnetic properties, *J. Magn. Magn. Mater.* **371** 43–8

[3] Park B J, Choi K-H, Nam K C, Ali A, Min J E, Son H, Uhm H S, Kim H-J, Jung J-S and Choi E H 2015 Photodynamic Anticancer Activities of Multifunctional Cobalt Ferrite Nanoparticles in Various Cancer Cells, *J. Biomed. Nanotechnol.* **11** 226–35

[4] Yardımcı F S, Şenel M and Baykal A 2012 Amperometric hydrogen peroxide biosensor based on cobalt ferrite–chitosan nanocomposite, *Mater. Sci. Eng. C* **32** 269–75

[5] Rao G S N, Kumar S A, Rao K, Rao B P, Gupta A, Caltun O, Dumitru I and Kim C 2007 *Doped Cobalt Ferrites for Stress Sensor Applications*, 2007 2nd IEEE International Conference on Nano/Micro Engineered and Molecular Systems (IEEE), pp 1186–9

[6] Craciunescu C M, Kishi Y, De Graef M, Lograsso T A and Wuttig M R 2002 Cobalt-base ferromagnetic shape memory alloys ed C S Lynch pp 235–44

[7] Li X, Liu X, Lin C, Zhang H, Zhou Z, Fan G and Ma J 2019 Cobalt ferrite nanoparticles supported on drinking water treatment residuals: An efficient magnetic heterogeneous catalyst to activate peroxymonosulfate for the degradation of atrazine, *Chem. Eng. J.* **367** 208–18

[8] Caltun O, Dumitru I, Feder M, Lupu N and Chiriac H 2008 Substituted cobalt ferrites for sensors applications, *J. Magn. Magn. Mater.* **320** e869–73

[9] Hashemi M, Mohandes F, Salavati-Niasari M and Esmaeily A S 2015 Preparation of cobalt ferrite micro/nanoparticles by solid-state thermal decomposition of a novel single-source precursor, *J. Mater. Sci. Mater. Electron.* **26** 6860–7

[10] Qiu B, Deng Y, Du M, Xing M and Zhang J 2016 Ultradispersed Cobalt Ferrite Nanoparticles Assembled in Graphene Aerogel for Continuous Photo-Fenton Reaction and Enhanced Lithium Storage Performance, *Sci. Rep.* **6** 29099

[11] Minak G, Fragassa C and de Camargo F V 2017 *A brief review on determinant aspects in energy efficient solar car design and manufacturing*, International Conference on Sustainable Design and Manufacturing, Bologna, Italy, April 26 – 28, pp. 847-856

[12] de Camargo F V, Giacometti M and Pavlovic A 2017 *Increasing the energy efficiency in solar vehicles by using composite materials in the front suspension*, International Conference on Sustainable Design and Manufacturing, Bologna, Italy, April 26 – 28, pp. 801-811

[13] Minak G, Brugo T M, Fragassa C, Pavlovic A, de Camargo F V and Zavatta N 2019 Structural Design and Manufacturing of a Cruiser Class Solar Vehicle, *Journal of Visual Experiments* **143** e58525, doi:10.3791/58525

[14] Da Dalt S, Takimi A S, Volkmer T M, Sousa V C and Bergmann C P 2011 Magnetic and Mössbauer behavior of the nanostructured MgFe2O4 spinel obtained at low temperature, *Powder Technol.* **210** 103–8

[15] Zampiva R Y S, Kaufmann Junior C G, Pinto J S, Panta P C, Alves A K and Bergmann C P 2017 3D CNT macrostructure synthesis catalyzed by MgFe2O4nanoparticles—A study of surface area and spinel inversion influence, *Appl. Surf. Sci.* **422** 321–30

[16] Dinka P and Mukasayan A 2005 In situ preparation of Oxide-based Supported Catalysts by Solution Combustion Synthesis, *J. Phys. Chem. B* **109** 21627–33

[17] Giorgini L, Fragassa C, Zattini G and Pavlovic A 2016 Acid aging effects on surfaces of PTFE gaskets investigated by Fourier Transform Infrared Spectroscopy, *Tribology in Industry* **38**(3) 286-296.

[18] Fragassa C, Giorgini L, Pavlovic A and Zattini G 2016 Acid aging effects on surfaces of PTFE gaskets investigated by Thermal Analysis, *Tribology in Industry* **38**(4) 435-444

[19] Zattini G, Pavlovic A, Fragassa C and Giorgini L 2017 Characterization of solid residues emerging during ultrasonic cleaning treatments of tyre moulds, *FME Transactions* **45**(3) 339-347,

[20] Jain S R, Adiga K C and Pai Verneker V R 1981 A new approach to thermochemical calculations of condensed fuel-oxidizer mixtures, *Combust. Flame* **40** 71–9

[21] Jovanović R, Rašuo B, Stefanović P, Cvetinović D and Swiatkowski B 2013 Numerical investigation of pulverized coal jet flame characteristics under different oxy-fuel conditions, *Int. J. Heat Mass Transf.* **58** 654–62

[22] Effenberg G and Ilyenko S 2008 *Ternary Alloy Systems Phase Diagrams, Crystallographic and Thermo– dynamic Data* (German: Springer Berlin Heidelberg)

IOP Conf. Series: Materials Science and Engineering **659** (2019) 012080 doi:10.1088/1757-899X/659/1/012080

First assessment on suspension parameter optimization for a solar – powered vehicle

S Sorrentino[1], A De Felice[1], P Grosso[1] and G Minak[2]

[1]University of Modena and Reggio Emilia, Department of Engineering Enzo Ferrari, Via Vivarelli 10, 41125 Modena, Italy

[2]University of Bologna, Department of Industrial Engineering, Viale Risorgimento 2, 40136, Bologna, Italy

E-mail: pasquale.grosso@unimore.it

Abstract. Optimization of suspension parameters with respect to comfort and road holding is a challenging issue for solar–powered cars, due to in–wheel electric engines on very light vehicles, carrying payloads which can exceed their total mass. The solar–powered car considered in this study was designed and manufactured for racing by the University of Bologna; with a mass of 300 kg and a payload of 320 kg due to four occupants, using 5 m^2 of monocrystalline silicon photovoltaic panel on the roof, 64 kg of lithium–ion batteries and two electric engines coupled directly to the rear wheels, it can achieve either a range of 600 km at cruising speed, or velocity peaks of 120 km/h. In this contribution, equivalent vertical stiffness and equivalent damping coefficients are optimized for both axles, achieving results that in terms of comfort and road holding are comparable to those of standard passenger cars.

1. Introduction

The interest in solar–powered vehicles arose as a topic of study mainly developed by academic institutions [1,2] with the aim of promoting sustainable mobility [3], including new standards and regulations [4]. The main challenges in designing such cutting–edge vehicles [5,6] consist in electrical systems, solar array design, structural materials [7] and mechanical subsystems such as suspensions [8]. Focusing on suspension systems, the load due to passengers, electric batteries and solar panels can be even higher than the total weight of the remaining parts of the vehicle, making the choice of suspension stiffness and weight distribution quite challenging in order to get good performances in terms of vehicle dynamics [9,10]. These technical demands become extreme in the case of the so-called 'cruiser' category, i.e. multi–passenger vehicles recently introduced in solar–powered car competitions. In fact, compared to more traditional single–seater solar cars, cruisers present a total weight from four to five times higher, as well as higher center of gravity [11].

The solar–powered car considered in this study and displayed in Figure 1 was designed and manufactured for racing. It is equipped with a solar–electric powertrain developed for efficiently transporting four passengers weighing 80 kg each. With a monocoque, structural and non–structural parts in carbon fiber reinforced polymer (CFRP), a roll–cage in titanium alloy among other distinct technical features, it represents one of the lightest multi–occupant solar cars ever built: a vehicle with total mass of 320 kg allows to transport 320 kg due to four occupants. Using 5 m^2 of monocrystalline silicon photovoltaic panel on the roof, 64 kg of lithium–ion batteries, two electric engines coupled directly to the rear wheels and further solutions for optimal energy control, either a range of 600 km at cruising speed, or velocity peaks of 120 km/h can be achieved [11].

Figure 1. Solar–powered car under investigation [11].

However, this vehicle still needs further mechanical improvements, for increasing the average speed of the race vehicle up to 75 km/h, a necessary condition to successfully take part in further competitions. In particular, the current suspension system (as shown in Figure 2) has to be designed considering the necessity to provide adequate levels of handling, road holding, comfort and vibration control in a process of design review and optimisation as done for more conventional vehicles [12,13].

Figure 2. Suspension system under investigation [11].

The focus of the present contribution is a first assessment on optimization of the main parameters affecting comfort and road holding, which are the equivalent stiffness and damping of the suspensions [14] for this multi-occupant solar–powered car.

To this purpose, in section 2 the equivalent vertical stiffness of the front and rear axles are tuned in order to fit the basic requirements for comfort in terms of natural frequencies and mode shapes of bounce and pitch [15]; the optimization of suspension damping [16] is then addressed in section 3, for reaching a good compromise between comfort and road holding, aimed at achieving results comparable to those of standard passenger cars.

2. Tuning of suspension stiffness

Two fundamental parameters in car suspension design are first considered, the dynamic index r_1 (related to mass distribution) and the stiffness index r_2 (related to the equivalent vertical stiffness of the two axles) [14], which are defined by:

$$r_1 = \frac{J_y}{m_s a_1 a_2}, \quad r_2 = \frac{k_1 a_1}{k_2 a_2} \tag{1}$$

where m_s is the total sprung mass, J_y is the lateral moment of inertia with respect to the center of gravity, a_1 and a_2 are the front and rear partial wheelbases (longitudinal distances of the center of gravity from the front and rear axles respectively, where $a_1 + a_2 = l$ is the wheelbase), while k_1 and k_2

are the equivalent vertical stiffness of the front and rear axles, respectively. Usually in road cars r_1 ranges between 0.90 and 0.97, while r_2 is set between 0.90 and 0.95 for respecting current standards in suspension design [14]. A 2–dofs model for free vertical oscillations is adopted, according to:

$$\mathbf{M\ddot{s}}+\mathbf{Ks}=\mathbf{0},\quad \mathbf{M}=\begin{bmatrix} m_s & 0 \\ 0 & J_y \end{bmatrix},\quad \mathbf{K}=\begin{bmatrix} K_{11} & K_{12} \\ K_{21} & K_{22} \end{bmatrix}=\begin{bmatrix} k_1+k_2 & k_1a_1-k_2a_2 \\ k_1a_1-k_2a_2 & k_1a_1^2+k_2a_2^2 \end{bmatrix},\quad \mathbf{s}=\begin{Bmatrix} z_G \\ \vartheta \end{Bmatrix} \tag{2}$$

where z_G is the vertical displacement of the center of gravity of the sprung mass and ϑ is the pitch angle, yielding the natural angular frequencies of the so–called bounce and pitch modes as:

$$\omega_{1,2}=\sqrt{\frac{m_sK_{22}+J_yK_{11}\mp\sqrt{(m_sK_{22}-J_yK_{11})^2+4m_sJ_yK_{12}K_{21}}}{2m_sJ_y}} \tag{3}$$

while the nodes of the two mode shapes are given by the following longitudinal coordinates (positive values backwards with respect to the center of gravity):

$$x_{01}=\frac{\omega_1^2J_y-K_{22}}{K_{21}},\quad x_{02}=\frac{\omega_2^2J_y-K_{22}}{K_{21}} \tag{4}$$

Assuming at a first stage r_1 = 1, then the nodes in equations (4) are located on the vertical of the two axles; moreover, if $r_2 < 1$ then the two natural angular frequencies are simply given by:

$$\omega_1=\sqrt{\frac{k_1l}{m_s(l-a_1)}}=\sqrt{\frac{g}{\Delta z_1}},\quad \omega_2=\frac{1}{\sqrt{r_2}}\omega_1>\omega_1,\quad \Delta z_2=r_2\Delta z_1<\Delta z_1 \tag{5}$$

where Δz_1 and Δz_2 are the static deflections at the front and rear axles, respectively. An upper bound for the largest static deflection produces a lower bound for the first natural frequency.
Setting r_2 = 0.95 and Δz_1 = 150 mm (design upper limit for the car under study) yields Δz_2 = 142.5 mm, f_1 = 1.29 Hz, f_2 = 1.32 Hz. The effect of reducing r_2 would simply be that of reducing Δz_2 and increasing f_2, according to equations (5). The (realistic) effect of reducing r_1 is addressed in section 4.

3. Tuning of suspension damping

The damping distribution is set to give a damping matrix proportional to the stiffness matrix in equations (2), i.e. $\mathbf{C} = \beta\mathbf{K}$, with damping coefficients $c_1 = \beta k_1$ and $c_2 = \beta k_2$ (as a consequence, the mode shapes are the same of the related undamped model).

For selection of the damping coefficients c, a different 2–dofs model is adopted, usually called Quarter car model, but herein referred to forced vibration of the whole vehicle [6], according to:

$$\mathbf{M\ddot{x}}+\mathbf{C\dot{x}}+\mathbf{Kx}=\mathbf{f},\quad \mathbf{M}=\begin{bmatrix} m_s & 0 \\ 0 & m_n \end{bmatrix},\quad \mathbf{C}=\begin{bmatrix} c & -c \\ -c & c \end{bmatrix},\quad \mathbf{K}=\begin{bmatrix} k & -k \\ -k & k+p \end{bmatrix},\quad \mathbf{f}=\begin{Bmatrix} 0 \\ ph \end{Bmatrix},\quad \mathbf{x}=\begin{Bmatrix} z \\ y \end{Bmatrix} \tag{6}$$

where m_n is the total unsprung mass, $c = \beta k$, $k = k_1 + k_2$ (hence $c_1 = \beta k_1$, $c_2 = \beta k_2$), p is the total vertical stiffness of the tyres, h is the displacement imposed at the ground by uneven surface (assumed harmonic, with angular frequency ω), z and y are the displacements of m_s and m_n, respectively. Notice that in this model the longitudinal position of the center of gravity, and therefore a_1, a_2, are not influential. The two natural frequencies of the undamped model described by equations (6) are:

$$\omega_{1,2}^{(i)}=\frac{1}{2}\left[\frac{p+k}{m_n}+\frac{k}{m_s}\mp\sqrt{\left(\frac{p+k}{m_n}-\frac{k}{m_s}\right)^2+\frac{4k^2}{m_nm_s}}\right] \tag{7}$$

According to the following data (design parameters for the car under study): m_s = 584 kg, m_n = 59.32 kg, k = 39203 N/m (due to $r_1 = r_2 = 0.95$, a_1 = 1421 mm, k_1 = 18748 N/m, k_2 = 20455 N/m), p = 3.6×10^5 N/m, then equation (7) yields f_1 = 1.24 Hz, f_2 = 13.06 Hz (with $f_2/f_1 \approx 10$, a typical value for road cars). As suggested by Bourcier de Carbon [16], a choice for an 'optimal' damping coefficient c is:

$$c_o = \sqrt{\frac{m_s k}{2}}\sqrt{\frac{p+2k}{p}} \tag{8}$$

in this case yielding c_o = 3734 Ns/m (which, introducing $\mathbf{C} = \beta\mathbf{K}$ in equations 2, gives modal damping ratios in the usual range of 0.38 to 0.40).

Equation (8) provides a value close to what can be considered the best choice for coefficient c in terms of comfort optimization, which has to be increased for road holding optimization [14]. To this purpose the following frequency response functions are considered:

1) receptance R_Z between amplitude Z (m_s vertical displacement) and amplitude H of ground harmonic input ($\omega^2 R_Z$ yielding a measure of m_s vertical acceleration, to be minimized for comfort optimization):

$$R_Z = \frac{|Z|}{H} = p\sqrt{\frac{k^2 + c^2\omega^2}{h_1^2(\omega) + c^2\omega^2 h_2^2(\omega)}}, \quad \begin{cases} h_1(\omega) = m_s m_n \omega^4 - [(p+k)m_s + k m_n]\omega^2 + pk \\ h_2(\omega) = p - (m_s + m_n)\omega^2 \end{cases} \tag{9}$$

2) receptance R_Y between amplitude Y (m_n vertical displacement), and amplitude H:

$$R_Y = \frac{|Y|}{H} = p\sqrt{\frac{(k - m_s\omega^2)^2 + c^2\omega^2}{h_1^2(\omega) + c^2\omega^2 h_2^2(\omega)}} \tag{10}$$

3) receptance $R_{Z\text{-}Y}$ between relative amplitude $Z - Y$, and amplitude H:

$$R_{Z-Y} = \frac{|Z - Y|}{H} = \frac{p\, m_s \omega^2}{\sqrt{h_1^2(\omega) + c^2\omega^2 h_2^2(\omega)}} \tag{11}$$

4) transmissibility T between ground vertical force amplitude N, and force pH:

$$T = \frac{N}{pH} = \omega^2\sqrt{\frac{\left[m_s m_n \omega^2 - k(m_s + m_n)\right]^2 + c^2\omega^2 (m_s + m_n)^2}{h_1^2(\omega) + c^2\omega^2 h_2^2(\omega)}} \tag{12}$$

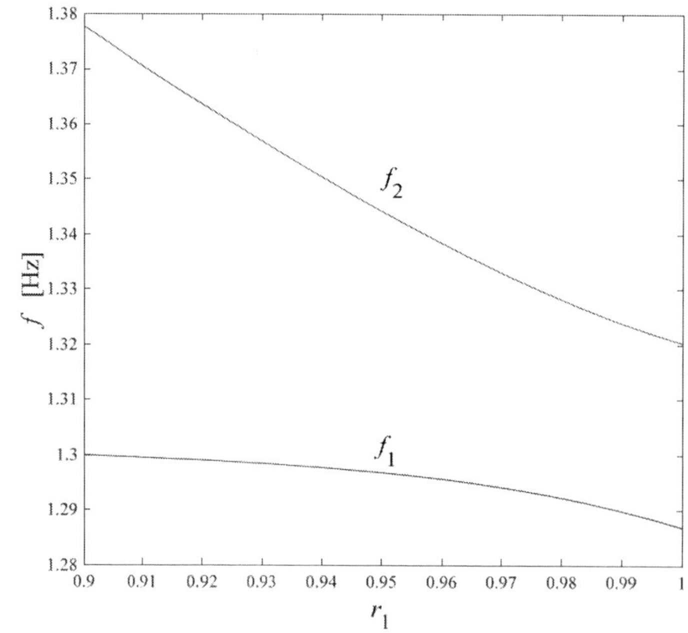

Figure 3. Effect of dynamic index r_1 (setting the stiffness index $r_2 = 0.95$) on natural frequencies of bounce (f_1) and pitch (f_2).

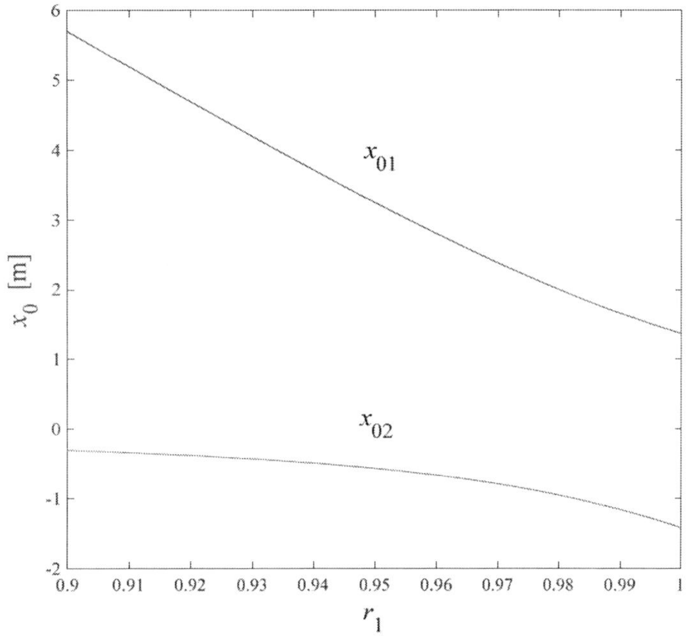

Figure 4. Effect of dynamic index r_1 (setting the stiffness index $r_2 = 0.95$) on the longitudinal position of nodes x_{01} (bounce mode) and x_{02} (pitch mode).

4. Results and discussion

The effects of reducing r_1 (from 1 to 0.9, for realistic mass distributions) can be studied by means of equations (2) to (4), with the following data (design parameters for the car under study): $m_s = 584$ kg, $l = 2792$ mm, $a_1 = 1421$ mm.

As shown in figures 3 and 4 for $r_2 = 0.95$, reducing r_1 from 1 to 0.9 produces a moderate increase of both natural frequencies (however without exceeding the upper bound of 1.50 Hz) together with a large shift of the nodal points x_{01} and x_{02} (x_{01} increases, meaning that the modal shape of the first mode, with lower natural frequency, is getting closer to that of pure bounce; while x_{02} reduces, meaning that the modal shape of the second mode, with higher natural frequency, is getting closer to that of pure pitch). Having the second nodal point (x_{02}) close to the center of gravity, and a bit ahead of it, helps to enhance the comfort of driver and passengers on front seats [6,7]. Realistic variations of a_1 do not affect significantly the results reported in figures 3 and 4 (1.0 Hz $< f_1 < f_2 <$ 1.5 Hz, mode at lower natural frequency f_1 = bounce, mode at higher natural frequency f_2 = pitch).

Therefore, with the adopted design parameters (mass distribution, partial wheelbases), it is possible in any case to fulfil the basic comfort requirements for free vibration, that is: 1) natural frequencies of bounce and pitch modes falling in the range 1.0 – 1.5 Hz; 2) pitch mode with its node located at about the front seats [15], as shown in figures 3 and 4. Slight variations of mass distribution (due to different arrangements of the battery pack) would result in small corrections to equivalent vertical stiffness parameters (in index r_2, equation 1), without affecting comfort.

Regarding forced vibration, in road cars the main interest is minimizing the amplitude of vertical acceleration of the sprung mass [14,16], for optimizing comfort.

As shown in Figure 5, c_o computed with equation (8) and $r_1 = r_2 = 0.95$ is a very good choice for improving comfort in the specific case of the car under study, since it yields a frequency response function with amplitudes comparable to those of standard passenger cars [14]. However, as shown in figure 6, for improving significantly road holding, c should be moderately increased (at least to $1.3c_o$), thus affecting comfort. Notice that increasing c would also have the advantageous effect of reducing the maximum amplitude of relative displacement between sprung and unsprung masses, as displayed in Figure 7.

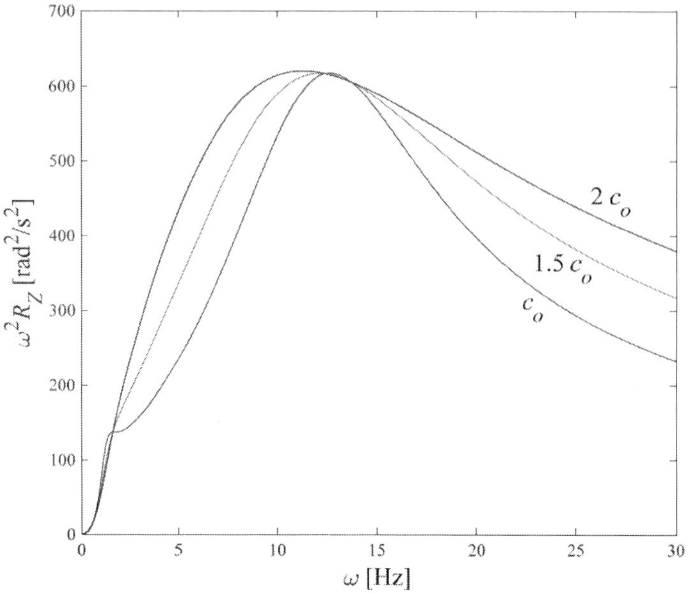

Figure 5. Amplification factor for the sprung mass m_s vertical acceleration ($\omega^2 R_Z$) as a function of frequency, for different values of suspension damping.

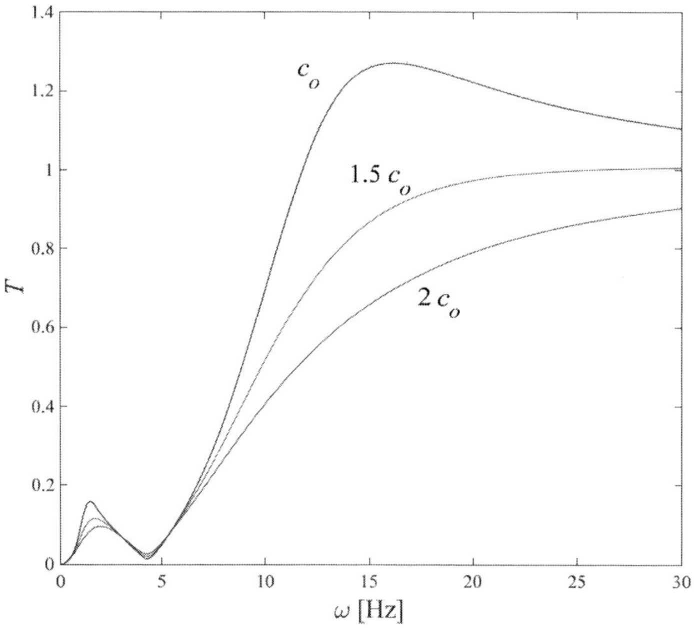

Figure 6. Transmissibility between ground vertical force amplitude N and force pH as a function of frequency (equation 12), for different values of suspension damping.

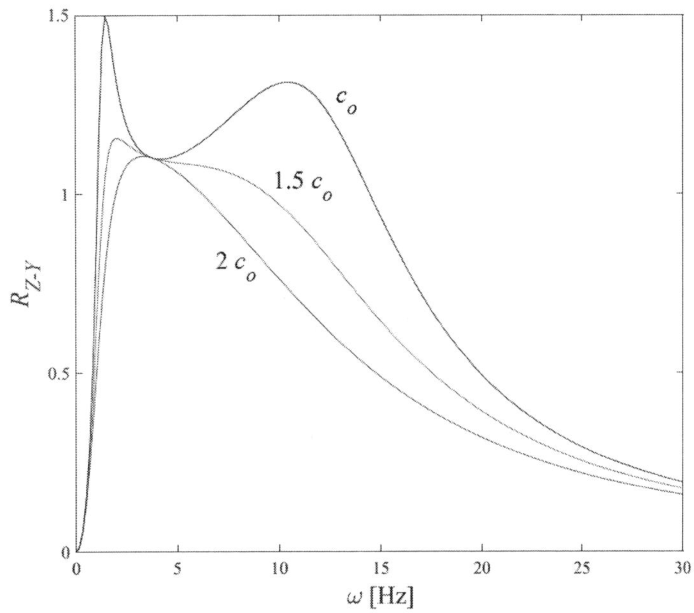

Figure 7. Amplification factor between relative amplitude $Z - Y$, and amplitude H as a function of frequency (equation 11), for different values of suspension damping.

5. Conclusions

In this contribution a preliminary study has been presented regarding the optimization of the suspension equivalent stiffness and damping parameters in the case of the Italian solar–powered multi-occupant car. This vehicle was designed and manufactured by the University of Bologna, together with the support of other Italian institutions, with the scope to compete in exclusive solar challenges all over the world. And it won in 2018 its first competition, the American Solar Challenge, running for 1,700km with 4 passengers and a cruising speed of, approximately, 65 km/h. But racing rules never stop to pretend more with the declared scope to foster researches and results on the solar mobility. Thus, an average speed higher than 75 km/h is now requested to take part to the next international competition, the World Solar Challenge, October 2019. This target also means a desired maximum speed of around 120-130km/h for an extremely light quadricycle. Nevertheless, this performance cannot be achieved without a relevant redesign and optimization in terms of vehicle dynamic, where this study represents a first fundamental step.

The equivalent vertical stiffness of the front and rear axles has been tuned in order to fit the basic requirements for comfort in terms of natural frequencies and mode shapes of bounce and pitch. The suspension damping has then been optimized for reaching a good compromise between comfort and road holding, aimed at achieving results comparable to those of standard passenger cars.

Acknowledgment

The present paper was presented inside the '*Toward a Sustainable Mobility*' special session as part of the '*Two Seats for a Solar Car*' research project, an action funded by the Italian Ministry of Foreign Affairs and International Cooperation within the Executive Programme of Cooperation in the field of Science and Technology between the Italian Republic and the Republic of Serbia.

References

[1] Wamborikar Y S and Sinha A *Solar powered vehicle*, World Congress on Engineering and Computer Science, October 2010, Vol. 2, pp. 20-22

[2] Rizzo G, Arsie I and Sorrentino, M, *Solar energy for cars: perspectives, opportunities and problems*. In GTAA Meeting, 2010, May, pp. 1-6

[3] Zacharof N, Fontaras G, Ciuffo B et al. 2016 *Review of in use factors affecting the fuel consumption and CO2 emissions of passenger cars*. European Union, EUR 27819 EN, doi:10.2790/140640

[4] Pavlovic A and Fragassa C 2015 General considerations on regulations and safety requirements for quadricycles, *International Journal for Quality Research* **9**(4) 657-674

[5] Fenton J and Hodkinson R 2001 *Lightweight electric/hybrid vehicle design*, Butterworth–Heinemann

[6] Thacher E F 2015 *A Solar Car Primer: A Guide to the Design and Construction of Solar–Powered Racing Vehicles*, Springer

[7] de Camargo F V, Fragassa C, Pavlovic A and Martignani M 2017 Analysis of the Suspension Design Evolution in Solar Cars, *FME Transactions* **45**(3) 394–404

[8] Odabaşi V, Maglio S, Martini A and Sorrentino S 2019 Static Stress Analysis of Suspension Systems for a Solar–Powered Car, *FME Transactions* **47**(1) 70–75

[9] Arsie I, Rizzo G and Sorrentino M 2006 Optimal design and dynamic simulation of a hybrid solar vehicle. *SAE Technical Paper* No. 2006-01-2997.

[10] Minak G, Fragassa C and de Camargo F V 2017 *A brief review on determinant aspects in energy efficient solar car design and manufacturing*, International Conference on Sustainable Design and Manufacturing, Bologna, Italy, April 26 – 28, pp. 847-856

[11] Minak G, Brugo T M, Fragassa C, Pavlovic A, de Camargo F V and Zavatta N 2019 Structural Design and Manufacturing of a Cruiser Class Solar Vehicle, *Journal of Visualized Experiments* **143** 1–15

[12] Martini A, Bellani G and Fragassa C 2018 Numerical assessment of a new hydro-pneumatic suspension system for motorcycles, *International Journal of Automotive and Mechanical Engineering* **15**(2) 5308-5325

[13] Leonelli L, Cattabriga S and Sorrentino S 2018 Driveline instability of racing motorcycles in straight braking manoeuvre, *Proceedings of the Institution of Mechanical Engineers, Part C: Journal of Mechanical Engineering Science* **232**(17) 3045-3061

[14] Guiggiani M 2014 *The Science of Vehicle Dynamics*, Springer

[15] Olley M 1934 Independent wheel suspension – its whys and wherefores, *SAE Transactions* **34**(3) 73–81

[16] Bourcier de Carbon C 1950 *Theorie mathématique et réalisation pratique de la suspension amortie des véhicules terrestres*, Troisieme Congres Technique de l'Automobile, Paris, France

Synthesis of $Bi_2Fe_xNbO_7$ films applied as a catalyst for hydrogen production using visible-light photo-electrolysis

A R A Scharnberg[1], A Pavlovic[2] and A K Alves[1]

[1] Laboratory of Ceramic Materials, Department of Materials, Federal University of Rio Grande do Sul - UFRGS, Osvaldo Aranha 99, Porto Alegre, RS 90035-190, Brazil
[2] Interdepartmental Center for Industrial Research on Advanced Mechanics and Materials, University of Bologna, Viale del Risorgimento 2, 40136, Bologna, Italy

E-mail: allan.ambiental@outlook.com

Abstract. The seek for feasible sustainable mobility alternatives is a major concern of the society nowadays. By its turn, the production of hydrogen represents one of the main lines of study on clean energy, since hydrogen presents the possibility of storage and association with other renewable energy sources. In this regard, photo-electrolysis is a promising option for hydrogen generation. This process optimizes the electrolysis of water by using external energy to increase the potential of a photo-electrode, which must be a material that absorbs sunlight (usually a semiconductor activated by solar radiation), promoting the generation of H_2 at the cathode. Titanium dioxide (TiO_2) is the most applied semiconductor in photocatalytic applications. However, the band gap of this material limits its activity only under UV light, disregarding about 90% of incident solar radiation. Researchers explore catalysts that can be activated under visible light, such as bismuth-based mixed oxide semiconductors, which have attracted interest because of their excellent stability, visible light absorption, and photocatalytic properties. This research aims to develop and characterize photo-anodes based on bismuth, niobium and iron oxides ($Bi_2Fe_xNbO_7$), for production of hydrogen via photo-electrolysis of water. The films were produced by the sol-gel process and deposited under a conductive glass slide by dip-coating. The effect of the iron concentration was evaluated by UV-Vis spectroscopy and SEM analysis, in order to estimate its hydrogen production potential.

1. Introduction

Since the beginning of time, the advances of our civilization are confused with technological advances [1]. Before the industrial revolution, the main source of energy was biomass, like firewood. After that, with the invention of the steam engine, the natural coal gradually replaced the firewood. After World War II, due to the development of the automobile industry and the emergence of multinational oil companies, oil consumption has grown exponentially: since the beginning of the 21st century, oil, coal and natural gas are the three main sources of global supply of energy, accounting for over 80% of all consumption [2]. Accompanying the trends of economic, technological and population growth, global energy demand has been growing continuously. As most of this energy matrix is composed of fossil fuels, this growth has given rise to a series of environmental problems, mainly due to the emission of greenhouse gases (GHG).

Carbon dioxide emissions from fossil fuels have increased the average concentration of CO_2 in the atmosphere from 270 to 400 ppm in the last 110 years and is estimated that the level reaches 600 ppm by 2050 [3]. This phenomenon can lead to glacier melting, rising and acidification of the ocean, coral extinction, epidemics of temperature-making diseases, tropical storms and extreme snowstorms [4]. Besides, fossil reserves are finite [5], subject to geopolitical conflicts, and there are more noble applications for them, such as the petrochemical industry. Due to these facts, there is a need to develop new technologies for energy generation.

In recent decades there has been a significant development of non-fossil energy sources, such as solar, wind, geothermal and nuclear [2], which has contributed to the reduction of GHG emissions. The challenge is to find a source of energy that is abundant, emission-free and sustainable. Representing a major cause of pollutant emissions, research on urban mobility has seek to take advantage of the solar energy to produce innovative, safe, lightweight and zero-emission vehicles [6–8], although the energy efficiency of these vehicles, using a photovoltaic technology, still has room for improvement.

One of the most interesting proposals to attain that goal is the photo-electrolysis, which also consists in the usage of the solar energy, but to produce hydrogen, since it presents the possibility of storage and association with other renewable sources of energy [3]. In order for hydrogen to be considered a truly emissions-free fuel, its only byproduct must be water, and its entire production process must be carried out cleanly. The most commonly used process for hydrogen production today is from the reform of methane to steam, which cannot be considered a clean process.

Photo-electrolysis is a process similar to electrolysis, but it makes direct use of light by the conversion of solar light into electrical current and then the transformation of water into useful chemical energy (H_2). A photo-electrolytic cell (PEC) is composed of a semiconductor device that absorbs solar energy and generates the voltage necessary to separate the water molecules as shown in figure 1. Photo-electrolysis integrates the generation of solar energy and electrolysis of water in a single photo-anode and is considered the most promising renewable method for hydrogen production [9]. Figure 2 demonstrate de reaction mechanism in the surface of the semiconductor photo-anode.

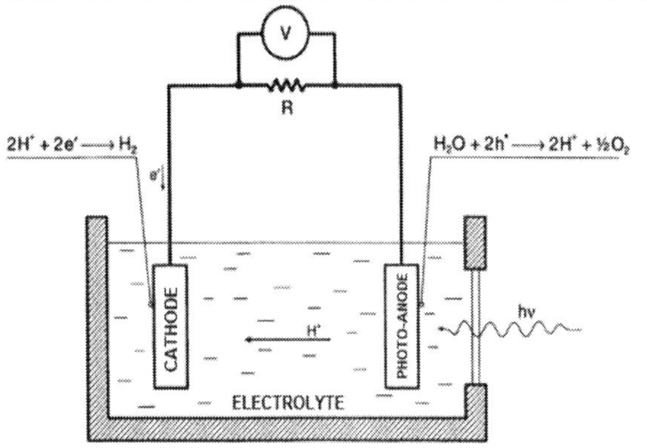

Figure 1. Structure of a photo-electrochemical cell (PEC) for water photo-electrolysis. Adapted from Bak et al, 2002 [10].

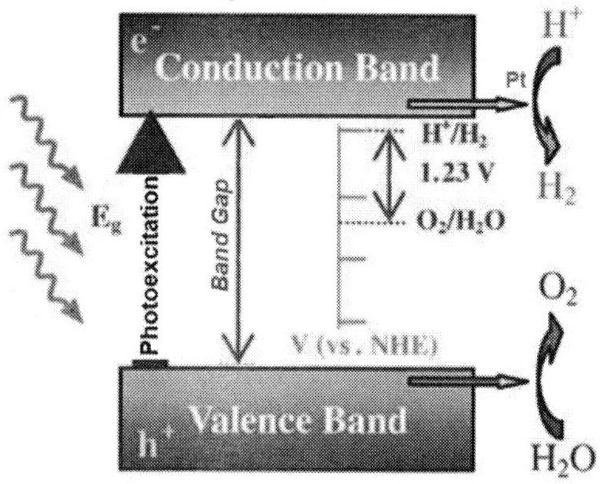

Figure 2. The principle diagram for photo-catalytic water splitting on a semiconductor. Adapted from Zhu et al, 2009 [11].

This technique has begun in 1972 when companies [12] used titanium oxide as the electrode for the production of hydrogen in a photo-electrochemical cell. Several studies have pointed out oxides with possible relevance for use in this process; many of them present difficulties as cost, obtaining and toxicity. The semiconductor must have a short band gap value to absorb a large part of visible light irradiation [13]. Therefore, ideal semiconductor materials for PEC application must have a low band gap, be stable in operation, permit electron mobility, do not show high electron/hole recombination capacity, be affordable and available [14].

Researchers explore catalysts that can be activated under visible light, such as coupled semiconductors, doped semiconductors and mixed metal oxide semiconductors, such as bismuth-based mixed oxide semiconductors, which have attracted interest because of their excellent stability, visible light absorption and photocatalytic properties. These include, among others, bismuth oxide (Bi_2O_3), bismuth niobate (Bi_3NbO_7), bismuth ferrite ($Bi_2Fe_4O_9$), perovskite bismuth ferrite (BiFeO), and mixed iron, bismuth and niobium oxides (Bi_2FeNbO_7) [11,15–19]. In this research, with the intention of developing photocatalysts based on bismuth, iron and niobium oxides for hydrogen production via photo-electrolysis of water, thin films were synthesized and characterized regarding their morphology and optical properties.

2. Experimental details

In order to develop and characterize photo-anodes based on bismuth, niobium and iron ($Bi_2Fe_xNbO_7$) for hydrogen production by photo-electrolysis of water, films were synthesized by the sol-gel method and deposited by dip coating on fluorine-doped SnO_2 (FTO) glass substrates. Four compositions were proposed, one without iron (Bi_2NbO_7) and with 0.8, 1.0 and 1.2 mole of iron ($Bi_2Fe_{0.8}NbO_7$; Bi_2FeNbO_7 and $Bi_2Fe_{1.2}NbO_7$), in order to evaluate its influence of iron concentration on the photocatalytic activity of the films.

The films were prepared in three steps: a) dissolution of the stoichiometric amounts of metal precursor ($Fe(NO_3)_3.9H_2O$/Synth) in 5 ml of Acetylacetone (Sigma-Aldrich) under magnetic stirring for 15 minutes. Parallel to this process, stoichiometric quantities of ($Bi(NO_3)_3.5H_2O$/Vetec) were solubilized in 5 ml of Acetylacetone under magnetic stirring for 15 minutes. Niobium chloride ($NbCl_5$) was solubilized in 4 ml of hydrochloric acid under magnetic stirring for 5 minutes. 2) After total solubilisation of both, the solution containing bismuth was homogenized, with agitation, with the solution containing iron. 3) Then, the niobium containing solution was titrated to the above, kept under constant stirring until complete homogenization, forming a uniform sol with orange-yellow coloration.

For the preparation of the films by the dip coating process, a universal machine of mechanical tests was used. FTO glass slides (15mm×25mm) were used as substrate for deposition of the sol and support for the thin films. They were previously cleaned with ethanol in the bath for 30 minutes. Five immersions were performed for each slide, with a 5 minute interval between each to dry the deposited material. The settling speed was set at 5 cm min^{-1}, as it conferred good uniformity to the deposited layers at each stage. After the dip coating process, the samples were dried at room temperature for 15 hours. Subsequently, they were heat-treated at 600°C. A slow heating rate (2°C min^{-1}) and 1h maximum temperature was used to produce thin films with good crystallinity.

2.1. Characterizations

In the area of catalytic materials for photo-electrolysis, it is essential to obtain information about the band gap value of the material, thus confirming the minimum energy size for the water splitting reaction, as well as the capacity of the material to absorb solar radiation. A Cary 5000 spectrometer was used to measurement. For the band gap calculation, the Kubelka-Munk method was utilized in the diffuse reflectance mode. The Kubelka-Munk Function is described by Myrick et al. 2011 [20] (equation 1), where *k* is the abortion coefficient and *s* is the scattering factor.

$$F(R) = (1\text{-}R)^2/2R = k/s \quad (1)$$

The band gap energy (Eg, eV) was obtained by extrapolation of the rising part of the curve to the x-axis (λg, nm), with calculation using: Eg = 1240/λg. Once the band gap is determined, it is possible to find the equivalent wavelength value (λeq), which is the minimum required energy for semiconductor activation related to the solar spectrum.

So it is possible to estimate the maximum wavelength that still contains sufficient energy to photo-excite electrons in the material, and thus estimate the percentage of the solar spectrum that the semiconductor is able to resort, and consequently its solar energy usage capacity.

In order to evaluate the morphology of the films, Scanning Electron Microscopy (SEM) studies were performed with a Carl Zeiss EVO MA10 instrument operated at 20 kV. Samples were gold-coated before starting the analysis and images were obtained in the secondary electron mode.

3. Results and discussion

3.1. Optical properties

The band gap analysis was performed with the purpose of determining the photo-activation energy of the films. The Kubelka-Munk Method was used with diffuse reflectance values. Figure 3 shows the diffuse reflectance spectrum of the samples and their band gap value (Eg).

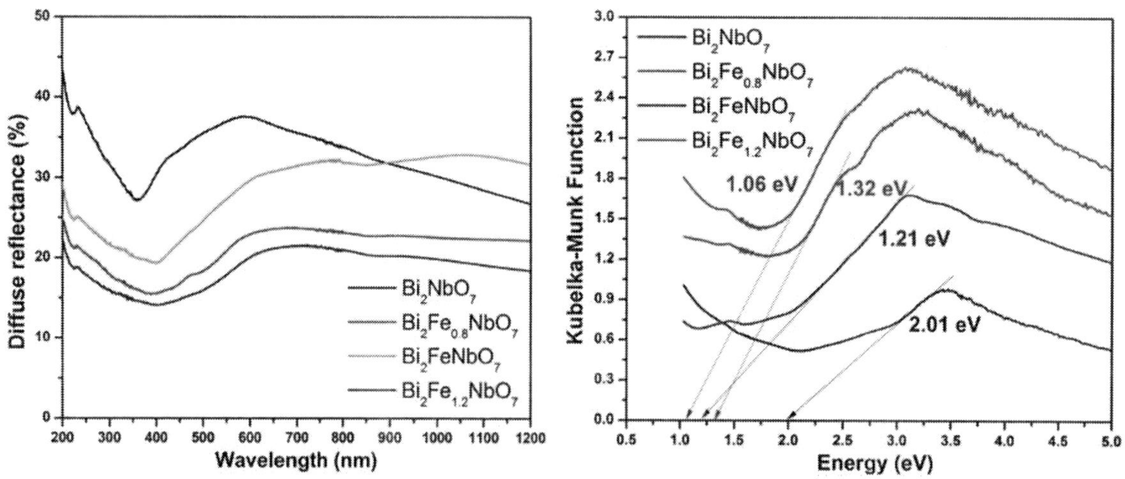

Figure 3. Diffuse reflectance and Kubelka-Munk function of the films.

All films presented semiconductor behaviour with Eg values lower than 2.01 eV. It is possible to observe a decreasing in the band gap value due the increasing of iron concentration. A linear regression was performed, identifying a first order polynomial with coefficient of determination (R^2) of 0.994, confirming a very strong relation between the increase of the amount of iron added and the decrease of the band gap value. One of the most consolidated explanations for this phenomenon claims that the adding of iron, which is a conductive material, produces intermediate energy levels between the valence and conduction band [21]. Then the material is able to absorb a larger part of the solar spectrum, since it requires photons with less energy to promote electrons from its valence band to conduction band. This phenomenon is known as red shift [22], and is accompanied by a change in the yellowish white colour of the film without iron to reddish in the highest concentration.

This higher light absorption capacity suggests a higher photo-catalytic activity. In addition, due to the effect of electrons and electron-holes in the electrical conductivity, lower values of band gap energy, due to the iron presence, could improve the electronic conductivity of the photo-electrodes, thus leading to a better electrical conductivity and ion transfer [23], which are desirable properties for photo-electrodes aiming water splitting. Table 1 summarizes the computed band gap values (Eg) and equivalent wavelengths (λeq) that are capable of providing enough energy for the activation of the photo-electrodes.

Table 1. Energy band gap values (Eg), Equivalent wavelengths (λ_{eq}) and solar spectrum usage percentage of the photo-electrodes.

$Bi_2Fe_xNbO_7$ x =	Band Gap (eV)	Equivalent wavelength (λ_{eq}) (nm)	Solar spectrum percentage (%) (AM1.5 Global) [24]
0	2.01	617	35.95
0.8	1.32	939	70.24
1.0	1.21	1025	75.36
1.2	1.06	1170	82.16

The radiation emitted by the sun is like that of a black body of 5,800 K. As it passes through the atmosphere, parts of sunlight are dispersed and/or absorbed. Chemical substances interact with sunlight and absorb certain wavelengths, altering the amount of light of those frequencies that reach the surface of the earth [25]. ASTM G-173 [24] formulates tables with values of energy quantity per wavelength (W m^{-2} nm^{-1}) that reaches the surface in normal plane (AM1.5 Direct), inclined plane (AM1.5 Global) and the irradiation before the atmospheric filter (AM0). Figure 4 shows the spectrum of solar radiation and equivalent wavelength (λeq) calculated from the optical band gap of the films.

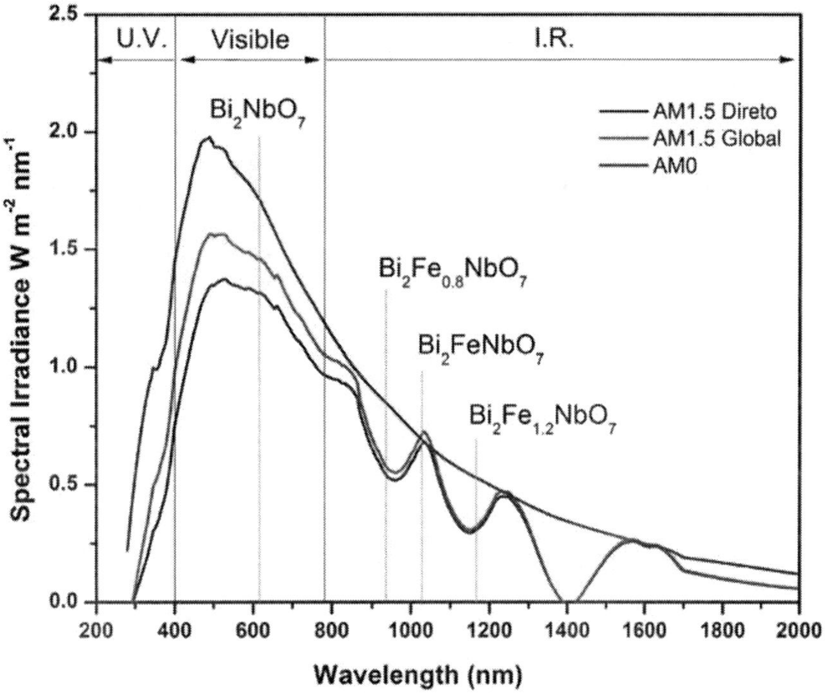

Figure 4. Spectral solar irradiance. Adapted from ASTM standard G-173-03 [24].

The lines corresponding to the equivalent wavelength (λeq) are represented for each film. It is important to emphasize that this is the maximum wavelength that still has enough energy to promote electrons from the valence band to the conduction band. Thus, the percentage of photons that have enough energy to activate the material can be obtained by the charge of the energy of all wavelengths smaller than the λeq. As shown in table 1, the films produced in this work can potentially be photo-excited by up to 82% of the solar spectrum that reaches the surface of the Earth, thus being a promising candidate for use as a photo-electrode for the hydrogen production by photo-electrolysis.

IOP Conf. Series: Materials Science and Engineering **659** (2019) 012081 doi:10.1088/1757-899X/659/1/012081

3.2. SEM studies

The morphology of mixed oxide films of $Bi_2Fe_xNbO_7$ was analyzed by Scanning Electron Microscope (SEM), and the results are shown in Figure 5. The micrographs on top show that the surface of the films is rough and with cracks distributed in all the analyzed area, characteristic from this kind of films [16,18]. Furthermore, these cracks are composed of particles with irregular size and shape. The film that does not contain iron (figure 5a, b) show fewer cracks and nanopores distributed by the shape, probably due to the exhaust gases during the heat treatment. This may be beneficial to the photo-electrolysis reaction by increasing the surface area of the film and thus the electrode-electrolyte interface [14].

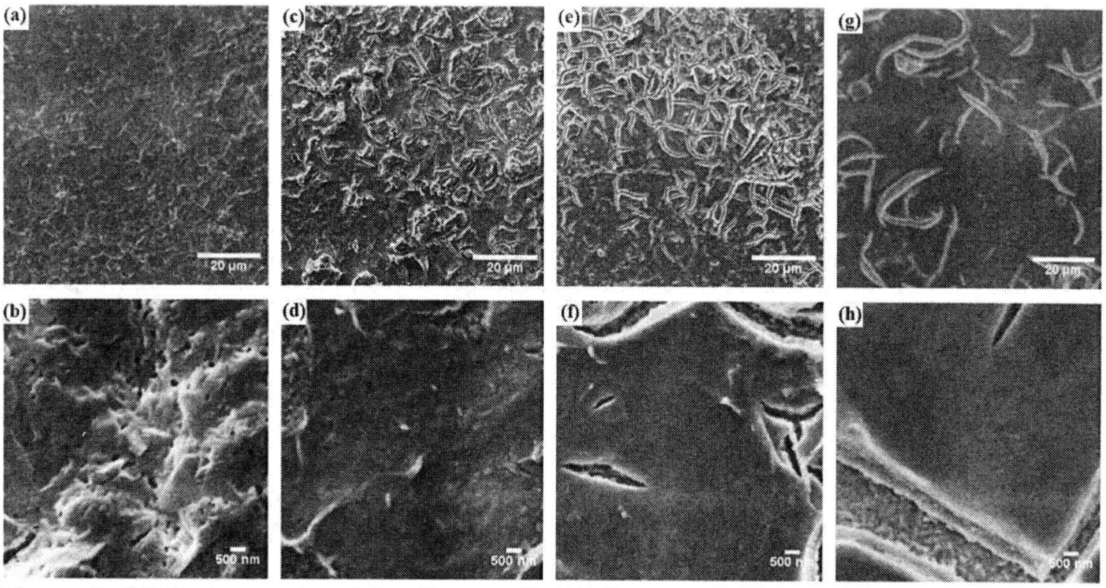

Figure 5. SEM micrographs of the films: **a, b)** Bi_2NbO_7; **c, d)** $Bi_2Fe_{0.8}NbO_7$; **e, f)** Bi_2FeNbO_7; **g, h)** $Bi_2Fe_{1.2}NbO_7$.

The films containing iron show progressive bigger cracks, it may have happened due the major amount of nitrates, its precursor. It is possible to observe that the increase in the concentration of iron causes a decrease in the amount of cracks and an increase in its size, on the other hand, the film surface is more homogeneous. In some cracks, it is possible to see the conductive layer of FTO.

Films deposited under substrates often develop tensile stresses large enough to cause cracking. A film heat treated at a high temperature and then cooled develops tensile stresses when the thermal expansion coefficient of the film exceeds that of the substrate. This is usually the case for oxides deposited on ceramic substrates, like FTO glass [26,27]. The cracked surface can be beneficial in the process of photo-electrolysis, mainly due to the greater surface area, directionality in the transport of electrons and the availability of electron-holes (h^+) for water oxidation [28]. However, if necessary, it is possible to manage the presence of cracks by changing the parameters of the heat treatment.

4. Conclusions

$Bi_2Fe_xNbO_7$ films were synthesized by an easy sol-gel route followed by heat treatment at relatively low temperature (600°C). The films exhibit a low band gap value and consequently an intense absorption of visible light. The increment of iron leads to a linear decrease in the band gap value. SEM studies show that the surface presents some cracks, possible due the presence of nitrates and residual tensile stresses. Further analyses should be performed to better understand their properties, however, based on data from UV-Vis spectroscopy and morphology studies, the photo-electrodes show high photo-catalytic activity under visible and even infrared light, which places them as potential candidates for photo-anodes, being able to overcome the barrier that the low absorption of visible light imposes to the processes hydrogen production by photo-electrolysis.

Acknowledgment

This work was financially supported by the Brazilian National Council for the Improvement of Higher Education (CAPES). The present paper was presented inside the '*Toward a Sustainable Mobility*' special session as part of the '*Two Seats for a Solar Car*' research project, an action funded by the Italian Ministry of Foreign Affairs and International Cooperation within the Executive Programme of Cooperation in the field of Science and Technology between the Italian Republic and the Republic of Serbia.

References

[1] Harari Y N 2015 Sapiens - A Brief History of Humankind *An Animal of No Significance,* Harper
[2] Fuqiang Zhang L C 2019 Review and outlook of world energy development, *Non-Fossil Energy Development in China* (Elsevier) pp 1–36
[3] U.S. Energy Information Administration (EIA) 2016 International Energy Outlook, *U.S. Energy Inf. Adm.*
[4] Letcher T M (Trevor M . *Managing Global Warming : an Interface of Technology and Human Issues*
[5] United Nations department for economic and social Affairs 2019 *World Economic situation and prospects 2019* (UNITED NATIONS)
[6] Minak G, Brugo TM, Fragassa C, Pavlovic A, de Camargo FV and Zavatta N. 2019 Structural Design and Manufacturing of a Cruiser Class Solar Vehicle*, J. Vis. Exp.* **143** *e58525*
[7] Minak G, Fragassa C and de Camargo F V 2017 A brief review on determinant aspects in energy efficient solar car design and manufacturing, International Conference on Sustainable Design and Manufacturing, Bologna, Italy, April 26 – 28, pp. 847-856
[8] *de Camargo F V, Giacometti M and Pavlovic A 2017 Increasing the energy efficiency in solar vehicles by using composite materials in the front suspension, International Conference on Sustainable Design and Manufacturing, Bologna, Italy, April 26 – 28, pp. 801-811*
[9] dos Santos K G, Eckert C T, De Rossi E, Bariccatti R A, Frigo E P, Lindino C A and Alves H J 2017 Hydrogen production in the electrolysis of water in Brazil, a review *Renew. Sustain. Energy Rev.* **68** 563–71
[10] Bak T, Nowotny J, Rekas M and Sorrell C 2002 Photo-electrochemical hydrogen generation from water using solar energy. Materials-related aspects, *Int. J. Hydrogen Energy* **27** 991–1022
[11] Zhu J and Zäch M 2009 Nanostructured materials for photocatalytic hydrogen production, *Curr. Opin. Colloid Interface Sci.* **14** 260–9
[12] Fujishima A and Honda K 1972 Electrochemical photolysis of water at a semiconductor electrode, *Nature* **238** 37–8
[13] Jang J S, Kim H G and Lee J S 2012 Heterojunction semiconductors: A strategy to develop efficient photocatalytic materials for visible light water splitting, *Catal. Today* **185** 270–7
[14] Ahmad H, Kamarudin S K, Minggu L J and Kassim M 2015 Hydrogen from photo-catalytic water splitting process: A review, *Renew. Sustain. Energy Rev.* **43** 599–610
[15] Ropero-Vega J L, Meléndez A M, Pedraza-Avella J A, Candal R J and Niño-Gómez M E 2014 Mixed oxide semiconductors based on bismuth for photoelectrochemical applications, *J. Solid State Electrochem.* **18** 1963–71
[16] Rosas-Barrera K L, Ropero-Vega J L, Pedraza-Avella J A, Niño-Gomez M E, Pedraza-Rosas J E and Laverde-Cataño D A 2011 Photocatalytic degradation of methyl orange using Bi_2MNbO_7 (M=Al, Fe, Ga, In) semiconductor film on stainless steel, *Catal. Today* **166** 135–9

[17] Ropero-Vega J L, Pedraza-Avella J A and Niño-Gómez M E 2015 Hydrogen production by photoelectrolysis of aqueous solutions of phenol using mixed oxide semiconductor films of Bi–Nb–M–O (M = Al, Fe, Ga, In) as photoanodes, *Catal. Today* **252** 150–6

[18] Ropero-Vega J L, Rosas-Barrera K L, Pedraza-Avella J A, Laverde-Cataño D A, Pedraza-Rosas J E and Niño-Gómez M E 2010 Photophysical and photocatalytic properties of Bi_2MNbO_7 (M = Al, In, Ga, Fe) thin films prepared by dip-coating, *Mater. Sci. Eng. B* **174** 196–9

[19] Bencina M, Valant M, Pitcher M W and Fanetti M 2014 Intensive visible-light photoactivity of Bi- and Fe-containing pyrochlore nanoparticles, *Nanoscale* **6** 745–8

[20] Myrick M L, Simcock M N, Baranowski M, Brooke H, Morgan S L and McCutcheon J N 2011 The Kubelka-Munk Diffuse Reflectance Formula Revisited *Appl. Spectrosc. Rev.* **46** 140–65

[21] Werapun U and Pechwang J 2019 Synthesis and Antimicrobial Activity of Fe:TiO_2 Particles, *J. Nano Res.* **56** 28–38

[22] Kumaravel V, Mathew S, Bartlett J and Pillai S C 2019 Photocatalytic hydrogen production using metal doped TiO_2: A review of recent advances, *Appl. Catal. B Environ.* **244** 1021–64

[23] Yousefipour K, Sarraf-Mamoory R and Yourdkhani A 2019 Iron-doping as an effective strategy to enhance supercapacitive properties of nickel molybdate, *Electrochim. Acta* **296** 608–16

[24] A.S.T.M. International 2012 ASTM G173-03(2012) *Stand. Tables Ref. Sol. Spectr. Irradiances Direct Norm. Hemispherical 37° Tilted Surf.*

[25] Lean J 2000 *Evolution of the Sun's Spectral Irradiance Since the Maunder Minimum* vol 27

[26] Xia Z C and Hutchinson J W 2000 Crack patterns in thin film, *J. Mech. Phys. Solids* **48** 1107–31

[27] Scharnberg A R A, Priebbnow A V, Arcaro S, Da Silva R M, Dos Santos P A M, Basegio T M and Rodriguez A A L 2019 Evaluation of the addition of soda-lime glass and yerba mate wastes in ceramic matrix, *Cerâmica* **65** 63–9

[28] Miklos D B, Remy C, Jekel M, Linden K G, Drewes J E and Hübner U 2018 Evaluation of advanced oxidation processes for water and wastewater treatment – A critical review, *Water Res.* **139** 118–31

Comparing the accuracy of 3D slicer software in printed end-use parts

M Šljivic[1], A Pavlovic[2], M Kraišnik[3] and J Ilić[1]

[1] University of Banja Luka, Faculty of Mechanical Engineering, S.Stepanovića 75, 78000 Banja Luka, Republic of Srpska, Bosnia Herzegovina
[2] University of Bologna, Faculty of Industrial Engineering, viale Risorgimento 2, 40133, Bologna, Italy
[3] University of East Sarajevo, Faculty of Mechanical Engineering, Lukavica 71126, Republika Srpska, Bosnia Herzegovina

E-mail: milan.sljivic@unibl.rs

Abstract. This study aims to compare the accuracy offered by 3D Slicer Software in printing end-use parts inside a Fused Deposition Modeling process of Additive Manufacturing. The purpose, in particular, is to investigate the surface quality and the dimensional stability of the manufactured parts comparing the effect of selecting a different 3D Slicer tool among *Simplify3D, Cura* and *Slic3r 3D*. With this scope, parts were produced using these process tools while results were analysed in terms of accuracy, production time and consumption of material. Results, graphically and visually presented, show significant differences in the dimensional and surface accuracy with an optimum outcome offered by the *Simplify3D* as best 3D slicer tool. The *Simplify3D* slicer has essential advantages in printed end-use parts because creates the 3D models with significantly better accuracy and quality support.

1. Introduction

The Fused Deposition Modeling (FDM) [1] is a relatively new technique for Additive Manufacturing (AM) [2], perfectly integrable as a process step in a modern and flexible manufacturing system [3]. This process is based on a technological evolution of the Rapid Prototyping (RP) process [4] that has evolved and expanded with the scope to produce end-use parts and final products [5, 6] embracing from medical applications [7] to architecture [8], aerospace [9] or automotive[10].

The FDM has achieved a large increase in all spheres of applications nowadays, from the professional one to the amateur domain. In this additive manufacturing process, wax or plastics are extruded, layer by layer, through a nozzle which follows the cross-section of a part, forming the part's geometry [11]. When a layer is done, the platform descends and the nozzle starts with the inflexion of the following layer [12]. This technology can use different types of materials including ABS [13,14], polyamide [13], polylactide [15], polyethene and polypropylene (including their reinforcements [16]), considering also unexpected materials as melted wax such as wood thermoplastics [17], advanced metals and ceramics [18]. The process can also benefit of 3D slicer software to transform a blueprint/model that it creates for printing and transfers to the machine, determines the speed of extrusion, optimizes the tool path for printing the object and controls the orientation of an object and the formation of layers [19-20]. Commercial FDM printers were developed and established by Stratasys [21]. The RepRap project [22] was started in 2005 in developed low-cost do-it-yourself FDM printers. Based on the RepRap, other printers were developed such as Makerbot [23], Ultimaker [24]) also allowing low-cost 3D printers to possibly produce end-use parts and reproduce parts of themselves. In the integration with a 3D printer, a good 3D slicer software is very important. If a good

IOP Conf. Series: Materials Science and Engineering **659** (2019) 012082 doi:10.1088/1757-899X/659/1/012082

slicer tool is used, it will surely get better results, even from a mediocre 3D printer. Otherwise, any valid 3D slicer tool will create a geometry based on an STL-file successfully, describing coordinates, printed speed, nozzle and bed temperatures, advanced dimensional accuracy, support requirements and other accuracy variables (further detailed in [25]). An FDM process with a commercial printer was used to create several rapid prototyping models. The application of the appropriate slicer software for processing has a direct impact on the quality of the printed parts, as in the case shown in Figure 1.

With this work, we want to showcase a subset of slicing tools to compare the accuracy offered by 3D slicer tools for parts production were the used upgraded low-cost FDM printer. The following 3D slicer software was compared: *Cura* (Free), *Slic3r* (Free) and *Simplify3D* ($150) [26]; examining the differences in terms of print quality as well as the selection of optimal processing parameters in the case of a low-cost/easy accessible FDM printer. The aspect of the accuracy of final products is not trivial for the use of such a technology on real market-oriented applications (as reported in [27-29]).

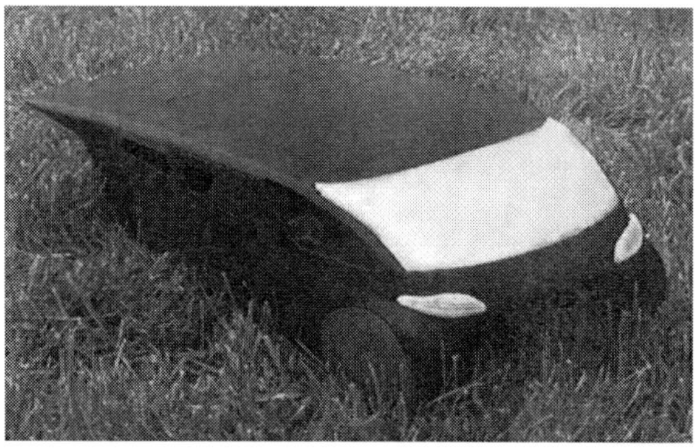

Figure 1. RP in-scale model of a Solar Car [30].

For this purpose, we selected two final parts that we printed on the same printer and processed the selected slicer tools under the same criteria. The aim is to examine the impact of slicing tools on the accuracy of 3D printed parts and the difference between slicer tools versus the print quality that is being achieved.

2. Experimental Setup

For the testing, two end-use parts were used: the *Phone Holder* (figure 2) and the *Pendant* (figure 3). The selected elements can have wide use in the demand of the customer because the design can be customized at the customer's request to change and adapt to a wider extent. In the dimensional accuracy and quality testing experiment, they are well-representing requirements that are required from parts for the free market as in the manufacturing of the RP model in the automotive industry. After executing additive manufacturing, the final parts were used to test the models in comparing the accuracy of the 3D Slicer software.

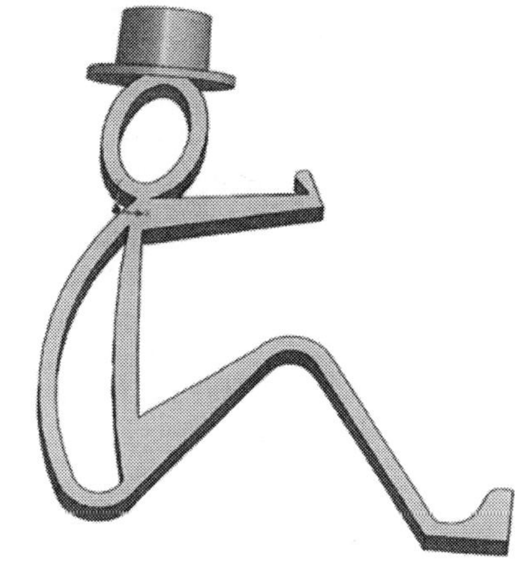

Figure 2. The CAD model of the Phone Holder (in SolidWorks software).

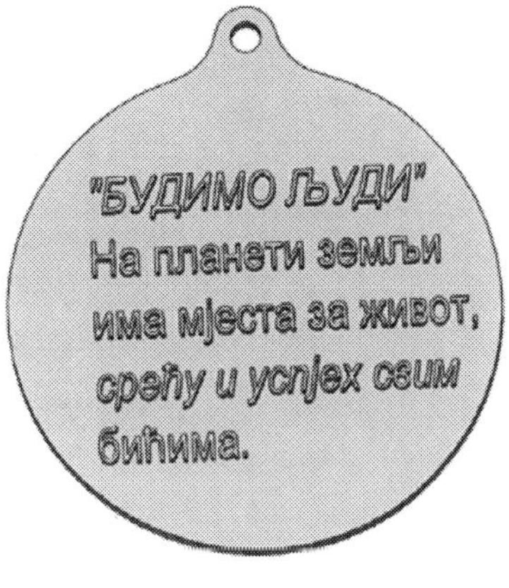

Figure 3. The CAD model of the Pendant (in SolidWorks software).

2.1. FDM printing technology

The printer used in this research was a low-cost 3D printer (*Infitary M508*) based on FDM technology as shown in figure 4.

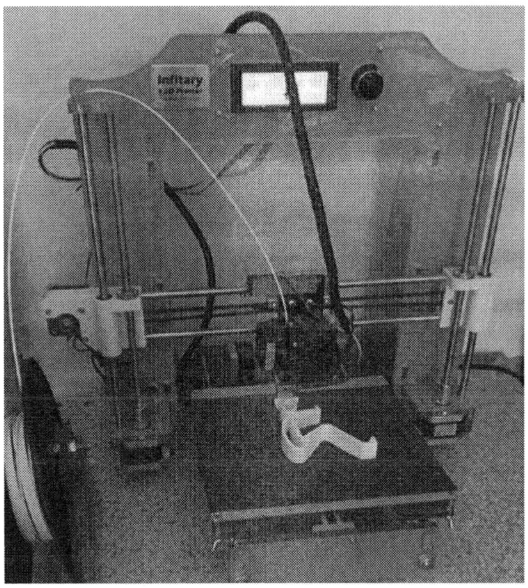

Figure 4. Low-cost 3D printer (*Infitary M508*) based on FDM technology.

The main technical characteristics of the desktop 3D printer used during the investigation including the independent variables are listed in table 1.

Table 1. Main technical characteristics of the desktop 3D printer.

Technical characteristics	3D printer *Infitary M508*
Printing technology	Fused Deposition Modeling
Standard nozzle	0.4mm
Best printing size	200x200x180mm
Support off-line print	Micro SD card printing

Print speed	30 – 120mm/s
Extruder head temperature	185 – 260 ^{0}C
Hot bed temperature	0 – 100 ^{0}C
Print head number	One
Printing materials diameter	1.75mm
Printing materials	PLA, ABS, HIPS etc.
XY axis positioning accuracy	0.012mm
Z axis positioning accuracy	0.004mm
Support file format	STL, G-code, OBJ

Using the selected low-cost FDM 3D printer, the manufacturing can be made in-house, which would enable time and cost savings.

2.2. Used material

Materials used to fabricate custom-parts were Polylactic Acid (PLA), a biodegradable thermoplastic derived from corn starch and sugar cane. PLA is environmentally friendly, safe to use and it is now the most advised used material for 3D printers. All parts are printed with PLA filament of the diameter 1.75 mm and machine nozzle size 0,4 mm. PLA has high mechanical strength and also good properties [31,32], able to expand its potential applications. The manufacturing temperature of PLA can be up to 230 ^{0}C, and the print bed temperature usually used is 50 ^{0}C.

2.3. Slicing software used

It is very important to have the right slicing software because, with the appropriate setting, the best quality print is possible. The slicing software that converts digital 3D models based on the settings to the 3D printer and to create a part. The slicer calculates how much material it will need and how long it will take to do it. In conclusion, all processed information is generated into a *GCode* file that is sent to the printer. For the comparison of the accuracy of slicer tools, three of the most popular following packages (as of February 2019 [26]) were selected:

- *Cura*, free open-source software developed by *Ultimaker* includes all settings for 3D printer slicing application [33]. The basic settings menu has possibilities the view mode Layers, Transparent, Overhang and Normal which are distributed in different tabs as and the following operations: rotate, scale and mirror.
- *Slic3r*, free open-source software, developed by Alessandro Ranellucci supported by several contributors [34]. Editing the model is performed via a four-tab interface: *Platter, Print Settings, Filament Settings* and *Printer Settings*.
- *Simplify3D*, one of the most common slicing software on the market. The software allows a huge variety of options to settings offering: extruders, layer control, various infill methods, temperature and cooling settings, import of *STL, OBJ* or *3MF* files, scale, rotate and repair 3D model. The software is compatible with many different printers, and it provides very good speed and control.

2.4. Slicing software configuration

For every part (the pendant and the holder) the same slicing software configuration was used to set the print parameters. The basic settings menu of *Cura* (version *15.04*) printed the pendant and the phone holder as shown in figure 5. Identically, the setting print parameters were used by slicer software *Cura, Simplify3D* and *Slic3r* (version *1.3.0*) for printing parts.

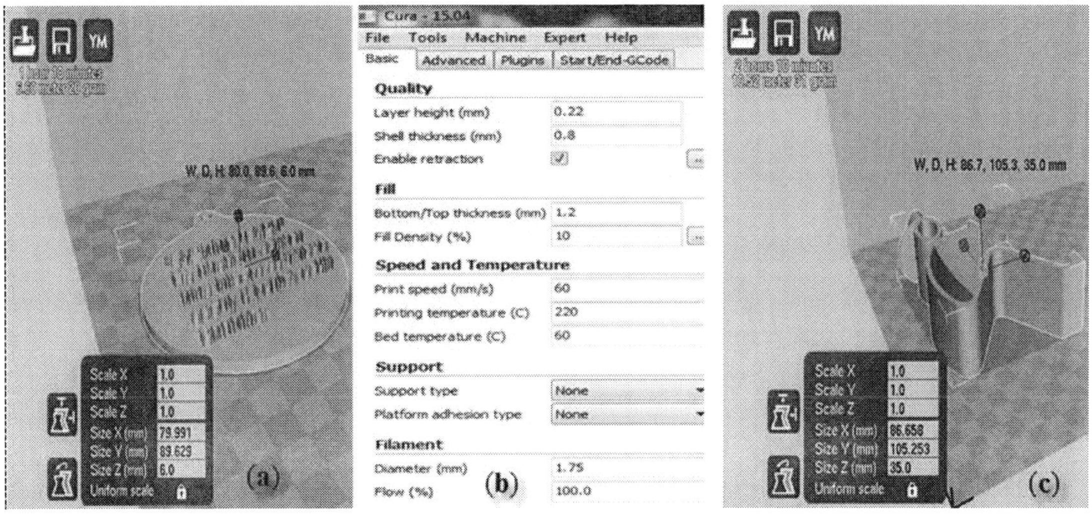

Figure 5. Basic settings of *Cura* (ver. 15.04) (**b**) as used to print the pendant (**a**) and the phone holder (**c**).

Two end-use parts were the set with the same print parameters used in *Cura, Simplify3D* and *Slic3r* as listed in table 2.

Table 2. 3D slicer settings print parameters adopted.

3D slicer settings print parameters	*Cura,Slic3r, Simplify*
Quality: Layer height	0.22 mm
Shell thickness	0.8 mm
Nozzle size	0.4 mm
Fill: Bottom/Top thickness	1.2 mm
Fill Density	10 %
Fill Pattern	Rectilinear
Print speed	60 mm/s
Extruder head temperature	220 ^{0}C
Bed temperature	60 ^{0}C
PLA filament: diameter	1.75 mm
Support structure build	None
Platform adhesion type	None
Flow compensation	100 %
Print time: (Pendant-*Cura*)	1h 18min
Print time: (Holder-*Cura*)	2h 10min
Print time: (Pendant-*Slic3r)*	1h 21min
Print time: (Holder-*Slic3r*)	2h 15min
Print time: (Pendant-*Simplify)*	58 minutes
Print time: (Holder- *Simplify)*	2h 5min

All parts have been printed using a low-cost 3D printer (*Infitary M508*), as represented in figure 4, without support structures and also without raft which improved the print platform adhesion.

2.5 Metric used for accuracy of slicing software and measuring print quality

For deviations between the printed part and the original part, the following metrics were used for measuring part deviations: very small deviation (<0.2 mm), small deviation (0.2–1 mm), rough deviation (1–2 mm) in the normal surface. Missing material in the shortest perpendicular: very small gap (<1 mm), small gap (1–2 mm), large gap (2–8 mm). Furthermore, the metric used for measuring text print quality was: Perfect, Readable, Partially readable, Not readable.

3. Results and discussion

The dimensional accuracy measurement of each printed part has been done using the *Digital Vernier Calliper* that offers a calliper precision of 0.01 mm. In Figure 6 (I) and (II) the locations used for measures are shown. Results of dimensional accuracy concerning the use of different slicing tools are listed in table 3. In particular, 5 samples for each slicer tool were considered, calculating the average value, as reported in columns under the label *'absolute'* (mm). For each of these mean values, the deviation in respect to the reference value has been calculated in mm and percentage. Observations on the achieved results for the accuracy of the printed parts analyzed were made for each used slicer tool. Figure 7 shows the pendants as done by printing the same geometry using three different slicer tools: *Slic3r, Cura* and *Simplify3D.*

3.1 Observed results for Slic3r software

The measurement results show that the dimensional deviation between the 3D model and the value of the printed parts for the indicated labels (a) to (j) (table 3) was in the range from 0.10 mm to 0.61 mm and in percentage 1.34% to 6.72%. According to the metric for measuring parts deviations in the surface normal, it belonged to the category of "small deviation" (0.2-1 mm). The measurement of the shortest perpendicular, label (g) and (h) (figure 6), was realized with a deviation from 0.12 mm to 0.20 mm, which belongs to the class: "very small gap (<1 mm)". As regards the readability as a quality indicator for the size and details of the visibility of the letters that were printed, the Slic3r tool did not become the required quality, as shown in Figure 7. Print time for the Pendant: 1 hour 21 minutes, and of the Holder: 2 hours 15 minutes.

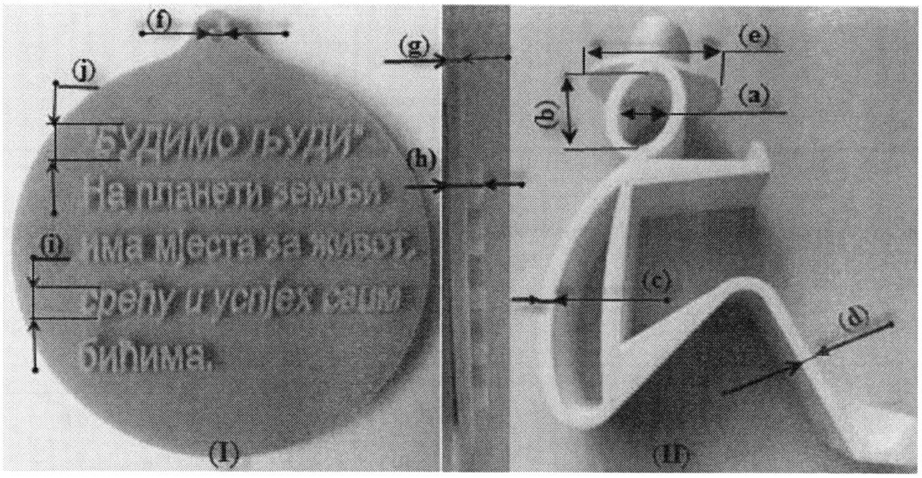

Figure 6. (I) Locations measured with the calliper by the pendant and (II) locations measured with the calliper by the phone holder.

Table 3. Results of the measurement dimensional accuracy by the pendant, figure 6 (I), and the phone holder, figure 6 (II) with different slicing tools.

	Reference	*Slic3r*			*CURA*			*Simplify3D*		
Label	Model [mm]	absolute [mm]	deviation [mm]	[%]	absolute [mm]	deviation [mm]	[%]	absolute [mm]	deviation [mm]	[%]
(a)	12.00	11.63	0.37	3.08%	11.77	0.23	1.91%	11.86	0.14	1.16%
(b)	21.00	20.61	0.61	2.90%	20.66	0.34	1.62%	20.96	0.04	0.19%
(c)	3.00	3.10	0.10	3.33%	3.05	0.05	1.66%	3.02	0.02	0.66%

(d)	3.00	3.19	0.19	6.33%	3.13	0.13	4.33%	3.12	0.12	4.00%
(e)	26.00	25.65	0.35	1.34%	25.68	0.32	1.23%	25.80	0.20	0.77%
(f)	5.00	4.81	0.19	3.80%	4.81	0.19	3.80%	4.85	0.15	3.00%
(g)	3.00	2.88	0.12	4.00%	2.92	0.08	2.66%	2.99	0.01	0.33%
(h)	6.00	5.80	0.20	3.33%	6.22	0.22	3.66%	5.85	0.15	2.50%
(i)	4.11	4.34	0.23	5.59%	4.29	0.18	4.37%	4.25	0.14	3.40%
(j)	5.65	6.03	0.38	6.72%	6.01	0.36	6.37%	5.93	0.28	4.95%

3.2 Observed results for Cura (version 15.04) software

It is possible to say that better dimensional accuracy respect to the use of *Slic3r* slicer tool was achieved when using *Cura 3D*. The dimensional deviation between the 3D model and the printed parts, as measured in the normal direction in respect to the surface, ranged from 0.05 mm to 0.36 mm (from 1.23% to 6.37%). These results fall inside the class of: "small deviation" (0.2-1 mm). The dimensional accuracy in the shortest perpendicular was achieved in deviation from 0.08 mm to 0.22 mm, label (g) and (h) shown in figure 6, which belongs to the class: "very small gap" (<1 mm). The pendant with complex letters done printed with good quality achieved a high level of detail, shown in figure 7. Print time for the Pendant: 1 hour 18 minutes, and of the Holder: 2 hours 10 minutes.

3.3 Observed results for Simplify3D software

The *Simplify3D* slicer software has achieved the best quality and dimensional accuracy of printed parts. The dimensional deviation in the surface normal was done in the range from 0.02 mm to 0.28 mm and in the percentage 0.19% to 4.95%. These results belong to the class: "small deviation" (0.2-1 mm). The deviation in the shortest perpendicular also was done with the best accuracy: from 0.01 mm to 0.15 mm, and these results belong to the very small class (<1 mm). The visually analyzed quality and accuracy of the printed complete letters are readable and close to the original shape, shown in figure 7. Print time for the Pendant: 58 minutes, and of the Holder: 2 hours 05 minutes.

Figure 7. Photograph of the pendants manufactured by printing from three different slicer tools: *Slic3r*, *Cura* and *Simplify3D*.

4. Conclusions

For quality test comparison, the corresponding parts that were printed on a low-cost 3D printer were selected and processed with three 3D slicer software: *Slic3r* and *Cura* (open source and free) and *Simplify3D* (powerful and professional). The comparison of the precision in the printed parts showed significant differences in the quality achieved with the three tools used.

According to the metric for measuring parts deviations in terms of dimensional accuracy in the surface normal, the *Slic3r* software has done from 0.10 mm to 0.61 mm which belongs to the class from small deviation. But if visibility is taken into consideration as an indicator of surface and letter quality, then the Slic3r software has not achieved a satisfactory quality level.

The *Cura* (version 15.04) software achieved better dimensional accuracy than the *Slic3r* slicer tool. The dimensional deviation between the 3D model and the printed parts was done from 0.05 mm to 0.36 mm, which is also in the range of small deviation. The surface and complex letters of parts, *Cura* was done with good quality and achieved a high level of detail.

The *Simplify3D* slicer software has achieved the best quality and dimensional accuracy of printed parts. The dimensional deviation was in the range from 0.02 mm to 0.28 mm. The visually analyzed quality of surface and letters of the printed parts are corresponding and close to the original shape.
Thus, it is possible to conclude that the slicer *Cura* gave good, but not necessarily accurate results. *Simplify3D* slicer has essential advantages because it has many detailed settings of the 3D model and also creates better quality support. The recommended software is firstly the Simplify slicer, then the free *Cura* slicer for processing not only low-cost but also semi-professional and professional 3D printers.

Acknowledgment
This research has been supported by the Italian Ministry of Foreign Affairs and International Cooperation thought the Joint Research Projects as Particular Relevance, with a project named '*Two Seats for a Solar Car*' within the Executive Programme of Cooperation between Italy and Serbia in the field of Science and Technology.

References
[1] Grimm T 2003 Fused deposition modeling: a technology evaluation, *Time-compression technologies* **11**(2) 1-6
[2] Huang S H, Liu P, Mokasdar A and Hou L 2013 Additive manufacturing and its societal impact: a literature review, *Int. J. Adv. Manuf. Technol.* **67**(5–8) 1191–1203
[3] Lukic L, Djapic M, Fragassa C, Petrovic A and Pavlovic A 2018 Optimization Model for Machining Processes Design in Flexible Manufacturing Systems, *Journal of Advanced Manufacturing Systems* **17**(2) 137-153
[4] Cooper K 2001 *Rapid prototyping technology: selection and application*. CRC press
[5] Jamieson R and Hacker H 1995 Direct slicing of CAD models for rapid prototyping, *Rapid Prototyping Journal*, **1**(2) 4-12
[6] Na J K and Oneida E K 2018 Nondestructive evaluation method for standardization of fused filament fabrication based additive manufacturing, *Additive Manufacturing* **24** 154-165
[7] Martelli N et al. 2016 Advantages and disadvantages of 3-dimensional printing in surgery: a systematic review, *Surgery* **159**(6) 1485-1500
[8] Ryder G, Ion B, Green G, Harrison D and Wood B 2002 Rapid design and manufacture tools in architecture, *Automation in construction* **11**(3) 279-290
[9] Klippstein H, Diaz De Cerio S, Hassanin H, Zweiri Y and Seneviratne L 2018 Fused deposition modeling for unmanned aerial vehicles (UAVS): A review, *Advanced Engineering Materials* **20**(2) 1700552
[10] Ilardo R and Williams C B 2010 Design and manufacture of a Formula SAE intake system using fused deposition modeling and fiber-reinforced composite materials, *Rapid Prototyping Journal* **16**(3) 174-179.
[11] Mohmed O A, Masood S H and Bhowmik J L 2015 Optimization of fused deposition modeling process parameters: a review of current research and future prospects, *Adv. Manuf.* **3** 42–53
[12] Baumann F, Bugdayci H, Grunert J, Keller F and Roller D 2016 Influence of slicing tools on quality of 3D printed parts, *Computer-aided design & applications* **13**(1) 14–31
[13] Pavlovic A, Šljivić M, Kraisnik M, Ilić J and Anić J 2017 Polymers in Additive Manufacturing: The Case of a Water Pump Impeller, *FME Transactions* **45**(3) 354-359.
[14] Sljivic M, Pavlovic A, Stanojevic M and Fragassa C 2016 Combining additive manufacturing and vacuum casting for an efficient manufacturing of safety glasses, *FME Transactions* **44**(4) 393-397
[15] Fragassa C, Minak G and Poodts E 2010 *Mechanical characterisation of photopolymer resins for rapid prototyping*. Proc. of the 27th Danubia-Adria Symposium on Advances in

Experimental Mechanics, 22-25 September, Wroclaw, Poland. Code 125161, pp. 55-56
[16] Zhong W, Li F, Zhang Z, Song L and Li Z 2001 Short fiber reinforced composites for fused deposition modeling, *Materials Science and Engineering: A* **301**(2) 125-130
[17] Zivkovic I, Pavlovic A and Fragassa C 2018 Improvements in wood thermoplastic matrix composite materials properties by physical and chemical treatments, *International Journal for Quality Research* **10**(1) 205–218
[18] Bellini A and Güçeri S 2003 Mechanical characterization of parts fabricated using fused deposition modeling, *Rapid Prototyping Journal* **9**(4) 252-264
[19] Mueller B 2012 Additive manufacturing technologies – Rapid prototyping to direct digital manufacturing, *Assembly Automation* **32**(2), doi: 10.1108/aa.2012.03332baa.010
[20] Tomiyama T, Gu P, Jin Y et al. 2009 Design Methodologies: Industrial and Educational Applications, *CIRP Annals - Manufacturing Technology* **58**(2) 543–565
[21] Stratasys Direct Inc. 2015. Fused Deposition Modeling (FDM) Design Guidelines. Stratasys Direct, Inc. https://www.stratasysdirect.com/resources/fused-deposition-modeling/
[22] RepRap.org Community Project: Welcome to RepRap.org, http://reprap.org, accessed on April 2019
[23] Makerbot Industries LLC: MakerBot | 3D Printers | 3D Printing, http://www.makerbot.com, accessed on April 2019
[24] Ultimaker BV: Home, https://www.ultimaker.com accessed on June 2019.
[25] Boschetto A and Bottini L 2014 Accuracy prediction in fused deposition modeling, *The International Journal of Advanced Manufacturing Technology*, **73**(5-8) 913-928
[26] Locker A 2019 *Best 3D Slicer Software for 3D Printers in 2019 (Most are Free)*, https://all3dp.com/1/best-3d-slicer-software-3d-printer/ (accessed on Feb. 2019)
[27] Alsoufi S M and Elsayed E A 2017 Warping Deformation of Desktop 3D Printed Parts Manufactured by Open Source Fused Deposition Modeling (FDM) System, *International Journal of Mechanical & Mechatronics Engineering IJMME-IJENS 17 No:04*
[28] Shazreen K, Ibrahim M and Saude N 2017 Dimensional accuracy analysis for desktop 3d printers for fused deposition modeling (FDM), *International Journal of Mechanical And Production Engineering* **5**(6)
[29] Galantuccia L M, Bodib I, Kacanib J and Lavecchia F 2015 Analysis of dimensional performance for a 3D open-source printer based on fused deposition modeling technique, *Procedia CIRP* **28** 82-87
[30] Minak G, Brugo T et al. 2019 Structural Design and Manufacturing of a Cruiser Class Solar Vehicle, *Journal of Visual Experiments* **143** e58525, doi:10.3791/58525.
[31] Armillotta A 2006 Assessment of surface quality on textured FDM prototypes, *Rapid Prototyping Journal* **12**(1) 35–41
[32] Šljivić M, Pavlovic A, Ilić J, Stanojević M and Todorović S 2017 Comparing the Accuracy of Professional and Consumer Grade 3D printers in Complex Models Production, *FME Transactions* **45**(3) 348-353
[33] Ultimaker BV: Software downloads, http://software.ultimaker.com/, accessed on June 2019
[34] Slic3r: Software downloads, http://slic3r.org/, accessed on June 2019

IOP Conf. Series: Materials Science and Engineering **659** (2019) 012083 doi:10.1088/1757-899X/659/1/012083

Sequential deposition method of $TiO_2/CH_3NH_3PbI_3$ films for solar cell application

A E R T P Oliveira[1], F Bonatto[1], A K Alves[1] and C Fragassa[2]

[1] Post-Graduation Program in Mining, Metallurgy and Materials Engineering (PPGE3M), Federal University of Rio Grande do Sul, Osvaldo Aranha 99, 90035-190, Porto Alegre, Brazil

[2] InterDepartmental Centre for Industrial Research on Advanced Applications in Mechanical Engineering & Materials Technology, Alma Mater Studiorum University of Bologna, viale Risorgimento 2, 40136 Bologna, Italy

E-mail: anne.targino@ufrgs.br

Abstract. Seeking to study innovative solar cell compositions with the goal to reach the highest energy efficiency level attainable, the aim of this study was to develop a route to obtain a solar cell composed by hybrid perovskite ($CH_3NH_3PbI_3$) using a sequential deposition method through the techniques of spin-coating and immersion. Initially, the deposition of PbI_2 thin film of was performed on a FTO/glass substrate coated with TiO_2, which was subsequently converted into perovskite crystals through spin coating using a CH_3NH_3I solution. The influence of the PbI_2 layer thickness on the formation of $CH_3NH_3PbI_3$ crystals was evaluated. The hydrophilic characteristic of TiO_2 affects the distribution of the crystals nucleation sites, since PbI_2 possesses a non-polar liquid characteristic. The characterization of the perovskite thin films showed that thickness affects directly the bandgap and the surface morphology, revealing the presence of dendritic structures and acicular crystals. Both growth and coverage increased for thinner layers of PbI_2. It was also possible to observe an increased uniformity in the film for smaller PbI_2 layers.

1. Introduction

The development of sustainable energy sources has become critical, since most of the global energy still relies on fossil matter, whereas a scenario of an industrial sector considerably dependent on coal and the urban mobility that almost entirely leans on polluting fuels such as gasoline and diesel can be verified. Within this context, solar energy is increasingly becoming an attractive alternative, causing a major increase in research and development of more efficient and inexpensive photovoltaic devices either for industrial plants or, more recently, solar-powered zero-emission vehicles [1-3]. Aiming to explore novel and more energy-efficient manufacturing alternatives for solar cells, recent literature have focused on perfecting techniques of deposition of different thin films over the solar cell substrate to allow a higher absoption of light and, consequently, enhanced levels of output energy.

The use of organic-inorganic perovskite halides as an absorbable layer in thin film solar cells started in 2009, resulting in a PV device with 3.1% efficiency [4]. In recent years, the rapid evolution of the device has led to a modification of its structure, with a large increase in efficiency, reaching values above 23.7% [5]. This organometallic material has been widely explored due to the singular characteristics of the perovskite material such as excellent and adjustable optical properties by controlling chemical compositions [6], ambipolar load transport [7], and very long diffusion lengths [8].

Deposition of perovskite halide thin films can be performed by vapor phase deposition, solution-gel chemistry approaches, one step deposition, and two-step solution synthesis [9-11]. Usually, perovskite is synthesized by combining two precursors, an organic and other inorganic, which can be combined by different deposition routes [12]. Most deposition methods are based on the same principle: the combination of an organic component iodide methylammonium (MAI), with an inorganic component, such as iodide or lead chloride (PbI_2 or $PbCl_2$), to produce a perovskite ($CH_3NH_3PbI_3$ or $CH_3NH_3PbI_{3-x}Cl_x$, respectively) [13].

In this work, we have compared the velocity of deposition of halide perovskite ($CH_3NH_3Pb_2I_3$) using a simple two-step method of deposition, over a mesoporous TiO_2 layer deposited in F doped SnO_2, to access morphological and optical properties for future use in solar cells.

2. Materials and methods

2.1. Preparation of precursor solutions

For the preparation of the $CH_3NH_3PbI_3$ films it was used glass covered with FTO (Fluorine doped Tin Oxide) with measurements of 2.5 cm long and 1 cm wide. The slides were immersed in acetone and placed on the ultrasound over 5 minutes for cleaning. They were washed with distilled water and then subjected to a temperature of 400 °C for 30 min to eliminate any organic matter that was present. A TiO_2 paste was prepared for the deposition of TiO_2 by spin coating. The paste was obtained by grinding 50 g of TiO_2-P25 (Evonik), 20 g of titanium isopropoxide (97 %, Sigma-Aldrich), 2.5 g of carboxi methyl cellulose (Sigma-Aldrich), 80 mL of terpineol (Sigma-Aldrich), 4 mL of acetyl acetonate (Sigma-Aldrich) and 5 mL of ethanol (99 %, Zeppelin) for a period of 12 h, so the paste was fully homogenized. For the preparation of PbI_2 solution, it was used a two-steps methodology: 1 M solution was obtained dissolving PbI_2 in anhydrous N, N-Dimethylmethanamide under agitation at a temperature of 70 °C. The precursor solution was ready for use when the salt was completely dissolved and with a translucent yellow color. For the second stage of two-step deposition, a solution was prepared composed of 10 mg/mL CH_3NH_3I in anhydrous isopropanol, agitated for a period of 10 min, in which all iodine methylammonium was dissolved.

2.2. Deposition in thin film

The TiO_2 mesoporous layer was made from a dissolution of the TiO_2 paste in ethanol with a proportion of 2:7. The solution was deposited at 3000 rpm per 20 s. The obtained film was dried at 125 °C for 10 min and then sintered 450 °C for 30 min. The perovskite layer was prepared using sequential two steps method describe by Burschka et al [11], but the second step was modified using the spin coating technique. The PbI_2 solution deposition was performed by spin coating, with different speeds for obtaining different thicknesses: 2000, 3000 and 4000 rpm, with two consecutive depositions. After this step, a 10-minute drying was made on a heating plate at 70 °C. Subsequently, each sample was covered with a solution of MAI in a spin coating process using a speed of 2000 rpm, followed by drying the film at 100 °C for 30 min. The film samples were named as PV-2, PV-3, PV-4 according to their respective deposition speeds: 2000, 3000 or 4000 rpm. The whole experiment was carried out at ambient condition (room temperature and open atmosphere).

3. Results and Discussion

For morphological characterization, scanning electron microscopy (SEM) was used to analyze the layer thickness influence of the PbI_2 films. The comparisons of the results obtained under the effect of the variation of deposition parameters are presented in Figure 1. The analyses revealed that the samples are homogeneous (without cracks) and with uniform appearance throughout the whole surface. It can be seen in Figure 1 (a), (c) and (e) that as the PbI_2 deposition velocity was increased, there was a growth of crystals. The hydrophilic character of TiO_2 [14,15] possibly affects the distribution of the nucleation sites of the crystals originating from the PbI_2. Figure 1 (e) revealed the presence of dendritic structures and needle-shaped crystals. There is the formation of perovskite grains for the speeds of 2000 and 3000 rpm (Figure 1 a to d). When 4000 rpm was used (Figure 1 f) there is no presence of these grains, and only dendritic structures are formed. This behavior of dendritic formation can be explained by the rapid cooling of the PbI_2 solution due to its higher velocity of deposition. It is also possible to observe the mesoporous layer of TiO_2 below the grains and dendrites of formed perovskites.

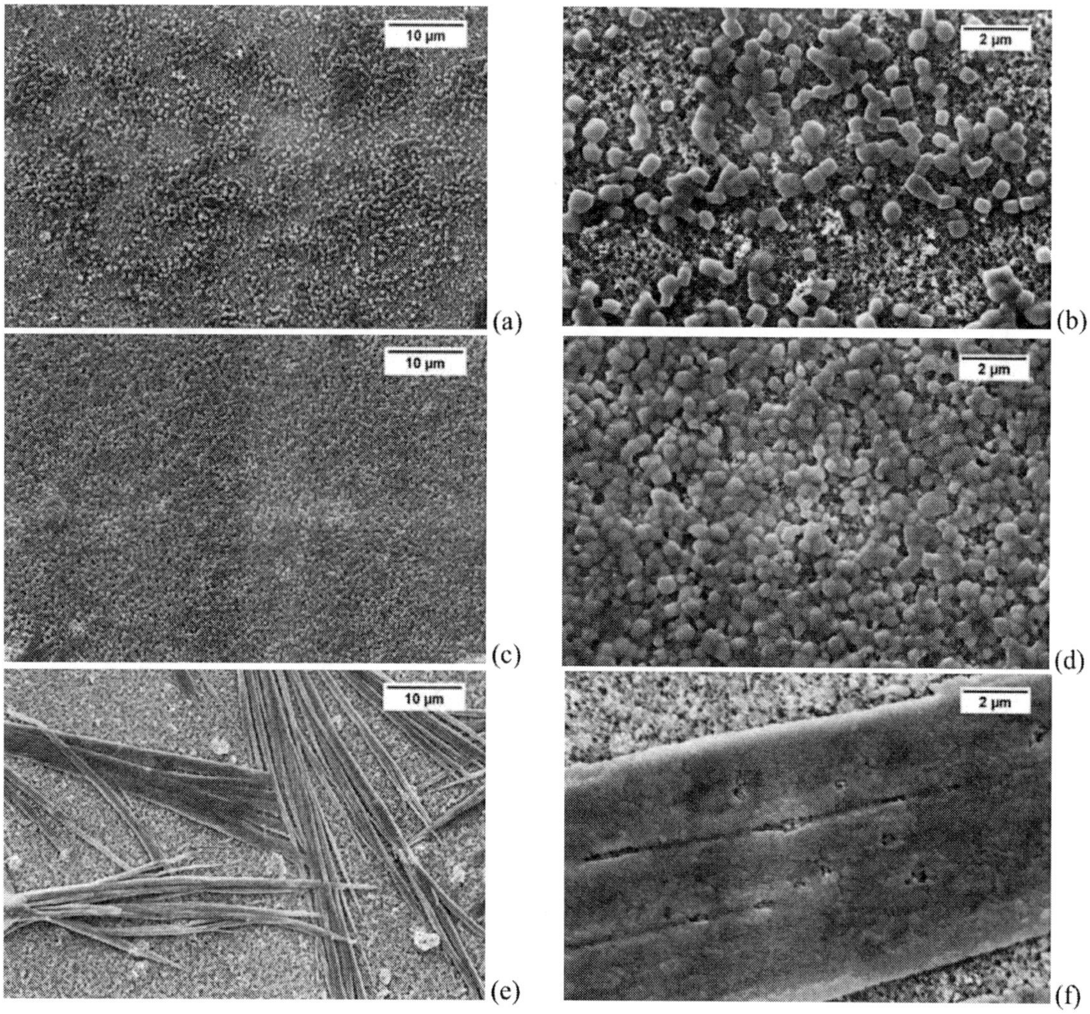

Figure 1. SEM images of $CH_3NH_3PbI_3/TiO_2$ films. Deposition velocity: (a)-(b) 2000 rpm, (c)-(d) 3000 rpm and (e)-(f) 4000 rpm.

The optical bands for $CH_3NH_3PbI_3/TiO_2/FTO$ films were determined by diffuse reflectance measurements (Figures 2 a, b and c). The measure of the extinction coefficient α , which is proportional to F(R), is calculated using the reflectance data according to the equation Kubelka-Munk [16,17], $F(R) \sim \alpha = (1-R)^2/2R$, where R is the percentage of reflected light, A is the energy of the incident photon ($h\nu$).The energy of the optical band gap (Eg) are related to the transformed Kubelka-Munk function, alpha ($h\nu$) = B ($h\nu$ – Eg)^n, where B is the absorption constant, Eg is the band gap energy and n is the power index that is related to the optical absorption, which in this case is related to a indirect transition and have the value of 1/2. According to other studies, the band gap of the TiO_2 determined based on the indirect transition is between 3.17 and 3.28 eV [18], due to the higher presence of the anatase phase. The $CH_3NH_3PbI_3$ bandgap is equivalent to 1.5 eV [6], which is formed between the orbital Pb and I. The calculated values for the optical band gaps are 3.15; 3.3 and 2.7 eV respectively, for PV-4, PV-3, PV-2 samples. It was observed that for higher deposition speeds, the bandgap is directly related to the presence of TiO_2, because the perovskite layer is thinner. For the deposition in lower velocity, the band gap was estimated between the values 1.5 and 3.2 eV due to the greater presence of perovskite.

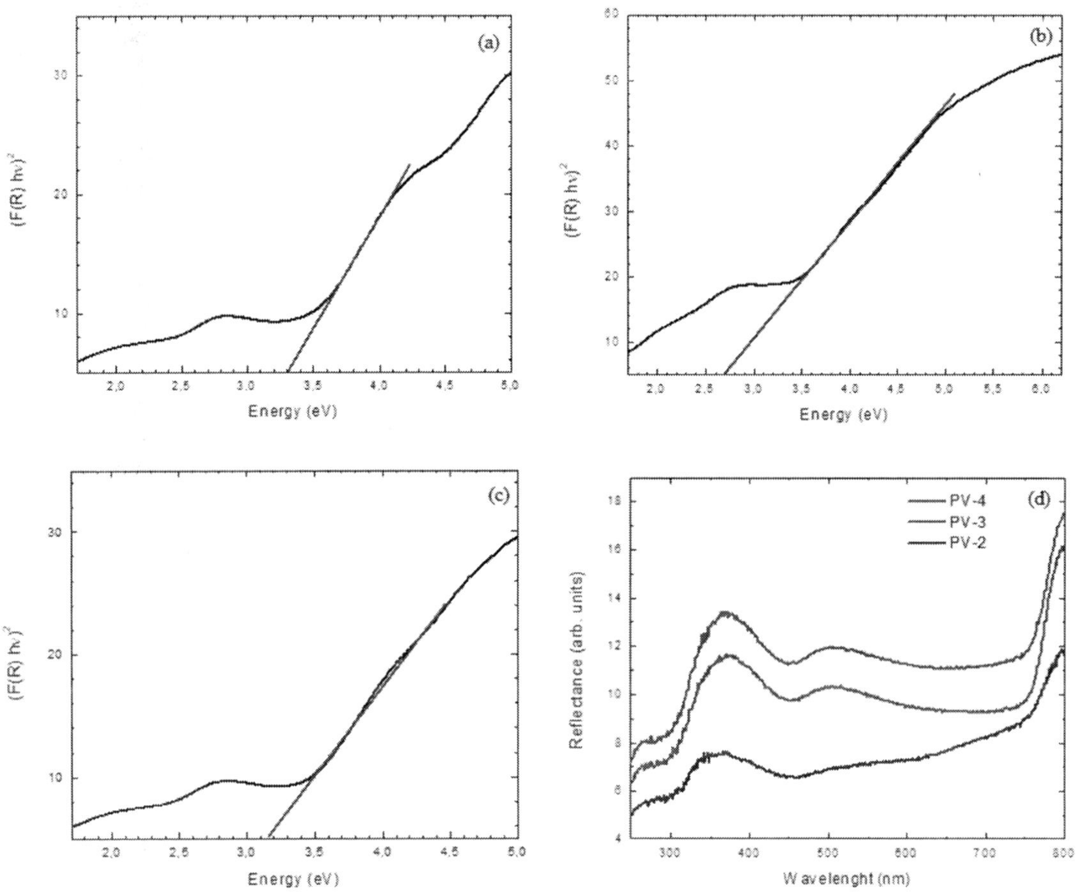

Figure 2. Transformed Kubelka-Munk spectrum of the $CH_3NH_3PbI_3$-sensitized TiO_2 film, PbI_2 depositions conditions: (a) 2000 rpm, (b) 3000 rpm, (c) 4000 rpm; (d) diffuse reflectance spectrum of the $CH_3NH_3PbI_3$-sensitized TiO_2 film.

The UV-vis absorption spectra of $CH_3NH_3PbI_3$ films deposited on TiO_2 were depicted in Figure 3a. As can be seen in the diagram, the absorption spectra in the UV region have prominent peaks at 280 nm for the PV-2 , for PV-3 around 282 nm and for PV-4 around 281 nm indicating a perovskite formation [19]. The film prepared at 3000 rpm presents a greater absorption compared to the other films corroborating the results of SEM where it is possible to observe a greater formation of perovskite. In the other films, there is no significant difference in the absorption spectrum.

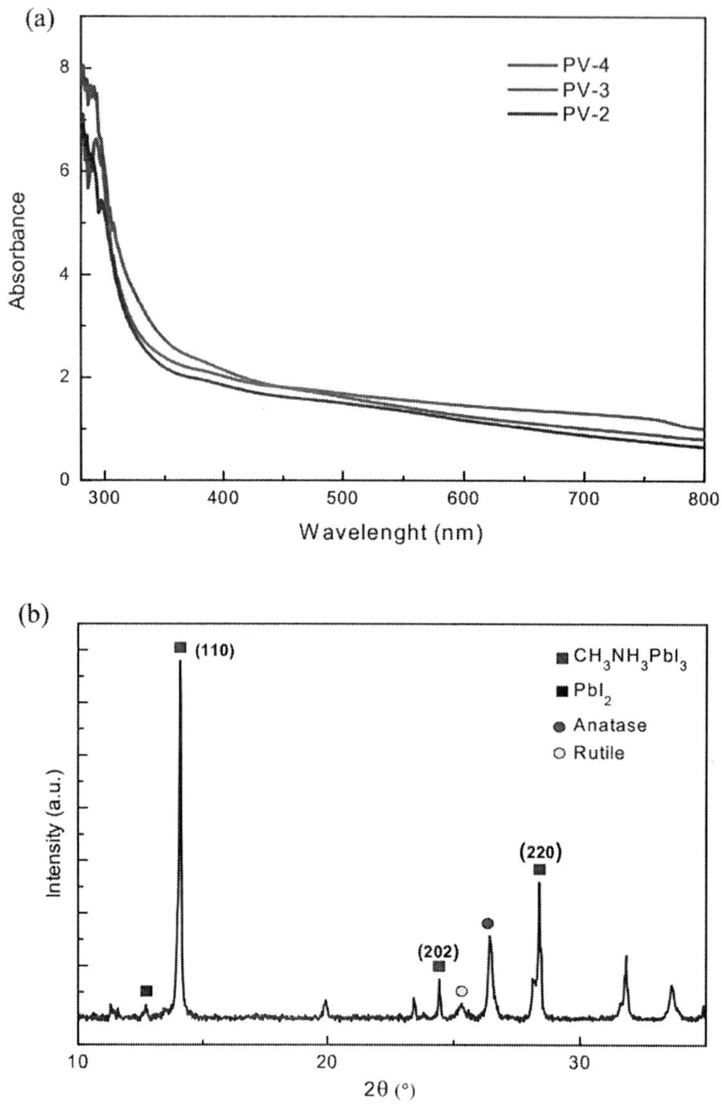

Figure 3. (a) UV-Vis absorption spectra of $CH_3NH_3PbI_3/TiO_2$ films. (b) XRD pattern of TiO_2/ $CH_3NH_3PbI_3$ thin film on FTO-glass.

The X-ray diffractogram for PV-3 is shown in Figure 3b. The diffractograms of the other samples were very similar to the PV-3 spectra. It can be noted that the diffraction signal related to the formation of perovskite $CH_3NH_3PbI_3$ in $2\theta = 14°$ is the most intense, and confirms the high crystallinity of the synthesized film. In addition, a weak signal at $2\theta = 12.4$ °, related to the precursor PbI_2, can be seen, indicating that the majority of PbI_2 precursor has been transformed into perovskite. Usually, the presence of PbI_2 is related to the degradation of the perovskite in the presence of moisture [20]. The XRD results indicates that the perovskite film has a good stability because only a residual presence of the precursor was identified. Finally, the perovskite planes (202) and (220), in $2\theta = 24.4°$ and $2\theta = 28.3°$, respectively, can be identified, in addition to the anatase peaks at $2\theta = 25.2°$ and rutile at $2\theta = 26.4°$, belonging to TiO_2 film from the substrate [21]. To calculate the average grain size of the perovskite, the Scherrer equation was used, considering the value of fwhm for the peak associated with the plane (110). Analysis of this data provided an average grain size of approximately 91 ± 3 nm.

4. Conclusion

In this study, the influence of the thickness of PbI_2 layers on $FTO/TiO_2/CH_3NH_3PbI_3$ films was investigated. The surface morphology of the films showed that with the increase of deposition velocity, there is a variation in the formation of perovkskite crystals, which can be explained by the rapid cooling of the PbI_2 solution in the FTO substrate. The PV-3 sample was the one with the better homogenous distribution of perovskite. There was a variation in the optical band gap for the lower velocity layer due to the greater presence of deposited material (PbI_2). The sample PV-3 had a more expressive absorption due to the greater amount of perovskite formed. Little crystallographic variation was observed, confirming the perovskite formation for all samples. The thin films have been synthesized successfully indicating that it is possible to employ this film in a future use for solar cells. The methodology allows fast production of thin films with lower cost due to their possible production in ambient conditions.

Acknowledgment

The authors would like to thank the Coordination for the Improvement of Higher Education Personnel – CAPES, the National Council for Scientific and Technological Development – CNPq and the Foundation of Support to Research of the State of Rio Grande do Sul – FAPERGS for the financial support. The present paper was presented inside the '*Toward a Sustainable Mobility*' special session as part of the '*Two Seats for a Solar Car*' research project, an action funded by the Italian Ministry of Foreign Affairs and International Cooperation within the Executive Programme of Cooperation in the field of Science and Technology between the Italian Republic and the Republic of Serbia.

References

[1] Minak G, Brugo T M, Fragassa C, Pavlovic A, de Camargo F V and Zavatta N 2019 Structural Design and Manufacturing of a Cruiser Class Solar Vehicle, *Journal of Visual Experiments* **143** e58525, doi:10.3791/58525

[2] Minak G, Fragassa C and de Camargo F V 2017 *A brief review on determinant aspects in energy efficient solar car design and manufacturing*, International Conference on Sustainable Design and Manufacturing, Bologna, Italy, April 26 – 28, pp. 847-856

[3] de Camargo F V, Giacometti M and Pavlovic A 2017 *Increasing the energy efficiency in solar vehicles by using composite materials in the front suspension*, International Conference on Sustainable Design and Manufacturing, Bologna, Italy, April 26 – 28, pp. 801-811

[4] Noh M F M, The C H, Daik, Lim E L, Yap C C, Ibrahim M A, Ludin N A, Yusoff A R M, Jang J and Teridi M A M 2018 The architecture of the electron transport layer for a perovskite solar cell, *J. Mater. Chem.* C **6** 682

[5] Jena A K, Kulkarni A and Miyasaka T 2019 Halide perovskite photovoltaics: background, status, and future prospects, *Chem. Rev.* **119** 3036-3103

[6] Noh J H, Im S H, Heo J H, Mandal T N and Seok S I 2013 Chemical Management for Colorful, Efficient, and Stable Inorganic-Organic Hybrid Nanostructured Solar Cells, *Nano Letters* **13**(4) 1764-1769

[7] Heo J H, Im S H, Noh J H, Mandal T N, Lim C S, Chang J A, Lee Y H, Kim H J, Sarkar A, Nazeeruddin M K, Grätzel M and Seok S I 2013 Efficient inorganic–organic hybrid heterojunction solar cells containing perovskite compound and polymeric hole conductor, *Nat Photonics* **7** 487–492

[8] Xing G, Mathews N, Sun S, Lim S S, Lam Y M, Gratzel M, Mhaisalkar S and Sum T C 2013 Long-range balanced electron- and hole-transport lengths in organic-inorganic $CH_3NH_3PbI_3$, *Science* **342** 344

[9] Lee M M, Teuscher J, Miyasaka T, Murakami T N and Snaith H J 2012 Efficient hybrid solar cells based on meso-superstructured organometal halide perovskites, *Science* **338**(6107) 643-647
[10] Chen Q, Zhou H, Hong Z, Luo S, Duan H S, Wang H H, Liu Y, Li G and Yang Y 2014 Planar heterojunction perovskite solar cells via vapor-assisted solution process, *J. Am. Chem. Soc.* **136**(2) 622-625
[11] Burschka J, Pellet N, Moon S J, Humphry-Baker R, Gao P, Nazeeruddin M K and M. Grätzel 2013 Sequential deposition as a route to high-performance perovskite-sensitized solar cells, *Nature* **499** (7458) 316–319
[12] Matthews P D, Lewis D J and O'Brien P 2017 Updating the Road Map to Metal-Halide Perovskites for Photovoltaics, *J. Mater. Chem.* A **5**(33) 17135–17150
[13] Stranks S D, Nayak P K, Zhang W, Stergiopoulos T and Snaith H J 2015 Formation of Thin Films of Organic–Inorganic Perovskites for High-Efficiency Solar Cells, *Angewandte. Chem. Int.* **5** 3240
[14] Sakai N, Fujishima A, Watanabe T and Hashimoto K 2003 Quantitative Evaluation of the Photoinduced Hydrophilic Conversion Properties of TiO_2 Thin Film Surfaces by the Reciprocal of Contact Angle, *The Journal of Physical Chemistry* B **107**(4) 1028–1035
[15] Jesus M A M L, Timòb G, Agustín-Sáenz C, Braceras I, Cornelli M and Ferreira A M 2018 Anti-soiling coatings for solar cell cover glass: Climate and surface properties influence, *Solar Energy Materials and Solar Cells* **185**(2018) 517–523
[16] Kortum G F A 1969 *Reflectance Spectroscopy: Principles, Methods, Applications*, Springer
[17] Simmons L E 1975 Diffuse reflectance spectroscopy: a comparison of the theories, *Appllied Optics* **14**(6) 1380–1386
[18] Lin H, Huang C P, Li W, Ni C, Ismat S S and Tseng Y-H 2006 Size dependency of nanocrystalline TiO2 on its optical property and photocatalytic reactivity exemplified by 2-chlorophenol, *Applied Catalysis B: Environmental* 68 1-11
[19] Eze V O and Mori T 2016 Organic-Inorganic Hybrid Perovskite Solar Cells Using Hole Transport Layer Based on α-Naphthyl Diamine Derivative, *Journal of Photopolymer Science and Technology* **29**(4) 581–586
[20] Mokhtar M Z, Chen M, Whittaker E, Hamilton B, Aristidou N, Ramadan S, Gholinia A, Haque S A, O'Brien P and Saunders B R 2017 $CH_3NH_3PbI_3$ films prepared by combining 1- and 2-step deposition: how crystal growth conditions affect properties, *Phys. Chem. Chem. Phys.* **19** 7204 -7214
[21] Jeong B, Hwang I, Cho S H, Kim E H, Cha S, Lee J, Kang H S, Cho S M, Choi H, Parl 2016 Solvent-assisted gel printing for micropatterning thin organic–inorganic hybrid perovskite films, *ACS Nano* **10**(9) 9026-9035

New method for modeling the topographical property of metals and its application in robot laser hardening with overlapping

M Babič[1]*, G Lesiuk [2] and L Giorgini[3]

[1] Faculty of Information Studies, 8000 Novo Mesto, Slovenia
[2] Department of Mechanics, Materials Science and Engineering, Wroclaw University of Science and Technology, 50-370 Wroclaw, Poland
[3] Department of Industrial Chemistry, University of Bologna, 40136 Bologna, Italy

E-mail: babicster@gmail.com

Abstract. Robot laser surface hardening is the part necessary for hardening instantly absorbs light energy, turning it into thermodynamic, due to which the temperature rises and austenite forms, then, after instant cooling, the martensite microsubstance is obtained. Due to the technology of heat treatment, it is possible to obtain other high-strength layers. The combination of laser hardening technology with the advantages of high cooling rate, uniform temperature distribution, minimum thermal stress, high strength, wide processing capabilities of various parts, etc. made this machine advanced in the processing of molding tools. In the laser a flat semiconductor is used as a heat source. The laser, in turn, has a high photoelectric conversion efficiency, has a short wavelength, low energy consumption and other advantages. In this article, we present the technology of robot laser hardening with overlapping and new method for predicting the topographical property of overlapping hardening process.

1. Introduction

Surface hardening through additional surface treatment processes represents an extremely competitive strategy to efficiently offer properties materials that they would not otherwise have [1,2]. This general concept is true for many materials, even belonging to families for which the idea of surface hardening is not obvious, as for example in the case of composites [3,4] or cermets [5].

Otherwise, it is precisely in metals and their alloys, and, in particular, in steel alloys that surface processes for hardening maybe reach the highest levels in terms of the variability and choice of treatments, technological advances, performance achieved by the materials, and so on [6-9].

The laser hardening technology [10], the core of one of these metal treatments, involves the study of principles of light energy in physics. Laser is an acronym for Light Amplification by Stimulated Emission of Radiation. The laser hardening [11-13] can be used as a metal surface treatment process, complementary to conventional flame and induction hardening or other processes [14]. A high-power laser beam is applied to heat a metal surface rapidly and selectively to produce hardened case depths of up to 1.5 mm with the hardness values of up to 65 HRC. It is used exclusively on ferrous materials suitable for hardening including, steels and cast iron with a carbon content of more than 0.2 %. Laser beam hardening is employed to locally improve the surface properties of components.

Use of this treatment can increase wear and fatigue resistance in parts of steel and cast iron [15,16], respect their traditional alloys [17]. Through a locally restricted heat treatment arises a minimum heat input, thereby minimized distortion. The associated high heating and cooling rates result in fine microstructures with good mechanical properties. During the laser skin hardening, the material (carbonaceous material) is heated up for a short time above austenitizing temperature and is transubstantiated by fast cooling down into the martensite structure. Different tool steels are widely introduced in industrial applications [18] due to good performance, a wide range of mechanical properties, machinability and wear cheapness.

With the laser hardening surface of the material, we can significantly improve their wear properties (Fig. 1 and 2). Heat is generated by absorbing the laser radiation on the surface and the material is quenched by heat transportation inside. The surface may not melt up.

In brief, robot laser hardening have many advantages:

- laser is source of energy with outstanding characteristics (contactless methods, controlled input of energy, high capacity, constant process, precise positioning),
- lower costs for additional machining,
- no use of cooling agents or chemicals,
- high flexibility,
- the process can be automated and integrated in the production process,
- superior wear resistance of hardened surface,
- selective hardening of complex geometrical shapes.

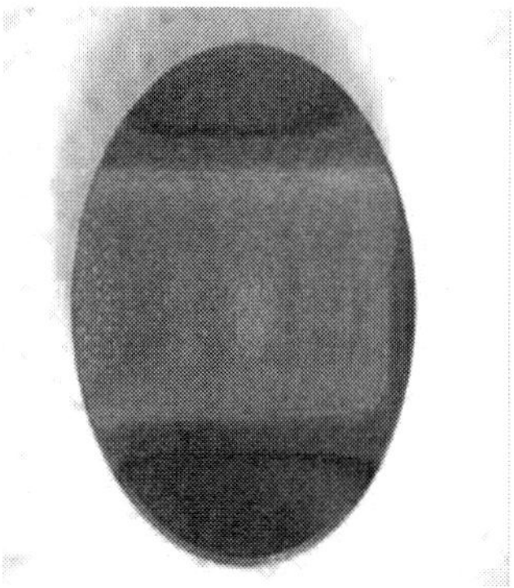

Figure 1. Hardened specimens.

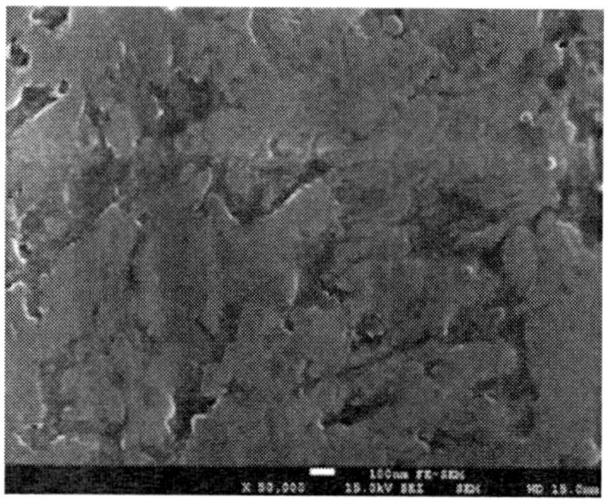

Figure 2. Microstructure of hardened specimens.

On the other side, surface treatments, including laser hardening, can be surely benefit from the adoption of intelligent systems and their modern tools for an optimal processing.

The term intelligent systems [19-21] is subject to the state of development of modern computer systems and their application inside the manufacturing processes. However, there are some characteristic features of intelligent systems that are independent of technological development. Intelligent systems process data electronically and can interpret information and their meaning from data and data records and link them to each other on a topic-specific basis. They are capable of learning, have artificial intelligence, and have cognitive abilities. Other characteristic features are their adaptability to various tasks and environmental conditions, their fault tolerance and safety. Intelligent systems are networked and can access other systems: knowledge-based systems, knowledge bases, expert systems and other software solutions.

In this work, in line with a methodological approach dealing with the intelligent systems, we will use the artificial intelligent techniques to investigate and optimise the robot laser hardening as done with success in the past on different materials and process technologies [22-24].

2. Materials preparation and experimental method

We made patterns of a standard label on the materials according to DIN standard GGG 70 and GGG 70 L. We hardened an ordinary tool steel with the laser at different temperature T ∈ [1000, 1400] ° C with steps 100 ° C and different speed v ∈ [2, 5] mm/s. We repeated this process in the way to produce specimens hardened with overlapping (Fig. 3).

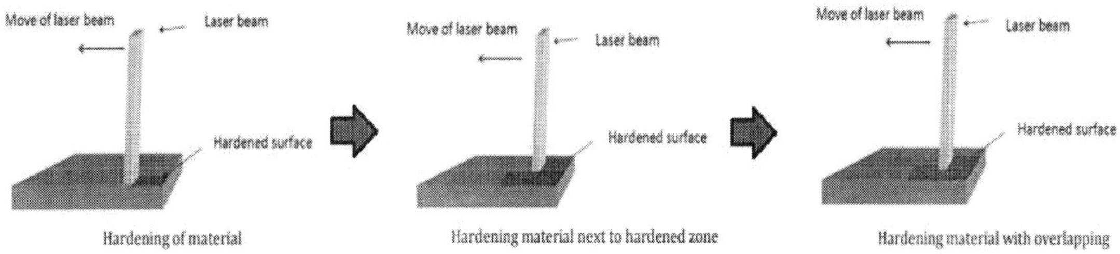

Figure 3. Process of robot laser hardening with overlapping.

In all these attempts we made pictures of the microstructure. We made recordings of hardened surface area. Each pattern was etching and polished before recording images by microscope. First, we made recordings using an optical microscope and then with an electron microscope. Images were made by field emission scanning electron microscope JSM-7600F JEOL company. To model the results, we used intelligent system methods, i.e. the neural network and support vector machine methods. In general, it can be said that the neural networks NN [25] have a large appeal to many researchers due to their great closeness to the structure of the brain, a characteristic not shared by more traditional systems. Support vector machines [26] (SVM) are supervised learning models with associated learning algorithms that analyze data used for classification and regression analysis. Hybrid evolutionary computation [27] is a generic, flexible, robust, and versatile method for solving complex global optimization problems and can also be used in practical applications. We present a new intelligent hybrid systems model in Fig. 4.

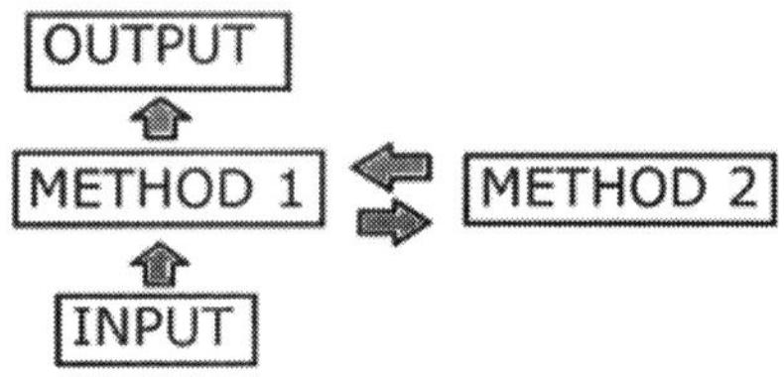

Figure 4. Hybrid system method of machine learning.

3. Results and discussion

In Table 1, the parameters of hardened specimens that impact on hardness are presented. We mark specimens from S1 to S20. Parameter X1 presents the parameter of temperature in degree of Celsius [C], X2 presents the speed of hardening [mm/s] and X3 presents base hardness of hardened specimens. The last parameter Y presents hardness of hardened specimens. Table 2 presents experimental and prediction hardness of laser hardened robot specimens. In Table 2 present symbol S name of specimens, E experimental data, NN prediction with neural network, SVM prediction with support vector machine and H prediction with new hybrid system of machine learning.

Table 1. Parameters of robot laser hardened specimens.

S	X1	X2	X3	Y
S1	1000	2	34.0	60.0
S2	1000	3	34.0	58.7
S3	1000	4	34.0	56.0
S4	1000	5	34.0	56.5
S5	1400	2	34.0	58.0
S6	1400	3	34.0	57.8
S7	1400	4	34.0	58.1
S8	1400	5	34.0	58.2
S9	1000	2	60.0	57.4
S10	1000	3	58.7	56.1
S11	1000	4	56.0	53.8
S12	1000	5	56.5	56.0
S13	1400	2	58.0	55.3
S14	1400	3	57.8	57.2
S15	1400	4	58.1	57.8
S16	1400	5	58.2	58.0
S17	800	0	34.0	52.0
S18	1400	0	34.0	57.0
S19	2000	0	34.0	56.0
S20	950	0	34.0	58.0

Table 2. Experimental and predicted data.

E	NN	SVM	H
60.0	60.16710	59.87995	59.42609
58.7	58.16926	58.14569	58.48085
56.0	56.56679	56.48387	55.92964
56.5	56.30757	56.38787	56.42103
58.0	58.12300	58.48259	57.90567
57.8	57.88996	57.98160	57.97986
58.1	57.65343	57.53997	58.53496
58.2	58.44157	58.36040	58.18148
57.4	58.65890	57.56091	57.36191
56.1	57.86290	55.93925	55.98725
53.8	58.94343	53.85237	53.70487
56.0	56.52490	52.81287	39.39352
55.3	58.61613	58.55416	55.34749
57.2	57.74158	56.35732	57.11229
57.8	57.97462	52.35552	57.73809
58.0	56.31630	52.13819	57.95978
52.0	60.90898	61.26072	51.93134
57.0	60.61049	60.23867	57.50505
56.0	59.79442	58.61351	55.85109
58.0	60.91160	61.19265	57.53997

The measured and predicted surface hardness of the RLH specimens are shown in Fig. 5. The neural network model presents a 2.26 % deviation from the measured data. The support vector machine model presents a 2.57 % deviation from the measured data. The hybrid method of machine learning presents a 1.77 % deviation from the measured data.

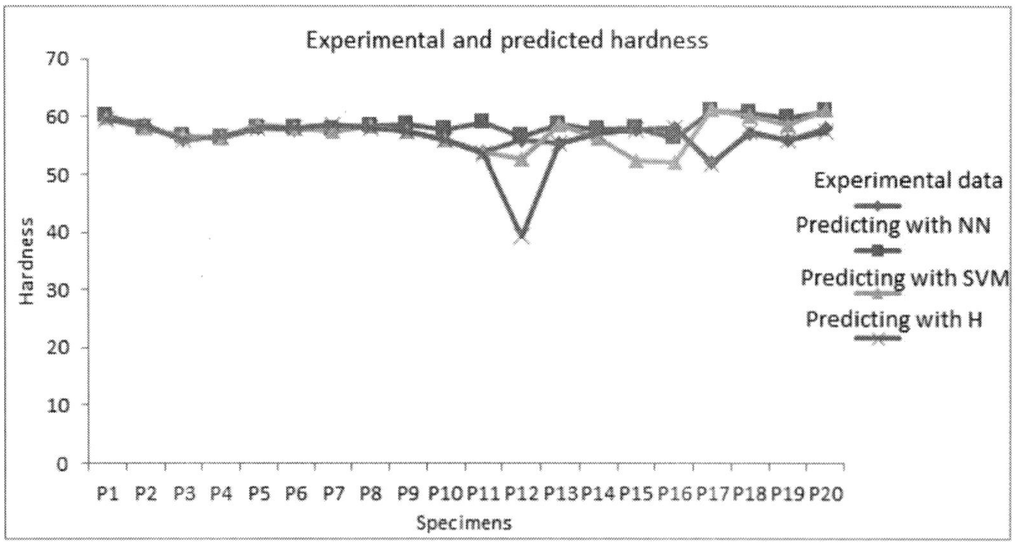

Figure 5. Experimental and predicted hardness.

4. Conclusion

The paper presents application of intelligent system methods in process of robot laser hardening with overlapping. We present hybrid method of intelligent system to predict topographical property of robot laser hardened specimens. The main findings can be summarized as follows:

1. For prediction of the hardness of hardened specimens we use neural network and support vector machine methods.
2. Predictions are in an acceptable accordance with experiments. The neural network and support vector machine models are almost equivalent in terms of accuracy. The hybrid method for machine learning leads to better results in predictions.
3. With the hybrid method of intelligent systems, we increase production of the process of laser hardening, because we decrease the time of the process and increase topographical properties of materials.
4. We describe the relationship between hardness and the parameters of the robot laser cell. This finding is important regarding certain alloys that are hard to mix, because they have different melting temperatures; however, such alloys have better technical characteristics. By varying different parameters (e.g., temperature), robot laser cells produce different patterns.
5. The present research can be easily extended with the scope of involving different material properties with valuable results in the case of cast irons (as done in [28])

References

[1] Davis J R 2002 *Surface hardening of steels: understanding the basics*, ASM international
[2] Lampman S 1991 Introduction to surface hardening of steels, *ASM Handbook* **4** 259-267.
[3] Fotouhi M, Saghafi H et al. 2017 Effect of PVDF nanofibers on the fracture behavior of composite laminates for high-speed woodworking machines, *P I Mech Eng C-J Mech* **231**(1) 31-43, doi: 10.1177/0954406216650711
[4] Zivkovic I, Pavlovic A and Fragassa C 2016 Improvements in wood thermoplastic composite materials properties by physical and chemical treatments, *International Journal of Quality Research* **10**(1) 205-218
[5] Guaglianoni W C, Cunha M A et al. 2018 Synthesis, Characterization and Application by HVOF of a WCCoCr/NiCr Nanocomposite as Protective Coating Against Erosive Wear, *Tribology in Industry* **40**(3) 477-487, doi: 10.24874/ti2018.40.03.13
[6] Béjar M A and Henríquez R 2009 Surface hardening of steel by plasma-electrolysis boronizing, *Mat & Des* **30**(5) 1726-1728.
[7] Chou, Y K 2002 Surface hardening of AISI 4340 steel by machining: a preliminary investigation, *J Mat Proc Techn* **124**(1-2) 171-177.
[8] Zou J X, Zhang K M et al. 2010 Mechanisms of hardening, wear and corrosion improvement of 316 L stainless steel by low energy high current pulsed electron beam surface treatment, *Thin Solid Films* **519**(4) 1404-1415.
[9] Sirin S Y, Sirin K and Kaluc E 2008 Effect of the ion nitriding surface hardening process on fatigue behavior of AISI 4340 steel, *Materials Characterization* **59**(4) 351-358.
[10] Totten G E 2006 *Steel heat treatment, 2nd edition.*
[11] Xu Z, Leong K H and Reed C B 2008 Nondestructive evaluation and real-time monitoring of laser surface hard-ening, *Journal of Materials Processing Technology* **206** (1-3) 120-125
[12] Dutta Majumdar J and Manna I 2011 Laser material processing, *International materials reviews* **56**(5-6), 341-388.
[13] Ion J. C. 2005 *Laser processing of Engineering Material - Principles, procedures and industrial applications,* Elsevier
[14] Santacruz G, Takimi A S, de Camargo F V, Bergmann C P and Fragassa C 2019 Comparative Study of Jet Slurry Erosion of Martensitic Stainless Steel with Tungsten Carbide HVOF Coating, *Metals* **9**(5) 600; doi:10.3390/met9050600
[15] Kennedy E, Byrne G, Collins D N 2004 A review of the use of high power diode lasers in surface hardening, *Journal of Materials Processing Technology* **155**, 1855-1860.
[16] Chen C H, Altstetter C J and Rigsbee, J M 1984 Laser processing of cast iron for enhanced erosion resistance, *Metallurgical Transactions A* **15**(4), 719-728.
[17] Fragassa C, Minak G and Pavlovic A 2016 Tribological aspects of cast iron investigated via

fracture toughness, *Tribology in Industry* **38**(1) 1-10

[18] Fragassa C, Lucisano G, Marinkovic D and Campana G 2019 A Practical Guideline for the Design and Use of Tools in Woodworking, *FME Transactions* **47**(2) 487-495, doi: 10.5937/fmet1903487F

[19] Serenko A and Dohan M 2011 Comparing the expert survey and citation impact journal ranking methods: Example from the field of Artificial Intelligence, *Journal of Informetrics* **5**(4) 629–649, doi:10.1016/j.joi.2011.06.002

[20] Lukic L, Djapic M, et al. 2018 Optimization Model for Machining Processes Design in Flexible Manufacturing Systems, *Journal of Advanced Manufacturing Systems* **17**(2) 137-153, doi: 10.1142/S0219686718500099

[21] Djapic M, Lukic L, et al. 2017 Multi-agent team for engineering: a machining plan in intelligent manufacturing systems, *International Journal of Machining and Machinability of Materials* **19**(6) 505-521, Doi: 10.1504/IJMMM.2017.10006404

[22] Babic M, Calì M, et al. 2018 Surface Roughness Evaluation in Hardened Materials by Pattern Recognition Using Network Theory, *International Journal on Interactive Design and Manufacturing* **13**(1) 211-219. Doi: 10.1007/s12008-018-0507-3

[23] Fragassa C, Babic M and Minak G 2019 Predicting the tensile behaviour of cast alloys by a pattern recognition analysis on experimental data, *Metals* 9, 557; doi:10.3390/met9050557

[24] Christopher C M L, Sasikumar T, et al. 2018 Neural network prediction of aluminum–silicon carbide tensile strength from acoustic emission rise angle data, *FME Transactions* **46**(2) 253-258. doi:10.5937/fmet1802253M.

[25] Ciresan D, Giusti A, Gambardella L and Schmidhuber J 2012 Deep Neural Networks Segment Neuronal Membranes in Electron Microscopy Images, *Advances in Neural Information Processing Systems* (NIPS 2012), Lake Tahoe

[26] Campbell C and Ying Y 2011 *Learning with Support Vector Machines,* Morgan & Claypool, ISBN 978-1-60845-616-1

[27] Senanayke S M N, Malik O A, Iskandar P and Zaheer D 2013 Anterior cruciate ligament recovery monitoring system using hybrid computational intelligent techniques, *International Journal of Hybrid Intelligent Systems* **10**(4) 2015-235

[27] Fragassa C, Babic M and Minak G 2019 Predicting the tensile behaviour of cast alloys by a pattern recognition analysis on experimental data, *Metals* **9**(557), doi:10.3390/met9050557

IOP Conf. Series: Materials Science and Engineering **659** (2019) 012085 doi:10.1088/1757-899X/659/1/012085

Piezoelectric PVDF sensor as a reliable device for strain/load monitoring of engineering structures

S Fotouhi[1], R Akrami[2], K Ferreira-Green[3], G A M Naser[3], M Fotouhi[3] and C Fragassa[4]

[1] Department of Mechanical Engineering, University of Tabriz, Tabriz, Iran
[2] University of Peyamnoor, Tehran, Iran
[3] Department of Design and Mathematics, University of the West of England, Bristol BS16 1QY, UK
[4] Department of Industrial Engineering, University of Bologna, Viale Risorgimento 2, 40136 Bologna, Italy

E-mail: mohammad.fotouhi@uwe.ac.uk

Abstract. This study investigates the potential of PVDF (polyvinyldiene fluoride) piezoelectric polymer material as a strain/load sensor for engineering structures. The PVDF sensor can be made in any shape/size and are flexible. In addition, the PVDF sensor is passive and offers the advantage of requiring no power to function. PVDF sensors were bonded to an aluminum specimen representative of an engineering structure and the voltage output of the PVDF sensors was found to vary linearly with the applied tensile load. This paper evaluates the possibility to make develop cheap, reliable and efficient sensors for structural health monitoring of engineering structures.

1. Introduction

Engineering structures, such as aircraft, wind turbines, marine vessels, buildings and offshore platforms, are experiencing a range of uncertain dynamic loadings during their life time, due to changes in operational and environmental conditions [1, 2].

These structures are usually over-engineered in order to prevent failure. This general situation is because there is not an appropriate and cost efficient devices to monitor the subjected loads, and therefore not possible to predict the fatigue life of these structures.

In order to have an accurate prediction about fatigue life of engineering structures, a Structural health monitoring (SHM) technology should be developed to understand the loads intensity and numbers.

A device, designed by NASA [3], showed the applicability in counting the strain-level for aircraft structures. This device was using a resistance strain gage that was used as the sensor with solid-state circuits and electro-mechanical counters. The device performance was compared against mechanical scratch guard. Unfortunately, the sensor was large (20.8 x 12.2 x 4.3 cm3 - without the counters) and had a high power consumption, requiring the provision of energy by the aircraft's auxiliary power supply.

Later, a patent was filed by Man Technologie in 1998 [4] that designed a portable device that can count and sort load levels. Again, a mechanical strain gauge was used, however, the results were not consistent over the rain-flow plots. The significantly-sized device was attached to the structure by screws and had its own power-supply.

PVDF film can be used for energy harvesting purpose as well as being developed as primary sensor due to its vast dynamic range, wide frequency coverage, high conversion efficiency, high elastic compliance, cost effectiveness etc. [5-7]. PVDF also shows excellent potential for embedding into structures because of its low Young's modulus when compared with modern materials [8]. This means that embedding such a sensor has little effect on the mechanical properties [9].

As a result, using PVDF as a sensor for the load-counting purpose can result in a compact, low-power consumption and wireless device that can be easily implemented in engineering structures. As a result, in another study [10], PVDF film was used to replace the strain gauge to monitor the load history of engineering structures. These results showed that PVDF can potentially be used to replace the strain gauge, however, no investigation was done about the reliability of the produced results measured by the PVDF sensor compared with the strain gauges.

This research is investigating the feasibility of replacing strain-gauges with Piezoelectric PVDF film, as a dynamic load gauge. The overall aim is to develop a monitoring-based method for fatigue life assessment with use of long-term monitoring data by PVDF.

2. Methodology and procedure

As shown in Figure 1, the experiment was coordinated by attaching the PVDF sensor attached to a piece of mild steel and subjecting it to cyclic tensile tests with different load levels. The examined sample was created by mounting the PVDF sensor 20mm from the midpoint of the steal which was cut to length of 300mm. This was followed by another sensor mounted on the other side of the mild steel to compare the results (see Figure 3).

The PVDF sensors were manufactured by TE Connectivity (part number 1-1002910-0) and the employed data acquisition system was Vishay Model 6100 scanner. Table 1 summarises some parameters and dimensions of the PVDF sensors. Table 2 shows characteristics of the steel substrate. The PVDF sensors was loaded in D33 direction where it has the highest piezoelectric co-efficient and therefore highest sensitivity.

The sample was cut from a longer piece with a length of 300mm, width of 25mm and a depth of 6mm. The sample was subjected to both tension and compression in order to evaluate the PVDF performance in both conditions. Different loads were applied in a sinusoidal wave pattern by varying the amplitude in order for the specimen to be strained, as the amplitude was varied the data acquisition device collected information for both the strain acquired and the voltages in respect to the amplitudes. The sinusoidal pattern was necessary as piezo electrics do not perform adequately when subjected to static loads or a constant loading condition but rather require a dynamic loading approach for voltage generation.

The machine was calibrated to 0 mean amplitude with no preload applied, and the specimen was subjected to tension under 2kn, 4kn, 6kn, 8kn and 10kn on different times. Readings were derived by the data acquisition device from the voltage signals obtained by the piezoelectric sensors and the strain signals acquired by the strain scanner.

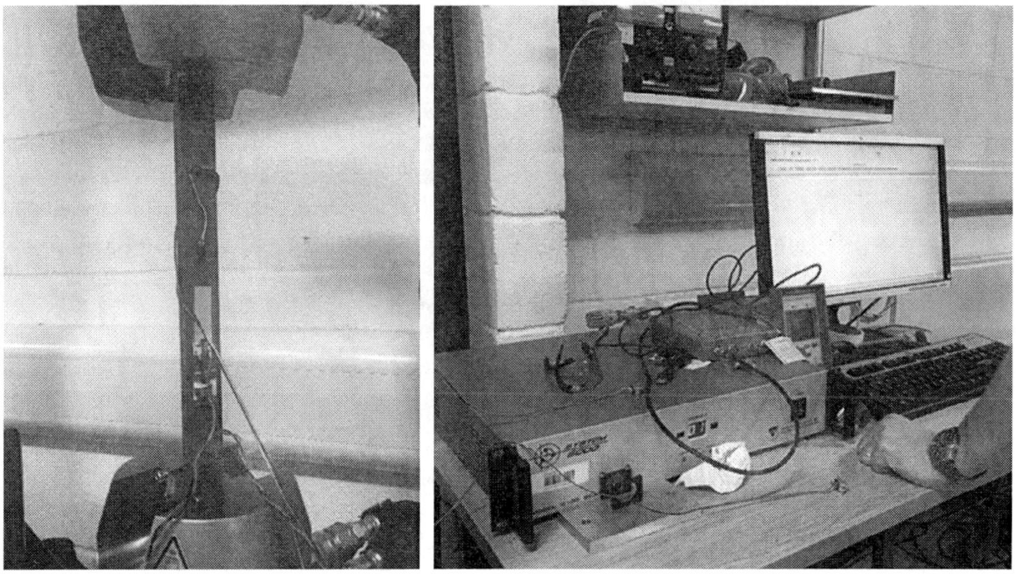

Figure 1. Experimental setup a) The tensile test and b) The data acquisition system.

Table 1. Dimensions and parameters of the PVDF sensors.

Parameter	Value	Units
Thickness	28	μm
Piezoelectric Coefficient D_{31}	23 x 10^{-12}	$(C/m^2)/(N/m^2)$
Piezoelectric Coefficient D_{33}	-33 x 10^{-12}	$(C/m^2)/(N/m^2)$
Piezoelectric Coefficient G_{31}	216 x 10^{-3}	$(m/m)/(C/m^2)$
Piezoelectric Coefficient G_{33}	-330 x 10^{-3}	$(m/m)/(C/m^2)$
Capacitance	1.38	NF
Young's Modulus	2 – 4 x 10^9	N/m^2
Electrodes Material	SILVER INK	
Electrode's Thickness	0.484	μm

Table 2. Characteristics of the beam.

Parameter	Value	Units
Mass Density	7850	Kg/m^3
Young's Modulus	210	GPa
Poisson Ratio	0.303	

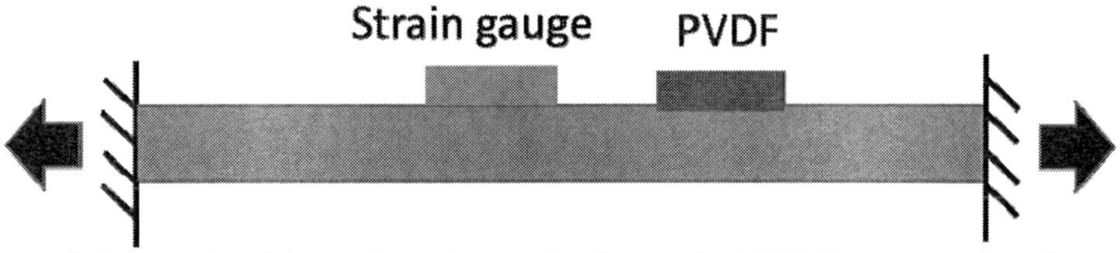

Figure 2. Schematic of the steel specimen with the attached PVDF sensors and strain gauges.

3. Results

Figure 3 shows the amplitude dependence of both the PVDF sensor according to the magnitude of the cyclic force, from 2kN to 10kN. As can be seen from Figure 3, there is a higher voltage by increasing the load level. This is because in the high load level, a higher displacement level is experienced by the piezo-electric material, compared with the low load level.

The ability for the PVDF sensors to detect cyclic load is examined by comparing the outputs of strain gauges to those of the PVDF as illustrated in Figure 4. As can be seen, the voltage results resemble a good correlation with strain gauge readings indicating PVDFs capability to monitor the load variation.

Results also show that PVDF was more sensitive to sense deformation than strain gauges giving it an advantage over strain gauge technology. As illustrated in Figure 5, the output voltage varied linearly with a peak amplitude of 0.11 volts being achieved for 2 kN of force and 0.39 volts for 10 kN.

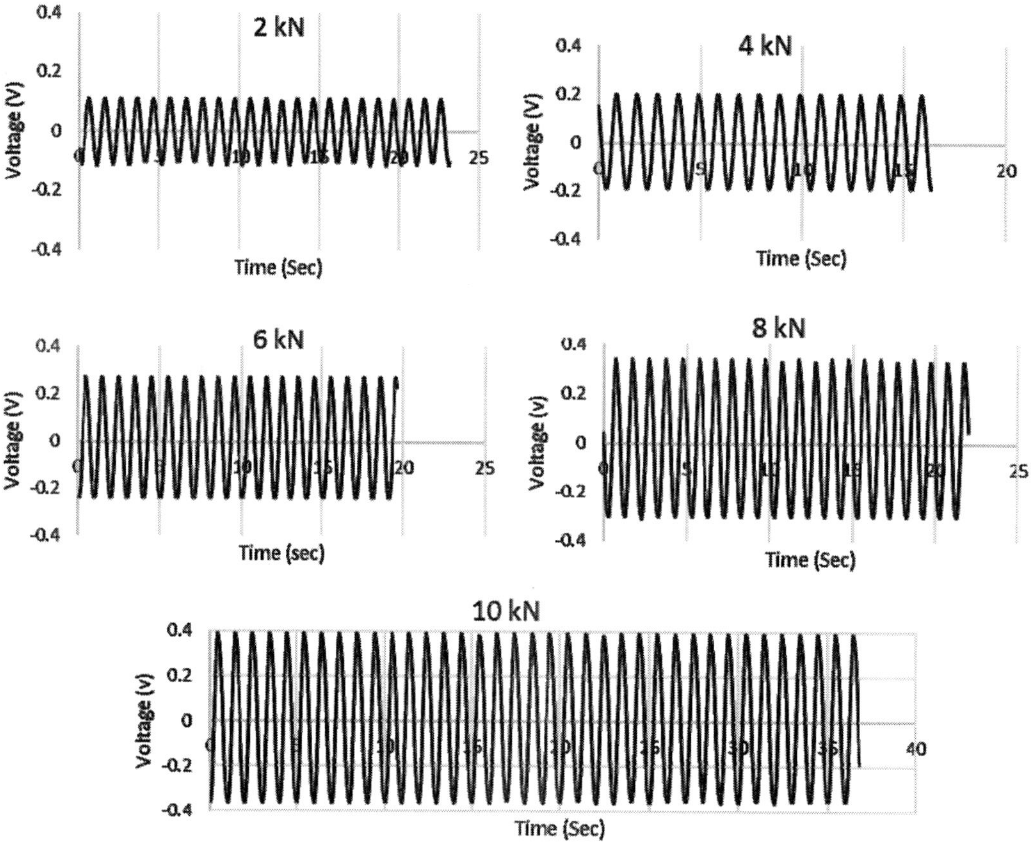

Figure 3. Voltage versus time results for the cyclic loads with different amplitudes.

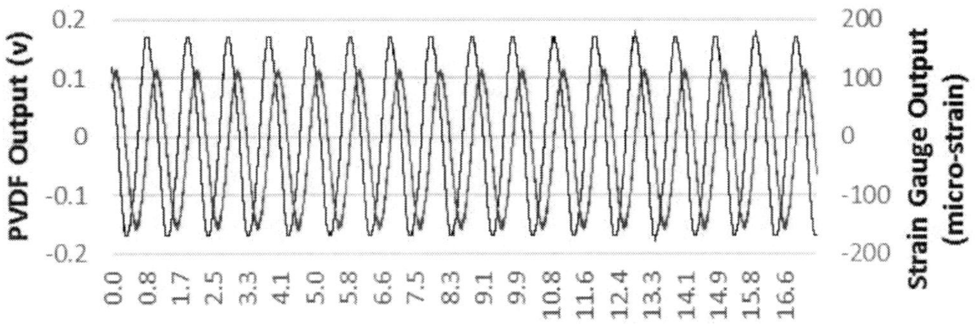

Figure 4: Comparison of PVDF and Strain Gauge vs. Time.

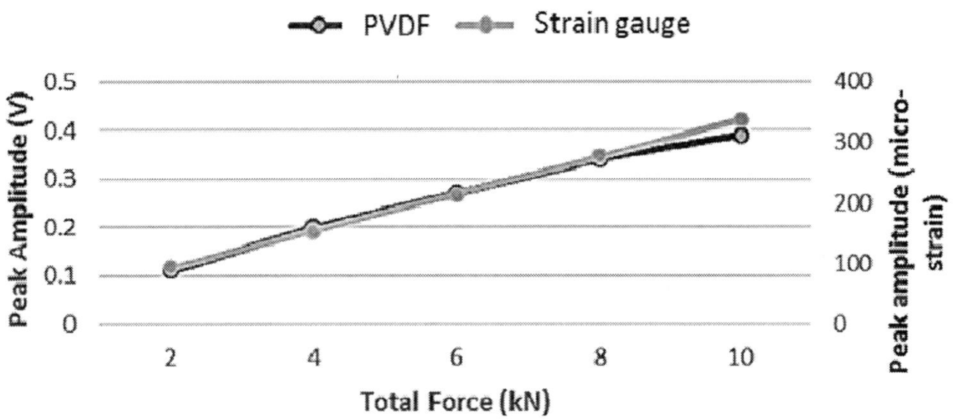

Figure 5: PVDF output dependence on amplitude compared with strain gauge.

4. Discussion

The PVDF sensors appear well matched to the task of load monitoring on structures. Even though a direct comparison of the strain gauge values and PVDF output cannot be made, however the trends in results between the two are undoubtedly visible.

This is especially obvious from Figure 5 which show the linear relationship between the load level and the voltage/strain. The sensitivity of the PVDF sensors appears to be adequate for the dynamic load monitoring purpose.

5. Conclusion

- Monitoring the load level of steel sample with PVDF sensors, the output of the sensors was found to vary linearly with applied load. A similar pattern was observed for the strain gauge measurements.
- It was concluded that the PVDF piezoelectric sensor is a good alternative to conventional strain gauges in order to measure dynamic loadings for structural health monitoring purposes.
- The PVDF sensors offer the advantage of needing no power to function and the sensors can be made to suit any size and geometry, therefore reducing the size and cost of the load monitoring device.

Acknowledgment
This research has been supported by the Italian Ministry of Foreign Affairs and International Cooperation thought the Joint Research Projects as Particular Relevance, with a project named 'Two Seats for a Solar Car' within the Executive Programme of Cooperation between Italy and Serbia in the field of Science and Technology.

References

[1] de Camargo F V, Guilherme C E M, Fragassa C and Pavlovic A 2016 Cyclic stress analysis of Polyester, Aramid, Polyethylene and Liquid Crystal Polymer yarns, *Acta Polytechnica* **56**(5) 402-408
[2] de Camargo F V 2019 Survey on Experimental and Numerical Approaches to Model Underwater Explosion, *Journal of Marine Science and Engineering* **7**(1) 15
[3] Pitts F L and Spencer J L 1973 Electronic Stain-Level, Counter, United States Patent: 3733424
[4] Friedho R and Kubsch G 1998 *Portable Device for Counting and Sorting Load Cycles in supporting, Structures with Various Loads*, United States Patent: 5,763,788
[5] Vodicka R and Galea S C 1998 *Use of PVDF Strain Sensors for Health Monitoring of Bonded Composite Patches*, Airframes and Engines Division Aeronautical and Maritime Research Laboratory, DSTO-TR-0684
[6] Chiu Y Y, Lin W Y, Wang H Y, Huang S B and Wu M H 2013 Development of a piezoelectric polyvinylidene fluoride (PVDF) polymer-based sensor patch for simultaneous heartbeat and respiration monitoring, *Sensors and Actuators A: Physical* **189**, 328-334
[7] Xin Y, Sun H, Tian H, Guo C, Li X, Wang C and Wang S 2016 The use of polyvinylidene fluoride (PVDF) films as sensors for vibration measurement: A brief review, *Ferroelectrics* **502**(1) 28-42
[8] Fotouhi M, Saghafi H, Brugo MT et al. 2017 Effect of PVDF nanofibers on the fracture behavior of composite laminates for high-speed woodworking machines. *Proc. Inst. Mech. Eng., Part C: Journal of Mechanical Engineering Science* **231**(1) 31-43
[9] Fotouhi M, Xiao B, Pozegic T et al. 2017 *High Performance Piezo Electric Nanocomposite Sensors*, ICCM-21, Xi'an, China
[10] De M, Pozegic T, Hamerton I and Fotouhi M 2018 *Development and evaluation of a novel piezoelectric PVDF sensor as a load spectrum counter*, ECCM 18

Innovation in solar vehicles: From concept to prototype in less than 24 months

S Maglio[1], M Lukovic[2], N Zavatta[3*] and A Leardini[3]

[1] Onda Solare Association, Castel San Pietro Terme, Bologna, Italy
[2] Industrial Design Department, University of Arts Belgrade, Belgrade, Serbia
[3] Interdepartmental Center for Industrial Research on Advanced Mechanics and Materials, Alma Mater Studiorum University of Bologna, Bologna, Italy

E-mail: nicola.zavatta2@unibo.it

Abstract. The present article describes the integrated path used for the conceptual, functional and constructive design of an exclusive multi-occupant solar-powered vehicle. The project was based on the massive implementation of concurrent engineering and quality tools, rarely used in such an integrated way, featuring a novel and attractive design, 3D CAD modelling, structural and fluid dynamics validations, in-scale rapid prototyping, functional tests, multi-objective optimization, parts manufacturing and assembly. Thanks to this approach, the solar car prototype presented high technological contents, especially in terms of materials, structures and processes. Furthermore, large CNC-machined multi-material molds and hybrid manufacturing solutions were adopted to speed up the production phase allowing to move from the initial concept to the final prototype within 24 months. Since June 2018, the solar vehicle is on the road transporting 4 people, weighing less than 300kg, reaching speeds of 120km/h and able to travel hundreds of kilometers with zero-emission and no fuel consumption.

1. Introduction

This work briefly summarizes the realization of the design and construction of one of the few multi-occupant solar vehicles in the world and its use to cross a large part of the United States, carried by a group of researchers from the University of Bologna and supported by partner institutions.

Everything started from the sporting passion for automotive competitions and from the previous experience in the field of solar vehicles, which led to the creation of a solar single-seater in recent years. With this vehicle, we took part in several races on different continents bringing back valuable successes. However, the category of single-seater solar vehicles did not seem to represent a real technological challenge as it appeared to be the category of multi-passenger vehicles. Known under the name of 'cruiser' car, this relatively recent category in the racing scene obliges designers to face particularly complex technical problems, linked to the specific needs of solar mobility.

Solar cars have to be imagined by the designers around the fundamental concept of energy efficiency [1,2]. This condition requires important technical and design choices that, combined with the race regulation requirements [3], lead to exclusive and highly innovative prototypes. Besides, everything was made possible only thanks to the use of concurrent engineering and its design tools, within a more general approach to Total Quality Management (TQM) [4] in automotive. This path is here briefly presented together with a list of references, able to provide full details on specific aspects.

IOP Conf. Series: Materials Science and Engineering **659** (2019) 012086 doi:10.1088/1757-899X/659/1/012086

2. Materials and Methods

Specifically, the following phases and tasks were considered:

2.1. Aesthetic Design

- International design competition.
- Comparison of style proposals.
- Realization of sketches (Figure 1).
- Definition of the final design and rendering (Figure 2b).
- European design patents (Figure 2a).

Figure 1. Aesthetic design: some proposals from the design competition (©Filip Relic).

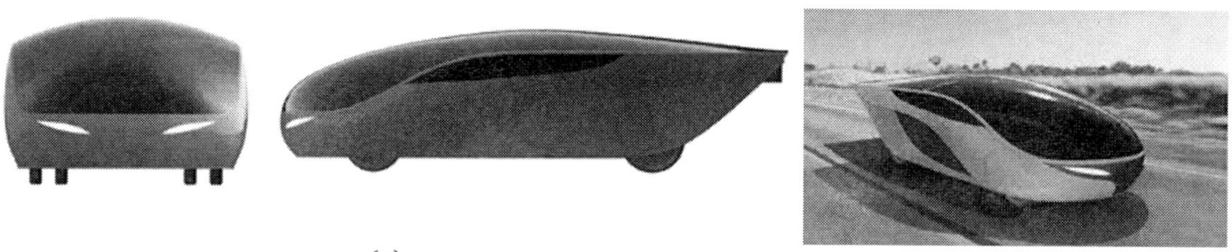

(a) (b)

Figure 2. Aesthetic final design (a) and patented version with rendering effect (b) (©Marko Lukovic).

2.2. Functional Design

- Conceptual design of solutions [5] (Figure 3a).
- Quality Function Deployment [6].
- Rapid prototypes (in SLS, scale 2:3) (Figure 3b).
- Wind Gallery investigation.
- Dedicated CFD algorithms [7,8].
- Safety and crash test simulation [9].
- Open-source platform for multi-objective optimization.

(a) (b)

Figure 3. Functional design: a) vehicle parts' design and b) rapid prototyping.

IOP Conf. Series: Materials Science and Engineering **659** (2019) 012086 doi:10.1088/1757-899X/659/1/012086

2.3. Construction Design

- Advanced modeling of composites [10-12].
- 3D detailed design of parts .
- FEM optimization [13,14].
- Real dimension prototypes for functional tests (in ABS and metals).

2.4. Material Design and Characterisation

- Massive use of light materials [15-17] with >90% in carbon fiber reinforced composites.
- Titanium use for structural reinforcement and safety cage.
- Use of hybrid composites, including natural reinforcements (flax and basalt [18,19]).
- Use of sandwiches structures and their optimisation [20,21].
- Use of natural/low VOC resins.
- Material testing [22,23].

2.5. Process Design

- Composite processing by pre-preg, manual stratification and autoclave forming [24].
- Communication protocols (from CAD to CAM) for hybrid processes.
- Realization of molds for car body (N. 5, in high-density foam) [25] (Figure 4).
- Realization of molds for parts and components (N. 53, in composites, aluminum and MDF).

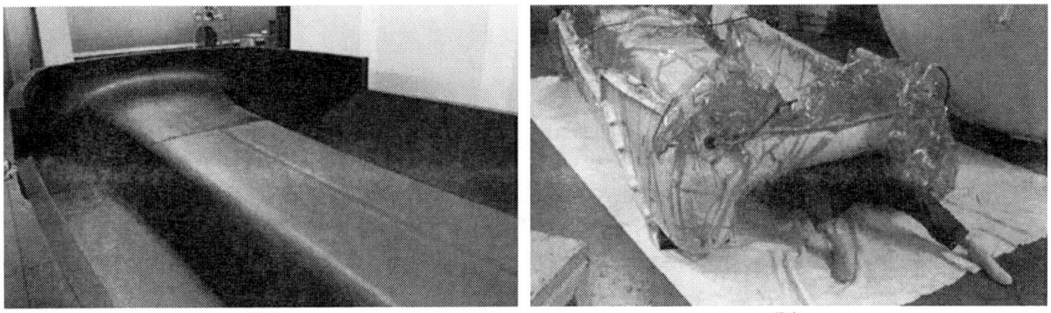

(a) **(b)**

Figure 4. Process design: a) mould in composite; b) use of mould for monocoque production.

2.6. Manufacturing and Assembly Design

- Structural parts (monocoque, frame, safety cage) manufacturing [26,27].
- Functional parts and kinematics (leaf springs [28], suspensions [29,30]) manufacturing.
- Non-structural (brakes, wheels [31], steering) manufacturing [32].
- Assembly (Figure 5).
- Standards and homologation [33].

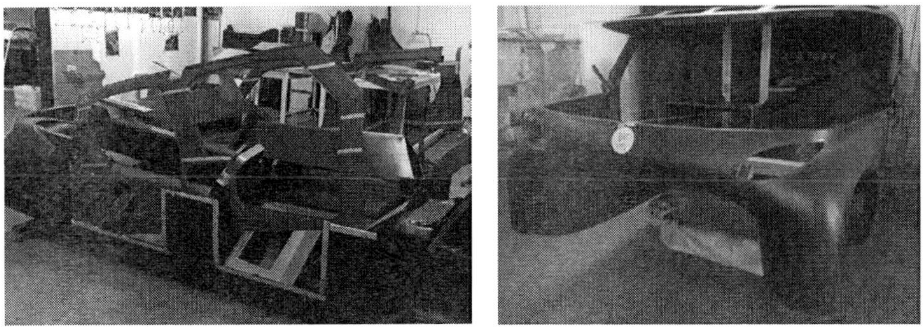

Figure 5. Manufacturing steps of the monocoque chassis.

2.7. Quality and Experimental Design

- Validation tests on parts [34].
- Laser scanning for quality control.
- Tests in supervised conditions (inside an airport).
- Tests on the roads and drivability [35].
- Tests during the competition (Figure 2).

A simplification of the design process together with the interactions between its main phases is represented in Figure 6 [36].

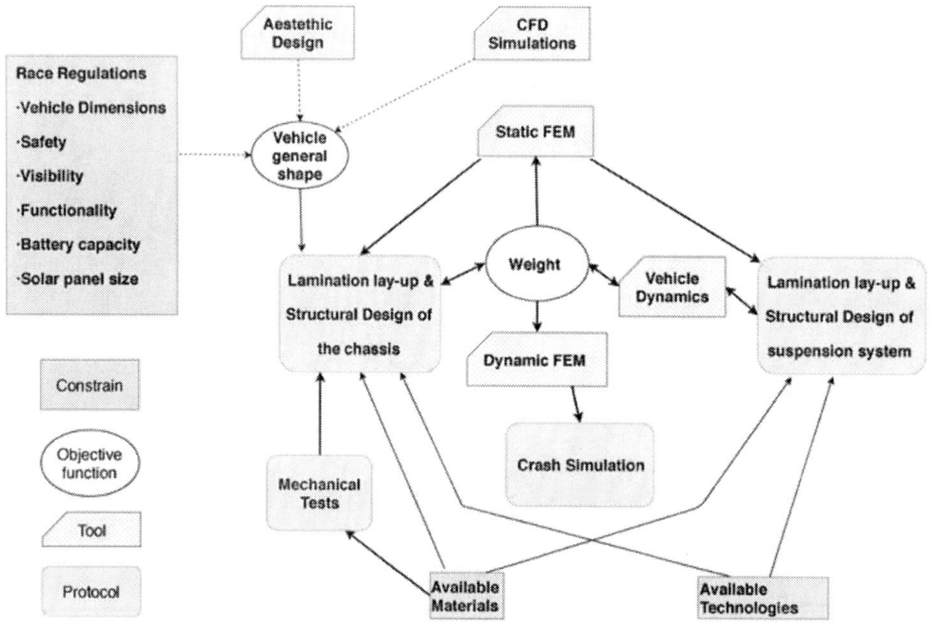

Figure 6. Design flow chart showing the interactions between phases during the design process [36].

3. Results

The result of this process is a 4-seater, electric solar-powered vehicle. Size 4.6x1.8x1.2m; empty mass (including batteries) of 230 kg and a maximum of 600 kg. Main structure and components in carbon fiber reinforced polymers (CFRP), with some parts in kevlar, titanium and ergal. Front surface of 1.60 m^2 with a 0.20 drag coefficient. Maximum expected speed 100 km/h, average consumption of 21 Wh/km (at 55 km/h), range of about 750 km (at 55 km/h). Photovoltaic panel in monocrystalline silicon, 5 m^2 with 326 SunPower cells, maximum efficiency 24% (at 25°C) and maximum panel power of 1.1kW. Converter with 98% efficiency. Battery pack: 83 kg positioned in the central tunnel, 1344 Samsung lithium ion cells with a nominal capacity of 3.4 Ah; rated voltage 48V, intensity 331.2 Ah, total energy 16.1 kWh; 2688 temperature sensors. Two synchronous motors with permanent magnets on the rear wheels: external rotor coupled directly to the wheel, and stator on the structure. Each 11kg motor has a nominal power of 1300 W and a maximum of 3000 W (nominal torque of 35 Nm and a maximum of 125 Nm); expected efficiency 97%.

These performances permitted to cross the United States, from the Missouri river to Portland, passing through the Rocky Mountains in less than 10 days, with an average speed of approx. 65 km/h, inside the American Solar Challenge 2018 (Figure 7).

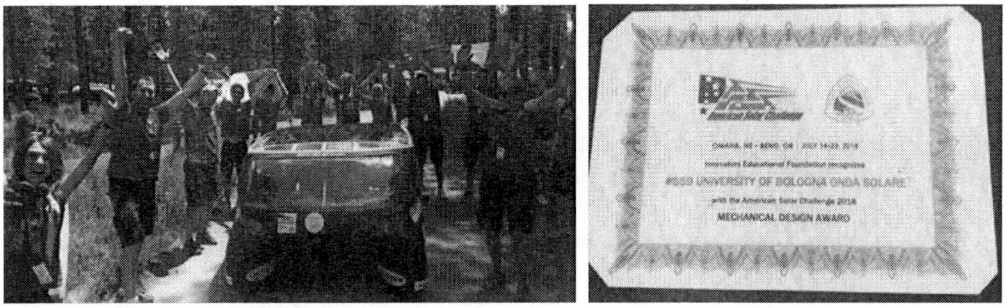

(a) (b)

Figure 7. The solar car as winner in its first race (a) and rewarded for the best mechanical design (b).

4. Discussion and Conclusions

As a synthesis, thanks to this design approach, it was possible to:

- design and build a four-wheeled light vehicle.
- Include it in the standard category of quadricycles.
- minimize the overall dimensions of space.
- maximize the space where to locate the solar panels.
- carry 4 passengers safely.
- balance weights and vehicle dynamics.
- provide adequate space for vehicle components such as steering and suspensions.
- provide adequate space for electrical components such as engines and battery.
- guarantee fast-access systems in case of maintenance.
- optimize the aerodynamic profile by reducing air resistance.
- guarantee a phenomenon of downforce in order to reduce the rolling resistance.
- correctly convey part of the air for cooling and ventilation.
- ensuring the internal liveability of the spaces and the installation of auxiliary devices.
- insert elements such as doors, air vents, lights, arrows, license plate in line with the standards.
- use shapes that can be built with available materials and technologies.
- provide for the possibility of registration of the vehicle.

Acknowledgment

The research is part of a larger action, managed by the University of Bologna (with references Cristiano Fragassa and Giangiacomo Minak), aiming at demonstrating the potentialities of the solar energy for a sustainable mobility. This publication has been supported by the Italian Ministry of Foreign Affairs and International Cooperation thought the Joint Research Projects as Particular Relevance, with a project named 'Two Seats for a Solar Car' within the Executive Programme of Cooperation between Italy and Serbia in the field of Science and Technology. Special thanks to Mauro Sassatelli and Ruggero Malossi (Onda Solare).

References

[1] Thacher E F 2015 *A Solar Car Primer, A Guide to the Design and Construction of Solar-Powered Racing Vehicles*. Springer

[2] Paterson G, Vijayaratnam P, Perera C and Doig G 2016 Design and development of the Sunswift eVe solar vehicle: a record-breaking electric car, *Journal of Automobile Engineering* **230** 1972-1986

[3] American Solar Challenge, *2018 Regulations - Revision B*, http://americansolarchallenge.org/ASC/wp-content/uploads/2017/09/ASC2018-Regs-External-Revision-B.pdf (accessed on Sep. 2017)

[4] Fragassa C, Pavlovic A and Massimo S 2014 Using a Total Quality Strategy in a new Practical Approach for Improving the Product Reliability in Automotive Industry, *International*

Journal for Quality Research **8**(3) 297–310
[5] Minak G, Fragassa C and de Camargo F V 2017 *A brief review on determinant aspects in energy efficient solar car design and manufacturing*, International Conference on Sustainable Design and Manufacturing, Bologna, Italy, April 26 – 28, pp. 847-856
[6] Joost W 2012 Reducing vehicle weight and improving U.S. energy efficiency using integrated computational materials engineering, *J of the Minerals, Met, and Mat Soc* **64** 1032-1038
[7] Kin W D, Kruger S, van Rensburg N J and Pretorius L 2013 *Numerical assessment of aerodynamic properties of a solar vehicle*, ASME International Mechanical Engineering Congress and Exposition, San Diego, United States of America, November 15 – 21
[8] de Kock JP, van Rensburg, N J, Kruger S and Laubscher R F 2014 *Aerodynamic optimization in a lightweight solar vehicle design*, ASME International Mechanical Engineering Congress and Exposition, Montreal, Canada, November 14 – 20, pp. 1-8
[9] De Camargo FV, Pavlovic A and Fragassa C 2018 Crash Safety and the Principles for an Impact Numerical Simulation in Composite Materials, *AIP Conference Proceedings* **1981**(1) 020032, doi:10.1063/1.5045894
[10] Boria S, Pavlovic A, Fragassa C and Santulli C 2016 Modeling of Falling Weight Impact Behavior of Hybrid Basalt/Flax Vinylester Composites, *Procedia Engineering* **167** 223–230
[11] Fragassa C, de Camargo, FV, Pavlovic A and Minak G 2019 Explicit Numerical Modeling Assessment of a Basalt Reinforced Composite For Low-Velocity Impact, *Composites Part B: Engineering* **163** 522-535, doi: 10.1016/j.compositesb.2019.01.013
[12] de Camargo F V and Pavlovic A 2017 Fracture Evaluation of the Falling Weight Impact Behaviour of a Basalt/Vinylester Composite Plate through a Multiphase Finite Element Model, *Key Engineering Materials* **754** 59-62, doi:10.4028/www.scientific.net/KEM.754.59.
[13] De Camargo F V, Fragassa C, Pavlovic A and Martignani M 2017 Analysis of the Suspension Design Evolution in Solar Cars, *FME Transactions* **45**(3) 394-404
[14] Shokrieh M M and Rezaei 2003 D Analysis and optimization of composite leaf spring, *Composite Structures* **60** 317-25
[15] Mathijsen D 2016 Redefining the motor car, *Reinforced Plastics* **60** 154-159
[16] Gay D 2014 *Composite Materials: Design and Applications*, CRC Press
[17] Maglio S, de Camargo F V and Rodrigues M R 2017 Benefits and Risks in the Use of Composite Materials in Solar Vehicles, *Key Engineering Materials* **754** 51-54, doi:10.4028/www.scientific.net/KEM.754.51.
[18] Fragassa C, Pavlovic A and Santulli C 2018 Mechanical and impact characterisation of flax and basalt fibre bio-vinylester composites and their hybrids, *Composites Part B: Engineering* **137** 247-259. doi: 10.1016/j.compositesb.2017.01.004
[19] Fragassa C, de Camargo F V, Pavlovic A and Minak G 2018 Experimental evaluation of static and dynamic properties of low styrene emission vinylester laminates reinforced by natural fibres, *Polymer Testing* **69** 437-449, doi: 10.1016/j.polymertesting.2018.05.050.
[20] Zenkert D 1995 *An Introduction to Sandwich Construction*, Engineering Materials Advisory Services Ltd, Sheffield, UK
[21] Poodts E, Panciroli R and Minak G 2013 Design rules for composite sandwich wakeboards, *Composites Part B: Engineering* **44**(1) 628-638
[22] ASTM D7264. Standard Test Method for Flexural Properties of Polymer Matrix Composite Materials. ASTM International. West Conshohocken, PA (2015)
[23] ASTM D2344. Standard Test Method for Short-Beam Strength of Polymer Matrix Composite Materials and Their Laminates. ASTM International. West Conshohocken, PA (2015)
[24] Betancur E, Fragassa C Coy J, Hincapie S and Osorio G 2017 *Aerodynamic effects of manufacturing tolerances on a solar car*, International Conference on Sustainable Design and Manufacturing, Bologna, Italy, April 26 – 28, pp. 868-876
[25] Fragassa C, Minak G and Sassatelli M 2019 Reducing Defects in Composite Monocoque Frames, *FME Transactions* **47**(1) 48-53, doi: 10.5937/fmet1901048F
[26] Betancur E, Mejía-Gutiérrez R, Osorio-Gómez G and Arbelaez A 2017 Design of structural parts for a racing solar car, *Lecture Notes in Mechanical Engineering* 25-32

[27] Liu Q, Lin Y, Zong Z, Sun G and Li Q 2013 Lightweight design of carbon twill weave fabric composite body structure for electric vehicle, *Composite Structures* **97** 231-238
[28] Wood K 2014 Composite leaf springs: Saving weight in production, *Composites Technology* **19**(1) 34-39
[29] Hurter W S, van Rensburg N J, Madyira D M and Oosthuizen GA 2014 *Static analysis of advanced composites for the optimal design of an experimental lightweight solar vehicle suspension system*, ASME International Mechanical Engineering Congress and Exposition, Montreal, Canada, November 14 – 20
[30] de Camargo F V, Giacometti M and Pavlovic A 2017 *Increasing the energy efficiency in solar vehicles by using composite materials in the front suspension*, International Conference on Sustainable Design and Manufacturing, Bologna, Italy, April 26 – 28, pp. 801-811
[31] Rondina F, Taddia S, Mazzocchetti L, Donati L, Minak G, Rosenberg P, Bedeschi A and Dolcini E 2018 Development of full carbon wheels for sport cars with high-volume technology, *Composite Structures* **192** 368-378
[32] Fragassa C, de Camargo, FV and Giorgini L 2018 Quality Assessment of Carbon Fiber Automotive Parts when Using Polyurethane Foam as Pattern Material, *AIP Conference Proceedings* **1981**(1) 020088, doi:10.1063/1.5045950
[33] Pavlovic A and Fragassa C 2015 General considerations on regulations and safety requirements for quadricycles, *International Journal for Quality Research* **9**(4) 657–674
[34] Sodena P D, Kaddourb A S and Hinton M J 2004 Recommendations for designers and researchers resulting from the world-wide failure exercise, *Composites Science and Technology* **64** 589-604
[35] Martini A, Bellani G, Fragassa C 2018 Numerical assessment of a new hydro-pneumatic suspension system for motorcycles, *International Journal of Automotive and Mechanical Engineering* **15**(2) 5308-5325
[36] Minak G, Brugo T M, Fragassa C, Pavlovic A, de Camargo F V and Zavatta N 2019 Structural Design and Manufacturing of a Cruiser Class Solar Vehicle, *Journal of Visual Experiments* **143** e58525, doi:10.3791/58525

Testing methods and equipment for palletized products

A Greco[1], A Renzini[1], M Vaccari[1] and F V De Camargo[2]

[1] Technology Lab Robopac, Aetna Group Spa, Via Cá Bianca, 1260, 40024 Castel San Pietro (BO), Italy
[2] Department of Industrial Engineering, Alma Mater Studiorum University of Bologna, Viale Risorgimento 2, 40136 Bologna, Italy

E-mail: mvaccari@robopac.com

Abstract. The ability to present cutting-edge solutions for packaging in the field of palletized products has to be considered as the result of the investments in research infrastructures and advances in applied researches. In the present article, main equipment and all the other testing facilities used to optimize palletized product are detailed. Tests are implemented in accordance with standard and no standard procedures up to fully innovative methods. This approach is able to provide advice both on the most appropriate palletizing scheme to pack the load, either on the best packaging methods able to conserve intact the product to the consumer. For customer it represents a great added value that allows to experience first-hand palletizing solutions, packaging and stability of the load, reducing production costs and optimizing efficiency.

1. Introduction

Over the years, the packaging industry has become increasingly important. At the beginning, the essential aspect was to guarantee a safe, functional and aesthetically appreciated product packaging. Later, and rather recently, the packaging has covered additional aspects such as environmental sustainability of transport and the integrity of the goods transported (as in [1]). In this regard it is usual to differentiate a 'primary packaging', the one intended to contain the product in contact with it (e.g. bottles, blisters, etc.), from a 'secondary packaging', used to collect and protect the goods during transport (e.g. [2]). For this second form of packaging it is common to group primary containers (e.g. bottles, blisters, etc.) in bundles, consolidated through heat-shrink material, and then in pallets [3]. These pallets are wrapped in prestressed films of plastic material that stabilize the package [4].

This secondary packaging process is characterized by a significant number of parameters, such as, for example, the type and thickness of the film, the number of layers, the strength of pre-stress and so on. This 'recipe' should be optimized in consideration of the type of product to be packaged (geometry and materials, e.g. glass bottles vs cardboard boxes) as well as the type of transport envisaged (e.g. distance, means of transport, time and place of storage) (as done in [5]). Only thanks to such an approach toward the optimization, it would be possible to guarantee an efficient packaging and a competitive product, fully in line with an approach oriented to the Total Quality Management [6]. Otherwise, the risks are of various types: from the waste of plastic film for an oversized packaging to damage to the product due to insufficient packaging. The related effects are not only of an economic nature, but also of safety (e.g. unforeseen fall of goods), and of environmental sustainability due to the waste of material and energy in the case of unusability of the product.

At the same time it is not at all easy to arrive at the determination and best packaging conditions as it is not always clear how each of the process parameters impact on the final packaging. Furthermore, the loads (mechanical, environmental, etc.) to which a package must withstand are not always known.

There are numerous studies that attempt to clarify both the aforementioned aspects in order to identify the most suitable package for each planned journey.

Transport by ship [7], by train [8], by truck [9] are just some of the different options that identify the specific load spectra to which a package is subjected during transport. In the various studies the attempt is to categorize these solicitations around some bases of reference, such as, for example, the geographical area (e.g. Europe [10], America [11], Asia [12-14]) or the type of route travelled (highway vs secondary roads [15, 16]).This is joined by all that is present although less intuitive. This relates, for example, to secondary handling (e.g. inside the depots, in the ports [17], in the loading and unloading phases [18]), including shocks (expected or not) [19, 20], storage inside or outdoors (changes in temperature, humidity [21] and irradiation with modification of the properties of the film.

This problem is so worthy of attention that innumerable standards have arisen. Among many of others it is interesting to mention: EUMOS 40509 for evaluating the load unit rigidity [22]; the ASTM D4003 as test method for horizontal impacts of containers [23]; ASTM D5276 for drop tests on containers [24]; the ISTA list on procedures for simulating the effect of different mean of transportations [25,26].

2. Aims and Scope

Regardless of the characteristics of the packaging considered to be the most suitable [27] and the stresses expected during the trip or their theoretical models (as in [28]), there will always be a time when all these considerations must be tested through an experimental approach.

In other words, we will always need a laboratory equipped to test packages. This goal is not obvious in terms of feasibility, especially considering how the packaging subject to analysis can be very large and heavy, and must be subjected to rather significant impulsive loads, capable of reproducing stresses similar to those present during transport, including events such as shock (as done in [29]].

Some regulations help simplifying real loads by considering them equivalent or less burdensome than those provided through standard tests. Several studies and even our experience suggest that these tests are not to be considered equivalent, nor more rigorous [30, 31].

Globalized transport, multimodality, speed, the quantity of goods around and the many other conditions present in a modern business toward a flexible manufacturing system [32, 33], let us believe that standard tests are rather a basis from which to start.

In short, a pallet must pass these tests undamaged, but this is not a sufficient condition to ensure that packaging is adequate for real. The perfect test equipment would be that able to solicit the pack with loading profiles similar to those found in the field (e.g. through experimental measurements).

This study describes our testing laboratory for palletized pallets through the description of the two testing machines, traditionally used for performing standard tests. In particular, it is possible to state that the two most used and effective ways to test the products that are packaged by means of the wrapping machines are the vibration platforms and the acceleration sledge.

Added to the, the study reports the description of a new testing machine, specifically designed and equipped to provide loads to pallets with a view to maximum experimental flexibility [34]. It consists of a commercial hexapod platform, with 6 degree of freedom (*DoF*), powered by a specific control system and a special loading platform, able to strictly fix the pallet during the test.

3. Vibrating platform

This platform is used to verify the stabilization effectiveness of the palletized load by vibrating the table at a configurable frequency (figure 1). The machine was developed to verify that these standards are satisfied: ASTM D5415-D5495 ruling standard test method for evaluating load containment performance of stretch wrap films by vibration testing [35]; ISO 2247 on Packaging complete filled transport packages and unit loads. Vibration tests at fixed low frequency [36].

The platform presents a rigid part, the basement, that is leaning on the ground through four feet. This part makes a rigid body able to bear the inertial stresses induced by the vibrating plane (as in [37]), faithfully providing a test setup that reproduces this harmful mechanism for the merchandise [38-39].

The operation principle of this platform is based on the rotation of two eccentric shafts with horizontal axis, moved by an asynchronous motor, that connect the moving part with the fixed one. The resulting components of the uniform circular motion are oscillatory harmonic with fixed amplitude and frequency proportional to motor speed.

Figure 1. Vibrating platform.

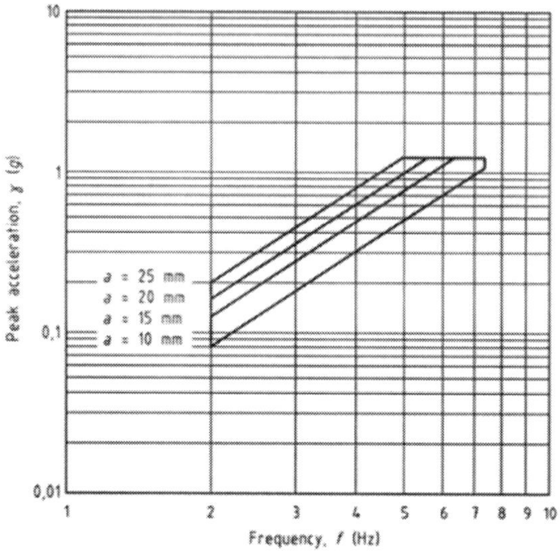

Figure 2. Graph acceleration-frequency (ISO 2247)

The test indicated in the standards consists of a table vibration at a fixed acceleration (g) for a certain period of time (10 min according to ASTM) and looking for load movements, wrapping material tearing or splitting, or other load integrity losses. If the load withstands to this test, it must be repeated at a higher frequency. Since the vertical displacement of the table is fixed, the relation between acceleration and frequency can be seen in figure 2 where $\gamma(g)$ is the peak of acceleration, expressed in terms of acceleration due to the gravity g ; a is the peak amplitude, expressed in millimetres (shifting in the diagram from 10 to 25 mm); f is the frequency, expressed in herz.

The displacement of the platform is fixed at 25 mm, so the correct line that connects acceleration to frequency must be followed. It is really helpful to know the vibration frequency of the load because every type of product has an own natural frequency and it can be seen resonant behavior also at very low frequency (0-5 Hertz range).

In figure 3, it can be seen, as example, the results of a test performed on the vibrating platform at 2.5 Hertz for 10 minutes, on a palletized load wrapped with a specific program, reported in Table 1. The related wrapping program consisted in that case of 25 revolutions on film, characterized by ~340g of weight, and a medium pre-stretch of 260%.

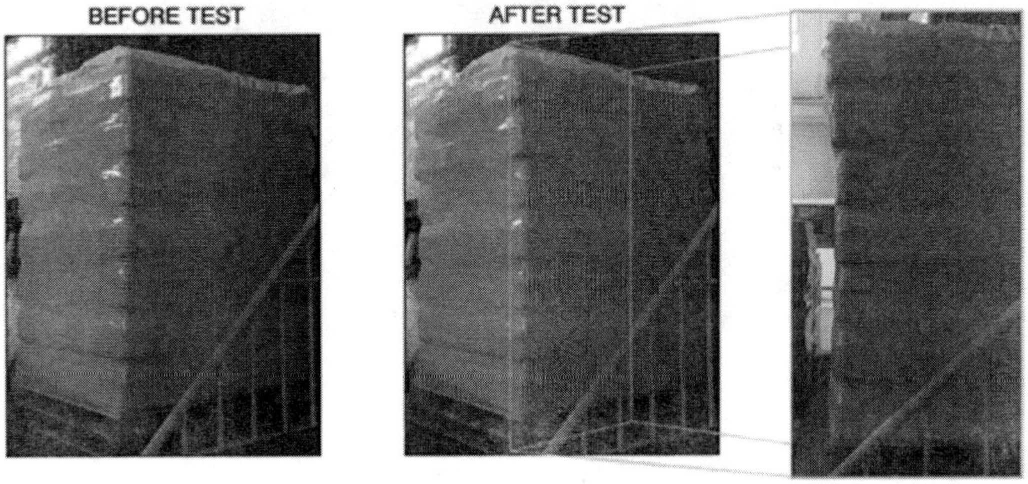

Figure 3. Results on wrapping load after test on vibrating platform.

4. Acceleration sledge

The acceleration sledge is used to verify the effectiveness of palletized load stabilization by means of an acceleration imposed to the load along the advancement direction. It is composed by a fixed base and a translating plane, on which the pallet is placed, and its motion law is configurable in acceleration (figure 4).

The supporting base of the sledge is a welded metal structure with rectangular base and it is configured in order to create a rigid body able to bear inertial stresses induced by the motion of the translating plane. The absolute value of these dynamic stresses increases with the increase of the acceleration imposed to the moving part and of the weight of unit load submitted to the test.

Also the moving part is a welded metal structure on which are mounted two asynchronous motors with planetary gearboxes. The value and the variability of the loads moved impose the use of a motion transmission system as rigid as possible, so it can be guaranteed the tracking of the trajectories imposed to the motors and it allows the repeatability of the tests.

This rigidity is entrusted to rack and pinion pair; the pinion is collocated on the output motor shaft, while the rack is bolted to the base, where are also bolted two linear ball guide on which slides the moving part.

The moving part that acts as a support for the load is delimited in front and behind, along advancement direction, from two panels, tilting at will.

The acceleration curve that controls the movement of the sledge is a modified trapezoidal type and it is characterized by the following phases:

- Acceleration ramp, configurable in duration. In this phase the acceleration is brought from 0 to the maximum selected value.
- Constant acceleration part, configurable in duration and absolute value.
- Deceleration ramp, configurable in duration. The acceleration passes from maximum value to minimum value (deceleration) passing through 0.
- Constant deceleration part, configurable in absolute value.

The control software verifies that the combination of parameters chosen by the operator does not lead to overcome the limits of stroke and maximum velocity of the system, and eventually forces him to modify the parameters.

Figure 4. Acceleration sledge.

Thanks to this application it is possible to evaluate load deformation when it is submitted to a certain pre-set horizontal acceleration, in particular it can be seen the tilt value of the load wall according to standard EUMOS 40509.

This standard is relevant in order to quantify the effectiveness of the wrapping during transport to prevent deformation or relative movement of the products, with a resulting elastic or permanent deformation of the load. The cause of these deformations is the horizontal inertial force that characterizes the transport, mostly road and railway transport.

Load rigidity is expressed as the maximum acceleration that the platform can reach without causing deformations or not acceptable movements of the load. This value is expressed in k·g where g is the acceleration gravity.

Whatever the value of acceleration imposed to the sledge, the standard imposed a constant acceleration part of at least 0.3 seconds, to reach from rest condition in maximum 0.05 seconds.
The standard imposes also some restrictions on the lateral panels, in particular the panel behind the load must be collocated at least 60 mm from the body and must have an inclination of at least 14 degrees with respect to vertical direction, creating a sufficient space for a certain deformation without contact between the two bodies.

In order to guarantee the rigidity level k the load must report deformations lower to the limits at the end of the test with acceleration $k{\cdot}g$, where these limits are defined in the following cases:

- Permanent deformation of every product, along horizontal direction, must not exceed 5% of total height of the load. Anyway, if the load is lower than 1200 mm, the limit is fixed at 60 mm. Furthermore the deformation related to the 200 lower mm must be less than 40 mm;
- Elastic deformation of the load must not exceed, along horizontal plane, 10% of the total height of the load. In case of tilt of the load it must be considered the tilt angle of the lower face in order to determine the real load deformation;
- If the composition of the load includes different layers of product, the maximum vertical space that forms between two successive layers must not exceed 2% of the total height of the load;
- The products must not be damaged at the end of the test.

In figure 5, it can be seen also in this case the results of a test on the acceleration sledge, with the same load of the previous machine, wrapped with the same program.

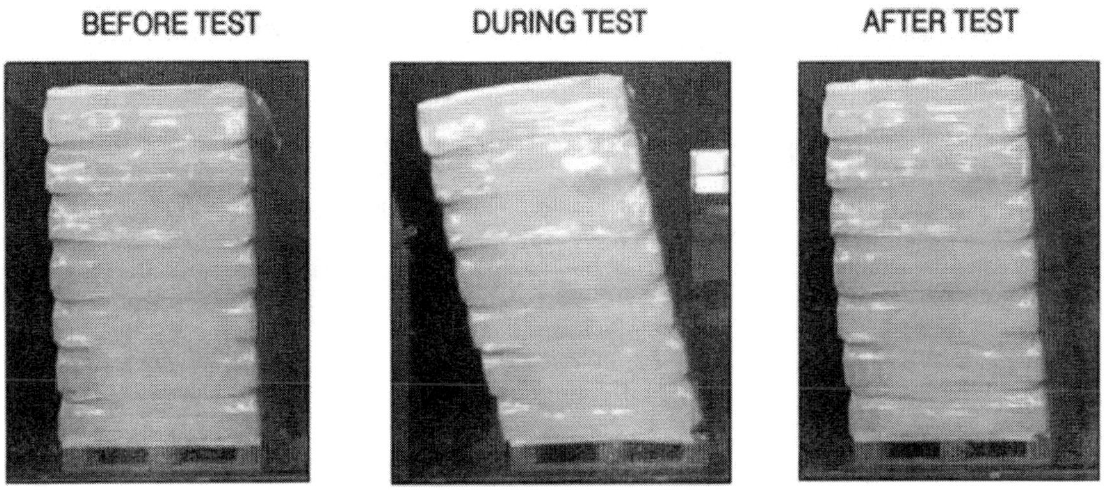

Figure 5. Results on wrapping load during and after test on acceleration sledge.

5. Hexapod platform

The machine that is also called motion platform is able to move a body through its six degrees of freedom by means of six actuators that move the legs of the system. The system configuration is reported in figure 6. The system is defined kinematic parallel robot because it is constituted by elementary kinematic chains that sustain in parallel the same end effector. It differs from a serial kinematic robot that is characterized by the absence of closed path within its kinematic chain.

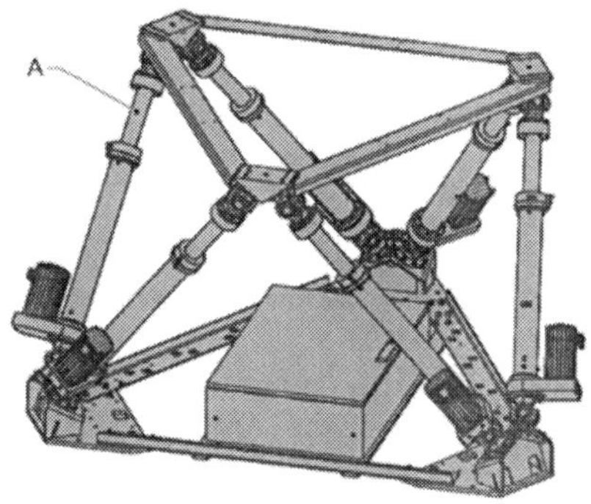

Figure 6. Mechanical assembly.

The hexapod parallel kinematic machines with extensible legs are also indicated with the name Gough-Stewart platform; the first realized a hexapod for the handling of tires to execute dynamic tests while the second published a relevant article about the use of Gough hexapod as simulator [40].

The advantages that the parallel kinematic configuration presents, with respect to the serial one, are:

- High dynamic performance (in particular if it can be used slender legs);
- Configuration modularity (economic saving);
- High rigidity and precision.

Otherwise the negative aspects of the parallel configuration are:

- Difficulty in realizing universal or spherical joints sufficiently rigid and precise;
- Bulky configuration with respect to the volume of work;

- Volume of work of strange shape;
- Isotropic behavior within volume of work;
- Risk of collision between legs;
- Limited possibility to incline the end effector.

The considered model is called MB-E-6DOF-2x-xx00KG, where MB means Motion Base, E indicates the use of electric actuators, 6DOF expresses the number of actionable degrees of freedom, 2x indicates the stroke of the actuators and depends on the performance type chosen for this model, as well as the payload, indicated as xx00KG.

The floor frame is attached to the ground by means of three separated floor plates, connected to the adjacent through a joint block that supports two joints on which are mounted two electric actuators. These joint blocks are bolted on the ground and are interconnected by floor beams (figure 7).

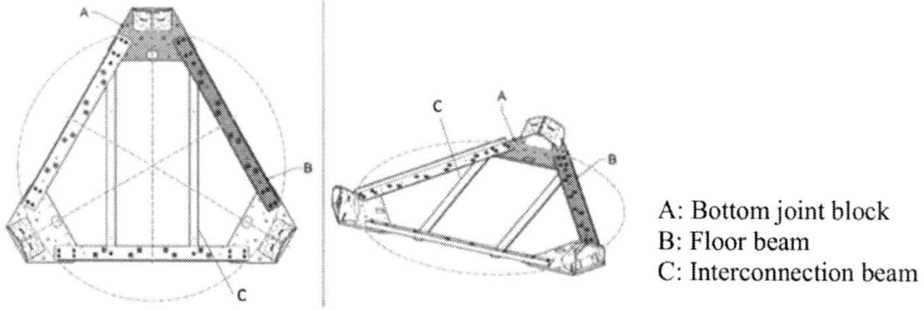

Figure 7. Floor frame assembly.

The moving frame is generally customer-supplied and interfaces with three top joint blocks, each supporting two electric actuators. So overall there are six actuators between floor and moving frames.

The beams that are represented in figure 8 support the top joint blocks but they are meant only for transportation and correct interface with the moving frame. After the installation they were removed because they reduce the payload.

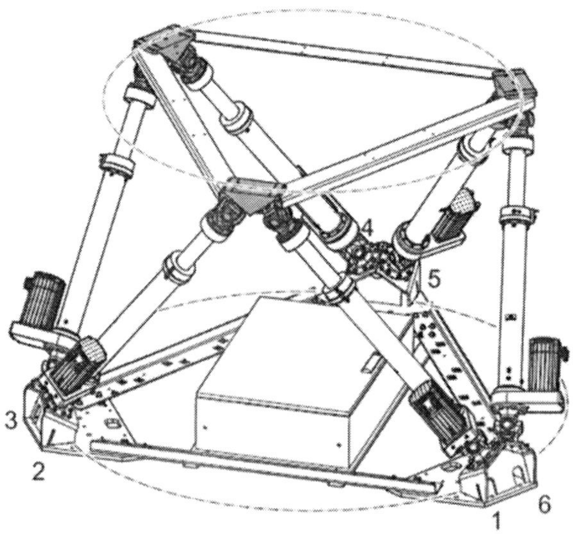

Figure 8. Mounting the top joint block to the frame.

Each electric actuator is attached to the floor frame and to the moving frame by a dedicated joint (figure 9). The motion base electric actuator lower joints are bolted onto the bottom joint blocks. The electric actuator upper joints are bolted onto the top joint blocks. The lower and upper joints are universal joints that allow rotation in two degrees of freedom.

The servo motor peak torque ensures that sufficient acceleration or deceleration is possible in every position. The mechanical brake locks the electrical actuator in some failure modes or when the simulator is not in use. Finally the possibility to include magnetorheological fluid devices for an active control and further functionalities was investigated (as proposed by [41]).

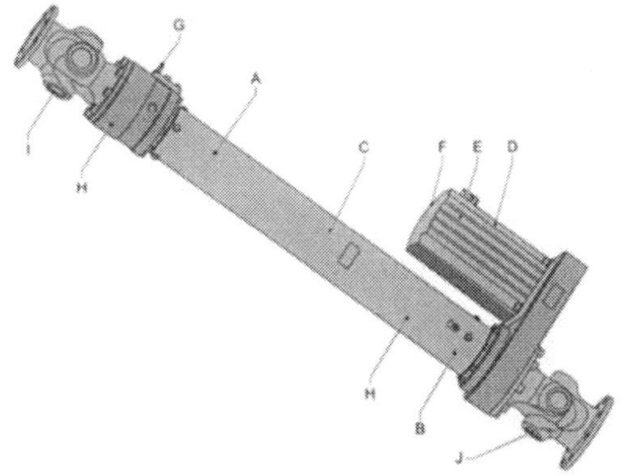

A: Stiff structure of steel telescopic tubes
B: Low friction sliding bearings
C: Ball screw with long lead
D: Brushless DC servo motor
E: Mechanical brake
F: Resolver for control loop feedback
G: Home switch
H: Buffer to absorb kinetic energy
I: Upper joint
J: Lower joint

Figure 9. Electric actuator.

5.1. Design of pallet fastening

The moving frame, used as pallet fastening, was designed to be compatible with machine supports and can be seen in figure 10.

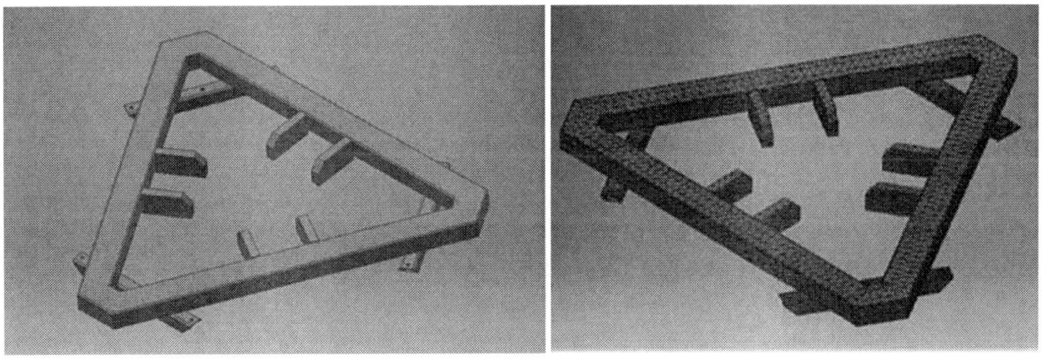

Figure 10. Design and FEM validation of the moving frame.

To simulate the pallet motion it is designed a fastening system that comprehends a wood loader that has to be fixed to the moving frame, on which the palletized load is placed.

The wood loader is fixed to the frame through three perforated sheets that are connected to the tubular elements placed in the lower part of the frame with some threaded rods. These rods are closed from both sides with a nut preceded by a washer and a spring-washer (figure 11).

Once the loader is fixed we had to constrain the palletized load. It was limited along its sides from four angular components connected to the loader with four plates placed below it. Also in this case the connection is performed by means of threated rods closed by nuts. Finally to constrain the vertical movement of the pallet to the platform movement the loader presents holes in different points (where it can be seen the washers) to allow the insertion of wood screws that fix the pallet to the loader.

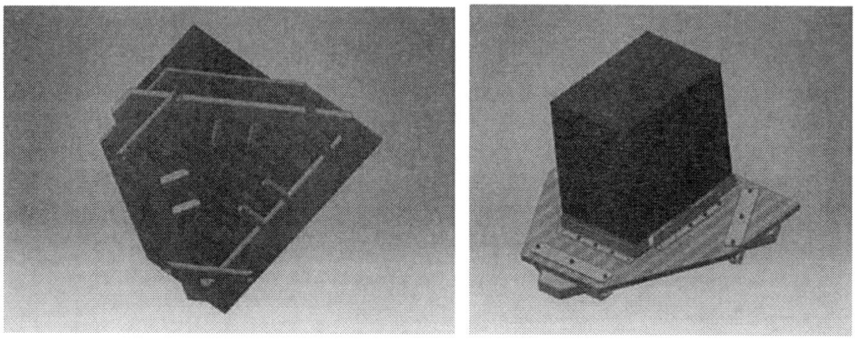

Figure 11. Fastening system of the pallet.

As it is important to assure that the pallet is correctly fixed to the platform is equally fundamental to know that the designed moving frame does not have a natural frequency in the frequency range excited by the system. In this case the platform start to vibrate and does not transmit the correct stresses between hexapod and pallet (the ones acquired in the tests).

To compute the first natural frequency of the steel component that was designed and that is used as moving frame of the system we used the finite element method (FEM) by means of integrated function in Inventor. Inventor is not a specific software for the finite element analysis but allows us, through a preliminary analysis, to estimate the natural frequencies of the body. After the setting of an automatic mesh and the definition of physical properties of the material we run a modal analysis that compute the natural frequencies of the body and the modal shapes. The results show that the first natural frequency of the system is around 90 Hertz. Considering that the data coming from our tests are used in the system after a filtering at 50 Hertz we can conclude that the platform does not influence the vibrations transmitted to the hexapod, allowing to correctly replicate them on the pallet and so on the load placed on it.

5.2. First use of the hexapod platform

The configuration shown in figure 12 was physically implemented and represented, to the knowledge of the authors, the first 'transport simulator for packages'. During the validation session, performed in MOOG Inc, Holland, after an initial performing test, it was programmed with transport test data acquired by means of our measurement system. The system proved to be able to return results (in terms of movements and accelerations) compatible with reality and readable from a dataset.

Figure 12. Moog hexapod configured for test simulation.

6. Conclusions

The success of the first simulations and the potentiality to replicate every type of transport or critical event thanks to the developed measurement system combined with Moog hexapod has led the laboratory to represent a first-class test environmental for this special products. Therefore this machine is added to the current one, able to study by standard tests the effect of transport on the stability due to wrapping of the palletized load. Meanwhile an instrumented pallet will be used to extend the available data related to transports, for every kind of product, transport, road and other parameters and configurations required by the customer.

Acknowledgment
The authors wish to acknowledge the assistance and encouragement from colleagues from the Department of Industrial Engineering at the University of Bologna.

References

[1] Berardinelli A, Donati V, et al. 2003 Effects of Transport Vibrations on Quality Indices of Shell Eggs, *Biosystems Engineering* **86**(4) 495-502
[2] Zhou R, Su S, Yan L and Li Y 2007 Effect of transport vibration levels on mechanical damage and physiological responses of Huanghua pears (Pyrus pyrifoliaNakai, cv. Huanghua), *Postharvest Biology and Technology* **46**(1) 20-28
[3] Wang Z W and Fang K 2016 Dynamic Performance of Stacked Packaging Units, *Packaging Technology and Science* **29**(10) 491-511
[4] Singh J, Cernokus E, Saha K and Roy S 2014 The effect of stretch wrap pre-stretch on unitized load containment *Packaging Technology and Science* **27**(12) 944-961
[5] Singh S P, Joneson E, Singh J and Grewal G 2007 Dynamic analysis of less-than-truckload shipments and test method to simulate this environment, *Packaging Technology and Science* **21**(8) 453-466
[6] Fragassa C, Pavlovic A and Massimo S 2014 Using a Total Quality Strategy in a new Practical Approach for Improving the Product Reliability in Automotive Industry, *International Journal for Quality Research* **8**(3) 297–310
[7] Borozc P, Singh P and Singh J 2015 Evaluation of Distribution Environment in LTL Shipment between Central Europe and South Africa, *Journal of Applied Packaging Research* **7**(2) n. 3
[8] Borozc P and Singh P 2016 Measurement and analysis of vibration levels in rail transport in central Europe, *Packaging Technology and Science,* DOI: 10.1002/pts.2225
[9] Jarimpos B, Singh S P and Saengnil W 2005 Measurement and analysis of truck transport vibration levels and damage to packaged tangerines during transit, *Packaging Technology and Science* **18**(4) 179-188
[10] Manuel G R M, Singh S P and Ballester V A 2008 Measurement and Analysis of Vibration Levels for Truck Transport in Spain as a Function of Payload. Suspension and Speed, *Packaging Technology and Science* **21**(8) 439-451
[11] Rissi G O, Singh S P, Burgess G and Singh J 2008 Measurement and Analysis of Truck Transport Environment in Brazil, *Packaging Technology and Science* **21**(4) 231-246
[12] Fei L, Yutaka I, Takeo S and Takaaki S 2008 Analysis of Shock and Vibration in Truck Transport in Japan, *Packaging Technology and Science* **21**(8) 479-489
[13] Singh S P, Sandhu A P S, Singh J and Joneson E 2007 Measurement and analysis of truck and rail shipping environment in India, *Packaging Technology and Science* **28**(6) 381-392
[14] Zhou R, Yan L, Li B and Xie J 2015 Measurement of Truck Transport Vibration Levels in China as a Function of Road Conditions, Truck Speed and Load Level, *Packaging Technology and Science* **28**(11) 949-957
[15] Lak M A, Degrande G and Lombaert G 2011 The effect of road unevenness on the dynamic vehicle response and ground-borne vibrations due to road traffic, *Soil dynamics and earthquake engineering* **31**(10) 1357-1377
[16] Beno P, Krilek J, Kovac J, et al. 2018 The Analysis of the New Conception Transportation Cableway System Based on the Tractor Equipment, *FME Transactions* **46**(1) 17-22
[17] Jovanović J, Fragassa C, Krivkapić Z and Vujović A 2019 Environmental Management Systems and Balanced Scorecard: An Integrated Analysis in the Marine Transport, *Journal of Marine Science and Engineering* **7**(119) 1-25, doi:10.3390/jmse7040119
[18] Martins G H A and Dell R 2008 Solving the pallet loading problem, *European Journal of*

Operational Research **184**(2) 429-440
[19] Blumer, T and Guadagnini D 2011 *Shock transmissibility of a palletized load caused by forklift truck handling*, Industrial Technology, California Polytechnic State of University
[20] Lepine J, Rouillard V and Sek M 2016 On the use of machine learning to detect shocks in road vehicle vibration signals, *Packaging Technology and Science*, DOI:10.1002/pts.2202
[21] Yui-Ping E and Wang Z W 2010 Effect of relative humidity on energy absorption properties of honeycomb paperboards, *Packaging Technology and Science* **23**(8) 471-483
[22] EUMOS 40509 (2012): Test method for load unit rigidity
[23] ASTM D4003 (2015). Standard Test Methods for Programmable Horizontal Impact Test for Shipping Containers and Systems, American Society of Testing and Materials
[24] ASTM D5276 (2009). Standard test method for drop test of loaded containers by free fall, American Society of Testing and Materials
[25] ISTA 3B. (2012) Packaged-Products for Less-Than-Truckload (LTL) Shipment. Ista 3 Series General Simulation Performance Test Procedure, 1-11
[26] ISTA 3E. (2009) Unitized Load Containment of Same Product. Ista 3 Series General Simulation Performance Test Procedure, 1-13
[27] Zhang Q, Saito K and Nagaoka K 2016 Proposal of hybrid damping package design, *Journal of applied packaging research* **8**(3) n. 6
[28] Zhong C, Saito K and Kawaguchi K 2013 Improvement of equivalent drop theory for transport packaging, *Packaging Technology and Science* **2**(2) 67-81
[29] Bruscella B 1997 *The analysis and simulation of the spectral and statistical properties of road roughness for package performance testing*, Department of Mechanical Engineering, Victoria University of Technology. Melbourne, Australia
[30] Fragassa C, Macaluso I, Vaccari M and Lucisano G 2017 Measuring the mechanical and climatic conditions encountered by palletized products in handling and transport, *FME Transactions* **45**(3) 382-393
[31] Ge C 2000 Model of accelerated vibration test *Packaging Technology and Science* **13**(1) 7-11
[32] Djapic M, Lukic L, et al. 2017 Multi-agent team for engineering: a machining plan in intelligent manufacturing systems, *International Journal of Machining and Machinability of Materials* **19**(6) 505-521. Doi: 10.1504/IJMMM.2017.10006404
[33] Lukic L., Djapic M, et al. 2018 Optimization Model for Machining Processes Design in Flexible Manufacturing Systems, *Journal of Advanced Manufacturing Systems*, **17**(2) 137-153
[34] Bernard C, Laspalas A, et al. 2011 Transport vibration laboratory simulation: On the necessity of multiaxis testing, *Packaging Technology and Science* **24**(1) 1-14
[35] ASTM D5415-D5495 Ruling standard test method for evaluating load containment performance of stretch wrap films by vibration testing
[36] ISO 2247 (2000) Packaging — complete, filled transport packages and unit loads — vibration tests at fixed low frequency
[37] Fragassa C 2017 Material selection in machine design: the change of cast iron for improving the high-quality in woodworking, *Proc. Institution of Mechanical Engineers, Part C: Journal of Mechanical Engineering Science* **231**(1) 18-30
[38] de Camargo F V, Queiroz J P T P, Silveira A C F, Pin T F P, Vaccari M 2018 Retrospect of the Influence of Transportation in the Delivered Quality of Palletized Product, *Int J for Quality Research* **12**(3) 773-788
[39] Queiroz J P T P, Silveira A C F, de Camargo F V and Vaccari M 2018 A brief review on palletized products transportation technology, *FME Transactions* **46**(1) 101-107
[40] Stewart D 1965 A Platform with Six Degrees of Freedom, *Proceedings of the Institution of Mechanical Engineers* **180**(1.15) 371–386. doi:10.1243/pime_proc_1965_180_029_02
[41] Fragassa C, Berardi L and Balsamini G 2016 Magnetorheological fluid devices: an advanced solution for an active control on the wood manufacturing process, *FME Transactions* **44**(4) 333-339

AUTHOR INDEX